	IB	IIB	IIIA	IVA	VA	VIA	VIIA	
								4.00260
			5 B Boron 10.81	6 C Carbon 12.011	7 N Nitrogen 14.0067	8 O Oxygen 15.9994	9 F Fluorine 18.998403	10 Ne Neon 20.179
			13 Al Aluminum 26.98154	14 Si Silicon 28.0855	15 P Phosphorus 30.97376	16 S Sulfur 32.06	17 Cl Chlorine 35.453	18 Ar Argon 39.948
28 Ni Nickel 58.70	29 Cu Copper 63.546	30 Zn Zinc 65.38	31 Ga Gallium 69.72	32 Ge Germanium 72.59	33 As Arsenic 74.9216	34 Se Selenium 78.96	35 Br Bromine 79.904	36 Kr Krypton 83.80
46 Pd Palladium 106.4	47 Ag Silver 107.868	48 Cd Cadmium 112.41	49 In Indium 114.82	50 Sn Tin 118.69	51 Sb Antimony 121.75	52 Te Tellurium 127.60	53 I Iodine 126.9045	54 Xe Xenon 131.30
78 Pt Platinum 195.09	79 Au Gold 196.9665	80 Hg Mercury 200.59	81 Tl Thallium 204.37	82 Pb Lead 207.2	83 Bi Bismuth 208.9804	84 Po Polonium (209)	85 At Astatine (210)	86 Rn Radon (222)

4 Gd adolinium 57.25	65 Tb Terbium 158.9254	66 Dy Dysprosium 162.50	67 Ho Holmium 164.9304	68 Er Erbium 167.26	69 Tm Thulium 168.9342	70 Yb Ytterbium 173.04	71 Lu Lutetium 174.97
6 Cm urium 247)	97 Bk Berkelium (247)	98 Cf Californium (251)	99 Es Einsteinium (254)	100 Fm Fermium (257)	101 Md Mendelevium (258)	102 No Nobelium (255)	103 Lr Lawrencium (260)

*Name not officially assigned

The values of atomic mass shown in parentheses are estimated. In most cases the values represent the mass number of the most stable isotope.

Metals

Nonmetals

Metalloids

Noble gases

101540

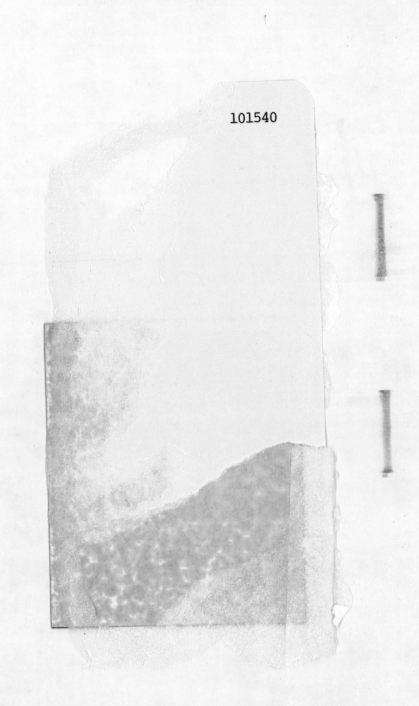

THE CHEMICAL BASIS OF LIFE

THE CHEMICAL BASIS OF LIFE

General, Organic, and Biological Chemistry for the Health Sciences

George H. Schmid, Ph.D.
Professor, Department of Chemistry,
University of Toronto,
Toronto, Ontario, Canada

Little, Brown and Company Boston

First Edition

Library of Congress Catalog Card No. 81-83245

ISBN 0-316-77371-9

Printed in the United States of America

HAL

NURSING ACQUISITIONS EDITOR: Julie Stillman
NURSING DEVELOPMENTAL EDITOR: Ann West
MANAGING BOOK EDITOR: Katharine Tsioulcas
EDITORIAL ASSISTANT: Priscilla Hurdle
EDITORIAL ASSISTANT: Nancy Mimeles Carey

In Memoriam

MAITA

Everything wonderful must end it is said.

But the end is not at hand,

It is only a brief parting until we meet again.

CONTENTS

PREFACE

I have written this text for use by students preparing for careers in the health sciences, such as nursing, medical laboratory technology, dietetics, health and physical education, and dental hygiene. It also provides the necessary chemistry background for students in community health, home economics, and the liberal arts. No previous course in the sciences is assumed.

The central theme of this text is that living systems are practical examples of the chemical principles that are discovered in the chemical laboratory. To emphasize this point, examples from living systems are used extensively to illustrate basic chemical principles. For example, breathing is presented as a practical example of the gas laws. Acid-base balance in the blood is used as an example of the basic principles of acid-base chemistry. Metabolic diseases and the action of drugs, vitamins, and hormones are discussed as examples of chemical reactions of organic compounds. A variety of other examples related to living systems are given throughout the text, in the examples, and in the exercises.

The topics are arranged in the following order. The first ten chapters cover the basic principles of general chemistry. The next nine chapters present the major functional groups of organic compounds; each of these nine chapters contains a section on naming, physical properties, and biologically important chemical reactions. Several chapters include optional sections (which can be omitted without loss of continuity) on the mechanisms of organic reactions.

Once the biologically important reactions of organic compounds containing only one functional group have been covered, polyfunctional compounds are introduced in Chapter 20. This prepares the student for the following four chapters, which introduce the major types of biologically important compounds: carbohydrates, lipids, proteins, and enzymes. Finally, these compounds and the chemical reactions learned in Chapters 11 through 19 are brought together and are applied to metabolism in the last five chapters. Extensive cross-references in these chapters emphasize the important point that *most of the chemical reactions of living systems are examples of the reactions of the major functional groups of organic compounds.*

Numerous study aids are provided for the student. New terms are italicized when they are first defined. These terms and their definitions, new concepts, any new chemical reactions, and a summary are given at the end of each chapter. Exercises are placed at appropriate places in the text and at the end of each chapter. The answers to all in-chapter exercises appear at the back of the book. Frequent cross-references throughout the text help the student tie together related concepts and facts found in different sections of the text. The Appendix provides a review of basic mathematics.

A study guide is available for this text. It includes detailed solutions to the exercises in the text as well as additional exercises. An accompanying laboratory manual and instructor's reference manual are also available.

G. H. S.

ACKNOWLEDGMENTS

In writing this book I have benefited greatly from the comments and suggestions of many colleagues and students. I wish to express my appreciation to Sara Chambers, Professor M. Bersohn, and Professor R. Kluger for detailed and thoughtful reviews of various sections of the book. Special thanks are due Snezana Dalipi and Wai-Fun Chan for their tireless efforts through the various stages of proofreading and to Claudette Garipy and Ann Young for their excellent work in typing the manuscript. I have been aided by the many detailed suggestions received from reviewers: Geoffrey Davies, Northeastern University; Sam Milosevich, University of Wisconsin—Milwaukee; Richard G. Pflanzer, Indiana University School of Medicine, Indianapolis; and Marlene Spero, Loretto Heights College.

I am especially grateful to Coleen Dean for her constructive criticism, invaluable advice, and moral support.

Converting the author's manuscript into the final textbook required the dedication of a number of people at Little, Brown and Company. I am indebted to editors Ann West and Julie Stillman for their continued enthusiasm, advice, and attention to detail, to Clif Gaskill for expertly solving the multitude of production problems, and to Katharine Tsioulcas, Janet Olsen, Priscilla Hurdle, and Nancy Mimeles Carey for their outstanding copyediting.

G. H. S.

TO THE STUDENT

This text is written to help you learn chemistry. It will help you by presenting chemistry in a clear, logical, and orderly way. The sections within each chapter are organized so that one subject leads smoothly and logically to another. Your progress in learning chemistry depends in large part on your mastering each concept or principle as it is presented. To help you do this, a number of study aids are provided.

New terms are italicized when they are first defined. In addition, all new terms and their definitions are given at the end of each chapter, together with a list of the new concepts and any new chemical reactions. Each chapter has a summary at the end. Read the chapter summary after you have finished studying the chapter to be sure you have understood all the essential points. You may also find it helpful to read the summary to get an overview before studying a chapter.

The most important study aids in the text are the exercises. There are exercises within many sections of every chapter. Do all the exercises pertaining to a section to test your understanding of the material before going on to the next section. Additional exercises are given at the end of each chapter. Solve these as well. The best way to understand and learn to apply a chemical principle is to solve a number of exercises that require the application of that chemical principle. It is not sufficient simply to follow an example and its worked-out solution in the text or in class. You must be able to solve the exercise—in the chapter and at the end of the chapter—yourself. Only then can you be confident of your knowledge of the material in that chapter. The answers to the in-chapter exercises are given at the end of the text. Detailed solutions to all the exercises are given in the study guide, which offers additional exercises as well.

Anyone learning a new subject finds it difficult to see interrelationships and general principles. The text contains frequent cross-references to help you tie together related concepts and facts found in different sections of the text. The Appendix presents a brief mathematical review. If you have any doubts about your ability to handle the mathematical skills needed in this text, you should read the Appendix and do the Appendix exercises. The extensive index at the end of the book should also prove useful.

In this text I have tried to help you understand chemistry and show you the importance of chemistry to all living systems. How successful I have been will be determined by you, the student who uses this text. Your comments and suggestions are always welcome.

G. H. S.

THE CHEMICAL BASIS OF LIFE

CHEMISTRY, METHODS, AND MEASUREMENTS 1

1.1 CHEMISTRY AND LIFE

Humans are only one of a wide variety of forms of life inhabiting the earth. People have always been fascinated by the different forms of life they see around them. Over the years, humans have considered themselves to be a unique form of life. But is this view justified? What is the difference between humans and other forms of life? In our view, the difference is usually one of size, shape, or function. While there are obvious physical differences, is there really a difference in the chemistry of the various forms of life on earth? Today we believe that *humans, rather than being unique, are one of many forms of life that obey certain basic principles of chemistry.* This idea has slowly evolved during the past century from the work of many scientists in many fields. Let us briefly examine how this idea developed.

The view that humans are a unique form of life was first seriously challenged about 100 years ago by the theory of evolution. This theory states that the various forms of life are not constant, but, rather, continually evolve. That is, they continually change to create slightly different forms, some of which are better adapted to survival. On the basis of the theory of evolution, we can reach the following conclusions: Life existed on earth several billion years ago in an extremely simple form. This form possibly resembled the bacterium, which is the simplest form of life now existing. If life evolved from such a simple form, then the essence of the living state is found in very small organisms. Furthermore, this indicates that the *basic principles of the living state are the same in all living forms.*

About the same time that the theory of evolution was proposed, other scientists were independently reaching a similar conclusion. These people were examining the structures of thin sections of many plants and animals under the microscope. As a result of this work, it was proposed that all larger plants and animals are made up of small fundamental units called *cells*. This does not mean that all cells are the same. On the contrary, there are many different classes of cells, each one having a particular function. But each cell is a fundamental unit containing everything needed to reproduce itself. In this way, all living systems can be reduced to a common denominator—the cell.

While these ideas were being developed, chemists were busy studying the chemical composition of things obtained from living systems. From this work it was discovered that there is no chemical way to distinguish between something formed in a cell and the same thing formed in a chemical laboratory.

On the basis of countless experiments in the fields of medicine, physiology, biology, and chemistry, scientists believe today that *life is a series of complex chemical reactions*. All these chemical reactions, no matter how complex, obey the laws of science. No new laws of science are yet needed to account for the chemical reactions involved in the process of life. Therefore, there is no special chemistry of the living system, and cells, the basic unit of living systems, must obey the laws of chemistry.

One inescapable conclusion is reached from this view of living systems. *To understand the process of life, you must understand the principles of chemistry.* But how can such an abstract knowledge of chemistry enable us to understand the chemistry of the complicated reactions found in living systems? The answer is the following. Before we can study the chemistry of complex molecules, we must learn the basic rules of how atoms combine to form simple molecules. Furthermore, we must learn certain principles of chemistry that have important applications to living systems. Once we understand these basic principles of chemistry, we can then examine the structure and chemical reactions of small molecules, which serve as models for the giant molecules found in living systems. We can then apply this knowledge to study the chemistry of the molecules of living systems. Let us start toward our goal by learning how scientists seek the solution to a problem in a logical way.

1.2 THE SCIENTIFIC METHOD

It requires a certain amount of mental discipline and logical thought to solve any problem. Scientists attack a problem by means of a series of logical

steps called the *scientific method.* The first step of this method is to gather and organize all the information related to the problem. These are the *facts*. For example, if we have a balloon filled with air and we measure the size of the balloon as the temperature of the surroundings changes, we find that as the temperature increases the balloon expands; conversely, the balloon contracts when the temperature is lowered. These are experimental facts that can be verified by any skilled observer. As more and more facts accumulate the scientist makes a *hypothesis,* which is an attempt to correlate the observed facts. Thus, we can state as a hypothesis that the volume of a balloon varies as the temperature of its surroundings changes. This hypothesis usually suggests more experiments, which provide more facts. These new facts either strengthen the hypothesis or discredit it. If repeated experiments confirm the hypothesis, a *law* can be formulated. A law is simply a statement of our universal experience of the behavior of nature. It does not explain why nature behaves as it does, but it does present an adequate description of all the experimental data. In this way, scientists have arrived at a law that states that the volume of a gas is directly proportional to its absolute temperature, other factors being held constant. This is Charles' law, as we will learn in Chapter 6.

After various laws are formulated, it is usually possible to propose a *theory*. A theory is actually a model or picture that exists in the human mind to help us understand the way in which nature behaves. For example, as our knowledge of the behavior of gases increased, it became possible to propose a theory, called the kinetic molecular theory, which explains the gas laws in terms of the motion and properties of minute particles. Scientists have successfully used this theory to predict many facts concerning the behavior of materials. Because it is useful, this theory is still retained by scientists.

By this scientific method the seemingly diverse experimental facts of science in general and chemistry in particular are molded into categories that help us in our understanding of nature. The task of learning about our world would become hopeless without these categories to relate facts.

As we categorize the observations of science, a careful distinction must be made between fact and theory. An experimental fact is invariant with time. For example, the color of pure gold at room temperature was the same yesterday as it is today, and will be the same at any time in the future. To answer the question—What gives gold its characteristic color?—requires a theory that is the product of human imagination. Depending on our scientific sophistication, these theories can and will change with time.

There is only one problem with the scientific method. It does not take into account accidental discoveries. Major scientific discoveries are often made in this way. The vulcanization of rubber and the antibiotic activity of penicillin are two examples of such discoveries. The scientific method is a logical way of attempting to solve a problem, but intuition and imagination often play an important part.

The scientific method is based on collecting facts. To obtain facts, we must make measurements. Furthermore, the measurements that we make must be done in a way that can be repeated worldwide. This requires an international system of measurement, as we will learn next.

1.3 THE INTERNATIONAL SYSTEM OF UNITS (SI)

The first international system of measurement was the metric system. It was adopted by the United States and 16 other nations in 1875. This is a decimal system, which means that larger and smaller parts of the basic unit are obtained by multiplying or dividing by ten. The metric system was extensively revised in 1960 and was given a new name, "International System of Units." Its official abbreviation is SI.* SI is still a decimal system, and many of the units are the same in SI and the metric system.

There are fractions and multiples of each basic unit of SI. Each has a

* This abbreviation is obtained from its French name, "Le *S*ystème *I*nternationale d'Unités."

Table 1-1. SI Prefixes

Prefix	Symbol	Meaning			
tera-	T	$10 \times 10 \times 10 \times 10 \times 10 \times 10 \times 10 \times 10 \times 10 \times 10 \times 10 \times 10$	$= 1{,}000{,}000{,}000{,}000$	$= 10^{12}$*	$\times$ basic unit
giga-	G	$10 \times 10 \times 10 \times 10 \times 10 \times 10 \times 10 \times 10 \times 10$	$= 1{,}000{,}000{,}000$	$= 10^{9}$	$\times$ basic unit
mega-	M	$10 \times 10 \times 10 \times 10 \times 10 \times 10$	$= 1{,}000{,}000$	$= 10^{6}$	$\times$ basic unit
kilo-	k	$10 \times 10 \times 10$	$= 1000$	$= 10^{3}$	$\times$ basic unit
					basic unit
deci-	d	$\frac{1}{10}$	$= 0.1$	$= 10^{-1}$	$\times$ basic unit
centi-	c	$\frac{1}{10} \times \frac{1}{10}$	$= 0.01$	$= 10^{-2}$	$\times$ basic unit
milli-	m	$\frac{1}{10} \times \frac{1}{10} \times \frac{1}{10}$	$= 0.001$	$= 10^{-3}$	$\times$ basic unit
micro-	μ	$\frac{1}{10} \times \frac{1}{10} \times \frac{1}{10} \times \frac{1}{10} \times \frac{1}{10} \times \frac{1}{10}$	$= 0.000001$	$= 10^{-6}$	$\times$ basic unit
nano-	n	$\frac{1}{10} \times \frac{1}{10} \times \frac{1}{10} \times \frac{1}{10} \times \frac{1}{10} \times \frac{1}{10} \times \frac{1}{10} \times \frac{1}{10} \times \frac{1}{10}$	$= 0.000000001$	$= 10^{-9}$	$\times$ basic unit
pico-	p	$\frac{1}{10} \times \frac{1}{10} \times \frac{1}{10} \times \frac{1}{10} \times \frac{1}{10} \times \frac{1}{10} \times \frac{1}{10} \times \frac{1}{10} \times \frac{1}{10} \times \frac{1}{10} \times \frac{1}{10} \times \frac{1}{10}$	$= 0.000000000001$	$= 10^{-12}$	$\times$ basic unit

*The use of exponents is reviewed in Appendix A-5.

characteristic name obtained by adding a prefix to the basic name of the unit. The prefix tells us how many times the basic unit has been divided or multiplied by ten. The names of the most commonly used prefixes are given in Table 1-1.

Let us now learn the SI units of distance, mass, and volume and learn how to use these prefixes.

Length

The basic unit of length or distance in SI is the *meter*. The most commonly used fractions of the meter are the centimeter and the millimeter. The kilometer is the most common multiple of the meter. Notice that the name of each of these units tells us the relationship between it and the basic unit. For example, the word "centimeter" is made up of two parts, the prefix centi-, indicating 1/100, and the basic unit, meter. Therefore, a centimeter is one one-hundredth of a meter. Similarly, a kilometer is one thousand meters. Other fractions and multiples of the meter are given in Table 1-2. With SI units we can easily estimate the size of the unit. A millimeter is a small unit that is convenient for measuring short distances. A kilometer, on the other hand, is a large unit that is commonly used to measure distances between cities.

Each SI unit is given a symbol. It is m for the meter. The symbols for the fractions and multiples are obtained by placing the symbol of the prefix before the symbol of the basic unit. Thus, km is the symbol for kilometer, cm is the symbol for centimeter, and so forth.

Table 1-2. Fractions and Multiples of the Meter

Unit	Symbol	Relationship to the Meter
terameter	Tm	10^{12} m
gigameter	Gm	10^{9} m
megameter	Mm	10^{6} m
kilometer	km	10^{3} m
meter	m	m
decimeter	dm	10^{-1} m
centimeter	cm	10^{-2} m
millimeter	mm	10^{-3} m
micrometer	μm	10^{-6} m
nanometer	nm	10^{-9} m
picometer	pm	10^{-12} m

Mass

The terms *mass* and *weight* are used interchangeably in everyday usage. However, this is incorrect, because there is a difference between the two terms. Mass is defined as the amount of material in a certain object compared to a reference standard mass. Weight is the measurement of the attraction of the earth's gravity for the mass of the object. The mass of an object is the same on the moon as on the earth, but it weighs less because of the lower gravity of the moon.

The SI unit of mass is the kilogram (kg). The most commonly used fractions are the gram (g), the milligram (mg), and the microgram (μg).

*EXERCISE 1-1 Construct a table similar to Table 1-2 indicating the relationship among Tg, Gg, Mg, kg, g, dg, cg, mg, μg, ng, and pg.

Volume

The SI unit of volume is the cubic meter. Its symbol is m^3. A cubic meter is a very large unit and consequently it is not used very often by chemists. The liter is the most commonly used unit of volume. The liter, whose symbol is L, is defined in terms of SI units as 0.001 m^3, or 1000 L = 1 m^3. Commonly used fractions of the liter are the milliliter (mL) and the centiliter (cL). One milliliter is identical to one cubic centimeter, whose symbol is cm^3. Although the old symbol for cubic centimeter (cc) should no longer be used, it is still used by some people, particularly those who are not familiar with the SI units. The various SI units of volume are summarized in Table 1-3.

Table 1-3. Common SI Units of Volume

1 cubic meter (m^3) = 1000 liters (L)
1 liter (L) = 1000 milliliters (mL)
1 milliliter (mL) = 1000 microliters (μL)

EXERCISE 1-2 Construct a Table similar to Table 1-2 indicating the relationship among TL, GL, ML, kL, L, dL, cL, mL, μL, nL, and pL.

Despite attempts to standardize the units of measurement in the world, systems of measurement other than SI are commonly used. In the United States, for example, a system that originated in England several centuries

* The answers for the exercises in this chapter begin on page 857. NOTE: The *answers* to all the end-of-chapter exercises are provided in *Instructor's Reference Manual for Schmid's The Chemical Basis of Life.* The *solutions* to *all* the text (in-chapter and end-of-chapter) exercises may be found in *Study Guide for Schmid's The Chemical Basis of Life.*

Table 1-4. Equivalent United States and SI Units

	U.S. Unit	Abbreviations for U.S. Unit	SI Unit
Length	1 inch	in.	2.54 cm
	1 yard	yd	0.914 m
	1 mile	mi	1.60 km
Mass	1 pound	lb	454 g
	1 ounce	oz	28.4 g
	1 dram	dr	3.88 g
	1 grain	gr	0.0648 g
Volume	1 pint	pt	0.473 L
	1 quart	qt	0.946 L
	1 gallon	gal	3.78 L

ago is widely used. People who are accustomed to the use of feet, pounds, and gallons as measures of length, mass, and volume, respectively, often find it difficult to use SI units. One way to solve this problem is to convert from one system of units to another. In the next section we will learn how to do this.

1.4 CONVERTING SYSTEMS OF UNITS

To convert from the system of units used in the United States to SI units, we must know the equivalents of the two systems. Table 1-4 lists the most widely used equivalents.

To make use of the information in Table 1-4, we must convert the equivalents into conversion factors. A conversion factor is simply the ratio of one unit divided by another. The relationship between any two units can be expressed as two conversion factors. For example, let us write the two conversion factors expressing the relationship between inches and centimeters. According to Table 1-4, 1 in. = 2.54 cm. We simply divided both sides of the equation by 1 in. to obtain one conversion factor.*

$$\frac{1\ \cancel{\text{in.}}}{1\ \cancel{\text{in.}}} = \frac{2.54\ \text{cm}}{1\ \text{in.}}$$

$$1 = \frac{2.54\ \text{cm}}{1\ \text{in.}}$$

* A review of basic mathematical operations is given in the Appendix.

This conversion factor tells us that there are 2.54 cm per 1 in. The other conversion factor can be obtained by dividing both sides of the equation by 2.54 cm:

$$\frac{1\text{ in.}}{2.54\text{ cm}} = \frac{\cancel{2.54\text{ cm}}}{\cancel{2.54\text{ cm}}}$$

$$\frac{1\text{ in.}}{2.54\text{ cm}} = 1$$

This conversion factor tells us that there is 1 in. per 2.54 cm. Each conversion factor contains the two different units. This is an important part of the ratio, as we will soon learn.

EXERCISE 1-3 Write two conversion factors expressing the relationship between each of the following units:
(a) pound and g (b) gallon and L (c) km and miles (d) ounce and g (e) pint and L (f) yards and m (g) grain and g

To convert from one set of units to another, multiply the unit you wish to convert by the conversion factor that allows you to cancel the unwanted unit. The following two examples illustrate this method.

Example 1-1 Convert 5.00 in. to cm.

Solution:

The first step is to write the quantity that you want to convert, *including its units:*

5.00 in.

Next, multiply this quantity by the conversion factor that allows you to cancel all unwanted units:

$$5.00\,\cancel{\text{in.}} \times \frac{2.54\text{ cm}}{1\,\cancel{\text{in.}}} = 12.7\text{ cm}$$

The inches on top cancel the inches on the bottom, leaving cm as the desired unit.

What happens if we multiply by the other conversion factor?

$$5.00\text{ in.} \times \frac{1\text{ in.}}{2.54\text{ cm}} = 1.97\,\frac{\text{in.}^2}{\text{cm}}$$

The unit in.^{2}/cm makes no sense, so we know that we have made a mistake. Therefore, the first conversion factor is the correct one to use. If you use this method all the time, you will not make a mistake.

Example 1-2 Convert 2.20 lb to g.
Solution:
From Table 1-4, 1 lb = 454 g.
The conversion factors are $1 = \frac{454\text{ g}}{1\text{ lb}}$ and $\frac{1\text{ lb}}{454\text{ g}} = 1$.
Using the first one gives the correct conversion:

$$2.20\,\cancel{\text{lb}} \times \frac{454\text{ g}}{1\,\cancel{\text{lb}}} = 999\text{ g}$$

EXERCISE 1-4 Convert each of the following quantities into its equivalent in the specified units:
(a) 1.00 m^3 to gal (b) 75.0 cm to in. (c) 25.0 oz to g
(d) 325 yd to m (e) 5.30 qt to L (f) 25.0 mi to km
(g) 5.00 kg to lb (h) 16.0 pt to L

Another major difference between measurements in the United States and the rest of the world is the scale used to measure temperature, as we will learn in the next section.

1.5 TEMPERATURE

We determine how hot or cold an object is by measuring its temperature. We measure temperature by means of thermometers. The principle of a thermometer is based on the fact that most substances expand when heated and contract when cooled.

The most common type of thermometer is the mercury thermometer, illustrated in Figure 1-1. It consists of a tube of uniform and very small bore (a capillary tube) that is sealed at its upper end and is enlarged into a spherical or cylindrical bulb at its lower end. This bulb is filled with mercury. When it is heated, the mercury expands and rises in the tube. Because of the very small bore of the tube, even small increases in the volume of the mercury will cause it to rise appreciably in the tube. Thermometers using other liquids such as alcohol can also be constructed.

To express the temperature of an object, the thermometer must have some kind of scale. Three scales of temperature are in use today. In the United States, temperatures are measured in degrees Fahrenheit (°F). In

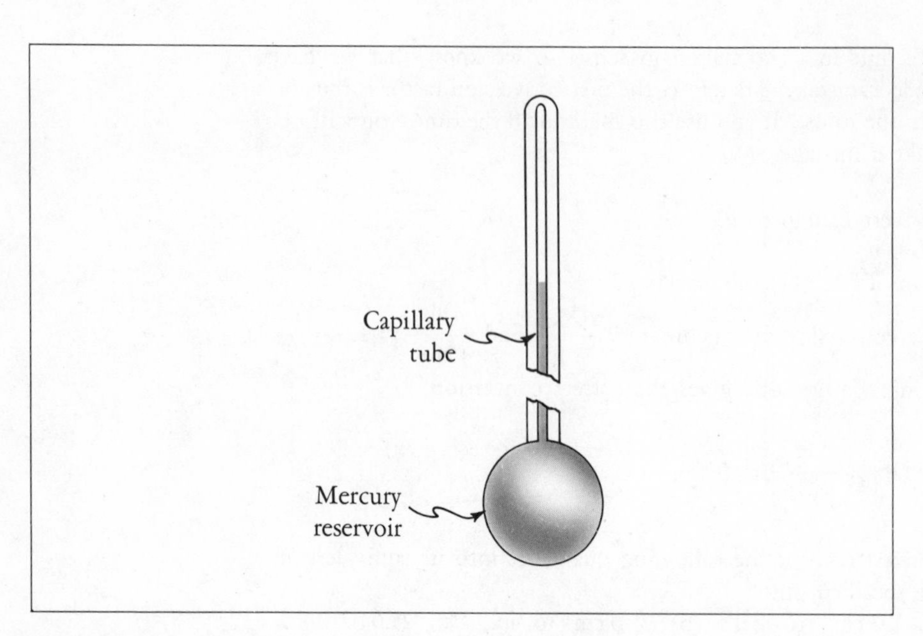

Fig. 1-1. A mercury thermometer.

science, however, the international temperature scales are the Celsius scale and the Kelvin scale.

The Celsius scale (abbreviated °C) uses the freezing point of water to define 0° and the boiling point of water as 100°. The scale between these two points is divided into 100 equal degrees. A comparison between the Celsius and Fahrenheit scales is shown in Figure 1-2.

The SI unit of temperature is the Kelvin (K).* A degree is identical on the Kelvin and Celsius scales. The only difference between them is that zero is set at different points on the two scales. On the Kelvin scale, zero is the coldest temperature possible, called absolute zero. The zero on the Kelvin scale is equal to −273.15° C. To convert from the Celsius to the Kelvin scale, simply add 273.15° to the Celsius temperature:

$$K = {}^\circ C + 273.15^\circ \qquad \text{or} \qquad {}^\circ C = K - 273.15^\circ$$

Notice that there are no negative temperatures on the Kelvin scale.

EXERCISE 1-5 Convert each of the following to Kelvin:
(a) 100° C (b) −20° C (c) 36.9° C (d) 350° C

EXERCISE 1-6 Convert each of the following to degrees Celsius:
(a) 298 K (b) 10 K (c) 150 K (d) 1500 K

* Notice that the abbreviation for Kelvin *does not* have a ° as superscript.

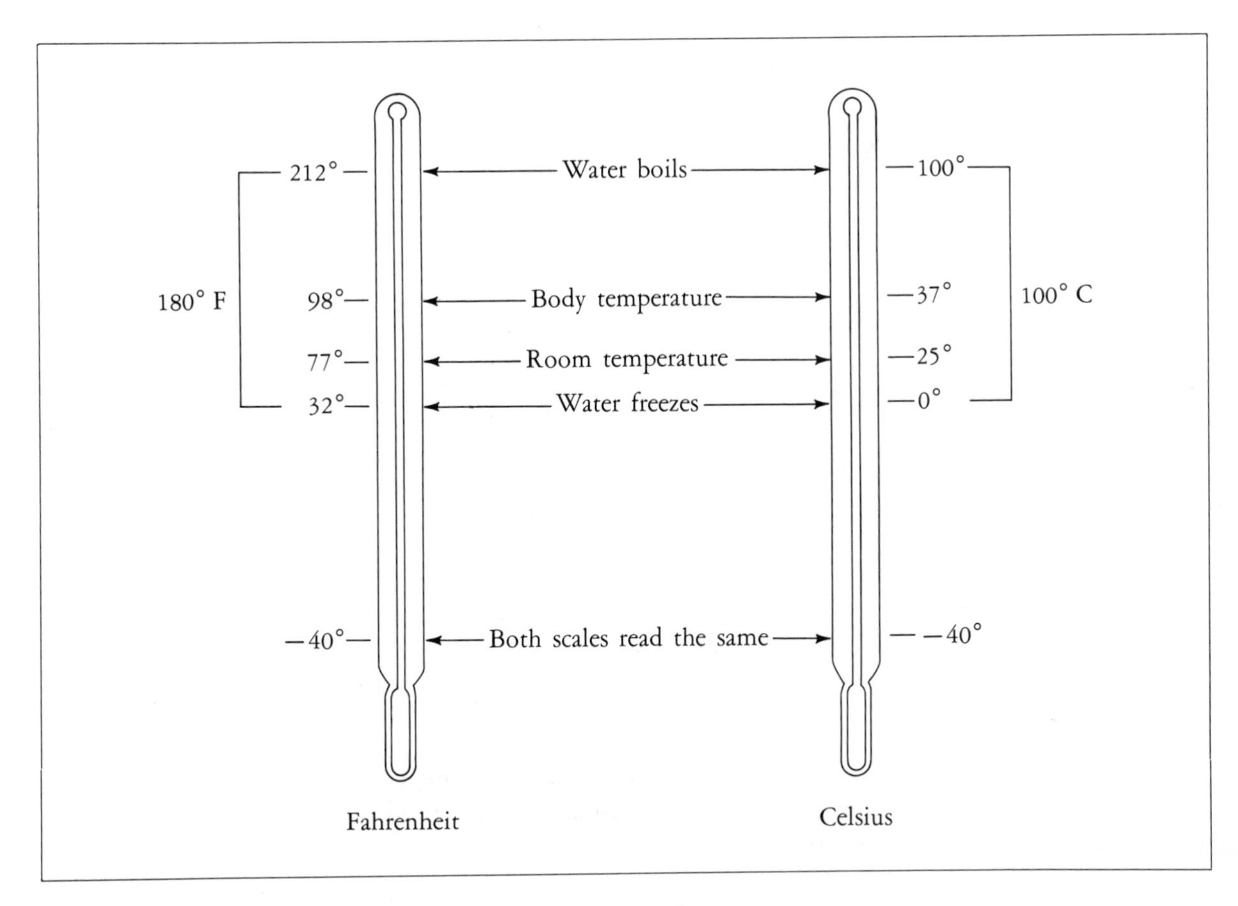

Fig. 1-2. A comparison of the Fahrenheit and Celsius temperature scales.

As shown in Figure 1-2, there are 180 degrees between the freezing point and the boiling point of water on the Fahrenheit scale, whereas there are only 100 degrees between these two points on the Celsius scale. Therefore, there are 9° F for every 5° C. This fact is the origin of the factors 9/5 and 5/9 in the following formulas, which are used to convert degrees Fahrenheit to degrees Celsius and vice versa.

$$°F = \frac{9}{5}\,°C + 32° \qquad °C = \frac{5}{9}\,(°F - 32°)$$

The following examples show how to use these formulas to convert a temperature from one scale to another.

Example 1-3 Convert 40° C to °F.

Solution:
This can be done by substituting into the following formula:

$$°F = \frac{9}{5}\ °C + 32°$$

$$°F = \frac{9}{5}\ (40°) + 32° = 9(8°) + 32° = 104°$$

Example 1-4 Convert 92° F to °C.
Solution:
This can be done by substituting into the following formula.

$$°C = \frac{5}{9}\ (°F - 32°)$$

$$°C = \frac{5}{9}\ (92° - 32°) = \frac{5}{9}\ (60°) = 33°$$

EXERCISE 1-7 Make the following conversions:
(a) 95° F to °C (b) −10° C to °F (c) 150° C to °F
(d) 0° F to °C (e) −50° F to K (f) 75 K to °F

The fundamental units are those that measure mass, volume, length, and temperature. Combinations of these fundamental units are needed, as we will learn in the next section.

1.6 DENSITY AND SPECIFIC GRAVITY

One often hears the statement that lead is heavier than aluminum. Such a statement is not always correct, because it does not take into account the relative sizes of the samples of the two materials. If the two samples are the same size, it is true that the one made of lead weighs more than the one made of aluminum. We express this difference in weight by saying that lead has a higher density than aluminum. *Density* is defined as the ratio of the mass of a material to its volume.

$$\text{Density} = \frac{\text{mass of sample}}{\text{volume of sample}}$$

Densities are usually expressed as g/cm^3 or g/mL. Thus, the densities of lead and aluminum are 11.3 g/cm^3 and 2.7 g/cm^3, respectively. This means

Table 1-5. Densities of Solids

Solid	Density (g/cm^3 at 25° C)
Aluminum	2.70
Copper	8.96
Gold	19.3
Lead	11.3
Mercury	13.6
Silver	10.5
Salt	2.16
Sucrose (table sugar)	1.57

that 1 cm^3 of lead weighs 11.3 g, whereas the same amount of aluminum weighs only 2.7 g. The densities of several solid materials are given in Table 1-5.

The density of any material can be calculated from its mass and volume, as illustrated in the following example.

Example 1-5 What is the density of a solid that weighs 175 g and has a volume of 100 cm^3?

Solution:

According to the definition,

$$\text{Density} = \frac{175\ \text{g}}{100\ \text{cm}^3} = 1.75\ \text{g/cm}^3$$

The mass or volume of a sample can be calculated if we know its density. This is illustrated in the following examples.

Example 1-6 The density of blood plasma is 1.026 g/mL at 25° C. What is the weight of 500 mL of blood plasma?

Solution:

For this problem, we must convert the 500 mL to g. The density gives the relationship between mass and volume. Because 1 mL weighs 1.026 g, it follows that $500\ \cancel{\text{mL}} \times 1.026\ \frac{\text{g}}{\cancel{\text{mL}}} = 513\ \text{g}$.

Example 1-7 A recipe calls for 1000 g of milk. The density of milk is 1.035 g/mL. What is the minimum size container into which you can pour the milk?

Solution:

Here mass must be converted to volume. From the density, we

know that 1 mL of milk weighs 1.035 g. We can express this as an equation, 1 mL = 1.035 g, which can be made into a conversion factor, $\frac{1 \text{ mL}}{1.035 \text{ g}} = 1$.

Therefore, $1000 \cancel{\text{g}} \times \frac{1 \text{ mL}}{1.035 \cancel{\text{g}}} = 966 \text{ mL}$.

Therefore, a 1-L container is the minimum size needed to hold 1000 g of milk.

EXERCISE 1-8 Convert the volume of each of the following materials into its corresponding mass:
(a) 1.00 m^3 of lead (b) 100 mL of salt (c) 2.30 qt of sucrose

EXERCISE 1-9 Convert the mass of each of the following materials into its corresponding volume:
(a) 25.0 g of mercury (b) 1.50 kg of aluminum (c) 10.0 lb of sucrose

Specific gravity is closely related to density. *Specific gravity* is the ratio of the density of any material to the density of water at the same temperature.

$$\text{Specific gravity} = \frac{\text{density of material}}{\text{density of water}}$$

According to this definition, the specific gravity of water is 1.0. Gold has a specific gravity of 19.3. This means that its density is 19.3 times greater than that of water.

Notice that specific gravity is a ratio of two values. The units cancel in this ratio and the value of specific gravity has no units. This is shown in the following example.

Example 1-8 A sample of moon rock weighs 125 g and occupies a volume of 50.0 mL at 25° C. What is the specific gravity of this sample?
Solution:
We must determine the density first.

$$\text{Density of moon rock} = \frac{125 \text{ g}}{50.0 \text{ mL}} = 2.50 \text{ g/mL}$$

The density of water at 25° C is 1.00 g/mL. Therefore,

$$\text{Specific gravity of moon rock} = \frac{2.50 \text{ g/}\cancel{\text{mL}}}{1.00 \text{ g/}\cancel{\text{mL}}} = 2.50$$

Notice that the specific gravity of a material is numerically the same as its density at 25° C because the density of water is 1.00 g/mL at 25° C.

EXERCISE 1-10 Calculate the specific gravity of each of the following: (a) 100 mL of ethyl alcohol weighing 79.0 g at 25° C (b) 1000 mL of seawater weighing 1.025 kg at 25° C (c) silver at 25° C (d) blood plasma at 25° C (see example 1-6)

The temperature has been given in all calculations because density and specific gravity vary with temperature. The reason for this is that the volume of a sample changes with a change in temperature. As a result, the density and specific gravity also change. Such changes are particularly important for liquids and gases. Therefore, comparisons of densities and specific gravities should be made at the same temperature.

Specific gravities of liquids are easily measured by an instrument called a *hydrometer*. The hydrometer is placed in the liquid whose specific gravity is to be measured. The density of the liquid determines how far the hydrometer sinks into the liquid. A scale, placed on the stem of the hydrometer, is calibrated in such a way that the specific gravity can be read directly at the intersection of the scale and the surface of the liquid, as shown in Figure 1-3.

The range of specific gravities that can be measured by any one hydrometer is quite limited. Therefore, different hydrometers are used to measure the

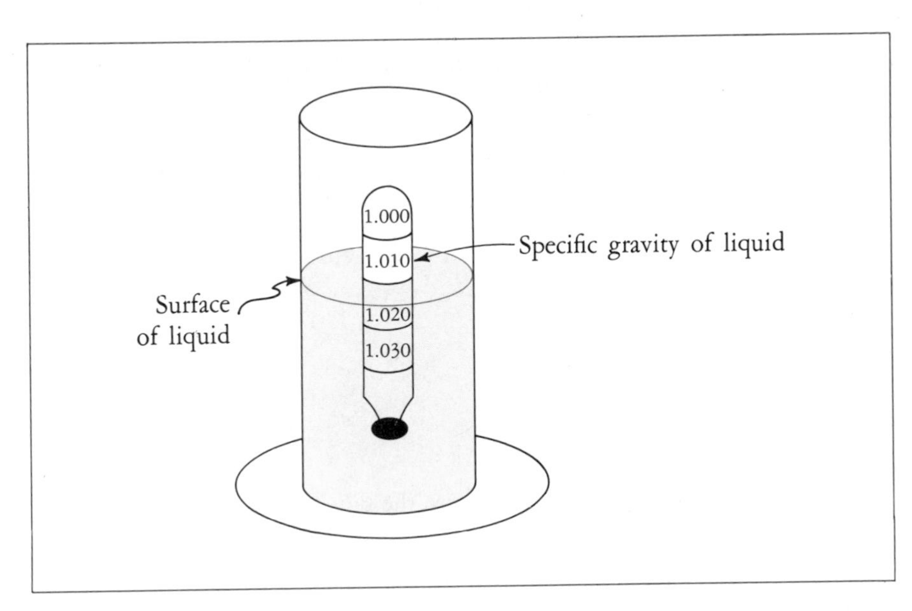

Fig. 1-3. A hydrometer floating in a liquid whose specific gravity is to be determined. The specific gravity of the liquid is read directly from the scale at the intersection of the scale and the surface of the liquid.

specific gravities of battery acid, antifreeze, and urine. In each case, the specific gravity tells us something about the composition of the liquid. For example, a sample of urine from a healthy person has a specific gravity between 1.018 and 1.025 at 25° C. The urine from persons suffering from diabetes contains large quantities of sugar. As a result, the specific gravity of their urine is higher than normal. Thus, the result of taking the specific gravity of a sample of urine, a simple and easy test, serves as a warning of a medical problem.

1.7 SUMMARY

The idea that humans are one of many forms of life that obey the laws of chemistry has developed slowly over the past few centuries. This conclusion was reached as a result of the work of many scientists in many fields. Scientists reach such conclusions because new facts are continually being obtained. These facts are then correlated in the form of a hypothesis. A hypothesis usually suggests more experiments, which provide more facts. These new facts either strengthen the hypothesis or discredit it. If repeated experiments confirm the hypothesis, it becomes a law. It is usually possible to propose a theory after various laws have been formulated.

Scientists must make reliable and reproducible measurements in order to obtain facts. An international system of units, called SI, has been adopted for this purpose. In SI, the meter is the basic unit of length, the kilogram is the basic unit of mass, and the cubic meter is the basic unit of volume. Other units such as density and specific gravity are combinations of these three basic units.

Various kinds of measurements are made on matter. The definition and classification of the various forms of matter are the topics of the next chapter.

REVIEW OF TERMS AND CONCEPTS

Terms

CUBIC METER (m^3) The basic unit of volume in SI.

DEGREE CELSIUS (°C) A measure of temperature that equals 1/100 the difference between the normal freezing point and the boiling point of water: 1° C = 9/5° F.

DEGREE FAHRENHEIT (°F) A measure of temperature that equals 1/180 the difference between the normal freezing point and the boiling point of water: 1° F = $5/9^\circ$ C.

DENSITY The ratio of the mass of an object to its volume.

FACTS All information related to a problem that can be repeatedly verified.

HYDROMETER An instrument used to measure specific gravity.

HYPOTHESIS An attempt to correlate observed facts.

KELVIN A measure of temperature that uses absolute zero as its zero.

KILOGRAM (kg) The basic unit of mass in SI.

LAW A concise statement that summarizes a number of facts.

METER (m) The basic unit of length in SI.

SCIENTIFIC METHOD A way of solving a problem by a series of logical steps.

SI The international system of units.

SPECIFIC GRAVITY The ratio of the density of any object to the density of water at the same temperature.

THEORY A proposed explanation of why or how something happens the way it does.

Concepts

1. All forms of life obey the laws of chemistry.
2. The scientific method is a logical, methodical, and idealized way of approaching a scientific problem.
3. The international system of units (SI) has been adopted to provide a standard system of measurement that can be repeated anywhere in the world.
4. Units are handled like numbers during conversions from one system of units to another. By using conversion factors, units can be canceled so that only the desired unit remains.

EXERCISES

1-11 Convert each of the following quantities into its equivalent in the units specified:
(a) 1 g to kg (b) 25 μL to L (c) 5 mm to km (d) 6 in. to cm (e) 1.0 L to pt (f) 16 km to mi (g) 5.0 lb to kg (h) 22 gal to L (i) 100 m to yd (j) 5.0 ft to cm (k) 25 g to oz (l) 5.0 L to qt

1-12 Make the following conversions:
(a) body temperature (98° F) to °C (b) −10° F to °C (c) 75 K to °C (d) 10 K to °F (e) −130° F to K

1-13 Which quantity is the greater in each of the following pairs? (a) 1 yd or 1 m (b) 1 L or 1 qt (c) 1 cm or 1 in. (d) 40° C or 80° F (e) 300 K or 40° C (f) 10 dL or 100 mL (g) 1 qt or 100 mL (h) 100 cm or 1000 mm

1-14 The volume of a 5.00-in. cube of gold is 125 in.^{3} Calculate its weight in g, kg, and lb.

1-15 You are given a sample of shiny metal that is supposed to be pure silver. You determine that the sample weighs 112 g and has a volume of 15.0 mL. Is it pure silver?

1-16 How large a volume will 1.00 lb of each of the following substances occupy?
(a) blood (b) urine (c) lead (d) aluminum

1-17 A person takes two tablets of aspirin three times per day. Each tablet contains 5 grains of aspirin. How many g of aspirin has the person taken each day?

MATTER AND ENERGY 2

Chapter 1 was a discussion of the scientific method and SI, the international system of units of measurement. This system is used to measure all forms of *matter*. But what is matter? Scientists define matter as anything that has mass and occupies space. This definition is so broad that it includes almost everything in the world. Can the various forms of matter be defined and classified more precisely? The answer is yes. How this is done is one of the two major topics of this chapter. The other topic is *energy*. There are many forms of energy. How they are related and their importance in bringing about changes in matter will be discussed in this chapter. Let us start by examining the properties of matter.

2.1 PROPERTIES OF MATTER

A particular kind of matter, such as water, gold, silver, salt, or sugar, is called a substance. Every substance has a characteristic set of properties that makes it different from all other substances. These properties give it a unique identity. For example, sugar is an odorless, white solid that has a sweet taste. When heated, sugar melts and turns brown. Salt is also an odorless, white solid but it has a unique taste and finally melts when heated above red heat. These properties of sugar and salt serve to distinguish them. Each has a characteristic set of properties that identifies it. The properties of a substance that are characteristic of that substance are called its *intrinsic properties*. These properties do not depend on the size or shape of the substance.

*EXERCISE 2-1 Give two intrinsic properties of each of the following substances:
(a) water (b) gold (c) silver (d) oxygen (e) glass

Some intrinsic properties are more useful than others in describing matter. It is difficult to give numerical values to such intrinsic properties as odor and taste. Properties such as melting points, boiling points, and densities are more useful because they are easily measured and expressed in numbers. For example, water can be identified as a clear, colorless liquid. But these properties do not distinguish water from alcohol, another clear, colorless liquid. More specific properties of water are its boiling point (100° C), melting point (0° C), and density (1.0 g/mL at 4° C). No other substance has exactly this set of properties. This set is unique to water and nothing else. These intrinsic properties are called *physical properties*.

Many substances react, either alone or with other matter, to form new materials. These reactions are called the *chemical properties* of a substance. For example, the rusting of iron and the burning of wood are chemical properties of these substances. The major emphasis in this book is to present and discuss the chemical properties of substances that are of importance to living systems.

A substance is identified by its properties. Once a substance is identified, it can be classified. The classifications of matter are discussed in the next section.

2.2 CLASSIFYING MATTER

Matter is classified as either a mixture or a pure substance. A *mixture* contains a number of different substances more or less mixed together. A mixture has no unique set of properties. Rather, it has the properties of all the substances that are a part of it. For example, air is a gaseous mixture made up of nitrogen, oxygen, argon, water vapor, and carbon dioxide. Each of these substances retains its own physical properties in the mixture.

A mixture is either heterogeneous or homogeneous. The parts of a *heterogeneous mixture* are visibly different. For example, a mixture of sand and gravel is a heterogeneous mixture because the individual pieces of sand and gravel are clearly visible. The parts of a *homogeneous mixture* cannot be detected even with a microscope. For example, a salt and water solution is a homogeneous mixture because the presence of the salt cannot be seen.

* The answers for the exercises in this chapter begin on page 858.

EXERCISE 2-2 Classify each of the following mixtures as either homogeneous or heterogeneous:
(a) alcohol and water (b) a piece of granite (c) brass (d) steel

A mixture can be separated into its parts by using the differences in the physical properties of the parts. For example, when the temperature of air is lowered, water vapor separates as liquid (rain) or solid water (snow). When the air is cooled still further, solid carbon dioxide (dry ice) forms. Finally, at even lower temperatures, the rest of the air becomes liquid. Thus, the differences in the physical properties of the substances in the mixture allow us to separate them.

Pure substances are classified as either elements or compounds. These pure substances are defined in the next section in terms of the composition of matter.

2.3 COMPOSITION OF MATTER

Humans have wondered about the composition of matter since history began. What makes up matter? The ancient Greeks gave an answer to this question that is still accepted today. They reached this answer by reasoning as follows. If a large quantity of matter, such as a bar of silver, is cut in half, both halves are still silver. The two halves can be cut into still smaller pieces. These pieces can be cut still smaller until minute particles are formed that cannot be cut any further without losing the properties of silver. These tiny particles are called *atoms*.

Atoms are the fundamental units of elements. An *element* is a substance that contains only one kind of atom. There are only 106 elements, so there are only 106 different kinds of atoms. Many elements are quite familiar to us. The precious metals gold, silver, and platinum and the less precious but still useful metals copper, aluminum, and iron are elements. Mercury, a liquid, and helium, a gas, are other examples of elements.

Atoms combine to form *molecules*. They combine according to certain well-defined principles, which will be discussed in Chapter 5. A substance that contains only one kind of molecule is called a *pure compound*. A molecule is the smallest particle that has the properties of a pure compound.

The relationship between atoms and molecules is similar to that between the letters of the alphabet and words. There are 26 letters in the English alphabet and these can be combined according to definite rules to form millions of words. Similarly, the 106 atoms combine according to definite rules to form millions of molecules. These relatively few atoms are the fundamental units of all matter.

Matter, made up of atoms or molecules, exists in three different states: gaseous, liquid, and solid. These states have been referred to briefly in the previous sections. Let us consider them in more detail.

2.4 STATES OF MATTER

All the elements and most simple compounds can exist as a gas, a liquid, or a solid. These are the three states of matter.

A gas has no shape or volume of its own. It takes the shape and volume of its container. A gas can easily be compressed or expanded as its container changes in size.

A liquid also has no shape of its own. However, it does have a specific volume, which conforms to the shape of its container. A liquid can be compressed only under high pressure.

A solid has a fixed volume and a fixed shape even when it is not in a container. A solid is more difficult to compress than a liquid. Very high pressure must be used to decrease its volume.

Water is a familiar example of a compound that exists in the three states. Liquid water is the most common form. Yet when the temperature is lowered, water freezes to form ice, the solid form of water. When water evaporates or is heated to its boiling point, it exists as water vapor, the gaseous form of water.

Not all substances can exist in three states. Many large and complex molecules exist only in the solid or liquid state because they are unstable when heated to their melting or boiling point. For example, sugar decomposes instead of melting when heated.

EXERCISE 2-3 Explain what change of state occurs in each of the following: (a) boiling water (b) heating solder (c) warming dry ice

Matter, such as water, can be changed from one state to another. Changes of state are examples of one of two kinds of change that occur in nature.

2.5 PHYSICAL AND CHEMICAL CHANGES

Everything in the world undergoes change. Animals are born, mature, and eventually die; iron rusts; mountains erode; water in lakes and oceans evaporates. These changes can be classified as either physical or chemical.

Chemical changes result in the disappearance of one or more substances and the formation of new ones. An example of a chemical change is the burning of the natural gas propane. Propane and some oxygen from the air disappear to form carbon dioxide and water vapor. Both these substances have different physical properties that distinguish them from propane and oxygen. Chemical changes are usually called chemical reactions.

No new substances are formed in a *physical change*. For example, the freezing of water is a common example of a physical change. The substance is still water before and after the change. No new substance is formed. Notice, however, that a physical change often results in the change of some intrinsic properties. For example, the density of ice is 0.917 g/mL, which is different from that of liquid water.

It is easy to define chemical and physical changes. It is more difficult to identify these changes in nature. Some are obvious. Cutting matter is a physical change. The rusting of iron is a chemical reaction. Other changes are not so obvious. Is taste a physical or a chemical change? The current view is that taste is a chemical reaction between the substance and the taste buds. The only way to establish the kind of change is to analyze the substances chemically before and after the change. This is not always easy to do. Therefore, it is not possible to classify a change simply by looking at it.

EXERCISE 2-4 Classify each of the following as either a chemical reaction or a physical change:
(a) boiling water (b) frying an egg (c) sanding a piece of wood (d) burning wood

During chemical reactions and physical changes, energy is either released or absorbed. We get the energy we need for our world from chemical reactions. Our bodies get the energy they need from the food we eat. But what is energy? We learn the answer to this question in the next section.

2.6 ENERGY AND ITS TRANSFER

The word energy is used to denote activity. Thus, a person who is active in business, social, or political organizations is often described as having lots of energy. We also speak of the energy that can be obtained from petroleum. But we cannot see, taste, or smell energy. Unlike matter, energy does not occupy space; yet we can feel its effects. Energy is not a thing but is more like a characteristic of a substance.

All matter has energy. This energy has many forms. Potential, kinetic, chemical, atomic, and radiant energy are common types. An object has *potential energy* because of its position. For example, water at the top of a waterfall has potential energy as a result of its position above the surface of the earth. An object has *kinetic energy* as a result of its motion. A moving automobile, a thrown ball, and an airplane in flight all have kinetic energy. *Chemical energy* is the energy stored in the molecule as a result of the kinds and positions of its atoms. *Atomic* or *nuclear energy* is associated with the structure of atoms. *Radiant energy* is the energy of light.

The energy of any object changes whenever it undergoes a chemical reaction or physical change. This energy change occurs by transferring energy. For example, part of the energy of wood is transferred to the surrounding air when it is burned. The energy is transferred by several methods. Four of the most common are work, heat, sound, and light. These are all visible signs of a transfer of energy.

Historically, one of the first persons to realize that energy can be transferred was Benjamin Thompson, Count Rumford, who in 1798 was in charge of the boring of cannons for the Holy Roman Empire. A large amount of heat is liberated during the boring of a cannon. Where does all this heat come from? At that time, heat was believed to be a substance that was contained in all matter. Cutting matter was supposed to release it. According to this theory, the amount of heat released during the boring of a cannon should depend on the size of the cuttings. Thompson showed that there was no relationship between the amount of heat released and the size of the cuttings. Furthermore, the amount of heat that could be generated by a blunt borer seemed to be inexhaustible. On the basis of his experiments, Thompson concluded that heat was not a substance. Rather, the energy needed to bore the cannon was being transferred or converted into heat by friction between the borer and the cannon.

It is important to realize that, although a substance may contain energy, it never contains heat or work. Heat and work are evident only when the energy of the material changes and a transfer of energy occurs. The terms heat content and work content of substances are often used. These terms mean that the change in energy of a substance will occur in the form of heat or work.

Many examples of the transfer of energy are available to us every day. The chemical energy of the fuel burned in an engine does work to make the engine run. Part of the energy is also transferred as heat and sound. Everyone has seen fireworks. The noise, light, and heat given off by an explosion represent another example of the transfer of energy. Here the energy of the chemicals that make up the fireworks is transferred to the surroundings as sound, light, and heat.

Chemical reactions involved in our body also demonstrate the transfer of energy. The food we eat provides the energy that allows our bodies to carry out the normal work of contraction and motion. Food also provides us with heat to maintain a constant body temperature. Food undergoes a series of chemical reactions in our bodies that result in the transfer of part of the chemical energy to muscles to do work and to the surroundings as heat.

EXERCISE 2-5 What kind of energy is contained in each of the following? (a) a thrown stone (b) fuel oil (c) a stone resting at the edge of a cliff

EXERCISE 2-6 What happens to the energy during the changes in each of the following substances?
(a) coal is burned (b) a book falls off a table (c) a nail is hit by a hammer (d) you jog

From these simple and isolated examples, we can move to the utilization and transfer of energy in living systems.

2.7 ENERGY AND LIFE

The sun is the source of all our energy. Sunlight transfers this energy to the earth, where it is stored for future use as chemical energy in the molecules of plants. This transfer occurs by a process called *photosynthesis*. This process is a series of complex chemical reactions that convert sunlight, carbon dioxide, and water to complex molecules. Some of the energy of carbon dioxide, water, and sunlight is transferred into the chemical energy of the complex molecules at the same time. This chemical energy can be used in a number of ways. Chemical reactions in the plant transfer a part of this energy into work and heat to allow the plant to live and grow. The plant can be used as food for humans and animals. The molecules of the plant, when digested, undergo more chemical reactions, which transfer energy wherever needed by animals and humans. In this way, the energy of the sun is transferred to humans and animals, who cannot get their energy directly by photosynthesis. Energy transfer, not only among the sun, plants, and animals, but also among the various parts of living organisms, is absolutely essential to life. The compounds involved in the transfer of energy in living systems are discussed in Chapter 26.

The transfer of energy does not stop with humans. After death, plants, animals, and humans decay. Any remaining energy is transferred into other forms outside the animal world. Although this energy is lost in a biological sense, it is never really lost. For example, the energy available to us in coal is the remains of the energy of plants that died long ago and were transformed into coal by a geological process.

The important point to remember is that the energy obtained from the sun is never lost. The energy simply appears in another form. This form may be useless to us. It may not heat our homes and it may not do useful work for us, but it is still there somewhere in the universe. Such a statement implies a conservation of energy. How can we be so sure that the energy of the universe is conserved? Let us examine this question next.

2.8 CONSERVATION OF MASS-ENERGY

The idea that energy is neither created nor destroyed had its beginnings in the seventeenth century. Careful experimental observations during the eighteenth and nineteenth centuries led scientists to the conclusion that the total energy of the universe is constant. This is a statement of the law of conservation of energy.

This law was carried a step further in 1905, when Albert Einstein proposed that mass and energy are related according to the equation $E = mc^2$. This equation tells us that the loss of a quantity of mass (m) times the speed of light squared (c^2) produces an amount of energy equivalent to E. Because mass and energy can be transformed into each other and therefore are the same, we must conclude that there is a law of conservation of mass-energy:

The total mass-energy in the universe is constant.

We can illustrate this law by tracing what happens to the mass and energy contained in a mixture of 100 g of coal and enough oxygen to burn it completely. This mixture contains a certain quantity of energy. As we burn the mixture, energy is transferred; both heat and light are given off. The rest of the original energy is transferred to the chemical products. The sum of all the energy contained in the heat, light, and chemical products must be equal to the amount of energy originally contained in the mixture of coal and oxygen.

This mixture of coal and oxygen also contains mass: 100 g of coal and 266 g of oxygen. The mass of the products formed by burning the coal weighs 366 g. This is exactly the mass of the original coal and oxygen. No mass is lost. Most of the chemical reactions in the world occur without loss of mass. The major exceptions are nuclear reactions, which are discussed in Chapter 4.

The units used to measure mass were given in Chapter 1. The units that are used to measure changes in energy are given in the next section.

2.9 UNITS OF ENERGY

Heat is one of the most easily measured forms of energy. The most common unit of measurement of heat is the *calorie*. Its symbol is cal. A calorie is defined as the amount of heat needed to raise the temperature of 1 g of water from 14.5° to 15.5° C. Although this is the exact definition of a calorie, approximately 1 calorie is needed to raise the temperature of 1 g of water by 1° C at any temperature between 0 and 70° C. Most chemical reactions release several thousand calories; as a result, quantities of heat are usually reported as kilocalories (1 kcal = 1000 cal).

The SI unit of energy is the *joule*. Its symbol is J. There are exactly 4.184 J in 1 cal. This unit is not as widely used as the calorie. However, as use of SI units becomes more widespread, the joule will eventually replace the calorie as the basic unit of heat.

Be careful in using the calorie as a unit, because there are actually two calories in use: the one defined above, and the Calorie (with a capital C), used in nutrition. One Calorie is equal to 1 kcal. The calorie (1000 cal = 1 kcal) is the basic unit of heat used in this book.

EXERCISE 2-7 How much heat, in cal and kcal, is needed to raise the temperature of 1000 g of water from 1° to 25° C?

The measurement of the heat absorbed or lost in chemical reactions or physical changes that are important in the human body lead to a better understanding of how the body carries out many of its functions. For example, the body temperature of a healthy human is constant. To maintain this constant temperature, the body must transfer heat to its surroundings. Some of the ways the body does this are discussed in the next section.

2.10 THE BODY AND HEAT TRANSFER

The human body at rest gets its energy by means of a series of complex chemical reactions called *metabolism*. The body gets its heat from a part of this energy. The temperature of the body must stay fairly constant to function properly. Either too much or too little heat can be fatal. The body gets rid of excess heat by transferring it to the surroundings in a number of ways.

Evaporation of water from the skin is one way the body loses heat. This process takes advantage of the fact that heat is needed to transform liquid water into its vapor. The heat needed to carry out this physical change is called the heat of vaporization. It is defined as the number of calories needed to change 1 g of substance from the liquid to the vapor state at its normal boiling point. The heats of vaporization of a number of compounds used in medicine are given in Table 2-1. Notice that the heat of vaporization of water is higher than that of the other compounds. This means that it takes more heat to vaporize 1 g of water than 1 g of most other compounds. Evaporation of perspiration is an efficient method of transferring body heat and is particularly important during physical exercise.

Ethyl and isopropyl alcohols have fairly high heats of vaporization. This is the reason these compounds are placed on the skin to reduce fever. The excess heat caused by the fever is transferred to the alcohol, which vaporizes. Similarly, the skin feels cool when either diethyl ether or ethyl chloride is applied. The vaporization of these compounds removes heat from the skin, making it feel cooler.

EXERCISE 2-8 Calculate the heat, in kcal, needed to vaporize 100 g of each of the following compounds at their normal boiling points:
(a) water (b) isopropyl alcohol (c) diethyl ether
(d) chloroform

The body also transfers heat to its surroundings by *radiation*. The body is like a hot water radiator used to heat a room. Both give off heat. The heat radiated from the body accounts for much of its heat loss, particularly

Table 2-1. Heats of Vaporization of Various Compounds

Compound	Heat of Vaporization (cal/g)
Water	540
Rubbing alcohol (isopropyl alcohol)	159
Diethyl ether	84
Ethyl chloride	93
Chloroform	59
Ethyl alcohol	204

Table 2-2. Specific Heats of Common Substances

Substance	Specific Heat (cal/g × °C)
Water	1.00
Ethyl alcohol	0.581
Copper	0.0949
Silver	0.0557
Aluminum	0.217
Gold	0.0298

during cold weather. Much heat is lost from an uncovered head. This loss can be greatly reduced simply by wearing a hat.

Heat is transferred from one substance to another substance that is colder. This is called heat *conduction*. This is another way that heat can be transferred to or from the body. For example, when an ice pack is placed on the skin, heat is transferred from the skin to the ice and the skin becomes cool.

The thin layer of air next to the skin is warmed by the transfer of heat from the body. If a breeze replaces this layer of air with colder air, the body transfers heat to warm the colder air. As a result, the body loses heat. We wear clothes to trap a layer of warmed air next to our skin to minimize this kind of heat loss.

Water is the most abundant compound in the body. Body water can act internally to control body temperature because it can absorb a fairly large amount of heat with relatively little change in temperature. The amount of heat needed to raise the temperature of 1 g of a substance by 1° Celsius is called its *specific heat*. The specific heats of several common substances are given in Table 2-2. Water has a higher specific heat than most compounds. For example, it takes 10 times as much heat to raise the temperature of a quantity of water by 1° than it does to increase the temperature of the same amount of copper by 1°. This means that water in the body can absorb a fairly large amount of heat without changing temperature. In this way, water acts as an internal temperature regulator.

EXERCISE 2-9 Calculate the amount of heat, in kcal, needed to raise the temperature of 100 g of each of the following substances from 37 to 47° C:

(a) water (b) copper (c) ethyl alcohol (d) aluminum

2.11 SUMMARY

Matter is the term given to anything that has mass and occupies space. A particular kind of matter, called a substance, has a characteristic set of chemical and physical properties that distinguish it from any other substance. A substance is either pure or a mixture. A mixture has no unique properties, but has the properties of all the substances that are a part of it. Pure substances are either elements or compounds. There are 106 elements, each made up of a different kind of atom. An atom is the smallest unit with the characteristic properties of an element. Atoms combine to form molecules. A molecule is the smallest unit with the characteristic properties of a compound.

Gas, solid, and liquid are the three states of matter. A change of state of a substance is one example of a physical change. In such a change, no new substance is formed. A chemical change results in the formation of one or more new substances.

All matter has energy, which exists in various forms, such as kinetic, potential, chemical, radiant, and atomic energy. The energy of a substance changes whenever it undergoes a chemical or physical change. The substance loses energy by transferring it to the surroundings. Transferring energy from the surroundings to the substance results in a gain of energy by the substance. Work, heat, sound, and light are four visible signs of energy transfer.

Such transfers of energy are absolutely essential to life. The energy of the sun is transferred to various molecules in plants by a process called photosynthesis. The energy stored in these molecules is used by the plants for their own purposes. Many plants are food for humans and animals. After digestion, the molecules of food undergo chemical reactions that transfer energy to wherever it is needed. In this way, the energy of the sun is transferred to compounds that provide the energy for humans.

Everything in the world is either matter or energy. Furthermore, the mass of a substance and its energy are related by the equation $E = mc^2$. This means that mass and energy can be transformed into one another. One of the fundamental conclusions reached from the study of matter and energy and their interrelationship is that the total amount of mass and energy in the universe is constant. This is the law of conservation of mass-energy.

The human body gets its energy from a series of complex chemical reactions called metabolism. A part of this energy is transferred as heat. A constant body temperature is maintained by transferring excess heat to the surroundings through evaporation of perspiration, radiation, and conduction. In addition, water in the human body acts as an internal thermostat.

The fundamental building block of matter is the atom. In the next chapter we will learn about the structure of the atoms most often encountered in living systems.

REVIEW OF TERMS AND CONCEPTS

Terms

ATOM The smallest particle that has all the properties of an element.

ATOMIC ENERGY Energy released by a change in the structure of an atom.

CALORIE The amount of heat needed to raise the temperature of 1 g of water from 14.5° to 15.5° C.

CHEMICAL CHANGE A change that results in the disappearance of one or more substances and the formation of new ones.

CHEMICAL ENERGY The energy stored in a molecule as a result of the kinds and positions of its atoms.

CHEMICAL PROPERTIES The characteristic chemical reactions of a substance.

COMPOUND A substance composed of molecules.

ELEMENT A substance that contains only one kind of atom.

ENERGY The capacity for doing work.

GAS A substance that has no shape or volume, but takes the shape and volume of its container.

HEAT OF VAPORIZATION The number of calories needed to change 1 g of a substance from liquid to vapor at its normal boiling point.

HETEROGENEOUS MIXTURE A mixture whose individual parts are clearly visible.

HOMOGENEOUS MIXTURE A mixture whose individual parts cannot be detected, even with a microscope.

JOULE The SI unit of energy which is equal to 4.184 calories.

KINETIC ENERGY The energy of an object as a result of its motion.

LIQUID A substance that has a specific volume but conforms to the shape of its container.

METABOLISM The complex chemical reactions that furnish energy to a living organism.

MIXTURE A number of substances more or less mixed together.

MOLECULE The smallest particle that has the properties of a compound.

PHOTOSYNTHESIS A series of chemical reactions that transfer the energy of sunlight, carbon dioxide, and water into the chemical energy of molecules of a plant.

PHYSICAL CHANGE A change in which no new substance is formed.

PHYSICAL PROPERTIES Properties of a substance that are easily measured and expressed as numbers.

RADIANT ENERGY The energy of light.

SOLID A substance that has a fixed shape and volume even when it is not in a container.

SPECIFIC HEAT The amount of heat needed to raise the temperature of 1 g of a substance by 1° C.

SUBSTANCE Any particular kind of matter.

Concepts

1. All matter is composed of atoms. An element contains only one kind of atom. Atoms combine to form molecules. A molecule is the fundamental particle of a compound.
2. Matter exists in three states: gas, liquid, and solid.
3. Matter undergoes two types of change: physical and chemical.
4. The energy of matter changes whenever matter undergoes a chemical or physical change.
5. The change in energy of matter is visible because the energy is transferred as heat, work, light, or sound.
6. The transfer of the energy of the sun to compounds in plants by photosynthesis furnishes the energy for life on earth.
7. The total of mass and energy in the universe is constant.

EXERCISES 2-10 Classify each of the following as either a physical or a chemical property:
(a) The melting point of iron is 1535° C. (b) Brass tarnishes. (c) Hydrogen gas burns. (d) The specific heat of aluminum is 0.215 cal/g.

2-11 What difference in properties of each of the following pairs of substances could you use to distinguish easily between them? (a) water and milk (b) salt and sugar (c) a piece of aluminum and the same size piece of iron (d) water and vinegar

2-12 What is the difference between an atom and a molecule?

2-13 Classify each of the following changes as either chemical or physical:
(a) Rain falls from a cloud. (b) A nail is driven into a board.

(c) A tree grows. (d) Grass is cut. (e) Food decays. (f) Food is eaten.

2-14 Natural gas is often used as a fuel in heating homes. Explain what happens to the energy of the natural gas as it is burned.

2-15 Why is it incorrect to say that an object contains heat?

2-16 Why is a hot water bottle placed in a cold bed?

2-17 Explain the following statement: Matter and energy are only different forms of a single reality.

2-18 Give an example of a process in which a part of the chemical energy of an object is transferred mostly into
(a) heat (b) light (c) kinetic energy.

2-19 The burning of 1 g of coal releases 7.84 kcal of heat. What quantity of water (in g, kg, and mL) can you heat from 25° to 100° C if all the heat released in burning 1 kg of coal is completely transferred to the water?

2-20 Which of the compounds ethyl alcohol, chloroform, and diethyl ether, when applied to the skin, would be the most effective in reducing fever?

ATOMS 3

The atom was identified in Chapter 2 as the smallest representative sample of an element. Only 106 different kinds of atoms are known to exist. Our first goal in this chapter is to learn what makes each of these 106 kinds of atoms different. Our second goal is to learn how the structure of the atoms is responsible for the differences and similarities in their chemical and physical properties.

Before we can reach these goals, we must learn the names of the various elements. It turns out that chemists have given the elements not only specific names, but also characteristic symbols.

3.1 ELEMENTS: THEIR NAMES AND SYMBOLS

Each element has both a symbol and a name. The symbol is made up of one or two letters taken from the name of the element. The first letter of the symbol is often the first letter of the name of the element. For example, the letter C is the symbol for the element carbon. The letters N and F are the symbols for the elements nitrogen and fluorine, respectively. Other elements have more complicated symbols that are derived from their Latin names. The two letters Na represent the element sodium, whose Latin name is natrium. The first letter of the symbol is always a capital. If there is a second letter, it is always lower case. Table 3-1 lists the symbols for several of the most common elements. These symbols should be learned. They are the shorthand used by chemists whenever they want to indicate the presence of

Table 3-1. Names and Symbols for a Number of Elements

Element	Symbol	Element	Symbol	Element	Symbol
Aluminum	Al	Gold	Au	Oxygen	O
Argon	Ar	Helium	He	Phosphorus	P
Arsenic	As	Hydrogen	H	Potassium	K
Barium	Ba	Iodine	I	Platinum	Pt
Boron	B	Iron	Fe	Selenium	Se
Bromine	Br	Lead	Pb	Silicon	Si
Calcium	Ca	Lithium	Li	Silver	Ag
Carbon	C	Magnesium	Mg	Sodium	Na
Chlorine	Cl	Manganese	Mn	Sulfur	S
Chromium	Cr	Mercury	Hg	Titanium	Ti
Cobalt	Co	Neon	Ne	Uranium	U
Copper	Cu	Nickel	Ni	Vanadium	V
Fluorine	F	Nitrogen	N	Zinc	Zn

any of the atoms of these elements. These symbols are used throughout this book.

The 106 elements can be listed in a number of ways. One way is to list them alphabetically, as in Table 3-1. Another and more useful way is to arrange the elements in the periodic table.

3.2 PERIODIC TABLE

Chemists of the nineteenth century noticed that many elements have similar chemical properties. Numerous attempts were made to arrange the elements that were known at that time into some systematic fashion to emphasize these similarities. The most successful arrangement was arrived at independently by two chemists, Dmitri Mendeleev and Lothar Meyer. They grouped the elements with similar properties together into families and designed the first periodic table of the elements. The modern periodic table, shown in Figure 3-1, is a direct descendant of the original Meyer-Mendeleev arrangement of the elements. (See the endpapers of this book for the periodic table shown in more detail.)

The periodic table consists of a number of columns called *groups*. The horizontal rows are called *periods*. *The elements in each group have similar chemical and physical properties*. For example, the elements at the far right of the periodic table, called the noble gases, are the least reactive of all the elements. Special names are also given to the elements in other groups. The

Period ↓ \ Group →	IA	IIA	IIIB	IVB	VB	VIB	VIIB	VIIIB			IB	IIB	IIIA	IVA	VA	VIA	VIIA	VIIIA
1	H 1																	He 2
2	Li 3	Be 4											B 5	C 6	N 7	O 8	F 9	Ne 10
3	Na 11	Mg 12											Al 13	Si 14	P 15	S 16	Cl 17	Ar 18
4	K 19	Ca 20	Sc 21	Ti 22	V 23	Cr 24	Mn 25	Fe 26	Co 27	Ni 28	Cu 29	Zn 30	Ga 31	Ge 32	As 33	Se 34	Br 35	Kr 36
5	Rb 37	Sr 38	Y 39	Zr 40	Nb 41	Mo 42	Tc 43	Ru 44	Rh 45	Pd 46	Ag 47	Cd 48	In 49	Sn 50	Sb 51	Te 52	I 53	Xe 54
6	Cs 55	Ba 56	La 57	Hf 72	Ta 73	W 74	Re 75	Os 76	Ir 77	Pt 78	Au 79	Hg 80	Tl 81	Pb 82	Bi 83	Po 84	At 85	Rn 86
7	Fr 87	Ra 88	Ac 89	Rf 104	Ha 105	106												

Nonmetals

Lanthanide series	Ce 58	Pr 59	Nd 60	Pm 61	Sm 62	Eu 63	Gd 64	Tb 65	Dy 66	Ho 67	Er 68	Tm 69	Yb 70	Lu 71
Actinide series	Th 90	Pa 91	U 92	Np 93	Pu 94	Am 95	Cm 96	Bk 97	Cf 98	Es 99	Fm 100	Md 101	No 102	Lr 103

H 1	Symbol Atomic number

Fig. 3-1. The periodic table of the elements.

Table 3-2. The Most Abundant Elements

Element	Weight Percent	Element	Weight Percent
Oxygen	49.5	Calcium	3.4
Silicon	25.7	Sodium	2.6
Aluminum	7.5	Potassium	2.4
Iron	4.7	Magnesium	1.9

elements in group IA are called the alkali metals. Those in group IIA are called the alkaline earth metals. The transition metals are the elements located in groups IIIB to IIB. The halogens are the elements in group VIIA. All the members of a group have similar physical and chemical properties.

The elements are divided into three classes based on their physical properties. Elements that show a metallic luster when polished, are capable of being drawn out into wire, can be hammered into sheets, and are good conductors of heat and electricity are classed as *metals*. Elements that do not have these properties are classed as *nonmetals*. A class between these two is called the *metalloids,* or borderline elements. Elements of the three classes are shown in the periodic table in Figure 3-1. There is a zigzag line on the right side of the periodic table that separates the nonmetals on the right from the metals on the left. Near this line are the metalloids. These elements, such as silicon (Si) and germanium (Ge), have some properties that are similar to those of nonmetals and some that are similar to those of metals.

Only 90 of the 106 elements are found in nature. The others are prepared in the laboratory by instruments and techniques discussed in Chapter 4. A very few elements, such as gold, silver, copper, and platinum, exist naturally in a pure state. The rest are found in nature as parts of compounds; that is, their atoms are combined with other atoms to form compounds. Of the 90 elements found in nature, only eight make up 98 percent of all the compounds on earth. These elements and their percentages are given in Table 3-2. Notice that the elements carbon, hydrogen, and nitrogen do not appear in this table. These elements make up the compounds found in living systems, yet each one accounts for less than 1 percent of all the elements on earth.

Most of the natural elements had been discovered by the end of the nineteenth century. Furthermore, the idea that the elements are made up of atoms was completely accepted. Beginning in 1895, however, a number of subatomic particles were discovered. Their discovery overturned the easily

Table 3-3. Three Important Subatomic Particles

Name	Charge	Mass (amu)	Symbol
Electron	-1	$1/1837$	e^-
Proton	$+1$	1.007	H^+, p^+, p
Neutron	0	1.004	n

understood world of indestructible atoms and laid the foundation for our modern ideas about atoms and chemical reactions. What are these subatomic particles? What relationship do they have to an atom?

3.3 PARTS OF ATOMS

New laboratory tools and techniques have led to the discovery of many new subatomic particles during the past century. We need to consider only three of these particles. With these three, we can construct a model of the atom that satisfactorily accounts for the chemical properties of all 106 atoms. These three particles are the *electron,* the *proton,* and the *neutron.* Each of these particles has characteristic properties.

The electron is the particle responsible for electric current. It is a negatively charged particle, which is defined as having one unit of negative charge. The symbol e^- is used to represent an electron. The superscript minus sign indicates its negative charge.

The proton is a positively charged particle. Its charge is equal to but opposite that of the electron. Therefore, a proton has one unit of positive charge. The symbol H^+ represents a proton.* The plus sign placed as a superscript indicates its positive charge.

The neutron is a particle that has no charge. For this reason, it escaped detection until 1932. The symbol n represents a neutron.

All three of these particles have mass. But their masses are so small that it is difficult to compare them with familiar objects. For example, the mass of a proton is 1.672×10^{-24} g. Such masses are so small and so inconvenient to use that a new unit, the atomic mass unit (amu), has been adopted. In this unit, a proton has a mass of 1.007 amu. The mass of a neutron is slightly less, 1.004 amu. The mass of an electron is still less, only $1/1837$ the mass of a proton. The characteristics of these three particles are summarized in Table 3-3.

* Other symbols such as p and p^+ are sometimes used to represent a proton. In this text only the symbol H^+ is used to represent a proton.

The discovery of these particles was a major step toward determining the structure of atoms. But how are these particles arranged in an atom?

3.4 PUTTING THE PARTS TOGETHER

All atoms have two things in common. First, they all have a *nucleus,* which contains the neutrons and protons. Second, their electrons occupy the space outside the nucleus. But here the similarity ends. The differences in atoms of different elements are due to the fact that the atoms have different numbers of protons in their nuclei. For example, the nucleus of the helium atom contains two protons, whereas the nucleus of an oxygen atom contains eight protons. The nucleus of an atom has a positive charge because the nucleus contains protons. The charge of a nucleus is equal to the number of protons it contains. Thus, the helium nucleus has a charge of $+2$, whereas the oxygen nucleus has a charge of $+8$.

An atom is electrically neutral. Therefore, the charge created by the protons in the nucleus must be balanced exactly by an equal number of electrons outside the nucleus. For example, the helium atom has two protons in its nucleus. To balance the charge of $+2$ created by these protons, it must have two electrons outside the nucleus. Similarly, an oxygen atom has eight protons in its nucleus and it must have eight electrons outside the nucleus.

Remember that neutrons, protons, and electrons are all very small particles. The atom that contains these particles is also very small. A uranium atom, one of the largest, has a diameter of 2.8×10^{-8} cm. This is so small that a length of 1 cm corresponds to 36 million uranium atoms placed side to side!

Atoms are small, but the nucleus is even smaller; it occupies only a small part of the total volume of the atom. If one could magnify the size of an atom so that its nucleus was the size of a baseball, the atom would be a sphere approximately 16 km in diameter. Two important points are made by this picture of an atom. First, the most massive particles of the atom are concentrated in a very small volume of space. Second, most of the atom is empty space.

The number of protons in the nucleus is an important property of an atom. This number determines not only the number of electrons outside the nucleus, but also the atomic number.

3.5 ATOMIC NUMBERS

At the turn of the century it was discovered that the number of positive charges on the nuclei increases by one from atom to atom as one moves from one group to the next in any period of the periodic table. For example, the positive charge on the hydrogen nucleus is +1; on helium, +2; on lithium, +3. The positive charge is due to the protons in the nucleus. Therefore, each element has one more proton in its nucleus than the element just before it in the periodic table. This fact allows us to give numbers, called *atomic numbers,* to each element. *The atomic number of an element is equal to the number of protons in its nucleus.* The atomic number of hydrogen is 1; of helium, 2; of lithium, 3. The atomic number of each element is given in the periodic table (it is the whole number placed directly below the symbol of the element in Figure 3-1).

*EXERCISE 3-1 Fill in the blank spaces in the following table:

Atom	A	D	G	H
Number of protons	5			10
Number of electrons	5		3	
Atomic number		8		

EXERCISE 3-2 What is the atomic number of each of the following elements?
(a) C (b) Cl (c) Fe (d) Sn (e) Au (f) Ag (g) U

All the atoms of an element have the same atomic number. The importance of this statement became clear with the discovery of isotopes.

3.6 ISOTOPES

So far, we have said nothing about the neutrons in the nucleus. It turns out that atoms of the same element can have different numbers of neutrons in their nuclei. For example, most hydrogen atoms contain only one proton in their nuclei. A few hydrogen atoms called deuterium atoms have nuclei that contain one proton and one neutron. The nuclei of still other hydrogen atoms, called tritium atoms, have one proton and two neutrons. All are atoms of hydrogen because their nuclei contain one proton (all have an atomic number of one). But the atoms differ in the number of neutrons in their nuclei. Atoms whose nuclei have the same number of protons but different numbers of neutrons are called *isotopes.*

Hydrogen is not the only element that has isotopes. Many other elements have two or more isotopes.

* The answers for the exercises in this chapter begin on page 859.

EXERCISE 3-3 Which of the following pairs of atoms are isotopes?

Atom	Number of Protons	Number of Neutrons	Number of Electrons
(a) E	4	5	4
L	5	5	5
(b) M	9	9	9
Q	9	10	9
(c) R	11	12	11
T	10	10	10

Isotopes of an element have the same number of protons, but different numbers of neutrons. Consequently, isotopes have different mass numbers and relative atomic masses.

3.7 MASS NUMBERS AND RELATIVE ATOMIC MASSES

The mass number of an atom is the sum of the number of protons and neutrons in its nucleus. The mass number of hydrogen is 1; that of deuterium is 2. The mass number identifies a particular isotope of an element. This information can be added to the symbol of the element. Thus, a specific isotope of an element is designated by writing the symbol of the

Table 3-4. Naturally Occurring Isotopes of Carbon, Oxygen, and Sulfur

Symbol of Isotope	Number of Protons	Number of Neutrons	Atomic Number	Mass Number
$^{12}_{6}C$	6	6	6	12
$^{13}_{6}C$	6	7	6	13
$^{16}_{8}O$	8	8	8	16
$^{17}_{8}O$	8	9	8	17
$^{18}_{8}O$	8	10	8	18
$^{32}_{16}S$	16	16	16	32
$^{33}_{16}S$	16	17	16	33
$^{34}_{16}S$	16	18	16	34
$^{36}_{16}S$	16	20	16	36

element with the atomic mass number placed as a superscript to the left. Usually, the atomic number is also added as a subscript to the left. In this way we distinguish between isotopes of the same element. The three isotopes of hydrogen are designated as follows:

$^{1}_{1}H$ $\quad$ $^{2}_{1}H$ $\quad$ $^{3}_{1}H$

Mass number (number of protons and neutrons)
← Symbol of the element
Atomic number

The mass number, atomic number, numbers of protons and neutrons, and the symbols for the isotopes of carbon, oxygen, and sulfur are given in Table 3-4.

EXERCISE 3-4 Determine the mass number and atomic number of each isotope in the following table:

Element	Number of Protons	Number of Neutrons
(a) Chlorine	17	18
(b) Chlorine	17	20
(c) Nitrogen	7	7
(d) Nitrogen	7	8

EXERCISE 3-5 Write the symbols that identify each isotope in exercise 3-4.

Isotopes differ in mass because they have different numbers of neutrons in their nuclei. But what about the masses of different elements? Are they also different? Atoms of different elements differ in their numbers of protons and neutrons. Therefore, it seems logical to suppose that they do have different masses. The masses of different atoms are determined by comparing them to a standard mass. For atoms, the standard mass, adopted by international agreement in 1961, is one isotope of carbon, $^{12}_{6}C$, called carbon-12. One atomic unit (amu) is defined as one-twelfth the mass of a carbon-12 atom. This arbitrarily sets the atomic mass of $^{12}_{6}C$ as exactly 12 amu.

The masses of other atoms are obtained by comparing them to carbon-12. This is done experimentally by means of an instrument called a mass spectrograph. The masses of all isotopes relative to carbon-12 can be determined by using this instrument. Values of the masses of selected isotopes are given in Table 3-5. Notice that the atomic mass of an isotope is very close to, but

Table 3-5. Isotopic Masses and Average Atomic Masses

Element	Isotope	Mass Number	Isotope Mass (amu)	Natural Abundance (%)	Average Atomic Mass of Elements (amu)
Nitrogen					14.0067
	$^{14}_{7}N$	14	14.003	99.63	
	$^{15}_{7}N$	15	15.000	0.37	
Boron					10.811
	$^{10}_{5}B$	10	10.013	19.60	
	$^{11}_{5}B$	11	11.009	80.40	
Sulfur					32.064
	$^{32}_{16}S$	32	31.972	95.00	
	$^{33}_{16}S$	33	32.971	0.76	
	$^{34}_{16}S$	34	33.967	4.22	
	$^{36}_{16}S$	36	35.967	0.014	
Magnesium					24.312
	$^{24}_{12}Mg$	24	23.985	78.70	
	$^{25}_{12}Mg$	25	24.986	10.13	
	$^{26}_{12}Mg$	26	25.983	11.17	

not exactly equal to, a whole number. In fact, the relative mass of an isotope is almost the same as its mass number.

What about the atomic masses of the elements? Are they whole numbers? The answer is no. This can be seen from the values of the atomic masses of several elements given in Table 3-5. The reason for this is that an element is made up of a number of isotopes. Therefore, the mass of an element is a weighted average of the masses of all its isotopes. For example, it is known that carbon is made up of 98.89 percent $^{12}_{6}C$ and 1.11 percent $^{13}_{6}C$. The average mass of carbon is the weighted average of the masses of these two isotopes:

$$\text{Average atomic mass of C} = (0.9889)(12.00) + (0.0111)(13.00) = 12.01$$

This is the way all the atomic masses of the elements are obtained.

The terms *mass* and *weight* are often used interchangeably (see Section 1.3). As a result, the atomic mass of an element is also called its atomic weight.

The composition of the nucleus determines many properties of an atom.

The number of protons in the nucleus determines the atomic number and the number of electrons outside the nucleus. Isotopes are atoms that have the same number of protons but different numbers of neutrons in their nuclei. Finally, the mass number of an atom is determined by the number of neutrons and protons in its nucleus.

But we are interested in the chemical properties of atoms. The arrangement of electrons outside the nucleus determines these properties.

3.8 ELECTRON ARRANGEMENT

The electrons occupy most of the space in an atom, but they do not have unlimited freedom to go anywhere within the atom. Electrons are located in certain well-defined shells in the space around the nucleus. This view of the arrangement of electrons in atoms has been well developed during the past 80 years. This modern theory describes the arrangement of electrons in mathematical terms. We shall ignore the mathematics and concentrate on the energy and arrangement of the electrons as predicted by this theory.

There seems to be a problem with our model of the atom. It is well known that a positive charge and a negative charge are attracted to each other. Why, then, do the electrons and protons not combine and destroy the atom? The reason is that there is a balance of forces. One force is the attraction between the positively charged nucleus and the negatively charged electron. An equal and opposing force is created by the motion of the electron. As a result, electrons stay outside the nucleus. Electrons stay with an atom unless the atom undergoes a chemical reaction. Then an atom gains or loses one or more electrons, as we will learn in Chapter 5.

The various shells in which electrons are located are given numbers. The first is shell number 1 (also sometimes called the K shell). This shell is nearest the nucleus. An electron in this shell has the lowest energy of any electron in the atom. Succeeding shells are numbered 2 (L shell), 3 (M shell), 4 (N shell), and so forth. The energy increases in each succeeding shell. We will consider here only the first four shells.

Each shell, except the first, has subshells. Shell 2 has 4 subshells, shell 3 has 9 subshells, and shell 4 has 16 subshells. These subshells are regions in space where the electron spends most of its time. Such regions are called *atomic orbitals*. Each of these atomic orbitals is given a symbol to distinguish it from the others. The shells, number of atomic orbitals, and their symbols are given in Table 3-6.

The symbol for each atomic orbital contains a number, a letter, and sometimes a subscript. The number indicates the shell to which the orbital

Table 3-6. Shells and Their Atomic Orbitals

Shell	Number of Atomic Orbitals	Symbols of Atomic Orbitals
1	1	1s
2	4	2s $2p_x$ $2p_y$ $2p_z$
3	9	3s $3p_x$ $3p_y$ $3p_z$ Five 3d orbitals*
4	16	4s $4p_x$ $4p_y$ $4p_z$ Five 4d and seven 4f orbitals*

* Each of the 3d, 4d, and 4f atomic orbitals also have a subscript. However, we need not concern ourselves with their complete symbol.

belongs. Thus, the 2s atomic orbital belongs to shell 2. The letter indicates the type of atomic orbital. Both the 2s and the 3s are s-type atomic orbitals, even though they belong to different shells. Finally, the entire symbol indicates the region in space where the electron is usually found.

One of the most important properties of an atomic orbital is its energy. The relative energy levels of the atomic orbitals of the first four shells are given in Figure 3-2. Notice several important features of this figure. First, all atomic orbitals with the same number and the same letter have the same energy. Thus, all three 2p atomic orbitals have the same energy. Second, there is a general increase in the energy of the atomic orbitals within any shell. For example, the 2p atomic orbitals are higher in energy than the 2s atomic orbital. Third, the energy levels of the shells are closer together the farther the shells are from the nucleus. In fact, there is an overlap of the energies of the atomic orbitals in shells 3 and 4. As a result, the 4s atomic orbital is lower in energy than the 3d orbitals. The importance of this feature of the energy levels of atomic orbitals will soon be clear.

The concept of atomic orbitals and their relative energy levels is extremely important to our understanding of the arrangement of electrons in atoms. Using the relative energy levels of these orbitals (Figure 3-2) and the rule that each atomic orbital can contain a maximum of two electrons, we can describe the arrangement of electrons in any atom.

Let us start with hydrogen, the simplest atom. It has only one electron, and that electron occupies the 1s atomic orbital, the orbital of lowest energy. This describes the electron arrangement of the hydrogen atom: one electron occupies the region in space described by the 1s atomic orbital.

The helium atom has two electrons, both of which occupy the 1s atomic orbital. The lithium atom has three electrons. Two electrons occupy the 1s atomic orbital, and the remaining electron must occupy the next lowest atomic orbital, the 2s. Beryllium, with four electrons, has two electrons in

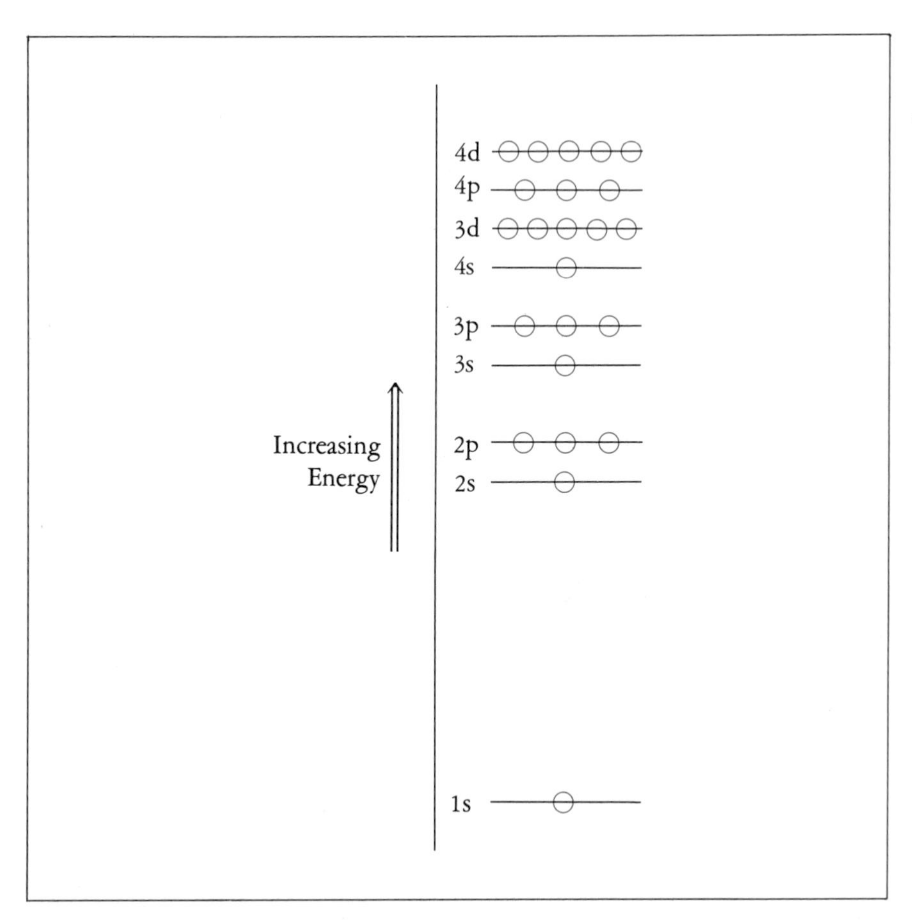

Fig. 3-2. Relative energy levels of the atomic orbitals of the first four shells. Each colored circle represents an atomic orbital that can be filled by one or two electrons.

the 1s atomic orbital and two electrons in the 2s atomic orbital. This kind of analysis can be used to describe the electron arrangement of all the atoms. The electron arrangements of the first 20 elements are given in Table 3-7.

Several important points should be noted about the electron arrangements of the elements given in Table 3-7. First, whenever there is more than one atomic orbital of the same energy, one electron occupies each atomic orbital of equal energy first before two electrons are placed in any one atomic orbital. For example, a carbon atom has two electrons in the 1s atomic orbital, two in the 2s atomic orbital, and one electron in each of two different 2p atomic orbitals. Second, the 4s atomic orbital is lower in energy than the 3d atomic orbitals. Therefore, it is occupied before the 3d atomic orbitals. The electron arrangement of potassium shows this. After the 3p atomic orbitals are filled, the next electron occupies the 4s atomic orbital.

Table 3-7. Electron Arrangement of the First Twenty Elements

	Atomic Orbital									
Element	1s	2s	$2p_x$	$2p_y$	$2p_z$	3s	$3p_x$	$3p_y$	$3p_z$	4s
H	x									
He	xx									
Li	xx	x								
Be	xx	xx								
B	xx	xx	x							
C	xx	xx	x	x						
N	xx	xx	x	x	x					
O	xx	xx	xx	x	x					
F	xx	xx	xx	xx	x					
Ne	xx	xx	xx	xx	xx					
Na	xx	xx	xx	xx	xx	x				
Mg	xx	xx	xx	xx	xx	xx				
Al	xx	xx	xx	xx	xx	xx	x			
Si	xx	xx	xx	xx	xx	xx	x	x		
P	xx	xx	xx	xx	xx	xx	x	x	x	
S	xx	xx	xx	xx	xx	xx	xx	x	x	
Cl	xx	xx	xx	xx	xx	xx	xx	xx	x	
Ar	xx	xx	xx	xx	xx	xx	xx	xx	xx	
K	xx	xx	xx	xx	xx	xx	xx	xx	xx	x
Ca	xx	xx	xx	xx	xx	xx	xx	xx	xx	xx

The electron arrangement of any of the first 20 elements can be determined by following a few simple steps. First, use the atomic number to determine the number of electrons in the atom. Starting with the 1s atomic orbitals, place the electrons into atomic orbitals of increasing energy until there are no more electrons. As an example, let us determine the electron arrangement of the sodium atom. The atomic number of sodium is 11, so it has 11 electrons. We place two electrons in each of the 1s, 2s, and three 2p atomic orbitals, for a total of 10 electrons. We then place the remaining electron in the 3s atomic orbital, and our description of the electron arrangement of the sodium atom is complete.

It is convenient to summarize the electron arrangement of an atom. We can do this by writing the symbol of the element followed by the symbols of the filled atomic orbitals. If the atomic orbital contains two electrons, the superscript 2 is added. The symbol of an atomic orbital without a superscript means that it contains one electron. In this way, we can summarize the electron arrangements of the lithium atom as follows:

Li : $1s^2\ 2s$ ← two electrons in 1s atomic orbital; one electron in 2s atomic orbital

Electron arrangement

EXERCISE 3-6 Summarize the electron arrangement of an atom of each of the following elements:
(a) Be (b) C (c) F (d) Na

Although this is an accurate method of summarizing the electron arrangement of an element, it can be quite complex. For example, the electron arrangement of potassium is summarized as follows:

$$\text{K}\ 1s^2\ 2s^2\ 2p_x{}^2\ 2p_y{}^2\ 2p_z{}^2\ 3s^2\ 3p_x{}^2\ 3p_y{}^2\ 3p_z{}^2\ 4s$$

A more convenient representation can be made by considering only the total number of electrons in each shell. In this way, the electron arrangement of potassium is represented by the following shorthand notation:

	K:2:8:8:1
Shell	1 2 3 4

The colons separate the electrons in each shell. We shall use this shorthand notation in this text.

EXERCISE 3-7 Using the shorthand notation, represent the electron arrangement of the following elements:
(a) B (b) N (c) Al (d) P (e) Cl

EXERCISE 3-8 What is the maximum number of electrons that can occupy each of the first four shells?

The most important feature of Table 3-7 is that the elements in the same group of the periodic table have similar electron arrangements. For example, all the noble gases except helium have eight electrons in their outermost

Table 3-8. Electron Arrangement of Atoms of Several Groups

Group	Element	Atomic Number	Electron Configuration
IA Alkali metals	Li	3	2:1
	Na	11	2:8:1
	K	19	2:8:8:1
IIA Alkaline-earth metals	Be	4	2:2
	Mg	12	2:8:2
	Ca	20	2:8:8:2
VIA	O	8	2:6
	S	16	2:8:6
VIIA Halogens	F	9	2:7
	Cl	17	2:8:7
VIIIA Noble gases	He	2	2
	Ne	10	2:8
	Ar	18	2:8:8

shell. The electron arrangement of the outermost shell for any noble gas, helium excepted, is $s^2\ p_x{}^2\ p_y{}^2\ p_z{}^2$. The alkali metals are another example. They all contain only one electron in their outermost shell. This similarity in electron arrangement of the atoms in several groups is shown in Table 3-8.

The modern theory that describes the detailed structure of atoms predicts a periodic relationship in their electron arrangements. The periodic relationship of elemental properties was discovered experimentally more than 100 years ago. This fact is summarized in the periodic table (see Section 3.2). But now we can understand why the elements have a periodic relationship. *The similarities in the electron arrangements of the outermost shell are responsible for the similarities in the properties of the elements of the same group.*

The predictions of the atomic theory agree with experimental observations of the elements. For this reason, scientists accept this theory of the detailed structure of atoms.

EXERCISE 3-9 How many electrons does each of the following atoms have in its outermost shell?
(a) N (b) C (c) Cl (d) K (e) Ne (f) Si

Let us turn our attention to three important periodic properties of the elements.

3.9 IONIZATION ENERGY, ELECTRON AFFINITY, AND ELECTRONEGATIVITY: THREE PERIODIC PROPERTIES OF ATOMS

Many of the properties of elements show periodic behavior. We will consider only three of their properties: ionization energy, electron affinity, and electronegativity. These three properties are important because they help us understand the kinds of bonds that an atom forms with another atom. Let us examine each of these properties in some detail.

The electrons of an atom remain in atomic orbitals unless something causes them to leave. In Section 3.8 we learned that an atom gains or loses an electron in a chemical reaction. Adding energy to an atom is another way to remove an electron from the atom. The amount of energy needed to remove the first electron from an atom is called its *first ionization energy*. The lower the ionization energy, the easier it is to remove an electron from the atom.

A graph of the ionization energies of the atoms and their atomic numbers reveals a periodic relationship, as shown in Figure 3-3. Starting with lithium, the values of the first ionization energy of the elements of the first period generally increase to reach a maximum at neon. Then there is a sudden decrease at sodium. The values then begin to increase again to reach another maximum at argon. From this graph, we can conclude that the alkali metals (lithium, sodium, and potassium) lose an electron most easily. As we proceed along a period, it becomes increasingly difficult for an element to lose an electron. The noble gases are the most reluctant of all the elements to lose an electron.

Electron affinity is the energy released when an atom gains an electron. It is a measure of the ability of an atom to attract an electron. The higher the value of the electron affinity, the greater the ability of an atom to gain an electron. The electron affinities of the first 20 elements are plotted against their atomic numbers in Figure 3-4. As in the graph of the ionization energies, a periodic relationship is apparent.

Starting with helium, the values of the electron affinities of the elements in the first period increase and decrease slightly until oxygen and fluorine are reached. These two elements have the highest electron affinities in the first period. Then there is a sudden decrease at neon. The values in the second period parallel those in the first row. The values of the elements in groups VIA and VIIA (sulfur and chlorine, respectively) are again the highest in the period. As before, there is a sudden decrease when argon is reached. From this graph, we conclude that the noble gases are the most reluctant to accept an electron, and the halogens are the most willing of all the elements to accept an electron.

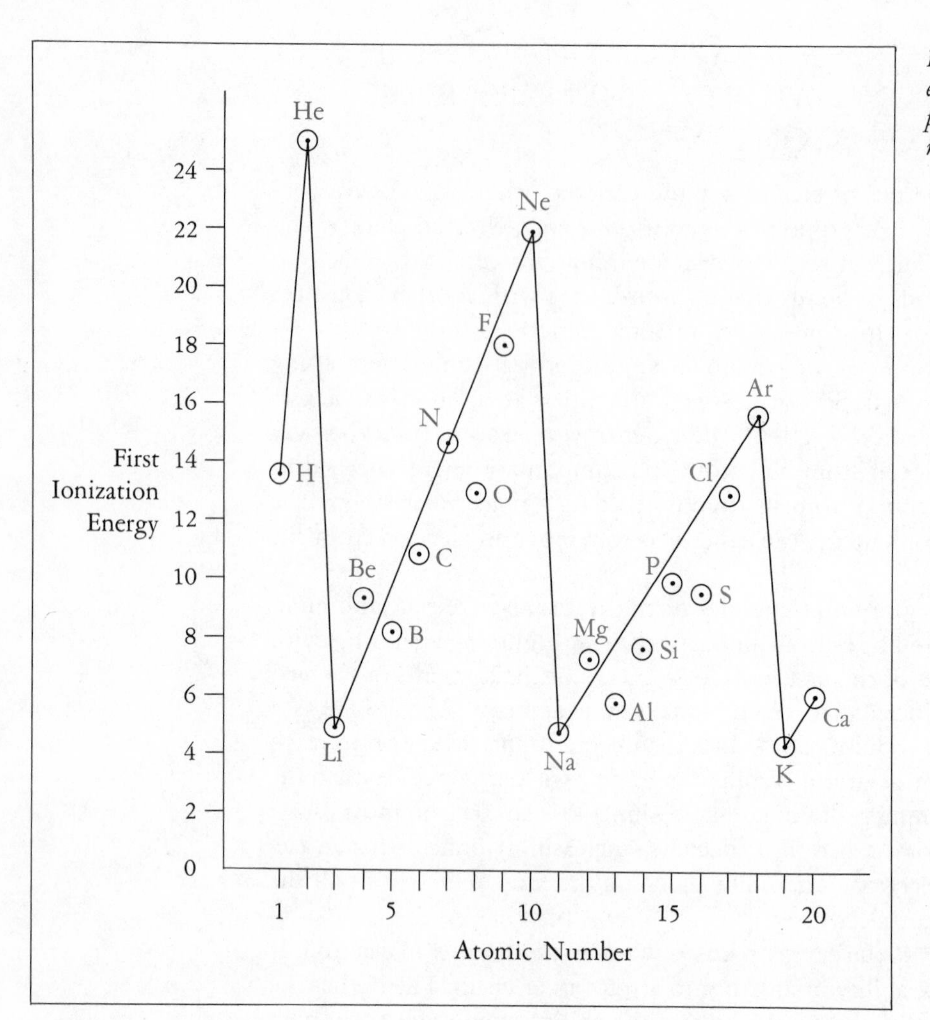

Fig. 3-3. The first ionization energies of the first 20 elements plotted against their atomic numbers.

There is an important relationship between the electron arrangement of atoms and the values of their ionization energies and electron affinities. The noble gases have high ionization energies and low electron affinities. This means that the atoms of these elements are reluctant either to lose or to gain an electron. The electron arrangement of the noble gases is particularly stable. This arrangement is so stable that many other atoms either lose or gain electrons to achieve this arrangement. For example, a sodium atom easily loses an electron to form a positively charged sodium ion. A sodium ion has 11 protons but only 10 electrons. Therefore, it has a net positive charge of +1. We represent this fact by writing the symbol of the element with a plus added as a superscript, Na^+. The electron arrangement of a

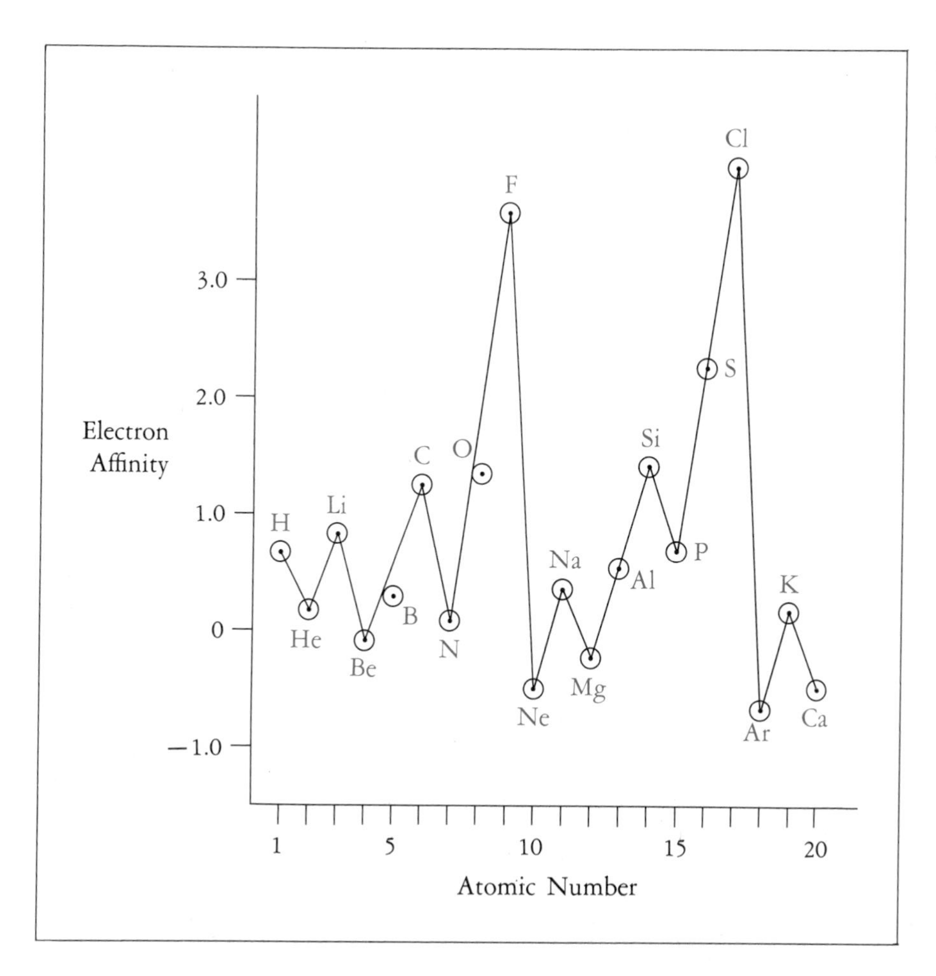

Fig. 3-4. The electron affinities of the first 20 elements plotted against their atomic numbers.

sodium ion is identical to that of the noble gas neon, as shown in Figure 3-5. This is the reason the alkali metals have such low ionization energies. They easily lose an electron to form the electron arrangement of the preceding noble gas.

Atoms can gain electrons to form the noble gas electron arrangement. For example, a fluorine atom easily gains an electron to form a fluoride ion. A fluoride ion has 9 protons but 10 electrons. Therefore, it has a net negative charge of −1. We represent this by writing the symbol of the element with a minus added as a superscript, F^-. The electron arrangement of the fluoride ion is also identical to that of the noble gas neon, as shown in Figure 3-5. This is the reason the halogens have such high electron affinities. They achieve the electron arrangement of the adjacent noble gases by adding an electron.

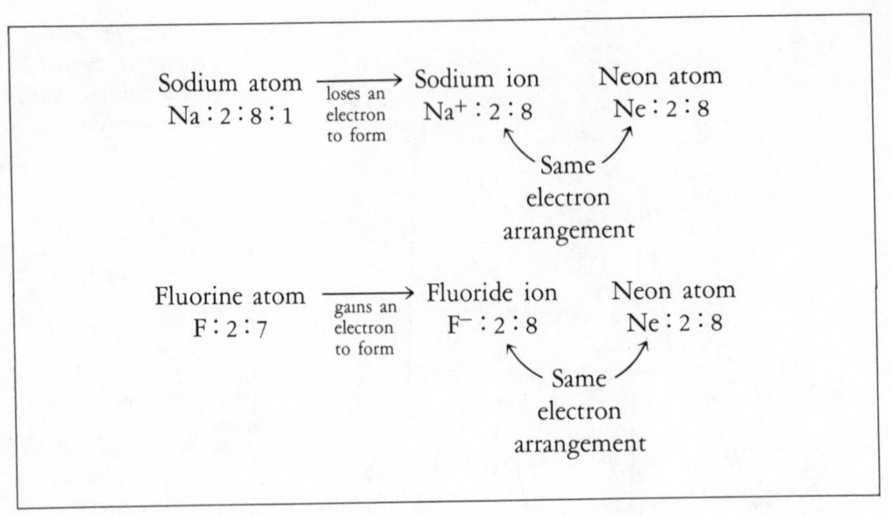

Fig. 3-5. A sodium atom loses an electron to form a sodium ion, which has the same electron arrangement as the neon atom. A chlorine atom gains an electron to form a chloride ion, which also has the same electron arrangement as the neon atom.

Electron affinity and ionization potential are measures of two different processes. One is the loss of an electron from an atom, and the other is a gain of an electron. To indicate which of these two processes is dominant in an atom, the idea of *electronegativity* was developed. Electronegativity is the tendency of an atom to attract electrons to itself. The electronegativity of an atom was originally calculated as the average of its electron affinity and its ionization potential. Since then, other scales of electronegativity have been proposed. The scale in most common use today is the one proposed by Linus Pauling. It assigns a value of 4 to fluorine and compares all other atoms to this standard value. For example, sodium, an atom with a relatively small attraction for electrons, has an electronegativity value of only 0.9. Carbon, whose electron attraction is moderate, has a value of 2.5. The higher the value of the electronegativity of an atom, the greater the tendency to attract electrons to itself.

A graph of the Pauling electronegativity values of the atoms and their atomic numbers has a periodic relationship, as shown in Figure 3-6. Starting with lithium, the values of electronegativity of the elements of the first period increase to reach a maximum at fluorine. Then there is a sudden decrease at sodium, followed by an increase until another maximum is reached at chlorine. In general, there is a decrease in the electronegativity with descent within a particular group. For example, the electronegativities of the halogens decrease in the order fluorine, chlorine, bromine, and iodine.

The importance of electronegativity is that we can tell the kind of bond formed between two atoms simply by looking at the difference in their electronegativities. We will learn about this in Chapter 5.

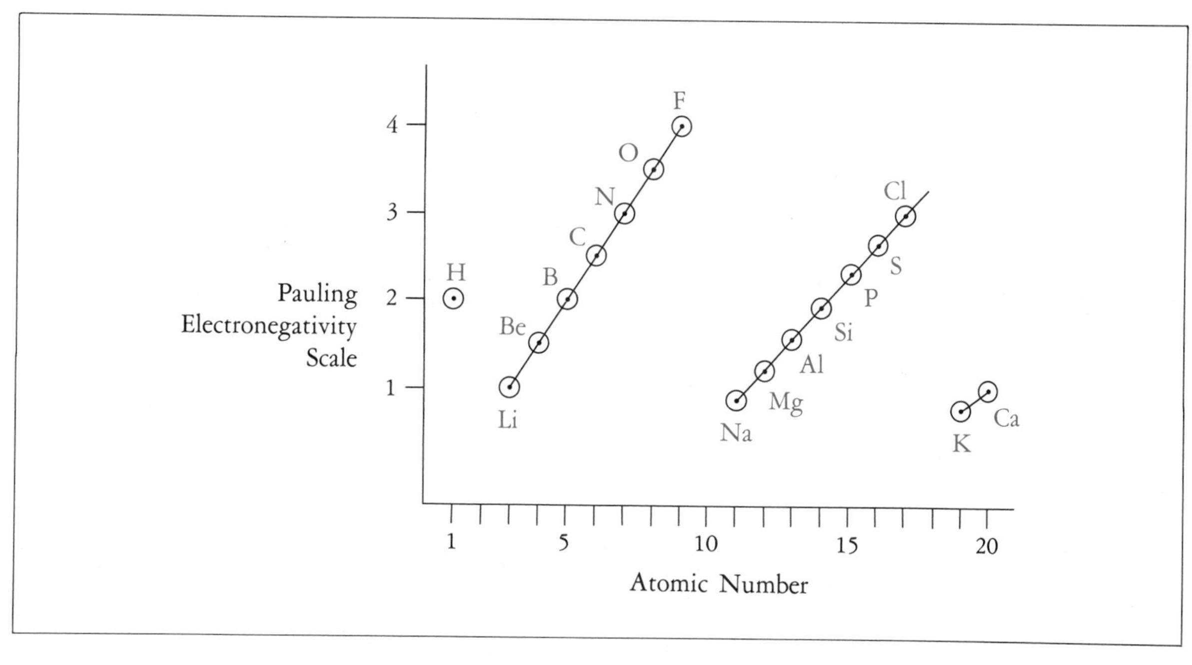

Fig. 3-6. The electronegativities of the first 20 elements plotted against their atomic numbers.

3.10 SUMMARY

The atom is the smallest representative sample of an element. Each of the 106 known atoms has a unique name and a distinctive symbol. Three subatomic particles—proton, neutron, and electron—are used to construct a model of the atom. Protons have a positive charge, electrons have a negative charge, and neutrons are electrically neutral. The nucleus of an atom contains the protons and neutrons, and the electrons occupy the space outside the nucleus. Because an atom contains an equal number of protons and electrons, it is electrically neutral. The number of protons in the nucleus of an atom determines its atomic number. Not all the atoms of an element have the same number of neutrons in their nuclei. Atoms that have the same number of protons but different numbers of neutrons in their nuclei are called isotopes. A particular isotope of an element is identified by its mass number, which is the sum of the number of protons and neutrons in its nucleus. The atomic masses of the elements are given in atomic mass units (amu). One amu is one-twelfth the mass of the $^{12}_{6}C$ isotope of carbon taken as the international standard of mass. The atomic mass of an element is the average of the masses of all its isotopes.

The modern theory of the structure of the atom places the electrons of an atom in various shells about the nucleus. Each shell, except the first, has subshells called atomic orbitals. Each atomic orbital contains a maximum of two electrons. The arrangement of electrons in these atomic orbitals predicts a periodic relationship that is identical to that found in the periodic table. The similarities in the properties of elements of the same group are due to the similarities in the electron arrangements of the outermost shell. The electron arrangement of the noble gases is particularly stable. Consequently, the atoms of many elements gain or lose electrons to form such an electron arrangement.

REVIEW OF TERMS AND CONCEPTS

Terms

ATOMIC MASS NUMBER (amu) A mass equal to one-twelfth the mass of a carbon-12 isotope of carbon.

ATOMIC NUMBER The number of protons in the nucleus of an atom.

ATOMIC ORBITAL A region in space near the nucleus where one or two electrons spend most of their time.

ELECTRON A subatomic particle that has a negative charge and a mass of $1/1837$ amu.

ELECTRONEGATIVITY The tendency of an atom to attract electrons to itself.

ELECTRON AFFINITY The energy released when an atom gains an electron.

GROUPS The elements in one column of the periodic table.

IONIZATION ENERGY The energy needed to remove an electron (usually, the first) from an atom.

ISOTOPES Atoms that have the same number of protons but different numbers of neutrons in their nuclei.

MASS NUMBER The sum of the number of protons and neutrons in the nucleus of an atom.

NEUTRON A subatomic particle that has no charge and a mass of 1.004 amu.

PERIODS The rows of the periodic table.

PERIODIC TABLE An arrangement of the elements that groups them according to the similarities in their properties.

PROTON A subatomic particle that has a positive charge and a mass of 1.007 amu.

Concepts

1. The periodic table emphasizes the similarities in the properties of the elements.
2. Three subatomic particles—neutrons, protons, and electrons—are sufficient to construct a model that satisfactorily explains the properties of atoms.
3. An atom is made up of a nucleus, which contains the neutrons and protons, and regions outside the nucleus where the electrons are found.
4. An atom is electrically neutral; consequently, it contains an equal number of protons and electrons.
5. Atomic orbitals are the regions outside the nucleus where the electrons are found.
6. Atoms of elements in the same group have similar arrangements of electrons.
7. Atoms of the noble gases have particularly stable electron arrangements. Many atoms lose or gain electrons to achieve this electron arrangement.

EXERCISES

3-10 Give the name of the element corresponding to each of the following symbols:
(a) N (b) C (c) O (d) H (e) S (f) Ag (g) Au (h) P (i) K

3-11 Give the symbol of the element corresponding to each of the following names:
(a) chlorine (b) calcium (c) helium (d) neon (e) iron (f) lead (g) iodine

3-12 Using the periodic table, give the name and symbol of two elements that belong to the same group as each of the following elements:
(a) F (b) Ca (c) Se (d) K (e) Kr

3-13 What are two similarities in the atomic structures of different atoms? What are two differences?

3-14 Give the number of electrons in each atom of the following elements:
(a) H (b) C (c) N (d) O (e) F (f) S

3-15 Which of the following atoms are isotopes?

Atom	Atomic Number	Number of Neutrons
V	6	6
X	16	16
Y	5	5
Z	6	7
AA	5	6
DD	16	20

3-16 Give the name and symbol of each of the atoms in exercise 3-15.

3-17 How many neutrons are there in each of the following atoms?
(a) ${}^{12}_{6}C$ (b) ${}^{2}_{1}H$ (c) ${}^{37}_{17}Cl$ (d) ${}^{15}_{7}N$ (e) ${}^{18}_{8}O$

3-18 The element silicon is made up of 92.21 percent ${}^{28}_{14}Si$, 4.70 percent ${}^{29}_{14}Si$, and 3.09 percent ${}^{30}_{14}Si$. What is the atomic mass of silicon?

3-19 Represent the electron arrangement of each of the following elements:
(a) H (b) Ne (c) O (d) Na (e) S

3-20 In terms of the electron arrangement of atoms, explain why there is a periodic arrangement of the atoms.

RADIOACTIVITY AND NUCLEAR CHEMISTRY 4

We learned in Chapter 3 that there are isotopes of almost all the elements. Most of these isotopes are stable, but some are unstable. The nuclei of unstable isotopes undergo spontaneous nuclear reactions that cause particles and energy, called *nuclear radiation,* to be given off. The emission of these particles and energy by an isotope is called *radioactivity*. Only a few isotopes found in nature are radioactive. The first example of a naturally occurring radioactive substance was discovered by Henri Becquerel in 1896. He found that uranium ore gave off penetrating radiation that darkened photographic film without exposing the film to light. Since then, other scientists have found many other radioactive elements. More than 50 naturally occurring radioactive isotopes are now known. In addition, scientists have been able to make many radioactive isotopes not found in nature.

The discovery of radioactive isotopes has greatly affected our lives. The awesome power of nuclear weapons, the promise of abundant energy, radiation therapy, and contamination of the environment by nuclear waste products are all results of the properties of the tiny nuclei of these radioactive isotopes. Given all the controversy, any educated person must know something about radioactivity. In this chapter, we will learn how to detect radioactive isotopes, how these isotopes are used in medicine, how radioactive isotopes that are not found in nature can be made in the laboratory, and how energy is obtained from nuclear reactions. Let us start by examining the types of radiation emitted by radioactive isotopes.

4.1 TYPES OF RADIATION

Early in the twentieth century, it was discovered that naturally occurring isotopes emit three kinds of radiation. At that time, scientists did not understand them, so their discoverers named them simply *alpha* (α), *beta* (β), and *gamma* (γ). Since then, scientists have discovered the identities and properties of these types of radiation. Each has characteristic properties that determine how it affects living systems.

Alpha radiation is a stream of particles moving at about one-tenth the speed of light. Each particle is the nucleus of a helium atom that contains two protons and two neutrons and has a charge of +2. Alpha particles are relatively large and heavy, so they cannot travel very far without colliding with other particles. As a result, these particles do little damage to internal organs because they cannot penetrate the skin. However, if a substance that emits alpha particles gets inside the body by being inhaled or swallowed, the alpha particles can then damage internal organs.

Beta radiation is also a stream of particles, but the particles are electrons. The electrons are produced within the nucleus by the transformation of a neutron into a proton and an electron. The proton stays in the nucleus and the electron is emitted. An electron is smaller than a helium nucleus (alpha particle), travels much faster, and can penetrate the skin to a depth of a few centimeters. Exposure to beta radiation causes the skin to appear burned. Damage to internal organs occurs when a substance that emits beta particles gets into the body.

Gamma radiation is not a particle, but a form of energy similar to light waves, radio waves, or x-rays. This radiation has high energy and can penetrate deep within the body and cause serious damage. Gamma radiation usually occurs along with alpha and beta radiation.

Two less common but still important types of nuclear radiation are *neutrons* and *positrons*. We learned in Chapter 3 that the neutron is one of the

Table 4-1. Properties of Various Forms of Nuclear Radiation

Type of Radiation	Composition	Symbol	Mass (amu)	Electrical Charge	Approximate Penetration of Skin (cm)
Alpha	Helium nucleus	α, ${}^{4}_{2}He$	4	+2	0.01
Beta	Electron	β, ${}^{0}_{-1}e$	$1/_{1,837}$	−1	1
Gamma	Energy	γ	0	0	100
Neutron	Neutron	n, ${}^{1}_{0}n$	1	0	10
Positron	Positron	β^{+}, ${}^{0}_{1}e$	$1/_{1,846}$	+1	1

three particles that make up atoms. A positron has about the same mass as an electron but has a positive charge. The properties and symbols of these various forms of radiation are summarized in Table 4-1.

The emission of radiation from radioactive isotopes is also called ionizing radiation, for reasons that we will learn in the next section.

4.2 IONIZING RADIATION

The radiation from radioactive isotopes and x-rays can form ions in matter by knocking electrons off the atoms and molecules in its path. For this reason, it is called *ionizing radiation*. The chief effects of radiation on living systems are due to these ionization reactions. Repeated exposure to low levels of radiation seems to have a number of major effects on health. Among them are cancer (carcinogenic effects), damage to the fetus, and genetic damage.

It has been known for many years that radiation causes cancer. Skin cancer, bone cancer, leukemia, and other cancers are products of exposure to radiation. Even at very low levels of exposure, there is danger from radiation. For example, x-rays used in diagnosis are not completely free of potential harm to a patient. Persons who administer x-rays must take precautions to avoid exposure, because the effect is cumulative. Exposure to high levels of radiation kills cells. Use is made of this fact to treat cancer. Cancer cells are exposed to high-energy x-rays or gamma radiation to destroy these cells or to retard the spread of cancer.

Fetuses and small children are particularly sensitive to radiation. Ionizing radiation affects them more strongly than adults. The effects of radiation are widespread. Damage occurs to the brain, eyes, bones, and other organs.

The genetic risk of exposure to radiation is more difficult to determine because the genetic damage may not be seen for several generations. Genetic damage is caused by damage to the genes in the nuclei of cells. The damage to the structure of the gene may cause death or a variety of physical defects in following generations.

Clearly, exposure to radiation is dangerous. But is there any level of exposure below which radiation has no effect? According to one theory, called the threshold theory, no damage occurs below a certain level of radiation, called the threshold value. Opposed to this is the linear theory. According to the linear theory, the risk of damage is proportional to exposure, even down to very low levels of radiation. The current view is a compromise of these two theories: There is a risk of damage even at low levels of radiation, but the risk is extremely small.

The dangers of ionizing radiation are compounded by the fact that this radiation cannot be detected by the human body. We cannot see, feel, or smell ionizing radiation. Therefore, a person can be exposed to lethal levels of radiation without knowing it until it is too late. But we do have methods of detecting ionization radiation. Several of these methods are given in the next section.

4.3 DETECTING IONIZING RADIATION

The methods of detecting ionizing radiation all make use of the fact that radiation disturbs the electronic environment of the atoms and molecules that it encounters. The three methods of detecting ionizing radiation most frequently used are the photographic method, use of scintillation counters, and use of the Geiger counter.

Becquerel discovered that uranium ore was radioactive because of its effect on photographic film, as we learned in the introduction to this chapter. Photographic film and paper shielded from light are exposed by ionizing radiation. This exposure is detected by developing the film or paper in the usual way. This fact is used to provide an individual detector for persons working near sources of radioactivity. Each person wears a badge containing a piece of film. The film is changed at regular intervals and developed. A certain level of radiation will cause the film to be exposed, warning the wearer of potential danger.

Scintillation counters are instruments that contain a surface coated with a special substance that gives off flashes of light when hit by ionizing radiation. The invisible ionizing radiation strikes the surface and some of its

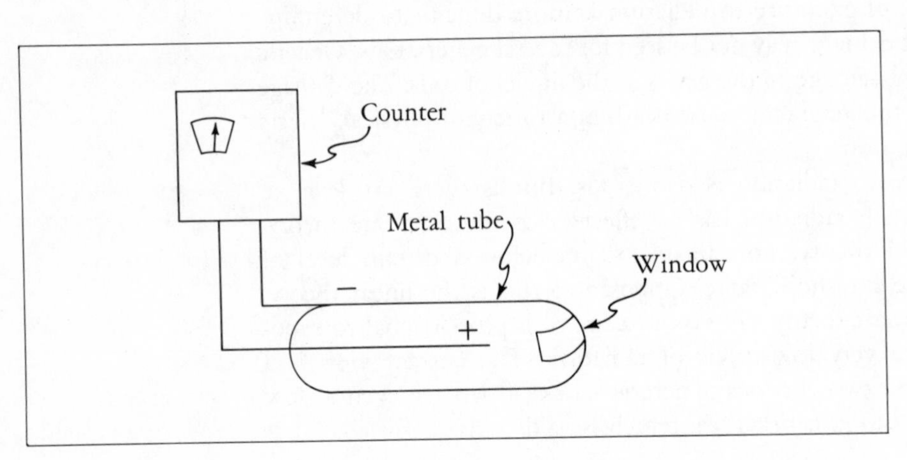

Fig. 4-1. A Geiger counter.

energy is transformed into visible light. Electronic devices magnify and record these flashes.

The most common instrument used to detect and measure ionizing radiation is the Geiger counter, shown schematically in Figure 4-1. The detecting part of the instrument is a metal tube. It contains a gas, a wire down the center, and a window at one end. The window is made of a thin material to allow alpha and beta particles to enter. A large potential difference is maintained between the metal walls of the tube and the central wire. When ionizing radiation enters the tube, it forms ions. This causes a pulse of electricity to flow from the wire to the metal walls of the tube. This pulse is counted by an electronic device that produces either a meter reading or an audible clicking sound. Small, portable Geiger counters are in common use to detect sources of radioactivity in the environment. When we detect ionizing radiation, it usually means that a nuclear reaction has taken place.

4.4 NUCLEAR REACTIONS

Alchemists in the Middle Ages dreamed of turning one chemical element into another. The atomic theory seemed to shatter this dream because the atom assumed the role of the stable and indivisible unit of matter. However, with the discovery of radioactivity, scientists soon realized that atoms do change from one kind to another when they emit nuclear radiation. This change occurs during a nuclear reaction whenever the nucleus of an isotope emits alpha or beta particles. When this happens, the nucleus gains or loses positive charges and its atomic number is changed. A change in atomic number means that one element has changed to another (see Section 3.5).

An example of a change of one element into another is the nuclear reaction of one isotope of carbon, ${}^{14}_{6}C$ (carbon-14). When ${}^{14}_{6}C$ emits a beta particle, it is changed into an isotope of nitrogen, ${}^{14}_{7}N$. We can represent this nuclear reaction by means of the following nuclear equation.

$$ {}^{14}_{6}C \longrightarrow {}^{14}_{7}N + {}^{0}_{-1}\beta \qquad (4\text{-}1) $$

In this equation, we designate the particular isotopes of carbon and nitrogen in the way we learned in Section 3.7. By convention, the symbol of the starting material is placed on the left and the symbols of the products are placed on the right. The products and reactants are joined by an arrow pointing in the direction of the product.

Furthermore, nuclear equations must be balanced. This means that the sum of all the protons and neutrons is the same in the products as in the starting materials. No protons or neutrons are lost during a nuclear reaction. Consequently, the sum of the superscripts (mass numbers) and the subscripts (atomic numbers) of the particles on the left side of the equation must equal the same sum on the right side. We can check this for equation 4-1 as follows:

Sum of superscripts = 14 Sum of superscripts = 14 + 0 = 14

Mass number → $^{14}_{6}C \longrightarrow ^{14}_{7}N + ^{0}_{-1}\beta$ ← Atomic number

Sum of subscripts = 6 Sum of subscripts = 7 − 1 = 6

Clearly, equation 4-1 is balanced.

We can understand why one element is changed into another by examining a balanced nuclear equation. When an alpha or a beta particle is emitted by an isotope, the atomic number must change to compensate for this loss or gain of one or more protons. The result of this change is the formation of an isotope of a different element. The atomic number determines the element, and the mass number determines the particular isotope of that element.

*EXERCISE 4-1 Balance each of the following nuclear equations:

(a) $^{40}_{19}K \longrightarrow ^{40}Ca + ^{0}_{-1}\beta$

(b) $^{224}_{88}Ra \longrightarrow Rn + ^{4}_{2}He$

(c) $^{232}_{90}Th + ^{1}_{0}n \longrightarrow Th$

(d) $^{238}_{92}U \longrightarrow ^{234}Th + {}_{2}He$

EXERCISE 4-2 Fill in the missing mass number, atomic number, and symbol in each of the following equations:

(a) $^{24}_{11}Na \longrightarrow$ ____ $+ ^{0}_{-1}\beta$

(b) $^{9}_{4}Be + ^{4}_{2}He \longrightarrow$ ____ $+ ^{1}_{0}n$

(c) $^{210}_{84}Po \longrightarrow$ ____ $+ ^{4}_{2}He$

(d) $^{14}_{7}N + ^{1}_{0}n \longrightarrow ^{14}_{6}C +$ ____

* The answers for the exercises in this chapter begin on page 859.

One of the characteristic properties of a nuclear reaction is the time needed for it to occur. The half-life of an isotope is a measure of this time, as we will learn next.

4.5 HALF-LIFE

The breakup or decay of the nuclei of a particular radioactive isotope requires a certain amount of time to occur. Not all nuclei decay at the same time; rather, they decay over a period of time. The time needed for one-half of the original nuclei of an isotope to decay to other substances is called the *half-life* of the isotope. The symbol $t_{1/2}$ is used to indicate half-life.

Each radioactive isotope has a characteristic half-life. Values of the half-lives of naturally occurring isotopes range from milliseconds to several billion years. The values of the half-lives of several naturally occurring radioactive isotopes are given in Table 4-2.

The importance of the half-life of a radioactive isotope is that it tells us how long a sample of the isotope will exist. For example, $^{234}_{90}Th$ (thorium-234) has a relatively short half-life of 24.1 days. If we have a 1.00 g sample of thorium-234, 0.50 g of it will be left after 24.1 days because the other half will be turned into protactinium (Pa), according to the following equation:

$$^{234}_{90}\mathrm{Th} \xrightarrow{t_{1/2} = 24.1 \text{ days}} {}^{234}_{91}\mathrm{Pa} + {}^{0}_{-1}\beta$$

Table 4-2. Half-Lives of Some Naturally Occurring Radioactive Isotopes

Element	Isotope	Half-Life
Hydrogen	$^{3}_{1}H$ (Tritium)	12.3 yr
Carbon	$^{14}_{6}C$	5700 yr
Potassium	$^{40}_{19}K$	200 million yr
Radon	$^{222}_{86}Rn$	3.8 days
Radium	$^{226}_{88}Ra$	1600 yr
Thorium	$^{234}_{90}Th$	24.1 days
Thorium	$^{230}_{90}Th$	80,000 yr
Uranium	$^{235}_{92}U$	700 million yr
Uranium	$^{238}_{92}U$	4.5 billion yr
Polonium	$^{214}_{84}Po$	0.15 msec
Lead	$^{214}_{82}Pb$	26.8 min

In another 24.1 days there would be only 25 percent of the original thorium-234 (0.25 g), and after three half-lives (3×24.1 days, or 72.3 days) only 12.5 percent (0.125 g) would remain. Theoretically, the amount of thorium-234 present never reaches zero. But in practice, after 10 half-lives, or 241 days, the amount left is so small (9.8×10^{-4} g) that its presence cannot be detected by any of the methods given in Section 4.3. For all practical purposes, after 241 days, our sample of thorium-234 is all gone. It has been converted into another element.

The value of the half-life of a radioactive isotope is independent of sample size. Thus, whether the size of the sample of thorium-234 is 10 kg, 1 kg, or 10 g, half of it will decay in 24.1 days.

EXERCISE 4-3 The half-life of radium-223, $^{223}_{88}Ra$, is 11.4 days. How long will it be until only 25 percent of the sample of radium-223 is left?

EXERCISE 4-4 Starting with 500,000 atoms of element 106, and given $t_{1/2} =$ 0.90 second, how many atoms will be left after 3.6 seconds?

In addition to how long a radioactive isotope will last, we also want to know how much radiation it gives off. For this reason, we need units to measure the radiation given off by radioactive isotopes.

4.6 RADIATION DOSAGES

Radiation causes damage to all living systems. The amount of damage is directly related to the dose of radiation received. In this section, we will define several units of radiation damage. Let us start with the basic unit of radioactivity.

Curie and Becquerel

The curie, represented by the symbol Ci, is the level of radioactivity caused by 3.7×10^{10} radioactive disintegrations per second. This unit is a measure of the number of nuclei that decay per second. It is independent of the size of the radioactive sample.

One curie represents a large amount of radioactivity. As a result, more moderate units such as picocurie (10^{-12} Ci), microcurie (10^{-6} Ci), and millicurie (10^{-3} Ci) are generally used. The SI unit of radioactivity is the becquerel, whose symbol is Bq. One curie is equal to 3.7×10^{10} becquerel.

The curie and the becquerel are basic units of radioactivity. Unfortunately, equal numbers of curies of radiation given off by different isotopes do not always produce the same biological effect. To assess biological damage caused by radiation, we need units that can be compared. One such unit is the rad.

Rad

The rad, whose symbol is D, stands for *r*adiation *a*bsorbed *d*ose. It is a unit that expresses the amount of energy from the radiation actually absorbed by tissue or other material. One rad is equal to 2.4×10^{-3} cal of energy absorbed by 1 kg of tissue. This is a very small unit. Converted to heat, this energy would raise the temperature of tissue only a few thousandths of a degree. Because it is concentrated in a small particle, however, it causes great damage to tissue.

The SI unit of absorbed dose is the gray, whose symbol is Gy. A gray is defined as 1 J of energy absorbed per 1 kg of tissue. One hundred rads are equivalent to one gray.

The energy absorbed by living tissue is not the only factor that contributes to the biological hazards of radiation. For this reason, a dose of 1 rad from one source is not necessarily equal to a dose of 1 rad from another. The rem is the unit that takes all these differences into account.

Rem

One rem is 1 rad multiplied by a factor called the relative biological equivalent (RBE). Thus,

$$\text{Number of rem} = \text{RBE} \times \text{number of rad}$$

The factor RBE takes into account the differences in biological damage caused by different kinds of ionizing radiation of the same energy. For example, the RBE of an alpha particle is 10 times that of a beta particle. This means that alpha particles released in tissue cause 10 times more damage than beta particles of the same energy. The rem is a more accurate and comparable measure of biological damage caused by different ionizing radiation. For this reason, it is commonly used to describe allowable doses of radioactivity.

The object of defining units of radiation is to be able to determine how much radiation is dangerous to humans. One way of expressing this danger is by LD_{50} values.

LD_{50} Values

The ionizing radiation dosage at which injury to humans begins is unknown, as we learned in Section 4.2. The best we can do is to express the short-term exposure to ionizing radiation that is fatal to various forms of life. This is expressed as the 30-day LD_{50} value, defined as the dose, in rem, that is fatal to 50 percent of the population within 30 days.

The LD_{50} value for humans is estimated to be 500 rem. Thus, a person suddenly exposed to 500 rem of ionizing radiation would have a 50 percent chance of remaining alive after 30 days. Even after 30 days, the possibility of illness and death would still be great. The LD_{50} values of most mammals are in the range 250 to 1000 rem. Simpler forms of life such as insects can tolerate much more radiation. For example, some insects can tolerate more than 50,000 rem, and some microorganisms can tolerate even more.

We cannot escape ionizing radiation. There are small amounts of naturally occurring radioactive substances in the soil, in the food we eat, and in the water we drink. Streams of particles called *cosmic rays* enter our atmosphere from the sun and bombard us. These are all part of *natural background radiation*. The exposure we receive from these natural sources is very small. Each person receives about 100 mrem annually from natural background radiation.

The use of radioactive substances in medicine, industry, and nuclear power plants has increased our risk of exposure to radiation. Many of these radioactive substances are not found naturally but are made for a specific use. We will learn how this is done in the next section.

4.7 TRANSMUTATION

All the naturally occurring radioactive isotopes have been known for a long time, and their properties have been thoroughly investigated. We can purify these isotopes and obtain many of them in large quantities. Many are widely used in our society. Sometimes, naturally occurring radioactive isotopes do not have the properties that we need. For example, the half-life of the isotope or the type of particle emitted may make it unsuitable for our purposes. We cannot do anything about this because these are fundamental properties of a particular isotope. But we are not limited just to naturally occurring radioactive isotopes, because scientists can change one isotope into another not found in nature. Changing one element into another, either in nature or in the laboratory, is called *transmutation*. Scientists do this by means of bombardment reactions.

A *bombardment reaction* occurs when particles of atomic or subatomic size strike atoms of an element and change them into another element. For example, the reaction between an alpha particle and a beryllium atom forms a carbon atom and a neutron according to the following equation:

$$^{9}_{4}Be + ^{4}_{2}He \longrightarrow ^{12}_{6}C + ^{1}_{0}n$$

In this reaction, an alpha particle is the bombarding particle. It meets a beryllium atom, the target atom, head on with such force that they momentarily fuse. The unstable nuclear particle formed by this fusion ejects a neutron and leaves behind a carbon atom.

For a bombardment reaction to occur, the bombarding particles must have enough energy to overcome the highly charged nucleus of the target atom, especially when the target atom has a high atomic mass. Most particles emitted by natural radioactive isotopes do not have enough energy to do this, so their energies must be increased. This can be done by an *accelerator,* a device that increases the speed of a charged atomic or subatomic particle. Increasing the speed of a particle increases its energy and therefore increases its effect on the atom it hits.

One type of accelerator is a circular accelerator. The operation of a circular accelerator is based on two well-known laws of nature. First, a charged particle is repelled by another particle of like charge and is attracted by one of opposite charge. Second, a charged particle moves in a curved path when in a magnetic field. A *cyclotron* is a circular accelerator, and its major parts are shown in Figure 4-2.

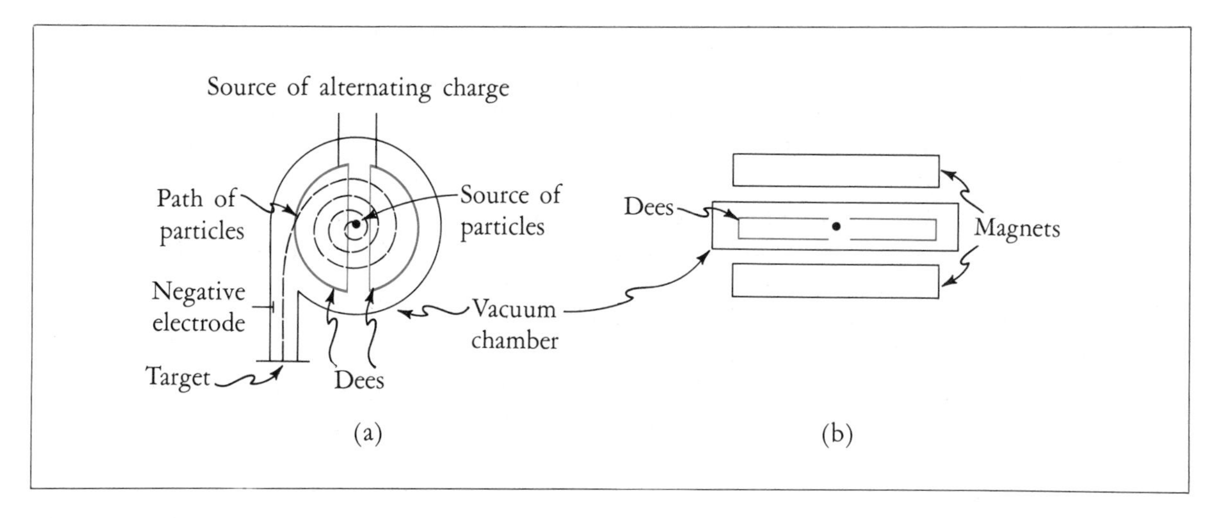

Fig. 4-2. A cyclotron: (a) top view; (b) side view.

A cyclotron consists of a hollow disc split into halves called dees. These dees are placed in a strong magnetic field. The electrical charges on the dees are opposite in sign and can be rapidly reversed. As a result, a positive particle, such as an alpha particle, is attracted to one, then the other, then back and forth between the dees as the electrical charge alternates. While the particle moves from one dee to another, it follows a curved path because the dees are placed in a magnetic field. As the speed of the particle increases, the radius of its spiral path becomes larger and larger. When the positively charged particles reach the outside of the dee, a negatively charged electrode attracts them and deflects them to the target. When the particles hit the target, the bombardment reaction occurs.

In this way, many radioactive isotopes are made that are useful in medicine. The radioactivity from isotopes made in this way is no different from that available from naturally occurring radioactive isotopes. These isotopes are useful because their chemical properties are such that they can get the radiation to where it is useful to diagnose or treat cancer and other diseases. We will learn about some of these isotopes prepared by scientists in the next section.

4.8 MEDICAL USES OF RADIOACTIVE ISOTOPES

Radiation from radioactive isotopes is used extensively in medicine to treat cancer. The radiation is directed at the cancer cells from either outside or inside the body, depending on the type of cancer. The most widely used external source of radiation is cobalt-60. This isotope of cobalt has a half-life of 5.3 years and is a powerful beta and gamma emitter. The sample of cobalt-60 is placed in a lead container with an opening that allows its radiation to be directed to the site of the cancer.

The use of radioactive isotopes internally to treat cancer is based on the fact that all isotopes of an element, radioactive or not, have identical electron arrangements and consequently identical chemical properties (see Section 3.8). For example, when taken internally, the iodide ion from sodium iodide is eventually concentrated in the thyroid gland. Use is made of this fact to treat thyroid cancer. Iodine-131, a radioactive isotope of iodine made by nuclear bombardment is used. A patient is given a drink of water containing sodium iodide. Some of the iodide ions are radioactive. When these ions are absorbed by the thyroid gland, radiation from their decay destroys the cancer cells. Other radioactive isotopes used to treat cancer are given in Table 4-3.

Radioactive isotopes are also used in medicine as *tracers* to diagnose

Table 4-3. Selected Radioactive Isotopes and Their Medical Uses

Isotope	Symbol	Half-Life	Emission	Use
Cobalt-60	${}^{60}_{27}Co$	5.3 yr	β, γ	Source of radiation for cancer therapy
Iodine-131	${}^{131}_{53}I$	8 days	β, γ	Treatment of thyroid cancer, diagnosis of thyroid gland malfunction
Iodine-123	${}^{123}_{53}I$	13.3 hr	β	Treatment of thyroid cancer
Phosphorus-32	${}^{32}_{15}P$	143 days	β	Treatment of leukemia
Sodium-24	${}^{24}_{11}Na$	15.0 hr	β, γ	Check of proper function of circulatory system
Iron-59	${}^{59}_{26}Fe$	45.6 days	β, γ	Determination of red blood cell formation and lifetime
Technetium-99	${}^{99}_{43}Tc$	6.1 hr	β	Brain, kidney, and lung scans

illnesses. The movement of these isotopes in the body can be followed or traced by means of the radiation given off. For example, sodium-24 in the form of sodium chloride (salt) is given to a patient in a water solution to locate blockages in the circulatory system. Several radioactive isotopes used as tracers are given in Table 4-3.

Radioactive isotopes used in humans must be chosen carefully. First, the half-life must be long enough for the isotope to do its job yet short enough that the isotope will disappear from the body without subjecting the person to unnecessary radiation. Second, no isotopes that emit alpha particles are used. The reason is that alpha particles are relatively large (see Section 4.1), and they cause a great deal of damage to all tissue and organs if an isotope emits them internally.

In addition to emitting particles, nuclear reactions also release energy. Certain nuclear reactions are particularly well suited to providing energy, as we will learn in the next section.

4.9 ENERGY AND NUCLEAR REACTIONS

The experiment that heralded the dawn of the nuclear age was carried out by a group of physicists in Italy in 1933. They bombarded uranium atoms with low-energy neutrons (called slow or thermal neutrons) and found that a number of elements of atomic masses lighter than uranium were formed. Scientists soon realized that these products were formed by splitting the nucleus of the uranium atom. The nuclear reaction that causes an atom to split into several smaller parts is called a *fission reaction*.

Further experiments provided additional facts about fission reactions. First, only a few isotopes are known to undergo fission reactions when bombarded by neutrons. Uranium-235 (U-235) is the only naturally occurring isotope that undergoes this reaction. Two other isotopes formed by nuclear bombardment, plutonium-239 and uranium-233, also undergo fission reactions. Second, a large amount of energy is released when a fission reaction occurs. Finally, fission reactions form neutrons and gamma radiation as well as other elements.

This last fact is particularly important because it allows a *chain reaction* to occur whenever enough uranium-235 atoms are present. A chain reaction is a series of repeated reactions in which one reaction produces a product that allows the series of reactions to continue. A chain reaction is shown in Figure 4-3. A slow neutron strikes an atom of uranium-235 and causes it to split into two nuclei of lighter mass with the emission of gamma radiation and two, three, or four neutrons. These neutrons can strike other nearby uranium-235 atoms and cause them to undergo fission reactions. Each fission reaction produces more neutrons, which cause more fission reactions, which produce more neutrons, and so forth. For a chain reaction to occur, enough fissionable nuclei must be present so that most of the neutrons will hit other fissionable nuclei. It is the emission of neutrons by a fission reaction that keeps the chain reaction going. The amount of material needed for a chain reaction to continue (to be self-sustaining) is called the *critical mass*.

The fission of uranium-235 can occur in more than 30 ways to produce about 200 different radioactive isotopes. These isotopes usually have atomic numbers between 30 and 64 and masses between 72 and 161. The following three equations are representative of the fission reactions of uranium-235:

$$^{1}_{0}n + ^{235}_{92}U \rightarrow ^{90}_{38}Sr + ^{143}_{54}Xe + 3\,^{1}_{0}n$$

$$^{1}_{0}n + ^{235}_{92}U \rightarrow ^{135}_{53}I + ^{97}_{39}Y + 4\,^{1}_{0}n$$

$$^{1}_{0}n + ^{235}_{92}U \rightarrow ^{103}_{42}Mo + ^{131}_{50}Sn + 2\,^{1}_{0}n$$

Notice that more neutrons are formed than are consumed in each of these fission reactions.

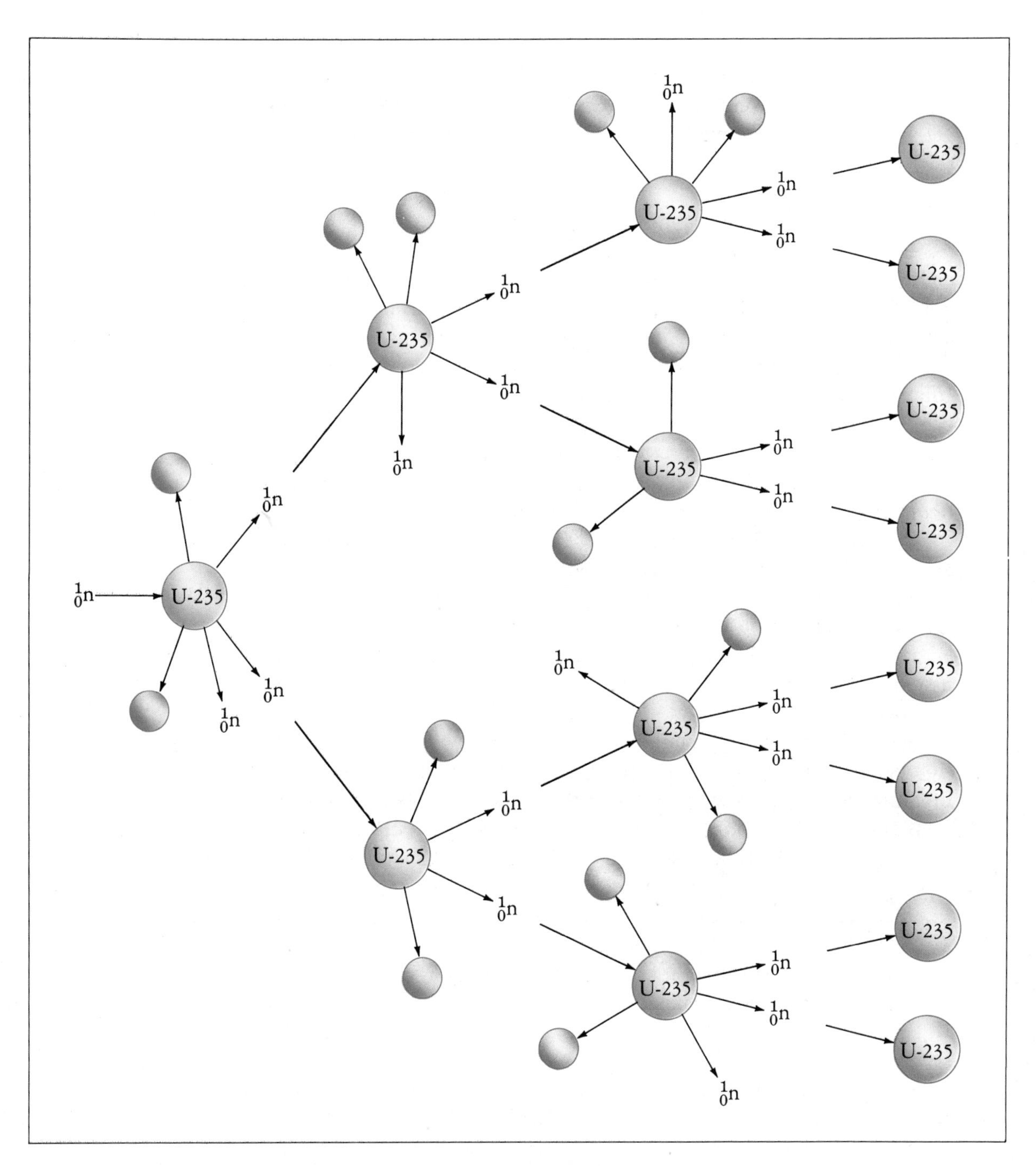

Fig. 4-3. A chain reaction involving the fission of uranium-235. Each act of fission produces more than one neutron, which keeps the chain reaction going.

An atomic explosion is an uncontrolled chain reaction. An atom bomb contains two separate quantities of fissionable material, each less than the critical mass. The atomic explosion is triggered by suddenly bringing together these subcritical masses. When this occurs, fission occurs so rapidly that the whole mass explodes, producing temperatures greater than 10,000,000° C.

Controlling the chain reaction allows the tremendous heat of the fission reaction to be released over a long period of time. Fission reactions are controlled by placing material that absorbs neutrons in the midst of the fissionable material. Carbon, boron, and cadmium absorb neutrons readily. These elements are made into rods that can be inserted into the fissionable material. The amount of fission and the amount of energy released can be controlled by the position of the rods in the fissionable material, called the core. The number of nuclear fission reactions is decreased by pushing the rods in and increased by pulling them out. All nuclear reactions are stopped by putting all the rods into the core. In this case, the rods absorb so many neutrons that not enough are present to cause fission reactions to occur.

The heat given off in a controlled nuclear reaction can be used to generate electricity. The method of generating electricity is the same in nuclear power plants and plants that burn fossil fuels. The heat is used to turn water into steam. The steam drives turbines that generate electricity. The difference is that the source of heat in a nuclear power plant is controlled nuclear fission rather than burning of fossil fuel. A diagram of a nuclear power plant is shown in Figure 4-4.

Less than 1 percent of uranium ore is uranium-235. The more abundant isotope, uranium-238, does not undergo fission reactions. To obtain enough uranium-235 for a chain reaction, this isotope must be separated from uranium-238. This cannot be done chemically, because both isotopes have identical chemical properties. Separation was originally accomplished in 1943 by converting uranium ore to the gas uranium hexafluoride. Uranium hexafluoride containing uranium-235 moves slightly more rapidly through air than does uranium hexafluoride containing uranium-238. Therefore, when passed through a long tube containing thousands of tiny consecutive pinholes, the uranium hexafluoride containing uranium-235 travels faster and reaches the end of the tube first. In this way, scientists were able to separate the two isotopes of uranium.

This method of separating uranium-235 from uranium-238 is long and tedious. Another source of fissionable material was made available by the discovery that bombarding uranium-238 with neutrons forms one isotope of plutonium, plutonium-239, that undergoes fission. Because uranium and

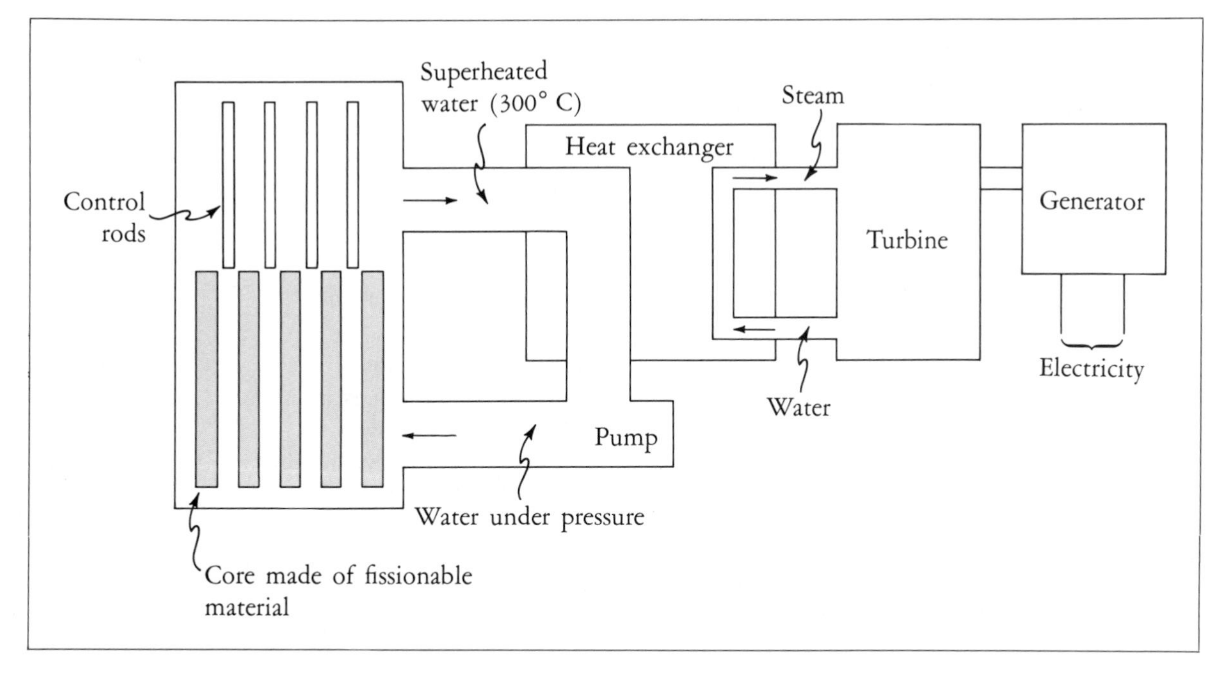

Fig. 4-4. A pressure nuclear power plant. Water under pressure is circulated through the core, where it is heated to about 300° C. The water does not boil because it is under pressure. This superheated water is used to form steam to turn a turbine that generates electricity.

plutonium are different elements, they can be separated chemically. This is a much easier way to obtain fissionable material. Today, plutonium-239 is the fissionable material in most nuclear power plants.

Although nuclear power plants promise unlimited energy, they have several disadvantages. The major one is that the products of fission are highly radioactive. As a result, their disposal is a serious problem. Several of these radioactive isotopes are particularly dangerous because they have long half-lives and they are absorbed and become part of plants and animals. Several of these isotopes are listed in Table 4-4. Strontium-90 is particularly dangerous to people because it belongs to the same group (Group IIA) as calcium and therefore behaves chemically like calcium. Both calcium and strontium are incorporated into the bones of humans and animals, where they stay for a long time. Because of its long half-life (27.7 years), strontium-90 remains as a source of radiation in the body for many years.

Another kind of nuclear reaction may allow us to obtain energy without forming dangerous radioactive products. These reactions, called *nuclear fusion reactions,* involve fusing two or more nuclei to make a heavier nucleus.

Table 4-4. Some Dangerous Radioactive Elements Produced by Nuclear Fission

Isotope	Half-Life	Radiation	Danger
Iodine-131	8 days	β, γ	Concentrates in thyroid and causes thyroid cancer
Krypton-85	10.8 yr	β, γ	Inhalation exposes lungs to radiation
Xenon-133	5.3 days	β, γ	
Strontium-90	27.7 yr	β	Concentrates in bone and causes bone cancer
Cesium-137	30 yr	β, γ	Acts like sodium and potassium because it is in the same group; travels throughout the body, thereby exposing it to radiation
Radium-226	1,602 yr	α, γ	Similar to strontium-90

The following is one example of a fusion reaction:

$$^{2}_{1}H + ^{2}_{1}H \longrightarrow ^{4}_{2}He$$

In this reaction, two nuclei of deuterium, a stable isotope of hydrogen (see Section 3.6), are forced together to form a helium nucleus, which is also stable. To accomplish this reaction, a temperature of at least 10,000,000° C is needed to provide the energy to force the two positive nuclei close enough to fuse.

Such fusion reactions occur in two places in the world. The first is the sun, whose heat and light are produced by fusion reactions. The other is in a hydrogen bomb. The high temperature needed to carry out the fusion reaction is furnished by the explosion of an atomic bomb (fission bomb). The atomic bomb is set off first and provides the energy to trigger the hydrogen bomb.

The control of fusion reactions to produce energy in a power plant involves such great technical difficulties that it will be many decades before this goal is realized.

4.10 SUMMARY

The nuclei of certain isotopes are unstable. These isotopes undergo spontaneous nuclear reactions that cause nuclear radiation to be given off. Such isotopes are radioactive. Five kinds of radiation are given off: alpha, beta, gamma, neutron, and positron. This radiation from radioactive isotopes and x-rays forms ions in matter by knocking electrons off the atoms in its path. Ionizing radiation can cause cancer, damage to the fetus, and genetic damage. This radiation is particularly dangerous because it cannot be seen or felt. It is detected by photographic film, scintillation counters, and Geiger counters. Two important characteristics of radioactive isotopes are their half-lives and the amount of radiation they give off.

The emission of alpha and beta radiation causes one element to change into another element. This change occurs whenever the nucleus of an atom gains or loses positive charge. As a result, its atomic number is changed. Scientists can make radioactive isotopes not found in nature by bombardment reactions. In these reactions, atomic or subatomic particles strike atoms and permanently change them. Some of these isotopes formed by bombardment reactions are used in medicine as tracers; others are used to treat cancer.

Only three isotopes, uranium-235, uranium-233, and plutonium-239, undergo fission reactions when bombarded by neutrons. In these reactions, the nucleus is split, giving off two, three, or four neutrons, a great deal of energy, and several elements of atomic mass lighter than the original element. The release of neutrons during a fission reaction is important because it keeps the chain reaction going. An atomic explosion is an uncontrolled fission chain reaction. By inserting rods made of carbon, boron, or cadmium into the fissionable material, a fission chain reaction can be controlled. As a result, the heat, released over a long period of time, can be used to generate electricity in nuclear power plants. The products of fission reactions are highly radioactive and dangerous to humans.

It is possible that nuclear energy can be obtained from fusion reactions that do not form radioactive products. In these reactions, the nuclei of two or more elements are fused to form an element of heavier mass. Such a fusion reaction, fusing two deuterium nuclei to form a helium nucleus, occurs in the sun and in a hydrogen bomb.

REVIEW OF TERMS AND CONCEPTS

Terms

ACCELERATOR A device that increases the speed of a charged particle to increase its effect on an atom it strikes.

ALPHA RADIATION Helium nuclei given off by a radioactive isotope.

BACKGROUND RADIATION Radiation received by plants, animals, and humans from natural sources.

BECQUEREL The SI unit that is a measure of the number of nuclei that decay per second.

BETA RADIATION Electrons given off by a radioactive isotope.

BOMBARDMENT REACTION A reaction that occurs when particles of atomic or subatomic size strike atoms and change one element into another element.

CHAIN REACTION A series of repeated reactions in which one reaction produces a product that allows another identical reaction to occur.

COSMIC RAYS Particles that enter our atmosphere from the sun.

CRITICAL MASS The amount of fissionable material that allows a chain reaction to be self-sustaining.

CURIE A unit of radioactivity: 1 curie (Ci) $= 3.7 \times 10^{10}$ disintegrations per second.

CYCLOTRON A circular accelerator of charged particles.

FISSION REACTION A nuclear reaction that splits a large nucleus into two intermediate-sized nuclei, neutrons, and energy.

FUSION REACTION A nuclear reaction in which the nuclei of two or more elements are fused to form an element of heavier mass.

GAMMA RADIATION A form of energy given off in a nuclear reaction.

GRAY The SI unit of absorbed dose of radiation.

HALF-LIFE The time needed for one half of the original sample of a material to disappear.

IONIZING RADIATION Any radiation that forms ions when it interacts with matter.

LD_{50} VALUE The dose of radiation, in rem, that is fatal within 30 days to 50 percent of those exposed.

NUCLEAR RADIATION Particles and energy given off by a nuclear reaction.

POSITRON A particle that has about the same mass as an electron but a positive charge.

RAD A measure of nuclear radiation energy absorbed by tissue or other material.

RADIOACTIVITY The emission of nuclear radiation.

REM A measure of biological damage caused by nuclear radiation.

TRACER A substance whose movement can be followed or traced by means of the radiation it gives off.

TRANSMUTATION The changing of one element into another.

Concepts

1. Most of the isotopes of the elements are stable. Only certain isotopes have nuclei that are unstable and consequently radioactive.
2. The five major types of nuclear radiation are alpha, beta, gamma, neutron, and positron.
3. Nuclear radiation is dangerous to humans. It can cause short-term damage such as cancer and long-term damage such as genetic damage.
4. Only three isotopes are known to undergo fission reactions.
5. A critical mass of fissionable material undergoes a chain reaction that gives off a tremendous amount of energy. When controlled, the release of this energy can be used to generate electricity in nuclear power plants.

EXERCISES

4-5 Give one physical property of each of the five kinds of nuclear radiation that distinguishes it from all the others.

4-6 Give three methods of detecting nuclear radiation.

4-7 Write equations for the emission of beta radiation from each of the following nuclei:
(a) hydrogen-3 (b) technetium-99 (c) cobalt-60
(d) strontium-90

4-8 Write equations for the emission of alpha radiation from each of the following nuclei:
(a) plutonium-239 (b) bismuth-214 (c) uranium-238
(d) polonium-210

4-9 Zirconium-85 emits beta radiation to form isotope X. Isotope X emits beta radiation to form the stable isotope Y. Identify isotopes X and Y.

4-10 A nuclear explosion releases 7 kg of radioactive plutonium-239, whose half-life is 24,000 years. How many kg of plutonium-239 remain in the environment from this explosion after 96,000 years?

4-11 Give an example of a transmutation reaction.

4-12 Which three isotopes are capable of undergoing fission reactions?

4-13 Why is it impossible to separate chemically uranium-235 and uranium-238, whereas uranium-238 and plutonium-239 can be separated chemically?

4-14 What is the difference between a fission reaction and a fusion reaction?

CHEMICAL BONDS 5

Molecules were identified in Chapter 2 as the smallest representative sample of a compound. Molecules are made up of atoms joined together. The links that join these atoms are called chemical bonds. The properties of a molecule are determined by its chemical bonds. The number of bonds per atom and the direction and strength of these bonds are all important in determining the shape and chemical reactions of a molecule. Chemical bonds are like mortar between bricks. They establish the structure and function of the molecule.

In this chapter we will learn about the principal kinds of chemical bonds. At the root of almost all chemical bonds is the fact that the noble gas electron arrangement of atoms is particularly stable.

5.1 OCTET RULE

In Chapter 3 we learned that the noble gases neither gain nor lose electrons easily. Consequently, they are relatively unreactive. This resistance to change is due to their electron arrangements. All the noble gases except helium have an octet of electrons (eight electrons) in their outer electron shell. Atoms of other elements gain or lose electrons from their outer shell to achieve the noble gas electron arrangement. This suggests that an octet of electrons in the outershell is a particularly stable arrangement in nature. *Atoms that do not have an octet of electrons in their outer shell will gain or lose electrons to achieve this electron arrangement.* This is the octet rule.

An atom can gain an octet of electrons in either of two ways. One way is transfer of outer shell electrons from an atom of one element to an atom of another element. This process forms ionic bonds. The second way is the sharing of outer shell electrons by two or more atoms. This process forms covalent bonds. We will examine these two kinds of bonds in more detail.

To understand how ionic bonds are formed, we must learn more about ions.

5.2 IONS

Positive ions, called *cations,* are formed when atoms lose electrons (see Section 3.9). Negative ions, called *anions,* are formed when atoms gain electrons (see Section 3.9).

The charge on an ion is determined by the number of electrons it must gain or lose to achieve the electron arrangement of the nearest noble gas. For example, atoms of the elements in group IA (Li, Na, K, Rb, and Cs) all lose one electron to form cations with one positive charge. Atoms of the elements in group IIA (Be, Mg, Ca, Sr, Ba, and Ra) must lose two electrons to achieve a noble gas electron arrangement. Consequently, these alkaline earth elements form cations with two positive charges. Finally, atoms of elements in group IIIA (Al, Ga, In, and Tl) lose three electrons to form cations with three positive charges. All these cations have a stable noble gas electron arrangement in their outer shell. The electron arrangements of representative examples of these cations are given in Table 5-1.

Other atoms achieve noble gas electron arrangements by adding electrons to form anions. Atoms of elements in group VIIA (F, Cl, Br, and I) add one electron to form anions with one negative charge. Atoms of elements in group VIA (O, S, Se, and Te) add two electrons to form anions with two

Table 5-1. Noble Gas Electron Arrangement of Ions

Element	Electron Arrangement of Element	No. of Electrons Gained or Lost to Form Ion	Electron Arrangement of Ion	Electron Arrangement of Nearest Noble Gas
Potassium	K : 2 : 8 : 8 : 1	Lose 1	K^{+1} : 2 : 8 : 8	Ar : 2 : 8 : 8
Magnesium	Mg : 2 : 8 : 2	Lose 2	Mg^{+2} : 2 : 8	Ne : 2 : 8
Aluminum	Al : 2 : 8 : 3	Lose 3	Al^{+3} : 2 : 8	Ne : 2 : 8
Chlorine	Cl : 2 : 8 : 7	Gain 1	Cl^{-1} : 2 : 8 : 8	Ar : 2 : 8 : 8
Oxygen	O : 2 : 6	Gain 2	O^{-2} : 2 : 8	Ne : 2 : 8

negative charges. All these anions have a stable noble gas electron arrangement in their outer shell, as shown in Table 5-1.

Notice that atoms of elements in groups at the far left of the periodic table lose electrons to form cations. This tendency is indicated by the fact that they all have relatively low ionization energies (see Section 3.9). In contrast, atoms of elements in groups VIA and VIIA gain electrons to form anions. This tendency is indicated by the fact that they all have relatively high electron affinities (see Section 3.9).

*EXERCISE 5-1 Write the electron arrangement and give the charge of the ion formed by the loss or gain of electrons from each of the following elements:
(a) Ca (b) S (c) Li (d) F (e) Al

Each cation and anion is given a unique name and symbol. The names of cations formed from a single atom are obtained from the name of the parent element. The cation formed from the element sodium is called a sodium ion. The names of anions formed from a single atom are obtained by changing the ending of the name of the parent element to -ide. The anion formed from a chlorine atom is called a chloride ion. The names of the most common cations and anions are given in Table 5-2.

The symbols of ions are the same as the symbols of the elements from which they are formed except that the charge is included as a superscript to the right of the symbol. The symbols of the most common ions are given in Table 5-2.

EXERCISE 5-2 Does each of the following symbols represent an ion or an element?
(a) Na (b) Cl^- (c) H^+ (d) Cu (e) Li^+ (f) O^{-2}
(g) Mg^{+2}

Included in Table 5-2 are several cations of the transition metals found in groups VIIIB, IB, and IIB. The simple octet rule cannot predict the electron arrangement of these cations. In fact, several of these metals form more than one cation. We shall not go into the reasons why these metals form such cations, which are beyond the scope of this book; however, the names and symbols of these cations should be learned, because they are important in living systems.

A distinction must be made between the two different ions formed by

* The answers for the exercises in this chapter begin on page 859.

Table 5-2. Names, Symbols, and Ionic Charges of the Most Common Ions

Symbol of Element	Symbol of Ion	Name of Ion	Ionic Charge
H	H^+	Hydrogen ion (proton)	+1
Group IA			
Li	Li^+	Lithium ion	+1
Na	Na^+	Sodium ion	+1
K	K^+	Potassium ion	+1
Group IIA			
Mg	Mg^{+2}	Magnesium ion	+2
Ca	Ca^{+2}	Calcium ion	+2
Ba	Ba^{+2}	Barium ion	+2
Group IIIA			
Al	Al^{+3}	Aluminum ion	+3
Transition metals			
Cu	Cu^+	Cuprous ion	+1
Cu	Cu^{+2}	Cupric ion	+2
Fe	Fe^{+2}	Ferrous ion	+2
Fe	Fe^{+3}	Ferric ion	+3
Hg	Hg^+	Mercurous ion	+1
Hg	Hg^{+2}	Mercuric ion	+2
Ag	Ag^+	Silver ion	+1
Mn	Mn^{+2}	Manganous ion	+2
Group VIA			
O	O^{-2}	Oxide ion	−2
S	S^{-2}	Sulfide ion	−2
Group VIIA			
F	F^-	Fluoride ion	−1
Cl	Cl^-	Chloride ion	−1
Br	Br^-	Bromide ion	−1
I	I^-	Iodide ion	−1

metals such as iron, copper, and mercury. One method is to use a different suffix for each ion. For example, Cu^+ is called a cuprous ion, while Cu^{+2} is called a cupric ion. In general, the ending -ic is used to name the ion with the most positive charge and the ending -ous is given to the ion with the least positive charge. Another method is to use a roman numeral after the symbol of the element. The ion Cu^+ is written Cu(I), whereas Cu^{+2} is written Cu(II).

The names and symbols of ions resemble those of the elements from which they are obtained. However, the physical and chemical properties of ions and elements are completely different. Iron is a typical example. Elemental iron is the major component of steel. Like most metals, iron shows a metallic luster and is a good conductor of heat and electricity. Ferric ions, on the other hand, exist only in combination with anions in compounds. These compounds have none of the properties associated with the metal.

This difference between elemental iron and ferrous ions is clearly illustrated by the clinical treatment of anemia. People who are anemic are deficient in ferrous ions. These people are given a compound that contains ferrous ions and an anion, as we will learn in Section 5.4, to cure anemia. They are not given small pieces of iron metal! The body recognizes that the element iron (Fe) and ferrous ions (Fe^{+2}) are entirely different.

Cations and anions do not exist alone. Anions are attracted to cations and vice versa. This attraction between ions of different charges forms ionic bonds.

5.3 IONIC BONDS

Ionic bonds are formed by a strong attraction between negative and positive ions. Anions and cations are formed in many chemical reactions by the transfer of one or more electrons from one element to another.

The transfer of electrons between atoms can be pictured by using an electron dot representation of the outer shell electron arrangement of atoms. In these representations, the symbol of the atom is surrounded by the number of dots equal to the number of electrons in the outer shell of the atom. Sodium atoms, for example, are represented as $Na\cdot$; beryllium atoms, as $Be\colon$; chlorine atoms, as $\colon\overset{\cdot\cdot}{\underset{\cdot\cdot}{Cl}}\cdot$; oxygen atoms, as $\colon\overset{\cdot\cdot}{\underset{\cdot\cdot}{O}}$; and so forth. It makes no difference where the dots are placed or how they are arranged. Only the total number of dots is important. Thus, all the following represent the chlorine atom and its outer-shell electron arrangement:

$$\colon\overset{\cdot\cdot}{\underset{\cdot\cdot}{Cl}}\cdot \qquad \cdot\overset{\cdot\cdot}{\underset{\cdot\cdot}{Cl}}\colon \qquad \colon\overset{\cdot}{\underset{\cdot\cdot}{Cl}}\colon \qquad \colon\overset{\cdot\cdot}{\underset{\cdot}{Cl}}\colon$$

EXERCISE 5-3 Write an electron dot representation for atoms of each of the following elements:
(a) K (b) Mg (c) S (d) F (e) Al (f) Li

We can use electron dot representations to show the transfer of an electron in the reaction between sodium and chlorine atoms. In this reaction, a sodium atom loses an electron to a chlorine atom with the result that two ions are formed, a sodium ion and a chloride ion. This electron transfer can be represented as follows:

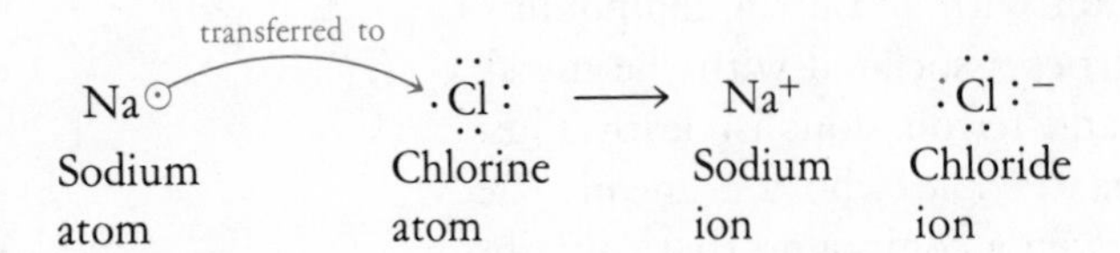

Notice that the transfer of an electron from a sodium atom to a chlorine atom forms one positive ion and one negative ion. The attraction between these two ions binds them together. We call such an attraction between ions of opposite charge an ionic bond.

Other atoms of elements in group IA can combine with atoms of elements in group VIIA by transferring one electron to form compounds with ionic bonds. All alkali metals give up one electron, and all halogens accept only one electron. Therefore, the ratio of cations to anions in these compounds is always one to one.

Elements of group IIA, the alkaline earth metals, lose two electrons to form cations with two positive charges. Halogen atoms can accept these electrons, but two halogen atoms are needed to accept the two electrons from each alkaline earth atom. The double charge of the cation is exactly balanced by two singly charged anions. Therefore, there are two anions to each cation in these compounds, and the ratio of cations to anions is one to two. An example of such a compound is the one formed by the combination of beryllium and fluorine atoms. We can represent this transfer of electrons as follows:

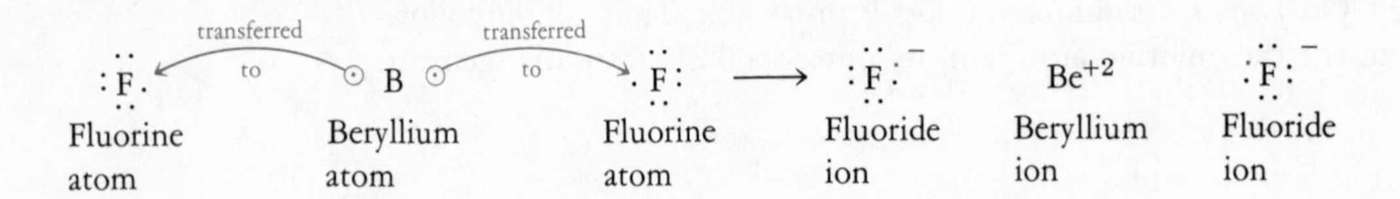

Again, the ions of opposite charge are attracted to each other and form an ionic bond.

The composition of a compound is indicated by its *chemical formula*. The two-to-one ratio of anions (F^-) to cations (Be^{+2}) is indicated by the chemical formula BeF_2. By convention, the symbol of the element that forms the

Table 5-3. Relationship Between Chemical Formula and Ratio of Cations to Anions

Chemical Formula	Ratio of Cations to Anions (Cations : Anions)
NaCl	1 : 1
$MgCl_2$	1 : 2
Na_2O	2 : 1
BeO	1 : 1

cation is given first, followed by the symbol of the element that forms the anion. The subscript 2 indicates that there are two F^- ions for each Ca^{+2} ion in the compound. When no subscript is found next to the symbol of an element, the number one is understood. The relationship between the chemical formula and the numbers of cations and anions is given in Table 5-3.

EXERCISE 5-4 Indicate the charge on each ion and the ratio of cations to anions in each of the following compounds:
(a) LiCl (b) MgF_2 (c) Li_2S (d) MgO (e) Al_2O_3 (f) $CaCl_2$

The cations of the alkaline earth metals can also combine with anions that have two negative charges, such as oxygen or sulfur anions. In these cases, the double charge of the cations is exactly balanced by the double charge of the anions. Therefore, the ratio of cations to anions is one to one in these compounds.

Many compounds with ionic bonds are formed by various combinations of cations and anions. But the combination must occur in such a way that a neutral compound is formed; that is, the charge on the cations must be exactly balanced by the charge on the anions. As a result, the chemical formulas of these compounds are determined by the charges on the cations and anions that combine to form the compound (Table 5-4).

The chemical formulas of most compounds made up of cations and anions can be obtained from Table 5-4. The letter M in Table 5-4 represents any metal that can form a cation having one, two, or three positive charges. The letter X represents any element or group (see Section 5.8) that can form an anion having one or two negative charges. From Table 5-4, we can determine that any cation having three positive charges (M^{+3}) will combine with any anion having two negative charges (X^{-2}) to form a compound

Table 5-4. Molecular Formula of Neutral Ionic Compounds Formed by Combining Cations and Anions[a]

Group IA M^{+1}	Group IIA M^{+2}	Group IIIA M^{+3}	
MX	MX_2	MX_3	Group VIIA X^{-1}
M_2X	MX	M_2X_3	Group VIA X^{-2}

[a] The letter M represents any metal that can form a cation having one, two, or three positive charges. The letter X represents any element or group that can form an anion having one or two negative charges.

with the chemical formula M_2X_3. An example of such a compound is Al_2O_3.

EXERCISE 5-5 Write the chemical formula of the compound formed by combination of the following ions:
(a) K^+, O^{-2} (b) Ca^{+2}, Cl^- (c) Fe^{+2}, Br^- (d) Cu^{+2}, S^{-2}
(e) Fe^{+3}, O^{-2}

Compounds made up of ions held together by ionic bonds are called ionic compounds. Because of their ionic bonds, such compounds have characteristic properties.

5.4 STRUCTURE, PROPERTIES, AND NAMES OF IONIC COMPOUNDS

Ionic compounds are composed of cations and anions held together by ionic bonds. These compounds are usually called salts and they are solids at room temperature. A visible sample of an ionic compound contains billions of cations and anions. These ions arrange themselves in an orderly fashion as a result of the fact that each cation repels all other positively charged ions and attracts anions. Similarly, each anion attracts positively charged ions and repels all other anions. As a result, there is a regular three-dimensional pattern of alternating cations and anions. Such an arrangement of ions is called a *crystal lattice*. The crystal lattice of common table salt, NaCl, is shown in Figure 5-1.

Each sodium ion in the crystal lattice of NaCl is surrounded by six chloride ions. Likewise, each chloride ion is surrounded by six sodium ions.

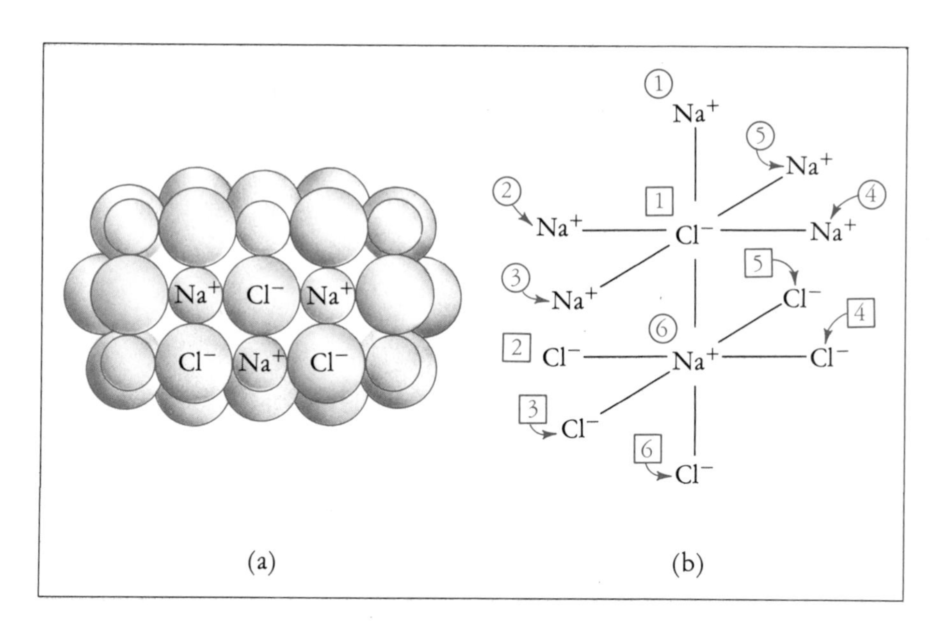

Fig. 5-1. Two representations of the crystal lattice of NaCl.

This is shown in Figure 5-1b for specific sodium and chloride ions. For clarity, each of the six nearest neighbors is numbered.

One question comes to mind as we look at the crystal lattice of NaCl in Figure 5-1. Where is the molecule of NaCl? The answer is that no molecule of NaCl exists! The reason is that we cannot pick out at random any adjacent pair of sodium and chloride ions and call them a molecule. The ionic compound NaCl is a regular arrangement of ions. Therefore, *chemists regard ionic compounds as an orderly pattern of oppositely charged ions.* This means that there is no such thing as a molecule of an ionic compound. Thus, the compound NaCl exists, but there is no molecule of NaCl.

The orderly arrangement of ions in ionic compounds gives them characteristic properties. Ionic compounds generally have high melting points, are soluble in water, and conduct electricity when melted or dissolved in water. Table salt, NaCl, has all these properties: it melts at 804° C; its solubility in water is familiar to everyone; it conducts electricity when melted or dissolved in water. These properties are characteristic of salts.

Ionic compounds are given names to indicate their composition. The name is made up of two words. The first word is the name of the cation, and the second word is the name of the anion. This method of naming salts is illustrated by the following examples:

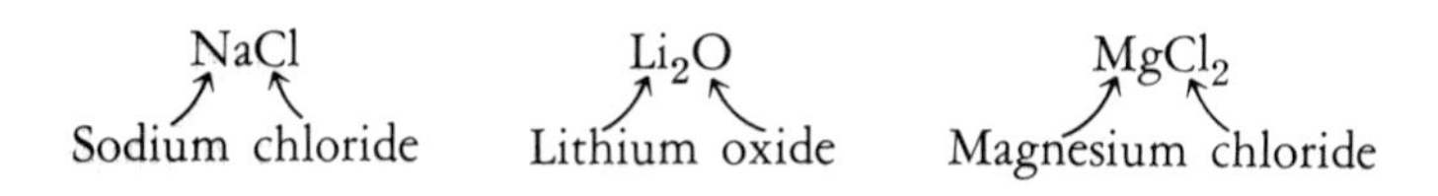

Table 5-5. Several Salts and Their Medicinal Uses

Name	Formula	Uses
Sodium bicarbonate[a] (baking soda)	$NaHCO_3$	Antacid
Silver nitrate[a]	$AgNO_3$	Germicide and antiseptic
Stannous fluoride	SnF_2	Added to toothpaste to decrease incidence of dental caries
Calcium sulfate[a] (plaster of Paris)	$(CaSO_4)_2 \cdot H_2O$	Plaster casts
Magnesium sulfate[a] (epsom salts)	$MgSO_4 \cdot 7H_2O$	Cathartic
Calcium carbonate[a]	$CaCO_3$	Antacid
Potassium permanganate[a]	$KMnO_4$	Cauterizing agent, antiseptic
Zinc oxide	ZnO	Base powder for calamine lotion
Ferrous sulfate[a]	$FeSO_4$	Prescribed for simple iron deficiency anemia
Zinc sulfate[a]	$ZnSO_4$	Used to treat skin conditions such as eczema
Barium sulfate[a]	$BaSO_4$	Important ingredient of a "barium cocktail": provides the contrast material for gastrointestinal radiographs
Mercurous chloride (calomel)	Hg_2Cl_2	Cathartic

[a] These salts contain polyatomic anions whose names are given in Table 5-7.

EXERCISE 5-6 Name each of the following compounds:
(a) LiBr (b) NaI (c) CaF_2 (d) $FeCl_3$ (e) Al_2O_3
(f) CuS (g) $HgCl_2$

Many of these salts have important medicinal applications, as shown in Table 5-5.

Atoms need not completely lose one or more electrons to form a chemical bond. The atoms of many elements share electrons to form covalent bonds.

5.5 COVALENT BONDS

Covalent bonds are formed by sharing electrons to achieve a noble gas electron arrangement for each atom. A pair of electrons shared between two atoms satisfies the needs of each of the two atoms. As an example, two hydrogen atoms combine to form the bond in a hydrogen molecule by sharing the two available electrons. We can represent this as follows:

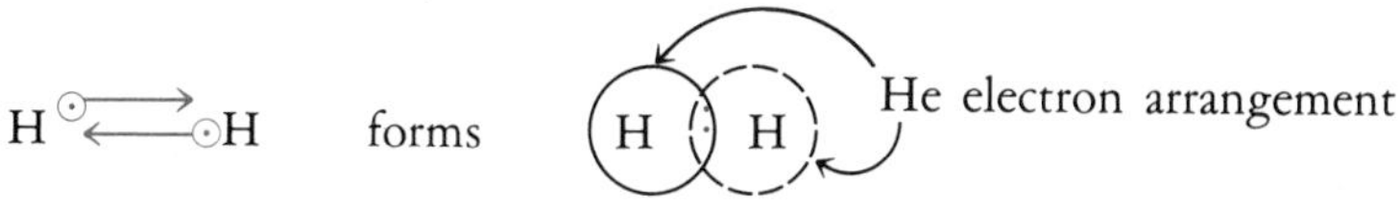

H⊙ ⇄ ⊙H forms (H : H) He electron arrangement

The two dots between the two symbols for the element hydrogen indicate that the pair of electrons is shared. Notice that only one pair of electrons is shared between the two atoms. These two electrons satisfy the requirement of a noble gas electron arrangement of *both* hydrogen atoms. As a result, the two atoms join together to form a very stable hydrogen molecule. This shared pair of electrons is called a covalent bond.

Although two dots (:) between two symbols for the elements can be used to represent the pair of electrons in a covalent bond, it is simpler to represent them by a dash (—). The structure of the hydrogen molecule is usually written in this way as H—H.

Other examples of covalent bonds formed by sharing an electron pair are those in chlorine and fluorine.

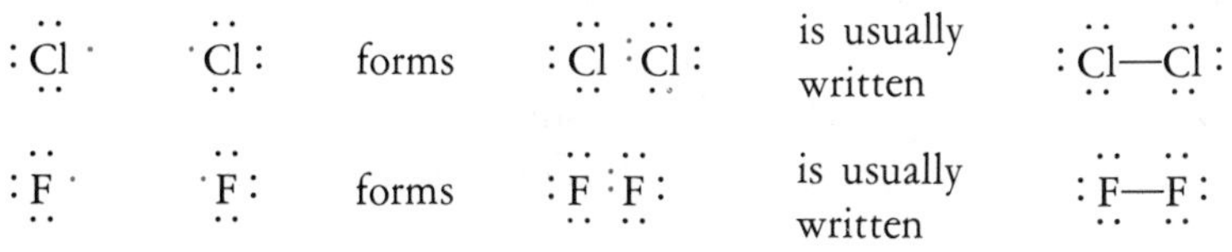

:C̤̈l· ·C̤̈l: forms :C̤̈l:C̤̈l: is usually written :C̤̈l—C̤̈l:

:F̤̈· ·F̤̈: forms :F̤̈:F̤̈: is usually written :F̤̈—F̤̈:

Two atoms can share more than one electron pair to attain the noble gas electron arrangement. Two covalent bonds between two atoms are called a *double bond*. The following is an example of a molecule that contains a double bond:

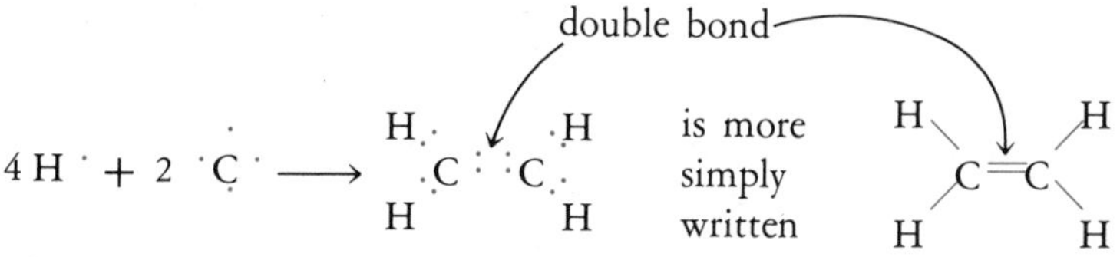

4 H· + 2 ·Ċ̣· ⟶ H:C::C:H (with H above and below each C) is more simply written $H_2C{=}CH_2$ (double bond)

Dashes, but two this time (=), replace the dots between the symbols for the two atoms.

Three electron pairs can be shared between two atoms to form a *triple bond.* The following are examples of molecules containing a triple bond:

2 H· + 2 ·Ċ· ⟶ H:C:::C:H is more simply written H—C≡C—H

triple bond

:Ṅ· + ·Ṅ: ⟶ :N:::N: is more simply written :N≡N:

Three parallel dashes (≡) are used, instead of the dots to represent a triple bond.

EXERCISE 5-7 Draw a circle around the noble gas electron arrangement for each atom in the following compounds:

(a) :Ï:Ï: (b) :N:::N: (c) H:C:::C:H

(d) H.C::C.H (H and H on each carbon)

The tendency to gain a noble gas electron arrangement by sharing electrons is so strong in certain elements that they exist as *diatomic molecules.* The elements hydrogen (H_2), oxygen (O_2), nitrogen (N_2), fluorine (F_2), chlorine (Cl_2), bromine (Br_2), and iodine (I_2) all exist as molecules made up of two atoms of the element.

Such diatomic molecules are extremely stable. Energy, usually as heat, is liberated when a covalent bond is formed. In the case of the bond between two hydrogen atoms, this amounts to 52 kcal/g H_2. Conversely, to break a covalent bond into two atoms requires energy. Therefore, a molecule is a more stable arrangement of the nuclei and electrons than are the isolated atoms. This is true because in the molecule each electron of the bond is attracted by two positive nuclei, whereas in the separate atoms, each electron is attracted by only one positive nucleus. However, this attractive force in a molecule is opposed by repulsion between the two positive nuclei and between the two electrons. In a covalent bond a balance is attained between these two opposing forces and the atoms remain bonded to one another.

Electrons are shared equally only between atoms of the same element. Two atoms of different elements can also share electrons, but the sharing is unequal. Bonds formed by the unequal sharing of a pair of electrons are called polar covalent bonds.

5.6 POLAR COVALENT BONDS

The ability to attract electrons differs from atom to atom. The electronegativity of an atom (see Section 3.9) is a measure of this ability. The greater the electronegativity of an atom, the more likely it is to attract electrons. The type of bond formed between two atoms of different elements depends on the difference in their electronegativities. Two atoms of widely differing electronegativity, such as sodium and chlorine, form an ionic bond. Two atoms of the same element have identical electronegativities and form a covalent bond. Between these two extremes are bonds formed between atoms of intermediate differences in electronegativity. The electron pair is still shared between the atoms, but the sharing is unequal. This unequal sharing results in a *polar covalent bond.*

An example of a polar covalent bond is found in a molecule of HCl (hydrogen chloride). The hydrogen and chlorine atoms share a pair of electrons. Because chlorine is more electronegative than hydrogen, the electron pair is drawn toward chlorine, giving it a partial negative charge and leaving a partial positive charge on the hydrogen. We indicate a partial charge on an atom by the symbol δ, delta. The symbol $\delta+$ indicates a partial positive charge; $\delta-$ indicates a partial negative charge. These symbols are often added to the structural formula of the molecule to indicate an unequal sharing of the electron pair as follows:

$$\overset{\delta+}{\mathrm{H}} - \overset{\delta-}{\mathrm{Cl}}$$

In a polar covalent bond, the more electronegative atom acquires a partial negative charge, and its less electronegative partner acquires a partial positive charge.

Based on the electronegativities of the elements given in Section 3.9, we can predict that the atom on the right in each of the following bonds is partially negative with respect to its partner:

$$\overset{\delta+}{\mathrm{C}} - \overset{\delta-}{\mathrm{N}} \qquad \overset{\delta+}{\mathrm{C}} - \overset{\delta-}{\mathrm{F}} \qquad \overset{\delta+}{\mathrm{Li}} - \overset{\delta-}{\mathrm{C}}$$

$$\overset{\delta+}{\mathrm{C}} - \overset{\delta-}{\mathrm{O}} \qquad \overset{\delta+}{\mathrm{H}} - \overset{\delta-}{\mathrm{O}}$$

$$\overset{\delta+}{\mathrm{C}} - \overset{\delta-}{\mathrm{Cl}} \qquad \overset{\delta+}{\mathrm{H}} - \overset{\delta-}{\mathrm{N}}$$

The electronegativities of carbon and hydrogen are nearly the same. Consequently, the electron pair in the carbon-hydrogen bond, so prevalent in organic chemistry, is shared about equally between the two atoms.

Although the sharing of electrons in a polar covalent bond is unequal, the sharing still satisfies the requirements of a noble gas electron arrangement for the two atoms forming the bond. Furthermore, the noble gas electron arrangement for an atom can be attained by sharing electrons with more than one other atom. It is this fact that allows us to determine the number of bonds that atoms are capable of forming.

5.7 BONDING CAPACITIES OF ATOMS

Atoms of the elements hydrogen, carbon, nitrogen, oxygen, sulfur, and the halogens form polar covalent bonds with atoms of other elements. The number of bonds that any of these atoms can form is determined by the number of electrons it needs to gain a noble gas electron arrangement. A carbon atom, for example, has four electrons in its outer shell. It must get four more electrons to gain a noble gas arrangement. It can do this by forming four bonds with atoms of other elements or other carbon atoms. Thus, one carbon atom combines with four hydrogen atoms to form the molecule CH_4. We can picture this process as follows:

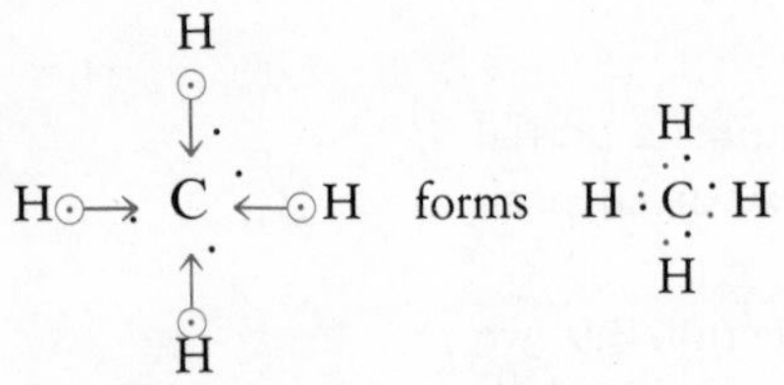

```
       H
       ⊙
       ↓.
H⊙→  C  ←⊙H    forms    H
      .                H:C:H
       ↑.                H
       ⊙
       H
```

In this way, we draw an electron-dot structure of the molecule CH_4. As before, each pair of dots representing a bond can be replaced by a dash to form the following *structural formula:*

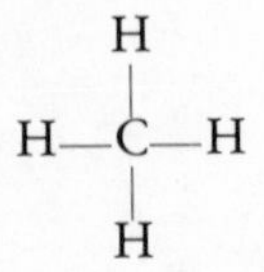

```
   H
   |
H—C—H
   |
   H
```

Table 5-6. Number of Polar Covalent Bonds Formed by the Atoms of Elements Important to Living Systems

Element	H	C	N	O	S	Cl
Atomic no.	1	6	7	8	16	17
Total no. of electrons	1	6	7	8	16	17
No. of outer shell electrons	1	4	5	6	6	7
No. of electrons needed to form nearest noble gas electron structure	1	4	3	2	2	1
No. of bonds formed	1	4	3	2	2	1

In a similar manner, oxygen, which has six electrons in its outer shell, forms two bonds to gain a noble gas electron arrangement. The oxygen atom in a molecule of water, H_2O, forms bonds with two hydrogen atoms as follows:

$$H\odot\!\!\rightarrow \ddot{\underset{\cdot\cdot}{O}}\!\leftarrow\!\odot H \quad \text{forms} \quad H\!:\!\ddot{\underset{\cdot\cdot}{O}}\!:\!H \quad \text{is more simply written} \quad H-\ddot{\underset{\cdot\cdot}{O}}-H$$

The number of bonds formed by atoms of the elements important to living systems is given in Table 5-6. Structural formulas are very useful representations of molecules because we know from them the way atoms in molecules are joined together. Thus, we know from the structural formula of water that the atoms are joined H—O—H and not H—H—O. This information is important because it allows us to establish a relationship between the structure of a molecule and its chemical properties. Such relationships are particularly important for organic compounds, as we will learn in Chapter 11.

EXERCISE 5-8 Circle the noble gas electron arrangement for each atom in the following structural formulas:

(a) $:\ddot{\underset{\cdot\cdot}{I}}-\ddot{\underset{\cdot\cdot}{I}}:$ (b) $H-\overset{H}{\underset{H}{\overset{|}{\underset{|}{N^{+}}}}}-H$ (c) $:\ddot{\underset{\cdot\cdot}{Cl}}-\overset{:\ddot{Cl}:}{\underset{:\underset{\cdot\cdot}{Cl}:}{\overset{|}{\underset{|}{C}}}}-\ddot{\underset{\cdot\cdot}{Cl}}:$

(d) $H-\ddot{\underset{\cdot\cdot}{S}}-H$ (e) $\begin{matrix}H\\ \ \ \diagdown\\ \ \ \ C=\ddot{O}:\\ \ \ \diagup\\ H\end{matrix}$

EXERCISE 5-9 Write the structural formula for each of the following compounds:
(a) CF_4 (b) $ClCH_3$ (c) ICl (d) PH_3

Molecules are electrically neutral, stable arrangements of atoms. There also exist stable arrangements of atoms that are not electrically neutral. These groups of atoms are called polyatomic ions.

5.8 POLYATOMIC IONS

Polyatomic ions are groups of two or more atoms held together by polar covalent bonds. Even though they have an electrical charge, these polyatomic ions are stable. In fact, these ions undergo most chemical reactions as a unit. The major polyatomic ions are listed in Table 5-7. Because of their importance, the names, formulas, and charges of these ions must be memorized.

Polyatomic cations and anions combine in exactly the same way as simple

Table 5-7. Important Polyatomic Ions

Chemical Formula	Electron-Dot Structure	Charge	Name
NH_4^+	$\left[\begin{matrix} & H & \\ H & :\ddot{\underset{..}{N}}: & H \\ & H & \end{matrix}\right]^+$	+1	Ammonium
OH^-	$:\ddot{\underset{..}{O}}:H^-$	−1	Hydroxide
HCO_3^-	$\left[\begin{matrix} & & :O: & \\ & & :: & \\ H & :\ddot{\underset{..}{O}}: & C & :\ddot{\underset{..}{O}}: \end{matrix}\right]^-$	−1	Bicarbonate
CO_3^{-2}	$\left[\begin{matrix} & :O: & \\ & :: & \\ :\ddot{\underset{..}{O}}: & C & :\ddot{\underset{..}{O}}: \end{matrix}\right]^{-2}$	−2	Carbonate
NO_3^-	$\left[\begin{matrix} & :O: & \\ & :: & \\ :\ddot{\underset{..}{O}}: & N & :\ddot{\underset{..}{O}}: \end{matrix}\right]^-$	−1	Nitrate
NO_2^-	$\left[\ddot{\underset{..}{O}}::\ddot{N}:\ddot{\underset{..}{O}}:\right]^-$	−1	Nitrite

Table 5-7 (*Continued*)

Chemical Formula	Electron-Dot Structure	Charge	Name
$H_2PO_4^-$	$\left[\begin{array}{c} :\underset{\cdot\cdot}{O}: \\ H:\underset{\cdot\cdot}{\ddot{O}}:\underset{\cdot\cdot}{\ddot{P}}:\underset{\cdot\cdot}{\ddot{O}}:H \\ :\underset{\cdot\cdot}{O}: \end{array}\right]_{a}^{-}$	−1	Dihydrogen phosphate
HPO_4^{-2}	$\left[\begin{array}{c} :\underset{\cdot\cdot}{O}: \\ :\underset{\cdot\cdot}{\ddot{O}}:\underset{\cdot\cdot}{\ddot{P}}:\underset{\cdot\cdot}{\ddot{O}}:H \\ :\underset{\cdot\cdot}{O}: \end{array}\right]_{a}^{-2}$	−2	Hydrogen phosphate
PO_4^{-3}	$\left[\begin{array}{c} :\underset{\cdot\cdot}{O}: \\ :\underset{\cdot\cdot}{\ddot{O}}:\underset{\cdot\cdot}{\ddot{P}}:\underset{\cdot\cdot}{\ddot{O}}: \\ :\underset{\cdot\cdot}{O}: \end{array}\right]_{a}^{-3}$	−3	Phosphate
HSO_4^-	$\left[\begin{array}{c} :\underset{\cdot\cdot}{O}: \\ H:\underset{\cdot\cdot}{\ddot{O}}:\underset{\cdot\cdot}{\ddot{S}}:\underset{\cdot\cdot}{\ddot{O}}: \\ :\overset{\cdot\cdot}{O}: \end{array}\right]_{a}^{-}$	−1	Hydrogen sulfate
SO_4^{-2}	$\left[\begin{array}{c} :\underset{\cdot\cdot}{O}: \\ :\underset{\cdot\cdot}{\ddot{O}}:\underset{\cdot\cdot}{\ddot{S}}:\underset{\cdot\cdot}{\ddot{O}}: \\ :\overset{\cdot\cdot}{O}: \end{array}\right]_{a}^{-2}$	−2	Sulfate

[a] In these ions, P and S do not obey the octet rule. The reason for this is beyond the scope of this text.

ions; that is, they combine so that the compound formed is electrically neutral. This is illustrated by the compound formed by combining calcium and nitrate ions. Two nitrate ions are needed for each calcium ion to balance the charges. Therefore, the chemical formula is $Ca(NO_3)_2$. Notice the use of parentheses around the nitrate ion. The subscript 2 refers to everything within the parentheses and tells us that two nitrate ions are needed for each calcium ion.

EXERCISE 5-10 Write the chemical formula of the compound formed by combining the following pairs of ions:
(a) Li^+ and OH^- (b) Na^+ and HCO_3^- (c) Ca^{+2} and CO_3^{-2} (d) NH_4^+ and HPO_4^{-2} (e) Ba^{+2} and PO_4^{-3}

The compounds containing these polyatomic ions are named by combining the name of the cation followed by the name of the anion, as shown by the following examples:

$NaHCO_3$ — Sodium bicarbonate $Ca(NO_3)_2$ — Calcium nitrate $(NH_4)_2HPO_4$ — Ammonium hydrogen phosphate

EXERCISE 5-11 Name each of the following compounds:
(a) $NaNO_2$ (b) $Ba_3(PO_4)_2$ (c) K_2HPO_4 (d) $LiHSO_4$
(e) $(NH_4)_2SO_4$

Polyatomic ions have an electrical charge because the number of electrons and the number of protons in the cluster of atoms are not equal. This fact can be confirmed by analyzing the electron arrangement of the hydroxide ion. The electron-dot structure of the ion is $:\ddot{\underset{..}{O}}:H^-$. Both hydrogen and oxygen have gained a noble gas electron arrangement. Such an arrangement requires one more electron than the total number of protons in the nuclei of both the oxygen and the hydrogen. We can verify this statement by counting the total number of electrons and protons in the ion.

Total number of protons = number of protons in nucleus of hydrogen + number of protons in nucleus of oxygen = 1 + 8 = 9

Total positive charge in ion = Total number of protons = +9

Total number of electrons = Number of electrons in outer shell of both atoms + Number of electrons in inner shells of each atom = 8 + 2 = 10

Total negative charge in ion = Total number of electrons = −10

Charge of ion = Total negative charge in ion + Total positive charge in ion = −10 + 9 = −1

The number of electrons in the outer shell of each atom is obtained from the number of dots in the electron-dot structure of the ion. In this example, the number is eight. In addition, the oxygen atom has two electrons in the first shell that are not shown in the electron-dot representation because they are not involved in bond formation.

The reason for the stability of the hydroxide ion is now clear. Both the hydrogen atom and the oxygen atom have gained a noble gas electron arrangement. In gaining this structure, the total number of electrons is one more than the total number of protons in the nuclei of the two atoms. As a result, the ion has a charge of -1.

EXERCISE 5-12 Determine the number of protons, electrons, and charge of the following ions:
(a) NH_4^+ (b) CO_3^{-2}

An electron-dot structure is given for each polyatomic ion in Table 5-7. More than one electron-dot structure or structural formula can be written for several of these ions. Does this mean that there is more than one structure for these ions? The answer is no! The resonance theory has been proposed to explain why certain ions and molecules *appear* to have more than one structural formula.

5.9 RESONANCE THEORY

More than one structural formula can be written to represent certain molecules and ions. The carbonate ion, CO_3^{-2}, is one such example. The three following structural formulas can be written for the carbonate ion:

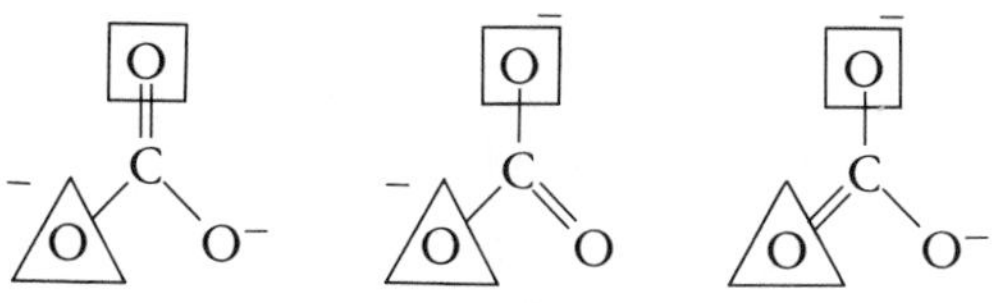

In these three structural formulas, the positions of the atoms have not changed. Only the locations of the bonding electrons have changed. In each of the structures, one of the three carbon-oxygen bonds is written as a double bond and the others are written as single bonds. These structures indicate that the three carbon-oxygen bonds of the carbonate ion are different (one is a double bond, and the other two are single bonds). However, this structure has not been confirmed by experiments. In fact, the reverse is true. Many experiments have been carried out on the carbonate ion and they all lead to the same conclusion: the three carbon-oxygen bonds of the carbonate ion are identical. We conclude, therefore, that these three structural formulas do not correctly represent the real structure of the carbonate ion. *It must be emphasized that the problem is not with the carbonate ion, but with our methods of representing on paper the bonds in molecules and ions.* How do we solve this problem?

The *resonance theory* was developed in the 1930s to solve this problem. According to this theory, the bonding in certain molecules or ions, such as the carbonate ion, is best described by writing two or more structural formulas that differ only in the positions of their electrons. The electron structure of the real molecule or ion is regarded as a *hybrid* of these structures. These structural formulas are known as *contributing* or *resonance* structures. We indicate that the real molecule or ion is a hybrid by interconnecting the contributing structures with a double-headed arrow. The hybrid structure of the carbonate ion is represented by the three contributing structures:

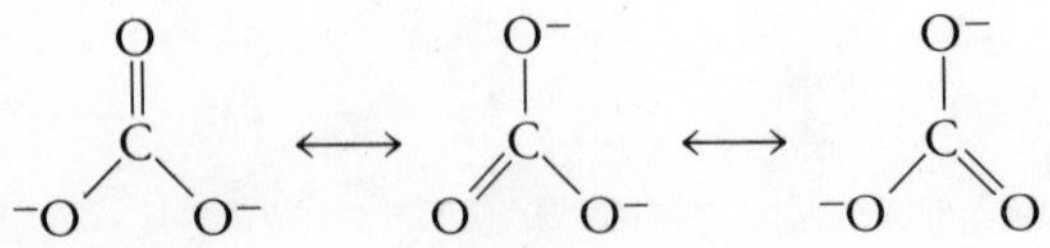

5-1 ⟷ *5-2* ⟷ *5-3*

This method of describing an object is not unique to chemistry. Consider the following analogy. Suppose that you are writing a letter and you wish to describe a friend. One way to do this would be to describe the friend in terms of fictional characters. For example, by describing your friend as a hybrid of James Bond and Sherlock Holmes, you provide the reader with a description of certain characteristics of your friend. Although James Bond and Sherlock Holmes are not real persons, they can be used to describe a real live person. Similarly, the structures *5-1*, *5-2*, and *5-3* are fictitious, but they too can be used to describe the real structure of the carbonate ion.

Because the contributing structures do not represent real molecules or ions, it is not surprising then that the hybrid structure has chemical and physical properties different from those expected of the contributing structures. We will learn about some examples of hybrid structures in Chapter 13.

It is impractical to continue to write the contributing structures of a molecule or ion whose real structure is a hybrid. How then do we write its structure? Chemists do not agree on the answer to this question. Sometimes the structure of the hybrid is written using dotted lines to indicate that certain electrons are not simply localized between two nuclei. Structure *5-4* represents the carbonate ion in this way:

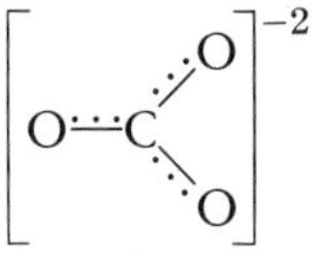

5-4

The major disadvantage of these symbols is that we lose count of the electrons in the molecule or ion. An alternative method is to use just one contributing structure to represent the molecule or ion. It is understood that this one symbol actually represents a hybrid structure. In this book, we will use one contributing structure to represent each molecule or ion whose real structure is a hybrid.

EXERCISE 5-13 Write the contributing structures to the hybrid of the NO_3^- ion.

5-10 SUMMARY

Atoms are joined together by chemical bonds to form molecules. Chemical bonds are formed because of the desire of many elements to gain a noble gas electron arrangement. The atoms of many elements gain this electron arrangement in one of two ways.

One way is to transfer one or more electrons to another atom to form ions of opposite charge. The attraction between these ions forms an ionic bond. Compounds that are made up of ions are called ionic compounds and they are electrically neutral. Consequently, the total number of charges of the cations are exactly balanced by the total number of charges of the anions. Ionic compounds are usually solids at room temperature, have high melting points, and conduct electricity when melted or dissolved in water. The ions of an ionic compound are arranged in an orderly fashion in the solid. As a result, there is no such thing as a molecule of an ionic compound.

A second way that atoms gain a noble gas electron arrangement is to share outer shell electrons. This sharing forms a covalent bond. Electrons are shared equally only between atoms of the same element. Atoms of different electronegativities can share electrons, but the sharing is unequal. Bonds formed by the unequal sharing of electrons are called polar covalent bonds. Many atoms gain a noble gas electron arrangement by sharing electrons with more than one atom. The number of covalent bonds that an atom of an element forms is determined by the number of electrons it needs to gain a noble gas electron arrangement.

Molecules are electrically neutral, stable arrangements of atoms. Certain groups of atoms, called polyatomic ions, are not electrically neutral. The atoms of these ions all have noble gas electron arrangements. As a result, they are stable and undergo most chemical reactions as a unit. However, their total number of electrons is not equal to their total number of protons. Consequently, the cluster of atoms has an electrical charge. More than one structural formula can be written for several of these polyatomic ions, although the ions actually have only one structure. The resonance theory has been proposed to explain why these ions and certain molecules appear to have more than one structure.

REVIEW OF TERMS AND CONCEPTS

Terms

ANION An atom or group of atoms that has a negative charge.

CATION An atom or group of atoms that has a positive charge.

CHEMICAL FORMULA A formula that shows the ratios of atoms of all the elements in a compound.

COVALENT BOND A bond formed between identical atoms by the sharing of a pair of outer shell electrons.

CRYSTAL LATTICE A regular three-dimensional pattern of alternating cations and anions.

DIATOMIC MOLECULE A molecule that contains two atoms.

DOUBLE BOND Two electron pairs shared between two atoms.

IONIC BOND A bond formed between elements of widely different electronegativities by the *transfer* of outer shell electrons from one atom to the other.

POLAR COVALENT BOND A bond formed by the unequal sharing of a pair of outer shell electrons between two atoms of intermediate differences in electronegativity.

POLYATOMIC IONS A stable cluster of atoms that has an electrical charge.

RESONANCE THEORY A theory that describes the bonding in certain ions or molecules as a hybrid of two or more contributing structures.

STRUCTURAL FORMULA A representation of a molecule that indicates specifically the way atoms are joined together and the kinds of bonds by which they are joined.

TRIPLE BOND Three pairs of electrons shared between two atoms.

Concepts

1. Atoms are joined together by chemical bonds to form molecules.
2. Atoms that do not have a noble gas electron arrangement in their outer shell will gain or lose electrons to achieve this arrangement.
3. Ionic bonds are formed by the mutual attraction between cations and anions.
4. Ionic compounds are made up of an orderly pattern of oppositely charged ions. There is no such thing as a molecule of an ionic compound.
5. Ions are chemically very different from the elements of which they are formed.
6. Covalent bonds are formed by sharing electrons between atoms. When the sharing is unequal, polar covalent bonds are formed.
7. Atoms that form covalent or polar covalent bonds have a characteristic bonding capacity that is determined by the number of electrons needed to gain a noble gas electron arrangement.
8. Atoms form molecules because molecules are a more stable arrangement of the nuclei and electrons than are the isolated atoms.

EXERCISES 5-14 The electron-dot representations of three hypothetical elements, X, Y, and Z, are as follows:

$$\dot{X} \qquad \cdot\underset{\cdot\cdot}{\ddot{Y}}: \qquad \cdot\underset{\cdot}{\dot{Z}}\cdot$$

(a) To which group of the periodic table does each element belong?

(b) Write electron-dot representations for the ions formed from elements X and Y. Does element Z easily form an ion?

(c) Write electron-dot formulas for each of the following compounds:
(i) X_2 (ii) Y_2 (iii) XY (iv) ZX_4 (v) ZY_4

(d) Identify each bond in exercise 5-14c as ionic, covalent, or polar covalent.

5-15 What is the difference between the electron structure of a sodium ion and that of the element sodium? Is the ion or the element essential to the health of people?

5-16 Write the chemical formula of the compound formed by the combination of the following ions, and name all the compounds.
(a) Na^+, S^{-2} (b) Li^+, HPO_4^{-2} (c) NH_4^+, CO_3^{-2}
(d) Al^{+3}, SO_4^{-2} (e) Ca^{+2}, PO_4^{-3}

5-17 Why is it incorrect to say that a molecule of lithium fluoride has ionic bonds?

5-18 Name each of the following compounds:
(a) $Al(NO_3)_3$ (b) $(NH_4)_2S$ (c) KOH (d) $CaHPO_4$
(e) $NaHCO_3$ (f) Na_2CO_3 (g) $LiHSO_4$

5-19 What is the difference between an oxygen atom and the element oxygen, which is a part of the air we breathe?

5-20 Indicate which atom in the following bonds has a partial positive charge ($\delta+$) and which has a partial negative charge ($\delta-$).
(a) —O—H (b) —C—N (c) —C—O (d) —C—Br

5-21 How many covalent or polar covalent bonds can be formed by atoms of each of the following elements?
(a) C (b) H (c) N (d) O (e) F (f) S (g) Si (h) P
(i) I

INTERMOLECULAR ATTRACTIONS AND STATES OF MATTER

6

We learned in Chapter 5 how atoms bond together to form molecules. One characteristic of molecules is that they are attracted to one another. They tend to stick or cling together because of these intermolecular attractions. A compound will exist at room temperature as a gas, solid, or liquid, depending on the strength of these intermolecular attractions. These attractions are weak in a gas, stronger in a liquid, and strongest in a solid.

In this chapter, we will identify the various types of intermolecular attractions and examine how they are related to the physical state of a compound. One factor that determines the strength of intermolecular attraction is the shape of the molecules. Let us begin by learning the shapes of three important molecules.

6.1 THE SHAPES OF WATER, AMMONIA, AND METHANE MOLECULES

The shape of a molecule is described in terms of its bond lengths and bond angles. A *bond length* is the distance between the two nuclei of the atoms forming the bond. The length of a bond depends on the kinds of atoms bonded together. For example, the length of a carbon-hydrogen bond is different from that of a carbon-carbon single bond. The length of a carbon-carbon single bond differs from that of a carbon-carbon double bond. Thus, bond lengths are determined both by the kinds of atoms forming the bond and by the kind of bond that holds the atoms together. A *bond angle* is the angle between any two bonds to the same atom. The bond length and bond angle of a water molecule (H_2O) are shown in the following structural formula:

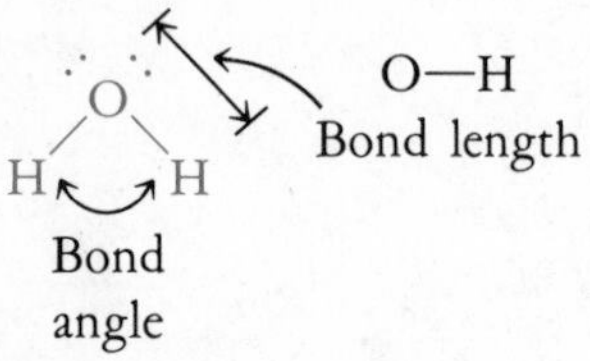

O—H
Bond length

Bond
angle

The bond angles in most simple molecules are determined by the *total* number of electron pairs around the central atom. All these electron pairs are negatively charged. Because like charges repel each other, it is reasonable to expect that these pairs of electrons will get as far away as possible from each other. This idea can be used to explain the shape of simple molecules such as methane (CH_4), ammonia (NH_3), and water (H_2O).

Consider first a molecule of methane. It has four hydrogens bonded to a central carbon atom. The maximal separation of the electrons in these four bonds can be achieved by pointing them toward the four corners of a regular tetrahedron. In such an arrangement, the carbon is located at the center of a regular tetrahedron and a hydrogen atom is located at each corner, as shown in Figure 6-1a.* The carbon atom is called a *tetrahedral carbon*. Each H—C—H bond angle is 109.5 degrees.

A molecule of ammonia (NH_3) also has four electron pairs around the central nitrogen atom. Three pairs form bonds between the nitrogen atom and three hydrogen atoms. The remaining pair, called a nonbonding electron pair, or a lone pair, does not form a bond with any atom. The maximal separation of these four pairs of electrons can be achieved again by pointing them toward the corners of a regular tetrahedron. The nitrogen atom is at the center of the tetrahedron; three hydrogens occupy three corners, and the nonbonding pair occupies the fourth corner, as shown in Figure 6-1b.

The shape of the ammonia molecule is determined by the positions in space of its four atoms, so the molecule is a trigonal pyramid, as shown in Figure 6-1c. This arrangement is like a tripod. Three legs extend from the

* Most people have difficulty visualizing the three-dimensional structure of molecules drawn on a page. The only way to solve this problem is to make a three-dimensional model of the molecule. Many inexpensive sets of molecular models are available commercially for this purpose. We will repeatedly discuss the shapes of molecules in this book because their shape is an important factor in determining the chemical reactions of molecules in living systems. For this reason, the reader is urged to obtain a set of molecular models. It is a good investment.

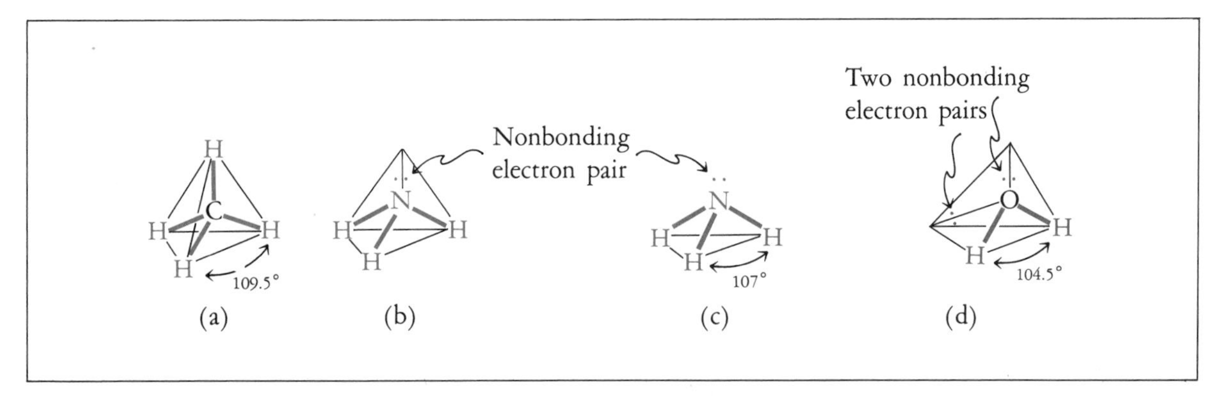

Fig. 6-1. The shapes of methane (CH_4), ammonia (NH_3), and water (H_2O) molecules.

nitrogen atom. A hydrogen atom is located at the bottom of each leg, and the nonbonding pair of electrons is at the top of the tripod. The H—N—H bond angle is 107 degrees. Although the nonbonding electron pair does not form a bond, it plays an important role in the chemical reactions of ammonia, as we will learn in Chapter 9.

A molecule of water (H_2O) also has four pairs of electrons around its central oxygen atom. Two pairs form bonds with two hydrogen atoms, whereas two pairs are nonbonding. Again, these four electron pairs occupy the corners of a tetrahedron with the oxygen atom at the center. Two hydrogens occupy two corners and the nonbonding pairs occupy the remaining two corners as shown in Figure 6-1d. The water molecule has a V shape with a H—O—H bond angle of 104.5 degrees.

*EXERCISE 6-1 Predict the shape and the H—N—H bond angle of the ammonium ion, NH_4^+.

The shape of a molecule is one factor that determines whether a molecule is polar.

6.2 POLAR MOLECULES

We learned in Section 5.6 that the unequal sharing of electrons in the polar covalent bond of HCl results in a partial negative charge on the chlorine atom and a partial positive charge on the hydrogen atom. As a result, the HCl molecule has a partial positive charge at one end and a partial negative

* The answers for the exercises in this chapter begin on page 860.

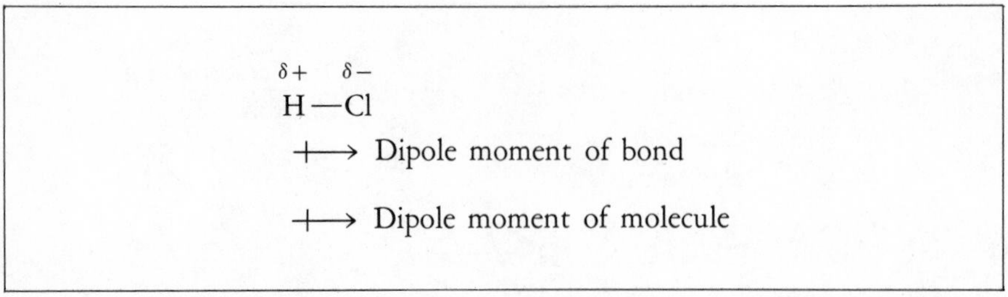

$\delta+$ $\delta-$
H—Cl

+⟶ Dipole moment of bond

+⟶ Dipole moment of molecule

Fig. 6-2. The bond dipole moment and the molecular dipole moment of HCl. The arrow indicating the dipole moment points toward the center of negative charge.

charge at the other end, making it like a small magnet. This separation of positive and negative charges within a bond or a molecule defines its *dipole moment*. Any molecule that has a dipole moment is a *polar molecule*.

In the case of the HCl molecule, the bond dipole moment is the same as its molecular dipole moment. Thus, any diatomic molecule with a polar covalent bond has a molecular dipole moment and it is a polar molecule. We can symbolize the dipole moment of a bond or molecule by an arrow pointing away from the center of positive charge toward the center of negative charge. This is shown for HCl in Figure 6-2.

Triatomic or polyatomic molecules that have polar covalent bonds may or may not have a molecular dipole moment, depending on their shape. Consider, for example, molecules of water and carbon dioxide (CO_2). Each O—H bond of the water molecule has a dipole moment. Because the water molecule is V shaped, the two O—H bond dipole moments reinforce each other, as shown in Figure 6-3. The molecule has a resultant dipole moment, and water is a polar molecule.

Carbon dioxide is also a triatomic molecule. Its C═O bonds are polar, but the molecule does not have a dipole moment. The reason is that the CO_2 molecule is linear. The dipole moments of the two C═O bonds are in opposite directions, as shown in Figure 6-3. The bond dipole moments cancel each other, so the molecule has no resultant molecular dipole moment. Consequently, carbon dioxide is a nonpolar molecule.

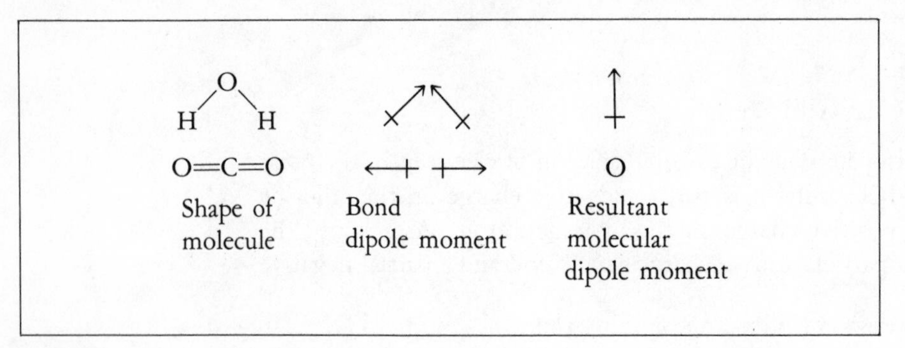

Fig. 6-3. The effect of molecular shape on the resultant molecular dipole moment.

Many molecules have polar bonds but no resultant molecular dipole moment. Molecules such as CH_4 and CCl_4 are examples in which the dipole moments of the bonds cancel each other. Such molecules are nonpolar.

EXERCISE 6-2 Which of the following molecules are polar?
(a) HF (b) O_2 (c) NH_3 (d) F_2 (e) CF_4

The separation of charge in polar molecules causes attractions between adjacent molecules. We will learn about these intermolecular attractions next.

6.3 INTERMOLECULAR ATTRACTIONS

Polar molecules have permanent dipole moments. These dipole moments cause an attraction called *dipolar attraction* between adjacent molecules. The positive charge at one end of a molecule attracts the negative charge at the end of an adjacent molecule. This attraction causes the molecules to orient themselves as shown in Figure 6-4 for iodine monochloride (ICl). The attraction between the partial positive charge on the iodine atom and the partial negative charge on the chlorine atom of an adjacent molecule causes the molecules to arrange themselves like many small magnets.

The strongest and an especially important kind of intermolecular attraction is the *hydrogen bond.* This attraction occurs between a hydrogen atom covalently bonded to a highly electronegative atom such as oxygen, nitrogen, or fluorine and an electronegative atom on an adjacent molecule. In these bonds, the hydrogen atom has a partial positive charge and the electro-

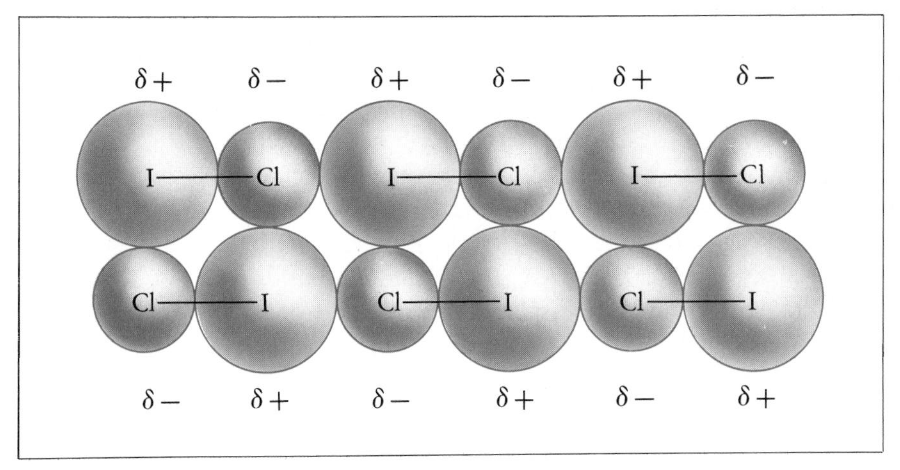

Fig. 6-4. The orientation of iodine monochloride (ICl) molecules caused by the attraction of one end of a molecule to the oppositely charged end of another molecule.

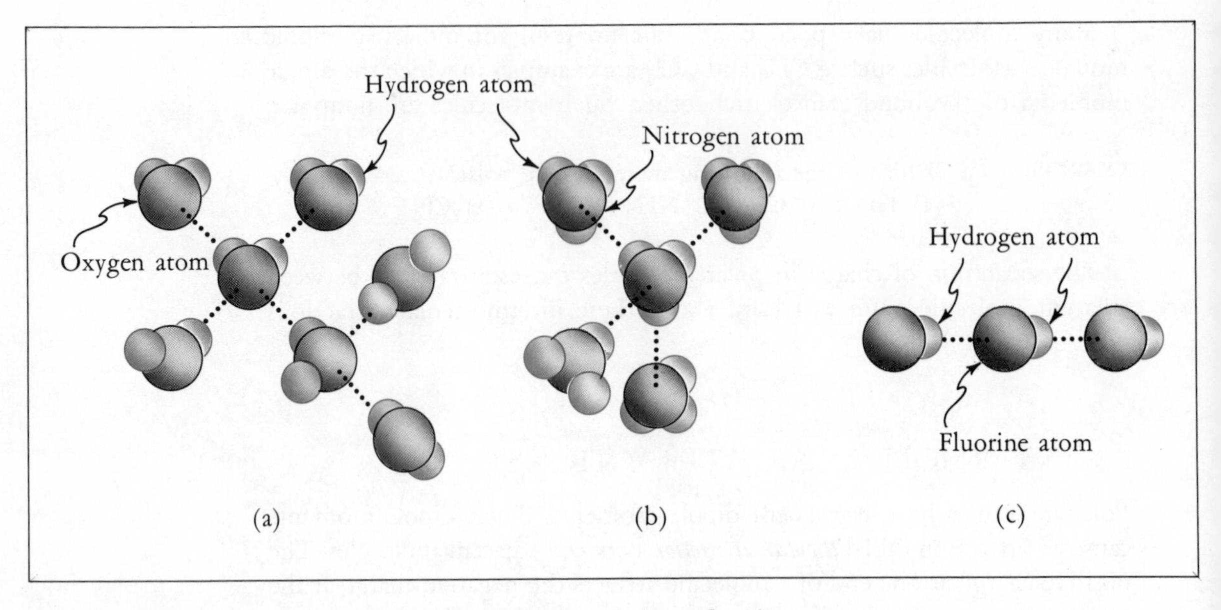

Fig. 6-5. Hydrogen bonds in (a) water, (b) ammonia, and (c) hydrogen fluoride.

negative atom has a partial negative charge. The specific attraction between these atoms is responsible for the hydrogen bond. Examples of hydrogen bonding are found in water, ammonia, and hydrogen fluoride. The electrostatic attractions between the atoms forming a hydrogen bond are usually indicated by a dotted line (···) joining the atoms, as shown in Figure 6-5. For hydrogen bonding to be important, both electronegative elements (the one covalently bonded to hydrogen and the one to which the hydrogen is attracted) must be oxygen, nitrogen, or fluorine. Only hydrogen bonded to these three highly electronegative elements is sufficiently positive, and only these three elements are sufficiently negative for such an attraction to exist.

Although hydrogen bonds are much weaker than covalent bonds, they are of enormous importance in ordering the arrangement of molecules in both solutions and crystals. As we will learn later, they play a key role in determining the shapes of many large molecules in living systems.

In addition to these relatively strong intermolecular attractions, there are weak interactions between nonpolar molecules. These attractions are caused by a pulsing of the electrons back and forth between the nuclei of a bond. At any one instant, the pair of electrons may be displaced toward one atom. Microseconds later, the electrons may be displaced toward the opposite end of the bond. This pulsating of the electrons gives rise to a momentary dipole moment in a nonpolar molecule. This polarization of the molecule, no matter how brief, causes a dipole moment to be formed in an adjacent molecule. These molecules now both have dipole moments and they are attracted to each other. These attractions are called *van der Waals* or *hydrophobic attractions*. They are weak for simple nonpolar molecules like hydrogen or oxygen. However, the attraction among many large nonpolar molecules in living systems is the sum of a great number of these weak van der Waals or hydrophobic attractions. As a result, the total attraction is strong enough to hold the molecules together. We will learn in Chapters 23, 24, and 25 how important these attractions are to the properties of many compounds in living systems.

EXERCISE 6-3 What kind of intermolecular attraction would be the most important for the following compounds?
(a) H_2 (b) BrCl (c) HCl (d) NH_3 (e) H_2O

All these intermolecular attractions, whether weak or strong, are important in determining the physical properties of matter. This will become clear as we examine in more detail the three states of matter. Let us start with solids.

6.4 SOLIDS

Solid substances are very familiar to us. We know from our everyday experiences that a solid is rigid. It retains its shape even when not in a container. Perhaps less commonly known is that solids are difficult to compress. Very high pressures must be applied to compress a solid.

We can explain these and other properties of solids by examining the details of their structure. The structure of the crystalline solid NaCl was presented in Section 5.4. It consists of sodium and chloride ions arranged in a crystal lattice. The attractions between the ions of unlike charges is so great that they are held rigidly in place. These attractions are called *electrostatic attractions* and are much stronger than any of the intermolecular attractions mentioned in Section 6.3. The ions probably vibrate about a central location, but movement from place to place is difficult.

Many solids are made up of molecules that are arranged in a regular three-dimensional pattern. The attractions between the molecules are not as great as those between the ions in an ionic compound. However, the attraction is strong enough that the molecules are held close together, and movement of individual molecules is difficult.

All metals except mercury are solids at room temperature. Metals consist of atoms arranged in a regular three-dimensional pattern. These atoms are different from isolated single atoms. They are held together by *metallic bonds* that are also much stronger than any of the intermolecular attractions mentioned previously. A pure crystal of a metal is regarded as a huge molecule made up of millions of atoms held together by metallic bonds.

Solid crystalline substances, therefore, are made up of ions, molecules, or atoms arranged in a regular three-dimensional pattern. The rigidity of this pattern accounts for the rigidity of the solid. The particles are held close together by various attractions. A great deal of pressure is needed to push them closer together, so solids cannot easily be compressed.

The melting point is another physical property that is determined by the strength of attractions between the ions or molecules in the solid. The *melting point* of a solid is the temperature at which it melts, that is, changes from the solid state to the liquid state. In general, the stronger the kind of attraction, the higher the melting point. For example, ionic compounds such as NaCl have very strong electrostatic attractions and have high melting points (500 to 1000° C). Compounds made up of molecules with polar covalent bonds have lower melting points because their intermolecular attractions are weaker. Simple nonpolar molecules have very low melting points. The melting point of hydrogen, for example, is −259° C.

The number of calories needed to change 1 g of a solid to 1 g of liquid at its melting point is called its *heat of fusion*. The heats of fusion of a number of substances are given in Table 6-1. The values of the heats of fusion also reflect the strength of the attractions between the ions or molecules in the solids. Ionic compounds have the highest heats of fusion and nonpolar compounds have the lowest heats of fusion.

Not all solids have their particles arranged in regular patterns. In some solids, the particles are in no recognizable pattern. These are called *amorphous solids*. The difference in arrangement of particles between crystalline and amorphous solids is shown in Figure 6-6.

Table 6-1. Heats of Fusion

Substance	Heat of Fusion (cal/g)
NaCl	123.5
H_2O	79.7
NH_3	83.9
Ag	25.0
Animal fat	17.6
HCl	14.0
O_2	3.3

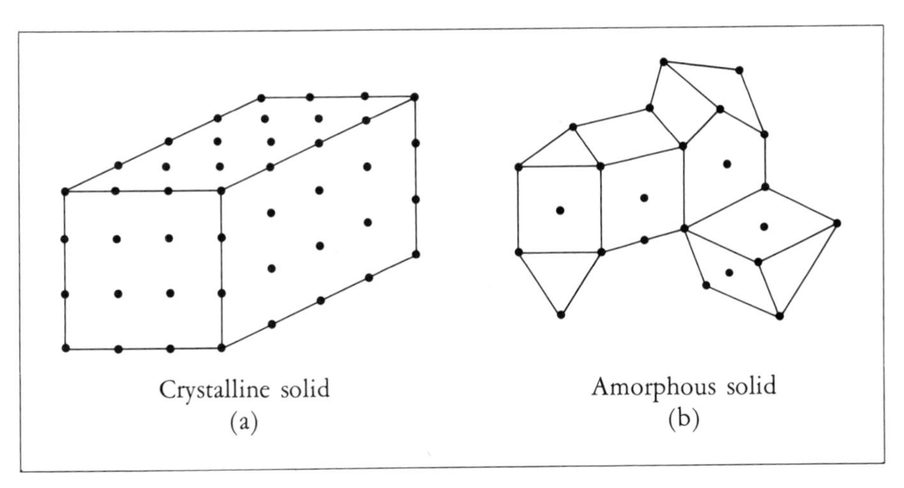

Fig. 6-6. (a) A part of the regular pattern of particles in a crystalline solid. (b) A part of the irregular distribution of particles in an amorphous solid.

The dots in Figure 6-6 represent the locations of the particles, and the solid lines represent the edges of the solid.

The physical properties of liquids are different from those of solids because of differences in the strengths of their intermolecular attractions. We shall learn about these differences in the next section.

6.5 LIQUIDS

Liquids, like solids, are quite familiar to us. Liquids are fluids that take the shape of their containers. They seek their own level under the influence of gravity. Like solids, they are difficult to compress. Unlike solids, liquids evaporate when left in an open container.

A molecular picture of the liquid state is provided by the *kinetic molecular theory*. According to this theory, the attraction between molecules in liquids is less than the attraction between the same molecules in the solid. As a result, the molecules of liquid are in constant motion. They are attracted to one another, but they are not rigidly held. They do not have the fixed, ordered rearrangement that is characteristic of crystalline solids. They slide over one another without much difficulty and move from place to place. This accounts for the fact that liquids are fluid. The molecules of a liquid are still close together, and they cannot be compressed much more.

The attraction between molecules of a liquid is sufficiently weak that occasionally a molecule near the surface of the liquid escapes. It escapes because it has enough energy to overcome the attractions of other molecules in the liquid. This process is called evaporation and is represented schematically in Figure 6-7.

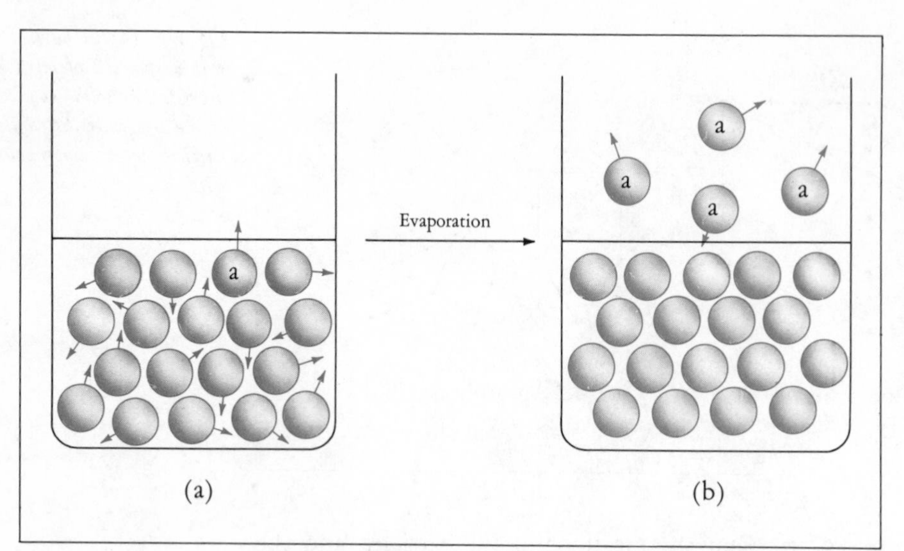

Fig. 6-7. Evaporation. (a) Molecules in the liquid are in constant motion. A few molecules like the one labeled a *are near and moving toward the surface. Molecule* a *has enough energy to escape from the liquid. (b) Other molecules have escaped from the liquid and are now part of the vapor.*

Molecules that escape from the liquid take their energy with them. This decreases the total energy of the remaining molecules in the liquid, so the liquid and its surroundings become cooler. The body uses evaporation as a method of losing excess heat, as we learned in Section 2.10.

The term *vapor* is given to gaseous molecules of a substance at a temperature at which the substance is a liquid. For example, water is a liquid at 25° C and normal atmospheric pressure. However, there are molecules of water above the surface of the liquid. These are molecules of water vapor. They mix with molecules of air above the surface of the liquid. If the container is open to the air, the vapor molecules are free to escape and evaporation continues until all the liquid is gone.

We know from experience that a liquid placed in a closed container will not evaporate. What happens to the molecules that leave the liquid and form the vapor? These molecules stay near the surface of the liquid because they cannot leave the container. Some of them collide with the surface of the liquid and are recaptured. Eventually, the number of molecules leaving the surface of the liquid per second is exactly equal to the number captured by the liquid per second. When this occurs, the liquid and vapor are in *dynamic equilibrium*. This means that some molecules return to the liquid and an equal number escape into the vapor, but the total number of molecules in the vapor remains constant. It is important to realize that, even though it looks as though nothing is happening, there is constant movement of molecules into and out of the liquid.

We know that there are molecules of vapor because we can measure the pressure they create. This pressure caused by the molecules in the vapor

Table 6-2. Vapor Pressure of Liquids at 25° C

Liquid	Vapor Pressure (mm Hg)
H_2O	23.8
Ethyl alcohol	97.3
CCl_4	91.0
Ether	614

above a liquid is called the *vapor pressure* of that liquid. The values of the vapor pressure of various liquids are given in Table 6-2. The vapor pressure of a liquid at a particular temperature is a measure of the attraction between molecules in the liquid. A polar liquid such as water has a lower vapor pressure at 25° C than does CCl_4, a nonpolar liquid, because the intermolecular attractions in water are greater than those in CCl_4. Liquids that have high vapor pressures at room temperature evaporate quickly. They are called *volatile liquids*.

We know from experience that a liquid evaporates more quickly on a hot day than a cold day. This occurs because the number of molecules in the vapor increases with increasing temperature. Another way of saying this is that the vapor pressure of a liquid increases as the temperature of the liquid increases. As a liquid is heated, a temperature will be eventually reached at which its vapor pressure equals atmospheric pressure. This temperature is the *boiling point* of the liquid.

Different liquids have different boiling points. This is a reflection of the difference in the strength of their intermolecular attractions. For example, ICl and Br_2 have approximately the same mass, yet the boiling point of ICl (97° C) is higher than that of Br_2 (59° C). This difference of 38° C between the boiling points of the two compounds reflects the extra energy in the form of heat that must be put into the liquid to overcome the attraction between the polar ICl molecules to vaporize them. The boiling point of a liquid is one of its characteristic physical properties.

The number of calories needed to vaporize 1 g of liquid at its boiling point is called the heat of vaporization. The heats of vaporization of a number of liquids are given in Chapter 2 (Table 2-1). The heats of vaporization are also a measure of the strength of intermolecular attractions. The higher the heat of vaporization, the greater the attraction between the molecules of the liquid.

EXERCISE 6-4 How are liquids and solids similar? How do they differ?

Converting a liquid to a gas requires enough energy to overcome most of the attractions between molecules in the liquid. Consequently, intermolecular attractions are small in a gas, as we will learn in the next section.

6.6 GASES

Gases are the least familiar form of matter to most people. Yet air is a gas and it is around us all the time. Using air and our experiences with it as an example, we can deduce some of the properties of gases. Gases are the least dense form of matter. This is evident from the light weight of a balloon filled with air. A gas uniformly fills a closed container such as a balloon. The gas rushes out if the balloon is punctured, indicating that a gas is mobile. Gases are easily compressed. For example, we can put a quantity of air into a tire that is two or three times greater than the volume of the tire. These properties of gases are explained by the kinetic molecular theory.

According to this theory, a gas is made up of molecules that are widely separated from each other in otherwise empty space. There is plenty of room to push these molecules closer together, so gases can easily be compressed. Gas molecules move at high speeds in straight but random paths. Thus, gases are mobile. They quickly fill a container or escape through an opening. Gas molecules collide but do not stick together because the attractions between them are weak.

Gas molecules collide not only with each other but also with the walls of the container. The result of these collisions is that gases exert a pressure. The air around us exerts a pressure on us. We know this because we can measure it, as we will learn in the next section.

6.7 UNITS OF PRESSURE

The earth is surrounded by a large atmosphere of air and we live at the bottom of it. This large mass of air exerts a pressure on us similar to the pressure exerted by water on a scuba diver. We do not consciously feel this pressure, but we can demonstrate its presence and actually measure it by means of the apparatus shown in Figure 6-8. A glass tube closed at one end is filled with mercury and the open end is then inverted into a container of mercury. The level of mercury in the tube falls until its height is approximately 760 mm above the surface of the mercury in the container. Why does the level of the mercury stop at 760 mm? Why doesn't the mercury all fall out? The reason is that the pressure at the surface of the container

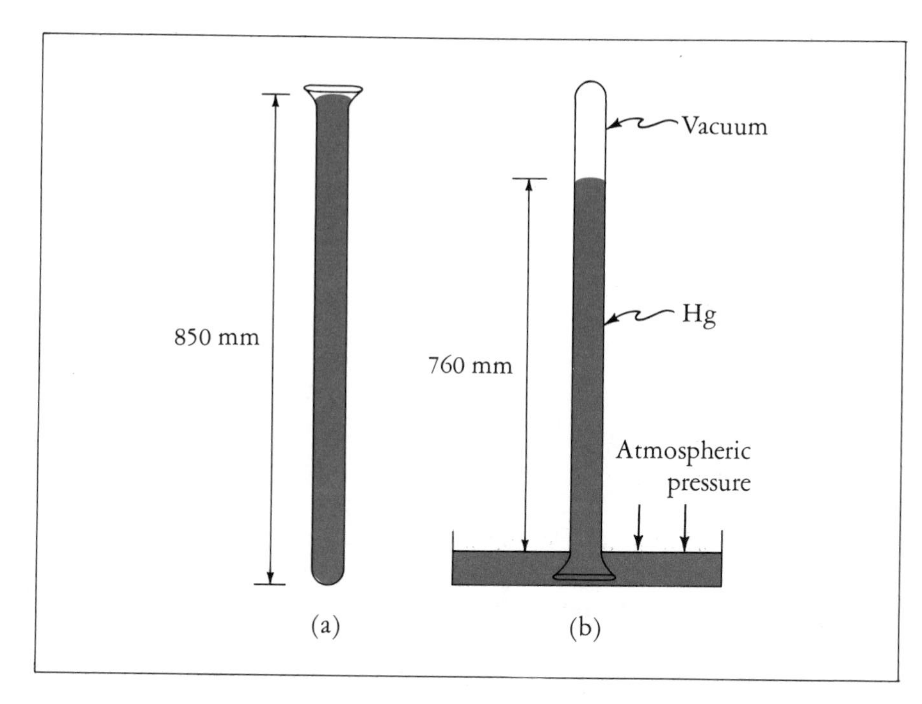

Fig. 6-8. A barometer. (a) A glass tube filled with mercury. (b) The glass tube inverted into a pan of mercury. The level of mercury in the tube falls until the pressure of 760 mm of Hg is exactly balanced by the atmospheric pressure at 0°C and sea level.

caused by the mercury in the tube is exactly balanced by the atmospheric pressure. No more mercury will flow in or out of the tube because there is a balance of the two opposing pressures. The height of the liquid column remaining in the tube is a measure of the atmospheric pressure. Thus, the atmospheric pressure will support a column of mercury 760 mm high. The apparatus in Figure 6-8 is a common scientific instrument called a barometer.

The standard measure of atmospheric pressure is defined as the pressure that supports a column of mercury 760 mm high at 0° C and sea level. Thus, one atmosphere (1 atm) is equal to 760 mm Hg. The unit mm Hg is also called a torr. One atmosphere is equal to 760 torr. Several other units of pressure are widely used. Atmospheric pressure is usually given in most weather reports as inches of mercury. One atmosphere is equal to 29.92 in. Hg. Engineers generally use pounds of air per square inch (psi). One atmosphere is equal to 14.7 psi. The SI unit of pressure is the pascal (Pa). One atmosphere is equal to 101,325 Pa. The pascal is not yet in widespread use in the United States. The units of pressure are summarized in Table 6-3. A substance at a pressure of 1 atm and 273 K is said to be at *standard temperature and pressure* (STP).

Often we must measure the pressure of a gas in a container. This can be done by means of an instrument called a manometer, which is similar but

Table 6-3. Units of Pressure

1 atmosphere (1 atm)	= 760 millimeters of mercury (mm Hg)
1 mm Hg	= 1 torr
1 atm	= 760 torr
1 atm	= 14.7 pounds per square inch (psi)
1 atm	= 29.92 inches of mercury (in. Hg)
1 atm	= 101,325 pascals (Pa)

not identical to a barometer. A manometer is shown in Figure 6-9. It consists of a glass U-tube partially filled with mercury. One side of the U-tube is connected to a container in which the pressure is to be measured; the other side is left open to the atmosphere. The gas in the container will exert a pressure against the mercury column. This pressure will be opposed by the atmospheric pressure. When the column of mercury becomes stationary, the pressure in the container is equal to the atmospheric pressure plus the height of the mercury column. With such an instrument, we can measure the pressure of most gases in a container.

We are most interested in the air around us. Life on earth would be impossible without it. Our lungs provide our bodies with the air we need. We must know certain physical properties of gases to understand how air gets into our lungs and eventually into cells of our bodies.

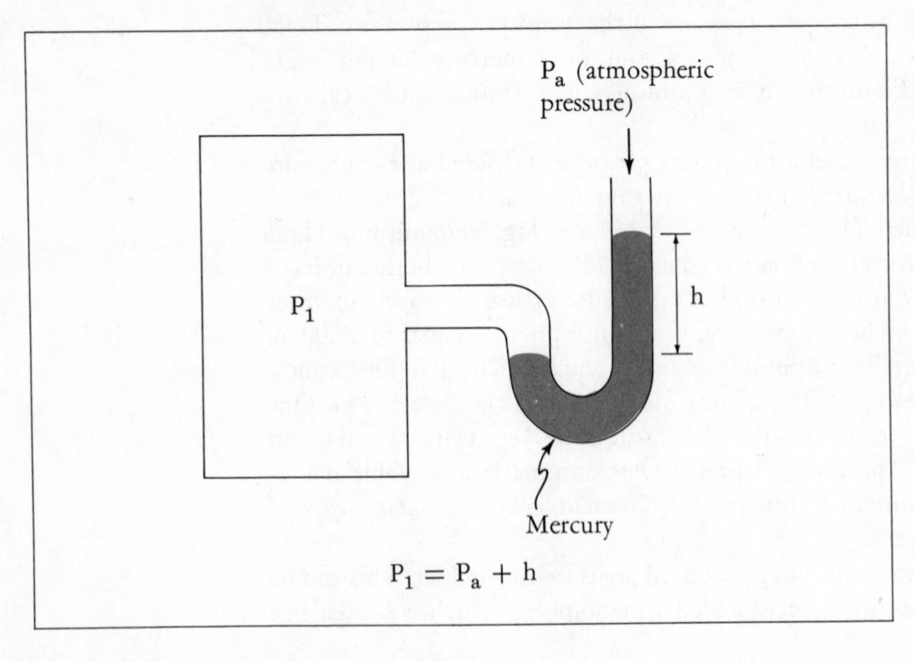

Fig. 6-9. The manometer. The pressure, P_1, inside the container is equal to the atmospheric pressure, P_a, plus the height, h, of the mercury column.

6.8 IDEAL GAS LAWS

The laws governing the effects of changes in temperature, pressure, and volume of a gas have been known for a long time. We will examine these laws in this section.

Boyle's Law

The first systematic study of the relationship between the pressure of a gas and the volume of a gas was made by Robert Boyle in 1662. He carried out his experiments by means of a simple J-tube (Figure 6-10). A quantity of air can be trapped in the closed end of the tube by carefully pouring mercury into the tube. The mercury not only serves to trap the air, but also serves as a manometer. The pressure of the gas is equal to the atmospheric pressure plus the height of the mercury column in the open end of the tube. The pressure of the gas can be increased systematically by adding various amounts of mercury into the open end; the corresponding change in the volume of the gas can be measured.

Boyle found that an increase in pressure caused the volume of the gas to decrease. This universal property of gases is called Boyle's law:

> The volume of a quantity of gas is inversely proportional to its pressure at a constant temperature.

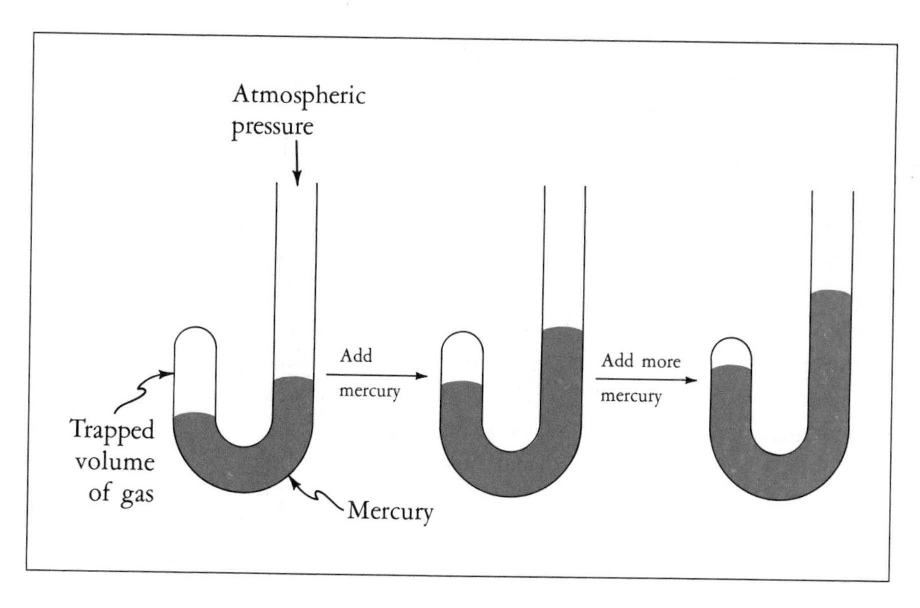

Fig. 6-10. The effect of pressure on volume. Adding more mercury to the J-tube increases the pressure on the trapped gas and causes its volume to decrease.

Boyle's law can be expressed mathematically as follows:

$$V \propto \frac{1}{P} \qquad \text{(quantity of gas and temperature constant)}$$

where V is the volume of gas and P is its pressure. The symbol $\propto$ means "is proportional to." This symbol can be replaced by an equals sign and a constant (K) to give the following equation:

$$V = \frac{1}{P} \times K$$

Rearranging, we obtain

$$P \times V = K \qquad \text{(quantity of gas and temperature constant)}$$

The following equation is a more useful way of expressing Boyle's law:

$$P_i \times V_i = P_f \times V_f \qquad \text{(quantity of gas and temperature constant)} \qquad (6\text{-}1)$$

where P_i and V_i are, respectively, the initial pressure and volume of the gas; P_f and V_f, the final pressure and volume of the gas. Equation 6-1 is true because both $P_i \times V_i$ and $P_f \times V_f$ are equal to the same constant, K, provided both the quantity of gas and the temperature are constant.

The mathematical form of Boyle's law given by equation 6-1 applies only to an ideal gas. There is no such thing as an ideal gas in nature, but this model of a gas can be used to explain the behavior of gases in terms of the behavior of molecules. One important feature of this model is that there is no attraction between molecules of the ideal gas. Consequently, the ideal gas remains a gas no matter what its temperature or pressure. Helium, a monoatomic gas, and hydrogen, a simple diatomic gas, are very nearly ideal gases because their intermolecular attractions are weak. The greater the size and complexity of the molecules of a gas, the less ideal is the behavior of the gas. Oxygen, carbon dioxide, and nitrogen, all gases important to living systems, are sufficiently small that they obey Boyle's law at pressures and temperatures found in nature.

The following example shows how equation 6-1 is used.

Example 6-1 The pressure of 100 mL of a gas is initially 760 mm Hg but is increased to 850 mm Hg. What is the final volume of the gas?

Solution:

First, identify each of the values given:

$V_i = 100\text{ mL}$ $\quad P_i = 760\text{ mm Hg}$

$V_f = ?$ $\quad P_f = 850\text{ mm Hg}$

Then insert these values into equation 6-1:

$$760\text{ mm Hg} \times 100\text{ mL} = 850\text{ mm Hg} \times V_f$$

$$V_f = \frac{760\ \cancel{\text{mm Hg}}}{850\ \cancel{\text{mm Hg}}} \times 100\text{ mL} = 89.4\text{ mL}$$

EXERCISE 6-5 The initial volume of a gas is 4.00 L. What is its final volume after the pressure is decreased from 760 mm Hg to 50 mm Hg?

EXERCISE 6-6 What pressure is needed to compress a gas, initially at 1 atm, from 2 L to 1 L?

The pressure of a gas also affects its movement. This fact can be demonstrated by the experiment shown in Figure 6-11. Two containers of air at different pressures are joined together by means of a tube containing a valve. When the valve is opened, gas from the container at the higher pressure will spontaneously flow into the other container. When the pressures of the gases in the two containers are equal, no further movement of air can be detected. We see examples of such movement of air every day in the wind. Wind is the movement of air from regions of high pressure to regions of

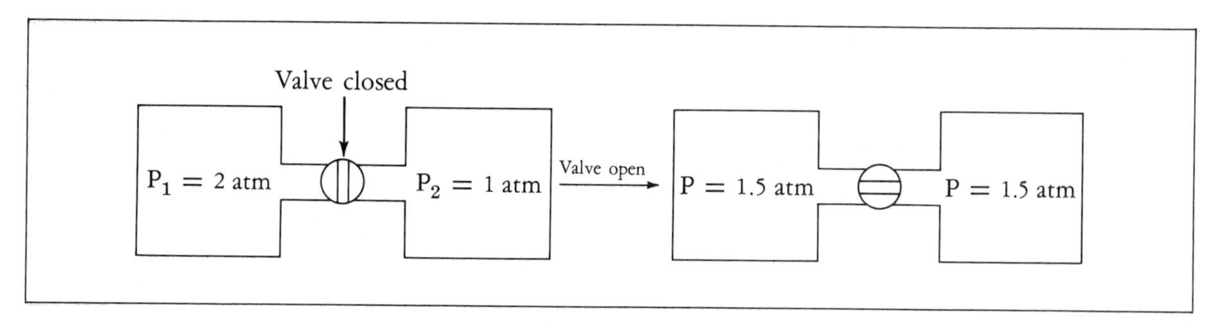

Fig. 6-11. When the valve is opened, air from the container at a higher pressure (P_1) flows into the container at a lower pressure (P_2) until the pressure in both containers is the same.

low pressure. Such movement of air also occurs in the lungs, as we will learn in Section 6.9.

In the original experiments carried out by Boyle, no attempt was made to keep the temperature constant, but because the experiments were carried out at room temperature, it probably varied only slightly. Boyle, however, did notice that warming the gas had a dramatic effect on the volume of the gas. Let us now examine exactly how temperature influences the volume of a gas, as did Jacques Charles nearly 100 years after the experiments of Boyle.

Charles' Law and the Kelvin Temperature Scale

Charles began an investigation of the relationship between the volume of a gas and its temperature at constant pressure in 1787. He found that the volume of a gas increases when the gas is heated and decreases when the gas is cooled.

The results of Charles' experiment form the basis of the absolute temperature scale (see Section 1.5). The idea behind this scale is as follows. We know that cooling a gas decreases its volume. Theoretically, if a gas were cooled to a low enough temperature, its volume could be reduced to zero. However, no gas or any sample of matter can have a volume of zero. Thus, there must be a minimal temperature to which all matter can be cooled. This temperature is defined as absolute zero, $-273.15°$ C. It is the coldest temperature possible and it is the zero point for the Kelvin scale of temperatures.

A quantitative relationship exists between the volume of a gas and its temperature *only* if the temperature is expressed in kelvins (K). It is important to remember this. All calculations involving gases must use temperatures in kelvins. This important condition is included in the statement of *Charles' law:*

> The volume of a quantity of gas at constant pressure is directly proportional to its temperature in kelvins (K).

Charles' law can be expressed mathematically as follows:

$V \propto T$ (quantity of gas and pressure constant)

where V is the volume of the gas and T is the temperature in kelvins. In the form of an equation, this becomes

$$V = C \times T \qquad \text{or}$$

$$\frac{V}{T} = C \qquad \text{(quantity of gas and pressure constant)}$$

The following equation is a more useful way to express Charles' law:

$$\frac{V_i}{T_i} = \frac{V_f}{T_f} \qquad \text{(quantity of gas and pressure constant)} \qquad (6\text{-}2)$$

where the subscripts i and f stand for the initial and final values, respectively, of gas volume and temperature in kelvins. Equation 6-2 is true because V_i/T_i and V_f/T_f are equal to the same constant, C, when the volume of gas and the pressure are both constant. Although equation 6-2 applies exactly only to an ideal gas, real gases such as oxygen, nitrogen, and carbon dioxide fit equation 6-2 well enough near atmospheric pressure and room temperature that we can apply it to these gases.

The use of equation 6-2 is shown in the following example:

Example 6-2 2.50 L of gas at 25° C is cooled to −15° C. What is the final volume of the gas?

Solution:

First, convert the temperatures to kelvins:

$$25^\circ\ C = 25 + 273 = 298\ K;$$
$$-15^\circ\ C = -15 + 273 = 258\ K$$

Then, identify each value given:

$$V_i = 2.50\ L \qquad V_f = ?$$
$$T_i = 298\ K \qquad T_f = 258\ K$$

Substitute the values in equation 6-2:

$$\frac{2.50\ L}{298\ K} = \frac{V_f}{258\ K}$$

$$V_f = \frac{258\ \cancel{K}}{298\ \cancel{K}} \times 2.50\ L = 2.16\ L$$

EXERCISE 6-7 What is the final volume of a gas, initially, at 100 mL, when the temperature is increased from 0° C to 75° C?

EXERCISE 6-8 To what temperature must 3.0 L of gas at 25° C be cooled so that its volume is decreased by half?

Aerosol cans are frequently used today as containers of a variety of products. They will explode violently if heated. The reason for this was discovered by Joseph Louis Gay-Lussac.

Gay-Lussac's Law

The pressure of a gas confined in a container, such as an aerosol can, will increase as the gas is heated. Gay-Lussac was the first person to discover the exact relationship between the temperature and pressure of a gas. His results are now given as *Gay-Lussac's law:*

> The pressure of a quantity of gas at constant volume is directly proportional to its temperature in kelvins.

Mathematically, this law is expressed in either of the following ways:

$$P \propto T \qquad \text{(quantity of gas and volume constant)}$$

$$\frac{P}{T} = \text{constant}$$

The following equation, however, is a more useful expression of Gay-Lussac's law:

$$\frac{P_i}{T_i} = \frac{P_f}{T_f} \qquad \text{(quantity of gas and volume constant)} \qquad (6\text{-}3)$$

The use of equation 6-3 is shown in the following example:

Example 6-3 An aerosol container holds a gas, originally at 25° C and 2.0 atm. The container is thrown into a fire whose temperature is 575° C. What is the final pressure of the gas?

Solution:
First, convert the temperatures to kelvins.

$25^\circ\ C = 25 + 273 = 298\ K$

$575^\circ\ C = 575 + 273 = 848\ K$

Then, identify each value given:

$P_i = 2.0\ atm \qquad P_f = ?$

$T_i = 298\ K \qquad T_f = 848\ K$

Substitute the values into equation 6-3:

$$\frac{2.0\ atm}{298\ K} = \frac{P_f}{848\ K}$$

$$P_f = \frac{848\ \cancel{K}}{298\ \cancel{K}} \times 2.0\ atm = 5.7\ atm$$

EXERCISE 6-9 What is the final pressure of a gas in a closed container originally at 0° C and 1 atm, if the temperature is doubled?

EXERCISE 6-10 What is the final temperature of a gas in a closed container, originally at 25° C and 453 mm Hg, if the pressure is doubled?

Boyle's, Charles', and Gay-Lussac's laws can be combined into one equation that tells us what happens when any or all of the three—temperature, pressure, and volume—are changed.

The Combined Gas Law

The combined gas law is expressed as follows:

$$\frac{PV}{T} = \text{constant} \qquad \text{(quantity of gas constant)}$$

where P is the pressure of the gas, V is its volume, and T is its temperature in kelvins. A given quantity of gas can be expanded, compressed, heated, cooled, or put under more or less pressure. During any of these changes, the quantity PV/T will remain constant. We can express this mathematically in the following way:

$$\frac{P_i \times V_i}{T_i} = \frac{P_f \times V_f}{T_f} \quad \text{(quantity of gas constant)} \qquad (6\text{-}4)$$

Again, the subscripts i and f stand for the initial and final values, respectively, of pressure, temperature, and volume. The use of equation 6-4 is shown by the following example.

Example 6-4 The pressure and temperature of 1.0 L of a gas are increased from 0° C and 1.0 atm to 100° C and 1520 mm Hg. What is the final volume?

Solution:

First, convert temperature to kelvins:

$$0^\circ \text{ C} = 0 + 273 = 273 \text{ K}$$

$$100^\circ \text{ C} = 100 + 273 = 373 \text{ K}$$

Next, convert pressure into the same units:

$$1520 \cancel{\text{mm Hg}} \times \frac{1.0 \text{ atm}}{760 \cancel{\text{mm Hg}}} = 2.0 \text{ atm}$$

Then, identify each value given

$V_i = 1.0$ L $\quad P_i = 1.0$ atm $\quad T_i = 273$ K

$V_f = ?$ $\quad P_f = 2.0$ atm $\quad T_f = 373$ K

Finally, substitute the values into equation 6-4:

$$\frac{1.0 \text{ atm} \times 1.0 \text{ L}}{273 \text{ K}} = \frac{2.0 \text{ atm} \times V_f}{373 \text{ K}}$$

$$V_f = \frac{1.0 \cancel{\text{atm}}}{2.0 \cancel{\text{atm}}} \times \frac{373 \cancel{\text{K}}}{273 \cancel{\text{K}}} \times 1.0 \text{ L} = 0.68 \text{ L}$$

EXERCISE 6-11 What is the volume of 1.0 L of gas, originally at −10° C and 733 mm Hg, when the temperature is increased to 50° C and the pressure is decreased to 350 mm Hg?

EXERCISE 6-12 Five liters of a gas originally at 100° C and 2 atm of pressure is compressed at 2.5 L by increasing the pressure to 4 atm. What is the final temperature of the gas?

Air was used as the gas in the experiments that led to the laws of Boyle, Charles, and Gay-Lussac. But air is a mixture of gases. How does this affect the gas laws? The answer to this question was provided by the work of John Dalton in 1801.

Dalton's Law of Partial Pressures

Dalton found that, in a mixture of two or more gases that do not react chemically, each gas behaves independently of the others. In particular, each gas makes a contribution to the total pressure. This contribution is equal to the pressure, called the *partial pressure,* that the gas would exert if it were in the same container alone (Figure 6-12). Each of the cylinders in Figure 6-12 has the same volume. We place enough nitrogen in the first cylinder so that the pressure is 1.0 atm. We place enough oxygen in the second cylinder so that its pressure is also 1.0 atm. We now mix the gases in a cylinder of the same size. We find that the total pressure of the gases is 2.0 atm, the sum of the partial pressures of the individual gases. This observation is called *Dalton's law of partial pressures:*

> The total pressure of a mixture of gases is equal to the sum of the partial pressures of the individual gases.

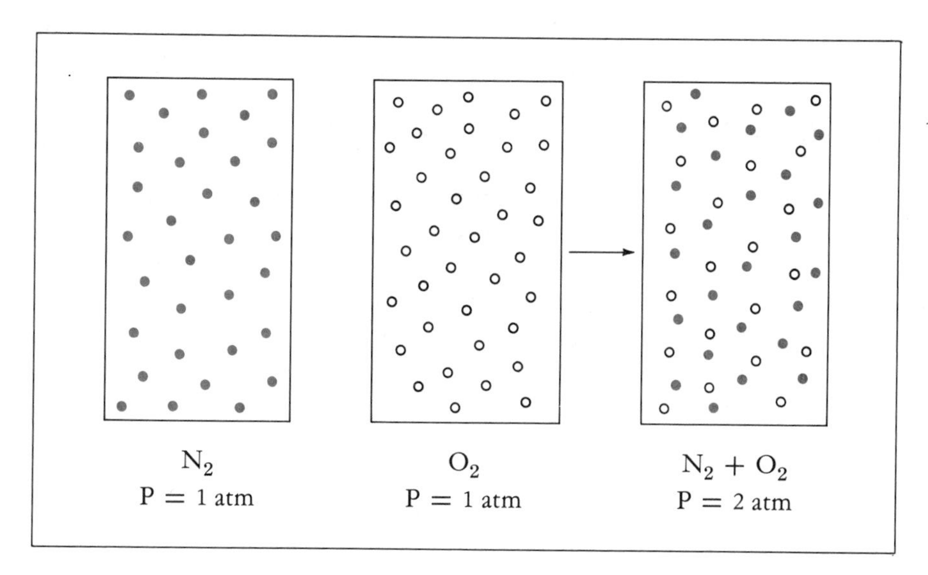

Fig. 6-12. An example of Dalton's law of partial pressures.

Dalton's law can be written as a mathematical equation:

$$P_T = P_1 + P_2 + P_3 + \ldots P_n \qquad (6\text{-}5)$$

where P_T is the total pressure and $P_1 \ldots P_n$ are the partial pressures of the n individual gases in the mixture.

The partial pressure of any gas in a mixture can be obtained if we know the total pressure and the composition of the mixture. This is expressed in the following equation:

$$P_1 = \frac{(\text{\% of gas 1 in mixture})}{100\%} \times P_T \qquad (6\text{-}6)$$

where P_1 is the partial pressure of gas 1 and P_T is the total pressure. The use of equation 6-6 is shown in the following example.

Example 6-5 A pressurized cylinder contains 30 percent oxygen and 70 percent helium at a total pressure of 5 atm. What is the partial pressure of oxygen (Po_2) in the cylinder?

Solution:

$$Po_2 = \frac{(\%\ O_2)}{100\%} \times P_T = \frac{30\%}{100\%} \times 5\ \text{atm} = 1.5\ \text{atm}$$

EXERCISE 6-13 Air is a mixture of 78 percent nitrogen, 21 percent oxygen, 0.93 percent argon, and 0.07 percent carbon dioxide. Calculate the partial pressure of each of these gases at 25° C and 760 mm Hg.

The fact that a gas in a mixture of gases behaves independently of the others accounts for the movement of a gas from one gas mixture to another. Such movement of gases, which is important in living systems, is represented schematically in Figure 6-13. The apparatus consists of two containers separated by a porous membrane. The partial pressures of oxygen and nitrogen are different in the two containers, but the total pressures are the same. We learned in Section 6.8 that gases move from regions of higher pressure to regions of lower pressure until the pressure in both regions becomes equal. The same thing happens in the apparatus in Figure 6-13. The partial pressure of oxygen is greater in one container than another. Over a period of time, oxygen moves to the container that has a lower

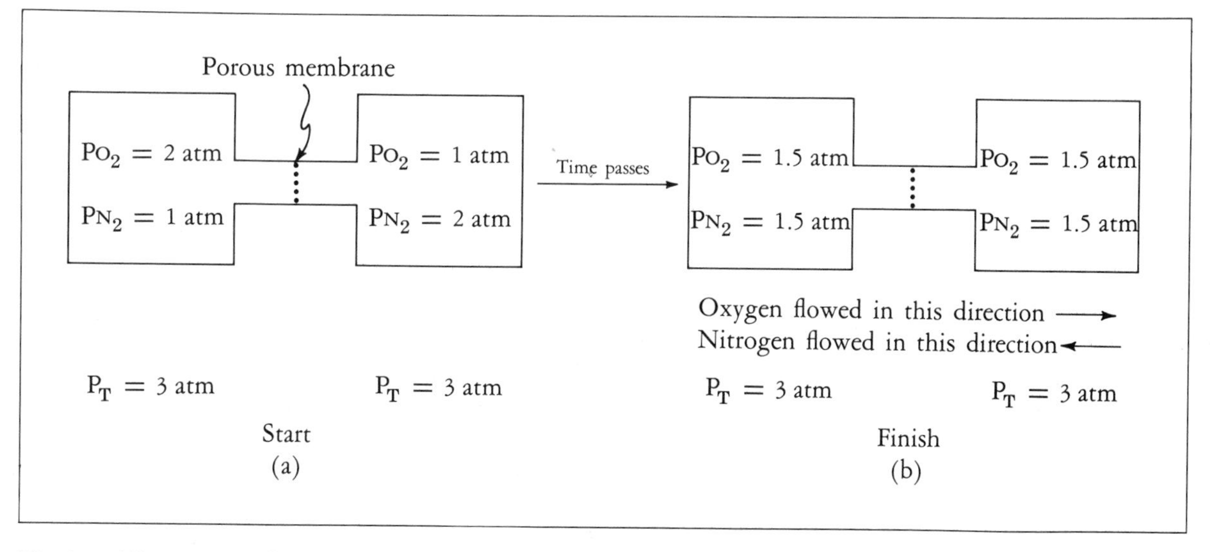

Fig. 6-13. The movement of gases between two mixtures. (a) At the start, the partial pressures of N_2 and O_2 are different in the two containers. (b) After some time, movement of both gases causes the partial pressures of the gases to be equal in the two containers.

partial pressure of oxygen. This flow of oxygen continues until its partial pressure is the same in both containers. Similarly, nitrogen flows from the container where the partial pressure of nitrogen is higher to the container where the partial pressure of nitrogen is lower.

The important point to remember is that the *movement of one gas in a mixture* from one region to another *is determined only by the difference in its own partial pressure in the two regions*. In the example given here, the two regions are made up of two gas mixtures. But the same principle applies to the movement of a gas from a mixture of gases into a liquid. In this case, the movement of a gas into the liquid depends on the difference between its partial pressure in the gas and in the liquid. During breathing, gases move into and out of a liquid, as we will learn in the next section.

6.9 GAS LAWS AND BREATHING

Breathing is a real-life application of the gas laws. When we inhale, our lungs expand and their volume increases. This causes the pressure of air within the lungs to decrease according to Boyle's law. As a result, the pressure of air in our lungs is slightly less than atmospheric pressure. This causes air to enter our lungs. The reverse occurs when we exhale. Our lungs contract and their volume decreases. This causes the pressure of air in our lungs to be slightly higher than atmospheric pressure and air leaves our lungs.

The oxygen in the air we breathe is needed by the body to convert food to energy, as we will learn in Chapter 26. In the process, carbon dioxide is produced. The body must move oxygen from the atmosphere into the cells and carbon dioxide in the reverse direction. Oxygen is absorbed and carbon dioxide is released from the body through the lungs. This transfer takes place across the thin porous membranes of alveoli in the lungs. The alveoli contain a fine network of capillaries where the oxygen and carbon dioxide are actually transferred between the lungs and the blood. The net direction of the movement of oxygen and carbon dioxide is determined by their

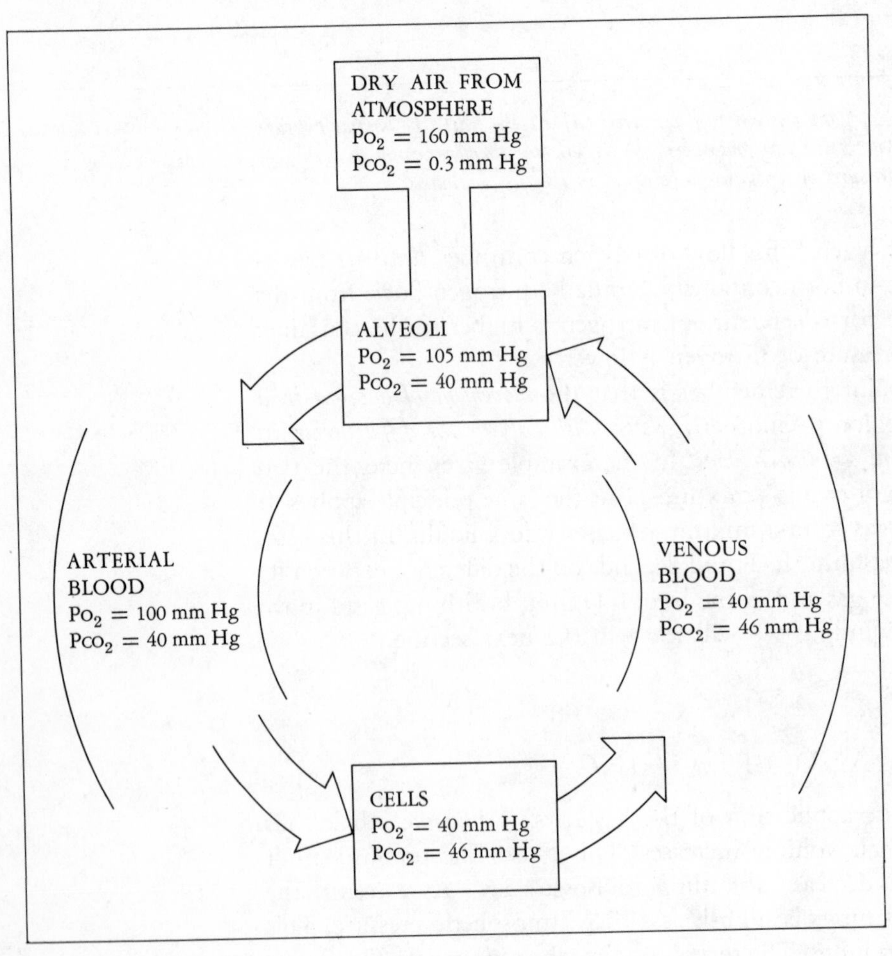

Fig. 6-14. Partial pressures of O_2 and CO_2 in the atmosphere, alveoli, and arterial and venous blood.

partial pressures* on either side of various membranes. Their movement can be understood with the aid of Figure 6-14, which gives their partial pressures in the lungs, the alveoli, and the arterial and venous blood.

Dry inhaled air has a partial pressure of oxygen of about 160 mm Hg and a partial pressure of carbon dioxide of about 0.3 mm Hg. Once the inhaled air enters the tracheobronchial tree, it picks up water vapor. As a result, when the air reaches the alveoli, the partial pressure of oxygen is decreased to 105 mm Hg. Venous blood returning to the lungs has a relatively low partial pressure of oxygen (40 mm Hg) and a relatively high partial pressure of carbon dioxide (46 mm Hg). As the venous blood passes through the capillaries in the alveoli, oxygen moves into the blood, the region of lower partial pressure. At the same time, carbon dioxide in the venous blood moves out into the lungs, the region of lower carbon dioxide partial pressure. This movement of gases through the membranes of the capillaries of the alveoli converts carbon dioxide–rich venous blood into oxygen-rich arterial blood that is circulated throughout the body.

Oxygen in the arterial blood is transferred to the cells by a similar process. The partial pressure of oxygen in the arterial blood is higher than its partial pressure in the cells. As a result, oxygen moves into the cells. Meanwhile, carbon dioxide moves from the cells into the arterial blood because its partial pressure is higher in the cells than in arterial blood. Venous blood, with its decreased partial pressure of oxygen and its increased partial pressure of carbon dioxide, returns to the lungs, where the cycle starts again.

Clearly, breathing is a very practical application of the gas laws.

6.10 SUMMARY

Polar covalent bonds have a dipole moment because one end is partially positive and the other end is partially negative. Any diatomic molecule with a polar covalent bond has a dipole moment and is a polar molecule. In diatomic molecules, the bond and molecular dipole moments are the same. In polyatomic molecules, the two dipole moments are not the same. The molecular dipole moment is made up of contributions from the individual bond dipole moments. These contributions depend on the shape of the molecule. If its shape is such that all bond dipole moments cancel each other, the molecule has no dipole moment and is nonpolar. An example is

* Called tension of a gas in medicine.

carbon dioxide. A molecule whose shape allows the bond dipole moments to reinforce each other has a molecular dipole moment and is polar. Water is an example of such a molecule. The shapes of simple molecules are determined by the fact that all pairs of electrons in the outer shell of the central atom tend to get as far away as possible from each other.

Polar molecules attract each other. There are three kinds of intermolecular attractions. Dipolar attractions are the attractions between the charged end of one molecule and the oppositely charged end of another molecule. The hydrogen bond is an attraction between a hydrogen atom covalently bonded to a highly electronegative atom such as oxygen, nitrogen, or fluorine and an electronegative atom on an adjacent molecule. Hydrophobic or van der Waals attractions are caused by pulsing of the electrons between the nuclei of a covalent bond. A momentary separation of charge occurs within the molecule that causes a weak attraction between adjacent molecules.

Modern theory pictures the three states of matter in terms of various attractions holding together atoms, molecules, or ions. The particles of a solid are held closely together by relatively strong intermolecular attractions. Weaker intermolecular attractions in liquids allow the molecules more movement but still hold them close together. Intermolecular attractions are weak in a gas. The molecules or atoms of a gas are widely separated from each other in otherwise empty space.

The behavior of gases in response to changes in temperature, pressure, and volume is described by the ideal gas laws. These laws apply exactly only to a nonexistent ideal gas, but in practice they apply quite well to many gases under conditions near room temperature and atmospheric pressure.

REVIEW OF TERMS AND CONCEPTS

Terms

AMORPHOUS SOLID A solid that has no clearly visible crystalline structure.

BOILING POINT The temperature at which the vapor pressure of the liquid equals the atmospheric pressure.

BOND ANGLE The angle between any two bonds to the same atom.

BOND LENGTH The distance between the two nuclei of atoms forming a bond.

DIPOLAR ATTRACTION Attractions between molecules that have dipole moments.

DIPOLE MOMENT A measure of separation of charge within a molecule.

ELECTROSTATIC ATTRACTIONS The attraction between ions of unlike charges.

HEAT OF FUSION The number of calories needed to change 1 g of a solid to 1 g of liquid at its melting point.

HYDROGEN BOND The attraction between a hydrogen atom covalently bonded to a highly electronegative atom and another electronegative atom on an adjacent molecule.

HYDROPHOBIC ATTRACTION A weak attraction between nonpolar molecules.

INTERMOLECULAR ATTRACTIONS Attractions between adjacent molecules.

MELTING POINT The temperature at which a solid changes to a liquid.

POLAR MOLECULE A molecule whose centers of positive and negative charge do not coincide.

STANDARD TEMPERATURE AND PRESSURE (STP) 273 K and 1 atm of pressure.

TETRAHEDRAL CARBON A carbon atom whose four bonds point toward the four corners of a tetrahedron.

VAPOR The gaseous molecules of a substance at a temperature at which the substance is a liquid.

VAPOR PRESSURE The pressure caused by the molecules in the vapor above a liquid.

VOLATILE LIQUID A liquid that evaporates quickly at room temperature.

Concepts

1. The shape of a molecule is an important factor in determining its intermolecular attractions.
2. The shapes of simple molecules are determined by the total number of electron pairs about the central atom. They get as far away as possible from one another.
3. The strengths of intermolecular attractions vary widely. They are strong in solids, weaker in liquids, and weakest in gases.
4. If the particles of a solid are molecules or atoms, they are held tightly together by strong interatomic or intermolecular attractions. If the particles of a solid are ions, these ions are held even more tightly together by electrostatic attraction.
5. Weaker intermolecular attractions in liquids allow the atoms or molecules more movement but still hold them close together.
6. The molecules or atoms of a gas are widely separated from each other in otherwise empty space.
7. The ideal gas laws describe the behavior of many real gases in response to changes in pressure, temperature, and volume under conditions near room temperature and atmospheric pressure.
8. Breathing is a practical application of the gas laws.

EXERCISES

6-14 List the three intermolecular attractions discussed in this chapter and give an example of each.

6-15 What is the difference between amorphous and crystalline solids?

6-16 Explain how a liquid evaporates.

6-17 The boiling point of HCl, $-85°$ C, is higher than that of F_2, $-188°$ C. Explain this 103° C difference in boiling points.

6-18 An aerosol can will withstand an internal pressure of 4 atm before it explodes. How hot can you heat a 100-mL aerosol can, originally at 25° C and 2 atm, before it explodes?

6-19 A balloon at sea level and 25° C containing 10 L of helium rises to an altitude where the temperature is $-30°$ C and the pressure is 100 mm Hg. What is the volume of the balloon?

6-20 At mile-high Denver, the atmospheric pressure is 632 mm Hg. Calculate the partial pressures of oxygen, nitrogen, carbon dioxide, and argon at this altitude.

6-21 If 10 L of oxygen and 10 L of nitrogen, both at 760 mm Hg, are compressed into a 10-L cylinder at constant temperature, what is the total pressure in the cylinder? What is the partial pressure of each gas in the cylinder?

CHEMICAL REACTIONS 7

Chemical reactions occur all around us. The burning of fuel, the rusting of iron, and the growth of plants and animals are all examples of chemical reactions. We represent these and other chemical reactions on paper by *chemical equations*. A chemical equation is a concise way of giving information about a chemical reaction: we know what elements or molecules react, how much of each one reacts, and the products of such a reaction. Clearly, a chemical equation contains a great deal of information. We shall learn how to get this information in this chapter.

The compounds that react in a chemical reaction are represented by their chemical formulas in the chemical equation. Let us start by reviewing the meaning of chemical formulas and how to write them.

7.1 CHEMICAL FORMULAS

We learned in Section 5.5 that a chemical formula is a symbol used to represent a chemical compound. The formula consists of a combination of letters and numbers that specify the numbers and kinds of atoms in a molecule. The letters used are the symbols of the atoms that make up the molecule; the numbers, which appear as subscripts, indicate how many of each atom are in the molecule. Usually the most metallic element is listed first, followed by the other elements in order of decreasing metallic property. For example, the chemical formula of water is H_2O. This means that a molecule of water contains two hydrogen atoms and one oxygen atom. The chemical formula of glucose, the compound produced in photosynthesis (Chapter 21), is $C_6H_{12}O_6$. We know from its chemical formula that a molecule of glucose contains 6 carbon atoms, 12 hydrogen atoms, and 6 oxygen atoms.

*EXERCISE 7-1 Give the number of atoms of each element in each of the following molecules:
(a) NH_3 (b) CCl_4 (c) Cl_2 (d) C_2H_6O (e) $C_{12}H_{26}$
(f) Na_2HPO_4

We can determine other facts about a compound from its chemical formula. For example, we can obtain its molecular mass by adding the masses of the individual atoms making up the molecule. This is shown in the following example.

Example 7-1 Calculate the molecular mass of water, H_2O.
Solution:
The molecular mass is calculated by adding the mass of each atom within the molecule. Therefore,

$$\begin{aligned}\text{Molecular mass of water} &= 2 \times \text{Atomic mass of hydrogen} + 1 \times \text{Atomic mass of oxygen} \\ &= 2 \times 1.01 \text{ amu} + 1 \times 16.00 \text{ amu} \\ &= 2.02 \text{ amu} + 16.00 \text{ amu} \\ &= 18.02 \text{ amu}\end{aligned}$$

Because the terms mass and weight are often used interchangeably (see Section 1.3), the molecular mass of a compound is usually called its *molecular weight*. The term molecular weight is in such common use that it will be used throughout this book.

We learned in Section 5.4 that there is no such thing as a molecule of an ionic compound. How then do we obtain the molecular weights of these compounds? Actually, they are calculated in exactly the same way as the weight of any other compound. But we give a different name to the sum of the atomic masses of the elements that make up an ionic compound. This sum is called the *formula weight*. Thus, the formula weight of sodium chloride is the sum of the masses of one sodium atom and one chlorine atom.

Several elements exist as diatomic molecules. Oxygen is a typical exam-

* The answers for the exercises in this chapter begin on page 861.

ple; it exists in nature as the diatomic molecule O_2. Its molecular weight is two times the mass of one oxygen atom. The atomic mass of one oxygen atom (16.00 amu) should not be confused with the molecular weight of the element oxygen, that is, oxygen gas, O_2 (32.00 amu).

EXERCISE 7-2 Determine the molecular weight or the formula weight of each of the following compounds:
(a) N_2 (b) Cl_2 (c) NH_3 (d) CCl_4 (e) CH_2Cl_2
(f) $C_6H_{12}O_6$ (g) $C_{12}H_{22}O_{11}$ (h) $CaCl_2$ (i) NaH_2PO_4
(j) Na_2CO_3

Another fact that we can determine about a compound from its chemical formula is the percentage of each element in the molecule. We can do this using the following equation:

$$\% \text{ Element} = \frac{\text{mass of element}}{\text{molecular weight of compound}} \times 100 \qquad (7\text{-}1)$$

The use of this equation is shown in the following example:

Example 7-2 Determine the percentages of hydrogen and oxygen in water.
Solution:
First, determine the mass of each element in the compound:

$$\text{Mass of hydrogen} = 2\ \cancel{\text{atoms H}} \times \frac{1.01 \text{ amu}}{\cancel{\text{atom H}}} = 2.02 \text{ amu}$$

$$\text{Mass of oxygen} = 1\ \cancel{\text{atom O}} \times \frac{16.00 \text{ amu}}{\cancel{\text{atom O}}} = 16.00 \text{ amu}$$

Then, determine the molecular weight of water:

Molecular weight of water = 18.02 amu

Finally, place the quantities into equation 7-1:

$$\% \text{ H} = \frac{2.02\ \cancel{\text{amu}}}{18.02\ \cancel{\text{amu}}} \times 100 = 11.2\%$$

$$\% \text{ O} = \frac{16.00\ \cancel{\text{amu}}}{18.02\ \cancel{\text{amu}}} \times 100 = 88.8\%$$

EXERCISE 7-3 Determine the percentage of each element in each of the following molecules:
(a) CO_2 (b) NH_3 (c) CH_2Cl_2 (d) $C_6H_{12}O_6$
(e) C_2H_5NO

Chemical formulas are an important part of chemical equations, as we will learn in the next section.

7.2 CHEMICAL EQUATIONS

A chemical equation summarizes our observations of what happens in a chemical reaction. In the world around us or in a laboratory, certain chemicals called *reactants* come together and react. As a result of this chemical reaction, new compounds called *products* are formed. We represent these facts in a chemical equation.

Chemical equations are written in a particular way. We can show how to write chemical equations by using as an example the reaction of carbon monoxide and oxygen to form carbon dioxide.

The correct symbols and chemical formulas are used to represent the reactants and products in all chemical equations. In our example, carbon dioxide is represented by the formula CO_2 and carbon monoxide, by CO. Oxygen gas, a diatomic molecule, is correctly represented by its molecular formula O_2. By convention, the molecular formulas of the reactants are placed on the left side of the equation and the products are placed on the right side. The products and reactants are separated by an arrow pointing in the direction of the products. The arrow means forms or yields. Using this notation, we can write the chemical equation that represents the reaction of carbon monoxide and oxygen as follows:

$$CO + O_2 \longrightarrow CO_2 \qquad (7\text{-}2)$$

We read this equation as, "Carbon monoxide plus oxygen forms carbon dioxide."

All chemical equations must conform to the laws of chemistry. We learned in Section 2.8 that mass is conserved in a chemical reaction. The chemical equation must represent this fact. Consequently, the number of atoms of each element must be equal on the two sides of the chemical equation. Let us count the number of atoms on each side of equation 7-2 to see whether mass is conserved.

Element	No. of Atoms on Right Side of Equation 7-2	No. of Atoms on Left Side of Equation 7-2
C	1	1
O	2	3

There are two oxygen atoms on the right side of the equation, but there are three oxygen atoms on the left side. Recalling that mass must be conserved, we correct this difference by balancing the equation. To balance the equation, we place numbers called coefficients before the chemical formulas of the reactants and products until the numbers of atoms on the two sides of the equation are equal. We can balance equation 7-2 by multiplying both CO_2 and CO by 2:

Coefficient

$$2\,CO + O_2 \longrightarrow 2\,CO_2 \quad (7\text{-}3)$$

We can check this by recounting the numbers of all atoms on both sides of the equation.

Element	No. of Atoms on Right Side of Equation 7-3	No. of Atoms on Left Side of Equation 7-3
C	2	2
O	4	4

We now have equal numbers of each atom on both sides of the equation, and equation 7-3 is balanced.

The coefficients in equation 7-3 are not the only set that can balance the equation. We can understand this by looking at equations 7-4, 7-5, and 7-6. Each is balanced, yet the coefficients are different.

$$CO + \tfrac{1}{2}\,O_2 \longrightarrow CO_2 \quad (7\text{-}4)$$

$$4\,CO + 2\,O_2 \longrightarrow 4\,CO_2 \quad (7\text{-}5)$$

$$6\,CO + 3\,O_2 \longrightarrow 6\,CO_2 \quad (7\text{-}6)$$

The coefficients are different, but the ratios of the coefficients are identical. We can illustrate this as follows:

Equation No.	Ratio CO/CO_2	Ratio O_2/CO_2
7-3	$2/2 = 1$	$1/2$
7-4	$1/1 = 1$	$1/2/1 = 1/2$
7-5	$4/4 = 1$	$2/4 = 1/2$
7-6	$6/6 = 1$	$3/6 = 1/2$

We conclude from this that the coefficients in front of the chemical formulas in a balanced equation are relative numbers. From these numbers we obtain the relative amounts of each compound involved in the reaction. Thus, one molecule of CO reacts to form one molecule of CO_2. Two molecules of CO react to form two molecules of CO_2 and so forth. No matter how many molecules of CO react, an equal number of CO_2 molecules are formed. We know this from the coefficients of the balanced equation. This is the reason why a balanced equation is so important.

The coefficients in equation 7-5 are all twice as large as those in equation 7-3; those in equation 7-4 are all half as large as those in equation 7-3. If all the coefficients of a balanced equation are multiplied by a constant factor, the new equation is still balanced. In general, we prefer to write the balanced equation with the smallest whole numbers as coefficients.

The balancing of chemical equations is usually a matter of trial and error as well as experience. Although the important skill to acquire now is to recognize a balanced chemical equation and its meaning, it is still useful to begin to balance equations. The following examples show how to balance a chemical equation.

Example 7-3 Balance the equation for the reaction of methane (CH_4) and oxygen (O_2) to form water (H_2O) and carbon dioxide (CO_2).

Solution:

First, place the chemical formulas of the reactants on the left side and the products on the right side. Separate reactants from products by one arrow pointing to the products.

$$CH_4 + O_2 \longrightarrow H_2O + CO_2$$

Note that there is one more oxygen atom on the right side than on the left side of the equation. To balance the oxygen atoms, multiply the chemical formula for oxygen and water by 2.

$$CH_4 + 2\,O_2 \longrightarrow 2\,H_2O + CO_2$$

This equation is balanced.

Example 7-4 Sodium metal (Na) reacts with water (H_2O) to form sodium hydroxide (NaOH) and hydrogen gas (H_2). Write a balanced chemical equation for the reaction. The unbalanced equation is

$$Na + H_2O \longrightarrow NaOH + H_2$$

Solution:
First, to balance the hydrogen atoms, multiply the molecular formulas of water and sodium hydroxide by 2:

$$Na + 2\,H_2O \longrightarrow 2\,NaOH + H_2$$

Note that this results in one sodium atom on the left side and two sodium atoms on the right side of the equation. To balance the sodium atoms, just multiply the atomic symbol for sodium by 2:

$$2\,Na + 2\,H_2O \longrightarrow 2\,NaOH + H_2$$

This is the final balanced equation.

It is important to understand that the subscripts of the chemical formulas can never be changed to balance an equation. *If you change the subscript, you change the chemical formula and it no longer represents the compound involved in the reaction.* For example, it might be tempting to balance the equation in Example 7-3 by changing the chemical formula of water to H_3O. But this formula for water is incorrect because it disagrees with the experimentally determined composition of water. *Remember: never change the subscripts of a chemical formula to balance a chemical equation.*

We can summarize the three requirements that must be met to write a balanced chemical equation:

1. We must know what reactants are consumed and what products are formed.
2. We must know the correct formulas of each reactant and product.
3. We must satisfy the law of conservation of mass by balancing the equation.

EXERCISE 7-4 Balance each of the following equations:

(a) $H_2 + Cl_2 \longrightarrow HCl$ (b) $C + H_2 \longrightarrow CH_4$

(c) $Al + O_2 \longrightarrow Al_2O_3$ (d) $PtO \longrightarrow Pt + O_2$

(e) $C_2H_2 + H_2 \longrightarrow C_2H_6$

Once we have learned to balance chemical equations, we can turn our attention to using balanced chemical equations to determine the relative weights of reactants and products involved in the reaction. To do this, we must learn about an amount of matter called the mole.

7.3 GRAM-MOLECULAR WEIGHTS AND MOLES

We learned in the preceding section that the relative numbers of atoms and molecules that react in a chemical reaction can be determined from a balanced chemical equation. But molecules are so small that we cannot see them or weigh them individually. How then do we relate these extremely small particles to a quantity of matter that we can see and weigh? We do this by relating the molecular weights of molecules to the actual weight of a sample by the concept of gram-molecular weights.

The gram-molecular weight of a compound is its molecular weight expressed in grams. The molecular weight of water, for example, is 18.0 amu and its gram-molecular weight is 18.0 g. Hydrogen (H_2) has a molecular weight of 2.02 amu. Its gram-molecular weight is 2.02 g. Thus, the gram-molecular weight of a compound is found by determining its molecular weight in amu and then converting this number to grams.

EXERCISE 7-5 Determine the gram-molecular weight of each of the following compounds:

(a) Cl_2 (b) CH_2Cl_2 (c) Al_2O_3 (d) NaCl (e) $C_2H_4O_2$

The gram-molecular weight of a compound is also the weight of 1 *mole* of that substance. Thus, 18.0 g of water is 1 mole of water. Similarly, 2.02 g of hydrogen gas equals 1 mole of hydrogen. We can express this relationship as follows:

$$\text{Molecular weight of compound in grams} = \text{Gram-molecular weight of compound} = \text{1 mole of compound}$$

One mole of a substance is an extremely important quantity of matter because it contains exactly 602,000,000,000,000,000,000,000 (6.02×10^{23}) particles (ions, atoms, or molecules). The number of particles in a mole was first determined by the scientist Amadeo Avogadro, so we call this number *Avogadro's number.*

The importance of the mole is that it serves to link the invisible particles such as ions, atoms, and molecules with the visible matter that we can weigh. We can show this as follows:

$$6.02 \times 10^{23} \text{ particles} = 1 \text{ mole of substance} = \text{molecular weight of substance in grams}$$

Knowing Avogadro's number means that we are, in fact, weighing atoms, ions, or molecules when we weigh a sample of matter. We can show this by re-examining the reaction of carbon monoxide and oxygen to form carbon dioxide:

$$2\,CO + O_2 \longrightarrow 2\,CO_2$$

We concluded in the preceding section that the number of molecules of carbon dioxide formed in this reaction is exactly equal to the number of molecules of carbon monoxide that react. If the number of molecules of carbon monoxide is equal to Avogadro's number, 6.02×10^{23}, then there is 1 mole of carbon monoxide. This mole of carbon monoxide, 6.02×10^{23} molecules, reacts to form 6.02×10^{23} molecules, or 1 mole, of carbon dioxide. One mole of carbon monoxide weighs 28.0 g and 1 mole of carbon dioxide weighs 44.0 g. Therefore, 28.0 g of carbon monoxide react to form 44.0 g of carbon dioxide. In this way we relate the weight of a substance to the number of molecules of the substance that react.

Carbon monoxide, oxygen, and carbon dioxide are all gases at room temperature and atmospheric pressure. It is easier to measure their volumes than to weigh them. Therefore, it is useful to know the volume of 1 mole of a gas at standard conditions. It has been determined that the volume of 1 mole of an ideal gas at STP (Section 6.7), called its *molar volume,* is exactly 22.4 L. Most of the gases in nature are nearly ideal at STP and their molar volumes are very close to 22.4 L. Because the molar volume contains exactly 1 mole of the gas, the weight of the molar volume is equal to the gram-molecular weight of the gas.

The mole is the quantity that we use to relate the number of particles of a particular substance to its weight. Therefore, it is essential that we learn how to determine the number of moles in a given quantity of matter. We can do this using the method of conversion factors (Section 1.4), as shown in the following examples.

Example 7-5 How many moles are there in 36.0 g of water?

Solution:

STEP 1. Determine the gram-molecular weight (gmw) of water. In example 7-1, the molecular weight of water was determined to be 18.02 amu. Therefore, the gmw of water is 18.02 g. Because the weight of water is given to three significant figures in this example, the gmw of water is needed to only three significant figures. Therefore, for this example, the gmw of water is 18.0 g.

STEP 2. Convert gmw to moles (mol*):

$$18.0\ \text{g}\ H_2O = 1\ \text{mol}\ H_2O$$

STEP 3. Set up the conversion factors relating g to mol (see Section 1.4):

$$\frac{18.0\ \text{g}\ H_2O}{1\ \text{mol}\ H_2O} = 1 \qquad \frac{1\ \text{mol}\ H_2O}{18.0\ \text{g}\ H_2O} = 1$$

STEP 4. Because we want to convert g to mol, we use the conversion factor on the right:

$$36.0\ \cancel{\text{g}\ H_2O} \times \frac{1\ \text{mol}\ H_2O}{18.0\ \cancel{\text{g}\ H_2O}} = 2\ \text{mol}\ H_2O$$

Example 7-6 How many moles are there in 16.0 L of oxygen at STP?

Solution:

STEP 1. From the molar volume at STP, we know that

$$1\ \text{mol}\ O_2 = 22.4\ \text{L}\ O_2\ \text{at STP}$$

STEP 2. Set up the conversion factors relating mol to L:

$$\frac{1\ \text{mol}\ O_2}{22.4\ \text{L}\ O_2} = 1 \qquad \frac{22.4\ \text{L}\ O_2}{1\ \text{mol}\ O_2} = 1$$

* Mol is an abbreviation for mole.

STEP 3. Use the conversion factor on the left to convert from L to mol:

$$16.0 \cancel{L\ O_2} \times \frac{1 \text{ mol } O_2}{22.4 \cancel{L\ O_2}} = 0.714 \text{ mol } O_2$$

EXERCISE 7-6 Determine the number of moles in the quantities of each of the following compounds.
(a) 1.50 L N_2 at STP (b) 25.0 g NaCl (c) 125 g H_3PO_4
(d) 1.00 kg $C_6H_{12}O_6$

Frequently we want to convert the number of moles of a substance into the equivalent number of grams. This can be done using the method of conversion factors, as shown in the following examples.

Example 7-7 How many grams are there in 0.500 mol of rubbing alcohol, C_3H_8O?

Solution:

STEP 1. To determine the gmw of C_3H_8O, first determine its molecular weight:

$$(3 \times \text{atomic mass C}) + (8 \times \text{atomic mass H}) + (\text{atomic mass O}) = 60.1 \text{ amu.}$$

The gmw of C_3H_8O is 60.1 g.

STEP 2. Convert gmw to moles:

$$60.1 \text{ g } C_3H_8O = 1 \text{ mol } C_3H_8O$$

STEP 3. Set up the conversion factors relating g to mol:

$$\frac{60.1 \text{ g } C_3H_8O}{1 \text{ mol } C_3H_8O} = 1 \qquad \frac{1 \text{ mol } C_3H_8O}{60.1 \text{ g } C_3H_8O} = 1$$

STEP 4. Use the conversion factor on the left to convert from mol to g:

$$0.500 \cancel{\text{mol } C_3H_8O} \times \frac{60.1 \text{ g } C_3H_8O}{1 \cancel{\text{mol } C_3H_8O}} = 30.1 \text{ g } C_3H_8O$$

Example 7-8 How many liters are there in 2.70 mol of CO_2 at STP?

Solution:

STEP 1. From the molar volume at STP, we know that

$$1 \text{ mol } CO_2 = 22.4 \text{ L } CO_2$$

STEP 2. Set up the conversion factors relating mol to L:

$$\frac{1 \text{ mol } CO_2}{22.4 \text{ L } CO_2} = 1 \qquad \frac{22.4 \text{ L } CO_2}{1 \text{ mol } CO_2} = 1$$

STEP 3. Use the conversion factor on the right to convert from mol to L:

$$2.70 \, \cancel{\text{mol } CO_2} \times \frac{22.4 \text{ L } CO_2}{1 \, \cancel{\text{mol } CO_2}} = 60.5 \text{ L } CO_2$$

EXERCISE 7-7 Determine the number of grams in the quantities of each of the following compounds.
(a) 5.25 mol He (b) 0.750 mol NaCl (c) 1.53 mol $CHCl_3$
(d) 2.54 mol C_2H_5Cl

EXERCISE 7-8 Determine the volume of the quantities of each of the following gases at STP.
(a) 5.25 mol He (b) 2.00 mol N_2 (c) 0.575 mol O_2
(d) 3.25 mol CO

Another method of converting from moles of a substance to its gram equivalent and vice versa is by the use of the following equation.

$$\text{Number of moles of compound} = \frac{\text{weight of sample of compound in grams}}{\text{gram-molecular weight of compound}} \tag{7-7}$$

Equation 7-7 contains three quantities: the number of moles of the compound, the weight of the sample of the compound in grams, and the gram-molecular weight of the compound. If we know any two of these three quantities, we can determine the third, as shown in the following examples.

Example 7-9 How many moles are there in 50.0 g of CCl_4?
Solution:
STEP 1. Relate the quantities given in the problem to those in equation 7-7.

The weight of the sample is 50.0 g of CCl_4. The gmw of the CCl_4, determined by the method given in example 7-1, is 154 g. Therefore 1 mol of CCl_4 = 154 g. Expressed another way, there are 154 g of CCl_4 per mol of CCl_4, or 154 g CCl_4/mol CCl_4.
STEP 2. Substitute the quantities into equation 7-7:

$$\text{Number of moles of } CCl_4 = \frac{50.0\ g\ \cancel{CCl_4}}{154\ g\ \cancel{CCl_4}/1\ mol\ CCl_4} = 0.325\ mol\ CCl_4$$

Example 7-10 How many grams are there in 0.275 mol of NaCl?
Solution:
STEP 1. Relate the quantities given in the problem to those in equation 7-7.
We wish to determine the number of grams of NaCl. There is 0.275 mol of NaCl. The gmw of NaCl is obtained by the method shown in example 7-1: 58.4 g NaCl/mol NaCl.
STEP 2. Substitute the quantities into equation 7-7:

$$0.275\ mol\ NaCl = \frac{g\ NaCl}{58.4\ g\ NaCl/mol\ NaCl}$$

$$g\ NaCl = 0.275\ \cancel{mol\ NaCl} \times \frac{58.4\ g\ NaCl}{\cancel{mol\ NaCl}} = 16.1\ g\ NaCl$$

EXERCISE 7-9 Convert each of the following quantities into moles:
(a) 25.0 g C_3H_8O (b) 72.8 g HCl (c) 100 g $CaCl_2$
(d) 250 g $C_4H_{10}O$

EXERCISE 7-10 Convert each of the following amounts into grams:
(a) 0.253 mol LiBr (b) 2.00 mol I_2 (c) 1.75 mol CH_2Cl_2
(d) 5.63 mol $C_6H_{12}O_6$

Converting grams to moles and vice versa is a skill that is needed to handle the weight relations in chemical reactions, as we will learn in the next section.

7.4 WEIGHT RELATIONS IN CHEMICAL REACTIONS

Chemists are not satisfied to know just that two elements or compounds react to form products. They also want to know how much of each reactant is consumed and how much product is formed. This is the quantitative side of chemistry. We obtain quantitative information about a chemical reaction from its balanced equation. The key is the number or coefficient placed before each chemical formula in the balanced chemical equation. We obtain the relative quantities of each chemical involved in the reaction from these coefficients. We briefly touched on this fact in Section 7.3. Now let us go more deeply into this subject.

Let us consider the fermentation of grains and fruits as an example. This chemical reaction has been known and used for centuries to make alcoholic beverages. In this reaction, glucose, $C_6H_{12}O_6$, is converted into ethyl alcohol, C_2H_6O, and carbon dioxide. The balanced equation is as follows:

$$C_6H_{12}O_6 \longrightarrow 2\,C_2H_6O + 2\,CO_2 \tag{7-8}$$

In terms of molecules, we know from this chemical equation that one molecule of glucose reacts to form two molecules of ethyl alcohol and two molecules of carbon dioxide. We can just as easily think in terms of moles. Thus, 1 mole of glucose (Avogadro's number, 6.02×10^{23}, of molecules of glucose) react to form 2 moles of ethyl alcohol (2 $\times$ Avogadro's number of molecules of ethyl alcohol) and 2 moles of carbon dioxide (2 $\times$ Avogadro's number of molecules of carbon dioxide). We know that 1 mole is equal to the gram-molecular weight of a compound (see Section 7.3). Therefore, we know that 180 g of glucose (1 $\times$ gmw of glucose) forms 92.2 g of ethyl alcohol (2 $\times$ gmw of ethyl alcohol) and 88.0 g of carbon dioxide (2 $\times$ gmw of carbon dioxide). These quantitative relations are summarized in Table 7-1.

The information in Table 7-1 shows us that the ratios of the numbers of molecules or moles of compounds involved in the reaction are available from the coefficients of a balanced equation. Thus, the ratio of molecules or moles of glucose to those of carbon dioxide is 1:2. Similarly, the ratio of molecules or moles of glucose to those of ethyl alcohol is 1:2. No matter

Table 7-1. Quantitative Relations in the Chemical Reaction
$C_6H_{12}O_6 \longrightarrow 2\,C_2H_6O + 2\,CO_2$

Quantity	$C_6H_{12}O_6$	C_2H_6O	CO_2
Molecules	1	2	2
Moles	1	2	2
Atomic mass units (amu)	180	92.2	88.0
Grams	180	92.2	88.0

how many molecules or moles of glucose we start with, twice as many molecules or moles of ethyl alcohol and carbon dioxide will be formed.

This theory is straightforward, but sometimes its application to real-life situations is a bit confusing. When we carry out a chemical reaction, we usually have a given quantity (in grams) of one of the reactants. We want to know how many grams of one or more of the products will be formed in the reaction. We can solve this problem using the following steps.

STEP 1. Obtain the balanced chemical equation for the chemical reaction.
STEP 2. Convert the given quantity from grams to moles.
STEP 3. Using the coefficients of the balanced equation, determine the ratio of moles of the reactant and the product we want.
STEP 4. Convert the moles of product into grams or any other unit that is needed.

The use of these four steps is shown in the following examples.

Example 7-11 How many grams of ethyl alcohol can be obtained from the fermentation of 1000 g of glucose?

Solution:

STEP 1. Equation 7-8 is the balanced equation for this chemical reaction.

STEP 2. Convert 1000 g of glucose to moles of glucose (see Section 7.3):

$$1000 \cancel{\text{g } C_6H_{12}O_6} \times \frac{1 \text{ mol } C_6H_{12}O_6}{180 \cancel{\text{g } C_6H_{12}O_6}} = 5.55 \text{ mol } C_6H_{12}O_6$$

STEP 3. Determine the ratio of moles C_2H_6O to moles $C_6H_{12}O_6$ from the balanced equation:

$$1 \text{ mol } C_6H_{12}O_6 = 2 \text{ mol } C_2H_6O \text{ or } \frac{2 \text{ mol } C_2H_6O}{1 \text{ mol } C_6H_{12}O_6}$$

The 5.55 mol of $C_6H_{12}O_6$ are equal to the following moles of C_2H_6O:

$$5.55 \cancel{\text{mol } C_6H_{12}O_6} \times \frac{2 \text{ mol } C_2H_6O}{1 \cancel{\text{mol } C_6H_{12}O_6}} = 11.1 \text{ mol } C_2H_6O$$

STEP 4. Convert moles of product to grams of product (see Section 7.3):

$$11.1\ \cancel{\text{mol } C_2H_6O} \times \frac{46.1\text{ g } C_2H_6O}{1\ \cancel{\text{mol } C_2H_6O}} = 512\text{ g } C_2H_6O$$

EXERCISE 7-11 How many g and how many mL of carbon dioxide can be obtained from the fermentation of 500 g of glucose at STP?

EXERCISE 7-12 The rusting of iron is the reaction of iron with oxygen to form ferric oxide (rust). How many g of ferric oxide are formed by the complete rusting of 25.0 g of iron?

Sometimes we want to carry out a reaction to produce a certain quantity of product. This is a variant of the problem solved in example 7-11 and is solved in a similar way, as shown by the following example.

Example 7-12 How many grams of glucose do we have to ferment to get 100 g of ethyl alcohol?

Solution:

STEP 1. Equation 7-8 is the balanced equation for this chemical reaction.

STEP 2. Convert 100 g of ethyl alcohol to moles of ethyl alcohol.

$$100\ \cancel{\text{g } C_2H_6O} \times \frac{1\text{ mol } C_2H_6O}{46.1\ \cancel{\text{g } C_2H_6O}} = 2.17\text{ mol } C_2H_6O$$

STEP 3. Relate the number of moles of ethyl alcohol to the number of moles of glucose in the balanced equation:

$$1\text{ mol } C_6H_{12}O_6 = 2\text{ mol } C_2H_6O$$

$$2.17\ \cancel{\text{mol } C_2H_6O} \times \frac{1\text{ mol } C_6H_{12}O_6}{2\ \cancel{\text{mol } C_2H_6O}} = 1.09\text{ mol } C_6H_{12}O_6$$

STEP 4. Convert moles of $C_6H_{12}O$ to grams of $C_6H_{12}O$:

$$1.09\ \cancel{\text{mol } C_6H_{12}O_6} \times \frac{180\text{ g } C_6H_{12}O_6}{1\ \cancel{\text{mol } C_6H_{12}O_6}} = 196\text{ g } C_6H_{12}O_6$$

EXERCISE 7-13 A spark causes a mixture of hydrogen and oxygen gases to react to form water. How many g of hydrogen gas are needed to form 25.0 g of water?

EXERCISE 7-14 Blood glucose reacts with oxygen in cells to form carbon dioxide and water. How many g of carbon dioxide are formed from 10.0 g of glucose?

So far we have been concerned only about the conservation of mass in a chemical reaction. But what about energy?

7.5 ENERGY AND CHEMICAL REACTIONS

We learned in Section 2.6 that a transfer of energy occurs in many chemical reactions. The burning of gasoline in the engine of a car is an example. The chemical energy of the gasoline is transferred into heat and work. The energy released in such a reaction is sometimes included as one of the products in the chemical equation. For example:

$$2\,C_8H_{18} + 25\,O_2 \longrightarrow 16\,CO_2 + 18\,H_2O + \text{energy}$$

When this energy is released as heat, the reaction is called an *exothermic reaction.*

The amount of energy released in a chemical reaction depends on the quantity of reactants used. For example, burning 1 mole of carbon (coal) releases 94.1 kcal of energy as heat. We can indicate this fact in the chemical equation as follows:

$$C + O_2 \longrightarrow CO_2 + 94.1\ \text{kcal} \qquad (7\text{-}9)$$

With this information, we can calculate the amount of heat released by the reaction of any quantity of reactant, just as we could for any other product. This is shown in the following example.

Example 7-13 How much heat is released by burning 10.0 kg of coal? Assume that the coal is pure carbon.

Solution:

STEP 1. Equation 7-9 is the balanced equation.

STEP 2. Convert 10.0 kg of carbon to moles of carbon:

$$10.0\,\cancel{\text{kg C}} \times \frac{1\ \text{mol C}}{12.0\,\cancel{\text{g C}}} \times \frac{1000\,\cancel{\text{g}}}{1\,\cancel{\text{kg}}} = 833\ \text{mol C}$$

STEP 3. Relate heat released in kcal to the number of moles of carbon in the balanced equation:

1 mol C gives off 94.1 kcal, or

$$1 \text{ mol C} = 94.1 \text{ kcal or } \frac{94.1 \text{ kcal}}{1 \text{ mol C}} = 1$$

STEP 4. Convert moles of carbon to kcal released:

$$833 \cancel{\text{mol C}} \times \frac{94.1 \text{ kcal}}{1 \cancel{\text{mol C}}} = 7.84 \times 10^4 \text{ kcal}$$

Not all chemical reactions release energy. Indeed, energy must be added to make some compounds react. For example, energy must be added to water to decompose it to hydrogen and oxygen according to the following equation:

$$2\,H_2O + 136 \text{ kcal} \longrightarrow 2\,H_2 + O_2$$

Thus, 68 kcal of heat (136 kcal/2 mol H_2O) must be added to decompose 1 mole of water. This energy must be supplied from the surroundings. A reaction that takes heat from its surroundings is called an *endothermic reaction*.

So far we have considered only reactions that proceed in one direction. But many reactions in nature are reversible; that is, they proceed in both directions. The reactants form products, but the products can also react to reform the reactants. An important example of a reversible reaction is the one between glucose and oxygen to form carbon dioxide and water according to the following equation:

$$C_6H_{12}O_6 + 6\,O_2 \xrightarrow{\text{Metabolism}} 6\,CO_2 + 6\,H_2O + \text{energy}$$

This reaction is one of several that are involved in the metabolism of glucose (see Chapter 27). The reaction occurs in living cells and provides the energy needed by the cells to function.

The reverse reaction occurs in the leaves of plants by a process called photosynthesis (see Section 27.8). The energy of sunlight converts carbon dioxide and water to glucose and oxygen according to the following equation:

$$6\,CO_2 + 6\,H_2O + \text{energy} \xrightarrow{\text{Photosynthesis}} C_6H_{12}O_6 + 6\,O_2$$

We can combine these two equations as follows:

$$C_6H_{12}O_6 + 6\,O_2 \underset{\text{Photosynthesis}}{\overset{\text{Metabolism}}{\rightleftarrows}} 6\,CO_2 + 6\,H_2O + \text{energy}$$

Two arrows pointing in opposite directions ($\rightleftarrows$) are the symbol used to indicate a reversible reaction. All life on our planet depends on this reversible reaction.

EXERCISE 7-15 Decide whether each of the following reactions is exothermic or endothermic:
(a) $2\,C + O_2 \longrightarrow 2\,CO + 52.8$ kcal
(b) $I_2 + Cl_2 + 4.20$ kcal $\longrightarrow 2\,ICl$
(c) $H_2 + Cl_2 \longrightarrow 2\,HCl + 22.1$ kcal
(d) $2\,C + 2\,H_2 + 12.5$ kcal $\longrightarrow C_2H_4$

EXERCISE 7-16 The reaction of methane and oxygen to form carbon dioxide and water releases 213 kcal/mol of methane. How much heat is released from 20.0 g of methane?

The energy released in a chemical reaction comes from the energy contained in the reactants. All compounds have a certain amount of chemical energy as a result of their chemical bonds and the arrangement of their atoms, but different compounds have different amounts of energy. When the products of a reaction have less energy than the reactants, there is a release of energy during the reaction. If the reverse occurs, that is, if the products have more energy than the reactants, then energy is taken from the surroundings. Thus, there is an energy balance in chemical reactions as required by the law of conservation of mass-energy. This is shown schematically in Figure 7-1.

The energy of the reactants in a chemical reaction can also be released as electrical energy. This occurs in what are called oxidation-reduction reactions, as we will learn next.

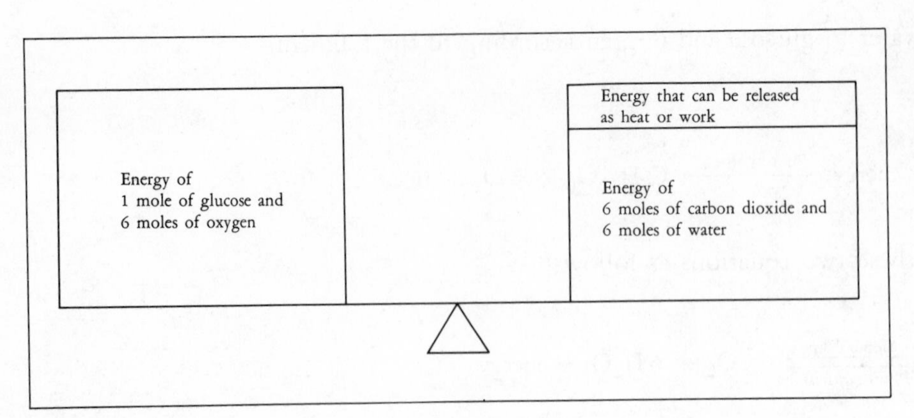

Fig. 7-1. Energy balance in the reaction of glucose and oxygen to form carbon dioxide, water, and energy.

7.6 OXIDATION-REDUCTION REACTIONS

We learned in Chapter 3 that certain atoms gain or lose electrons to achieve the particularly stable noble gas electron arrangement. This transfer of one or more electrons from one atom to another occurs in oxidation-reduction reactions. (Oxidation-reduction reactions are sometimes called redox reactions. The word redox is a contraction of the two words *red*uction-*ox*idation.) In such reactions, two processes occur at the same time. One atom loses one or more electrons, and another atom gains the same number of electrons. These two processes are given different names.

Oxidation is the loss of one or more electrons by an atom.
Reduction is the gain of one or more electrons by an atom.

The atom that loses electrons is oxidized, and the atom that gains electrons is reduced. Oxidation and reduction always occur together. One cannot happen without the other. The electrons lost by the atom that is oxidized are gained by the atom that is reduced.

The reaction of sodium metal and chlorine gas to form sodium chloride that we studied in Chapter 5 is an example of an oxidation-reduction reaction.

$$2\,Na + Cl_2 \longrightarrow 2\,NaCl$$

We can divide this equation into two half-reactions, one representing the oxidation part and the other representing the reduction part of the over-all reaction. Consider first the sodium metal. Each sodium atom loses an electron to form a sodium cation. We write the equation for this half-reaction as follows:

$$Na \longrightarrow Na^+ + e^-$$

Sodium is oxidized because it loses an electron. Chlorine gains an electron to form a chloride anion. We write the equation for this half-reaction as follows:

$$2\,e^- + Cl_2 \longrightarrow 2\,Cl^-$$

Chlorine is reduced because it gains an electron.

Both oxidation and reduction occur in the reaction of sodium metal and chlorine. The over-all reaction is the sum of the oxidation and reduction half-reactions. We can add the two half-reactions as follows:

$$\begin{array}{rcll} 2\,Na & \longrightarrow & 2\,Na^+ + 2\,e^- & \text{oxidation half-reaction} \\ 2\,e^- + Cl_2 & \longrightarrow & 2\,Cl^- & \text{reduction half-reaction} \\ \hline 2\,e^- + 2\,Na + Cl_2 & \longrightarrow & 2\,NaCl + 2\,e^- & \\ 2\,Na + Cl_2 & \longrightarrow & 2\,NaCl & \text{over-all equation} \end{array}$$

The sodium half-reaction is multiplied by 2 to balance the number of electrons on each side of the equation. The sodium and chloride ions are written together as sodium chloride. The electrons, which are common to both sides of the equation, are canceled to give the over-all chemical equation.

EXERCISE 7-17 Lithium metal reacts with bromine to form lithium bromide. Which element is oxidized? Which is reduced? Write the oxidation and reduction half-reactions and the over-all equation.

In the reaction of sodium metal and chlorine, chlorine causes the sodium atoms to lose electrons. Therefore, chlorine is called an oxidizing agent. In the same reaction, sodium atoms cause chlorine to gain electrons. Sodium is called a reducing agent. Notice that the substance (either atoms or molecules) oxidized is a reducing agent and the substance reduced is an oxidizing agent. We can summarize this as follows:

An oxidizing agent is an atom, molecule, or ion that is reduced (i.e., gains electrons) in a chemical reaction and at the same time causes another substance to be oxidized (i.e., to lose electrons).

A reducing agent is an atom, molecule, or ion that is oxidized (i.e., loses electrons) in a chemical reaction and at the same time causes another substance to be reduced (i.e., to gain electrons).

It is clear that electron transfer occurs in oxidation-reduction reactions. To keep track of which atoms gain or lose electrons, we assign them *oxidation numbers*. The oxidation number of an atom, given by a number and a positive or negative sign, is determined by the following set of rules.

1. Uncombined elements have an oxidation number of zero. For example, oxygen (O_2), chlorine (Cl_2), sodium metal (Na), and gold (Au) all have oxidation numbers of zero.
2. The oxidation number of a simple cation of groups IA, IIA, and IIIA and a simple anion of groups VIA and VIIA is equal to the charge on the ion. For example, Cl^- has an oxidation number of -1, and Na^+ has an oxidation number of $+1$.
3. Hydrogen in compounds has an oxidation number of $+1$ except when it is combined with a metal. When hydrogen is combined with a metal, its oxidation number is -1. In HCl the hydrogen has an oxidation number of $+1$. In lithium hydride, LiH, hydrogen has an oxidation number of -1 because it is combined with the metal lithium.
4. Oxygen has an oxidation number of -2 except in compounds called peroxides, where its oxidation number is -1. The most common example of a peroxide is hydrogen peroxide, H_2O_2.
5. The algebraic sum of all the oxidation numbers must be equal to zero in a compound. For example, in the compound sodium hydroxide, NaOH, the oxidation numbers are: sodium, $+1$; oxygen, -2; hydrogen, $+1$. Their sum is equal to zero.

Element	Sodium		Oxygen		Hydrogen	
Oxidation No.	$(+1)$	$+$	(-2)	$+$	$(+1)$	$= 0$

6. The algebraic sum of all the oxidation numbers must be equal to the charge of a polyatomic ion. For example, in the hydroxide ion, OH^-, the oxidation numbers are oxygen, -2; hydrogen, $+1$. Their sum is -1.

Element	Oxygen		Hydrogen	
Oxidation No.	(-2)	$+$	$(+1)$	$= -1$

We can determine the oxidation number of any element in a compound using these rules, as shown by the following examples:

Example 7-14 Determine the oxidation number of C in CO_2.
Solution:
STEP 1. The oxidation number of oxygen is -2 (rule 4).
STEP 2. There are two oxygen atoms for a total oxidation number of -4.
STEP 3. The sum of the oxidation numbers in a compound must be zero (rule 5). Therefore, the oxidation number of C is $+4$.

Example 7-15 Determine the oxidation number of N in NH_4^+.
Solution:
STEP 1. The oxidation number of hydrogen is $+1$ (rule 3).
STEP 2. There are four hydrogen atoms for a total oxidation number of $+4$.
STEP 3. The sum of the oxidation numbers must be equal to $+1$ for this polyatomic ion (rule 6). Therefore, we can write:

$$\underset{\text{of N}}{\text{Oxidation number}} + 4 \times \underset{\text{of H}}{\text{Oxidation number}} = +1$$

$$X + (+4) = (+1)$$

$$X = -3$$

The oxidation number of N is -3.

EXERCISE 7-18 Determine the oxidation number of each element in the following compounds or ions:
(a) CO (b) H_2O (c) CO_3^{-2} (d) $MgCl_2$ (e) Fe_2O_3
(f) CH_4 (g) CH_4O

An oxidation-reduction reaction can be identified from the fact that the oxidation numbers of the elements are different in the products and reactions. As an example, let us return to the reaction of sodium metal and chlorine. The oxidation numbers are given above the chemical symbols of the reactants and products in the following equation.

$$2\ Na^0 + Cl_2^0 \longrightarrow 2\ Na^{+1}Cl^{-1}$$

The oxidation number of sodium changes from zero to +1 in this reaction. An *increase* in the oxidation number of an element in a chemical reaction indicates that the element has been oxidized. Thus, sodium is oxidized. The oxidation number of chlorine changes from zero to −1 in this reaction. A *decrease* in the oxidation number indicates that the element has been reduced. In this way we can identify not only oxidation-reduction reactions but also which element is oxidized and which element is reduced. Let us do a few more examples.

Example 7-16 Determine whether the reaction of zinc metal and HCl to form hydrogen gas and zinc chloride, $ZnCl_2$, is an oxidation-reduction reaction.

Solution:

STEP 1. Write the balanced chemical equation:

$$Zn + 2\ HCl \longrightarrow ZnCl_2 + H_2$$

STEP 2. Determine the oxidation state of each element in the products and reactants:

$$Zn^0 + 2\ H^{+1}Cl^{-1} \longrightarrow Zn^{+2}Cl_2^{2(-1)} + H_2^0$$

STEP 3. Compare the oxidation states of the same element in the reactants and products:

Zn changes from zero to +2.

H changes from +1 to zero.

Cl does not change.

This is an oxidation-reduction reaction because the oxidation states of both zinc and hydrogen change.

Example 7-17 Which reagent is oxidized in the reaction in example 7-16? Which is reduced? Which reagent is the oxidizing agent? Which one is the reducing agent? Write the oxidation and reduction half-reactions.

Solution:

The oxidation number of an oxidized element increases. The oxidation number of zinc increases from zero to +2, so zinc is

oxidized. The oxidation number of hydrogen decreases from +1 to zero, so hydrogen is reduced.
Zinc metal is the reducing agent, and H^+ is the oxidizing agent. The two half-reactions are

$$Zn \longrightarrow Zn^{+2} + 2\,e^-$$

$$2\,H^{+1} + 2\,e^- \longrightarrow H_2$$

EXERCISE 7-19 Determine the oxidation number of each element in both the reactants and products of the following balanced equations. In each reduction identify the reagent that is oxidized, the reagent that is reduced, the reducing agent, and the oxidizing agent. Write the equation for the oxidation and reduction half-reactions.

(a) $4\,Al + 3\,O_2 \longrightarrow 2\,Al_2O_3$
(b) $H_2 + CuO \longrightarrow Cu + H_2O$
(c) $Zn + CuSO_4 \longrightarrow Cu + ZnSO_4$
(d) $CH_2O + H_2 \longrightarrow CH_4O$

There are many oxidation-reduction reactions in living systems. One of them supplies most of the energy required by a cell. This reaction is the oxidation of compounds obtained from food. The over-all equation for this reaction is the following:

$$2\,MH_2 + O_2 \longrightarrow 2\,M + 2\,H_2O \qquad (7\text{-}9)$$

The symbol MH_2 represents any of several different compounds obtained from food (Chapters 27, 28, and 29). Each molecule of MH_2 transfers two electrons to oxygen in this oxidation-reduction reaction. The compound MH_2 is oxidized, and oxygen gas is reduced to water. The two half-reactions for this oxidation-reduction are the following:

$$2\,MH_2 \longrightarrow 2\,M + 4\,H^{+1} + 4\,e^- \qquad \text{oxidation half-reaction}$$

$$4\,e^- + 4\,H^{+1} + O_2 \longrightarrow 2\,H_2O \qquad \text{reduction half-reaction}$$

Equation 7-9 represents the over-all chemical equation of the oxidation of compounds obtained from food by oxygen. This reaction actually occurs in living cells by a series of oxidation-reduction reactions that are called collectively *oxidative phosphorylation* (Chapter 27). In these reactions, the energy of food is transferred from MH_2 to special compounds that distribute the energy wherever it is needed. These compounds and some of the oxidation-reduction reactions that they undergo are discussed in Chapter 26.

Certain substances have a great effect on a chemical reaction but are not included in the chemical equation because they are not altered during the reaction. We will learn about these substances next.

7.7 CATALYSIS

Many chemical reactions involve substances that do not appear in the chemical equation for the over-all reaction. Although these substances are not usually altered, their presence causes an increase in the speed at which the reaction occurs. These substances are called *catalysts* and their effect is called *catalysis*. Catalysis is extremely important in many reactions in chemistry and biochemistry. Many of the reactions used in industry would not occur fast enough to be of practical importance without a catalyst. Among the most powerful catalysts are enzymes, compounds of high molecular weight that catalyze reactions in living systems. We will study enzymes in Chapter 25.

There are many kinds of catalysts. They can be as simple as a hydrogen ion (H^+) or as complex as an enzyme. Some catalysts are very specific and catalyze only one reaction. Others catalyze a wide variety of reactions. Some catalysts are soluble in the reaction mixture; they are called homogeneous catalysts. Others, called heterogeneous catalysts, are insoluble in the reaction mixture.

The reaction of sugar (sucrose) with water to form fructose and glucose is an example of a reaction that is catalyzed by an enzyme called invertase (see Chapter 21). We indicate that the reaction is catalyzed by placing either the name or the chemical formula of the catalyst above the arrow in the chemical equation as follows:

$$\underset{\text{Sugar (sucrose)}}{C_{12}H_{22}O_{11}} + H_2O \xrightarrow{\text{invertase}} \underset{\text{Glucose}}{C_6H_{12}O_6} + \underset{\text{Fructose}}{C_6H_{12}O_6}$$

We will encounter many examples of catalysis in future chapters.

7.8 SUMMARY

A chemical reaction is represented on paper by a chemical equation. Certain conventions are used in writing chemical equations. The reactants and products are represented by their correct chemical formulas. The chemical formulas of the reactants are placed on the left side and those of the products are placed on the right side of the chemical equation. The products and reactants are separated by an arrow pointing in the direction of the products.

A chemical equation must be balanced to obey the law of conservation of mass-energy. The number of atoms of each element must be equal on the two sides of a chemical equation. A chemical equation is balanced by placing numbers called coefficients in front of the chemical formulas of the reactants and products. The numbers are chosen so that the number of atoms of each element is the same on both sides of the equation. The subscripts of chemical formulas are never changed to balance a chemical equation.

From the coefficients of a balanced chemical equation, we can obtain the relative number of particles (atoms, molecules, or ions) involved in the reaction. To relate the number of these invisible particles to the amount of a substance that can be weighed, we use the concept of the mole. A mole is the molecular weight of a compound in grams and contains Avogadro's number of particles, 6.02×10^{23}. The mole bridges the gap between the invisible world of atoms, molecules, and ions and the visible world. Thus, when we weigh a sample of matter, we are actually counting the number of particles contained in that sample.

Energy is also a reactant or product in a chemical reaction. When energy is released as heat, the chemical reaction is called an exothermic reaction. A chemical reaction that takes energy from its surroundings is called an endothermic reaction. The energy of a chemical reaction can also be released as electrical energy. This occurs in oxidation-reduction reactions. Oxidation is the loss of one or more electrons by an atom. Reduction is the gain of one or more electrons by an atom. Oxidation and reduction always occur together in a chemical reaction. One of the most important oxidation-reduction reactions in living system supplies most of the energy required by a cell.

REVIEW OF TERMS AND CONCEPTS

Terms

AVOGADRO'S NUMBER 6.02×10^{23} particles.

CATALYSIS The effect of a catalyst on a chemical reaction.

CATALYST A substance that increases the speed at which a reaction occurs but is not altered during the reaction.

CHEMICAL EQUATION The representation of a chemical reaction.

COEFFICIENTS The numbers placed before chemical formulas to balance a chemical equation.

ENDOTHERMIC REACTION A chemical reaction that takes energy from its surroundings.

EXOTHERMIC REACTION A chemical reaction that releases energy.

FORMULA WEIGHT The molecular weight of an ionic compound in grams.

GRAM-MOLECULAR WEIGHT The molecular weight of a compound in grams.

MOLAR VOLUME The volume (22.4 L) occupied by 1 mole of a gas at STP.

MOLE The molecular or formula weight of a compound in grams (also equal to 6.02×10^{23} particles).

MOLECULAR WEIGHT The sum of the weight (actually, mass) of the individual atoms that make up a molecule.

OXIDATION The loss of one or more electrons by an atom.

OXIDIZING AGENT An atom, molecule, or ion that is reduced in a chemical reaction and at the same time causes another substance to be oxidized.

PRODUCTS Atoms, molecules, or ions formed as a result of a chemical reaction.

REACTANTS Atoms, molecules, or ions that undergo a chemical reaction.

REDUCING AGENT An atom, molecule, or ion that is oxidized in a chemical reaction and at the same time causes another substance to be reduced.

REDUCTION The gain of one or more electrons by an atom.

REVERSIBLE REACTION A chemical reaction that can occur in both directions.

Concepts

1. Two important facts are obtained from a balanced chemical equation. First, the products and reactants are identified. Second, the relative ratios of the chemicals that participate in the reaction are obtained.
2. The coefficients of the chemical formulas in a balanced chemical equation are the keys to determining the relative ratios of the chemicals involved in the reaction. These ratios can be interpreted in terms of moles of reactants and products.
3. The mole bridges the gap between the invisible world of individual atoms, molecules, and ions and the visible world of samples of matter that we can weigh.

4. Energy, usually in the form of heat, is often a product or reactant in a chemical reaction.
5. One or more electrons are transferred from one atom to another in an oxidation-reduction reaction.
6. The speed at which a chemical reaction occurs is increased by the presence of a catalyst.

EXERCISES 7-20 Determine the molecular weight of each of the following compounds:
(a) O_2 (b) $NaOH$ (c) H_3PO_4 (d) $C_4H_{10}O$
(e) C_2H_4NOCl

7-21 Why is it necessary to balance a chemical equation?

7-22 Balance each of the following equations:
(a) $C_3H_4 + H_2 \longrightarrow C_3H_8$
(b) $FeO + O_2 \longrightarrow Fe_2O_3$
(c) $CO + NO \longrightarrow N_2 + CO_2$
(d) $CaCO_3 + SO_2 + O_2 \longrightarrow CaSO_4 + CO_2$

7-23 Write balanced equations for each of the following reactions:
(a) Aluminum + bromine $\longrightarrow$ aluminium bromide
(b) Hydrogen + chlorine $\longrightarrow$ hydrogen chloride
(c) Ferric oxide + carbon monoxide $\longrightarrow$ iron + carbon dioxide
(d) Hydrogen sulfide + sulfur dioxide $\longrightarrow$ sulfur + water

7-24 Determine the gram-molecular weight of each of the following compounds:
(a) NO (b) C_6H_6 (c) Cu_3FeS_3 (d) CH_3HgCl
(e) $C_{10}H_7NO_2$

7-25 By definition, one mole of $^{12}_{6}C$ weighs 12.000 g. What is the weight of one atom of $^{12}_{6}C$?

7-26 How many moles are there in 25.0 g of each of the following compounds?
(a) N_2 (b) $LiBr$ (c) C_4H_{10} (d) H_2SO_4 (e) $CaHPO_4$

7-27 How many grams are there in 0.350 mol of each of the following compounds?
(a) I_2 (b) CCl_4 (c) C_4H_8 (d) $Ca(OH)_2$ (e) $C_6H_5NO_2$

7-28 Determine the weight, in grams, of 1 mole of
(a) chlorine atoms (b) chlorine molecules (c) chloride ions

7-29 A liquid fat has a composition of $C_{57}H_{68}O_6$. Its reaction with hydrogen, called hydrogenation, converts it to a solid that has a composition of $C_{57}H_{80}O_6$. How many moles of hydrogen are needed to hydrogenate 1 mole of liquid fat? How many grams of hydrogen are needed? How many liters of hydrogen at STP are needed? How many grams of solid fat are formed?

7-30 Sulfur dioxide produced by burning coal that contains sulfur in a furnace can be removed from the gases emitted in the chimney by treating it with carbon monoxide to form sulfur and carbon dioxide. How many moles of carbon monoxide are needed to react with 1.00 kg of sulfur dioxide? How many grams does this represent? How many grams of sulfur and carbon dioxide are formed?

7-31 Sulfur dioxide released into the atmosphere eventually reacts with oxygen to form toxic sulfur trioxide (SO_3). This sulfur trioxide reacts with water to form sulfuric acid (H_2SO_4), the acid of acid rain. Write the balanced equations for these two reactions. How many grams of sulfuric acid will be formed from 1.00 kg of sulfur dioxide?

7-32 Determine the oxidation number of each element in the following:
(a) C (b) CO (c) NO (d) NH_3 (e) CH_4O
(f) CH_2O

7-33 Which of the following are oxidation-reduction reactions?
(a) $HCl + NaOH \longrightarrow NaCl + H_2O$
(b) $2\,CO + 2\,NO \longrightarrow N_2 + 2\,CO_2$
(c) $NH_3 + HCl \longrightarrow NH_4Cl$
(d) $2\,H_2S + 3\,O_2 \longrightarrow 2\,SO_2 + 2\,H_2O$

AQUEOUS SOLUTIONS AND COLLOIDS 8

Solutions and colloids are essential to life. The solutions in living systems are aqueous solutions; that is, they are made with water. We will learn about aqueous solutions and colloids in this chapter. First, we will identify the types of solutions; then, we will learn various ways of expressing their concentrations. A molecular model of solutions will be presented and will be used to explain their biologically important properties. Finally, we will study colloids and learn how they differ from solutions.

8.1 TYPES OF SOLUTIONS

We are all familiar with the fact that sugar dissolves in water. The result of mixing a small amount of sugar in water is a homogeneous mixture called a *solution*. We define a solution as a homogeneous mixture of the molecules, atoms, or ions of two or more different substances. The substances that make up a solution are called its components. There is usually more of one component than the other components in the solution. The component present in excess is called the *solvent*. The other components are called the *solutes*. In a solution of sugar in water, water is the solvent and sugar is the solute.

The three states of matter can combine in nine different ways to form solutions containing two components. These are listed in Table 8-1.

Solutions that contain liquids as solvents are the types of solutions most familiar to us. Numerous examples of solutions containing solids in liquids,

Table 8-1. Types of Solutions

Solvent	Solute	Examples
Liquid	Liquid	Alcoholic beverages
Solid	Liquid	An amalgam such as mercury in silver
Gas	Liquid	—
Gas	Solid	—
Liquid	Solid	Salt water
Solid	Solid	Metal alloys such as brass or tin
Gas	Gas	Air
Liquid	Gas	Carbonated beverages (soda pop, beer, champagne)
Solid	Gas	Hydrogen gas in palladium metal

gases in liquids, and liquids in liquids are available from everyday experiences. Less familiar as solutions are those with solids as solvents, yet alloys and amalgams are important in many commercial products.

A given solution has a particular composition, but the composition can be varied by adding more of either component. For example, the sweetness of a solution of sugar and water varies with the amount of sugar dissolved. By contrast, the composition of a pure substance never varies.

The amount of a solute that dissolves in a solvent depends on several factors, as we will learn next.

8.2 SOLUBILITY

There is usually a limit to the amount of solute that can be dissolved in a solvent at a particular temperature. When this limit is reached, no more solute will dissolve in the solvent. When this happens, we say that the solvent is saturated with solute. We call such a solution a *saturated solution* of the solute in the solvent. Solubility is defined as the amount of solute that dissolves in a given quantity of solvent to form a saturated solution. The solubility of a solute in a particular solvent depends on a number of factors, such as the kind of solvent, the kind of solute, the temperature of the solvent, and the pressure above the solvent. Let us examine each of these factors in turn.

The results of our experiences in the world have led to the very general rule that "like dissolves like." By this we mean that a polar solvent such as

Table 8-2. Solubilities of Several Compounds in Water at 20° C and 100° C

Compound	Solubility at 20° C (g/100 mL)	Solubility at 100° C (g/100 mL)
NaCl	36.2	39.1
NH_4Br	97.1	146.0
KBr	59.4	102.0
NH_3	47.5	6.9
KNO_3	37.8	247.0
O_2	0.00434	0.00080
Li_2CO_3	1.33	0.725
$CaSO_4$	0.21	0.16

water is a good solvent for ionic compounds such as sodium chloride. Gasoline, a mixture of nonpolar organic compounds, is a good solvent for other nonpolar organic compounds such as greases and oils. It follows from this general rule that polar and nonpolar substances will not form solutions. An example is gasoline and water.

Sometimes there is no limit to the amount of one substance that can dissolve in another. This is particularly true for solutions of a liquid in a liquid. Some liquids are infinitely soluble in another liquid: any amount of one liquid will dissolve in any amount of another liquid. Ethyl alcohol and water provide an example of two liquids that are infinitely soluble in each other. Such a pair of liquids is said to be *completely miscible*. Other liquids are only slightly soluble in each other. Such liquids are said to be *partially miscible*. Liquids that are insoluble in each other are said to be *immiscible*. Thus, gasoline is immiscible with water.

The temperature of the solvent affects the solubility of a solute. In general, solutes are more soluble in hot than cold solvents. This is shown by the solubilities of several solutes in water at 20° C and 100° C listed in Table 8-2. The solubilities of several solids increase greatly with increasing temperature. Others increase only slightly, and some actually decrease.

Gases are other compounds whose solubilities in water decrease with increasing temperature. A familiar example is boiling water. The bubbles that form when water is heated are air escaping from solution because dissolved air is less soluble in water at higher temperatures. Boiled water has a characteristic flat taste that is due to the absence of dissolved gases.

The solubility of a gas is greatly affected by the pressure of that gas above the solution. In general, the solubility of any gas increases as the partial pressure of the gas above the solution is increased. Examples are the solubilities of oxygen and carbon dioxide in blood discussed in Section 6.9. Carbonated beverages are another example. These beverages contain the gas carbon dioxide dissolved in water and are bottled under high pressure. The amount of carbon dioxide in solution depends directly on the partial pressure of carbon dioxide above the liquid. When a bottle of carbonated beverage is opened, the partial pressure of carbon dioxide above the liquid decreases and the solubility of the carbon dioxide decreases. As a result, bubbles of carbon dioxide form and escape from the liquid. This is how a bottle of soft drink that is left open for a while goes flat. In contrast to gases, the solubilities of solids and liquids are practically unchanged with a change in pressure.

It is useful to know in general what substances dissolve in water and what factors affect the solubilities of gases, liquids, and solids in water. However, in the laboratory and in clinical work it is necessary to specify exactly the amount of solute in a solution. The methods devised to do this are given in the next section.

8.3 CONCENTRATIONS OF SOLUTIONS

We describe the relative amounts of solute and solvent in a solution by means of units of concentration. There are several such units, and we will examine the most commonly used ones.

Weight/Weight Percent

One way to specify the concentration of a solute in a solution is as a *percent by weight*. The concentration of the solute is given by the following equation (8-1):

$$\text{Percent by weight solute} = \frac{\text{weight of solute, in g}}{\text{weight of solute, in g} + \text{weight of solvent, in g}} \times 100 \quad (8\text{-}1)$$

The following example shows how to express the concentration of a solute in this unit.

Example 8-1 What is the percent by weight of sugar in a solution made by dissolving 10 g of sugar in 90 g of water?

Solution:

STEP 1. Relate the quantities in equation 8-1 to those given in the problem:

Solvent = water Solute = sugar

Weight of solvent = 90 g Weight of solute = 10 g

STEP 2. Place the quantities into equation 8-1:

$$\text{Percent by weight sugar} = \frac{10\text{ g}}{10\text{ g} + 90\text{ g}} \times 100 = \frac{10\not{\text{g}}}{100\not{\text{g}}} \times 100 = 10\%$$

*EXERCISE 8-1 Determine the percent by weight of the solute in each of the following solutions:
(a) 1.50 g of sodium chloride in 100 g of water
(b) 3.50 g of glucose in 250 g of water

Volume/Volume Percent

A convenient way of expressing the concentration of a liquid solute dissolved in a liquid is as a *percent by volume.* This unit of concentration is similar to percent by weight except that volumes in milliliters are used instead of weights in grams. The equation is as follows:

$$\text{Percent by volume solute} = \frac{\text{volume of solute, in mL}}{\text{total volume of solution, in mL}} \times 100 \qquad (8\text{-}2)$$

We can express the concentration of a solute in this unit, as shown in the following example.

Example 8-2 What is the percent by volume of ethyl alcohol in a solution made by diluting 10 mL of ethyl alcohol to 100 mL with water?
Solution:
STEP 1. Relate the quantities in equation 8-2 to those given in the problem:

Volume of solute = 10 mL

Total volume of solution = 100 mL

STEP 2. Place the quantities into equation 8-2:

$$\text{Percent by volume ethyl alcohol} = \frac{10\not{\text{mL}}}{100\not{\text{mL}}} \times 100 = 10\%$$

* The answers for the exercises in this chapter begin on page 862.

EXERCISE 8-2 Determine the percent by volume of the solute in each of the following solutions:
(a) 5.0 mL of rubbing alcohol diluted to 150 mL with water
(b) 15 mL of ethyl alcohol diluted to 500 mL with water

Weight/Volume Percent

This widely used method of expressing concentrations is a combination of weight and volume. The weight is usually that of the solid solute and the volume is that of the total solution. This unit is defined as follows:

$$\text{Percent by weight/volume solute} = \frac{\text{weight of solute, in g}}{\text{total volume of solution, in mL}} \times 100 \quad (8\text{-}3)$$

The use of this unit is shown in the following example:

Example 8-3 What is the percent by weight/volume of sodium chloride in a solution made by diluting 1.5 g of sodium chloride to 100 mL with water?

Solution:

STEP 1. Relate the quantities in equation 8-3 to those given in the problem:

Weight of solute = 1.5 g

Total volume of solution = 100 mL

STEP 2. Place the quantities into equation 8-3:

$$\text{Percent by weight/volume NaCl} = \frac{1.5\text{ g}}{100\text{ mL}} \times 100 = 1.5\%$$

Low concentrations of solute are often expressed in *milligrams per 100 mL*. This weight/volume percent unit is defined as follows:

$$\text{Milligrams per 100 mL} = \text{mg/100 mL} = \frac{\text{weight of solute, in mg}}{\text{100 mL of solution}} \quad (8\text{-}4)$$

This unit is often used to express the concentrations of solute in blood and urea, as shown in the following example:

Example 8-4 A 1-mL sample of blood plasma is found to contain 3.3 mg of sodium ions. Express this concentration in mg/100 mL.

Solution:

STEP 1. Express the amount of solute in mg per total volume of solution:

Solute = 3.3 mg

Solution = 1 mL of blood plasma

$$\frac{3.3 \text{ mg}}{1 \text{ mL}}$$

STEP 2. The definition of mg/100 mL is the weight of solute in 100 mL of solution. We know how many mg are in 1 mL of blood plasma. To find out the number of mg in 100 mL of blood plasma, we set up the following proportion:

$$\frac{3.3 \text{ mg}}{1 \text{ mL}} = \frac{X}{100 \text{ mL}}$$

STEP 3. Rearrange and solve for X.

$$X = \frac{3.3 \text{ mg} \times 100 \cancel{\text{mL}}}{1 \cancel{\text{mL}}} = 330 \text{ mg}$$

There are 330 mg of sodium ions in 100 mL of blood plasma. Therefore, according to equation 8-4,

$$\text{Milligrams per 100 mL} = \frac{330 \text{ mg}}{100 \text{ mL}}$$

This value is within the normal concentration range of sodium ions in blood plasma (320 to 350 mg/100 mL).

The unit mg/100 mL is sometimes called mg percent.

EXERCISE 8-3 Determine the percent by weight/volume of the solute in each of the following solutions:

(a) 5.00 g of sugar diluted to 500 mL with water
(b) 25.0 g of lithium bromide diluted to 1.00 L with water

EXERCISE 8-4 Determine the concentration of solute in each of the following solutions in mg/100 mL.

(a) 32.0 mg of sugar diluted to 10.0 mL with water
(b) 5.00 mL of solution that contains 1.00 g of sodium ion

Parts Per Million and Parts Per Billion

These units of concentration are widely used to report very small amounts of solute in a solution. The concentration of pollutants in water and air are usually reported in these units.

One part per million, abbreviated ppm, contains 1 part of solute per 1 million (10^6) parts of solution. By parts we mean any unit of measure such as grams, liters, or anything else we choose. For example, the concentration of solid pollutants in solid food is given in ppm expressed as mg of pollutant (the solute) in 1 million mg of solid food (the solution). Because 1 million mg is equivalent to 1 kg, ppm is usually defined as follows:

$$\text{ppm} = \frac{\text{weight of solute, in mg}}{\text{weight of solution, in kg} = 10^6 \text{ mg of solution}}$$

To express the concentrations of small quantities of solid solutes in water, the unit ppm is usually defined as mg of solute per liter of solution. This change from weight to volume of solvent can be made because 1 million mg (1 kg) of water occupies approximately 1 L. This definition of ppm is also frequently used even though the solution may weigh somewhat more or less than 1 kg.

Air pollution is measured in ppm on the basis of measurements of volume rather than weight. Thus, 1 ppm means that there is 1 μL of pollutant (the solute) per 1 million (10^6) μL (1 L) of air (the solution).

The sensitivity of analytical methods has improved so much that parts per billion, abbreviated ppb, has become a common unit of concentration. Its use and definition are similar to those of ppm. Thus, 1 ppb contains 1 part of solute per 1 billion (10^9) parts of solution. Again, the parts refer to weight or volume, depending on whether the solution is a gas, liquid, or solid.

The units ppm and ppb are extremely small. For example, 1 ppm is equivalent to 1 cent in 10,000 dollars, and 1 ppb is equivalent to 1 cent in 10 million dollars! These units are convenient means to express very small concentrations without using exponential notation. The use of these units is shown in the following example:

Example 8-5 The maximum Food and Drug Administration (FDA) tolerance of mercury in fish is 0.5 ppm. A 10-g sample of fish is found to contain 72 μg of mercury. Does the amount of mercury in the fish exceed the FDA maximum tolerance?

Solution:

STEP 1. Calculate the mg of mercury in 1 kg of this fish using the following proportion:

$$\frac{72\ \mu g}{10\ g} = \frac{X}{1\ kg}$$

STEP 2. Rearrange and solve for X:

$$X = \frac{72\ \cancel{\mu g}}{10\ \cancel{g}} \times 1\ \cancel{kg} \times \frac{10^3\ \cancel{g}}{1\ \cancel{kg}} \times \frac{1\ mg}{10^3\ \cancel{\mu g}} = 7.2\ mg$$

STEP 3. Determine ppm from its definition:

$$ppm = \frac{\text{weight of solute, in mg}}{\text{1 kg of solution}} = \frac{\text{7.2 mg of mercury}}{\text{1 kg of fish}} = 7.2\ ppm$$

The amount of mercury in the fish exceeds the maximum FDA tolerance. The fish is contaminated with mercury.

EXERCISE 8-5 Many fish are contaminated with the insecticide DDT. One such fish is the Coho salmon from Lake Michigan. A 20-g sample of Coho salmon was found to contain 0.010 mg of DDT. Express this amount of DDT in ppm.

Notice that the molecular weight of the solute is not needed when we express the concentration of the solute in any of the units of percent, ppm, or ppb. Thus, we can specify the concentration of a substance without knowing anything about its chemical composition.

Molar Concentrations (Molarity)

We introduced the concept of moles as a measure of quantities of matter in Section 7.3. We can extend this concept to solutions by expressing the amount of solute present in 1 L of solution in terms of moles. Molar concentration, or molarity, which is defined as the number of moles of solute per liter of solution, is designated by the capital letter M. This definition is given in the form of an equation, as follows:

$$M = \frac{\text{number of moles of solute}}{\text{number of liters of solution}} \qquad (8\text{-}5)$$

This method of expressing concentrations is shown in the following example:

Example 8-6 Determine the molar concentration of a solution that contains 25.0 g of glucose, $C_6H_{12}O_6$, in 500 mL of solution.

Solution:

STEP 1. Convert 25.0 g of glucose into moles:

Gram-molecular weight of glucose = 180 g/mol

$$\text{Number of moles} = \frac{25.0\ \cancel{g}}{180\ \cancel{g}}\ \text{mol} = 0.139\ \text{mol}$$

STEP 2. Convert volume of solution into L:

$$500\ \cancel{mL} \times \frac{1\ L}{1000\ \cancel{mL}} = 0.500\ L$$

STEP 3. Insert the calculated information into equation 8-5 and solve:

$$M = \frac{0.139\ \text{mol}}{0.500\ L} = 0.278\ M$$

The concentration of the solution is 0.278 molar, or 0.278 M.

EXERCISE 8-6 Determine the molar concentration of each of the following solutions:

(a) 40.0 g of NaOH made up to 1.0 L with water

(b) 250 mL of a solution that contains 5.40 g of NaCl

Molar concentration expresses the ratio of solute to solution. Two solutions that have the same molar concentrations have the same ratios of solute to solution even though the total volumes of the two solutions may be different. We can demonstrate this important fact by means of the three solutions of glucose in water shown in Figure 8-1. Solution A is made by adding water to 90 g of glucose to make exactly 1 L of solution; solution B, by adding water to 22.5 g of glucose to make exactly 250 mL of solution; solution C, by adding water to 0.450 g of glucose to make exactly 5 mL of solution. The molar concentration of all three solutions is 0.500 M. Although the molar concentrations are identical, each solution contains a

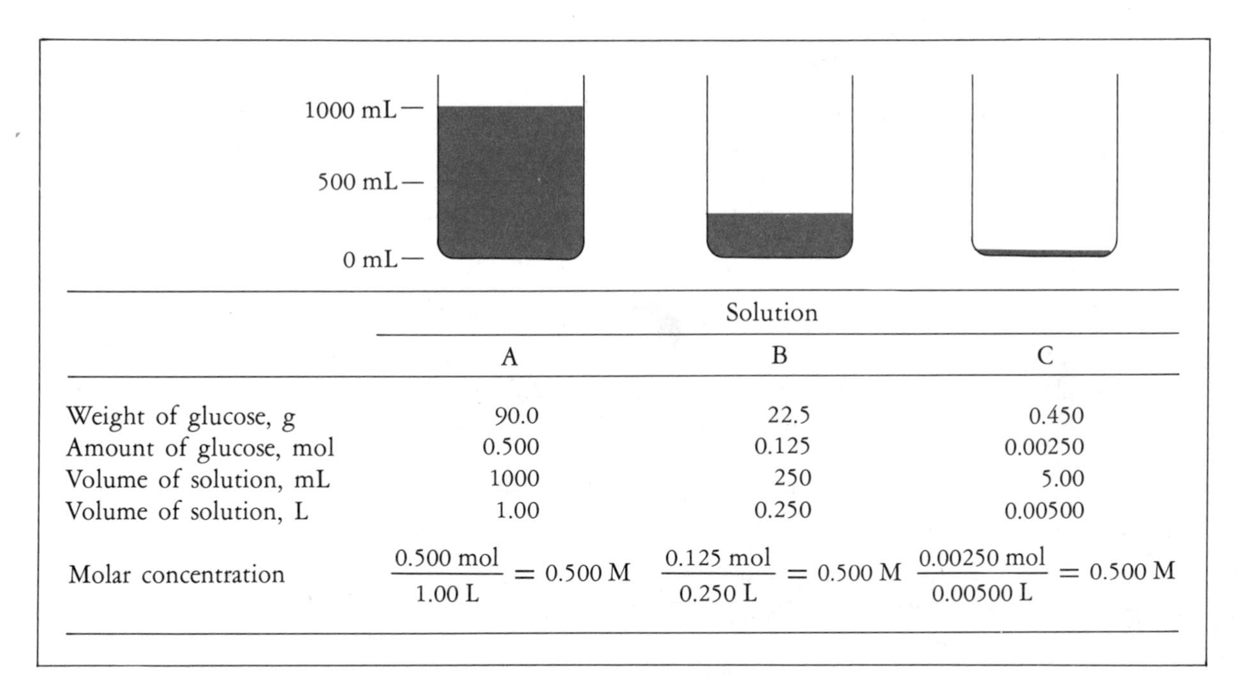

	Solution		
	A	B	C
Weight of glucose, g	90.0	22.5	0.450
Amount of glucose, mol	0.500	0.125	0.00250
Volume of solution, mL	1000	250	5.00
Volume of solution, L	1.00	0.250	0.00500
Molar concentration	$\frac{0.500 \text{ mol}}{1.00 \text{ L}} = 0.500 \text{ M}$	$\frac{0.125 \text{ mol}}{0.250 \text{ L}} = 0.500 \text{ M}$	$\frac{0.00250 \text{ mol}}{0.00500 \text{ L}} = 0.500 \text{ M}$

Fig. 8-1. Three different volumes of solutions of the same concentration.

different total volume and a different total number of moles of glucose. We know from the concentration 0.500 M only the ratio of moles of solute to liters of solution. We know nothing about the total volume of solution. Thus, a bottle labeled 0.500 M glucose may contain 10 L or as little as 1 mL. No matter how much solution there is in the bottle, every drop of it has a glucose concentration of 0.500 M.

The importance of molar concentration is that we can determine the weight of the solute contained in any volume of solution. This fact is shown in the following examples.

Example 8-7 A patient is fed intravenously 0.50 L of a 1.0 M glucose solution. How many grams of glucose has the patient received?

Solution:

STEP 1. We know that a 1.0 M solution contains 1.0 mole per liter of solution. We can determine the number of moles in 0.50 L by use of equation 8-5:

$$1.0 \text{ M} = \frac{1.0 \text{ mol}}{1.0 \text{ L}} = \frac{X}{0.5 \text{ L}}$$

STEP 2. Rearrange and solve for X:

$$X = \frac{1.0 \text{ mol}}{1.0 \cancel{L}} \times 0.50 \cancel{L} = 0.50 \text{ mol glucose}$$

STEP 3. Convert moles to grams:

$$0.50 \cancel{\text{mol glucose}} \times \frac{180 \text{ g glucose}}{1.0 \cancel{\text{mol glucose}}} = 90 \text{ g glucose}$$

Example 8-8 How many grams of glucose are needed to make 2.00 L of a 2.00 M solution?

Solution:

STEP 1. The number of moles needed to make 2.00 L is obtained by use of equation 8-5:

$$2.00 \text{ M} = \frac{2.00 \text{ mol glucose}}{1.00 \text{ L}} = \frac{X}{2.00 \text{ L}}$$

STEP 2. $X = \frac{2.00 \text{ mol glucose}}{1.00 \cancel{L}} \times 2.00 \cancel{L} = 4.00 \text{ mol glucose}$

$$4.00 \cancel{\text{mol glucose}} \times \frac{180 \text{ g glucose}}{1 \cancel{\text{mol glucose}}} = 720 \text{ g glucose}$$

EXERCISE 8-7 How many grams are needed to make each of the following solutions?

(a) 1.00 L of a 0.100 M NaCl solution
(b) 250 mL of a 1.50 M glucose solution
(c) 500 mL of a 0.150 M sucrose ($C_{12}H_{22}O_{11}$) solution

EXERCISE 8-8 How many grams of solute are there in each of the following quantities of solution?

(a) 25.0 mL of a 1.00 M LiBr solution
(b) 100 mL of a 0.500 M NaOH solution
(c) 250 mL of a 0.100 M $NaHCO_3$ solution

Milliequivalents Per Liter

This unit is used to express low concentrations of ions in body fluids. To use this unit, we must learn about *equivalents*. One equivalent of an ion, abbreviated Eq, is defined as 1 mole of that ion multiplied by the absolute value of its charge. For example, 1 mole of sodium ions contains one equivalent of sodium ions. One mole of chloride ions contains one equivalent of chloride ions. One mole of magnesium ions contains two equivalents of magnesium ions.

EXERCISE 8-9 How many equivalents are there of each underlined ion in the following quantities?

(a) 1.00 mol $\underline{Na}HCO_3$
(b) 0.150 mol $Na_2\underline{CO_3}$
(c) 2.00 mol $Li\underline{Br}$
(d) 0.350 mol $\underline{Fe}Cl_3$
(e) 1.50 mol $Mg\underline{Cl_2}$
(f) 0.250 mol $Ca_3\underline{(PO_4)_2}$

The unit milliequivalent per liter, abbreviated mEq/L, is defined as follows:

$$\text{Milliequivalent per liter (mEq/L)} = \frac{\text{number of milliequivalents of ion}}{\text{volume of solution, in L}} \qquad (8\text{-}6)$$

The use of this unit is shown in the following example:

Example 8-9 Express the concentrations of sodium and chloride ions in a 0.001 M NaCl solution in terms of milliequivalents per liter.

Solution:

STEP 1. A 0.001 M NaCl solution contains $\frac{0.001 \text{ mol NaCl}}{\text{L}}$.

$$\frac{0.001 \text{ mol NaCl}}{\text{L}} = \frac{0.001 \text{ mol Na}^+}{\text{L}}$$

and

$$\frac{0.001 \text{ mol Cl}^-}{\text{L}} \text{ (See Section 8.5.)}$$

STEP 2. $1 \text{ mol Na}^+ = 1 \text{ Eq Na}^+ \text{ or } 1 = \frac{1 \text{ Eq Na}^+}{1 \text{ mol Na}^+} = \frac{1 \text{ mEq Na}^+}{0.001 \text{ mol Na}^+}$

$$1 \text{ mol Cl}^- = 1 \text{ Eq Cl}^- \text{ or } 1 = \frac{1 \text{ Eq Cl}^-}{1 \text{ mol Cl}^-} = \frac{1 \text{ mEq Cl}^-}{0.001 \text{ mol Cl}^-}$$

STEP 3. $\frac{0.001 \text{ mol Na}^+}{\text{L}} \times \frac{1 \text{ mEq Na}^+}{0.001 \text{ mol Na}^+} = \frac{1 \text{ mEq Na}^+}{\text{L}}$

$$\frac{0.001 \text{ mol Cl}^-}{\text{L}} \times \frac{1 \text{ mEq Cl}^-}{0.001 \text{ mol Cl}^-} = \frac{1 \text{ mEq Cl}^-}{\text{L}}$$

EXERCISE 8-10 Determine the concentration of each ion, in mEq/L, in each of the following solutions:
(a) 0.001 M LiBr (b) 0.05 M Na_2CO_3

EXERCISE 8-11 If enough water is added to 25.0 g of $NaHCO_3$ to make 5.00 L of solution, what is the concentration of HCO_3^- ion, in mEq/L?

The various units of concentration are summarized in Table 8-3.

We have learned in this section how to express the concentration of a solution in several different units. The practical use of these units in clinical

Table 8-3. Summary of Concentration Units

Units	Definition
Weight/weight percent	$\frac{\text{Weight of solute, in g}}{\text{Weight of solute, in g + weight of solvent, in g}} \times 100$
Volume/volume percent	$\frac{\text{Volume of solute, in mL}}{\text{Total volume of solution, in mL}} \times 100$
Weight/volume percent	$\frac{\text{Weight of solute, in g}}{\text{Total volume of solution, in mL}} \times 100$
Milligram/100 mL	$\frac{\text{Weight of solute, in mg}}{\text{100 mL of solution}}$
Parts per million	$\frac{\text{Weight of solute, in mg}}{\text{Weight of solution, in kg}}$ or $\frac{\text{mg of solute}}{\text{L of solution}}$ or $\frac{\mu\text{L}}{\text{L}}$
Parts per billion	$\frac{\text{Weight of solute, in }\mu\text{g}}{\text{Weight of solution, in kg}}$ or $\frac{\mu\text{g of solute}}{\text{L of solution}}$ or $\frac{\mu\text{L}}{10^3\text{ L}}$
Molar concentration (molarity)	$\frac{\text{Number of moles of solute}}{\text{Number of liters of solution}}$
Milliequivalents per liter	$\frac{\text{Number of milliequivalents}}{\text{Volume of solution, in L}}$

work involves solutions made with water as a solvent. The properties of these aqueous solutions are vital to all living systems. To understand these properties and their importance to life, we must develop a molecular picture or model of solutions. A first step toward such a model is to recognize that substances that dissolve in water can be classified as either electrolytes or nonelectrolytes.

8.4 ELECTROLYTES AND NONELECTROLYTES

Electricity is the flow of electrons in a circuit from a battery or electrical generator along a wire back to the source. The electricity passing through the circuit can do work, such as running a motor or providing heat and light. The flow of electricity stops if the circuit is broken. An electrical circuit can also contain an aqueous solution, as shown in Figure 8-2. This circuit contains a battery, a beaker containing an aqueous solution, a light bulb, and two wires or plates called electrodes that are in the solution. All these items are connected by wires. The two electrodes are oppositely charged, just like the two poles of the battery. For electricity to flow through this circuit after all the connections have been made, the solution

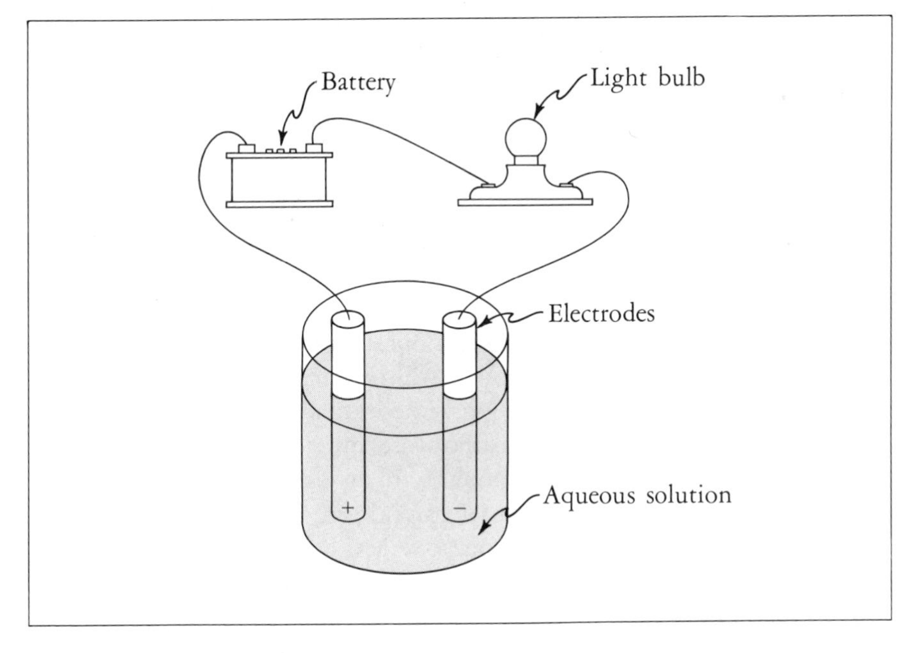

Fig. 8-2. An electrical circuit containing a battery, two electrodes, a light bulb, and an aqueous solution. Wires connect the battery, the electrodes, and the light bulb. The light bulb glows only if the aqueous solution conducts electricity.

Table 8-4. Some Electrolytes and Nonelectrolytes

Electrolytes		Nonelectrolytes	
$NaCl$	Sodium chloride	$C_{12}H_{22}O_{11}$	Sucrose
KI	Potassium iodide	C_2H_5OH	Ethyl alcohol
$LiBr$	Lithium bromide	C_3H_6O	Acetone
Na_2SO_4	Sodium sulfate	CH_4	Methane
KNO_3	Potassium nitrate	N_2	Nitrogen
$CaCl_2$	Calcium chloride	O_2	Oxygen
LiF	Lithium fluoride	CO	Carbon monoxide

must be able to conduct electricity. A glowing light bulb indicates that electricity is flowing through the circuit.

Aqueous solutions either conduct electricity or they do not. One that conducts electricity is called an *electrolytic solution;* one that does not is called a *nonelectrolytic solution.* A solute that forms an aqueous electrolytic solution is called an *electrolyte.* For example, the light bulb glows when an aqueous solution of sodium chloride is placed in the beaker. Thus, sodium chloride is an electrolyte. A solute that forms a nonelectrolytic solution is called a *nonelectrolyte.* Aqueous solutions of sugar and alcohol do not conduct electricity; both sugar and alcohol are classified as nonelectrolytes. Other examples of electrolytes and nonelectrolytes are given in Table 8-4.

How do we explain this difference between electrolytic and nonelectrolytic solutions? To answer this question, we must turn our attention to the model of solutions first proposed by Svante Arrhenius in 1887.

8.5 ARRHENIUS' THEORY OF ELECTROLYTES

Arrhenius proposed that molecules dissolve in water to form particles that mix completely with solvent molecules. Electrolytes and nonelectrolytes form different kinds of particles when they dissolve in water. All the electrolytes in Table 8-4 are compounds that contain ionic bonds. We learned in Section 5.4 that such compounds are solids at room temperature and contain ions arranged in a crystal lattice. When these compounds are dissolved in water, the ions are released and they distribute themselves uniformly in the water. In addition to salts, many compounds with polar covalent bonds also form ions when dissolved in water. When nonelectrolytes dissolve in water, neutral molecules rather than ions are released. Pictorial representa-

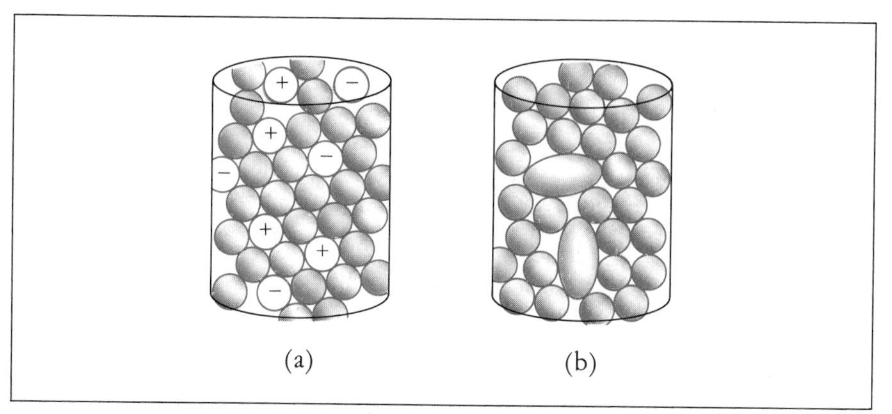

Fig. 8-3. A pictorial representation of (a) an aqueous solution of an electrolyte and (b) an aqueous solution of a nonelectrolyte. The open spheres represent water molecules. The spheres containing a plus or minus sign represent the ions; the spheroids represent neutral molecules.

tions of solutions of electrolytes and nonelectrolytes are shown in Figure 8-3. Water molecules surround ions. Such close association of water molecules with an ion is called *hydration*. We say that the ion is hydrated.

According to Arrhenius' model, an aqueous solution of sodium chloride contains an equal number of individual sodium and chloride ions, each surrounded by water molecules. Thus, 1 mole of sodium chloride forms 1 mole of sodium ions and 1 mole of chloride ions when dissolved in water. *Aqueous solutions of electrolytes are really solutions of hydrated ions.* The total number of ions formed per mole of electrolyte depends on the chemical formula of the electrolyte, as shown in Table 8-5.

Thus, 1 mole of calcium chloride dissolved in water forms 1 mole of hydrated calcium ions and 2 moles of hydrated chloride ions.

EXERCISE 8-12 Give the correct symbol and the number of moles of each ion formed when one mole of the following electrolytes is dissolved in water:
(a) HCl (b) NaOH (c) $CaCl_2$ (d) Na_2CO_3 (e) Li_3PO_4

Table 8-5. Number of Ions Formed per Mole of Electrolyte

Chemical Formula	Ions Formed in Aqueous Solution	Number of Ions in 1 Mol of Electrolyte
NaCl	Na^+ Cl^-	$2 \times 6.02 \times 10^{23}$
LiBr	Li^+ Br^-	$2 \times 6.02 \times 10^{23}$
KNO_3	K^+ NO_3^-	$2 \times 6.02 \times 10^{23}$
$CaCl_2$	Ca^{+2} Cl^- Cl^-	$3 \times 6.02 \times 10^{23}$
Na_2SO_4	Na^+ Na^+ SO_4^{-2}	$3 \times 6.02 \times 10^{23}$
Na_3PO_4	Na^+ Na^+ Na^+ PO_4^{-3}	$4 \times 6.02 \times 10^{23}$

We can use Arrhenius' model to explain how solutions of electrolytes conduct electricity. Let us return to the electrical circuit shown in Figure 8-2. One of the two electrodes has a positive charge; the other has a negative charge. The positive ions (cations) in an electrolytic solution are attracted to the negatively charged electrode and move freely in that direction. The negative ions (anions) are attracted to the positively charged electrode and move in that direction. This movement of ions in the solution results in a transport of electrical charge from one electrode to the other. The net effect is a flow of electrons through the solution. This is represented in Figure 8-4.

The situation is different in an aqueous solution of sugar, a nonelectrolyte. The sugar molecules, surrounded by water molecules, are neutral. When a pair of electrodes is placed in this solution, the sugar molecules are not attracted by either electrode. Consequently no electric current flows through the solution.

We now have a model of electrolytic and nonelectrolytic solutions that we have used to explain the difference in electrical conductivity between the two types of solution. This model of solutions has been used successfully to explain all the physical properties of solutions. Osmosis is one physical property of solutions that is vital to the life of any cell. For this reason, we will examine this physical property next.

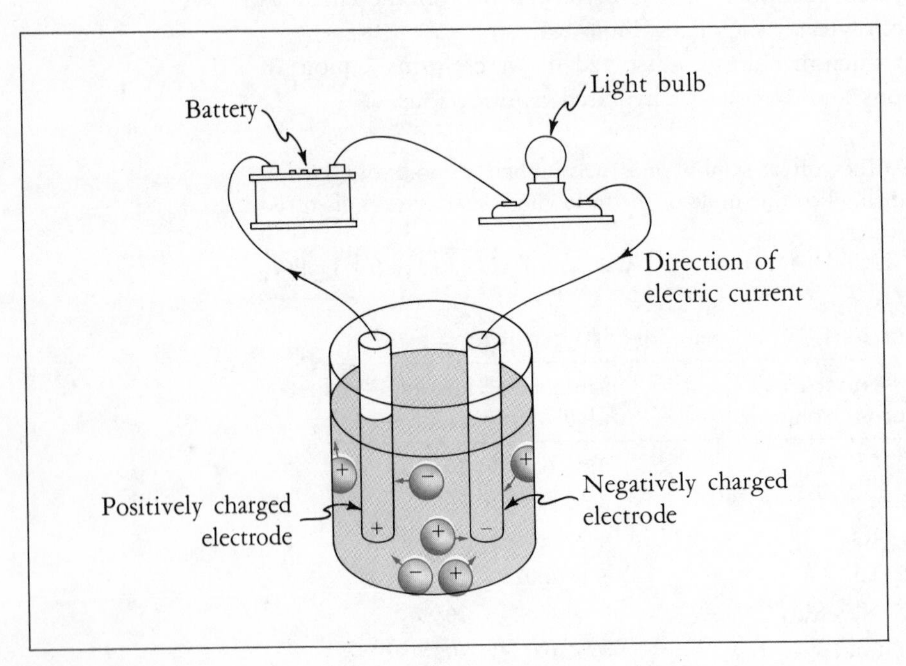

Fig. 8-4. The passage of an electric current through an electrolytic solution by the movement of ions. The spheres containing a plus or minus sign represent ions.

8.6 OSMOSIS AND OSMOTIC PRESSURE

Cells have limiting boundary membranes that are called plasma membranes. These membranes not only keep the cell intact but also allow the exchange of materials back and forth between the interior of the cell and its exterior surroundings. Dialysis and osmosis are two ways that such an exchange of materials occurs. Let us consider osmosis first.

Osmosis can be demonstrated by use of the U-tube shown in Figure 8-5. We place a thin membrane made of cellophane at the bottom of the U-tube. Pure water is placed in one arm of the U-tube, and an aqueous solution of glucose is placed in the other arm. We make sure that the heights of the columns in both arms of the U-tube are equal (Figure 8-5a). Several hours later, we find that the height of the column of glucose solution is greater than the height of the column of pure water (Figure 8-5b). For this change to occur, water must have passed through the membrane. Materials that allow only certain molecules to pass through are called *semipermeable membranes* or osmotic membranes. In our experiment, cellophane was the semipermeable membrane. In cells, the semipermeable membrane is the plasma membrane mentioned earlier.

We can now define osmosis. *Osmosis is the movement of water through an osmotic membrane from an aqueous solution that is less concentrated to one that is more concentrated.* This is a general phenomenon that occurs whenever an osmotic membrane separates two solutions of different concentrations.

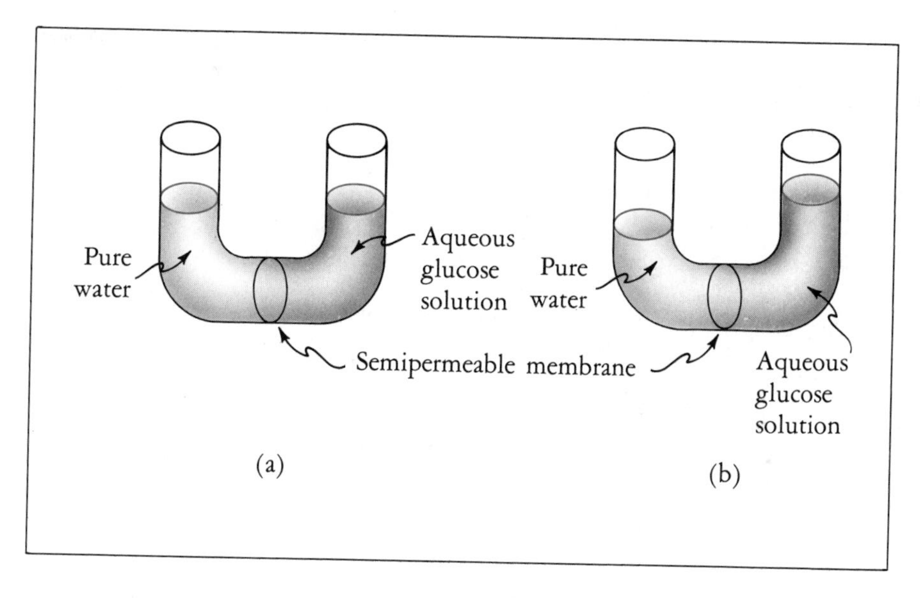

Fig. 8-5. A demonstration of osmosis. (a) The levels of the water and the aqueous glucose solution are the same at the beginning of the experiment. (b) After some time, the level of the glucose solution is higher than that of the pure water.

EXERCISE 8-13 Each of the following pairs of aqueous solutions is separated by an osmotic membrane. In which direction will the water move?
(a) water, 1 M NaCl (b) 1 M glucose, 0.5 M glucose
(c) 0.5 M NaBr, 1.0 M NaCl

We can prevent osmosis from occurring by applying pressure to the right arm of the U-tube in Figure 8-5. If we apply just the right amount of pressure, we can keep the heights of the columns in both arms equal and osmosis does not occur. The pressure needed to prevent osmosis is called the *osmotic pressure* of a solution. Notice that a high solute concentration means high osmotic pressure. Water moves from dilute to more concentrated solutions. The purpose of this movement of water is to make the concentrations of the solutions equal.

We must look at the structure of the osmotic membrane at the molecular level to understand osmosis. An osmotic membrane contains small holes. The size of these holes is an important property, which determines what kinds of molecules will pass through the membrane. Molecules larger than the holes will not pass through. The membrane therefore acts like a molecular sieve. Certain molecules pass through the membrane, and others do not. This selectivity of the membrane is responsible for osmosis, as we will learn from the diagram in Figure 8-6.

Figure 8-6 shows a molecular view of two aqueous glucose solutions of different concentrations separated by an osmotic membrane. The more con-

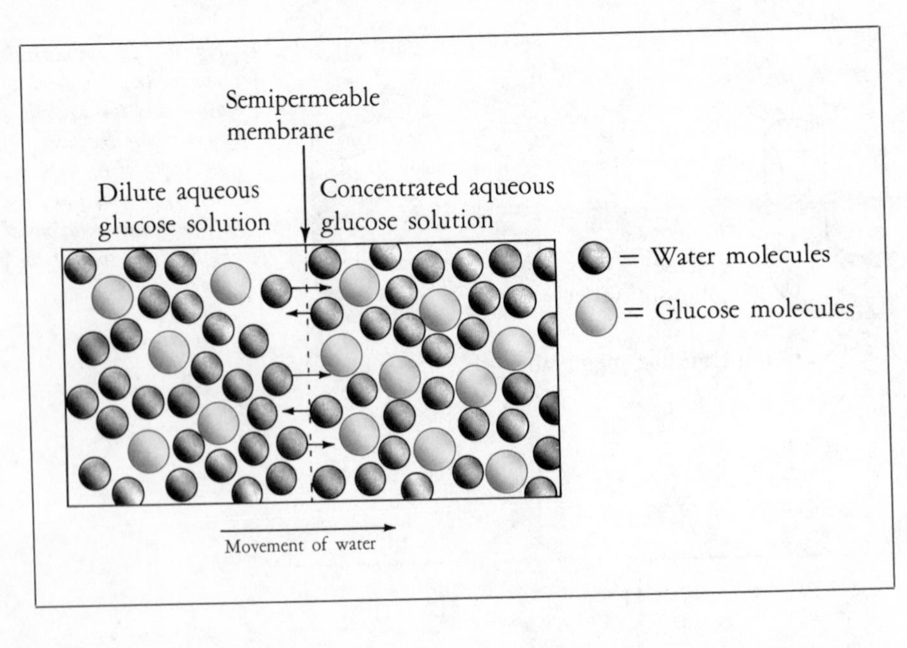

Fig. 8-6. A molecular view of osmosis.

centrated solution is in the right compartment. The holes in the osmotic membrane are large enough that water molecules can pass in both directions. But the holes are so small that glucose molecules cannot get through. All the molecules in both solutions are in continual motion, as we learned in Section 6.5. As a result of this motion, water molecules reach the membrane and collide with it. A water molecule that happens to find a hole in the membrane passes through it. The amount of water in the concentrated solution is less than that in the dilute solution, so the number of water molecules that collide with the membrane is smaller. As a result, more water molecules pass through the membrane from the dilute glucose solution to the more concentrated glucose solution. The result is a net movement of water into the more concentrated glucose solution. This is visible as an increase in its volume.

We can see in Figure 8-6 that water moves to the solution that has the greater number of dissolved particles (the more concentrated solution). This solution also has the higher osmotic pressure. We can conclude that the greater the number of particles, whether ions or molecules, in a solution, the greater its osmotic pressure. Any property of a solution that depends on the number of dissolved particles in the solvent is called a *colligative property*.

We can easily show that osmotic pressure is a colligative property. For example, if we measure the osmotic pressure of a 1 M aqueous sodium chloride solution, we find that it is exactly twice that of a 1 M aqueous glucose solution. The reason for this difference in osmotic pressure is that sodium chloride is an electrolyte, whereas glucose is a nonelectrolyte. An aqueous solution containing 1 mole of sodium chloride actually contains 1 mole of sodium ions and 1 mole of chloride ions, as we learned in Section 8.5. A 1 M solution of sodium chloride contains twice as many particles as an equal volume of a 1 M solution of glucose, a nonelectrolyte. As a result, its osmotic pressure is exactly twice that of a 1 M glucose solution.

EXERCISE 8-14 Which of each pair of solutions has the higher osmotic pressure? (a) 1 M LiBr, 1 M glucose (b) 1 M NaCl, 1 M Na_2CO_3 (c) 1 M Glucose, 0.6 M NaCl

The relative osmotic pressures of two solutions are extremely important in living systems. In fact, they are so important that special terms have been given to describe their relative osmotic pressure. Two solutions that have the same osmotic pressure are said to be *isotonic*. If one solution has a higher osmotic pressure than the other, it is said to be *hypertonic* with respect to the other. One of two solutions that has the lower osmotic pressure is said to be

Table 8-6. Examples of Terms Describing Relative Osmotic Pressures of Two Solutions

1. Isotonic
   A 1 M glucose solution and a 1 M urea (a nonelectrolyte) solution are isotonic.
2. Hypertonic
   A 1 M NaCl solution has a higher osmotic pressure than a 1 M glucose solution. Therefore, it is hypertonic compared to a 1 M glucose solution.
3. Hypotonic
   A 1 M NaCl solution has a lower osmotic pressure than a 2 M LiBr solution. Therefore, it is hypotonic compared to a 2 M LiBr solution.

hypotonic compared to the other. Examples of each of these terms are given in Table 8-6.

EXERCISE 8-15 For each pair of solutions in exercise 8-14, which solution is hypertonic? Which is hypotonic?

The plasma membranes of red blood cells behave as osmotic membranes. The cells contain an aqueous fluid made up of dissolved compounds. This fluid has an osmotic pressure determined by the concentration of dissolved molecules and ions in the fluid. Osmosis occurs when a red blood cell is placed in water. The solution inside the cell is hypertonic compared to pure water, so water enters the cell. So much water enters that the cell is ruptured. The rupture of red blood cells in this way is called *hemolysis*. We say that the cells are hemolyzed.

Osmosis also occurs when a red blood cell is placed in a concentrated saline (sodium chloride) solution. But in this case, the solution inside the cell is hypotonic compared to the saline solution and osmosis occurs in the reverse direction. Water leaves the cell and passes into the solution. This causes the red blood cell to shrivel and shrink. This process is called *crenation*.

A 0.95% saline solution is isotonic compared to the solution inside red blood cells. Consequently, red blood cells placed in such a solution undergo neither crenation nor hemolysis.

There is a very important practical reason for worrying about the osmotic pressure of the fluid inside a red blood cell compared to that of the cell's environment. Patients often must be fed intravenously. To prevent damage to their red blood cells, the concentration of the solution must be controlled so that neither hemolysis nor crenation occurs. Therefore, the concentration of the solution must match closely the concentration of all of the particles within the red blood cells. In other words, the solution to be given a patient intravenously must be isotonic with blood.

The fluids in living systems carry not only dissolved ions and molecules, but also larger particles called colloids.

8.7 COLLOIDS AND COLLOIDAL DISPERSIONS

We learned in Section 8.1 that solutions are homogeneous mixtures of solute and solvent molecules. The particles in a solution are the size of atoms and molecules. That is, their sizes range from 0.05 to 0.25 nm (remember that 1 nm $= 10^{-9}$ m, Section 1.3). Sometimes intermolecular attractions (Section 6.3) between molecules cause several hundred to several thousand of them to cluster together. The sizes of these clusters range from 1 to 100 nm. Matter containing particles of this size is called a *colloid.*

A uniform dispersion of a colloid in water is called a *colloidal dispersion.* This dispersion is similar to a solution in that the particles do not settle out on standing. However, a colloidal dispersion usually appears cloudy, and its particles are large enough to be photographed with the aid of an electron microscope. The colloid in a colloidal dispersion is called the *dispersed substance.* The continuous matter in which the colloid is dispersed is called the *dispersing substance.* The dispersed and dispersing substances can be liquids, solids, or gases. They can combine in nine different ways to form colloidal dispersions containing two components. Only eight of these nine possible combinations are known. A mixture of two gases cannot be a colloidal dispersion because the particles of a gas are individual molecules, as we learned in Section 6.6. As the molecules form clusters, the gas changes to a liquid. The eight types of colloidal dispersion are given in Table 8-7 with examples. Many compounds of high molecular weight in living systems

Table 8-7. Examples of Colloidal Dispersions

Dispersed Substance	Dispersing Substance	Example
Liquid	Gas	Fog, clouds
Solid	Gas	Smoke
Gas	Liquid	Foams, whipped cream
Liquid	Liquid	Milk, butter
Solid	Liquid	Paints, glue
Gas	Solid	Foam rubber, pumice
Liquid	Solid	Jellies, cheese
Solid	Solid	Colored glass, gems

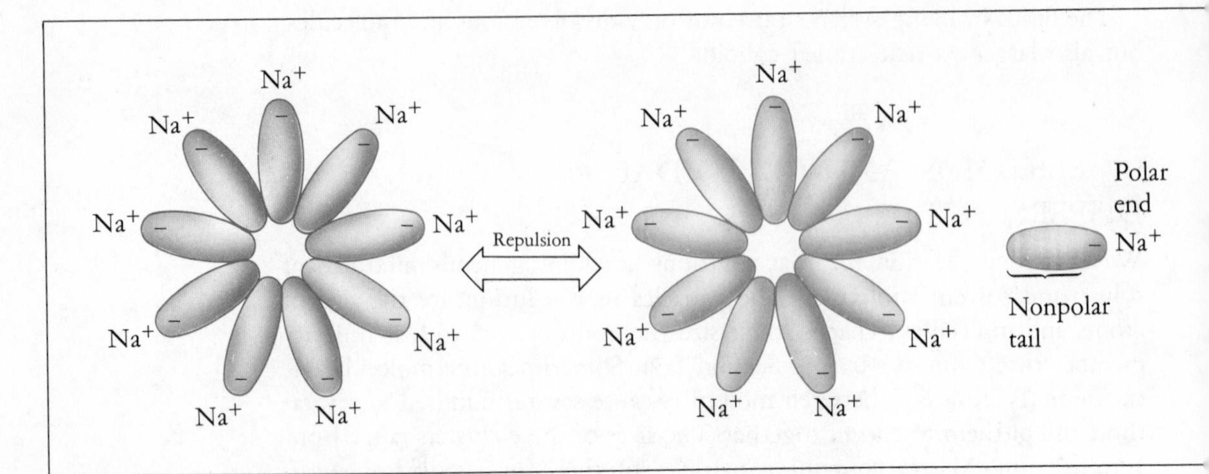

Fig. 8-7. Colloids formed by attractions between complex molecules. One end of each individual molecule has a negative charge (balanced by a sodium ion), and the other end is a long nonpolar tail. The long tails are held together by hydrophobic attractions. The negatively charged ends form the surface of a sphere. Adjacent colloids are repelled by their identical charges.

form colloidal dispersions rather than solutions in water. Starch (Section 21.7) and proteins (Chapter 23) are examples of such compounds.

If colloids are clusters of molecules, why don't the clusters increase in size until they get large enough to settle out? The reason is that the particles in the most stable colloidal dispersions all have the same electrical charge. These charges can be caused by adsorption of ions to the surface of the particles, or the large particles themselves can be charged. As a result, the particles repel each other and cannot form particles large enough to settle out. This repulsion between colloids in water is shown in Figure 8-7.

Other colloids are stabilized in water by the action of a third substance called an *emulsifying agent*. An example is a mixture of oil and water. Oil is immiscible with water. However, if we add soap (Section 22.7) to the mixture, the oil is emulsified by the soap. The soap is the emulsifying agent. The soap breaks up the oil into small drops. The soap molecules form a negatively charged layer on the surface of each oil drop. This causes the oil drops to repel each other, and they disperse throughout the water. Bile salts are another example of an emulsifying agent. These salts break up the fats we eat into small globules that can be more effectively digested.

The fluids of living systems are a complex mixture of colloids and dissolved ions and molecules. The behavior of these fluids in the body is vital to life. A particularly important property of these fluids is dialysis. Dialysis is similar to osmosis, as we will learn in the next section.

8.8 DIALYSIS AND LIVING SYSTEMS

We learned in Section 8.6 that an osmotic membrane allows water molecules, but not solute particles, to pass through. Cell membranes must be able to do more than this, because cells need to take in nutrients and discharge waste products. Membranes that allow small molecules and ions to pass while holding back large molecules and colloidal particles are called *dialyzing membranes.* Plasma membranes are examples of such membranes. The selective passage of small molecules and ions in either direction by a dialyzing membrane is called *dialysis.* Dialysis differs from osmosis in that osmotic membranes allow only solvent molecules to pass.

The process of dialysis is shown by the apparatus in Figure 8-8. The apparatus consists of a bag made of a dialyzing membrane such as an animal bladder. The bag contains a mixture of colloids and dissolved molecules and ions. The bag is placed in a container of pure water and water is continually passed through the membrane. The water carries the ions and molecules through the membrane, leaving the colloids behind. The ability of dialyzing membranes to allow the passage of only selected substances is extremely important to living systems.

The kidneys are an example of organs in the body that use dialysis to maintain the solute and electrolyte balance of the blood. The main purpose of the kidneys is to cleanse the blood by removing the waste products of

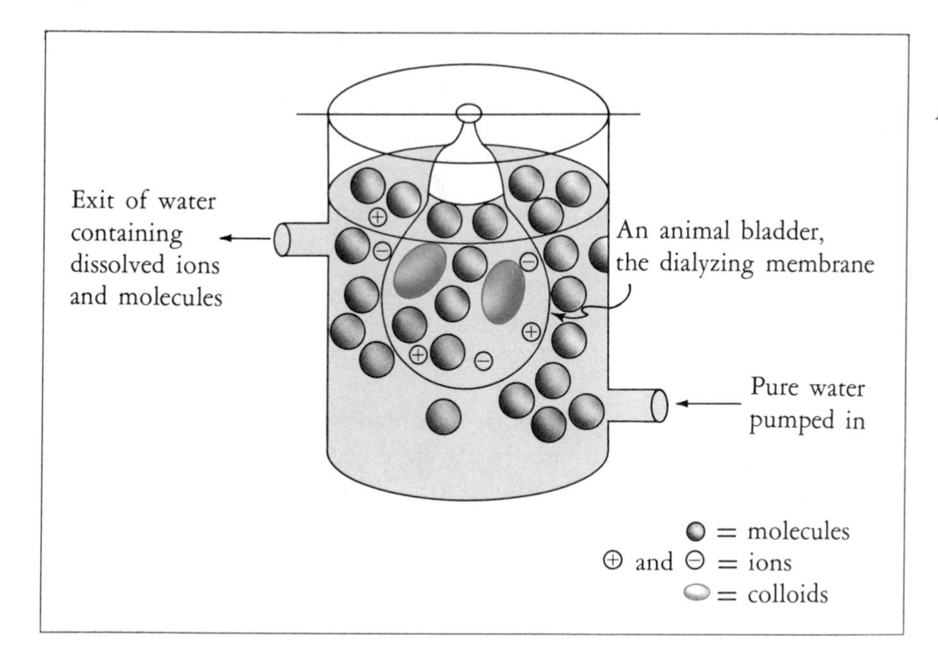

Fig. 8-8. Dialysis apparatus. Dissolved molecules and ions pass through the dialyzing membrane, but colloids do not.

metabolism and control the concentrations of electrolytes. The kidneys do this job very efficiently. Approximately 180 L of blood are purified daily in a 68-kg (150-lb) adult. Approximately 99 percent of the total volume processed is retained, and the remaining 1 percent is eliminated as urine. Part of the purification of blood occurs by dialysis.

In recent years, kidney machines have been built that purify the blood of patients with kidney failure. Each machine contains a series of tubes that are, in effect, dialyzing membranes. These tubes are chosen so that as the blood flows through, waste products pass through the membrane, but the larger cells and other molecules needed by the body do not. The machine thus removes many of the waste products of the blood and allows the patient to live a relatively normal life.

8.9 SUMMARY

Solutions are homogeneous mixtures of molecules, atoms, or ions of two or more different substances. The substances that make up a solution are called its components. The one component in excess is usually called the solvent, and the others are called solutes. The solubility of a solute in a particular solvent depends on many factors such as the kind of solute, the kind of solvent, the temperature of the solvent, and the pressure of solute above the surface of the solvent. The relative amounts of solute and solvent in a solution are expressed in units of concentration. The most common are percent by weight, volume, and weight/volume, ppm, ppb, molar concentration, and milliequivalents per liter.

Compounds that dissolve in water are either electrolytes or nonelectrolytes. Molecules of electrolytes contain at least one ionic or polar covalent bond. Most molecules of nonelectrolytes contain only relatively nonpolar bonds. Arrhenius' theory of electrolytes pictures aqueous solutions of electrolytes as individual ions each surrounded by a number of water molecules. Nonelectrolytes are pictured as individual neutral molecules, each surrounded by water molecules. Electrical conductivity, osmosis, and dialysis are explained by this theory.

Osmosis is the movement of water through an osmotic membrane from an aqueous solution that is less concentrated to one that is more concentrated. Osmotic pressure is the pressure that must be applied to a solution to prevent osmosis. Both osmosis and osmotic pressure are colligative properties; that is, they depend on the number of dissolved particles in the solution.

Molecules can cluster together to form particles of matter whose overall size is larger than the size of atoms or small molecules. These particles are called colloids. They are dispersed in water to form colloidal dispersions. Unlike solutions, which are transparent, colloidal dispersions are often cloudy. Most fluids in the body contain colloids as well as dissolved ions and molecules. Dialysis is a particularly important property of these fluids. This selective passage of molecules and ions in either direction through membranes is one method used by kidneys to purify blood.

REVIEW OF TERMS AND CONCEPTS

Terms

COLLIGATIVE PROPERTY Any property that depends on the number of dissolved particles in a solvent.

COLLOID Particles of matter between 1 and 100 nm in size.

COLLOIDAL DISPERSION A uniform dispersion of a colloid in water.

COMPLETELY MISCIBLE LIQUIDS Two liquids that are infinitely soluble in each other.

CRENATION The shriveling and shrinking in size of red blood cells caused by water leaving the cells.

DIALYSIS The selective passage of small molecules and ions in either direction through a dialyzing membrane.

DIALYZING MEMBRANE A membrane that allows small molecules and ions to pass through in either direction while holding back large molecules and colloids.

ELECTROLYTE A compound that forms ions when dissolved in water resulting in the formation of an electrolytic solution.

ELECTROLYTIC SOLUTION A solution that conducts electricity.

EMULSIFYING AGENT A compound or substance that stabilizes a colloidal dispersion.

HEMOLYSIS The rupture of red blood cells caused by water entering the cells.

HYDRATION The close association of water molecules with an ion.

HYPERTONIC SOLUTION A solution that has a higher osmotic pressure than another solution.

HYPOTONIC SOLUTION A solution that has a lower osmotic pressure than another solution.

ISOTONIC SOLUTIONS Two solutions that have the same osmotic pressure.

IMMISCIBLE LIQUIDS Two liquids that are insoluble in each other.

NONELECTROLYTE A compound that forms neutral molecules when dissolved in water resulting in the formation of a nonelectrolytic solution.

NONELECTROLYTIC SOLUTION A solution that does not conduct electricity.

OSMOSIS The movement of water through a membrane from a more dilute aqueous solution to a more concentrated one.

OSMOTIC PRESSURE The pressure needed to prevent osmosis.

PARTLY MISCIBLE LIQUID A liquid that is only slightly soluble in another liquid.

SATURATED SOLUTION A solution in which no more solute will dissolve in the solvent at a particular temperature.

SEMIPERMEABLE MEMBRANE A material that allows only certain molecules to pass through.

SOLUTE The component of a solution not present in excess.

SOLVENT The component of a solution present in excess.

SOLUTION A homogeneous mixture of the molecules, atoms, or ions of two or more different substances.

Concepts

1. Solutions are homogeneous mixtures of two or more substances.
2. The relative amounts of solute and solvent in a solution are expressed in terms of units of concentration.
3. Substances that form ions when dissolved in water are called electrolytes; those that form neutral molecules in solution are called nonelectrolytes.
4. Osmosis occurs between a dilute solution and a more concentrated solution through a semipermeable membrane.
5. Osmotic pressure is a colligative property.
6. Dialysis is one of the ways that dissolved ions and small molecules pass through the membranes of cells.

EXERCISES 8-16 Express the concentration of each of the following solutions as percent by weight/volume, M, and mEq/L:

(a) 12.0 g NaCl in 250 mL of solution
(b) 25.0 g $NaHCO_3$ in 500 mL of solution
(c) 50.0 g of Na_2CO_3 in 500 mL of solution
(d) 100 g of K_3PO_4 in 1.00 L of solution

8-17 How many grams of solute are needed to make each of the following solutions?
(a) 100 mL of 1.00 M $NaOH$
(b) 1.0 L of 0.030 M glucose
(c) 500 mL of 1.20 M KCl
(d) 500 mL of 0.500 M urea (CH_4N_2O)

8-18 How many grams of solute are there in each of the following quantities of solution?
(a) 50.0 mL of 0.100 M $MgCl_2$
(b) 100 mL of 1.00 M sucrose ($C_{12}H_{22}O_{11}$)
(c) 250 mL of 0.250 M $NaHCO_3$
(d) 1.50 L of 1.20 M KCl
(e) 0.750 L of 1.30 M $NaOH$
(f) 2.00 L of 0.750 M Na_2CO_3

8-19 What is the molar concentration of each ion formed when 0.50 mol of each of the following salts is dissolved in 1.00 L of solution?
(a) $NaBr$ (b) $NaHCO_3$ (c) Na_2CO_3 (d) $CaCl_2$
(e) K_2SO_4 (f) Na_3PO_4

8-20 Express the concentration of each ion in Exercise 8-19 in mEq/L.

8-21 Using the data in Table 8-2, determine how many grams of NaCl must be added to 250 mL of boiling water to make a saturated solution. If the solution is cooled to 20° C, how many grams of NaCl remain in solution?

8-22 A 1.00-mL sample of blood is found to contain 0.250 mg of calcium ions. What is the calcium ion concentration in M, mEq/L, and ppm?

8-23 Given equal volumes of the following solutions, decide which of each pair of solutions has the higher osmotic pressure.
(a) 1 M $LiBr$, 1 M Na_2CO_3
(b) 0.5 M glucose, 0.5 M $NaHCO_3$
(c) 1.0 M Na_3PO_4, 0.5 M Na_2CO_3

8-24 Two solutions are made up as follows:

VOLUME AND COMPONENTS	SOLUTION A	SOLUTION B
Volume composition	1.0 L	1.0 L
Electrolytes present	0.5 mol NaCl	0.5 mol KCl
Soluble compound present	0.01 mol sugar	0.01 mol sugar
Colloid present	0.005 mol starch	0.05 mol starch

Which solution has the higher osmotic pressure?

8-25 A person whose blood contains 0.0650 M alcohol (C_2H_5OH) is usually considered to be intoxicated. Express this concentration in ppm. If the total volume of blood in the person is 7.00 L, the body contains how many grams of alcohol?

8-26 Most people experience dizziness and headache when the concentration of carbon monoxide in the air reaches 100 ppm. Express this concentration in percent by volume.

8-27 The air that enters our lungs each day contains approximately 6.0×10^{26} molecules. How many of these molecules are sulfur dioxide on a day when the concentration of sulfur dioxide in the air is 0.14 ppm?

CHEMICAL REACTIONS IN AQUEOUS SOLUTIONS 9

We learned how to write and balance chemical equations in Chapter 7. In Chapter 8 we learned about several biologically important properties of aqueous solutions. In this chapter we will put this knowledge together and learn about several important kinds of chemical reactions in aqueous solutions.

Many of the chemical reactions that occur in nature take place between substances dissolved in water. Reactions between ions are particularly important in water and in many body fluids. Salts are compounds that release ions when dissolved in water. Other compounds called acids and bases also form ions in aqueous solutions. We will examine the reactions of the ions formed from these compounds in this chapter. To start, let us learn exactly which salts dissolve in water.

9.1 SOLUBILITIES OF SALTS IN WATER

We learned in Section 8.4 that many salts are electrolytes. Most salts that are electrolytes are quite soluble in water. The ions that are contained in the solid are released into the water when the salt is dissolved. But not all salts are soluble in water, as shown by the information given in Table 9-1. From the information in Table 9-1, we can make the following general statements about the solubilities of salts:

1. All salts containing ions of elements of group IA (Li^+, Na^+, K^+) are soluble in water no matter what the anion.

Table 9-1. Solubilities of Common Solid Salts in Water

Cation		Anion						
		Cl^- Chloride	NO_3^- Nitrate	CO_3^{-2} Carbonate	$C_2H_3O_2^-$ Acetate	SO_4^{-2} Sulfate	S^{-2} Sulfide	PO_4^{-3} Phosphate
Lithium	Li^+	S*	S	S	S	S	S	S
Sodium	Na^+	S	S	S	S	S	S	S
Potassium	K^+	S	S	S	S	S	S	S
Magnesium	Mg^{+2}	S	S	I	S	S	D	I
Calcium	Ca^{+2}	S	S	I	S	I	D	I
Barium	Ba^{+2}	S	S	I	S	I	D	I
Ammonium	NH_4^+	S	S	S	S	S	S	S
Ferrous	Fe^{+2}	S	S	I	S	S	I	I
Ferric	Fe^{+3}	S	S	—	—	SS	D	I
Cupric	Cu^{+2}	S	S	I	S	S	I	I
Zinc	Zn^{+2}	S	S	I	S	S	I	I
Silver	Ag^+	I	S	I	SS	I	I	I
Mercurous	Hg^+	I	S	I	SS	I	I	I
Mercuric	Hg^{+2}	S	S	—	S	D	I	I
Lead	Pb^{+2}	SS	S	I	S	I	I	I

*S = soluble in water; SS = slightly soluble in water; I = insoluble in water; D = decomposes in water; — = does not exist as salt.

2. All salts containing ammonium ions (NH_4^+) are soluble in water no matter what the anion.
3. All salts containing nitrate ions (NO_3^-) and acetate ions ($C_2H_3O_2^-$) are soluble in water no matter what the cation.
4. All salts containing chloride ions (Cl^-) are soluble in water except when the cations are Pb^{+2}, Ag^+, and Hg^+.
5. All salts containing sulfate ions (SO_4^{-2}) are soluble in water except when the cations are Ca^{+2}, Ba^{+2}, Fe^{+3}, Ag^+, Hg^+, and Pb^{+2}.
6. All salts containing sulfide ions (S^{-2}) are insoluble in water except when the cations are Li^+, Na^+, K^+, or NH_4^+.
7. Salts containing other combinations of ions are generally insoluble in water.

These rules of solubility are used to predict possible reactions between ions in solution, as we will learn next.

9.2 IONIC REACTIONS

An *ionic reaction* is a chemical reaction between ions or between ions and molecules. An ionic reaction occurs only if the product is one or more of the following:

1. a compound insoluble in water, called a precipitate
2. a gas
3. a compound that is soluble in water but does not exist as ions, called an un-ionized compound.

As an example, let us consider what happens when we dissolve equal molar (i.e., equimolar) quantities of lithium chloride and sodium nitrate in water. Will the following chemical reaction occur?

$$LiCl + NaNO_3 \longrightarrow LiNO_3 + NaCl \qquad (9\text{-}1)$$

The lithium and sodium ions simply exchange anions in this reaction. This type of reaction is called a double decomposition reaction.

We know from Table 9-1 that both lithium chloride and sodium nitrate are soluble in water and exist as ions in aqueous solution. The products, lithium nitrate and sodium chloride, also exist as ions in solution. We can indicate this fact by rewriting equation 9-1 in an ionic form, as follows:

$$Li^+ + Cl^- + Na^+ + NO_3^- \longrightarrow Li^+ + NO_3^- + Na^+ + Cl^-$$

This equation represents the true state of the four salts in an aqueous solution. But notice that the same four ions are on both sides of the chemical equation. This means that an aqueous solution of the two salts lithium chloride and sodium nitrate is simply a mixture of the four ions, Li^+, Na^+, NO_3^-, and Cl^-. No precipitate, no gas, no un-ionized compound is formed as product. Therefore, no chemical reaction has taken place.

Consider another example. Equimolar quantities of sodium chloride and silver nitrate are dissolved in water. Will the following reaction occur?

$$NaCl + AgNO_3 \longrightarrow AgCl + NaNO_3 \qquad (9\text{-}2)$$

The salts sodium chloride, silver nitrate, and sodium nitrate are all electrolytes. In contrast, silver chloride is insoluble in water (Table 9-1); it forms a precipitate. We can rewrite equation 9-2 as follows to indicate these facts:

$$Na^+ + Cl^- + Ag^+ + NO_3^- \longrightarrow AgCl\downarrow + Na^+ + NO_3^-$$

We indicate the fact that silver chloride forms a precipitate by placing an arrow pointing down (↓) after its chemical formula. The sodium and nitrate ions are common to both sides of the equation, and we may cancel them as follows:

$$\cancel{Na^+} + Cl^- + Ag^+ + \cancel{NO_3^-} \longrightarrow AgCl\downarrow + \cancel{Na^+} + \cancel{NO_3^-}$$

The part of the equation that remains, called the *net ionic equation,* is the following:

$$Cl^- + Ag^+ \longrightarrow AgCl\downarrow$$

Notice that this equation is balanced. Both the number of atoms of each element and the total charge are the same on both sides of the equation.

From this balanced net ionic equation, we know that any aqueous solution that contains silver ions and chloride ions will form a precipitate of silver chloride. We could have reacted lithium chloride, calcium chloride, or any water-soluble chloride ion containing salt with silver nitrate instead of sodium chloride and the net ionic equation would have been the same.

Notice the difference between a balanced net ionic equation and a balanced molecular equation. Only the ions that react and the products that are insoluble in water, a gas, or an un-ionized compound are included in a balanced net ionic equation. The complete chemical formulas of all reactants and products are included in the complete chemical equation, as we learned in Chapter 7. We will use both of these types of chemical equation in this book.

The balanced net ionic equation for any chemical reaction can be obtained by use of the following five steps.

Step 1. Determine the products of the reaction by exchanging the cations and anions of the reactants.

Step 2. Balance the complete chemical equation.

Step 3. Decide which of the reactants and which of the products are soluble and exist as ions in solution. If they all exist as ions, no reaction occurs and the solution is simply a collection of ions.

Step 4. Rewrite the chemical equation to separate the ions of all the electrolytes.

Step 5. Cancel all ions that appear on both sides of the equation. What remains is the net balanced ionic equation.

The use of these five steps is shown in the following example.

Example 9-1 Equimolar amounts of sodium sulfate and calcium chloride are dissolved in water. Will a chemical reaction occur? If so, write its net balanced ionic equation.

Solution:

STEP 1. Exchange partners to determine the products:

$$Na_2SO_4 + CaCl_2 \longrightarrow NaCl + CaSO_4$$

STEP 2. Balance the complete chemical equation:

$$Na_2SO_4 + CaCl_2 \longrightarrow 2\,NaCl + CaSO_4$$

STEP 3. Identify the salts that are electrolytes. From Table 9-1, Na_2SO_4, $CaCl_2$, and NaCl are all soluble in water and exist as ions in solution. Only $CaSO_4$ is insoluble in water.

STEP 4. Rewrite the chemical equation to indicate which salts are electrolytes and which are insoluble:

$$2\,Na^+ + SO_4{}^{-2} + Ca^{+2} + 2\,Cl^- \longrightarrow 2\,Na^+ + 2\,Cl^- + CaSO_4\downarrow$$

STEP 5. Cancel all ions that appear on both sides of the equation:

$$2\,\cancel{Na^+} + SO_4{}^{-2} + Ca^{+2} + 2\,\cancel{Cl^-} \longrightarrow 2\,\cancel{Na^+} + 2\,\cancel{Cl^-} + CaSO_4\downarrow$$

The balanced net ionic equation is the following:

$$SO_4{}^{-2} + Ca^{+2} \longrightarrow CaSO_4\downarrow$$

Because a precipitate is formed, a chemical reaction will take place.

*EXERCISE 9-1 Write the balanced net ionic equations for the reactions of $AgNO_3$ with $CaCl_2$, with LiCl, and with KCl. Verify the statement that the balanced net ionic equation is the same for all three reactions.

EXERCISE 9-2 Write the balanced net ionic equation for the reaction of each of the following pairs of salts. If no reaction occurs, state this fact as your answer.
(a) $Fe(NO_3)_2$, Na_2S (b) LiCl, KNO_3
(c) $Ba(NO_3)_2$, NaCl (d) Li_2S, $CuSO_4$
(e) NH_4NO_3, K_2CO_3 (f) $Hg(NO_3)_2$, $CaCl_2$
(g) $AgNO_3$, NH_4Cl (h) $Ba(NO_3)_2$, Na_2SO_4
(i) K_2CO_3, $C_2H_3O_2NH_4$

We learned in Section 8.8 that ions in body fluids are partly responsible for their osmotic pressure. Some of these ions have other functions in the body, as we will learn next.

9.3 IONS IN LIVING SYSTEMS

The most common cations and anions in the fluids of living systems are given in Table 9-2. Potassium and magnesium ions are found mostly in cellular fluids, whereas sodium and calcium ions are found in the fluid between cells, called *intercellular fluid.* Calcium ions are needed for healthy bones and teeth, for blood clotting, and for regulation of the heartbeat.

Table 9-2. The Most Common Cations and Anions in Body Fluids

Cations	Anions
Na^+	Cl^-
K^+	HCO_3^-
Mg^{+2}	$H_2PO_4^-$
Ca^{+2}	HPO_4^{-2}
Fe^{+2}	—
Fe^{+3}	—

Very small quantities, called trace amounts, of many metallic cations are also needed to maintain life. Hemoglobin (Section 13.6) contains iron in the form of ferrous ions. These ions play an important role in the transport of oxygen and carbon dioxide. Ferrous and ferric ions are also part of the cytochrome system that is involved in oxidative phosphorylation (Section 27.6). Other metallic cations such as cupric (Cu^{+2}), zinc (Zn^{+2}), cobalt (Co^{+2}), and manganous (Mn^{+2}) ions assist enzymes in their biological roles (See Section 25.3).

The metallic ions present in trace amounts in living systems usually exist as complex ions. A complex ion is made up of one or more metallic cations surrounded by other ions or molecules. These other ions or molecules contain nitrogen, oxygen, or sulfur atoms that form bonds with the metal-

* The answers for the exercises in this chapter begin on page 863.

lic cation. A simple example of a complex ion is the one formed between cupric ion and ammonia. Four ammonia molecules react with each cupric ion to form the complex ion shown in the following equation:

$$Cu^{+2} + 4\,NH_3 \longrightarrow Cu(NH_3)_4{}^{+2}$$

Many large molecules in living systems contain ammonia-like parts, as we will learn in Chapter 17. The nitrogen atoms contained in these molecules form complex ions with metallic cations. In this way, the metallic cation is bound to the molecule and becomes part of the living system. Several examples of metallic ion complexes of importance to life will be found in later chapters.

Sometimes the reaction of a metallic cation and a large molecule is poisonous to the living system. This is the case with the ions of mercury (Hg^{+2}) and lead (Pb^{+2}). Both are particularly poisonous to humans. Lead ions have a toxic effect on the kidneys and cause nerve damage. Mercuric ions cause extensive damage to the brain and nervous system. These ions react with the sulfur atoms of large molecules involved in many important functions of the body. The result is that the molecules are disrupted and are prevented from performing their normal functions. We will learn more about the effect of these poisons in Chapter 25.

Many of the reactions of ions and molecules in living systems are in a state of chemical equilibrium.

9.4 CHEMICAL EQUILIBRIUM

We encountered our first example of a reversible reaction in Section 7.5. There we learned that a chemical reaction is reversible when the products react to reform the reactants. Reversible reactions are usually in a state of chemical equilibrium. To understand chemical equilibrium, let us examine another example of a reversible reaction: the formation of ammonia from nitrogen and hydrogen. This reaction is the basis of the Haber process that is used worldwide to make ammonia for the manufacture of fertilizer. The forward reaction forms ammonia from hydrogen and nitrogen:

$$N_2 + 3\,H_2 \longrightarrow 2\,NH_3 \tag{9-3}$$

The reverse reaction forms nitrogen and hydrogen from ammonia:

$$2\,NH_3 \longrightarrow 3\,H_2 + N_2 \qquad (9\text{-}4)$$

The forward reaction occurs when a mixture of 3 moles of hydrogen and 1 mole of nitrogen are heated to 200° C and are subjected to a pressure of 30 atm. When there appears to be no further change in the reaction, we find that some hydrogen and nitrogen are still present. In fact, 32 percent of the volume of gases is still hydrogen and nitrogen; only 68 percent is ammonia. Further reaction does not increase the amount of ammonia formed.

The reverse reaction occurs when we start with pure ammonia. Under the same reaction conditions, ammonia decomposes to nitrogen and hydrogen. When there is no further apparent change, we find that ammonia is still present. Again, we find that 68 percent of the total amount of gas is ammonia, whereas 32 percent is nitrogen and hydrogen.

Regardless of whether we start with ammonia or with hydrogen and nitrogen, neither reaction goes to completion. Each reaction forms the same mixture of 68 percent ammonia and 32 percent hydrogen and nitrogen. Once this mixture is formed, however, the two reactions do not stop. It may appear to us that the forward and reverse reactions have stopped. In reality, both reactions still occur, but for every molecule of ammonia formed by the reaction of hydrogen and nitrogen, one is lost by decomposition to nitrogen and hydrogen. Thus, both forward and reverse reactions continue and the reaction system is in a state of *dynamic equilibrium.*

A chemical reaction is in a state of *equilibrium* when the amount of products formed per second by the forward reaction exactly equals the amount of products lost per second by the reverse reaction. Thus, equilibrium is reached when all spontaneous changes stop. Because an equilibrium reaction is also a reversible reaction, we use the same symbol, two arrows pointing in opposite directions ($\rightleftarrows$), for both, as shown in the following equation:

$$3\,H_2 + N_2 \rightleftarrows 2\,NH_3$$

The quantities of products and reactants present at equilibrium in any reaction are related to each other by the equilibrium constant expression. For the reaction of hydrogen, nitrogen, and ammonia, the equilibrium constant expression is

$$K = \frac{[NH_3]^2}{[H_2]^3[N_2]}$$

The symbol K is called the *equilibrium constant.* This constant is a ratio of the concentrations of the products to the concentrations of the reactants. Its numerical value depends on the particular chemical reaction, the temperature, and the pressure. (Equilibrium constants never have units. The reason for this is beyond the scope of this book. Just remember that the value of an equilibrium constant is a dimensionless quantity.) The general form of this ratio depends on the *balanced* equation for the equilibrium reaction. The molar concentrations of the products are placed in the numerator, and the molar concentrations of the reactants are placed in the denominator. The concentration of each reagent is raised to the power equal to its coefficient in the balanced equation. The following examples show this relationship between the balanced chemical equation and the equilibrium constant expression.

BALANCED CHEMICAL EQUATION	EQUILIBRIUM CONSTANT EXPRESSION
$2\ SO_2 + O_2 \rightleftharpoons 2\ SO_3$	$K = \frac{[SO_3]^2}{[SO_2]^2[O_2]}$
$3\ O_2 \rightleftharpoons 2\ O_3$ (Ozone)	$K = \frac{[O_3]^2}{[O_2]^3}$
$H_2O + CO \rightleftharpoons H_2 + CO_2$	$K = \frac{[H_2][CO_2]}{[H_2O][CO]}$

EXERCISE 9-3 Write the equilibrium constant expression for each of the following equilibrium reactions:

(a) $CO + 2\ H_2 \rightleftharpoons CH_3OH$
(b) $Cl_2 + 2\ HBr \rightleftharpoons Br_2 + 2\ HCl$
(c) $N_2 + O_2 \rightleftharpoons 2\ NO$
(d) $N_2O_4 \rightleftharpoons 2\ NO_2$
(e) $C_2H_4O_2 + C_2H_6O \rightleftharpoons H_2O + C_4H_8O_2$
(f) $C_2H_6O + HCl \rightleftharpoons C_2H_5Cl + H_2O$
(g) $Ag^+ + 2\ NH_3 \rightleftharpoons Ag(NH_3)_2^+$
(h) $2\ NH_3 + CO_2 \rightleftharpoons CN_2H_4O + H_2O$

We can tell whether the products or reactants are favored at equilibrium from the numerical value of the equilibrium constant of a reaction. When the equilibrium constant is greater than 10^2, most of the reactants have been converted to products. Thus, the *products are favored in an equilibrium reaction whose equilibrium constant is greater than* 10^2. Conversely, when the equilibrium constant is less than 10^{-2}, only a very small amount of product is formed. Therefore, *the reactants are favored in an equilibrium reaction whose equilibrium constant is less than* 10^{-2}. If the equilibrium constant is between 10^{-2} and 10^2, neither product nor reactant is greatly favored, and appreciable quantities of both are present at equilibrium.

EXERCISE 9-4 From the value of the equilibrium constant of each of the following chemical reactions, decide whether mostly products, mostly reactants, or both reactants and products are present at equilibrium.

REACTION	TEMPERATURE (° C)	K
(a) $SO_2 + NO_2 \rightleftharpoons SO_3 + NO$	375	5
(b) $H_2 + I_2 \rightleftharpoons 2\,HI$	425	55
(c) $2\,SO_2 + O_2 \rightleftharpoons 2\,SO_3$	400	397
(d) $N_2 + O_2 \rightleftharpoons 2\,NO$	37	1×10^{-12}

The equilibrium constant, K, for an equilibrium reaction will always have the same value at the same temperature and pressure. But what happens to an equilibrium reaction if we add more reactant? Does this change the equilibrium constant? These questions were answered by the French chemist Henri Louis Le Châtelier a century ago.

9.5 THE LE CHÂTELIER PRINCIPLE

A chemical reaction in equilibrium can be disturbed by an externally caused change such as addition of more reactants. The way the concentrations of both the reactants and the products are affected by such a change was first explained by H. L. Le Châtelier in 1884. His explanation, now called the Le Châtelier principle, is the following:

> If a system at equilibrium is disturbed by an externally applied stress, the system changes in such a way that this external stress is minimized.

Many examples of this principle are found in living systems. Let us examine two of them.

Oxygen needed by the body is carried in the bloodstream to the cells as part of the molecule oxyhemoglobin. This molecule is formed in the blood by the reaction between dissolved oxygen gas obtained from air and hemoglobin in the red blood cells. This reaction is reversible. In the cells, oxyhemoglobin releases its oxygen to reform hemoglobin. We can write the equation for this reversible reaction as follows:

$$\text{Hemoglobin} + 4\,O_2 \rightleftharpoons \text{Oxyhemoglobin} \qquad (9\text{-}5)$$

The chemical formulas and structures of hemoglobin and oxyhemoglobin are very complicated. We do not need to concern ourselves with their structures at this time. For our purposes, it is sufficient to use the symbol HHG for hemoglobin and the symbol $HG(O_2)_4^-$ for oxyhemoglobin. We can abbreviate equation 9-5 using these symbols as follows:

$$HHG + 4\,O_2 \rightleftharpoons HG(O_2)_4^- + H^+$$

The partial pressure of oxygen gas in the lungs is higher than anywhere else in the body, as we learned in Section 6.9. This increases the oxygen gas concentration in the blood entering the alveoli (the small, thin sacs of blood capillaries in the lungs). The presence of this additional oxygen disturbs the equilibrium in equation 9-5. This stress causes the reaction that forms $HG(O_2)_4^-$ to be favored until equilibrium is restored. The result is a reduction in the concentration of oxygen gas and formation of more oxyhemoglobin. This is the way that oxygen is transferred from the lungs to oxyhemoglobin and then to regions of the body that need oxygen. This application of the Le Châtelier principle is shown pictorially in Figure 9-1.

The important thing to understand about an equilibrium reaction is that the numerical value of the equilibrium constant does not change as long as the temperature remains constant. To relieve a stress caused by addition of more reactants or products, the concentrations of the reactants and products change to maintain the value of the equilibrium constant. We can show this by using another example, the enzyme-catalyzed reaction of glucose-1-phosphate (G-1-P) to glucose-6-phosphate (G-6-P) (Section 27.3). The chemical equation for this reaction and its equilibrium constant expression are as follows:

$$\text{Glucose-1-phosphate} \xrightleftharpoons{\text{Enzyme}} \text{Glucose-6-phosphate}$$

$$K = \frac{[\text{Glucose-6-phosphate}]}{[\text{Glucose-1-phosphate}]} = 20.0$$

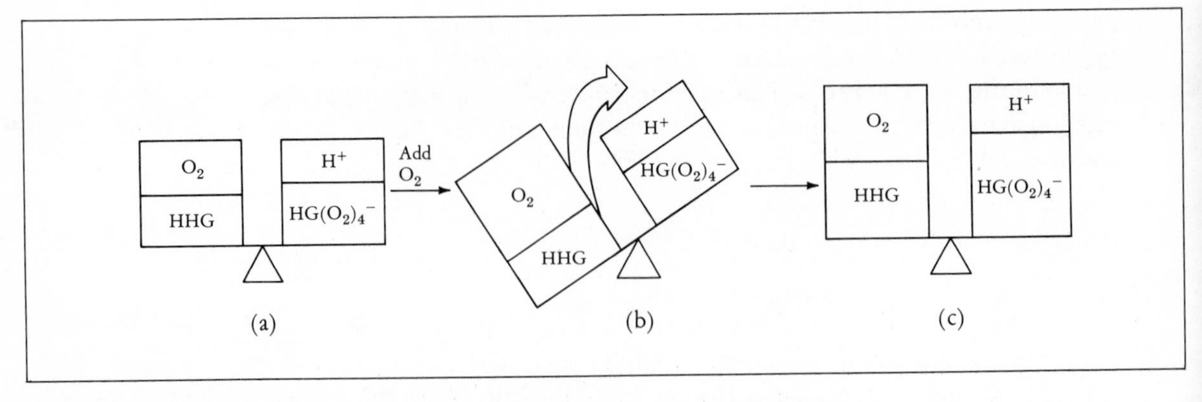

Fig. 9-1. A pictorial representation of the Le Châtelier principle. (a) The reaction between hemoglobin and oxyhemoglobin at equilibrium. (b) Addition of oxygen disturbs the equilibrium. (c) Oxygen and hemoglobin react to form more oxyhemoglobin and H^+ to re-establish the equilibrium.

We know from many experiments that the equilibrium constant for this reaction is 20.0 at body temperature (37° C).

Let us imagine that we have a solution in which the concentration of glucose-6-phosphate is 1.00 M and the concentration of glucose-1-phosphate is 0.0500 M. This is an equilibrium mixture of the two compounds. We can easily check this statement because the concentration of G-6-P divided by the concentration of G-1-P should equal 20.0. Thus,

$$\frac{[\text{G-6-P}]}{[\text{G-1-P}]} = \frac{1.00 \text{ M}}{0.0500 \text{ M}} = 20.0$$

This mixture is, indeed, at equilibrium. If we leave the solution undisturbed, the concentrations of the two reactants will not change.

Table 9-3. Effect of Added Reactant on the Composition of Equilibrium Mixtures

Initial Equilibrium Concentration, M			Reactant Added, M		New Equilibrium Concentration, M		
G-1-P*	G-6-P*	K	G-1-P	G-6-P	G-1-P	G-6-P	K
0.0500	1.00	20	0	0.500	0.0740	1.48	20
0.0250	0.500	20	0.0100	0	0.0260	0.509	20

*G-1-P means glucose-1-phosphate. G-6-P means glucose-6-phosphate.

What happens to the concentrations of the reactants in this solution if we add enough G-6-P to increase its concentration from 1.00 M to 1.50 M? We know that reaction will occur to convert G-6-P to G-1-P until the ratio of their concentrations is again equal to 20.0. In our example, equilibrium is re-established when the concentration of G-6-P reaches 1.48 M and that of G-1-P reaches 0.0740 M, as shown in Table 9-3. Thus, when a reaction at equilibrium is disturbed by changing the concentration of any reactant, chemical reaction occurs to change the concentrations of all the reactants until their ratio again equals the equilibrium constant.

EXERCISE 9-5 Predict the effect of the change indicated on each of the following equilibrium reactions:

(a) $CH_3CO_2H + CH_3OH \rightleftharpoons CH_3CO_2CH_3 + H_2O$
(i) add H_2O (ii) add CH_3CO_2H (iii) remove water

(b) $2\ NH_3 + CO_2 \rightleftharpoons CH_4N_2O + H_2O$
(i) add water (ii) remove CH_4N_2O (iii) lower pressure of CO_2

One very important and fundamental equilibrium in aqueous solutions is the one involved in the ionization of water.

9.6 IONIZATION OF WATER

We learned in Section 6.3 that molecules of liquid water are held together by hydrogen bonds. Within this three-dimensional structure of water molecules, there exist very small concentrations of *hydronium ions* (H_3O^+) and *hydroxide ions* (OH^-). These ions are formed from water molecules by the following reaction:

$$2\ H_2O \rightleftharpoons H_3O^+ + OH^- \qquad (9\text{-}6)$$

The formation of ions by the breaking of a polar covalent bond in a molecule is called *ionization*. Notice that in this process the electron pair of the broken bond stays with the more electronegative element.

The tendency for water to ionize is very small. This is shown by the fact that the concentration of hydronium ions in water at 25° C is only 1×10^{-7} M. Because a hydroxide ion is formed with each hydronium ion, the concentration of hydroxide ions is also 1×10^{-7} M. This is such a small concentration of ions that only one of every 550,000,000 water molecules is ionized. This proportion of ions is so small that we would not worry about the ionization were it not for the fact that the concentrations of hydronium and hydroxide ions change when certain substances dissolve in water. To understand how this occurs, we must begin with the equilibrium constant expression for the ionization of water. The equilibrium constant expression for equation 9-6 is the following:

$$K = \frac{[H_3O^+][OH^-]}{[H_2O]^2}$$

This equation can be simplified, because the concentration of water molecules is essentially constant. Thus, we can write

$$K[H_2O]^2 = K' = [H_3O^+][OH^-]$$

The ionization of water is often written simply as

$$H_2O \rightleftharpoons H^+ + OH^- \qquad (9\text{-}7)$$

Comparing equations 9-6 and 9-7 we see that

$$[H_3O^+] = [H^+]$$

This allows us to make a further simplification:

$$K' = K_w = [H^+][OH^-] \qquad (9\text{-}8)$$

The symbol K_w stands for the *ion product constant of water.* We can calculate its value by substituting the concentrations of hydronium and hydroxide ions into equation 9-8 to give

$$K_w = [H^+][OH^-] = [1 \times 10^{-7}\ M][1 \times 10^{-7}\ M] = 1 \times 10^{-14}*$$

* It appears from this equation that the units of K_w should be M^2. This is not true because equilibrium constants never have units, as we learned in Section 9-4.

We know from this equation that in pure water or in *any aqueous solution, both hydronium and hydroxide ions must be present and the product of their concentrations must be a constant equal to* 1×10^{-14}.

The relative concentrations of hydronium and hydroxide ions in aqueous solutions are determined by equation 9-8. If we know the concentration of one, we immediately know the concentration of the other. For example, if we add a substance to water that increases the concentration of hydronium ions from 1×10^{-7} M to 1×10^{-5} M, what happens to the concentration of hydroxide ions? Applying the Le Châtelier principle to equation 9-7, we know that reaction will occur to form more water. This results in a smaller concentration of hydroxide ions, as can be calculated using equation 9-8:

$$K_w = [H^+][OH^-] = 1 \times 10^{-14}$$

$$[H^+] = 1 \times 10^{-5}\ M$$

$$[OH^-] = \frac{1 \times 10^{-14}}{[H^+]} = \frac{1 \times 10^{-14}}{1 \times 10^{-5}} = 1 \times 10^{-9}\ M$$

EXERCISE 9-6 For each of the following solutions, calculate the concentration of hydroxide ions:
(a) $[H^+] = 1 \times 10^{-4}$ M (b) $[H^+] = 1 \times 10^{-7}$ M
(c) $[H^+] = 1 \times 10^{-10}$ M (d) $[H^+] = 1 \times 10^{-12}$ M

Because the product of the concentrations of hydronium and hydroxide ions must equal 1×10^{-14} at 25° C, it follows that as the concentration of one increases, the concentration of the other must decrease. This relationship between hydronium and hydroxide ions is shown in Table 9-4. All aqueous solutions contain both hydronium and hydroxide ions. However, their concentrations need not be equal. Only in a neutral solution is $[H^+] = [OH^-]$. An aqueous solution in which $[H^+]$ is greater than $[OH^-]$ is called an *acidic solution*. In a *basic solution*, $[OH^-]$ is greater than $[H^+]$.

EXERCISE 9-7 Decide whether each of the following solutions is acidic, basic, or neutral.
(a) $[OH^-] = 1 \times 10^{-12}$ M (b) $[H^+] = 1 \times 10^{-2}$ M
(c) $[OH^-] = 1 \times 10^{-3}$ M (d) $[OH^-] = 1 \times 10^{-6}$ M
(e) $[H^+] = 1 \times 10^{-7}$ M (f) $[OH^-] = 1 \times 10^{-7}$ M

Table 9-4. Relationship Between $[H^+]$ and $[OH^-]$

		$[H^+]$ (M)	$[OH^-]$ (M)		$[H^+][OH^-] = K_w$
Acidic solutions	Increasing $[H^+]$ ↑	1×10^{-1}	1×10^{-13}	↓	1×10^{-14}
		1×10^{-2}	1×10^{-12}		1×10^{-14}
		1×10^{-3}	1×10^{-11}		1×10^{-14}
		1×10^{-4}	1×10^{-10}		1×10^{-14}
		1×10^{-5}	1×10^{-9}		1×10^{-14}
		1×10^{-6}	1×10^{-8}		1×10^{-14}
Neutral solution		1×10^{-7}	1×10^{-7}		1×10^{-14}
Basic solutions		1×10^{-8}	1×10^{-6}		1×10^{-14}
		1×10^{-9}	1×10^{-5}		1×10^{-14}
		1×10^{-10}	1×10^{-4}		1×10^{-14}
		1×10^{-11}	1×10^{-3}		1×10^{-14}
		1×10^{-12}	1×10^{-2}	Increasing $[OH^-]$	1×10^{-14}
		1×10^{-13}	1×10^{-1}		1×10^{-14}

Certain compounds called acids dissolve in water and increase the concentration of hydronium ions relative to hydroxide ions. Other compounds called bases increase the concentration of hydroxide ions relative to hydronium ions. We will begin a study of these important kinds of compounds in the next section.

9.7 INTRODUCTION TO ACIDS AND BASES

Acids and bases have been known for centuries. The modern definitions of these compounds are due to the work of Arrhenius, the same scientist who proposed the theory of electrolytes (Section 8.5). Let us first consider acids. Arrhenius defined an acid as any compound that forms a proton, H^+ (also called a hydrogen ion), in aqueous solution. This definition is still acceptable, even though we know that protons do not exist alone in aqueous solutions. A proton in solution is hydrated; that is, it is surrounded by water molecules.

We learned in Section 9.6 that the ionization of water forms hydronium ions as well as hydroxide ions. The hydronium ion is our representation of a hydrated proton. Even the hydronium ion is a simplification, because a proton is hydrated by more than one water molecule. However, we repre-

Table 9-5. Common Strong and Weak Acids

Strong Acids	Weak Acids
HCl, HBr, $HClO_4$, HI HNO_3, H_2SO_4	$C_2H_3O_2H$, H_2CO_3, HF, HCN

sent a hydrated proton as a hydronium ion and ignore any further hydration. It is convenient to simplify an equation and represent the hydrated proton as H^+, as we did in equation 9-7. This is acceptable as long as we remember that it is a simplification of the real situation in an aqueous solution. The name proton or hydrogen ion is used as well as hydronium ion to designate the H^+ ion in water.

An acid is any compound that ionizes in water to form hydrogen ions. Certain acids ionize completely when dissolved in water. Such acids are called *strong acids.* Hydrogen chloride (HCl) is an example of a strong acid. When hydrogen chloride is bubbled into water, the molecules ionize to form hydrogen and chloride ions, according to the following equation:

HCl	$+ H_2O$	$\longrightarrow$	H_3O^+	$+$	Cl^-
Hydrogen chloride	Water		Hydrogen ion		Chloride ion

An aqueous solution of hydrogen chloride is a solution of hydrogen and chloride ions. Other compounds that are strong acids are listed in Table 9-5.

The acidic solution formed by dissolving hydrogen chloride in water is given another name. It is called hydrochloric acid. Acidic aqueous solutions of many compounds are given names different from those of the pure compound, as shown in Table 9-6. The names are usually derived from the

Table 9-6. Names of Aqueous Solutions of Acids

Compound	Name of Compound	Name of Aqueous Solution of Compound
HCl	Hydrogen chloride	Hydrochloric acid
HBr	Hydrogen bromide	Hydrobromic acid
H_2SO_4	Dihydrogen sulfate	Sulfuric acid*
HF	Hydrogen fluoride	Hydrofluoric acid
HCN	Hydrogen cyanide	Hydrocyanic acid

* In certain cases, extra letters are added for phonetic reasons.

name of the anion contained in the compound. If the name of the anion ends in *-ide* or *-ate,* then the name of the acidic solution becomes hydro_____ic or _____ic acid, respectively. Examples are given in Table 9-6.

Other acids ionize only slightly when dissolved in water. Such acids are called *weak acids.* An example is acetic acid ($C_2H_3O_2H$).* When dissolved in water, only approximately 2 percent of the molecules ionize to form hydrogen and acetate ions according to the following equation:

$$\underset{\text{Acetic acid}}{C_2H_3O_2H} + \underset{\text{Water}}{H_2O} \rightleftharpoons \underset{\text{Acetate ion}}{C_2H_3O_2^-} + \underset{\text{Hydrogen ion}}{H_3O^+} \tag{9-9}$$

Thus, an aqueous solution of acetic acid contains predominantly un-ionized molecules. Other compounds that are weak acids are listed in Table 9-5.

We sometimes indicate that ionization of a compound is not complete by double arrows of unequal length. This is shown in equation 9-9, where the longer arrow points toward the un-ionized molecules, indicating that the equilibrium lies in that direction.

Another way of classifying acids is by the number of hydrogen ions they form when dissolved in water. An acid such as hydrogen chloride that forms only one hydrogen ion per molecule is called a *monoprotic acid.* Sulfuric acid (H_2SO_4) is called a *diprotic acid* because both of its hydrogens form hydrogen ions when dissolved in water. These two hydrogen ions are formed in two separate steps, as shown by the following equations:

$$\underset{\text{Sulfuric acid}}{H_2SO_4} + \underset{\text{Water}}{H_2O} \rightleftharpoons \underset{\text{Bisulfate ion}}{HSO_4^-} + \underset{\text{Hydrogen ion}}{H_3O^+}$$

$$\underset{\text{Bisulfate ion}}{HSO_4^-} + \underset{\text{Water}}{H_2O} \rightleftharpoons \underset{\text{Sulfate ion}}{SO_4^{-2}} + \underset{\text{Hydrogen ion}}{H_3O^+}$$

The hydrogen atoms of an acidic molecule that are released to form hydrogen ions in water are called *acidic hydrogens.* In many acids such as

* The chemical formula of acetic acid is written in a number of ways. For the time being, we will write its chemical formula as $C_2H_3O_2H$. In Chapter 18 we will learn the structural formula of acetic acid and learn another way of writing its chemical formula.

HCl, H_2SO_4, and HNO_3, all the hydrogens in the compound are acidic. But this is not always the case. In acetic acid, for example, only one of the hydrogens is acidic. The formula of acetic acid is written $C_2H_3O_2H$ to emphasize this fact. Which hydrogens are acidic in molecules that contain many hydrogens will become clear when we examine the structures of organic compounds in Chapters 11 through 20.

EXERCISE 9-8 Write the chemical equation for the ionization of each of the following acids:
(a) HBr (b) HI (c) HNO_3 (d) $HClO_4$
(e) $C_2Cl_3O_2H$

EXERCISE 9-9 Phosphoric acid, H_3PO_4, can lose three protons. Write the equations representing the loss of the first, the second, and the third proton.

Let us turn our attention now to bases. According to Arrhenius' definition, a base is any compound that increases the hydroxide ion concentration in water. Bases are also called alkaline substances. Sodium hydroxide ($NaOH$) is a common example of a base. It is an ionic compound that exists as sodium ions (Na^+) and hydroxide ions (OH^-) in the solid state. When dissolved in water, hydrated sodium ions and hydroxide ions are formed, as shown in the following equation:

$$NaOH \longrightarrow Na^+ + OH^-$$

Other common bases are potassium hydroxide (KOH), calcium hydroxide ($Ca(OH)_2$), and magnesium hydroxide ($Mg(OH)_2$). These bases all exist as ions in aqueous solution and are all called *strong bases.*

Calcium hydroxide and magnesium hydroxide differ from sodium hydroxide and potassium hydroxide in that they are both only slightly soluble in water. Because of this fact, an aqueous solution of either calcium or magnesium hydroxide contains a relatively low concentration of hydroxide ions. Solutions of these two bases are weakly basic even though the two compounds are strong bases. This may seem confusing, but remember that a compound that is only slightly soluble has relatively few molecules dissolved in water. Only a small amount of magnesium hydroxide dissolves in water, but all of it forms ions because it is a strong base. So little dissolves that the concentration of hydroxide ion is low and the solution is only weakly basic. It is so weak that it does not harm either clothing or the human body. In fact, a slurry of magnesium hydroxide, called milk of magnesia, is taken as an antacid or purgative.

Many acids are in common use today. Some are in the body, and others are found in industrial and household products. Hydrochloric acid is produced in our stomachs to aid in the digestion of food. Industrially, an aqueous solution of hydrochloric acid is called muriatic acid. It is used to clean swimming pools, stonework, and toilet bowls. Vinegar is a dilute solution of acetic acid. Lactic acid accumulates in muscles during vigorous exercise (Chapter 27). Basic solutions are used as household cleaners because they remove fats and greases. For example, dilute solutions of sodium hydroxide are used as oven cleaners.

An acid reacts with a base in a characteristic reaction called neutralization.

9.8 NEUTRALIZATION

An acid reacts with a base to form water and a salt. For example, aqueous solutions of sodium hydroxide and hydrochloric acid react to form water and sodium chloride, according to the following equation:

$NaOH$ +	HCl	$\longrightarrow$	H_2O +	$NaCl$
Sodium hydroxide	Hydrochloric acid		Water	Sodium chloride

This is the most characteristic reaction of acids and bases. If equimolar concentrations of sodium hydroxide and hydrochloric acid are mixed, the resulting solution is neither acidic nor basic. It is neutral, and the reaction is called a *neutralization reaction.*

EXERCISE 9-10 Write the equation for the neutralization reaction of KOH and HBr.

Let us look at this neutralization reaction a bit more closely by writing its balanced net ionic equation. We learned how to do this in Section 9.2. Sodium hydroxide, hydrochloric acid, and sodium chloride all exist predominantly as ions in aqueous solution; water is mostly un-ionized. Thus, we can write the following equation:

$$Na^+ + OH^- + H^+ + Cl^- \longrightarrow Na^+ + Cl^- + H_2O$$

Canceling ions that appear on both sides of the equation, we obtain the following net ionic equation:

$$H^+ + OH^- \longrightarrow H_2O \qquad (9\text{-}10)$$

Notice that this is an example of an ionic reaction that goes to completion because an un-ionized molecule (H_2O) is formed.

Equation 9-10 is the net ionic equation for the neutralization reaction between *any* strong base and *any* strong acid. The only chemical reaction that occurs in a neutralization reaction between a strong acid and a strong base is the reaction of a hydrogen ion and a hydroxide ion to form water.

One mole of a monoprotic acid such as hydrochloric acid is completely neutralized by 1 mole of a strong base. Strong diprotic acids require 2 moles of base per mole of acid to be neutralized completely. The neutralization of sulfuric acid, for example, occurs by means of the following two reactions:

Reaction 1. $H_2SO_4 + NaOH \longrightarrow NaHSO_4 + H_2O$

Sulfuric acid, Sodium hydroxide, Sodium bisulfate, Water

Reaction 2. $NaHSO_4 + NaOH \longrightarrow Na_2SO_4 + H_2O$

Sodium bisulfate, Sodium hydroxide, Sodium sulfate, Water

Overall reaction $H_2SO_4 + 2\,NaOH \longrightarrow Na_2SO_4 + 2\,H_2O$

Again, the overall net ionic equation is $H^+ + OH^- \rightarrow H_2O$.

EXERCISE 9-11 Write the full equation and the net ionic equation for each of the following reactions. Verify the statement that the net ionic equation is the same for the neutralization reaction between any strong acid and any strong base.

(a) $KOH + HCl$ (b) $NaOH + HBr$
(c) $KOH + HNO_3$ (d) $NaOH + H_2SO_4$
(e) $KOH + HClO_4$ (f) $2\,KOH + H_2SO_4$

As in the case of all net ionic equations, the other ions are left unchanged in the solution. If the water evaporates, the solid crystals of the salt are left.

Acids and bases also react with many salts. Their reactions with salts of bicarbonate and carbonate ions are particularly important to living systems.

9.9 REACTIONS OF ACIDS AND BASES WITH CARBONIC ACID AND ITS SALTS

The reactions of bicarbonate and carbonate salts with acids and bases are important for controlling both the amount of carbon dioxide in the body and the acidity of the blood. Bicarbonate and carbonate salts are derived from a diprotic acid called carbonic acid, H_2CO_3. This acid is both weak and unstable.

Just like any acid, carbonic acid reacts with strong bases such as sodium hydroxide to form salts. Because it is a diprotic acid, 1 mole of carbonic acid can react with 2 moles of base. As a result, two salts can be formed in two separate reactions. Equimolar quantities of carbonic acid and base form sodium bicarbonate, according to the following equation:

$$\underset{\text{Carbonic acid}}{H_2CO_3} + \underset{\text{Sodium hydroxide}}{NaOH} \longrightarrow \underset{\text{Sodium bicarbonate}}{NaHCO_3} + \underset{\text{Water}}{H_2O} \qquad (9\text{-}11)$$

Sodium bicarbonate can react with more base to form sodium carbonate, as shown by the following equation:

$$\underset{\text{Sodium bicarbonate}}{NaHCO_3} + \underset{\text{Sodium hydroxide}}{NaOH} \longrightarrow \underset{\text{Sodium carbonate}}{Na_2CO_3} + \underset{\text{Water}}{H_2O} \qquad (9\text{-}12)$$

These two reactions are one way that excess base is neutralized in the body, as we will learn in the next chapter.

EXERCISE 9-12 Write the net ionic equations for equations 9-11 and 9-12.

Carbonic acid is unstable in water. Much of the acid decomposes to carbon dioxide and water, according to the following equation:

$$H_2CO_3 \rightleftharpoons H_2O + CO_2 \qquad (9\text{-}13)$$

However, some molecules of carbonic acid still remain in solution. Carbonic acid, water, and carbon dioxide are in equilibrium in water, but the equilibrium lies far to the right. Thus, when carbon dioxide is dissolved in water, a very small amount of carbonic acid is formed.

Carbonic acid can be formed in solution in a number of ways. One way, as we have just learned, is to dissolve carbon dioxide in water. Another way is to react a bicarbonate salt, such as sodium bicarbonate, with a strong acid such as hydrochloric acid. The full equation for this reaction is the following:

$NaHCO_3$ +	HCl	$\longrightarrow$	H_2CO_3 +	$NaCl$	(9-14)
Sodium bicarbonate	Hydrochloric acid		Carbonic acid	Sodium chloride	

But carbonic acid is unstable. It decomposes to carbon dioxide and water, as shown in equation 9-13. If we combine equations 9-13 and 9-14, we obtain the complete equation for the reaction of sodium bicarbonate and hydrochloric acid.

$$NaHCO_3 + HCl \longrightarrow NaCl + H_2O + CO_2\uparrow \qquad (9\text{-}15)$$

Let us again write the net ionic equation to focus our attention on exactly what happens in this reaction. The balanced net ionic equation for the reaction of *any* bicarbonate salt with *any* strong acid is the following:

$$HCO_3^- + H^+ \longrightarrow H_2CO_3 \rightleftarrows H_2O + CO_2\uparrow \qquad (9\text{-}16)$$

Thus, when any strong acid is added to a bicarbonate salt, carbon dioxide is released. Notice that equation 9-16 is an example of a chemical reaction that occurs because both a gas and an un-ionized molecule are formed.

Another way that carbonic acid can be formed is by the reaction of a carbonate salt, such as sodium carbonate, with a strong acid. This reaction occurs in two steps. The first is the reaction to form sodium bicarbonate as follows:

Na_2CO_3 +	HCl	$\longrightarrow$	$NaHCO_3$ +	$NaCl$	(9-17)
Sodium carbonate	Hydrochloric acid		Sodium bicarbonate	Sodium chloride	

Sodium bicarbonate reacts with more acid, as we learned in equation 9-15. Combining equations 9-15 and 9-17, we obtain the following overall equation for the reaction of sodium carbonate and hydrochloric acid:

$$Na_2CO_3 + 2\,HCl \longrightarrow 2\,NaCl + H_2O + CO_2\uparrow$$

Carbonate salts, like bicarbonate salts, react with strong acids to form carbon dioxide and water according to the following net ionic equation:

$$CO_3^{-2} + 2\,H^+ \longrightarrow H_2CO_3 \rightleftharpoons H_2O + CO_2\uparrow$$

This is the reaction that occurs when baking soda is added to acid. The fizzing is caused by the release of carbon dioxide.

Notice that sometimes bicarbonate salts react with acids while at other times they react with bases. The net ionic equations for these two reactions are as follows:

$$OH^- + HCO_3^- \longrightarrow CO_3^{-2} + H_2O$$

$$H^+ + HCO_3^- \longrightarrow H_2CO_3 \rightleftharpoons H_2O + CO_2\uparrow$$

Ions or compounds that react with both acids and bases are called *amphoteric substances*. This amphoteric behavior of bicarbonate ion is very important to its buffering action in the blood, as we will learn in the next chapter.

Unlike strong acids and strong bases, mixtures of equimolar aqueous solutions of weak acids and strong bases are not neutral. The reason for this will be examined in the next section.

9.10 AQUEOUS SOLUTIONS OF SALTS

We learned in Section 9.8 that the net ionic equation for the reaction of a strong acid and a strong base is $H^+ + OH^- \rightarrow H_2O$. This solution is neutral because the concentrations of hydrogen and hydroxide ions are equal. In contrast, the concentrations of hydrogen and hydroxide ions are not equal in the reaction of a strong base and a weak acid. Consider the reaction of equimolar amounts of acetic acid and sodium hydroxide as an example. The full and net ionic equations for this reaction are as follows:

$$C_2H_3O_2H + NaOH \rightleftharpoons C_2H_3O_2Na + H_2O \qquad (9\text{-}18)$$

$$C_2H_3O_2H + OH^- \rightleftharpoons C_2H_3O_2^- + H_2O \quad (9\text{-}19)$$

The net ionic equation (equation 9-19) is different from that of the reaction of a strong acid and a strong base. The difference is due to the fact that the weak acid, acetic acid, exists predominantly as un-ionized molecules in aqueous solution. Equations 9-18 and 9-19 represent an equilibrium reaction that lies far to the right. At equilibrium, there are small amounts of hydroxide ions, sodium ions, and un-ionized acetic acid in solution. As a result, the hydroxide ion concentration is larger than the hydrogen ion concentration and the solution is slightly basic.

The equilibrium reaction represented by equation 9-18 can be established also by dissolving sodium acetate in water. Sodium acetate will react with water to form the equilibrium concentrations of un-ionized acetic acid and sodium hydroxide. The same slightly basic solution can be made in either of two ways: by dissolving sodium acetate in water, or by mixing equimolar amounts of sodium hydroxide and acetic acid. This is a general rule. The same aqueous solution can be made by either dissolving a salt in water or by reacting equimolar amounts of the acid and base to form that particular salt.

We can classify salts according to the acid and base from which they are formed. Such a classification will help us decide whether an aqueous solution of a salt is acidic, basic, or neutral. For example, sodium chloride is classified as a salt of a strong acid (HCl) and a strong base (NaOH). Sodium acetate is classified as a salt of a weak acid (acetic acid) and a strong base (NaOH). We can determine which acid reacts with which base to form a particular salt by the following method. Simply add a hydroxide ion to the cation of the salt to determine the base. Add a proton to the anion of the salt to determine the acid. This method is shown in the following example.

Example 9-2 What acid and base react to form sodium sulfate (Na_2SO_4)?

Solution:

STEP 1. To determine the base, combine the cation of the salt, Na^+, with OH^-. The base is NaOH.

STEP 2. To determine the acid, combine the anion of the salt, SO_4^{-2}, with H^+. The acid is H_2SO_4.

STEP 3. Because NaOH is a strong base and H_2SO_4 is a strong acid, Na_2SO_4 is the salt of a strong acid and a strong base.

EXERCISE 9-13 Classify each of the following as either the salt of a strong acid and a strong base or the salt of a weak acid and a strong base.
(a) KCl (b) $NaNO_3$ (c) $C_2H_3O_2K$ (d) Na_2CO_3
(e) K_2SO_4 (f) NaBr (g) $Mg(ClO_4)_2$ (h) $CaCl_2$

We know that the reaction of a strong acid and a strong base forms a neutral solution. This same neutral solution can be made by dissolving the salt formed from this strong acid and strong base. In general, an aqueous solution of a salt of a strong acid and a strong base is neutral. We also know that an aqueous solution of sodium acetate is basic. Similarly, an aqueous solution of any salt of a weak acid and a strong base is basic.

EXERCISE 9-14 Will an aqueous solution of each of the salts in exercise 9-13 be basic or neutral?

So far, we have only identified compounds that are acids and bases and solutions that are basic, acidic, or neutral. We have said nothing about how we measure the acidity or basicity of these solutions. We will learn how to do this in the next chapter.

9.11 SUMMARY

Salts that dissolve in water exist predominantly as ions in aqueous solution. Reactions between ions occur whenever a gas, a precipitate, or an un-ionized molecule is formed as product. Ionic reactions are best represented by a balanced net ionic equation. Their advantage is that they include only the ions and molecules that actually take part in the reaction.

Many of the reactions of ions and molecules are reversible. Most reversible reactions can reach a state of equilibrium. The quantities of products and reactants present at equilibrium are related to each other by the equilibrium constant expression. This ratio defines the equilibrium constant of a particular reaction. The numerical value of an equilibrium depends on the reaction, the temperature, and the pressure. A large value of the equilibrium constant means that the products are favored. A small value of the equilibrium constant means that the reactants are favored. An equilibrium constant near one means that both the reactants and the products are present at equilibrium. Whenever an equilibrium is disturbed by an externally applied stress, the system changes to minimize this stress.

A particularly important equilibrium in aqueous solution is the ionization of water. In pure water, there exist small concentrations of hydrogen ions and hydroxide ions. The product of their concentrations is called the ion product of water. The concentrations of hydrogen and hydroxide ions are equal in pure water. In an acidic solution, the hydrogen ion concentration is larger than the hydroxide ion concentration; in a basic solution, the reverse is true.

Certain compounds, called acids, form hydrogen ions when dissolved in water. Other compounds, defined as bases by Arrhenius, form hydroxide ions when dissolved in water. Acids and bases can be strong or weak. Strong acids and bases exist predominantly as ions in solution; weak acids and bases exist predominantly as un-ionized molecules.

A neutralization reaction is the reaction of a strong base with a strong acid. The net ionic reaction for neutralization, $H^+ + OH^- \longrightarrow H_2O$, is the same for the reaction of all strong bases with all strong acids. All acids, whether strong or weak, react with strong bases to form a salt and water. The resulting solution may be acidic, basic, or neutral, depending on the salt formed. The salt of a strong acid and a strong base forms a neutral solution, whereas the salt of a weak acid and a strong base forms a basic solution. Thus, the same solution can be formed either by dissolving a salt in water or by mixing equimolar quantities of the acid and base that will react to form the same salt.

REVIEW OF TERMS AND CONCEPTS

Terms

ACIDIC HYDROGENS Hydrogen atoms in an acidic molecule, which are released in water to form hydrogen ions.

ACIDIC SOLUTION An aqueous solution in which the concentration of hydrogen ions is greater than the concentration of hydroxide ions.

AMPHOTERIC SUBSTANCE A compound or ion that reacts with both acids and bases.

BASIC SOLUTION An aqueous solution in which the concentration of hydroxide ions is greater than the concentration of hydrogen ions.

DIPROTIC ACID An acid that has two acidic protons.

EQUILIBRIUM The state reached when all spontaneous change stops.

EQUILIBRIUM CONSTANT The ratio of the equilibrium concentrations of products to the equilibrium concentrations of reactants.

INTERCELLULAR FLUID The fluid between cells.

ION PRODUCT CONSTANT OF WATER The product of the concentrations of hydrogen ions and hydroxide ions. Its symbol is K_w and at 25° C, it is equal to 1×10^{-14}.

IONIC REACTION A chemical reaction between ions or between ions and molecules.

IONIZATION The formation of ions by the breaking of a polar covalent bond in a molecule.

LE CHÂTELIER PRINCIPLE If a system at equilibrium is disturbed by an externally applied stress, the system changes in such a way that this external stress is minimized.

MONOPROTIC ACID An acid that has one acidic hydrogen.

NET IONIC EQUATION An equation that includes only the ions and un-ionized molecules actually involved in the chemical reaction.

NEUTRALIZATION The reaction between a strong base and a strong acid.

Concepts

1. Salts that dissolve in water exist as ions in solution.
2. All apparent changes in a reaction stop at equilibrium. Both the forward and reverse reactions actually continue, but the amount of product formed in the forward reaction is exactly balanced by the amount of product lost in the reverse reaction.
3. When a reaction at equilibrium is disturbed by a change in the concentration of any reactant, chemical reaction occurs to change the concentrations of reactants and products until their ratio again equals the equilibrium constant.
4. The product of the concentrations of hydrogen ions and hydroxide ions in water, K_w, is equal to 1×10^{-14}. An acidic solution contains a higher concentration of hydrogen ions than hydroxide ions. A basic solution contains a higher concentration of hydroxide ions than hydrogen ions.
5. An acid is a compound that forms hydrogen ions in aqueous solution. By the Arrhenius definition, a base is a compound that forms hydroxide ions in aqueous solution.
6. Any acid reacts with any base to form water and a salt. The resulting aqueous solution may be acidic, basic, or neutral, depending on the salt formed. A salt of a strong base and a strong acid forms a neutral solution; a basic solution is formed by dissolving a salt of a weak acid and a strong base in water.

EXERCISES 9-15 Which of the following salts are insoluble in water?
(a) $CaCl_2$ (b) $HgCl_2$ (c) $BaSO_4$ (d) PbS (e) $CaCO_3$

9-16 Write the balanced net ionic equation for the reaction of each of the following pairs of compounds. If no reaction occurs, state this.
(a) $Ca(NO_3)_2$, NaCl (b) $Ca(OH)_2$, H_2SO_4
(c) $KHCO_3$, HBr (d) Na_2S, $Pb(NO_3)_2$
(e) $CaCl_2$, NaOH (f) $AgNO_3$, LiCl (g) NaCl, LiBr
(h) $MgCl_2$, H_2SO_4 (i) $(C_2H_3O_2)_2Pb$, NaCl

(j) $Cu(NO_3)_2$, Li_2S (k) $Zn(NO_3)_2$, H_2SO_4
(l) $NaHCO_3$, HCl

9-17 Citric acid, $C_6H_8O_7$, is a weak acid present in citrus fruits. The equation for the ionization of citric acid is the following.

$$C_6H_8O_7 \rightleftharpoons C_6H_7O_7^- + H^+$$

Write the equilibrium constant expression for the ionization of citric acid.

9-18 The equation for the ionization of the weak acid HF is as follows:

$$HF \rightleftharpoons H^+ + F^-$$

Indicate the effect on the concentration of HF, H^+, and F^- at equilibrium of adding each of the following reagents:
(a) HF (b) NaF (c) H_3O^+ (d) OH^-

9-19 How does a hydronium ion differ from a proton?

9-20 Give one example each of a monoprotic acid and a diprotic acid.

9-21 Why is magnesium hydroxide classified as a strong base although its aqueous solution is only weakly basic?

9-22 Write the balanced net ionic equation for each of the following reactions:
(a) $NaOH + HClO_4$ (b) $KOH + H_2CO_3$
(c) $HCl + NaHCO_3$ (d) $KOH + NaHCO_3$
(e) $C_2H_3O_2H + KOH$ (f) $HCl + CaCO_3$
(g) $HNO_3 + NaHCO_3$ (h) $HClO_4 + K_2CO_3$

9-23 Classify each of the following as either the salt of a strong acid and a strong base or the salt of a weak acid and a strong base:
(a) $CaCO_3$ (b) $Mg(C_2H_3O_2)_2$ (c) KNO_3 (d) NaCN
(e) NaI (f) CaF_2

9-24 Decide whether an aqueous solution of each of the salts in exercise 9-23 is neutral or basic.

9-25 Give the chemical formulas of the acid and the base that must be mixed in equimolar quantities in water to form the same solution as that made by dissolving each of the following salts:
(a) $C_2H_3O_2Na$ (b) KNO_3 (c) KF (d) $NaHSO_4$

ACIDS AND BASES 10

Our bodies contain many different kinds of fluids. These fluids are solutions and colloidal dispersions whose properties we studied in Chapters 8 and 9. The concentrations of the various solutes in these solutions are delicately balanced in a healthy person. Particularly important is the balance between acids and bases in the blood. To understand how this balance is maintained, we must learn how to measure acidity, how to express it in logarithmic units, how to detect it, and finally, how to control it. Let us start by learning another way to express the concentrations of hydrogen and hydroxide ions in aqueous solutions.

10.1 THE pH SCALE

We used molar concentration units in Section 9.6 to express the concentration of hydrogen and hydroxide ions in water. Another way to state hydrogen ion concentration is pH. The pH of a solution is defined as follows:

$$pH = \log \frac{1}{[H^+]} = -\log [H^+] \qquad (10\text{-}1)$$

The symbol log stands for logarithm. The logarithm of a number is the power to which the number 10 must be raised to equal that number. For example, the number 1000 is equal to 10^3. The logarithm (log) of 1000 is

3; that is, 10 raised to the power 3 equals 1000. We can write equation 10-1 in the following form to show this relationship:

$$[H^+] = 1 \times 10^{-pH} \quad (10\text{-}2)$$

The pH of a solution is the negative power to which the number 10 must be raised to equal the molar concentration of hydrogen ions in the solution. For example, the hydrogen ion concentration of pure water at 25° C is 1×10^{-7} M. Its pH can be calculated as follows:

$$[H^+] = 1 \times 10^{-7}\ M = 1 \times 10^{-pH}$$

$$pH = 7$$

*EXERCISE 10-1 Calculate the pH corresponding to each of the following values of $[H^+]$:
(a) 1×10^{-4} (b) 1×10^{-2} (c) 1×10^{-6}
(d) 1×10^{-8} (e) 1×10^{-12}

The relationship between the hydrogen ion concentration of an aqueous solution and its pH is given in Table 10-1. The values of pH in Table 10-1 range from zero, for $[H^+] = 1$ M (1×10^0 M) to 14, for $[H^+] = 1 \times 10^{-14}$ M. Notice that the values of pH and molar concentrations of hydrogen ion are the inverse of each other: the higher the value of the pH of a solution, the lower its hydrogen ion concentration and vice versa. Thus, an acidic solution has a pH value less than 7, and a basic solution has a pH value greater than 7.

Most people are not familiar with logarithmic scales such as pH, and the use of such scales can be deceptive. For example, a change of one unit in pH from pH 1 to pH 2 may not seem like very much, but actually represents a change of a factor of ten in hydrogen ion concentration; a change of two pH units from pH 1 to pH 3 actually represents a 100-fold change in hydrogen ion concentration. This difference of one or two pH units is substantial. A solution with a pH of 1 will damage fabric and many metals, whereas a solution with a pH of 3 will cause little damage. Spilling a soft drink (pH of 2 to 4) on a shirt will not damage the fabric.

We learned in Section 9.6 that hydroxide and hydrogen ion concentra-

* The answers for the exercises in this chapter begin on page 864.

Table 10-1. Relationships Among pH, $[H^+]$, pOH, and $[OH^-]$

$[H^+]$ (M)	pH	pOH	$[OH^-]$ (M)	
$1 \times 10^0 = 1$	0	14	1×10^{-14}	↑
1×10^{-1}	1	13	1×10^{-13}	
1×10^{-2}	2	12	1×10^{-12}	Acidic solutions
1×10^{-3}	3	11	1×10^{-11}	
1×10^{-4}	4	10	1×10^{-10}	
1×10^{-5}	5	9	1×10^{-9}	
1×10^{-6}	6	8	1×10^{-8}	↓
1×10^{-7}	7	7	1×10^{-7}	Neutral solution
1×10^{-8}	8	6	1×10^{-6}	↑
1×10^{-9}	9	5	1×10^{-5}	
1×10^{-10}	10	4	1×10^{-4}	
1×10^{-11}	11	3	1×10^{-3}	Basic (alkaline) solutions
1×10^{-12}	12	2	1×10^{-2}	
1×10^{-13}	13	1	1×10^{-1}	
1×10^{-14}	14	0	$1 \times 10^0 = 1$	↓

tions are related by K_w, the ion product constant of water. If we know the molar concentration of either the hydroxide ion or the hydrogen ion, we can easily calculate the other. We can do the same with pH by defining the quantity pOH as follows:

$$[OH^-] = 1 \times 10^{-pOH} \qquad (10\text{-}3)$$

Substituting equations 10-2 and 10-3 into the expression for K_w, we obtain

$$(1 \times 10^{-pH})(1 \times 10^{-pOH}) = K_w = 1 \times 10^{-14}$$

Another way of writing this equation is the following:*

$$pH + pOH = 14$$

* This follows from the mathematical properties of exponents given in the Appendix (A.5).

Thus, if we know the pH of a solution, we can easily determine its pOH. For example, if the pH of a solution is 5, its pOH is calculated as follows:

$$pOH = 14 - pH = 14 - 5 = 9$$

The relationship between $[OH^-]$ and pOH is shown in Table 10-1.

The molar concentrations of all substances, and hydrogen ions in particular, are written in the following way:

$$[H^+] = n \times 10^m$$

The value of m is always a *whole* number that can be either positive or negative (e.g., 1, 5, or −10). Equation 10-2 can be used only if *the value of n is 1.* If n is any other number, equation 10-1 must be used to convert hydrogen ion concentration to pH. The use of equation 10-1 is shown in the following examples.

Example 10-1 Calculate the pH of a solution whose hydrogen ion concentration is 1.50×10^{-3}.

Solution:

STEP 1. According to equation 10-1

$$pH = \log \frac{1}{[H^+]}$$

STEP 2. Insert $[H^+]$ into equation 10-1:

$$pH = \log \frac{1}{1.50 \times 10^{-3}}$$

STEP 3. Carry out the calculation:*

$$pH = \log \frac{1}{1.50 \times 10^{-3}} = \log (6.66 \times 10^2)$$

$$pH = \log 6.66 + \log 10^2$$

$$pH = 0.82 + 2.00 = 2.82$$

* The use of logarithms is explained in the Appendix (A.6).

Frequently, we want to convert pH to hydrogen ion concentration. We can still use equation 10-1 for this purpose, as shown in the following example.

Example 10-2 Calculate the hydrogen ion concentration of a solution whose pH is 5.25.

Solution:

STEP 1. Rearrange equation 10-1 as follows:

$$pH = -\log [H^+]$$

$$\log [H^+] = -pH$$

STEP 2. Insert value of pH into this rearranged equation:

$$\log [H^+] = -5.25$$

STEP 3. Rewrite -5.25 as $-6 + 0.75$

$$\log [H^+] = -6 + 0.75$$

STEP 4. Take antilogs. From a log table, the antilog of 0.75 is 5.62. The antilog of -6 is 10^{-6}. Therefore

$$[H^+] = 5.62 \times 10^{-6}$$

EXERCISE 10-2 Calculate the pH of solutions with the following $[H^+]$ values: (a) 5.75×10^{-4} (b) 1.23×10^{-5} (c) 7.93×10^{-7} (d) 3.45×10^{-8} (e) 8.63×10^{-10}

EXERCISE 10-3 The pH of urine ranges from 4.9 to 8.3. What hydrogen ion concentrations correspond to this range of pH?

Acidic and basic solutions are common in the world. Many of them are body fluids; others are products that we use every day. A few of them were mentioned in Section 9.7. We can now specify the acidity of many of these solutions, as shown in Table 10-2. The various body fluids differ both in their acidity and in their range of acidities. Stomach acid is the most acidic, and it has a fairly narrow pH range (1.0 to 3.0). Blood plasma is slightly basic, and it has a very narrow pH range. If the pH of blood plasma changes to a value outside this range, the ability of the blood to transport oxygen is reduced. Therefore, maintaining the pH of blood plasma within a narrow range is important to human life. We will learn one way that the body

Table 10-2. pH Values of Some Common Solutions

Substance	pH	
Lime juice	2.0	
Stomach acid	1.0–3.0	
Soft drinks	2.0–4.0	
Wine	3.0–4.0	Acidic solutions
Tomato juice	4.0	
Beer	4.0–5.0	
Black coffee	5.0	
Milk	6.2–6.6	
Urine	4.8–7.0	
Pure water	7.0	Neutral
Urine	7.0–8.4	
Blood	7.35–7.45	
Baking soda	8.0	
Borax solution	9.0	Basic (alkaline) solutions
Milk of magnesia	10.0	
Household ammonia	12.0	
0.1 M lye (NaOH) solution	13.0	

maintains the pH of its blood plasma in Section 10.8. In contrast to blood plasma, urine has a wide pH range. It can be acidic, basic, or neutral. This wide range of the pH of urine is due to the fact that many acidic and basic substances are removed from the body through the urine to help maintain the pH of blood plasma.

A list of the pH values of various solutions given in Table 10-2 indicates that we can measure pH. How we do this is the subject of the next section.

10.2 MEASURING pH

The easiest, most common, and most accurate method of measuring pH in the laboratory is by means of an instrument called a *pH meter.* The pH of a solution is measured by a pair of special electrodes placed in the solution. The pH is read directly from the digital readout on the instrument. A commercially manufactured pH meter is shown in Figure 10-1.

The acidity of a solution can also be determined by an *indicator.* An indicator is an organic compound that changes color within a characteristic

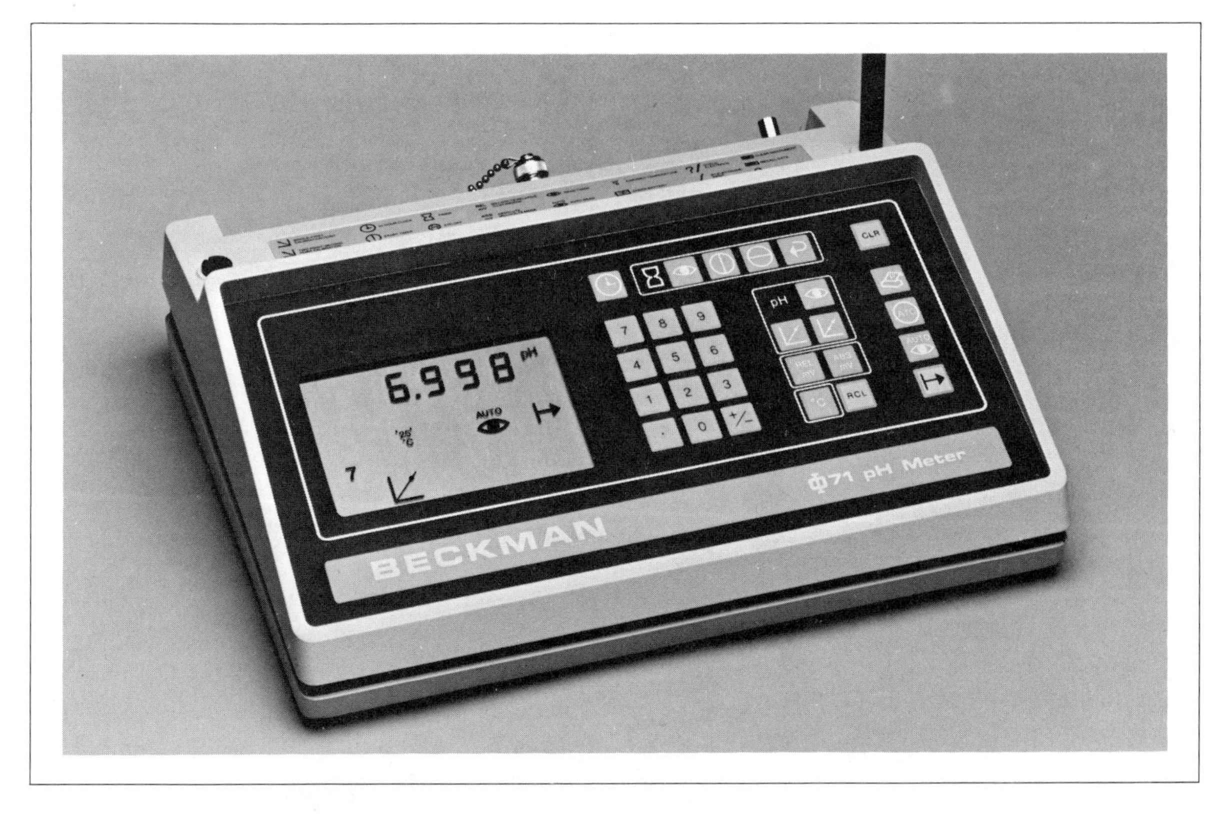

Fig. 10-1. A pH meter. Photograph courtesy of Beckman Instruments, Inc.

pH range. A well-known example is litmus. The color of litmus is red at pH values less than 5, but it is blue at pH values greater than 8.5. Litmus can be used in two ways. First, it can be absorbed by a piece of porous paper called litmus paper. When a drop of a solution is placed on the litmus paper, the color of the paper indicates whether the solution is acidic or basic. The litmus can also be added directly to the solution. The color of the solution indicates whether the solution is acidic or basic.

The pH meter is the more accurate of the two ways of measuring the pH of a solution. Indicators are most often used in acid-base titrations, as we will learn next.

10.3 ACID-BASE TITRATIONS

A titration is a method of determining the amount of acid or base in a solution. This method is based on the chemical reaction between an acid

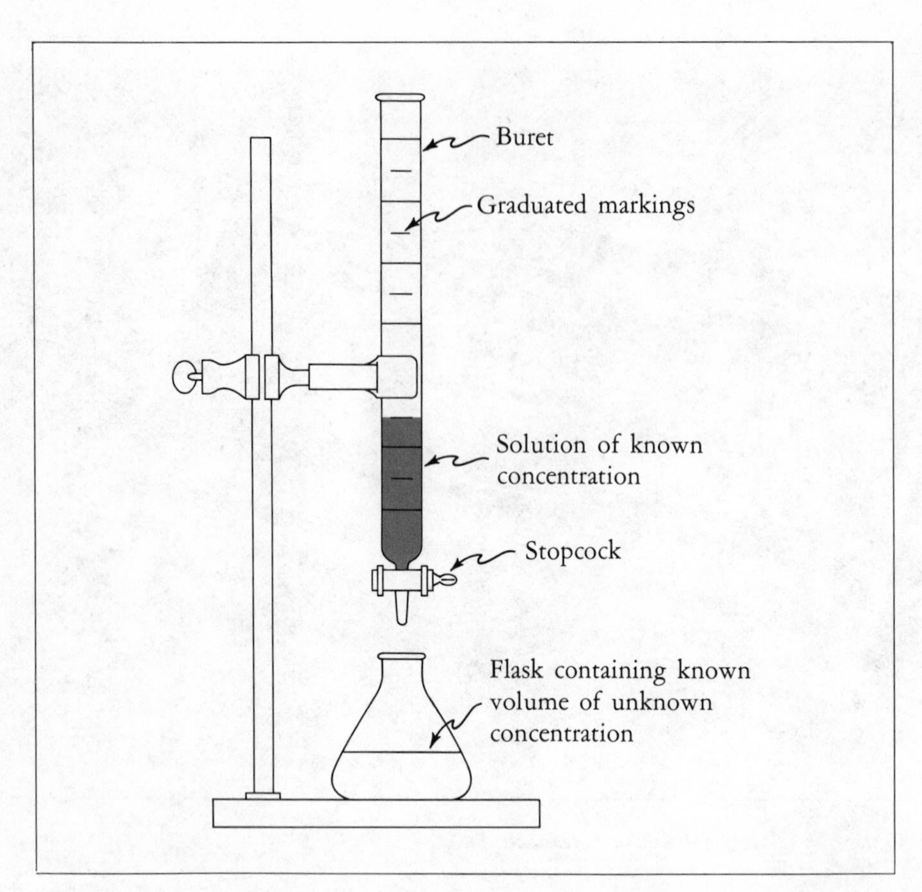

Fig. 10-2. Titration apparatus.

and a base. For example, if we want to know the amount of base in a certain volume of solution, we measure the amount of an acid of known concentration needed to react completely with the base. Conversely, if we have another solution and want to know the amount of acid it contains, we measure the amount of base needed to react with all the acid.

We carry out an acid-base titration with the apparatus shown in Figure 10-2. The apparatus consists of a buret filled with a solution (acid or base) of known concentration and a flask below the buret. A buret is a glass tube that has graduated markings to allow us to measure exactly the amount of solution added. The flow of liquid from the buret is controlled by turning the stopcock. We wish to determine the concentration of the acid or base contained in the flask.

Let us follow the steps in determining the concentration of a sodium hydroxide solution using this apparatus. First, we place a carefully measured volume of the sodium hydroxide solution into the flask. Next, we add an indicator to the solution in the flask. Then we place a solution of hydrochloric acid of known concentration into the buret. Finally, we add the acid solution by drops from the buret into the flask. We add acid just until the color of the indicator changes. The indicator changes color when the acid has reacted with all of the base in the solution. This tells us that we are at the end of the titration. This color change is called the *endpoint* or *equivalence point*.

We can appreciate the importance of the equivalence point by referring to the balanced chemical equation of the acid-base reaction. It is the following for the reaction of sodium hydroxide and hydrochloric acid, as we learned in Section 9.8:

$$NaOH + HCl \longrightarrow NaCl + H_2O$$

The molar ratio of sodium hydroxide to hydrochloric acid is 1:1 at the equivalence point. The number of moles of hydrochloric acid needed to react with the base can be calculated from the concentration of the hydrochloric acid and volume added. This is also equal to the number of moles of sodium hydroxide in solution. Because we know the volume of the basic solution placed in the flask, we can calculate its concentration. These calculations are similar to those we learned to do in Section 8.3. They are reviewed in the following example.

Example 10-3 It takes 35.2 mL of a 0.100 M HCl solution to neutralize exactly 25.0 mL of a NaOH solution. What is the concentration of this NaOH solution?

Solution:

STEP 1. Calculate the number of moles of acid used (see Section 8.3):

$$\text{moles HCl} = 0.0352\,\cancel{L} \times 0.100\,\frac{\text{mol}}{\cancel{L}} = 0.00352\ \text{mol}$$

STEP 2. At the equivalence point, mol HCl = mol NaOH:

$$0.00352\ \text{mol HCl} = 0.00352\ \text{mol NaOH}$$

STEP 3. M = mol/L

$$\text{M of NaOH} = \frac{0.00352\ \text{mol NaOH}}{0.025\ \text{L}} = 0.141\ \text{M}$$

EXERCISE 10-4 A 0.250 M HCl solution is used to titrate 50.0 mL each of three solutions of NaOH. Solution A requires 62.0 mL, Solution B requires 75.2 mL, and Solution C requires 125 mL. What is the concentration of NaOH in each of the solutions?

EXERCISE 10-5 How many mL of a 0.100 M HCl solution are needed to neutralize exactly 50 mL of a 0.0500 M NaOH solution?

The concentrations of acids and bases used for titrations are not usually expressed as molar concentrations. Instead, a unit called normality is used.

10.4 NORMALITY

We learned in Section 9.8 that 1 mole of hydrochloric acid neutralizes 1 mole of sodium hydroxide. We also know that 1 mole of sulfuric acid neutralizes *2* moles of sodium hydroxide. Therefore, 1 mole of sulfuric acid is equivalent to 2 moles of hydrochloric acid in an acid-base reaction. Because of this fact, we can define the gram-equivalent weight of an acid as follows:

The weight of an acid, in grams, that supplies 1 mole of hydrogen ions is defined as the gram-equivalent weight of that acid.

A similar definition can be given for bases:

The weight of a base, in grams, that supplies 1 mole of hydroxide ions or reacts with 1 mole of hydrogen ions is defined as the gram-equivalent weight of that base.

The gram-equivalent weight of an acid takes into account the fact that a diprotic acid supplies twice as many hydrogen ions per mole of acid as does a monoprotic acid. Thus, to determine the gram-equivalent weight of an acid or a base, we can use the following equations:

$$\text{Gram-equivalent weight of an acid} = \frac{\text{gram-molecular weight of acid}}{\text{number of acidic hydrogens per molecule}} \quad (10\text{-}4)$$

$$\text{Gram-equivalent weight of a base} = \frac{\text{gram-molecular weight of base}}{\text{number of hydroxide ions per molecule}} \qquad (10\text{-}5)$$

The use of these equations to determine the gram-equivalent weights of acids and bases is shown in the following examples.

Example 10-4 Determine the gram-equivalent weight of H_2CO_3.
Solution:
STEP 1. Determine the gram-molecular weight of H_2CO_3 (see Section 7.3):

Molecular weight = 62.0 amu

Gram-molecular weight = 62.0 g

STEP 2. Determine the number of acidic hydrogens in H_2CO_3. Note that H_2CO_3 is a diprotic acid; therefore, it contains two acidic hydrogens.
STEP 3. Substitute values from steps 1 and 2 into equation 10-4:

$$\text{Gram-equivalent weight of } H_2CO_3 = \frac{62.0\text{ g}}{2} = 31.0\text{ g}$$

Example 10-5 Determine the gram-equivalent weight of $Ca(OH)_2$.
Solution:
STEP 1. Molecular weight = 74.1 amu

Gram-molecular weight = 74.1 g

STEP 2. Two hydroxide ions per molecule of $Ca(OH)_2$.
STEP 3.

$$\text{Gram-equivalent weight of } Ca(OH)_2 = \frac{74.1\text{ g}}{2} = 37.0\text{ g}$$

EXERCISE 10-6 Calculate the gram-equivalent weight of each of the following compounds:
(a) HCl (b) NaOH (c) $Ba(OH)_2$ (d) H_2SO_4
(e) $Mg(OH)_2$ (f) H_3BO_3

Notice that the gram-equivalent weight of a monoprotic acid is the same as its gram-molecular weight. The gram-equivalent weight of a diprotic acid is half its gram-molecular weight.

The ratio of the gram-molecular weight of an acid or base to its gram-equivalent weight is the number of equivalents per mole. This concept of equivalents, introduced in Section 8.3, can be expressed by the following equation:

$$\text{Number of equivalents of acid or base per mole of acid or base} = \frac{\text{gram-molecular weight of acid or base}}{\text{gram-equivalent weight of acid or base}}$$

It follows that 1 mole of a monoprotic acid contains 1 equivalent of acid, whereas 1 mole of a diprotic acid contains 2 equivalents of acid.

EXERCISE 10-7 How many equivalents are there in 1 gram-molecular weight of each compound in exercise 10-6?

Frequently we must determine the number of equivalents in a weight of an acid or base. This can be done using the following equation:

$$\text{Number of equivalents in a sample} = \frac{\text{weight of sample, in grams}}{\text{gram-equivalent weight of acid or base}} \qquad (10\text{-}6)$$

This equation is very similar to the one given in Section 7.3 to determine the number of moles in a given amount of a compound. The use of equation 10-6 to determine the number of equivalents in a given weight of acid or base is shown in the following examples.

Example 10-6 How many equivalents are there in 24.0 g of H_2SO_4?

Solution:

STEP 1. Determine the gram-equivalent weight of H_2SO_4.

$$\text{Gram-molecular weight of } H_2SO_4 = 98.1 \text{ g}$$

$$\text{Gram-equivalent weight} = \frac{98.1 \text{ g}}{2} = 49.0 \text{ g/Eq } H_2SO_4$$

STEP 2. Substitute values into equation 10-6:

$$\text{Number of equivalents} = \frac{24.0 \not{g}}{49.0 \not{g}/\text{Eq } H_2SO_4} = 0.490 \text{ Eq } H_2SO_4$$

Example 10-7 How many equivalents are there in 52.0 g of $Al(OH)_3$?
Solution:
STEP 1. Gram-equivalent weight $= 26.0\ g/Eq\ Al(OH)_3$
STEP 2.

$$\text{Number of equivalents} = \frac{52.0\ \cancel{g}}{26.0\ \cancel{g}/Eq\ Al(OH)_3} = 2.00\ Eq\ Al(OH)_3$$

EXERCISE 10-8 How many equivalents are there in the specified quantity of each of the following acids or bases?
(a) 25.3 g HI (b) 62.5 g H_3BO_3 (c) 5.79 g NaOH
(d) 50.0 g $Mg(OH)_2$

The important point to realize about the idea of equivalents is that any number of equivalents of any base reacts with *exactly* the same number of equivalents of any acid. For example, 1.32 Eq of sodium hydroxide reacts with 1.32 Eq of boric acid. In this way, we do not have to worry about a balanced equation. The definition of equivalents takes care of this detail.

EXERCISE 10-9 How many equivalents of each of the following bases will react with 2.75 Eq of HCl?
(a) NaOH (b) $Ca(OH)_2$ (c) $Mg(OH)_2$ (d) $Al(OH)_3$

The concept of equivalents can be extended to solutions by expressing the amount of acid or base present in a liter of solution in terms of equivalents. *Normality* is defined as the number of equivalents of acid or base per liter of solution and is designated by the capital letter N. This definition is given in the form of an equation as follows:

$$N = \frac{\text{number of equivalents of acid or base}}{\text{volume of solution, in liters}} \qquad (10\text{-}7)$$

The use of normality to express the concentration of a base is shown in the following example.

Example 10-8 What is the normality of a solution made by adding enough water to 20.0 g of NaOH to make 500 mL of solution?
Solution:
STEP 1. Determine the gram-equivalent weight of NaOH.

Gram-molecular weight $= 40.0$ g

Gram-equivalent weight = 40.0 g/1 = 40.0 g/Eq NaOH

STEP 2. Determine the number of equivalents in 20.0 g of NaOH:

$$\text{Number of equivalents} = \frac{20.0\ \cancel{g}}{40.0\ \cancel{g}/\text{Eq}} = 0.500\ \text{Eq}$$

STEP 3. Substitute values into equation 10-7:

$$\text{N} = \frac{0.500\ \text{Eq}}{0.500\ \text{L}} = 1.00\ \text{N}$$

EXERCISE 10-10 Determine the normality of each of the following solutions:
(a) 45.0 g of KOH made up to 1.00 L with water
(b) 137 g of HCl made up to 1.50 L with water
(c) 84.3 g of H_2SO_4 made up to 500 mL with water
(d) 94.2 g of H_3BO_3 made up to 750 mL with water

Normality is most useful as a unit of concentration for the calculations involved in a titration. At the equivalence point of an acid-base titration, we know that

$$\text{Number of equivalents of base} = \text{number of equivalents of acid} \qquad (10\text{-}8)$$

Equation 10-7 can be rearranged to give the following equation:

$$\text{Number of equivalents of acid or base} = \text{normality of acid or base} \times \text{volume of solution, in liters} \qquad (10\text{-}9)$$

Substituting equation 10-9 into equation 10-8, we obtain

$$\text{Normality of base} \times \text{volume of base} = \text{normality of acid} \times \text{volume of acid}$$

This equation can be abbreviated further:

$$\text{N}_{\text{base}} \times \text{V}_{\text{base}} = \text{N}_{\text{acid}} \times \text{V}_{\text{acid}} \qquad (10\text{-}10)$$

We can use this equation to determine the concentration of an acidic or basic solution by titration, as shown in the following example.

Example 10-9 It takes 52.5 mL of a 0.200 N NaOH solution to neutralize 25.0 mL of an H_2SO_4 solution. What is the normality of the H_2SO_4 solution?

Solution:

STEP 1. Match the quantities in equation 10-10 with those given:

$$N_{base} = 0.200\ N \qquad V_{base} = 52.5\ mL$$

$$N_{acid} = \text{unknown} \qquad V_{acid} = 25.0\ mL$$

STEP 2. Substitute the quantities into equation 10-10.

$$0.200\ N \times 52.5\ mL = N_{acid} \times 25\ mL$$

STEP 3. Solve for N_{acid}:

$$N_{acid} = 0.200\ N \times \frac{52.5\ \cancel{mL}}{25.0\ \cancel{mL}} = 0.420\ N$$

Notice that the volumes in equation 10-10 do not need to be expressed in liters. Any unit of volume can be used provided that both volumes are measured in the same unit. Because burets are usually calibrated in milliliters, most calculations use this unit, as shown in the preceding example.

EXERCISE 10-11 The following volumes of a 0.250 N solution of H_2SO_4 are needed to neutralize 50.0 mL of various basic solutions. What is the normality of each basic solution?
(a) 43.5 mL (b) 22.4 mL (c) 35.2 mL

The advantage of normality as a unit of concentration of acids and bases is that we know directly the volume of solution that supplies 1 mole of hydrogen ions or hydroxide ions. We do not need to worry about the balanced equation for the acid-base reaction. This makes acid-base titration calculations easier and more convenient. Normality is widely used as the unit of concentration in acid-base titrations because of this fact.

The Arrhenius definition of acids and bases, given in Section 9.7, is satisfactory for most acids and bases used in titrations. Other compounds can be classified as acids or bases by using the more general Brønsted definition.

10.5 BRØNSTED ACIDS AND BASES

A more general definition of acids and bases was proposed by Johannes N. Brønsted in 1923. He defined an acid as any compound or ion that donates a proton and a base as any compound or ion that accepts a proton. All the substances that are acids according to the Arrhenius definition are also acids according to the Brønsted definition. For example, the Arrhenius acids HCl, H_2SO_4, and H_2CO_3 donate protons, so they are also Brønsted acids. All the Arrhenius bases are also Brønsted bases. Sodium hydroxide and potassium hydroxide, for example, are both proton acceptors and they are both Arrhenius and Brønsted bases.

The Brønsted definition is important because it allows us to classify as acids and bases many compounds and ions that do not fit the Arrhenius definition. Consider a monoprotic acid. It is a Brønsted acid. When it donates its proton, the remainder of the molecule is no longer an acid; it is now a base. Hydrogen chloride is an example of such a monoprotic acid. In water, it donates a proton to water to form a hydronium ion and a chloride ion. Hydrogen chloride is an acid according to both the Arrhenius and the Brønsted definitions. But the chloride ion that remains after the proton is donated is a base. Why is chloride ion a base? Because it fits the Brønsted definition. It can accept a proton to form hydrogen chloride. Notice that chloride ion is not a base according to the Arrhenius definition.

EXERCISE 10-12 Classify each of the following molecules or ions as a Brønsted acid or base:
(a) HI (b) F^- (c) H_2SO_4 (d) OH^- (e) CO_3^{-2}
(f) NO_3^- (g) ClO_4^-

When an acid donates a proton, the remainder of the molecule or ion is a base. These two parts of the same molecule are called a *conjugate acid-base* pair. To use hydrogen chloride again as an example, hydrogen chloride is the acid and chloride ion is its conjugate base. The two, HCl and Cl^-, are a conjugate acid-base pair. We say that hydrogen chloride is the conjugate acid of chloride ion.

There is another conjugate acid-base pair in an aqueous solution of hydrogen chloride. Hydrogen chloride donates a proton to water to form a hydronium ion. Because water accepts a proton, it is a Brønsted base. Furthermore, a hydronium ion can donate a proton, so it is a Brønsted acid. Therefore, water and hydronium ion are the second conjugate acid-base pair in the solution. The two conjugate acid-base pairs in an aqueous hydrogen chloride solution are shown in the following equation:

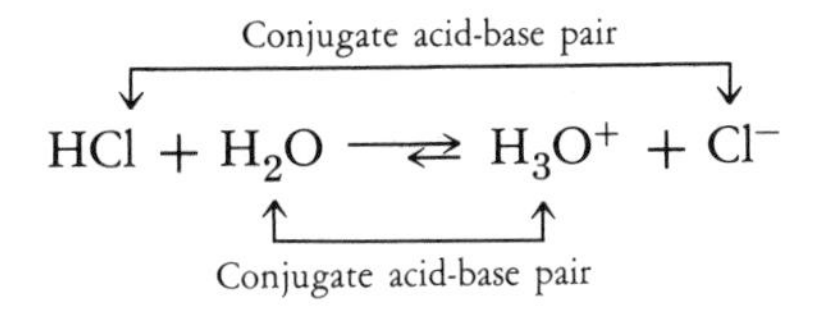

There is a simple relationship between the structure of an acid and its conjugate base. The structure of the conjugate base is that of the acid without its acidic hydrogen. For example, the conjugate base of hydrogen cyanide, HCN, is obtained by removing the acidic hydrogen. What is left, cyanide ion, CN^-, is the conjugate base of HCN. To obtain the structure of the conjugate acid of a base, just add a proton to the structure of the base. Thus, the conjugate acid of the base F^- is obtained by adding a proton to form HF.

EXERCISE 10-13 Write the structure of the conjugate acid or conjugate base of each of the following:
(a) OH^- (b) $HClO_4$ (c) Br^- (d) HI (e) NO_3^-
(f) HCO_3^-

EXERCISE 10-14 Identify the two conjugate acid-base pairs in each of the following equations:

(a) $HBr + H_2O \rightleftharpoons H_3O^+ + Br^-$

(b) $H_2SO_4 + H_2O \rightleftharpoons HSO_4^- + H_3O^+$

(c) $H_3O^+ + OH^- \rightleftharpoons H_2O + H_2O$

(d) $OH^- + C_2H_3O_2H \rightleftharpoons H_2O + C_2H_3O_2^-$

A particularly important Brønsted base is ammonia, NH_3. Ammonia reacts with a proton (or hydrogen ion) to form an ammonium ion according to the following equation:

Conjugate acid-base pair

$$NH_3 + H_3O^+ \rightleftharpoons H_2O + NH_4^+$$

Conjugate acid-base pair

Ammonia and ammonium ion are one conjugate acid-base pair; water and hydronium ion are the other. Ammonia is another molecule that does not fit the Arrhenius definition of a base. It reacts as a base because the nitrogen atom has an unshared pair of electrons (see Section 6.1). These electrons form the bond between the nitrogen and the proton, as shown in the following equation:

$$\mathrm{H{:}\overset{\displaystyle H}{\underset{\cdot\cdot}{\overset{\cdot\cdot}{N}}}{:}H + H{:}\overset{+}{\underset{\displaystyle (H)}{\overset{\cdot\cdot}{O}}}{:}H \rightleftharpoons H{:}\overset{\displaystyle H^{+}}{\underset{\displaystyle H}{N}}{:}H + H{:}\overset{\cdot\cdot}{\underset{\cdot\cdot}{O}}{:}H}$$

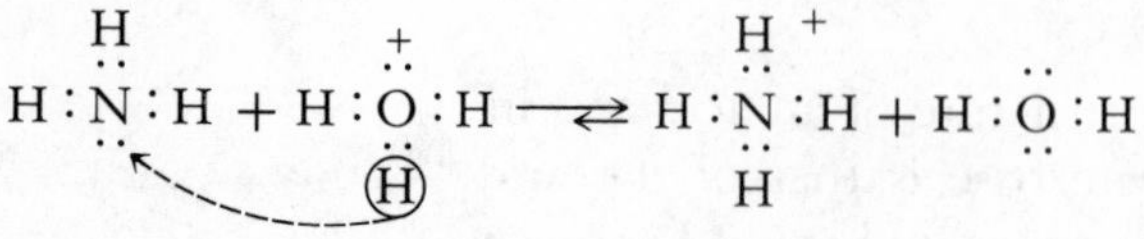

Many compounds in living systems have groups that are ammonia-like; others act like ammonium ions. These groups accept and donate hydrogen ions to control the acidity of body fluids. We will learn about these compounds called amines in Chapter 17.

Certain ions and molecules can be both a Brønsted acid and a Brønsted base. To see an example, let us re-examine the ionization of water (Section 9.6). In the process of ionization, one water molecule donates a proton to another molecule of water to form a hydronium ion and a hydroxide ion. One water molecule acts as an acid and the other acts as a base, as shown in the following equation:

Conjugate acid-base pair

$$\underset{\text{Acid}_1}{H_2O} + \underset{\text{Base}_2}{H_2O} \rightleftharpoons \underset{\text{Acid}_2}{H_3O^+} + \underset{\text{Base}_1}{OH^-}$$

Conjugate acid-base pair

Water acts as both a Brønsted acid and a Brønsted base in this reaction.

Many compounds and ions show such amphoteric behavior. The bicarbonate ion is an example of such an ion. It reacts as a base according to the following reaction:

Conjugate acid-base pair

$$\underset{\text{Base}_1}{HCO_3^-} + \underset{\text{Acid}_2}{H_3O^+} \rightleftharpoons \underset{\text{Base}_2}{H_2O} + \underset{\text{Acid}_1}{H_2CO_3}$$

Conjugate acid-base pair

It also reacts as an acid, according to the following reaction:

Conjugate acid-base pair

$$\underset{\text{Acid}_1}{HCO_3^-} + \underset{\text{Base}_2}{OH^-} \rightleftharpoons \underset{\text{Acid}_2}{H_2O} + \underset{\text{Base}_1}{CO_3^{-2}}$$

Conjugate acid-base pair

Other ions formed by the loss of one proton from a diprotic acid also show this behavior.

EXERCISE 10-15 Write the equations for the reactions of each of the following ions as both an acid and a base. Identify the conjugate acid-base pairs in each reaction.
(a) HSO_4^- (b) HSO_3^- (c) $H_2PO_4^-$

The compounds or ions involved in acid-base reactions can be divided into two conjugate acid-base pairs. This means that there are two acids and two bases in solution. For example, in an aqueous solution of hydrogen chloride, the two bases are chloride ion and water. We can regard these acid-base reactions as a competition between two bases for the proton. Thus, chloride ions and water compete for protons in the aqueous hydrogen chloride solution. Who wins? We can answer this question only by carrying out an experiment. The result of this experiment was given in Section 9.7, where we learned that hydrogen chloride is almost completely ionized in water. This means that water molecules capture most of the protons, and the equilibrium in the following equation lies far to the right:

$$H_2O + HCl \rightleftharpoons H_3O^+ + Cl^-$$

Therefore, we conclude that water is a stronger base than is chloride ion.

From experimental observations, we can arrive at the list of the relative strengths of Brønsted acids and bases given in Table 10-3. A strong acid is defined as a compound that readily donates a proton; a strong base readily accepts a proton. In contrast, a weak acid is a compound that holds its proton tightly, and a weak base reluctantly accepts a proton. The relative strengths of the acids and bases in Table 10-3 are arranged to show an important relationship. The strongest acid is located in the upper left corner. Its conjugate base turns out to be the weakest base! This is a general rule. *The stronger the acid, the weaker is its conjugate base. The weaker an acid, the stronger is its conjugate base.*

Table 10-3. Relative Strengths of Some Brønsted Acids and Their Conjugate Bases

Acid		Base	
Chemical Formula	Name	Chemical Formula	Name
$HClO_4$	Perchloric acid	ClO_4^-	Perchlorate ion
HBr	Hydrobromic acid	Br^-	Bromide ion
HCl	Hydrochloric acid	Cl^-	Chloride ion
H_2SO_4	Sulfuric acid	HSO_4^-	Bisulfate ion
HNO_3	Nitric acid	NO_3^-	Nitrate ion
H_3O^+	Hydronium ion	H_2O	Water
HSO_4^-	Bisulfate ion	SO_4^{-2}	Sulfate ion
H_3PO_4	Phosphoric acid	$H_2PO_4^-$	Dihydrogen phosphate ion
$C_2H_3O_2H$	Acetic acid	$C_2H_3O_2^-$	Acetate ion
H_2CO_3	Carbonic acid	HCO_3^-	Bicarbonate ion
NH_4^+	Ammonium ion	NH_3	Ammonia
HCO_3^-	Bicarbonate ion	CO_3^{-2}	Carbonate ion
H_2O	Water	OH^-	Hydroxide ion

Increasing Acid Strength ↑ (acids, from bottom to top); Increasing Base Strength ↓ (bases, from top to bottom)

Why are we interested in the relative strengths of Brønsted acids and bases? The reason is that acid-base reactions tend to produce the weaker of the possible acids and bases. We can predict the position of an acid-base equilibrium if we know the relative strengths of the acids and bases involved. Consider as an example the following reaction:

$$NH_3 + H_2O \rightleftharpoons NH_4^+ + OH^-$$

Does the equilibrium lie to the left or to the right? According to the information in Table 10-3, ammonium ion is a stronger acid than is water, and hydroxide ion is a stronger base than is ammonia. Because the weaker acid and base, ammonia and water, tend to form, we predict that the equilibrium lies to the left. From experimental observations, we know that this conclusion is correct. By simply memorizing the list of strong acids and strong bases given in Tables 9-5 and 10-3, it becomes possible to predict the position of the equilibrium in many acid-base reactions.

EXERCISE 10-16 Predict the position of the equilibrium in each of the following reactions:

(a) $C_2H_3O_2H + H_2O \rightleftharpoons H_3O^+ + C_2H_3O_2^-$

(b) $HCO_3^- + H_2O \rightleftharpoons CO_3^{-2} + H_3O^+$

(c) $C_2H_3O_2H + OH^- \rightleftharpoons H_2O + C_2H_3O_2^-$

(d) $HCN + H_2O \rightleftharpoons H_3O^+ + CN^-$

The information in Table 10-3 is only qualitative. Ionization constants are a way of quantitatively expressing the strengths of acids and bases.

10.6 IONIZATION CONSTANTS OF ACIDS AND BASES

We learned in Section 9.7 that strong acids are more completely ionized in solution than are weak acids. The degree of ionization of any acid is given by its ionization constant, K_a. The equilibrium constant for the ionization of an acid in water is defined as its ionization constant. This reaction for acetic acid is the following:

$$\underset{\text{Acetic acid}}{C_2H_3O_2H} + H_2O \rightleftharpoons H_3O^+ + \underset{\text{Acetate ion}}{C_2H_3O_2^-} \tag{10-11}$$

The equilibrium constant expression for equation 10-11 is

$$K = \frac{[H_3O^+][C_2H_3O_2^-]}{[H_2O][C_2H_3O_2H]} \tag{10-12}$$

We learned in Section 9.7 that $[H_3O^+]$ and $[H^+]$ are merely different symbols for a hydrogen ion in aqueous solution. Furthermore, the concentration of water is virtually constant. Therefore, we can rewrite equation 10-12 as follows:

$$K[H_2O] = K_a = \frac{[H^+][C_2H_3O_2^-]}{[C_2H_3O_2H]}$$

This is the ionization constant expression for acetic acid.

Diprotic acids ionize in two steps. Such acids have two ionization constants, one for each step. For example, the two equations and their ionization constant expressions for the ionization of carbonic acid are:

$$H_2CO_3 + H_2O \rightleftharpoons H_3O^+ + HCO_3^- \qquad K_{a_1} = \frac{[H^+][HCO_3^-]}{[H_2CO_3]}$$

$$HCO_3^- + H_2O \rightleftharpoons H_3O^+ + CO_3^{-2} \qquad K_{a_2} = \frac{[H^+][CO_3^{-2}]}{[HCO_3^-]}$$

EXERCISE 10-17 Write the equations and the ionization constant expression for the ionization of each of the following acids:
(a) HCN (b) HF (c) H_2SO_4 (d) H_2SO_3 (e) H_3PO_4

The values of K_a for several acids and their ionization reactions are given in Table 10-4. The acids that were classified as strong in Sections 9.7 and 10.5 all have large values of K_a. Weak acids have small values of K_a. Notice that the second ionization constant of a diprotic acid is smaller than the first. The third ionization constant of phosphoric acid (H_3PO_4) is even smaller. The value of K_{a_3} for HPO_4^{-2} is so small that almost no ionization occurs in water. Thus, the concentration of phosphate ion (PO_4^{-3}) is very low in an aqueous solution of phosphoric acid.

Table 10-4. Ionization Constants, K_a, of Acids at 25° C

Acid	Ionization Reaction	K_a	pK_a
HCl	$HCl \rightleftharpoons H^+ + Cl^-$	1×10^3	−3
H_2SO_4	$H_2SO_4 \rightleftharpoons H^+ + HSO_4^-$	1×10^2	−2
HSO_4^-	$HSO_4^- \rightleftharpoons H^+ + SO_4^{-2}$	1.20×10^{-2}	1.92
H_3PO_4	$H_3PO_4 \rightleftharpoons H^+ + H_2PO_4^-$	7.52×10^{-3}	2.12
$H_2PO_4^-$	$H_2PO_4^- \rightleftharpoons H^+ + HPO_4^{-2}$	6.23×10^{-8}	7.21
HPO_4^{-2}	$HPO_4^{-2} \rightleftharpoons H^+ + PO_4^{-3}$	2.2×10^{-13}	12.6
HF	$HF \rightleftharpoons H^+ + F^-$	3.53×10^{-4}	3.45
$C_2H_3O_2H$	$C_2H_3O_2H \rightleftharpoons H^+ + C_2H_3O_2^-$	1.75×10^{-5}	4.75
H_2CO_3	$H_2CO_3 \rightleftharpoons H^+ + HCO_3^-$	4.30×10^{-7}	6.36
HCO_3^-	$HCO_3^- \rightleftharpoons H^+ + CO_3^{-2}$	5.61×10^{-11}	10.3
HCN	$HCN \rightleftharpoons H^+ + CN^-$	4.93×10^{-10}	9.31

We learned in Section 10.1 that pH is another way of expressing hydrogen ion concentration. Ionization constants can also be expressed on a logarithmic scale. These values are called pK_a and they are defined as follows:

$$pK_a = -\log K_a$$

As with pH, pK_a and K_a are the inverse of each other: the larger the value of pK_a, the weaker the acid, and vice versa. The values of pK_a are included in Table 10-4.

The values of K_a and pK_a in Table 10-4 allow us to compare the strengths of acids as shown in the following example.

Example 10-10 Which is the stronger acid, HCN or HF?

Solution:

STEP 1. Find the values of K_a for each acid in Table 10-4.

For HF, $K_a = 3.53 \times 10^{-4}$

For HCN, $K_a = 4.93 \times 10^{-10}$

STEP 2. Decide which value of K_a is smaller.

4.93×10^{-10} is smaller than 3.53×10^{-4}

Therefore, HCN is the weaker acid.

EXERCISE 10-18 Which is the stronger acid in each of the following pairs of acids? (Consider only the first ionization constant for diprotic acids.)

(a) HCl, HF (b) H_2SO_4, H_2SO_3 (c) $C_2H_3O_2H$, H_2CO_3

We can get an idea of the hydrogen ion concentration in an aqueous solution of an acid from its ionization constant. A small value of K_a means that most of the molecules of the acid are un-ionized and the hydrogen ion concentration in solution is small. For example, the hydrogen ion concentration of a 1 M solution of acetic acid is approximately 10^{-3} M, a weakly acidic solution. Hydrochloric acid, on the other hand, has a large value of K_a and is completely ionized. The hydrogen ion concentration of a 1 M solution of hydrochloric acid is 1 M, a strongly acidic solution. Although there is a great difference in the pH of the two solutions, we must remember that both will react with 1 equivalent of base in an acid-base titration. Therefore, we must be careful to distinguish between the hydrogen ion concentration in an aqueous solution of an acid and the total acid present, which includes both hydrogen ions and un-ionized molecules.

The strengths of bases are expressed as base ionization constants, K_b. The equilibrium reaction and the equilibrium constant expression for the weak base ammonia in water are as follows:

$$NH_3 + H_2O \rightleftharpoons NH_4^+ + OH^-$$

$$K = \frac{[NH_4^+][OH^-]}{[H_2O][NH_3]}$$

Again, the concentration of water is constant in the solution, and we can write

$$K[H_2O] = K_b = \frac{[NH_4^+][OH^-]}{[NH_3]}$$

Values of K_b are given for several bases in Table 10-5. Weak bases have small values of K_b and strong bases have large values of K_b. For example, phosphate ion is a relatively strong base, whereas bicarbonate ion is a weak one.

The values of K_b can be expressed on a logarithmic scale. Thus, pK_b is defined as follows:

$$pK_b = -\log K_b$$

Table 10-5. Ionization Constants, K_b, of Some Bases at 25° C

Base	Ionization Reaction	K_b	pK_b
PO_4^{-3}	$PO_4^{-3} + H_2O \rightleftharpoons HPO_4^{-2} + OH^-$	4.52×10^{-2}	1.34
$Zn(OH)_2$	$Zn(OH)_2 \rightleftharpoons Zn(OH)^+ + OH^-$	9.62×10^{-4}	3.02
CO_3^{-2}	$CO_3^{-2} + H_2O \rightleftharpoons HCO_3^- + OH^-$	1.84×10^{-4}	3.74
AgOH	$AgOH \rightleftharpoons Ag^+ + OH^-$	1.10×10^{-4}	3.96
NH_3	$NH_3 + H_2O \rightleftharpoons NH_4^+ + OH^-$	1.77×10^{-5}	4.75
HCO_3^-	$HCO_3^- + H_2O \rightleftharpoons H_2CO_3 + OH^-$	2.30×10^{-8}	7.64

Values of pK_b are included for the bases in Table 10-5.

EXERCISE 10-19 Write the base ionization constant expression for each of the following bases:
(a) $AgOH$ (b) HCO_3^- (c) $Ba(OH)_2$

EXERCISE 10-20 Convert the following values of K_b to pK_b:
(a) 1.54×10^{-3} (b) 2.75×10^{-6} (c) 5.43×10^{-8}
(d) 9.45×10^{-10}

So far, we have been interested in the hydrogen ion concentrations of aqueous solutions of pure acids and bases. Most of the acidic or basic solutions in the body do not contain just pure acid or pure base. Many of these solutions are mixtures of a weak acid and a salt of its conjugate base. These are called buffer solutions and they have a very important property.

10.7 BUFFER SOLUTIONS

A buffer solution is a mixture of either a weak acid *plus* a salt of this weak acid or a weak base *plus* a salt of this base. Such a mixture reacts with both acids and bases, so small additions of either strong acids or strong bases cause little change in its pH.

A mixture of equimolar quantities of acetic acid and sodium acetate dissolved in water is an example of a buffer solution. The mixture has a large reservoir of both weak acid molecules, $C_2H_3O_2H$, and the conjugate base of the acid, the acetate ions of sodium acetate. Small amounts of a strong acid added to the buffer solution react with the conjugate base as follows:

$$H_3O^+ + C_2H_3O_2^- \rightleftharpoons C_2H_3O_2H + H_2O$$

The equilibrium lies to the right because water is the weaker base and acetic acid is the weaker acid. As a result, most of the added hydrogen ion is removed from the solution and the pH hardly changes.

Hydroxide ions added to the buffer solution react with molecules of acetic acid to form acetate ions and water as follows:

$$OH^- + C_2H_3O_2H \rightleftharpoons C_2H_3O_2^- + H_2O$$

Again, the equilibrium lies to the right. Consequently, most of the hydroxide ions are removed from solution and the pH is only changed slightly.

The pH of a buffer solution is determined by the pK_a of the weak acid and the log of the ratio of the concentration of the conjugate base to the concentration of the acid. The following equation, called the Henderson-Hasselbalch equation, expresses this relationship for the acetic acid–acetate buffer.

$$pH = pK_a + \log \frac{[C_2H_3O_2^-]}{[C_2H_3O_2H]}$$

A small change in the concentration of either acetate ion or acetic acid will hardly change the pH. This fact is shown pictorially in Figure 10-3. There is a balance between the equimolar concentrations of acetate ions and acetic acid, as shown in Figure 10-3a. A few drops of strong acid added to the buffer solution react with acetate ions to form acetic acid. But the amounts of acetic acid and acetate ions in solution are much larger than the amount of acid added. As a result, the changes in the concentrations of acetate ion and acetic acid are small, and the ratio $[C_2H_3O_2^-]/[C_2H_3O_2H]$ remains about the same. The balance is hardly disturbed and the pH remains almost constant, as shown in Figure 10-3b.

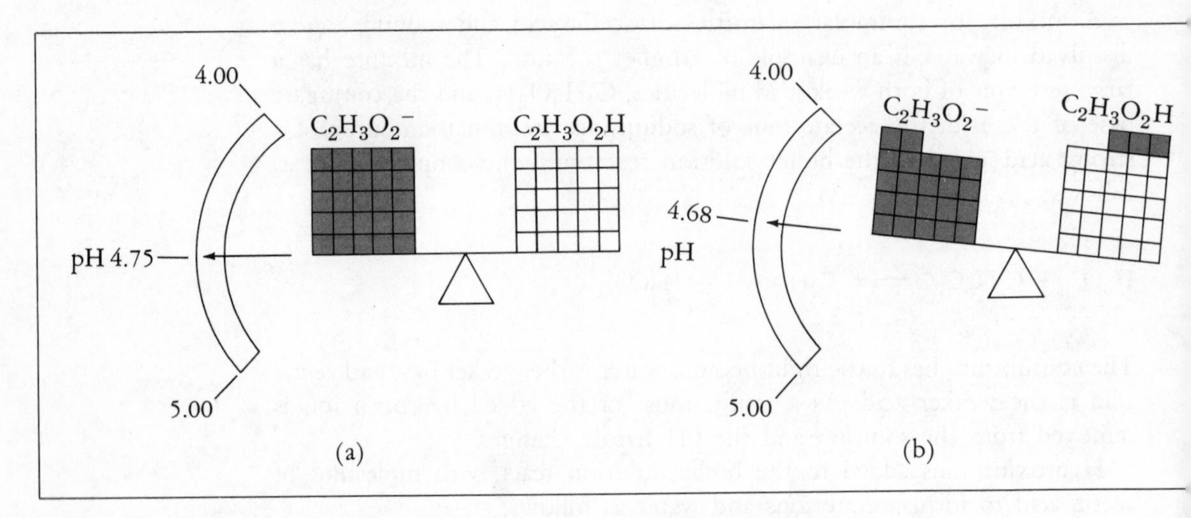

Fig. 10-3. The effect of adding small amounts of acid to an acetic acid–acetate ion buffer. (a) The pH of the original buffer containing equimolar quantities of acetic acid and acetate ion is 4.75. (b) After the addition of a small amount of acid, the pH is 4.68.

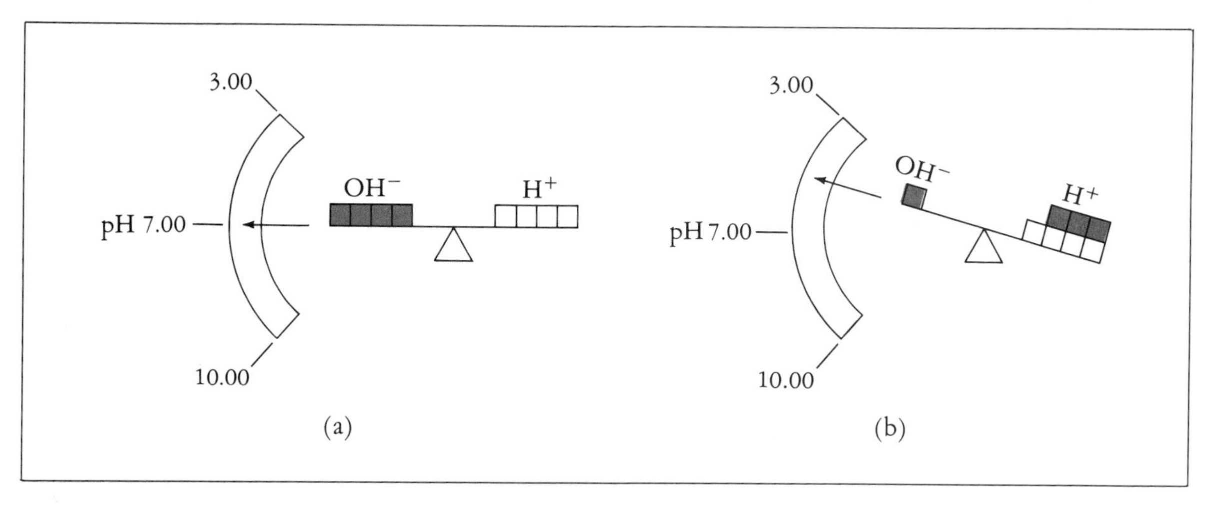

Fig. 10-4. The effect of adding a small amount of acid to pure water. (a) Pure water contains equal concentrations of OH^- and H^+. (b) A small amount of a strong acid reacts with almost all the hydroxide ions. As a result, the concentration of H^+ is much greater than the concentration of OH^-.

In contrast, addition of small amounts of acid to water cause a drastic change in pH. This is shown pictorially in Figure 10-4a. There is a balance between the small amounts of hydrogen and hydroxide ions in water, as we learned in Section 9.6. A few drops of strong acid added to water react with almost all the hydroxide ions. This decrease in hydroxide ions causes the solution to become strongly acidic, as shown in Figure 10-4b. Thus, addition of small amounts of a strong acid to an unbuffered solution causes drastic changes in its pH.

A buffer solution has a limited ability to react with acids and bases without drastically changing its pH. A solution acts as a buffer because it contains *both* members of a conjugate acid-base pair. Removal of one of these two by either a chemical or a physical process destroys the buffer action of the solution. For example, if enough strong acid is added to an acetic acid–acetate ion buffer solution, all at once or little by little, to react with all the acetate ion, the solution loses its ability to act as a buffer. The solution is no longer a buffer solution because it does not contain large reservoirs of both the weak acid and its conjugate base. Thus, continued addition of strong acids or bases to a buffer solution eventually exhausts its ability to act as a buffer.

Buffer solutions are important in the body, because they maintain the acid-base balance in the blood.

10.8 ACID-BASE BALANCE IN BLOOD

The pH of various body fluids is maintained by buffers. There are several different buffer systems in the body. Dihydrogen phosphate ($H_2PO_4^-$) and monohydrogen phosphate (HPO_4^{-2}) are one weak acid-base conjugate pair that acts as a buffer in the blood. Any acid reacts with monohydrogen phosphate according to the following equation:

$$HPO_4^{-2} + H_3O^+ \rightleftarrows H_2PO_4^- + H_2O$$

Dihydrogen phosphate is a weak acid (Table 10-3) that reacts with any base as follows:

$$H_2PO_4^- + OH^- \longrightarrow HPO_4^{-2} + H_2O$$

Another conjugate acid-base pair that acts as a buffer is carbonic acid–bicarbonate ion. Carbonic acid is formed by dissolving carbon dioxide in aqueous body fluids (see Section 9.9). It is a weak acid that ionizes to bicarbonate ion. The equation for these two equilibrium reactions is as follows:

$$CO_2 + H_2O \rightleftarrows H_2CO_3 \rightleftarrows HCO_3^- + H^+$$

Normally, in body fluids such as blood, there is 24 mEq/L of bicarbonate ion to 1.2 mEq/L of carbonic acid. The pH of the blood is within its normal range of 7.35 to 7.45 when this ratio, 20 parts bicarbonate ion to 1 part carbonic acid, is maintained. The pH of the blood becomes more acidic when the ratio $[HCO_3^-]/[H_2CO_3]$ becomes less than 20/1, say, 16/1, or 12/1. *The acidic condition of the blood signified by a pH less than 7.35 is called acidemia.* The pH of the blood becomes more basic when the ratio $[HCO_3^-]/[H_2CO_3]$ becomes greater than 20/1, say, 25/1, or 30/1. *The alkaline condition of the blood signified by a pH greater than 7.45 is called alkalemia.* Death occurs if the pH of the blood is more acidic than 6.8 or more basic than 7.8. The effect on pH of a change in the ratio $[HCO_3^-]/[H_2CO_3]$ is shown pictorially in Figure 10-5.

We learned in the previous section that all buffer solutions have a limited ability to withstand additions of strong acids or bases without changing their pH very much. As soon as either the acid or its conjugate base is used up, the solution loses its ability to act as a buffer. The buffers in the body are

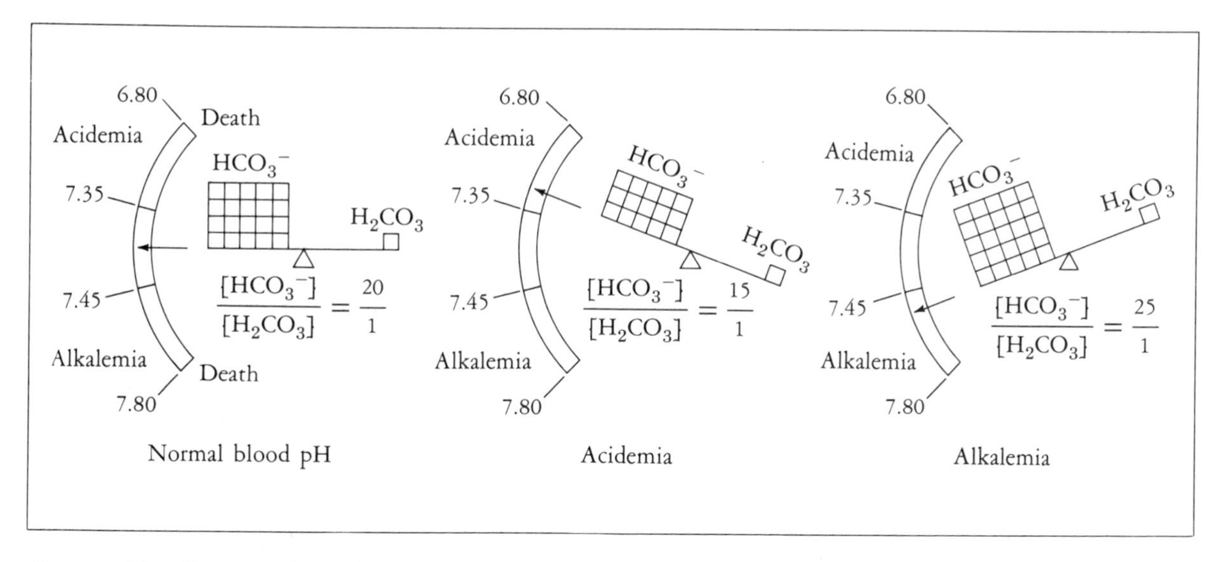

Fig. 10-5. The effect on blood pH of a change in the ratio of bicarbonate ion to carbonic acid.

no exception to this rule. But buffers in the body differ in one important respect from those in the laboratory. The body can replenish components of the buffer solution as they are used up or can remove from the body any excess component. As an example, let us consider how the body uses the carbonic acid–bicarbonate ion buffer system to cope with an increase in either the acid or the base concentration in the blood.

First, consider a patient who has an illness that causes an increase in the concentration of acidic products in the blood. *The physiologic processes causing acidemia are called acidosis.* The acidic products react with bicarbonate ions to produce carbonic acid. This causes a decrease in the ratio $[HCO_3^-]/[H_2CO_3]$. Unless something is done to return this ratio to normal, acidosis will occur. One of the functions of both the lungs and the kidneys is to maintain the pH of the blood by replenishing the buffer components that are used up or removing any excess components from the body. The circulation of air into and out of the lungs, called *ventilation,* produces the quickest response. An increase in the amount of carbonic acid in the blood causes a corresponding increase in the amount of carbon dioxide formed from the decomposition of carbonic acid. To lose this excess carbon dioxide, deeper and faster breathing, called *hyperventilation,* occurs. This causes a decrease in the acidity of the blood because the carbon dioxide formed is lost through the lungs. If this does not return the pH to normal, the kidneys can help by releasing more bicarbonate ion into the blood and removing hydrogen ions. In these ways, the body tries to return the

$[HCO_3^-]/[H_2CO_3]$ ratio to its normal value of 20 and maintain the acid-base balance in the blood.

Consider next a patient who has an illness that causes an increase in the concentration of basic products in the blood. *The physiologic processes causing alkalemia are called alkalosis.* The lungs and the kidneys are equally well equipped to handle an increase of basic products in the blood. These basic products react with carbonic acid to form bicarbonate ions. This time the ratio $[HCO_3^-]/[H_2CO_3]$ increases. The simplest way to prevent this ratio from increasing is to conserve the carbon dioxide in the body and use it to produce more carbonic acid. To do this, loss of carbon dioxide through the lungs is minimized by slower and shallow breathing, called *hypoventilation.* As before, the kidneys can help if needed. But this time, bicarbonate ions are removed and hydrogen ions are added to the blood. Thus, the lungs and kidneys can function to maintain the pH of the blood within its normal range of 7.35 to 7.45.

There is a growing awareness of the importance of acid-base balance in maintaining human health. The examples given in this section show that acid-base balance in the blood is governed by the same principles as acid-base balance in the laboratory.

10.9 SUMMARY

Hydrogen ion concentrations are expressed in either molar concentration units or pH. The pH scale ranges from zero to 14. A neutral solution has a pH of 7. An acidic solution has a pH value less than 7, and a basic (alkaline) solution has a pH value greater than 7. The total amount of acid or base in a solution can be determined by an acid-base titration. Calculations in acid-base titrations are greatly simplified when the concentrations of acids and bases are expressed in normality. The advantage of normality as a unit of concentration of acids and bases is that we know directly the volume that supplies 1 mole of hydrogen or hydroxide ions. We do not need to worry about the balanced equation for the acid-base reaction.

An acid is a proton donor, and a base is a proton acceptor according to the Brønsted definition. This general definition allows us to classify many compounds and ions as acids and bases. When a Brønsted acid loses a proton, the remainder of the molecule is a Brønsted base. These two parts of the same molecule are a conjugate acid-base pair. There is a relationship between these two. In general, the stronger the acid, the weaker is its conjugated base, and vice versa. We use this relationship to predict the equilibrium position of an acid-base reaction. An acid-base reaction occurs to produce the weaker of the possible acids and bases. The strengths of acids and bases are expressed quantitatively as ionization constants.

A buffer solution is a mixture of either a weak acid plus a salt of this acid or a weak base plus a salt of this base. Such a mixture reacts with both acids and bases, so small additions of either strong acids or strong bases cause little change in its pH. A buffer solution has a limited ability to react with acids and bases without drastically changing its pH. A solution acts as a buffer because it contains both members of a conjugate acid-base pair. Removal of one of these two by either a chemical or a physical process destroys the buffer action of the solution.

Buffer solutions are used by the body to maintain the pH of the blood. An important buffer is the carbonic acid–bicarbonate ion system. Buffers in the body differ from those in the laboratory in that they can replenish the components of the buffer solution as they are used up and they can remove any excess component from the body. One of the functions of the lungs and kidneys is to maintain the $[HCO_3^-]/[H_2CO_3]$ ratio.

REVIEW OF TERMS AND CONCEPTS

Terms

ACID IONIZATION CONSTANT The equilibrium constant that expresses the degree of ionization of an acid in water.

ACIDEMIA An acid condition of the blood signified by a pH of less than 7.35.

ACIDOSIS The physiologic processes causing acidemia.

ALKALEMIA An alkaline condition of the blood signified by a pH greater than 7.45.

ALKALOSIS The physiologic processes causing alkalemia.

BASE IONIZATION CONSTANT The equilibrium constant that expresses the degree of ionization of a base in water.

BRØNSTED ACID Any substance that donates a proton.

BRØNSTED BASE Any substance that accepts a proton.

BUFFER SOLUTION A solution that contains a mixture of either a weak acid plus a salt of that acid or a weak base plus a salt of that base.

CONJUGATE ACID-BASE PAIR An acid molecule (the acid) and the part of the molecule that remains after it loses a proton (the conjugate base).

ENDPOINT The change in color of an indicator that indicates the end of a titration.

EQUIVALENCE POINT The point at which equal numbers of equivalents of acid and base react.

GRAM-EQUIVALENT WEIGHT OF AN ACID The weight of an acid, in grams, that supplies 1 mole of hydrogen ions.

GRAM-EQUIVALENT WEIGHT OF A BASE The weight of a base, in grams, that supplies 1 mole of hydroxide ions or reacts with 1 mole of hydrogen ions.

HYPERVENTILATION Breathing that is faster and deeper than normal.

HYPOVENTILATION Breathing that is slower and more shallow than normal.

INDICATOR A substance that changes color when the pH of a solution changes within a characteristic range.

NORMALITY A unit of concentration of acids and bases that expresses equivalents of solute per liter of solution.

pH A method of expressing the concentration of hydrogen ions in aqueous solutions on a logarithmic scale.

pH METER An instrument that measures pH.

pOH A method of expressing the concentration of hydroxide ions in aqueous solutions on a logarithmic scale.

Concepts

1. Basic solutions have values of pH between 7 and 14; acidic solutions have values of pH between 0 and 7.
2. A titration is a method of determining the total amount of an acid or a base in a solution.
3. At the endpoint or equivalence point of a titration, the number of equivalents of acid is equal to the number of equivalents of base.
4. Acid-base reactions are a competition between two bases for a proton. The reaction occurs to form the weaker acid and base.
5. There is a difference between the pH of a solution containing a weak acid or a weak base and the total amount of acid or base it contains. The pH measures the concentration of hydrogen ions, whereas the total amount of acid (or base) includes concentrations of both hydrogen ions *and* any un-ionized molecules of acid (or base).
6. A buffer solution contains both members of a conjugate acid-base pair.
7. A buffer solution has a limited ability to react with strong acids and bases without drastically changing its pH.

EXERCISES 10-21 Which solution is the more acidic?

	(a)	(b)	(c)	(d)
Solution A	pH = 4	$[H^+] = 1 \times 10^{-3}$ M	$[H^+] = 1 \times 10^{-4}$ M	pOH = 5
Solution B	pH = 7	$[H^+] = 1 \times 10^{-12}$ M	pH = 5	pOH = 10

10-22 Calculate the pH corresponding to each of the following values of $[OH^-]$:
(a) 1×10^{-5} M (b) 3.28×10^{-8} M (c) 6.95×10^{-10} M
(d) 1.53×10^{-2} M

10-23 Why do you think that vinegar and citrus fruit juices are listed as first-aid treatment for people who have accidentally swallowed a solution of lye (NaOH)?

10-24 A 25.0 mL sample of gastric juices is titrated with a 0.121 N NaOH solution. It takes 33.1 mL of this solution to reach the endpoint. What is the acid concentration of this sample of gastric juices?

10-25 What is the difference between the total acidity of a solution and its pH?

10-26 Which solution is the more acidic?
(a) 0.1 N HCl or 1 N HCl (b) 1 N HCl or 1 N $C_2H_3O_2H$
(c) 1 N H_2SO_4 or 1 N HCl (d) 1 M H_2SO_4 or 1 N H_2SO_4
(e) 1 M $NaHSO_4$ or 1 M $NaHCO_3$
(f) 1 M NaCl or 1 M Na_2CO_3

10-27 Identify the two conjugate acid-base pairs in each of the following reactions:

(a) $HCO_3^- + H_3O^+ \rightleftharpoons H_2CO_3 + H_2O$

(b) $H_3PO_4 + H_2O \rightleftharpoons H_2PO_4^- + H_3O^+$

(c) $HF + H_2O \rightleftharpoons F^- + H_3O^+$

(d) $HSO_4^- + H_2O \rightleftharpoons H_2SO_4 + OH^-$

(e) $NH_4^+ + H_2O \rightleftharpoons NH_3 + H_3O^+$

(f) $HNO_3 + OH^- \rightleftharpoons H_2O + NO_3^-$

10-28 Predict the position of the equilibrium in each of the reactions in exercise 10-27.

10-29 List the following acids in order of increasing acid strength:
Acid A $K_a = 1 \times 10^{-9}$ Acid D $K_a = 6.5 \times 10^{-5}$
Acid B $K_a = 1 \times 10^{-3}$ Acid E $K_a = 2.5 \times 10^{-11}$
Acid C $K_a = 1.5 \times 10^{-9}$ Acid F $K_a = 8.9 \times 10^{-2}$

10-30 Convert the following values of K_a to pK_a:
(a) 1×10^{-4} (b) 7.32×10^{-6} (c) 1.75×10^{-10}
(d) 8.50×10^{-8}

10-31 Explain in your own words how the carbonic acid–bicarbonate ion buffer solution acts to maintain the pH of blood constant.

10-32 For each of the following weak acids, write the equation for its ionization, and its ionization constant expression.

(a) HCO_2H (Formic acid, found in bees and ants, is a monoprotic acid.)

(b) $C_6H_7O_6H$ (Ascorbic acid, found in fresh fruit, is a monoprotic acid.)

(c) $C_9H_7O_4H$ (Acetylsalicylic acid, ASA, the active ingredient in aspirin, is a monoprotic acid.)

(d) $C_3H_5O_3H$ (Lactic acid, a waste product formed when muscles are exercised, is a monoprotic acid.)

(e) $C_6H_5O_7H_3$ (Citric acid, found in citrus fruits, is a triprotic acid.)

10-33 Each of the weak acids given in exercise 10-32 and its conjugate base can form a buffer solution. Write the equations for the chemical reactions that occur when either (a) a strong acid or (b) a strong base is added to each buffer solution.

10-34 How much 1 M HCl solution must be added to 1 L of a 1 M acetic acid ($C_2H_3O_2H$)–1 M sodium acetate buffer solution to completely destroy its ability to act as a buffer?

10-35 The advertising for an antacid tablet claims that it neutralizes 47 times its weight in stomach acid. Stomach acid is HCl. If the active ingredient in the tablet weighs 0.50 g, how many mL of a 3 N HCl solution will it neutralize?

ALKANES AND CYCLOALKANES: AN INTRODUCTION TO ORGANIC CHEMISTRY

11

We learned in Chapter 1 that about 100 years ago chemists began to study the chemical composition of compounds obtained from living systems. Two important facts emerged from this work. First, no unique chemical elements were found in living systems; that is, a chemical element found in a cell is no different from the same element found in lifeless objects. Second, carbon, hydrogen, oxygen, and nitrogen are the predominant elements in living systems. The result of this last fact was the arbitrary division of chemistry into two categories—organic chemistry, the study of compounds obtained from living systems, and inorganic chemistry, the study of compounds obtained from minerals. Today we realize that this is an artificial distinction that has no biological basis. As a result, we now define organic chemistry as the chemistry of carbon and its chemical compounds. No reference to the source, living or lifeless, is made in the modern definition. In fact, there is no chemical way of distinguishing between the same compound formed in a cell or in a chemical laboratory.

Carbon atoms form four polar covalent bonds with atoms of other elements, as we learned in Section 5.7. Carbon atoms also form covalent bonds with other carbon atoms. In fact, covalent bonds link many carbon atoms together to form long chains and rings. As a result, carbon atoms bond together to form millions of compounds with different chemical and physical properties. We need some way to classify these compounds if we want to make sense out of all this information about them. Learning to classify organic compounds is the first goal of this chapter. The second goal is to learn the chemical and physical properties of alkanes and cycloalkanes, the simplest class of organic compounds. Finally, we will learn how to name alkanes and cycloalkanes. To start, let us learn how to write the structural formulas of compounds that contain more than one carbon atom.

11.1 STRUCTURAL FORMULAS

We have frequently used the chemical (or molecular) formulas of compounds in the preceding chapters. Such molecular formulas as H_2O, NaOH, and H_3PO_4 are shorthand notations that specify the number and kinds of atoms in a molecule. They enable us to calculate the molecular weight of a compound and they facilitate the writing and balancing of chemical equations. For organic compounds, the chemical formula loses some of its usefulness because we find that one chemical formula does not always represent just one compound. For example, three different organic compounds all have the same chemical formula, C_5H_{12}. These three compounds are *isomers*. They are different compounds that have the same number and kinds of atoms. Each of these has different physical and chemical properties that distinguish it from the others. Yet the chemical formulas of all three are identical. How do we account for this? The only reasonable explanation is that the arrangement of the atoms in the molecules must be different even though the total number and kinds of atoms are identical for each compound. Therefore, we need some way to specify how the atoms in a molecule are arranged. We use *structural formulas* to do this.

Structural formulas are constructed by writing the symbols for the elements and joining with dashes those that are bonded together. We learned how to construct such structural formulas for simple compounds such as HF, H_2O, and NH_3 in Section 5.7. We can extend this idea to carbon-containing compounds by using the normal rules of bonding for carbon, hydrogen, oxygen, nitrogen, and the halogens. For example:

CH_3Cl $\qquad$ C_2H_6 $\qquad$ CH_4O $\qquad$ CH_5N

```
   H            H  H           H              H
   |            |  |           |              |   H
 H—C—Cl      H—C—C—H       H—C—O—H        H—C—N/
   |            |  |           |              |   \
   H            H  H           H              H    H
```

Carbon can also form double and triple bonds with adjacent carbon or other elements. For example:

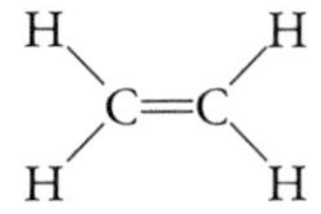

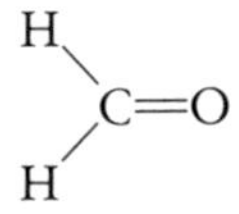

$$\begin{array}{ccc} H & & H \\ & C{=}C & \\ H & & H \end{array} \qquad \begin{array}{cc} H & \\ & C{=}O \\ H & \end{array}$$ Double bonds

$$H{-}C{\equiv}C{-}H \qquad H{-}C{\equiv}N$$ Triple bonds

These structural formulas provide us with information about the number and kinds of atoms in a molecule, and they also indicate specifically which atoms are bonded to which other atoms and by what kind of bonds. Consequently, structural formulas are of more value in organic chemistry than are the simple chemical formulas.

To conserve space, the structural formulas shown above are usually written in an abbreviated form. For example, the molecule C_5H_{12} can be written in the following ways:

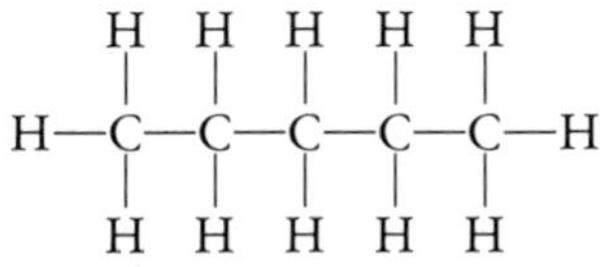

$$\begin{array}{ccccccccccc} & & H & & H & & H & & H & & H & \\ & & | & & | & & | & & | & & | & \\ H & - & C & - & C & - & C & - & C & - & C & -H \\ & & | & & | & & | & & | & & | & \\ & & H & & H & & H & & H & & H & \end{array}$$ Abbreviated, it becomes

$$CH_3{-}CH_2{-}CH_2{-}CH_2{-}CH_3$$ or

$$CH_3CH_2CH_2CH_2CH_3$$ or even

$$CH_3(CH_2)_3CH_3$$

Other examples are:

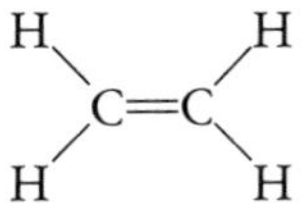

$$\begin{array}{ccc} H & & H \\ & C{=}C & \\ H & & H \end{array}$$ is equivalent to $H_2C{=}CH_2$

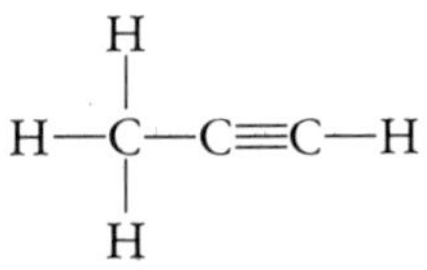

$$\begin{array}{ccc} & H & \\ & | & \\ H- & C & -C{\equiv}C{-}H \\ & | & \\ & H & \end{array}$$ is equivalent to $CH_3C{\equiv}CH$

Once we understand the concept of structural formulas, we are in a position to account for the existence of more than one compound with the same chemical formula.

11.2 STRUCTURAL ISOMERS

Compounds that have the same chemical formulas but differ in their structural formulas are called *structural isomers.*

Consider, for example, the chemical formula C_5H_{12}. If we arrange these atoms according to the ordinary rules of bonding, which require that each carbon atom form four bonds to its neighbors and each hydrogen atom bond to one other atom, we find that there are three—and only three—ways in which these atoms can be arranged.

```
                                  H                      H
                                  |                      |
                               H—C—H                  H—C—H
                                  |                      |
   H  H  H  H  H           H      |     H  H      H      |      H
   |  |  |  |  |           |      |     |  |      |      |      |
H—C—C—C—C—C—H         H—C————C————C—C—H    H—C————C————C—H     Full structure
   |  |  |  |  |           |      |     |  |      |      |      |
   H  H  H  H  H           H      H     H  H      H      |      H
                                                         |
                                                      H—C—H
                                                         |
                                                         H
A                          B                      C
```

A	B	C	
$CH_3CH_2CH_2CH_2CH_3$	$CH_3CHCH_2CH_3$ with CH_3 on the second carbon	CH_3CCH_3 with CH_3 above and below the central carbon	Condensed structure
C_5H_{12}	C_5H_{12}	C_5H_{12}	Chemical formula

Corresponding to these three possible arrangements, three and only three substances have been found with the chemical formula C_5H_{12}. Two are liquids with low boiling points (28° and 36° C, respectively), and the third is a gas at room temperature. These three compounds exhibit different physical and chemical properties because of the different arrangement of the atoms within the molecules. The structural formulas A, B, and C indicate to us these differences, whereas the chemical formula C_5H_{12} does not.

Be careful when you draw the structural formulas of isomers and you decide which ones are different. Structural formulas indicate only which atoms are bonded to one another, not their relative positions in space. Thus, as long as the atoms are bonded to one another, the formulas are identical no matter where their relative positions are on the paper. For this reason, the following structures are identical:

$$\begin{array}{l} CH_3 \\ | \\ CH_2CH_2CH_2CH_3 \end{array} \quad \text{or} \quad \begin{array}{l} CH_3CH_2CH_2 \\ \qquad\quad | \\ \qquad\quad CH_2CH_3 \end{array} \quad \text{is the same structural formula as} \quad CH_3CH_2CH_2CH_2CH_3$$

$$\begin{array}{l} CH_3CHCH_2CH_3 \\ \qquad | \\ \quad\; CH_3 \end{array} \quad \text{or} \quad \begin{array}{l} CH_3CH_2CHCH_3 \\ \qquad\qquad | \\ \qquad\quad\; CH_3 \end{array} \quad \text{is the same structural formula as} \quad \begin{array}{l} \quad\; CH_3 \\ \qquad | \\ CH_3CHCH_2CH_3 \end{array}$$

You can convince yourself of this fact by making models of these compounds.

*EXERCISE 11-1 Two different compounds are known to have the same chemical formula, C_4H_{10}. Draw the structural formula for each compound.

We have seen that structural formulas serve to explain the occurrence of isomers. This is only one use of structural formulas. Their most important function is to permit us to classify molecules in terms of certain *reactive parts* or *functional groups.* In so doing, we develop the structural theory of organic chemistry. What are functional groups and how do they help us to systematize the study of organic chemistry?

11.3 FUNCTIONAL GROUPS

As organic chemistry developed, it was soon realized that the chemical reactions of organic compounds occurred at specific sites in the molecules.

* The answers for the exercises in this chapter begin on page 865.

$$CH_3NH_2 + HCl \longrightarrow CH_3\overset{+}{N}H_3Cl^-$$

11.1

$$CH_3(CH_2)_5CH_2NH_2 + HCl \longrightarrow CH_3(CH_2)_5CH_2\overset{+}{N}H_3Cl^-$$

11.2

$$\begin{array}{l} CH_2CH_2 \\ | \\ CH_2CH_2 \end{array} \!\!>\! CHNH_2 + HCl \longrightarrow \begin{array}{l} CH_2CH_2 \\ | \\ CH_2CH_2 \end{array} \!\!>\! CH\overset{+}{N}H_3Cl^-$$

11.3

Fig. 11-1. The similarities of the chemical reactions of an —NH_2 *group bonded to carbon in three different compounds.*

These sites usually contained specific atoms or groups of atoms bonded to carbon, such as —Cl or —OH or —NH_2. These groups at which reactions occur are called *functional groups.* In general, these groups account for a very small part of the total organic molecule. The chemical reactions of these functional groups are found to be generally independent of the carbon skeleton. For example, the three organic compounds represented by structures *11.1, 11.2,* and *11.3* in Figure 11-1 all contain an —NH_2 group bonded to a carbon atom. All three of these compounds react in the same way with hydrochloric acid, as shown in Figure 11-1. Organic compounds that contain the same functional group undergo similar chemical reactions and have similar physical properties. In this way, the large number of reactions of organic compounds can be organized into classes, each having the characteristic reactions of that particular functional group. This is the way that we will organize the study of organic chemistry in this text. Some of the functional groups whose chemistry we will study are listed in Table 11-1.

Table 11-1. Functional Groups

Functional Group	Class of Compound	Example
$>C=C<$	Alkene	$H_2C=CH_2$
$-C\equiv C-$	Alkyne	$HC\equiv CH$
(benzene ring)	Aromatic	(benzene ring with CH_3)

Table 11-1. (*Continued*)

Functional Group	Class of Compound	Example
$>C{-}X$ (X = F, Cl, Br, or I)	Halide	CH_3Cl, CH_3Br
$>C{-}OH$	Alcohol	CH_3OH
$>C{-}O{-}C<$	Ether	H_3COCH_3
$>C{-}NH_2$	Amine	CH_3NH_2
$>C{=}N{-}OH$	Oxime	$(H)(CH_3)C{=}N{-}OH$
$-C(=O)OH$	Carboxylic acid	$CH_3C(=O)OH$
$-C(=O){-}O{-}C<$	Ester	$CH_3C(=O)OCH_3$
$-C(=O)NH_2$	Amide	$CH_3C(=O)NH_2$
$-C(=O)X$ (X = Cl, Br)	Acid halide	$CH_3C(=O)Cl$
$-C(=O)H$	Aldehyde	$CH_3C(=O)H$
$-C(=O)-$	Ketone	$CH_3C(=O)CH_3$
$-C{-}O{-}P(=O)(OH){-}OH$	Phosphate monoester	$CH_3OP(=O)(OH){-}OH$
$-C{-}O{-}P(=O)(OH){-}O{-}P(=O)(OH){-}OH$	Alkyl pyrophosphate	$CH_3OP(=O)(OH){-}O{-}P(=O)(OH){-}OH$
$-C{\equiv}N$ ($-CN$)	Nitrile	$CH_3C{\equiv}N$

EXERCISE 11-2 Circle the functional group in each of the following compounds. Then, arrange the compounds in groups, each group containing a different functional group.

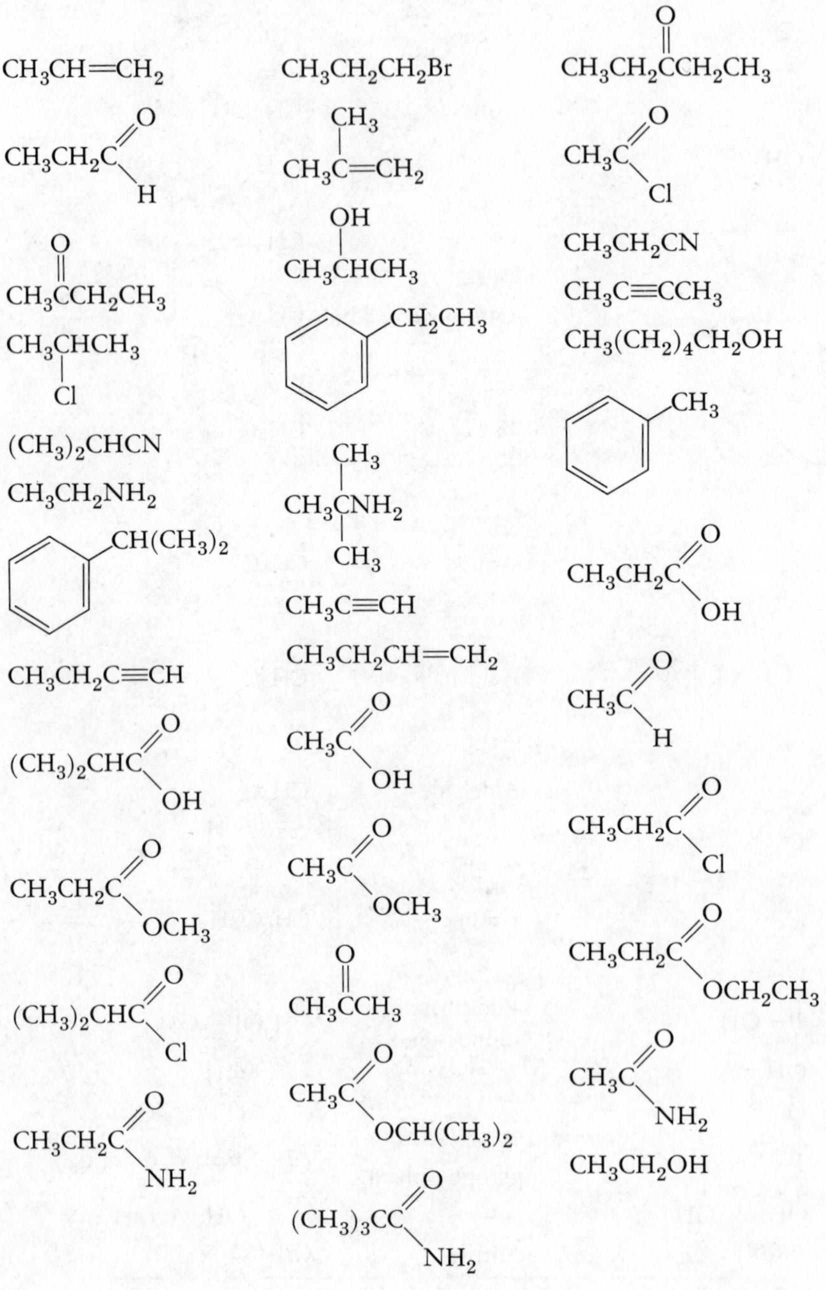

$CH_3CH{=}CH_2$

$CH_3CH_2C(=O)H$

$CH_3C(=O)CH_2CH_3$

$CH_3CH(Cl)CH_3$

$(CH_3)_2CHCN$

$CH_3CH_2NH_2$

$C_6H_5CH(CH_3)_2$

$CH_3CH_2C{\equiv}CH$

$(CH_3)_2CHC(=O)OH$

$CH_3CH_2C(=O)OCH_3$

$(CH_3)_2CHC(=O)Cl$

$CH_3CH_2C(=O)NH_2$

$CH_3CH_2CH_2Br$

$CH_3C(CH_3){=}CH_2$

$CH_3CH(OH)CH_3$

$C_6H_5CH_2CH_3$

$CH_3C(CH_3)_2NH_2$

$CH_3C{\equiv}CH$

$CH_3CH_2CH{=}CH_2$

$CH_3C(=O)OH$

$CH_3C(=O)OCH_3$

$CH_3C(=O)CH_3$

$CH_3C(=O)OCH(CH_3)_2$

$(CH_3)_3CC(=O)NH_2$

$CH_3CH_2C(=O)CH_2CH_3$

$CH_3C(=O)Cl$

CH_3CH_2CN

$CH_3C{\equiv}CCH_3$

$CH_3(CH_2)_4CH_2OH$

$C_6H_5CH_3$

$CH_3CH_2C(=O)OH$

$CH_3C(=O)H$

$CH_3CH_2C(=O)Cl$

$CH_3CH_2C(=O)OCH_2CH_3$

$CH_3C(=O)NH_2$

CH_3CH_2OH

Let us begin our study of organic chemistry by examining the class of compounds called hydrocarbons. These simple compounds are obtained from petroleum.

11.4 PETROLEUM

Petroleum is the major source of energy for our industrial society. Petroleum also furnishes most of the simple organic compounds used in the manufacture of synthetic materials. Almost all of the compounds available from petroleum contain only carbon and hydrogen. As a result, they are given the general name *hydrocarbons* (a contraction of the two words hydrogen and carbon).

Petroleum, while known to the ancients, was first obtained in large quantities in North America in the latter part of the nineteenth century. Its use during this century has transformed our civilization from an essentially agricultural one to a mechanized, industrial one. The petroleum and natural gas being removed from the ground today were formed as the result of a geological process that started millions of years ago when organic materials from plants and animals were buried with sediments from oceans and rivers. With the passing of time, the sediments were slowly subjected to great pressures and bacterial action, which formed natural gas and petroleum deposits.

The petroleum obtained directly from an oil well is a complex mixture of hydrocarbons. To be of commercial value, the crude oil must be refined, that is, separated into various fractions such as gasoline, kerosene, and fuel oil. This separation can be accomplished by a process called *distillation.* The crude oil is heated to boiling, and its vapors are passed into a fractionating column, which is a tall column containing perforated plates or irregular glass pieces. The purpose of this column is to separate the various compounds of the crude oil according to their boiling points. The components with lower boiling points are concentrated at the top of the column, and the components with higher boiling points are concentrated lower in the column. The various fractions obtained from the distillation of crude oil are listed in Table 11-2.

Even the fractions listed in Table 11-2 are complicated mixtures of hydrocarbons whose composition depends on the location of the oil well from which the crude oil was obtained. Over the years, much work has been done by scientists to identify the components of crude oil. As a result of this work, it has been found that all hydrocarbons can be placed into four classes:

Table 11-2. Distillation Fractions from Crude Oil

Fraction	Range of Boiling Points (°C)	Uses
Natural gas	Below 0	Fuel
Petroleum ethers and light naphtha	30–100	Solvents and degreasing solvents
Gasoline	30–200	Motor fuel
Kerosene	175–275	Jet fuel
Diesel	190–330	Diesel fuel
Fuel oil	230–360	Heating fuel
Lubricating and mineral oil	Above 350	Lubricants

1. *Alkanes* that contain only single bonds
2. *Alkenes* that contain one or more carbon-carbon double bonds
3. *Alkynes* that contain one or more carbon-carbon triple bonds
4. *Aromatic hydrocarbons* that contain six electrons in a six-member ring

Alkenes and alkynes are also known as *unsaturated hydrocarbons,* whereas alkanes are also known as *saturated hydrocarbons.*

Each one of these classes contains a large number of compounds. We will examine the alkanes in this chapter and the other classes in subsequent chapters.

11.5 ALKANES

It is possible to isolate and identify a large number of the alkanes contained in crude oil by careful distillation. The chemical formulas, structural formulas, melting points, and boiling points of all the alkanes that contain one to five carbon atoms are listed in Table 11-3.

From the information in Table 11-3, we find that the chemical formulas of alkanes correspond to the general formula C_nH_{2n+2}, where n is the number of carbon atoms in the molecule. By using this general formula, we can obtain the chemical formula (and, consequently, the molecular weight) of any alkane. For example, if we want the chemical formula of the alkane containing ten carbon atoms, we need only substitute n = 10 into the general chemical formula to arrive at the correct answer $C_{10}H_{22}$. Similarly, an alkane containing 12 carbon atoms has the chemical formula $C_{12}H_{26}$.

Table 11-3. Properties and Formulas of All Alkanes Containing One to Five Carbon Atoms

Molecular Formula	No. of Carbon Atoms	Structural Formula	Boiling Point (°C)	Melting Point (°C)
CH_4	1	CH_4	−162	−183
C_2H_6	2	CH_3CH_3	−89	−172
C_3H_8	3	$CH_3CH_2CH_3$	−42	−187
C_4H_{10}	4	$CH_3CH_2CH_2CH_3$	−0.5	−135
C_4H_{10}	4	CH_3CHCH_3 CH_3 (on the central C)	−10	−145
C_5H_{12}	5	$CH_3CH_2CH_2CH_2CH_3$	+36	−130
C_5H_{12}	5	$CH_3CHCH_2CH_3$ CH_3 (on the second C)	+28	−160
C_5H_{12}	5	CH_3 (above) CH_3CCH_3 CH_3 (below)	+9.5	−20

EXERCISE 11-3 Calculate the chemical formulas for the alkanes containing 15, 20, 25, and 50 carbon atoms.

It can be seen in Table 11-3 that the chemical formula of each alkane in the series differs from the preceding one by a $—CH_2—$ unit, called a methylene group. Thus,

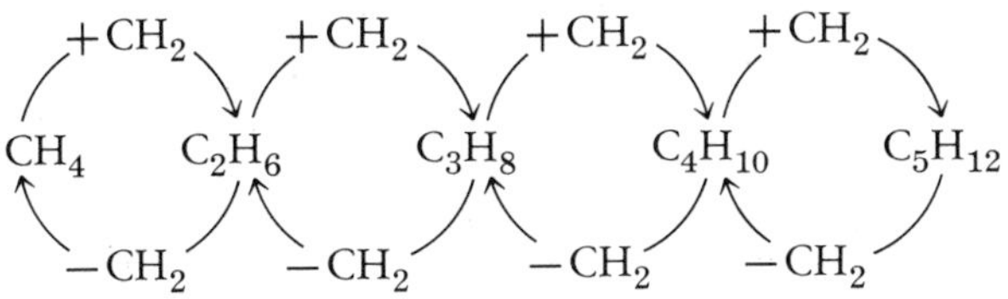

$$CH_4 \underset{-CH_2}{\overset{+CH_2}{\rightleftharpoons}} C_2H_6 \underset{-CH_2}{\overset{+CH_2}{\rightleftharpoons}} C_3H_8 \underset{-CH_2}{\overset{+CH_2}{\rightleftharpoons}} C_4H_{10} \underset{-CH_2}{\overset{+CH_2}{\rightleftharpoons}} C_5H_{12}$$

A series of compounds in which successive members differ by a $—CH_2—$ unit is called a *homologous series.* Members of such series are usually closely related in both physical and chemical properties. This is extremely important because it means that we do not have to investigate the properties of every single organic compound. Rather, the properties of representative members of homologous series are studied, and these properties are used to predict the behavior of all other members of that series.

Although we can identify an individual member of a homologous series by its structural formula, we eventually come to the point where we want to give it a distinctive name. To do this requires a simple yet clear system of naming organic compounds. Starting with the Geneva Convention in 1892, such a system has been developed over the years. This system is intended to be international, that is, independent of any one language. Minor revisions to the rules are made periodically, usually to provide consistent and internationally acceptable names for new classes of compounds. The most recent revision by the Convention, which is now called the International Union of Pure and Applied Chemistry (IUPAC), was made in 1957.

The IUPAC system is a set of rules by which any organic compound can be named in a logical manner. We will first learn the basic rules as they apply to the saturated hydrocarbons. In succeeding chapters we will learn how to modify these basic rules to indicate the presence of the various functional groups.

11.6 NAMING ALKANES AND CYCLOALKANES

The names of the first ten straight-chain alkanes are given in Table 11-4. The first member of this series is called methane. The next three are called ethane, propane, and butane, respectively. Other members are simply named by adding the ending *-ane* to the Greek name for the number of carbon atoms in the compound. Thus, *pentane* is formed by adding the ending *-ane* to the Greek word *penta,* meaning five. (The *a* of the prefix *penta-* is dropped when the prefix is placed before a vowel.) It is important to learn these

Table 11-4. Names of the First Ten Straight-Chain Alkanes

Structural Formula	No. of Carbon Atoms	Name
CH_4	1	Methane
CH_3CH_3	2	Ethane
$CH_3CH_2CH_3$	3	Propane
$CH_3(CH_2)_2CH_3$	4	Butane
$CH_3(CH_2)_3CH_3$	5	Pentane
$CH_3(CH_2)_4CH_3$	6	Hexane
$CH_3(CH_2)_5CH_3$	7	Heptane
$CH_3(CH_2)_6CH_3$	8	Octane
$CH_3(CH_2)_7CH_3$	9	Nonane
$CH_3(CH_2)_8CH_3$	10	Decane

names, because all other hydrocarbons are named as relatives of these simple straight-chain hydrocarbons.

Branched-chain alkanes, that is, alkanes whose carbons are not connected one after another, are named by using the following rules:

1. *Determine the longest continuous chain of carbon atoms in the molecule.* The name of the alkane corresponding to this number of carbon atoms, given in Table 11-4, is the parent name of the compound.
2. *Carbons (or functional groups) that have replaced one or more hydrogens on the parent chain are called substituents and are designated by numbers and prefixes to the parent name.* To indicate its location on the parent chain, we assign the substituent a number. To do this, we number the parent chain from one end to the other. The direction of numbering is chosen to give the lowest number possible to the substituent. Often, more than one substituent is present on the chain. These substituents can be identical or different. The following rules cover both possibilities:
   a. If the same substituent appears more than once, the additional prefixes *di-* (for two), *tri-* (for three), and *tetra-* (for four) are used.
   b. If two or more different substituents are present, they are listed in alphabetical order.
3. *If the substituent is a hydrocarbon group, this group is named by changing the* -ane *ending to* -yl. For example, methane becomes methyl (CH_3—), and ethane becomes ethyl (C_2H_5—). The names of the most important saturated hydrocarbon substituents, called alkyl groups, are given in Table 11-5.

Table 11-5. Names and Structures of Important Alkyl Groups

Chemical Formula	Name	Chemical Formula	Name
CH_3—	Methyl	$CH_3CH_2CH_2CH_2$—	*n*-Butyl
CH_3CH_2—	Ethyl	$CH_3CH_2CHCH_3$ |	*sec*-Butyl
$CH_3CH_2CH_2$—	Propyl	CH_3╲ $CHCH_2$— CH_3╱	*iso*-Butyl
CH_3CHCH_3 |	*iso*-Propyl		
CH_3 | CH_3CCH_2— | CH_3	*neo*-Pentyl	CH_3 | CH_3C— | CH_3	*t*-Butyl

When the name of an alkyl group has no prefix, the name stands for the straight chain alkyl group. For example, pentyl designates the substituent $CH_3CH_2CH_2CH_2CH_2$—. Other alkyl groups in Table 11-5 have prefixes that are abbreviated for convenience. The prefix *iso-* means that the substituent is branched one carbon from the end of the chain opposite the end where it is attached to the parent chain. The *sec-* (for secondary) means that the carbon of the alkyl group bonded to the parent chain has two additional carbons bonded to it. The *t-* (for tertiary) means that the carbon of the alkyl group bonded to the parent chain has three additional carbons bonded to it.

Let us name a few compounds using these rules.

Example 11-1 Name the compound whose structure is

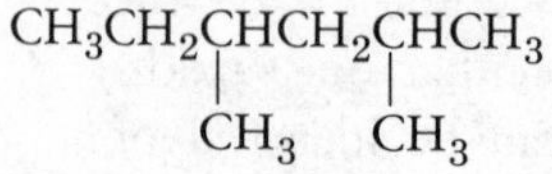

```
CH3CH2CHCH2CHCH3
       |    |
      CH3  CH3
```

Solution:

RULE 1. The longest continuous chain contains six carbon atoms. Therefore, the parent name is *hexane*. The compound is a substituted hexane.

RULE 2. The parent chain contains two substituents. To assign them numbers, we number the chain in both directions.

Numbering the longest chain from left to right gives the following:

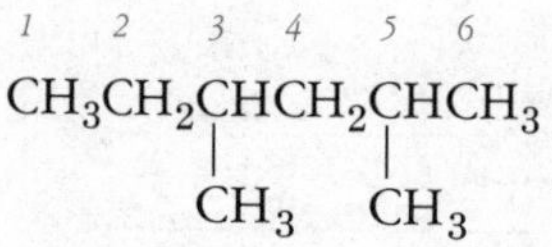

```
 1   2   3  4    5  6
CH3CH2CHCH2CHCH3
       |    |
      CH3  CH3
```

Numbering the longest chain from right to left gives the following:

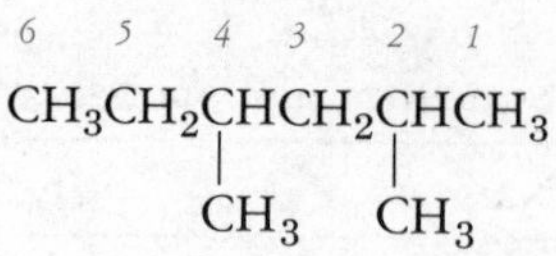

```
 6   5   4  3    2  1
CH3CH2CHCH2CHCH3
       |    |
      CH3  CH3
```

If we number from the right to left, as in the formula on the right, the lowest number (2) is given to the first substituent encountered. Thus, the two substituents are located at positions 2 and 4.

RULE 3. The two substituents both contain only one carbon atom. Therefore, they are both methyl groups. Using the names and numbers of the substituents as prefixes to the name of the parent chain, we obtain the IUPAC name for the compound, 2,4-dimethylhexane. Note that each part of the name refers to a specific feature of the molecule:

2,4- indicates the positions of both substituents on the parent chain

di indicates that two identical substituents are on the parent chain
methyl indicates the structure of the substituent
hexane indicates that the parent chain contains six carbon atoms

Example 11-2 Name the compound whose structure is

$$\begin{array}{lll} CH_3 & & CH_3 \\ | & & | \\ CH_2 & CH_3 & CH_2 \\ | & | & | \\ CH_3CHCH_2C & \!\!\!\text{———} & CH_2 \\ & | & \\ & CH_2CH_3 & \end{array}$$

Solution:

Numbering the longest chain in one direction gives:

$$\begin{array}{lll} {}^{8}CH_3 & & CH_3{}^{1} \\ | & & | \\ {}^{7}CH_2 & CH_3 & CH_2{}^{2} \\ | & | & | \\ CH_3CHCH_2C & \!\!\!\text{———} & CH_2{}^{3} \\ \;\;\;\;\;{}_{6}\;\;{}_{5} & |\,{}_{4} & \\ & CH_2CH_3 & \end{array}$$

Numbering the same chain in the opposite direction gives:

$$\begin{array}{lll} {}^{1}CH_3 & & CH_3{}^{8} \\ | & & | \\ {}^{2}CH_2 & CH_3 & CH_2{}^{7} \\ | & |\,{}^{5} & | \\ CH_3CHCH_2C & \!\!\!\text{———} & CH_2{}^{6} \\ \;\;\;\;\;{}_{3}\;\;{}_{4} & | & \\ & CH_2CH_3 & \end{array}$$

1. The name of the parent chain is octane.
2. There are three substituents (two methyl groups and one ethyl group).
3. Numbering from left to right gives the lowest number to the *first* substituent encountered.

The correct IUPAC name is 5-ethyl-3,5-dimethyloctane.

Note three important facts about the names derived for the compounds in examples 11-1 and 11-2.

1. In each case, the name is one word. This is a general rule for the names of alkanes. To obtain one word, the various parts, separated by numbers, are reconnected by dashes or commas. Commas are used between numbers, and dashes connect the numbers and the substituents.
2. The prefixes di-, tri-, tetra-, and so on do *not* count when one is deciding on the *alphabetical* order of the substituents. Thus, ethyl precedes dimethyl because e comes before m in the alphabet.
3. Remember that the structural formula of an alkane containing numer-

ous alkyl substituents can be written in a variety of ways, all of which are equivalent. The following example illustrates this fact:

$$\begin{array}{l} \quad CH_2CH_3 \\ \quad\ \ | \\ CH_3CHCH_2CHCH_3 \\ \quad | \\ \quad CHCH_3 \\ \quad | \\ \quad CH_3 \end{array} \quad \text{identical to} \quad \begin{array}{l} \qquad\qquad CH_3 \\ \qquad\qquad\ | \\ CH_3CHCH_2CHCH_2CH_3 \\ \quad | \\ \quad CHCH_3 \\ \quad | \\ \quad CH_3 \end{array} \quad \text{identical to} \quad \begin{array}{l} \quad CH_3 \qquad CH_3 \\ \quad | \qquad\quad\ \ | \\ CH_3CHCHCH_2CHCH_2CH \\ \qquad\ \ | \\ \qquad\ \ CH_3 \end{array}$$

Note that the longest carbon-containing chain is not necessarily the horizontal one as written.

EXERCISE 11-4 Give the IUPAC name for the following compounds:

$$\text{(a)}\ \begin{array}{l} \qquad\qquad\ CH_2CH_3 \\ \qquad\qquad\quad | \\ CH_3CH_2CHCHCH_2CH_3 \\ \qquad\qquad | \\ \qquad\quad CH_2CH_3 \end{array} \qquad \text{(b)}\ \begin{array}{l} \quad CH_3\ \ CH_2CH_2CH_3 \\ \quad\ | \qquad | \\ CH_3CHCHCHCH_2CH_3 \\ \qquad\quad | \\ \qquad\quad CH_3 \end{array}$$

$$\text{(c)}\ \begin{array}{l} \quad CH_3 \\ \quad | \\ CH_3CHCHCH_2CH_2CH_3 \\ \qquad\quad | \\ \qquad\quad CH_2 \\ \qquad\quad | \\ \ CH_3CH_2CHCH_3 \end{array}$$

EXERCISE 11-5 Draw the structural formulas corresponding to each of the following names:
(a) 2-methylbutane
(b) 2,2,4-trimethyloctane
(c) 2,3-dimethylpentane
(d) 4,4-diethyl-2,3,5-trimethylheptane

EXERCISE 11-6 The following names are *incorrect.* Give the reason why and give the correct IUPAC name.
(a) 1-methylbutane (b) 4,4-dimethylpentane
(c) 2-ethylhexane (d) 5-ethyl-6-methyl-4-propyloctane

In addition to forming long chains, saturated hydrocarbons can also exist as rings. Cyclic compounds of carbon containing only single bonds are called cycloalkanes. If the compounds contain only one ring, they have the general formula C_nH_{2n}. Note that cycloalkanes are *not* isomeric with alkanes.

Cycloalkanes are named by adding the prefix *cyclo-* to the name of the straight-chain hydrocarbon containing the same number of carbon atoms. For example:

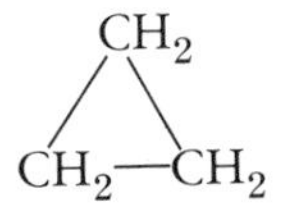

CH_2 / CH_2—CH_2 identical to

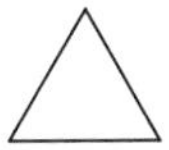

Cyclopropane

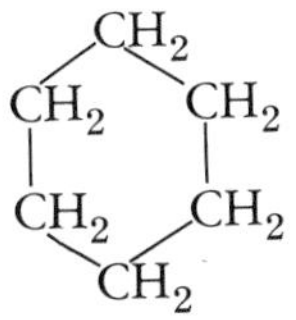

CH_2 CH_2 CH_2 CH_2 CH_2 CH_2 identical to

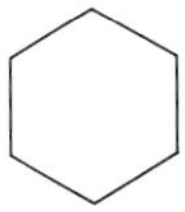

Cyclohexane

The structural formulas of cycloalkanes are usually abbreviated by drawing a polygon containing the appropriate number of carbon atoms. It is understood that there are a carbon atom and two hydrogen atoms at the intersection of two lines.

All the carbon atoms of a cycloalkane are equivalent. Therefore, no number prefix is needed for monosubstituted cycloalkanes. For example:

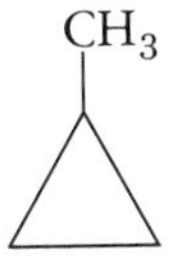

CH_3 — Methylcyclopropane

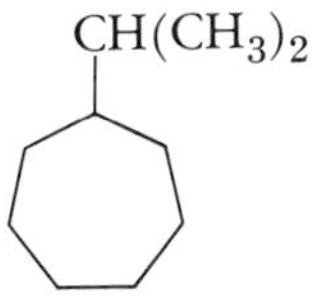

$CH(CH_3)_2$ — Isopropylcycloheptane

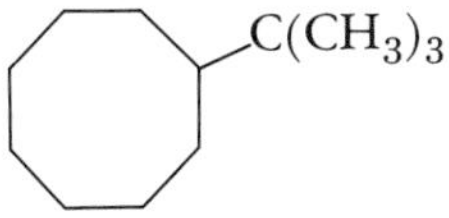

$C(CH_3)_3$ — *t*-Butylcyclooctane

However, numbers must be used if there is more than one substituent:

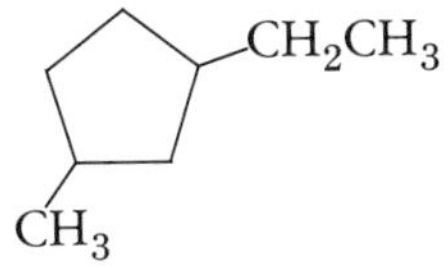

CH_2CH_3, CH_3 — 1-Ethyl-3-methylcyclopentane

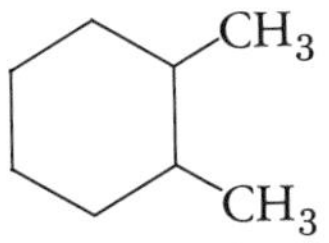

CH_3, CH_3 — 1,2-Dimethylcyclohexane

One substituent is given the number one. The direction of numbering around the ring is then chosen to give the lowest number to the second substitution.

When cycloalkanes are attached to a complex chain, it is usually better to consider the cycloalkane as a substituent. In this case, *cycloalkyl* is the name for a cycloalkane as a substituent. For example:

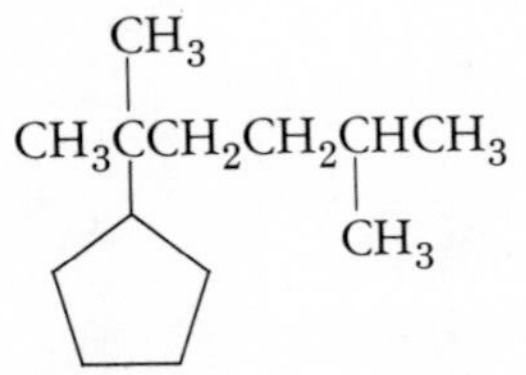

CH_3
$CH_3CCH_2CH_2CHCH_3$
CH_3

2-Cyclopentyl-2,5-dimethylhexane

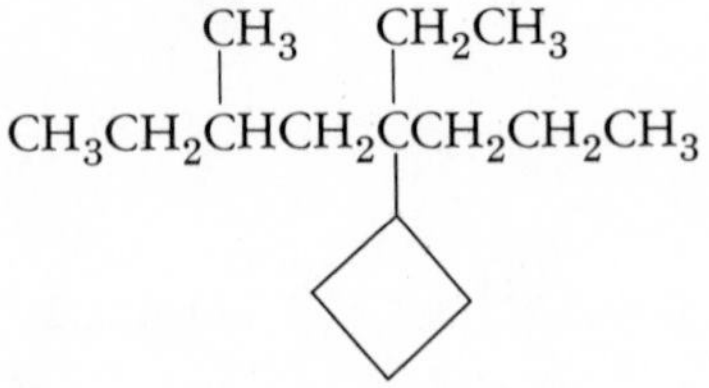

CH_3 CH_2CH_3
$CH_3CH_2CHCH_2CCH_2CH_2CH_3$

5-Cyclobutyl-5-ethyl-3-methyloctane

EXERCISE 11-7 Give the IUPAC name for each of the following compounds:

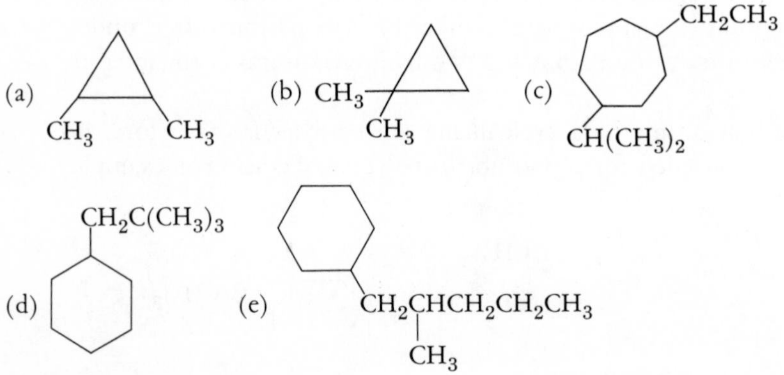

(a) CH_3 CH_3

(b) CH_3 CH_3

(c) CH_2CH_3 $CH(CH_3)_2$

(d) $CH_2C(CH_3)_3$

(e) $CH_2CHCH_2CH_2CH_3$
CH_3

EXERCISE 11-8 Draw the structural formulas corresponding to each of the following IUPAC names:
(a) *sec*-butylcyclobutane
(b) 1,3-di-*t*-butylcyclopentane
(c) 1-isobutyl-3-propylcyclononane
(d) hexamethylcyclopropane

A carbon atom is designated primary, secondary, or tertiary according to the number of carbon atoms bonded to it. A single carbon atom attached to only one other carbon is called a primary carbon and is sometimes designated as 1°. A carbon that is attached to two other carbons is called a secondary carbon (2°). A carbon attached to three other carbons is called a tertiary carbon (3°). A carbon attached to four other carbons is called a quaternary carbon (4°). An example of each is shown in Figure 11-2a.

Each hydrogen atom can be given the same designation of primary (1°),

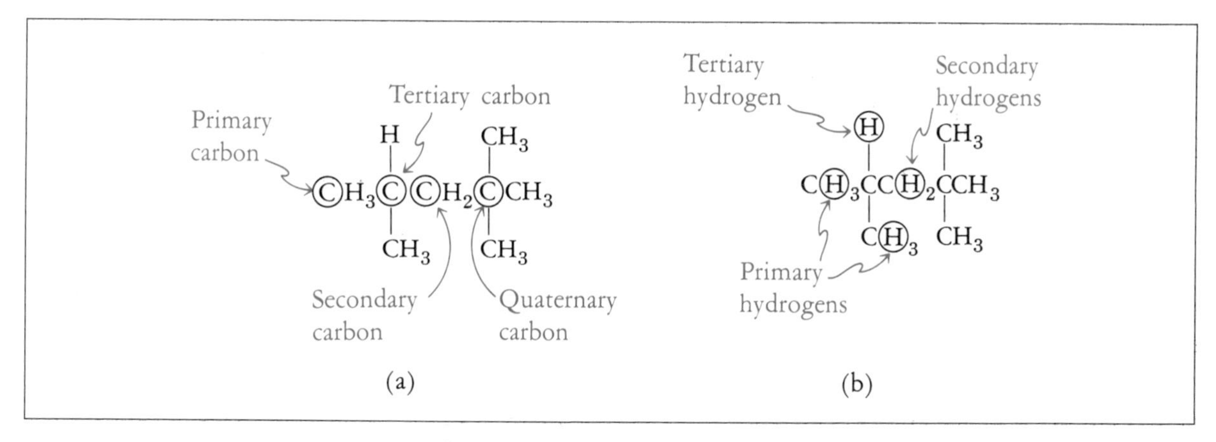

Fig. 11-2. Examples of primary (1°), secondary (2°), and tertiary (3°) carbons and hydrogens.

secondary (2°), or tertiary (3°) as the carbon atom to which it is bonded. Examples of such hydrogen atoms are shown in Figure 11-2b.

EXERCISE 11-9 In the following structural formulas, classify the carbons and hydrogens as 1°, 2°, or 3°.

(a) $(CH_3)_2CHCH_3$ (drawn as CH_3 and CH_3 both bonded to $CHCH_3$)

(b) $CH_3CH_2C(CH_3)_2$ with H on the C

(c) $CH_3CH_2CH_2CH_3$

(d) $(CH_3)_2CHC(CH_3)_2CH_3$ (drawn as CH_3 and CH_3 bonded to $CHCCH_3$, with CH_3 above and below the second C)

11.7 PHYSICAL PROPERTIES

In our everyday life we come into contact with alkanes. Natural gas or fuel oil may heat our homes. Gasoline is used as fuel for automobiles, and asphalt is used to pave our streets. From these daily observations, we realize that the physical state of alkanes can vary from gases to thick, viscous liquids and solids. In general, the boiling and melting points increase as the number of carbon atoms increases. Thus, alkanes containing fewer than five carbon atoms are gases, whereas those containing five to ten carbons are liquids. Alkanes containing 11 or more carbons are viscous liquids, whereas those containing more than 20 carbon atoms are waxlike solids. The melting and boiling points of the alkanes containing one to five carbon atoms are given in Table 11-3.

Alkanes contain only single covalent bonds, either carbon-carbon or carbon-hydrogen. Because the electronegativities of carbon and hydrogen are nearly the same, alkanes are *nonpolar* molecules. This nonpolar character accounts for the fact that alkanes have lower boiling points than most other organic compounds of comparable molecular weight. For example, the boiling point of ethane, CH_3CH_3 (molecular weights, 30 amu), is $-89°$ C, whereas that of methyl alcohol, CH_3OH (molecular weight, 32 amu) is 66° C. In addition, their nonpolarity makes alkanes insoluble in water.

11.8 THREE-DIMENSIONAL STRUCTURE OF ALKANES

All of the objects that we see around us exist in three dimensions. Therefore, it is not surprising to find that the molecules that make up all of these objects also occupy the three dimensions of space; that is, a molecule has a particular size and shape. We examined the shapes of a few simple molecules in Section 6.1. We learned that the four hydrogen atoms bonded to the central carbon of methane describe a unique arrangement in space called a tetrahedron. The four H—C—H bond angles are 109.5 degrees in the arrangement. It has been found that a value very near 109 degrees is characteristic of the bond angle between any two of the four atoms bonded to a carbon atom of an alkane. Such a bond angle allows the four atoms around the central carbon atom to be as far away as possible from one another. A carbon atom with a bond angle near 109 degrees is called a tetrahedral carbon.

Size and shape are two of several factors that affect the chemical and physical properties of molecules, as we learned in Chapter 6. To understand how the three-dimensional shape of a molecule affects its properties, we must visualize the arrangement of its atoms in space. Therefore, we must be able to draw its structure in three dimensions. How do we do this? The problem is to represent a three-dimensional object on a two-dimensional surface such as the page of a book. One way is to try to draw a three-dimensional representation of the molecule on the page, just as an artist paints a landscape on canvas. Such a representation is shown for methane in Figure 11-3. A solid line between atoms indicates that the bond is in the plane of the page. A dashed line indicates that the bond projects below the page, and a solid wedge indicates that the bond projects above the page.

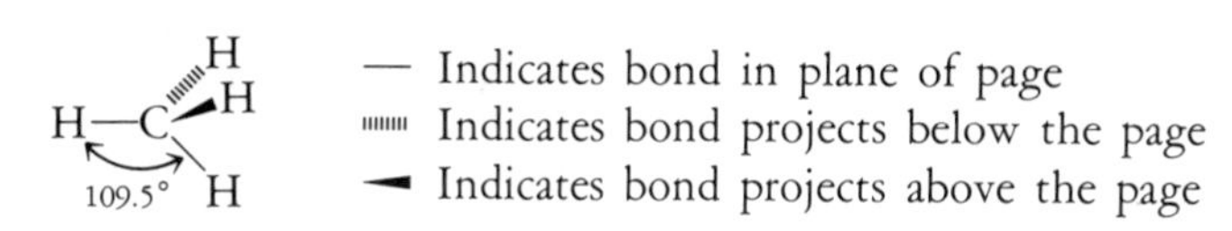

Fig. 11-3. A three-dimensional representation of methane.

It may be difficult for the reader to visualize the three-dimensional arrangement of the atoms of methane when viewed on the two-dimensional page of a book. There is only one remedy for this problem: construction of a three-dimensional model. Molecular models are available from most college bookstores. These permit the construction of three-dimensional models for most of the common classes of organic molecules.

11.9 CONFORMATIONAL ISOMERS

OPTIONAL SECTION

At room temperature, the atoms of a molecule are not stationary. Rather, the atoms forming a bond vibrate about an equilibrium position, and atoms and groups rotate about certain kinds of bonds. Let us examine one particular type of rotation within organic molecules that results in conformational isomers.

To show conformational isomers, it is necessary to introduce a projection method of representing the three-dimensional structure of molecules. This projection, called a Newman projection, is made by viewing the molecule along the carbon-carbon single bond, as shown in Figure 11-4. The front carbon atom (carbon 2) is designated on the page as a point. The bonds from this carbon are designated by three lines, 120 degrees apart, radiating from the point. The circle represents the back carbon (carbon 1), and the bonds to this carbon are also designated by three lines, 120 degrees apart, extending from the circle. This projection is summarized in Figure 11-5a. In Figure 11-5b this projection is applied to ethane. Notice that in the Newman projection the normal 109.5 degree H—C—H angle becomes 120 degrees. This occurs as a result of projecting a three-dimensional structure onto a planar surface. These conclusions should be verified with the aid of molecular models.

Let us turn our attention to the rotation of the carbon-carbon bond in ethane. When this bond is rotated, the positions of the three hydrogens on one carbon change relative to those on the other. This is shown in Figure 11-6 using Newman projections of ethane. Let us start with the hydrogens

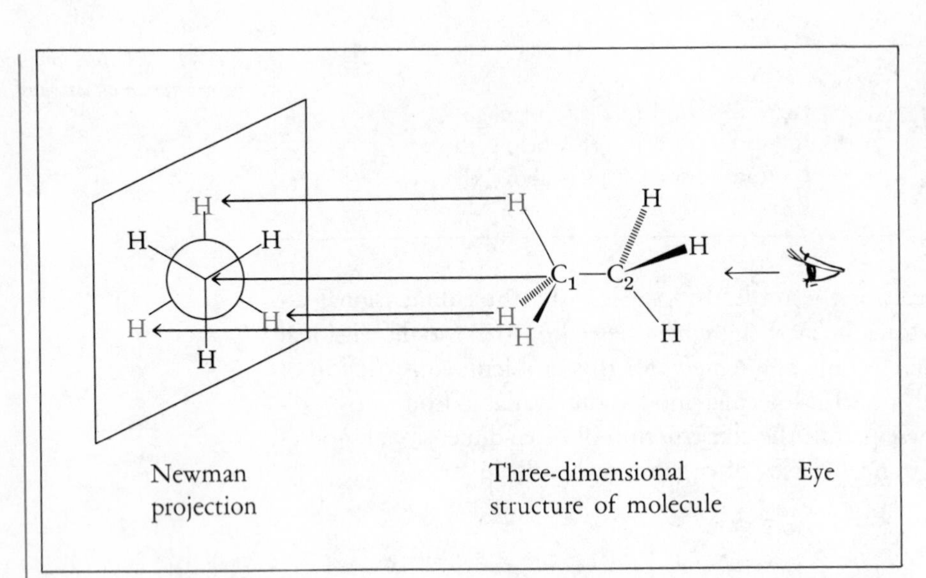

Fig. 11-4. The relationship between the three-dimensional representation of ethane and its projection onto a plane.

of ethane arranged as in Figure 11-6a. In this structure, each bond to a hydrogen on the front carbon bisects the angle between the two hydrogens of the rear carbon. If we rotate the front carbon 60 degrees clockwise relative to the rear carbon, the positions of the hydrogens on the front carbon are changed relative to those on the rear carbon. The new relative positions are illustrated in Figure 11-6b. When you look down the carbon-carbon axis, each bond to a hydrogen of the front carbon will eclipse a bond to a hydrogen of the rear carbon. By rotating the front carbon by another 60 degrees clockwise, the hydrogens are changed to relative positions identical to the starting positions (Figure 11-6c). These statements should be verified with the aid of molecular models.

These different arrangements of atoms achieved by rotation about single

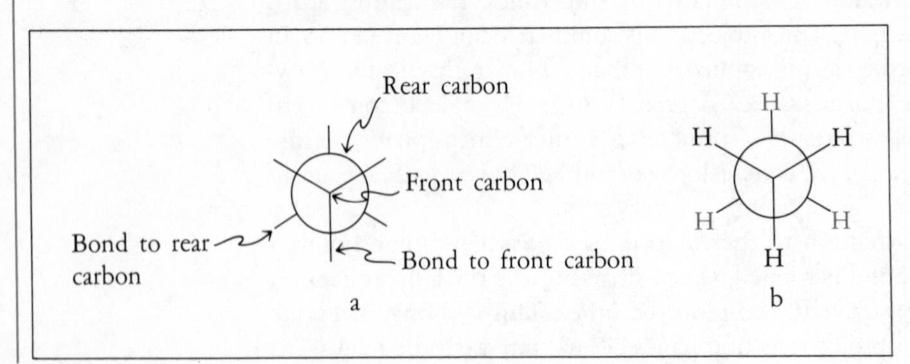

Fig. 11-5. The Newman projection (a) viewed along carbon-carbon bond, (b) applied to ethane.

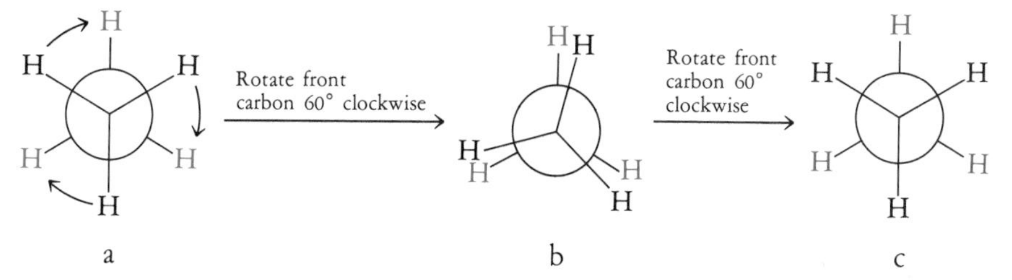

Fig. 11-6. Newman projections of ethane.

bonds are called *conformations.* The structures represented by Figures 11-6a and 11-6b are conformational isomers. They can be interconverted simply by rotating one carbon relative to the other. Three of these conformations have been given specific names. Figures 11-6a and 11-6c represent the *staggered conformation;* Figure 11-6b represents the *eclipsed conformation.* The conformations intermediate between staggered and eclipsed are called *skew conformations.*

There are more than three conformations of ethane. In fact, any rotation of one carbon relative to the other forms a new conformation of the molecule. Consequently, there are an infinite number of conformations of ethane. Particular attention is given to the staggered and eclipsed conformations of ethane because they represent the two extremes of the relative positions of hydrogens on adjacent carbons. In the staggered conformation the hydrogens on adjacent carbons are as far away from each other as possible, whereas in the eclipsed conformation the hydrogens on adjacent carbons are as close together as they can get. At room temperature ethane exists in all possible conformations.

Remember that conformational isomers can be interconverted by simple rotation about a single carbon-carbon bond. *No bonds need to be broken.* Because of this facile interconversion, it is impossible to isolate a single conformational isomer at room temperature.

EXERCISE 11-10 Draw the Newman projections for the staggered and eclipsed conformations of $Cl_3C—CCl_3$.

In cycloalkanes, the ring prevents free rotation about the carbon-carbon single bonds in the ring. However, cycloalkanes have characteristic three-dimensional structures.

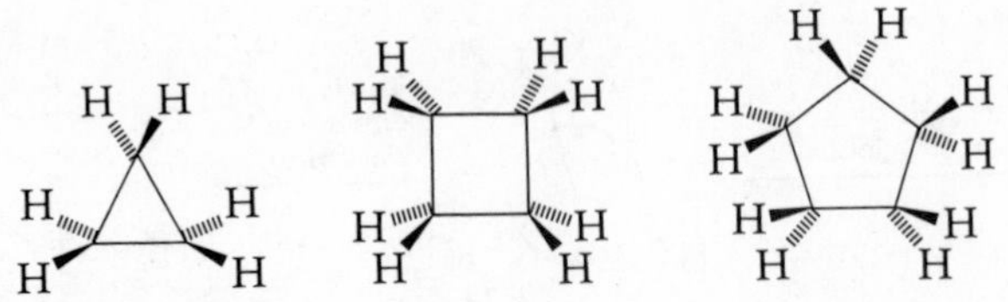

Cyclopropane Cyclobutane Cyclopentane

Fig. 11-7. Three-dimensional structure of cyclopropane, cyclobutane, and cyclopentane.

11.10 THREE-DIMENSIONAL STRUCTURE OF CYCLOALKANES

In this section, we will consider the three-dimensional structure of the cycloalkanes containing three, four, five, and six carbon atoms.

The three carbon atoms of cyclopropane define a plane. The carbon atoms of cyclobutane and cyclopentane are close enough to a plane that we can regard these rings as flat. The three-dimensional structures of these three rings are shown in Figure 11-7.

Because these rings are flat, it is possible to place substituents on adjacent carbons in two different positions relative to each other. For example 1,2-dimethylcyclopropane can exist as two compounds that differ in the arrangement of the methyl groups. In the *cis* isomer, both methyl groups are above (or below) the ring; in the *trans* isomer, one group is on each side of the ring.

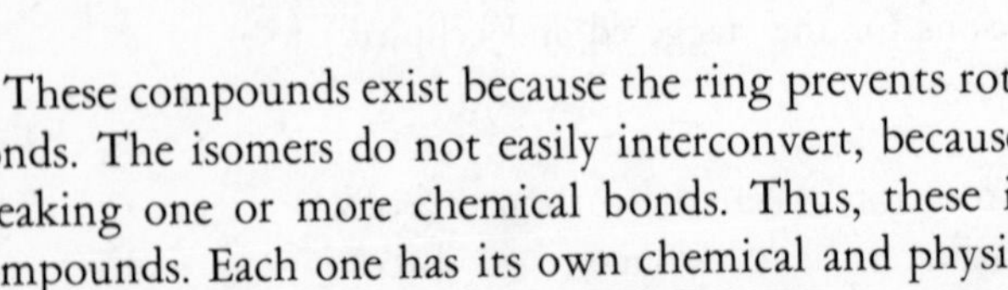

cis-1,2-Dimethylcyclopropane *trans*-1,2-Dimethylcyclopropane

These compounds exist because the ring prevents rotation about the ring bonds. The isomers do not easily interconvert, because this would require breaking one or more chemical bonds. Thus, these isomers are different compounds. Each one has its own chemical and physical properties. These compounds are *geometric isomers.* The following are examples of other cycloalkanes that can exist as geometric isomers:

cis-1,3-Dimethylcyclobutane

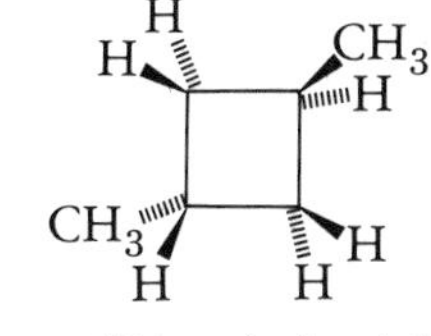

trans-1,3-Dimethylcyclobutane

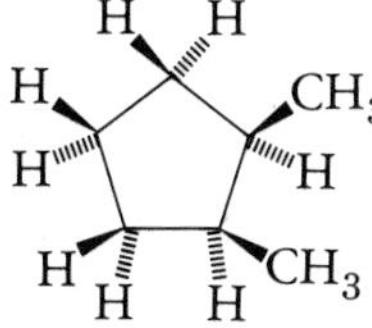

cis-1,2-Dimethylcyclopentane

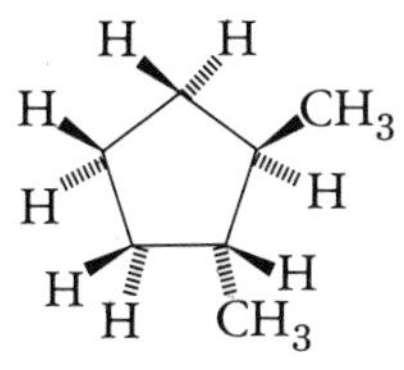

trans-1,2-Dimethylcyclopentane

Rings containing six or more carbon atoms are not flat. Cyclohexane is a typical example of a nonplanar ring. One of the ways that the six atoms of the cyclohexane ring can be arranged is shown in Figure 11-8. For simplicity, only the carbon atoms are shown. Because of its resemblance to a chair, this arrangement is called the chair form of cyclohexane.

When the hydrogens are included in this representation, it is apparent that there are two kinds of hydrogen. Six of them are perpendicular to the ring, as shown in Figure 11-9a. These hydrogens are located in the axial

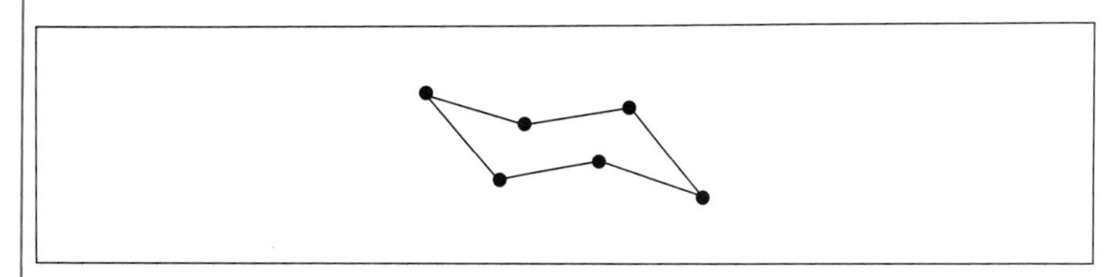

Fig. 11-8. The chair form of cyclohexane. The hydrogen atoms are omitted for clarity.

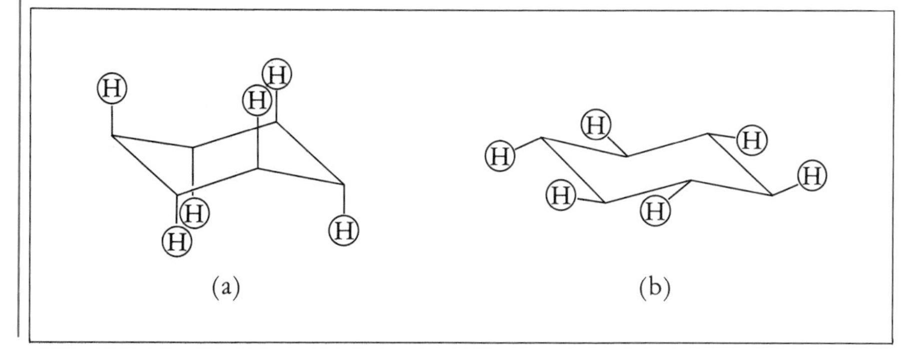

Fig. 11-9. The geometrically different hydrogens of cyclohexane: (a) the axial hydrogens; (b) the equatorial hydrogens.

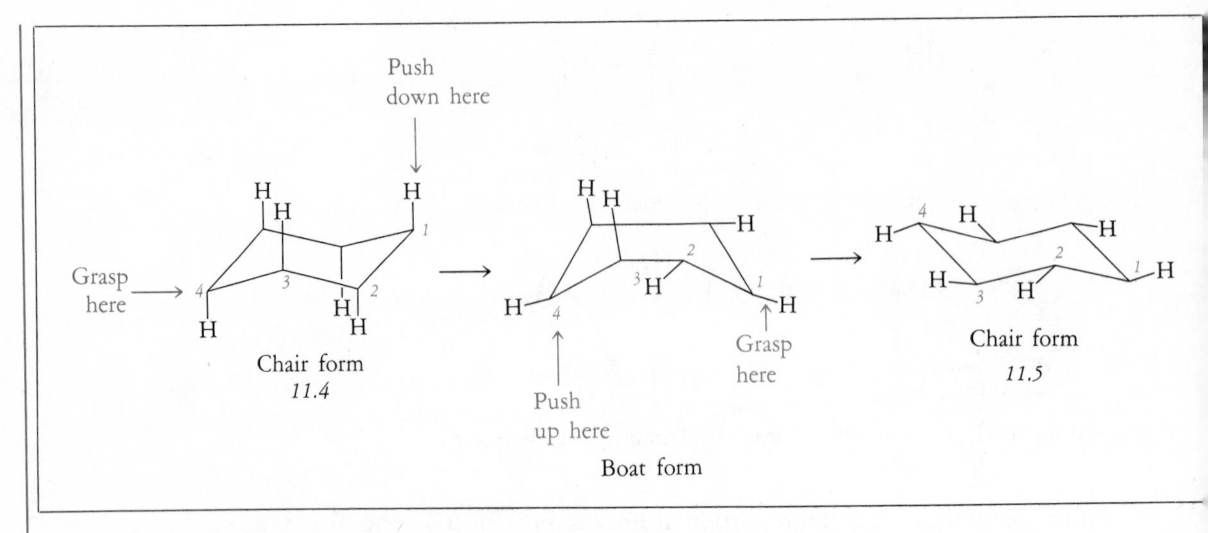

Fig. 11-10. The interchange of hydrogens from the axial to the equatorial positions. Some hydrogens have been omitted for simplicity.

positions and are called the axial hydrogens. The other six radiate outward from the ring, as shown in Figure 11-9b. These hydrogens are located in the plane of the ring and are called the equatorial hydrogens.

It has been found that the positions of these two sets of hydrogens can be interchanged by "flipping" the ring, as illustrated in Figure 11-10. In this way, the axial hydrogens of chair form *11.4* become the equatorial hydrogens of chair form *11.5*. This can easily be demonstrated with the aid of molecular models. Make a model of the chair form of cyclohexane. Position it so that it resembles chair form *11.4*. Grasp carbon 4 firmly and pull down carbon 1. This arrangement of the six ring carbons is called the boat form. Now grasp carbon 1 of the boat form and push up at carbon 4. This forms another chair form of cyclohexane, structure *11.5*. But the hydrogens in the axial positions of the original chair form *11.4* are now in the equatorial positions of chair form *11.5*. Notice that no bonds are broken in this process. This type of "flipping" of the ring occurs many times per second at room temperature and, as a result, all the hydrogens of cyclohexane are equivalent.

Disubstituted cyclohexanes can also exist as *cis-trans* isomers. Because the ring of cyclohexane is not flat, it is more difficult to visualize these isomers.

One chair form of each of the *cis-* and *trans-*1,2-dimethylcyclohexanes is illustrated in Figure 11-11. Notice that the two methyl groups of the *trans* isomer are both in the equatorial positions. In the *cis* isomer, one methyl group is in the axial position and the other is in the equatorial position.

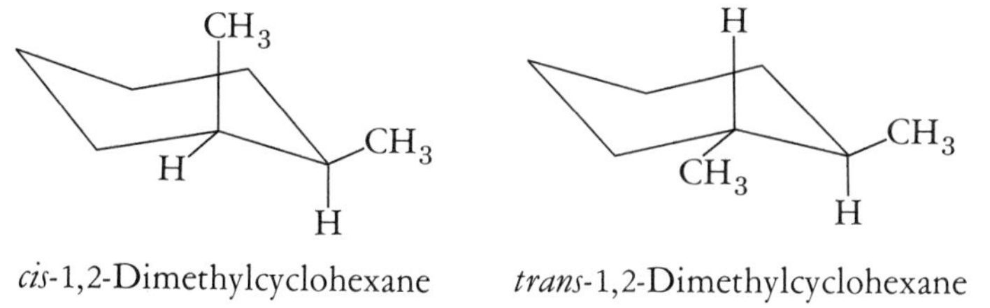

CH_3 H CH_3 H H CH_3 CH_3 H

cis-1,2-Dimethylcyclohexane *trans*-1,2-Dimethylcyclohexane

Fig. 11-11. cis- *and* trans-*1,2-Dimethylcyclohexane.*

These small differences in the shapes of molecules can have important effects on their reactivity in living systems, as we will learn in Chapter 21. In the meantime, let us turn our attention to the chemical reactions of alkanes and cycloalkanes.

11.11 CHEMICAL REACTIONS

The alkanes are the least reactive organic compounds that we will encounter. They do not react at room temperature with most acids (HCl, H_2SO_4), bases (aqueous NaOH or KOH), oxidizing reagents ($K_2Cr_2O_7$, $KMnO_4$), or reducing reagents.

Recognition of this fact makes predicting the reactions of organic molecules much easier, because all organic molecules can be divided into three parts, as illustrated in Figure 11-12. One part is alkane-like. The second part is the functional group (or groups), and the third part is the methylene group adjacent to the functional group. Because of the presence of the adjacent functional group, this methylene group is activated and often undergoes specific chemical reactions. The alkane-like part, however, is unreactive. This allows us to focus our attention on the reactions of the functional group and the methylene group adjacent to it. Thus, *the alkane-like parts of organic molecules serve as carriers of the functional group and rarely are involved in the chemical reaction.*

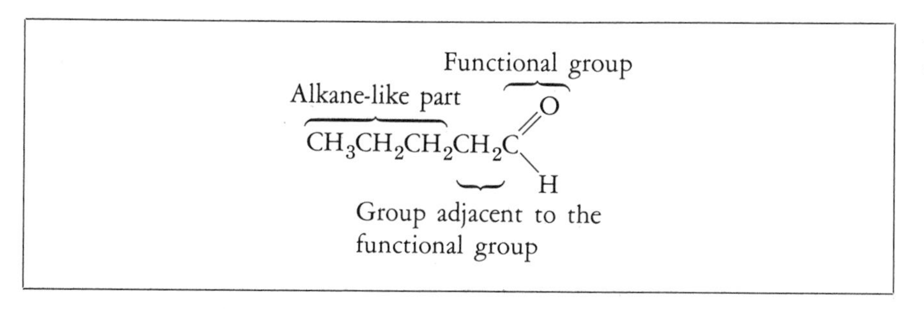

Fig. 11-12. The three parts of an organic molecule.

EXERCISE 11-11 In each of the following molecules, circle the three regions described in Section 11-11.

(a) $CH_3CH_2CH_2CH_2OH$ (b) $(CH_3)_2CHCO_2H$

(c) $(CH_3)_3CCH_2CH_2NH_2$

(d) 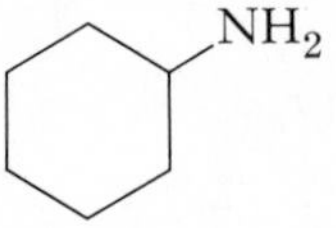 cyclohexyl–NH_2

(e) $CH_3CH_2CH_2\overset{O}{\overset{\|}{C}}CH_2CH_2CH_3$

Among the few reactions that alkanes undergo are their reactions with chlorine and oxygen. We will examine these reactions, because the first is used commercially to prepare chlorinated hydrocarbons and the second liberates heat, which accounts for the use of alkanes as fuels.

Chlorination

When a mixture of an alkane and chlorine gas is heated or exposed to light, a reaction quickly occurs. The products are chloroalkanes and hydrogen chloride. The overall reaction can be summarized as follows:

$$RH + Cl_2 \xrightarrow[\text{or } \Delta]{\text{energy as light}} RCl + HCl$$

The symbol RH represents the alkane or cycloalkane, where R is an alkyl or cycloalkyl group. This shorthand will be used frequently in this book to indicate that a chemical reaction is general for any R group. The symbol Δ indicates energy as heat. Specific examples of monochlorination are:

$$CH_4 + Cl_2 \xrightarrow{\text{energy}} CH_3Cl + HCl$$

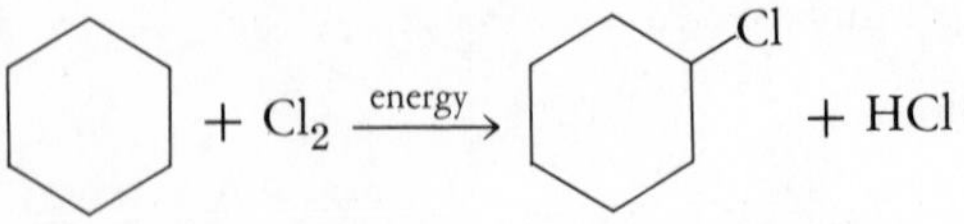

$$\text{cyclohexane} + Cl_2 \xrightarrow{\text{energy}} \text{chlorocyclohexane (C}_6\text{H}_{11}\text{Cl)} + HCl$$

Chlorination of an alkane is a *substitution reaction.* One or more hydrogen atoms of the alkane are substituted by a chlorine atom.

If an excess of chlorine gas is used, further chlorination of the product occurs. For example, the chlorination of methane can form a mixture of four products: methyl chloride (CH_3Cl), methylene chloride (CH_2Cl_2), chloroform ($CHCl_3$), and carbon tetrachloride (CCl_4). These mixtures of chloroalkanes are commercially important as solvents and are widely used in dry cleaning.

Oxidation

Hydrocarbons are used principally as fuels. In the presence of excess oxygen and a spark, alkanes burn to form carbon dioxide, water, and a large quantity of energy released as heat:

$$2\,C_2H_6 + 7\,O_2 \longrightarrow 4\,CO_2 + 6\,H_2O + \Delta$$

It is this energy, liberated in the reaction, that accounts for the use of alkanes as sources of heat and power.

If insufficient oxygen is available, partial oxidation of the alkanes occurs to form carbon monoxide or even elemental carbon:

$$2\,CH_4 + 3\,O_2 \longrightarrow 2\,CO + 4\,H_2O$$

$$CH_4 + O_2 \longrightarrow C + 2\,H_2O$$

Such incomplete oxidation often occurs in automobile engines. The result is carbon deposits on the cylinder head and the emission of poisonous carbon monoxide.

11.12 SUMMARY

We know exactly how the atoms in a molecule are arranged by looking at the structural formula of the molecule. We can distinguish different compounds, called structural isomers, that have the same chemical formula but different structural formulas. From the structural formula of a molecule, we can identify its functional group(s). These specific atoms or groups of atoms bonded to carbon undergo characteristic reactions. Organic compounds that contain the same functional group undergo similar chemical reactions. In this way, we can classify and systematically study the chemical and physical properties of organic compounds.

The first class of organic compounds that we study are those obtained from petroleum. Almost all of these compounds contain only the elements carbon and hydrogen, and they are given the general name hydrocarbons. Alkanes are hydrocarbons that contain only single bonds. They have the general formula C_nH_{2n+2}. The alkanes represent a homologous series; that is, successive members of the series differ by a methylene ($—CH_2—$) group. Alkanes are systematically named by the IUPAC system. In this way, each member of a homologous series can be given a distinctive name.

The carbon bond angle is 109.5 degrees. This is the characteristic bond angle of a carbon bonded to four atoms or groups. The four bonds of carbon point toward the corners of a tetrahedron. Cycloalkanes are alkanes that contain carbon atoms in a ring. They form a homologous series with the general formula C_nH_{2n}.

Alkanes and cycloalkanes are the least reactive types of organic compound. They do not react at room temperature with most acids, bases, or oxidizing agents or any reducing agent. Among the few reactions that alkanes undergo are their reactions with chlorine and oxygen. The first is used commercially to prepare degreasing solvents, and the second releases heat, which accounts for the use of alkanes as fuels.

REVIEW OF TERMS, CONCEPTS, AND REACTIONS

Terms

ALKANES Compounds containing only the two elements carbon and hydrogen and single covalent bonds.

ALKYL GROUP An alkane in which one hydrogen has been removed. For example, methane (CH_4) minus one hydrogen becomes a methyl group ($CH_3—$).

CONFORMATIONS The different arrangements of atoms in space achieved by rotation about single bonds.

CYCLOALKANES Alkanes in which the carbon atoms form a ring.

FUNCTIONAL GROUP An atom or group of atoms bonded to carbon where chemical reactions occur.

GEOMETRIC ISOMERS Molecules that differ in the three-dimensional arrangements of their atoms in space.

HOMOLOGOUS SERIES Compounds of the same type (containing the same functional group) that differ by a $—CH_2—$ group.

IUPAC International Union of Pure and Applied Chemistry. The international society that now governs the system for naming organic compounds.

METHYLENE GROUP A $—CH_2—$ unit of an organic compound.

PRIMARY CARBON A carbon atom bonded to only one other carbon.

PRIMARY HYDROGEN A hydrogen atom bonded to a primary carbon.

QUATERNARY CARBON A carbon atom bonded to four other carbons.

SATURATED HYDROCARBONS Hydrocarbons that contain only single bonds (another name for alkanes).

SECONDARY CARBON A carbon atom bonded to two other carbons.

SECONDARY HYDROGEN A hydrogen atom bonded to a secondary carbon.

STRUCTURAL FORMULA A chemical formula that indicates specifically which atoms in a molecule are bonded to which other atoms and by what kind of bonds.

STRUCTURAL ISOMERS Different compounds that have the same molecular formulas.

SUBSTITUTION REACTION A reaction in which a group or atom of a molecule is replaced by another atom or group.

TERTIARY CARBON A carbon atom bonded to three other carbons.

TERTIARY HYDROGEN A hydrogen atom bonded to a tertiary carbon.

TETRAHEDRAL CARBON A carbon atom whose four bonds are directed toward the corners of a tetrahedron.

UNSATURATED HYDROCARBONS Hydrocarbons that contain double or triple bonds (another name for alkenes and alkynes).

Concepts

1. Organic compounds that contain the same functional group undergo similar chemical reactions.
2. Petroleum supplies much of the energy and many of the simple organic compounds used in our society.
3. Most compounds in petroleum are hydrocarbons.
4. The general formula for an alkane is C_nH_{2n+2}, where n is equal to the number of carbon atoms. For a cycloalkane, the general formula is C_nH_{2n}.
5. The ending *-ane* in the IUPAC name signifies that the compound is an alkane.
6. Alkanes do not react at room temperature with most acids, bases, or oxidizing agents or any reducing agent. Similarly, the alkane-like parts of organic molecules are rarely involved in chemical reactions.

Reactions

1. Chlorination

$$RH + Cl_2 \xrightarrow[\text{or heat}]{\text{light}} RCl + HCl$$

2. Oxidation

$$2\,C_nH_{2n+2} + (3n + 1)\,O_2 \longrightarrow 2n\,CO_2 + 2(n + 1)\,H_2O + \Delta$$

EXERCISES 11-12 Give the IUPAC name for each of the following compounds:

(a) 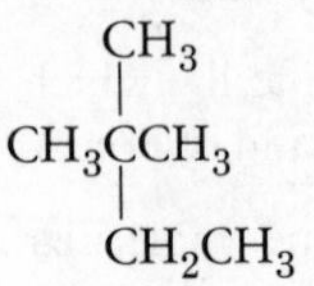 CH_3 above, $CH_3\underset{|}{\overset{|}{C}}CH_3$, CH_2CH_3 below

(b) CH_3 above, $CH_3\overset{|}{C}HCH_2CH_2CH_2CH_3$

(c) CH_2CH_3 above, $CH_3\overset{|}{C}H\underset{|}{C}HCH_2CH_3$, $CH_2CH_2CH_3$ below

(d) CH_3 and CH_2CH_3 above, $CH_3\overset{|}{C}H\underset{|}{C}HCH_2CH_2\overset{|}{C}HCH_2CH_3$, $CH_3CH_2CH_2$ below

(e) CH_3 CH_3 above, $CH_3\underset{|}{\overset{|}{C}}CH_2\underset{|}{\overset{|}{C}}CH_2CH_2CH_3$, CH_3 CH_2CH_3 below

(f) cyclobutane with $CH(CH_3)_2$ and CH_3

(g) cyclohexane with CH_3 and $C(CH_3)_3$

(h) cyclopentane with $-\overset{CH_3}{\overset{|}{C}}HCH_2CH_3$

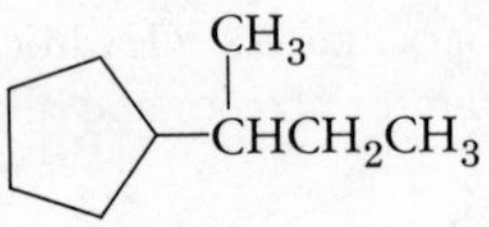

(i) 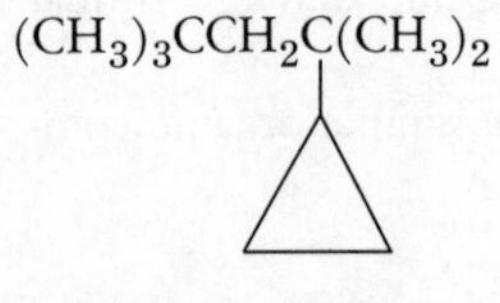 $(CH_3)_3CCH_2C(CH_3)_2$ with cyclopropane below

(j) 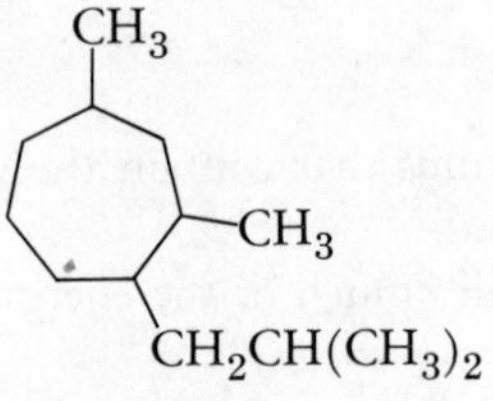 cycloheptane with CH_3, $-CH_3$, and $CH_2CH(CH_3)_2$

(k) 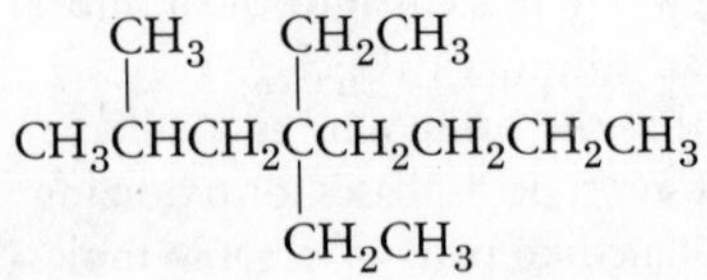 CH_3 and CH_2CH_3 above, $CH_3\overset{|}{C}HCH_2\underset{|}{\overset{|}{C}}CH_2CH_2CH_2CH_3$, CH_2CH_3 below

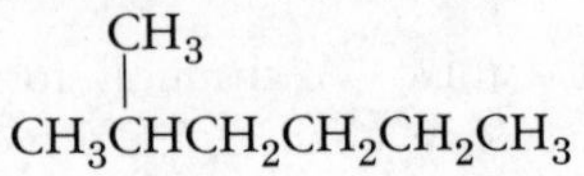

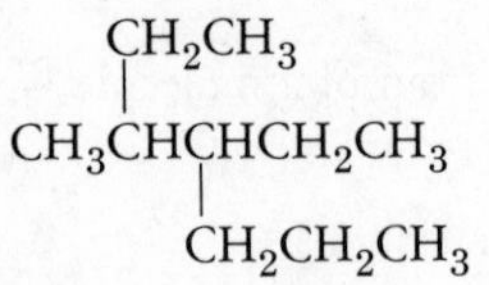

(l) 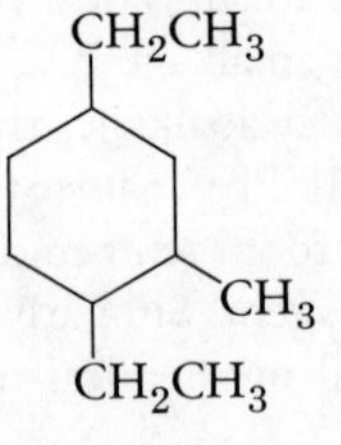 cyclohexane with CH_2CH_3, CH_3, and CH_2CH_3

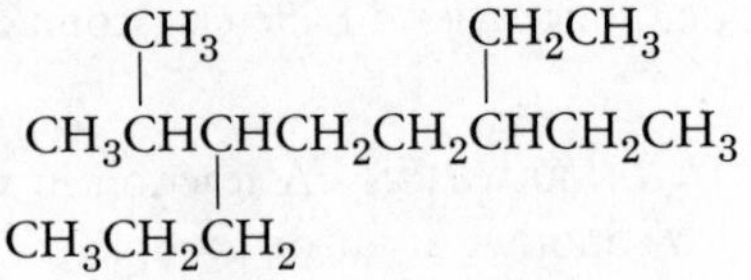

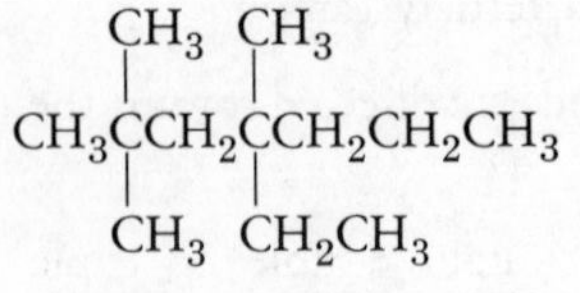

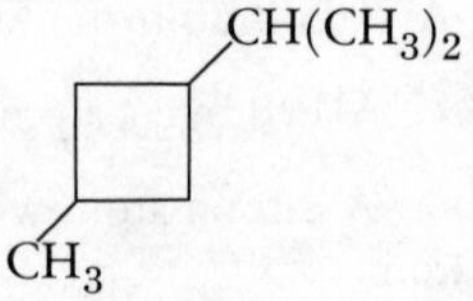

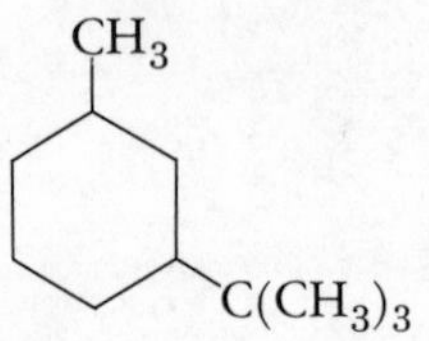

11-13 Draw the structural formulas corresponding to each of the following names:

(a) 1-*t*-butyl-4-ethylcyclohexane
(b) 2,2,3-trimethylbutane
(c) 1,4-diethylcycloheptane
(d) 2,5-dimethylhexane
(e) 4-ethyl-3-methylheptane
(f) 5-*t*-butyl-7-ethyl-3,3,5,8-tetramethyldecane

11-14 Among the structures in exercise 11-13, pick out and show specifically a compound that has

(a) a primary carbon
(b) a secondary carbon
(c) a tertiary carbon
(d) a quaternary carbon
(e) a *t*-butyl group
(f) two ethyl groups

11-15 Match the structural formula in list A with its correct name in list B.

A

(a) $(CH_3)_2CHCH_2CH_2CH_3$
(b) $(CH_3)_2CHCH_2C(CH_3)_3$
(c) $CH_3CHCH_2CHCH_3$ (with CH_3CH_2 on the first CH and CH_2CH_3 on the second CH)
(d)

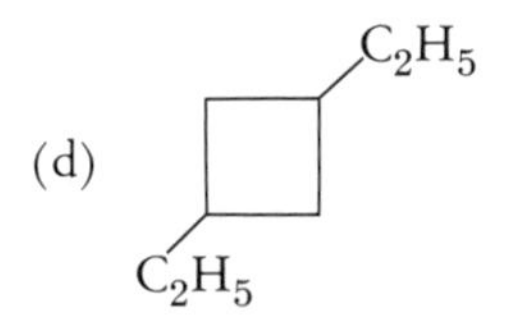

(e) $(CH_3)_2CHCHCH_2CH_2CH_3$ (with $CH_3CH_2CHCH_3$ on the second CH)
(f) cyclobutane with three CH_3 groups

CH_3
CH_3 CH_3

B

(a) 1,2,3-trimethylcyclobutane
(b) 3,5-dimethylheptane
(c) 2-methylpentane
(d) 3-methyl-4-isopropylheptane
(e) 2,2,4-trimethylpentane
(f) 1,3-diethylcyclobutane

ALKENES AND ALKYNES

12

In this chapter we will study the chemistry of the carbon-carbon double and triple bonds. Compounds that contain these functional groups are important both in industry and in living systems. We study alkenes and alkynes together because their chemical and physical properties are very similar. Let us start our study of these unsaturated hydrocarbons by learning to name alkenes.

12.1 NAMING ALKENES

Alkenes, especially complex ones, are best named by the logical IUPAC system. The system is similar to that for alkanes, with the following additions:

1. The longest continuous chain of carbon atoms *containing all the double bonds* is chosen for the parent name.
2. The parent name is obtained by substituting the ending *-ene* for the ending *-ane* of the corresponding alkane.
3. The chain is numbered to give the lowest numbers to the first carbon of the double bonds *regardless of substituent numbers.*
4. The suffix *-ene* is used for one carbon-carbon double bond, *-diene* for two, *triene* for three, and so forth.

Some examples of the use of these rules are as follows:

$$\overset{5}{C}H_3\overset{4}{C}H_2\overset{3}{C}H(CH_3)\overset{2}{C}H{=}\overset{1}{C}H_2$$

3-Methyl-1-pentene
(*not* 3-methyl-4-pentene)

$$\overset{1}{C}H_2{=}\overset{2}{C}(CH_3)\overset{3}{C}H_2\overset{4}{C}H{=}\overset{5}{C}(C(CH_3)_3)\overset{6}{C}H_2\overset{7}{C}H{=}\overset{8}{C}H\overset{9}{C}H_3$$

5-*t*-Butyl-2-methyl-1,4,7-nonatriene

$$\overset{1}{H}C(C_5H_9){=}\overset{2}{C}H\overset{3}{C}H_2\overset{4}{C}H_3$$

1-Cyclopentyl-1-butene

Because alkenes are named by selecting the longest continuous carbon chain that contains the double bonds, it is possible that this chain is not the longest continuous chain in the molecule. In the following example there is a continuous chain of seven carbons, but the longest chain that contains the double bond is a six-carbon chain. Therefore, the proper name for this compound is 2-ethyl-1-hexene:

$$CH_3CH_2\overset{2}{C}(={}_1CH_2)\overset{3}{C}H_2\overset{4}{C}H_2\overset{5}{C}H_2\overset{6}{C}H_3$$

2-Ethyl-1-hexene

Common names are often used for the smaller alkenes. The three most common are the following:

$CH_2{=}CH_2$ Ethylene $CH_3CH{=}CH_2$ Propylene $CH_3C(CH_3){=}CH_2$ Isobutylene

Common names are generally avoided with alkenes containing more than four carbons because of the large number of isomers possible.

*EXERCISE 12-1 Give the IUPAC names for ethylene, propylene, and isobutylene.

In naming alkenes, it is sometimes convenient to regard the double bond as a substituent. This most frequently occurs when the double bond is attached to a ring containing other substituents. In this case, the double bond is named as a substituent. Three of the most common substituents and their names are the following:

* The answers for the exercises in this chapter begin on page 868.

$CH_2{=}CH{-}$ Vinyl $CH_2{=}CHCH_2{-}$ Allyl $CH_3CH{=}CH{-}$ Propenyl

Examples of compounds in which a double bond is named as a substituent are as follows:

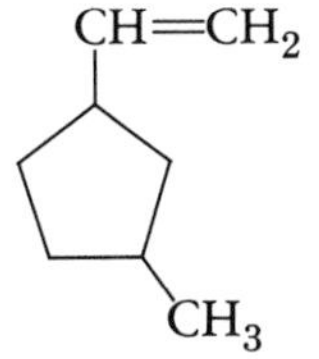

3-Methyl-1-vinylcyclopentane

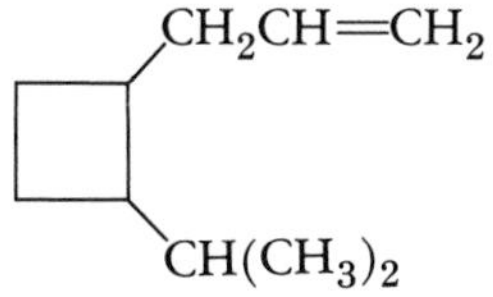

1-Allyl-2-isopropylcyclobutane

When naming cyclic alkenes, the direction of numbering the ring carbons is chosen so that the carbons of the double bond receive numbers 1 and 2 *and* the substituents receive the lowest numbers possible.

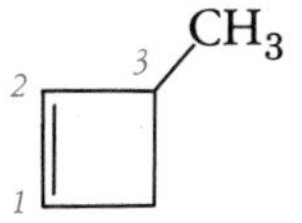

3-Methylcyclobutene

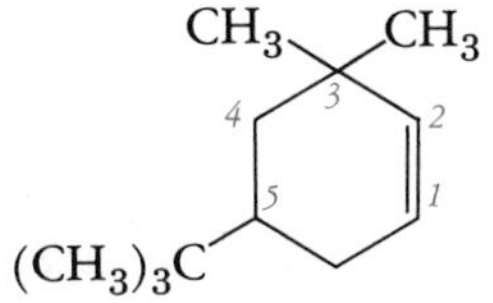

5-*t*-Butyl-3,3-dimethylcyclohexene

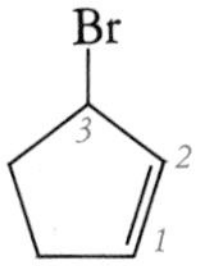

3-Bromocyclopentene

Notice that the number 1 indicating the position of the double bond in the cyclic alkene is understood and is not included in the name.

EXERCISE 12-2 Give the IUPAC name for each of the following compounds:

(a) $CH_2{=}CHCH_2CH(CH_3)_2$

(b) 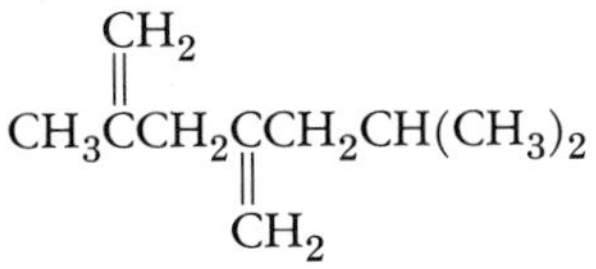 $CH_3C({=}CH_2)CH_2C({=}CH_2)CH_2CH(CH_3)_2$

(c)

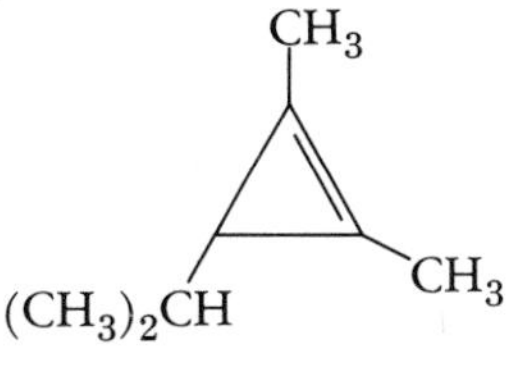

(d)

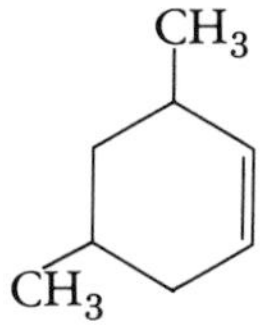

(e)

EXERCISE 12-3 Write the structure for each of the following names:
(a) 1,3-cyclohexadiene
(b) propenylcyclopropane
(c) 1,2-dimethylcyclopentene
(d) 3-*t*-butyl-4,4-dimethyl-1-pentene
(e) 2,5-dimethyl-2-hexene
(f) 2,3-dimethyl-2-pentene

Although the method of naming alkenes and alkanes is similar, there is an added feature to the naming of alkenes. This is caused by the fact that certain alkenes can exist as geometric isomers, as we will learn in the next section.

12.2 STRUCTURE OF ALKENES AND GEOMETRIC ISOMERS

We learned in Section 6.1 that the four hydrogens bonded to the carbon atom of methane are arranged in space in the shape of a tetrahedron. The carbon atoms of a double bond and the atoms bonded to them have a different shape. All of these atoms lie in a plane, as shown for ethylene in Figure 12-1.

The four hydrogens and the two carbon atoms of the double bond of ethylene lie in the same plane with a H—C—H bond angle of nearly 120 degrees. This is the characteristic shape of three atoms or groups bonded to a carbon atom. In this case two atoms are bonded by single bonds and the third atom is bonded by a double bond. This planar arrangement of the bonds about a carbon atom is called a trigonal arrangement and is shown in Figure 12-2. The carbon bond angles are approximately 120 degrees.

The double bond of an alkene is rigid. As a result, alkenes can form geometric isomers. This is shown in Figure 12-3 for 2-butene. The *trans* isomer has the two methyl groups across from each other, and the *cis* isomer has the groups adjacent to each other. The two compounds clearly have the same carbon skeleton, but they differ in the arrangement of this skeleton in

H H
C=C
H H

Fig. 12-1. All the atoms of ethylene lie in the plane of the page.

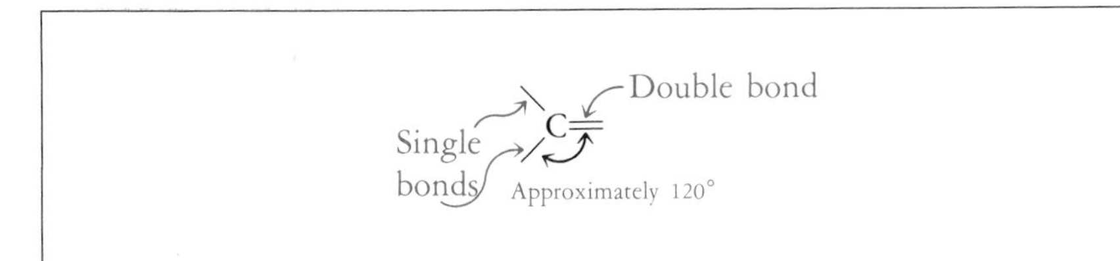

Fig. 12-2. The trigonal arrangement of three bonds to carbon (two single bonds and one double bond) all lie in the plane of the page.

$$\begin{array}{ccc} CH_3 & & H \\ & C{=}C & \\ H & & CH_3 \end{array} \qquad \begin{array}{ccc} CH_3 & & CH_3 \\ & C{=}C & \\ H & & H \end{array}$$

trans-2-Butene *cis*-2-Butene

Fig. 12-3. The cis *and* trans *isomers of 2-butene.*

space. These isomers cannot be interconverted without breaking the double bond, a process that requires a great deal of energy. As a result, *cis* and *trans* isomers of an alkene do not interconvert at room temperature. Each can be isolated and each has its own characteristic chemical and physical properties. This is shown in Table 12-1, where the melting and boiling points of a number of *cis-trans* geometric isomers are listed.

EXERCISE 12-4 Draw the structural formulas for the *cis* and *trans* isomers of the following alkenes:
(a) 2-pentene (b) 3-hexene (c) 4-methyl-2-pentene

The double bonds of all alkenes are rigid, yet not all alkenes can exist as geometric isomers. Why? To identify alkenes that cannot exist as *cis-trans* isomers, look at the atoms or groups bonded to the carbons of the double bond. *If either carbon of the double bond of an alkene is bonded to identical atoms or groups, geometric isomers cannot exist.* For example:

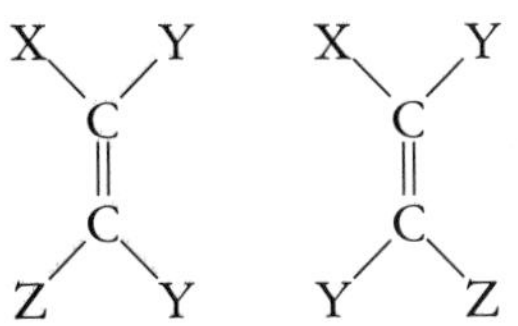

The two structures bonded to each carbon are different, so geometric isomers are possible.

$$\begin{array}{ccc} X & & Z \\ & C & \\ & \| & \\ & C & \\ Y & & Y \end{array} \quad \text{is identical to} \quad \begin{array}{ccc} Z & & X \\ & C & \\ & \| & \\ & C & \\ Y & & Y \end{array}$$

The two structures bonded to one of the carbons are identical, so no geometric isomers are possible.

Table 12-1. Examples of Geometric Isomers of Alkenes

Structure	Name	Boiling Point (°C)	Melting Point (°C)
CH_3 CH_3 C=C H H	*cis*-2-Butene	4	−139
CH_3 H C=C H CH_3	*trans*-2-Butene	1	−160
H H C=C Cl Cl	*cis*-1,2-Dichloroethene	−80	60
H Cl C=C Cl H	*trans*-1,2-Dichloroethene	−50	48
H H C=C HO_2C CO_2H	Maleic acid (*cis*-Butenedioic acid)	130	—
H CO_2H C=C HO_2C H	Fumaric acid (*trans*-Butenedioic acid)	287	—

EXERCISE 12-5 Which of the following alkenes can exist as geometric isomers?
(a) 2,3-dimethyl-2-butene (b) 2-methyl-2-butene
(c) 3-ethyl-3-hexene (d) 1-hexene
(e) 4,4-dimethyl-2-pentene (f) 2-methyl-1-butene

It is important to recognize the alkenes that can exist as *cis-trans* isomers, because the chemical reactivities of these isomers differ in living systems.

12.3 IMPORTANCE OF GEOMETRIC ISOMERS IN LIVING SYSTEMS

In the previous section we learned that geometric isomers are compounds that have the same carbon skeleton but differ in the arrangement of this skeleton in space and consequently have different physical properties. It has been found that many geometric isomers also have different biological activities. One example is the *pheromones.* These are chemical compounds secreted by a species that produce a response in others of the same or different species. For example, some compounds act as sex attractants for a particular species; others act as defenses against enemies, and still others provide various means of communication.

Much work has been done on compounds that are sex attractants for certain types of insects. Such compounds have great potential value as insect-controlling devices. The scent of a particular compound that is a sex attractant can be used to attract a destructive insect to a trap, where it can be eliminated. This method of insect control is valuable because it can be made specific for one type of insect and consequently does not harm the environment.

The following compounds are sex attractants for the insects specified:

$$\begin{array}{c} CH_3(CH_2)_2CH_2(H)C{=}C(H)CH_2(CH_2)_4CH_2O\overset{\displaystyle O}{\overset{\|}{C}}CH_3 \\ \text{Cabbage looper moth} \end{array}$$

$$\begin{array}{c} CH_3(H)C{=}C(H){-}(H)C{=}C(H)(CH_2)_6CH_2OH \\ \text{Codling moth} \end{array}$$

$$\begin{array}{c} CH_3CH_2CH_2(H)C{=}C(H){-}(H)C{=}C(H)(CH_2)_8CH_2OH \\ \textit{12-1} \\ \text{Silkworm moth} \end{array}$$

$$\begin{array}{c} CH_3CH_2(H)C{=}C(H)(CH_2)_9CH_2O\overset{\displaystyle O}{\overset{\|}{C}}CH_3 \\ \text{European corn borer} \end{array}$$

The importance of the geometry about the double bond for the effectiveness of these compounds as sex attractants is shown by the fact that changing the *cis* double bond in compound *12-1* to the *trans* isomer decreases by a factor of 100,000,000,000 the ability of the compound to excite a response in male silkworm moths!

Another example of a biologically active alkene is the synthetic female sex hormone diethylstilbestrol, commonly called DES. This compound was given to women during pregnancy to prevent miscarriages.

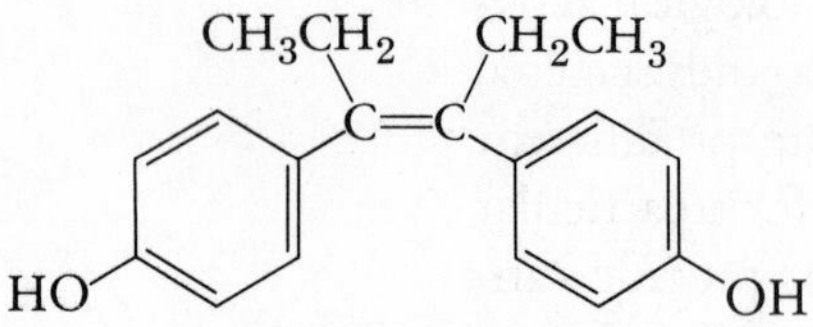

cis-Diethylstilbestrol

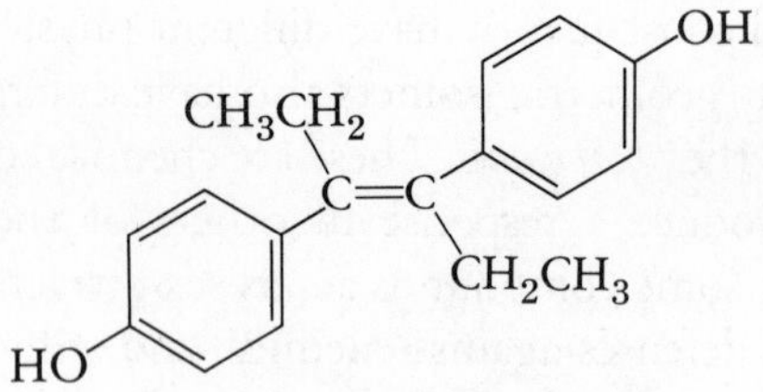

trans-Diethylstilbestrol

Only the *trans* isomer is active for this purpose. Unfortunately this drug has a serious side effect. An unusually large percentage of the daughters of the women who took DES during pregnancy developed cancer of the uterus. As a result, DES is banned for use by humans. However, it is still used as an additive in cattle and chicken feed to promote growth and accelerate gain in weight.

It is the aim of scientists to explain why these isomers have such different chemical reactivities. To do this, it is necessary to learn the characteristic reactions of simple alkenes. We will learn some of these reactions in the next section.

12.4 ADDITION REACTIONS OF ALKENES

Alkenes and alkanes undergo different types of reactions. Alkanes react by substitution, whereas *addition to the double bond* is the characteristic reaction of alkenes. This fact is illustrated by comparing the reactions of chlorine with ethane and ethylene.

$$CH_3CH_3 + Cl_2 \xrightarrow{\text{Light}} ClCH_2CH_3 + HCl \qquad \text{Substitution}$$

$$CH_2{=}CH_2 + Cl_2 \longrightarrow ClCH_2CH_2Cl \qquad \text{Addition}$$

Bromine, strong acids, water in the presence of an acid catalyst, and hydrogen in the presence of a metal catalyst also add to alkenes. In contrast to the low reactivity of the alkanes, alkenes readily undergo addition reactions at room temperature and require no light. Let us examine each of these reactions.

Addition of Hydrogen: Reduction of the Carbon-Carbon Double Bond

The addition of hydrogen to alkenes in the presence of a metal catalyst converts them to alkanes. In this reaction, one hydrogen adds to each carbon of the double bond, as shown in the following equation:

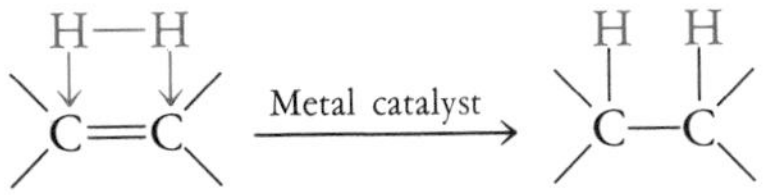

$$\text{H—H} + \text{>C=C<} \xrightarrow{\text{Metal catalyst}} \text{>C(H)—C(H)<}$$

This *hydrogenation* of the double bond is a reduction of the carbon-carbon double bond and is an important reaction in living systems as well as in the laboratory. Some specific examples are as follows:

$$\text{Cyclohexene} + H_2 \xrightarrow[\text{25° C}]{\text{Metal catalyst}} \text{Cyclohexane}$$

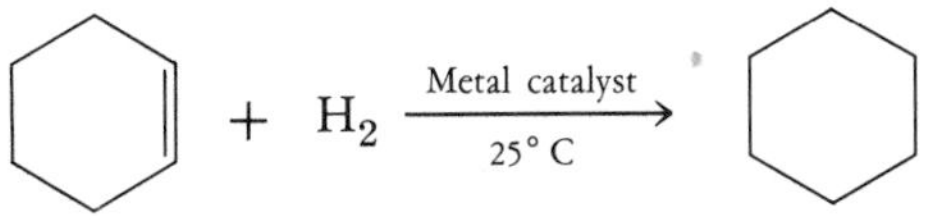

Cyclohexene Cyclohexane

$$CH_2{=}CHCH{=}CHCH_3 + 2\,H_2 \xrightarrow[\text{25° C}]{\text{Metal catalyst}} CH_3(CH_2)_3CH_3$$

1,3-Pentadiene Pentane

Hydrogenation of alkenes occurs extremely slowly at room temperatures but occurs readily in the presence of a metal catalyst such as finely divided platinum, nickel, or palladium.

Addition of Halogens

The addition of bromine and chlorine to alkenes occurs readily. Some specific examples are as follows:

$$CH_3CH{=}CH_2 + \text{Br—Br} \longrightarrow CH_3CH(Br)CH_2Br$$

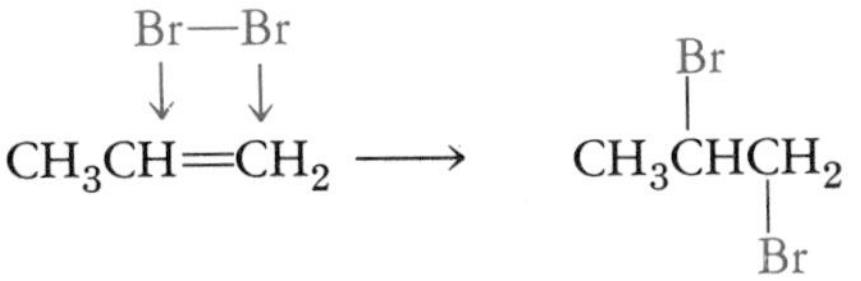

Propene 1,2-Dibromopropane

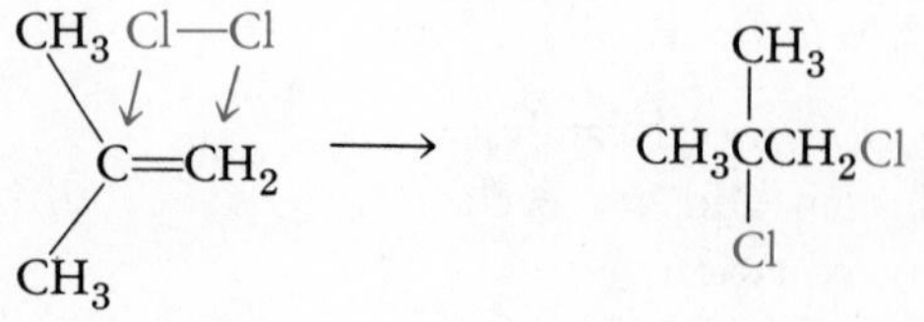

$$\mathrm{(CH_3)_2C{=}CH_2} + \mathrm{Cl{-}Cl} \longrightarrow \mathrm{CH_3C(CH_3)(Cl)CH_2Cl}$$

Methylpropene 1,2-Dichloro-2-methylpropane

Iodine usually does not react with alkenes but the interhalogens iodine monochloride (ICl) and iodine monobromide (IBr) are added readily.

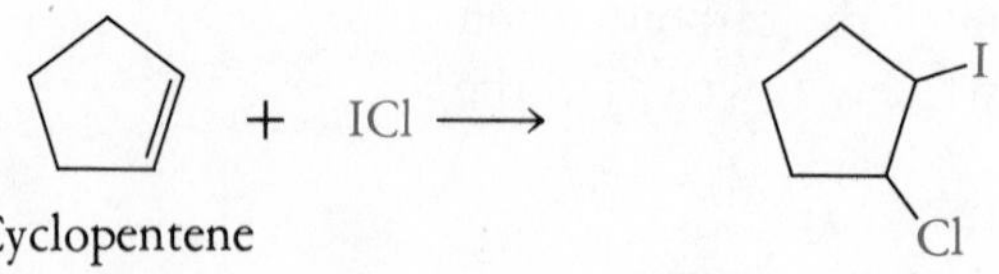

Cyclopentene + ICl ⟶ 1-Chloro-2-iodocyclopentane

The addition of bromine in carbon tetrachloride is a simple and useful test for the presence of a carbon-carbon unsaturated bond in a molecule. A solution of bromine in carbon tetrachloride has an orange-brown color, whereas alkenes and dibromoalkanes are colorless. The disappearance of the bromine color as the bromine solution is added by drops to a solution of a compound in carbon tetrachloride indicates the presence of a carbon-carbon unsaturated bond.

Addition of Acids

Acids such as sulfuric acid and the hydrogen halides are readily added to alkenes. Some specific examples are as follows:

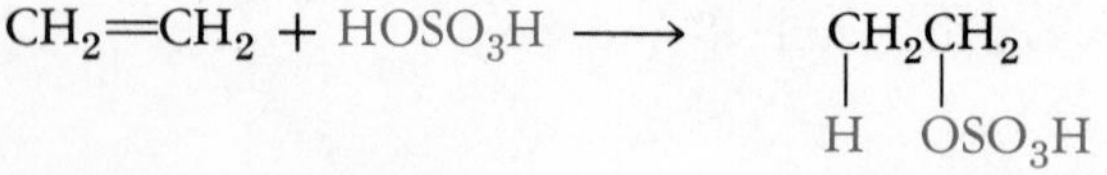

$$\mathrm{CH_2{=}CH_2} + \mathrm{HOSO_3H} \longrightarrow \mathrm{CH_2(H)CH_2(OSO_3H)}$$

Ethylene Sulfuric acid Ethyl hydrogen sulfate

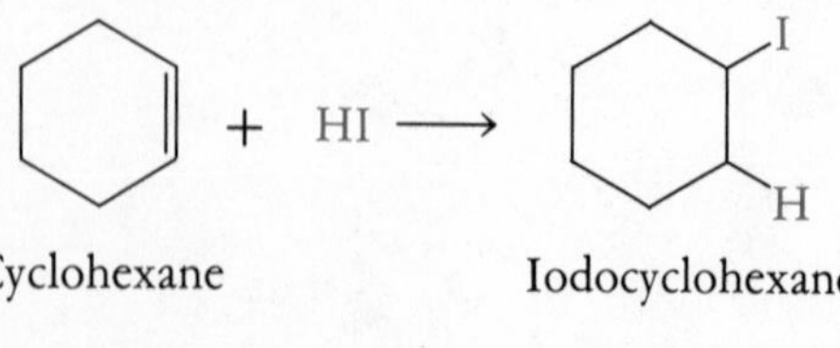

Cyclohexane + HI ⟶ Iodocyclohexane

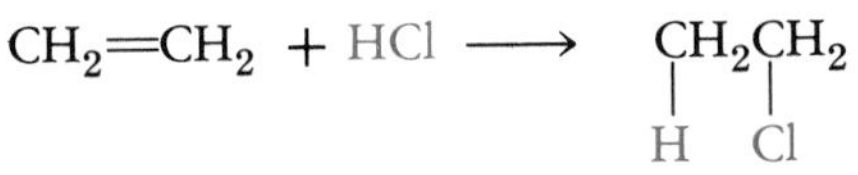

$$CH_2{=}CH_2 + HCl \longrightarrow \underset{\substack{| \quad |\\ H \quad Cl}}{CH_2CH_2}$$

Ethylene → Chloroethane

The addition of an unsymmetrical reagent such as an acid HX to an unsymmetrical alkene can form two isomeric products:

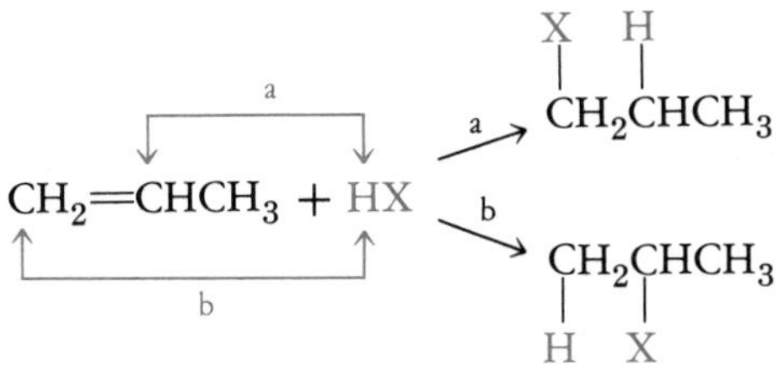

$$CH_2{=}CHCH_3 + HX \begin{cases} \xrightarrow{a} \overset{\substack{X \quad H\\ | \quad |}}{CH_2CHCH_3} \\ \xrightarrow{b} \underset{\substack{| \quad |\\ H \quad X}}{CH_2CHCH_3} \end{cases} \tag{12-1}$$

As illustrated in equation 12-1, the hydrogen of the acid could be added either to carbon 1 (path b) or to carbon 2 (path a) of propene to form the two products. Actually, only one product is obtained, the one formed when *the hydrogen of the acid is added to the carbon of the double bond containing the greatest number of hydrogens* (path b in equation 12-1). This rule is called the *Markownikoff rule.* The following examples illustrate application of this rule:

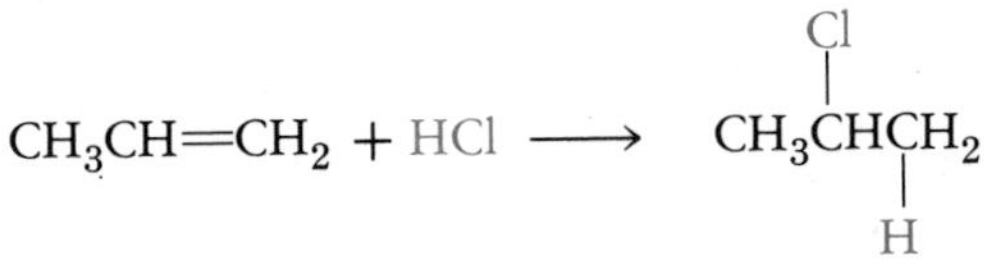

$$CH_3CH{=}CH_2 + HCl \longrightarrow CH_3\overset{\substack{Cl\\ |}}{CH}\underset{\substack{|\\ H}}{CH_2}$$

Propene → 2-Chloropropane

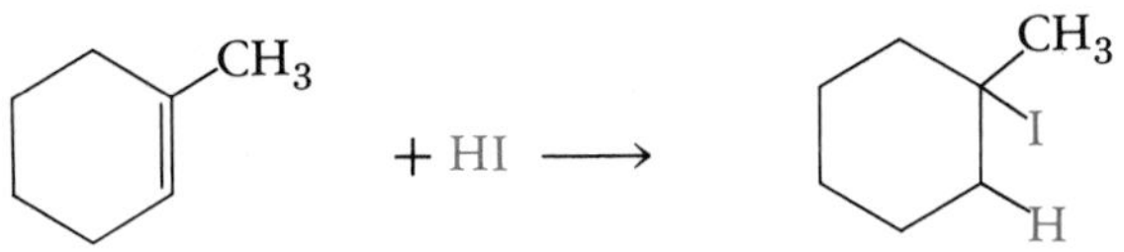

1-Methylcyclohexene + HI ⟶ 1-Iodo-1-methylcyclohexane

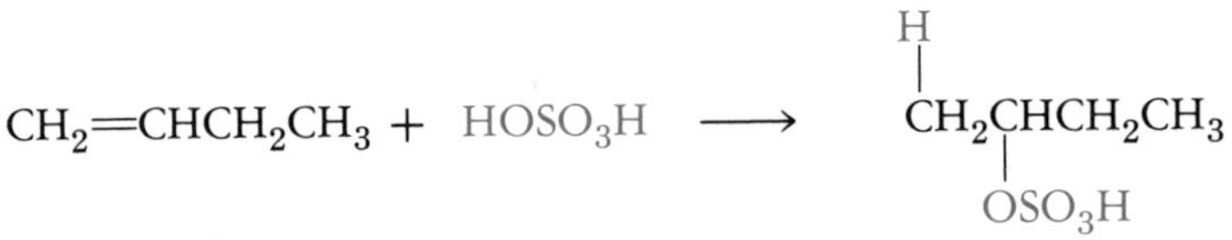

$$CH_2{=}CHCH_2CH_3 + HOSO_3H \longrightarrow \overset{\substack{H\\ |}}{CH_2}\underset{\substack{|\\ OSO_3H}}{CH}CH_2CH_3$$

1-Butene, Sulfuric acid → sec-Butyl hydrogen sulfate

This rule can be generalized to the addition to an alkene of any reagent XY that is polarized $X^{\delta+} - Y^{\delta-}$. (See Section 5.6 to determine which element acquires a partial positive charge.) The positive portion, $X^{\delta+}$, will be added to the carbon of the double bond containing the greatest number of hydrogens.

Addition of Water

The addition of water to alkenes, called *hydration,* requires the presence of a strong acid catalyst such as sulfuric or phosphoric acid. Some specific examples are as follows:

$$CH_2{=}C(CH_3)_2 + HOH \xrightarrow{H^+} CH_2(H)C(CH_3)_2(OH)$$

Methylpropene → Methyl-2-propanol

$$CH_3CH{=}CH_2 + HOH \xrightarrow{H^+} CH_3CH(OH)CH_2(H)$$

Propene → 2-Propanol

$$\text{1-methylcyclohexene} + HOH \xrightarrow{H^+} \text{cyclohexane with } CH_3, OH \text{ on C-1 and } H \text{ on C-2}$$

1-Methylcyclohexene → 1-Methylcyclohexanol

The hydration of alkenes follows the Markownikoff rule, as evident from the examples given above.

EXERCISE 12-6 Write the structural formula of the product of addition to 2-methyl-1-butene of the following reagents:
(a) Br_2 (b) H_2SO_4 (c) H_2O, H^+ (d) HCl (e) Cl_2
(f) HI (g) ICl (Hint: the molecule is polarized $I^{\delta+} - Cl^{\delta-}$.)

The addition of water to an alkene also occurs in living systems. Rather than use a strong acid catalyst, which would destroy the system, reactions in cells use enzymes, biological catalysts, that work efficiently at the temperature and pH of the cell. An example of such a reaction is the hydration of fumarate to malate that is catalyzed by the enzyme fumarase. This is an important reaction in the citric acid cycle (see Section 27.5).

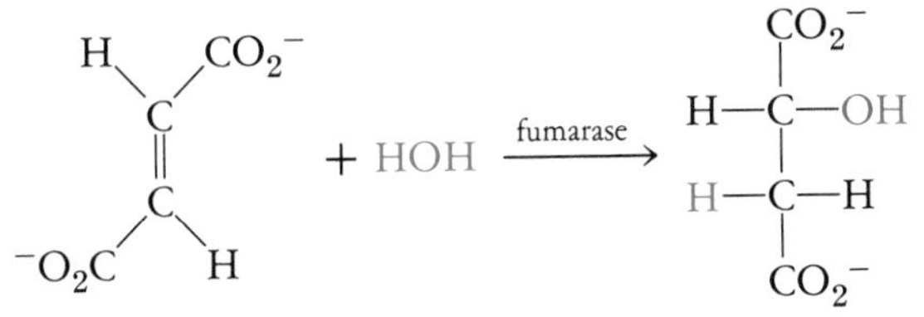

$$\underset{\text{Fumarate}}{\overset{H\diagdown\quad\diagup CO_2^-}{\underset{^-O_2C\diagup\quad\diagdown H}{C=C}}} + HOH \xrightarrow{\text{fumarase}} \underset{\text{Malate}}{H-\overset{\overset{CO_2^-}{|}}{C}-OH,\ H-\underset{\underset{CO_2^-}{|}}{C}-H}$$

12.5 REACTION MECHANISMS: HOW REACTIONS OCCUR

OPTIONAL SECTION

Organic chemists are not satisfied with simply identifying the products of a chemical reaction. They want to know more about the reaction. In particular, how does it occur? To answer this question, chemists propose a *reaction mechanism.* This is an attempt to picture how a reaction occurs. It involves a detailed description (the more detailed the description, the better) of the step, or steps, by which the reagents combine to form products. A reaction mechanism is proposed from the facts that are known about the reaction. To illustrate, let us examine the reaction mechanism proposed by organic chemists for the addition of acids to alkenes.

First, we must examine the facts. We know that the addition takes place with acids or by means of acid catalysis and follows the Markownikoff rule. From these and other facts, organic chemists have proposed the following reaction mechanism. An acid, such as hydrochloric acid (HCl), transfers a proton to an alkene, such as 2-butene, in the first step:

$$\text{Step 1.}\quad CH_3CH{=}CHCH_3 + H{-}Cl \longrightarrow CH_3\overset{\overset{H}{|}}{C}H\overset{+}{C}HCH_3 + \overset{-}{Cl} \qquad (12\text{-}2)$$

The electrons of one of the double bonds are used to form the new C—H bond. As a result of this proton transfer, the other carbon of the original double bond has only six electrons in its outer shell. Because it is deficient in electrons, it possesses a full positive charge. An ion in which carbon bears a positive charge is called a *carbonium ion.* A carbonium ion has an extremely short lifetime. It is one example of a *reactive intermediate.* A carbonium ion reacts rapidly with any reagent that can furnish a pair of electrons to neutralize its positive charge. An ion or reagent that seeks electrons is called an *electrophile* or *electrophilic reagent.* The carbonium ion finds the electrons it needs by combining with the chloride ion in the second and final step to form the product:

$$\text{Step 2.}\quad CH_3\overset{H}{\overset{|}{C}}H\overset{+}{C}HCH_3 + \overset{-}{Cl} \longrightarrow CH_3\overset{H}{\overset{|}{C}}H\overset{Cl}{\overset{|}{C}}HCH_3 \qquad (12\text{-}3)$$

These two steps represent a reaction mechanism for the addition of acids to alkenes.

Curved arrows have been used in equations 12-2 and 12-3 to indicate the movement of electron pairs during the reaction. According to convention, the pair of electrons at the foot of the arrow shifts to a new position at the head of the arrow. For example, in equation 12-2, an arrow indicates that one electron pair of the double bond is shifted to form a new C—H bond. During the reaction, the electrons of the H—Cl bond are also shifted to form a chloride ion. In equation 12-3, a pair of electrons from the outer shell of the chloride ion shift to form the C—Cl bond of the product. We use arrows in this way to keep track of the movement of electrons during the reaction.

How well does our proposed reaction mechanism, summarized by equations 12-2 and 12-3, agree with the observed Markownikoff rule? To answer this question, let us write the reaction mechanism for the addition of HCl to propene.

$$CH_2{=}CHCH_3 + H{-}Cl \nearrow \overset{+}{C}H_2\overset{H}{\overset{|}{C}}HCH_3 + \overset{-}{Cl}\ \ \text{12.2 Primary (1°) carbonium ion}$$

$$CH_2{=}CHCH_3 + H{-}Cl \searrow H{-}CH_2\overset{+}{C}HCH_3 + \overset{-}{Cl}\ \ \text{12.3 Secondary (2°) carbonium ion} \qquad (12\text{-}4)$$

A carbonium ion is formed in the first step (equation 12-4). But in this case, two isomeric carbonium ions can be formed. One is a primary carbonium ion, structure *12.2,* and the other, structure *12.3,* is a secondary one. Note that carbonium ions are designated primary, secondary, and tertiary in exactly the same way as carbon atoms (see Section 11.6). If the primary carbonium ion is formed, its reaction with chloride ion would form CH_3CH_2Cl in the second step (equation 12-5). But this product is not found! Therefore, only the secondary carbonium ion is formed. Why?

$$\begin{array}{l} CH_3\underset{\displaystyle |}{\overset{\displaystyle H}{C}}H\overset{+}{C}H_2 + \overset{-}{Cl} \not\longrightarrow CH_3CH_2CH_2Cl \\ CH_3\overset{+}{C}HCH_2{-}H + \overset{-}{Cl} \longrightarrow CH_3\overset{\displaystyle Cl}{\overset{|}{C}}HCH_3 \end{array} \qquad (12\text{-}5)$$

To answer this question, we must use a principle that has been developed by observing a large number of reactions. The principle is as follows: if a reaction can occur by two or more paths, it will follow the path that requires the least energy. This usually means the path leading to the more stable reactive intermediate. In terms of our reaction mechanism, this means that the reaction will follow the path that leads to the most stable carbonium ion. But which carbonium ion, the primary or the secondary one, is the more stable? Experimental results have shown that the order of carbonium ion stability is tertiary $>$ secondary $>$ primary. Because the secondary carbonium ion is more stable than the isomeric primary one, we would expect it to be formed. Its reaction in the second step with chloride ion forms the observed product, $CH_3CHClCH_3$. Thus, our mechanism is consistent with the observed Markownikoff rule of addition.

In summary, the reaction mechanism for the addition of acids to alkenes involves two steps. A carbonium ion is formed as a reactive intermediate in the first step. If two isomeric carbonium ions can be formed, the more stable one is formed. Finally, the carbonium ion reacts with an anion or other electron-rich reagent to form the product.

A word of caution: a reaction mechanism is a theory that is based on experimental facts. If new experimental facts are discovered that do not agree with the reaction mechanism, *then the reaction mechanism must be modified to agree with the facts.* Therefore, a reaction mechanism can vary with the passage of time. In contrast, experimental facts, if correctly obtained and accurately reported, do not change with time. For example, we know that HCl adds to propene to form $CH_3CHClCH_3$ as product. This is an experimental fact that was known many years ago. It is the same today and will be the same in the year 2077. The mechanism of this reaction, discussed above, is based on all the experimental facts known to date. However, it is possible that new facts will emerge during the next 50 years that will require major modifications, or even rejection, of this mechanism. It is for this reason that chemists are always interested in new facts about a particular reaction. So let us learn more facts about addition reactions of alkenes. In particular, we should learn the facts about polymerization, a reaction that is important both industrially and in living systems.

EXERCISE 12-7 Write a two-step mechanism for the addition of HCl to

(a) $(CH_3)_2C{=}CH_2$ (b) (cyclopentane ring)$={CH_2}$

12.6 POLYMERIZATION: ONE ALKENE ADDING TO ANOTHER

With the aid of a catalyst, an alkene can be made to add to another molecule of the same alkene. The first product of this addition reaction contains two molecules of the original alkene and is called a *dimer.* The reaction need not stop at this point. The dimer can add to another molecule of the alkene to form a *trimer.* This reaction can continue until a huge molecule is formed. This process is called *polymerization.* For example, ethylene polymerizes to form polyethylene with the aid of a special catalyst:

$$n\,CH_2{=}CH_2 \xrightarrow{\text{Catalyst}} \left[-CH_2CH_2-\right]_n$$

Ethylene (a monomer) — Polyethylene (a polymer) — n is a whole number that varies from several hundred to several thousand

The starting alkene is called a *monomer;* the large molecule formed as the final product is called a *polymer.* It is important to realize that a polymer is made up of many units of the monomer joined together. This can be seen by examining in more detail a section of the polyethylene polymer, as shown in Figure 12-4.

Most alkenes can be made to polymerize. A number of such polymers have found wide use in our society. Polyethylene, for example, is made into a material used to package food and clothes. Propylene, when polymerized, forms a stronger polymer than polyethylene:

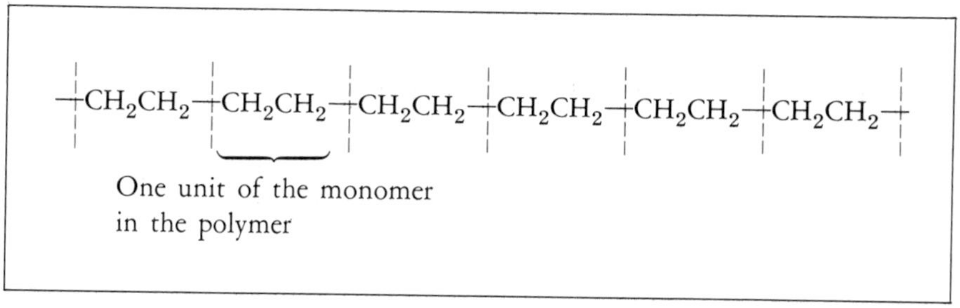

$$-\!\!\!|CH_2CH_2-\!\!\!|CH_2CH_2-\!\!\!|CH_2CH_2-\!\!\!|CH_2CH_2-\!\!\!|CH_2CH_2-\!\!\!|CH_2CH_2-\!\!\!|$$

One unit of the monomer in the polymer

Fig. 12-4. A portion of the polyethylene polymer. The dashed lines indicate the bonds linking the individual monomer units.

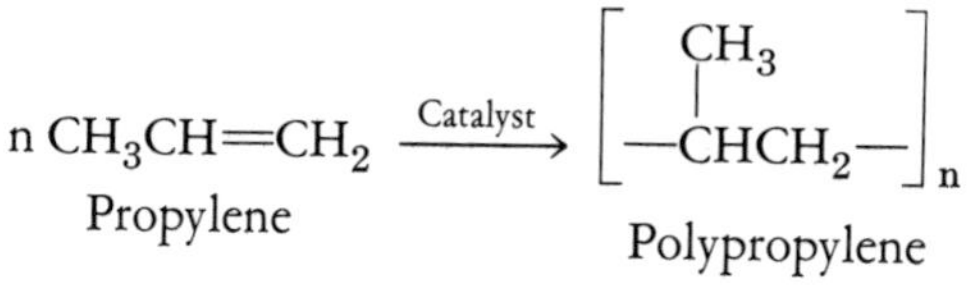

$$\underset{\text{Propylene}}{n\,CH_3CH{=}CH_2} \xrightarrow{\text{Catalyst}} \underset{\text{Polypropylene}}{\left[-\overset{\overset{CH_3}{|}}{C}HCH_2-\right]_n}$$

Polypropylene is widely used to make indoor-outdoor carpeting. Notice that both polyethylene and polypropylene are alkanes and are chemically inert. This behavior is expected for members of the alkane family (see Section 11.11).

Several halogen-containing monomers related to ethylene are of particular commercial interest. Vinyl chloride forms polyvinyl chloride (PVC) on polymerization:

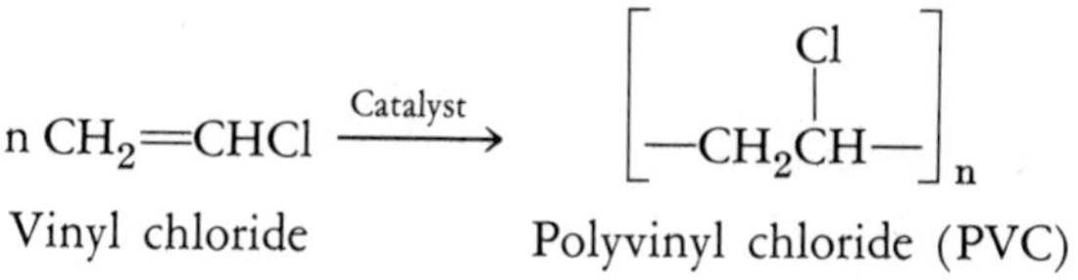

$$\underset{\text{Vinyl chloride}}{n\,CH_2{=}CHCl} \xrightarrow{\text{Catalyst}} \underset{\text{Polyvinyl chloride (PVC)}}{\left[-CH_2\overset{\overset{Cl}{|}}{C}H-\right]_n}$$

This polymer is widely used to make plastic bottles. Teflon, which is made from the monomer tetrafluoroethylene, is one of the most chemically inert of all organic substances. For this reason, it is widely used as a coating for cooking utensils.

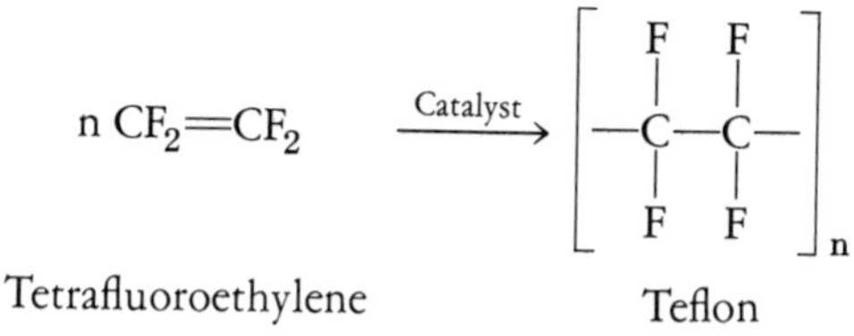

$$\underset{\text{Tetrafluoroethylene}}{n\,CF_2{=}CF_2} \xrightarrow{\text{Catalyst}} \underset{\text{Teflon}}{\left[-\underset{\underset{F}{|}}{\overset{\overset{F}{|}}{C}}-\underset{\underset{F}{|}}{\overset{\overset{F}{|}}{C}}-\right]_n}$$

Dienes can also polymerize. For example polybutadiene, a synthetic rubber, is made from 1,3,-butadiene:

$$n\ CH_2{=}CHCH{=}CH_2 \xrightarrow{\text{Catalyst}} \left[-CH_2CH{=}CHCH_2-\right]_n$$

1,3-Butadiene Polybutadiene

EXERCISE 12-8 Write a structure for polypropylene and polybutadiene similar to that given in Figure 12-4 for polyethylene. Clearly indicate each unit of monomer.

These are only a few examples of the many polymers that have been prepared commercially. Chemists have learned to prepare synthetic polymers with a wide variety of different chemical and physical properties. However, chemists still cannot match the wide variety of polymers formed by living systems. One kind of compound that living systems make is described in the next section.

12.7 POLYMERS FORMED BY LIVING SYSTEMS

Polymerization also occurs in living systems. A large number of compounds called *terpenes* are found in living systems. These compounds are all polymers made of a repeating five-carbon unit that is structurally related to isoprene. Notice that isoprene is a substituted butadiene.

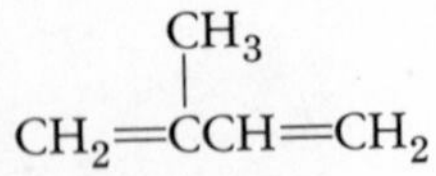

$$CH_2{=}C(CH_3)CH{=}CH_2$$

Isoprene

```
        C
        |
        C     C
       / \   /
      C    C
```

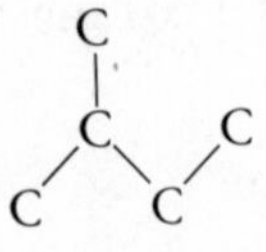

Isoprene unit

identical to

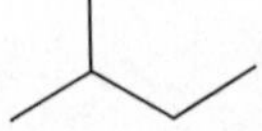

Natural rubber is an example of a polymer made by combining thousands of isoprene units.

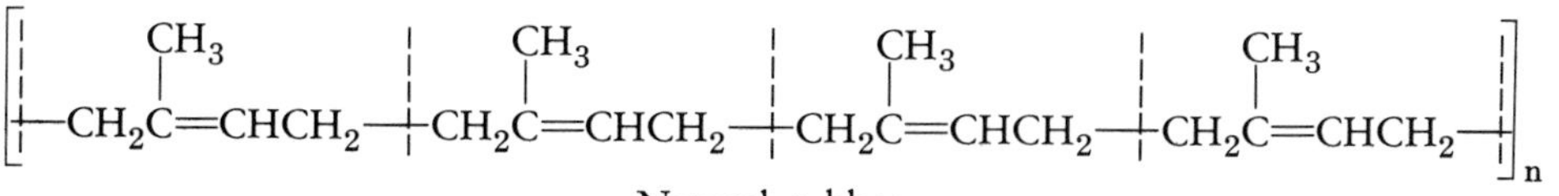

$$\left[-CH_2\overset{\overset{\displaystyle CH_3}{|}}{C}{=}CHCH_2- \Big| -CH_2\overset{\overset{\displaystyle CH_3}{|}}{C}{=}CHCH_2- \Big| -CH_2\overset{\overset{\displaystyle CH_3}{|}}{C}{=}CHCH_2- \Big| -CH_2\overset{\overset{\displaystyle CH_3}{|}}{C}{=}CHCH_2- \right]_n$$

Natural rubber

The dashed lines indicate where isoprene units are joined.

The number of isoprene units in a terpene varies greatly. One of the largest terpenes is natural rubber; the smallest terpenes contain only two units. Despite this difference in size, the isoprene units are usually connected to one another in a head-to-tail manner. For example:

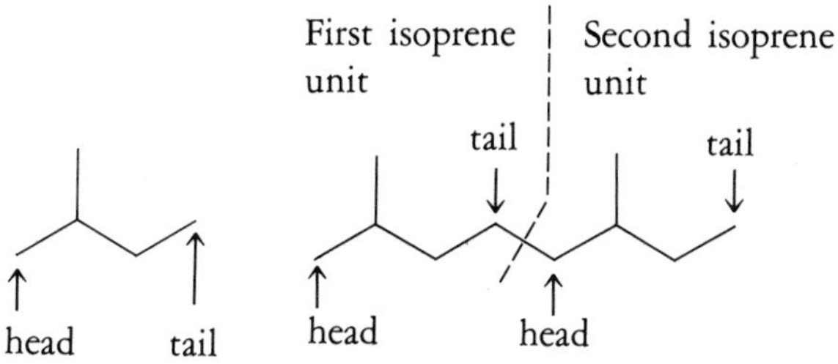

First isoprene unit | Second isoprene unit

tail ↓ tail ↓

↑ head ↑ tail ↑ head ↑ head

As with most general rules, exceptions are known. The following examples illustrate both the general rule and exceptions for some terpenes.

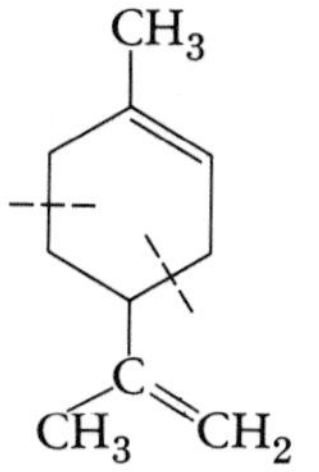

CH_3 CH_3 CH_2

Limonene
(Lemon oil)
2 Isoprene units

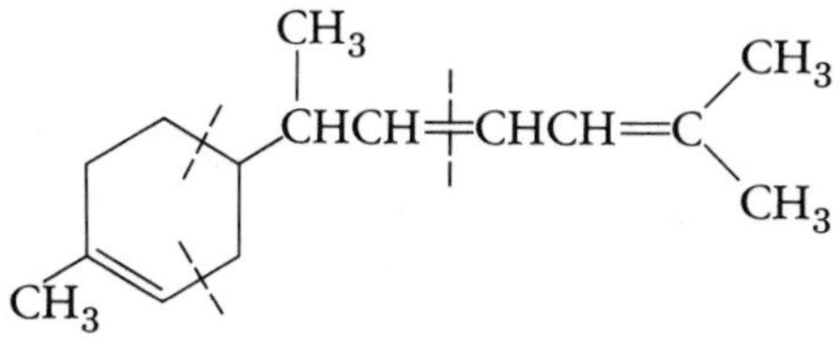

CH_3 $CHCH{=}CHCH{=}C(CH_3)_2$ CH_3

Zingibenene
(Oil of ginger)
3 Isoprene units

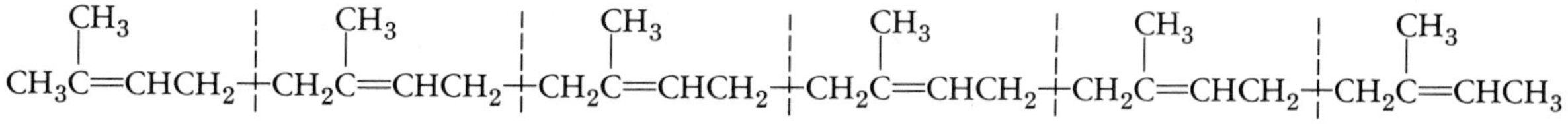

$$CH_3\overset{\overset{\displaystyle CH_3}{|}}{C}{=}CHCH_2 \Big| CH_2\overset{\overset{\displaystyle CH_3}{|}}{C}{=}CHCH_2 \Big| CH_2\overset{\overset{\displaystyle CH_3}{|}}{C}{=}CHCH_2 \Big| CH_2\overset{\overset{\displaystyle CH_3}{|}}{C}{=}CHCH_2 \Big| CH_2\overset{\overset{\displaystyle CH_3}{|}}{C}{=}CHCH_2 \Big| CH_2\overset{\overset{\displaystyle CH_3}{|}}{C}{=}CHCH_3$$

Squalene
6 Isoprene units

EXERCISE 12-9 How many isoprene units are there in each of the following terpenes? Used dashed lines to separate the isoprene units.

(a) 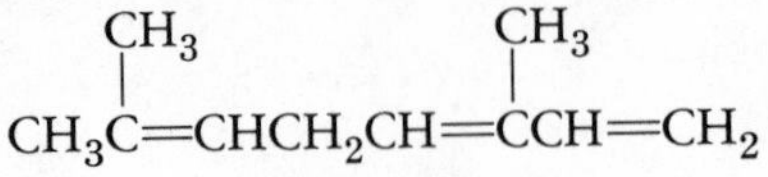 $CH_3C(CH_3){=}CHCH_2CH{=}C(CH_3)CH{=}CH_2$

(b) 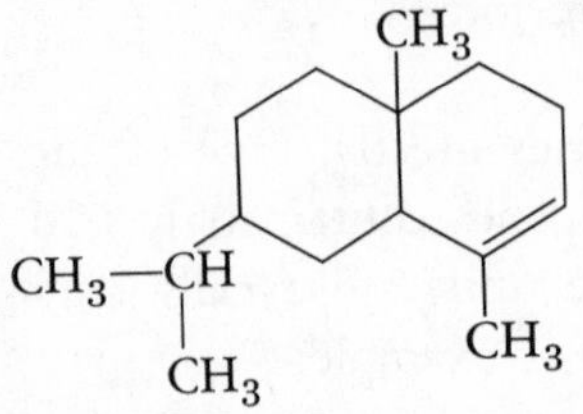 [bicyclic structure: decalin ring system with CH_3 at the ring fusion carbon, a CH_3 on a ring double-bond carbon, and a $CH_3{-}CH(CH_3)$ (isopropyl) group]

There is one important difference between the way terpenes and industrial polymers are formed. Polymers made commercially, like polyethylene, use monomers as the starting material. Living systems, on the other hand, *do not* use alkenes or dienes as starting materials. The compounds used by living systems usually require a number of steps to convert them into terpenes. Each of these various steps, including the polymerization step, requires the presence of an enzyme. Just how these reactions occur on an enzyme is not known for many systems. However scientists are slowly learning how enzymes carry out these chemical transformations. Some of their conclusions will be presented in future chapters.

12.8 OXIDATIONS OF CARBON-CARBON DOUBLE BONDS

Another typical reaction of alkenes is oxidation (review Section 7.6). The carbon-carbon double bond of an alkene reacts readily with a number of oxidizing reagents such as potassium permanganate ($KMnO_4$), peracids, and ozone (O_3). The product of the reaction depends on the reagent and the experimental conditions. Under mild oxidation, the carbon skeleton remains intact. Under more vigorous conditions, oxidation breaks both bonds of the carbon-carbon double bond.

A dilute aqueous basic solution of $KMnO_4$ is a mild oxidizing reagent. It reacts with a carbon-carbon double bond at room temperature to form a compound that contains a hydroxyl group (—OH) attached to each carbon of the original double bond. Such a compound is called a diol (or glycol). For example:

$$CH_3CH{=}CH_2 \xrightarrow[H_2O]{KMnO_4,\ KOH} CH_3\underset{\substack{|\\ OH}}{C}H{-}\underset{\substack{|\\ OH}}{C}H_2$$

Propene → 1,2-Propanediol

Notice that the original carbon skeleton remains intact at the end of the reaction.

Peracids, like peracetic acid (CH_3CO_3H) or perbenzoic acid ($C_6H_5CO_3H$), are other mild oxidizing reagents. These acids react with alkenes to form a three-member ring containing an oxygen atom. This three-member ring is called an *epoxide* (common name) or an *oxirane* (IUPAC). This reaction is called *epoxidation.*

$$>C{=}C< + C_6H_5C({=}O){-}O{-}OH \longrightarrow >C\overset{O}{—}C< + C_6H_5CO_2H$$

Alkene, Perbenzoic acid → An oxirane (an epoxide), Benzoic acid

The following is a specific example:

$$(CH_3)_2C{=}CH_2 + C_6H_5C({=}O){-}O{-}OH \longrightarrow (CH_3)_2C\overset{O}{—}CH_2 + C_6H_5CO_2H$$

Methylpropene, Perbenzoic acid → 1,1-Dimethyloxirane, Benzoic acid

EXERCISE 12-10 Write the structure of the product of the reaction of a dilute aqueous basic solution of potassium permanganate with the following alkenes:
(a) propene (b) cyclopentene (c) 1-methylcyclohexene
(d) 4,4-dimethyl-1-pentene

EXERCISE 12-11 Write the structure of the product of the reaction of perbenzoic acid with the alkenes in exercise 12-10.

A hot acidic solution of $KMnO_4$ is a strong oxidizing reagent that breaks both bonds of the double bond. The products formed depend on the structure of the original alkene, as illustrated by the following examples:

$$\mathrm{RCH{=}CH_2} \xrightarrow[\mathrm{KMnO_4}]{\text{Hot } \mathrm{H^+}} \underset{\text{Carboxylic acid}}{\mathrm{RC(=O)OH}} + \underset{\text{Carbon dioxide}}{\mathrm{O{=}C{=}O}}$$

$$\mathrm{R_2C{=}CHR'} \xrightarrow[\mathrm{KMnO_4}]{\text{Hot } \mathrm{H^+}} \underset{\text{Ketone}}{\mathrm{R_2C{=}O}} + \underset{\text{Carboxylic acid}}{\mathrm{R'C(=O)OH}}$$

The symbols R and R′ (and sometimes R″) indicate that the two or more alkyl groups in a molecule may be different. Notice that the original carbons of the double bond end up as the carbons of the C=O bonds of the products. The C=O bond can be part of a ketone, a carboxylic acid, or carbon dioxide, depending on the number of hydrogens on the original carbons of the double bond. If one of the two carbons is bonded to two hydrogens, it will end up as the carbon of carbon dioxide. If it is bonded to only one hydrogen, it will end up as the carbon of the carboxylic acid group. If it has no hydrogens bonded to it, it will end up as the carbon of a ketone group.

A strong oxidizing reagent found in nature is ozone (O_3). Ozone is produced by lightning during electrical storms and is also present in smog. Its reaction with alkenes, called *ozonolysis,* also breaks the double bond. The destructive action of ozone on living systems is due to this reaction. Ozone destroys the structure of many biologically important alkenes by breaking their double bonds. Once their structure is destroyed, their biological activity is also destroyed.

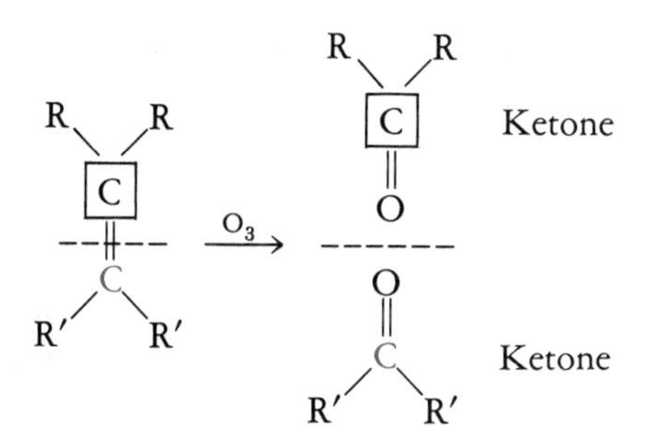

EXERCISE 12-12 Write the structure of the products of the reaction of a hot acidic solution of potassium permanganate with the alkenes in exercise 12-10.

The reader may have noticed that a number of the chemical equations in this section are not balanced. The number of such unbalanced equations will increase as we begin to study the chemical reactions of living systems. The reason for not balancing these equations is simple; we want to focus our attention on the organic portion of the reaction, not on the quantitative aspect of the reaction, as we did in Chapter 7. Therefore we often write only the organic reactant and product with an arrow between them and place the other reagents above the arrow. We ignore the other nonorganic products of the reaction. This greatly simplifies the chemical equation and allows us to focus our attention on the fate of the organic reactant.

12.9 ALKYNES

The other unsaturated hydrocarbons that we wish to study are the alkynes. These compounds all contain at least one carbon-carbon triple bond. Alkynes are not as common in nature as are alkenes. However, the simplest member of the series, acetylene (C_2H_2), is an important industrial chemical. Acetylene is also the name given to this class of compounds.

Acetylene is a linear molecule. All four of its atoms lie in a straight line, as shown in Figure 12-5. As a result, the carbon bond angle is 180 degrees.

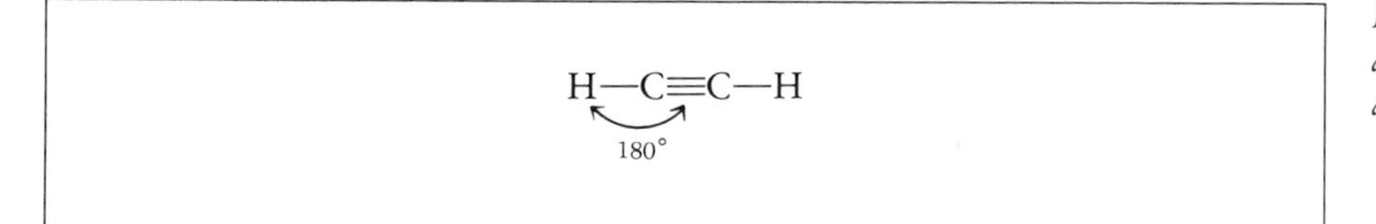

Fig. 12-5. The linear arrangement of the atoms of acetylene.

This is the characteristic bond angle of a carbon bonded to two groups or atoms (one by a single bond, the other by a triple bond).

Simple alkynes are often named as derivatives of acetylene. The names of the alkyl groups substituted for the hydrogens on acetylene precede the word acetylene. For example:

$HC{\equiv}CH$	$CH_3CH_2C{\equiv}CH$	$CH_3C{\equiv}CCH_3$	$(CH_3)_3CC{\equiv}CH$
Acetylene	Ethylacetylene	Dimethylacetylene	*t*-Butylacetylene

The IUPAC rules for naming alkynes are the same as those for alkenes, except that the ending is *-yne.* The following examples illustrate the rules:

$HC{\equiv}CH$	$CH_3C{\equiv}CH$	$CH_3C{\equiv}CCH_2CH_3$	$\begin{array}{l} CH_3\underset{\displaystyle CH_3}{\underset{\vert}{C}}HC{\equiv}CH \end{array}$
Ethyne	Propyne	2-Pentyne	3-Methyl-1-butyne

EXERCISE 12-13 Give the IUPAC name for the following compounds:

(a) $CH_3C{\equiv}CCH_3$ (b) $CH_3CH_2C{\equiv}CCH_2CH_3$

(c) $CH_3C{\equiv}C\underset{\displaystyle CH_3}{\underset{\vert}{C}}H\overset{\displaystyle CH_3}{\overset{\vert}{C}}HCH_2CH_3$ (d) $\text{(cyclopentyl)}{-}\overset{\displaystyle CH_3}{\overset{\vert}{C}}HC{\equiv}CH$

(e) $(CH_3)_3CC{\equiv}CH$ (f) $CH_3CH_2C{\equiv}CH$

EXERCISE 12-14 Draw a structural formula for each of the following names:

(a) cyclohexylacetylene (b) 4,4-dimethyl-2-pentyne

(c) 3-cyclopropylpropyne

(d) 4-*t*-butyl-6-methyl-5-isopropyl-2-octyne

EXERCISE 12-15 Give the IUPAC name for the following compounds:

(a) ethylacetylene (b) ethylmethylacetylene

(c) dimethylacetylene (d) *sec*-butylacetylene

(e) diisopropylacetylene

Most of the reactions of alkynes are similar to those of alkenes. The same reagents that add to carbon-carbon double bonds also add to carbon-carbon triple bonds. But it is possible to add two molecules of reagent to each alkyne. If sufficiently mild conditions are used, the addition can be stopped after addition of only one molecule to form a substituted alkene. Addition of a second molecule forms a substituted alkane. The following are a few specific examples:

$$CH_3CH_2C{\equiv}CH + Br_2 \longrightarrow CH_3CH_2C(Br){=}CHBr \xrightarrow[\Delta]{Br_2} CH_3CH_2CBr_2CHBr_2$$

1-Butyne — 1,2-Dibromo-1-butene — 1,1,2,2-Tetrabromobutane

$$CH_3C{\equiv}CH \xrightarrow[Pd]{H_2} CH_3CH{=}CH_2 \xrightarrow[\text{Metal catalyst}]{H_2} CH_3CH_2CH_3$$

Propyne — Propene — Propane

Bromination of propyne at 25°C leads to 1,2-dibromopropene. Bromine can be added to the alkene by carrying out the reaction at higher temperatures. An alkene such as propene can be prepared from the corresponding alkyne (in this case, propyne) by the use of a special palladium catalyst, which prevents the addition of hydrogen to propene.

Addition reactions to alkynes also follow the Markownikoff rule. The more electropositive part of the XY molecule is added to the triple-bonded carbon bearing the most hydrogens:

$$CH_3C{\equiv}CH + HCl \longrightarrow CH_3C(Cl){=}CH_2 \xrightarrow{HCl} CH_3CCl_2CH_3$$

One molecule of acetylene can also be added to another to form a dimer. The product is a useful industrial chemical because it can be chemically transformed to a variety of commercial products.

$$2\,HC{\equiv}CH \xrightarrow{\text{Catalyst}} HC{\equiv}CCH{=}CH_2$$

Acetylene — Butenyne (vinylacetylene)

One minor difference between alkenes and alkynes is the product formed by acid-catalyzed hydration. Alkenes form alcohols (see Section 12.4). Alkynes form compounds containing a C=O bond. This occurs because the first formed product rearranges to form a compound with a C=O. Once this product is formed, the reaction stops. Thus, alkynes add only one molecule of water:

$$CH_3C{\equiv}CH + H_2O \xrightarrow[H_2SO_4]{HgSO_4} \left[CH_3\overset{\overset{\displaystyle OH}{|}}{C}{=}\overset{\overset{\displaystyle H}{|}}{C}H \right] \xrightarrow{\text{rearranges}} CH_3\overset{\overset{\displaystyle O}{\|}}{C}{-}\underset{\underset{\displaystyle H}{|}}{\overset{\overset{\displaystyle H}{|}}{C}}H$$

Notice also that both $HgSO_4$ and H_2SO_4 are needed to catalyze the reaction.

Despite a few minor differences, the chemical reactions of alkynes are generally the same as those of alkenes. This is evident from the examples given in this chapter.

EXERCISE 12-16 Write the structure of the product of the addition of the following reagents to 1-butyne:
(a) 1 mole Cl_2 (b) H_2O, H_2SO_4, $HgSO_4$
(c) 1 mole HCl (d) 2 mole HCl (e) 1 mole H_2, Pd

12.10 SUMMARY

Alkenes and alkynes are both unsaturated hydrocarbons. Alkenes contain a double bond, whereas alkynes contain a triple bond. In the IUPAC system of naming organic compounds, the ending *-ene* is used to designate alkenes, and the ending *-yne* is used to designate alkynes. Certain alkenes can exist as *cis-trans* isomers that differ in the arrangement of their atoms in space. Each isomer has characteristic physical and chemical properties.

The chemical reactions of alkenes and alkynes are very similar. Both undergo additions to the double or triple bonds, their characteristic reaction. Among the reagents added are the halogens, acids, hydrogen, and water. The last two react with the aid of catalysts. Alkenes and alkynes also undergo polymerization. In this reaction, one unsaturated hydrocarbon adds to another.

The double bond of alkenes also reacts with oxidizing reagents. The products of the reaction depend on the oxidizing reagent used and the experimental conditions. Under mild oxidation, the carbon skeleton remains intact. Under more vigorous conditions, both of the double bonds are broken.

REVIEW OF TERMS, CONCEPTS, AND REACTIONS

Terms

ACETYLENES Another name for alkynes.

ALKENES Unsaturated hydrocarbons containing double bonds.

ALKYNES Unsaturated hydrocarbons containing triple bonds.

CARBONIUM ION An ion in which carbon bears a positive charge.

ELECTROPHILE An ion or molecule that seeks electrons (literally, "electron-loving").

EPOXIDATION The reaction of an alkene with a peracid to form a three-member oxygen-containing ring.

HYDRATION The addition of water, usually to a carbon-carbon double or triple bond.

HYDROGENATION The addition of hydrogen to a molecule.

MARKOWNIKOFF RULE When an unsymmetrical reagent is added to an unsymmetrical alkene or alkyne, the more positive portion of the reagent adds to the double-bonded carbon containing the greatest number of hydrogens.

MONOMER One molecule of an alkene.

POLYMER A huge molecule with a high molecular weight formed by combining a large number of monomers.

POLYMERIZATION A reaction that forms polymers.

REACTION MECHANISM A detailed description of the step, or steps, by which reagents combine to form products.

REACTIVE INTERMEDIATE An ion or molecule that has an extremely short lifetime.

Concepts

1. The major reactions of alkenes and alkynes are those of addition.
2. Alkenes and alkynes react more easily with acids, halogens, and oxidizing reagents than do alkanes.

3. Alkenes and alkynes add halogens, acids water, and hydrogen. The latter two additions require a catalyst.
4. An alkene can be added to another molecule of the same alkene with the aid of a catalyst.
5. The product of oxidation of an alkene depends on the oxidizing reagent and the experimental conditions. Under mild oxidation, the carbon skeleton remains intact. Under more vigorous conditions, oxidation breaks both bonds of the double bond.

Reactions

Reactions of Alkenes (see Section 12.4)

1. Addition of hydrogen (see p. 301)

$$\mathrm{RCH{=}CHR} + \mathrm{H_2} \xrightarrow[\text{catalyst } 25^\circ \text{ C}]{\text{Metal}} \underset{\substack{|\\ \mathrm{H}}}{\mathrm{RCH}}-\underset{\substack{|\\ \mathrm{H}}}{\mathrm{CHR}}$$

2. Addition of chlorine and bromine (see p. 301)

$$\mathrm{RCH{=}CHR} + \mathrm{X_2} \longrightarrow \underset{\substack{|\\ \mathrm{X}}}{\mathrm{RCH}}-\underset{\substack{|\\ \mathrm{X}}}{\mathrm{CHR}}$$

X = Cl or Br

3. Addition of acids (see p. 302)

$$\mathrm{RCH{=}CHR} + \mathrm{HX} \longrightarrow \underset{\substack{|\\ \mathrm{H}}}{\mathrm{RCH}}-\underset{\substack{|\\ \mathrm{X}}}{\mathrm{CHR}}$$

X = Cl, Br, I, or OSO_3H

4. Addition of water (hydration) (see p. 304)

$$\mathrm{RCH{=}CHR} + \mathrm{H_2O} \xrightarrow{\mathrm{H^+}} \underset{\substack{|\\ \mathrm{H}}}{\mathrm{RCH}}-\underset{\substack{|\\ \mathrm{OH}}}{\mathrm{CHR}}$$

5. Polymerization (see p. 308)

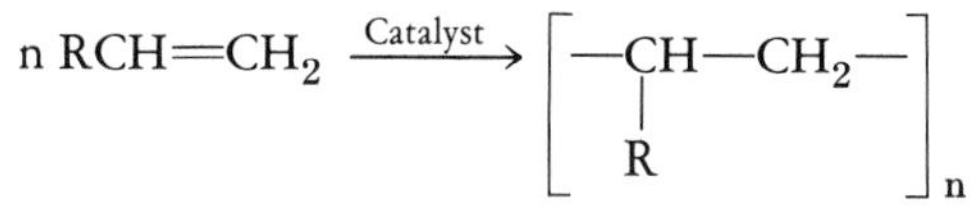

$$n\,RCH{=}CH_2 \xrightarrow{\text{Catalyst}} \left[\begin{array}{c}-\underset{\displaystyle R}{\underset{|}{C}}H-CH_2-\end{array}\right]_n$$

6. Oxidation (see p. 312)

a. Dilute aqueous basic $KMnO_4$

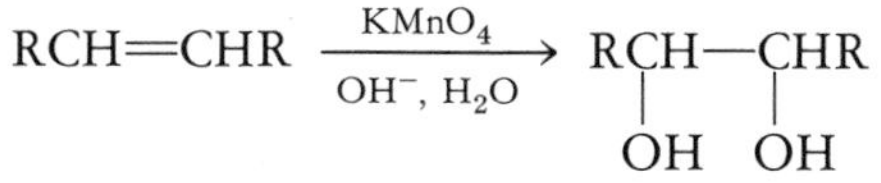

$$RCH{=}CHR \xrightarrow[OH^-,\ H_2O]{KMnO_4} \underset{\displaystyle OH}{\underset{|}{RCH}}-\underset{\displaystyle OH}{\underset{|}{CHR}}$$

b. Epoxidation

$$RCH{=}CHR + C_6H_5CO_3H \longrightarrow RCH\overset{O}{-}CHR + C_6H_5CO_2H$$

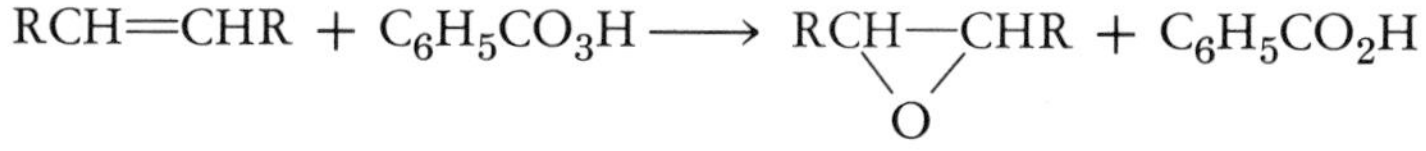

c. Hot aqueous acidic $KMnO_4$

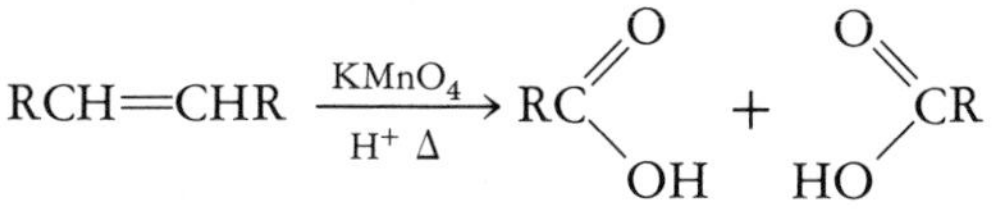

$$RCH{=}CHR \xrightarrow[H^+\ \Delta]{KMnO_4} RC(=O)OH + HO(O=)CR$$

d. Ozone

$$R_2C{=}CR'_2 \xrightarrow{O_3} R_2C{=}O + O{=}CR'_2$$

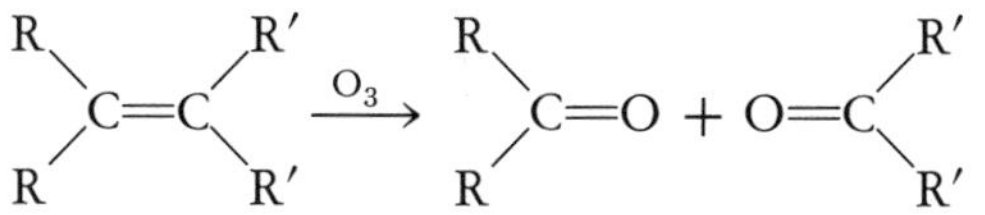

Reactions of alkynes (see Section 12.9)

1. Addition of chlorine and bromine

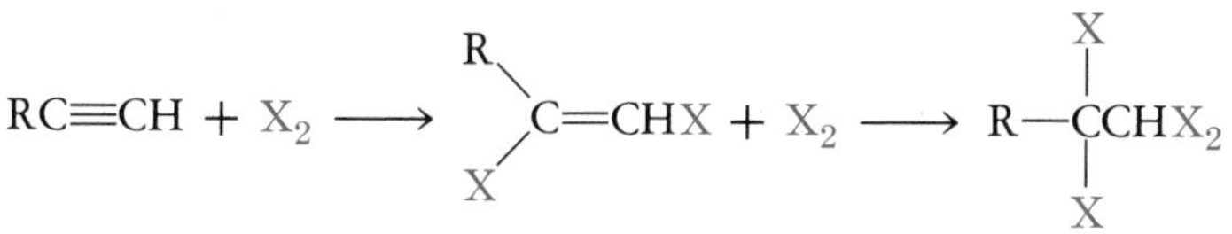

$$RC{\equiv}CH + X_2 \longrightarrow \underset{X}{\overset{R}{>}}C{=}CHX + X_2 \longrightarrow R-\underset{\displaystyle X}{\underset{|}{\overset{\displaystyle X}{\overset{|}{C}}}}CHX_2$$

X = Cl or Br

2. Addition of acids, HX

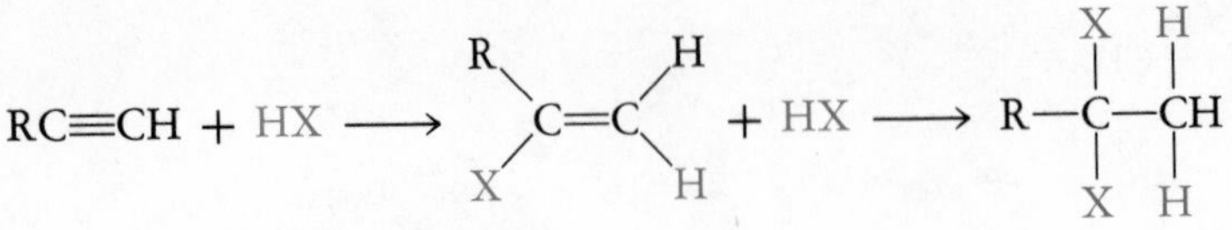

$$RC{\equiv}CH + HX \longrightarrow \underset{X}{\overset{R}{}}\!\!>C{=}C<\!\!\underset{H}{\overset{H}{}} + HX \longrightarrow R{-}\overset{X}{\underset{X}{C}}{-}\overset{H}{\underset{H}{C}}H$$

X = Cl, Br, I, or OSO_3H

3. Addition of water (hydration)

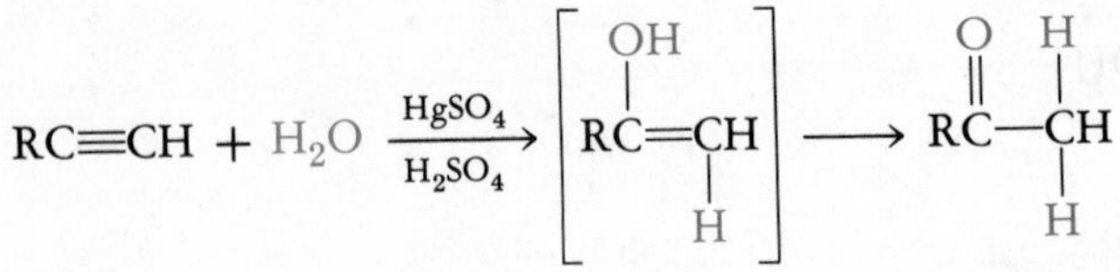

$$RC{\equiv}CH + H_2O \xrightarrow[H_2SO_4]{HgSO_4} \left[R\overset{OH}{C}{=}\underset{H}{C}H \right] \longrightarrow R\overset{O}{\overset{\|}{C}}{-}\overset{H}{\underset{H}{C}}H$$

4. Addition of hydrogen (hydrogenation)

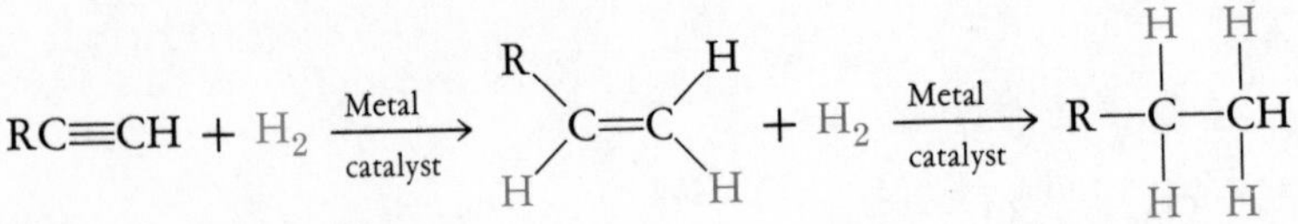

$$RC{\equiv}CH + H_2 \xrightarrow[catalyst]{Metal} \underset{H}{\overset{R}{}}\!\!>C{=}C<\!\!\underset{H}{\overset{H}{}} + H_2 \xrightarrow[catalyst]{Metal} R{-}\overset{H}{\underset{H}{C}}{-}\overset{H}{\underset{H}{C}}H$$

5. Self-addition

$$HC{\equiv}CH + HC{\equiv}CH \xrightarrow{Catalyst} HC{\equiv}CCH{=}CH_2$$

EXERCISES 12-17 Write structural formulas for all the alkenes with each of the following molecular formulas:

(a) C_3H_5Cl (b) $C_3H_4Cl_2$ (c) C_3H_4ClBr (d) C_4H_7Cl

12-18 For all the answers to exercise 12-17, indicate (a) which structures can exist as *cis-trans* isomers, and (b) which structures cannot exist as *cis-trans* isomers.

12-19 Give the IUPAC name for each of the following compounds:

(a) 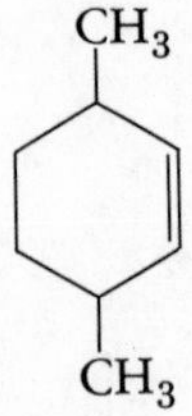 [cyclohexene ring with CH_3 at top and CH_3 at bottom]

(b) 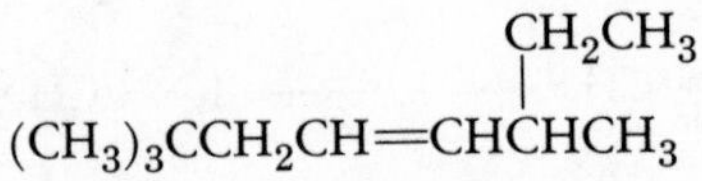 $(CH_3)_3CCH_2CH{=}CH\overset{CH_2CH_3}{\overset{|}{C}}HCH_3$

(c) 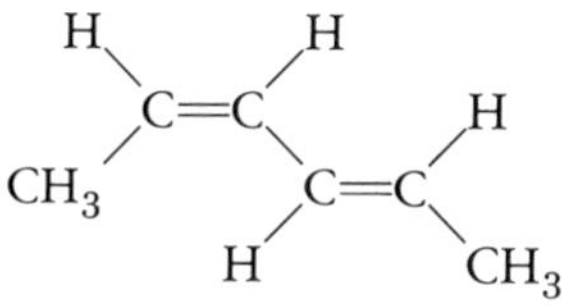 H, H, CH_3 / C=C / H / C=C / H, CH_3

(d) 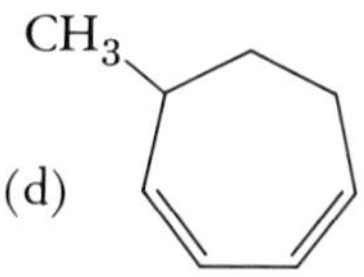 CH_3 (cycloheptadiene ring)

(e) $CH_2{=}CHCH_2CH(CH_3)_2$ (f) $H_2C{=}C{=}C(CH_3)_2$

(g) 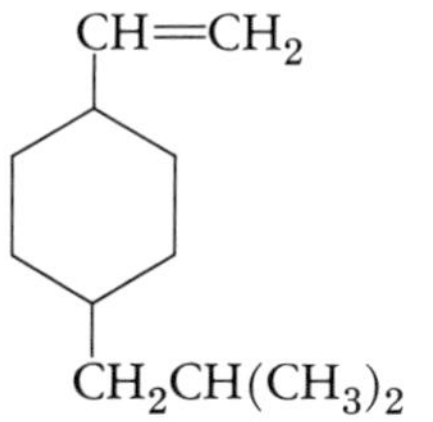 $CH{=}CH_2$ / (cyclohexane ring) / $CH_2CH(CH_3)_2$

(h) 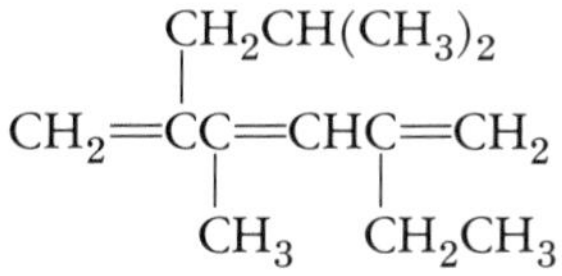 $CH_2{=}CC{=}CHC{=}CH_2$ with $CH_2CH(CH_3)_2$ above the second C, CH_3 below the second C, and CH_2CH_3 below the fourth C

12-20 Write structural formulas for each of the following names:
(a) 1,3-hexadiene (b) 2,3,5,5-tetramethyl-1-hexene
(c) allylcyclopentane (d) 3-methyl-5-vinylcyclohexene
(e) 3-cyclopentyl-1-butene (f) 2-methyl-1,3-pentadiene
(g) 2-methyl-1,3-cyclopentadiene (h) 1,3-hexadiyne
(i) 3-methyl-1-pentyne (j) 4-ethyl-3-isopropyl-1-heptyne
(k) 3-*sec*-butyl-3-*t*-butyl-1-heptyne

12-21 Write the equation for the reaction of 2-methyl-2-pentene with each of the following reagents:
(a) Br_2 (b) concentrated H_2SO_4 (c) HCl
(d) H_2, Pt (e) Cl_2 (f) $C_6H_5CO_3H$
(g) dilute basic aqueous solution of $KMnO_4$ (h) H_2O, H^+

12-22 Repeat exercise 12-21 using cyclopentene instead of 2-methyl-2-pentene.

12-23 Write the equation for the reaction of 1 mole of 1-pentyne with 1 mole of each of the following reagents:
(a) H_2, Pt (b) Br_2 (c) HCl (d) Cl_2

12-24 Repeat exercise 12-23 using 2 moles of each reagent per mole of 1-pentyne.

12-25 Write the structure of the organic product in each of the following reactions:
(a) cyclohexene + Cl_2 $\longrightarrow$
(b) 1-butene + HBr $\longrightarrow$
(c) cyclohexylacetylene + H_2O $\xrightarrow[H_2SO_4]{HgSO_4}$
(d) 1,3-cyclohexadiene + 2 H_2 $\xrightarrow{Pt}$
(e) 2,3-dimethyl-2-butene + $C_6H_5CO_3H$ $\longrightarrow$
(f) propene + H_2O $\xrightarrow{H_2SO_4}$
(g) 2-butyne + 2 HBr $\longrightarrow$
(h) 1-butene + $KMnO_4$ (dilute basic aqueous solution) $\longrightarrow$

12-26 Propene reacts with concentrated H_2SO_4 to form $\underset{\displaystyle|\atop\displaystyle OSO_3H}{CH_3CHCH_3}$ as product. Write a reaction mechanism for this reaction.

12-27 Hydrochloric acid (HCl) adds to 2-pentene to form both $CH_3\underset{\displaystyle|\atop\displaystyle Cl}{C}HCH_2CH_2CH_3$ and $CH_3CH_2\underset{\displaystyle|\atop\displaystyle Cl}{C}HCH_2CH_3$ as products. Write a reaction mechanism that accounts for this fact.

AROMATIC COMPOUNDS

13

Chemists were presented with an intriguing puzzle in 1825 with the discovery of the compound benzene. Chemical analysis of benzene revealed a chemical formula of C_6H_6. A formula C_nH_n indicates a high degree of unsaturation such as that found in alkynes. By analogy, it might be expected that benzene would be highly reactive and undergo addition reactions characteristic of alkenes and alkynes. Surprisingly, that is not the case. Benzene does not react readily with halogens, hydrogen halides, or other reagents that are usually added to alkenes and alkynes. It is also unaffected by reaction conditions under which alkenes are oxidized and reduced.

Benzene can be forced to react with most of these reagents by the use of catalysts or more vigorous conditions. *But these reactions of benzene are substitutions rather than additions.* For example, benzene reacts with bromine in the presence of ferric bromide ($FeBr_3$) as catalyst to form bromobenzene:

$$\underset{\text{Benzene}}{C_6H_6} + \underset{\text{Bromine}}{Br_2} \xrightarrow{FeBr_3} \underset{\text{Bromobenzene}}{C_6H_5Br} + \underset{\text{Hydrogen bromide}}{HBr} \qquad (13\text{-}1)$$

In this reaction, a bromine atom is substituted for one of the hydrogens of benzene. Clearly, benzene does not resemble alkenes and alkynes in its reactivity. By the end of the nineteenth century, it was clear to chemists that benzene was not related to any known alkene or alkyne.

Benzene has another characteristic that distinguishes it from most other organic compounds: it has a rather pleasant aroma. As a result, chemists of the nineteenth century classified it as an *aromatic hydrocarbon.* Through the years the meaning of the word *aromatic* has changed. We no longer use it to refer to the aroma of a compound, but rather, to the unusual chemical stability of a group of compounds of which benzene is the parent hydrocarbon.

The relative chemical inertness of benzene and other aromatic hydrocarbons and their tendency to undergo substitution rather than addition reactions were puzzling to scientists. It took almost 100 years for chemists to arrive at a satisfactory structure of the benzene molecule.

13.1 STRUCTURE OF BENZENE

In the latter part of the nineteenth century, chemists were able to deduce the structure of many compounds solely on the basis of their chemical reactions. Using this method, chemists arrived at a satisfactory description of the structure of benzene. The following reactions provided significant information about the structure of benzene.

The bromination of benzene forms one and only one isomer of bromobenzene (C_6H_5Br). Attempts to prepare and isolate more than one isomer of bromobenzene were unsuccessful. From this result, it was concluded that all of the hydrogens of benzene are equivalent. Further bromination of bromobenzene results in the formation and isolation of three and only three isomers of dibromobenzene ($C_6H_4Br_2$).

Although substitution reactions are typical of benzene, under certain conditions benzene can also undergo addition reactions. For example, under high pressure and temperature, 3 moles of hydrogen can be added to benzene in the presence of a platinum catalyst to form cyclohexane (C_6H_{12}). This result establishes that the basic structure of benzene is a six-member ring containing three units of unsaturation.

The last important piece of information was not obtained from chemical reactions. It was obtained early in the twentieth century after development of the x-ray method of determining the geometry of molecules. X-ray diffraction measurements of benzene show that all six bonds of the ring are equal in length. Therefore, the carbon skeleton of benzene is a regular hexagon with bond angles of 120 degrees.

Based on these facts, chemists proposed several descriptions of the structure of benzene. The first satisfactory one was provided by Linus Pauling in the 1930s using resonance theory. You will remember that according to this theory (Section 5.9), when a molecule can be written as two or more equivalent or nearly equivalent structures that differ only in the arrangement of electrons (without movement of any atoms), its structure does not correspond to any one of the *contributing structures.* Instead, the structure of the actual molecule is a *hybrid* of all the resonance contributors. The two major contributors to the hybrid structure of benzene are the following:

13-1 *13-2*

Again, neither structure *13-1* nor structure *13-2* represents the true structure of benzene. The real structure is a resonance hybrid that lies somewhere between the two. This representation of the benzene molecule means that the six-member ring does not contain three double bonds. Benzene is not cyclohexatriene and therefore the structure of benzene is different from that of typical alkenes. *The presence of six electrons in a six-member ring results in a unique structure with characteristic chemical and physical properties.*

What is so important about molecules or ions whose structures are hybrids of a number of contributing resonance structures? Such compounds or ions have increased stability. By this we mean that they have a lower energy content than similar molecules whose structures are not hybrids. This increased stability affects their chemical reactivity. For example, benzene is much less reactive than any alkene. This decrease in reactivity is due to the increased stability of benzene caused by resonance. Resonance theory was proposed to account for this increased stability of many organic molecules and ions. We will encounter more examples of resonance-stabilized molecules and ions in future chapters.

If the structure of benzene is neither 13-1 nor 13-2, how should we represent its structure on paper? There are two views. Some people prefer the following structural formula of benzene:

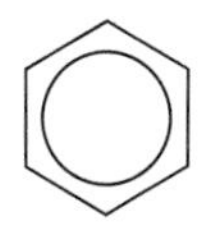

This representation conveys the aromatic nature of benzene. However, the major objection to this representation is that it is impossible to keep track of the electrons. The reader may well encounter this representation of benzene elsewhere. Others, including the author, prefer to represent benzene by drawing just one of the structures *13-1* or *13-2*. This representation will be used in this text. Neither structure *13-1* nor structure *13-2* alone conveys the aromatic nature of benzene. However, use of either structure is generally accepted by most chemists, who understand that it is a convenient symbol for representing the hybrid structure of benzene.

Using this representation, we can write the chemical equation of the bromination of benzene as follows:

$$C_6H_6 + Br_2 \xrightarrow{FeBr_3} C_6H_5Br + HBr$$

All the hydrogens are equivalent, so no matter which one is substituted, only one isomer of bromobenzene can be formed. Bromination of bromobenzene forms three, and only three, isomeric dibromobenzenes:

$$C_6H_5Br + Br_2 \xrightarrow{FeBr_3} o\text{-}C_6H_4Br_2 + m\text{-}C_6H_4Br_2 + p\text{-}C_6H_4Br_2 + HBr \quad (13\text{-}2)$$

Now that we can write the structures of aromatic compounds, we must learn how to name them. Unfortunately, the method used to name aromatic compounds is not very systematic.

13.2 NAMING AROMATIC COMPOUNDS

Many aromatic compounds were isolated and given nonsystematic names years before the IUPAC rules were developed. Because these names had become so widespread, they were simply incorporated into the IUPAC rules. No new IUPAC rules are needed, but it is necessary to learn a number of nonsystematic names.

When only one group is attached to benzene, the compound is named by placing the name of the group as a prefix to the word benzene. For example:

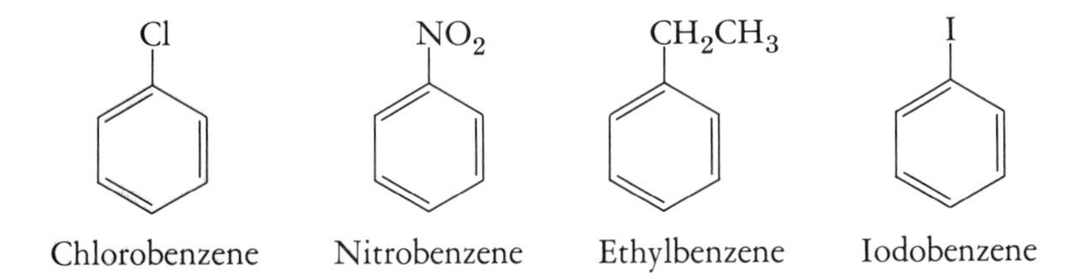

The names of the most common groups placed as prefixes are the following:

—F	fluoro-	—I	iodo-
—Cl	chloro-	$—NO_2$	nitro-
—Br	bromo-	—NO	nitroso-

The names of other monosubstituted benzenes show no resemblance to the name of the attached group. For example, methylbenzene is *always* known as toluene; aminobenzene, as aniline; hydroxybenzene, as phenol. The most important of these compounds are the following:

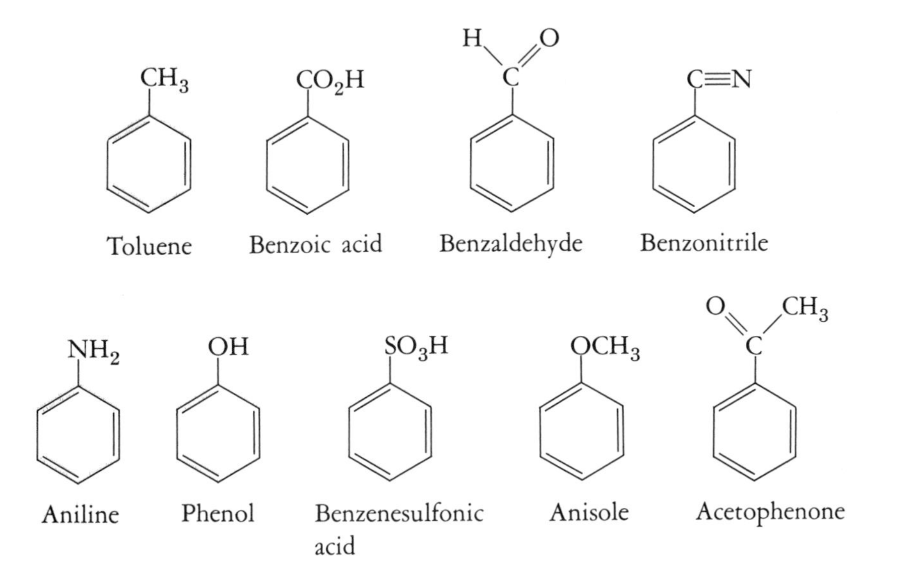

If the group attached to a benzene ring is a complex one, it is often more convenient to designate the benzene ring as a substituent on the group. In this case, the benzene ring with one hydrogen removed is called a *phenyl* group:

C_6H_5— or

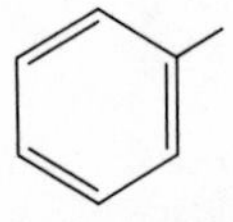

Phenyl group

The following are examples of this way of naming aromatic compounds:

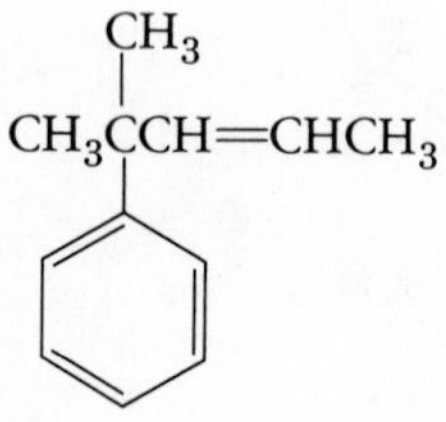

CH_3
$CH_3CCH{=}CHCH_3$

C_6H_5 C_6H_5

4-Methyl-4-phenyl-2-pentene 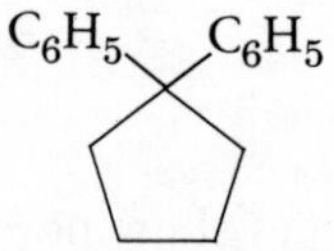 1,1-Diphenylcyclopentane

The common name *benzyl* is given to the following group:

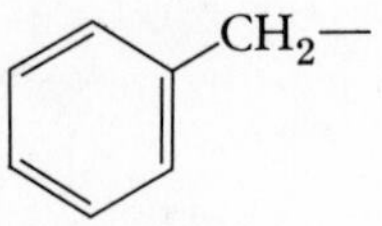

CH_2—

Benzyl group

For example:

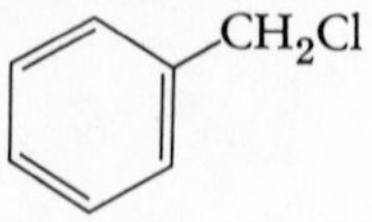

CH_2Cl CH_2Br

Benzyl chloride 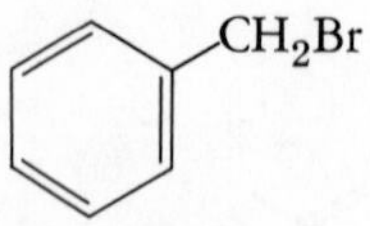 Benzyl bromide

*EXERCISE 13-1 Give the IUPAC name for each of the following compounds:

(a) 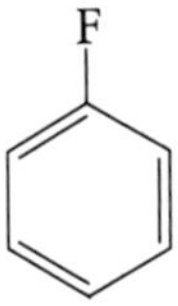 F on benzene ring (b) $C(CH_3)_3$ on benzene ring (c) $CH_3CHC{\equiv}CH$ on benzene ring (d) 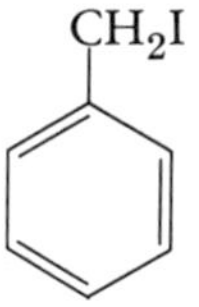 CH_2I on benzene ring (e) 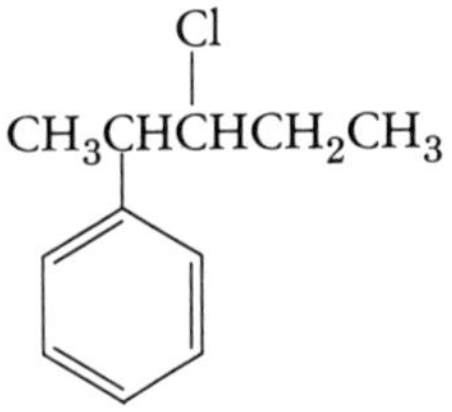 $CH_3CHCHCH_2CH_3$ with Cl, on benzene ring

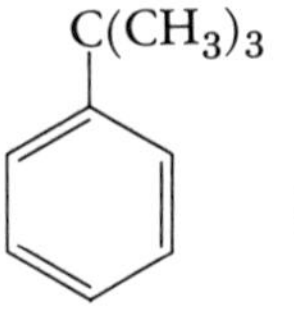

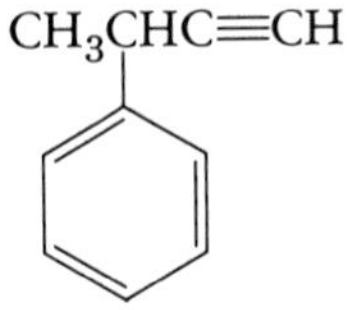

If two groups are attached to the benzene ring, the name must not only tell what groups are present, but also where they are located. We can differentiate the three possible isomers of a disubstituted benzene in two ways. The first is by use of the prefixes *ortho-*, *meta-*, and *para-*, which are abbreviated *o*, *m*, and *p*. The second is to number the carbons of the benzene ring. For example:

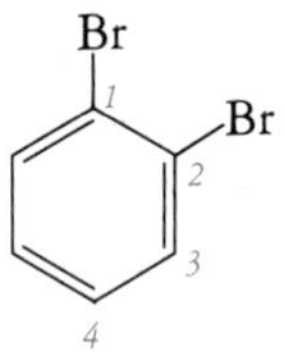

o-Dibromobenzene	*m*-Dibromobenzene	*p*-Dibromobenzene
1,2-Dibromobenzene	1,3-Dibromobenzene	1,4-Dibromobenzene

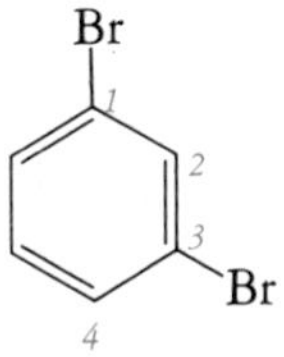

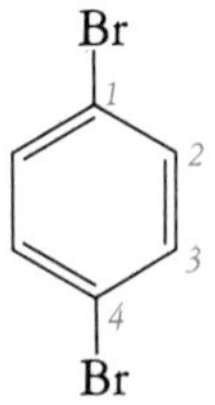

If the two groups are different and one of the groups is the kind that gives a special name to the molecule, then it is named as a derivative of that special compound. For example:

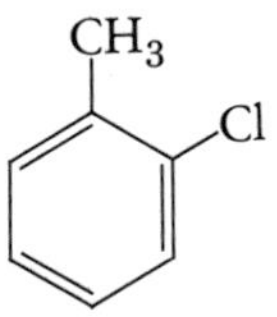

o-Chlorotoluene	*p*-Nitrobenzoic acid	*m*-Bromoaniline
2-Chlorotoluene	4-Nitrobenzoic acid	3-Bromoaniline

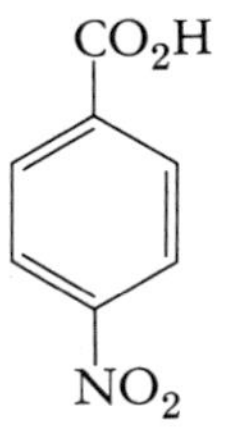

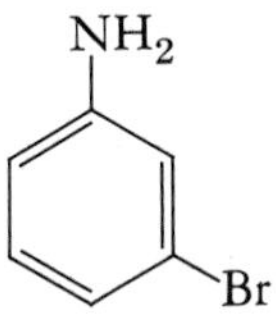

* The answers for the exercises in this chapter begin on page 871.

If neither is a group that gives a special name to the molecule, then the names of the two groups are placed successively followed by benzene. For example:

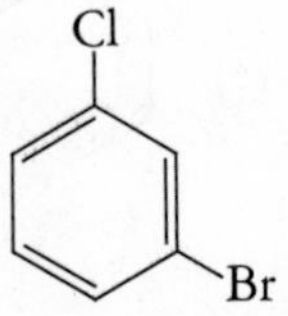

Cl, Br — *m*-Bromochlorobenzene
3-Bromochlorobenzene

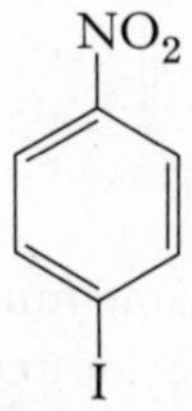

NO_2, I — *p*-Iodonitrobenzene
4-Iodonitrobenzene

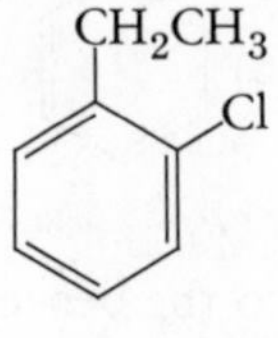

CH_2CH_3, Cl — *o*-Chloroethylbenzene
2-Chloroethylbenzene

Note that the number 1 is not included in the name. It is understood that the substituent without a number is bonded to the 1 position.

EXERCISE 13-2 Give the IUPAC name for each of the following compounds:

(a) 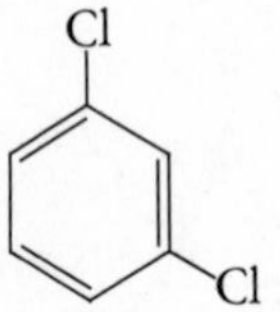 Cl, Cl
(b) 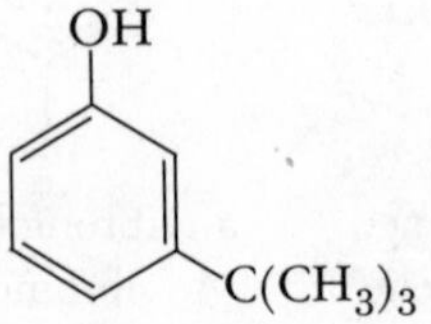 OH, $C(CH_3)_3$
(c) 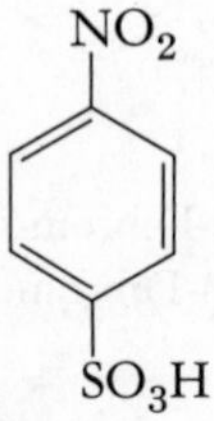 NO_2, SO_3H
(d) 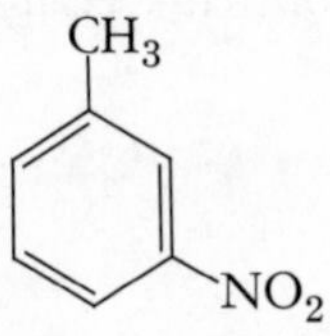 CH_3, NO_2
(e) 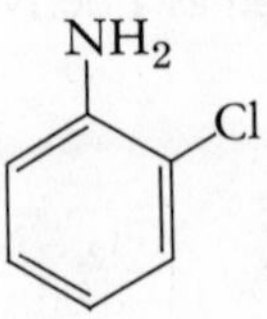 NH_2, Cl
(f) 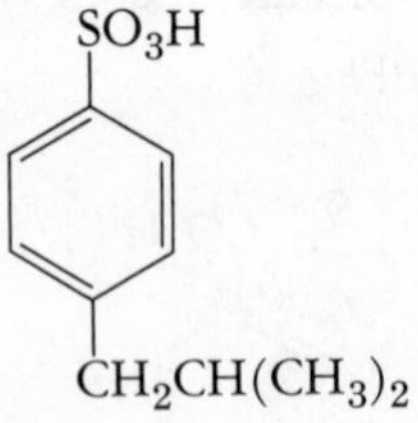 SO_3H, $CH_2CH(CH_3)_2$
(g) 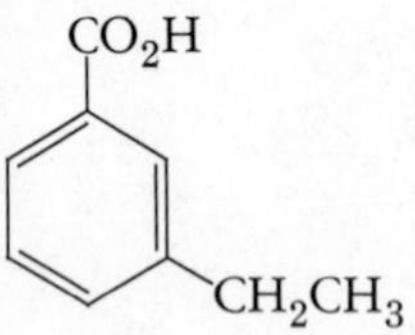 CO_2H, CH_2CH_3
(h) 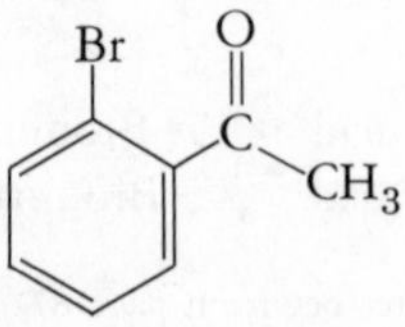 Br, $C(=O)CH_3$

(i) 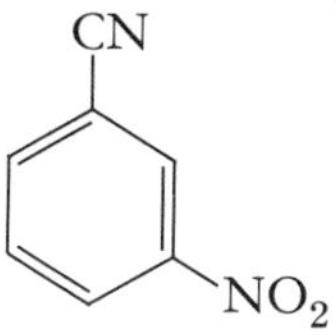 [structure: benzene ring with CN and NO_2 in meta positions]

(j) 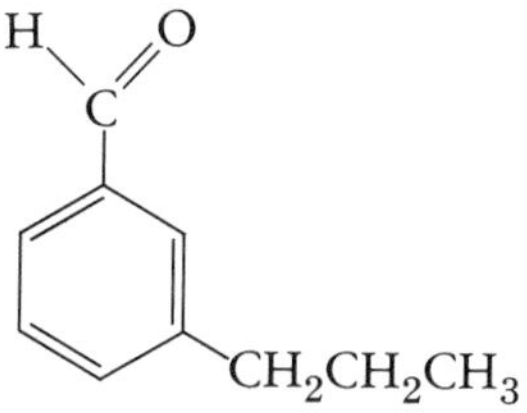 [structure: benzene ring with CHO (H–C=O) and $CH_2CH_2CH_3$ in meta positions]

EXERCISE 13-3 Write the structure for each of the following compounds:
(a) *p*-chlorofluorobenzene (b) 3-propylbenzoic acid
(c) *o*-fluorotoluene (d) 3-nitroaniline
(e) *m-t*-butylnitrobenzene (f) 2-*sec*-butylaniline
(g) 4-toluenesulfonic acid (h) 4-phenyltoluene
(i) *p*-fluorophenol (j) *m*-nitroethylbenzene
(k) 3-chlorobenzaldehyde (l) *p*-isobutylanisole
(m) 4-ethylacetophenone

Several disubstituted benzenes have been given names that, again, give no indication of the kind of groups attached to the ring. The following are several of the most common examples:

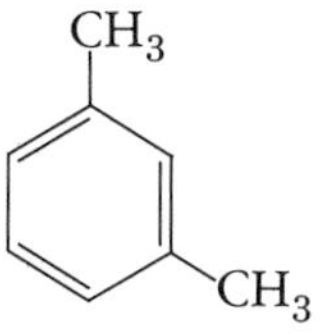

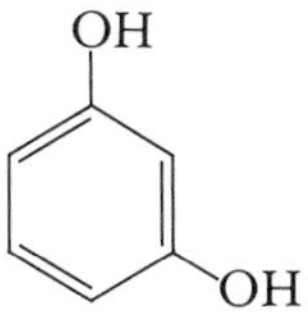

m-Xylene (CH_3, CH_3) Catechol (OH, OH) Resorcinol (OH, OH) Hydroquinone (OH, OH) *p*-Cresol (OH, CH_3)

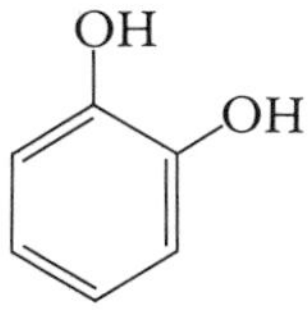

EXERCISE 13-4 Write the structure of the following compounds:
(a) *o*-xylene (b) *p*-xylene (c) *o*-cresol (d) *m*-cresol

Now that we have learned to name compounds that contain a benzene ring, let us turn our attention to some of their chemical reactions.

13.3 SUBSTITUTION REACTIONS OF AROMATIC COMPOUNDS

The most common reaction of the benzene ring involves substituting one or more of the ring hydrogens by other groups or atoms. *Bromination, chlorination, nitration, alkylation, acylation,* and *sulfonation* are typical examples of such *substitution* reactions:

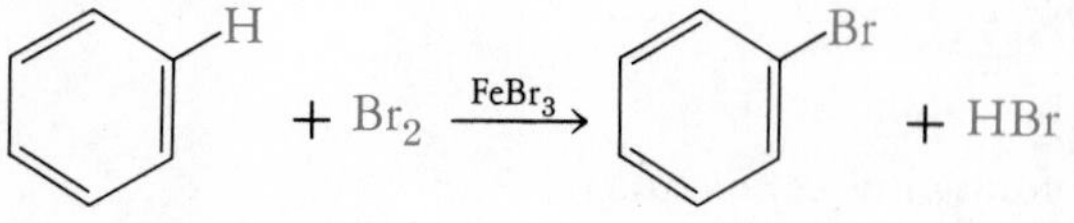

Bromination

$$C_6H_5{-}H + Br_2 \xrightarrow{FeBr_3} C_6H_5{-}Br + HBr$$

Chlorination

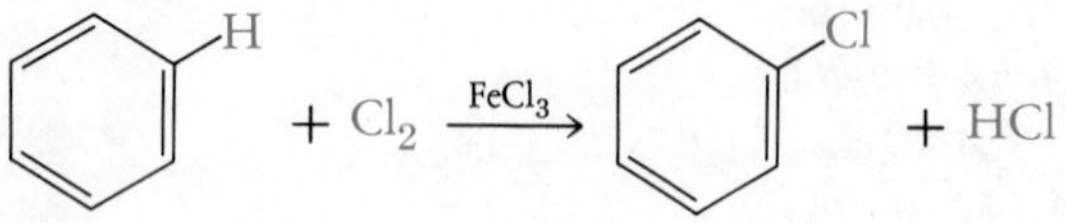

$$C_6H_5{-}H + Cl_2 \xrightarrow{FeCl_3} C_6H_5{-}Cl + HCl$$

Nitration

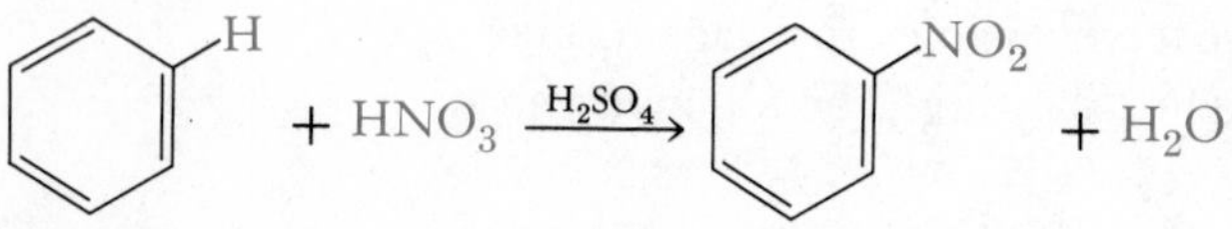

$$C_6H_5{-}H + HNO_3 \xrightarrow{H_2SO_4} C_6H_5{-}NO_2 + H_2O$$

Sulfonation

$$C_6H_5{-}H + H_2SO_4 \xrightarrow{SO_3} C_6H_5{-}SO_3H + H_2O \qquad (13\text{-}3)$$

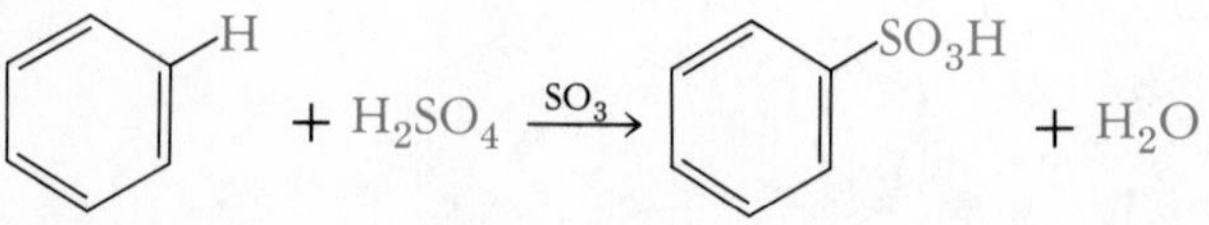

Friedel-Crafts alkylation

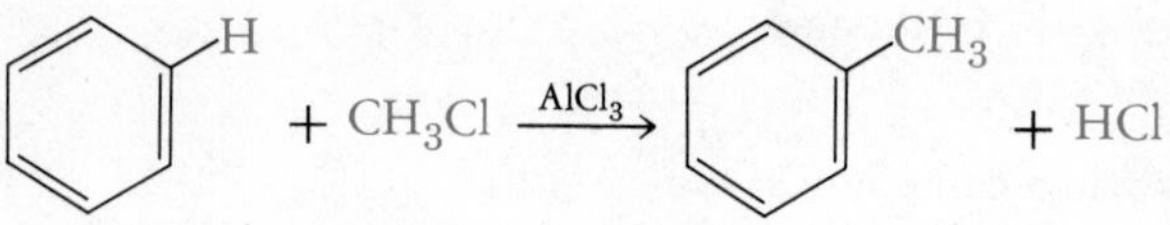

$$C_6H_5{-}H + CH_3Cl \xrightarrow{AlCl_3} C_6H_5{-}CH_3 + HCl$$

Friedel-Crafts acylation

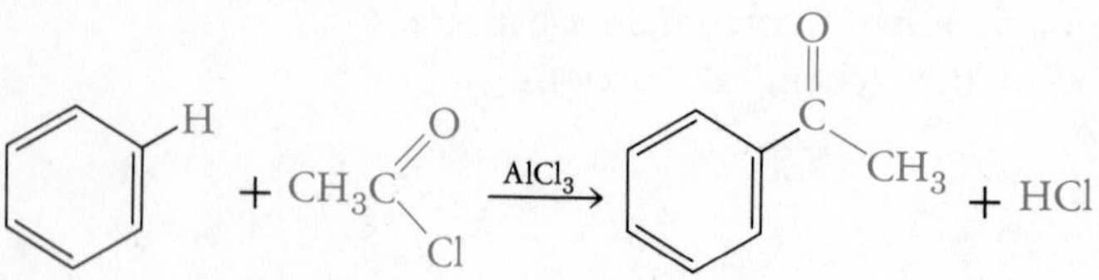

$$C_6H_5{-}H + CH_3C(=O)Cl \xrightarrow{AlCl_3} C_6H_5{-}C(=O){-}CH_3 + HCl$$

The experimental conditions and kinds of catalysts vary depending on the particular reaction. For example, nitration occurs more easily than sulfonation, as shown by the fact that sulfuric acid is used as a solvent for nitration.

For best results, sulfonation is carried out with fuming sulfuric acid. This more reactive reagent contains sulfur trioxide (SO_3) dissolved in concentrated sulfuric acid. In these reactions, it is possible to substitute one or more hydrogens as desired, by choosing the proper experimental conditions.

The substitution reactions shown in equation 13-3 use benzene as an example. But benzene is not the only aromatic compound that can undergo this type of reaction. Toluene, nitrobenzene, chlorobenzene, and many other aromatic compounds also undergo substitution reactions. These compounds differ from benzene in that they already have a substituent on the benzene ring. Consequently, the remaining five hydrogens are not equivalent. When these hydrogens undergo substitution reactions, isomers can be formed depending on which hydrogen is replaced. But which one will be replaced, the one in the *ortho,* the *meta,* or the *para* position? To answer this question, let us compare the products formed by the mononitration of toluene and nitrobenzene.

The nitration of toluene forms the three isomeric nitrotoluenes in the following percentages:

CH_3 + HNO_3 $\xrightarrow{H_2SO_4}$ CH_3 NO_2 + CH_3 NO_2 + CH_3 NO_2

58% 4% 38%

Together the *ortho* and *para* isomers account for 96 percent of the product. Very little *meta* isomer is formed. The methyl group of toluene therefore directs the nitro group into the *ortho* and *para* positions.

The nitration of nitrobenzene forms all three isomeric dinitrobenzenes in the following percentages:

NO_2 + HNO_3 $\xrightarrow{H_2SO_4}$ NO_2 NO_2 + NO_2 NO_2 + NO_2 NO_2

6% 93% 1%

The major product is the *meta* isomer. In this case, the nitro group on the ring directs the second nitro group into the *meta* position.

By analyzing the reaction products of a great many substitution reactions, it has been found that every group can be put into one of two classes: either *ortho-para* directing or *meta* directing. The most common groups and their directing effects are listed in Table 13-1.

The directing influence of a given group is the same, either predominantly *ortho-para* or predominantly *meta,* whatever the substitution reaction. The actual percentages of the various isomers may vary slightly from reaction to reaction. The following substitution reactions of toluene illustrate this point:

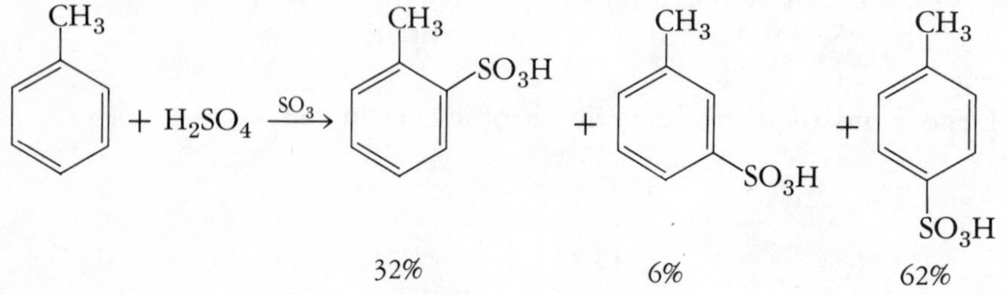

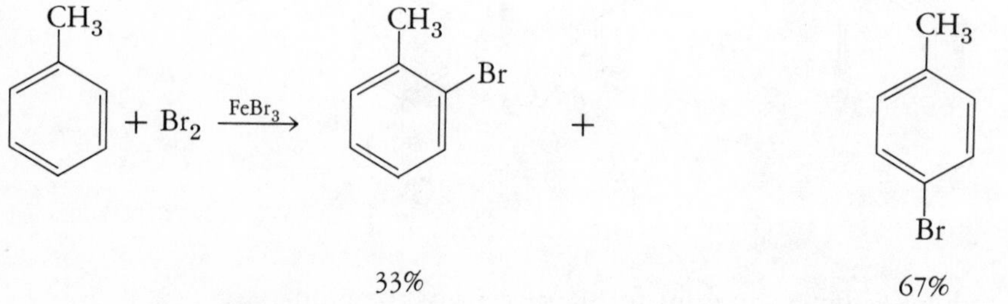

CH_3 + H_2SO_4 $\xrightarrow{SO_3}$ CH_3, SO_3H (32%) + CH_3, SO_3H (6%) + CH_3, SO_3H (62%)

CH_3 + Br_2 $\xrightarrow{FeBr_3}$ CH_3, Br (33%) + CH_3, Br (67%)

Thus, nitration, bromination, and sulfonation of toluene all form predominantly the *ortho* and *para* isomers. Other substitution reactions of toluene would give similar results.

EXERCISE 13-5 Write the structures of the principal products expected from mononitration of each of the following compounds:

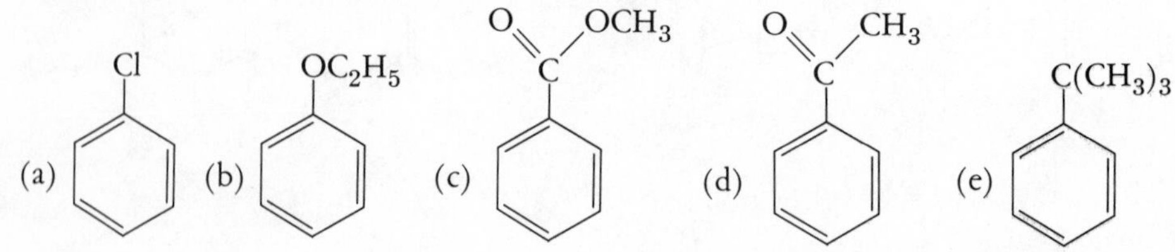

(a) Cl (b) OC_2H_5 (c) O, OCH_3, C (d) O, CH_3, C (e) $C(CH_3)_3$

Table 13-1. Directing Effect of Substituents[a]

Predominant Directing	
o, p	*m*
$—NH_2$, —NHR, $—NR_2$	$—NO_2$
$—NH_2$, —NHAr, $—N(Ar)_2$	—CN
—OH, —OR, —OAr	$—CO_2H$, —CHO, —C(=O)R, $—CO_2R$, $—CO_2Ar$
—NHC(=O)R, —NHC(=O)Ar	$—SO_3H$
Alkyl groups	$—\overset{+}{N}H_3$, $—\overset{+}{N}H_2R$, $—\overset{+}{N}HR_2$, $—\overset{+}{N}R_3$
Halogens	$—\overset{+}{N}H_3$, $—\overset{+}{N}H_2Ar$, $—\overset{+}{N}H(Ar)_2$, $—\overset{+}{N}(Ar)_3$

[a] R = alkyl groups; Ar = aryl groups.

EXERCISE 13-6 Write the structures of the principal products of monochlorination of each of the compounds in exercise 13-5.

In addition to controlling the direction of substitution, the group attached to the benzene ring can make the ring more or less susceptible than benzene to substitution reactions. For example, benzene reacts completely with fuming sulfuric acid at room temperature within 25 minutes, whereas toluene reacts within 2 minutes under the same reaction conditions. The methyl group of toluene makes the ring more reactive than the unsubstituted benzene. The methyl group is therefore an activating group. On the other hand, nitrobenzene has been found to react more slowly than benzene. The nitro group makes the ring less reactive than is unsubstituted benzene and is called a deactivating group.

Experiments have shown that all groups can be classified as either activating or deactivating. In general, groups that direct *ortho* and *para* are also activating, whereas groups that direct *meta* are also deactivating. The halogens are the exception to this rule. They direct *ortho* and *para* but are deactivating. The effects on product and reactivity of most of the common groups are summarized in Table 13-2.

EXERCISE 13-7 For each of the compounds in exercise 13-5, tell whether bromination will occur faster or slower than with benzene.

In summary, the results of many experiments clearly establish that in aromatic substitution reactions, the group already on the benzene ring has

Table 13-2. Directive Influence and Effect on Reactivity of Substituents[a]

I. Activating and *ortho* and *para* directing

A. Strongly activating

$-NH_2$, $-NHR$, $-NR_2$, $-NHAr$, $-N(Ar)_2$

$-OH$, $-OR$, $-OAr$

B. Moderately activating

$-NHC(=O)R$, $-NHC(=O)Ar$

C. Weakly activating

Alkyl groups

II. Strongly deactivating and *meta* directing

$-NO_2$

$-CN$

$-CO_2H$, $-CHO$, $-C(=O)R$, $-C(=O)Ar$, $-CO_2R$, $-CO_2Ar$

$-SO_3H$

$-\overset{+}{N}H_3$, $-\overset{+}{N}H_2R$, $-\overset{+}{N}HR_2$, $-\overset{+}{N}R_3$, $-\overset{+}{N}H_2Ar$, $-\overset{+}{N}H(Ar)_2$, $-\overset{+}{N}(Ar)_3$

III. Deactivating and *ortho* and *para* directing

Halogens

[a] R = alkyl groups; Ar = aryl groups.

two effects. It directs the second group to specific positions, and it influences the reactivity of the ring toward further substitution.

13.4 MECHANISM OF ELECTROPHILIC AROMATIC SUBSTITUTION REACTIONS

OPTIONAL SECTION

The general mechanism that has been proposed for aromatic substitution reactions involves the following three steps:

Step 1. Formation of the electrophile:

$$EY + \text{Catalyst} \longrightarrow \overset{+}{E}\,\overset{-}{Y}\ \text{Catalyst}$$

The symbol EY is a general representation of a molecule that contains a

polar covalent bond in which the part Y is the more electronegative. The catalyst enhances this polarity and makes the part E an electrophile.
Step 2. Reaction of the electrophile with the benzene ring:

$$+ \overset{+}{E}\ \overset{-}{Y}\ \text{Catalyst} \longrightarrow \quad \text{H}\ \overset{-}{Y}\ \text{Catalyst}$$

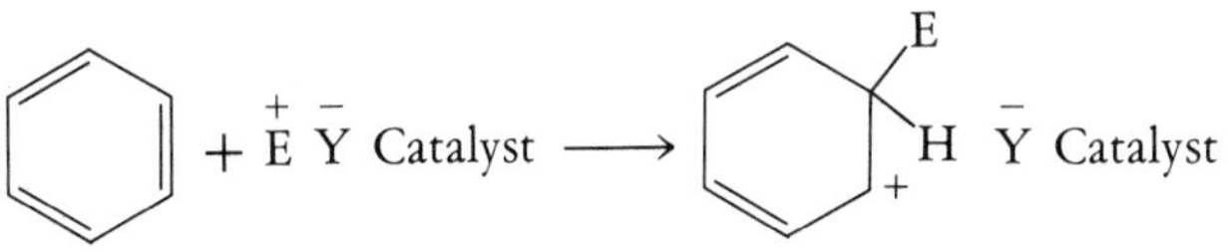

Step 3. Loss of a proton to form the product:

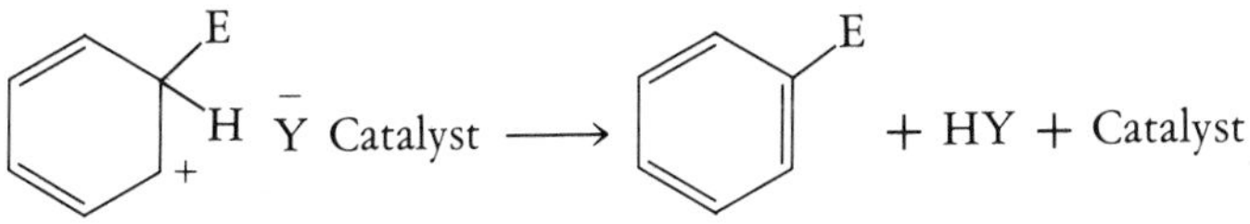

$$\text{H}\ \overset{-}{Y}\ \text{Catalyst} \longrightarrow \quad + \text{HY} + \text{Catalyst}$$

The key step is the reaction of the electrophile with the benzene ring. To emphasize this point, the reaction is called an electrophilic aromatic substitution. Notice that the catalyst assists in the formation of the electrophile in step 1 and is reformed in step 3. A specific catalyst is needed to generate the strong electrophile for each type of reaction discussed in Section 13.3. Let us apply this general mechanism to several specific reactions.

Bromination

Bromine reacts with $FeBr_3$ to form the electrophile Br^+ in the first step, as follows:

Step 1. $Br_2 + FeBr_3 \longrightarrow Br^+\ FeBr_4^-$

Bromine by itself is not a strong enough electrophile to react with a benzene ring, so the catalyst $FeBr_3$ is needed to increase its electrophilicity.
The electrophile adds to the benzene ring in the second step to form the carbonium ion *13-3:*

Step 2. 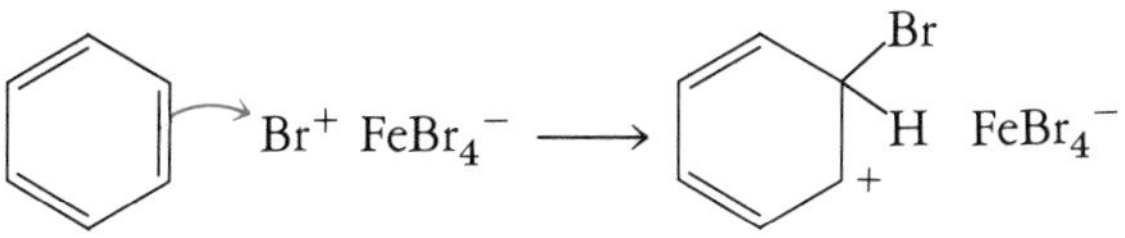 $\quad Br^+\ FeBr_4^- \longrightarrow \quad H\ FeBr_4^-$

13-3

The C—Br bond is formed by taking two electrons from the aromatic ring. Notice that this step is similar to the first step in the reaction mechanism of additions of acids to alkenes (see Section 12.5). The carbonium ion 13-3 is a resonance hybrid with three contributing structures. This indicates that the positive charge is delocalized to the carbons *ortho* and *para* to the one bonded to bromine:

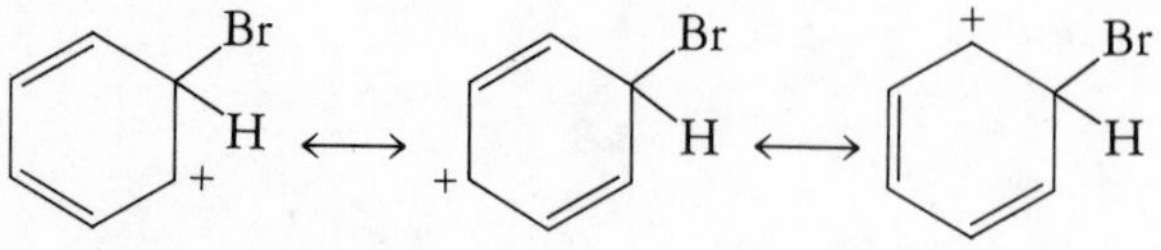

Br H + ⟷ + Br H ⟷ + Br H

13-3

In the third and final step, the carbonium ion loses a proton to form the products hydrogen bromide and bromobenzene, and reforms the catalyst.

Step 3. Br H + $FeBr_4^-$ ⟶ Br + HBr + $FeBr_3$

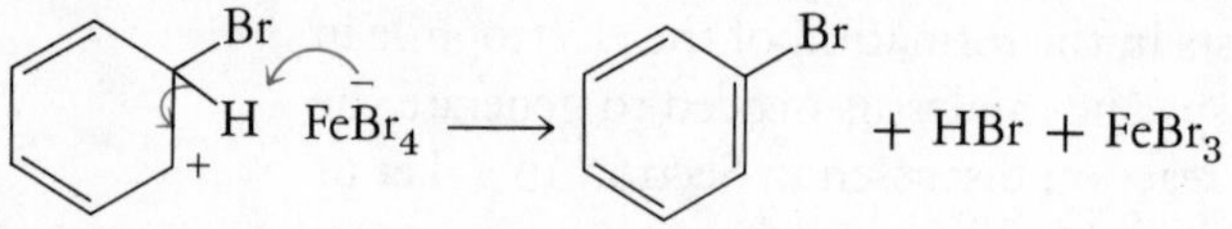

A proton is lost in the final step to reform a stable aromatic system, because this is the path that requires the least energy.

EXERCISE 13-8 Write the reaction mechanism for the chlorination of benzene with $FeCl_3$ as catalyst.

Nitration

The electrophile in the nitration reaction is the *nitronium ion,* $O{=}\overset{+}{N}{=}O$. The nitronium ion is formed by the reaction of nitric acid with concentrated sulfuric acid, which is the catalyst:

Step 1. $HO{-}\overset{+}{N}(=O)O^-$ + H_2SO_4 ⇌ $O{=}\overset{+}{N}{=}O$ + H_2O + HSO_4^-

Reaction of the nitronium ion with the benzene ring occurs in the second step:

Step 2. $\text{C}_6\text{H}_6 + \text{O}{=}\overset{+}{\text{N}}{=}\text{O} \longrightarrow$ [resonance structures of the $\text{C}_6\text{H}_6\text{NO}_2^+$ intermediate, with $-\overset{+}{\text{N}}(\text{O}^-){=}\text{O}$ and H on the same ring carbon and the positive charge in the ring] $\longleftrightarrow \cdots \longleftrightarrow \cdots$

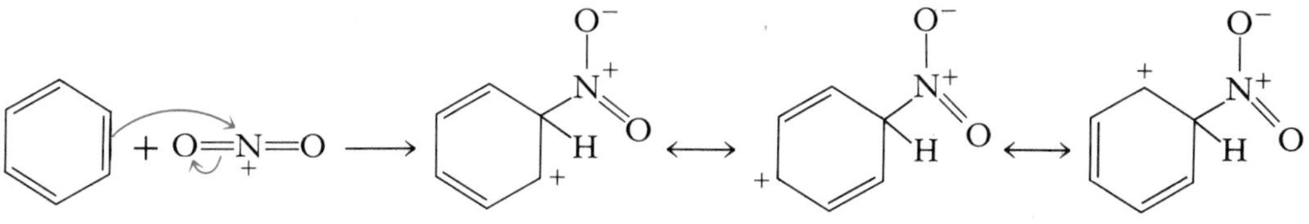

Loss of a proton in the third step forms the product:

Step 3. [$\text{C}_6\text{H}_6\text{NO}_2^+$ intermediate] $+ \overset{-}{\text{HSO}_4} \longrightarrow \text{C}_6\text{H}_5\overset{+}{\text{N}}(\text{O}^-){=}\text{O} + \text{H}_2\text{SO}_4$

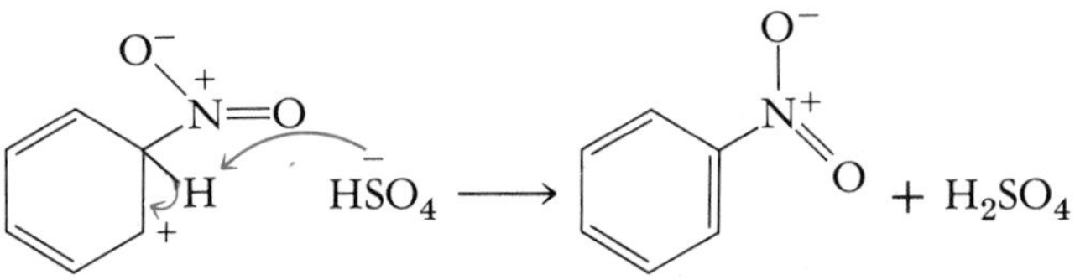

Friedel-Craft Acylation and Alkylation Reactions

Aluminum salts such as $AlCl_3$ and $AlBr_3$ are used as catalysts in these reactions. They react with alkyl halides, such as CH_3Cl, and acyl halides, such as $CH_3C(=O)Cl$, to form salt-like compounds in which the C—Cl bond is strongly polarized.

Step 1. $$CH_3Cl + AlCl_3 \longrightarrow \overset{\delta+}{CH_3}\text{—}\overset{\delta-}{ClAlCl_3}$$

$$CH_3C(=O)Cl + AlCl_3 \longrightarrow CH_3\overset{\delta+}{C}(=O)\text{—}\underset{\delta-}{ClAlCl_3}$$

In this way the carbon atom acquires a partial positive charge and becomes a strong electrophile. It reacts with a benzene ring in the second step:

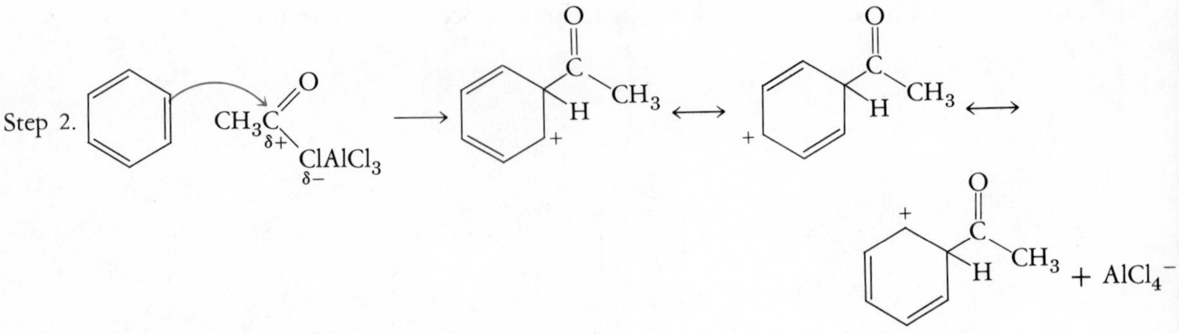

Step 2. $CH_3\overset{\delta+}{C}(=O)-\overset{\delta-}{ClAlCl_3}$ $\longrightarrow$ (C(=O)CH$_3$, H, +) $\longleftrightarrow$ (C(=O)CH$_3$, H, +) $\longleftrightarrow$ (C(=O)CH$_3$, H, +) $+ AlCl_4^-$

Loss of a proton in the third step forms the product:

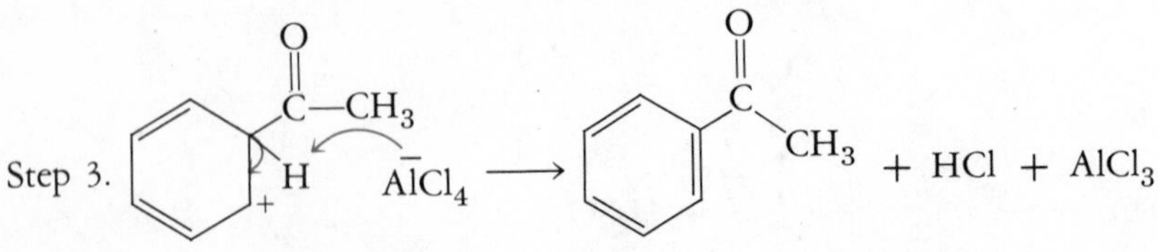

Step 3. (C(=O)—CH$_3$, H, +) $\overset{-}{Al}Cl_4$ $\longrightarrow$ (C(=O)CH$_3$) $+ HCl + AlCl_3$

EXERCISE 13-9 Write a mechanism for the reaction of CH_3Br and benzene with $AlBr_3$ as catalyst to form toluene as product.

We have focused our attention on the substitution reactions of the benzene ring. However, benzene can also undergo other reactions, as we will learn in the next section.

13.5 OTHER REACTIONS OF AROMATIC COMPOUNDS

Substitution is their most common type of reaction, but aromatic compounds can also undergo a few other reactions. We are already familiar with one such nonsubstitution reaction, the hydrogenation of benzene to cyclohexane. Another example of such an addition reaction is the *light-catalyzed* addition of 3 moles of chlorine to benzene:

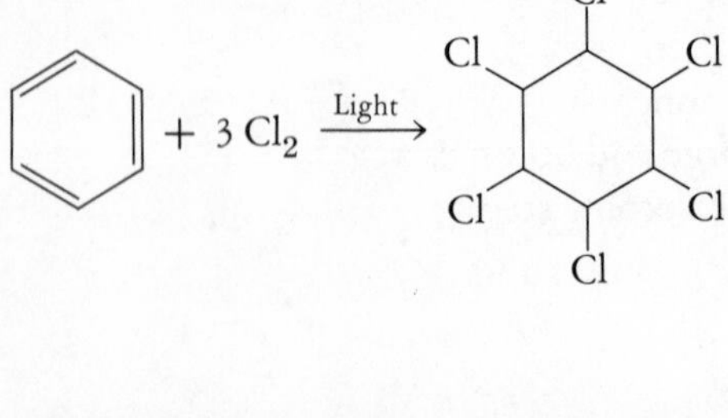

$$C_6H_6 + 3\,Cl_2 \xrightarrow{\text{Light}} C_6H_6Cl_6$$

1,2,3,4,5,6-Hexachlorocyclohexane

(13-4)

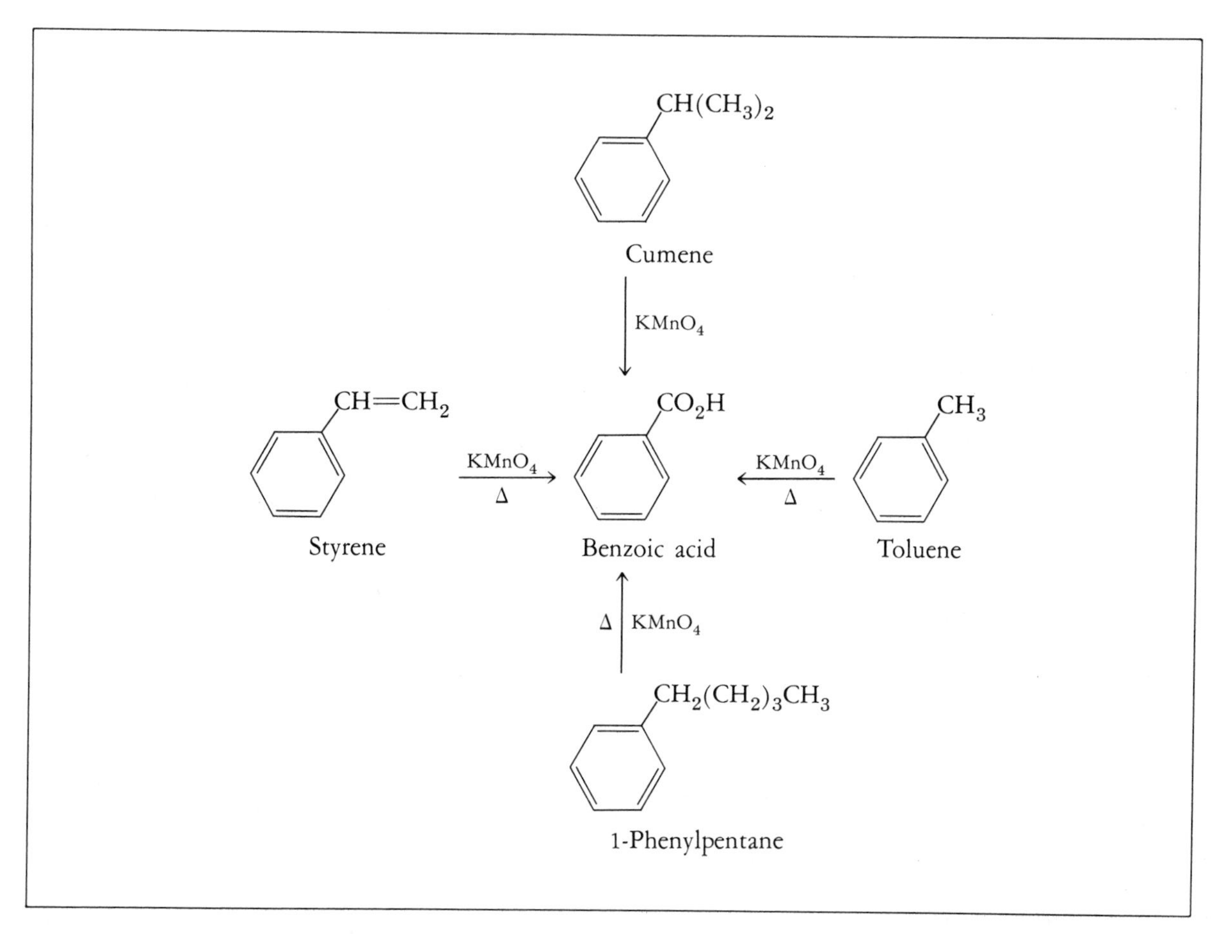

Fig. 13-1. The formation of benzoic acid by the oxidation of four monoalkylbenzenes.

Notice that light is used as a catalyst to cause addition. The reaction conditions are very different from those used in the substitution reaction of chlorine (Cl_2 and $FeCl_3$). The product of addition, 1,2,3,4,5,6-hexachlorocyclohexane, more commonly known as benzene hexachloride (BHC), is a potent insecticide.

The reaction of alkylbenzenes with strong oxidizing reagents demonstrates the remarkable stability of the benzene ring. When heated with aqueous alkaline or acidic potassium permanganate, any monoalkylbenzene is oxidized to benzoic acid. This fact is illustrated in Figure 13-1. If more than one alkyl group is present, both are oxidized to the $—CO_2H$ group. For example:

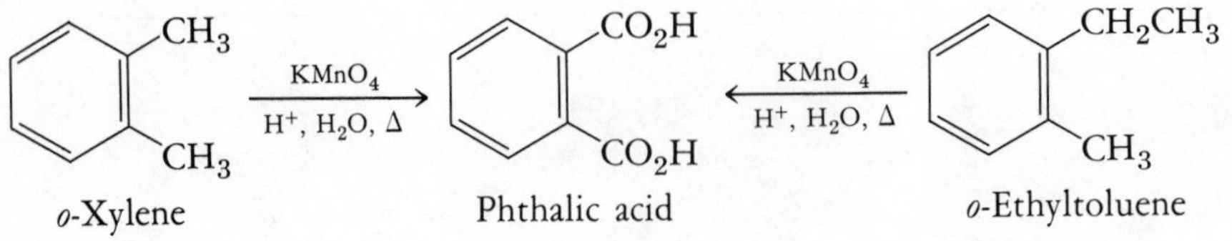

$$\text{o-Xylene (ring with } CH_3, CH_3) \xrightarrow[H^+, H_2O, \Delta]{KMnO_4} \text{Phthalic acid (ring with } CO_2H, CO_2H) \xleftarrow[H^+, H_2O, \Delta]{KMnO_4} \text{o-Ethyltoluene (ring with } CH_2CH_3, CH_3)$$

o-Xylene Phthalic acid *o*-Ethyltoluene

Remember that, although benzene is relatively inert toward oxidizing agents, it can be oxidized under vigorous conditions like any organic compound. For example, when heated hot enough in air, benzene burns to form carbon dioxide and water.

To get the benzene ring to react, it is clear that vigorous conditions (catalysts, heat, and/or light) must be used. In general, reactions at other functional groups require less vigorous conditions. In practical terms, this means that the benzene ring is usually unaffected by conditions that cause reactions elsewhere in the molecule. Thus, benzene, like the alkane portion of a molecule, remains intact during most reactions. This is an important point to remember, because it is true for both living and nonliving systems.

What kinds of compounds in living systems contain a benzene ring? In the next section, we will give a few examples.

13.6 AROMATIC COMPOUNDS IN NATURE

Derivatives of Benzene

Benzene is toxic to humans. It causes severe liver damage and is a possible carcinogen (causes cancer). Although benzene itself is not found in nature, many compounds that have a benzene ring in their basic structure are biologically important. Such compounds are called *derivatives* of benzene. They usually differ in the kind and number of substituents attached to the basic structure. For example, phenylalanine, a compound essential for life, contains a benzene ring and is a derivative of benzene:

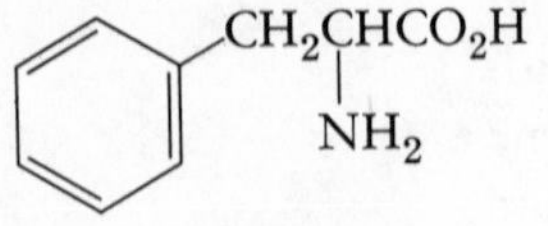

$$C_6H_5-CH_2CH(NH_2)CO_2H$$

Phenylalanine

This compound cannot be made in the human body and must be obtained from the food we eat. However, plants can make phenylalanine. They prepare it from the nonaromatic compound shikimic acid in several steps:

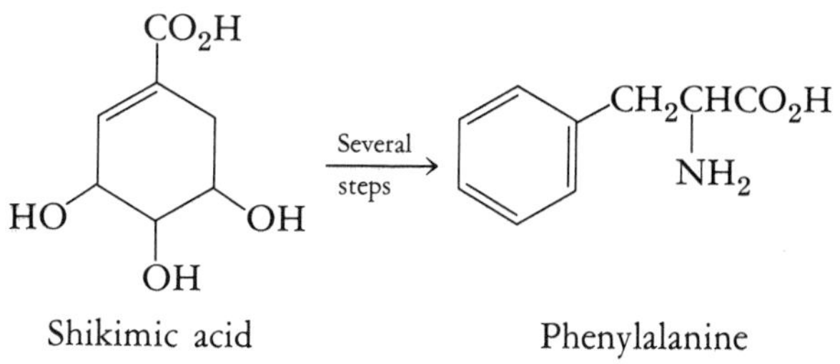

Shikimic acid $\xrightarrow{\text{Several steps}}$ Phenylalanine

In the human body, phenylalanine can be used to prepare other important compounds such as the hormones epinephrine and dopa:

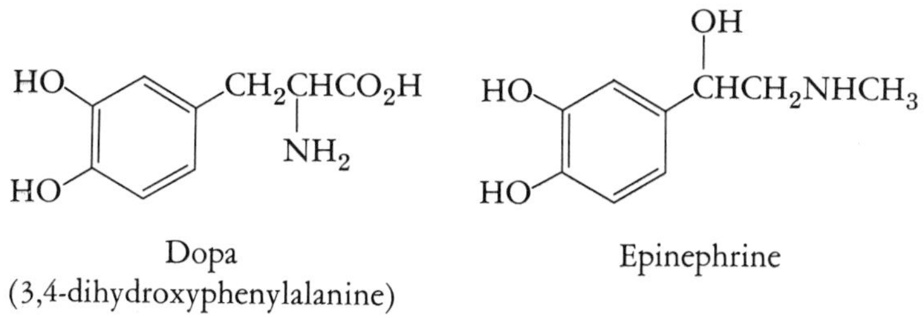

Dopa (3,4-dihydroxyphenylalanine)

Epinephrine

Vanillin and oil of wintergreen, two important flavoring materials, are two more examples of compounds that contain a benzene ring:

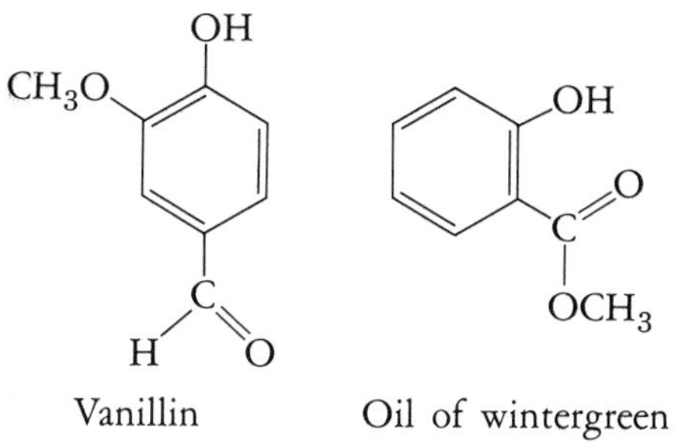

Vanillin

Oil of wintergreen

There are many other examples of biologically important derivatives of benzene. We will encounter other examples in future chapters. In living systems, the benzene rings of these compounds also undergo substitution reactions. One example is the transformation of phenylalanine into tyrosine. In this reaction, the *para* hydrogen of phenylalanine is replaced by an —OH group:

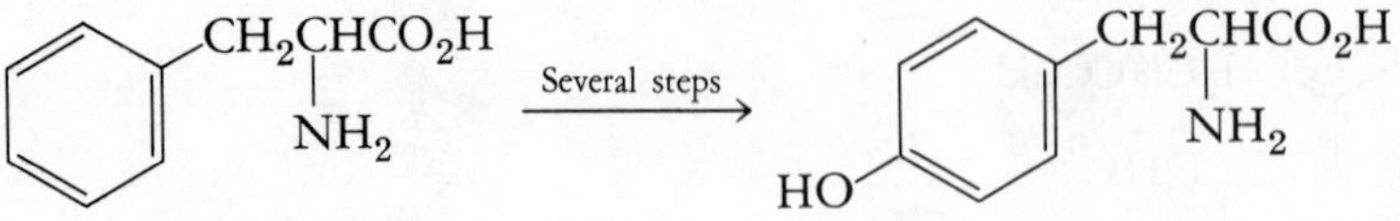

$$C_6H_5CH_2CH(NH_2)CO_2H \xrightarrow{\text{Several steps}} HO{-}C_6H_4CH_2CH(NH_2)CO_2H$$

Phenylalanine → Tyrosine

This simple looking reaction actually requires several steps and is catalyzed by enzymes (see Section 29.6).

Table 13-3. Some Aromatic Heterocyclic Compounds

Structure	Name	Boiling Point (°C)	Melting Point (°C)
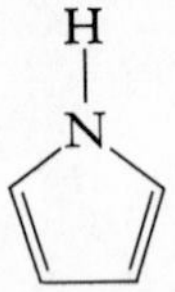	Pyrrole	130	−24
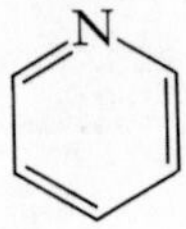	Pyridine	115	−42
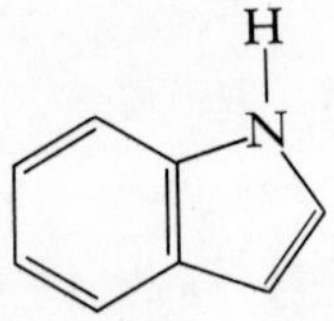	Indole	254	53
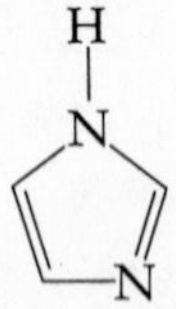	Imidazole	257	91
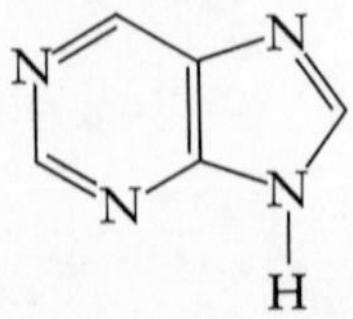	Purine	—	217

Another group of compounds that undergo substitution reactions are the aromatic heterocyclic compounds. Many of their derivatives are biologically important. Let us examine a few of the more important ones.

Aromatic Heterocyclic Compounds

Another group of compounds are known that resemble benzene in that they are highly unsaturated yet undergo electrophilic substitution reactions. Unlike benzene, they contain at least one element other than carbon or hydrogen, usually nitrogen. These compounds, called *aromatic heterocyclic compounds,* contain a ring (or rings) in which one or more of the atoms of the ring are elements other than carbon. The term hetero- indicates that at least one carbon atom of a cyclic hydrocarbon has been replaced by another element. These other elements are called *heteroatoms.*

These compounds resemble benzene in another way. There are six electrons in the ring containing the heteroatom (or atoms). It is this feature, six electrons in a ring, that is responsible for the similarity of their reactions to those of benzene. The aromatic heterocyclic compounds that we will study are given in Table 13-3. Although these compounds are not usually found in nature, many of their derivatives are biologically important. Let us start by learning some of the important derivatives of pyrrole.

The most important derivatives of pyrrole in living systems are the *porphyrins.* These are compounds derived by joining four pyrrole rings together by means of a CH (methine) bridge. The simplest member of this series is known as *porphin.* The most important chemical property of porphyrins is their ability to bind metal ions in the space between the four nitrogen atoms. This system is of great importance because it forms part of the

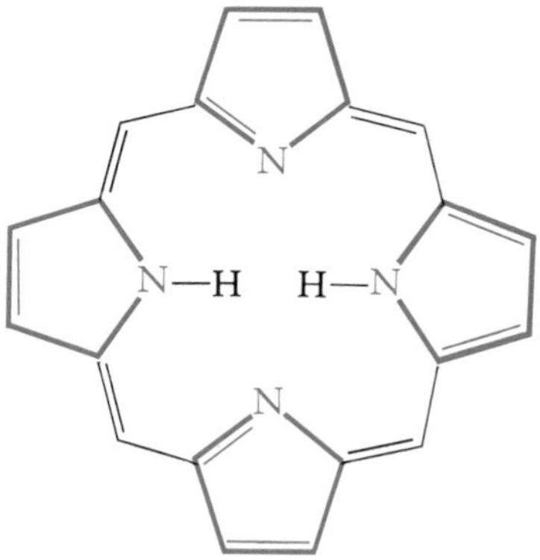

N
N—H H—N
N

Porphin

protein *hemoglobin,* which is responsible for carrying oxygen to the tissues from the lungs via the bloodstream as well as for returning carbon dioxide to the lungs from the tissues. Hemoglobin is composed of two parts. One part is a protein portion called *globin.* The other part is a porphyrin part called *heme* that does the actual transporting of oxygen and carbon dioxide. Heme contains an iron atom in the ferrous oxidation state in the space between the nitrogen atoms. Its structure is the following:

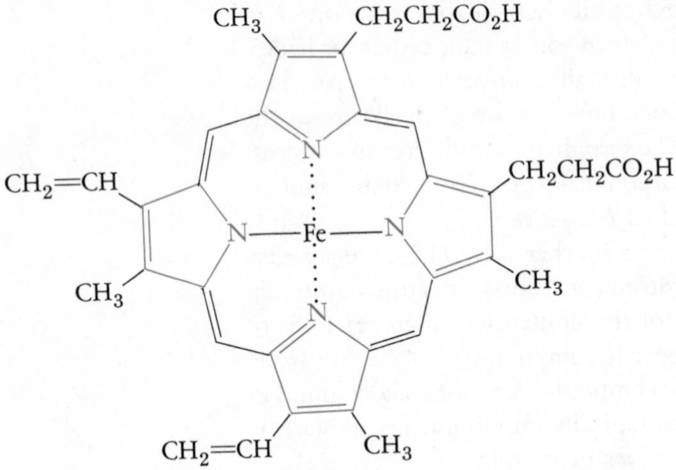

CH_3 $CH_2CH_2CO_2H$
$CH_2{=}CH$ N $CH_2CH_2CO_2H$
N—Fe—N
CH_3 CH_3
N
$CH_2{=}CH$ CH_3

Heme

Other porphyrin-containing compounds are the chlorophylls, which are pigments in all photosynthetic cells. The structure of one of these compounds, chlorophyll a, is the following:

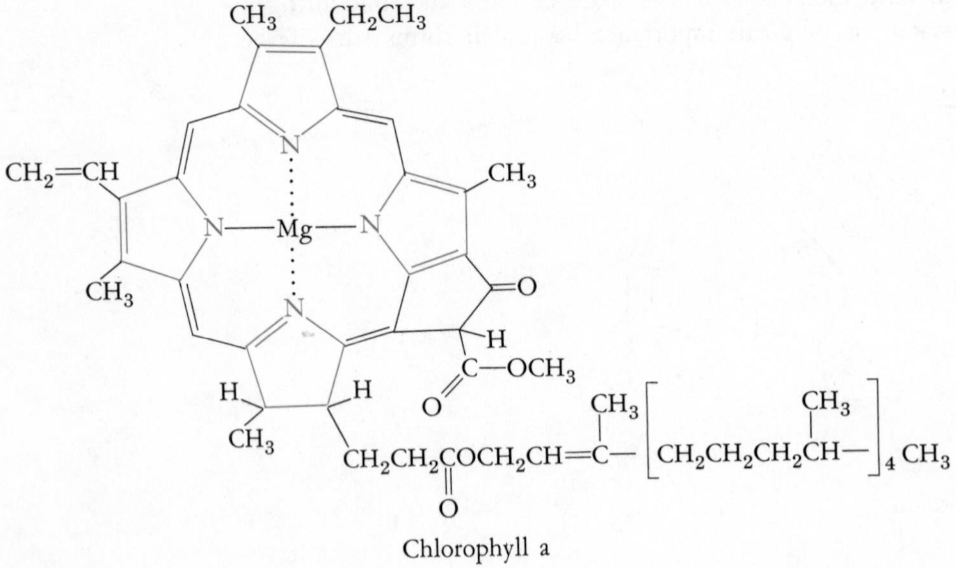

CH_3 CH_2CH_3
N
$CH_2{=}CH$ CH_3
N—Mg—N
CH_3 =O
N
H
C—OCH_3
H H O
CH_3 CH_3 CH_3
CH_3 $CH_2CH_2\underset{\|}{C}OCH_2CH{=}\overset{|}{C}{-}\left[CH_2CH_2CH_2\overset{|}{C}H{-}\right]_4 CH_3$
O

Chlorophyll a

Pyridine can be regarded as benzene in which one carbon has been replaced by a nitrogen atom:

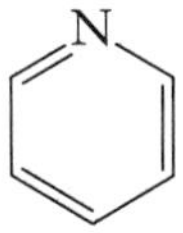

Pyridine

Many derivatives of pyridine are found in nature. The following are a few examples that are biologically important:

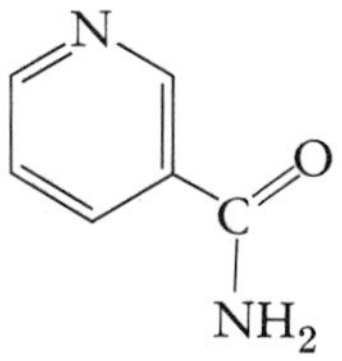

Nicotinamide

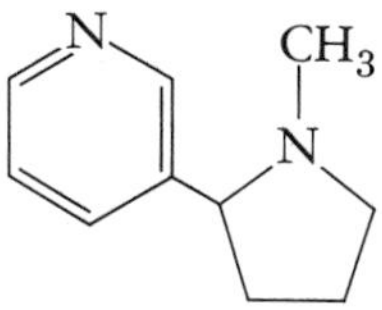

Nicotine

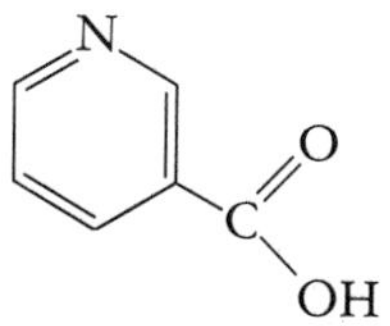

Nicotinic acid

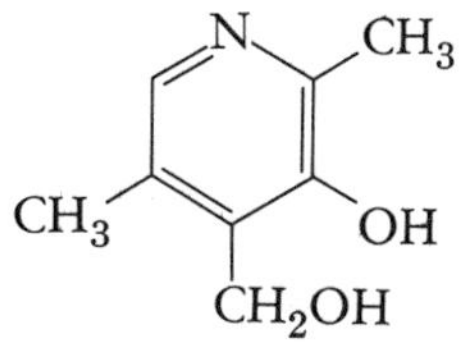

Pyridoxine (Vitamin B_6)

Nicotinamide is a reactive part of a more complicated molecule called NAD (*n*icotinamide *a*denine *d*inucleotide), which plays a key role in many important oxidation-reduction reactions in cells. Among them are respiration, photosynthesis, and the process of vision.

Indole is an aromatic heterocyclic compound containing a pyrrole ring and a benzene ring with one side in common. The single most important

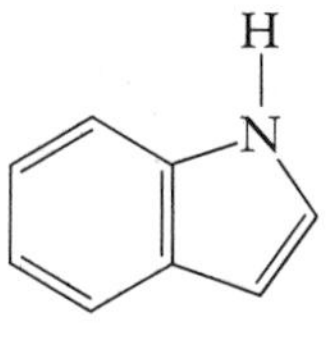

Indole

derivative of indole in living systems is the amino acid *tryptophan.* It is an essential amino acid for mammals; that is, mammals cannot manufacture tryptophan, but must get it from foods. Many important compounds are prepared from tryptophan in living systems. One of them is serotonin, an important hormone that affects blood pressure and is a factor in the func-

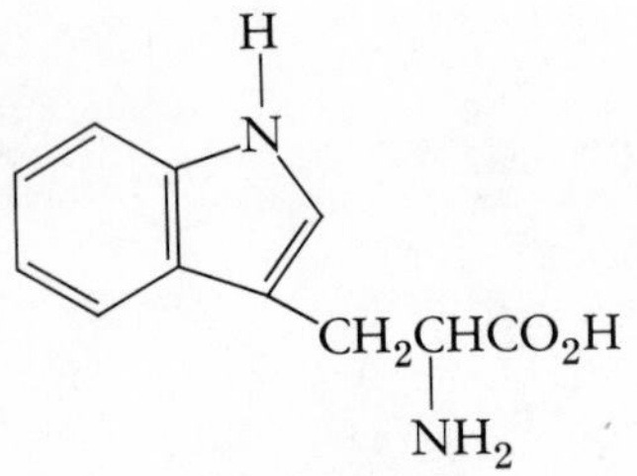

Tryptophan

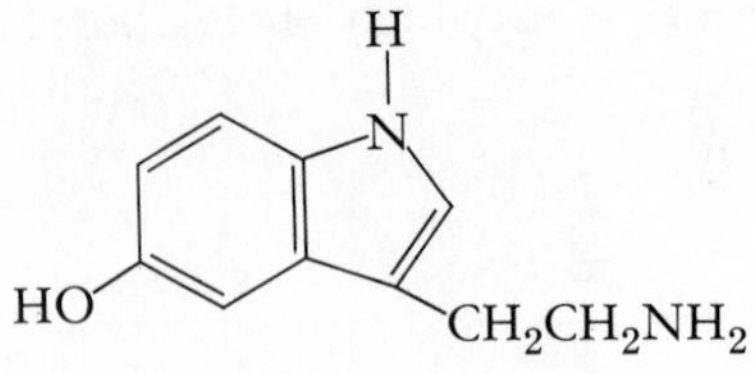

Serotonin
(5-Hydroxytryptamine)

tioning of the brain. More than 500 different compounds that have been isolated from various plants are known to contain the indole ring system. Two structures of these *indole alkaloids* are as follows:

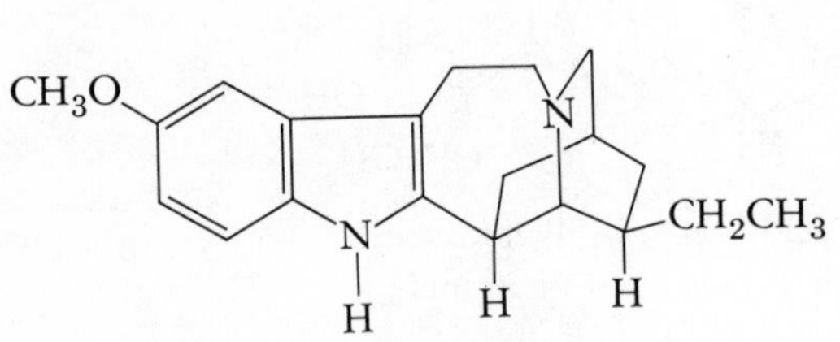

Ibogaine

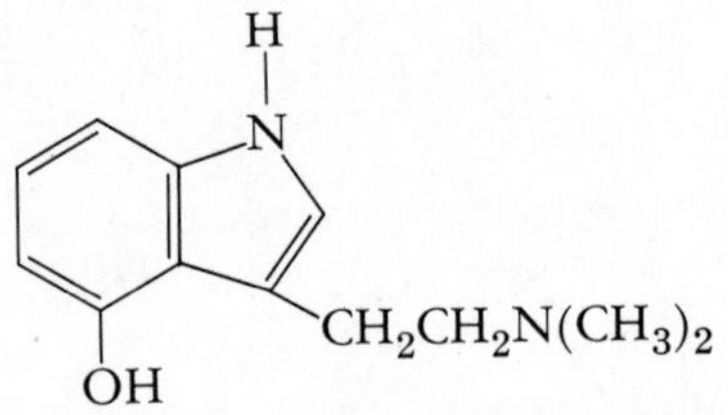

Psilocin

EXERCISE 13-10 In the structures of tryptophan, serotonin, ibogaine, and psilocin, circle the indole ring system.

Imidazole is a compound that contains two nitrogen atoms in a five-member ring:

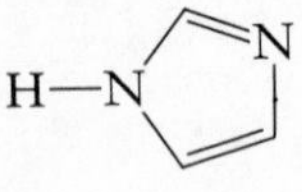

The most important derivative of imidazole is the amino acid histidine. In living systems, enzymes transform histidine to the pharmacologically active

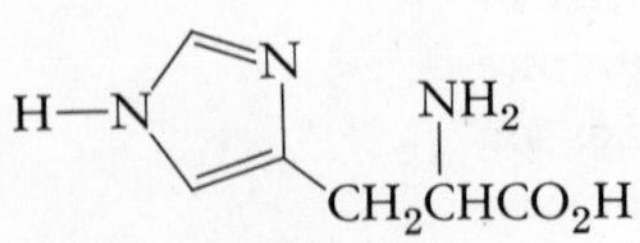

Histidine

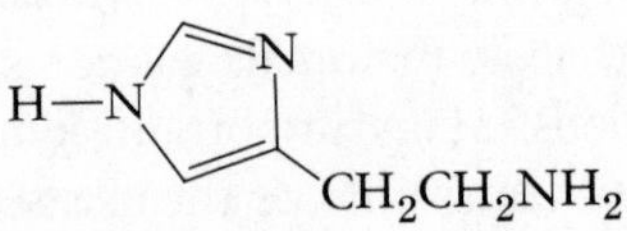

Histamine

histamine, which causes dilation of capillaries and stimulation of muscles. An excess of histamine is associated with allergic reactions. Drugs called antihistamines are compounds that counteract the biological effects of histamines.

Purine is a compound that contains two heteroaromatic rings that have one side in common. A derivative of purine with an $—NH_2$ (amine) substituent in the 6 position is called adenine. It is part of *a*denine *trip*hosphate (ATP), the compound that is involved in energy transfer in living systems. We will learn more about ATP in Chapter 26. Caffeine, the stimulant in coffee and tea, is also a purine derivative.

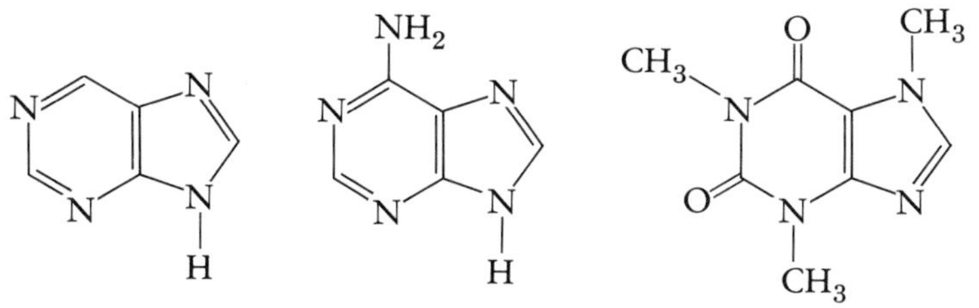

Purine Adenine Caffeine

It is clear from these few examples that aromatic heterocyclic compounds are important to living systems.

13.7 SUMMARY

Benzene has a high degree of unsaturation yet is surprisingly unreactive. It does not undergo the usual addition reactions of alkenes and alkynes. It does react under more vigorous conditions (heat or catalysts) to form products of substitution. This difference in chemical reactivity is due to the structure of benzene. A benzene ring does not contain three isolated double bonds. Rather, the six electrons are associated with all six ring carbons. This is represented in resonance theory by a hybrid structure of benzene made up of two equivalent contributing resonance structures. Six electrons in a ring result in a special structure with characteristic physical and chemical properties.

Bromination, chlorination, nitration, sulfonation, Friedel-Crafts alkylation and acylation are typical substitution reactions of benzene and its derivatives. When these reactions occur with benzene derivatives, the group already on the benzene ring has two effects: it directs the incoming electrophile to specific positions, and it influences the reactivity of the ring toward further substitution. Thus, a substituent may be classed as either activating and *ortho-para* directing or deactivating and *meta* directing. The halogens are the exception; they are deactivating yet *ortho-para* directing.

Under even more vigorous conditions, addition to the benzene ring can occur. Two such reactions are hydrogenation, with a metal catalyst and high temperature and pressure, and light-catalyzed chlorination. In general, the conditions required for substitution or addition reactions of a benzene ring are more vigorous than those needed for reactions of other functional groups. Thus, the benzene ring is like the alkane portion of a molecule in that it remains unaffected during most of the reactions of other functional groups.

Other compounds are known that resemble benzene in that they are highly unsaturated yet undergo predominantly substitution reactions. Many of them contain heteroatoms in place of carbon atoms in the ring. The derivatives of several of these aromatic heterocyclic compounds are important to the chemistry of living systems.

REVIEW OF TERMS, CONCEPTS, AND REACTIONS

Terms

ACTIVATING GROUP A substituent on a benzene ring that causes a substitution reaction to occur more rapidly than it would occur with benzene.

AROMATIC COMPOUNDS Benzene and compounds that resemble benzene in their chemical behavior.

DEACTIVATING GROUP A substituent on a benzene ring that causes a substitution reaction to occur more slowly than it would occur with benzene.

DERIVATIVES Compounds that have the same basic structure and differ only in the number and position of substituents.

HETEROATOMS Atoms of another element that replace carbon in a cyclic hydrocarbon.

META DIRECTOR A substituent on a benzene ring that causes a substitution reaction to occur predominantly at a position *meta* to itself.

ORTHO-PARA DIRECTOR A substituent on a benzene ring that causes a substitution reaction to occur predominantly at a position *ortho* or *para* to itself.

Concepts

1. Benzene and aromatic hydrocarbons are unlike alkenes and alkynes in their reactivity. The aromatic compounds are less reactive and undergo substitution reactions rather than addition reactions characteristic of alkenes and alkynes.
2. The structure of benzene and other aromatic compounds can be described in terms of resonance theory as a hybrid of several contributing resonance structures.
3. Aromatic compounds, with their six electrons in a five- or six-member ring, have characteristic chemical and physical properties.
4. Many derivatives of aromatic heterocyclic compounds are biologically important.

Reactions

Substitution reactions (see Section 13.3)

1. Halogenation

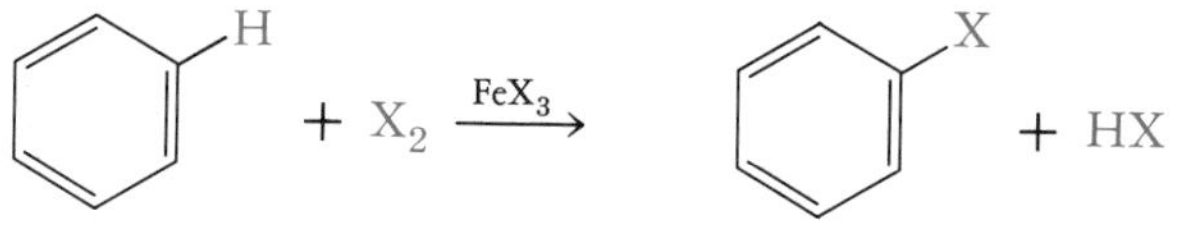

$$C_6H_5H + X_2 \xrightarrow{FeX_3} C_6H_5X + HX$$

X = Br or Cl

2. Nitration

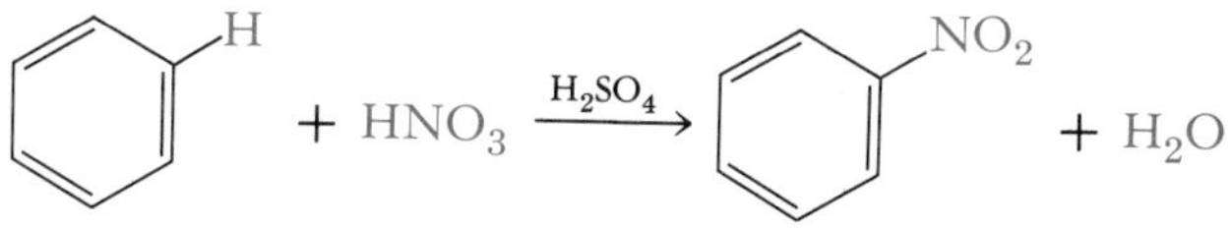

$$C_6H_5H + HNO_3 \xrightarrow{H_2SO_4} C_6H_5NO_2 + H_2O$$

3. Sulfonation

$$C_6H_5H + H_2SO_4 \xrightarrow{SO_3} C_6H_5SO_3H + H_2O$$

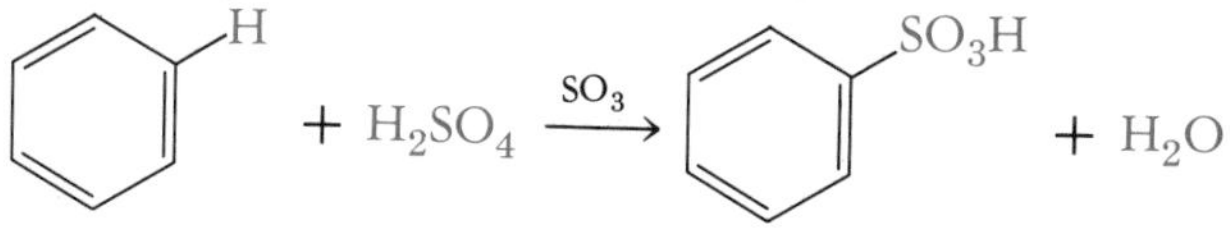

4. Friedel-Crafts alkylation

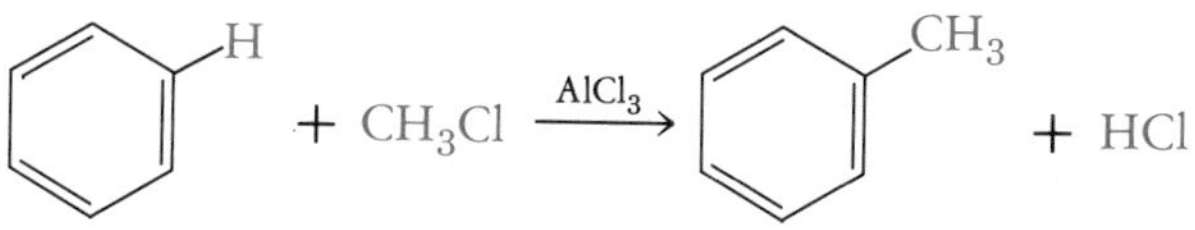

$$C_6H_5H + CH_3Cl \xrightarrow{AlCl_3} C_6H_5CH_3 + HCl$$

5. Friedel-Crafts acylation

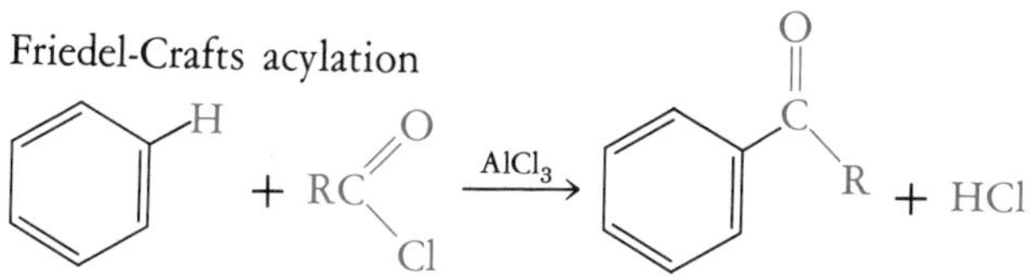

$$C_6H_5H + RC(=O)Cl \xrightarrow{AlCl_3} C_6H_5C(=O)R + HCl$$

Addition reactions (see Section 13.5)

1. Hydrogenation

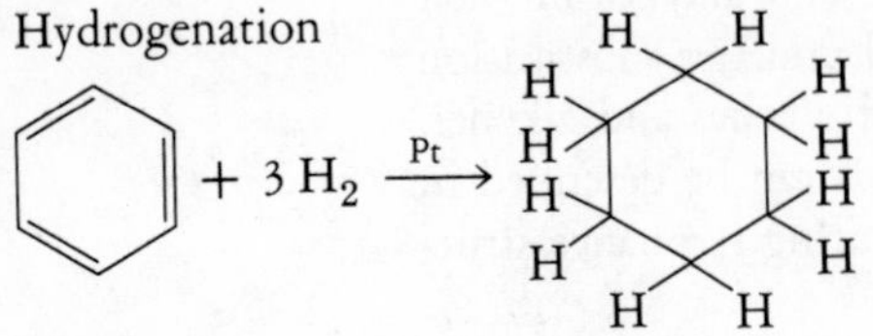

$$C_6H_6 + 3\,H_2 \xrightarrow{Pt} C_6H_{12}$$

2. Chlorination

$$C_6H_6 + Cl_2 \xrightarrow{Light} C_6H_6Cl_6$$

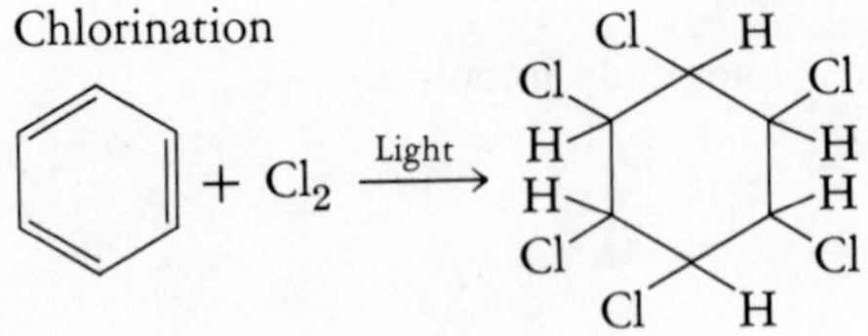

Oxidation reactions (see Section 13.5)

$$C_6H_5R \xrightarrow[KMnO_4]{H^+,\ H_2O,\ \Delta} C_6H_5CO_2H$$

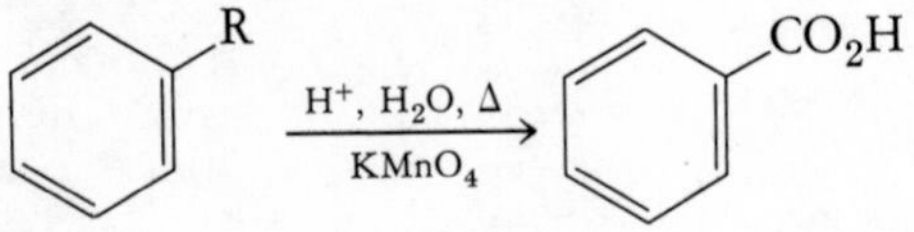

R = any alkyl group

EXERCISES 13-11 Give the IUPAC name for each of the following compounds:

(a) 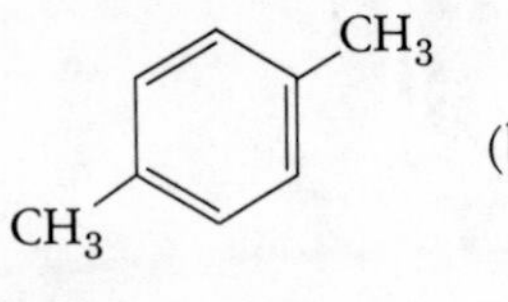 CH_3, CH_3 (p-disubstituted benzene)

(b) 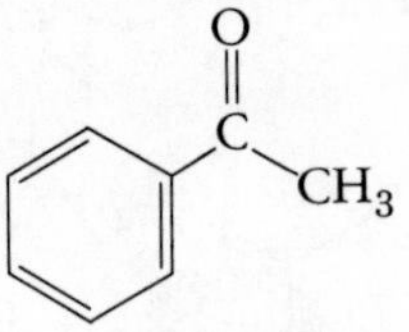 $C_6H_5C(=O)CH_3$

(c) $C_6H_5SO_3H$

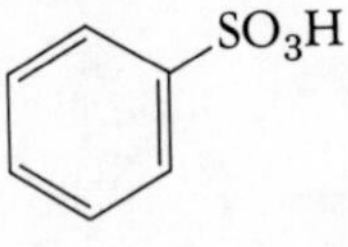

(d) 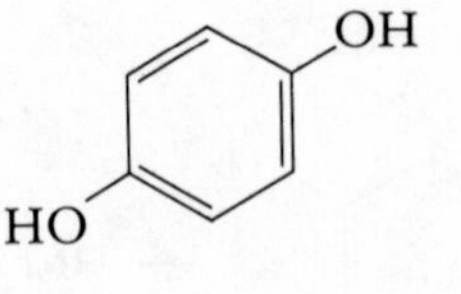 OH, HO (p-disubstituted benzene)

(e) 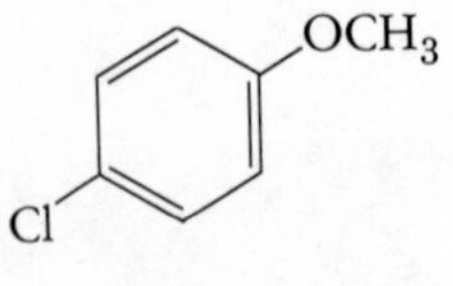 OCH_3, Cl (p-disubstituted benzene)

(f) 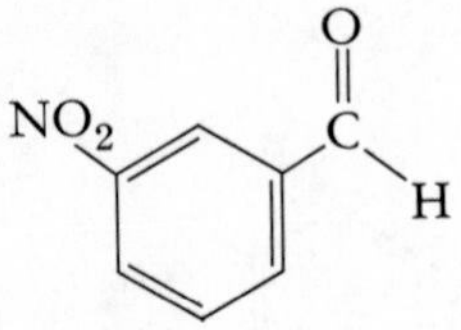 NO_2, $C(=O)H$ (m-disubstituted benzene)

(g) 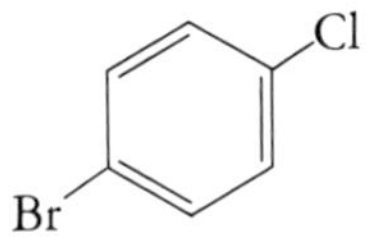 Cl, Br

(h) 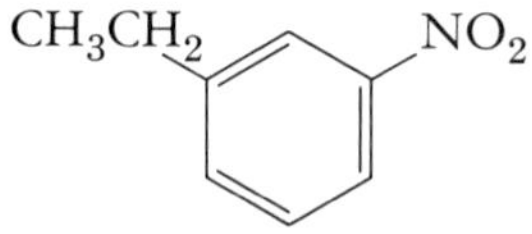 CH_3CH_2, NO_2

(i) 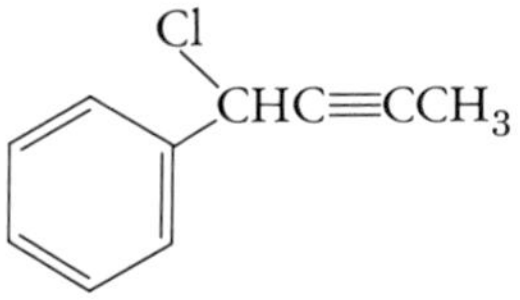 Cl, $CHC{\equiv}CCH_3$

(j) 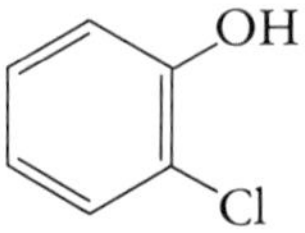 OH, Cl

(k) 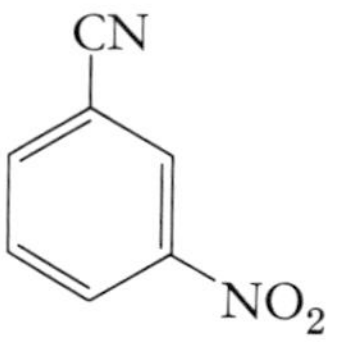 CN, NO_2

(l) 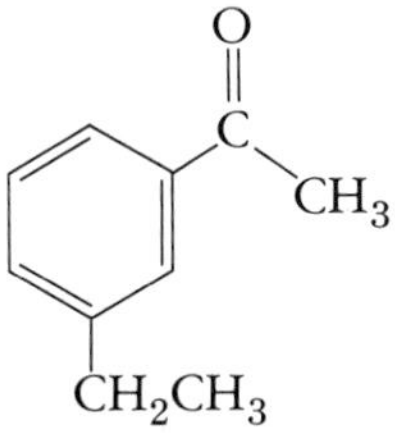 O, C, CH_3, CH_2CH_3

13-12 Write the structures and give the IUPAC names for all of the isomers of the following compounds:

(a) triethylbenzenes (b) dichlorofluorobenzenes

(c) bromochlorotoluenes

13-13 Write the structural formula of the *major* monosubstituted product in each of the following reactions. For each reaction, indicate whether the reaction will occur faster or slower than it would with benzene itself.

(a) toluene + Cl_2 $\xrightarrow{FeCl_3}$

(b) chlorobenzene + HNO_3 $\xrightarrow{H_2SO_4}$

(c) benzonitrile + H_2SO_4 $\xrightarrow{SO_3}$

(d) acetophenone + Br_2 $\xrightarrow{FeBr_3}$

(e) nitrobenzene + HNO_3 $\xrightarrow{H_2SO_4}$

(f) anisole + CH_3Cl $\xrightarrow{AlCl_3}$

(g) ethylbenzene + HNO_3 $\xrightarrow{H_2SO_4}$

(h) phenol + H_2SO_4 $\xrightarrow{SO_3}$

13-14 For each of the following pairs of compounds, indicate which one will react faster to form a monosubstituted product.

(a) benzene and toluene + $HNO_3 \xrightarrow{H_2SO_4}$

(b) anisole and benzene + $H_2SO_4 \xrightarrow{SO_3}$

(c) chlorobenzene and nitrobenzene + $Cl_2 \xrightarrow{FeCl_3}$

(d) toluene and chlorobenzene + $CH_3Cl \xrightarrow{AlCl_3}$

(e) phenol and toluene + $Br_2 \xrightarrow{FeBr_3}$

13-15 Give the structure and name of the product for each of the following reactions.

(a) *t*-butylbenzene + $3H_2 \xrightarrow[\text{Pressure}]{\text{Pt}, \Delta}$

(b) toluene + $KMnO_4 \xrightarrow[\text{NaOH}]{\Delta}$

(c) *p*-xylene + $KMnO_4 \xrightarrow[\text{NaOH}]{\Delta}$

(d) benzene + 3 $Cl_2 \xrightarrow{\text{Light}}$

(e) *p*-chloroethylbenzene + $KMnO_4 \xrightarrow[\text{NaOH}]{\Delta}$

STEREOISOMERS OF CARBON COMPOUNDS

14

We introduced the concept of isomers in Section 11.1. Isomers are different compounds that have the same number and kinds of atoms. But there are different kinds of isomers. We have already identified several of them. Structural isomers, introduced in Section 11.2, differ in their structural formulas. Geometric isomers, introduced in Section 12.2, differ in the arrangements of their atoms in space. Actually, geometric isomers are specific examples of a larger class of isomers called stereoisomers. In this chapter, we will learn about other kinds of stereoisomers. Let us start by learning about leucine, a chiral compound.

14.1 CHIRAL COMPOUNDS

One of the characteristics of isomers is that they have different physical properties. This is shown in Table 14-1 for the structural isomers of chemical formula C_7H_7Br and the geometric isomers of 3-hexene. From this observation we conclude that if *all* properties of two substances are identical, then the two substances must have identical molecular structures. This fact is fundamental to the study of organic chemistry. Yet a number of compounds were discovered during the nineteenth century that seemed to contradict this fact. The problem with these compounds can be illustrated by the example of leucine.

The chemical formula of leucine is $C_6H_{13}O_2N$, and it has the following structural formula:

```
CH3    CH3
   \  /
    CH
    |
    CH2
    |
 H—C—CO2H
    |
    NH2
```

Leucine

Two isomers of leucine exist. Both have the same structural formula and the same melting and boiling points. But one tastes bitter and the other tastes sweet. Because not *all* their properties are the same, the individual molecules of these two isomers of leucine must differ in some way. From the structural formula, it seems difficult to understand how leucine could possibly exist as two isomers. However, if we draw the structural formula of leucine in three dimensions, we can see the difference.

The three-dimensional structure of leucine is shown in Figure 14-1. Remember that in such a representation a solid line between atoms indicates that the bond is in the plane of the page. A dashed line means that the bond projects below the plane of the page, and a solid wedge indicates that the bond projects above the page (see Section 11.8). We can write two such structures for leucine: *14-1* and *14-2*. These two structures are *not* identical. In structure *14-1* the $—NH_2$ group projects below the page, whereas in structure *14-2* this group projects above the page. The positions of the $—CO_2H$ group are also reversed. The two structures *14-1* and *14-2* differ only in the way the four groups bonded to the carbon are oriented in space. Consequently, they are stereoisomers. These two structures represent the two isomers of leucine.

Table 14-1. Several Isomers and Their Characteristic Melting and Boiling Points and Densities

		Physical Constants		
Type of Isomer	Compound	Melting Point (°C)	Boiling Point (°C)	Density at 25°C (g/mL)
Structural	Benzyl bromide	−3.9	201	1.4380
	o-Bromotoluene	−26	181	1.4222
	m-Bromotoluene	−40	184	1.4019
	p-Bromotoluene	28	184	1.3898
Geometric	*trans*-3-Hexene	−113	67	0.6770
	cis-3-Hexene	−135	67	0.6796

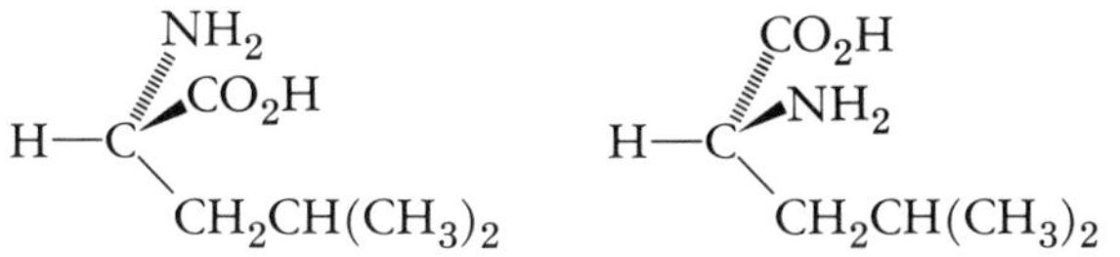

NH_2, CO_2H, H—C, $CH_2CH(CH_3)_2$

14-1

CO_2H, NH_2, H—C, $CH_2CH(CH_3)_2$

14-2

Fig. 14-1. Two three-dimensional structures of leucine.

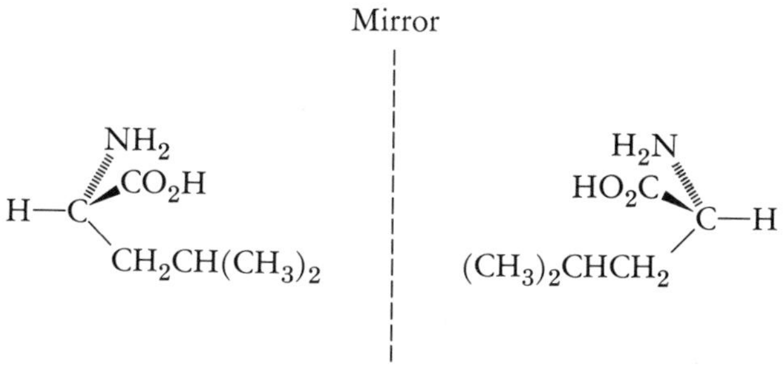

Mirror

NH_2, CO_2H, H—C, $CH_2CH(CH_3)_2$

H_2N, HO_2C, C—H, $(CH_3)_2CHCH_2$

Fig. 14-2. The mirror image relationship of the two stereoisomers of leucine.

Stereoisomers *14-1* and *14-2* are related in an important way. They are mirror images of each other, as shown in Figure 14-2. Furthermore, *these mirror images cannot be superimposed.* Stereoisomers that are mirror images but cannot be superimposed are called *enantiomers.*

It is important to be convinced that these two enantiomers of leucine are mirror images and cannot be superimposed. This can be done by building two three-dimensional molecular models, one of each enantiomer, and trying to superimpose them.

If you superimpose any two groups or atoms, you will find that the other two cannot be superimposed. This is shown in Figure 14-3. The hydrogens and the $—NH_2$ groups of the two enantiomers can be superimposed, but the $—CO_2H$ and the alkyl groups cannot be superimposed. No matter how you turn the models around, all four groups cannot be superimposed unless you break bonds to the central carbon and interchange the positions of two groups. Clearly, the two enantiomers *14-1* and *14-2* are different and are not easily interconverted.

It is not unusual that an object cannot be superimposed on its mirror image. Many objects show this relationship. Hands are an example. One hand is the mirror image of the other, yet the two hands cannot be superimposed. You can easily test this by using your own hands.

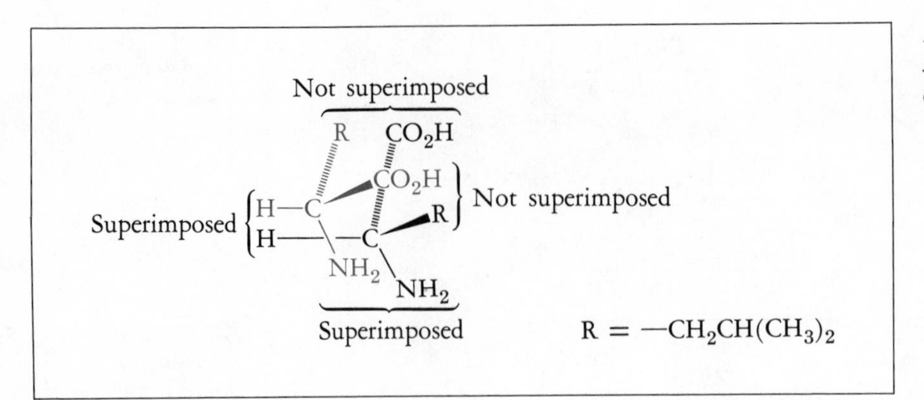

Fig. 14-3. An attempt to superimpose the two enantiomers of leucine.

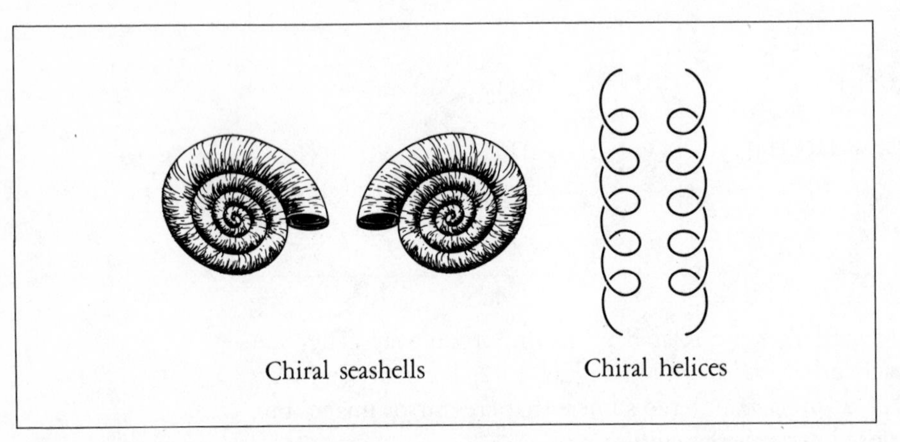

Fig. 14-4. Two objects that cannot be superimposed on their mirror images.

*EXERCISE 14-1 Which of the following objects cannot be superimposed on its mirror image?
(a) a hammer (b) a pencil (c) one shoe (d) one glove

Compounds that are not identical to their mirror images are said to be *chiral.* This word is derived from the Greek word *cheir,* meaning hand. A molecule that is chiral is like one hand. It has a partner, the other hand, that is a mirror image. This mirror image cannot be superimposed on the original. Clearly, an enantiomer is chiral. Two other examples of chiral substances are given in Figure 14-4.

Remember that all objects have a mirror image. However, many objects can be superimposed on their mirror image. Such objects are called *achiral. Enantiomers are not possible for achiral compounds.* But how do we distinguish

* The answers for the exercises in this chapter begin on page 872.

between chiral and achiral molecules? We will learn to do this in the next section.

14.2 RECOGNIZING CHIRAL COMPOUNDS

The only foolproof way to recognize a chiral molecule is to determine whether it can be superimposed on its mirror image. This method is not easy. It requires a well-developed ability to visualize objects in three dimensions. Most people find that it is easiest to build a three-dimensional model of both the molecule and its mirror image and then try to superimpose them.

EXERCISE 14-2 Draw a three-dimensional representation for each of the following compounds and their mirror images. Decide which compounds are chiral. Make a molecular model of each compound and its mirror image. Try to superimpose these models to verify your answers.
(a) CH_4 (b) CH_3Cl (c) CH_2FCl (d) CHIFCl (e) CCl_4

A shortcut can often be used. Most chiral molecules contain one or more carbon atoms that are bonded to four different groups or atoms. Each of the following compounds has one such carbon atom, indicated by an asterisk:

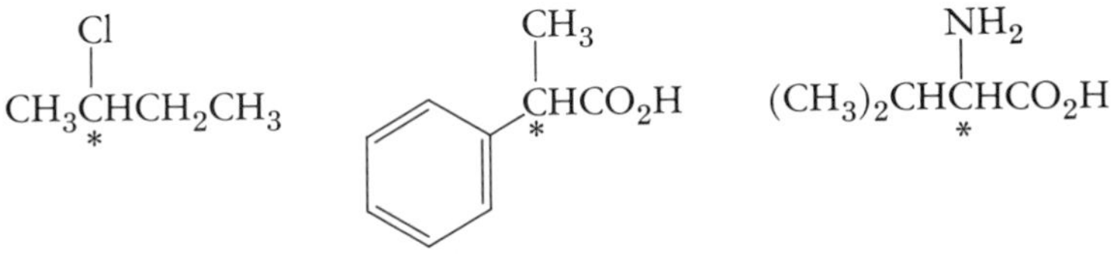

Cl
|
$CH_3\underset{*}{C}HCH_2CH_3$

CH_3
|
$C_6H_5\underset{*}{C}HCO_2H$

NH_2
|
$(CH_3)_2CH\underset{*}{C}HCO_2H$

A carbon atom that is bonded to four different groups or atoms is called a *chiral atom* or a *chiral center*. Another name for a chiral atom is an asymmetric atom.

Like many shortcuts, this one can lead us astray. Some compounds contain chiral centers yet can be superimposed on their mirror images. As a result, they are achiral. We will encounter an example of this kind of compound in Section 14.7. Such compounds can be recognized quite easily.

Despite this exception, the concept of a chiral carbon is still useful. With a bit of experience, you will be able to recognize chiral molecules.

Let us now learn how to detect chiral molecules experimentally.

14.3 OPTICAL ACTIVITY

In Section 14.1 we learned that the two enantiomers of leucine taste different. Other pairs of enantiomers do not necessarily show this difference to our sense of taste. But the enantiomers of leucine do show a difference in another property that is common to all enantiomers. This property is the ability to rotate plane-polarized light. This is a property of *all* chiral molecules. What is plane-polarized light, and how can we use it to detect chiral molecules?

Classic physics describes the motion of light in terms of waves, as illustrated in Figure 14-5. In ordinary light, the rise and fall of the waves occurs equally in all directions perpendicular to the direction of travel of the light, as shown in Figure 14-5. When placed in a beam of light, certain materials affect the light in an unusual way. One such material, called a polarizing filter, blocks all the waves except those moving parallel to the axis of the filter. This effect is illustrated in Figure 14-6. The filtered light is now *plane-polarized light.* There are several materials that can be used as filters. One is a Nicol prism. Another is a material known by the trade name Polaroid; this material is used in the lenses of sunglasses to filter out some of the sunlight that would otherwise enter the eyes.

If we look at an object through two polarizing filters, one behind the other, as shown in Figure 14-7, we observe an interesting effect. The amount of light reaching our eyes depends on the relative orientations of the axes of the two filters. If the axes of the two filters are parallel, the amount of light is unaffected. If the axes are arranged in any other orientation, some of the

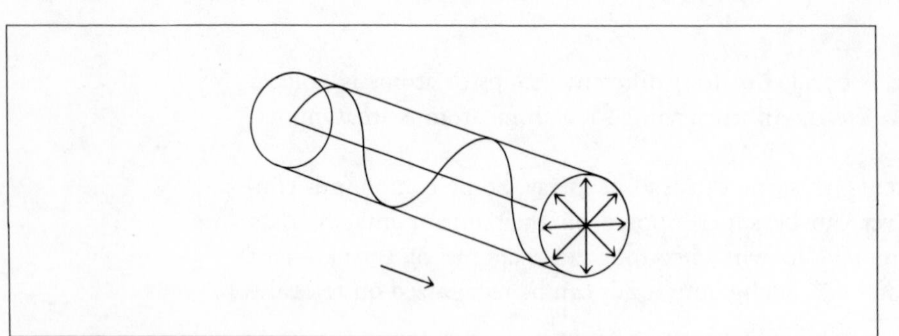

Fig. 14-5. The wave nature of a light beam. For simplicity, only one light wave is shown.

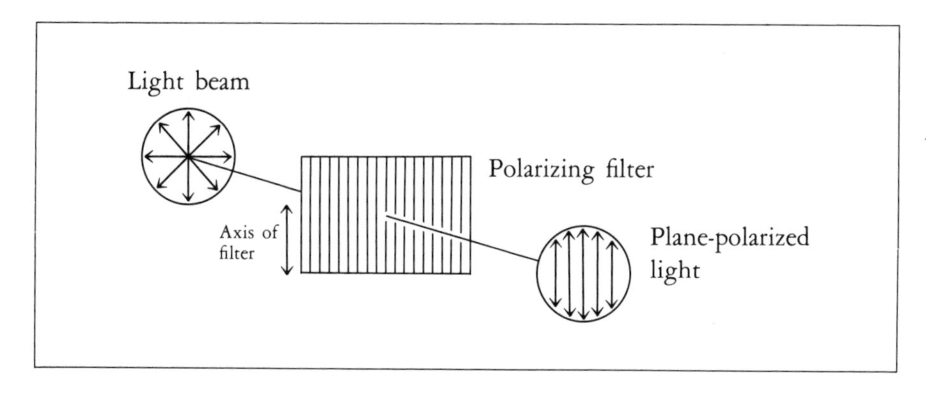

Fig. 14-6. When light passes through a polarizing filter, the light becomes plane-polarized.

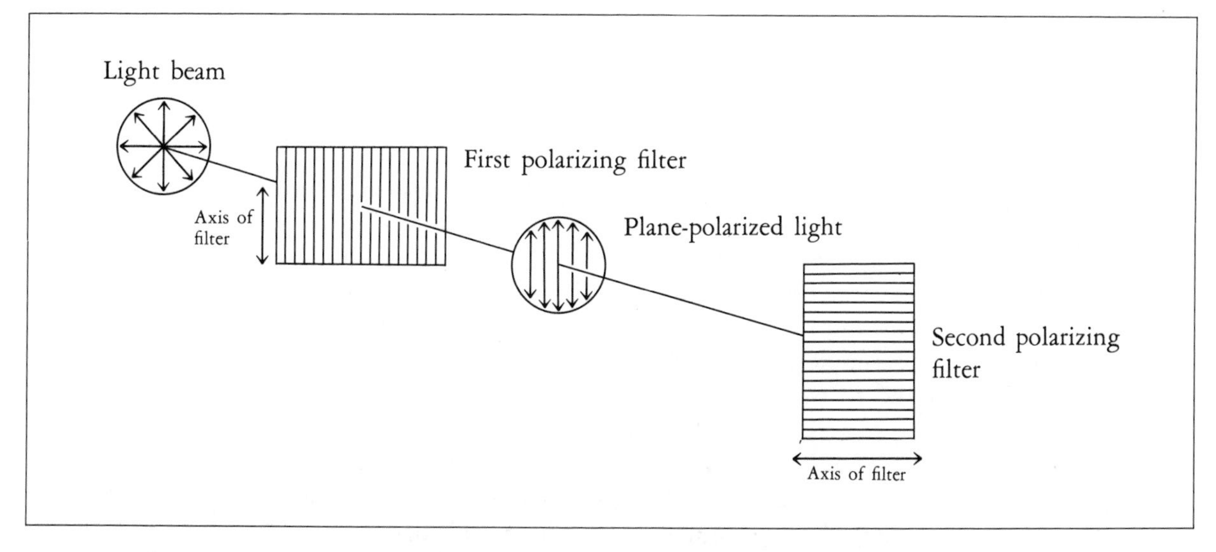

Fig. 14-7. Effect of two polarizing filters on a light beam. The first polarizing filter produces a beam of plane-polarized light. When this beam encounters a second polarizing filter, the amount of light that passes through the filter depends on the relative orientation of the axes of the two filters. When they are at right angles, no light gets through.

light is blocked by the second filter and we see less light. When the axes of the two filters are at 90 degrees to each other, no light reaches our eyes.

We can use this effect to detect chiral compounds with an instrument called a *polarimeter,* illustrated in Figure 14-8. The working parts of a polarimeter consist of a light source; a polarizing filter (usually a Nicol prism); a sample tube; another polarizing filter (again, a Nicol prism), which can be rotated (the analyzer); a scale capable of measuring degrees; and an eyepiece. When the substance in the sample tube is achiral, the plane-polarized light emerges from the tube unchanged.

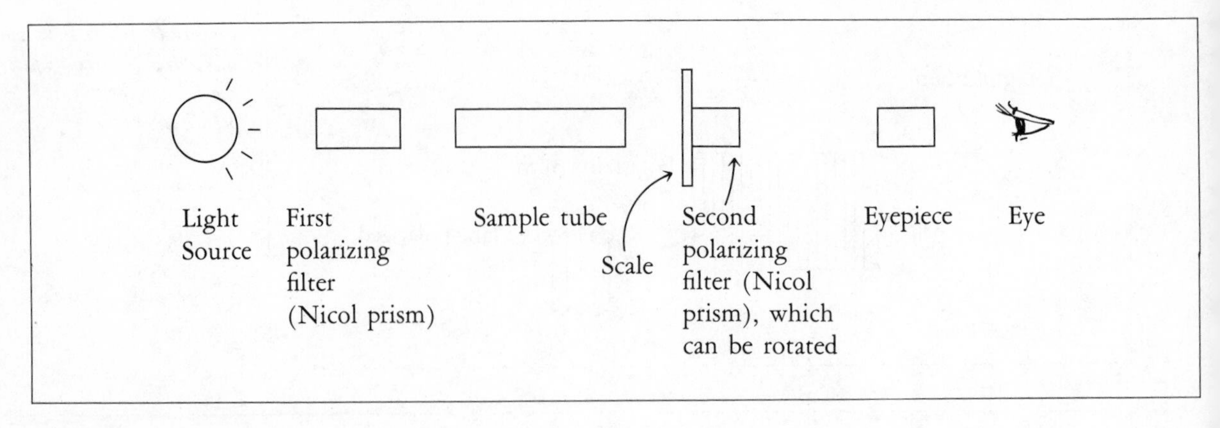

Fig. 14-8. Schematic representation of a polarimeter.

If a solution of one enantiomer of a chiral molecule is placed in the sample tube, it will rotate the plane-polarized light. That is, the polarized light that emerges from the sample tube is no longer parallel to the light that entered the tube. As a result, the emerging light is not parallel to the axis of the analyzer and the intensity of light reaching the eyepiece is reduced. The original intensity of light can be restored by rotating the analyzer, left to right, until its axis is parallel to the plane-polarized light emerging from the sample tube. Any substance that will rotate plane-polarized light is said to be *optically active.*

A compound that rotates plane-polarized light to the right when viewed by an observer looking toward the light source is called *dextrorotatory.* If the rotation is to the left, the compound is called *levorotatory.* The name of an enantiomer can include the direction in which it rotates plane-polarized light by addition of the prefix (+) or (−). Thus, (+)-leucine is dextrorotatory, and (−)-leucine is levorotatory.

The angle by which the analyzer must be rotated to restore the maximal intensity of light is called the optical rotation of a compound. This angle depends on four factors: the temperature of the solution, the light source, the concentration of the solution, and the length of the sample tube. To take these factors into account, a new quantity called the *specific rotation* of a compound, $[\alpha]$, is defined as follows:

$$[\alpha]_\lambda^t = \frac{100\,\alpha}{c\,l}$$

where α = observed optical rotation in degrees, including the sign; t = temperature of the sample, in degrees centigrade; λ = wavelength of

Table 14-2. Some Physical Properties of Selected Enantiomers

Compound	Chemical Formula	Melting Point (°C)	Boiling Point (°C)	Density at 25° C (g/mL)	Refractive Index	$[\alpha]_D^{25}$ (degrees)
(−)-Leucine	$(CH_3)_2CHCH_2CH(NH_2)CO_2H$	293	—	1.293	—	−10.8
(+)-Leucine		293	—	1.293	—	+10.8
(−)-2-Methyl-1-butanol	$CH_3CH_2CH(CH_3)CH_2OH$	—	129	0.819	1.411	−5.76
(+)-2-Methyl-1-butanol		—	129	0.819	1.411	+5.76
(−)-Mandelic acid	$C_6H_5CH(OH)CO_2H$	133.8	—	1.341	—	−155.5
(+)-Mandelic acid		133.8	—	1.341	—	+155.5

light source; l = length of sample tube, in decimeters; c = concentration of the compound, expressed as g/100 mL of solution.

The specific rotation is a physical constant of a compound, just like boiling and melting points. The specific rotations and some physical constants for selected pairs of enantiomers are given in Table 14-2. Notice an important fact about the information given in Table 14-2. The physical properties of enantiomers are identical except for the signs of their specific rotations. For example, the specific rotation of (+)-leucine is +10.8°, and the value for (−)-leucine is exactly the same but in the opposite direction (−10.8°). *Enantiomers rotate plane-polarized light by equal amounts but in opposite directions.* For most enantiomers, this is the only difference in their physical properties. Because these compounds rotate plane-polarized light, they were originally called *optical isomers,* a term that is still used today.

Before studying other kinds of stereoisomers, we must recognize an important difference between a chiral substance and one that is optically active.

14.4 DIFFERENCE BETWEEN CHIRALITY AND OPTICAL ACTIVITY

In the previous section, we concentrated on the relationship between chiral substances and their ability to rotate plane-polarized light. However, we must stress a very important difference between these two concepts. On the one hand, we know that a chiral molecule cannot be superimposed on its mirror image. On the other hand, an optically active compound rotates plane-polarized light. A chemical compound either is chiral or is not. In Section 14.2 we learned how to recognize some of these chiral compounds. However, individual samples of such compounds are not necessarily optically active! How can this possibly be true? To answer this question, we will consider a sample of a compound that is a mixture of equal amounts of both enantiomers. The specific rotations of the enantiomers are equal, but in opposite directions. Consequently, the effect of one enantiomer on the plane-polarized light is exactly canceled by the effect of the other enantiomer and the sample will be optically inactive. This means that the optical activity of a sample of a compound depends on the relative proportions of the two enantiomers. As long as there is more of one enantiomer than the other, the sample will be optically active. The largest value of its specific rotation will be obtained when only one of the two enantiomers is present.

A mixture of equal amounts of the two enantiomers is called a *racemic mixture.* Such a mixture is optically inactive. When reactions are carried out in the laboratory, racemic mixtures are usually formed. For example, the following light-catalyzed reaction of chlorine and butane forms 2-chlorobutane as one of the products:

$$CH_3CH_2CH_2CH_3 + Cl_2 \xrightarrow{\text{light}} \underset{CH_3}{\overset{CH_2CH_3}{H\!-\!C}}\!-\!Cl + Cl\!-\!\underset{CH_3}{\overset{CH_2CH_3}{C\!-\!H}} \quad (+\text{ Other isomers})$$

2-Chlorobutane contains a chiral center. Yet the sample prepared in this way is optically inactive because equal amounts of the dextrorotatory and levorotatory isomers are formed.

In living systems, the chemical reactions usually produce only one enantiomer of a compound. This is an important difference between reactions that occur in the laboratory and those that occur in living systems.

EXERCISE 14-3 Isopropyl alcohol $CH_3\underset{OH}{\underset{|}{C}}HCH_3$ and 2-butanol $CH_3\underset{OH}{\underset{|}{C}}HCH_2CH_3$ are prepared industrially and can be purchased from any chemical supplier. Samples of these two compounds obtained in this way are optically inactive. Explain.

Stereoisomers are clearly stable compounds that can be handled and stored in bottles. What names do we write on the labels to distinguish a bottle containing one stereoisomer from a bottle containing another? To answer this question, let us learn to name stereoisomers.

14.5 NAMING STEREOISOMERS

We have learned that the structures of compounds can be expressed by the use of names. For example, methylpropane represents a unique chemical structure. We would like to extend the use of names to enantiomers to designate their three-dimensional structures. But to do this requires that we know their *absolute configuration,* that is, the actual orientation in space of the atoms of the molecule. How can we do this? The only bit of experimental information that we have is the sign of the specific rotation of the enantiomers. From this it is impossible to know anything about a molecule's absolute configuration. The D,L-system of naming stereoisomers is one way to solve this problem.

The idea of this system is to relate the absolute configuration of every chiral molecule to one or the other of the enantiomers of glyceraldehyde, $HOCH_2CH(OH)CHO$:

$$HOCH_2\underset{\displaystyle OH}{\underset{|}{C}}HCHO$$

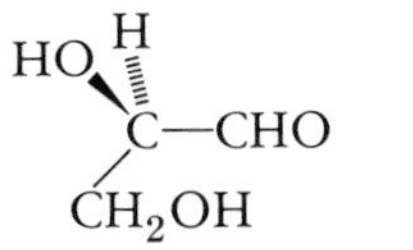

D-(+)-Glyceraldehyde *14-3*

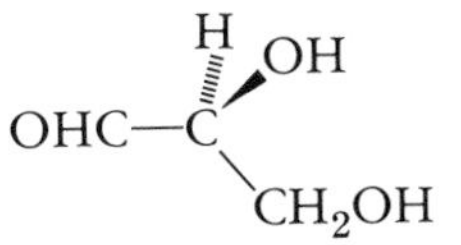

L-(−)-Glyceraldehyde *14-4*

The absolute configuration represented by structure *14-3* is given to (+)-glyceraldehyde.* The symbol D is given to this structure and all chiral molecules related to it. Structure *14-4* represents (−)-glyceraldehyde and is designated by the symbol L. The configuration of other stereoisomers are related to these two by means of chemical reactions. For example, it is known that (+)-glyceraldehyde reacts with an aqueous bromine solution to form (−)-glyceric acid, $HOCH_2\underset{|}{\overset{OH}{C}}HCO_2H$. Because no bonds to the chiral carbon are broken, and because we know the absolute configuration of (+)-glyceraldehyde, we can write structure *14-5* to represent the absolute configuration of (−)-glyceric acid:

* These absolute configurations were originally assigned *arbitrarily*. However, in 1949, it was found by the use of x-ray crystallography that these assignments were correct. Thus, the original assignment was a lucky choice.

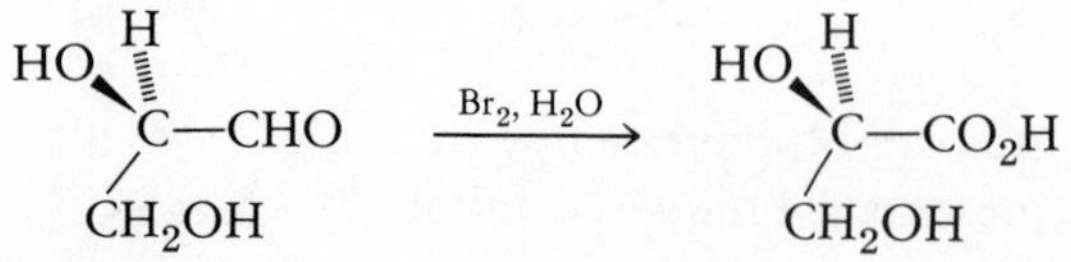

$$\underset{\text{D-(+)-Glyceraldehyde}}{\text{HO}\overset{\text{H}}{\text{C}}\text{—CHO},\ \text{CH}_2\text{OH}} \xrightarrow{Br_2,\ H_2O} \underset{\text{D-(−)-Glyceric acid}}{\text{HO}\overset{\text{H}}{\text{C}}\text{—CO}_2\text{H},\ \text{CH}_2\text{OH}}$$

14-5

Because one of the enantiomers of glyceric acid is obtained by a chemical reaction from one enantiomer (D) of glyceraldehyde, we know that they both have the same *relative configuration.* Therefore, it is a member of the D family and is called D-(−)-glyceric acid.

Notice that, although both (+)-glyceraldehyde and (−)-glyceric acid belong to the same D family, one is dextrorotatory and the other is levorotatory. This is clear proof that the sign of the optical rotation of a chiral molecule is not related to its three-dimensional structure. Many other compounds have been related to D- and L-glyceraldehyde by one or more chemical reactions. This method has been widely used to designate the absolute configuration of amino acids and carbohydrates, as we will learn in Chapters 21 and 23.

EXERCISE 14-4 For each of the following reactions, draw the three-dimensional structures of the reactant and the product and decide whether the product belongs to the D or L family of stereoisomers.

(a) 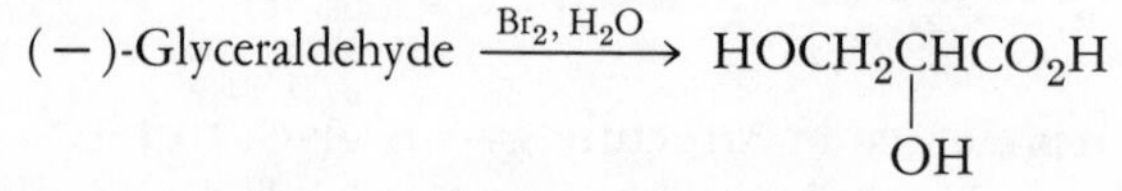 (−)-Glyceraldehyde $\xrightarrow{Br_2,\ H_2O}$ $HOCH_2CH(OH)CO_2H$

(b) 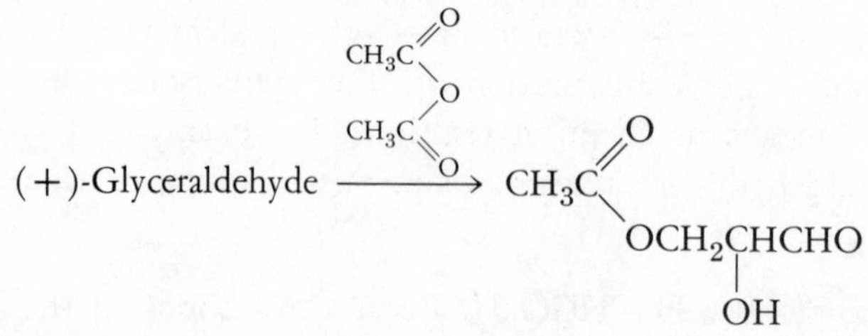 (+)-Glyceraldehyde $\xrightarrow{(CH_3CO)_2O}$ $CH_3C(=O)OCH_2CH(OH)CHO$

(c) 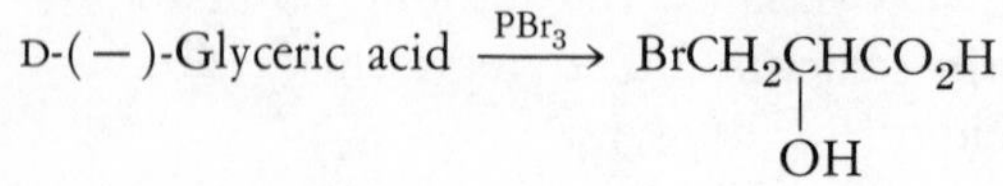 D-(−)-Glyceric acid $\xrightarrow{PBr_3}$ $BrCH_2CH(OH)CO_2H$

So far we have concentrated our attention on compounds that contain only one chiral center. However, most chiral compounds in the world contain more than one chiral center. Therefore, let us examine briefly the relationship between the stereoisomers of compounds containing two chiral centers.

14.6 DIASTEREOMERS

Compounds that contain two or more chiral centers can exist as more than two stereoisomers. The general rule is that the *maximal* number of stereoisomers is 2^n, where n is the number of chiral centers. For a compound containing two chiral centers, we should find a maximum of four stereoisomers ($n = 2$; $2^2 = 4$).

EXERCISE 14-5 Glucose has the following structural formula:

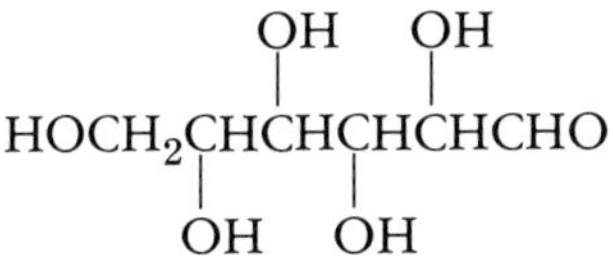

$$HOCH_2CH(OH)CH(OH)CH(OH)CH(OH)CHO$$

How many chiral carbon atoms does it contain? Place an asterisk next to each one. How many stereoisomers are possible?

A specific example of a compound containing two chiral centers is the carbohydrate 2,3,4-trihydroxybutyraldehyde.

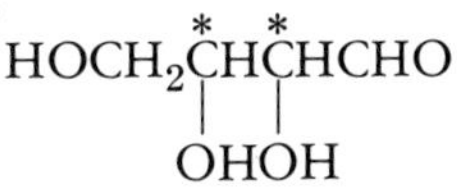

$$HOCH_2\overset{*}{C}H(OH)\overset{*}{C}H(OH)CHO$$

Asterisks mark the two chiral carbons. The four stereoisomers are represented by structures *14-6, 14-7, 14-8,* and *14-9.* Each has a unique configuration.

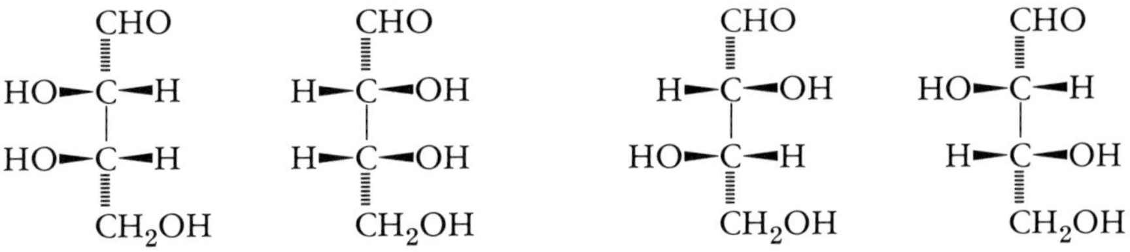

	14-6	*14-7*	*14-8*	*14-9*
top	CHO	CHO	CHO	CHO
C2	HO—C—H	H—C—OH	H—C—OH	HO—C—H
C3	HO—C—H	H—C—OH	HO—C—H	H—C—OH
bottom	CH_2OH	CH_2OH	CH_2OH	CH_2OH
name	Erythrose	Erythrose	Threose	Threose

Note that structures *14-6* and *14-7* are mirror images and therefore are a pair of enantiomers. This pair of enantiomers has been given the common name erythrose. Similarly, structures *14-8* and *14-9* are mirror images and are a second pair of enantiomers, which have been given the common name threose.

Careful examination of the structures *14-6* through *14-9* confirms that neither structure *14-6* nor structure *14-7* is a mirror image of structure *14-8* or structure *14-9*. Furthermore, neither structure *14-6* nor structure *14-7* can be superimposed on structure *14-8* or structure *14-9*. Clearly, the stereoisomers represented by structures *14-6* and *14-8* are *not* enantiomers. What, then, is their relationship? Any two of the stereoisomers represented by structures *14-6* to *14-9* that are not mirror images are called *diastereomers*. Stereoisomers represented by structures *14-6* and *14-8* are diastereomers, as are structures *14-7* and *14-9*.

EXERCISE 14-6 Write stereochemical representations of all stereoisomers of 2,3-dibromopentane ($CH_3CH_2\underset{Br}{\underset{|}{C}}H\underset{Br}{\underset{|}{C}}HCH_3$). Indicate which stereoisomers are enantiomers and which are diastereomers.

One of the important differences between diastereomers is their difference in physical properties. For example, the compound represented by structure *14-6* is a liquid, whereas the one represented by structure *14-9* is a solid with a melting point of 132° C. These two stereoisomers also have different solubilities in ethyl alcohol. The erythrose isomers are very soluble, whereas the threose isomers are only slightly soluble. This difference can be put to practical use: it can be used to separate enantiomers, as we will learn in Section 14.8.

Not all compounds that contain two chiral centers exist as four stereoisomers. Some of these compounds have a special feature that decreases the total number of stereoisomers. We will learn about these compounds in the next section.

14.7 MESO STEREOISOMERS

Certain molecules that contain two chiral carbon atoms do not exist as four stereoisomers. Tartaric acid, $HO_2C\underset{OH}{\underset{|}{C}}H\overset{OH}{\overset{|}{C}}HCO_2H$, is a classic example of such a compound. Because tartaric acid has two chiral atoms, four stereoisomers would be expected. However, only three stereoisomers have ever been isolated. Why? To answer this, let us examine the structures of the stereoisomers of tartaric acid. Structures *14-10* and *14-11* cannot be superimposed on their mirror images and are one pair of enantiomers.

CO_2H H—C—OH HO—C—H CO_2H	CO_2H HO—C—H H—C—OH CO_2H	CO_2H H—C—OH H—C—OH CO_2H	CO_2H HO—C—H HO—C—H CO_2H
L-Tartaric acid	D-Tartaric acid	Identical	
14-10	*14-11*	*14-12*	*14-13*

At first glance, it appears that another pair of enantiomers, structures *14-12* and *14-13,* can be drawn. However, structures *14-12* and *14-13* are mirror images that *can be superimposed* and therefore are identical. This can be confirmed by rotating structure *14-12* by 180 degrees in the plane of the page. The result is the same structure as *14-13:*

CO_2H H—C—OH H—C—OH CO_2H	Rotate by 180° →	CO_2H HO—C—H HO—C—H CO_2H	identical to *14-13*
14-12			

Again, molecular models can be used to prove that structures *14-12* and *14-13* are identical.

The term *meso* is given to an isomer that contains chiral atoms but can be superimposed on its mirror image. Because of this, a meso stereoisomer is optically inactive. Therefore, there are three stereoisomers of tartaric acid: a meso isomer and a pair of enantiomers. The fact that the four substituents on both chiral carbon atoms of tartaric acid are identical is the structural feature that is responsible for the existence of a meso stereoisomer. This is a general rule. The meso stereoisomer then is a compound with chiral centers that can never be optically active because it can be superimposed on its mirror image. Look out for this kind of compound. They occur often in nature.

EXERCISE 14-7 Which of the following compounds can exist as a meso stereoisomer?

(a) $CH_3CH(Cl)CH(Cl)CH_3$ (b) $CH_3CH(Cl)CH(Br)CH_3$ (c) $HOCH_2CH(OH)CH(OH)CH_2OH$

EXERCISE 14-8 Write stereochemical representations for all the stereoisomers of the compounds in exercise 14-7.

In this chapter, we have given the physical properties of a number of pure enantiomers. How were these properties obtained? How is one enantiomer separated from its stereoisomer? In the next section, we will examine several methods for doing this.

14.8 RESOLUTION OF RACEMIC MIXTURES

The process of separating a pair of enantiomers into the pure stereoisomers is called *resolution.* Let us examine three methods of resolving racemic mixtures.

Mechanical Separation

This method was used by Louis Pasteur in 1848 to separate the stereoisomers of a crystalline tartaric acid salt. He noticed that the salts crystallized into two forms that were mirror images of each other: some were "left-handed" crystals and others were "right-handed" crystals. By using a pair of tweezers and a magnifying glass, he was able to separate the two kinds of crystals. Separate solutions of the two had the same specific rotation, but differed in the sign of rotation. Thus, Pasteur resolved the mixture by a mechanical separation of the two crystal forms. This is a rather special method, which cannot be applied to all racemic mixtures. A better and more general method involves the difference in physical properties of diastereomers.

Resolution Using Diastereomers

In Section 14.6, we learned that diastereomers have different physical properties. This fact provides us with a method of resolving a racemic mixture. The idea is simple. We react a racemic mixture with an optically active reagent. The product is a mixture of two diastereomers. We then pick one property, such as their solubility in ethyl alcohol, that is different for the two diastereomers. We use this property to separate one diastereomer from the other. The separated diastereomers are each reacted to reconvert them to the optically active reagent and the separated enantiomers. Let us apply this method to the resolution of benzedrine, *14-14,* a compound that is used as a stimulant.

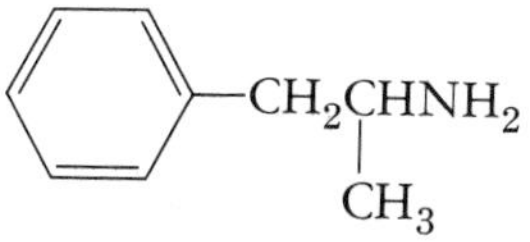

$$C_6H_5{-}CH_2\underset{\displaystyle CH_3}{\underset{|}{C}}HNH_2$$

Benzedrine
14-14

The functional group of benzedrine is an amine group. It is a base and reacts with organic and inorganic acids to form salts (see Chapter 17). An organic acid can be used that is a pure enantiomer. Such acids are available from natural sources. (+)-Tartaric acid is one such acid. The acid reacts with racemic benzedrine to form two diastereomeric salts, as shown diagrammatically in Figure 14-9. Once the diastereomers are separated, each can be reacted with an aqueous solution of sodium hydroxide to form one enantiomer of benzedrine. In this way, the two enantiomers of many racemic mixtures can be separated. This is the most general technique for resolving racemic mixtures. Another technique is to use an enzyme.

Enzymatic Resolution

This method involves giving a living system a racemic mixture. The system reacts with only one enantiomer and leaves the other one behind. This method was first reported by Pasteur. He found that a yeast used in wine making reacted with (+)-tartaric acid. When he gave the yeast racemic tartaric acid, he found that only the dextrorotatory isomer was consumed. The levorotatory isomer was isolated at the end of the reaction.

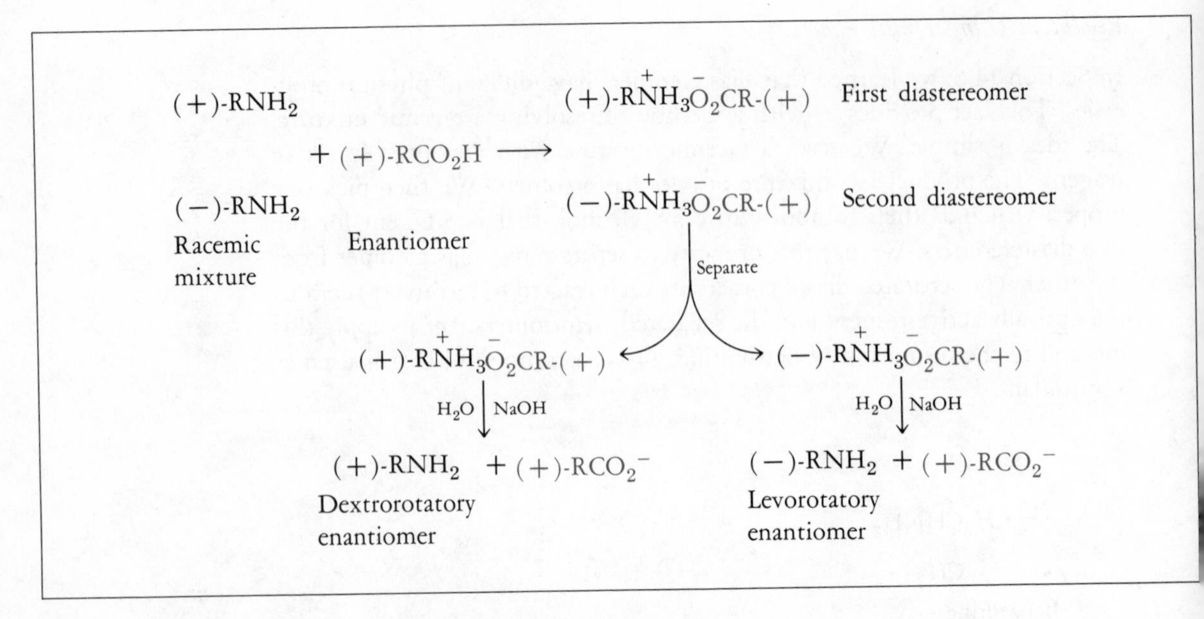

Fig. 14-9. A schematic representation of the resolution of racemic benzedrine.

This is not a general method for resolving racemic mixtures, but it does give us an insight into how living systems react with chiral molecules.

14.9 CHIRAL COMPOUNDS AND LIVING SYSTEMS

We have stressed that the physical properties of enantiomers are identical except for the direction in which they rotate plane-polarized light. But what about their chemical properties? We have already noted several differences. For example, the two enantiomers of leucine taste different. Taste is a chemical reaction, and there is clearly a difference. In many other chemical reactions, there are no differences in the reactivity of the two enantiomers. What determines the difference in the chemical reactivity of a pair of enantiomers?

It has been found that *a pair of enantiomers will react differently only with chiral reagents.* When they react with achiral reagents, no difference in their chemical reactivity is observed. A similar situation occurs every time anyone puts on socks and shoes. A pair of socks is achiral, because one sock can be superimposed on the other one, its mirror image. Either sock will fit your left foot. Either sock will also fit your right foot. Both socks are the same to

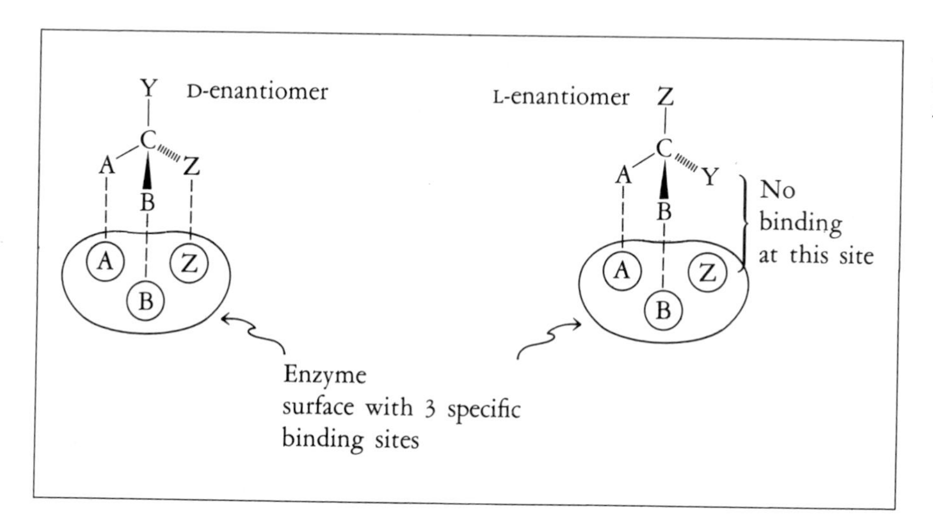

Fig. 14-10. A schematic representation of the stereospecificity of an enzyme.

your feet. In chemical terms, we can say that either member of a pair of enantiomers (one foot) is affected the same way by an achiral reagent (a sock). In contrast, a left (or right) shoe is chiral. The left shoe, for example, fits only the left foot. Therefore, a chiral reagent affects one member of a pair of enantiomers differently from the other.

Because living systems usually react with only one of a pair of enantiomers, the living system is also chiral. The part of the living system that is chiral is the enzyme. Consequently, an enzyme has the ability to react with only one of a pair of enantiomers. This ability of an enzyme is called its *stereospecificity.*

We can illustrate the stereospecificity of an enzyme in an oversimplified manner in the following way. The enzyme reacts with a compound, called the *substrate,* to form a complex. The enzyme is chiral and therefore forms a complex with only one enantiomer of the substrate. The chemical reaction occurs in the complex. Once the reaction has occurred, the complex breaks apart and the enzyme is freed.

This interaction between an enzyme and the substrate, CABZY, is illustrated schematically in Figure 14-10. In this example, the enzyme has three binding sites, one specific for A, another for B, and a third for Z. The binding sites are arranged on the surface as shown in Figure 14-10. To form an enzyme-substrate complex, all three sites on the enzyme must bind with the corresponding groups or atoms of the substrate. Only the D-enantiomer can bind to the surface with all three groups attached to the correct binding sites. The L-enantiomer can bind to a maximum of two sites. In this way, the enzyme can distinguish between the two enantiomers.

It is simple to explain in general terms why the chiral environment of a molecule is so important to its reactions in biological systems. However, the manner in which these interactions are accomplished with such high precision and efficiency is the subject of much investigation. Slowly, our understanding of these processes is growing. A few of the conclusions will be presented in later chapters.

14.10 SUMMARY

In this chapter, we have learned about stereoisomers that contain one or more chiral centers. Chiral molecules that cannot be superimposed on their mirror images are called enantiomers. Any two stereoisomers that are not mirror images are called diastereomers. Molecules that contain chiral centers yet can be superimposed on their mirror images are called meso stereoisomers.

The physical properties of both members of a pair of enantiomers are identical. Only the direction in which each enantiomer rotates plane-polarized light is different. One rotates plane-polarized light to the left, whereas the other rotates it by an equal amount to the right. The physical properties of diastereomers are different. It is this fact that makes it possible to resolve many racemic mixtures.

The chemical properties of enantiomers differ only when the enantiomers react with chiral reagents. Enzymes are chiral reagents, so living systems can react with only one of a pair of enantiomers. From the examples given in this chapter, it is clear that the three-dimensional structures of molecules are important in many aspects of our lives.

REVIEW OF TERMS AND CONCEPTS

Terms

ABSOLUTE CONFIGURATION The actual orientation in space of the groups or atoms of a molecule.

CHIRAL CARBON ATOMS A carbon atom bonded to four different atoms or groups.

CONFIGURATION The arrangement in space of the atoms of a molecule.

DEXTROROTATORY ISOMER A substance that rotates the plane of polarized light to the right.

DIASTEREOMERS Stereoisomers that are optically active but are not mirror images.

ENANTIOMERS Stereoisomers that are mirror images but cannot be superimposed.

ENZYME STEREOSPECIFICITY The ability of an enzyme to react with only one of a pair of enantiomers.

LEVOROTATORY ISOMER A substance that rotates the plane of polarized light to the left.

MESO STEREOISOMER A molecule that contains chiral centers but is optically inactive.

OPTICALLY ACTIVE COMPOUND A compound capable of rotating the plane of polarized light.

POLARIZED LIGHT Light in which all rays vibrate in a single plane.

RACEMIC MIXTURE A mixture containing equal amounts of a pair of enantiomers. As a result, racemic mixtures are optically inactive.

RESOLUTION The process of separating the enantiomers of a racemic mixture.

STEREOISOMERS Isomers that have identical structural formulas but differ in the spatial arrangement (configuration) of their atoms.

Concepts

1. A molecule consisting of a carbon atom bonded to four different groups or atoms can exist as two enantiomers that cannot be superimposed.
2. Enantiomers have the same physical properties except that they rotate the plane of polarized light in opposite directions to the same degree.
3. The chirality of a molecule is responsible for its optical activity.
4. The configurations of optically active compounds can be designated by relating them to D- and L-glyceraldehyde.
5. Compounds containing n chiral carbon atoms can have a *maximum* of 2^n stereoisomers.
6. Diastereomers have different physical properties. It is this fact that is the basis of the process of resolution.
7. Molecules containing chiral carbon atoms are not necessarily optically active (for example, meso compounds).
8. A pair of enantiomers will react differently only with chiral reagents. When they react with achiral reagents, no difference in their chemical reactivity is observed.
9. The chiral environment of an enzyme is responsible for its ability to react with only one of a pair of enantiomers.

EXERCISES 14-9 Give examples of each of the following types of isomer:
(a) structural (b) geometric (c) enantiomer
(d) diastereomer

14-10 Which of the following compounds contain(s) a chiral carbon?

(a) $CH_3CH(Cl)CH_2CH_3$ (b) $CH_3CH(OH)-C_6H_5$

(c) $CHBr(CO_2H)CO_2H$ (d) $CHCl_2CO_2H$ (e) p-dibromobenzene ($Br-C_6H_4-Br$)

14-11 Draw a three-dimensional representation of each of the compounds in exercise 14-10 and confirm your answer.

14-12 The reaction of 1-butene and HCl forms $CH_3CH(Cl)CH_2CH_3$ as product. Is the product formed in this reaction optically active? Explain.

14-13 For each of the following compounds, place an asterisk beside each chiral carbon; draw three-dimensional structures for all the stereoisomers; and identify them as meso stereoisomers, enantiomers, or diastereomers.

(a) $C_6H_5CH(NH_2)CH(OH)CH_3$ (b) $CH_3CH(OH)CH(OH)CH_3$ (c) $CH_3CH(Cl)CH(OH)CH_3$

14-14 Explain how you would separate a racemic mixture of $CH_3CH(NH_2)CH_2CH_3$ into its pure enantiomers.

14-15 The product of the reaction of the inorganic acid HCl with benzedrine (compound *14-14*) is the salt

$C_6H_5CH_2CH(CH_3)\overset{+}{N}H_3\overset{-}{Cl}$

It is impossible to separate this salt into its pure enantiomers, yet the product of the reaction of benzedrine with one enantiomer of the organic acid, tartaric acid, can be separated easily. Explain.

ORGANIC HALIDES 15

Compounds that contain one or more halogen atoms in addition to carbon and hydrogen are called organic halides. These compounds are rare in nature. Most are prepared commercially or in the laboratory and are used widely in the world for a variety of purposes. Although organic halides are extremely useful, their continued use presents a pollution hazard. One of the purposes of this chapter is to make you aware of the kinds of organic halides used today and some of the problems associated with their continued use. The study of these compounds also has another aim. Organic halides undergo two types of reaction, *nucleophilic substitution* and *elimination,* that occur in living cells. Thus, a study of these reactions provides the foundation that we need to understand the biologically related reactions of organic halides.

Let us start by examining the occurrence of these compounds in the world.

15.1 OCCURRENCE

Thyroxine and triiodothyronine are examples of organic halides found in living systems. Both compounds are secreted by the thyroid gland to influence the rate of metabolic activity of practically every tissue in the body.

Tyrian purple is obtained from sea snails, and certain *Streptomycetes* produce antibiotics such as chloramphenicol and chlortetracycline. Very few other naturally occurring organic halides are known.

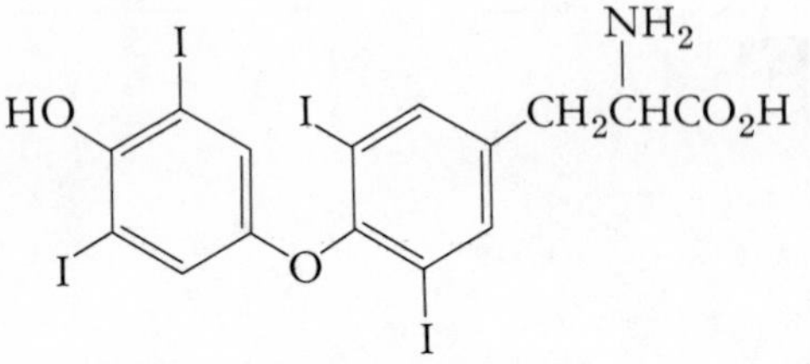

Thyroxine

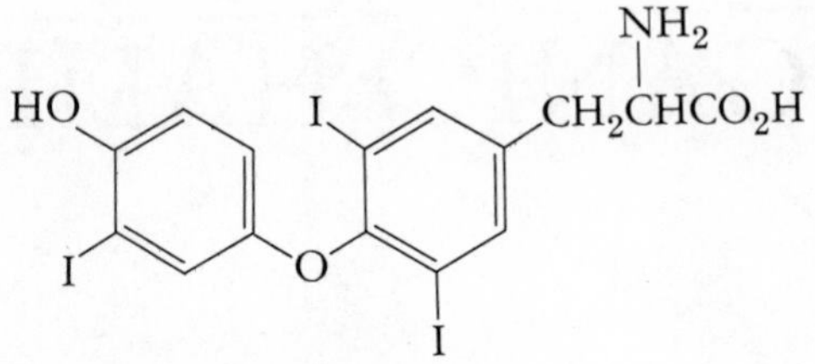

Triiodothyronine

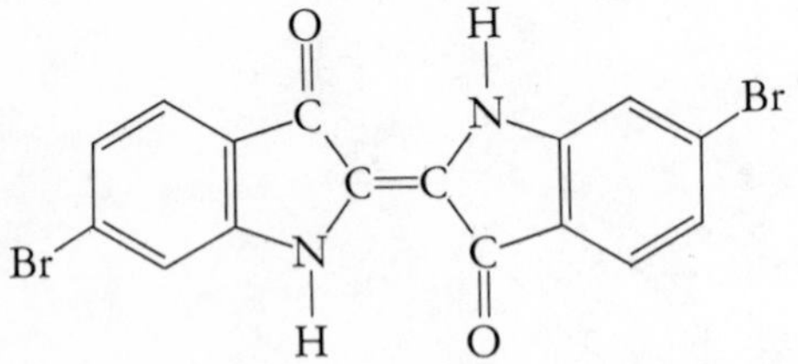

Tyrian purple

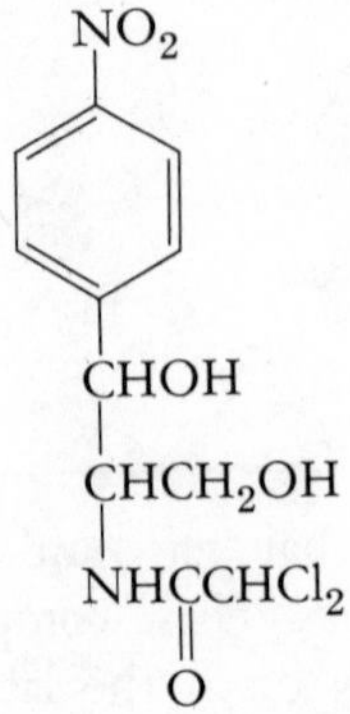

Chloramphenicol

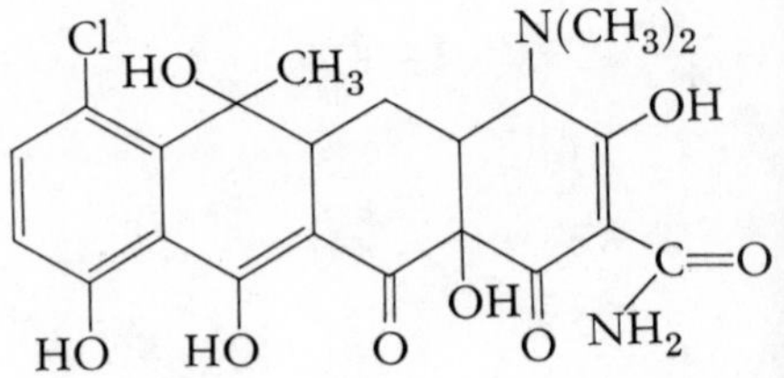

Chlortetracycline (Aureomycin®)

Human ingenuity has produced many simple organic halides that have found wide commercial use. Carbon tetrachloride, trichloroethylene, and tetrachloroethylene are used to dry-clean fabrics. These compounds, as well as isomeric mixtures of chlorinated saturated hydrocarbons (see Section 11.11), are used to degrease metals. Chloroform was used for many years as a general anesthetic for surgery. It is no longer used for this purpose because prolonged exposure to the vapor of most organic chlorides results in extensive liver damage.

CCl_4 Carbon tetrachloride

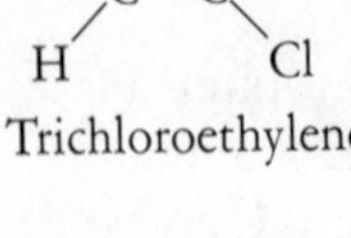

Trichloroethylene

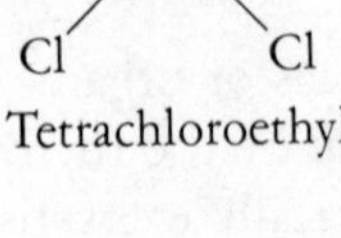

Tetrachloroethylene

$CHCl_3$ Chloroform

Two other organic halides are widely used as anesthetics. Ethyl chloride,

CH_3CH_2Cl, is a fast-acting local anesthetic. It is a gas at room temperature unless kept under pressure. When sprayed on the skin, it evaporates and cools the skin and the nerve endings. Halothane ($CHClBrCF_3$), when inhaled, is an anesthetic. Its chief advantage over other anesthetics such as ether is that it is nonflammable and nonexplosive.

Certain simple organic compounds that contain both chlorine and fluorine are used as propellants in aerosol containers. The best known of these are the Freons manufactured by Dupont. Two examples are Freon-11 and Freon-12:

CCl_3F	CCl_2F_2
Freon-11	Freon-12

Freons are useful as propellants and refrigerants because they are nontoxic, nonflammable, odorless, and noncorrosive.

Several widely used polymers are organic halides. Two examples, Teflon and polyvinyl chloride (PVC), were mentioned in Section 12.6. Another example of a polyhalogenated polymer is the wrap material called Saran Wrap®. In this case, the alkene vinylidene chloride ($CH_2{=}CCl_2$) is polymerized to produce a clear, inert plastic useful in packaging.

$$n\,CH_2{=}CCl_2 \xrightarrow{\text{Catalyst}} \left[-CH_2\overset{\overset{\displaystyle Cl}{|}}{\underset{\underset{\displaystyle Cl}{|}}{C}}- \right]_n$$

Vinylidene chloride — Polyvinylidene chloride

Organic halides have attracted the most public attention because of their use as insecticides. Many organic halides have been used as insecticides during the past two decades. Perhaps the best known of these compounds is DDT (1,1,1-trichloro-2-bis(4-chlorophenyl)ethane). Several compounds of similar structure such as perthane are also insecticides. Chlordane and the structurally related aldrin are also potent insecticides. A compound widely used in place of moth balls (naphthalene) is *p*-dichlorobenzene. Other organic halides such as 2,4,5-trichlorophenoxyacetic acid (2,4,5-T) and pentachlorophenol (PCP) are herbicides. They have been used to control the

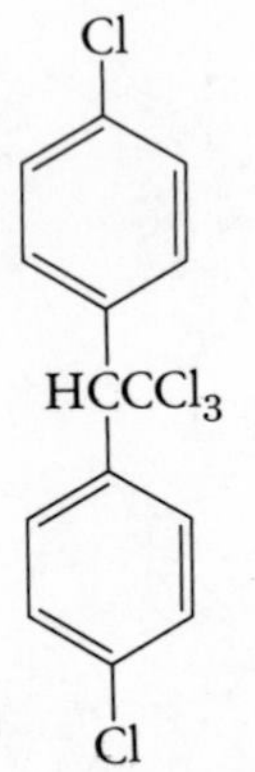

1,1,1-Trichloro-2-bis (4-chlorophenyl)ethane (DDT)

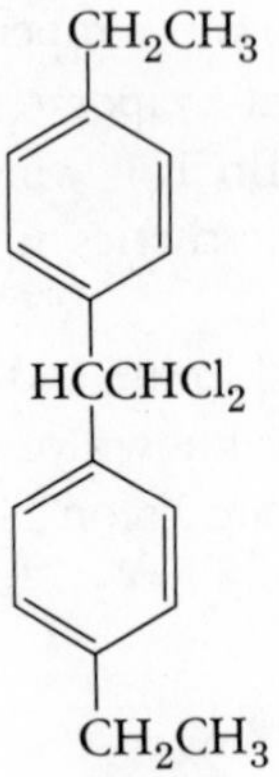

Perthane

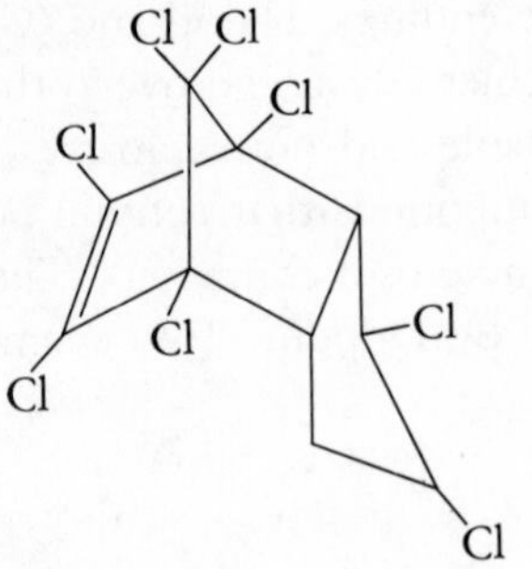

Chlordane

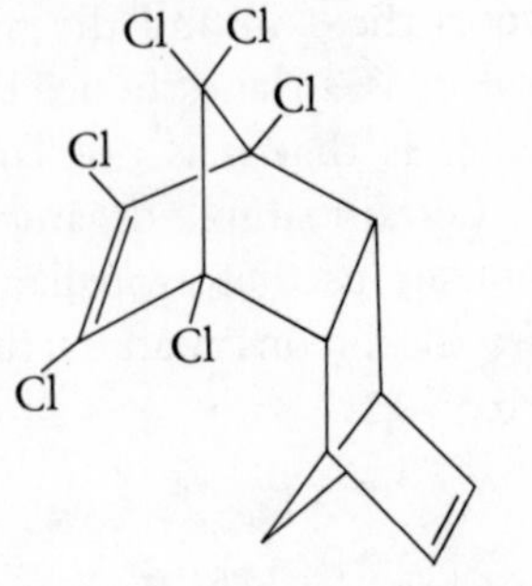

Aldrin

growth of weeds in lawns and gardens. Pentachlorophenol is also used as a wood preservative.

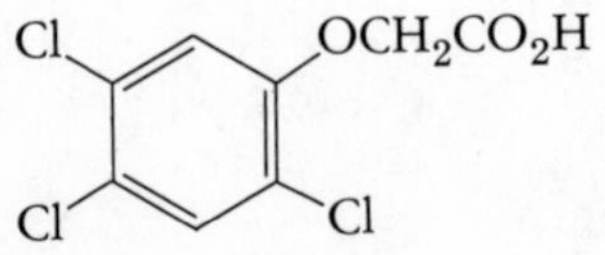

2,4,5-Trichlorophenoxyacetic acid (2,4,5-T)

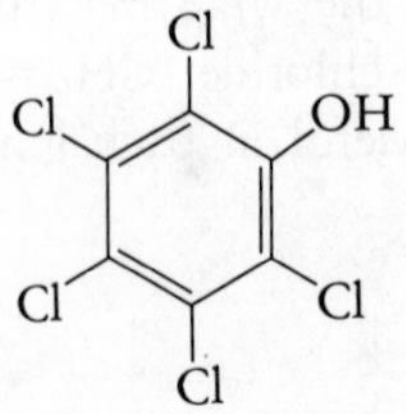

Pentachlorophenol (PCP)

These insecticides have been invaluable in controlling many diseases that are carried by insects, but there are problems associated with using them. After several years of use, the insects develop resistance to the insecticide. As a result, larger and larger amounts of insecticide must be used to produce the desired effect; but in larger amounts these compounds are toxic to animals and humans.

Two examples illustrate this fact. In 1976 at Sevesco, Italy, an industrial accident released an estimated 2 to 10 pounds of 2,3,7,8-tetrachlorodibenzo-*p*-dioxin (TCDD) into the air. As a result, vegetation in the area was destroyed, animals died, people became ill, and more than 800 people were evacuated from the most contaminated area.

In the United States, serious illness has affected workers in plants manufacturing the insecticide Kepone®. The symptoms of this poisoning include tremors and jitters followed by a slow loss of motion. The mechanism by which Kepone affects the body is unknown, but the evidence so far indicates that the central nervous system is attacked.

There is a real danger in the short term to persons coming into contact with large quantities of these insecticides. There is also a long-term danger to all life. This danger arises from the fact that organic halides are soluble in hydrocarbon-like portions of cells but are essentially insoluble in water. Therefore, the usual biological processes by which organic materials in nature are decomposed are unable to affect these insecticides. Consequently, the insecticides decompose slowly and persist in the environment for many years. Eventually, they are ingested by a variety of plants and microorganisms. Once they enter a living system, they tend to concentrate in the hydrocarbon-like regions of the tissue.

There is now clear evidence that the concentrations of these insecticides increase in species along the food chain. A species high on the food chain has a larger concentration of insecticide than does a species lower on the chain. For example, the concentration of an insecticide in lake water after the surrounding region is sprayed may reach 0.01 ppm. This concentration of insecticide presents no immediate danger to plants, animals, or humans. After a time, the concentration of the insecticide in the plankton of the lake may reach 7 ppm. Fish feeding on these microorganisms are found to contain concentrations of the insecticide as high as 1000 ppm. As a result of this accumulation, the concentration of the insecticide is now dangerous to animal and human life. The birds who eat these fish usually die directly or lay eggs whose shells are too thin to survive incubation. This has resulted in the near extinction of several species of birds. Humans are not immune to this process, because we are at the top of the food chain. For this reason, the use of many chlorinated hydrocarbons as insecticides has been severely restricted.

The use of organic halides as propellants in aerosol containers poses a potential environmental hazard. These compounds are lighter than air and tend to concentrate many miles above the surface of the earth. At high altitude they undergo a reaction catalyzed by sunlight that some scientists fear will deplete the atmosphere of its protective ozone layer. If this occurs, more ultraviolet light will strike the surface of the earth, with serious consequences for all life.

Organic halides have proved to be a mixed blessing. Their widespread use has made life easier, more comfortable, and more enjoyable for many people.

However, their persistence in the environment may yet prove to be the bane of human existence on earth. Only time will tell.

Although the number of different kinds of organic halides made and used is very large, they can all be classified into five types, as we will learn in the next section.

15.2 CLASSIFYING ORGANIC HALIDES

Most organic halides are classified as aromatic, aliphatic, or vinyl. Aromatic organic halides are called *aryl halides.* In these compounds, the halogen is attached to a carbon that is part of the ring of an aromatic compound. Bromobenzene is an example of an aryl halide. Some of the reactions of aryl halides were given in Chapter 13. Others will be discussed in Section 15.11. Aliphatic organic halides are called *alkyl halides.* In these compounds, the halogen is attached to a carbon in an alkyl group. *Vinyl halides* are compounds in which the halogen is bonded to a carbon that is part of a carbon-carbon double bond.

$C{=}C{-}Cl$ — A vinyl chloride

C_6H_5Br — An aryl bromide

CH_3CH_2F — An alkyl fluoride

Alkyl halides are classified further according to their structure. This classification is based on the type of carbon to which the halogen is attached. The halogen of a *primary alkyl halide* is bonded to a primary carbon. (To review the definitions of primary (1°), secondary (2°), and tertiary (3°) carbons, see Section 11.6.) The following are examples of primary alkyl halides (arrows identify primary carbon atoms):

CH_3CH_2Br $C_6H_5CH_2Cl$ $CH_3C(CH_3)_2CH_2I$

Primary (1°) alkyl halides

The halogen of a *secondary alkyl halide* is bonded to a secondary carbon, and that of a *tertiary alkyl halide* is bonded to a tertiary carbon. The following are examples of secondary and tertiary alkyl halides:

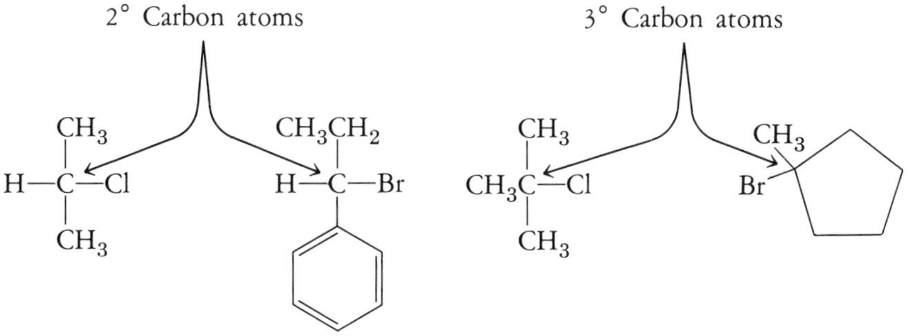

2° Carbon atoms

3° Carbon atoms

$$H-\overset{CH_3}{\underset{CH_3}{C}}-Cl \qquad H-\overset{CH_3CH_2}{\underset{C_6H_5}{C}}-Br \qquad CH_3\overset{CH_3}{\underset{CH_3}{C}}-Cl \qquad \text{1-bromo-1-methylcyclopentane}$$

Secondary (2°) alkyl halides

Tertiary (3°) alkyl halides

In Chapter 13 we learned how to name aryl halides. Let us now turn our attention to naming alkyl halides.

15.3 NAMING ALKYL HALIDES

We have encountered alkyl halides in previous chapters and in a few instances have given them IUPAC names without explanation. The reason for this is that alkyl halides are named by a straightforward extension of the IUPAC rules. The halogens are considered as substituents on the parent chain. The name of the halogen is placed as a prefix to the parent name. The following prefixes, introduced in Section 13.2, are used for the halogens: fluoro- for —F, chloro- for —Cl, bromo- for —Br, and iodo- for —I.

The rules that apply to alkenes and compounds with alkyl branches apply here. The only additional rule needed is that when a halogen and a carbon-carbon double or triple bond are in the same molecule, the numbering of the parent chain is chosen to give the lowest number to the double or triple bond. The following examples illustrate the rules:

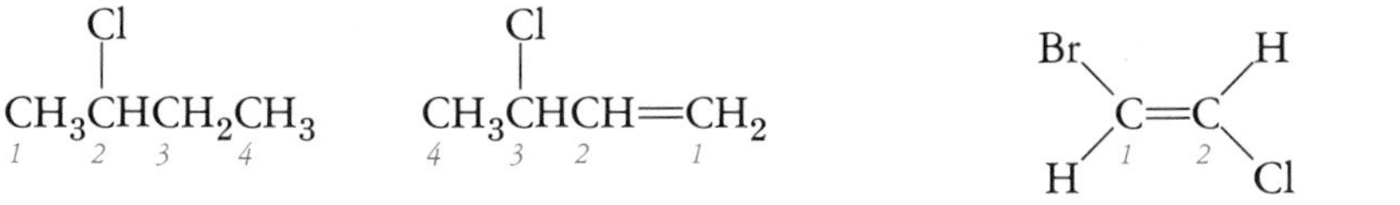

$$\underset{1\ \ 2\ \ 3\ \ 4}{CH_3\overset{Cl}{C}HCH_2CH_3} \qquad \underset{4\ \ 3\ \ 2\ \ 1}{CH_3\overset{Cl}{C}HCH{=}CH_2} \qquad \text{(Br)(H)}\underset{1}{C}{=}\underset{2}{C}\text{(H)(Cl)}$$

2-Chlorobutane

3-Chloro-1-butene

trans-1-Bromo-2-chloroethene

1,2-Diiodocyclopropene

$$\begin{array}{c} \,CH_3\ \ Br \\ ICH_2\overset{|}{\underset{|}{C}}{-}\overset{|}{\underset{|}{C}}CH_2CH_2Cl \\ Cl\ \ \ CH_3 \end{array}$$

3-Bromo-2,5-dichloro-1-iodo-2,3-dimethylpentane

*EXERCISE 15-1 Give the IUPAC name for each of the following compounds:

(a) $CH_3CH(CH_3)CH(Br)CH_3$ (b) $(CH_3)_2C(Cl)CH(Br)CH_2CH_3$ (c) $CH_3CH_2C(CH_3)(I)CH_2CH(I)CH_2CH(CH_2CH_3)CH_3$

(d) $C_6H_5CH(Cl)CH_2CH_3$ (e) cyclohexane ring with CH_3, Cl, F, Cl, CH_3, F substituents (f) $ClHC{=}CHC(CH_3)_3$

(g) $CH_3CH{=}CHCH(Cl)CH_3$ (h) $H_2C{=}C(CH_2CH(CH_3)_2)CH_2CH(Cl)CH_2CH_3$ (i) 3-chlorocyclohexene ring with Cl

EXERCISE 15-2 Write the structure for each of the following compounds:
(a) 2-bromohexane (b) 1-chlorocyclohexene
(c) bromocyclopentane (d) 2,5-diiodoheptane
(e) 3-chloropropene (f) 5-chloro-2,3-dimethylhexane
(g) 1-bromo-2-methyl-3-phenylpropane
(h) 1-chloro-1-methylcyclobutane
(i) 1-bromo-5-*t*-butylcyclohexene
(j) 4-bromo-1-pentene

Common names are often used for alkyl halides containing five or fewer carbon atoms. By this method, the name of the alkyl group to which the halogen is attached is given first. The name of the halogen, as halide,

* The answers for the exercises in this chapter begin on page 874.

follows as a separate word. The following examples illustrate this method of naming alkyl halides:

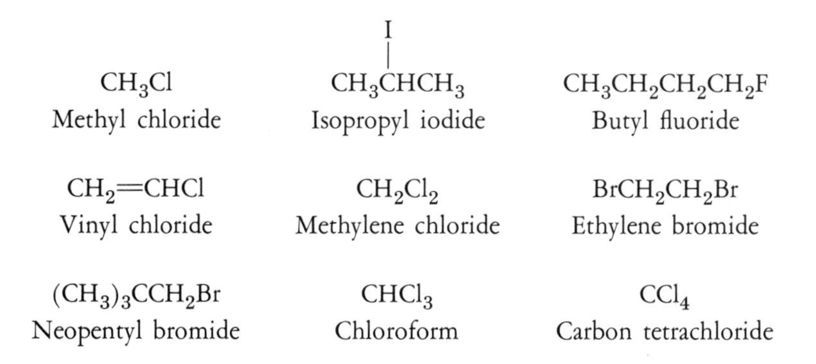

The last two examples, chloroform and carbon tetrachloride, are exceptions to the method of using common names for alkyl halides.

EXERCISE 15-3 Give the IUPAC name for each of the following compounds:
(a) propyl bromide (b) *sec*-butyl iodide
(c) isopropyl chloride (d) neopentyl fluoride
(e) ethyl chloride (f) isobutyl bromide
(g) chloroform (h) *t*-butyl iodide
(i) carbon tetrachloride (j) ethylene chloride
(k) vinyl chloride (l) methylene chloride

EXERCISE 15-4 Classify the compounds in exercises 15-1, 15-2, and 15-3 as primary, secondary, or tertiary alkyl halides or vinyl halides.

Now that we know how to name alkyl and vinyl halides, let us turn our attention to their physical properties.

15.4 PHYSICAL PROPERTIES

Alkyl halides containing one or two carbon atoms are either gases or liquids with low boiling points. As mentioned previously, organic halides are insoluble in water but are soluble in organic solvents. Because the masses of the halogen atoms are all much greater than that of hydrogen, the halogenated hydrocarbons are more dense than the corresponding saturated hydrocarbon. They are often even more dense than water. Thus, most of the alkyl

Table 15-1. Boiling Point and Density of Selected Organic Halides

Compound	Chloride X = Cl		Bromide X = Br		Iodide X = I	
	Boiling Point (°C)	Density (g/mL)	Boiling Point (°C)	Density (g/mL)	Boiling Point (°C)	Density (g/mL)
CH_3X	−24	Gas	5	Gas	43	2.28
CH_3CH_2X	13	Gas	38	1.44	72	1.93
$CH_3CH_2CH_2X$	47	0.89	71	1.34	102	1.75
$CH_3CH_2CH_2CH_2X$	79	0.88	102	1.28	130	1.62
$(CH_3)_2CHCH_2X$	69	0.88	91	1.26	120	1.61
$(CH_3)_3CX$	51	0.84	73	1.22	100	1.57
$CH_2{=}CHX$	−14	Gas	16	1.52	56	2.03
C_6H_5—X (benzene ring—X)	132	1.11	156	1.52	189	1.82

chlorides and all the bromides and iodides listed in Table 15-1 sink when added to a container of water.

Although rare in nature, a wide variety of organic halides are readily available from industrial sources. Consequently, chemists have studied their reactions in great detail. Organic halides undergo two types of reaction: substitution and elimination. These types of reaction also occur in living systems. The information obtained from the study of organic halides can be directly applied to reactions in living systems. Let us examine the first of these reactions, substitution.

15.5 SUBSTITUTION REACTIONS OF ALKYL HALIDES

In general, a substitution reaction involves the substitution of one atom or group bonded to carbon for another atom or group (see Sections 11.11 and 13.3 for examples). In the case of alkyl halides, the reaction involves *substituting another atom or group for the halogen.* For example, hydroxide ion reacts with methyl bromide to form methyl alcohol by substitution of the hydroxy group for the bromine atom, as shown in the following equation:

$$\underset{\text{Hydroxide ion}}{HO^-} + \underset{\text{Methyl bromide}}{CH_3{-}Br} \longrightarrow \underset{\text{Methyl alcohol}}{CH_3OH} + \underset{\text{Bromide ion}}{Br^-}$$

This type of reaction is also called a *displacement reaction,* because the bromine can be regarded as being displaced by the hydroxide ion. A variety of different groups can displace the halogen, as illustrated in the following equations:

$$CH_3O^- + CH_3{-}X \longrightarrow CH_3OCH_3 + X^-$$
$$HS^- + CH_3{-}X \longrightarrow CH_3SH + X^-$$
$$CN^- + CH_3{-}X \longrightarrow CH_3CN + X^-$$
$$N_3^- + CH_3{-}X \longrightarrow CH_3N_3 + X^-$$
$$H_3N\!: + CH_3{-}X \longrightarrow CH_3\overset{+}{N}H_3 + X^-$$
$$CH_3\ddot{N}H_2 + CH_3{-}X \longrightarrow CH_3\overset{+}{N}H_2CH_3 + X^-$$

These reactions, which appear to be different, are actually quite similar. The similarity is in the type of reagent that displaces the halogen. In the examples given, all possess an unshared pair of electrons. Such reagents seek a positive charge or a nucleus, and they are classified as *nucleophiles* (nucleus-loving). The relatively positive carbon of the carbon-halogen bond attracts these nucleophiles. It is this carbon where the nucleophile displaces the halogen, as illustrated in Figure 15-1. Because the halogen is displaced, it is called the *leaving group.* Because the nucleophile is the attacking group, we can call these reactions *nucleophilic displacement* or *nucleophilic substitution* reactions.

Alkyl halides have been used to illustrate nucleophilic substitution reactions, but such reactions can occur with other classes of compounds. The following are two examples of nucleophilic substitution reactions:

$$\overset{\downarrow}{C}H_3CH_2\overset{+}{S}(CH_3)_2 + :NH_3 \longrightarrow CH_3CH_2\overset{+}{N}H_3 + S(CH_3)_2$$

$$\overset{\downarrow}{C}H_3CH_2\overset{+}{N}_2 + OH^- \longrightarrow CH_3CH_2OH + N_2$$

Nucleophilic substitution reactions all have the following common characteristics. One of the reagents is a nucleophile. The other reagent contains a

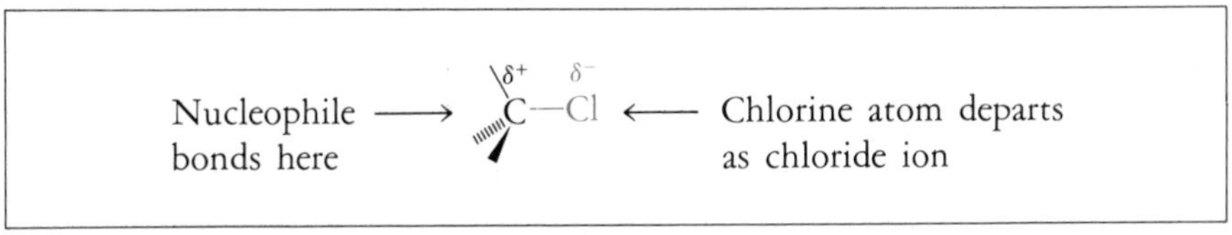

Nucleophile bonds here $\longrightarrow$ $\overset{\delta+}{C}{-}\overset{\delta-}{Cl}$ $\longleftarrow$ Chlorine atom departs as chloride ion

Fig. 15-1. The site of substitution in a nucleophilic displacement reaction.

carbon atom bonded to four other atoms or groups, one of which is strongly electron-attracting. This group is the one displaced (the leaving group) by the nucleophile. The carbon atom to which the leaving group is bonded is the reaction site. In the preceding two examples, an arrow indicates the site of nucleophilic substitution and the leaving group is indicated in color. It is important to recognize this type of reaction because it occurs frequently in biological systems.

EXERCISE 15-5 Which of the following reagents are nucleophiles?

(a) NH_2^- (b) H^+ (c) 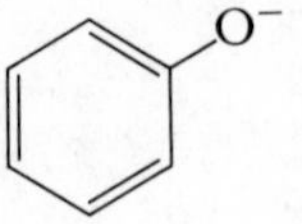 $C_6H_5O^-$ (benzene ring bearing O^-) (d) Na^+ (e) I^- (f) 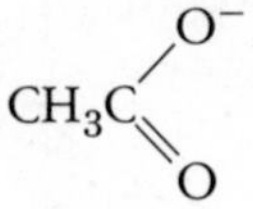 $CH_3C(=O)O^-$

EXERCISE 15-6 For each of the following reactions, indicate the nucleophile, the leaving group, and the site of nucleophilic substitution:

(a) $CH_3O^- + C_6H_5CH_2Br \longrightarrow C_6H_5CH_2OCH_3 + Br^-$

(b) $CH_3CH_2Cl + CH_3S^- \longrightarrow CH_3CH_2SCH_3 + Cl^-$

(c) $CH_3CH_2Br + I^- \longrightarrow CH_3CH_2I + Br^-$

(d) $CH_3CH_2CH_2OSO_2C_6H_5 + N_3^- \longrightarrow CH_3CH_2CH_2N_3 + {}^-OSO_2C_6H_5$

EXERCISE 15-7 Write the structure of the product in each of the following reactions:

(a) 1-chloroethane + $CN^- \longrightarrow$

(b) methyl chloride + $CH_3O^- \longrightarrow$

(c) 1-bromobutane + $CH_3C(=O)O^- \longrightarrow$

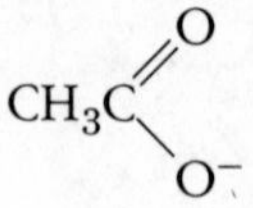

(d) isobutyl iodide + $C_6H_5O^-$ (benzene ring bearing O^-) $\longrightarrow$

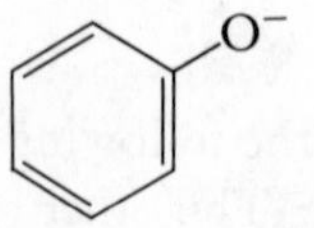

Nucleophilic substitution reactions occur most readily with primary alkyl halides. The reaction of secondary and tertiary alkyl halides with nucleophiles is complicated by the fact that they undergo both substitution and elimination reactions. This is discussed in the next section.

15.6 ELIMINATION REACTIONS OF ALKYL HALIDES

Alkyl halides react with bases to form alkenes by the elimination of a molecule of hydrogen halide. This reaction, called an elimination reaction, involves removal of the halogen together with the hydrogen atom from a carbon adjacent to the one bonded to the halogen. The following reaction of *t*-butyl chloride with hydroxide ion is an example of such a reaction:

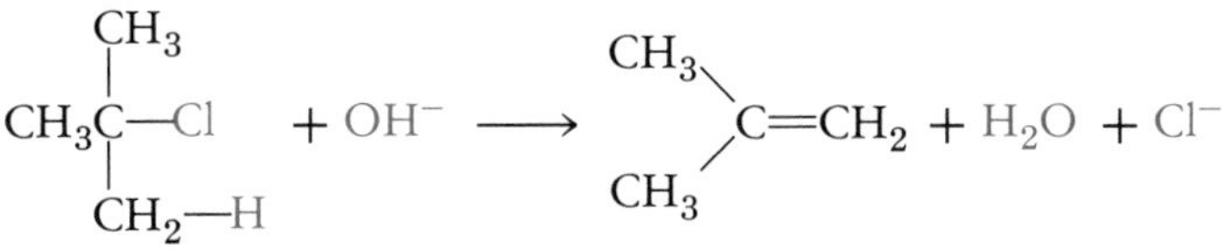

$$\begin{array}{c} CH_3 \\ | \\ CH_3C{-}Cl \\ | \\ CH_2{-}H \end{array} + OH^- \longrightarrow \begin{array}{c} CH_3 \diagdown \\ \quad C{=}CH_2 \\ CH_3 \diagup \end{array} + H_2O + Cl^-$$

Often, the elimination reaction of an alkyl halide can form a mixture of isomeric alkenes. For example, 2-chlorobutane can eliminate a hydrogen from either carbon 1 or carbon 3 to yield both 1-butene and 2-butene. The alkene with the most substituted double bond is usually the chief product, as shown in equation 15-1:

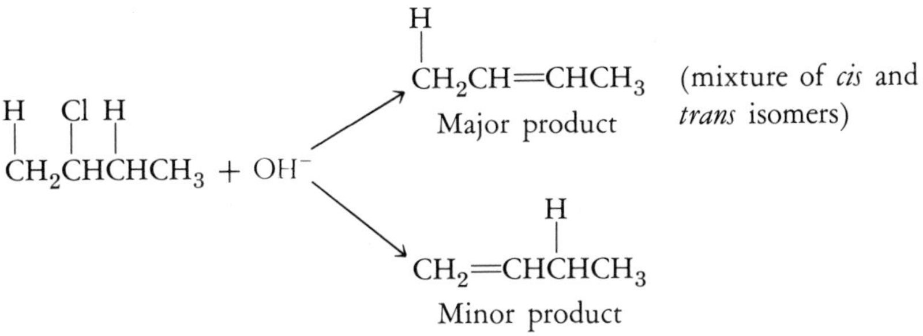

$$\begin{array}{c} H\ \ Cl\ H \\ |\ \ \ |\ \ \ | \\ CH_2CHCHCH_3 \end{array} + OH^- \begin{array}{l} \nearrow \begin{array}{c} H \\ | \\ CH_2CH{=}CHCH_3 \\ \text{Major product} \end{array} \quad \text{(mixture of } cis \text{ and } trans \text{ isomers)} \\ \searrow \begin{array}{c} H \\ | \\ CH_2{=}CHCHCH_3 \\ \text{Minor product} \end{array} \end{array}$$

EXERCISE 15-8 For each of the following compounds, indicate which hydrogens would be lost in an elimination reaction:

(a) CH_3CH_2Br (b) 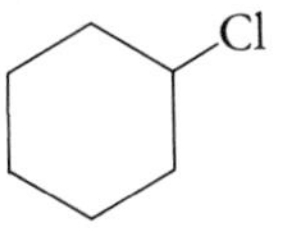 cyclohexane ring bearing Cl (c) 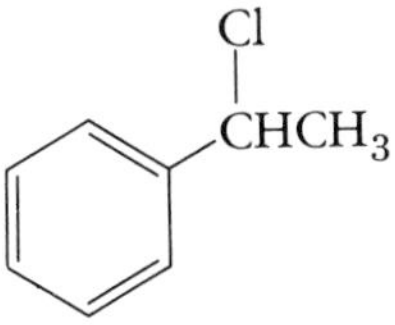 benzene ring bearing $\begin{array}{c} Cl \\ | \\ CHCH_3 \end{array}$

(d) 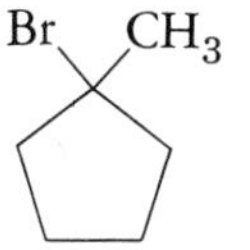 cyclopentane ring bearing Br and CH_3 on the same carbon (e) 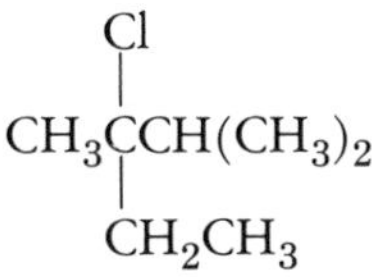 $\begin{array}{c} Cl \\ | \\ CH_3CCH(CH_3)_2 \\ | \\ CH_2CH_3 \end{array}$

EXERCISE 15-9 For the hydrogens that can be lost in the compounds in exercise 15-8, indicate which are equivalent.

EXERCISE 15-10 What is the maximal number of isomeric alkenes that can be obtained by an elimination reaction of an alkyl halide?

Notice that these elimination reactions of alkyl halides occur with the same reagent, hydroxide ion, that was used to illustrate a nucleophilic substitution reaction in Section 15.5. The hydroxide ion reacts with alkyl halides to form products of both elimination and substitution. The reason for this is that hydroxide ion is not only a base but also a nucleophile. In general, the nucleophiles given in Section 15.5 are also bases, so *both substitution and elimination reactions can occur when an alkyl halide reacts with a nucleophile.* This presents us with a problem. How can we determine which reaction will predominate? Analysis of experimental results has shown that several factors determine the course of the reaction. Two of the most important are the structure of the alkyl halide and the base strength of the nucleophile. First, consider the structure of the alkyl halide. Tertiary alkyl halides generally form the largest amounts of products from elimination reactions. Secondary alkyl halides produce less, and primary alkyl halides produce the least. The reverse is true for nucleophilic substitution reactions. Primary alkyl halides form the most product, and tertiary alkyl halides form none or only small amounts of the products of substitution reaction. We can summarize this as follows:

$$\xrightarrow{\text{Increasing amounts of products of elimination}}$$

Alkyl halide structure	Primary	Secondary	Tertiary

$$\xleftarrow{\text{Increasing amounts of products of substitution}}$$

The second factor is the base strength of the nucleophile. What is this? We have established that a hydroxide ion can react as both a base and a nucleophile. As a base it has an affinity for a proton; as a nucleophile it has an affinity for a carbon nucleus. These are two different characteristics of the same reagent. We can illustrate this by the following reactions:

$OH^- + H^+ \longrightarrow HOH$ Hydroxide ion acting as a base

$OH^- + CH_3Cl \longrightarrow HOCH_3 + Cl^-$ Hydroxide ion acting as a nucleophile

In general, if the reagent is a better base than a nucleophile, elimination will be the preferred reaction with alkyl halides. Conversely, if it is a better nucleophile than a base, substitution reactions will predominate. In this way we can establish the relative basicities and nucleophilicities of reagents. Thus, we can arrange the anions presented in Section 15.5 into three classes as follows:

Strong Base, Weak Nucleophile	Moderate Base and Nucleophile	Weak Base, Strong Nucleophile
NH_2^-	Cl^-, CN^-	I^-, N_3^-
CH_3O^-	$CH_3CO_2^-$, OH^-(aqueous)	$C_6H_5O^-$
OH^-(alcohol)		$C_6H_5S^-$

By combining these two factors, we can make the following generalizations:

1. Tertiary alkyl halides react with nucleophiles to form substantial amounts of the products of elimination. As the base strength of the nucleophile increases, so does the amount of elimination product.
2. Secondary alkyl halides react with nucleophiles to form products of both substitution and elimination. The proportion of each depends on the basicity of the nucleophile. With strong bases, elimination products predominate; with weak bases, substitution products predominate.
3. The products of the reaction of primary alkyl halides with nucleophiles depend on the basicity of the nucleophile. With strong bases, elimination products are formed. With weak bases, substitution products are formed.

For example, sodium amide ($NaNH_2$), a strong base, reacts with primary, secondary, and tertiary alkyl halides to form products of elimination. Sodium cyanide (NaCN), a base of moderate strength, reacts with tertiary alkyl halides to form predominantly products of elimination; with secondary alkyl halides to form a mixture of elimination and substitution; with primary alkyl halides to give predominantly products of substitution. Sodium iodide (NaI), a relatively weak base, reacts with tertiary alkyl halides to form substantial amounts of the products of elimination, whereas with primary and secondary alkyl halides it forms products of substitution.

15.7 HOW NUCLEOPHILIC SUBSTITUTION REACTIONS OCCUR

OPTIONAL SECTION

The generalizations given in the preceding section have been arrived at by studying, comparing, and analyzing many reactions of alkyl halides. However, scientists are not satisfied with simply listing the chemical reactions of functional groups. Their intellectual curiosity demands knowing how a particular reaction occurs. Much work, time, and money has been devoted to learning how substitution and elimination reactions occur. We will now summarize some of the conclusions that scientists have reached about how these reactions occur.

Two mechanisms have been proposed for nucleophilic substitution reactions of alkyl halides. One is a one-step mechanism and the other is a two-step mechanism.

In the one-step mechanism, the nucleophile attacks the carbon of the alkyl halide and the leaving group departs in a single step. That is, a bond between the nucleophile and the carbon is formed as the bond between the same carbon and the leaving group is broken. This is called the S_N2 mechanism. The S refers to substitution; N, to nucleophilic; 2, to bimolecular. The term bimolecular means that the two compounds, an alkyl halide and a nucleophile, must be present for the reaction to occur. We can picture this mechanism as shown in Figure 15-2.

Notice that when a reaction occurs by this mechanism, the configuration about the carbon is inverted. We can actually detect this change in configuration experimentally by starting with an optically active compound. When hydroxide ion reacts with optically active (+)-2-bromooctane, the alcohol product is found to be optically active. However, it has the inverted configuration. If the hydroxy group simply occupied the same position as the bromine atom in the original molecule, the alcohol would have the same configuration, not the inverted configuration as observed. This is shown in

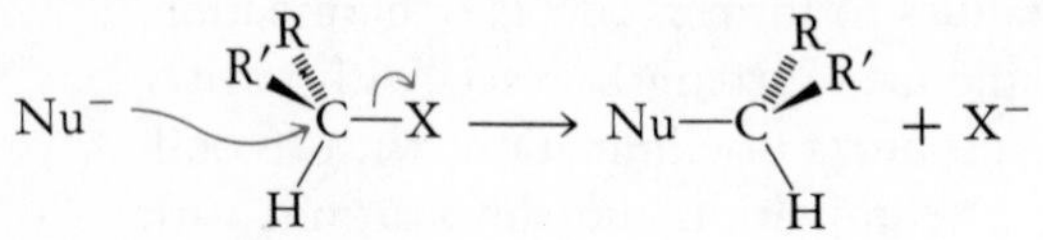

$$Nu^- + R'RC(H)—X \longrightarrow Nu—C(H)RR' + X^-$$

X = leaving group
Nu = nucleophile
R and R′ = different alkyl groups

Fig. 15-2. The S_N2 mechanism.

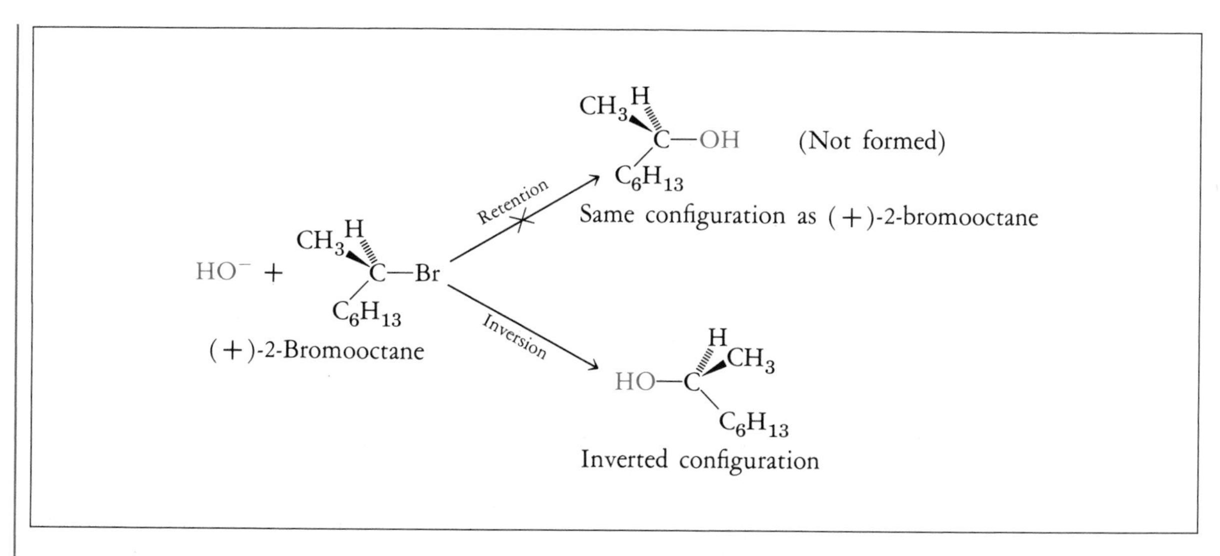

Fig. 15-3. The inversion of configuration at a chiral carbon during a reaction that occurs by an S_N2 mechanism.

Figure 15-3. The simplest way to explain this experimental result is by means of the S_N2 mechanism. All primary and most secondary alkyl halides undergo nucleophilic substitution reactions by an S_N2 mechanism.

The second mechanism involves two steps. In the first step, ionization of the alkyl halide occurs slowly to form a carbonium ion. In the second step, the carbonium ion reacts rapidly with a nucleophile to form the product. This is called the S_N1 mechanism.

Again, S and N refer to substitution and nucleophilic, and 1 refers to unimolecular. The term unimolecular means that the reaction occurs without the assistance of any other compound. All tertiary alkyl halides undergo nucleophilic displacement reactions by this mechanism because they form very stable carbonium ions. (Review Section 12.5 for carbonium ion stabilities.) We can picture this mechanism as shown in Figure 15-4.

An optically active alkyl halide that reacts by an S_N1 mechanism usually forms optically inactive products. The reason for this is shown in Figure 15-4. The positive carbon of the carbonium ion is bonded to only three groups. It is therefore a trigonal carbon atom (see Section 12.2). All the bonds to such a carbon lie in one plane with 120 degree bond angles. Because of this planar arrangement of the groups bonded to the positive carbon, the nucleophile can attack from both sides (paths a and b in Figure 15-4). The result is a mixture of equal amounts of the two enantiomers, that is, a racemic mixture that is optically inactive.

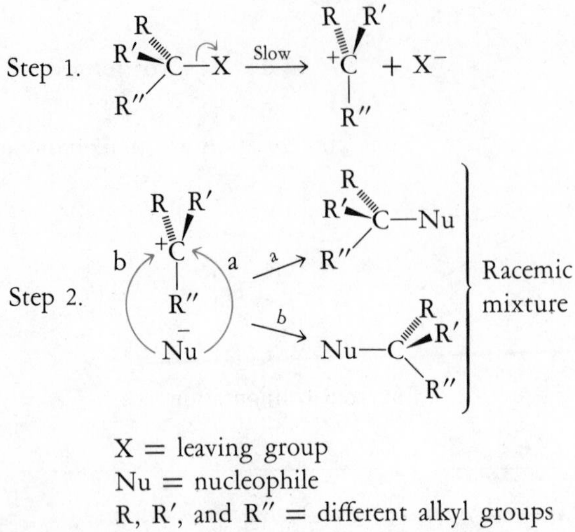

Fig. 15-4. The S_N1 mechanism.

Step 1. R, R', R''–C–X $\xrightarrow{\text{Slow}}$ R, R', R''–$\overset{+}{C}$ + X^-

Step 2. $\overset{+}{C}(R)(R')(R'')$ + $\overset{-}{Nu}$ —a→ R, R', R''–C–Nu ; —b→ Nu–C–R, R', R'' (Racemic mixture)

X = leaving group
Nu = nucleophile
R, R′, and R″ = different alkyl groups

In summary, the stereochemistry of the products is one of the major differences between S_N1 and S_N2 mechanisms. The S_N1 mechanism forms racemic products, whereas the S_N2 mechanism forms products of inverted configuration. In general, reactions of tertiary alkyl halides occur by an S_N1 mechanism, whereas reactions of all primary and most secondary alkyl halides occur by an S_N2 mechanism.

EXERCISE 15-11 Draw the configuration of the product of each of the following reactions:

(a) I^- + $(CH_3)(H)(CH_3CH_2)C{-}Br \longrightarrow CH_3\underset{|}{C}HCH_2CH_3$ (I) + Br^-

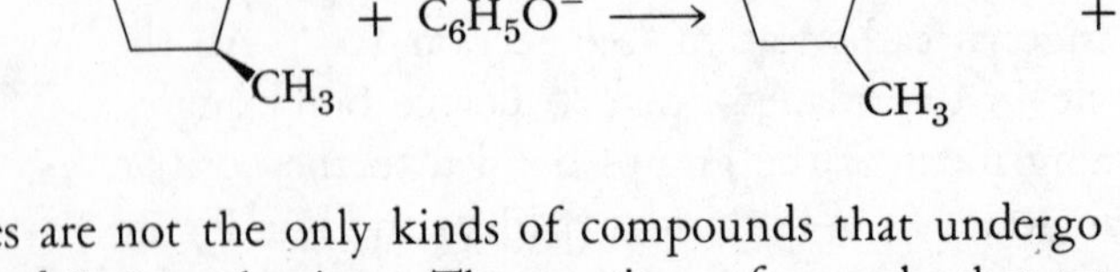

(b) 1-bromo-2-methylcyclopentane (Br, CH_3) + $C_6H_5O^- \longrightarrow$ 1-phenoxy-2-methylcyclopentane (OC_6H_5, CH_3) + Br^-

Alkyl halides are not the only kinds of compounds that undergo reactions by S_N1 and S_N2 mechanisms. The reactions of several other types of organic compounds occur by these two mechanisms. Before we encounter them, however, let us learn how elimination reactions occur.

15.8 HOW ELIMINATION REACTIONS OCCUR

Two mechanisms have also been proposed for the elimination reactions of alkyl halides. Again, one involves one step and the other involves two steps. Let us begin with the two-step mechanism, because it resembles the S_N1 mechanism.

The two-step mechanism begins exactly like the S_N1 mechanism. The alkyl halide ionizes to a carbonium ion in the first step. But in this mechanism, the carbonium ion rapidly loses a proton to a base in the second step to form an alkene rather than reacting with a nucleophile. This is the E1 mechanism. The E refers to elimination, and 1 means unimolecular, as in S_N1. We can picture this mechanism as shown in Figure 15-5.

The other mechanism involves a single step. The base attacks and removes a hydrogen from a carbon adjacent to the one bonded to the halogen. While this occurs, the halogen leaves and a new carbon-carbon double bond is formed. This is called a bimolecular elimination mechanism, or E2 for short. We can picture this mechanism as shown in Figure 15-6.

Like substitution reactions, eliminations occur with compounds other

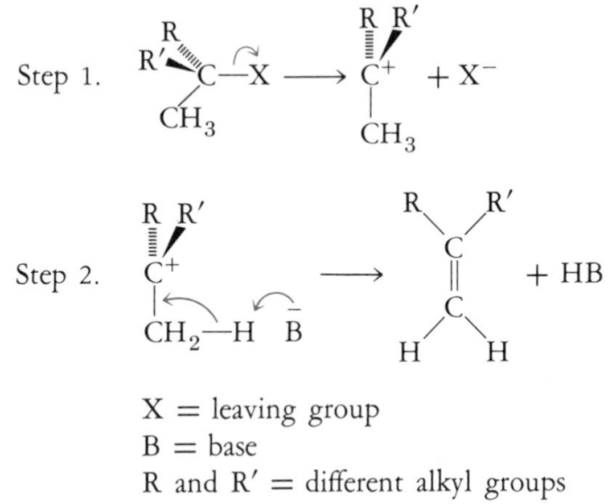

Step 1. $RR'(CH_3)C{-}X \longrightarrow RR'C^{+}(CH_3) + X^-$

Step 2. $RR'C^{+}{-}CH_2{-}H \;\; \bar{B} \longrightarrow RR'C{=}CH_2 + HB$

X = leaving group
B = base
R and R′ = different alkyl groups

Fig. 15-5. The E1 mechanism.

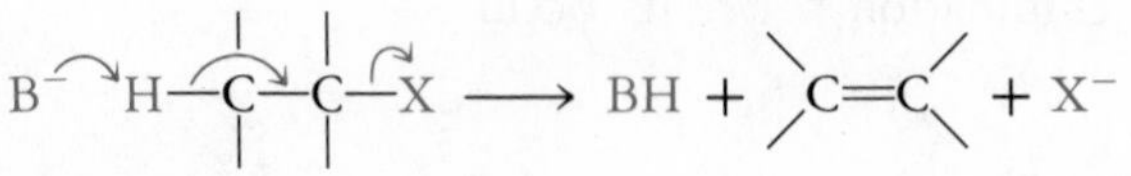

$$\mathrm{B^- \; H{-}\overset{|}{\underset{|}{C}}{-}\overset{|}{\underset{|}{C}}{-}X \longrightarrow BH + \;\rangle C{=}C\langle\; + X^-}$$

X = leaving group
B = base

Fig. 15-6. The E2 mechanism.

than alkyl halides. The truth of this statement is clearly demonstrated by the examples of elimination and substitution reactions in living systems discussed in the next section.

15.9 SUBSTITUTION AND ELIMINATION REACTIONS IN LIVING SYSTEMS

Substitution and elimination reactions occur in the cells of plants and animals. Because alkyl halides do not usually exist in living systems, groups or atoms other than halides are eliminated or displaced. The most important of these leaving groups in biological systems are the phosphates $-O-\overset{O}{\overset{\|}{\underset{O^-}{\underset{|}{P}}}}-O^-$, the pyrophosphates $-O-\overset{O}{\overset{\|}{\underset{O^-}{\underset{|}{P}}}}-O-\overset{O}{\overset{\|}{\underset{O^-}{\underset{|}{P}}}}-O^-$, the sulfonium ion $-\overset{+}{S}R_2$, and the quaternary ammonium ion $-\overset{+}{N}R_3$ (see Section 17-3).

Methylation is one of the more common reactions in living systems. It involves the transfer of a methyl group from one molecule to another. This is done by using a compound called S-adenosylmethionine as the methyl carrier. Its structure is as follows:

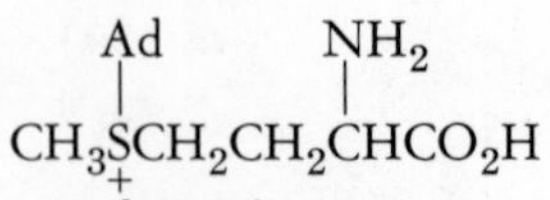

$$\underset{\text{S-adenosylmethionine}}{CH_3\overset{\overset{\text{Ad}}{|}}{\underset{+}{S}}CH_2CH_2\overset{\overset{NH_2}{|}}{C}HCO_2H}$$

The large and complicated adenosyl group is abbreviated Ad. Because this part of the molecule is not involved in the reaction, we can ignore its structure for now. The important part of this molecule in the methylation

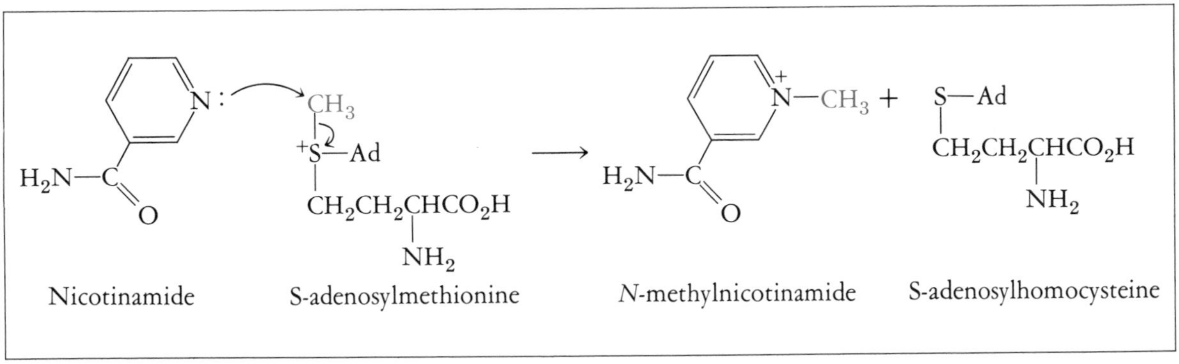

$H_2N{-}C(=O)$ pyridine $N\!:$ + $CH_3{-}S^+({-}Ad)CH_2CH_2CH(NH_2)CO_2H$ $\longrightarrow$ $H_2N{-}C(=O)$ pyridine $N^+{-}CH_3$ + $Ad{-}S{-}CH_2CH_2CH(NH_2)CO_2H$

Nicotinamide S-adenosylmethionine *N*-methylnicotinamide S-adenosylhomocysteine

Fig. 15-7. The transfer of a methyl group from S-adenosylmethionine to nicotinamide by a nucleophilic displacement reaction.

reaction is the methyl group bonded to the sulfur. It is transferred by a nucleophilic displacement reaction, as shown in Figure 15-7. In this reaction, the ring nitrogen of nicotinamide acts as the nucleophile, and the leaving group is S-adenosylhomocysteine. Forty different compounds in living systems react with S-adenosylmethionine by such nucleophilic displacement reactions.

EXERCISE 15-12 Norepinephrine $(HO)_2C_6H_3{-}CH(OH)CH_2NH_2$ reacts with S-adenosylmethionine to form epinephrine (also called adrenaline) $(HO)_2C_6H_3{-}CH(OH)CH_2\overset{+}{N}H_2CH_3$. Identify the site of nucleophilic substitution, the nucleophile, and the leaving group.

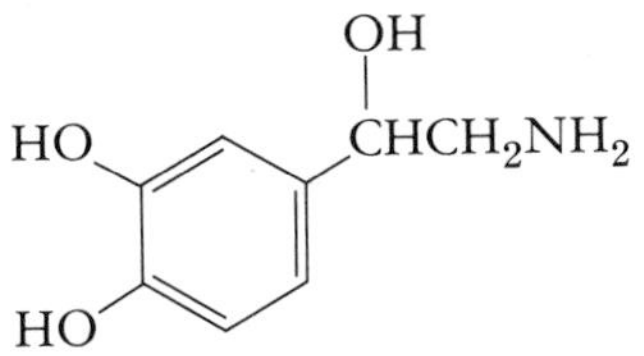

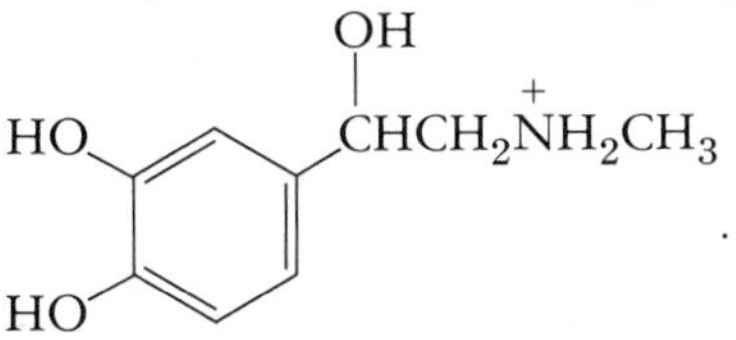

All living systems have a way of getting rid of toxic compounds. Unfortunately, most of the methods of disposal cannot handle water-insoluble compounds that contain halogens. Some living systems solve this problem by replacing the halogen atoms by an —OH group. This increases the water solubility of the molecule and allows the system to dispose of it. The following enzyme-catalyzed reaction is one such example:

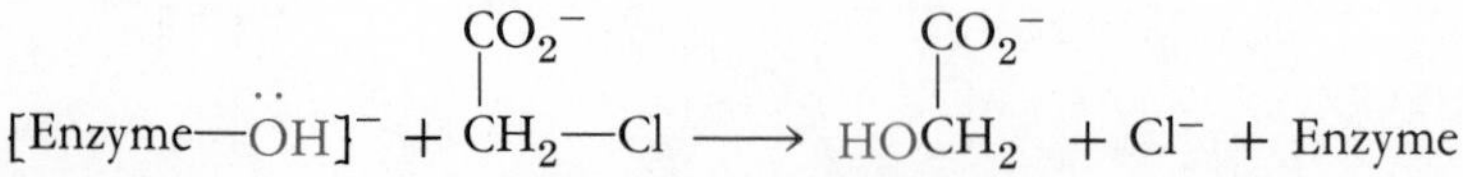

$$[\text{Enzyme}{-}\ddot{\text{O}}\text{H}]^- + \underset{}{\overset{\displaystyle CO_2^-}{\overset{|}{CH_2}}}{-}Cl \longrightarrow \overset{\displaystyle CO_2^-}{\overset{|}{HOCH_2}} + Cl^- + \text{Enzyme}$$

An —OH group bound to an enzyme displaces the halogen on a carbon in a typical substitution reaction. Certain living systems can also remove toxic chlorinated compounds by eliminating hydrogen chloride (HCl) from a molecule. The following is such a reaction that occurs with DDT:

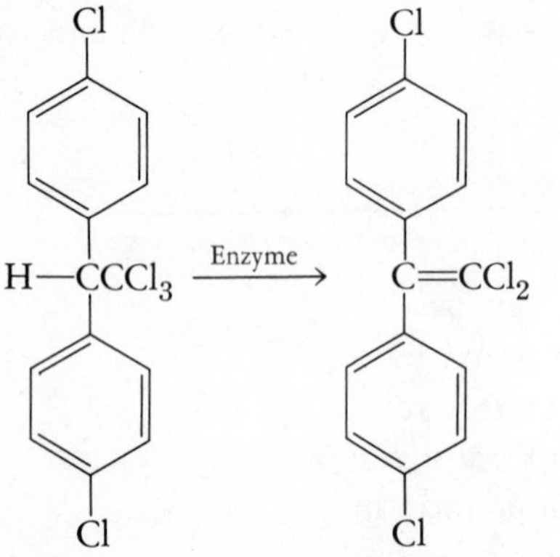

$$(p\text{-}ClC_6H_4)_2\text{C(H)}{-}CCl_3 \xrightarrow{\text{Enzyme}} (p\text{-}ClC_6H_4)_2C{=}CCl_2$$

DDT DDE

However, in this case, the technique is not successful. The product of this elimination reaction, *d*ichlorodiphenyl*d*ichloro*e*thylene (DDE) is as toxic as DDT.

Enzyme-catalyzed dehydrations are the usual elimination reactions found in living systems. They often occur as the reverse of an enzyme-catalyzed hydration reaction. For example, it was mentioned in Chapter 12 that the enzyme fumarase catalyzes the hydration of fumarate to malate. This reaction is reversible. Under certain conditions, the same enzyme can dehydrate malate to form fumarate.

Another dehydration reaction occurs in the conversion of citrate to isocitrate. Citrate is dehydrated by the enzyme aconitase, which catalyzes the reversible interconversion of citrate and isocitrate via the enzyme-bound intermediate *cis*-aconitate.

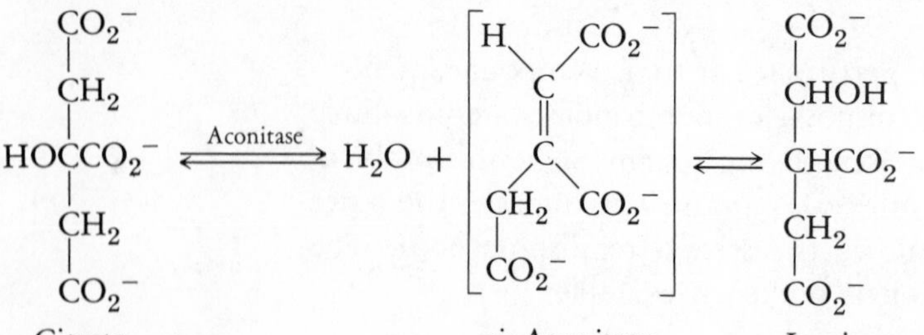

$$\underset{\text{Citrate}}{CO_2^-{-}CH_2{-}HOCCO_2^-{-}CH_2{-}CO_2^-} \underset{}{\overset{\text{Aconitase}}{\rightleftharpoons}} H_2O + \underset{\textit{cis}\text{-Aconitate}}{\left[\begin{matrix}H & & CO_2^- \\ & C & \\ & \| & \\ & C & \\ CH_2 & & CO_2^- \\ | & & \\ CO_2^- & & \end{matrix}\right]} \rightleftharpoons \underset{\text{Isocitrate}}{CO_2^-{-}CHOH{-}CHCO_2^-{-}CH_2{-}CO_2^-}$$

Alkyl halides react with many metals to form organometallic compounds. These compounds, although relatively small in number, are very important in living systems. Let us briefly examine some of these compounds.

15.10 ORGANOMETALLIC COMPOUNDS

Organometallic compounds contain a metal atom bonded to one or more carbons. Many of these compounds are widely used. For example, tetraethyl lead is an anti-knock additive in gasoline; ethylmercuric chloride is a fungicide; merthiolate is an antiseptic. Many organometallic compounds are toxic. For example, the toxicity of dimethylmercury is a problem because it is involved in the process by which mercury, a pollutant in the environment, enters the life cycle.

$(CH_3CH_2)_4Pb$	$CO_2^- Na^+$ / $SHgCH_2CH_3$ (on benzene ring)	CH_3CH_2HgCl	CH_3HgCH_3
Tetraethyl lead	Merthiolate	Ethylmercuric chloride	Dimethylmercury

Organic chemists have prepared many organometallic compounds in the laboratory. They are usually prepared by the reaction of an alkyl halide with a metal. The following are two specific examples:

$$CH_3I + Mg \longrightarrow CH_3MgI \qquad \text{(Grignard reaction)}$$

$$CH_3(CH_2)_2CH_2Br + 2\,Li \longrightarrow \underset{\text{Organolithium compounds}}{CH_3(CH_2)_2CH_2Li} + LiBr$$

These are important reactions of alkyl halides. The organometallic compounds formed in this way are useful *synthetic intermediates*. That is, they can undergo further reactions to make other types of compounds. Such organometallic compounds do not exist in living systems.

A number of organometallic compounds are involved in the reactions of living systems. Chlorophyll (see Section 13.6) contains a magnesium atom. This compound is one of the major light-absorbing pigments in most green cells and is involved in photosynthesis. Another organometallic compound is heme (see Section 13.6), which contains an iron atom. Heme is involved in carrying oxygen in the respiratory chain.

The metal atom may seem like only a small part of a large and complex organic molecule, but its presence is vital. Without it, the molecule cannot carry out its job. Clearly, these organometallic compounds are necessary to life.

So far we have been concerned mainly with the reactions of alkyl halides. But how about aryl and vinyl halides? Do they undergo substitution reactions? We will learn the answer to this question in the next section.

15.11 SUBSTITUTION REACTIONS OF ARYL AND VINYL HALIDES

It is important to recognize the difference between the substitution reactions of alkyl halides and aryl halides. Nucleophilic substitution reactions occur easily with primary and secondary alkyl halides. In contrast, aryl and vinyl halides undergo nucleophilic substitution reactions only with extreme difficulty. Under the reaction conditions that cause nucleophilic displacement reactions to occur with alkyl halides, no reaction occurs with aryl or vinyl halides. For example:

Cl (chlorobenzene) or $CH_2{=}CHCl$ $\xrightarrow{CH_3O^-,\ HS^-,\ CN^-,\ N_3^-,\ I^-}$ No reaction

Nucleophilic substitution reactions can be made to occur with aryl halides in two ways. The first is to increase the severity of the reaction conditions. Reacting chlorobenzene with sodium hydroxide at high temperature (350° C) and high pressure (1500 psi) forms phenol. In this way, phenol has been prepared commercially by the Dow process for many years:

$$C_6H_5Cl + NaOH \xrightarrow[\text{Pressure}]{\Delta} C_6H_5OH + NaCl \quad \text{Dow process}$$

The second way is to modify the structure of the aryl halide. If the aromatic ring contains, in addition to the halogen, certain other properly placed groups, aryl halides do undergo nucleophilic substitution reactions. Thus, when nitro- ($—NO_2$), nitroso- ($—NO$), cyano- ($—CN$), sulfonate ($—SO_3H$), or carboxylate ($—COOH$) groups are placed *ortho* or *para* to the halogen, nucleophilic substitution reactions will occur. Some specific examples are the following:

Cl, NO_2 $\xrightarrow[160^\circ C]{\text{Aqueous NaOH}}$ OH, NO_2

Cl, NO_2, NO_2 $\xrightarrow[130^\circ C]{\text{Aqueous } Na_2CO_3}$ OH, NO_2, NO_2

NO_2, Cl, NO_2, NO_2 $\xrightarrow{\text{Warm } H_2O}$ NO_2, OH, NO_2, NO_2

Note that as the number of nitro- groups increases, nucleophilic substitution occurs with greater ease.

An aryl halide of this type, 2,4-dinitrofluorobenzene, is a particularly useful reagent for determining the end amino acid of a peptide or protein, as we will learn in Chapter 23.

15.12 SUMMARY

Alkyl halides are rarely found in nature. However, alkyl halides made industrially are widely used. Their continued use has created problems for the public. Alkyl halides undergo two major types of reaction: substitution and elimination. These types of reaction occur often in living systems. Organometallic compounds are rare in nature but play important roles in living systems. Numerous organometallic compounds prepared in the laboratory are widely used as synthetic intermediates. Most aryl and vinyl halides, unlike primary and secondary alkyl halides, undergo nucleophilic substitution reactions with difficulty.

REVIEW OF TERMS, CONCEPTS, AND REACTIONS

Terms

ALKYL HALIDE A compound containing a halogen bonded to an alkyl group.

ARYL HALIDE A compound containing a halogen bonded to a carbon that is part of the ring of an aromatic compound.

DISPLACEMENT REACTION Another term for substitution reaction.

ELIMINATION REACTION A reaction in which two atoms or groups bonded to adjacent carbon atoms are removed, and a double bond forms between these carbons.

NUCLEOPHILE A group or atom that seeks a positive charge or a nucleus.

ORGANOMETALLIC COMPOUNDS Organic compounds that contain a covalent bond between a metal atom and one or more carbon atoms.

PRIMARY ALKYL HALIDE A compound in which the halogen is bonded to a primary carbon.

SECONDARY ALKYL HALIDE A compound in which the halogen is bonded to a secondary carbon.

SYNTHETIC INTERMEDIATES Compounds that are used to prepare other compounds.

TERTIARY ALKYL HALIDE A compound in which the halogen is bonded to a tertiary carbon.

VINYL HALIDE A compound containing a halogen bonded to a carbon that is part of a carbon-carbon double bond.

Concepts

1. Both substitution and elimination reactions occur when alkyl halides react with nucleophiles.
2. Tertiary alkyl halides react with nucleophiles to form substantial amounts of the products of elimination.
3. Secondary alkyl halides react with nucleophiles to form products of both substitution and elimination. The proportion of each depends on the basicity of the nucleophile. With strong bases, elimination products predominate; with weak bases, substitution products predominate.
4. The products of the reaction of primary alkyl halides with nucleophiles depend on the basicity of the nucleophiles. With strong bases, elimination products are formed. With weak bases, substitution products are formed.

5. Aryl and vinyl halides usually undergo nucleophilic substitution reactions with great difficulty. Such reactions can be forced by increasing the severity of the reaction conditions or, for aryl halides, by modifying the structure of the aryl halide.

Reactions

Reactions of alkyl halides

1. Nucleophilic substitution reaction (for specific examples, see Section 15.5)

$$R{-}X + Nu^- \longrightarrow R{-}Nu + X^-$$

X = leaving group
Nu = nucleophile
R = alkyl group

2. Elimination reaction (for specific examples, see Section 15.6)

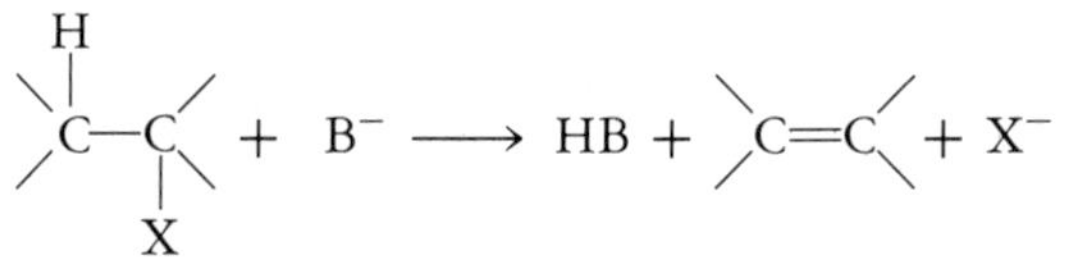

$$\overset{H}{\overset{|}{>C}}-\underset{\underset{X}{|}}{C}< + B^- \longrightarrow HB + >C{=}C< + X^-$$

X = leaving group (e.g., Cl or Br)
B = base

3. Reactions with metals (for specific examples, see Section 15.10)
   (a) $RX + Mg \longrightarrow RMgX$
   (b) $RX + 2\,Li \longrightarrow RLi + LiX$
   X = Cl, Br, or I
   R = alkyl group

Reactions of aryl halides

1. Nucleophilic substitution reaction (for specific examples, see Section 15.11)

$$ArX + Nu^- \longrightarrow ArNu + X^-$$

Ar = aryl group
X = Cl, Br, or F
Nu^- = nucleophile

The aryl (Ar) group must contain groups such as $-NO_2$, $-NO$, $-CN$, $-SO_3H$, or $-CO_2H$ *ortho* and/or *para* to X.

2. Electrophilic substitution reaction (see Chapter 13)
3. Reaction with metals
   (a) $ArX + Mg \longrightarrow ArMgX$
   (b) $ArX + 2Li \longrightarrow ArLi + LiX$
   Ar = aryl group
   X = Cl or Br

EXERCISES 15-13 Write the structure for each of the following compounds:
(a) 2-fluorobutane (b) isobutyl chloride
(c) 1-chloro-2,2-dimethylpentane
(d) 2,4-dicyanochlorobenzene
(e) 1-chloro-1-isopropylcyclopentane
(f) *cis*-4,4-dichloro-2-pentene
(g) 1,4-diiodo-2-butyne
(h) 1-bromo-4-chloro-3-methylpentane
(i) 1-chloro-1-sec-butylcyclohexane
(j) 1-chloro-2,3,3-trimethylbutane

15-14 Give the IUPAC name for each of the following compounds:

(a) $CF_2{=}CF_2$ (b) $ClCH_2C{\equiv}CCCl_2CH_3$ (c) 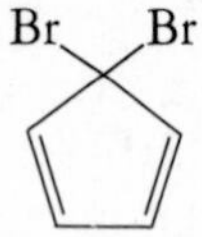 [cyclopentadiene ring with two Br on the same carbon]

(d) 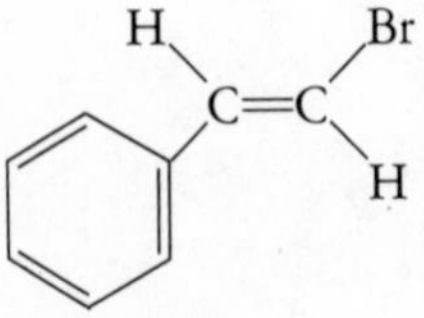 [benzene ring–CH=CHBr: H and Br on the double bond, H on the Br-bearing carbon] (e) $(CH_3)_3CCH_2F$ (f) $(CH_3CH_2)_3CCl$

(g) 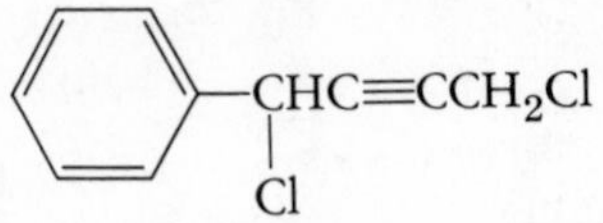 [benzene ring]$-\underset{\displaystyle Cl}{CH}C{\equiv}CCH_2Cl$ (h) 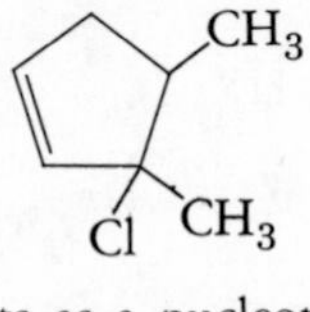 [cyclopentene ring with CH_3 on one carbon and Cl and CH_3 on the adjacent carbon]

15-15 Classify each of the following reagents as a nucleophile or an electrophile:
(a) NO_3^- (b) H^+ (c) ClO_4^- (d) HSO_3^- (e) Ag^+
(f) Hg^{+2}

15-16 Identify each of the following as an addition, elimination, or nucleophilic substitution reaction:
(a) $CH_3CH_2Br + NH_2^- \longrightarrow CH_2{=}CH_2 + NH_3 + Br^-$

(b) $CH_3CH{=}CH_2 + Br_2 \longrightarrow CH_3\underset{\displaystyle Br}{C}HCH_2Br$

(c) $CH_3CH_2\overset{+}{S}(C_6H_5)_2 + Br^- \longrightarrow CH_3CH_2Br + S(C_6H_5)_2$

(d) $CH_3O-\overset{O}{\overset{\|}{\underset{O^-}{\underset{|}{P}}}}-O-\overset{O}{\overset{\|}{\underset{O^-}{\underset{|}{P}}}}-O^- + N_3^- \longrightarrow CH_3N_3 + {}^-O-\overset{O}{\overset{\|}{\underset{O^-}{\underset{|}{P}}}}-O-\overset{O}{\overset{\|}{\underset{O^-}{\underset{|}{P}}}}-O^-$

(e) $(CH_3)_3CBr + OH^- \longrightarrow CH_3\underset{CH_3}{\underset{|}{C}}{=}CH_2 + HOH + Br^-$

(f) 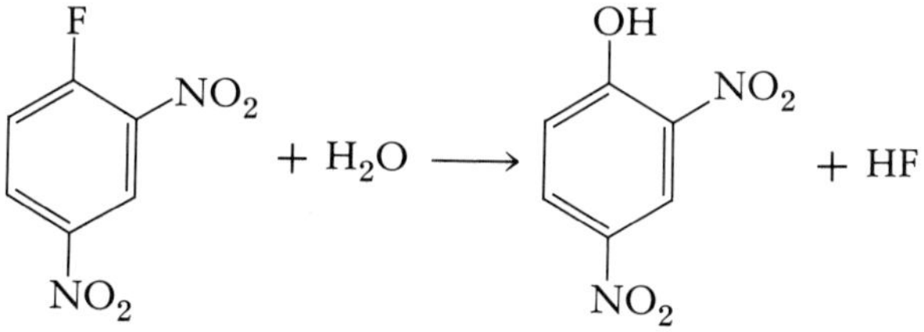 2,4-dinitrofluorobenzene (F, NO_2, NO_2) $+ H_2O \longrightarrow$ 2,4-dinitrophenol (OH, NO_2, NO_2) $+ HF$

15-17 Write the structure of the product of each of the following reactions:

(a) 1-chlorobutane $+ I^- \longrightarrow$

(b) 2,4-dinitrobromobenzene $+ CH_3O^- \longrightarrow$

(c) propyl chloride $+ CN^- \longrightarrow$

(d) *t*-butyl bromide $+ NH_2^- \longrightarrow$

(e) isopropyl bromide $+ Mg \longrightarrow$

(f) 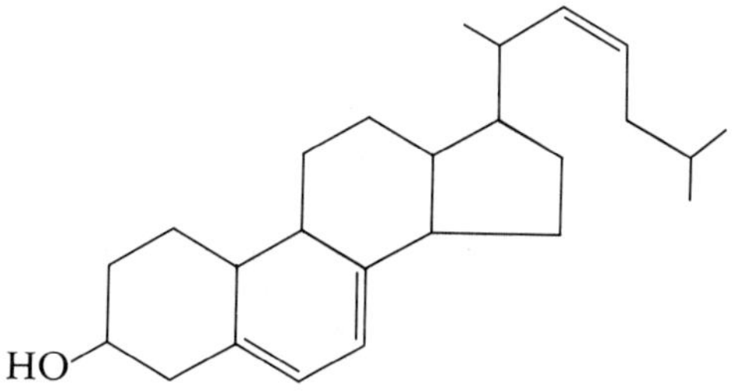 Ergosterol (HO) $+ {}^+\overset{Ad}{\overset{|}{\underset{CH_3}{\underset{|}{S}}}}CH_2CH_2\overset{NH_2}{\overset{|}{C}}HCO_2H \longrightarrow$

Ergosterol

15-18 Optically active 2-chloropentane reacts with NaOH in water to form inverted 2-pentanol ($CH_3\underset{OH}{\underset{|}{C}}HCH_2CH_2CH_3$), 1-pentene, and a mixture of *cis*- and *trans*-2-pentene. Write the mechanism for this reaction.

15-19 Predict whether the major products of the following reactions are elimination products or substitution products or both. Write the structure(s) of the product(s).

(a) 1-bromobutane $+$ NaOH $\xrightarrow{H_2O}$

(b) 2-chloro-2-methylpentane $+$ NaCN $\longrightarrow$

(c) bromocyclopentane + NaI $\xrightarrow{\text{Acetone}}$

(d) 2-bromopentane + $NaOCH_3$ $\xrightarrow{CH_3OH}$

(e) 1-chloropropane + NaSH $\longrightarrow$

(f) isobutyl chloride + $C_6H_5O^-$ $\longrightarrow$

(g) 3-chlorohexane + $NaNH_2$ $\xrightarrow{\text{Ether}}$

ALCOHOLS, PHENOLS, ETHERS, AND THIOLS

16

To many people alcohol means an intoxicating drink. However, to organic chemists, the word alcohol refers to a class of compounds that contain an —OH group, called a *hydroxyl* or *hydroxy* group, bonded to an alkyl group. One specific example of this type of compound is ethyl alcohol, CH_3CH_2OH, the intoxicating ingredient of many drinks. There are many more examples of alcohols, none of which has this effect on humans. In this chapter we will study alcohols and three related compounds: phenols, ethers, and thiols.

Alcohols can be viewed as organic analogues of water in which one hydrogen is replaced by an alkyl group. If one hydrogen is replaced by an aryl group, the compound is classed as a *phenol.* Replacing both hydrogens by alkyl or aryl groups forms a class of compounds called *ethers.* This relationship is shown in Figure 16-1. Regarding water, alcohols, phenols, and ethers in this way emphasizes their structural similarities and makes it easier to understand their physical and chemical properties.

Let us start our study of these important classes of compounds by learning how to classify and name alcohols.

16.1 CLASSIFYING AND NAMING ALCOHOLS

There are several similarities in the methods of classifying and naming alcohols and alkyl halides. Alcohols, like alkyl halides, are classified accord-

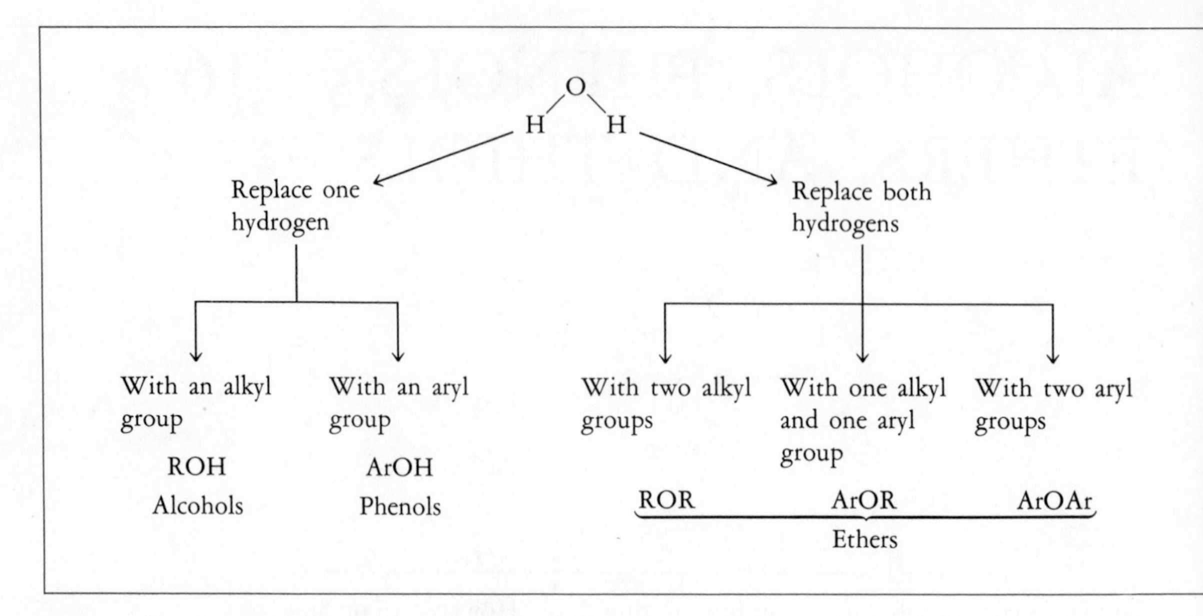

Fig. 16-1. Structural similarities of water, alcohols, phenols, and ethers.

ing to their structure. Thus, a *primary alcohol* is a compound in which the hydroxy group is bonded to a primary carbon. In a *secondary alcohol* the hydroxy group is bonded to a secondary carbon. In a *tertiary alcohol* the hydroxy group is bonded to a tertiary carbon. The following are examples of primary, secondary, and tertiary alcohols:

1° Carbon atom	2° Carbon atom	3° Carbon atom
RCH_2OH	$R'{-}\underset{OH}{\overset{H}{C}}{-}R$	$R'{-}\underset{OH}{\overset{R}{C}}{-}R''$
Primary alcohol	Secondary alcohol	Tertiary alcohol

As for alkyl halides, common names are often used for alcohols containing five or fewer carbon atoms. The name consists of two words. The first word is the name of the alkyl group (see Section 11.6 for the names of alkyl groups) to which the hydroxy group is bonded, and the second is the word alcohol. The following examples illustrate the application of common names to alcohols:

CH_3OH	CH_3CH_2OH	$CH_3CH_2CH_2CH_2OH$
Methyl alcohol	Ethyl alcohol	Butyl alcohol

$CH_2{=}CHCH_2OH$	$(CH_3)_3CCH_2OH$	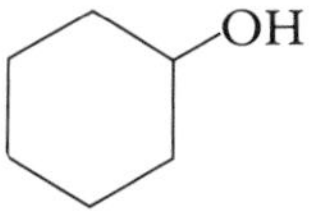 cyclohexane–OH
Allyl alcohol	Neopentyl alcohol	Cyclohexyl alcohol

*EXERCISE 16-1 Give a common name for each of the following compounds:
(a) $CH_3CH_2CH_2OH$ (b) $(CH_3)_2CHCH_2OH$ (c) $CH_3CH(OH)CH_3$

(d) 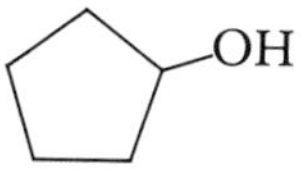 cyclopentane–OH (e) $(CH_3)_3COH$ (f) $CH_3CH_2CH(OH)CH_3$

(g) $CH_3(CH_2)_3CH_2OH$ (h) 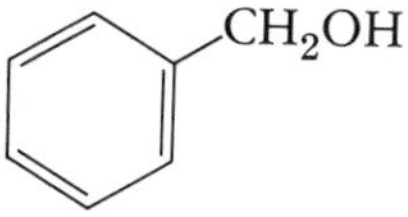 benzene–CH_2OH

Alcohols containing more than five carbons are best named by the IUPAC rules. The following additions to the rules set out in Chapter 11 are needed to deal with alcohols:

1. The longest continuous chain of carbon atoms that contains the hydroxy group is taken as the parent chain.
2. The parent name is obtained by substituting the ending *-ol* for the ending *-e* of the corresponding alkane.
3. The chain is numbered to give the lowest number to the carbon bonded to the hydroxyl group.
4. The suffix *-ol* is used for one hydroxy group; *-diol,* for two; *-triol,* for three, and so forth.

Some examples of the use of these rules are as follows:

* The answers for the exercises in this chapter begin on page 876.

CH_3CH_2OH	OH (cyclohexane ring)	1 2 3 4 $CH_3CHCH_2CHCH_3$ (OH on C-2; CH_2CH_3 on C-4, numbered 5 6)	3 2 1 $CH_2{=}CHCH_2OH$
Ethanol	Cyclohexanol	4-Methyl-2-hexanol	2-Propen-1-ol

Notice that when a double bond (or a triple bond) is part of the parent chain, the hydroxy group is given the lowest number.

EXERCISE 16-2 Classify the alcohols in exercise 16-1 as 1°, 2°, or 3°.

EXERCISE 16-3 Give the IUPAC name for the compounds in exercise 16-1.

EXERCISE 16-4 Write the structure for each of the following compounds:
(a) 2,2-dimethyl-1-propanol (b) 2-methyl-1-butanol
(c) cyclopropanol (d) 2-chloro-1-propanol
(e) 1-methylcyclohexanol (f) 1,2-ethanediol
(g) 3-methyl-3-buten-2-ol (h) 2-propyn-1-ol
(i) 1,2,3-propanetriol (j) 1-phenylpropanol
(k) 2-cyclohexenol

EXERCISE 16-5 Give the IUPAC name for each of the following compounds:

(a) CH_3OH (b) OH on a cyclobutane ring (c) CH_3CHCH_2OH with OH on C-2

(d) $ClCH_2CH_2OH$ (e) benzene ring–$CHCH{=}CH_2$ with OH on the CH

(f) cyclohexane ring with OH and Cl (g) $HOCHCH_2CH_2Br$ attached to a cyclopentane ring (h) OH on a cyclopentene ring

(i) $ClCH_2$(H)C=C(H)CH_2OH (j) $CH_3CHCH_2CCH_2OH$ with I on C-2 (numbered from the left) and two CH_3 on the quaternary carbon

(k) $HOCH_2CHCHCH_2OH$ with HO and OH on the two central carbons

16.2 PHYSICAL PROPERTIES

The relationship between the structure of a molecule and its physical properties is illustrated by comparing alcohols, phenols, and ethers. Alcohols and phenols have boiling points considerably higher than those of alkanes, aromatic hydrocarbons, or alkyl or aryl halides of similar molecular weight. This is evident from the physical properties of selected compounds listed in Table 16-1. Water also has an abnormally high boiling point. Hydrogen bonding is the explanation for this property of water (see Section 6.3 for an explanation of hydrogen bonding). Alcohols and phenols are similar to water in that they contain a polar O—H bond that can form hydrogen bonds with neighboring molecules. To vaporize an alcohol or a phenol, energy in the form of heat must be added to overcome this intermolecular attraction. As a result, these compounds have higher boiling points than anticipated from their molecular weights.

Ethers, in contrast, do not have an electron-deficient hydrogen to form hydrogen bonds. Therefore, hydrogen bonding cannot occur between ether molecules in the liquid state. As a result, the boiling points of ethers are similar to those of alkanes of similar molecular weight. For example, diethyl ether, $C_2H_5OC_2H_5$ (molecular weight, 74 amu), has almost the same boiling point (35° C) as pentane (molecular weight, 72 amu; boiling point, 36° C).

The hydroxy group of an alcohol or a phenol can also form hydrogen bonds with water. This interaction results in much greater solubility of

Table 16-1. Physical Properties of Some Alcohols, Phenols, Ethers, and Reference Compounds

Compound	Molecular Weight (amu)	Boiling Point (°C)	Melting Point (°C)	Water Solubility
CH_3CH_2OH	46	78	−117	Completely miscible
CH_3OH	32	65	−98	Completely miscible
CH_3CH_2Cl	64.5	13	−138	Insoluble
$CH_3CH_2CH_3$	44	−43	−188	Insoluble
$CH_2{=}CHCH_3$	42	−48	−185	Insoluble
C_6H_6	78	80	5	Insoluble
C_6H_5Cl	112.5	132	−45	Insoluble
C_6H_5OH	94	182	41	Slightly soluble
$CH_3CH_2OCH_2CH_3$	74	35	−116	Slightly soluble
n-C_5H_{12}	72	36	−130	Insoluble

alcohols and phenols in water than that of the corresponding alkanes, aromatic hydrocarbons, alkyl halides, or aryl halides. The presence of hydroxy groups in many molecules of biological importance makes these compounds water soluble. As a result, they are able to move about in the aqueous environment of living systems. This explains why converting toxic alkyl halides to alcohols as discussed in Section 15.9 makes it possible to get rid of them.

EXERCISE 16-6 Would you expect the two isomers ethanol and dimethyl ether (CH_3OCH_3) to have the same boiling points and solubility in water? Explain.

In the next section we will learn two methods of preparing alcohols that can be used in the laboratory or in living systems.

16.3 PREPARING ALCOHOLS

There are several ways to prepare alcohols in the laboratory. However, we will learn only the two methods that are similar to those used for preparing alcohols in living systems. We are already familiar with one of these methods: the hydration of alkenes.

Hydration of Alkenes

The acid-catalyzed addition of water to an alkene forms an alcohol as a product. This reaction, which is discussed in Section 12.4, adds water to the double bond according to the Markownikoff rule:

$$CH_3CH{=}CH_2 + HOH \xrightarrow{H^+} CH_3\underset{\displaystyle OH}{\underset{|}{C}}H\overset{\displaystyle H}{\overset{|}{C}}H_2$$

Examples of this reaction in living systems are presented in Sections 12.4, 27.5, and 28.3.

Reduction of Carbonyl Compounds

Alcohols can be prepared by adding hydrogen to the carbon-oxygen double bond of certain carbonyl compounds. This reaction is called the reduction of carbonyl compounds. The reductions of aldehydes and ketones are examples of these reactions:

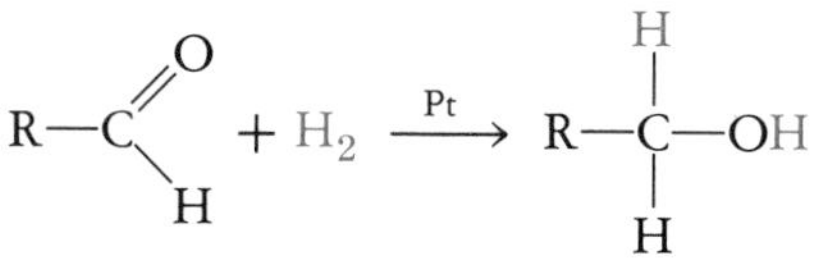

$$\underset{\text{Aldehyde}}{R-C(=O)-H} + H_2 \xrightarrow{Pt} \underset{1^\circ\ \text{Alcohol}}{R-CH_2-OH}$$

$$\underset{\text{Ketone}}{R-C(=O)-R'} + H_2 \xrightarrow{Pt} \underset{2^\circ\ \text{Alcohol}}{R-CH(R')-OH}$$

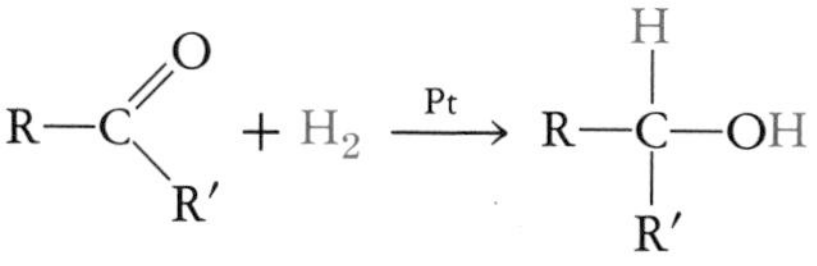

This reaction is similar to the metal-catalyzed addition of hydrogen to the carbon-carbon double bond of an alkene (Section 12.4). In both reactions, a hydrogen atom becomes bonded to each atom of the double bond:

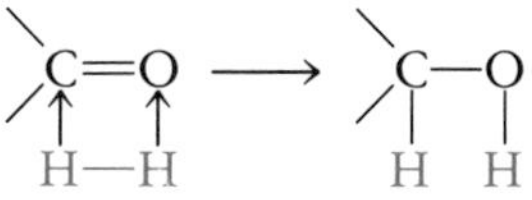

$$>C{=}O \ (+\ H{-}H) \longrightarrow >C(H){-}O{-}H$$

Another reagent that reduces carbonyl compounds is lithium aluminum hydride ($LiAlH_4$):

Aldehydes

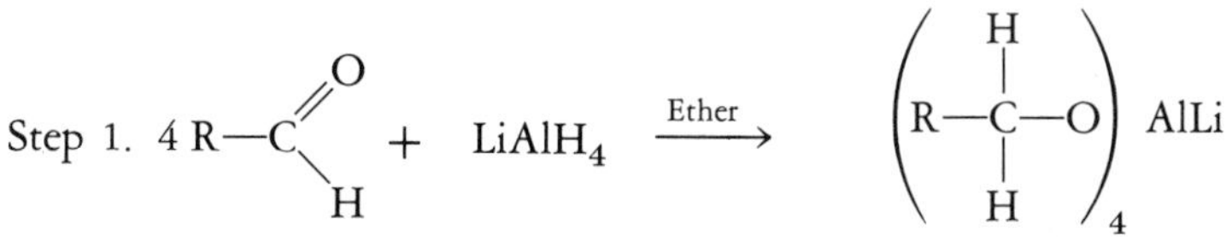

Step 1. $$\underset{\text{Aldehyde}}{4\,R-C(=O)-H} + \underset{\text{Lithium aluminum hydride}}{LiAlH_4} \xrightarrow{\text{Ether}} \underset{\text{Lithium aluminum alkoxide}}{(R-CH_2-O)_4AlLi}$$

Step 2. $$\underset{\text{Lithium aluminum alkoxide}}{(R-CH_2-O)_4AlLi} + 4\,H_2O \longrightarrow \underset{1^\circ\ \text{Alcohol}}{4\,RCH_2OH} + LiOH + Al(OH)_3$$

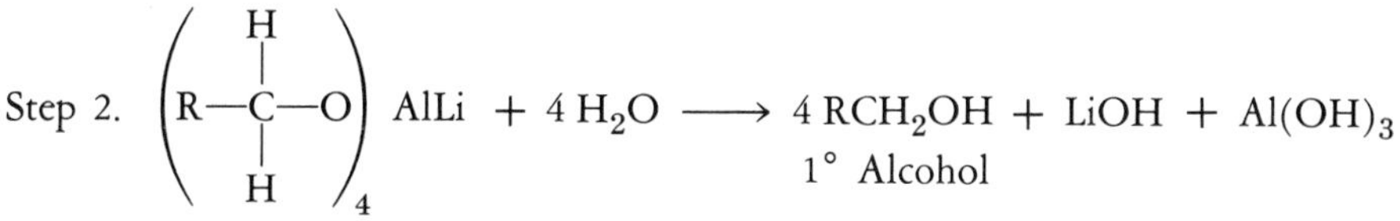

Ketones

Step 1. $4\,R{-}\underset{\underset{O}{\|}}{C}{-}R + LiAlH_4 \xrightarrow{\text{Ether}} (R_2CH{-}O)_4AlLi$

Ketone — Lithium aluminum hydride — Lithium aluminum alkoxide

Step 2. $(R_2CH{-}O)_4AlLi + 4\,H_2O \longrightarrow R_2CHOH + LiOH + Al(OH)_3$

Lithium aluminum alkoxide — 2° Alcohol

The first product formed is actually a lithium aluminum alkoxide, which forms the alcohol on reaction with water in the second step.

In both of these reactions, the reducing agents, hydrogen and the platinum catalyst and lithium aluminum hydride, reduce aldehydes to primary alcohols and ketones to secondary alcohols.

EXERCISE 16-7 Write the structure of the major product of each of the following reactions:

(a) $CH_3\overset{O}{\overset{\|}{C}}H + H_2 \xrightarrow{Pt}$

(b) $CH_3CH_2CH{=}CH_2 + H_2O \xrightarrow{H^+}$

(c) $CH_3\underset{\underset{O}{\|}}{C}CH_3 + LiAlH_4 \longrightarrow \xrightarrow{H_2O}$

(d) 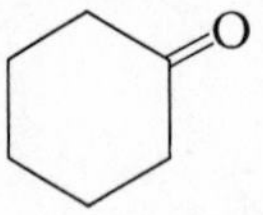 cyclohexanone $+ H_2 \xrightarrow{Pt}$

(e) 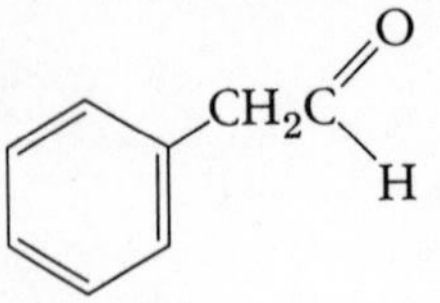 $C_6H_5CH_2\overset{O}{\overset{\|}{C}}H + LiAlH_4 \longrightarrow \xrightarrow{H_2O}$

As we will learn in the next section, two of the reactions of alcohols are simply the reverse of these reactions.

16.4 REACTIONS OF ALCOHOLS

Two of the reactions of alcohols are simply the reverse of the methods of preparing alcohols that we learned in the preceding section.

Dehydration

Alcohols react with concentrated sulfuric acid (H_2SO_4) to eliminate a molecule of water to form an alkene, as shown by the following equation:

One of these two hydrogens is removed

$$\text{R—}\underset{\text{H}}{\overset{\text{H}}{\underset{|}{\overset{|}{\text{C}}}}}\text{—CH}_2\text{—OH} \xrightleftharpoons{H_2SO_4} \begin{matrix}\text{R} \\ \text{H}\end{matrix}\!\!>\!\text{C=CH}_2 + \text{HOH}$$

This dehydration reaction is the reverse of the addition of water to alkenes discussed in Sections 12.4 and 16.3. These two reactions actually form the same equilibrium mixture of alkene, aqueous acid, and alcohol. By choosing the correct experimental conditions, the equilibrium shown in equation 16-1 may be shifted in either direction to form the alkene or the alcohol as the major product. For example, the use of concentrated sulfuric acid removes water and shifts the equilibrium to the left, forming alkene as the product. If only a small amount of acid is used in a large quantity of water, the equilibrium is shifted to the right and the alcohol is the major product.

$$\text{RCH=CH}_2 + \text{HOH} \xrightleftharpoons{H^+} \text{R}\underset{\text{OH}}{\underset{|}{\text{C}}}\text{H}\overset{\text{H}}{\overset{|}{\text{C}}}\text{H}_2 \qquad (16\text{-}1)$$

In general, the ease of dehydration of alcohols is tertiary > secondary > primary. This order is shown by the following examples:

$$\text{CH}_3\underset{\text{CH}_3}{\underset{|}{\overset{\text{CH}_3}{\overset{|}{\text{C}}}}}\text{OH} \xrightarrow[25^\circ\,C]{H_2SO_4} \begin{matrix}\text{CH}_3 \quad \text{CH}_3 \\ \diagdown\;\diagup \\ \text{C} \\ \| \\ \text{CH}_2\end{matrix} + \text{H}_2\text{O}$$

3° Alcohol

$$\text{CH}_3\underset{\text{OH}}{\underset{|}{\text{C}}}\text{HCH}_3 \xrightarrow[100^\circ\,C]{H_2SO_4} \text{CH}_2\text{=CHCH}_3 + \text{H}_2\text{O}$$

2° Alcohol

$$\underset{1^\circ \text{ Alcohol}}{CH_3CH_2OH} \xrightarrow[150^\circ C]{H_2SO_4} CH_2{=}CH_2 + H_2O$$

Thus, *t*-butyl alcohol is easily dehydrated at room temperature by concentrated sulfuric acid. Progressively higher temperatures are needed to dehydrate isopropyl alcohol and ethyl alcohol.

Ester Formation

Alcohols react with carboxylic acids in the presence of a strong acid catalyst to form a class of compounds called esters. The general equation for this reaction is the following:

$$\underset{\text{Alcohol}}{ROH} + \underset{\text{Carboxylic acid}}{R'C(=O)OH} \underset{}{\overset{H^+}{\rightleftharpoons}} \underset{\text{Ester}}{R'C(=O)OR} + HOH$$

The reactions of this biologically important class of compounds will be discussed in Section 18.5.

Alcohols also react with mineral acids to form compounds that are called esters of inorganic acids. In these compounds the acidic hydrogen of the inorganic acid is replaced by an alkyl group. Alkyl phosphates are an important class of such compounds in living systems. Their relation to phosphoric acid is clear from the following example:

Phosphoric acid $\quad H{-}O{-}P(=O)(OH){-}OH$

Alkyl phosphate $\quad R{-}O{-}P(=O)(OH){-}OH$

At the pH of living systems, the phosphate group is ionized. Thus, the structure of alkyl phosphates in living systems is best represented as follows:

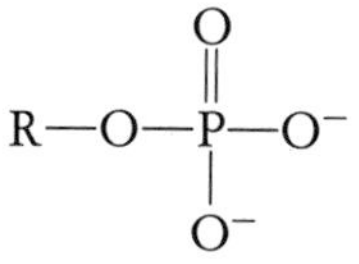

$$\mathrm{R{-}O{-}\overset{\overset{\displaystyle O}{\|}}{\underset{\underset{\displaystyle O^-}{|}}{P}}{-}O^-}$$

The phosphate group, as mentioned in Section 15.9, is important in many biological reactions because it is a good leaving group. Thus, alkyl phosphates undergo the usual nucleophilic substitution reactions. For example, alkyl phosphates react with water to form an alcohol and hydrogen phosphate:

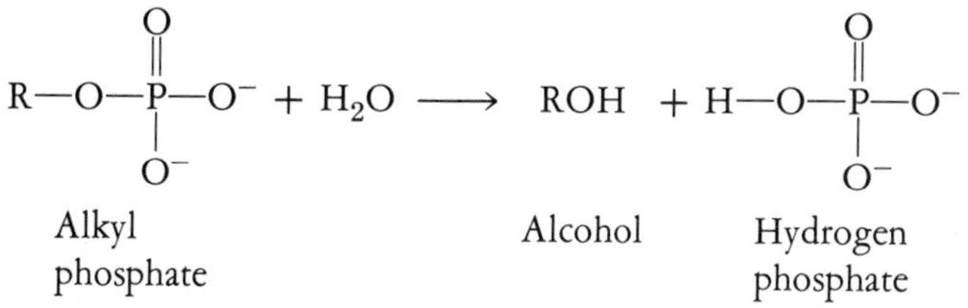

$$\underset{\text{Alkyl phosphate}}{\mathrm{R{-}O{-}\overset{\overset{\displaystyle O}{\|}}{\underset{\underset{\displaystyle O^-}{|}}{P}}{-}O^-}} + H_2O \longrightarrow \underset{\text{Alcohol}}{\mathrm{ROH}} + \underset{\text{Hydrogen phosphate}}{\mathrm{H{-}O{-}\overset{\overset{\displaystyle O}{\|}}{\underset{\underset{\displaystyle O^-}{|}}{P}}{-}O^-}}$$

We will encounter this reaction many times in Chapters 27, 28, and 29.

Oxidation

The products of oxidation of alcohols depend on the structure of the alcohol. Primary alcohols are first oxidized to aldehydes, which are further oxidized to carboxylic acids. Secondary alcohols form ketones on oxidation, and tertiary alcohols are not easily oxidized.

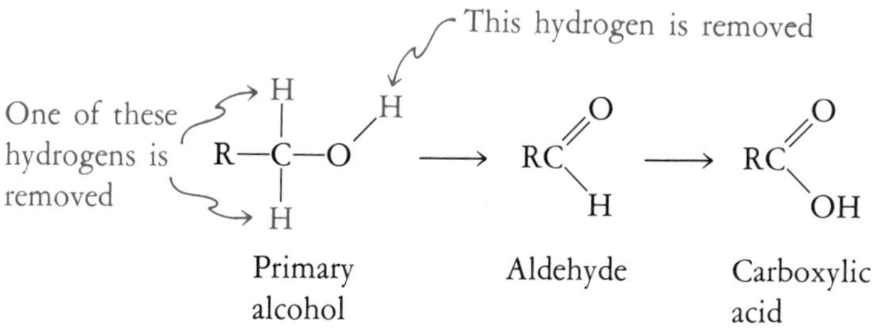

$$\underset{\text{Primary alcohol}}{\mathrm{R{-}\overset{\overset{\displaystyle H}{|}}{\underset{\underset{\displaystyle H}{|}}{C}}{-}O{-}H}} \longrightarrow \underset{\text{Aldehyde}}{\mathrm{RC({=}O)H}} \longrightarrow \underset{\text{Carboxylic acid}}{\mathrm{RC({=}O)OH}}$$

One of these hydrogens is removed (the two H atoms on carbon); This hydrogen is removed (the H on oxygen)

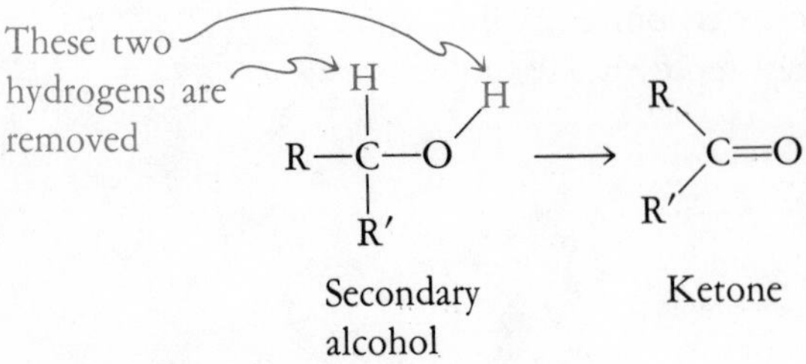

These two hydrogens are removed

$$\text{R—}\overset{\text{H}}{\underset{\text{R}'}{\text{C}}}\text{—O—H} \longrightarrow \begin{matrix}\text{R}\\ \text{R}'\end{matrix}\!\!>\text{C=O}$$

Secondary alcohol → Ketone

Notice that these reactions are the reverse of the reduction of carbonyl compounds to form alcohols (Section 16.3). For example, reduction of a ketone forms a secondary alcohol, whereas the oxidation of the same alcohol forms the ketone:

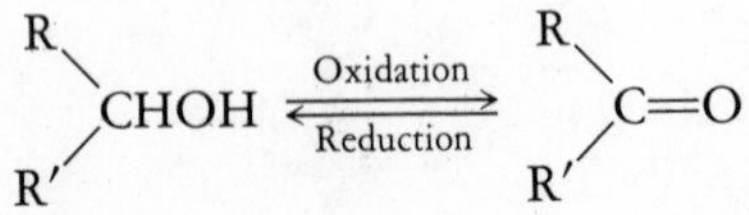

$$\begin{matrix}\text{R}\\ \text{R}'\end{matrix}\!\!>\text{CHOH} \underset{\text{Reduction}}{\overset{\text{Oxidation}}{\rightleftharpoons}} \begin{matrix}\text{R}\\ \text{R}'\end{matrix}\!\!>\text{C=O}$$

The usual reagents used in the laboratory for the oxidation of alcohols are hot acidic potassium dichromate ($K_2Cr_2O_7$) or a hot alkaline solution of potassium permanganate ($KMnO_4$). For example:

$$CH_3CH_2CH_2OH + KMnO_4 \xrightarrow[\Delta]{KOH} CH_3CH_2CO_2H$$

$$CH_3\underset{\text{OH}}{\underset{|}{C}}HCH_3 + K_2Cr_2O_7 \xrightarrow[\Delta]{H_2SO_4} CH_3\underset{\text{O}}{\underset{\|}{C}}CH_3$$

Tertiary alcohols are not easily oxidized. If the oxidizing agent is acidic, dehydration of the alcohol may occur and the resulting alkene will be oxidized.

EXERCISE 16-8 Write the structure of the product in each of the following reactions:

(a) Cyclohexanol (cyclohexane ring bearing OH) + concentrated H_2SO_4 $\longrightarrow$

(b) $CH_3CH_2OH + CH_3CO_2H \overset{H^+}{\rightleftharpoons}$

(c) $CH_3CH(OH)CH_2CH_3 + K_2Cr_2O_7 \xrightarrow[\Delta]{H^+}$

(d) $CH_3CH_2CH_2CH_2OH + KMnO_4 \xrightarrow[\Delta]{OH^-}$

Oxidation of primary and secondary alcohols in living systems occurs under milder conditions. These oxidations are catalyzed by enzymes, as we will learn in the next section.

16.5 OXIDATION OF ALCOHOLS IN LIVING SYSTEMS

The oxidation of alcohols is an important reaction in living systems. Enzymes called *dehydrogenases* catalyze these reactions. One example is the oxidation of malate to oxaloacetate, which occurs in the citric acid cycle (Section 27.5):

$$\underset{\text{Malate}}{H{-}C(CO_2^-)(CH_2CO_2^-){-}OH} \xrightarrow{\text{Dehydrogenase}} \underset{\text{Oxaloacetate}}{(CO_2^-)C(=O)(CH_2CO_2^-)}$$

This reaction is known to require the presence of nicotinamide, a derivative of pyridine (Section 13.6). Because the nicotinamide is linked to the enzyme, the enzyme is called pyridine-linked dehydrogenase. It is known that the pyridine ring accepts one of the hydrogens from the alcohol. With this knowledge, we can write a more complete equation for the oxidation of malate, as follows:

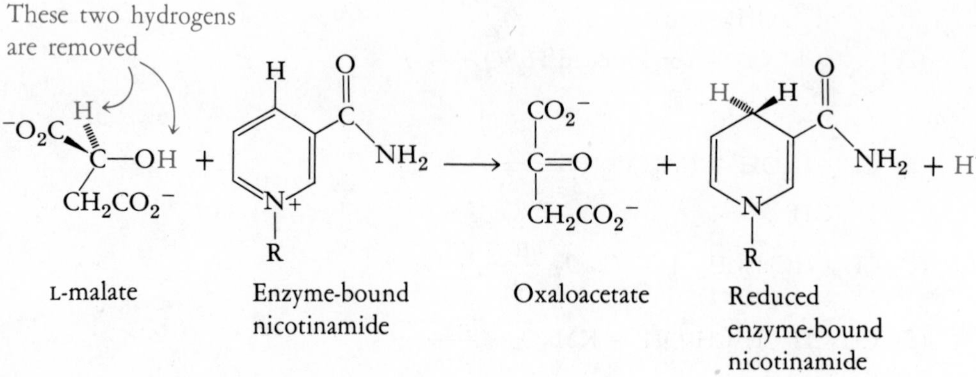

These two hydrogens are removed

L-malate + Enzyme-bound nicotinamide ⟶ Oxaloacetate + Reduced enzyme-bound nicotinamide $+ H^+$

One hydrogen is transferred to the 4 position of the pyridine ring, and the other ends up as a hydrogen ion that adds to a basic site of the enzyme.

This reaction is very stereospecific. Only L-malate ion is oxidized. The other enantiomer, D-malate ion, does not react with this enzyme. This is an example of the stereospecificity of enzymes that was first mentioned in Section 14.9. Here, a chiral reagent, the enzyme, distinguishes between chiral compounds.

16.6 PHENOLS

Phenols are compounds that contain a hydroxy group bonded to a benzene ring. The following are examples of phenols:

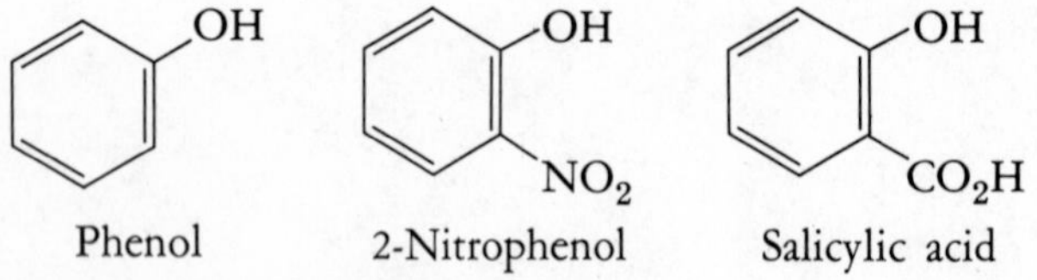

Phenol   2-Nitrophenol   Salicylic acid

Phenols undergo electrophilic aromatic substitution reactions as discussed in Section 13.3. The hydroxy group is strongly activating and *ortho-para* directing. For example, phenol itself reacts with bromine without catalyst to form 2,4,6-tribromophenol:

$$\text{Phenol} + 3\,Br_2 \longrightarrow \text{2,4,6-Tribromophenol} + 3\,HBr$$

Phenols differ from alcohols in one important way. Phenols are much stronger acids than are alcohols. This fact is evident from their acid ionization constants (see Section 10.6):

$$\text{Phenol} + H_2O \rightleftharpoons C_6H_5O^- + H_3O^+ \quad K_a = 1 \times 10^{-10}$$

$$\text{Cyclohexanol} + H_2O \rightleftharpoons C_6H_{11}O^- + H_3O^+ \quad K_a \approx 1 \times 10^{-16}$$

Thus, phenol is approximately 1 million times stronger as an acid than is cyclohexanol. In practical terms, this means that a dilute aqueous solution of phenol is acidic, whereas a solution of cyclohexanol is not. Furthermore, phenol reacts with bases such as hydroxide ion, whereas cyclohexanol does not:

$$\text{Phenol} + NaOH \longrightarrow \text{Sodium phenoxide } (C_6H_5O^-\,Na^+) + H_2O$$

$$\text{Cyclohexanol} + NaOH \longrightarrow \text{No reaction}$$

Phenol, or carbolic acid as it is sometimes called, has antiseptic properties in dilute solution. In fact, all phenolic compounds appear to have germicidal properties. Several commercial germicides contain phenols. Lysol® contains *o*-phenylphenol, and *n*-hexylresorcinol is the active ingredient in Sucrets® lozenges and several mouthwashes.

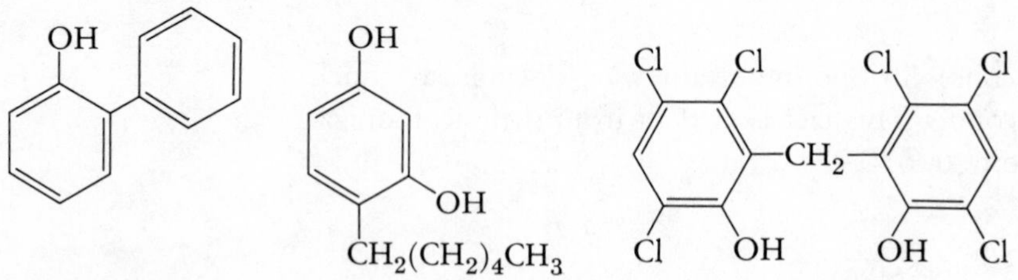

OH

OH

OH

$CH_2(CH_2)_4CH_3$

Cl Cl Cl Cl

$-CH_2-$

Cl OH OH Cl

o-Phenylphenol   *n*-Hexylresorcinol   Hexachlorophene

Hexachlorophene was used in germicidal soaps, some toothpastes, and deodorants until it was discovered to have undesirable side effects.

Aromatic 1,2- and 1,4-dihydroxy compounds are phenols that undergo an important oxidation-reduction reaction. For example, hydroquinone is easily oxidized to quinone. This reaction is reversible, because quinone is easily reduced to hydroquinone:

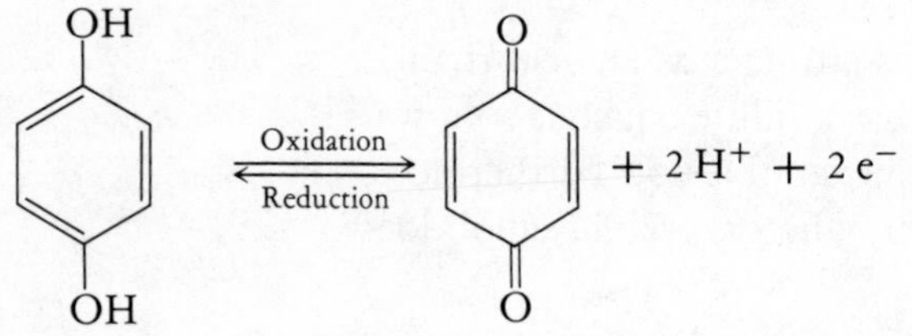

OH

O

Oxidation

Reduction

$+ 2\,H^+ + 2\,e^-$

OH

O

Hydroquinone   Quinone

One compound that can oxidize hydroquinone is the light-activated silver bromide of exposed photographic film. Consequently, the developing solution of many black and white photographic films is an alkaline solution of hydroquinone.

Hydroquinones and quinones are important in the respiratory systems of living systems. When hydroquinone is oxidized, it loses two electrons. In effect, it transfers two electrons from itself to another molecule. It is this property that makes hydroquinones important in biological systems. A number of hydroquinones are involved in transferring electrons to molecular oxygen in the respiratory system. One such compound is ubiquinone (also known as coenzyme Q):

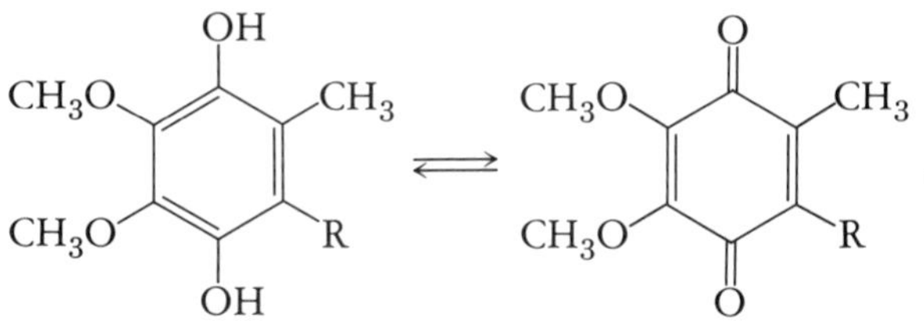

CH_3O, CH_3O, OH, OH, CH_3, R $\rightleftharpoons$ CH_3O, CH_3O, O, O, CH_3, R $+ 2\,H^+ + 2\,e^-$ $R = (CH_2CH{=}C(CH_3)CH_2)_6H$

Reduced form of ubiquinone

Oxidized form of ubiquinone

EXERCISE 16-9 Write the structure of the product in each of the following reactions:

(a) 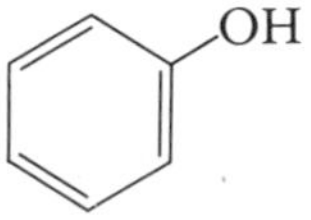 C_6H_5OH + KOH $\longrightarrow$

(b) 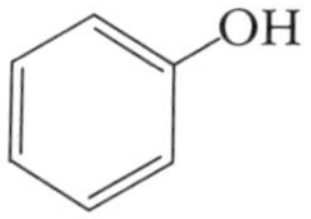 C_6H_5OH + Cl_2 $\longrightarrow$

(c) 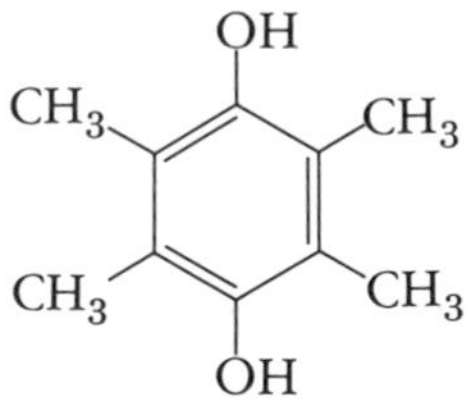 OH, OH, CH_3, CH_3, CH_3, CH_3 + Oxidizing agent $\longrightarrow$

(d) $CH_3CH_2CH_2CH_2OH$ + KOH $\longrightarrow$

(e) 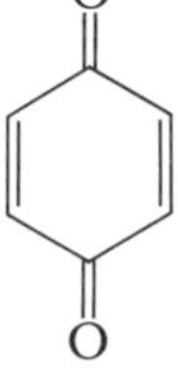 O, O + Reducing agent $\longrightarrow$

16.7 ETHERS

Ethers are compounds that contain an oxygen atom bonded to two alkyl groups, two aryl groups, or one aryl and one alkyl group. Ethers, like most other classes of organic compounds, can be named either by a common method or by IUPAC rules. The simpler ethers are usually known by their common names. By this method, the word ether is used as a root, and the names of the two groups bonded to the oxygen are prefixed alphabetically. The following examples illustrate this method:

$CH_3CH_3OCH_2CH_3$ — Diethyl ether

$C_6H_5OCH(CH_3)_2$ — Isopropyl phenyl ether

$C_6H_{11}OCH_3$ — Cyclohexyl methyl ether

Notice that the individual parts of the name are separated by a space.

To name ethers by the IUPAC rules, the more complicated group attached to the oxygen is chosen as the parent. By complicated, we mean the group that is the most branched or the most substituted or has the longest chain. The other group and the oxygen are considered as substituents on this chain. For example:

$\underbrace{CH_3O}_{\text{Substituent}}\underbrace{CH_2CH{=}CH_2}_{\text{Parent chain}}$

$\underbrace{CH_3CH_2O}_{\text{Substituent}}\underbrace{CH(CH_3)CH_2CH_3}_{\text{Parent chain}}$

The substituent is named as a prefix, and its name is constructed by replacing the *-yl* of the alkyl name or phenyl by *-oxy*. In this way, $CH_3O—$ is called methoxy; $C_6H_5O—$, phenoxy; $CH_3CH_2O—$, ethoxy, and so forth. The following examples illustrate the application of these rules:

$C_6H_5O\overset{4}{C}H_2\overset{3}{C}H(OCH_3)\overset{2}{C}H{=}\overset{1}{C}H_2$

3-Methoxy-4-phenoxy-1-butene

$CH_3CH_2O\overset{4}{C}H(\overset{5}{C}H_3)\overset{3}{C}H_2\overset{2}{C}H_2\overset{1}{C}H_2Cl$

1-Chloro-4-ethoxypentane

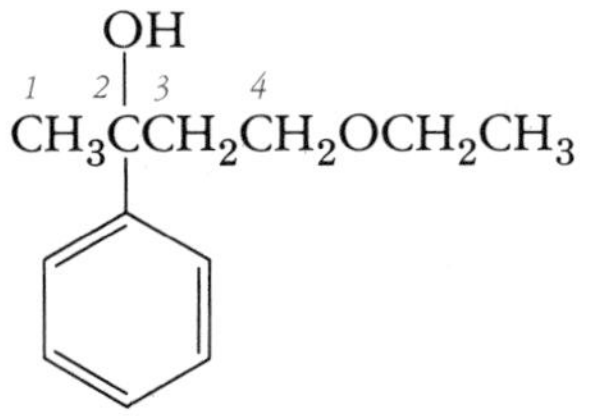

$CH_3OCH_2CH_3$

Methoxyethane

OH (on C-2), numbering 1 2 3 4

$CH_3CCH_2CH_2OCH_2CH_3$ (with phenyl on C-2)

4-Ethoxy-2-phenyl-2-butanol

The following cyclic ethers have widely used common names:

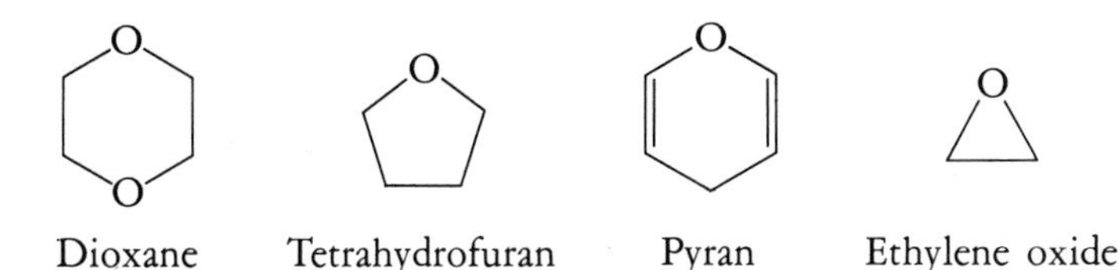

Dioxane Tetrahydrofuran Pyran Ethylene oxide

EXERCISE 16-10 Give a common name for each of the following:

(a) CH_3OCH_3 (b) $CH_3OCH_2CH_3$

(c) $CH_3CH_2OCH(CH_3)_2$ (d) 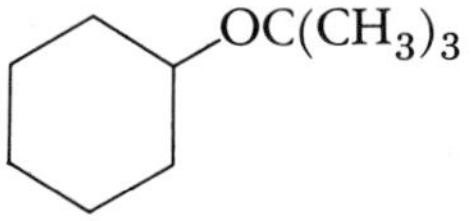 cyclohexyl–$OC(CH_3)_3$

(e) 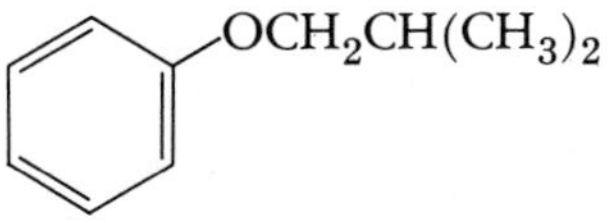 phenyl–$OCH_2CH(CH_3)_2$ (f) 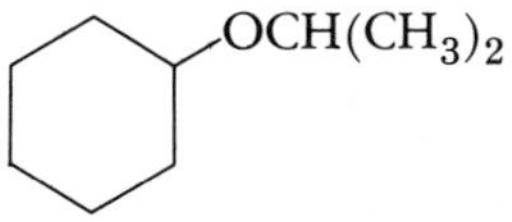 cyclohexyl–$OCH(CH_3)_2$

(g) $CH_3CH_2OCH{=}CH_2$ (h) $CH_3OCH_2CH{=}CH_2$

(i) $(CH_3)_3COC(CH_3)_3$

EXERCISE 16-11 Give the IUPAC name for each structure in exercise 16-10.

EXERCISE 16-12 Write the structure for each of the following:
(a) 2-methoxypentane (b) 1-ethoxy-1-methylcyclopentane
(c) diphenyl ether (d) 2,3-diethoxybutane
(e) 2-chloro-1-ethoxy-1-phenylpropane
(f) 4-chloro-1-propoxy-1-butyne
(g) 2-isopropoxy-2,3-dimethylbutane
(h) 2,4-dimethoxytoluene (i) cyclopropoxybenzene

Ethers are quite unreactive. For example, they do not react with bases, most oxidizing reagents, or metals such as sodium or magnesium.

$$CH_3CH_2OCH_2CH_3 \xrightarrow{NaOH,\ K_2Cr_2O_7,\ Na} \text{No reaction}$$

However, ethers do react with acids. The products of the reaction with hydroiodic acid (HI) are an alkyl iodide and an alcohol. For example:

$$\underset{\text{Diethyl ether}}{CH_3CH_2OCH_2CH_3} + \underset{\text{Hydroiodic acid}}{HI} \xrightarrow{\Delta} \underset{\text{Ethyl iodide}}{CH_3CH_2I} + \underset{\text{Ethyl alcohol}}{CH_3CH_2OH}$$

A similar reaction occurs with hydrobromic acid (HBr).

In Section 12.8 we learned how to make epoxides, which are special kinds of ethers. These compounds are special because they have a three-member ring that contains an oxygen atom. To form such a ring, the C—C—C and C—O—C bond angles must be greatly distorted from their normal values. As a result, the bonds in the ring are strained. To relieve this strain, epoxides react with many reagents to open the ring. For example:

$$HOCH_2CH_2NH_2 \xleftarrow{NH_3} \overset{O}{CH_2\text{—}CH_2} \xrightarrow{HBr} HOCH_2CH_2Br$$

$$\overset{O}{CH_2\text{—}CH_2} \xrightarrow{H^+,\ H_2O} HOCH_2CH_2OH$$

Thus, epoxides, although classified as ethers, are far more reactive than are typical ethers.

Ethers are not the most common compounds in the world. Yet most people are familiar with diethyl ether, or ether as it is commonly called. Ether has been used as a general anesthetic. It is easy to administer, has little effect on the rate of respiration, blood pressure, or pulse rate, and is an excellent muscular relaxant. However, it does have a number of disadvantages. When mixed with air, diethyl ether is highly explosive. Furthermore, it irritates the membranes of the respiratory tract and it has an aftereffect of nausea. Because these disadvantages outweigh its advantages, diethyl ether is rarely used as a general anesthetic today. It has been replaced by other compounds such as halothane (Section 15.1), which are effective as general anesthetics and are not explosive.

One example of an epoxide in nature is squalene 2,3-epoxide. This compound is formed in a series of reactions that leads to the biosynthesis of cholesterol.

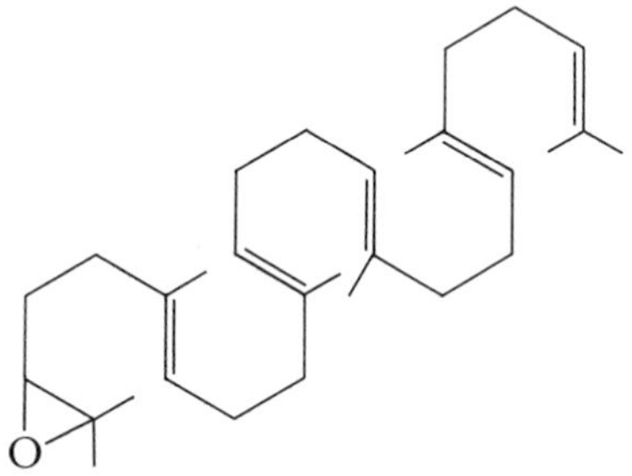

O

Squalene 2,3-epoxide

EXERCISE 16-13 Write the structure of the product in each of the following reactions:

(a) $CH_3OCH_3 + HBr \longrightarrow$ (b) $CH_2\overset{O}{\frown}CH_2 + HBr \longrightarrow$

(c) $CH_2\overset{O}{\frown}CH_2 + H_2S \xrightarrow{H^+}$

16.8 THIOLS

Thiols are the sulfur analogues of alcohols and contain the $-\overset{|}{\underset{|}{C}}-SH$ functional group. The IUPAC names of thiols are formed by adding the suffix *-thiol, -dithiol,* and so forth to the name of the parent hydrocarbon. Common names are obtained by first naming the alkyl group followed by the word *mercaptan.* It is sometimes necessary to name the functional group by the prefix *mercapto-*. The following examples illustrate these rules:

CH_3CH_2SH
Ethanethiol
(ethyl mercaptan)

$\overset{3}{C}H_3\overset{2}{C}H(SH)\overset{1}{C}H_3$
2-Propanethiol
(isopropyl mercaptan)

$HSCH_2CH_2SH$
Ethanedithiol

$\overset{3}{C}H_3\overset{2}{C}H(SH)\overset{1}{C}H_2OH$
2-Mercapto-1-propanol

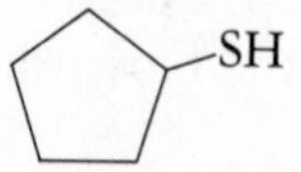

Cyclopentanethiol

Thiols are more volatile than are the corresponding alcohols and have a very disagreeable odor. Various thiols are found in nature. A number of thiols are responsible for the odor of a skunk; 1-propanethiol is released when an onion is peeled, and thiols are responsible for the odor of garlic. Thiols are highly reactive and are present in small quantities in living systems, where they play key roles.

Thiols, like alcohols, react with carboxylic acids to form compounds called *thioesters,* which have the following structure:

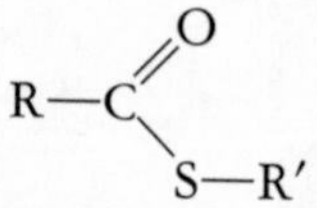

$$R{-}C({=}O){-}S{-}R'$$

Notice that the structure of these compounds resembles that of esters. The difference is that one oxygen has been replaced by a sulfur atom. The reactions of these biologically important compounds will be discussed in Section 18.6.

Many of the reactions of thiols are similar to those of alcohols; the most significant and biologically important difference between the two is the ease with which thiols are oxidized. A variety of mild oxidizing reagents such as oxygen (in the presence of metal catalysts), iodine, and hydrogen peroxide form disulfides from thiols. For example:

$$\underset{\text{Thiol}}{4\,RCH_2SH} + O_2 \xrightarrow{Fe} \underset{\text{Disulfide}}{2\,RCH_2SSCH_2R} + 2\,H_2O$$

Notice that the oxidation of a thiol occurs at the sulfur atom, whereas oxidation of an alcohol occurs at the carbon atom.

This oxidation of thiols to disulfides is important in the biological activity of proteins (Chapter 23). One of the amino acids in proteins is cysteine, which contains a thiol group. Cysteine is oxidized by atmospheric oxygen in the presence of iron salts to cystine:

$$\underset{\text{Cysteine}}{\underset{\displaystyle NH_2}{\overset{\displaystyle CO_2H}{HCCH_2SH}}} + \underset{\text{Cysteine}}{\underset{\displaystyle NH_2}{\overset{\displaystyle CO_2H}{HSCH_2CH}}} \underset{\text{Reduction}}{\overset{\text{Oxidation}}{\rightleftharpoons}} \underset{\text{Cystine}}{\underset{\displaystyle NH_2 \qquad\qquad NH_2}{\overset{\displaystyle CO_2H \qquad\qquad CO_2H}{HCCH_2S{-}SCH_2CH}}}$$

Cystine is important in protein structure because its disulfide bond serves as a covalent cross-link between two polypeptide chains. This reaction is reversible. Many mild reducing agents break the disulfide bond of cystine to form two molecules of cysteine.

The reversibility of the cysteine-cystine reaction is the basis of curling hair by a permanent. Hair is about 15 percent cystine. The hair to be curled is first treated with a mild reducing agent to break the disulfide bonds in the cystine units of the hair protein. The hair is then curled, set, and treated with a mild oxidizing agent to reform the disulfide bonds from different cysteine units.

16.9 SUMMARY

In this chapter we have studied three classes of oxygen-containing compounds—alcohols, phenols, and ethers—and the sulfur analogue of alcohols.

Alcohols contain an —OH group bonded to an alkyl group. Three important reactions of alcohols are dehydration, oxidation, and formation of esters with acids. These reactions occur in both the laboratory and in living systems. In living systems, these reactions are highly stereospecific. For example, the enzyme-catalyzed oxidation of malate ion occurs only with the L-enantiomer.

Phenols contain an —OH group bonded to an aromatic ring. Unlike alcohols, phenols are acids. Aromatic 1,2- or 1,4-dihydroxy compounds, called hydroquinones, are oxidized to quinones. This reaction, which is reversible, is important in the respiratory system.

Ethers have the general structure ROR, ROAr, or ArOAr where Ar represents an aryl group. They are quite unreactive except with hydroiodic acid (HI) and hydrobromic acid (HBr). In contrast, the three-member oxygen-containing ring of epoxides is easily opened by both acids and bases.

Thiols are the sulfur analogues of alcohols. Many of the reactions of alcohols and thiols are similar, but thiols are more easily oxidized. Their oxidation to disulfides is important in the biological activity of proteins.

REVIEW OF TERMS, CONCEPTS, AND REACTIONS

Terms

ALCOHOLS Compounds containing a hydroxy group bonded to the carbon of an alkyl group.

DEHYDROGENASE An enzyme that catalyzes the oxidation of alcohols to carbonyl compounds.

ETHERS Compounds containing an oxygen atom bonded to two aryl groups, two alkyl groups, or one of each.

HYDROXY GROUP The —OH functional group.

PHENOLS Compounds containing a hydroxy group bonded to the ring carbon of an aromatic compound.

THIOLS Compounds containing a —SH functional group bonded to the carbon of an alkyl group; the sulfur analogues of alcohols.

Concepts

1. Alcohols, like alkyl halides, are classified as primary, secondary, or tertiary, according to their structure.
2. Alcohols and phenols have higher boiling points and water solubility than other classes of compounds of similar molecular weight because of their ability to form hydrogen bonds.
3. The enzyme-catalyzed oxidation of alcohols is very stereospecific. Only one of a pair of enantiomers is oxidized.
4. Phenols are stronger acids than are alcohols.
5. Ethers are generally quite unreactive compounds. Epoxides are an exception. They undergo ring-opening reactions with acids and bases.
6. Thiols are the sulfur analogues of alcohols. They react with a variety of mild oxidizing agents to form disulfides.

Reactions

Preparation of alcohols (see Section 16.3)

1. Hydration of alkenes (see p. 414 and Section 12.4)

$$RCH{=}CH_2 + H_2O \xrightarrow{H^+} \underset{\displaystyle \overset{|}{OH}}{RCHCH_3}$$

2. Reduction of carbonyl compounds (see p. 414)

$$R{-}C\begin{matrix}{/\!\!/}O \\ \diagdown H\end{matrix} + H_2 \xrightarrow{Pt} RCH_2OH \qquad (1° \text{ alcohol})$$

$$R{-}C\begin{matrix}{/\!\!/}O \\ \diagdown R'\end{matrix} + H_2 \xrightarrow{Pt} \underset{\displaystyle \overset{|}{OH}}{RCHR'} \qquad (2° \text{ alcohol})$$

Reactions of alcohols (see Section 16.4)

1. Dehydration (see Section 16.4)

$$RCH_2CH_2OH \xrightarrow[H_2SO_4]{\text{concentrated}} RCH{=}CH_2 + H_2O$$

2. With organic acids (see p. 418 and Section 18.5)

$$ROH + R'C\begin{matrix}{/\!\!/}O \\ \diagdown OH\end{matrix} \overset{H^+}{\rightleftharpoons} R'C\begin{matrix}{/\!\!/}O \\ \diagdown OR\end{matrix} + H_2O$$

3. Oxidation (see p. 419)

$$RCH_2OH \xrightarrow[\text{Oxidation}]{} RC\begin{matrix}{/\!\!/}O \\ \diagdown H\end{matrix} \xrightarrow[\text{Oxidation}]{} RC\begin{matrix}{/\!\!/}O \\ \diagdown OH\end{matrix}$$

$$\underset{\displaystyle \overset{|}{OH}}{RCHR} \xrightarrow[\text{Oxidation}]{} R{-}\underset{\displaystyle \overset{\|}{O}}{C}{-}R$$

Preparing phenol (see Section 13.5)

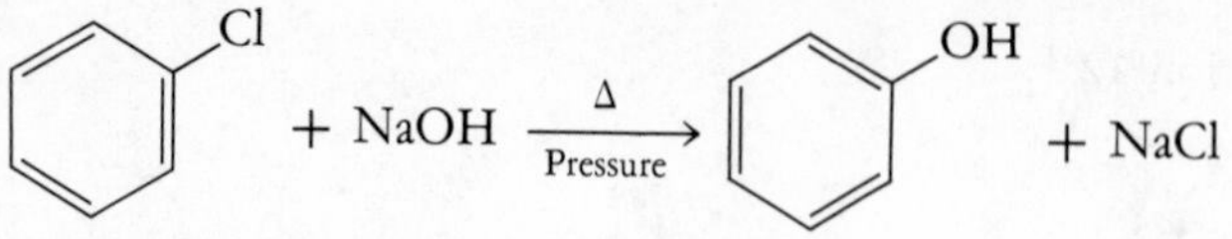

$$C_6H_5Cl + NaOH \xrightarrow[\text{Pressure}]{\Delta} C_6H_5OH + NaCl$$

Reactions of phenols (see Section 16.6)

1. Reaction with bases

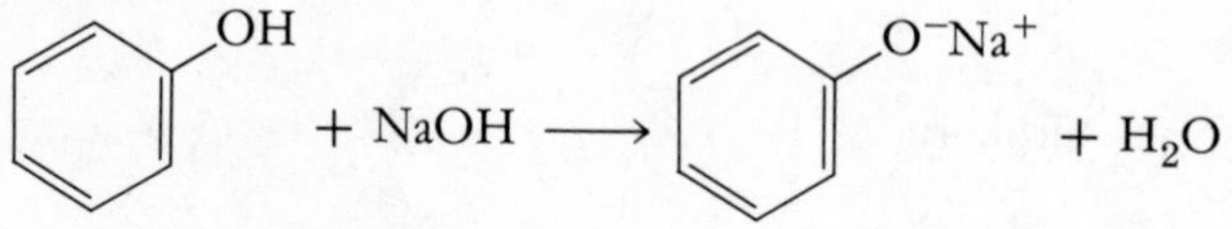

$$C_6H_5OH + NaOH \longrightarrow C_6H_5O^-Na^+ + H_2O$$

2. Aromatic substitution reactions (see Section 13.3)
3. Oxidation-reduction reactions

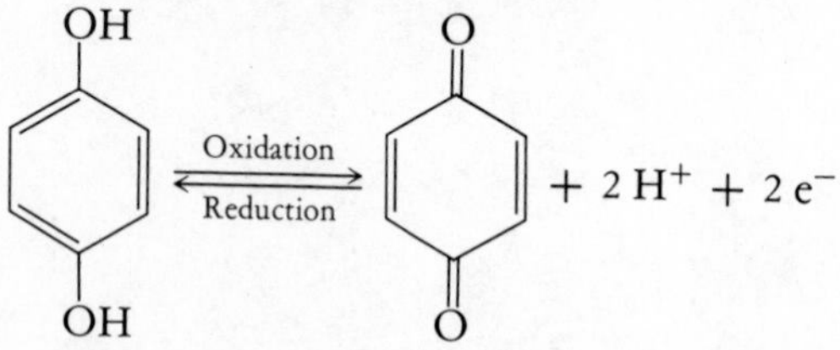

$$HO{-}C_6H_4{-}OH \underset{\text{Reduction}}{\overset{\text{Oxidation}}{\rightleftharpoons}} O{=}C_6H_4{=}O + 2\,H^+ + 2\,e^-$$

Reactions of ethers (see Section 16.7)

$$ROR + HX \longrightarrow RX + ROH$$

R = alkyl group
X = Br or I

Reactions of epoxides (see Section 16.7)

1. With acids

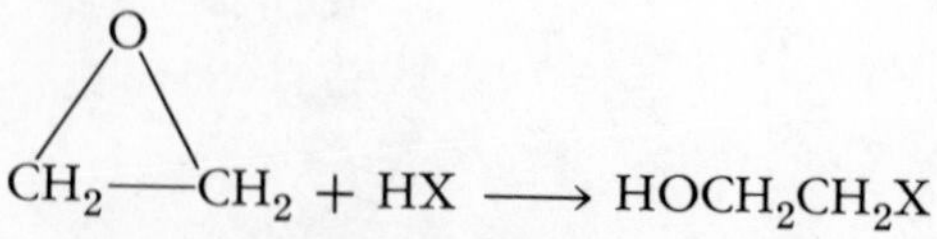

$$\underset{\text{(epoxide: O bridging } CH_2{-}CH_2)}{CH_2{-}CH_2} + HX \longrightarrow HOCH_2CH_2X$$

X = Cl, Br, or I

2. With bases

$$\overset{O}{\widehat{CH_2—CH_2}} + NH_3 \longrightarrow HOCH_2CH_2NH_2$$

Oxidation-reduction reactions of thiols (see Section 16.8)

$$4\,RSH + O_2 \underset{\text{Reduction}}{\overset{\text{Oxidation}}{\rightleftharpoons}} 2\,RSSR + 2\,H_2O$$

EXERCISES 16-14 Write the structure of a compound that is an example of each of the following:
(a) a primary alcohol (b) a cyclic secondary alcohol
(c) a tertiary alcohol (d) a phenol (e) a cyclic ether
(f) an aryl ether (g) a thiol (h) a disulfide
(i) a hydroquinone (j) a quinone

16-15 Give the IUPAC name for each of the following compounds:

(a) $CH_3CH(OH)CH(CH_3)CH_2CH_3$ (b) benzene ring bearing OH and NO_2 (c) cyclohexadiene ring bearing OH

(d) benzene ring bearing OCH_3, with Cl para to each other (e) $CH_3CH_2CH_2CH_2SH$

(f) cyclohexane ring bearing OCH_2CH_3 (g) phenyl–$C(OH)(CH_3)CH_2CH_2CH_2CH_3$

(h) $CH_3CH_2OC(CH_3)_2CH{=}CH_2$ (i) benzene ring bearing OH, $OCH(CH_3)_2$, and $CH_2CH_2CH_2CH_3$

16-16 Write the structure for each of the following compounds:
(a) 2-octanol (b) 1,3-propanediol (c) butyl ethyl ether
(d) 2-ethyl-1-pentanol (e) 3-ethyl-2-methoxypentane
(f) 2-methyl-2-butanethiol (g) 1,2-diethoxyethane
(h) 4-chloro-1-pentanol (i) 4-methoxy-2-methyl-2-butanol
(j) 1-cyclohexylethanol (k) 3,3-dimethyl-2-butanol

16-17 Write the structure of the product for each of the following reactions:
(a) methylpropene + $H_2O \xrightarrow{H^+}$
(b) cyclopentanol + concentrated $H_2SO_4 \longrightarrow$
(c) 4-nitrophenol + NaOH $\longrightarrow$
(d) diisopropyl ether + HI $\longrightarrow$
(e) ethanethiol + $O_2 \xrightarrow{Fe}$
(f) oxirane + HI $\longrightarrow$

16-18 Identify each of the following as an oxidation or a reduction reaction:

(a) $CH_3C(=O)CH_3 \longrightarrow CH_3CH(OH)CH_3$ (b) $CH_3SSCH_3 \longrightarrow 2\ CH_3SH$

(c) 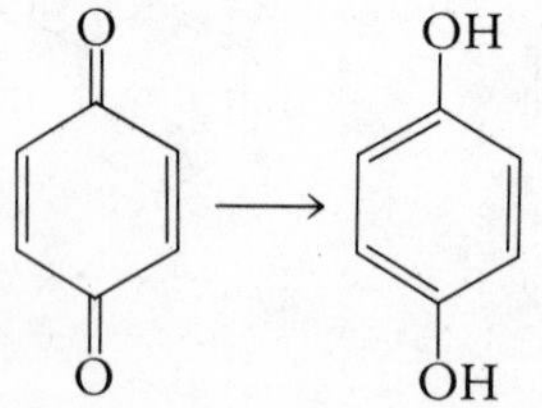 1,4-benzoquinone (ring with O at top and O at bottom) $\longrightarrow$ benzene ring with OH at top and OH at bottom

(d) $CH_3CH_2OH \longrightarrow CH_3C(=O)OH$

(e) 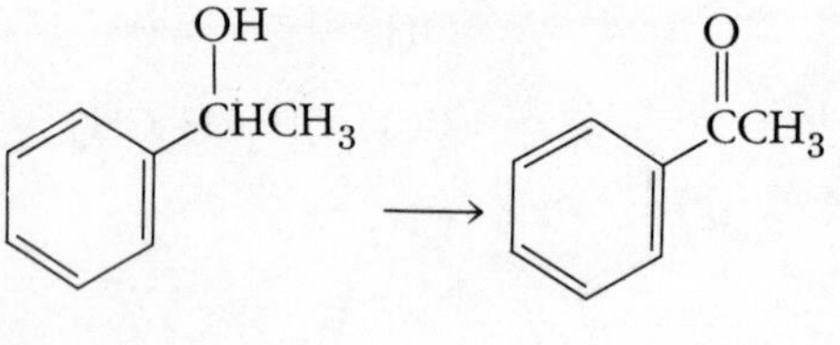 benzene ring–CH(OH)CH$_3$ $\longrightarrow$ benzene ring–C(=O)CH$_3$

AMINES

17

Amines are organic derivatives of ammonia in which one, two, or all three hydrogens of ammonia are replaced by alkyl or aryl groups. The following are examples of amines:

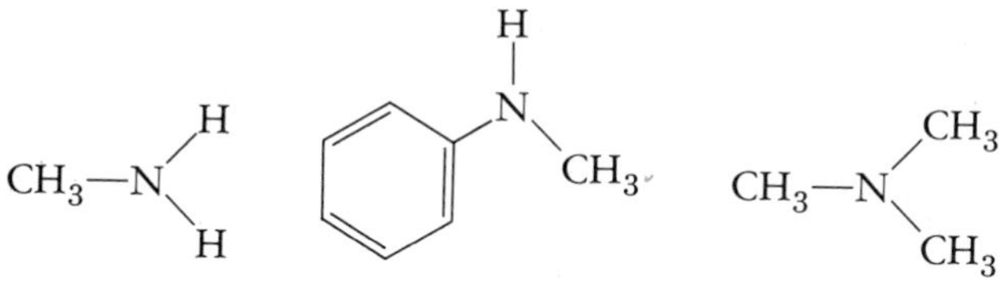

Methylamine: $CH_3{-}NH_2$ | *N*-methylaniline: $C_6H_5{-}NH{-}CH_3$ | Trimethylamine: $(CH_3)_3N$

In this chapter we will study the biologically important reactions of amines. As in previous chapters, let us start by classifying and naming amines.

17.1 CLASSIFYING AND NAMING AMINES

Amines can be classified as primary, secondary, or tertiary, depending on whether one, two, or three hydrogen atoms of ammonia have been replaced by organic groups. Like the nitrogen atom of ammonia, the nitrogen atom of an amine possesses an unshared pair of electrons:

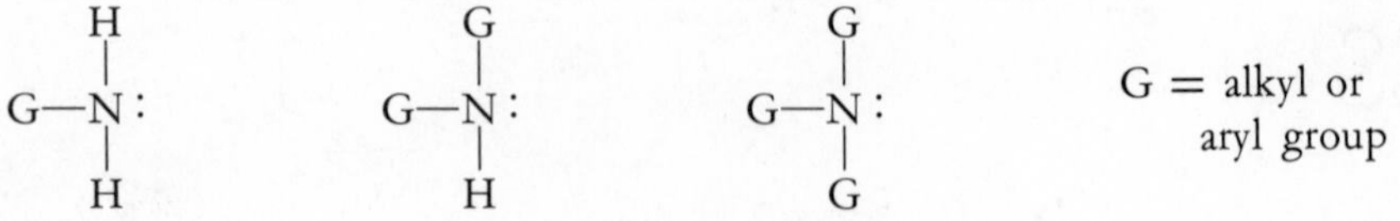

$$\begin{array}{c} H \\ | \\ G{-}\ddot{N}{:} \\ | \\ H \end{array} \qquad \begin{array}{c} G \\ | \\ G{-}N{:} \\ | \\ H \end{array} \qquad \begin{array}{c} G \\ | \\ G{-}N{:} \\ | \\ G \end{array}$$

Primary amine Secondary amine Tertiary amine

G = alkyl or aryl group

The groups bonded to the nitrogen in secondary and tertiary amines may be identical or different from one another. In some examples the groups bonded to the nitrogen are joined together to form a ring. For example:

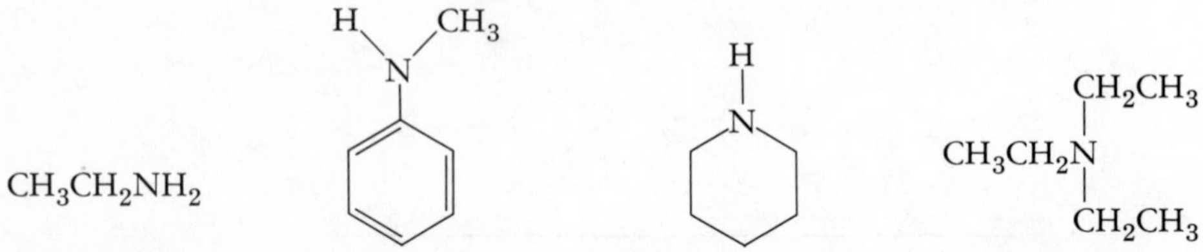

$CH_3CH_2NH_2$

Ethylamine (primary amine)

N-methylaniline (secondary amine)

Piperidine (secondary amine)

$(CH_3CH_2)_3N$

Triethylamine (tertiary amine)

Be careful not to confuse primary, secondary, and tertiary carbon atoms with primary, secondary, and tertiary amines. This distinction is clearly shown by comparing the structures of *t*-butylamine and trimethylamine:

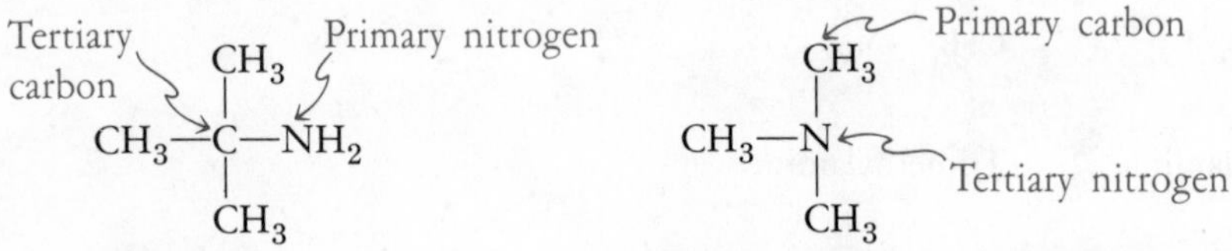

Tertiary carbon — Primary nitrogen

$$\begin{array}{c} CH_3 \\ | \\ CH_3{-}C{-}NH_2 \\ | \\ CH_3 \end{array}$$

t-Butylamine (primary amine)

Primary carbon — Tertiary nitrogen

$$\begin{array}{c} CH_3 \\ | \\ CH_3{-}N \\ | \\ CH_3 \end{array}$$

Trimethylamine (tertiary amine)

As in the case of alcohols and alkyl halides, common names are often used for simple primary aliphatic amines. The name is obtained by placing the name of the attached alkyl groups as a prefix to the word *amine.* The following primary amines are named in this way:

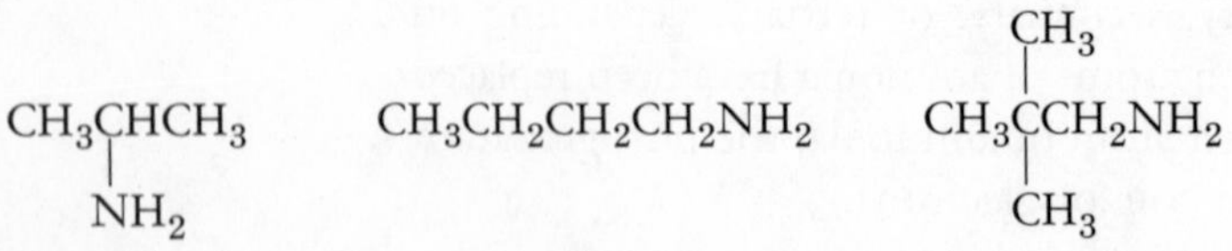

$$\begin{array}{c} CH_3CHCH_3 \\ | \\ NH_2 \end{array}$$

Isopropylamine

$CH_3CH_2CH_2CH_2NH_2$

Butylamine

$$\begin{array}{c} CH_3 \\ | \\ CH_3CCH_2NH_2 \\ | \\ CH_3 \end{array}$$

Neopentylamine

The prefixes *di-* and *tri-* are used for secondary and tertiary amines containing identical substituents. For example:

$(CH_3)_2NH$ — Dimethylamine

$(CH_3CH_2CH_2CH_2)_3N$ — Tributylamine

Notice that the name is *one* word.

The rules for obtaining the systematic name of primary amines are similar to those for alcohols (Section 16.1). The ending *-e* of the name of the longest hydrocarbon chain is changed to *-amine*. Some examples are as follows:

$\overset{3}{C}H_3\overset{2}{C}H_2\overset{1}{C}H_2NH_2$

1-Propanamine (common name is propylamine)

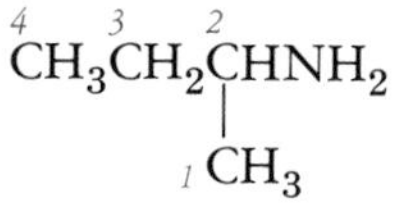

$\overset{4}{C}H_3\overset{3}{C}H_2\overset{2}{C}HNH_2$, with ${}_1CH_3$ on C-2

2-Butanamine (common name is *sec*-butylamine)

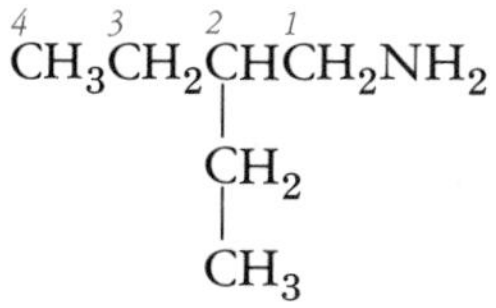

$\overset{4}{C}H_3\overset{3}{C}H_2\overset{2}{C}H\overset{1}{C}H_2NH_2$, with CH_2CH_3 on C-2

2-Ethyl-1-butanamine

Secondary and tertiary amines are named as *N*-substituted amines. The *N* is included to indicate that the substituent is on the nitrogen atom. In general, the group that is the most branched or the most substituted or has the longest chain is chosen as the parent chain. For example:

$\overset{3}{C}H_3\overset{2}{C}H_2\overset{1}{C}H_2NHCH_3$

N-methyl-1-propanamine (common name is *N*-methylpropylamine)

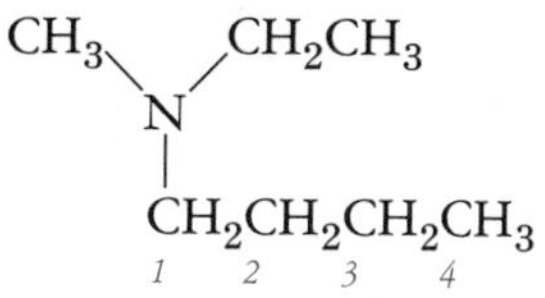

$(CH_3)(CH_3CH_2)N\overset{1}{C}H_2\overset{2}{C}H_2\overset{3}{C}H_2\overset{4}{C}H_3$

N-ethyl-*N*-methyl-1-butanamine (common name is *N*-ethyl-*N*-methylbutylamine)

Aromatic amines are usually named as derivatives of *aniline*. For example:

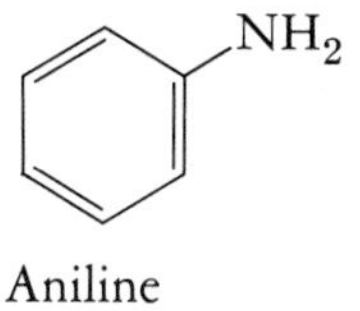

$C_6H_5NH_2$

Aniline (primary amine)

$C_6H_5NHCH_2CH_3$

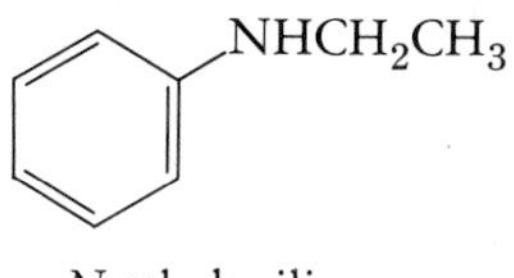

N-ethylaniline (secondary amine)

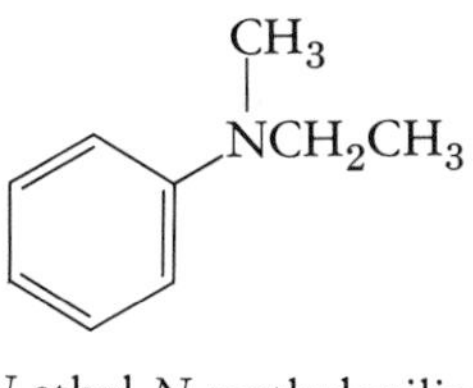

$C_6H_5N(CH_3)CH_2CH_3$

N-ethyl-*N*-methylaniline (tertiary amine)

*EXERCISE 17-1 Give a common name for each of the following compounds:

(a) $(CH_3)_3CNH_2$ (b) $(CH_3CH_2)_2NH$

(c) $CH_3CH_2CH_2NHCH_3$ (d) $(CH_3)_3N$

(e) 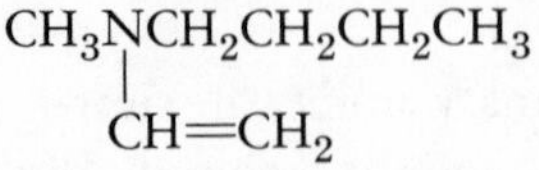 $CH_3NCH_2CH_2CH_2CH_3$ with $CH{=}CH_2$ on N (f) 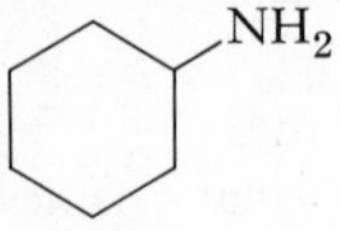 cyclohexane ring bearing NH_2

(g) 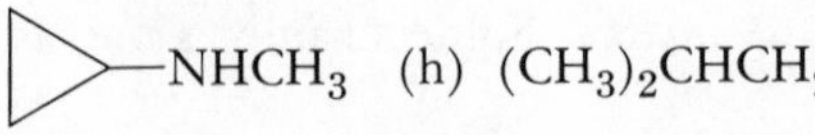 cyclopropane ring bearing $-NHCH_3$ (h) $(CH_3)_2CHCH_2NH_2$

(i) $CH_2{=}CHCH_2NH_2$ (j) 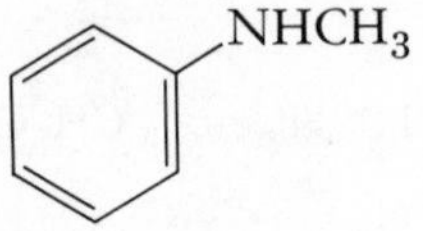 benzene ring bearing $NHCH_3$

(k) 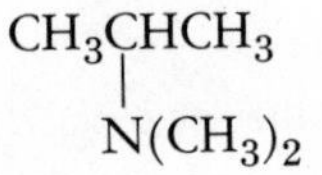 CH_3CHCH_3 with $N(CH_3)_2$ on the middle carbon (l) 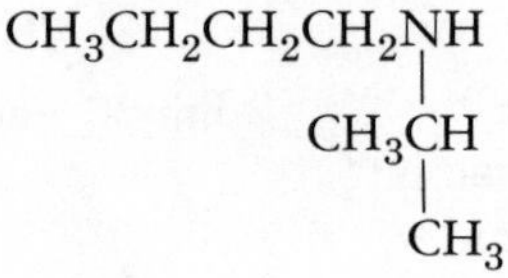 $CH_3CH_2CH_2CH_2NH$ with CH_3CH on N, and CH_3 on that CH

EXERCISE 17-2 Classify the amines in exercise 17-1 as 1°, 2°, or 3° amines.

EXERCISE 17-3 Give the IUPAC name for each compound in exercise 17-1.

EXERCISE 17-4 Write the structure for each of the following compounds:
(a) 1-butanamine (b) *N*-ethyl-3-heptanamine
(c) tri-*t*-butylamine (d) *N,N*-diethylaniline
(e) *N*-propyl-1-hexanamine

Now that we know how to name amines, let us turn our attention to their physical properties.

17.2 PHYSICAL PROPERTIES

The low-molecular-weight aliphatic amines resemble ammonia in many of their physical properties. They are gases that are readily soluble in water to give basic solutions, and their odors are very similar to that of ammonia.

* The answers for the exercises in this chapter begin on page 878.

Table 17-1. Physical Properties of Ammonia and Representative Amines

Name	Structure	Melting Point (°C)	Boiling Point (°C)
Ammonia	NH_3	−78	−33
Methylamine	CH_3NH_2	−94	−6
Ethylamine	$C_2H_5NH_2$	−81	17
Propylamine	*n*-$C_3H_7NH_2$	−83	49
Butylamine	*n*-$C_4H_9NH_2$	−49	77
t-Butylamine	*t*-$C_4H_9NH_2$	−67	44
Dimethylamine	$(CH_3)_2NH$	−93	7
Trimethylamine	$(CH_3)_3N$	−117	3
Aniline	C_6H_5—NH_2 (benzene ring—NH_2)	−6	184

The higher alkylamines have less pungent, more fishlike odors. Aromatic amines are generally very toxic.

The boiling points of the amines, like that of ammonia, are higher than those of nonpolar compounds of the same molecular weight. The reason for this is that amines, like ammonia and water, can form intermolecular hydrogen bonds (see Section 6.3). This also accounts for the high water solubility of amines containing fewer than six carbon atoms. The hydrogen bond N—H••••N formed by an amine is not as strong as that formed by an alcohol or a carboxylic acid group. Consequently, the boiling points of amines are lower than those of alcohols or carboxylic acids of similar molecular weight. For example, methyl alcohol has a boiling point (65° C) higher than that of methylamine (−6° C). Some of the properties of representative amines are listed in Table 17-1.

17.3 SUBSTITUTED AMMONIUM IONS

There are many compounds in which four atoms or groups are bonded to nitrogen. These compounds are actually salts. They contain an anion such as a halide ion and a positively charged nitrogen-containing cation. Let us focus our attention on these cations. Such a cation can be regarded as an ammonium ion in which one, two, three, or all four hydrogens are replaced by any combination of alkyl or aryl groups. The structural types of these ions are as follows:

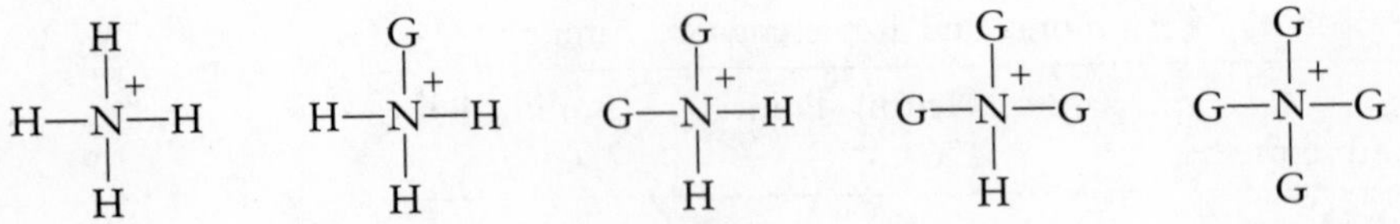

$$\underset{\text{Ammonium ion}}{H-\overset{\overset{H}{|}}{\underset{\underset{H}{|}}{N^+}}-H} \quad \underset{\text{Primary ammonium ion}}{H-\overset{\overset{G}{|}}{\underset{\underset{H}{|}}{N^+}}-H} \quad \underset{\text{Secondary ammonium ion}}{G-\overset{\overset{G}{|}}{\underset{\underset{H}{|}}{N^+}}-H} \quad \underset{\text{Tertiary ammonium ion}}{G-\overset{\overset{G}{|}}{\underset{\underset{H}{|}}{N^+}}-G} \quad \underset{\text{Quaternary ammonium ion}}{G-\overset{\overset{G}{|}}{\underset{\underset{G}{|}}{N^+}}-G}$$

G = an alkyl or aryl group

They are named as substituted ammonium ions. For example:

$CH_3\overset{+}{N}H_3\ Br^-$ Methylammonium bromide

$CH_3\overset{+}{N}H_2C_2H_5\ Cl^-$ Ethylmethylammonium chloride

$(CH_3)_4\overset{+}{N}\ I^-$ Tetramethylammonium iodide

If one of the groups bonded to nitrogen is a benzene ring, the compound is named as a substituted anilinium ion. For example:

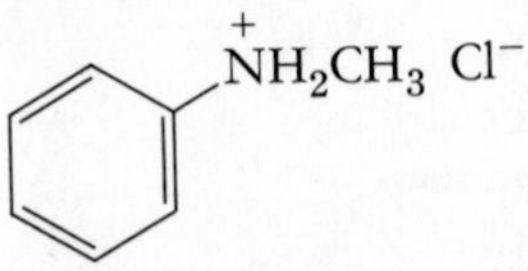

$C_6H_5\overset{+}{N}H_2CH_3\ Cl^-$ *N*-methylanilinium chloride

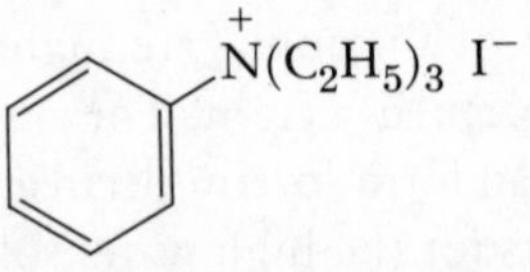

$C_6H_5\overset{+}{N}(C_2H_5)_3\ I^-$ *N,N,N*-triethylanilinium iodide

EXERCISE 17-5 Give a name to each of the following compounds:

(a) $CH_3CH_2CH_2CH_2\overset{+}{N}H_3\ Br^-$ (b) 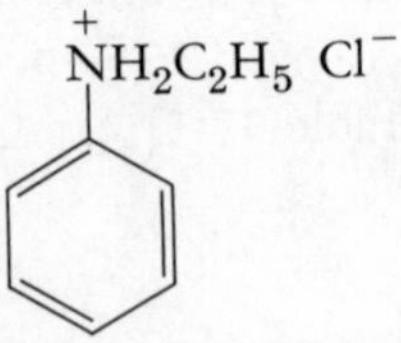 $C_6H_5\overset{+}{N}H_2C_2H_5\ Cl^-$

(c) $(CH_3CH_2CH_2)_4\overset{+}{N}\ I^-$ (d) $CH_3\overset{+}{N}H_2CH(CH_3)_2\ Br^-$

Quaternary ammonium compounds are important industrially and biologically. Compounds in which one alkyl group is a long saturated carbon chain, such as hexadecyltrimethylammonium chloride, are used as detergents and germicides.

$$CH_3(CH_2)_{14}CH_2\overset{+}{N}(CH_3)_3 \quad Cl^-$$

Hexadecyltrimethylammonium chloride

A biologically important quaternary ammonium compound is acetylcholine, which is present in nerve cells and is involved in the transmission of nerve impulses in the body (see Section 25.8).

$$CH_3\overset{O}{\overset{\|}{C}}OCH_2CH_2\overset{+}{N}(CH_3)_3 \quad OH^-$$

Acetylcholine

All primary, secondary, and tertiary ammonium ions are acids that readily lose a proton to a strong base such as hydroxide ion. For example:

$$\underset{\substack{\text{acid}\\ \text{Dimethyl}\\ \text{ammonium ion}}}{(CH_3)_2\overset{+}{N}H_2} + \underset{\text{base}}{OH^-} \longrightarrow \underset{\substack{\text{base}\\ \text{Dimethyl}\\ \text{amine}}}{(CH_3)_2NH} + \underset{\text{acid}}{H_2O}$$

This is another example of an acid-base reaction that occurs to form the weaker acid and base (see Section 10.5).

Quaternary ammonium ions do not have a hydrogen bonded to nitrogen, so they are not acids and they do not undergo this reaction. Instead, when heated with a strong base, they undergo the following elimination reaction:

$$R\underset{\substack{|\\ H}}{C}HCH_2{-}\overset{+}{N}(CH_3)_3 \xrightarrow[OH^-]{\Delta} RCH{=}CH_2 + N(CH_3)_3 + HOH$$

This reaction is similar to the elimination reactions of secondary and tertiary halides (see Section 15.6).

In living systems, certain compounds eliminate ammonia by a similar reaction. The following are two examples:

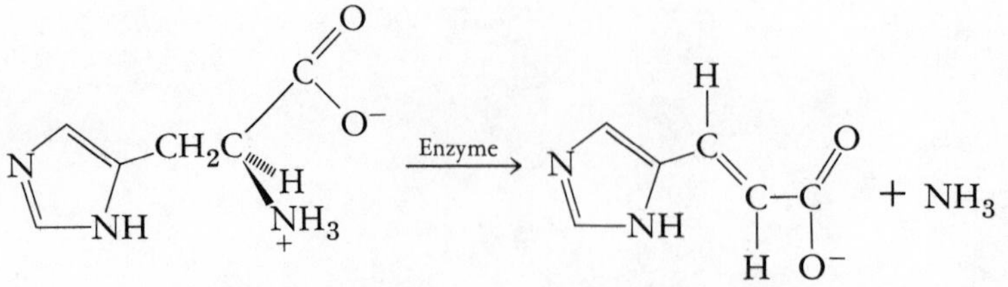

$$\text{L-histidine} \xrightarrow{\text{Enzyme}} \text{Urocanate} + NH_3$$

L-histidine   Urocanate

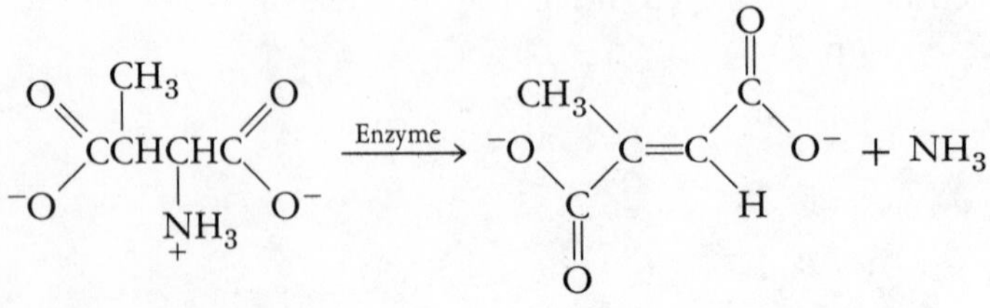

$$\beta\text{-Methylaspartate} \xrightarrow{\text{Enzyme}} \text{Mesaconate} + NH_3$$

β-Methylaspartate   Mesaconate

In these reactions, only the *trans* isomer of the alkene is formed. These examples again illustrate the stereospecificity of enzyme-catalyzed reactions.

EXERCISE 17-6 Write the structure of the product of each of the following reactions:

(a) 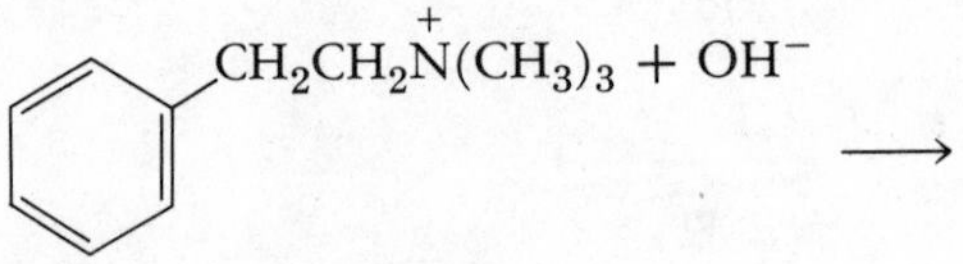 $C_6H_5CH_2CH_2\overset{+}{N}(CH_3)_3 + OH^- \longrightarrow$

(b) 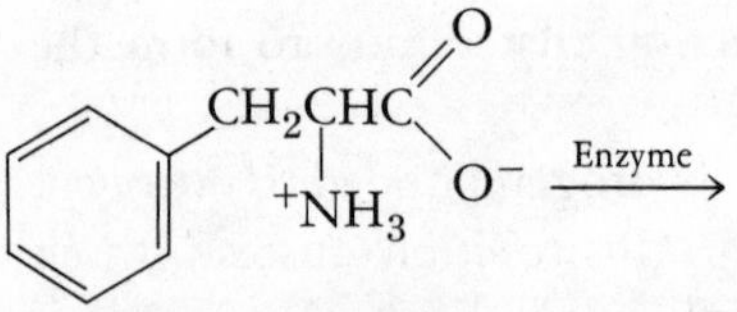 $C_6H_5CH_2CH(\overset{+}{N}H_3)CO_2^- \xrightarrow{\text{Enzyme}}$

(c) $(CH_3CH_2CH_2CH_2)_4\overset{+}{N} + OH^- \longrightarrow$

In living systems there are many more reactions that make amines than reactions that eliminate ammonia. Let us now learn the types of reactions that living systems use to make amines.

17.4 PREPARING AMINES

Amines can be prepared in living systems by a nucleophilic substitution reaction. The important features of this reaction are well illustrated by a similar laboratory method of preparing amines, the reaction of ammonia or one of its derivatives with an alkyl halide. We will consider two such reactions, the alkylation of ammonia and amines and the Gabriel synthesis.

Alkylation of Ammonia and Amines

Ammonia and amines react with alkyl halides to form alkylammonium halides:

$$H_3N\colon + \underset{\text{Alkyl halide}}{RX} \longrightarrow \underset{\text{Alkyl ammonium halide}}{H_3\overset{+}{N}R\ X^-} \qquad (17\text{-}1)$$

This is another example of a nucleophilic displacement reaction of alkyl halides discussed in Section 15.5. As would be expected, primary alkyl halides react to form more product of substitution than do secondary alkyl halides, and tertiary alkyl halides form mostly products of elimination.

Unfortunately, the reaction is more complicated than is indicated by equation 17-1. The complication is caused by the fact that some of the unreacted ammonia can react with the product alkylammonium ion as well as with the alkyl halide. This reaction, shown in equation 17-2, is a transfer of a proton to ammonia to form the ammonium ion and an alkylamine:

$$H_3N + R\overset{+}{N}H_3 \longrightarrow \overset{+}{N}H_4 + RNH_2 \qquad (17\text{-}2)$$

Once formed, this alkylamine can undergo further alkylation to yield dialkylamine, trialkylamine, and tetraalkylammonium halide. The reaction occurs in stages so that the major product formed can be controlled by choosing carefully the relative concentrations of the two reactants. If a large excess of ammonia (say, 30 to 40 times) is used, the primary amine, RNH_2, will be the major product. If one or more moles of alkyl halide are used per mole of ammonia, the quaternary ammonium halide will be the major product.

Gabriel Synthesis

A method of preparing primary amines that eliminates many of the problems encountered in the alkylation of ammonia is the Gabriel synthesis.

This reaction is similar to the alkylation reaction in that it is also a substitution reaction. But it is different because it makes use of the fact that placing a negative charge on nitrogen increases its nucleophilicity. The nucleophile is the phthalimide ion, which is formed by the reaction of phthalimide with a strong base such as hydroxide ion:

$$\text{Phthalimide (C}_6\text{H}_4(\text{C=O})_2\text{NH)} + OH^- \longrightarrow \text{Phthalimide ion (C}_6\text{H}_4(\text{C=O})_2\text{N}^-) + H_2O$$

Phthalimide   Phthalimide ion

Reaction of this nucleophile with an alkyl halide results in the substitution of the halide by the phthalimide ion to form a substituted phthalimide. This is a typical nucleophilic substitution reaction. Best results are obtained with primary halides, as we learned in Section 15.5. The substituted phthalimide can be hydrolyzed by aqueous hydrochloric acid to form the alkylammonium halide and phthalic acid:

$$\text{C}_6\text{H}_4(\text{C=O})_2\text{N}^- + CH_3Br \longrightarrow \text{C}_6\text{H}_4(\text{C=O})_2\text{NCH}_3 + Br^-$$

Phthalimide ion   *N*-methylphthalimide

$$\text{C}_6\text{H}_4(\text{C=O})_2\text{NCH}_3 + HCl \xrightarrow{H_2O} \text{C}_6\text{H}_4(CO_2H)_2 + CH_3\overset{+}{N}H_3\ Cl^-$$

N-methylphthalimide   Phthalic acid   Methylammonium chloride

$$CH_3\overset{+}{N}H_3\ Cl^- + OH^- \longrightarrow CH_3NH_2 + H_2O + Cl^-$$

Methylammonium chloride   Methylamine

The free amine can be liberated by treatment with a base such as hydroxide ion. The Gabriel synthesis is a specific method for preparing primary amines. Although this method appears to be very complicated, it is relatively simple in the laboratory and yields a very pure product.

EXERCISE 17-7 Write the structure of the product of each of the following reactions:

(a) NH_3 + excess $CH_3I \longrightarrow$
(b) $CH_3CH_2CH_2CH_2NH_2$ + large excess of $CH_3I \longrightarrow$
(c) $(CH_3)_3N + CH_3CH_2I \longrightarrow$

EXERCISE 17-8 Write the equations for the preparation of (a) ethylamine and (b) cyclohexylamine by means of the Gabriel synthesis.

17.5 PREPARING AMINES IN LIVING SYSTEMS

The preparation of amines by a nucleophilic displacement reaction involving ammonia or its derivatives also occurs in nature. Such reactions in living systems do not involve alkyl halides for the reasons given in Chapter 15. Rather, the molecules contain phosphate or pyrophosphate as leaving groups. One such example of the preparation of an amine is the biosynthesis of the purine ribonucleotides. This particular step is the enzyme-catalyzed reaction of glutamine with 5-phospho-α-D-ribose-1-pyrophosphate to form 5-phospho-β-D-ribosylamine:

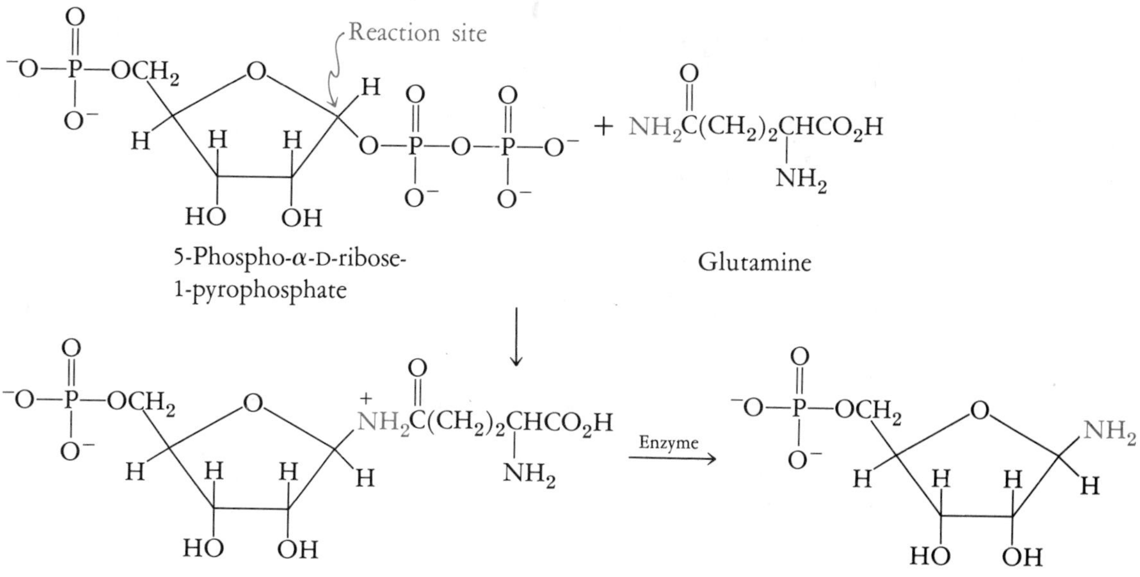

Reaction site

$NH_2\overset{O}{\overset{\|}{C}}(CH_2)_2\underset{NH_2}{CHCO_2H}$

5-Phospho-α-D-ribose-1-pyrophosphate + Glutamine

$\downarrow$

17-1 $\xrightarrow{\text{Enzyme}}$ 5-Phospho-β-D-ribosylamine

In this biological nucleophilic displacement reaction, the pyrophosphate group is the leaving group, and the amide nitrogen of glutamine is the nucleophile. The displacement occurs at the site indicated to form compound *17-1,* which is rapidly hydrolyzed to the product.

Although this reaction occurs between two rather complicated molecules, it is still an example of a nucleophilic displacement reaction that we learned in Chapter 15. Now that we know how amines are prepared, let us learn some of their reactions.

17.6 REACTIONS OF AMINES

Amines, like ammonia, have an unshared pair of electrons on the nitrogen atom. It is these electrons that are responsible for their two major reactions: basicity and nucleophilicity.

Basicity of Amines

We learned in Section 10.5 that, according to the Brønsted definition, a base is any compound or ion that accepts a proton. Ammonia and all three classes of amines are bases according to this definition because they react with protons, as shown by the following equations:

$$NH_3 + H_3O^+ \rightleftharpoons \overset{+}{N}H_4 + H_2O$$

$$RNH_2 + H_3O^+ \rightleftharpoons R\overset{+}{N}H_3 + H_2O$$

$$R_2NH + H_3O^+ \rightleftharpoons R_2\overset{+}{N}H_2 + H_2O$$

$$R_3N + H_3O^+ \rightleftharpoons R_3\overset{+}{N}H + H_2O$$

In each of these equations, the ammonium ion and the amine are a conjugate acid-base pair (see Section 10.5). Amines are bases, and the corresponding ammonium ions are their conjugate acids.

EXERCISE 17-9 Write the structure of the conjugate acid or base of each of the following:
(a) H_3O^+ (b) $CH_3CO_2^-$ (c) OH^- (d) Cl^-
(e) HNO_3 (f) $CH_3\overset{+}{N}H_3$ (g) HCl (h) C_6H_5OH
(i) SO_4^{-2} (j) CF_3CO_2H (k) $CH_3\overset{\overset{O}{\|}}{C}CH_3$

EXERCISE 17-10 Identify the conjugate acid-base pairs in the following equations:

(a) $HI + H_2O \longrightarrow H_3O^+ + I^-$

(b) $H_2O + NH_2^- \longrightarrow NH_3 + OH^-$

(c) $HNO_3 + CH_3NH_2 \longrightarrow CH_3\overset{+}{N}H_3 + NO_3^-$

(d) $CH_3CHO + OH^- \longrightarrow \overset{-}{C}H_2CHO + H_2O$

In Section 10.6, we expressed the ionization constants of bases in terms of K_b. However, it is now common in the biological sciences to *express the strength of a base in terms of the ionization constant, K_a, of its conjugate acid.* For amines this means we compare the ability of a substituted ammonium ion to donate a proton to water according to the following equation:

$$R_2\overset{+}{N}H_2 + H_2O \rightleftharpoons R_2NH + H_3O^+$$

Conjugate acid of the amine (labels $R_2\overset{+}{N}H_2$); Amine (labels R_2NH)

The ionization constant K_a is defined as follows:

$$K_a = \frac{[R_2NH][H_3O^+]}{[R_2\overset{+}{N}H_2]}$$

The K_a and pK_a for the conjugate acids of ammonia and several of its derivatives are given in Table 17-2.

In Table 17-2, the weakest acid is found in the upper left corner. Because a weak acid has a strong conjugate base, the strongest base is found in the upper right corner. In terms of the numerical value of K_a, this means that the smaller the value of K_a, the weaker the acid. For bases we must examine the K_a of the conjugate acid. The smaller the K_a of its conjugate acid, the stronger the base.

To illustrate this reasoning, let us arrange the following bases in order of increasing strength:

NH_3, CH_3NH_2, pyridine (C_5H_5N), aniline (C_6H_5—NH_2)

The K_a values of their conjugate acids are:

$\overset{+}{N}H_4$, 6.3×10^{-10}; $CH_3\overset{+}{N}H_3$, 2.3×10^{-11}; $C_5H_5\overset{+}{N}H$, 5.0×10^{-6}; $C_6H_5\overset{+}{N}H_3$, 2.5×10^{-5}

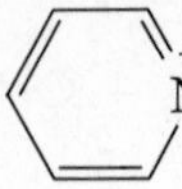

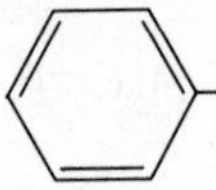

Therefore, the relative strengths of the bases are as follows:

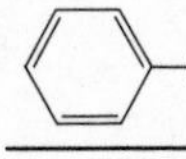

$C_6H_5NH_2 < C_5H_5N < NH_3 < CH_3NH_2$

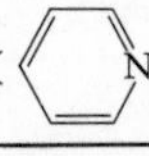

Increasing base strength →

At first, this method of expressing base strengths may seem a bit confusing. After a bit of experience, however, you should have no trouble using this method.

EXERCISE 17-11 Using the values of K_a in Table 17-2, arrange the following acids in order of increasing strength:

$p\text{-}CH_3C_6H_4\overset{+}{N}H_3$ $\quad \overset{+}{N}H_4 \quad CH_3C(=O)\overset{+}{N}H_3 \quad NH_3$

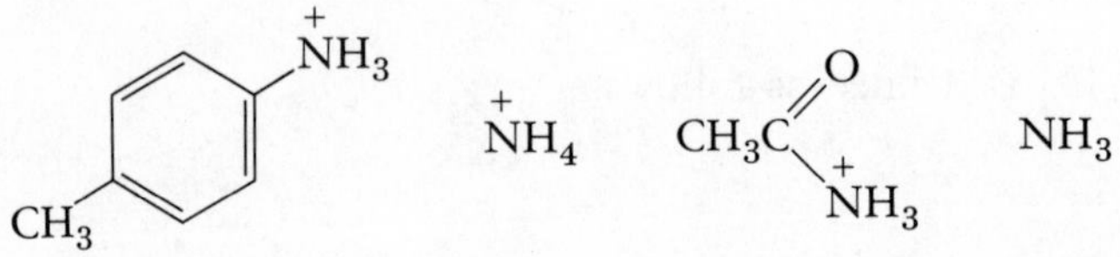

EXERCISE 17-12 Using the values of K_a in Table 17-2, arrange the following bases in order of increasing strength:

$C_5H_5N \quad C_6H_5NH_2 \quad NH_3 \quad CH_3NH_2$

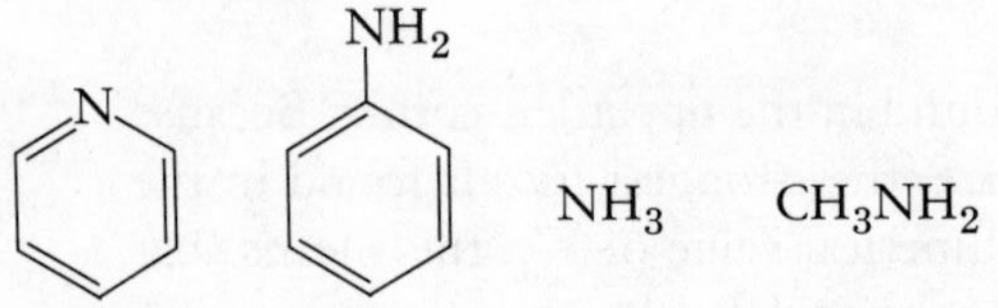

A word of caution: Do not confuse the K_a of ammonia with the K_a of ammonium ion! Ammonia is a very weak acid whose K_a is 10^{-36}. This simply means that ammonia does not easily give up a proton to form the amide ion, NH_2^-. In contrast, ammonium ion gives up a proton much more easily to form ammonia.

From the information in Table 17-2, we can reach several conclusions about the structure of an amine and its base strength. If we substitute a hydrogen of ammonia with a group that we identified as deactivating in

Table 17-2. Strengths of Conjugate Acids of Ammonia and Several of its Derivatives

Conjugate Acid	K_a	pK_a	Conjugate Base
NH_3	10^{-36}	36	NH_2^-
$(CH_3)_2\overset{+}{N}H_2$	2.0×10^{-11}	10.7	$(CH_3)_2NH$
$CH_3\overset{+}{N}H_3$	2.3×10^{-11}	10.6	CH_3NH_2
$\overset{+}{N}H_4$	6.3×10^{-10}	9.2	NH_3
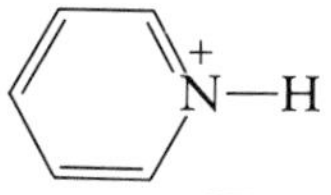 pyridinium ($C_5H_5\overset{+}{N}$—H)	5.0×10^{-6}	5.2	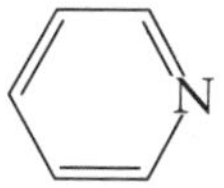 pyridine (C_5H_5N)
 CH_3—C_6H_4—$\overset{+}{N}H_3$	8.3×10^{-6}	5.1	 CH_3—C_6H_4—NH_2
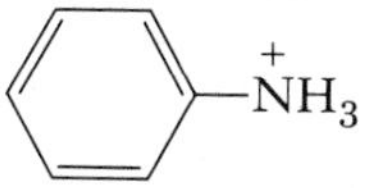 C_6H_5—$\overset{+}{N}H_3$	2.5×10^{-5}	4.6	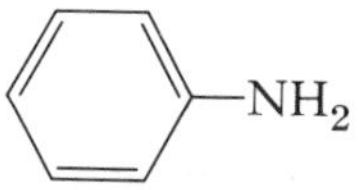 C_6H_5—NH_2
 NO_2—C_6H_4—$\overset{+}{N}H_3$	10	1.0	 NO_2—C_6H_4—NH_2
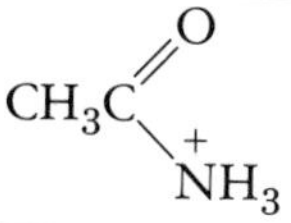 $CH_3C(=O)\overset{+}{N}H_3$	0.50	0.30	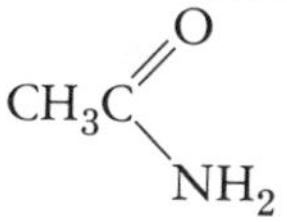 $CH_3C(=O)NH_2$

Increasing acid strength (downward) — Increasing base strength (upward)

electrophilic aromatic substitution reactions (see Table 13-2 and Section 13.3), the basicity of the resulting compound is *decreased* compared to ammonia. For example, $CH_3C(=O)NH_2$ is a much weaker base than is ammonia. The same effect is evident in ring-substituted anilines: 4-nitroaniline is a much weaker base than aniline. Conversely, activating substituents increase the basicity of amines. For example, CH_3NH_2 is a stronger base than is ammonia, and CH_3—C_6H_4—NH_2 is a stronger base than aniline.

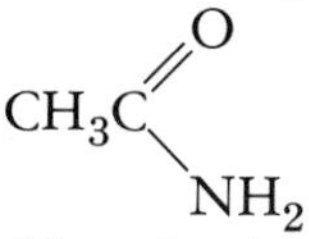

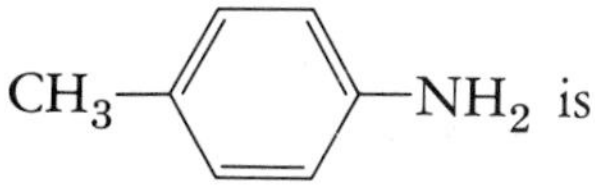

EXERCISE 17-13 Decide which one of each pair of amines is the more basic.

(a) 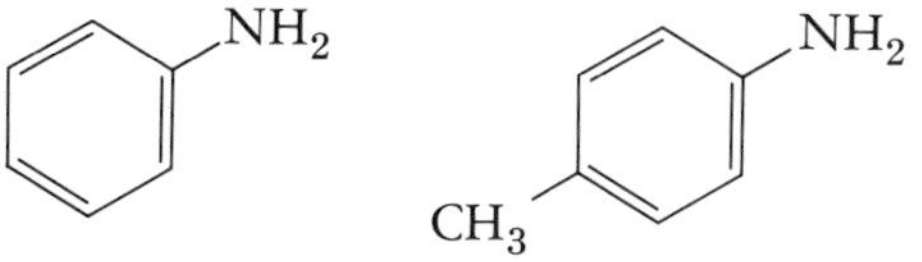 $C_6H_5NH_2$ (aniline)   4-$CH_3C_6H_4NH_2$

(b) CH_3NH_2 $\quad CH_3C(=O)NH_2$ $\quad$ (c) $ClNH_2$ $\quad NH_3$

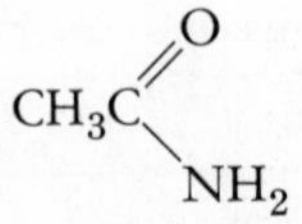

(d) 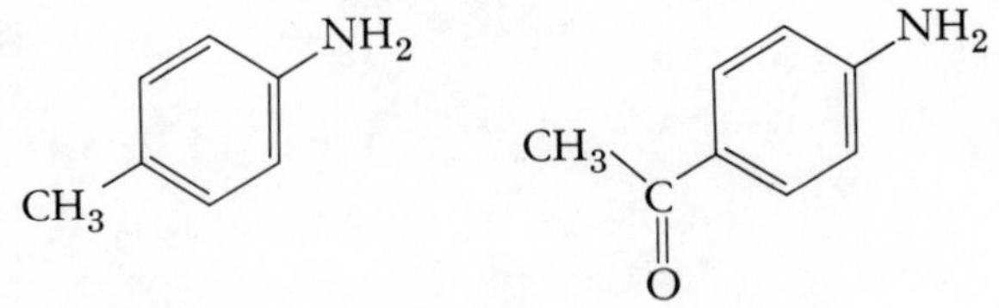 4-methylaniline ($CH_3C_6H_4NH_2$) $\quad$ 4-aminoacetophenone ($CH_3C(=O)C_6H_4NH_2$)

(e) $(CH_3)_2NH$ $\quad NH_3$

Nucleophilicity of Amines

The preparations of amines discussed in Sections 17.4 and 17.5 demonstrate the nucleophilicity of ammonia and its derivatives. In these reactions, the ammonia or ammonia derivative acts as a nucleophile toward a saturated carbon, that is, a carbon bonded to four atoms or groups. Ammonia, primary amines, and secondary amines also act as nucleophiles toward the carbon of a carbonyl group ($>C{=}O$). For example:

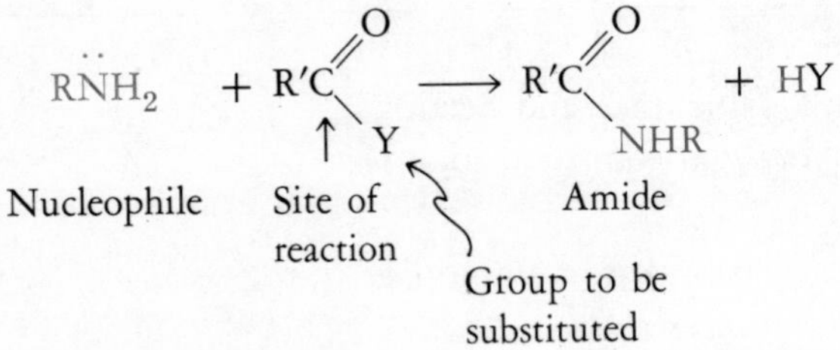

$$\underset{\text{Nucleophile}}{R\ddot{N}H_2} + \underset{\text{Site of reaction (C); Group to be substituted (Y)}}{R'C(=O)Y} \longrightarrow \underset{\text{Amide}}{R'C(=O)NHR} + HY$$

In this reaction, the nitrogen replaces group Y on the carbon of the carbonyl group to form an amide. Only ammonia, primary amines, and secondary amines react in this way. Tertiary amines do not form amides.

Specific examples of these reactions are given in the following equations:

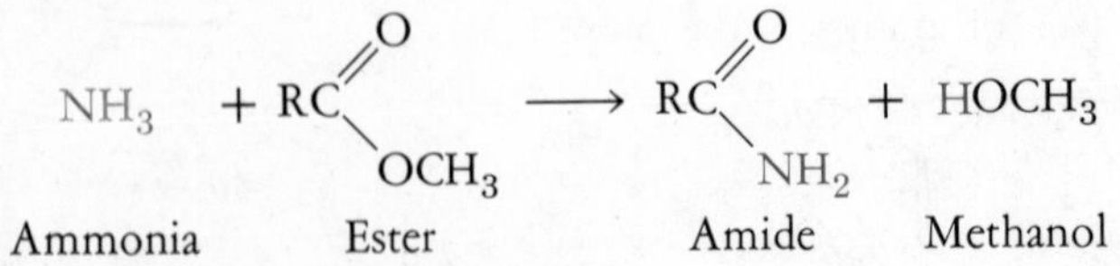

$$\underset{\text{Ammonia}}{NH_3} + \underset{\text{Ester}}{RC(=O)OCH_3} \longrightarrow \underset{\text{Amide}}{RC(=O)NH_2} + \underset{\text{Methanol}}{HOCH_3}$$

$$CH_3NH_2 + RC(=O)SCH_3 \longrightarrow RC(=O)NHCH_3 + HSCH_3$$

Methylamine Thioester An *N*-methyl amide Methanethiol

$$(CH_3)_2NH + (RC(=O))_2O \longrightarrow RC(=O)N(CH_3)_2 + RC(=O)OH$$

Dimethylamine Anhydride An *N*,*N*-dimethyl amide Carboxylic acid

The net result of all these reactions is the replacement of one hydrogen on the nitrogen of ammonia, a primary amine, or a secondary amine by a $R{-}C(=O){-}$ group, called an acyl group (see section 18.1).

In living systems, amines are converted into amides by a similar enzyme-catalyzed reaction. The reaction occurs between the amine and a compound called acetyl coenzyme A, usually designated $CH_3C(=O)SCoA$. The abbreviation CoA represents a complex structure (see Section 26.5). However, the reactivity of the acetyl part does not depend on the details of the CoA part and we can ignore it for the moment. The important point is that acetyl CoA is a thioester. Consequently, the SCoA portion is readily replaced by ammonia, a primary amine, or a secondary amine. For example:

$$RNH_2 + CH_3C(=O)SCoA \xrightarrow{\text{Enzyme}} CH_3C(=O)NHR + HSCoA$$

Amine Acetyl coenzyme A Amide Coenzyme A

In this case, the amine replaces the SCoA group.

In addition to reacting with the carbonyl groups of compounds such as anhydrides, esters, and thioester, ammonia and primary and secondary amines also react with sulfonic acid halides to form sulfonamides. Specific examples are the reaction of benzenesulfonyl chloride with ethylamine and dimethylamine:

$$C_6H_5SO_2Cl + NH_2C_2H_5 \xrightarrow{NaOH} C_6H_5SO_2NHC_2H_5 + NaCl + H_2O$$

Benzenesulfonyl chloride — Ethylamine — *N*-ethylbenzenesulfonamide

$$C_6H_5SO_2Cl + NH(CH_3)_2 \xrightarrow{NaOH} C_6H_5SO_2N(CH_3)_2 + NaCl + H_2O$$

Benzenesulfonyl chloride — Dimethylamine — *N,N*-dimethylbenzenesulfonamide

In these reactions the amine group replaces the chlorine of benzenesulfonyl chloride. Again, tertiary amines do not react to form sulfonamides.

Sulfonamides are biologically important compounds. They were one of the first antimetabolites developed to combat disease. The first one of the class, popularly called sulfa drugs, was sulfanilamide.

NH_2–C_6H_4–SO_2NH_2 Sulfanilamide

NH_2–C_6H_4–SO_2NH–(thiazolyl: C, N, CH, CH, S) Sulfathiazole

NH_2–C_6H_4–SO_2NH–(2-pyrimidinyl) Sulfadiazine

These sulfonamides are believed to act by taking the place in the body of 4-aminobenzoic acid, which is essential for the continued growth of the invading bacteria (see Section 25.9). Although sulfanilamide was found to be effective against a multitude of infections, it did have one major disadvantage. It is only slightly soluble in water, and damage to the kidney may occur by its accumulation in that organ during excretion. In an attempt to overcome this difficulty as well as other problems, other derivatives such as sulfathiazole and sulfadiazine were synthesized.

EXERCISE 17-14 Write the structure of the product formed in each of the following reactions:

(a) 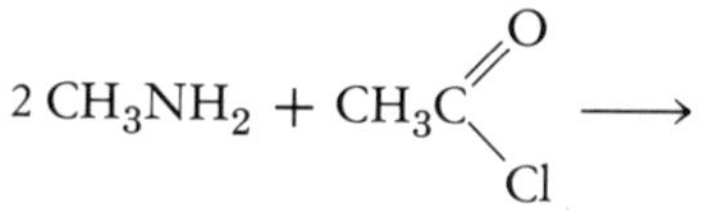 $2\,CH_3NH_2 + CH_3C(=O)Cl \longrightarrow$

(b) 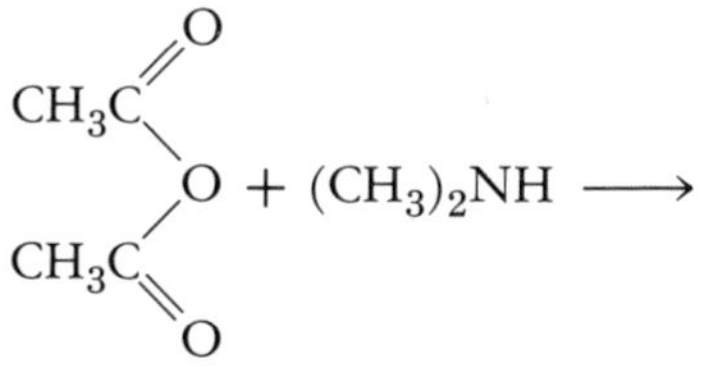 $(CH_3C(=O))_2O + (CH_3)_2NH \longrightarrow$

(c) 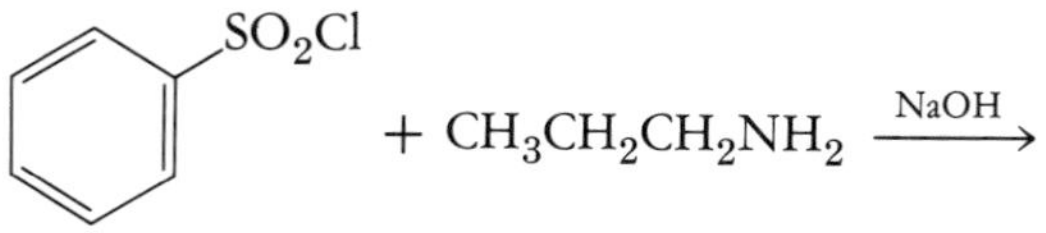 $C_6H_5SO_2Cl + CH_3CH_2CH_2NH_2 \xrightarrow{NaOH}$

(d) 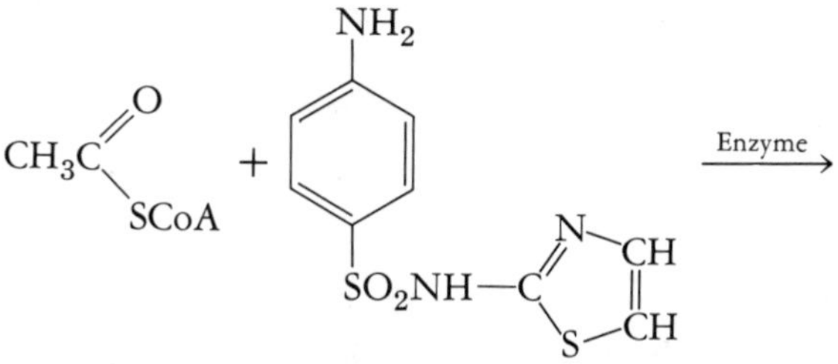 $CH_3C(=O)SCoA + H_2N{-}C_6H_4{-}SO_2NH{-}C_3H_2NS$ (2-thiazolyl: C, N, CH, CH, S ring) $\xrightarrow{Enzyme}$

17.7 OXIDATIVE DEALKYLATION OF AMINES

In Section 17.4, we learned how to alkylate ammonia and amines. Furthermore, we learned in Section 17.5 that this type of reaction also occurs in living systems. The reverse reaction is called *dealkylation.* Dealkylation is the removal of an alkyl group from an amine and is an important reaction in mammals. When a methyl group is removed, the reaction is called *demethylation.* It occurs in the liver in the presence of oxygen and is catalyzed by enzymes. The overall demethylation reaction is the following:

$$>N{-}CH_3 + 3/2\,O_2 \longrightarrow\ >N{-}H + CO_2 + H_2O$$

Oxidative demethylation reactions have been observed within many living organisms. The following examples illustrate some of these reactions:

Nicotine $+ \frac{3}{2} O_2 \xrightarrow{\text{Enzyme}}$ (N-demethylated nicotine, N–H) $+ CO_2 + H_2O$

Nicotine

N-methylaniline ($C_6H_5NHCH_3$) $+ \frac{3}{2} O_2 \xrightarrow{\text{Enzyme}}$ Aniline ($C_6H_5NH_2$) $+ CO_2 + H_2O$

N-methylaniline Aniline

Epinephrine ($(HO)_2C_6H_3CH(OH)CH_2NHCH_3$) $+ \frac{3}{2} O_2 \xrightarrow{\text{Enzyme}}$ Norepinephrine ($(HO)_2C_6H_3CH(OH)CH_2NH_2$) $+ CO_2 + H_2O$

Epinephrine Norepinephrine

The last example is the reverse of the reaction between norepinephrine and *S*-adenosyl methionine to form epinephrine (see Exercise 15.12). Thus, certain living systems have pathways for both alkylating and dealkylating amines.

Oxidative dealkylation is one of a number of reactions that break down or degrade compounds in living systems. In most cases, these reactions occur with the usual compounds needed for the system to live. Sometimes demethylation is used to degrade foreign compounds such as drugs. For example, removal of the *N*-methyl groups of morphine and codeine is known to occur in humans. The purpose of this reaction is to alter the structure of the drug so that it is no longer biologically active.

Clearly, living systems can alkylate and dealkylate amines to produce exactly the right amine needed for a particular reaction.

17.8 SUMMARY

Amines are widely found in nature. One way to prepare them in living systems is by an enzyme-catalyzed nucleophilic displacement reaction. Amines are also nucleophiles toward the carbon of carbonyl groups. Amides are formed as products in this reaction. Amines are weak bases. Their ability to accept a proton from other compounds is one of their principal functions in living systems.

Oxidative dealkylation of amines has been observed in many mammals. It is one method of degrading an amine. Finally, living systems can prepare amines as well as alkylate and oxidatively dealkylate them to produce exactly the right amine needed for a particular reaction.

REVIEW OF TERMS, CONCEPTS, AND REACTIONS

Terms

ACYLATION A reaction that results in the replacement of a hydrogen bonded to the nitrogen of ammonia or an amine by an acyl group.

ALKYLATION A reaction that results in the replacement of a hydrogen bonded to the nitrogen of ammonia or an amine by an alkyl group.

AMINE An organic derivative of ammonia in which one or more hydrogen atoms are replaced by alkyl or aryl groups.

OXIDATIVE DEALKYLATION A reaction that results in the replacement of an alkyl group bonded to the nitrogen of an amine by a hydrogen.

QUATERNARY AMMONIUM ION An organic derivative of an ammonium ion in which all four hydrogens are replaced by alkyl or aryl groups.

Concepts

1. Amines may be regarded as derivatives of ammonia.
2. Amines are classified as primary, secondary, or tertiary according to their structure.
3. Amines, like ammonia, possess an unshared pair of electrons on the nitrogen atom, which accounts for their basicity and nucleophilicity.

Reactions

Preparation of Amines (see Section 17.4)

1. Alkylation of ammonia and amines (see p. 445)

$$NH_3 \xrightarrow{RX} \underset{\text{Primary amine}}{RNH_2} \xrightarrow{RX} \underset{\text{Secondary amine}}{R_2NH} \xrightarrow{RX} \underset{\text{Tertiary amine}}{R_3N} \xrightarrow{RX} \underset{\text{Quaternary ammonium halide}}{R_4\overset{+}{N}\overset{-}{X}}$$

2. Gabriel synthesis (see p. 445)

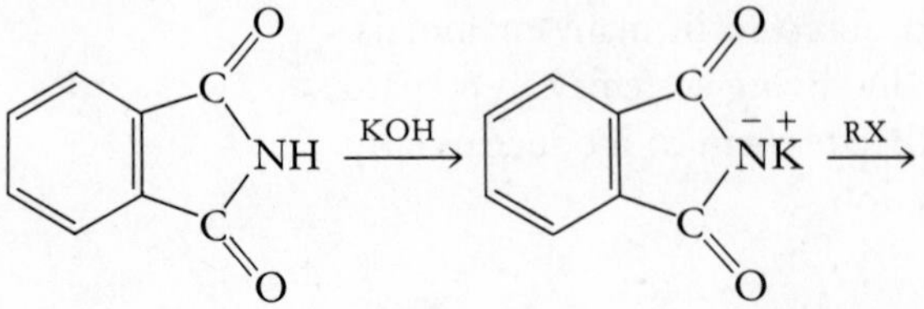

$$\text{Phthalimide (C}_6\text{H}_4(\text{CO})_2\text{NH)} \xrightarrow{\text{KOH}} \text{C}_6\text{H}_4(\text{CO})_2\overset{-}{\text{N}}\overset{+}{\text{K}} \xrightarrow{\text{RX}}$$

RX must be an
alkyl halide, preferably
primary or secondary

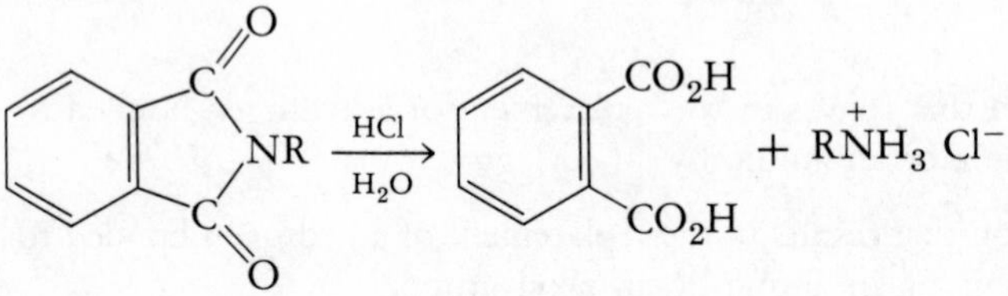

$$\text{C}_6\text{H}_4(\text{CO})_2\text{NR} \xrightarrow[\text{H}_2\text{O}]{\text{HCl}} \text{C}_6\text{H}_4(\text{CO}_2\text{H})_2 + \text{R}\overset{+}{\text{N}}\text{H}_3\ \text{Cl}^-$$

Reaction of Amines (see Section 17.6)

1. Reaction with acids to form salts (see p. 448)

$$H^+ + RNH_2 \longrightarrow R\overset{+}{N}H_3$$

$$H^+ + R_2NH \longrightarrow R_2\overset{+}{N}H_2$$

$$H^+ + R_3N \longrightarrow R_3\overset{+}{N}H$$

2. Acylation: amide formation (see p. 452)

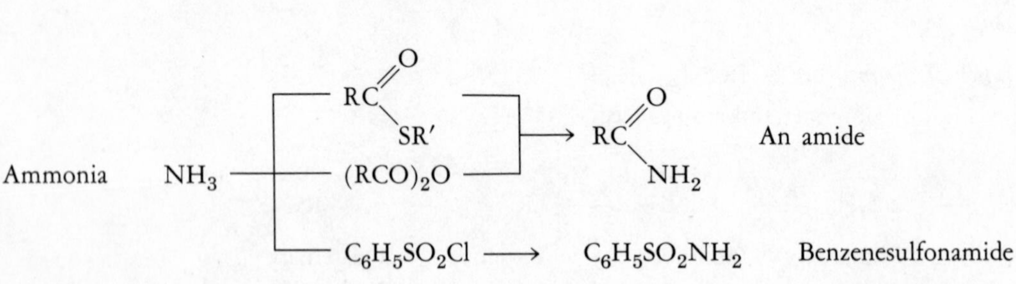

Ammonia NH_3 + $RC(=O)SR'$ or $(RCO)_2O$ $\longrightarrow$ $RC(=O)NH_2$ An amide

NH_3 + $C_6H_5SO_2Cl \longrightarrow C_6H_5SO_2NH_2$ Benzenesulfonamide

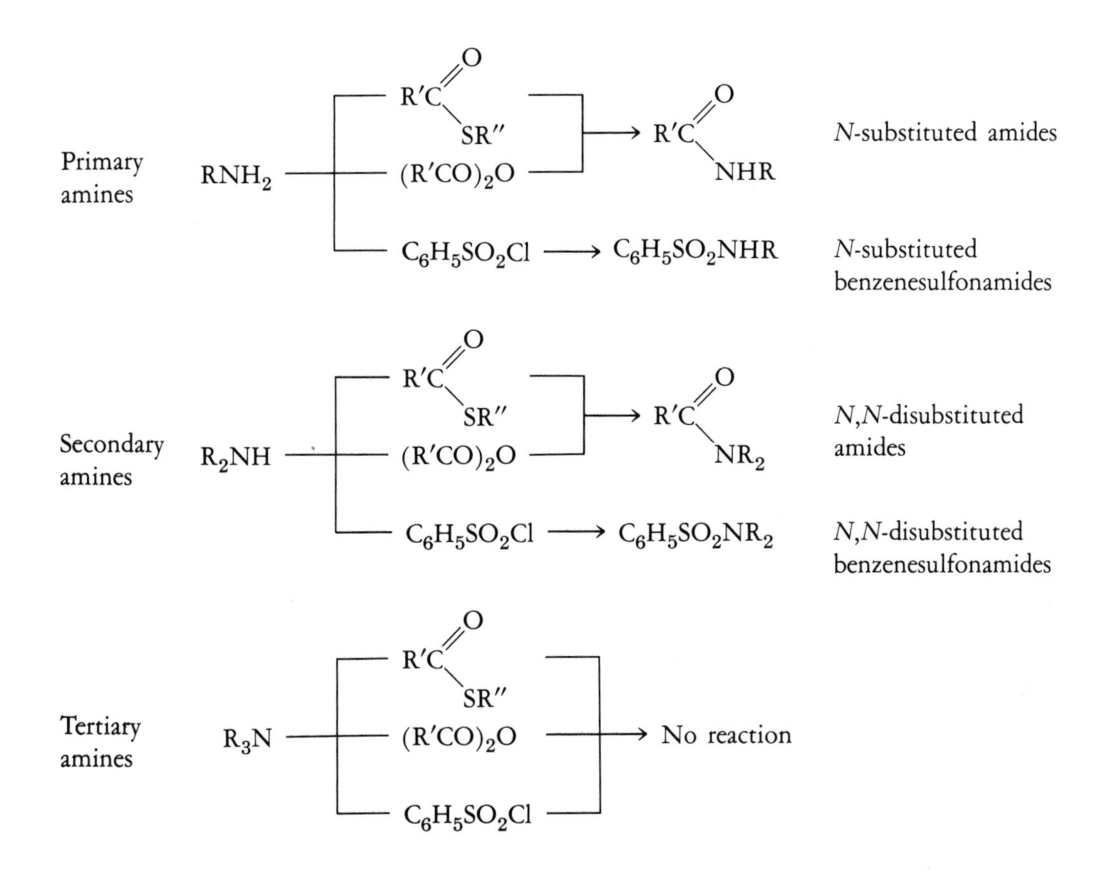

Oxidation demethylation (see Section 17.7)

$$\rangle N{-}CH_3 + 3/2\ O_2 \xrightarrow{\text{Enzyme}} \rangle N{-}H + CO_2 + H_2O$$

EXERCISES 17-15 Write the structure of the compound that is an example of each of the following:

(a) a primary aliphatic amine (b) a primary aromatic amine
(c) a cyclic secondary amine (d) a tertiary amine
(e) a quaternary ammonium ion
(f) a secondary ammonium ion

17-16 Write the structural formulas of, and name and classify as primary, secondary, or tertiary amines, all the isomers of the following molecular formulas:

(a) C_3H_9N (b) $C_4H_{11}N$

17-17 Give the IUPAC name for each of the following compounds:

(a) $CH_3(CH_2)_7NH_2$ (b) $(CH_3CH_2)_4\overset{+}{N}\ OH^-$

(c) $C_6H_5CH(NH_2)CH_2CH_3$ (benzene ring bearing $CHCH_2CH_3$ with NH_2 on the CH) (d) $BrCH_2CH_2CH_2CH_2\overset{+}{N}H_3\ Cl^-$

(e) cyclopropyl–$NCH_2CH_2CH_3$ with a second cyclopropyl group on N (f) cyclopentyl–$N(CH_3)_2$

(g) benzene ring with NH_2 and NO_2 (meta) (h) benzene ring with $N(CH_3)_2$, Br, Br (3,5-positions) (i) six-membered ring containing N bearing CH_3

17-18 Write the structure for each of the following compounds:
(a) 2-phenylethylamine (b) diphenylamine
(c) *N*-isopropyl-1-butanamine (d) *N,N*-dibutyl-1-hexanamine
(e) 4-aminobenzoic acid (f) 4-chlorobenzylamine
(g) neopentylamine (h) *N*-ethyl-*N*-methyl-1-pentanamine
(i) tetrapentylammonium iodide
(j) 2,6-dibromo-4-isopropylaniline

17-19 Write equations for the reaction of 1-butanamine with each of the following reagents:

(a) nitric acid (b) $CH_3C(=O)\text{–}O\text{–}C(=O)CH_3$

(c) benzenesulfonyl chloride and base (d) excess methyl iodide

(e) $CH_3C(=O)SCH_3$ (f) $CH_3CH_2C(=O)OCH_3$

17-20 Repeat exercise 17-19 with diethylamine instead of 1-butanamine.

17-21 Repeat exercise 17-19 with triethylamine instead of 1-butanamine.

17-22 Repeat exercise 17-19 with aniline instead of 1-butanamine.

17-23 Write the structure of the major organic product for each of the following reactions:

(a) methyl iodide + *N*,*N*-dimethylaniline $\longrightarrow$

(b) tetraisopropylammonium chloride + OH^- $\xrightarrow{\Delta}$

(c) *N*-methylcyclohexylamine + O_2 $\xrightarrow{\text{Enzyme}}$

(d) 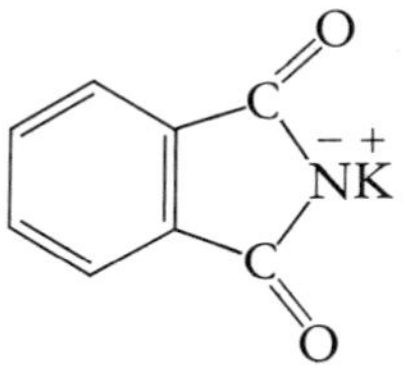 $C_6H_4(CO)_2\overset{-}{N}\overset{+}{K}$ + 1-bromopropane $\longrightarrow$ A $\xrightarrow[H_2O]{H^+}$ B

CARBOXYLIC ACIDS AND THEIR DERIVATIVES

18

Carboxylic acids are widely found in nature. Formic acid, the simplest carboxylic acid, is responsible for the sting of certain ants. Acetic acid gives vinegar its pungent odor and taste. Butyric acid contributes to the strong odor of rancid butter. Lactic acid is formed when milk sours. Citric acid is found in cells, in blood serum, and in urine, although it is more familiar to most people as the ingredient that gives citrus fruits their tart taste. Long straight-chain carboxylic acids, such as stearic acid, are called fatty acids because they are obtained by the hydrolysis of vegetable or animal fats.

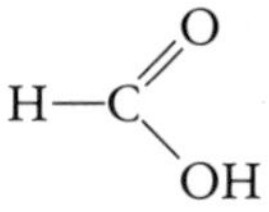

$H{-}C(=O)OH$ Formic acid

$CH_3C(=O)OH$ Acetic acid

$CH_3CH_2CH_2C(=O)OH$ Butyric acid

$CH_3CH(OH)C(=O)OH$ Lactic acid

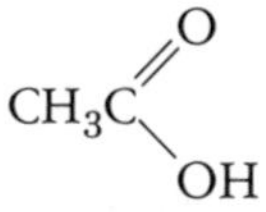

$HO{-}C(CH_2C(=O)OH)_2{-}CO_2H$ Citric acid

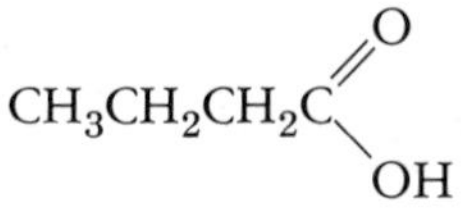

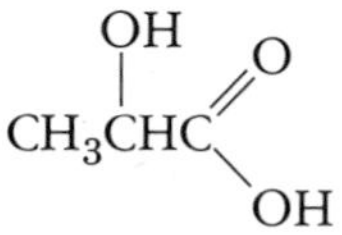

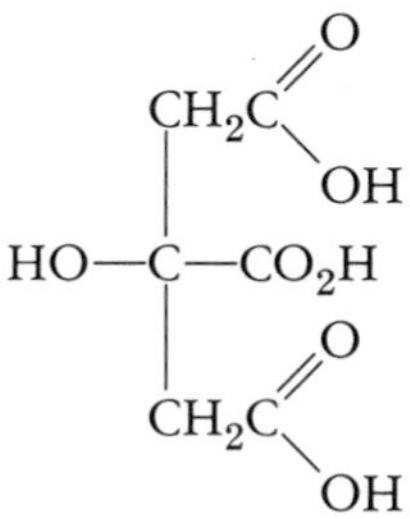

$CH_3(CH_2)_{16}C(=O)OH$ Stearic acid

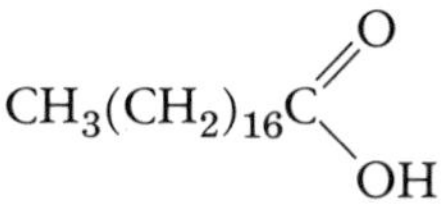

Each of these compounds contains the $-\overset{\displaystyle O}{\overset{\|}{C}}-OH$ group. This functional group is called a *carboxylic acid group* or a *carboxyl group.* It is a combination of the carbonyl $\left(>C{=}O\right)$ and the electronegative hydroxy ($-OH$) groups. These two groups combine to form a functional group with properties different from those of either the carbonyl or the hydroxy group alone. The carboxyl group is often written $-COOH$ or $-CO_2H$ for convenience.

Other classes of compounds are known that have electronegative groups bonded to the carbon of the carbonyl group. If an $-OR$ or $-OAr$ group is bonded to the carbon of the carbonyl group, the compound is called an *ester.* If a group containing sulfur is bonded to the carbonyl carbon, the compound is called a *thioester.* An *amide* contains an amino ($-NH_2$) or a substituted amino group ($-NHR$ or $-NR_2$) bonded to the carbonyl carbon. If a phosphate $\left(-O-\overset{\displaystyle O}{\overset{\|}{P}}(O^-)-O^-\right)$ or a carboxylate group $\left(-O-\overset{\displaystyle O}{\overset{\|}{C}}-R\right)$ is bonded to the carbonyl carbon, the compound is called an *acyl phosphate* or a *carboxylic acid anhydride,* respectively. The general formulas for these classes of compounds are given in Table 18-1. These compounds are called

Table 18-1. Structural Formulas of Carboxylic Acids and Their Derivatives

Formula	Name
$RC(=O)OH$	Carboxylic acid
$RC(=O)OR'$	Ester
$RC(=O)SR'$	Thioester
$RC(=O)NH_2$	Amide
$RC(=O)NHR'$	*N*-substituted amide
$RC(=O)NR'_2$	*N,N*-disubstituted amide
$RC(=O)-O-P(=O)(O^-)-O^-$	Acyl phosphate
$RC(=O)-O-C(=O)R$	Carboxylic acid anhydride

R and R′ can be either alkyl or aryl groups in this table.

derivatives of carboxylic acids because they can be hydrolyzed to the carboxylic acid according to the following equation:

$$\mathrm{RC}(=\mathrm{O})\mathrm{X} + \mathrm{H_2O} \longrightarrow \mathrm{RC}(=\mathrm{O})\mathrm{OH} + \mathrm{HX}$$

Carboxylic acid derivative — Carboxylic acid

$X = OR, OAr, SR, NH_2, NHR', NR'_2, OPO_3^{-2}, O_2CR$

The major type of reaction that carboxylic acids and their derivatives undergo in living systems is *substitution reaction.* In this reaction, the group or atom bonded to the carbonyl carbon is replaced by another group. The general reaction is the following:

$$\mathrm{RC}(=\mathrm{O})\mathrm{X} + \mathrm{HY} \longrightarrow \mathrm{RC}(=\mathrm{O})\mathrm{Y} + \mathrm{HX}$$

In this chapter, we will examine four of these substitution reactions that are important in living systems. Finally, we will examine the structure and substitution reactions of several derivatives of phosphoric acid that are important in living systems.

Let us start our study of carboxylic acids and their derivatives by learning how to name carboxylic acids. Their names are important because they are the roots of the common names of carboxylic acid derivatives.

18.1 NAMING CARBOXYLIC ACIDS

Most of the lower-molecular-weight carboxylic acids have been available for centuries from natural sources. As a result, they were given names long before the IUPAC rules were developed. Many of these common names are still used today. The common and IUPAC names of some carboxylic acids are given in Table 18-2.

Most of the common names are derived from a Latin or Greek name indicating a natural source of the acid. For example, formic acid received its

Table 18-2. Names of Representative Carboxylic Acids

Structural Formula	Common Name	IUPAC Name
HCO_2H	Formic acid	Methanoic acid
CH_3CO_2H	Acetic acid	Ethanoic acid
$CH_3CH_2CO_2H$	Propionic acid	Propanoic acid
$CH_3CH_2CH_2CO_2H$	Butyric acid	Butanoic acid
$(CH_3)_2CHCO_2H$	Isobutyric acid	2-Methylpropanoic acid
$CH_3(CH_2)_3CO_2H$	Valeric acid	Pentanoic acid
$CH_3(CH_2)_4CO_2H$	Caproic acid	Hexanoic acid
$CH_3(CH_2)_{16}CO_2H$	Stearic acid	Octadecanoic acid
$C_6H_5CO_2H$	Benzoic acid	Benzoic acid
$CH_2{=}CHCO_2H$	Acrylic acid	Propenoic acid
HO_2CCO_2H	Oxalic acid	Ethanedioic acid
$HO_2CCH_2CO_2H$	Malonic acid	Propanedioic acid
$HO_2CCH_2CH_2CO_2H$	Succinic acid	Butanedioic acid
$HO_2CCH_2CH_2CH_2CO_2H$	Glutaric acid	Pentanedioic acid
$(HO_2C)(H)C{=}C(H)(CO_2H)$ (cis)	Maleic acid	*cis*-Butenedioic acid
$(HO_2C)(H)C{=}C(CO_2H)(H)$ (trans)	Fumaric acid	*trans*-Butenedioic acid

name from the Latin word *formica* meaning ant, and acetic acid is named after the Latin word *acetum* meaning vinegar. Stearic acid receives its name from the Greek word *stear* meaning tallow, which indicates the source of this fatty acid. These common names are widely used for carboxylic acids containing four or fewer carbon atoms.

Common names are also used to name substituted low-molecular-weight carboxylic acids. The location of the substituent on the chain is designated by the letters α, β, γ, or δ, which refer to the following positions:

$$\overset{\delta}{C}-\overset{\gamma}{C}-\overset{\beta}{C}-\overset{\alpha}{C}-CO_2H$$

The following examples illustrate this method of naming carboxylic acids:

$$\underset{\displaystyle NH_2}{CH_3CHCO_2H}$$
α-Aminopropionic acid

$$BrCH_2CH_2CO_2H$$
β-Bromopropionic acid

$$\underset{\displaystyle CH_3}{CH_3CHCH_2CH_2CO_2H}$$
γ-Methylvaleric acid

According to the IUPAC rules, the parent name of a carboxylic acid is obtained by determining the longest continuous chain of carbon atoms containing the carboxyl functional group. The *-e* ending of the corresponding alkane name is replaced by the suffix *-oic acid.* The chain is numbered so that the carbon of the carboxylic group is given the number 1. Because the carboxyl group *must* be at the end of the chain, the number 1 is never included in the name. The following examples illustrate the IUPAC rules for naming carboxylic acids:

$$\overset{4}{C}H_3\overset{3}{C}H(Cl)\overset{2}{C}H_2\overset{1}{C}O_2H$$
3-Chlorobutanoic acid

$$\overset{5}{C}H_3\overset{4}{C}H(CH_3)\overset{3}{C}H_2\overset{2}{C}H(CH_3)\overset{1}{C}O_2H$$
2,4-Dimethylpentanoic acid

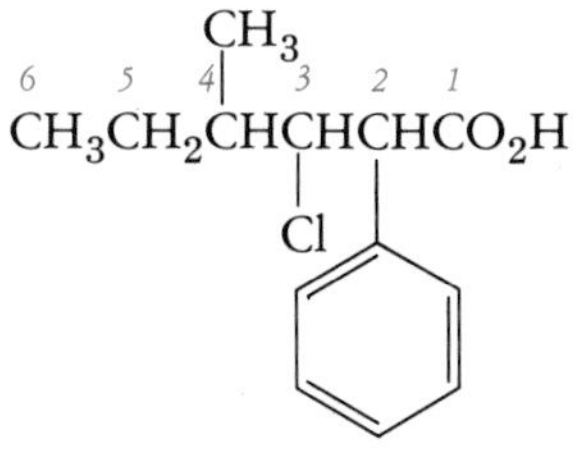

$$\overset{6}{C}H_3\overset{5}{C}H_2\overset{4}{C}H(CH_3)\overset{3}{C}H(Cl)\overset{2}{C}H(C_6H_5)\overset{1}{C}O_2H$$
3-Chloro-4-methyl-2-phenylhexanoic acid

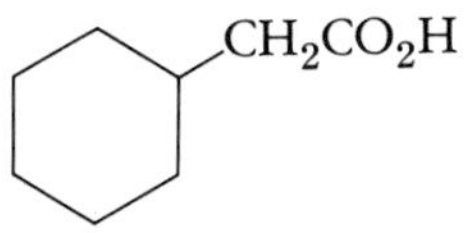

$$C_6H_{11}CH_2CO_2H$$
Cyclohexylethanoic acid

Many compounds contain two or more carboxyl groups. The common names of six simple dicarboxylic acids are given in Table 18-2. The IUPAC name of a dicarboxylic acid is obtained by adding the suffix *-dioic acid* to the name of the longest carbon chain that contains both carboxyl groups. For example:

$$HO_2\overset{3}{C}\overset{2}{C}H(Br)\overset{1}{C}O_2H$$
2-Bromopropanedioic acid

$$HO_2\overset{4}{C}\overset{3}{C}H_2\overset{2}{C}H(CH_3)\overset{1}{C}O_2H$$
2-Methylbutanedioic acid

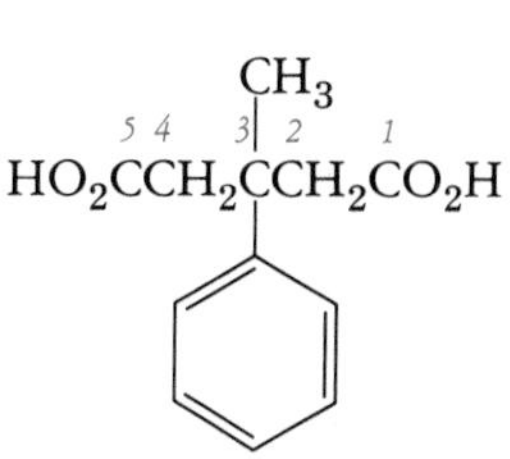

$$HO_2\overset{5}{C}\overset{4}{C}H_2\overset{3}{C}(CH_3)(C_6H_5)\overset{2}{C}H_2\overset{1}{C}O_2H$$
3-Methyl-3-phenylpentanedioic acid

Notice that the α-position of the common name is the 2-position of the IUPAC name; the β-position is the 3-position, and so forth. For example:

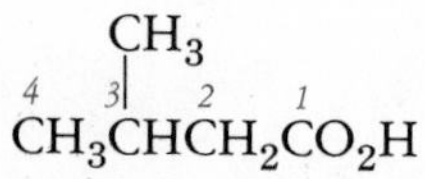

$CH_3CH(CH_3)CH_2CO_2H$ (carbons numbered 4, 3, 2, 1)

3-Methylbutanoic acid
β-Methylbutyric acid

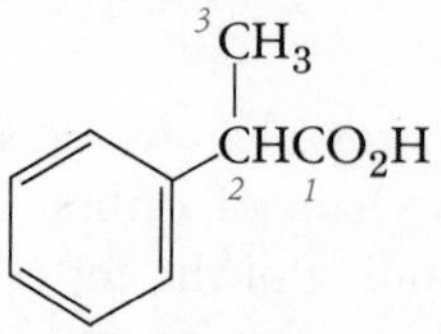

$C_6H_5CH(CH_3)CO_2H$ (carbons numbered 3, 2, 1)

2-Phenylpropanoic acid
α-Phenylpropionic acid

In certain molecules, particularly cyclic ones, it is easier to name the carboxyl group as a substituent. In these cases, the parent hydrocarbon name may be used, followed by *carboxylic acid.* For example:

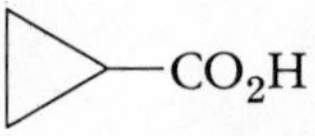

Cyclopropyl–CO_2H

Cyclopropanecarboxylic acid

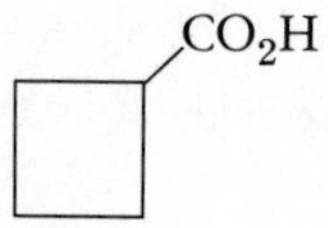

Cyclobutyl–CO_2H

Cyclobutanecarboxylic acid

The names of carboxylate ions (RCO_2^-) are formed by changing the *-ic acid* suffix to *-ate.* Salts of carboxylic acids are named by adding the name of the cation as a prefix to the name of the carboxylate ion. For example:

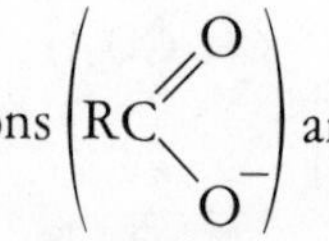

$CH_3CO_2^-NH_4^+$

Ammonium acetate
Ammonium ethanoate

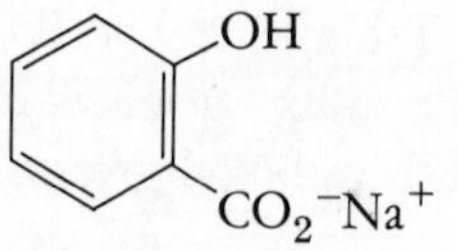

$o\text{-}HOC_6H_4CO_2^-Na^+$

Sodium salicylate

$CH_3CH(CH_3)CH_2CO_2^-K^+$

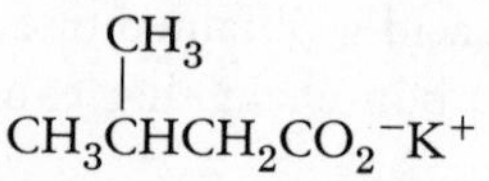

Potassium isovalerate
Potassium 3-methylbutanoate

The structure $-\overset{|}{\underset{|}{C}}-C(=O)-$ occurs in all carboxylic acids and their derivatives. Therefore, it is given the general name *acyl group.* A particular acyl group is named by replacing the *-ic acid* ending of the acid name (either common or IUPAC) by *-yl.* For example:

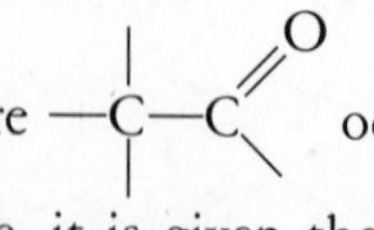

$CH_3C(=O)-$ Acetyl group (ethanoyl group)

$C_6H_5C(=O)-$ Benzoyl group

$CH_3CH_2CH_2C(=O)-$ Butanoyl group (butyryl group)

*EXERCISE 18-1 Give a name to each of the following compounds:

(a) $CH_3CH_2CO_2H$ (b) $(CH_3)_2CHCH_2CO_2H$

(c) $BrC(CH_3)_2CO_2H$ (d) cyclopentane–CO_2H (e) cyclohexane–$CO_2^-Na^+$

(f) $ClCH_2(H)C{=}C(H)CO_2H$ (g) $(CH_3)_2C{=}CHCH(C_6H_5)CH_2CO_2H$

(h) $HO_2CC(CH_3)_2CO_2H$ (i) cyclohexene–CO_2H

EXERCISE 18-2 Write the structure for each of the following compounds:
(a) butyric acid (b) propenoic acid
(c) cycloheptanecarboxylic acid (d) sodium propanoate
(e) *cis*-butenedioic acid (f) 2,4-dimethylpentanedioic acid
(g) 2,5-dimethylhexanoic acid (h) 8-nonenoic acid
(i) 3-chlorocyclobutanecarboxylic acid

EXERCISE 18-3 Give a name to each of the following acyl groups:

(a) $CH_3CH_2C(=O)-$ (b) $(CH_3)_2CHC(=O)-$ (c) $H-C(=O)-$

(d) $BrCH_2C(=O)-$ (e) $CH_2{=}CHC(=O)-$ (f) $CH_3(CH_2)_6C(=O)-$

* The answers for the exercises in this chapter begin on page 879.

18.2 PHYSICAL PROPERTIES

The effect of hydrogen bonding on the physical properties of compounds is well illustrated by carboxylic acids and their derivatives. As in the case of water and alcohols, carboxylic acids and amides have higher boiling points than hydrocarbons of similar molecular weight. This is illustrated in Table 18-3, where the boiling and melting points of some compounds of similar molecular weight are compared.

The higher-than-expected melting points of pentanoic acid and pentanamide, which is actually a solid at room temperature, are due to intermolecular hydrogen bonding (Section 6.3). This hydrogen bond is formed by the attraction of the oxygen of the carbon-oxygen double bond for the hydrogen of the —OH and $—NH_2$ groups. As the hydrogens attached to the nitrogen of an amide group are progressively replaced by alkyl groups, the ability to form hydrogen bonds decreases, causing a decrease in the melting points. The melting points of the isomers, pentanamide, *N*-methyl butanamide, and *N,N*-dimethylpropanamide illustrate this phenomenon.

Methyl butanoate is unable to form intermolecular hydrogen bonds, so its boiling point and melting point do not differ greatly from those of heptane. Acetic anhydride is an exception: its boiling point is high even though it is incapable of forming intramolecular hydrogen bonds. Acetic anhydride is a polar molecule (see Section 6.2), so there are strong intermolecular attractions that result in a higher-than-expected boiling point.

Hydrogen bonding also occurs between the hydrogens of the —COOH and $—CONH_2$ groups and the oxygen atom of water. As a result, carbox-

Table 18-3. Physical Properties of Some Compounds of Similar Molecular Weight

Structure	Name	Molecular Weight (amu)	Boiling Point (°C)	Melting Point (°C)
$CH_3(CH_2)_5CH_3$	Heptane	100	98	−91
$CH_3(CH_2)_3CO_2H$	Pentanoic acid	102	187	−34
$CH_3(CH_2)_3CONH_2$	Pentanamide	101	—	101
$CH_3(CH_2)_2CONHCH_3$	*N*-methyl butanamide	101	—	75
$CH_3CH_2CON(CH_3)_2$	*N,N*-dimethyl propanamide	101	191	−20
$CH_3(CH_2)_2CO_2CH_3$	Methyl butanoate	102	103	−95
$CH_3CO(SCH_2CH_3)$	S-Ethyl ethanethiolate	104	116	—
$(CH_3CO)_2O$	Acetic anhydride	102	140	−73

ylic acids and amides containing four or fewer carbon atoms are completely miscible with water. Esters are not very soluble in water, and acid anhydrides react slowly with water.

Carboxylic acids, thioesters, and anhydrides generally all have disagreeable odors. Esters, on the other hand, have pleasant odors that occur in natural perfumes and flavors. The occurrence in nature of some of the esters is included in Table 18-5.

EXERCISE 18-4 Explain why one compound of each pair has a higher boiling point (bp) than the other.

(a) $CH_3C(=O)CH_3$ (bp, 56° C) and $(CH_3)_2CHOH$ (bp, 83° C)

(b) $CH_3CH_2C(=O)NH_2$ (bp, 213° C) and $H{-}C(=O)N(CH_3)_2$ (bp, 155° C)

(c) $CH_3(CH_2)_3CO_2H$ (bp, 187° C) and $CH_3CH_2C(=O)OCH_2CH_3$ (bp, 99° C)

18.3 PREPARING CARBOXYLIC ACIDS

We have already learned in Section 16.4 that the vigorous oxidation of primary alcohols forms carboxylic acids. For example:

$$RCH_2OH \xrightarrow[KMnO_4]{H^+, H_2O} RC(=O)OH$$

The reverse of this reaction can be accomplished. Thus, a carboxylic acid can form a primary alcohol by reaction with lithium aluminum hydride ($LiAlH_4$), a strong reducing agent. For example:

$$RC(=O)OH \xrightarrow{LiAlH_4} RCH_2OH$$

In this way, primary alcohols and monocarboxylic acids are readily interconverted:

$$RCH_2OH \underset{\text{Reduction}}{\overset{\text{Oxidation}}{\rightleftharpoons}} RC(=O)OH$$

Vigorous oxidation of alkenes results in the breaking of the double bond, as we learned in Section 12.8. For example:

$$\underset{\text{2-Pentene}}{CH_3CH_2CH{=}CHCH_3} \xrightarrow[H^+,\, H_2O]{K_2Cr_2O_7} \underset{\text{Propionic acid}}{CH_3CH_2C(=O)OH} + \underset{\text{Acetic acid}}{CH_3C(=O)OH}$$

This reaction is particularly useful for forming dicarboxylic acids from cycloalkenes. For example:

$$\text{Cyclopentene} \xrightarrow[H^+,\, H_2O]{KMnO_4} \underset{\text{Glutaric acid}}{HO_2C(CH_2)_3CO_2H}$$

EXERCISE 18-5 Write the structure and name the carboxylic acid (or acids) formed by the oxidation of the following compounds:
(a) 1-propanol (b) 1,4-butanediol (c) 3-hexene
(d) 3-heptene (e) cyclohexene (f) 1-phenylpentane

Aromatic carboxylic acids can be prepared by the oxidation of an alkyl side chain on an aromatic ring, as discussed in Section 13.5.

$$C_6H_5CH_2CH_2CH_3 \xrightarrow[H^+,\, H_2O]{KMnO_4} C_6H_5CO_2H + 2\,CO_2$$

Carbons of the side chain not directly attached to the aromatic ring are oxidized to carbon dioxide.

As their names indicate, carboxylic acids are acids. This is their most important function in living systems. Let us examine this property of carboxylic acids.

18.4 ACIDITY OF CARBOXYLIC ACIDS

We learned in Section 10.5 that when an acid such as hydrochloric acid (HCl) is dissolved in water, it donates a proton to water to form a hydronium ion and a chloride ion:

$$HCl + H_2O \longrightarrow \overset{+}{H_3O} + \overset{-}{Cl}$$

Hydronium ion   Chloride ion

Carboxylic acids such as acetic acid also react in this way:

$$CH_3CO_2H + H_2O \leftrightarrows CH_3CO_2^- + H_3O^+$$

Acetic acid   Acetate ion (carboxylate ion)

Unlike hydrochloric acid, which is almost completely converted to hydronium and chloride ions, carboxylic acids protonate water only to a small extent. For example, only about 1 percent of a 0.1 M acetic acid solution is converted to acetate and hydronium ions. As a result, carboxylic acids are weak acids compared to inorganic acids such as hydrochloric acid (HCl), nitric acid (HNO_3), and sulfuric acid (H_2SO_4).

We express quantitatively the extent to which any acid donates a proton to water by means of its ionization constant, K_a (see Section 10.6). The values of K_a and pK_a for a number of carboxylic acids are given in Table 18-4. Remember: the larger the value of K_a or the smaller the value of pK_a, the stronger the acid.

We can reach certain conclusions about the acidity of carboxylic acids from the information given in Table 18-4. It is clear that all carboxylic acids are stronger acids than phenol. Furthermore, the strengths of unsubstituted aliphatic carboxylic acids are all about the same: $K_a \approx 10^{-5}$. Placing sub-

Table 18-4. K_a and pK_a Values of Some Acids in Water at 25° C

Acid Name	Structural Formula	K_a	pK_a
Formic acid	HCO_2H	1.8×10^{-4}	3.7
Acetic acid	CH_3CO_2H	1.8×10^{-5}	4.7
Propionic acid	$CH_3CH_2CO_2H$	1.4×10^{-5}	4.9
Butyric acid	$CH_3CH_2CH_2CO_2H$	1.6×10^{-5}	4.8
Isobutyric acid	$(CH_3)_2CHCO_2H$	1.7×10^{-5}	4.8
Chloroacetic acid	$ClCH_2CO_2H$	1.4×10^{-3}	2.9
Benzoic acid	$C_6H_5CO_2H$	6.6×10^{-5}	4.2
4-Nitrobenzoic acid	$4\text{-}NO_2C_6H_4CO_2H$	4.0×10^{-4}	3.4
Phenol	C_6H_5OH	1.0×10^{-10}	10

stituents on the α carbon affects the strength of aliphatic carboxylic acids. Substituents in the *para* position affect the strength of aromatic carboxylic acids. When an electronegative group or atom is placed in this position, the acid strength *increases*. For example, chloroacetic acid ($ClCH_2CO_2H$), with a $K_a = 1.4 \times 10^{-3}$, is a *stronger* acid than acetic acid (CH_3CO_2H), with a $K_a = 1.8 \times 10^{-5}$; 4-nitrobenzoic acid ($4\text{-}NO_2C_6H_4CO_2H$), with a $K_a = 4.0 \times 10^{-4}$, is a *stronger* acid than benzoic acid ($C_6H_5CO_2H$), with a $K_a = 6.6 \times 10^{-5}$. We can reach one further conclusion from the information in Table 18-4: the pH of living cells is buffered at approximately 7 (the pH of different cells varies slightly). At a pH of 7, any carboxylic acid with a K_a greater than about 10^{-5} exists predominantly in cells as the carboxylate ion.

Clearly, the structure of a carboxylic acid determines its acid strength. Although there is wide variation in their acid strengths, carboxylic acids undergo the one reaction typical of all acids, namely, the reaction with strong bases to form salts.

All carboxylic acids react completely with aqueous solutions of strong bases such as sodium hydroxide (NaOH) to form salts. These same salts and carbon dioxide are formed when carboxylic acids react with aqueous solutions of sodium bicarbonate ($NaHCO_3$) or sodium carbonate (Na_2CO_3):

$$RCO_2H + NaOH \longrightarrow RCO_2^-Na^+ + H_2O$$

$$RCO_2H + NaHCO_3 \longrightarrow RCO_2^-Na^+ + H_2O + CO_2$$

$$2\,RCO_2H + Na_2CO_3 \longrightarrow 2\,RCO_2^-Na^+ + H_2O + CO_2$$

The reaction with sodium bicarbonate serves as a simple general test to distinguish carboxylic acids from most other organic compounds that are not strong enough acids to react with this compound. The evolution of carbon dioxide serves to indicate that reaction with an acid has occurred.

The salts formed by the reactions of equivalent amounts of carboxylic acid and base may be isolated by removing the water by evaporation. When the solid formed in this way reacts with strong acids such as HCl, HNO_3, or H_2SO_4, the carboxylic acid is reformed:

$$RCO_2^-Na^+ + HCl \longrightarrow RCO_2H + NaCl$$

Carboxylic acids and their salts are easily interconverted:

$$\underset{\text{Carboxylic acid}}{RCO_2H} \underset{H^+ \text{(Acid)}}{\overset{OH^- \text{(Base)}}{\rightleftharpoons}} \underset{\text{Carboxylate ion (salt)}}{RCO_2^-}$$

Aqueous base readily converts carboxylic acids to their salts, and strong acids convert the salt back into the carboxylic acid.

EXERCISE 18-6 Write the structure and give the name for the organic product of each of the following reactions:
(a) $(CH_3)_2CHCO_2H + KOH \longrightarrow$
(b) $CH_3CH_2CH_2CO_2H + (CH_3)_2NH \longrightarrow$
(c) $CH_3CH_2CO_2^-K^+ + H_2SO_4 \longrightarrow$

The ability to act as an acid is one of the most important reactions of carboxylic acids in nature. Now let us turn our attention to the derivatives of carboxylic acids that also play vital roles in the chemistry of living systems.

18.5 ESTERS

Esters occur widely in nature. They are responsible for the odor of many natural perfumes and flavors. The boiling points and characteristic odors of selected esters are given in Table 18-5.

Table 18-5. Boiling Points and Characteristic Odors of Selected Esters

Name	Structure	Odor	Boiling Point (°C)
Pentyl acetate	$CH_3CO_2CH_2(CH_2)_3CH_3$	Bananas	148
Octyl acetate	$CH_3CO_2CH_2(CH_2)_6CH_3$	Oranges	210
Methyl butyrate	$CH_3CH_2CH_2CO_2CH_3$	Apples	103
Butyl butyrate	$CH_3CH_2CH_2CO_2CH_2CH_2CH_2CH_3$	Pineapples	167
Methyl salicylate	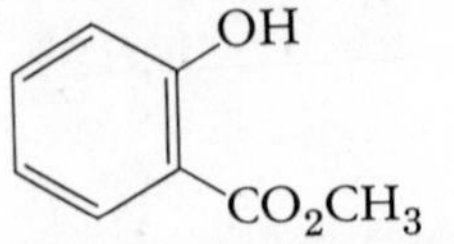 benzene ring with OH and CO_2CH_3 (ortho)	Wintergreen	223
3-Methylbutyl acetate	$CH_3CO_2CH_2CH_2CH(CH_3)_2$	Pears	146

Naming Esters

Esters are made of two parts, the carboxyl part and an alkyl or aryl group:

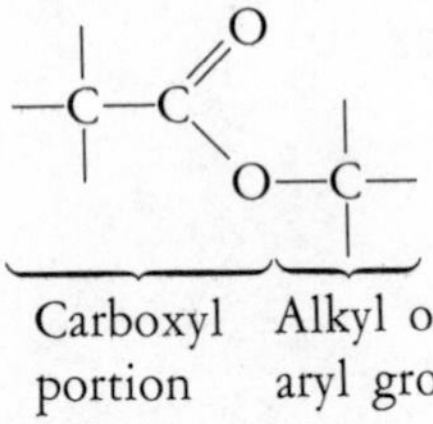

Carboxyl portion | Alkyl or aryl group

The carboxyl portion is named by replacing the *-ic acid* suffix of the name (either common or IUPAC) of the corresponding acid by *-ate*. The name of the alkyl or aryl group is placed as a separate word before the name of the carboxyl portion. For example:

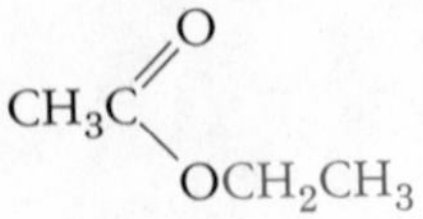

$CH_3C(=O)OCH_2CH_3$

Ethyl acetate (Ethyl ethanoate)

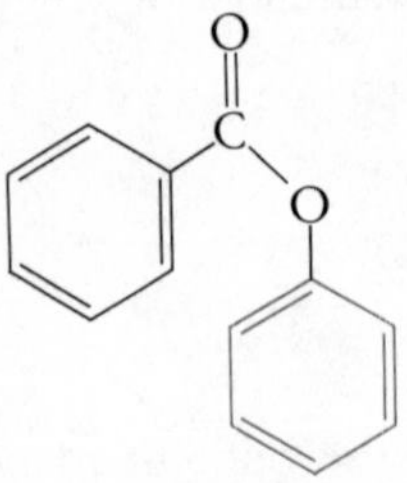

Phenyl benzoate

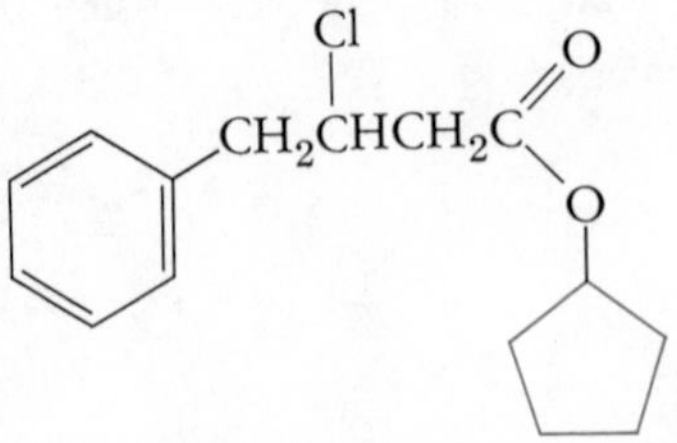

$CH_2CHClCH_2C(=O)O$

Cyclopentyl 3-chloro-4-phenylbutanoate

Notice the similarity to the method for naming esters and salts of carboxylic acids described in Section 18.1.

EXERCISE 18-7 Give a name to each of the following compounds:

(a) $HC(=O)OCH_2CH_2CH_3$ (b) $CH_3C(=O)OCH(CH_3)_2$

(c) $CH_3CH_2C(=O)OCH_2CH_3$ (d) $CH_3C(=O)O$–cyclohexyl

(e) CH_3(H)C=C(H)C(=O)OCH_3 (f) $(CH_3)_2CHCH_2C(=O)O$–phenyl

EXERCISE 18-8 Write the structure for each of the following compounds:
(a) methyl butanoate (b) hexyl valerate
(c) propyl benzoate (d) phenyl propanoate
(e) 4-chlorophenyl 2-methylpentanoate
(f) isobutyl propenoate (g) benzyl acetate

Preparation

When a phenol or a primary or secondary alcohol is heated with a carboxylic acid in the presence of a mineral acid as a catalyst, an equilibrium is established with the ester and water:

$$\underset{\text{Carboxylic acid}}{RC(=O)OH} + \underset{\substack{1^\circ \text{ or } 2^\circ \\ \text{Alcohol} \\ \text{or phenol}}}{R'OH} \overset{H^+}{\rightleftharpoons} \underset{\text{Ester}}{RC(=O)OR'} + \underset{\text{Water}}{H_2O}$$

Notice that in this reaction the —OH group of the carboxylic acid is replaced by the —OR′ group.

When this esterification reaction reaches equilibrium, appreciable quantities of both reactants and products are present. To make this method useful for preparing esters, it is necessary to increase the equilibrium concentration of the product. This can be done in two ways. The first way is to use an excess of either the alcohol or the carboxylic acid, whichever is cheaper. The second way is to remove the water as it is formed. One way to remove the water is simply to distill it from the reaction mixture. This works only if the carboxylic acid, alcohol, and ester all boil at temperatures higher than 100° C. A second method involves the use of a dehydrating agent that is inert in all of the reagents except water, which it absorbs.

EXERCISE 18-9 Write the equation for the preparation of each of the following esters by the acid-catalyzed reaction of a carboxylic acid and an alcohol:
(a) ethyl acetate (b) isopropyl benzoate
(c) ethyl 3-phenylpropanoate

Reactions

When an ester is heated with aqueous acid, the equilibrium between ester, water, alcohol, and carboxylic acid, discussed on page 477, is re-established. If the purpose of this reaction is to recover either the carboxylic acid or the alcohol, we are faced once again with the problem of increasing the equilibrium concentration of the product.

We can avoid this problem by carrying out the hydrolysis of an ester in an alkaline solution. This is an irreversible reaction that forms the salt of the carboxylic acid and an alcohol as products:

$$\underset{\text{Ester}}{RC(=O)OR'} + \underset{\text{Sodium hydroxide}}{NaOH} \xrightarrow{H_2O} \underset{\text{Sodium carboxylate}}{RC(=O)O^-Na^+} + \underset{\text{Alcohol}}{R'OH}$$

The carboxylic acid can be recovered by adding a mineral acid to the salt (see Section 18.4).

The conversion of an ester to an alcohol and the salt of a carboxylic acid by an aqueous base is called *saponification.* This name refers to the fact that soaps are formed by the alkaline hydrolysis of naturally occurring fats and oils. For example:

$$3\,NaOH + \begin{array}{l} \quad\;\; CH_2OCR \\ R'COCH \\ \quad\;\; CH_2OCR'' \end{array} \xrightarrow{H_2O} \begin{array}{l} CH_2OH \\ HOCH \\ CH_2OH \end{array} + \begin{array}{l} RCO_2^-Na^+ \\ R'CO_2^-Na^+ \\ R''CO_2^-Na^+ \end{array}$$

Fat or Oil — Glycerol — Soap

Esters, like carboxylic acids, can be reduced to alcohols by reducing agents such as lithium aluminum hydride ($LiAlH_4$) or hydrogen and a metal catalyst. For example:

$$CH_3CH_2C(=O)OCH_3 + H_2 \xrightarrow{Pt} CH_3CH_2CH_2OH + CH_3OH$$

Methyl propanoate — 1-Propanol — Methanol

Notice that two alcohols are formed in this reaction. One is formed by reduction of the acyl portion, and the other comes from the alkoxy (—OR) portion of the ester.

There are a number of reactions in which the alkoxy group of the ester is replaced by another group. One such reaction, called *transesterification,* is the acid-catalyzed reaction of an ester and an alcohol. For example:

$$CH_3C(=O)OCH_3 + CH_3CH_2OH \overset{H^+}{\rightleftharpoons} CH_3C(=O)OCH_2CH_3 + CH_3OH$$

Methyl acetate — Ethanol — Ethyl acetate — Methanol

In this reaction the $—OCH_3$ group of the original ester is replaced by an $—OCH_2CH_3$ group from ethanol.

A related reaction occurs by treating an ester with ammonia or an amine. This is called an *ammonolysis reaction.* In this case, an amide or an *N*-substituted amide is the product. For example:

$$CH_3CH_2C(=O)OCH_3 + NH_3 \longrightarrow CH_3CH_2C(=O)NH_2 + CH_3OH$$

Methyl propanoate — Ammonia — Propanamide — Methanol

$$CH_3CH_2C(=O)OCH_2CH_3 + CH_3CH_2NH_2 \longrightarrow CH_3CH_2C(=O)NHCH_2CH_3 + CH_3CH_2OH$$

Ethyl propanoate — Ethylamine — *N*-Ethyl propanamide — Ethanol

In these reactions, the alkoxy group bonded to the carbonyl carbon of the original ester is replaced by either a different alkoxy group or a group containing a nitrogen atom. These reactions occur widely in nature. In Section 18.6, we will learn that thioesters undergo similar reactions.

Esters of Salicylic Acid

Salicylic acid contains both a carboxyl group and a phenol group. Both of these functional groups can form esters, as shown in the following equations:

$$\text{Salicylic acid} + CH_3OH \overset{H^+}{\rightleftharpoons} \text{Methyl salicylate} + H_2O$$

Salicylic acid — Methanol — Methyl salicylate (oil of wintergreen)

$$\text{Salicylic acid} + CH_3C(=O)OH \overset{H^+}{\rightleftharpoons} \text{Acetylsalicylic acid} + H_2O$$

Salicylic acid — Acetic acid — Acetylsalicylic acid (aspirin)

Salicylic acid has both pain-reducing (analgesic) and fever-reducing (antipyretic) properties. Unfortunately, it irritates the lining of the stomach. Methyl salicylate and acetylsalicylic acid are two derivatives of salicylic acid that reduce this undesirable side effect while retaining medicinal properties.

Methyl salicylate is an oil that is found in several plants. It has the aroma and flavor of wintergreen. It is used for flavoring candy and as a perfume. It is also used in liniments because it has the ability to penetrate the skin. Once under the skin, methyl salicylate is hydrolyzed to salicylic acid, which relieves the soreness.

Acetylsalicylic acid, better known as aspirin or sometimes ASA, is the most widely used drug in the world. More than 25 million pounds of aspirin are produced annually in the United States. Although aspirin does irritate the lining of the stomach, its effect is much less than that of salicylic acid. Aspirin passes through the stomach and into the small intestines. There it is hydrolyzed to salicylic acid, which is absorbed by the body. The sodium salt of aspirin, sodium acetyl salicylate, is the active ingredient in Alka Seltzer®.

EXERCISE 18-10 Write the structure and give the names of the products formed by the saponification of each of the following esters:

(a) $CH_3CO_2CH_2CH_3$ (b) $(CH_3)_2CHCO_2CH_3$

(c) $CH_3CO_2CH_2CH_2O_2CCH_2CH_3$

(d) C_6H_5–$CO_2C(CH_3)_3$ (benzene ring bearing $CO_2C(CH_3)_3$)

EXERCISE 18-11 Write the structures of the products formed in the following reactions:

(a) methyl butanoate + ammonia $\longrightarrow$

(b) ethyl benzoate + isopropyl alcohol $\xrightarrow{H^+}$

(c) isopropyl *cis*-2-butenoate + isopropylamine $\longrightarrow$

18.6 THIOESTERS

Thioesters are compounds containing an $—S—\overset{|}{\underset{|}{C}}—$ group bonded to the carbonyl carbon. The reactions of thioesters are similar to those of esters, but they occur more rapidly. The general equations for the reactions of thioesters with water, alcohols, amines, and other thiols are as follows:

$$\underset{\text{Ester}}{CH_3C(=O)OR'} + \underset{\text{Thiol}}{RCH_2CH_2SH} \xleftarrow{R'OH} \underset{\text{Thioester}}{CH_3C(=O)SCH_2CH_2R} \xrightarrow{R'NH_2} \underset{\text{Amide}}{CH_3C(=O)NHR'} + \underset{\text{Thiol}}{RCH_2CH_2SH}$$

$$\underset{\text{Thioester}}{CH_3C(=O)SCH_2CH_2R} \xrightarrow{H_2O} \underset{\text{Carboxylic acid}}{CH_3C(=O)OH} + \underset{\text{Thiol}}{RCH_2CH_2SH}$$

$$\underset{\text{Thioester}}{CH_3C(=O)SCH_2CH_2R} \xrightarrow{R'SH} \underset{\text{Thioester}}{CH_3C(=O)SR'} + \underset{\text{Thiol}}{RCH_2CH_2SH}$$

Many examples of these reactions of thioesters are found in living systems, as we will learn in the next section.

EXERCISE 18-12 Write the structure of the products formed by the reaction of $(CH_3)_2CHC(=O)SCH_2CH_2CH_3$ with each of the following reagents:
(a) H_2O (b) $CH_3CH_2NH_2$ (c) CH_3SH
(d) $(CH_3)_2CHOH$

18.7 PREPARATION AND HYDROLYSIS OF ESTERS IN LIVING SYSTEMS

Because it is catalyzed by enzymes, esterification in biological systems is a rapid and efficient reaction. One of the most thoroughly studied esterification reactions is that involving acetyl coenzyme A. As we learned in Section 17.6, the important point about acetyl coenzyme A is that it is a thioester. Consequently, the —SCoA portion is readily replaced by alcohols or other nucleophiles to yield acetylated derivatives. For example:

$$\underset{\text{Choline}}{(CH_3)_3\overset{+}{N}CH_2CH_2OH} + \underset{\text{Acetyl coenzyme A}}{CH_3C(=O)SCoA} \xrightarrow{\text{Enzyme}} \underset{\text{Acetylcholine}}{(CH_3)_3\overset{+}{N}CH_2CH_2OC(=O)CH_3} + \underset{\text{Coenzyme A}}{HSCoA}$$

This is a general reaction by which many groups replace the —SCoA portion of an acetyl coenzyme A molecule to form a carboxylic acid derivative.

In living systems, the hydrolysis of esters is also catalyzed by enzymes. An example is the hydrolysis of fats by a specific enzyme into their constituent fatty acids and glycerol. These fatty acids are then biologically oxidized to provide energy for the cell, as we will learn in Chapter 28.

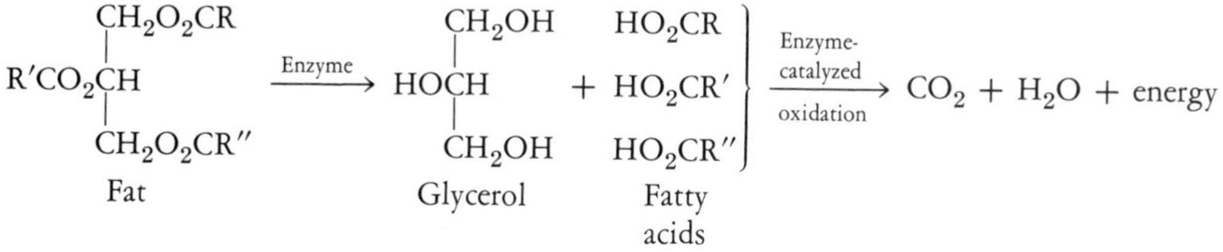

$$\underset{\text{Fat}}{\begin{array}{c} CH_2O_2CR \\ | \\ R'CO_2CH \\ | \\ CH_2O_2CR'' \end{array}} \xrightarrow{\text{Enzyme}} \underset{\text{Glycerol}}{\begin{array}{c} CH_2OH \\ | \\ HOCH \\ | \\ CH_2OH \end{array}} + \underset{\text{Fatty acids}}{\left.\begin{array}{c} HO_2CR \\ HO_2CR' \\ HO_2CR'' \end{array}\right\}} \xrightarrow[\text{oxidation}]{\text{Enzyme-catalyzed}} CO_2 + H_2O + \text{energy}$$

Unlike the acid-catalyzed reaction of preparing esters in the laboratory, which is a reversible reaction, the preparation and hydrolysis of esters in living systems occur by two separate paths. The hydrolysis of acetylcholine is an example. Acetylcholine is not hydrolyzed by the same enzyme that catalyzes its formation from choline and acetyl coenzyme A. Instead, acetylcholine reacts with a hydroxy group on the enzyme to form choline and the acetylated enzyme. This is an example of a transesterification reaction:

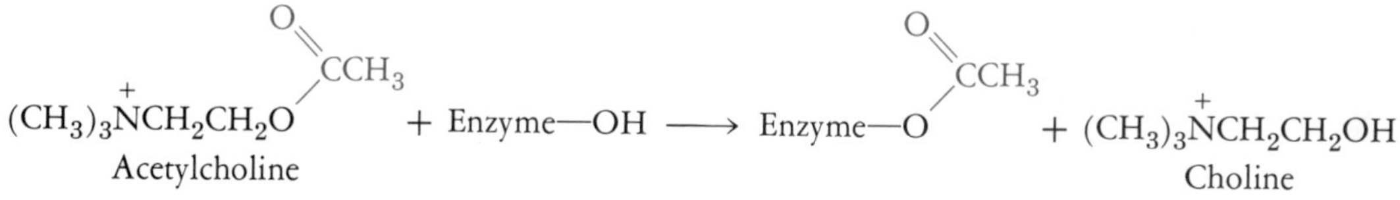

$$\underset{\text{Acetylcholine}}{(CH_3)_3\overset{+}{N}CH_2CH_2OC(=O)CH_3} + \text{Enzyme—OH} \longrightarrow \text{Enzyme—OC(=O)CH}_3 + \underset{\text{Choline}}{(CH_3)_3\overset{+}{N}CH_2CH_2OH}$$

EXERCISE 18-13 Write the structure of the product in each of the following reactions:

(a) 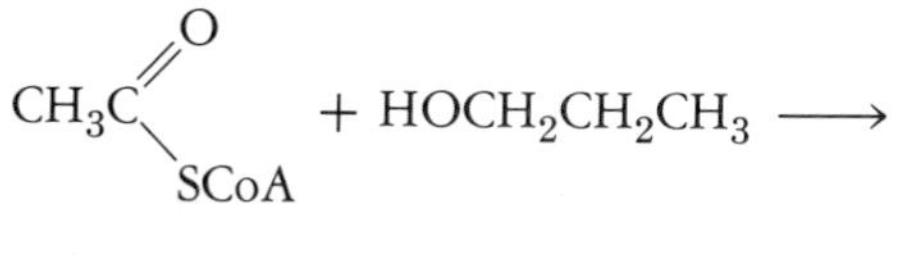 $CH_3C(=O)SCoA + HOCH_2CH_2CH_3 \longrightarrow$

(b) 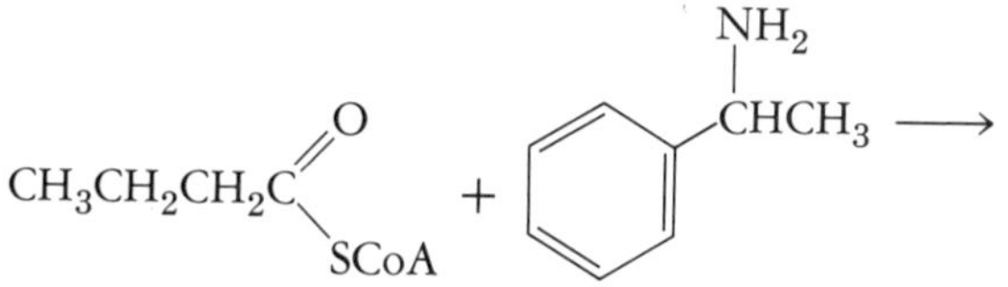 $CH_3CH_2CH_2C(=O)SCoA + C_6H_5CH(NH_2)CH_3 \longrightarrow$

Many of the enzymes that catalyze the hydrolysis of esters also catalyze the hydrolysis of amides. In fact, several reactions of amides resemble those of esters and thioesters.

18.8 AMIDES

Amides can be regarded as carboxylic acids in which the hydroxy group is replaced by an $-NH_2$, $-NHR$, or $-NR_2$ group. The names of amides emphasize this relationship, because they are named as derivatives of the carboxylic acids. Thus, the ending *-oic acid* or *-ic acid* is replaced by *amide*. For example:

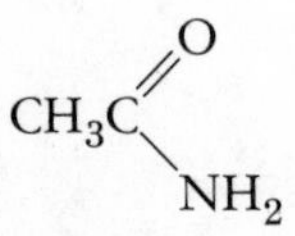

$CH_3C(=O)NH_2$

Acetamide
(related to acetic acid)

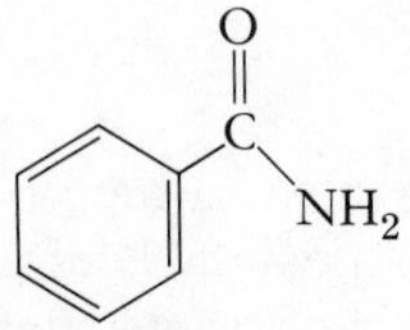

$C_6H_5C(=O)NH_2$

Benzamide
(related to benzoic acid)

To name substituted amides, the substituents on nitrogen are designated as prefixes in which a capital *N* is placed before the name of each substituent. For example:

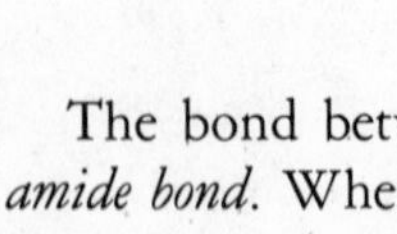

$CH_3C(=O)NHCH_3$

N-Methylacetamide

$CH_3CH_2C(=O)N(CH_3)CH_2CH_3$

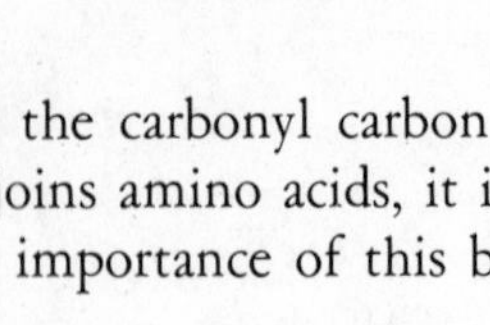

N-Ethyl-*N*-methylpropanamide

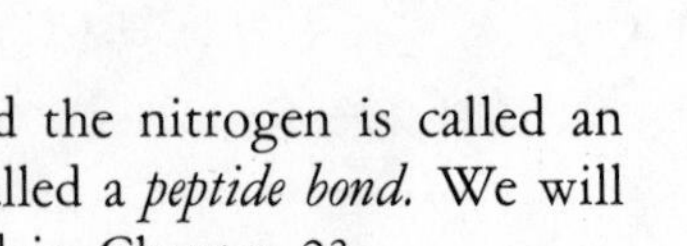

$CH_3CH_2CH_2C(=O)N(CH_2CH_3)_2$

N,N-diethylbutanamide

The bond between the carbonyl carbon and the nitrogen is called an *amide bond*. When it joins amino acids, it is called a *peptide bond*. We will learn more about the importance of this bond in Chapter 23.

EXERCISE 18-14 Give a name to each of the following compounds:

(a) $CH_3CH_2C(=O)NH_2$ (b) $H{-}C(=O)NHCH_3$

(c) $C_6H_5C(=O)NH{-}C_5H_9$ (cyclopentyl) (d) $CH_3CH_2CH_2C(=O)NHCH_2CH(CH_3)_2$

(e) $CH_3CH(Cl)C(=O)NHCH_3$ (f) $(CH_3)_2CHC(=O)N(CH_2CH_2CH_3)_2$

EXERCISE 18-15 Write the structure for each of the following compounds:
(a) butyramide (b) 3-methylpentanamide
(c) *N*-methylpentanamide (d) *N,N*-dimethylbenzamide
(e) *N*-propylpropenamide

In Section 17.6 we learned that there is a relationship between the structure of an amine and its base strength. Thus, substitution for a hydrogen of ammonia by a group that is deactivating in electrophilic aromatic substitution reactions decreases the basicity of the amine. We can apply this idea to amides, because they can be viewed as ammonia in which a hydrogen atom is substituted by an acyl group. The acyl group is deactivating, as we learned in Section 13.3. Therefore, amides should be less basic than ammonia. In fact, their basicity is decreased so much that *amides are neutral compounds.*

Amides are prepared by the reaction of ammonia or amines with a variety of acid derivatives such as esters (Section 18.5) and acid anhydrides (Section 18.9). Their major reaction of biological interest is hydrolysis to a carboxylic acid and an amine or ammonia. For example:

$$C_6H_5C(=O)NH_2 + H_2O \xrightarrow{H^+} C_6H_5C(=O)OH + NH_4^+$$

Benzamide Benzoic acid Ammonium ion

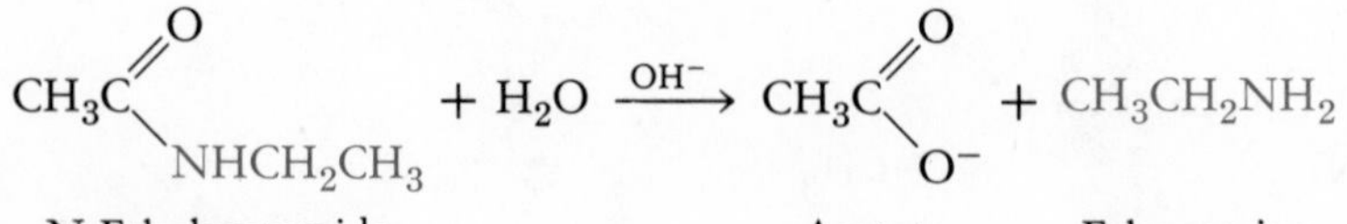

$$\underset{\text{N-Ethylacetamide}}{CH_3C(=O)NHCH_2CH_3} + H_2O \xrightarrow{OH^-} \underset{\text{Acetate ion}}{CH_3C(=O)O^-} + \underset{\text{Ethanamine}}{CH_3CH_2NH_2}$$

Notice that in these reactions the $—NH_2$ group of the amide is replaced by an —OH group. Hydrolysis of amides is catalyzed in the laboratory by either acids or bases; in living systems, enzymes are the catalysts. Carboxylic acid anhydrides undergo similar reactions.

EXERCISE 18-16 Write the structures of the products of acid hydrolysis of each amide in exercises 18-14 and 18-15.

18.9 CARBOXYLIC ACID ANHYDRIDES

A carboxylic acid anhydride can be regarded as a carboxylic acid whose —OH group has been substituted by a carboxylate group. They are named by replacing the word *acid* of the corresponding common or IUPAC name of the carboxylic acid with *anhydride*. For example:

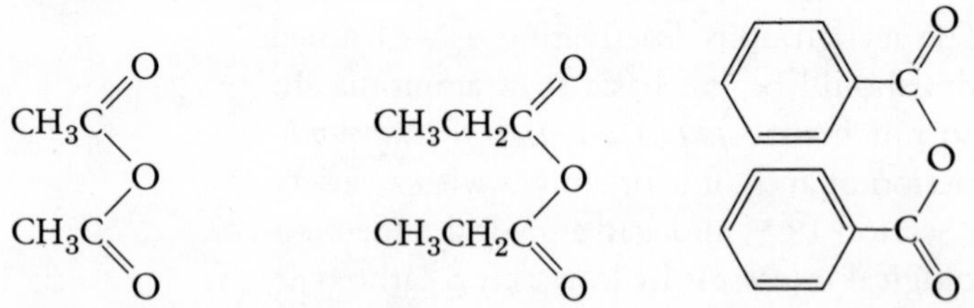

$$CH_3C(=O)-O-C(=O)CH_3 \qquad CH_3CH_2C(=O)-O-C(=O)CH_2CH_3 \qquad C_6H_5C(=O)-O-C(=O)C_6H_5$$

Acetic anhydride (Ethanoic anhydride) — Propionic anhydride (Propanoic anhydride) — Benzoic anhydride

Anhydrides react with water, amines, alcohols, and thiols. All of these reactions involve substituting an —OH, —NHR, —OR, or —SR group for the $—O\overset{\overset{\displaystyle O}{\|}}{C}R$ group, as shown by the following reactions:

$$CH_3C(=O)OH + CH_3C(=O)NHR$$

$$\uparrow NH_2R$$

$$CH_3C(=O)OR + CH_3C(=O)OH \xleftarrow{ROH} (CH_3C(=O))_2O \xrightarrow{H_2O} CH_3C(=O)OH + CH_3C(=O)OH$$

$$\downarrow RSH$$

$$CH_3C(=O)SR + CH_3C(=O)OH$$

Clearly, most of the reactions of carboxylic acids and their derivatives are substitution reactions. However, there is another way of looking at these same reactions, as we will learn in the next section.

18.10 ACYL TRANSFER REACTIONS

Many of the substitution reactions of carboxylic acids and their derivatives in living systems are called *acyl transfer reactions.* This is just another name for the same type of reaction. We can show this by using an example. Consider the following general equation:

Transfer to

$$RCH_2C(=O)OCH_2R' + HSCH_2R'' \longrightarrow RCH_2C(=O)SCH_2R'' + R'CH_2OH$$

The way that we have looked at this reaction so far is to focus our attention on the group bonded to the carbonyl carbon. When viewed in this way, we consider it a substitution reaction. The $—OCH_2R'$ group is substituted by a $—SCH_2R''$ group. If we focus our attention on the fate of the acyl group, we can regard the reaction as an acyl transfer reaction. That is, the acyl group is transferred from the oxygen atom of the ester to the sulfur atom of the thiol.

In most carboxylic acid derivatives in living systems, the group bonded to the carbonyl carbon is much larger and more complex than the acyl group. In the reaction of these compounds, it is easier to focus our attention on the smaller part of the molecule and consider the reaction as an acyl transfer. A typical example is the reaction of acetyl coenzyme A with an acyl carrier protein (ACP). The equation for this reaction is as follows:

$$CH_3C(=O)SCoA + HSACP \rightleftharpoons CH_3C(=O)SACP + HSCoA$$

Acetyl coenzyme A ACP Acetyl ACP Coenzyme A

Both the coenzyme A portion of acetyl coenzyme A and acyl carrier protein are large and complicated. Therefore, it is easier to focus on the transfer of the acyl group, which in this case is an acetyl group. This view is emphasized by the name given to the enzyme that catalyzes this reaction: *ACP-acyltransferase* (see Section 25.2 for naming enzymes). Remember that the reactions are the same. The name given them depends on how we look at them!

EXERCISE 18-17 Give an example of an acetyl transfer reaction in which the acetyl group is transferred from a nitrogen atom to an oxygen atom.

18.11 MECHANISM OF REACTIONS OF CARBOXYLIC ACIDS AND THEIR DERIVATIVES

OPTIONAL SECTION

The proposed mechanism of the reactions of carboxylic acids and their derivatives is based on the fact that a carbonyl group is polarized. In this bond the carbon, which is more electropositive, has a partial positive charge and the oxygen has a partial negative charge. The charge distribution is often represented as follows:

$$\overset{\delta+}{C}=\overset{\delta-}{O} \quad \left[C=O \longleftrightarrow \overset{+}{C}-\overset{-}{O} \right]$$

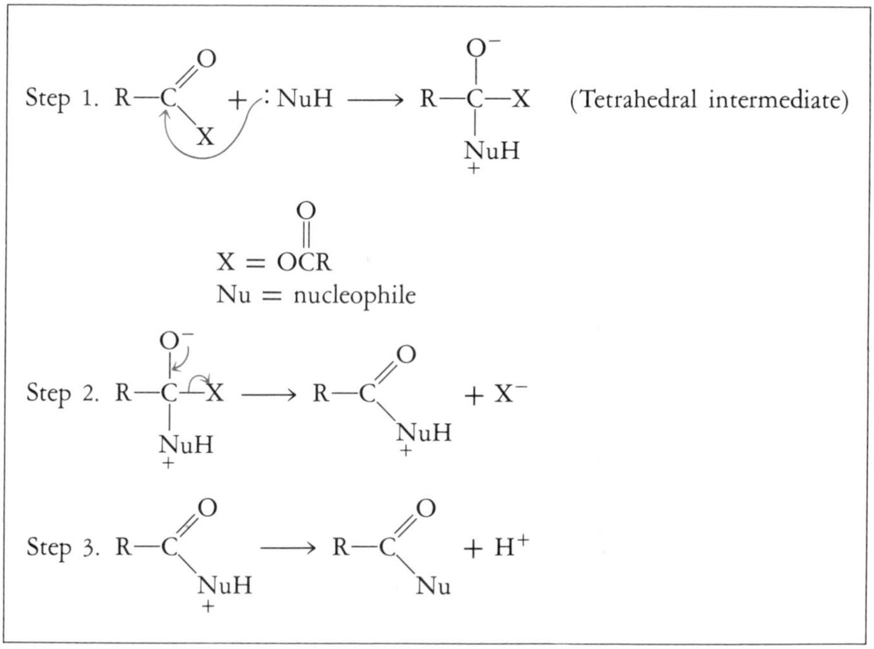

Step 1. $R-C(=O)-X + :NuH \longrightarrow R-C(O^-)(X)-\overset{+}{N}uH$ (Tetrahedral intermediate)

$X = -O\overset{O}{\overset{\|}{C}}R$
Nu = nucleophile

Step 2. $R-C(O^-)(X)-\overset{+}{N}uH \longrightarrow R-C(=O)-\overset{+}{N}uH + X^-$

Step 3. $R-C(=O)-\overset{+}{N}uH \longrightarrow R-C(=O)-Nu + H^+$

Fig. 18-1. General mechanism proposed for the reaction of acid anhydrides and nucleophiles.

The positive charge on the carbonyl carbon is increased when a strong electronegative group, such as the $-O\overset{O}{\overset{\|}{C}}R$ group is bonded to it. As a result, a nucleophile readily adds to this carbon to form a tetrahedral intermediate. During the addition, one of the two pairs of electrons of the carbonyl group moves to the oxygen atom. Because of its electronegativity, it can easily support the charge that it acquires. This is the first step of the mechanism given in Figure 18-1. The group X is expelled from the intermediate in the second step. Finally, a proton is lost to form the observed product. This is the general mechanism proposed for the reaction of acid anhydrides with water, alcohols, and amines. This general mechanism is applied to a specific reaction in the following example:

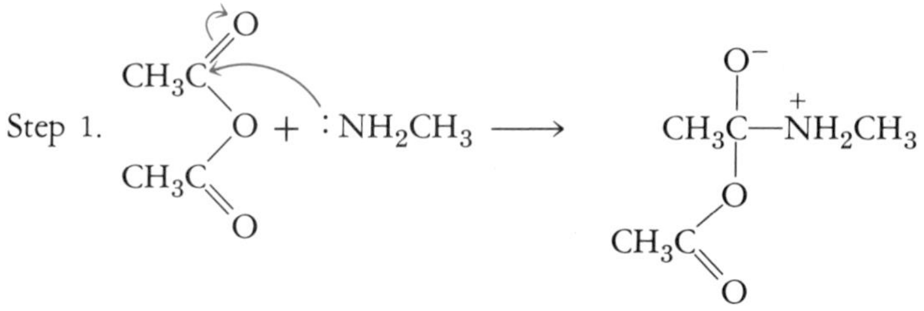

Step 1. $(CH_3C(=O))_2O + :NH_2CH_3 \longrightarrow CH_3C(O^-)(\overset{+}{N}H_2CH_3)-O-C(=O)CH_3$

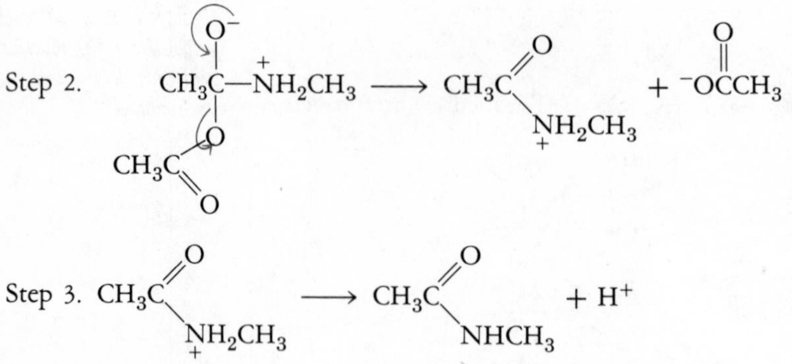

Step 2. $CH_3C(O^-)(OC(=O)CH_3)-\overset{+}{N}H_2CH_3 \longrightarrow CH_3C(=O)\overset{+}{N}H_2CH_3 + {}^-OC(=O)CH_3$

Step 3. $CH_3C(=O)\overset{+}{N}H_2CH_3 \longrightarrow CH_3C(=O)NHCH_3 + H^+$

EXERCISE 18-18 Write the mechanism of the reaction of acetic anhydride and each of the following compounds:
(a) water (b) dimethylamine

The nitrogen of an amide or the oxygen of the alkoxy group of an ester is less electronegative, so the positive charge on the carbonyl carbon is less in

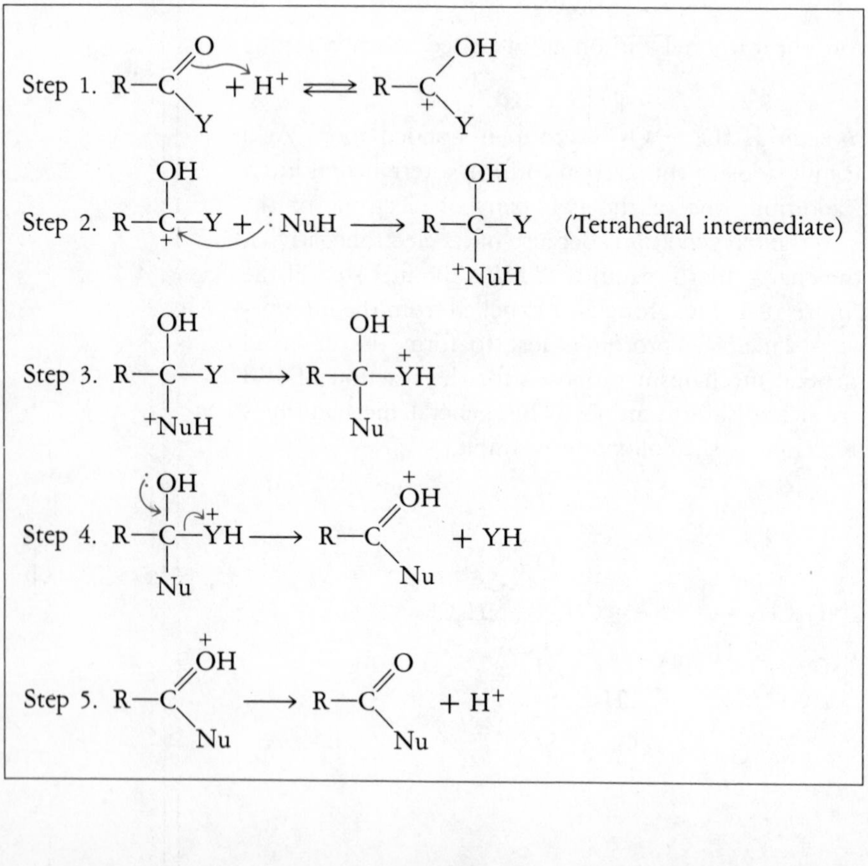

Step 1. $R-C(=O)Y + H^+ \rightleftharpoons R-\overset{+}{C}(OH)Y$

Step 2. $R-\overset{+}{C}(OH)-Y + :NuH \longrightarrow R-C(OH)(Y)-{}^+NuH$ (Tetrahedral intermediate)

Step 3. $R-C(OH)(Y)-{}^+NuH \longrightarrow R-C(OH)(Nu)-\overset{+}{Y}H$

Step 4. $R-C(OH)(Nu)-\overset{+}{Y}H \longrightarrow R-C(=\overset{+}{O}H)Nu + YH$

Step 5. $R-C(=\overset{+}{O}H)Nu \longrightarrow R-C(=O)Nu + H^+$

Fig. 18-2. General mechanism proposed for the acid-catalyzed reactions of amides or esters and nucleophiles.

these compounds than in acid anhydrides. As a result, amides and esters react very slowly with nucleophiles. An acid catalyst is used to speed up the reaction. The proton adds to the carbonyl oxygen and increases the positive charge on the carbonyl carbon. We can represent this reaction as follows:

$$\gt C{=}O + H^+ \rightleftharpoons \left[\gt C{=}OH^+ \longleftrightarrow \gt \overset{+}{C}{-}OH \right]$$

By increasing the positive charge on the carbon, the carbonyl group becomes more attractive to a nucleophile, which bonds to the carbonyl carbon to form a tetrahedral intermediate. These are the first two steps of the mechanism given in Figure 18-2.

Before the tetrahedral intermediate shown in Figure 18-2 can decompose, it must transfer a proton to the group Y. Once this occurs, the group Y is lost. Finally, a proton is lost in the last step to form the product. This general mechanism is applied to a specific specific reaction in the following example:

Step 1. $CH_3C(=O)NHCH_3 + H^+ \rightleftharpoons CH_3\overset{+}{C}(OH){-}NHCH_3$

Step 2. $CH_3\overset{+}{C}(OH){-}NHCH_3 + OH_2 \longrightarrow CH_3C(OH)(\overset{+}{O}H_2){-}NHCH_3$

Step 3. $CH_3C(OH)(\overset{+}{O}H_2){-}NHCH_3 \longrightarrow CH_3C(OH)(HO){-}\overset{+}{N}H(H)CH_3$

Step 4. $CH_3C(\ddot{O}H)(OH){-}\overset{+}{N}H(H)CH_3 \longrightarrow CH_3C(=\overset{+}{O}H)OH + NH_2CH_3$

Step 5. $CH_3C(=\overset{+}{O}H)OH \longrightarrow CH_3C(=O)OH + H^+$

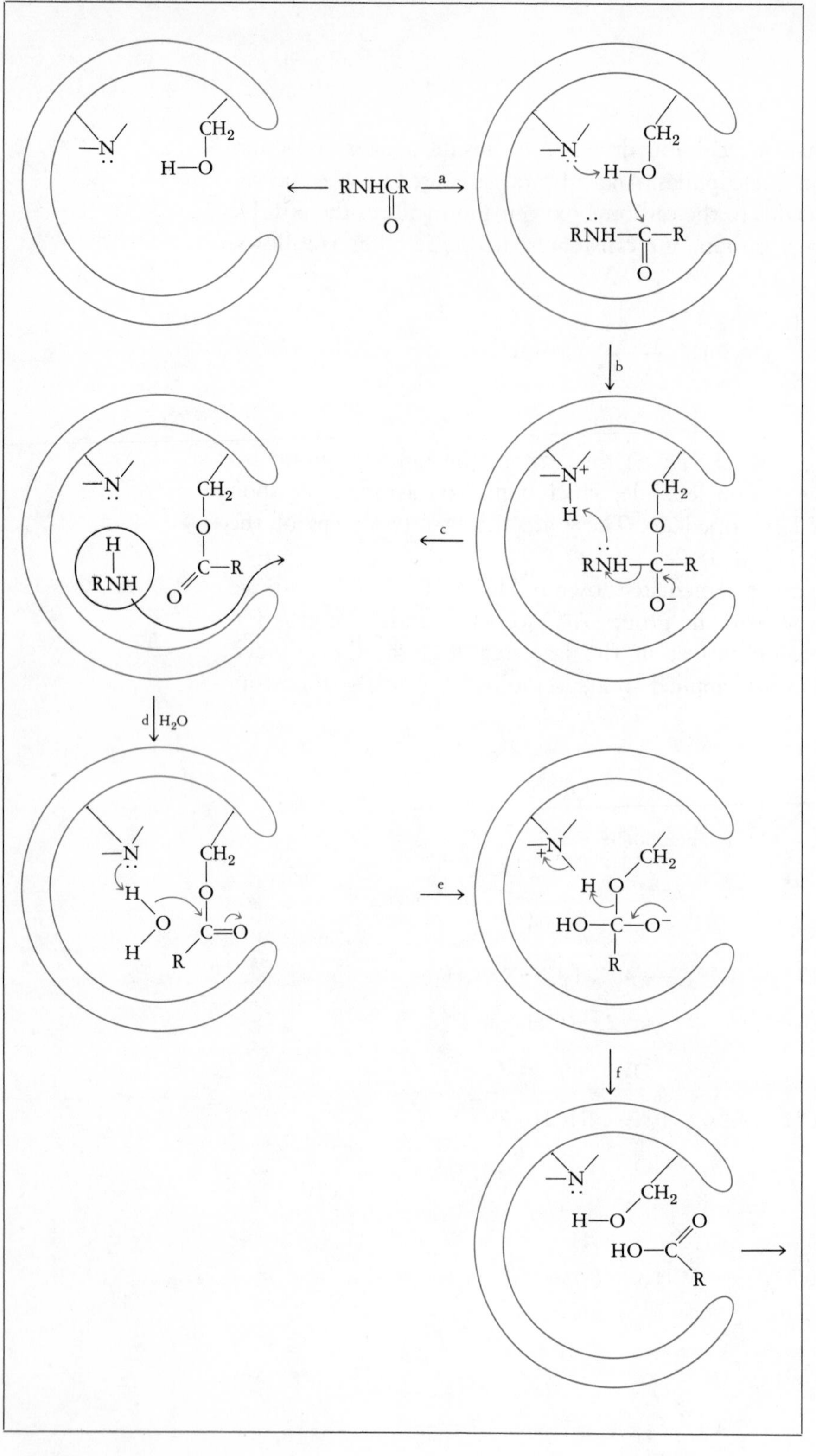

Fig. 18-3. A schematic representation of the enzyme-catalyzed hydrolysis of an amide. The amide enters the cavity (a). A hydroxy group bonded to the enzyme attacks the carbonyl group, forming a tetrahedral intermediate. The hydrogen of the hydroxy group is transferred to the enzyme-bound nitrogen base (b). The tetrahedral intermediate decomposes, forming an acyl-enzyme ester and the free amine (c). The amine leaves the cavity and is replaced by a water molecule (d). The water molecule attacks the carbonyl carbon to form another tetrahedral intermediate (e). This tetrahedral intermediate decomposes to form the carboxylic acid, and the enzyme is restored to catalyze another reaction (f).

EXERCISE 18-19 Write a mechanism for the acid-catalyzed reaction of methyl propanoate with:
(a) water (b) ethanol (c) ammonia

It is proposed that a tetrahedral intermediate is also involved in the enzyme-catalyzed hydrolysis of amides. A schematic representation of this proposed mechanism is given in Figure 18-3. The reaction occurs in a cavity of the enzyme. Specific groups within the cavity, a hydroxy group and a nitrogen base, are involved in this reaction. The hydroxy group is the nucleophile, and the nitrogen base transfers a proton among various sites during the reaction.

Although there are many similarities between the mechanisms in Figures 18-1, 18-2, and 18-3, there is one major difference. Enzyme-catalyzed hydrolysis of amides occurs about 1 billion times faster than the acid-catalyzed reaction! This is a general characteristic of the reactions in living systems. They occur much faster than similar reactions carried out in nonliving systems.

18.12 PHOSPHORIC ACID AND ITS DERIVATIVES

Phosphoric acid, like carboxylic acids, can form esters, amides, and anhydrides. A greater variety of such compounds can be formed with phosphoric acid because it has three replaceable hydrogens. Examples of some of the phosphoric acid derivatives found in living systems are shown in Figure 18-4.

These compounds are quite soluble in the aqueous environment of living systems because at the pH of the cell they exist predominantly as ions. Usually one or more of the acidic hydrogens are dissociated. These compounds undergo the same general types of reaction as the corresponding carboxylic acid derivative, as shown by the following examples:

$$\mathrm{HO{-}\overset{\overset{\displaystyle O}{\|}}{\underset{\underset{\displaystyle O^-}{|}}{P}}{-}O{-}\overset{\overset{\displaystyle O}{\|}}{\underset{\underset{\displaystyle O^-}{|}}{P}}{-}OH} + H_2O \longrightarrow 2\,\mathrm{HO{-}\overset{\overset{\displaystyle O}{\|}}{\underset{\underset{\displaystyle O^-}{|}}{P}}{-}OH}$$

Pyrophosphate Hydrogen phosphate

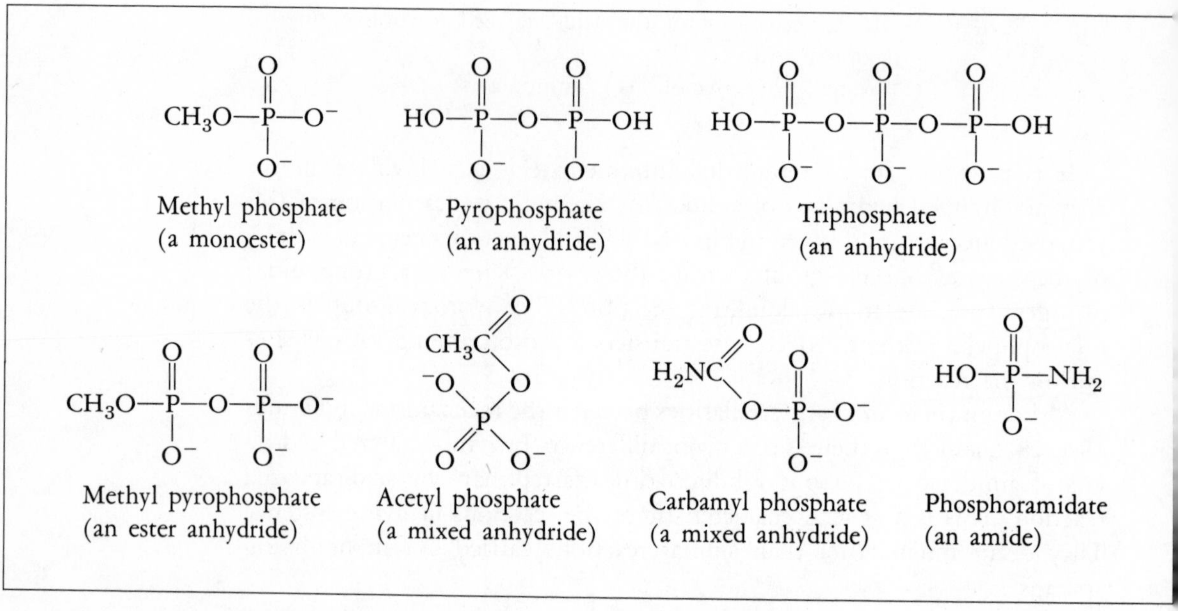

$CH_3O-P(=O)(O^-)-O^-$

Methyl phosphate
(a monoester)

$HO-P(=O)(O^-)-O-P(=O)(O^-)-OH$

Pyrophosphate
(an anhydride)

$HO-P(=O)(O^-)-O-P(=O)(O^-)-O-P(=O)(O^-)-OH$

Triphosphate
(an anhydride)

$CH_3O-P(=O)(O^-)-O-P(=O)(O^-)-O^-$

Methyl pyrophosphate
(an ester anhydride)

$CH_3C(=O)-O-P(=O)(O^-)-O^-$

Acetyl phosphate
(a mixed anhydride)

$H_2NC(=O)-O-P(=O)(O^-)-O^-$

Carbamyl phosphate
(a mixed anhydride)

$HO-P(=O)(O^-)-NH_2$

Phosphoramidate
(an amide)

Fig. 18-4. Structures of phosphoric acid derivatives.

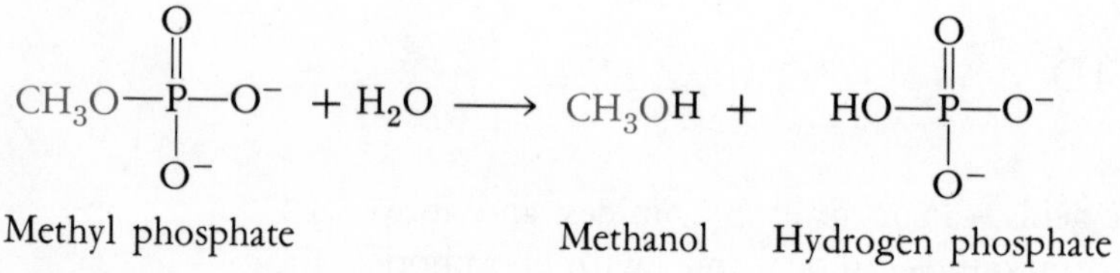

$$CH_3O-P(=O)(O^-)-O^- + H_2O \longrightarrow CH_3OH + HO-P(=O)(O^-)-O^-$$

Methyl phosphate — Methanol — Hydrogen phosphate

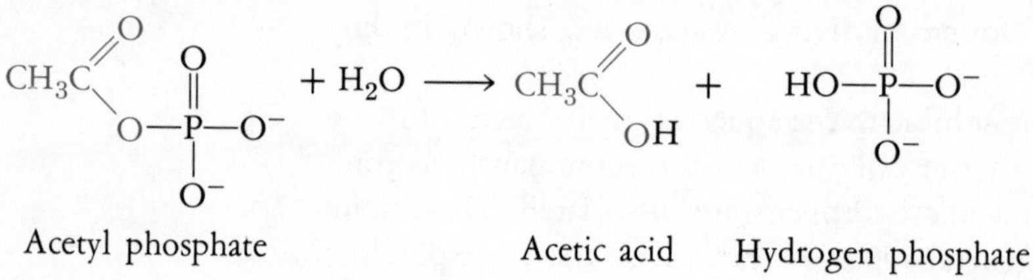

$$CH_3C(=O)-O-P(=O)(O^-)-O^- + H_2O \longrightarrow CH_3C(=O)OH + HO-P(=O)(O^-)-O^-$$

Acetyl phosphate — Acetic acid — Hydrogen phosphate

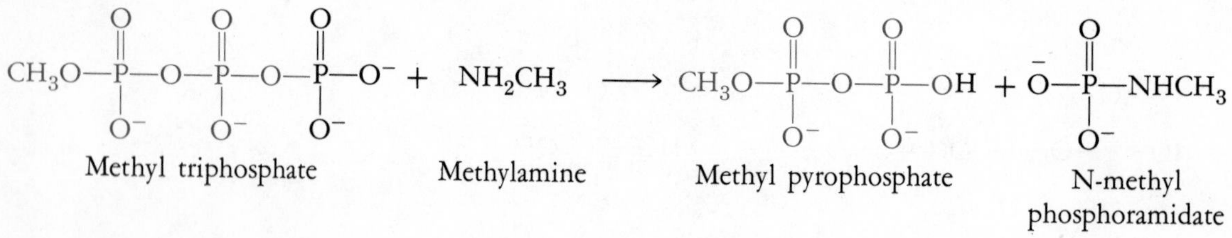

$$CH_3O-P(=O)(O^-)-O-P(=O)(O^-)-O-P(=O)(O^-)-O^- + NH_2CH_3 \longrightarrow CH_3O-P(=O)(O^-)-O-P(=O)(O^-)-OH + {}^-O-P(=O)(O^-)-NHCH_3$$

Methyl triphosphate — Methylamine — Methyl pyrophosphate — N-methyl phosphoramidate

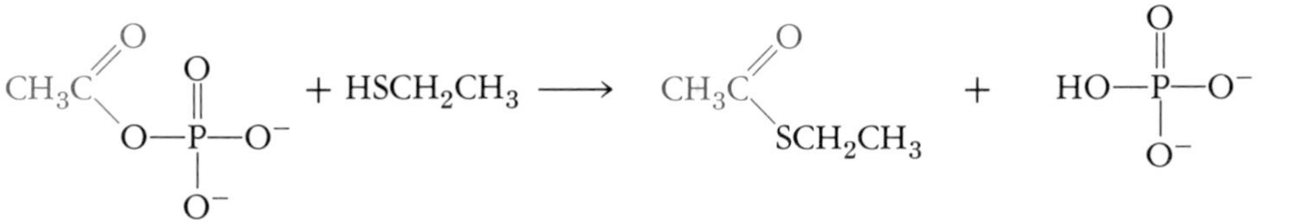

$$CH_3C(=O)-O-P(=O)(O^-)-O^- + HSCH_2CH_3 \longrightarrow CH_3C(=O)SCH_2CH_3 + HO-P(=O)(O^-)-O^-$$

Acetyl phosphate   Ethanethiol   S-Ethyl ethanethiolate   Hydrogen phosphate

We will encounter several examples of these compounds and their reactions in Chapters 27, 28, and 29.

18.13 SUMMARY

Carboxylic acids are compounds that contain a carboxyl functional group $\left(-C(=O)OH\right)$. Other classes of compounds are also known that have electronegative groups bonded to the carbonyl carbon. Esters $\left(-C(=O)OR\right)$, thioesters $\left(-C(=O)SR\right)$, amides $\left(-C(=O)NH_2\right)$, and carboxylic acid anhydrides $\left(-C(=O)-O-C(=O)-\right)$ are examples of such compounds. They are called derivatives of carboxylic acids because they can be hydrolyzed to carboxylic acids.

The major reaction that derivatives of carboxylic acids undergo in living systems is a substitution reaction. In this reaction, the group bonded to the carbonyl carbon is substituted by another group. Four important examples of this reaction are found in living systems. They are hydrolysis, reactions with alcohols, reactions with amines, and reactions with thiols.

The most important reaction of carboxylic acids is their acidity. The strength of an acid is quantitatively expressed as K_a, the dissociation constant. Carboxylic acids are weak acids. They generally have $K_a \approx 10^{-5}$. However, at the buffered pH of cells, most carboxylic acids exist predominantly as the carboxylate ion. All carboxylic acids react completely with strong bases to form salts. Treating these salts with strong inorganic acid re-forms the carboxylic acid.

REVIEW OF TERMS, CONCEPTS, AND REACTIONS

Terms

ACYL GROUP The general name of the structural unit $-\overset{|}{\underset{|}{C}}-C(=O)-$

ACYL PHOSPHATES A class of compound that contains the $-C(=O)-O-\overset{O}{\overset{\|}{P}}(-O^-)-O^-$ functional group; also called a mixed anhydride.

AMIDES Compounds that contain the $-C(=O)-\underset{|}{N}<$ functional group.

AMMONOLYSIS Reaction of an ester with ammonia to form an amide.

CARBONYL GROUP A carbon-oxygen double bond.

CARBOXYL GROUP The $-C(=O)-OH$ functional group.

CARBOXYLATE ION General name for compounds that contain the $-C(=O)-O^-$ functional group.

CARBOXYLIC ACID ANHYDRIDES Compounds that contain the $-\underset{\underset{O}{\|}}{C}-O-\underset{\underset{O}{\|}}{C}-$ functional group.

ESTERS Compounds that contain the $-C(=O)-O-\overset{|}{\underset{|}{C}}-$ functional group.

SAPONIFICATION The hydrolysis of an ester by an aqueous base.

THIOESTERS Compounds that contain the $-C(=O)-S-\overset{|}{\underset{|}{C}}-$ functional group.

Concepts

1. Esters, thioesters, amides, and anhydrides are derivatives of carboxylic acids that undergo four reactions that are important in living systems: hydrolysis, and reactions with amines, alcohols, and thiols.

2. Most carboxylic acids are weak acids. They react with bases to form salts.
3. Amides, unlike amines, are neutral compounds.
4. Substitution and acyl transfer are two ways of looking at the same type of reaction.

Reactions

Carboxylic Acids (see Sections 18.3, 18.4, and 18.5)

1. Preparation by Oxidation
   a. Of alcohols
   $$RCH_2OH \longrightarrow RCO_2H$$
   b. Of alkenes
   $$RCH{=}CHR' \longrightarrow RCO_2H + HO_2CR'$$
   c. Of aromatic compounds
   $$C_6H_5CH_2R \longrightarrow C_6H_5CO_2H + CO_2 + H_2O$$
2. Reactions
   a. With bases
   $$RCO_2H + NaOH \longrightarrow RCO_2^- Na^+ + H_2O$$
   b. With alcohols
   $$RCO_2H + R'OH \xrightleftharpoons{H^+} RC(=O)OR' + H_2O$$

Reactions of Carboxylic Acid Anhydrides (see Section 18.9)

$$RC(=O)X + H_2O \longrightarrow RC(=O)OH + HX$$

$$RC(=O)X + R'OH \longrightarrow RC(=O)OR' + HX$$

$$X = \overset{O}{\overset{\|}{O}}CR$$

$$RC(=O)X + NHY_2 \longrightarrow RC(=O)NY_2 + HX$$

Y = H, alkyl group, or aryl group

$$RC(=O)X + HSR' \longrightarrow RC(=O)SR' + HX$$

Esters (see Sections 18.5 and 18.9)

1. Preparation

a. $RCO_2H + R'OH \xrightleftharpoons{H^+} RC(=O)OR' + H_2O$

b. $(RC(=O))_2O + R'OH \longrightarrow RC(=O)OR' + RCO_2H$

2. Reactions

a. $RCO_2R' + NaOH \xrightarrow{H_2O} RC(=O)O^- \, Na^+ + R'OH$

b. $RCO_2R' + R''CH_2OH \xrightleftharpoons{H^+} RC(=O)OCH_2R'' + R'OH$

c. $RCO_2R' + NH_2R'' \longrightarrow RC(=O)NHR'' + R'OH$

d. $RCO_2R' + HSR'' \longrightarrow RC(=O)SR'' + R'OH$

Amides (see Sections 17.6, 18.5, and 18.8)

1. Preparation from carboxylic acid anhydrides and esters

$$RC(=O)X + NHY_2 \longrightarrow RC(=O)NY_2 + HX$$

$X = OC(=O)R$ or OR' $\quad$ $Y = H$, alkyl group, or aryl group

2. Hydrolysis

$$RC(=O)NY_2 + H_2O \xrightarrow[\text{or enzyme}]{\text{Acid, base,}} RC(=O)OH + HNY_2$$

Reactions of thioesters (see Section 18.6)

a. $$RC(=O)SCH_2R' + NH_2R'' \longrightarrow RC(=O)NHR'' + HSCH_2R'$$

b. $$RC(=O)SCH_2R' + HOR'' \longrightarrow RC(=O)OR'' + HSCH_2R'$$

c. $$RC(=O)SCH_2R' + H_2O \longrightarrow RC(=O)OH + HSCH_2R'$$

d. $$RC(=O)SCH_2R' + R''SH \longrightarrow RC(=O)SR'' + HSCH_2R'$$

Reactions of Phosphoric Acid Derivatives (see Section 18.12)

a. $$RO{-}P(=O)(O^-){-}O{-}P(=O)(O^-){-}O{-}P(=O)(O^-){-}O^- + NH_2R'' \longrightarrow RO{-}P(=O)(O^-){-}O{-}P(=O)(O^-){-}OH + {}^-O{-}P(=O)(O^-){-}NHR''$$

b. $$HO{-}P(=O)(O^-){-}O{-}P(=O)(O^-){-}OH + H_2O \longrightarrow 2\,HO{-}P(=O)(O^-){-}OH$$

c. $$RC(=O){-}O{-}P(=O)(O^-){-}O^- + H_2O \longrightarrow RC(=O)OH + HO{-}P(=O)(O^-){-}O^-$$

d. $$RC(=O){-}O{-}P(=O)(O^-){-}O^- + HSR' \longrightarrow RC(=O)SR' + HO{-}P(=O)(O^-){-}O^-$$

e. $$RCH_2O{-}P(=O)(O^-){-}O^- + H_2O \longrightarrow RCH_2OH + HO{-}P(=O)(O^-){-}O^-$$

EXERCISES 18-20 Give an example of each of the following:
(a) a thioester (b) a carboxylic acid (c) an amide
(d) an acyl phosphate (e) an acyl transfer reaction
(f) an anhydride (g) an ester (h) an *N*-substituted amide
(i) the salt of a carboxylic acid

18-21 Give a name to each of the following compounds:

(a) 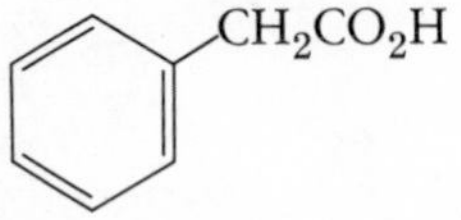 $C_6H_5CH_2CO_2H$ (b) $(CH_3)_2CHC(=O)OCH_3$

(c) $(CH_3)_2CHCH_2C(=O)OCH_2CH_3$ (d) 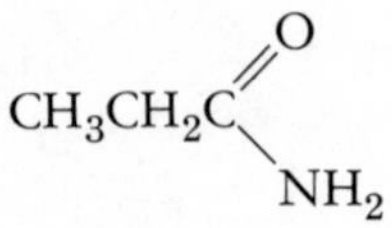 $CH_3CH_2C(=O)NH_2$

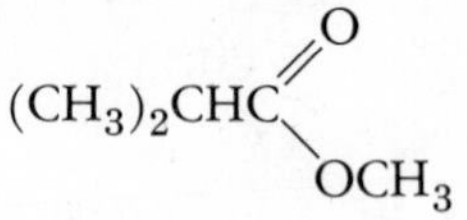

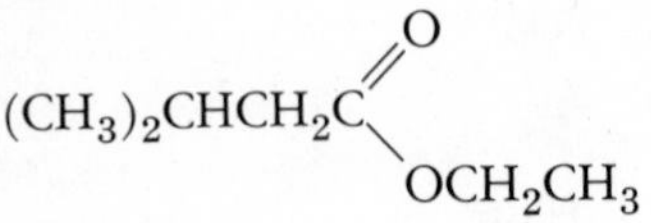

(e) 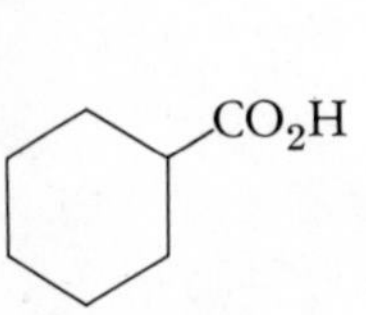 $C_6H_{11}CO_2H$ (f) 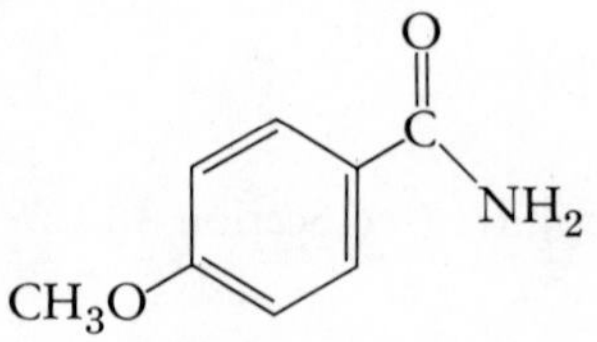 p-$CH_3OC_6H_4C(=O)NH_2$

(g) 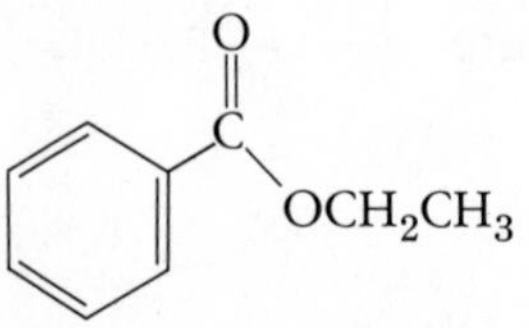 $C_6H_5C(=O)OCH_2CH_3$ (h) 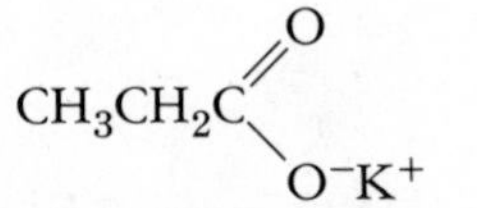 $CH_3CH_2C(=O)O^-K^+$

(i) 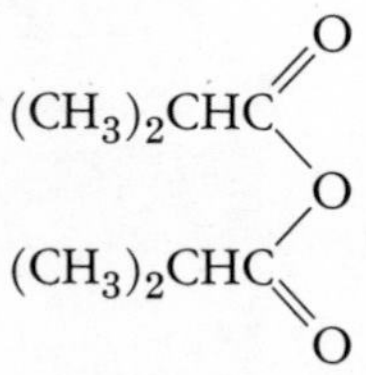 $(CH_3)_2CHC(=O)OC(=O)CHC(CH_3)_2$

18-22 Write the structure for each of the following compounds:
(a) 2-chloropropanamide (b) *p*-nitrophenyl acetate
(c) *N,N*-dimethylpropanamide (d) benzoic anhydride
(e) trimethylacetic acid (f) sodium benzoate
(g) 2-ethyloctanoic acid (h) α-aminobutyric acid

18-23 Write an equation for the reaction of each of the following reagents with acetic acid:
(a) CH_3CH_2OH, H^+ (b) aqueous NaOH

18-24 Write the structure(s) and give the name(s) of the major organic product(s) of the following reactions:

(a) 1-butene + $KMnO_4 \xrightarrow[H_2O]{H^+}$

(b) toluene + $K_2Cr_2O_7 \xrightarrow[H_2O]{H^+}$

(c) pentanoic acid + NaOH $\xrightarrow{H_2O}$

(d) propionic acid + $LiAlH_4 \longrightarrow$

(e) acetic acid + ethanol $\xrightarrow{H^+}$

(f) sodium hexanoate + HCl $\longrightarrow$

(g) 1-pentanol + $KMnO_4 \xrightarrow[H_2O]{H^+}$

(h) acetyl CoA + 1-propanol $\xrightarrow{\text{Enzyme}}$

(i) ethyl butyrate + 2-propanol $\xrightarrow{H^+}$

(j) *t*-butyl propanoate + NaOH $\xrightarrow{H_2O}$

(k) 2-methylbutanamide + NaOH $\xrightarrow{H_2O}$

(l) acetic anhydride + $CH_3OH \longrightarrow$

(m) ethyl benzoate + $NH_3 \longrightarrow$

ALDEHYDES AND KETONES

19

Aldehydes and ketones are two more classes of compounds that contain the carbonyl group. An aldehyde contains a carbonyl group whose carbon is bonded to one hydrogen and either an alkyl or an aryl group. For example:

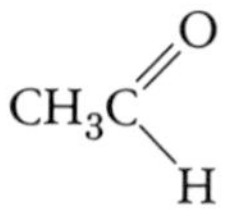

CH_3CHO

Acetaldehyde
(an aliphatic aldehyde)

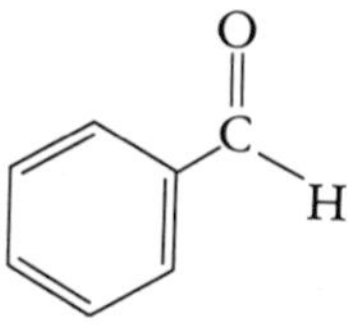

C_6H_5CHO

Benzaldehyde
(an aromatic aldehyde)

The aldehyde group is often written RCHO for convenience.

A ketone contains a carbonyl group whose carbon is bonded to two alkyl groups, two aryl groups, or one alkyl and one aryl group. The groups need not be identical. For example:

$CH_3\overset{O}{\overset{\|}{C}}CH_2CH_3$

2-Butanone
(an aliphatic ketone)

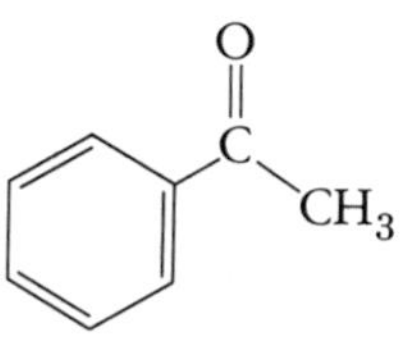

$C_6H_5COCH_3$

Acetophenone
(an aryl alkyl ketone)

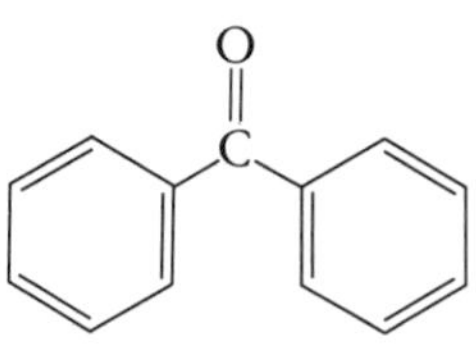

$C_6H_5COC_6H_5$

Benzophenone
(an aromatic ketone)

In this chapter, we will study the general chemistry of aldehydes and ketones with particular attention to their two major reactions: addition to the carbonyl group and condensation reactions. Many examples of these two reactions are found in both the laboratory and living systems. Let us start, as usual, by learning to name aldehydes and ketones.

19.1 NAMING ALDEHYDES AND KETONES

Aldehydes and ketones, like most other types of organic compounds, are known by either common or IUPAC names. Common names are almost always used for aldehydes of five carbons or less. They are obtained from the name of the carboxylic acid formed by oxidation of the aldehyde. The *-ic* or *-oic acid* ending of the name of the carboxylic acid is replaced by *-aldehyde.*

Thus, the aldehyde that yields formic acid when oxidized is called formaldehyde, the aldehyde that yields acetic acid when oxidized is called acetaldehyde, and so forth. The following are more examples of the common method of naming aldehydes:

CH_3CH_2CHO — Propionaldehyde

C_6H_5CHO — Benzaldehyde

$CH_3CH_2CH_2CHO$ — Butyraldehyde

As with carboxylic acids, the position of substituents in the aldehyde can be indicated in the common name by Greek letters starting with α at the position adjacent to the carbonyl group:

$$\overset{\delta}{C}—\overset{\gamma}{C}—\overset{\beta}{C}—\overset{\alpha}{C}—C(=O)H$$

For example:

$CH_3CH(Cl)CHO$ — α-Chloropropionaldehyde

$BrCH_2CH_2CH_2CHO$ — γ-Bromobutyraldehyde

$CH_3OC(CH_3)_2CHO$ — α-Methoxyisobutyraldehyde

The IUPAC names of aldehydes are formed by replacing the *-e* of the name of the longest carbon chain containing the aldehyde group with *-al.* Because the aldehyde group, like the carboxyl group, must always be at the end of the chain, it is assigned the number 1. However, the number is never included in the name. In the following examples, the common names, where applicable, are given in parentheses:

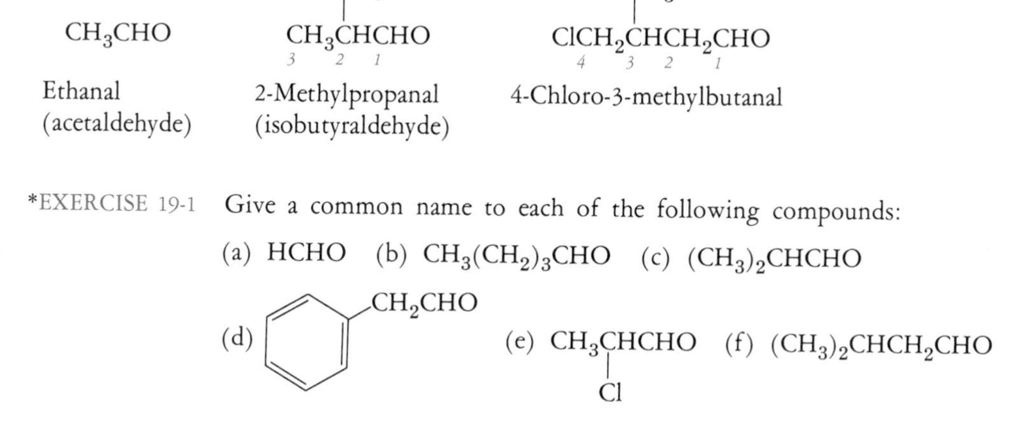

CH_3CHO — Ethanal (acetaldehyde)

$\overset{3}{CH_3}\overset{2}{CH}(CH_3)\overset{1}{CHO}$ — 2-Methylpropanal (isobutyraldehyde)

$\overset{4}{ClCH_2}\overset{3}{CH}(CH_3)\overset{2}{CH_2}\overset{1}{CHO}$ — 4-Chloro-3-methylbutanal

*EXERCISE 19-1 Give a common name to each of the following compounds:

(a) $HCHO$ (b) $CH_3(CH_2)_3CHO$ (c) $(CH_3)_2CHCHO$

(d) $C_6H_5CH_2CHO$ (e) $CH_3CH(Cl)CHO$ (f) $(CH_3)_2CHCH_2CHO$

EXERCISE 19-2 Give the IUPAC name for each of the structures in exercise 19-1.

EXERCISE 19-3 Give the IUPAC name for each of the following compounds:

(a) $CH_3(CH_2)_5CHO$ (b) $CH_3(H)C{=}C(H)CH_2CHO$ (CH$_3$ and CH$_2$CHO trans)

(c) $CH_3CH(CH_2CH_3)CH(Br)CHO$ (d) $H(Br)C{=}C(H)CHO$ (Br and CHO trans)

(e) $C_6H_{11}CH_2CH(Br)CHO$ (cyclohexyl) (f) $CH_3C{\equiv}CCHO$

(g) 2,4-dichloro-substituted benzene ring bearing CHO (CHO with Cl ortho, and Cl para to CHO)

* The answers for the exercises in this chapter begin on page 883.

EXERCISE 19-4 Write the structural formula corresponding to each of the following names:
(a) valeraldehyde (b) propanal (c) 3-methylpentanal
(d) 2-methyl-3-phenylpropenal (e) *trans*-2-butenal
(f) trimethylacetaldehyde (g) α,γ-dimethylcaproaldehyde
(h) *p*-tolualdehyde

EXERCISE 19-5 Indicate which of the compounds in exercises 19-1, 19-3, and 19-4 are capable of existing as
(a) enantiomers (b) diastereomers (c) *cis* and *trans* isomers

The common names of ketones are formed by placing the names of the groups attached to the carbonyl group as prefixes to the word ketone. The names are written as separate words. For example:

$CH_3CH_2\overset{\overset{O}{\parallel}}{C}CH_3$	$CH_3CH_2\overset{\overset{O}{\parallel}}{C}CH_2CH_3$	$CH_3CH_2\overset{\overset{O}{\parallel}}{C}CH{=}CH_2$	$C_6H_5CH_2\overset{\overset{O}{\parallel}}{C}CH_3$
Ethyl methyl ketone	Diethyl ketone	Ethyl vinyl ketone	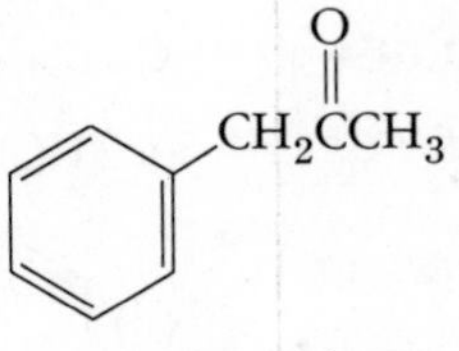 Benzyl methyl ketone

The following ketones are known generally by their trivial names:

$CH_3\overset{\overset{O}{\parallel}}{C}CH_3$	$C_6H_5\overset{\overset{O}{\parallel}}{C}CH_3$	$C_6H_5\overset{\overset{O}{\parallel}}{C}C_6H_5$
Acetone	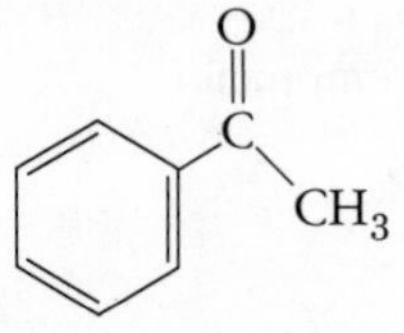 Acetophenone	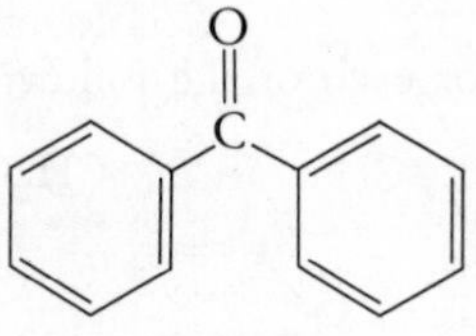 Benzophenone

The IUPAC names of ketones are derived from the name of the longest carbon chain that contains the carbonyl group by replacing the *-e* with the suffix *-one.* The location of the carbonyl group is designated by the lowest number placed just before the parent name. If another suffix is present, the number indicating the position of the carbonyl group is placed just before its suffix. For example:

$\overset{5}{C}H_3\overset{4}{C}H_2\overset{3}{C}H_2\overset{\overset{O}{\parallel}}{\underset{2}{C}}\overset{1}{C}H_3$	$(\overset{5}{C}H_3)_2\overset{4}{C}{=}\overset{3}{C}H\overset{\overset{O}{\parallel}}{\underset{2}{C}}\overset{1}{C}H_3$	$(CH_2)_5C{=}O$ (ring)
2-Pentanone	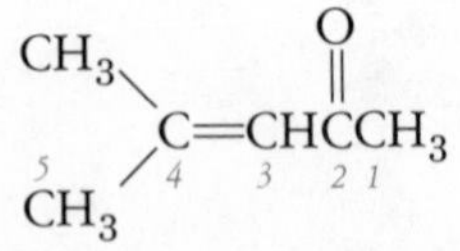 4-Methyl-3-penten-2-one	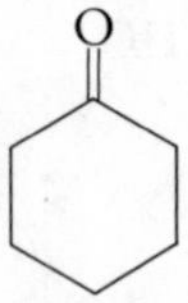 Cyclohexanone

EXERCISE 19-6 Give a common name to each of the following compounds:

(a) $CH_3CH_2CH_2\overset{O}{\overset{\|}{C}}CH_2CH_3$ (b) $(CH_3)_2CHCH_2\overset{O}{\overset{\|}{C}}CH_3$

(c) $(CH_3)_2CH\overset{O}{\overset{\|}{C}}CH_2-C_6H_5$

(d) $CH_3CH_2CH_2\overset{O}{\overset{\|}{C}}CH_2CH_2CH_3$ (e) $(CH_3)_2CH\overset{O}{\overset{\|}{C}}CH(CH_3)_2$

(f) $CH_3CH_2\underset{CH_3}{\underset{|}{C}}H\overset{O}{\overset{\|}{C}}CH{=}CH_2$

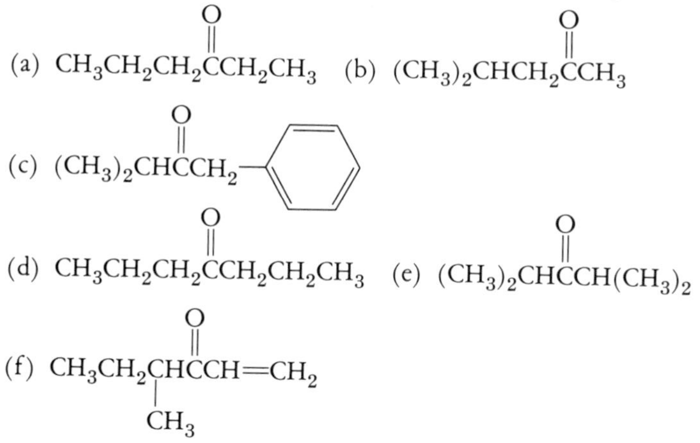

EXERCISE 19-7 Give the IUPAC name for each of the structures in exercise 19-6.

EXERCISE 19-8 Give the IUPAC name for each of the following compounds:

(a) $ClCH_2\overset{O}{\overset{\|}{C}}CH_3$ (b) $CH_2{=}CH\overset{O}{\overset{\|}{C}}CH{=}CH_2$ (c) $CH_2{=}CHCH_2\overset{O}{\overset{\|}{C}}CH_3$

(d) $HC{\equiv}CCH_2\underset{CH_3}{\underset{|}{C}}HCH_2\overset{O}{\overset{\|}{C}}CH_3$ (e) $C_6H_5(H)C{=}C(H)\overset{O}{\overset{\|}{C}}CH_3$

(f) cyclobutanone ring (g) cyclopentanone ring with Cl on each carbon adjacent to C=O (h) cyclohexenone ring

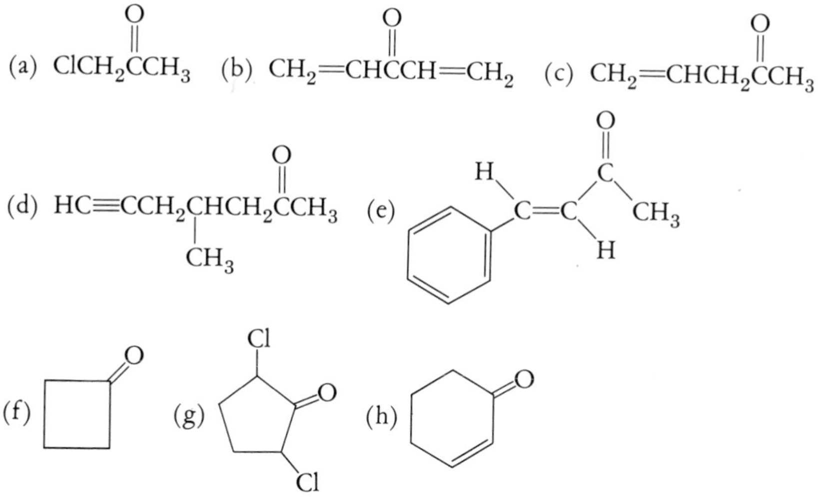

EXERCISE 19-9 Write the structural formula corresponding to each of the following names:

(a) butanone (b) 3-methyl-2-butanone
(c) *cis*-3-penten-2-one (d) cycloheptanone
(e) divinyl ketone (f) benzyl phenyl ketone
(g) 3-methylcyclohexanone (h) 3-octanone
(i) 4-phenyl-2-butanone

Now that we have learned to name aldehydes and ketones, let us examine some of their physical properties.

19.2 PHYSICAL PROPERTIES

The melting and boiling points and water solubilities of some representative aldehydes and ketones are given in Table 19-1. Most simple aldehydes and ketones are polar molecules. Strong intermolecular attractions in these compounds account for their boiling points, which are intermediate between those of hydrocarbons and alcohols of similar molecular weight.

Many aldehydes have pleasant odors. In contrast, the low-molecular-weight aliphatic aldehydes have sharp unpleasant odors. One familiar example is the aqueous solution of formaldehyde used for the preservation of biological specimens. In general, even ketones of low molecular weight have rather pleasant odors.

19.3 PREPARING ALDEHYDES AND KETONES

Alcohols can be oxidized, as we learned in Section 16.4. Primary alcohols are oxidized first to aldehydes, which are then further oxidized to carboxylic acids. Secondary alcohols are oxidized to ketones. This latter reaction is a general method of preparing ketones. The following are specific examples of this method:

$$\underset{\text{2-Butanol}}{CH_3\overset{\overset{\displaystyle OH}{|}}{C}HCH_2CH_3} \xrightarrow[H^+,\,H_2O]{K_2Cr_2O_7} \underset{\text{Butanone}}{CH_3\overset{\overset{\displaystyle O}{\|}}{C}CH_2CH_3}$$

$$\underset{\text{Cyclohexanol}}{C_6H_{11}OH} \xrightarrow[H^+,\,H_2O]{KMnO_4} \underset{\text{Cyclohexanone}}{C_6H_{10}{=}O}$$

To use the oxidization of primary alcohols as a method of preparing aldehydes, it is necessary to stop the reaction at the intermediate aldehyde stage. This can be done by removing the aldehyde from the reaction mix-

Table 19-1. Physical Properties of Some Aldehydes and Ketones

Structure	Common Name	IUPAC Name	Melting Point (°C)	Boiling Point (°C)	Solubility (g compound/100 g H_2O at 20° C)
Aldehydes					
HCHO	Formaldehyde	Methanal	−92	−21	Miscible
CH_3CHO	Acetaldehyde	Ethanal	−121	20	Miscible
CH_3CH_2CHO	Propionaldehyde	Propanal	−81	49	16
$CH_3(CH_2)_2CHO$	Butyraldehyde	Butanal	−99	76	7
$(CH_3)_2CHCHO$	Isobutyraldehyde	2-Methylpropanal	−66	64	11
$CH_3(CH_2)_3CHO$	Valeraldehyde	Pentanal	−91	103	Slightly soluble
$CH_2{=}CHCHO$	Acrolein	Propenal	−87	53	
H CHO C=C CH_3 H	Crotonaldehyde	*trans*-2-Butenal	−76	104	18
CHO	Benzaldehyde	Benzaldehyde	−26	178	0.3
Ketones					
CH_3COCH_3	Dimethyl ketone (acetone)	Propanone	−94	56	Miscible
$CH_3CH_2COCH_3$	Methyl ethyl ketone	Butanone	−86	80	26
$CH_3(CH_2)_2COCH_3$	Methyl propyl ketone	2-Pentanone	−78	102	6
$CH_3CH_2COCH_2CH_3$	Diethyl ketone	3-Pentanone	−41	101	5
$CH_3(CH_2)_3COCH_3$	Methyl butyl ketone	2-Hexanone	−35	150	2
=O	Cyclohexanone	Cyclohexanone	−45	157	5
O C CH_3	Methyl phenyl ketone (acetophenone)	1-Phenylethanone	21	202	Insoluble
O C	Diphenyl ketone (benzophenone)	Diphenylmethanone	48	306	Insoluble

ture as soon as it is formed. This can be done quite easily by distillation, because aldehydes have lower boiling points than that of the starting alcohol. By carrying out the oxidation at a temperature slightly above the boiling point of the aldehyde, it can be removed by distillation as it forms and before it can be further oxidized. The preparation of propionaldehyde from the oxidation of 1-propanol by chromic oxide (CrO_3) in acid solution is an example of this method:

$$CH_3CH_2CH_2OH \xrightarrow[65^\circ\,C,\ H_2O]{CrO_3,\ H^+} CH_3CH_2CHO$$

1-Propanol (boiling point, 97° C) — Propanaldehyde (boiling point, 49° C)

EXERCISE 19-10 Write the structure of the alcohol that forms each of the following aldehydes and ketones on oxidation:
(a) acetaldehyde (b) propanone (c) benzaldehyde
(d) methyl ethyl ketone

The fact that aldehydes are easily oxidized to carboxylic acids is used as a method of testing for the presence of an aldehyde, as we will learn in the next section.

19.4 TESTS FOR ALDEHYDES

Most of the reactions of aldehydes and ketones are very similar, but their ability to be oxidized differs greatly. Ketones are oxidized only under the most vigorous conditions, whereas aldehydes are easily oxidized. In fact, aldehydes are slowly oxidized by air.

This ease of oxidation is the basis of several specific tests for aldehydes. One of these is the *Tollens' test,* which consists of adding the Tollens' reagent, an alkaline solution of di(ammine) silver(I) ion, $Ag(NH_3)_2^+$, to the aldehyde. The aldehyde is oxidized to the carboxylate ion, and the silver ion is reduced to silver metal:

$$2\,Ag(NH_3)_2^+ + RCHO + 3\,OH^- \longrightarrow RCO_2^- + 2\,Ag\downarrow + 4\,NH_3 + 2\,H_2O$$

The silver metal coats the glass surface and forms a silver mirror, which serves to indicate that the reaction has occurred. For this reason, this reaction is commonly called the *silver mirror test for aldehydes.*

Another reagent often used to test for aldehydes is the *Fehling's solution.* This is an alkaline solution of cupric ion (Cu^{+2}) complexed with tartrate ion. The aldehyde is oxidized to the carboxylate ion and the cupric ion is reduced to cupric oxide (Cu_2O), which precipitates as a brick-red solid.

Because alcohols, alkenes, ketones, and carboxylic acids and their derivatives are not oxidized by either the Tollens' or the Fehling's reagent, these tests are highly specific for aldehydes.

Although they differ in their ease of oxidation, aldehydes and ketones undergo many of the same reactions, as we will learn in the next few sections.

19.5 ADDITION REACTIONS OF ALDEHYDES AND KETONES

Addition to the carbonyl group is one of the two major reactions of aldehydes and ketones (we will study the other reaction in Section 19.8):

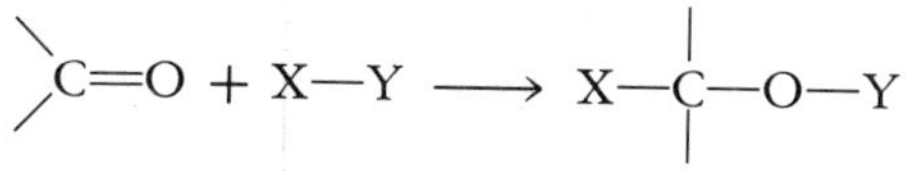

$$>C{=}O + X{-}Y \longrightarrow X{-}\overset{|}{\underset{|}{C}}{-}O{-}Y$$

While this reaction appears to be similar to the addition reactions of the carbon-carbon double bond, there is an important difference that is due to the difference in the polarity of the two double bonds. The carbon-oxygen double bond is made up of two atoms of different electronegativities. Consequently, the bonding electrons are not shared equally and the carbonyl group is polarized so that the carbon is slightly positive and the oxygen is slightly negative:

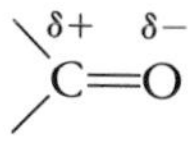

$$>\overset{\delta+}{C}{=}\overset{\delta-}{O}$$

As a result, the carbon of the carbonyl group forms bonds with the more nucleophilic part of the reagent X—Y. Thus,

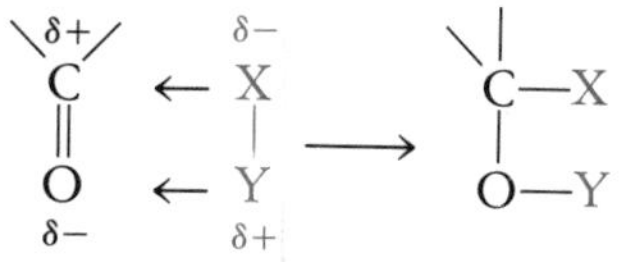

$$\begin{matrix} \overset{\delta+}{C} & \leftarrow & \overset{\delta-}{X} \\ \| & & | \\ \underset{\delta-}{O} & \leftarrow & \underset{\delta+}{Y} \end{matrix} \longrightarrow \begin{matrix} C{-}X \\ | \\ O{-}Y \end{matrix}$$

Many reagents are added to the carbonyl group of aldehydes and ketones in this way. Those of biological interest are water, hydrogen, alcohols, and ammonia and its derivatives. All of these reactions, though they appear to be different, are examples of the characteristic addition reactions of the carbonyl group of aldehydes and ketones.

Addition of Water

Most aldehydes react with water to form an equilibrium mixture of the aldehyde and an aldehyde hydrate:

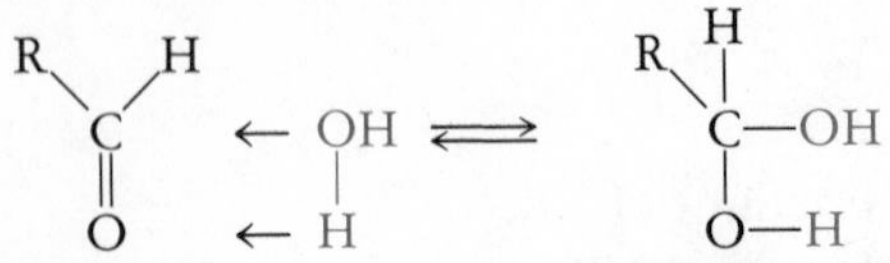

$$\mathrm{RCH{=}O} + \mathrm{H{-}OH} \rightleftharpoons \mathrm{RCH(OH)(O{-}H)}$$

Aldehyde — Aldehyde hydrate

These aldehyde hydrates are 1,1-diols. They are usually too unstable to isolate and purify because they easily lose water to reform the aldehyde. Most ketones are less readily hydrated.

In certain special cases, the 1,1-diol is sufficiently stable that it can be isolated. In these compounds, the carbonyl carbon is bonded to one or more highly electronegative groups. Hexafluoroacetone is one of the few ketones that form stable hydrates. Chloral hydrate is another example of a stable hydrate. It is a crystalline compound that is used in veterinary medicine as an anesthetic for animals.

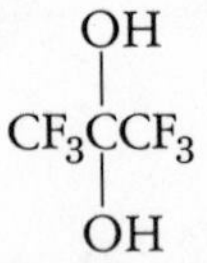

$$CF_3C(OH)_2CF_3 \qquad CCl_3CH(OH)_2$$

Hexafluoroacetone hydrate 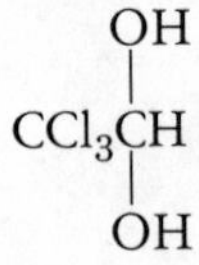 — Chloral hydrate

Addition of Alcohols

Aldehydes react with alcohols in the presence of an acid catalyst to form a *hemiacetal* as product:

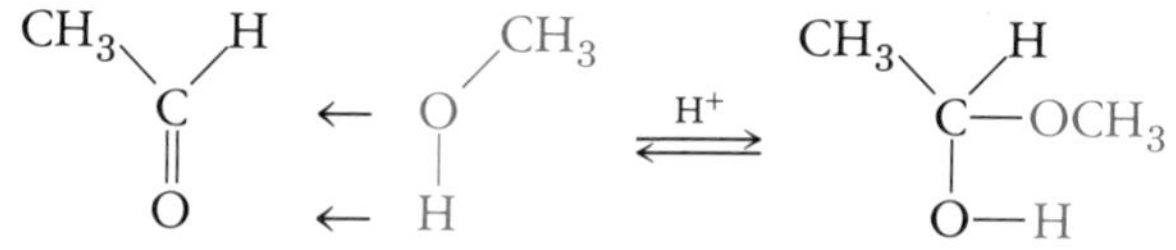

$$CH_3CH{=}O + CH_3OH \overset{H^+}{\rightleftharpoons} CH_3CH(OH)(OCH_3)$$

Acetaldehyde Methanol Hemiacetal

A hemiacetal contains an alkoxy and an alcohol group, both bonded to the original carbon of the carbonyl group. In the presence of excess alcohol, a hemiacetal can react to form an *acetal* and water.

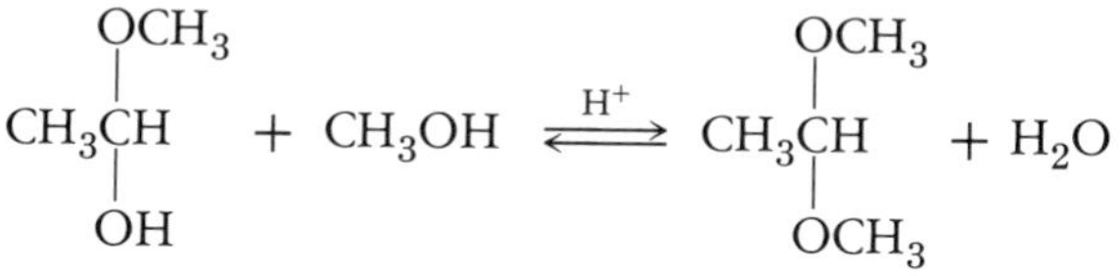

$$CH_3CH(OCH_3)(OH) + CH_3OH \overset{H^+}{\rightleftharpoons} CH_3CH(OCH_3)(OCH_3) + H_2O$$

Hemiacetal Methanol Acetal

These two steps can be summarized by the following equation:

$$CH_3CHO + 2\,CH_3OH \overset{H^+}{\rightleftharpoons} CH_3CH(OCH_3)_2 + H_2O$$

An acetal contains two alkoxy groups bonded to the same carbon. Thus, an acetal resembles an ether, and its reactions are similar to those of ethers (see Section 16.7). Acetals are stable and unreactive to aqueous base, but are cleaved by aqueous acid to their aldehyde and alcohol components. The reaction of aldehydes and alcohols to form hemiacetals and acetals is readily reversible. The reversibility of this reaction is an important feature of the reactions of carbohydrates that contain a hemiacetal functional group (see Chapter 21).

In addition to forming hemiacetals between two different molecules, such a reaction can occur *within* a molecule. Thus, a molecule that contains a hydroxy and an aldehyde group in the proper positions can form a five- or six-member cyclic hemiacetal. For example:

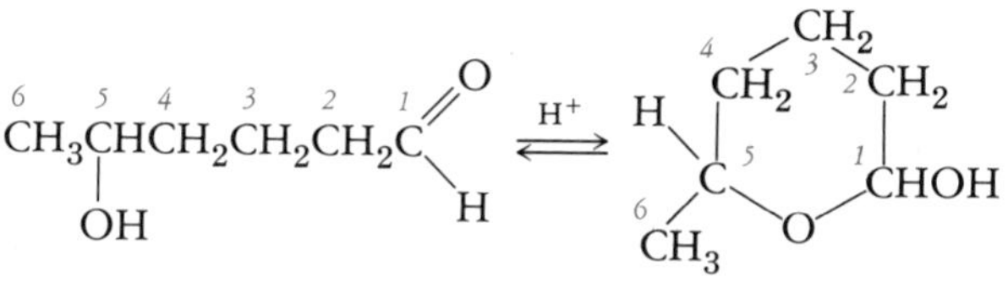

$$\overset{6}{C}H_3\overset{5}{C}H(OH)\overset{4}{C}H_2\overset{3}{C}H_2\overset{2}{C}H_2\overset{1}{C}HO \overset{H^+}{\rightleftharpoons} \text{cyclic hemiacetal (C-1 CHOH bonded through O to C-5 bearing H and } \overset{6}{C}H_3\text{; ring } CH_2\text{ at C-2, C-3, C-4)}$$

Carbohydrates (see Chapter 21) are compounds that form cyclic hemiacetals.

Ketones react with alcohols in a similar manner to form *hemiketals* and *ketals.* For example:

$$\underset{\text{Acetone}}{(CH_3)_2C{=}O} + \underset{\text{Methanol}}{CH_3OH} \overset{H^+}{\rightleftharpoons} \underset{\text{Hemiketal}}{(CH_3)_2C(OCH_3)(OH)} + \underset{\text{Methanol}}{CH_3OH} \overset{H^+}{\rightleftharpoons} \underset{\text{Ketal}}{(CH_3)_2C(OCH_3)_2} + H_2O$$

If a diol is used as the alcohol component, the acetal or the ketal will have a cyclic structure. For example:

$$(CH_3)_2C{=}O + HOCH_2{-}CH_2{-}CH_2OH \overset{H^+}{\rightleftharpoons} \left[(CH_3)_2C(OH){-}OCH_2{-}CH_2{-}CH_2OH\right] \overset{H^+}{\rightleftharpoons} (CH_3)_2C\langle{}^{OCH_2}_{OCH_2}\rangle CH_2 + H_2O$$

Ketals, like acetals, do not react with aqueous base but are readily cleaved by aqueous acid.

EXERCISE 19-11 Write the structure of the hemiacetal and acetal formed by the reaction of ethanol with the following aldehydes:
(a) acetaldehyde (b) benzaldehyde (c) propenaldehyde
(d) butyraldehyde

EXERCISE 19-12 Write the structure of the hemiketal and ketal formed by the reaction of ethanol with the following ketones:
(a) acetone (b) cyclopentanone (c) acetophenone
(d) 2-pentanone

EXERCISE 19-13 Repeat exercises 19-11 and 19-12 using 1,2-ethanediol instead of ethanol.

Addition of Ammonia and Its Derivatives: Schiff Bases

Ammonia adds to the carbonyl group of aldehydes and ketones. The initial product, in which the nitrogen is bonded to the carbon and one hydrogen is bonded to the oxygen, is unstable and spontaneously loses water to form an *imine:*

$$\underset{\text{Ketone or aldehyde}}{R(R')C{=}O} + \underset{\text{Ammonia}}{NH_2{-}H} \longrightarrow \left[R(R')C(OH)NH_2 \right] \xrightarrow{-H_2O} \underset{\text{An imine}}{R(R')C{=}NH}$$

Compounds having the general structures $RHC{=}NR'$ and $R_2C{=}NR'$ are called *aldimines* and *ketimines*, respectively. These compounds are also generally called *Schiff bases.* The Schiff bases of ammonia are usually unstable and undergo further reaction, but many derivatives of ammonia form stable Schiff bases with aldehydes and ketones. Some of these are listed in Table 19-2.

The net result of the addition is to convert a carbon-oxygen double bond into a carbon-nitrogen double bond. Oximes, hydrazones, and semicarba-

Table 19-2. Derivatives of Ammonia that React with Aldehydes and Ketones to Form Schiff Bases

Derivative of Ammonia	Structure	Product	Product Name
Primary amine	RNH_2	$>C{=}NR$	*N*-substituted imine
Secondary amine	R_2NH	$=C{-}NR_2$	Enamine
Hydroxylamine	NH_2OH	$>C{=}N{-}OH$	Oxime
Hydrazine	H_2NNH_2	$>C{=}NNH_2$	Hydrazone
Phenylhydrazine	$C_6H_5NHNH_2$	$>C{=}NNHC_6H_5$	Phenylhydrazone
2,4-dinitrophenyl-hydrazine	$NO_2{-}C_6H_3(NO_2){-}NHNH_2$	$>C{=}NNH{-}C_6H_3(NO_2){-}NO_2$	2,4-Dinitrophenyl-hydrazone
Semicarbazide	$NH_2\overset{O}{\overset{\parallel}{C}}NHNH_2$	$>C{=}NNH\overset{O}{\overset{\parallel}{C}}NH_2$	Semicarbazone

zones are often crystalline solids whose characteristic melting points can be used to identify a particular aldehyde or ketone.

EXERCISE 19-14 Write the equation for the reaction of 2-pentanone with the following reagents:
(a) phenylhydrazine (b) hydroxylamine
(c) 2,4-dinitrophenylhydrazine (d) semicarbazide

Reduction of Aldehydes and Ketones

The reduction of aldehydes and ketones forms alcohols (see Section 16.3). Primary alcohols are formed from aldehydes, whereas secondary alcohols are formed from ketones:

$$\underset{\text{Aldehyde}}{RCHO} \underset{\text{Oxidation}}{\overset{\text{Reduction}}{\rightleftharpoons}} \underset{\text{Primary alcohol}}{RCH_2OH} \qquad \underset{\text{Ketone}}{R\overset{O}{\overset{\|}{C}}R} \underset{\text{Oxidation}}{\overset{\text{Reduction}}{\rightleftharpoons}} \underset{\text{Secondary alcohol}}{R\overset{OH}{\overset{|}{C}}HR}$$

Notice that these reactions are the reverse of the preparation of aldehydes and ketones by the *oxidation* of primary and secondary alcohols (see Section 19.3).

Reduction of aldehydes and ketones is usually carried out in the laboratory by the use of two general reagents: metallic hydrides or catalytic hydrogenation.

Lithium aluminum hydride ($LiAlH_4$) and sodium borohydride ($NaBH_4$) are metallic hydrides that reduce aldehydes and ketones to alcohols. For example:

$$\underset{\text{Acetone}}{(CH_3)_2C{=}O} \xrightarrow{NaBH_4} \underset{\text{2-Propanol}}{(CH_3)_2CHOH}$$

Aluminum and boron are both less electronegative than is hydrogen (see p. 55). Consequently, the Al—H and B—H bonds of AlH_4^- and BH_4^- are polarized so that hydrogen acquires a partial negative charge. We can represent this as follows:

$$-\overset{|}{\underset{|}{Al}}{}^{\delta+}-H^{\delta-}$$ Polarization of the Al—H Bond

The hydrogen is the more electronegative part of the bond. Thus, the reduction by lithium aluminum hydride and sodium borohydride can be regarded as a transfer of a hydride ion (H : $^-$) from the metal hydride to the carbonyl carbon to form the alkoxide ion. Again, the more electronegative part of the reactant bonds to the carbon of the carbonyl group. Once the reduction with $NaBH_4$ is complete, the reaction mixture is treated with water, which hydrolyzes the salt to the alcohol and boric acid:

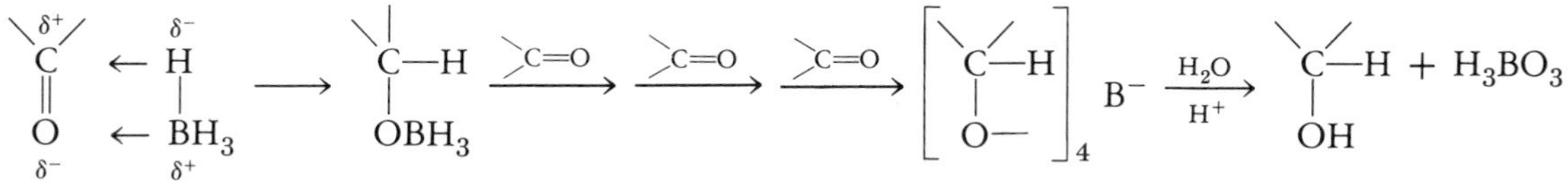

$$\overset{\delta+}{C}{=}\overset{\delta-}{O} \leftarrow \overset{\delta-}{H}{-}\overset{\delta+}{BH_3} \longrightarrow \text{C—H}(\text{OBH}_3) \xrightarrow{\text{>C=O}} \xrightarrow{\text{>C=O}} \xrightarrow{\text{>C=O}} \left[\text{C(H)—O—}\right]_4 B^- \xrightarrow[H^+]{H_2O} \text{C(H)—OH} + H_3BO_3$$

As we will learn in the next section, the reduction of carbonyl compounds in living systems also occurs by a hydride ion transfer. But metallic hydrides are not involved in this reaction, because they react with water. For example:

$$2\ LiAlH_4 + 5\ H_2O \longrightarrow 8\ H_2 + 2\ LiOH + Al_2O_3$$

Consequently, the transfer of a hydride ion in living systems occurs by an enzyme-catalyzed reaction between two organic compounds.

As we learned in Section 16.3, the carbonyl group of aldehydes and ketones is readily reduced with hydrogen and a metal catalyst such as nickel or platinum. For example:

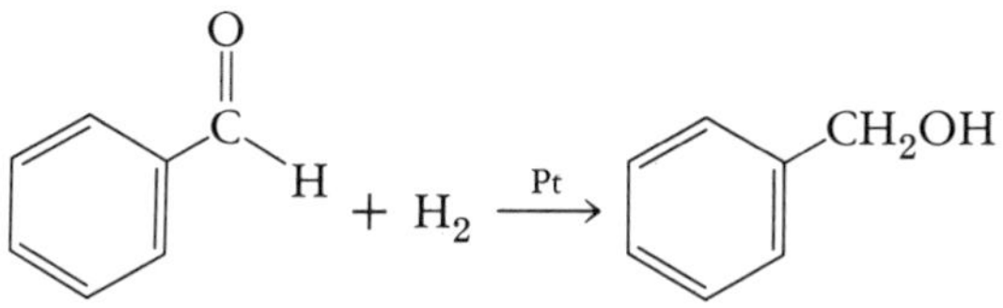

$$C_6H_5{-}CHO + H_2 \xrightarrow{Pt} C_6H_5{-}CH_2OH$$

Let us now examine several examples of addition reactions of aldehydes and ketones in living systems.

19.6 ADDITION REACTIONS OF ALDEHYDES AND KETONES IN LIVING SYSTEMS

One of the simplest addition reactions of a carbonyl group in living systems is the enzyme-catalyzed hydration of carbon dioxide to bicarbonate ion according to the following equation:

$$O{=}C{=}O + H_2O \xrightarrow{\text{Enzyme}} O{=}C\begin{matrix} O^- \\ \\ OH \end{matrix} + H^+$$

The enzyme that catalyzes this reaction is widely distributed in mammals. It is especially active in tissues that are involved in respiration, such as red blood cells.

The formation of Schiff bases is an important reaction in the formation of many compounds in living systems. One example is the transamination reaction. In such reactions, the α-amino group of an amino acid is transferred to the α carbon of an α ketoacid. For example:

$$\underset{\substack{\text{Alanine} \\ \text{(an amino acid)}}}{\begin{matrix} CH_3 \\ | \\ CHNH_2 \\ | \\ CO_2H \end{matrix}} + \underset{\substack{\alpha\text{-Ketoglutaric acid} \\ \text{(an } \alpha\text{-keto acid)}}}{\begin{matrix} CO_2H \\ | \\ O{=}C \\ | \\ CH_2 \\ | \\ CH_2CO_2H \end{matrix}} \overset{\text{Enzymes}}{\rightleftharpoons} \underset{\substack{\text{Pyruvic} \\ \text{acid}}}{\begin{matrix} CH_3 \\ | \\ C{=}O \\ | \\ CO_2H \end{matrix}} + \underset{\substack{\text{Glutamic} \\ \text{acid}}}{\begin{matrix} CO_2H \\ | \\ NH_2CH \\ | \\ CH_2 \\ | \\ CH_2CO_2H \end{matrix}}$$

Such reactions are catalyzed by enzymes and are readily reversible. It is by these reactions that amino acids are synthesized in living systems (see Chapter 29).

The first step in this transformation is the reaction of the amino group of the amino acid with pyridoxal phosphate, the reactive part of the enzyme, to form a Schiff base:

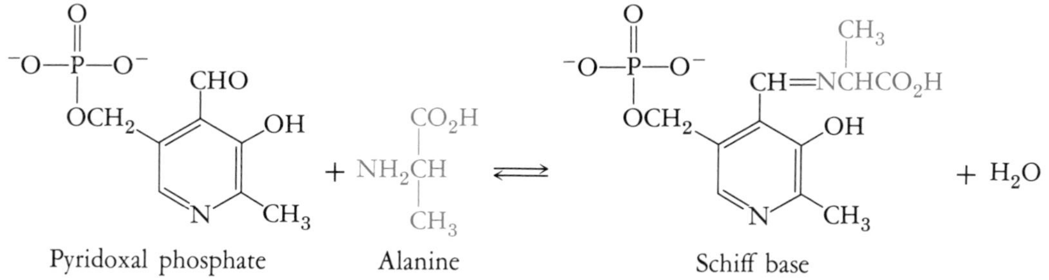

Pyridoxal phosphate + Alanine ⇌ Schiff base + H_2O

This Schiff base is the amino group carrier between the amino acid and the keto acid. The remaining steps in this reaction are discussed in Section 29.3.

The enzyme-catalyzed reduction of aldehydes and ketones occurs frequently in biological reactions. In these reactions, the same enzyme frequently catalyzes the oxidation of alcohols to aldehydes and ketones. The oxidation of L-malate is an example of such an oxidation-reduction reaction:

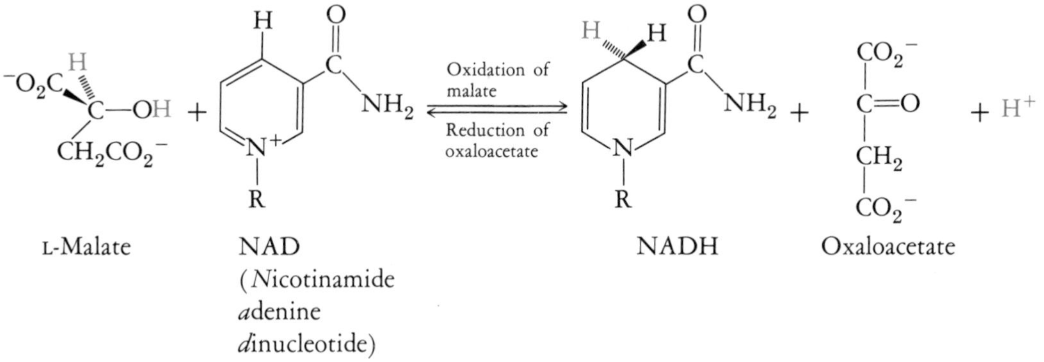

L-Malate + NAD (*N*icotinamide *a*denine *d*inucleotide) ⇌ NADH + Oxaloacetate + H^+

(Forward: Oxidation of malate; reverse: Reduction of oxaloacetate)

We have seen this example before in Section 16.5. There we concentrated on the oxidation reaction. You will recall that the oxidation is very stereospecific: only L-malate is oxidized. The reduction of oxaloacetate is also very stereospecific. Only L-malate is formed in this enzyme-catalyzed reduction. Because of this fact, the reduction is believed to be the reverse of the oxidation reaction. Thus, a hydride ion is transferred from the pyridine ring to the carbonyl carbon of oxaloacetate. Notice that this is the biological equivalent of the metallic hydride reduction of carbonyl groups carried out in the laboratory.

19.7 MECHANISM OF ADDITION REACTIONS OF ALDEHYDES AND KETONES

OPTIONAL SECTION

The mechanism proposed for these reactions is similar in many respects to that proposed for the reactions of esters, amides, and carboxylic acid anhydrides (review Section 18.11). As before, nucleophiles add to the positively charged carbonyl carbon of the aldehyde or ketone to form a tetrahedral intermediate. During the addition, one of the two pairs of electrons of the carbonyl group moves to the more electronegative oxygen. This is the first step of the mechanism given in Figure 19-1. In the second step, a proton is transferred from water to the oxygen atom to form the observed product. This general mechanism is applied to the lithium aluminum hydride ($LiAlH_4$) reduction of acetone as follows:

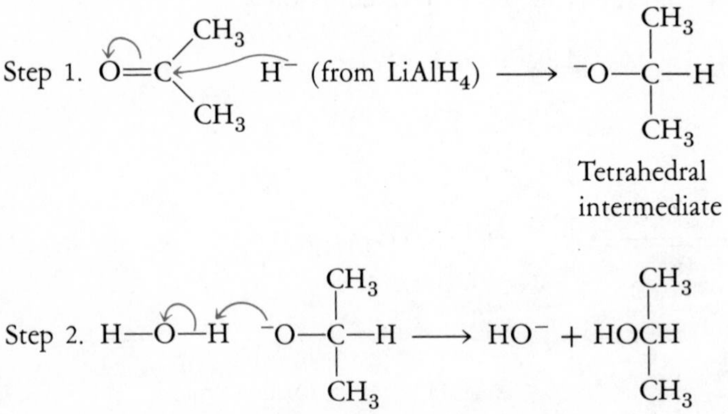

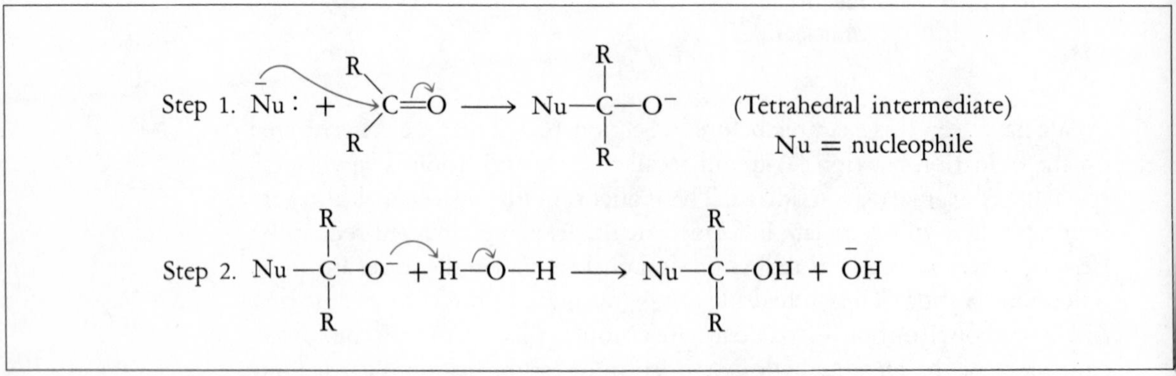

Step 1. $O{=}C(CH_3)_2 + H^-$ (from $LiAlH_4$) $\longrightarrow$ $^-O{-}C(CH_3)_2{-}H$

Tetrahedral intermediate

Step 2. $H{-}O{-}H \;\; ^-O{-}C(CH_3)_2{-}H \longrightarrow HO^- + HOCH(CH_3)_2$

Step 1. $\bar{N}u: + R_2C{=}O \longrightarrow Nu{-}CR_2{-}O^-$ (Tetrahedral intermediate)

Nu = nucleophile

Step 2. $Nu{-}CR_2{-}O^- + H{-}O{-}H \longrightarrow Nu{-}CR_2{-}OH + \bar{O}H$

Fig. 19-1. General mechanism proposed for the reaction of aldehydes or ketones and nucleophiles.

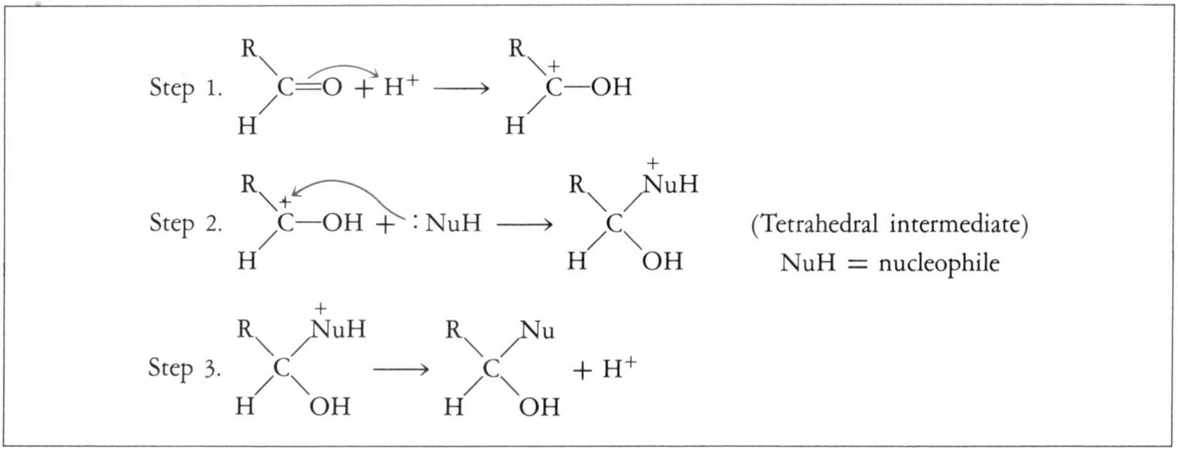

Step 1. $$\begin{matrix} R \\ & \backslash \\ & & C{=}O + H^+ \longrightarrow \\ & / \\ H \end{matrix} \quad \begin{matrix} R \\ & \backslash^{+} \\ & & C{-}OH \\ & / \\ H \end{matrix}$$

Step 2. $$\begin{matrix} R \\ & \backslash^{+} \\ & & C{-}OH + :NuH \longrightarrow \\ & / \\ H \end{matrix} \quad \begin{matrix} R & & \overset{+}{N}uH \\ & C & \\ H & & OH \end{matrix}$$

(Tetrahedral intermediate)
NuH = nucleophile

Step 3. $$\begin{matrix} R & & \overset{+}{N}uH \\ & C & \\ H & & OH \end{matrix} \longrightarrow \begin{matrix} R & & Nu \\ & C & \\ H & & OH \end{matrix} + H^+$$

Fig. 19-2. General mechanism proposed for the acid-catalyzed reaction of aldehydes or ketones and nucleophiles.

In this reaction, a proton is obtained from water, which is added after the first step is completed.

Acids catalyze the addition of relatively weak nucleophiles to aldehydes and ketones. Again, a proton adds to the carbonyl oxygen. This makes the carbonyl carbon more positive and therefore more susceptible to attack by nucleophiles. The general acid-catalyzed mechanism is given in Figure 19-2. The nucleophile adds to the protonated carbonyl group to form a tetrahedral intermediate. Loss of a proton from this intermediate forms the product and regenerates the acid catalyst. This general mechanism is applied to the reaction of methanol and acetaldehyde as follows:

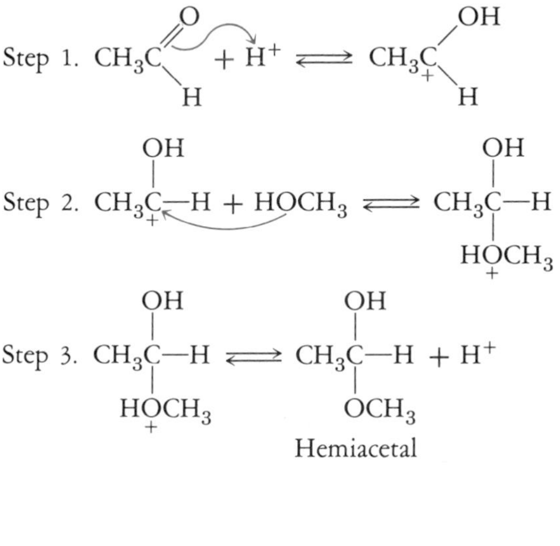

Step 1. $$CH_3\overset{O}{\overset{\|}{C}}{-}H + H^+ \rightleftharpoons CH_3\overset{+}{C}(OH){-}H$$

Step 2. $$CH_3\overset{+}{C}(OH){-}H + HOCH_3 \rightleftharpoons CH_3\overset{OH}{\underset{H\overset{+}{O}CH_3}{C}}{-}H$$

Step 3. $$CH_3\overset{OH}{\underset{H\overset{+}{O}CH_3}{C}}{-}H \rightleftharpoons CH_3\overset{OH}{\underset{OCH_3}{C}}{-}H + H^+$$

Hemiacetal

The hemiacetal reacts further to form an acetal by the following series of reactions:

$$CH_3CH(\ddot{O}H)(OCH_3) + H^+ \rightleftharpoons CH_3CH(\overset{+}{O}H_2)(OCH_3)$$

$$CH_3CH(\overset{+}{O}H_2)(OCH_3) \rightleftharpoons CH_3\overset{+}{C}H(OCH_3) + H_2O$$

$$CH_3\overset{+}{C}H(OCH_3) + H\ddot{O}CH_3 \rightleftharpoons CH_3CH(H\overset{+}{O}CH_3)(OCH_3)$$

$$CH_3CH(H\overset{+}{O}CH_3)(OCH_3) \rightleftharpoons CH_3CH(OCH_3)(OCH_3) + H^+$$

Other products of the addition reaction of aldehydes and ketones undergo further reactions. An example is the initial product of the reaction of methylamine and acetone which reacts further to form a Schiff-base by the loss of water to form a carbon-nitrogen double bond:

$$CH_3C(OH)(NHCH_3)CH_3 \longrightarrow CH_3C(=NCH_3)CH_3 + H_2O$$

We have learned how the polarity of the carbonyl group is responsible for nucleophilic addition reactions. This feature of the carbonyl group is also responsible for the other reaction of aldehydes and ketones, as we will learn in the next few sections.

EXERCISE 19-15 Write a mechanism for the reaction of acetaldehyde and the following reagents:
(a) H_2O, H^+ (b) $HOCH_2CH_2OH$, H^+ (c) H^- of $LiAlH_4$
(d) CH_3NH_2

19.8 CONDENSATION REACTIONS

In condensation reactions, two compounds are joined together (or condensed) to form one larger compound. In this section, we will learn about two condensation reactions: the aldol condensation and the Claisen condensation. We will also learn that these two condensation reactions are simply examples of addition reactions of aldehydes and ketones and substitution reactions of esters (see Section 18.5). Let us start by learning about the aldol condensation.

An aldol condensation reaction is a reaction in which the carbonyl carbon of one molecule forms a bond with the α carbon of another carbonyl-containing molecule. For example:

$$>C{=}O \;+\; H{-}CH_2C({=}O)R \longrightarrow >C(OH){-}CH_2C({=}O)R$$

Carbonyl carbon — α Carbon — Aldol condensation product

Aldehydes and ketones undergo this type of reaction. For example, acetaldehyde reacts in an aqueous basic solution to form a condensation product called aldol:

$$CH_3CHO \;+\; H{-}CH_2CHO \underset{}{\overset{OH^-}{\rightleftharpoons}} CH_3CH(OH){-}CH_2CHO$$

Acetaldehyde — Aldol

The name aldol is derived from the structure of the product, which is both an *ald*ehyde and an alcoh*ol*. Notice that this reaction is reversible. The reverse reaction is called a *retro-aldol condensation.*

Most aldehydes and many ketones that have α hydrogens undergo the aldol condensation. All the products of an aldol condensation reaction have a common structural feature, a β-hydroxy carbonyl skeleton:

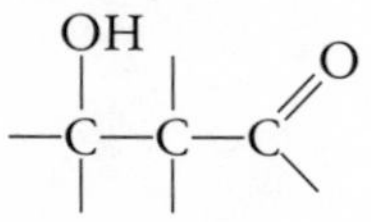

$$-\overset{\overset{\text{OH}}{|}}{\underset{|}{C}}-\overset{|}{\underset{|}{C}}-C\overset{\diagup\!\!\diagup O}{\diagdown}$$

This structure is the result of the way the two pieces join together. The α carbon of one aldehyde molecule becomes bonded to what was the original carbonyl carbon of the second aldehyde molecule. For example:

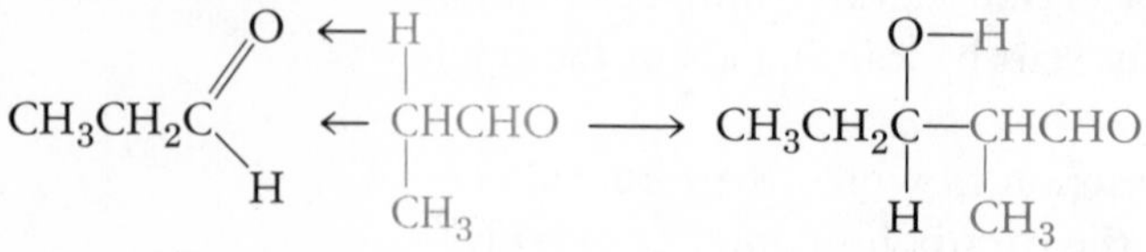

$$CH_3CH_2\overset{O}{\overset{\|}{C}}H + \overset{H}{\overset{|}{\underset{\underset{CH_3}{|}}{C}}}HCHO \longrightarrow CH_3CH_2\overset{\overset{O-H}{|}}{\underset{\underset{H}{|}}{C}}-\underset{\underset{CH_3}{|}}{C}HCHO$$

Notice that this reaction is another example of an addition to a carbonyl group.

The aldol condensation is an equilibrium. The equilibrium constant for formation of the products is favorable for most aldehydes. However, for ketones, the reaction is much less favorable. For example, the condensation of acetone under basic conditions produces only a small percentage of the product at equilibrium:

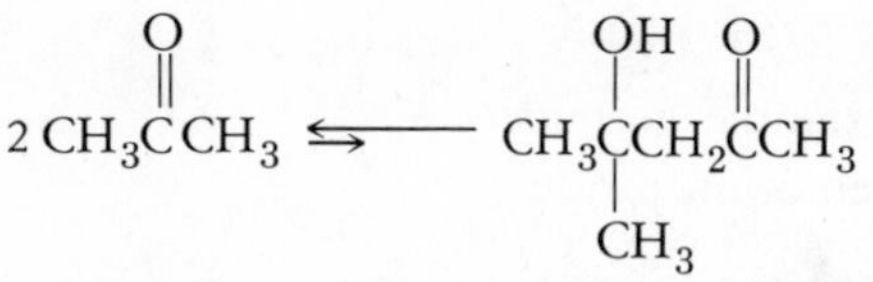

$$2\,CH_3\overset{O}{\overset{\|}{C}}CH_3 \rightleftharpoons CH_3\overset{\overset{OH}{|}}{\underset{\underset{CH_3}{|}}{C}}CH_2\overset{O}{\overset{\|}{C}}CH_3$$

This fact makes it possible to use aldol condensation reactions between different pairs of reagents to prepare other molecules. The following reaction is an example of such a mixed aldol condensation reaction:

$$C_6H_5\overset{O}{\overset{\|}{C}}H + CH_3\overset{O}{\overset{\|}{C}}CH_3 \xrightarrow{OH^-} C_6H_5\overset{\overset{OH}{|}}{\underset{\underset{H}{|}}{C}}CH_2\overset{O}{\overset{\|}{C}}CH_3$$

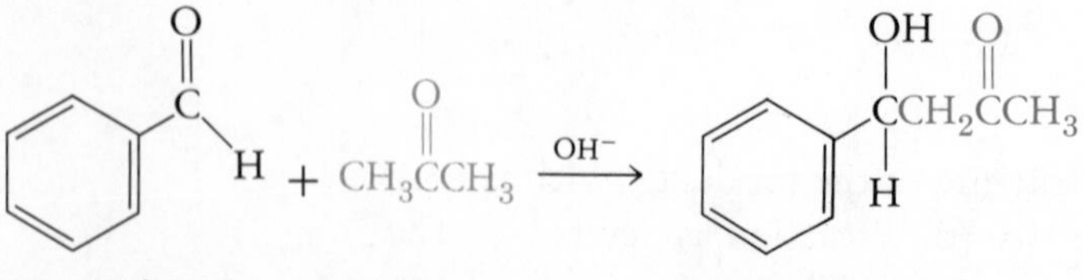

Benzaldehyde | Acetone | Mixed aldol condensation product

In this reaction, benzaldehyde does not undergo an aldol condensation because it does not have any α hydrogens. Acetone does not form a condensation product because the equilibrium is unfavorable for that reaction. Consequently, only the mixed condensation product is formed.

In contrast, a mixed aldol condensation reaction between two different aldehydes with α hydrogens will form four different products. For example:

$$CH_3CH_2CHO + CH_3CHO \xrightarrow{OH^-} CH_3CH(OH)CH_2CHO + CH_3CH(OH)CH(CH_3)CHO + CH_3CH_2CH(OH)CH_2CHO + CH_3CH_2CH(OH)CH(CH_3)CHO$$

Such reactions are not useful for preparing a specific compound. In general, mixed aldol condensation reactions are successful only when the α carbon of a ketone adds to the carbonyl group of an aldehyde that does not have α hydrogens.

EXERCISE 19-16 Write the structure of the aldol condensation product formed from the following compounds:
(a) pentanal (b) isobutyraldehyde (c) α-phenylacetaldehyde
(d) acetone and formaldehyde

Esters also undergo condensation reactions when treated with base. For example:

$$CH_3C(=O)OCH_2CH_3 + CH_3C(=O)OCH_2CH_3 \xrightarrow{NaOCH_2CH_3} CH_3C(=O)CH_2C(=O)OCH_2CH_3 + CH_3CH_2OH$$

This reaction is called a *Claisen condensation.* Notice that in this reaction the $—OCH_2CH_3$ group of one ester molecule is replaced by the α carbon of another ester molecule. This is another example of the typical substitution reactions of esters (see Section 18.5).

The overall result of a Claisen condensation is the transfer of an acyl group from one ester molecule to another (see Section 18.10):

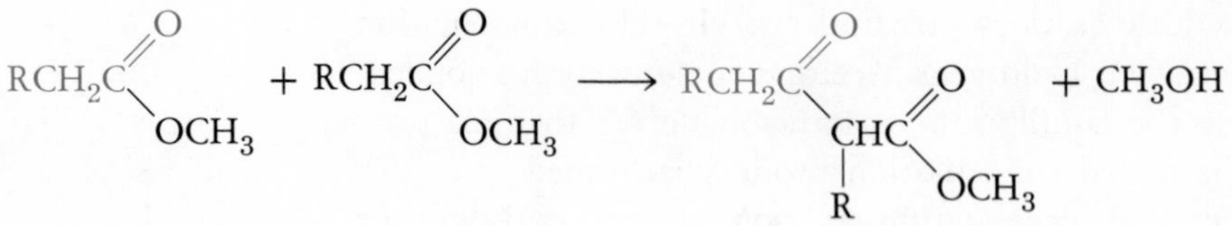

$$RCH_2C(=O)OCH_3 + RCH_2C(=O)OCH_3 \longrightarrow RCH_2C(=O)CH(R)C(=O)OCH_3 + CH_3OH$$

As a result, the products of Claisen condensations have a common structural feature, a β-ketocarbonyl skeleton:

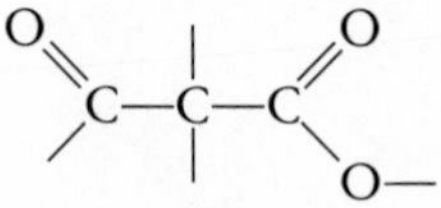

$$\text{>C(=O)}-\overset{|}{\underset{|}{C}}-C(=O)-O-$$

EXERCISE 19-17 Write the structure of the Claisen condensation product formed from each of the following esters:

(a) 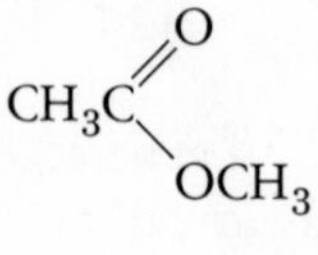 $CH_3C(=O)OCH_3$ (b) 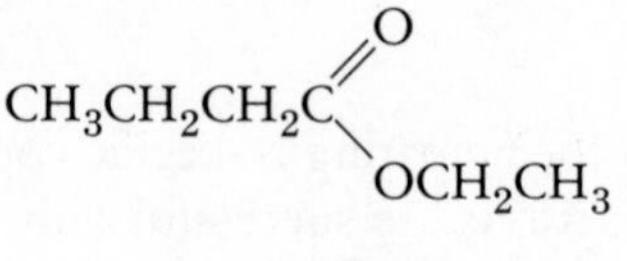 $CH_3CH_2CH_2C(=O)OCH_2CH_3$

In the next section, we will learn that both aldol and Claisen condensation reactions occur in living systems.

19.9 CONDENSATION REACTIONS IN LIVING SYSTEMS

Carbohydrates are prepared in living systems by an enzyme-catalyzed aldol condensation. A specific example is the preparation of D-fructose 1,6-diphosphate by the condensation of D-glyceraldehyde 3-phosphate and 1,3-dihydroxyacetone phosphate:

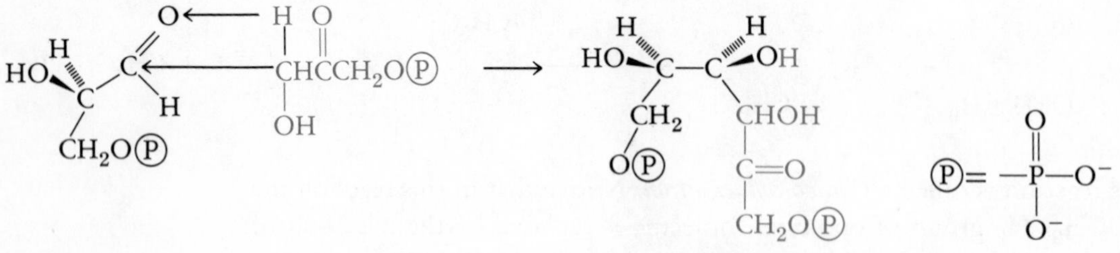

$$\text{HO–CH(CH}_2\text{O}Ⓟ\text{)–CH=O} + \text{HOCH–C(=O)–CH}_2\text{O}Ⓟ \longrightarrow \text{Ⓟ}\text{OCH}_2\text{–C(H)(OH)–C(H)(OH)–CHOH–C=O–CH}_2\text{O}Ⓟ$$

D-Glyceraldehyde 3-phosphate

1,3-Dihydroxyacetone-phosphate

19-1

D-Fructose 1,6-diphosphate

$$Ⓟ = -\overset{O}{\overset{\|}{P}}-O^- \text{ (with } O^- \text{ below P)}$$

This condensation reaction again shows the stereospecificity of enzyme-catalyzed reactions. Only the product with the stereochemistry shown in structure *19-1* is formed.

The acetyl group of acetyl coenzyme A reacts with many carbonyl groups in an aldol-like condensation. An example is the enzyme-catalyzed preparation of citrate by the reaction of acetyl coenzyme A and oxaloacetate:

$$\underset{\text{Oxaloacetate}}{{}^-O_2C\text{—}CH_2\text{—}C(=O)\text{—}CO_2^-} + \underset{\text{Acetyl coenzyme A}}{H\text{—}CH_2C(=O)SCoA} \xrightarrow{\text{Enzyme}} {}^-O_2C\text{—}CH_2\text{—}C(OH)(CO_2^-)\text{—}CH_2C(=O)SCoA \xrightarrow{H_2O} \underset{\text{Citrate}}{{}^-O_2C\text{—}CH_2C(OH)(CO_2^-)CH_2CO_2^-}$$

The acetyl group of acetyl coenzyme A also undergoes Claisen condensations. An example is the formation of acetoacetyl coenzyme A by an enzyme-catalyzed reaction of acetyl coenzyme A:

$$\underset{\text{Acetyl coenzyme A}}{CH_3C(=O)\text{—}S\text{—}CoA} + \underset{\text{Acetyl coenzyme A}}{H\text{—}CH_2C(=O)SCoA} \xrightarrow{\text{Enzyme}} \underset{\text{Acetoacetyl coenzyme A}}{CH_3C(=O)CH_2C(=O)SCoA} + \underset{\text{Coenzyme A}}{H\text{—}SCoA}$$

The last two examples demonstrate the dual role of the acetyl group of acetyl coenzyme A. In the first case, the α carbon bonds to a carbonyl carbon; in the second case, the acetyl group is transferred to an α carbon of another molecule.

In the next section we will learn how it is possible for this functional group to have a dual role.

EXERCISE 19-18 Write the structure of the product of the enzyme-catalyzed reaction of $HOCH_2COCH_2O$Ⓟ and L-glyceraldehyde 3-phosphate.

19.10 ACIDITY OF α HYDROGENS

In Chapter 11, we learned that saturated hydrocarbons are generally unreactive compounds. Therefore, it is not surprising to learn that the C—H bond is not acidic in these compounds. For example, K_a is 10^{-40} for methane (CH_4). Thus, there is little tendency for methane to ionize to form a proton and the ion $^-CH_3$ in which carbon bears a negative charge. Such anions are called *carbanions*.

However, aldehydes, ketones, and esters do react with strong bases to form carbanions. The presence of a carbonyl group makes the C—H bond α to it much more acidic than usual. The reason for this is that the carbanion can be stabilized by resonance. For example:

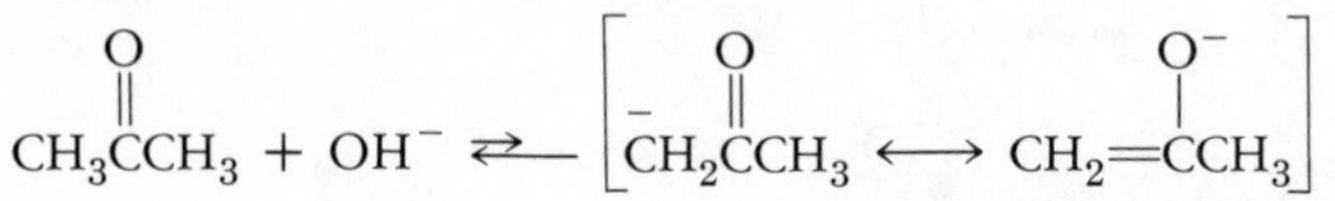

$$CH_3\overset{\overset{\displaystyle O}{\|}}{C}CH_3 + OH^- \rightleftarrows \left[\overset{-}{C}H_2\overset{\overset{\displaystyle O}{\|}}{C}CH_3 \longleftrightarrow CH_2{=}\overset{\overset{\displaystyle O^-}{|}}{C}CH_3\right]$$

Resonance-stabilized
enolate anion

This enolate anion is a resonance hybrid of two structures. One structure has the negative charge on carbon, whereas the other has the negative charge on oxygen. As we learned in Chapter 5, molecules or ions that can be written as a resonance hybrid of two or more structures are stabilized. This stabilization of the enolate anion is the reason why the hydrogen α to a carbonyl group is lost to a strong base.

EXERCISE 19-19 Indicate which hydrogens of the following compounds can be removed by an aqueous sodium hydroxide solution:

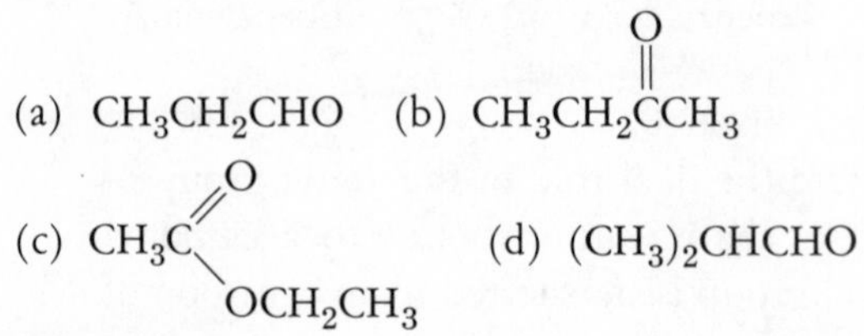

(a) CH_3CH_2CHO (b) $CH_3CH_2\overset{\overset{\displaystyle O}{\|}}{C}CH_3$

(c) $CH_3C(=O)OCH_2CH_3$ (d) $(CH_3)_2CHCHO$

It must be emphasized that in aqueous solution, these compounds show no acidic properties, because they are very weak acids. It is only in the presence of strong bases such as sodium hydroxide or sodium ethoxide ($NaOC_2H_5$) that a proton can be removed. Even then, in the equilibrium between the enolate anion and the neutral organic molecule, appreciable quantities of each are present. The formation of an enolate anion is the first step in the base-catalyzed condensation reactions of aldehydes, ketones, esters, and diethyl malonate (see p. 547).

EXERCISE 19-20 Write the equation for the equilibrium reaction that forms an enolate anion between sodium hydroxide and each of the compounds in exercise 19-19.

19.11 MECHANISM OF CONDENSATION REACTIONS

OPTIONAL SECTION

The general mechanism proposed for the aldol condensation is given in Figure 19-3. The first step is the formation of the enolate anion. Because at equilibrium appreciable quantities of both the enolate anion and the aldehyde are present, the enolate anion adds to the carbonyl carbon of another aldehyde molecule in the second step. Thus, the aldehyde plays two roles in this reaction: it reacts with base to form the enolate anion, and it contains the carbonyl group to which the enolate anion adds. In the final step, a proton is transferred from water to the oxygen atom and the hydroxide ion catalyst is reformed. Notice that this mechanism is very similar to that given in Figure 19-1. This general mechanism is applied to the base-catalyzed aldol condensation of acetaldehyde in the following series of reactions:

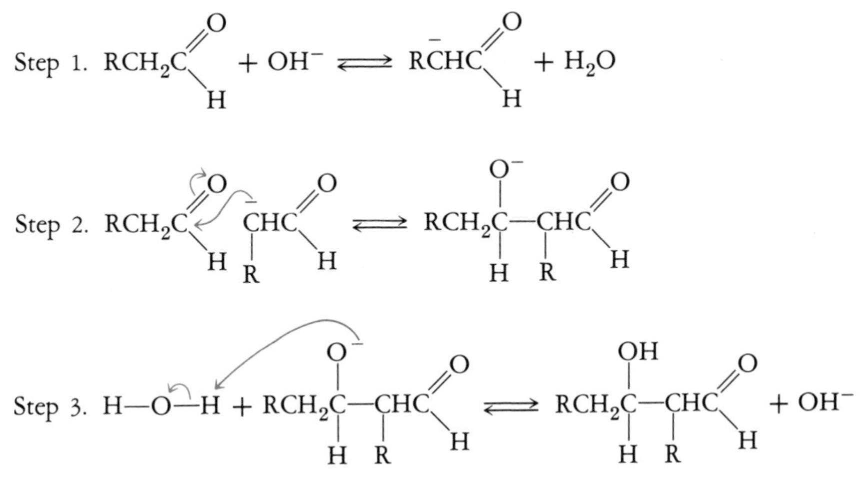

Step 1. $RCH_2CHO + OH^- \rightleftharpoons R\bar{C}HCHO + H_2O$

Step 2. $RCH_2CHO + \bar{C}H(R)CHO \rightleftharpoons RCH_2C(H)(O^-)-CH(R)CHO$

Step 3. $H-O-H + RCH_2C(H)(O^-)-CH(R)CHO \rightleftharpoons RCH_2C(H)(OH)-CH(R)CHO + OH^-$

Fig. 19-3. General mechanism proposed for the base-catalyzed aldol condensation.

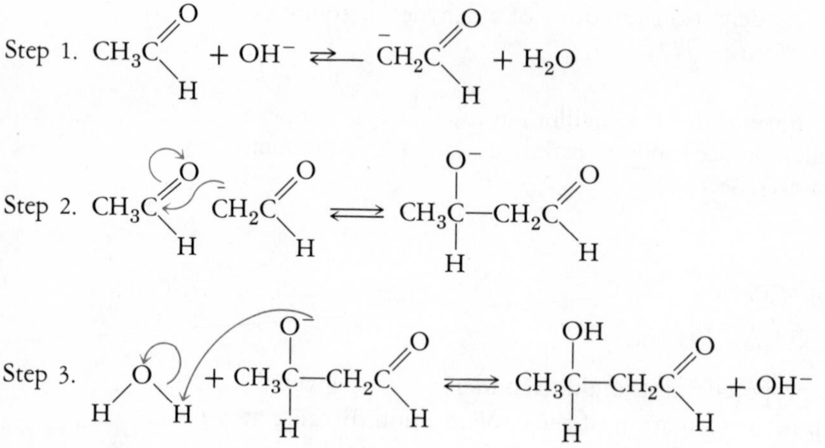

Step 1. $$CH_3CH{=}O + OH^- \rightleftharpoons \overset{-}{C}H_2CH{=}O + H_2O$$

Step 2. $$CH_3CH{=}O + \overset{-}{C}H_2CH{=}O \rightleftharpoons CH_3\underset{H}{\overset{O^-}{C}}{-}CH_2CH{=}O$$

Step 3. $$H_2O + CH_3\underset{H}{\overset{O^-}{C}}{-}CH_2CH{=}O \rightleftharpoons CH_3\underset{H}{\overset{OH}{C}}{-}CH_2CH{=}O + OH^-$$

EXERCISE 19-21 Write the mechanism for the base-catalyzed aldol condensation of propanal.

The proposed mechanism of the Claisen condensation is very similar. It is illustrated for ethyl acetate in Figure 19-4. Again, the first step is formation of an enolate anion. In the second step the enolate anion adds to the carbonyl carbon to form a tetrahedral intermediate that decomposes to the

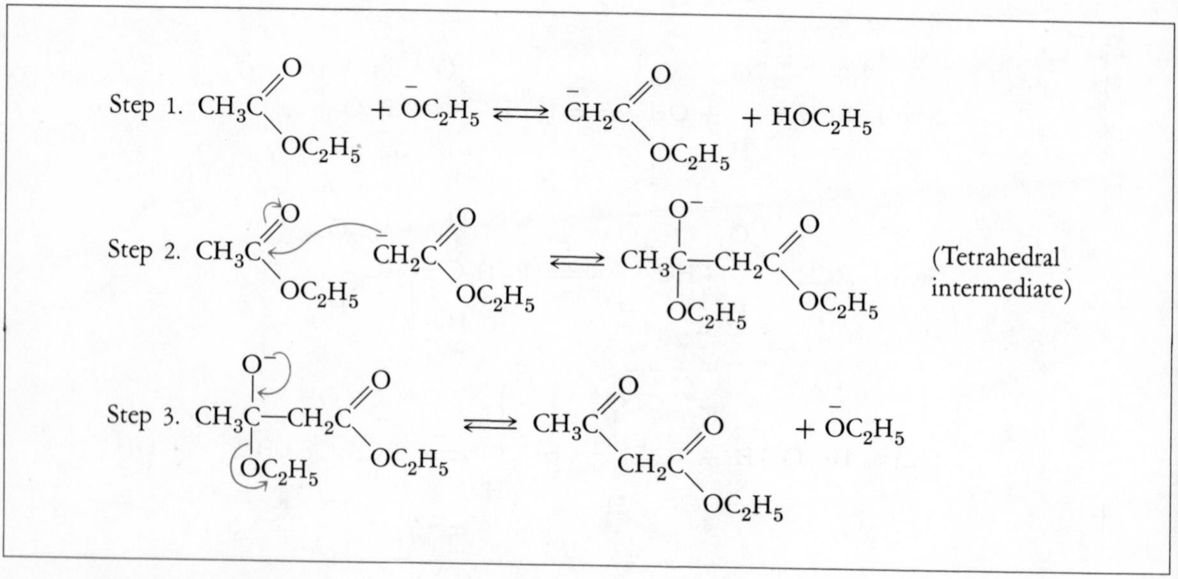

Step 1. $$CH_3C({=}O)OC_2H_5 + \overset{-}{O}C_2H_5 \rightleftharpoons \overset{-}{C}H_2C({=}O)OC_2H_5 + HOC_2H_5$$

Step 2. $$CH_3C({=}O)OC_2H_5 + \overset{-}{C}H_2C({=}O)OC_2H_5 \rightleftharpoons CH_3\underset{OC_2H_5}{\overset{O^-}{C}}{-}CH_2C({=}O)OC_2H_5$$ (Tetrahedral intermediate)

Step 3. $$CH_3\underset{OC_2H_5}{\overset{O^-}{C}}{-}CH_2C({=}O)OC_2H_5 \rightleftharpoons CH_3C({=}O)CH_2C({=}O)OC_2H_5 + \overset{-}{O}C_2H_5$$

Fig. 19-4. The proposed mechanism for the Claisen condensation of ethyl acetate.

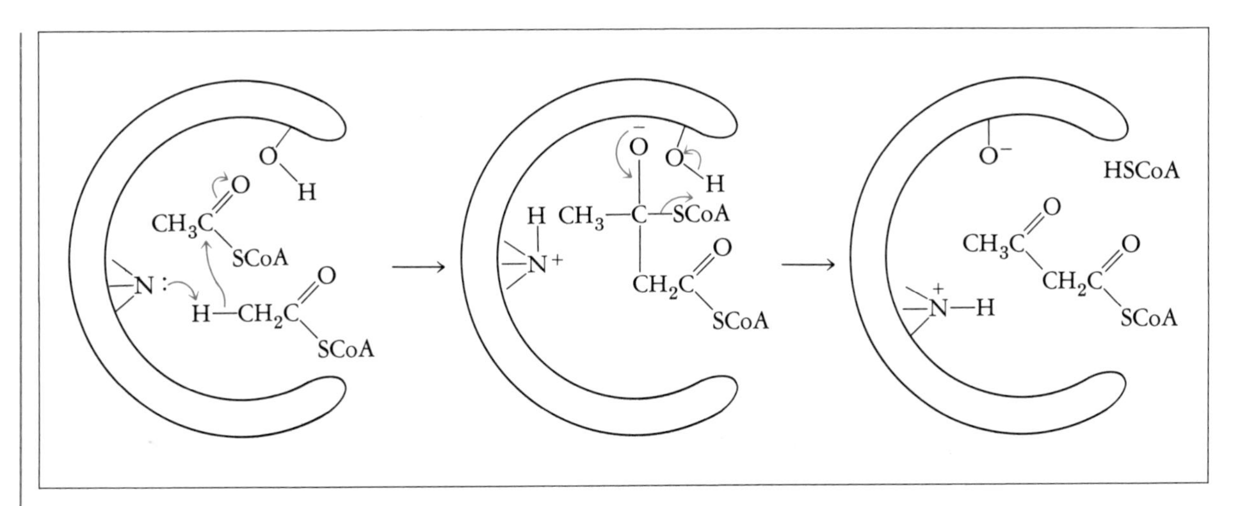

Fig. 19-5. A schematic representation of the mechanism of the enzyme-catalyzed condensation of acetyl coenzyme A.

product and reforms the $C_2H_5O^-$ catalyst in the third step. This mechanism is similar to that proposed for the reaction of carboxylic acid anhydrides and nucleophiles given in Figure 18-1 (Section 18.11).

The self-condensation of acetyl coenzyme A is the exact counterpart of the self-condensation of ethyl acetate. Because of this similarity, it has been suggested that the formation of the enolate anion of acetyl coenzyme A occurs at mildly basic sites on the enzyme. A proposed mechanism for this reaction is illustrated in Figure 19-5.

19.12 SUMMARY

Aldehydes and ketones both contain carbonyl groups. In an aldehyde, the carbonyl carbon is bonded to one hydrogen and either an alkyl or aryl group. In a ketone, the carbonyl carbon is bonded to two carbon-containing groups, either alkyl or aryl groups.

Aldehydes and ketones differ in their ease of oxidation. Aldehydes are easily oxidized to carboxylic acids, whereas ketones are oxidized only with difficulty. This difference in reactivity is used as a chemical means to distinguish between these two classes of compounds.

Aldehydes and ketones undergo the same two general reactions. One is the addition to the carbonyl group. Addition of water, alcohols, hydrogen, and ammonia and its derivatives are examples of addition reactions also found in living systems. The second type of reaction is a condensation in which the α-carbon of one carbonyl-containing compound bonds to the carbonyl carbon of another molecule. Aldehydes, ketones, and esters undergo this reaction, which is used in the laboratory and in living systems to build a larger molecule from two smaller ones.

In condensation reactions, the nucleophile is the enolate anion formed by a strong base removing the proton α to the carbonyl group. The carbonyl group serves two functions in condensation reactions. It serves as the site where a nucleophile adds, and it makes its α hydrogens more acidic than usual for a C—H bond.

Many of the reactions of aldehydes and ketones in the laboratory are similar to their reactions in living systems.

REVIEW OF TERMS, CONCEPTS, AND REACTIONS

Terms

ACETAL A compound that contains the $R'\text{—}\overset{\overset{\displaystyle H}{|}}{\underset{\underset{\displaystyle OR}{|}}{C}}\text{—}OR$ functional group.*

ALDEHYDE A compound that contains a carbonyl group whose carbon is bonded to one hydrogen and either an alkyl or aryl group.

CARBANION A compound that contains a carbon atom bearing a negative charge and bonded to three groups or atoms.

CONDENSATION REACTION A reaction in which two compounds are joined to form one larger compound.

FEHLING'S SOLUTION An alkaline solution of cupric ion complexed with tartrate ion (a test reagent for aldehydes).

HEMIACETAL A compound that contains the $R'\text{—}\overset{\overset{\displaystyle H}{|}}{\underset{\underset{\displaystyle OH}{|}}{C}}\text{—}OR$ functional group.*

HEMIKETAL A compound that contains the $R'\text{—}\overset{\overset{\displaystyle OH}{|}}{\underset{\underset{\displaystyle R''}{|}}{C}}\text{—}OR$ functional group.*

* R′ and R″ can be an alkyl or aryl group; R is an alkyl group.

KETAL A compound that contains the $R'-\underset{\underset{R''}{|}}{\overset{\overset{OR}{|}}{C}}-OR$ functional group.*

KETONE A compound that contains a carbonyl group, whose carbon is bonded to two alkyl or aryl groups.

SCHIFF BASE A compound that contains the $\begin{matrix} R' & & R' \\ & C{=}N & \\ R'' & & \end{matrix}$ functional group.*

TOLLENS' REAGENT An alkaline solution of $Ag(NH_3)_2^+$ used as a test for aldehydes.

Concepts

1. Aldehydes and ketones differ in their ease of oxidation. Aldehydes are easily oxidized by mild oxidizing agents, whereas ketones require more vigorous conditions.
2. Aldehydes and ketones undergo two types of reaction: addition to the carbonyl group, and condensation reactions.
3. The hydrogen α to the carbonyl group is more acidic than most C—H bonds. The reason for this is that the enolate anion formed by loss of an α hydrogen of an aldehyde or ketone is stabilized by resonance.

Reactions

Preparation of Aldehydes and Ketones (see Section 19.3)

1. $RCH_2OH \xrightarrow[\text{Oxidation}]{} RCHO$

2. $\underset{\underset{OH}{|}}{RCHR} \xrightarrow[\text{Oxidation}]{} \underset{\underset{O}{\|}}{RCR}$

Oxidation of Aldehydes (see Section 19.4)

$$2\,Ag(NH_3)_2^+ + RCHO + 3\,OH^- \longrightarrow RCO_2^- + 2\,Ag\downarrow + 4\,NH_3 + 2\,H_2O$$

*R′ and R″ can be an alkyl or aryl group; R is an alkyl group.

Addition Reactions (see Section 19.5)

1. Water

$$CCl_3CHO + H_2O \rightleftharpoons CCl_3CH(OH)_2$$

2. Alcohols

$$RCHO + R'OH \overset{H^+}{\rightleftharpoons} \underset{\text{Hemiacetal}}{RCH(OH)(OR')} + R'OH \overset{H^+}{\rightleftharpoons} \underset{\text{Acetal}}{RCH(OR')_2} + H_2O$$

$$RC(=O)R'' + R'OH \overset{H^+}{\rightleftharpoons} \underset{\text{Hemiketal}}{RC(OH)(OR')R''} + R'OH \overset{H^+}{\rightleftharpoons} \underset{\text{Ketal}}{RC(OR')_2R''} + H_2O$$

3. Ammonia and its derivatives

$$RCHO + NH_2R' \longrightarrow RCH{=}NR' + H_2O$$

$$RC(=O)R'' + NH_2R' \longrightarrow (R'')(R)C{=}NR' + H_2O$$

4. Hydrogen

$$RCHO \xrightarrow{LiAlH_4} RCH_2OH$$

$$R_2C{=}O \xrightarrow{LiAlH_4} R_2CHOH$$

$$RCHO \xrightarrow[\text{Metal catalyst}]{H_2} RCH_2OH$$

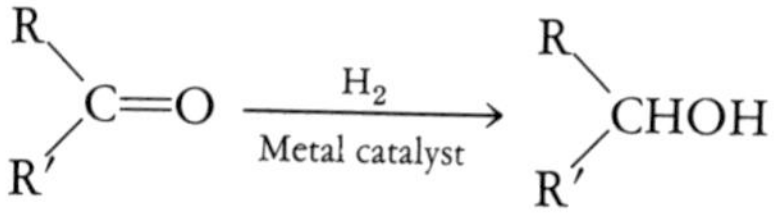

$$R(R')C{=}O \xrightarrow[\text{Metal catalyst}]{H_2} R(R')CHOH$$

Condensation Reactions (see Section 19.8)

1. Aldol condensation

$$2\,CH_3CHO \underset{}{\overset{OH^-}{\rightleftharpoons}} CH_3CH(OH)CH_2CHO$$

2. Claisen condensation

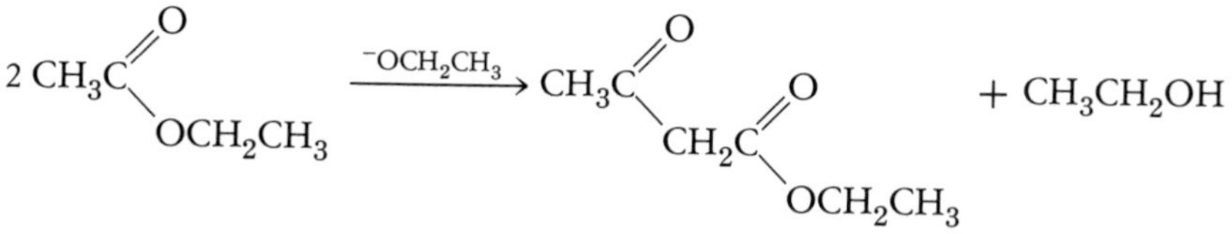

$$2\,CH_3C(=O)OCH_2CH_3 \xrightarrow{^-OCH_2CH_3} CH_3C(=O)CH_2C(=O)OCH_2CH_3 + CH_3CH_2OH$$

EXERCISES 19-22 Write the structural formulas and give the common and IUPAC names for all carbonyl compounds of formula $C_5H_{10}O$.

19-23 Write the structural formula of each of the following compounds:
(a) acetophenone (b) formaldehyde (c) acetone
(d) 3-hexenal (e) benzyl isopropyl ketone
(f) methoxyacetaldehyde (g) β-methylbutyraldehyde
(h) methyl *p*-tolyl ketone (i) 5-methyl-3-hexanone
(j) 3-methyl-2-butenal (k) 3,4-dimethoxyacetophenone

19-24 Give the IUPAC name for each of the following compounds:

(a) $CH_3CH_2CH_2CHO$ (b) $CH_3CH_2C(=O)CH_2CH_3$

(c) 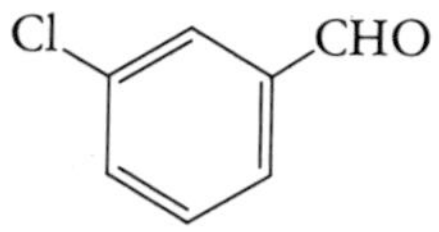 benzene ring with Cl and CHO substituents (meta): Cl–C_6H_4–CHO

(d) 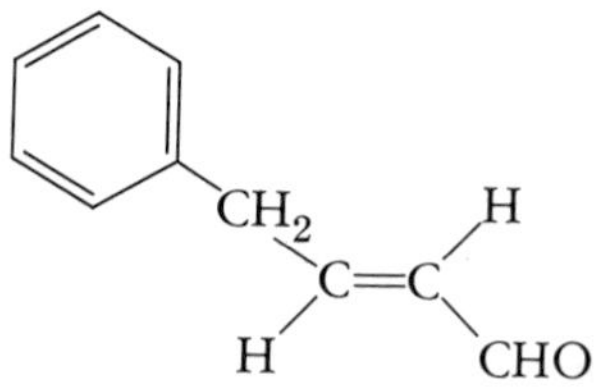 $C_6H_5CH_2$–CH=CH–CHO (trans; H on each alkene carbon)

(e) 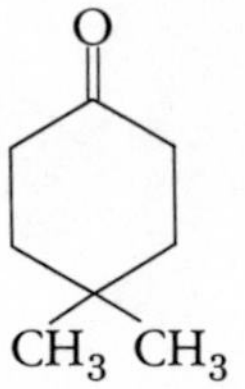 O; CH$_3$ CH$_3$ (f) 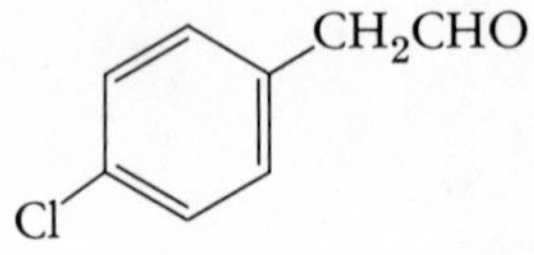 CH$_2$CHO; Cl

(g) 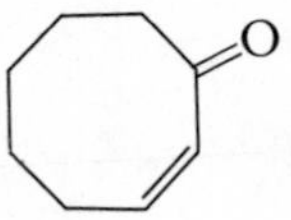 O (h) 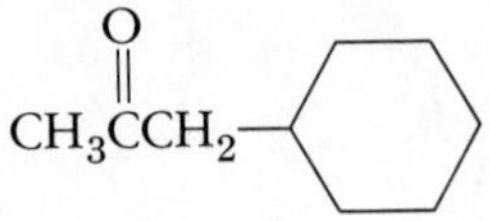 $CH_3\overset{O}{\overset{\|}{C}}CH_2$—

(i) 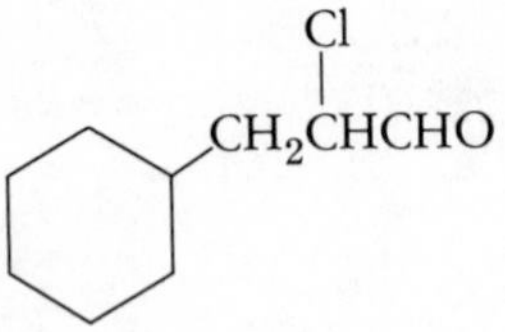 Cl; CH$_2$CHCHO

19-25 Give an example of each of the following:
(a) an acetal (b) a ketal (c) an oxime (d) a hemiketal
(e) a phenylhydrazone (f) a cyclic ketal
(g) a Claisen condensation (h) a hemiacetal
(i) an aldol condensation

19-26 Write the equation for the reaction, if any, of acetaldehyde with each of the following reagents:
(a) Tollens' reagent (b) H_2, Pt (c) semicarbazide
(d) $NaBH_4$, then H_2O (e) Ethyl alcohol, H^+
(f) hydroxylamine (g) aqueous NaOH solution
(h) $LiAlH_4$, then H_2O (i) CrO_3, H_2SO_4, H_2O
(j) 2,4-dinitrophenylhydrazine (k) Fehling's solution

19-27 Repeat exercise 19-26 using cyclopentanone instead of acetaldehyde.

19-28 Write the structure of the major organic product of each of the following reactions:

(a) 2-pentanol + CrO_3 $\xrightarrow[H_2O]{H_2SO_4}$

(b) $HOCH_2\overset{O}{\overset{\|}{C}}CH_2O$Ⓟ + O=C(H)$\overset{OH}{\overset{|}{C}}H\underset{OH}{\underset{|}{C}}HCH_2O$Ⓟ $\xrightarrow{\text{Enzyme}}$

(c) Pentanal + $Ag(NH_3)_2{}^+$ $\longrightarrow$

(d) 4-hydroxyhexanal $\xrightarrow{H^+}$

(e) cyclohexanone + 1,2-ethanediol $\xrightarrow{H^+}$

(f) isobutyraldehyde + 1-propanol (excess) $\xrightarrow{H^+}$

(g) 2-pentanone + methylamine $\longrightarrow$

19-29 Write a mechanism for each of the following reactions:

(a) $$CH_3\overset{\overset{\displaystyle O}{\|}}{C}CH_3 + H_2C{=}O \xrightarrow{OH^-} CH_3\overset{\overset{\displaystyle O}{\|}}{C}CH_2CH_2OH$$

(b) $$2\ CH_3C(=O)OCH_3 \xrightarrow{CH_3O^-} CH_3OC(=O)CH_2C(=O)OCH_3$$

19-30 1,2-Ethanedithiol ($HSCH_2CH_2SH$) reacts with acetaldehyde according to the following reaction:

$$CH_3C(=O)H + HSCH_2CH_2SH \overset{H^+}{\rightleftharpoons} CH_3CH\begin{matrix} S-CH_2 \\ | \\ S-CH_2 \end{matrix}$$

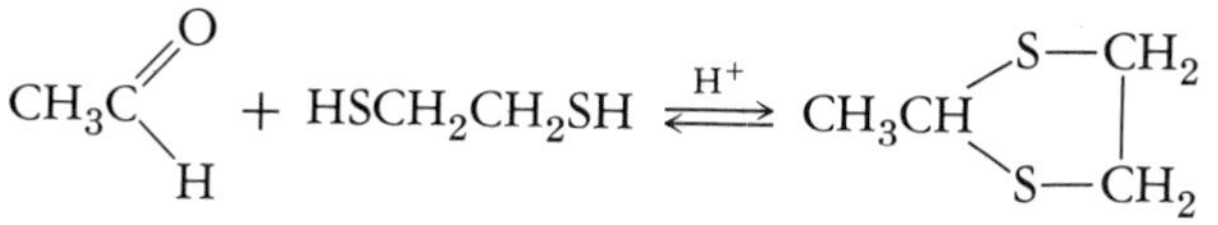

Propose a mechanism for this reaction.

20 REACTIONS OF POLYFUNCTIONAL COMPOUNDS

In the preceding chapters we learned the chemical reactions of the most important functional groups in living systems. The compounds used as examples usually contained only one functional group. However, most compounds in nature are more complicated. They are *polyfunctional,* that is, they contain more than one functional group. How does the presence of one functional group in a compound affect the reactions of another functional group? The answer to this question depends on the relative positions of the functional groups.

It is possible that the functional groups in a compound are so far from each other that they are completely independent. For example, lanosterol contains different functional groups that are widely separated.

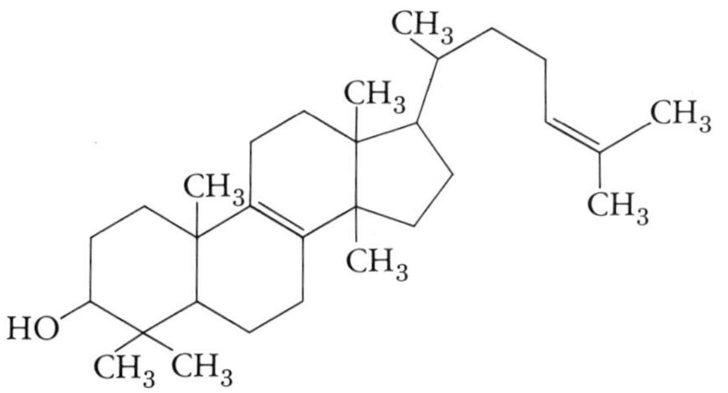

CH_3 CH_3 CH_3 CH_3 CH_3 CH_3 HO CH_3 CH_3

Lanosterol

Because they are so far apart, the two double bonds and the hydroxy group undergo their usual reactions independently of each other.

A second possibility is that one functional group may influence the reactivity of another functional group without changing its reactions. An example is the effect of the structure of a carboxylic acid on its acidity. We learned in Section 18.4 that carboxylic acids with electronegative substituents (in reality, functional groups) adjacent to the carboxylic acid functional group are stronger acids than are those lacking such substituents. In this case, the substituent does not change the reactions of the carboxylic acid. It is still an acid, but its strength is changed.

A third possibility is that the two functional groups together can be considered a new functional group that undergoes its own characteristic reactions. In Section 20.2, we will introduce such a new functional group, the β-dicarbonyl group.

Finally, it is possible that a compound will contain two functional groups that can react with each other. For example, a carboxylic acid and an alcohol are two functional groups that react to form an ester. If these two functional groups are correctly positioned in the same molecule, they can react with each other. Such *intramolecular reactions* form rings. If they are not positioned correctly for intramolecular reactions, then they usually react with functional groups in another molecule. These *intermolecular reactions* lead to dimers, trimers, and sometimes polymers.

In this chapter, we will learn the important intermolecular and intramolecular reactions of polyfunctional compounds and also learn the chemistry of the β-dicarbonyl functional group. But first let us learn to name polyfunctional compounds.

20.1 NAMING POLYFUNCTIONAL COMPOUNDS

A number of compounds have been encountered so far that contain two different functional groups. Chloroacetic acid and 3-hydroxybutanal (aldol) are two examples:

$$ClCH_2CO_2H \qquad CH_3\underset{\displaystyle OH}{\underset{|}{C}}HCH_2CHO$$

Chloroacetic acid　　3-Hydroxybutanal

In these names, one functional group is designated by a prefix; the other, by a suffix. How do we establish whether a functional group is denoted by a

Table 20-1. Order of Priority of Functional Groups

Functional Group	Formula	Name as Prefix	Name as Suffix	
Ammonium	$R_4\overset{+}{N}$	—	ammonium	
Carboxylic acid	$—CO_2H$	Carboxy-	-oic acid	
Esters	$—CO_2R$	Alkyloxycarbonyl- or aryloxycarbonyl-	Alkyl -oate or aryl -oate	Decreasing order of priority ↓
Acid halide	—COX (X = halogen)	Haloformyl-	-oyl halide	
Amide	$—CONH_2$	Amido-	-amide	
Aldehyde	—CHO	Aldo- or oxo-	-al	
Nitrile	—C≡N	Cyano-	-nitrile	
Ketone	—CO—	Keto- or oxo-	-one	
Alcohol	—OH	Hydroxy-	-ol	
Thiol	—SH	Mercapto-	-thiol	
Amine	$—NH_2$	Amino-	-amine	
Double bond	>C=C<	—	-ene	
Triple bond	—C≡C—	—	-yne	
Halogen	—X	Halo-	—	
Alkyl groups	—R	Alkyl	—	
Alkoxy	—OR	Alkoxy-	—	
Alkylthio	—SR	Alkylthio-	—	
Epoxy	C—C bridged by O (three-membered ring)	Epoxy-	—	
Nitro	$—NO_2$	Nitro-	—	

prefix or a suffix? Simply by establishing an order of priority for the various functional groups. The functional group of highest priority in the molecule is designated by a suffix and receives the lowest number. The other functional groups are usually denoted by prefixes. The priorities of the principal functional groups and their names as suffixes and prefixes are given in Table 20-1.

Using this list of priorities and the usual IUPAC rules, we can name many polyfunctional molecules. Let us name a few compounds to illustrate the use of the priorities given in Table 20-1. We will start with the following compound:

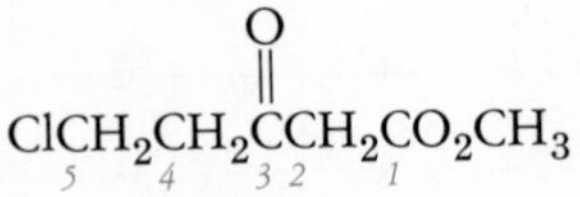

$$\underset{5}{Cl}\underset{4}{CH_2}CH_2\overset{\overset{\displaystyle O}{\|}}{\underset{3}{C}}\underset{2}{CH_2}\underset{1}{CO_2}CH_3$$

This molecule contains three functional groups: an ester group, a carbonyl group, and a chlorine atom. Their order of priority, according to Table 20-1, is $-CO_2CH_3 > \rangle C{=}O > -Cl$. Therefore, the compound is named as a substituted methyl pentanoate, and the carboxyl carbon is numbered 1. The carbonyl group is designated by the prefix keto- and the chlorine, by the prefix chloro-. The complete name is methyl 5-chloro-3-ketopentanoate.

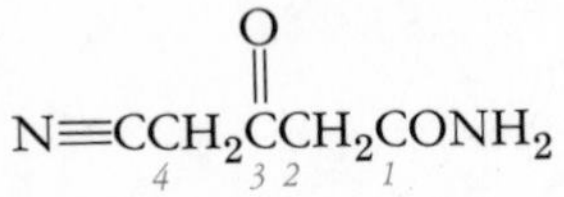

$$N{\equiv}\underset{4}{C}CH_2\overset{\overset{\displaystyle O}{\|}}{\underset{3}{C}}\underset{2}{CH_2}\underset{1}{C}ONH_2$$

This compound contains three functional groups whose order of priority is $-CONH_2 > -C{\equiv}N > \rangle C{=}O$. Therefore, the compound is named as a substituted butanamide. The complete name is 4-cyano-3-ketobutanamide. The following are other examples:

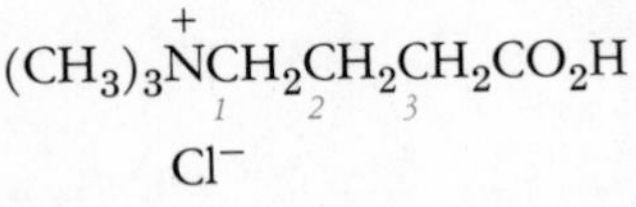

$$(CH_3)_3\overset{+}{N}\underset{1}{CH_2}\underset{2}{CH_2}\underset{3}{CH_2}CO_2H \quad Cl^-$$

3-Carboxypropyltrimethylammonium chloride

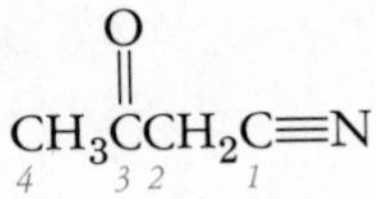

$$\underset{4}{CH_3}\overset{\overset{\displaystyle O}{\|}}{\underset{3}{C}}\underset{2}{CH_2}\underset{1}{C}{\equiv}N$$

3-Ketobutanenitrile

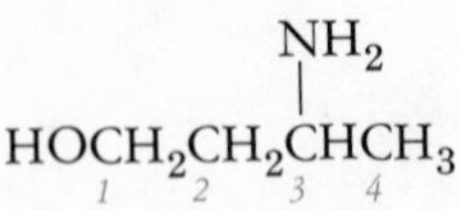

$$HO\underset{1}{CH_2}\underset{2}{CH_2}\overset{\overset{\displaystyle NH_2}{|}}{\underset{3}{C}}H\underset{4}{CH_3}$$

3-Amino-1-butanol

$$HS\underset{5}{CH_2}\overset{\overset{\displaystyle CH_3}{|}}{\underset{4}{C}}H\underset{3}{CH_2}\overset{\overset{\displaystyle C\equiv N}{|}}{\underset{2}{C}}H\underset{1}{C}OCl$$

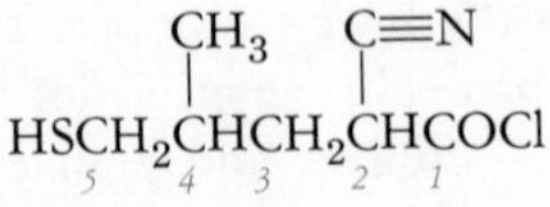

2-Cyano-5-mercapto-4-methylpentanoyl chloride

Notice that in Table 20-1 certain functional groups are designated only by prefixes and others are designated only by suffixes. Carbon-carbon double and triple bonds are designated only by suffixes, whereas halogens and alkyl, alkoxy, alkylthio, epoxy, and nitro groups are designated only by prefixes. Consequently, a molecule containing a double or triple bond and another functional group of higher priority must contain two suffixes in its name. This is an exception to the general rule that the name of a compound can contain only one suffix. For example:

$$\overset{4}{H_2N}\overset{3}{CH_2}\overset{2}{CH}{=}\overset{1}{CH}CHO$$
4-Amino-2-butenal

$$HO\overset{5}{CH_2}\overset{\overset{CH_3}{|}}{\overset{4}{C}H}\overset{3}{C}{\equiv}\overset{2}{C}\overset{1}{C}O_2H$$
5-Hydroxy-4-methyl-2-pentynoic acid

*EXERCISE 20-1 Write the structural formula for each of the following compounds:
(a) 1-buten-3-yne (b) 4-amino-2-pentanol
(c) 3-hydroxybutanamide (d) 3-penten-1-thiol
(e) 3-ketohexanenitrile
(f) 6-chloro-4-keto-3-methylhexanoic acid
(g) 4-amino-2-butanone (h) 2,3-epoxybutanoyl chloride
(i) 4-methoxy-2-butenal

EXERCISE 20-2 Give the IUPAC name for each of the following compounds:

(a) $HOCH_2\overset{\overset{O}{\|}}{C}CH_2CH_2OH$ (b) $CH_3(H)C{=}C(H)C(=O)CH_2CH_3$

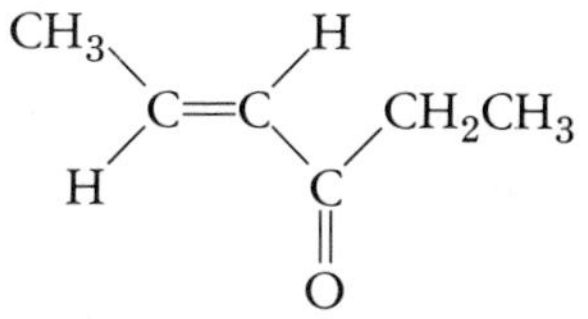

(c) $(CH_3)_2C{=}CHCHO$ (d) $BrCH_2(H)C{=}C(H)CH_2OH$

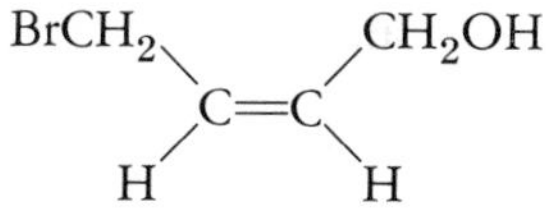

(e) $HOCH_2C{\equiv}CCH_2CONH_2$ (f) [cyclohexenone ring with OCH_3 substituent]

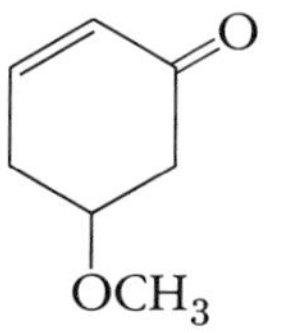

* The answers for the exercises in this chapter begin on page 888.

20.2 THE β-DICARBONYL FUNCTIONAL GROUP

A β-dicarbonyl functional group consists of two carbonyl groups separated by a single carbon atom:

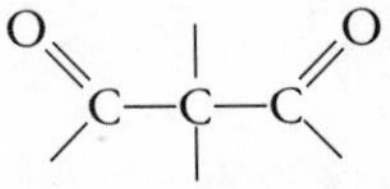

$$\underset{O\diagdown}{}\!\!\!\!\!\!\diagup C-\overset{|}{\underset{|}{C}}-C\!\overset{\diagup\!\!\!\diagup O}{\diagdown}$$

Either or both of the carbonyl groups can be part of a carboxylic acid, ester, aldehyde, or ketone functional group. The following compounds all contain this new polyfunctional group:

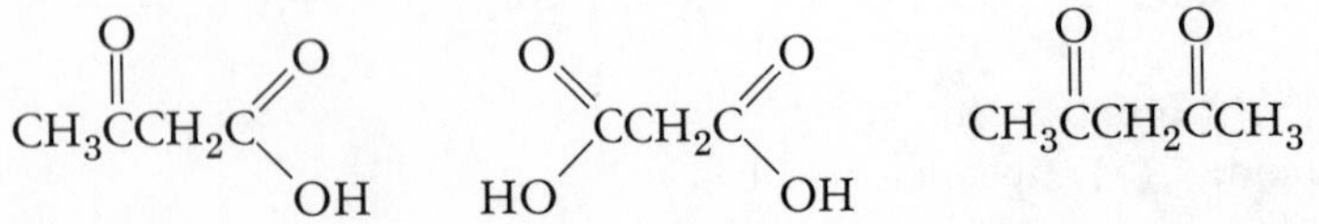

$$\underset{\text{Acetoacetic acid}}{CH_3\overset{O}{\overset{\|}{C}}CH_2\overset{O}{\overset{\|}{C}}OH} \qquad \underset{\text{Malonic acid}}{HO\overset{O}{\overset{\|}{C}}CH_2\overset{O}{\overset{\|}{C}}OH} \qquad \underset{\text{Acetylacetone}}{CH_3\overset{O}{\overset{\|}{C}}CH_2\overset{O}{\overset{\|}{C}}CH_3}$$

$$\underset{\text{Diethyl malonate}}{CH_3CH_2O\overset{O}{\overset{\|}{C}}CH_2\overset{O}{\overset{\|}{C}}OCH_2CH_3} \qquad \underset{\text{Acetylacetaldehyde}}{CH_3\overset{O}{\overset{\|}{C}}CH_2\overset{O}{\overset{\|}{C}}H} \qquad \underset{\text{Ethyl acetoacetate}}{CH_3\overset{O}{\overset{\|}{C}}CH_2\overset{O}{\overset{\|}{C}}OCH_2CH_3}$$

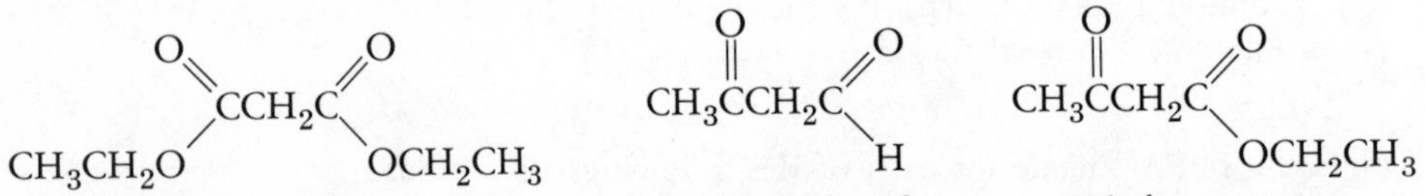

Compounds that contain the β-dicarbonyl group exist as an equilibrium mixture of two isomers. One isomer contains the two carbonyl groups and is called the *keto* isomer. The other isomer contains an —OH group bonded to the carbon of the carbon-carbon double bond and is called the *enol* isomer. The word enol comes from a combination of the IUPAC endings for an alkene (-ene) and an alcohol (-ol). For example, acetylacetone is an equilibrium mixture of 70 percent of the enol isomer and 30 percent of the keto isomer:

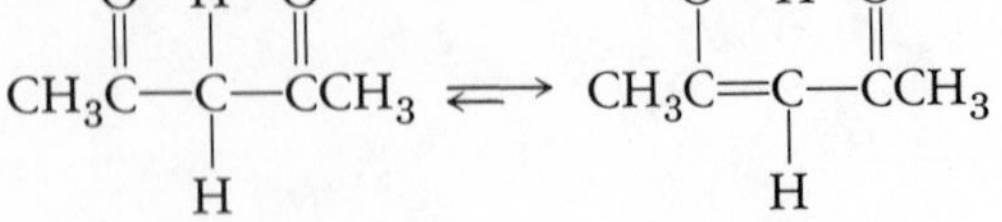

$$\underset{\substack{30\% \\ \text{Keto isomer of} \\ \text{acetylacetone}}}{CH_3\overset{O}{\overset{\|}{C}}-\overset{H}{\overset{|}{\underset{H}{\underset{|}{C}}}}-\overset{O}{\overset{\|}{C}}CH_3} \rightleftharpoons \underset{\substack{70\% \\ \text{Enol isomer of} \\ \text{acetylacetone}}}{CH_3\overset{O-H}{\overset{|}{C}}=\underset{H}{\underset{|}{C}}-\overset{O}{\overset{\|}{C}}CH_3}$$

This isomerization is called *tautomerism.* The two isomers are also called *tautomers.* By careful work, chemists have been able to isolate the pure enol isomer of several β-dicarbonyl compounds.

Tautomerism also occurs in simple ketones and aldehydes. However, in these cases, the keto isomer predominates at equilibrium. For example, more than 99 percent of acetaldehyde exists as the keto isomer:

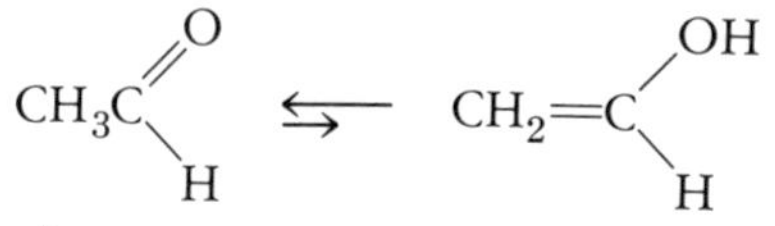

$$CH_3C(=O)H \rightleftharpoons CH_2{=}C(OH)H$$

$> 99\%$ Keto isomer of acetaldehyde

$< 1\%$ Enol isomer of acetaldehyde

EXERCISE 20-3 Write the structure of the enol isomer of each of the following compounds:

(a) 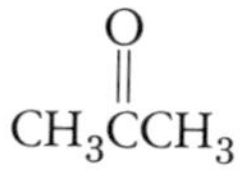 $CH_3\overset{O}{\overset{\|}{C}}CH_3$ (b) 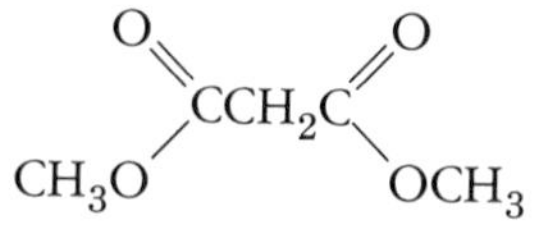 $CH_3O(O{=})CCH_2C({=}O)OCH_3$

Compounds that contain a β-dicarbonyl functional group undergo a number of characteristic reactions. Decarboxylation, alkylation, and acylation are reactions that occur both in the laboratory and in living systems.

Decarboxylation of β-Ketoacids

When a monofunctional carboxylic acid such as acetic acid is heated, it loses carbon dioxide only with great difficulty; e.g., when sodium acetate and soda lime (a mixture of sodium hydroxide and calcium oxide) are heated above 300° C, methane is formed in poor yield:

$$CH_3CO_2Na \xrightarrow[>300^\circ C]{\text{Soda lime}} CH_4 + CO_2$$

Sodium acetate — Methane

Such a reaction is called *decarboxylation,* meaning loss of carbon dioxide.

In contrast, the decarboxylation of 1,3-dicarboxylic acids is more easily accomplished. For example, when malonic acid is heated to only 140° C, acetic acid and carbon dioxide are formed in good yield:

$$\underset{\text{Malonic acid}}{HO_2CCH_2CO_2H} \xrightarrow{140^\circ\text{ C}} \underset{\text{Acetic acid}}{HO_2CCH_3} + CO_2$$

β-Keto carboxylic acids also undergo such decarboxylation reactions. For example:

$$\underset{\text{Acetoacetic acid}}{CH_3\overset{\overset{\displaystyle O}{\|}}{C}CH_2C(=O)OH} \xrightarrow{100^\circ\text{ C}} \underset{\text{Acetone}}{CH_3\overset{\overset{\displaystyle O}{\|}}{C}CH_3} + CO_2$$

In fact, this is a general reaction for any carboxylic acid that contains electronegative atoms or groups bonded to the α carbon. The following compounds all easily undergo decarboxylation reactions:

$$\underset{\text{Nitroacetic acid}}{O_2NCH_2CO_2H} \xrightarrow{50^\circ\text{ C}} \underset{\text{Nitromethane}}{O_2NCH_3} + CO_2$$

$$\underset{\text{Cyanoacetic acid}}{N{\equiv}CCH_2CO_2H} \xrightarrow{150^\circ\text{ C}} \underset{\text{Acetonitrile}}{N{\equiv}CCH_3} + CO_2$$

$$\underset{\text{Trichloroacetic acid}}{Cl_3CCO_2H} \xrightarrow{150^\circ\text{ C}} \underset{\text{Chloroform}}{Cl_3CH} + CO_2$$

$$\underset{\text{4-Chlorobenzoic acid}}{4\text{-}ClC_6H_4CO_2H} \xrightarrow{170^\circ\text{ C}} \underset{\text{Chlorobenzene}}{C_6H_5Cl} + CO_2$$

EXERCISE 20-4 Write the structures of the products formed by thermal decarboxylation of the following compounds:

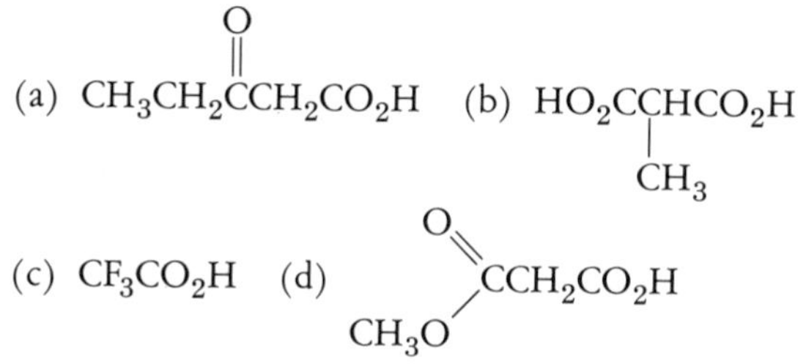

$$\text{(a) } CH_3CH_2\overset{\overset{O}{\|}}{C}CH_2CO_2H \quad \text{(b) } HO_2C\underset{\underset{CH_3}{|}}{C}HCO_2H$$

$$\text{(c) } CF_3CO_2H \quad \text{(d) } CH_3O{-}\overset{O}{\overset{\|}{C}}CH_2CO_2H$$

Alkylation and Acylation

In Chapter 19, we learned that the α hydrogens of aldehydes and ketones are sufficiently acidic to be removed by strong bases. The hydrogens bonded to the central carbon of the β-dicarbonyl functional group are even more acidic because they are located adjacent to two carbonyl groups. Consequently, they can be removed by a strong base. An example is diethyl malonate, which undergoes the following reaction when treated by a strong base such as sodium ethoxide in ethanol:

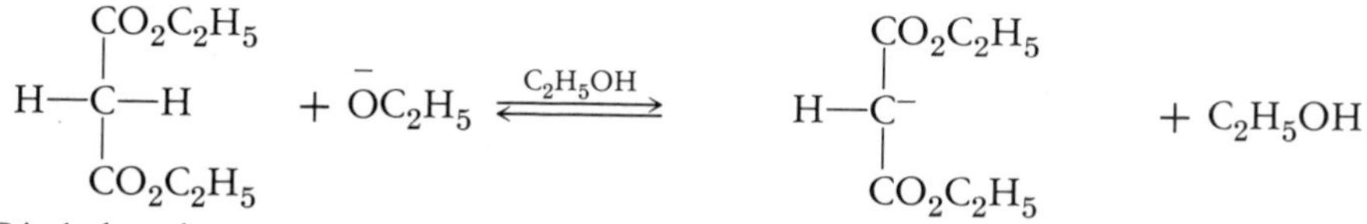

$$\underset{\text{Diethyl malonate}}{H{-}\underset{\underset{CO_2C_2H_5}{|}}{\overset{\overset{CO_2C_2H_5}{|}}{C}}{-}H} + \bar{O}C_2H_5 \underset{}{\overset{C_2H_5OH}{\rightleftharpoons}} \underset{\text{Anion of diethyl malonate}}{H{-}\underset{\underset{CO_2C_2H_5}{|}}{\overset{\overset{CO_2C_2H_5}{|}}{C^-}}} + C_2H_5OH$$

The anion of diethyl malonate is a hybrid of the following three structures:

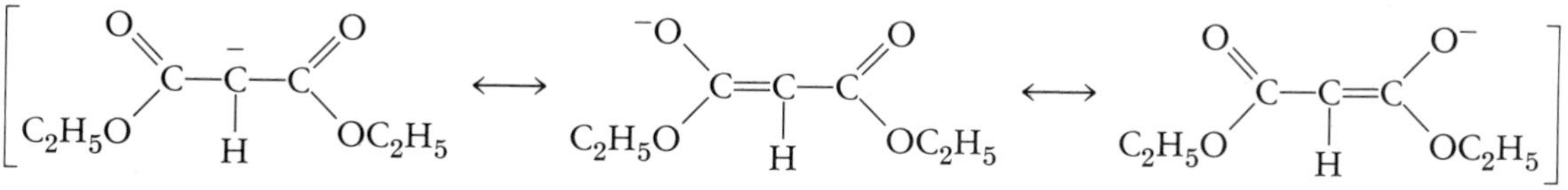

$$\left[C_2H_5O{-}\overset{\overset{O}{\|}}{C}{-}\underset{\underset{H}{|}}{\bar{C}}{-}\overset{\overset{O}{\|}}{C}{-}OC_2H_5 \longleftrightarrow C_2H_5O{-}\overset{\overset{^-O}{|}}{C}{=}\underset{\underset{H}{|}}{C}{-}\overset{\overset{O}{\|}}{C}{-}OC_2H_5 \longleftrightarrow C_2H_5O{-}\overset{\overset{O}{\|}}{C}{-}\underset{\underset{H}{|}}{C}{=}\overset{\overset{O^-}{|}}{C}{-}OC_2H_5 \right]$$

It is a resonance-stabilized enolate anion which is the reason the hydrogens α to the carbonyl groups are removed by a strong base.

EXERCISE 20-5 Write the structure of the enolate anion formed from each of the following compounds:

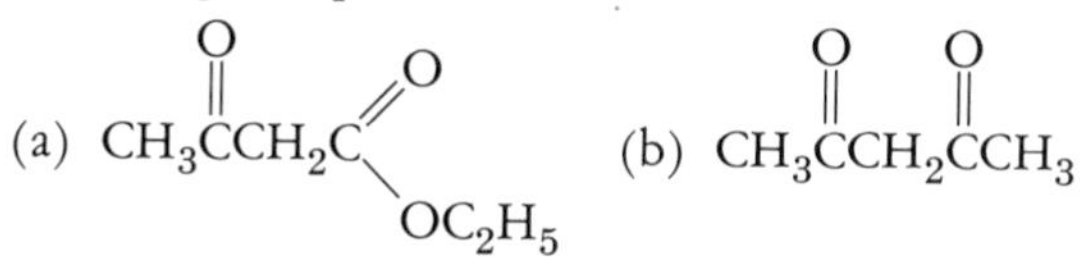

$$\text{(a) } CH_3\overset{\overset{O}{\|}}{C}CH_2\overset{\overset{O}{\|}}{C}{-}OC_2H_5 \qquad \text{(b) } CH_3\overset{\overset{O}{\|}}{C}CH_2\overset{\overset{O}{\|}}{C}CH_3$$

The enolate anion of diethyl malonate is a nucleophile. As a result, it can displace a halide from alkyl halides or be substituted for the group or atom bonded to the carbonyl group of carboxylic acid derivatives to form a substituted diethyl malonate. These are general reactions that are widely used in the laboratory and in living systems to form carboxylic acids and methyl ketones. Let us examine a few examples of these reactions.

The enolate anion of diethyl malonate reacts with a primary or secondary alkyl halide to form an alkylmalonic ester. For example:

$$\underset{\text{Anion of diethyl malonate}}{\mathrm{H{-}\overset{\displaystyle CO_2C_2H_5}{\underset{\displaystyle CO_2C_2H_5}{C^-}}}} + \underset{\text{Alkyl bromide}}{\overset{\displaystyle R}{CH_2}{-}Br} \longrightarrow \underset{\text{Alkylmalonic ester}}{\mathrm{H\overset{\displaystyle CO_2C_2H_5}{\underset{\displaystyle CO_2C_2H_5}{C}}CH_2R}} + Br^-$$

This is another example of a nucleophilic displacement reaction (see Section 15.5).

When the product of this reaction is treated with an aqueous solution of a strong acid, the ester groups are hydrolyzed to form an alkylmalonic acid. When heated, the alkylmalonic acid loses carbon dioxide to form a carboxylic acid. Notice that the final carboxylic acid contains two more carbon atoms than the starting alkyl halide.

$$\underset{\text{Alkyl malonic ester}}{\overset{\displaystyle CO_2C_2H_5}{\underset{\displaystyle CO_2C_2H_5}{CHCH_2R}}} \xrightarrow[H^+]{H_2O} \underset{\text{Alkyl malonic acid}}{\overset{\displaystyle CO_2H}{\underset{\displaystyle CO_2H}{CHCH_2R}}} \xrightarrow{\Delta} \underset{\text{Carboxylic acid}}{\underset{\displaystyle CO_2H}{CH_2CH_2R}} + CO_2$$

These two steps are usually combined in the laboratory. By heating the aqueous acid solution of diethyl alkylmalonate, both hydrolysis of the ester and decarboxylation occur in a single step.

The following is a specific example of the preparation of a carboxylic acid by this method:

$$\overset{\displaystyle CO_2C_2H_5}{\underset{\displaystyle CO_2C_2H_5}{CH_2}} + \bar{O}C_2H_5 \longrightarrow \overset{\displaystyle CO_2C_2H_5}{\underset{\displaystyle CO_2C_2H_5}{CH^-}} + C_2H_5OH$$

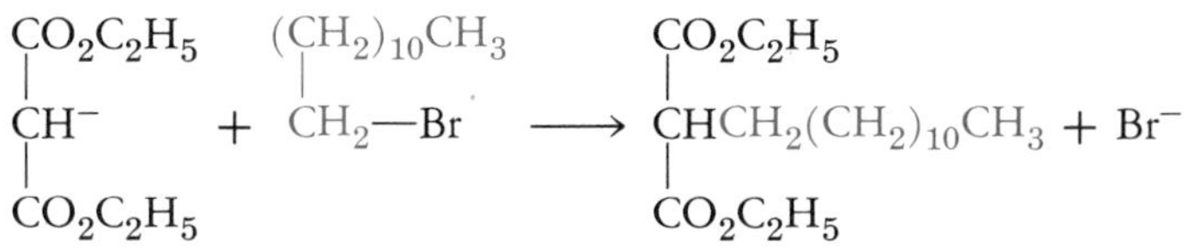

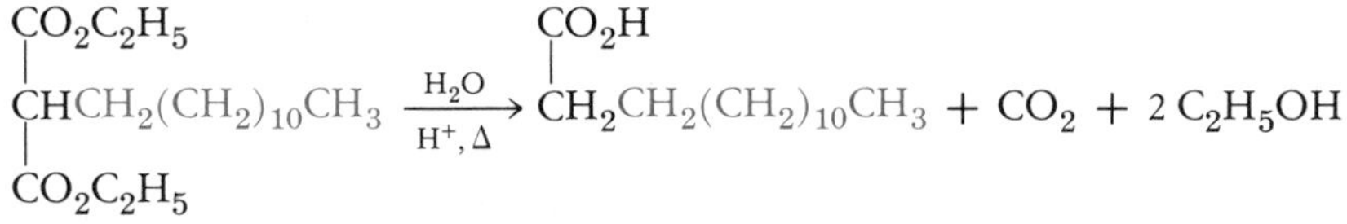

$$\begin{array}{l} CO_2C_2H_5 \\ | \\ CH^- \\ | \\ CO_2C_2H_5 \end{array} + \begin{array}{l} (CH_2)_{10}CH_3 \\ | \\ CH_2{-}Br \end{array} \longrightarrow \begin{array}{l} CO_2C_2H_5 \\ | \\ CHCH_2(CH_2)_{10}CH_3 \\ | \\ CO_2C_2H_5 \end{array} + Br^-$$

$$\begin{array}{l} CO_2C_2H_5 \\ | \\ CHCH_2(CH_2)_{10}CH_3 \\ | \\ CO_2C_2H_5 \end{array} \xrightarrow[H^+, \Delta]{H_2O} \begin{array}{l} CO_2H \\ | \\ CH_2CH_2(CH_2)_{10}CH_3 \end{array} + CO_2 + 2\ C_2H_5OH$$

In Chapter 18 we learned that a group or an atom bonded to the carbonyl carbon of a carboxylic acid derivative can be replaced by a nucleophile such as ammonia or an alcohol to form amides and esters, respectively. In a similar manner, the enolate anion of diethyl malonate will replace such groups to form diethyl acylmalonate as product. For example:

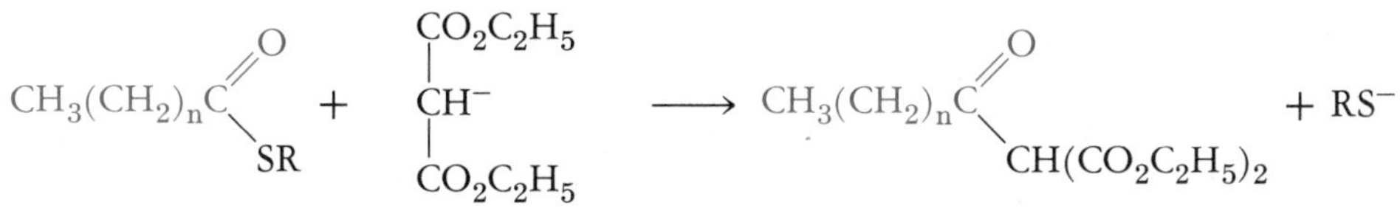

$$\underset{\text{Thioester}}{CH_3(CH_2)_nC(=O)SR} + \underset{\text{Anion of diethyl malonate}}{\begin{array}{l} CO_2C_2H_5 \\ | \\ CH^- \\ | \\ CO_2C_2H_5 \end{array}} \longrightarrow \underset{\text{Diethyl acylmalonate}}{CH_3(CH_2)_nC(=O)CH(CO_2C_2H_5)_2} + RS^-$$

Again, acid-catalyzed hydrolysis of the product forms an acylmalonic acid that loses carbon dioxide on heating:

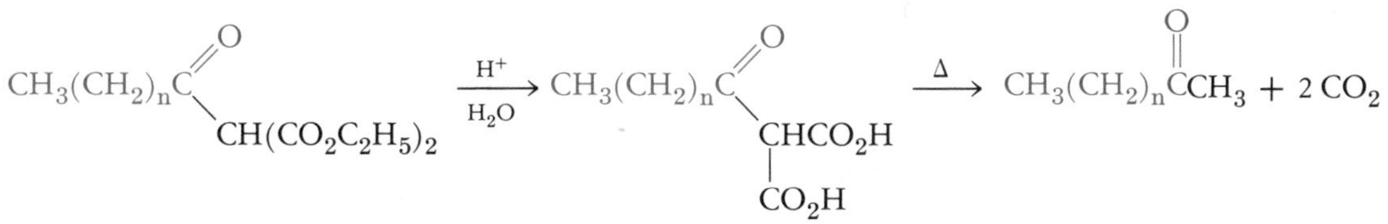

$$\underset{\text{Diethyl acylmalonate}}{CH_3(CH_2)_nC(=O)CH(CO_2C_2H_5)_2} \xrightarrow[H_2O]{H^+} \underset{\text{Acylmalonic acid}}{CH_3(CH_2)_nC(=O)\begin{array}{l} CHCO_2H \\ | \\ CO_2H \end{array}} \xrightarrow{\Delta} \underset{\text{Ketone}}{CH_3(CH_2)_n\overset{O}{\overset{\|}{C}}CH_3} + 2\ CO_2$$

Notice that in the decarboxylation step two molecules of carbon dioxide are lost to form a methyl ketone.

EXERCISE 20-6 Write the structure of the product of each of the following reactions:

(a) $CH_2(CO_2C_2H_5)_2 + \bar{O}C_2H_5 \longrightarrow$ Compound A

(b) Compound A + $CH_3(CH_2)_6CH_2Cl \longrightarrow$ Compound B

(c) Compound B + H_2O, $H^+ \xrightarrow{\Delta}$

(d) Compound A + $CH_3CH{=}CHCH_2CH_2C(=O)Cl \longrightarrow$ Compound C

(e) Compound C + $H_2O \xrightarrow[\Delta]{H^+}$

In the laboratory, diethyl malonate is a useful compound that we can use as a starting material to prepare a wide variety of carboxylic acids and methyl ketones. As we will learn in the next section, living systems use similar reactions of a derivative of malonic acid for many similar purposes.

20.3 THE β-DICARBONYL FUNCTIONAL GROUP IN LIVING SYSTEMS

Malonyl coenzyme A is an important compound in living systems because it is involved in the preparation of fatty acids (see Chapter 28). It is formed in an enzyme-catalyzed condensation (see Section 19.8) reaction between carbon dioxide and acetyl coenzyme A.

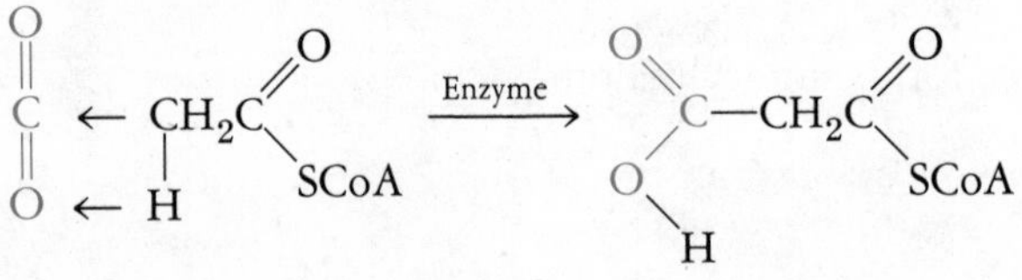

$$O{=}C{=}O + H{-}CH_2C(=O)SCoA \xrightarrow{\text{Enzyme}} HO{-}C(=O){-}CH_2C(=O)SCoA$$

Acetyl coenzyme A — Malonyl coenzyme A

Malonyl coenzyme A reacts with an acyl coenzyme A to add a two-carbon piece to the acyl group:

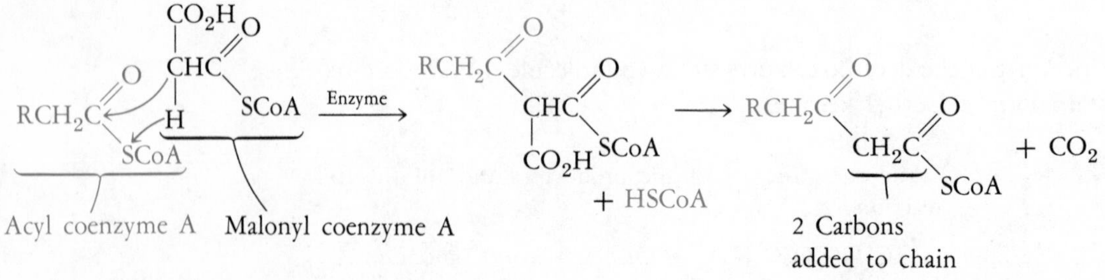

$$RCH_2C(=O)SCoA + H{-}CH(CO_2H)C(=O)SCoA \xrightarrow{\text{Enzyme}} RCH_2C(=O){-}CH(CO_2H)C(=O)SCoA + HSCoA \longrightarrow RCH_2C(=O){-}CH_2C(=O)SCoA + CO_2$$

Acyl coenzyme A — Malonyl coenzyme A — 2 Carbons added to chain

This is an example of the acylation of a β-dicarbonyl compound (see Sections 18.10 and 20.2). The following is a specific example:

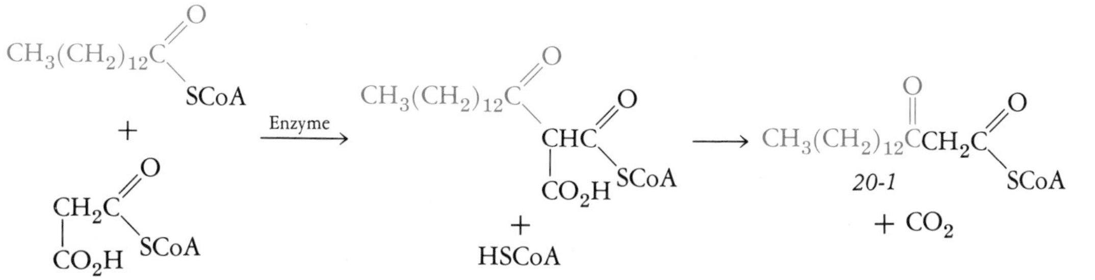

$$CH_3(CH_2)_{12}C(=O)SCoA + CH_2(CO_2H)C(=O)SCoA \xrightarrow{\text{Enzyme}} CH_3(CH_2)_{12}C(=O)CH(CO_2H)C(=O)SCoA + HSCoA \longrightarrow \underset{\textit{20-1}}{CH_3(CH_2)_{12}C(=O)CH_2C(=O)SCoA} + CO_2$$

This acylated product *20-1* undergoes a number of further reactions that transform it into a fatty acid. This requires the reduction of the β-keto group to a $-CH_2$ group, which occurs by means of the enzyme-catalyzed reactions given in Figure 20-1.

It must be emphasized that none of the reactions in the scheme in Figure 20-1 is new. They have all been presented in previous chapters. The first reaction is simply the reduction of a carbonyl group to form an alcohol (see Sections 16.3 and 19.5). Dehydration of the alcohol to an alkene is the second reaction (see Section 16.4), and addition of hydrogen to the double

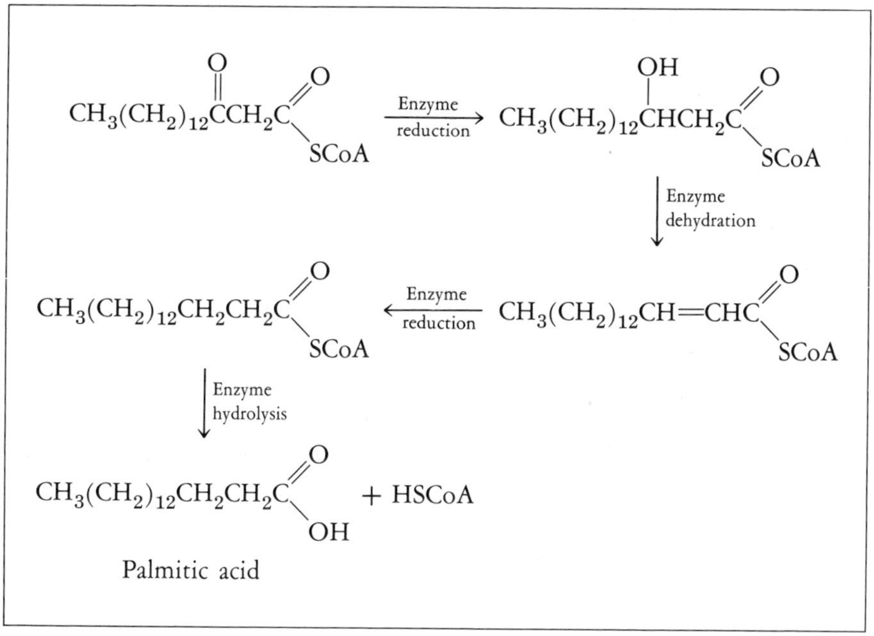

$$CH_3(CH_2)_{12}C(=O)CH_2C(=O)SCoA \xrightarrow[\text{reduction}]{\text{Enzyme}} CH_3(CH_2)_{12}CH(OH)CH_2C(=O)SCoA$$

$$CH_3(CH_2)_{12}CH(OH)CH_2C(=O)SCoA \xrightarrow[\text{dehydration}]{\text{Enzyme}} CH_3(CH_2)_{12}CH{=}CHC(=O)SCoA$$

$$CH_3(CH_2)_{12}CH{=}CHC(=O)SCoA \xrightarrow[\text{reduction}]{\text{Enzyme}} CH_3(CH_2)_{12}CH_2CH_2C(=O)SCoA$$

$$CH_3(CH_2)_{12}CH_2CH_2C(=O)SCoA \xrightarrow[\text{hydrolysis}]{\text{Enzyme}} \underset{\text{Palmitic acid}}{CH_3(CH_2)_{12}CH_2CH_2C(=O)OH} + HSCoA$$

Fig. 20-1. The preparation of a fatty acid from an acyl coenzyme A.

bond is the third reaction (see Section 12.4). Finally, hydrolysis of a thioester forms the product (see Section 18.6). This scheme illustrates the similarities in the types of reactions that occur in the laboratory and in living systems.

Compound *20-1* can also be hydrolyzed to a β-keto acid, which then decarboxylates to form a methyl ketone. For example:

$$\underset{\textit{20-1}}{CH_3(CH_2)_{12}\overset{O}{\overset{\|}{C}}CH_2\overset{O}{\overset{\|}{C}}SCoA} \xrightarrow[\text{Enzyme}]{H_2O} CH_3(CH_2)_{12}\overset{O}{\overset{\|}{C}}CH_2\overset{O}{\overset{\|}{C}}OH + HSCoA \longrightarrow CH_3(CH_2)_{12}\overset{O}{\overset{\|}{C}}CH_3 + CO_2$$

Persons with diabetes have acetone in their blood. It is formed from acetoacetic acid by a similar reaction (see Section 30.6):

$$\underset{\text{Acetoacetic acid}}{CH_3\overset{O}{\overset{\|}{C}}CH_2\overset{O}{\overset{\|}{C}}OH} \xrightarrow{\text{Enzyme}} \underset{\text{Acetone}}{CH_3\overset{O}{\overset{\|}{C}}CH_3} + CO_2$$

Clearly, decarboxylation of β-keto carboxylic acids and the acylation of β-dicarbonyl compounds are important reactions in living systems. In the next section, we will learn several important ring-forming reactions.

20.4 INTRAMOLECULAR REACTIONS

In the introduction to this chapter, we noted that molecules sometimes contain two functional groups that can react with each other. If the two groups have the proper spatial relationship, their intramolecular reaction usually forms cyclic products whose physical and chemical properties are often different from those of their noncyclic counterparts. In general, the two groups react with each other only when they can form five- or six-member rings. In this section, we will examine the cyclic products formed by the interaction of several functional groups of biological interest.

Dicarboxylic Acids

The formation of a five- or six-member anhydride ring is readily achieved if a dicarboxylic acid has carboxyl groups in the proper positions. The following are some examples:

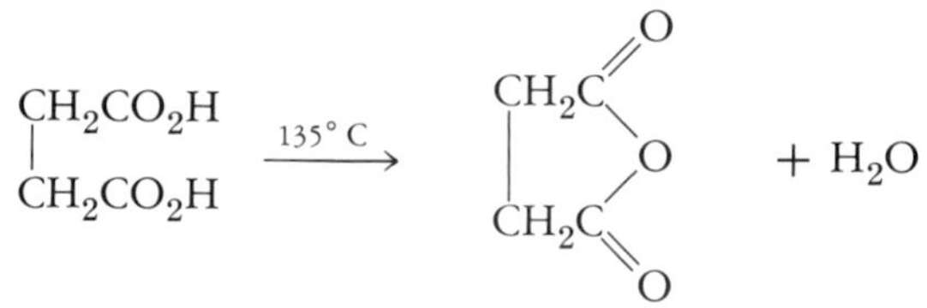

$$\begin{array}{l} CH_2CO_2H \\ | \\ CH_2CO_2H \end{array} \xrightarrow{135^\circ C} \text{Succinic anhydride} + H_2O$$

Succinic acid   Succinic anhydride

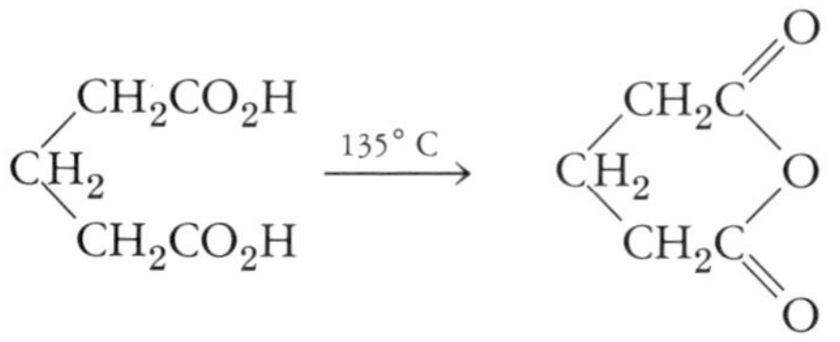

$$CH_2(CH_2CO_2H)_2 \xrightarrow{135^\circ C} \text{Glutaric anhydride}$$

Glutaric acid   Glutaric anhydride

EXERCISE 20-7 Write the structure of the product formed when each of the following dicarboxylic acids is heated to 135° C:

(a) 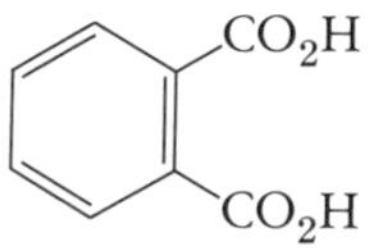 Phthalic acid   (b) 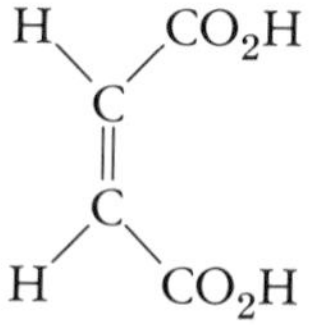 Maleic acid

Dicarboxylic acids containing fewer than four carbons do not form cyclic anhydrides. As we learned in Section 20.2, malonic acid loses carbon dioxide to form acetic acid when heated. Oxalic acid undergoes a similar reaction to form formic acid:

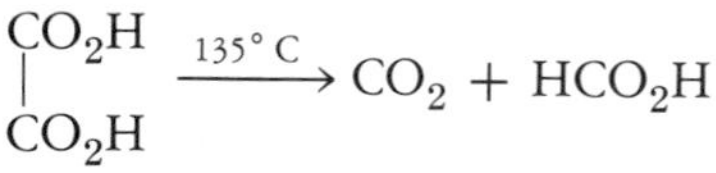

$$\begin{array}{l} CO_2H \\ | \\ CO_2H \end{array} \xrightarrow{135^\circ C} CO_2 + HCO_2H$$

Oxalic acid   Formic acid

Dicarboxylic acids whose two carboxyl groups are separated by more than three carbon atoms do not form cyclic anhydrides. As we will learn in Section 20.5, they form polymeric anhydrides.

Cyclic anhydrides undergo the same types of reactions as do linear anhydrides (see Section 18.9). For example:

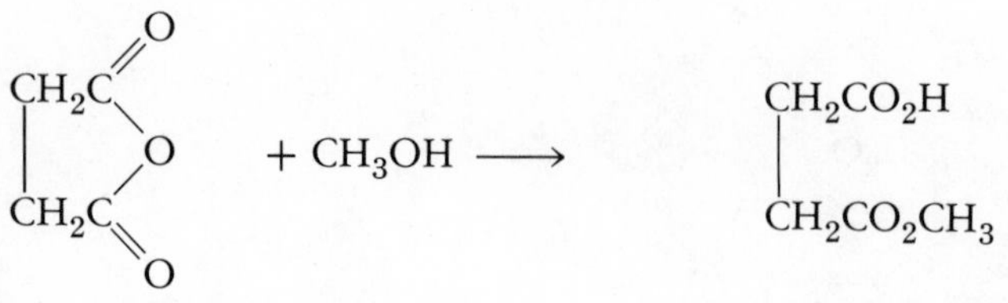

Succinic anhydride 3-Methoxycarbonylpropanoic acid

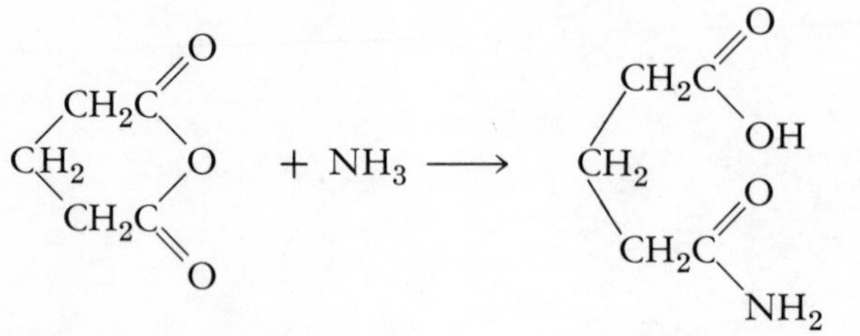

Glutaric anhydride 4-Amidobutanoic acid

Hydroxy Acids

Hydroxy acids are important biological compounds. Some of the more important low-molecular-weight hydroxy acids are listed in Table 20-2.

EXERCISE 20-8 Which of the compounds listed in Table 20-2 are capable of existing as
(a) enantiomers (b) diastereomers

Because carboxylic acids and alcohols react to form esters, it might be expected that the same reaction would occur with hydroxy acids. Indeed, hydroxy acids with suitably located hydroxy and carboxyl groups do react to form cyclic esters called *lactones*. The γ- and δ-hydroxy acids react to give the five- and six-member rings of γ and δ lactones so readily that it is difficult to isolate the free acids.

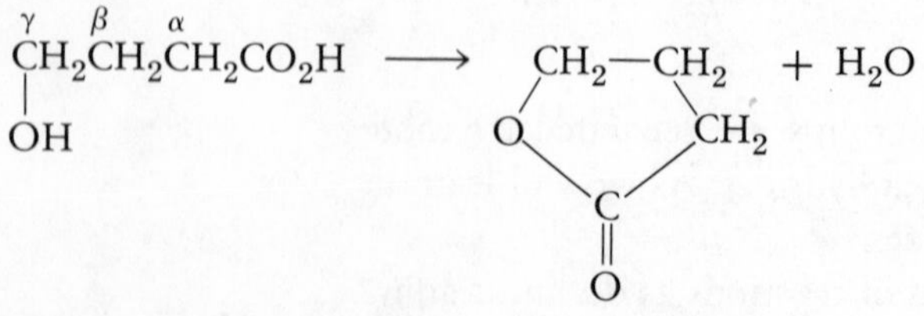

γ-Hydroxybutyric acid γ-Butyrolactone

Table 20-2. Some Biologically Important Hydroxy Acids

Name	Structure	Melting Point (°C)	Boiling Point (°C)
Glycolic acid	$HOCH_2CO_2H$	80	dec.*
Lactic acid	$CH_3\underset{OH}{\underset{\vert}{C}}HCO_2H$	18	dec.
Glyceric acid	$HOCH_2\underset{OH}{\underset{\vert}{C}}HCO_2H$	Syrup	dec.
β-Hydroxybutyric acid	$CH_3\underset{OH}{\underset{\vert}{C}}HCH_2CO_2H$	50	dec.
Malic acid	$HO_2CCH_2\underset{OH}{\underset{\vert}{C}}HCO_2H$	100	dec.
Tartaric acid	$HO_2C\underset{OH}{\underset{\vert}{C}}H-\underset{OH}{\underset{\vert}{C}}HCO_2H$	206	—
Mevalonic acid	$HOCH_2CH_2\overset{CH_3}{\overset{\vert}{\underset{OH}{\underset{\vert}{C}}}}CH_2CO_2H$	Oil	dec.
Citric acid	$HO_2CCH_2\overset{OH}{\overset{\vert}{\underset{CO_2H}{\underset{\vert}{C}}}}CH_2CO_2H$	153	dec.
Salicylic acid	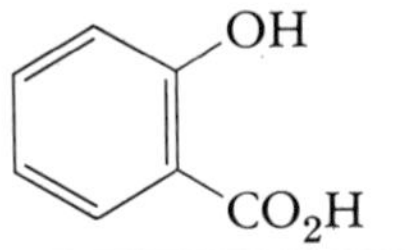 OH CO_2H	159	—

* Decomposes before boiling

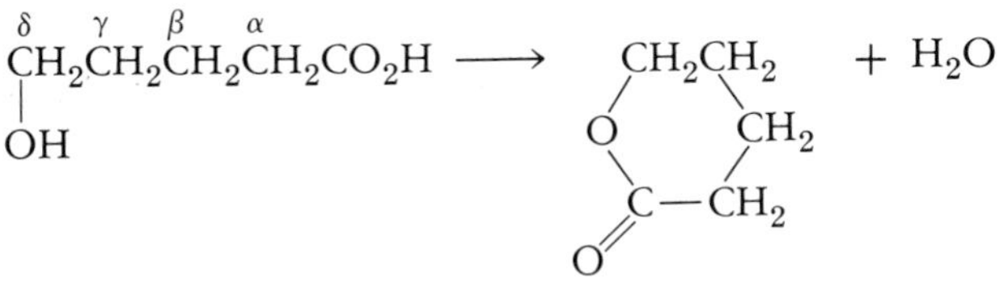

$$\underset{OH}{\underset{\vert}{\overset{\delta}{C}}}H_2\overset{\gamma}{C}H_2\overset{\beta}{C}H_2\overset{\alpha}{C}H_2CO_2H \longrightarrow \begin{array}{c} CH_2CH_2 \\ O \quad\quad CH_2 \\ C-CH_2 \\ O \end{array} + H_2O$$

δ-Hydroxyvaleric acid $\qquad$ δ-Valerolactone

The α-hydroxy acids react differently from γ- or δ-hydroxy acids in that they undergo a bimolecular esterification to form *lactides:*

$$\begin{array}{c} CH_3CHC(=O)OH \\ | \\ OH \\ HO \quad OH \\ CCHCH_3 \\ \| \\ O \end{array} \longrightarrow \begin{array}{c} CH_3CHC(=O) \\ O \quad O \\ C{-}CHCH_3 \\ \| \\ O \end{array} + 2\,H_2O$$

α-Hydroxy acid　　　　Lactide

β-Hydroxy acids under acidic or basic conditions dehydrate to form α-β-unsaturated carboxylic acids rather than a β lactone.

$$\underset{\beta\text{-Hydroxybutyric acid}}{CH_3\overset{OH}{\overset{|}{C}}HCH_2CO_2H} \xrightarrow[\Delta]{H^+ \text{ or } OH^-} \underset{\text{Crotonic acid}}{CH_3CH{=}CHCO_2H} + H_2O$$

Hydroxy Aldehydes and Ketones

The carbonyl group of aldehydes and ketones can add an alcohol to form hemiacetals and hemiketals, respectively (Section 19.5). When the carbonyl and the hydroxy groups are suitably located in the same molecule, the formation of cyclic hemiacetals and hemiketals occurs. For example:

$$\begin{array}{c} CH_2CH_2 \\ CH_2 \quad O{-}H \\ HC{=}O \end{array} \rightleftarrows \begin{array}{c} CH_2CH_2 \\ CH_2 \quad O \\ CH \\ | \\ O{-}H \end{array}$$

4-Hydroxybutyraldehyde　　　　Cyclic hemiacetal

$$\begin{array}{c} O \\ \| \\ CH_2CCH_3 \\ CH_2 \quad OH \\ CH_2CH_2 \end{array} \rightleftarrows \begin{array}{c} OH \\ | \\ CH_2CCH_3 \\ CH_2 \quad O \\ CH_2CH_2 \end{array}$$

6-Hydroxy-2-hexanone　　　　Cyclic hemiketal

As in the case of noncyclic acetals and ketals, this is an equilibrium reaction. The equilibrium is toward the cyclic product only for five- and six-member cyclic hemiacetals and hemiketals.

The α-hydroxy ketones react with mild oxidizing agents such as Fehling's solution and Tollens' reagent (see Section 19.4) to form diketones:

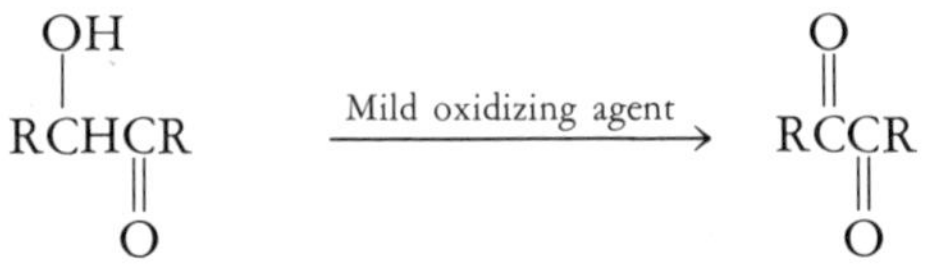

$$\underset{\alpha\text{-Hydroxy ketone}}{\mathrm{R\overset{OH}{\overset{|}{C}}H\underset{\underset{O}{\|}}{C}R}} \xrightarrow{\text{Mild oxidizing agent}} \underset{\text{Diketone}}{\mathrm{R\overset{O}{\overset{\|}{C}}\underset{\underset{O}{\|}}{C}R}}$$

We will encounter specific examples of this reaction in Section 21.5.

When the two functional groups are not correctly placed to form a ring, polymers are formed.

20.5 INTERMOLECULAR REACTIONS: POLYMER FORMATION

When the two functional groups in a molecule are not properly located for an intramolecular reaction to occur, an intermolecular reaction usually occurs instead. For example:

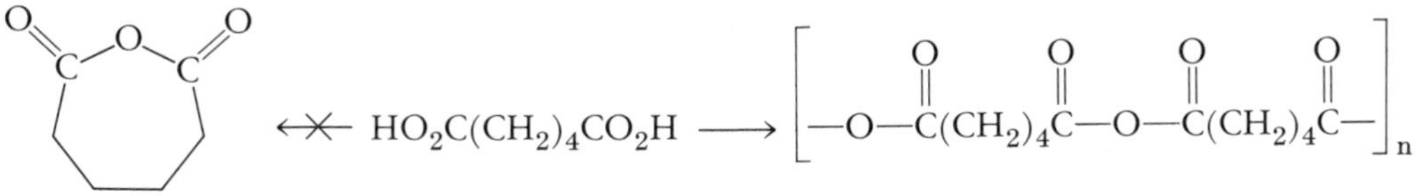

$$\underset{\text{Adipic anhydride}}{\text{(cyclic anhydride)}} \not\leftrightarrow \underset{\text{Adipic acid}}{HO_2C(CH_2)_4CO_2H} \longrightarrow \underset{\text{Adipic anhydride polymer}}{\left[-O-\overset{O}{\overset{\|}{C}}(CH_2)_4\overset{O}{\overset{\|}{C}}-O-\overset{O}{\overset{\|}{C}}(CH_2)_4\overset{O}{\overset{\|}{C}}-\right]_n}$$

Adipic acid does not form a cyclic anhydride because of the difficulty in getting the ends together to form a ring. Instead, carboxylic acid groups from adjacent molecules react to form a polymeric anhydride. These polymers, like those of alkenes that we studied in Section 12.6 are made up of a single monomer that is continually repeated. Polymers containing a single repeating monomer can be generalized by the structure:

$$(-A-A-A-A-A-)_n$$

A polymer may contain more than one kind of monomeric unit. In these cases, the structure of the polymer increases in complexity. Even with two monomeric units present in equal amounts, many combinations are possible. For example:

$$(\text{—A—B—A—B—})_n \qquad (\text{—A—A—B—B—A—A—B—B—})_n \qquad (\text{—A—A—B—A—B—B—})_n$$

If either A or B is chiral, more combinations are possible.

Many biologically important molecules are polymers. Carbohydrates, deoxyribonucleic acid (DNA), and proteins are all naturally occurring polymers. Their complex structures are due to the fact that they contain numerous monomers. For DNA, there are four different monomeric nucleotides. In proteins, 20 different monomeric amino acids are combined in a definite sequence in the polymer. The structure, properties, and reactions of these macromolecules will be examined in later chapters. In this chapter, we will examine some of the simpler natural and industrial polymers containing no more than two monomers.

Polymers can be classified according to their method of preparation. When monomeric units are added together to form a polymer, the compound is called an *addition polymer.* When the polymer is formed by a reaction that results in the elimination of a molecule of water or alcohol, it is called a *condensation polymer.* Let us examine each type of polymer briefly.

Addition Polymers

The monomeric units of an addition polymer are usually alkenes. For example:

$$n\ \text{RCH}{=}\text{CH}_2 \xrightarrow{\text{Catalyst}} \left[\text{—}\overset{\overset{\displaystyle \text{R}}{|}}{\text{C}}\text{H—CH}_2\text{—} \right]_n$$

This addition reaction was first presented in Section 12.6.

1,3-Dienes can also be polymerized. The polymers result by addition to the 1 and 4 positions of the diene, forming a polymer with double bonds:

$$n\ \underset{\underset{4}{\uparrow}}{\text{CH}_2}{=}\text{CH—}\overset{\overset{\displaystyle \text{R}}{|}}{\text{C}}{=}\underset{\underset{1}{\uparrow}}{\text{CH}_2} \xrightarrow{\text{Catalyst}} \left[\text{—CH}_2\text{CH}{=}\overset{\overset{\displaystyle \text{R}}{|}}{\text{C}}\text{CH}_2\text{—} \right]_n$$

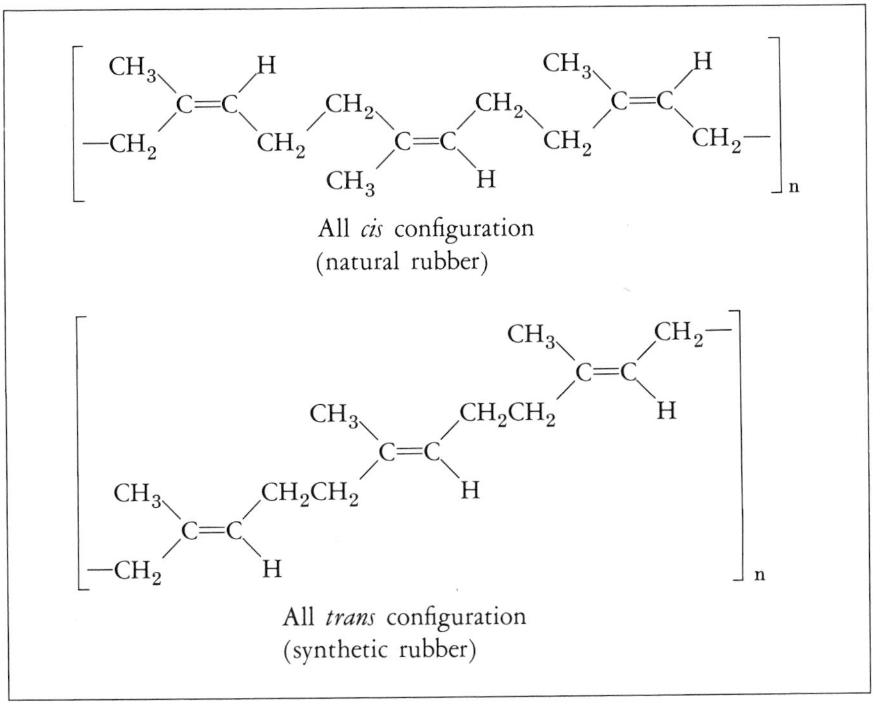

$$\left[\begin{array}{l} CH_3 \quad H \qquad\qquad\qquad CH_3 \quad H \\ \quad C{=}C \quad CH_2 \quad CH_2 \quad C{=}C \\ -CH_2 \quad CH_2 \quad C{=}C \quad CH_2 \quad CH_2- \\ \qquad\qquad CH_3 \quad H \end{array}\right]_n$$

All *cis* configuration
(natural rubber)

$$\left[\begin{array}{l} \qquad\qquad\qquad CH_3 \quad CH_2- \\ \qquad\qquad\qquad\quad C{=}C \\ \qquad CH_3 \quad CH_2CH_2 \quad H \\ \qquad\quad C{=}C \\ CH_3 \quad CH_2CH_2 \quad H \\ \quad C{=}C \\ -CH_2 \quad H \end{array}\right]_n$$

All *trans* configuration
(synthetic rubber)

Fig. 20-2. The cis *and* trans *configurations of the double bond of polyisoprene.*

For 1,3-dienes such as isoprene, $CH_2{=}\overset{\displaystyle CH_3}{\overset{|}{C}}{-}CH{=}CH_2$, the double bonds in the polymer can exist in either the *cis* or *trans* isomer configuration, as shown in Figure 20-2. Natural rubber is *cis*-polyisoprene. The *trans* isomer can be formed industrially and is known as *gutta-percha* rubber. A *cis* polymerization of isoprene can also be achieved by the use of special catalysts. The *cis* and *trans* polyisoprenes differ greatly in physical properties. The *trans* polymer is hard and brittle, whereas the *cis* polymer is elastic.

Polymers formed from a single monomer are known as *homopolymers*. If two monomers are used, then *copolymers* are formed. An example is the copolymerization of 1,3-butadiene and styrene to a copolymer called GRS that is useful for automobile tires:

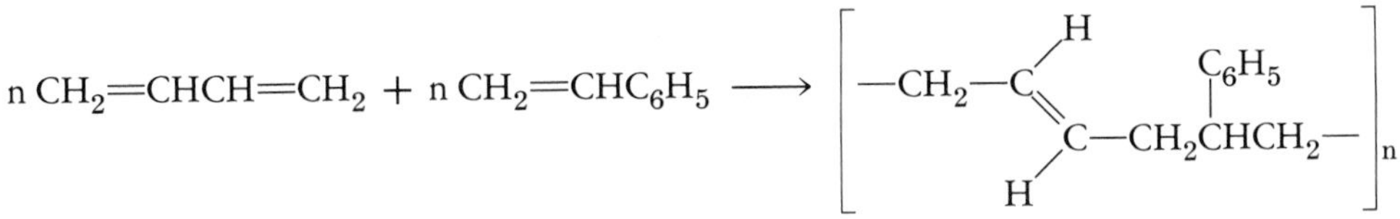

$$\underset{\text{1,3-Butadiene}}{n\,CH_2{=}CHCH{=}CH_2} + \underset{\text{Styrene}}{n\,CH_2{=}CHC_6H_5} \longrightarrow \underset{\text{GRS rubber (a copolymer)}}{\left[-CH_2{-}\overset{H}{C}{=}\underset{H}{C}{-}CH_2\overset{C_6H_5}{\overset{|}{C}}HCH_2-\right]_n}$$

Table 20-3. Some Industrially Important Polymers Prepared from Alkenes

Monomers	Polymer Name	Use
$CH_2{=}CClCH{=}CH_2 + CH_2{=}CClCH{=}CH_2$	Polychloroprene (Neoprene)	Oil-resistant rubber
$CH_2{=}C(CH_3)CH{=}CH_2 + CH_2{=}C(CH_3)CH{=}CH_2$	*trans*-Polyisoprene	Hard synthetic rubber
$CH_2{=}C(CH_3)CH{=}CH_2 + CH_2{=}C(CH_3)CH{=}CH_2$	*cis*-Polyisoprene	Natural rubber
$CH_2{=}CHCH{=}CH_2 + C_6H_5CH{=}CH_2$	GRS	Automobile tires
$CH_2{=}CF_2 + CH_2{=}CCl_2$	Viton	Chemically resistant rubber
$C_6H_5CH{=}CH_2 + C_6H_5CH{=}CH_2$	Polystyrene	Insulation
$CH_2{=}CHCN + CH_2{=}CHCN$	Polyacrylonitrile (Orlon)	Fibers

This GRS synthetic rubber is longer wearing than natural rubber and has largely replaced natural rubber for tires.

Some of the industrially important polymers prepared from alkenes are listed in Table 20-3.

Condensation Polymers

One of the most important condensation polymers is Nylon 66, which is made into fibers used in the manufacture of various fabrics. It can be prepared by the reaction of adipic acid and 1,6-diaminohexane:

$$\underset{\text{1,6-Diaminohexane}}{n\ NH_2(CH_2)_6NH_2} + \underset{\text{Adipic acid}}{n\ HO_2C(CH_2)_4CO_2H} \longrightarrow \underset{\substack{\text{Nylon-66} \\ \text{(a polyamide)}}}{\left[-\overset{O}{\overset{\|}{C}}(CH_2)_4\overset{O}{\overset{\|}{C}}-NH(CH_2)_6NH-\right]_n} + 2n-1\ H_2O$$

The number 66 refers to the fact that the polymer is made from a six-carbon diamine and a six-carbon dicarboxylic acid. Nylon 66 contains many amide groups and is a polyamide. As we will learn in Chapter 23, proteins are also polyamides.

EXERCISE 20-9 In polymerization, 6-aminohexanoic acid, $NH_2(CH_2)_5CO_2H$, forms a polyamide. What is the structure of the polyamide?

Condensation polymers can also be formed by the reaction of a dicarboxylic acid and a diol. Dacron is an example of such a polyester formed by the reaction of terephthalic acid and 1,2-ethanediol:

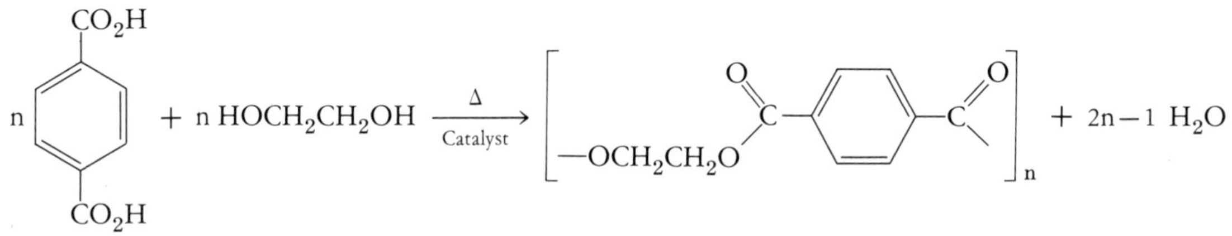

$$n\ HO_2C{-}C_6H_4{-}CO_2H + n\ HOCH_2CH_2OH \xrightarrow[\text{Catalyst}]{\Delta} \left[-OCH_2CH_2O{-}\overset{O}{\overset{\|}{C}}{-}C_6H_4{-}\overset{O}{\overset{\|}{C}}-\right]_n + 2n-1\ H_2O$$

Terephthalic acid; 1,2-Ethanediol (Ethylene glycol); Dacron (a polyester)

As we will learn in future chapters, most of the polymers in living systems are condensation polymers.

EXERCISE 20-10 Examine the structure of each of the following polymers. Tell whether the polymer is an addition or condensation polymer and write the structure of the monomer(s).

(a) $\left[-CH_2CCl_2CH_2CCl_2CH_2CCl_2-\right]_n$

(b) $\left[-\underset{\overset{\|}{O}}{C}CH_2(CH_2)_4NH\underset{\overset{\|}{O}}{C}CH_2(CH_2)_4NH-\right]_n$

(c) $\left[-CH_2\underset{}{\overset{Cl}{\overset{|}{C}}}{=}CHCH_2CH_2\overset{Cl}{\overset{|}{C}}{=}CHCH_2-\right]_n$

(d) $\left[-CF_2CF_2CF_2CF_2CF_2CF_2-\right]_n$

EXERCISE 20-11 It has proved to be impossible to prepare a polyamide by heating 1,6-diaminohexane and glutaric acid. Explain.

20.6 PREDICTING THE REACTIONS OF POLYFUNCTIONAL COMPOUNDS

In previous chapters, we have learned many reactions of the biologically important functional groups. One reason why we study organic chemistry is to use this knowledge to predict the chemical reactions of many compounds that are new to us. You have been predicting the reactions of new compounds by doing some of the exercises included in previous chapters, so this is nothing new. The compounds in these exercises usually contained only one functional group. But now we must predict the reactions of polyfunctional compounds. How do we do this? First, we must know the reactions of the individual functional groups presented in the previous chapters. In addition, we must consider the reactions of the reagents with all functional groups in the molecule. Frequently, a reagent will react with more than one functional group. We must also recognize when intermolecular or intramolecular reactions can occur. To illustrate this, let us predict the chemical reactions of several polyfunctional compounds.

Citronellal is a terpene. What chemical reactions does it undergo?

CH_3
CH
CH_2 CH_2
CH_2 CHO
CH_2
C
CH_3 CH_2

Citronellal

To answer this question, let us first identify the functional groups in the molecule. Citronellal contains an aldehyde group and a carbon-carbon double bond attached to a saturated hydrocarbon framework (the $—CH_2$ and $—CH_3$ groups). What reactions do each of these functional groups undergo? The saturated hydrocarbon framework is generally inert (see Chapter 11). Because it is rarely involved in chemical reactions, we can ignore its presence. The aldehyde group undergoes nucleophilic addition reactions, oxidation reactions, and condensation reactions. The carbon-carbon double bond undergoes electrophilic addition reactions and oxidation reactions.

It would be tempting to list these as all the chemical reactions of citronellal. After all, the reactions of the entire molecule should be the sum of the reactions of the individual parts. But before we reach this conclusion, we must ask two questions. Will the two functional groups react with each other? The answer to this question is no. An aldehyde group and a carbon-carbon double bond do not react with each other. Will any reagent react with more than one functional group? The answer to this question is yes. There are two reagents that will react with both functional groups of citronellal. The first is hydrogen (with a metal catalyst), which reduces both the aldehyde group (see Section 19.5) and the carbon-carbon double bond (see Section 12.4). The other is potassium permanganate ($KMnO_4$), or any strong oxidizing reagent, which will oxidize both functional groups. These reactions are as follows:

CH_3 CO_2H CH_3 O ← H^+ / $KMnO_4$, Δ, H_2O — CH_3 CHO CH_3 CH_2 — H_2, Pt → CH_3 CH_2OH CH_3 CH_3

H_2O, NaOH ↓ $KMnO_4$

CH_3 CO_2H CH_3 OH CH_2 OH

Notice that if we choose the proper reagent, reaction occurs at only one functional group. Tollens' reagent (see Section 19.4) oxidizes only the aldehyde group, and lithium aluminum hydride (see Section 19.5) reduces only the aldehyde group:

CH_3 CO_2H CH_3 CH_2 ← Tollens' reagent — CH_3 CHO CH_3 CH_2 — $LiAlH_4$ → CH_3 CH_2OH CH_3 CH_2

One of the challenges of organic chemistry is to find reagents that will react with one specific functional group in the presence of others. As we will learn in future chapters, enzymes in living systems frequently do this rapidly and stereospecifically. This is one of the major differences between reactions in the laboratory and in living systems.

The other reactions that we would predict for citronellal are the following:

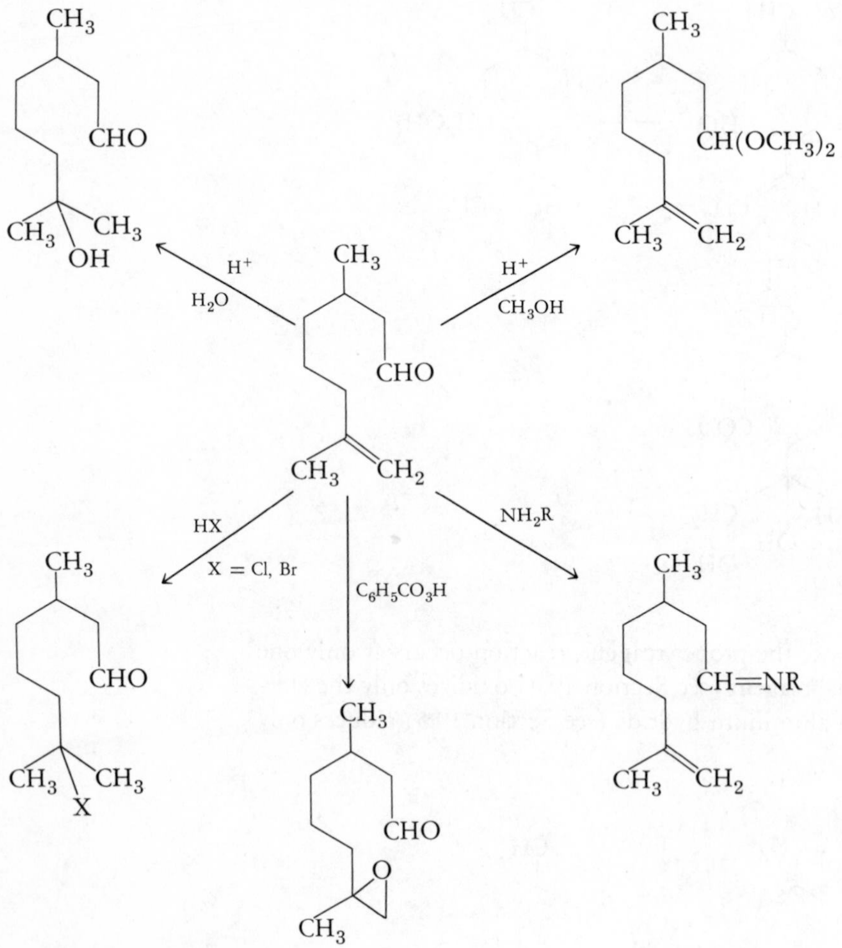

CH_3 CHO CH_3 CH_3 OH

CH_3 $CH(OCH_3)_2$ CH_3 CH_2

H^+ H_2O CH_3 H^+ CH_3OH

CHO

CH_3 CH_2

HX NH_2R

CH_3 X = Cl, Br CH_3

$C_6H_5CO_3H$

CHO CH=NR

CH_3

CH_3 CH_3 X CH_3 CH_2

CHO

O

CH_3

Using the same approach, we can predict the following reactions of 4-hydroxypentanamide:

$$CH_3CH(OH)CH_2CH_2C(=O)NH_2 \xrightarrow{\text{Oxidation}} CH_3C(=O)CH_2CH_2C(=O)NH_2$$

$$CH_3CH(OH)CH_2CH_2C(=O)NH_2 \xrightarrow{(CH_3CO)_2O} CH_3CH(OCCH_3)CH_2CH_2C(=O)NH_2 \quad (\text{OCCH}_3 = \text{O–C(=O)–CH}_3)$$

$$CH_3CH(OH)CH_2CH_2C(=O)NH_2 \xrightarrow{H^+,\ H_2O} \underbrace{CH_3CHCH_2CH_2C(=O)O}_{\text{ring: O–CH(CH}_3\text{)–CH}_2\text{–CH}_2\text{–C(=O)}} + \overset{+}{N}H_4$$

Notice that the acid-catalyzed hydrolysis of 4-hydroxypentanamide forms a lactone. This compound is formed because the carboxylic acid formed as the product of the acid-catalyzed hydrolysis of the amide group reacts with the hydroxy group.

Finally, we can predict that *trans*-3-(4′-hydroxyphenyl)-2-propenyl acetate will undergo the reactions shown in Figure 20-3.

Again, we find that several reagents react with more than one functional group. Hydrogen and a metal catalyst react with both the carbon-carbon double bond (see Section 12.4) and the ester group (see Section 18.5). An aqueous sodium hydroxide solution saponifies the ester group (see Section 18.5) and removes the acidic phenol proton (see Section 16.6). An aqueous dilute alkaline solution of potassium permanganate oxidizes the double bond to a glycol (see Section 12.8) and also saponifies the ester (see Section 18.5). Bromine adds to the double bond (see Section 12.4) and replaces the hydrogens *ortho* to the —OH group of the phenol (see Section 16.6). Finally, an aqueous solution of hydrochloric acid adds to the double bond (see Section 12.4) and hydrolyzes the ester group (see Section 18.5).

From these three examples, we can conclude that, in general, the reactions of a polyfunctional molecule are the sum of the reactions of its individual functional groups. Therefore, we have reached one of our goals. We can predict many of the biologically important reactions of polyfunctional molecules by applying our knowledge of the reactions of various functional groups. Consequently, there is nothing mysterious about the chemistry of polyfunctional molecules.

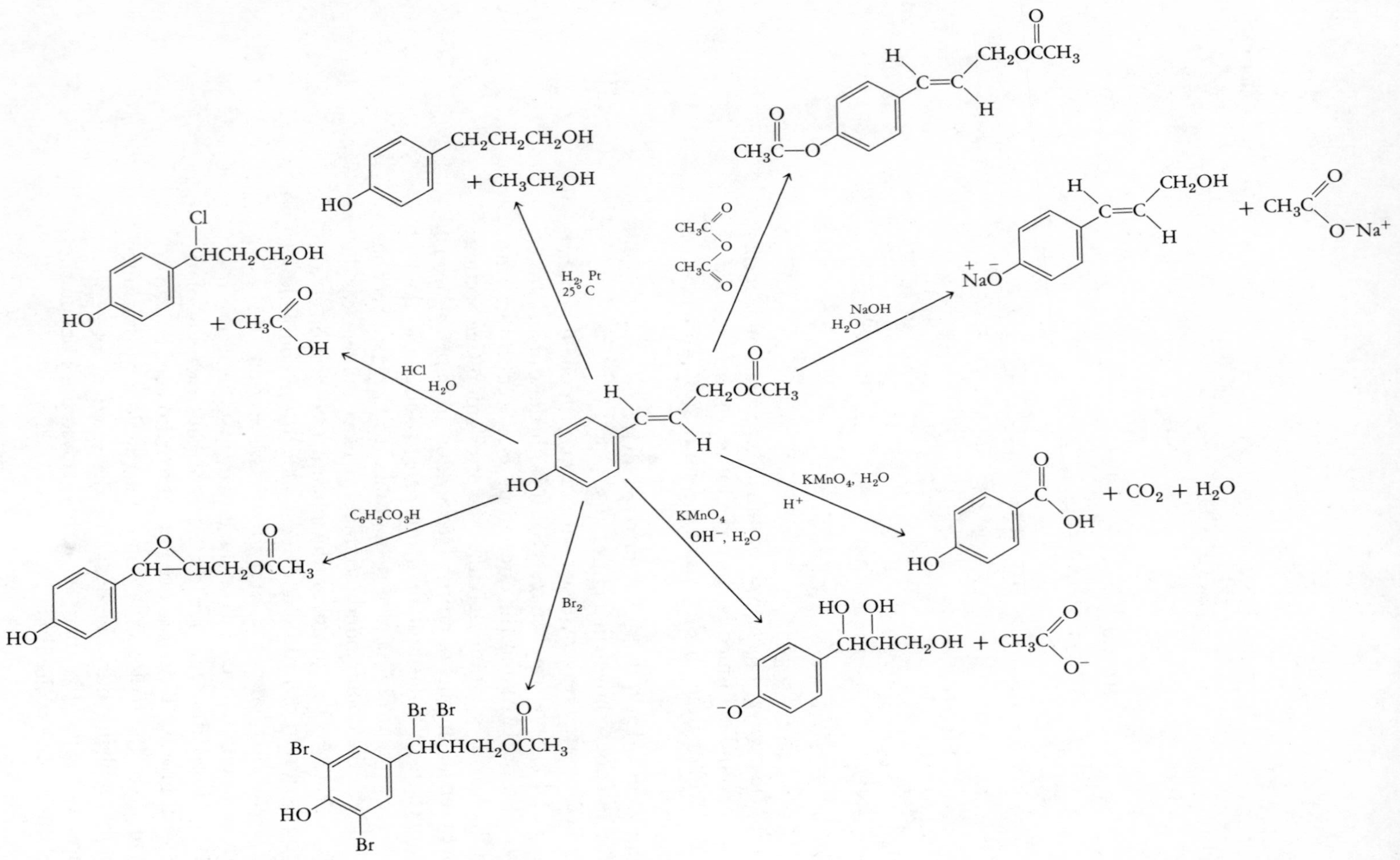

Fig. 20-3. Reactions of trans-*3-(4′-hydroxyphenyl)-2-propenyl acetate.*

We are now prepared to focus our attention on some rather complicated polyfunctional molecules. These include carbohydrates, lipids, and proteins. All of these compounds are vital to any living system. We study their chemistry in the next few chapters.

20.7 SUMMARY

In general, the biologically important chemical reactions of a polyfunctional molecule are the sum of the reactions of the individual functional groups. This is especially true when the functional groups are so far apart that they are completely independent, as is often the case. When two functional groups are close together, it is still possible that they react independently. But it is also possible that they make up a new functional group with its own characteristic reactions. An example is the β-dicarbonyl functional group.

Finally, certain functional groups in a molecule can react with each other. They undergo either intermolecular or intramolecular reactions. Intramolecular reactions occur to form five- or six-member rings, whereas intermolecular reactions occur to form polymers. A number of compounds vital to all living systems are polyfunctional molecules that form biologically important polymers.

REVIEW OF TERMS, CONCEPTS, AND REACTIONS

Terms

ADDITION POLYMER A polymer formed when monomer units are added together.

CONDENSATION POLYMER A polymer formed by a reaction that results in the elimination of a molecule of water or alcohol.

DECARBOXYLATION Loss of carbon dioxide from a molecule.

β-DICARBONYL FUNCTIONAL GROUP A molecule containing two carbonyl groups separated by a single carbon atom.

INTERMOLECULAR REACTIONS Reactions occurring between two molecules.

INTRAMOLECULAR REACTIONS Reactions occurring within the same molecule.

LACTONES Cyclic esters.

TAUTOMERS Isomers whose arrangements of atoms differ greatly but are in equilibrium.

Concepts

1. The reactions of polyfunctional molecules depend on the relative positions of the functional groups. If the functional groups are far enough apart, they can react independently of each other or can react with each other intermolecularly to form polymers. If the functional groups are close together, they can also react independently of each other, can influence the reactivity of one another without changing the reaction, can form a new multifunctional group, or can react with each other intramolecularly to form a ring.
2. The reactions of polyfunctional molecules are the sum of the reactions of the individual functional groups.

Reactions

β-Dicarbonyl Functional Groups (see Section 20.2)

1. Thermal decarboxylation

$$RC(=O)CH_2CO_2H \xrightarrow{\Delta} RC(=O)CH_3 + CO_2$$

2. Malonic ester synthesis

$$C_2H_5O_2CCH_2CO_2C_2H_5 \xrightarrow{OH^-} C_2H_5O_2C\bar{C}HCO_2C_2H_5$$

$$C_2H_5O_2C\bar{C}HCO_2C_2H_5 + RX \longrightarrow C_2H_5O_2CCH(R)CO_2C_2H_5$$

Intramolecular reactions (see Section 20.4)

1. Dicarboxylic acids

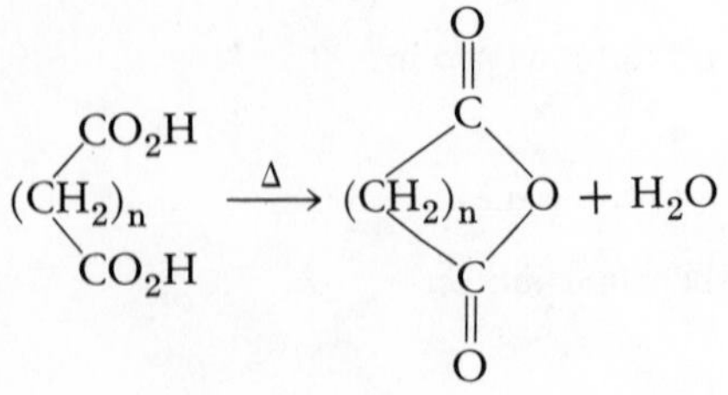

$$(CH_2)_n(CO_2H)_2 \xrightarrow{\Delta} (CH_2)_n(C=O)_2O + H_2O$$

n = 2 or 3

2. Hydroxy acids

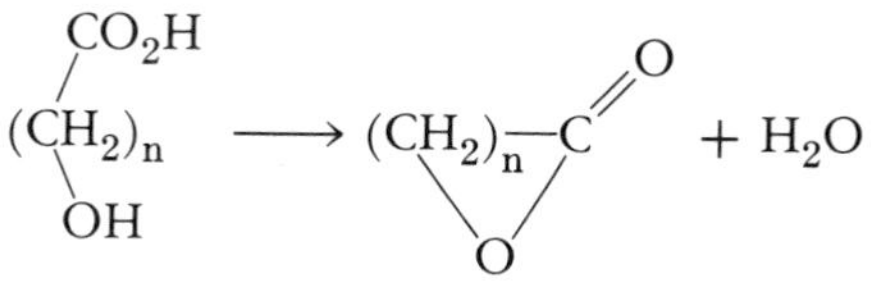

$$\underset{\text{OH}}{\overset{\text{CO}_2\text{H}}{(\text{CH}_2)_n}} \longrightarrow (\text{CH}_2)_n\text{—C(=O)—O (ring)} + H_2O$$

n = 3 or 4

3. Hydroxy aldehydes and ketones

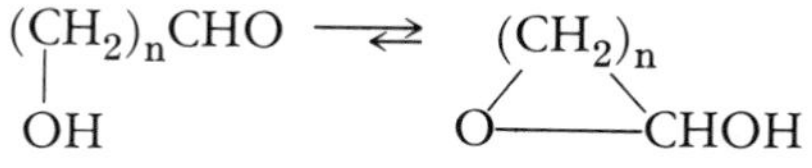

$$\underset{\text{OH}}{(\text{CH}_2)_n\text{CHO}} \rightleftarrows \overset{(\text{CH}_2)_n}{\text{O———CHOH}}$$

n = 3 or 4

Intermolecular reactions (see Section 20.5)

1. Addition polymers

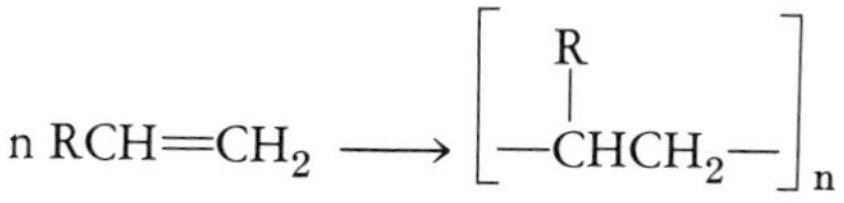

$$n\ RCH{=}CH_2 \longrightarrow \left[\text{—}\overset{\text{R}}{\text{CH}}\text{CH}_2\text{—}\right]_n$$

2. Condensation polymers
   (a) Polyamides

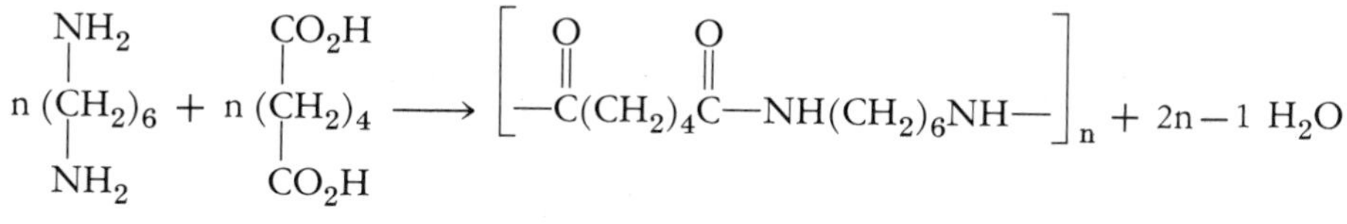

$$n\ \underset{\text{NH}_2}{\overset{\text{NH}_2}{(\text{CH}_2)_6}} + n\ \underset{\text{CO}_2\text{H}}{\overset{\text{CO}_2\text{H}}{(\text{CH}_2)_4}} \longrightarrow \left[\text{—}\overset{\text{O}}{\overset{\|}{\text{C}}}(\text{CH}_2)_4\overset{\text{O}}{\overset{\|}{\text{C}}}\text{—NH}(\text{CH}_2)_6\text{NH—}\right]_n + 2n-1\ H_2O$$

   (b) Polyesters

$$n\ \underset{\text{CO}_2\text{CH}_3}{\overset{\text{CO}_2\text{CH}_3}{\text{C}_6\text{H}_4}} + n\ HOCH_2CH_2OH \longrightarrow \left[\text{—OCH}_2\text{CH}_2\text{O}\overset{\text{O}}{\overset{\|}{\text{C}}}\text{—C}_6\text{H}_4\text{—}\overset{\text{O}}{\overset{\|}{\text{C}}}\text{—}\right]_n + 2n-1\ CH_3OH$$

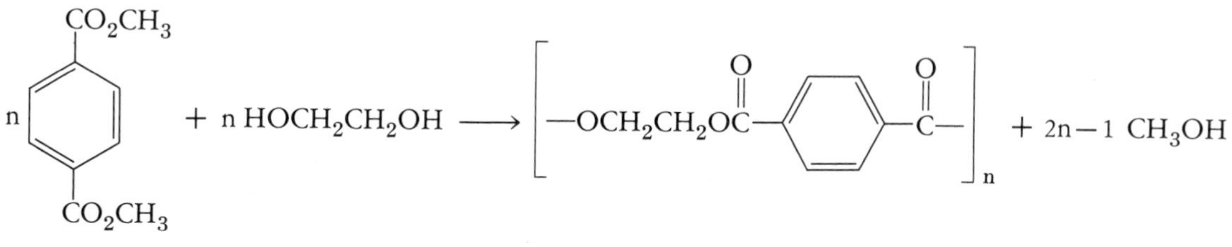

EXERCISES 20-12 Predict four reactions for each of the following compounds:

(a) [structure: estradiol — steroid skeleton with aromatic A ring bearing HO, angular CH_3 and OH on five-membered ring]

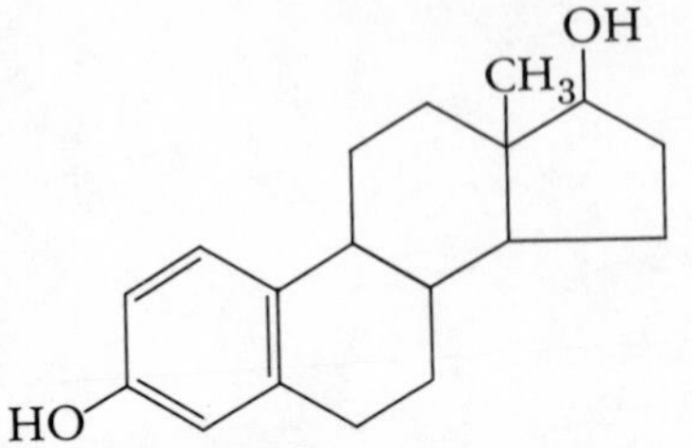

Estradiol (female sex hormone)

(b) $(CH_3CH_2)_2NCH_2CH_2OC(=O)-C_6H_4-NH_2$

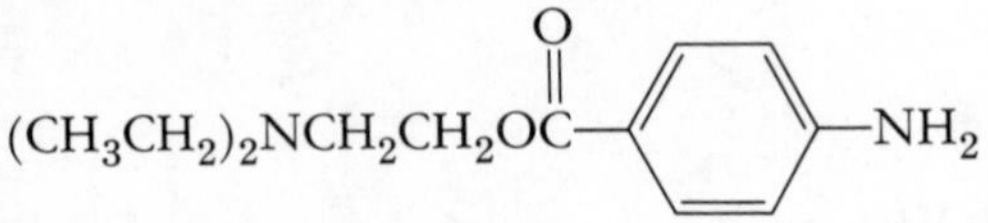

Procaine (a pain deadener)

(c) [structure: benzene ring with three OH groups (HO, OH, OH) and CO_2H]

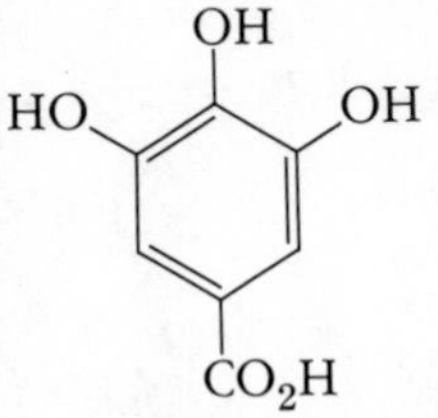

Gallic acid

20-13 Give the IUPAC name for each of the following compounds:

(a) $HSCH_2CH_2CH(CH_3)OH$ (b) 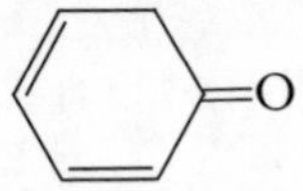 [structure: cyclohexadienone ring with =O]

(c) $N{\equiv}CCH{=}CHCO_2CH_3$ (d) $(CH_3)_2NCH{=}CHCH_2OH$

(e) $NH_2C(=O)CH_2C{\equiv}CCHO$ (f) $CCl_3CH_2CH(OH)CH(OH)CH_3$

20-14 Write the structural formula for each of the following compounds:

(a) 3,3,3-trichloro-1-cyclopropylpropene
(b) methyl 2-mercapto-4-oxobutanoate
(c) 5-chloro-5-methylcyclopentadiene
(d) 1-chloroformyl-3-chloro-3-cyano-4-hydroxycyclohexene
(e) 4-methoxy-3-butenamide
(f) 4-ketopentanal

20-15 Write the structure of the major organic product (or products) of the following reactions.

(a) 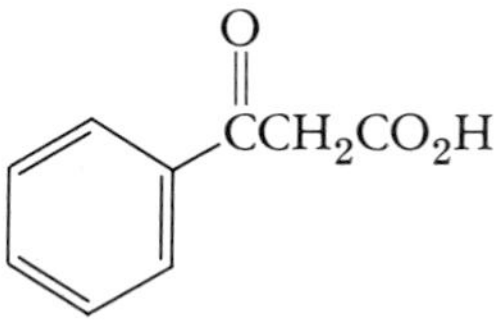 $C_6H_5CCH_2CO_2H$ (with C=O on the first C) $\xrightarrow{100^\circ C}$

(b) $CH_3\overset{O}{\overset{\|}{C}}CH_2CO_2C_2H_5 \xrightarrow[100^\circ C]{H^+, H_2O}$

(c) $C_2H_5O_2CCH_2CO_2C_2H_5 \xrightarrow{NaOC_2H_5} \xrightarrow{C_2H_5Br}$

(d) $HO_2CCH_2C(CH_3)_2CH_2CO_2H \xrightarrow{130^\circ C}$

(e) 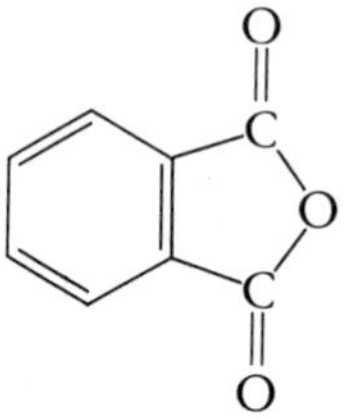 Phthalic anhydride $+ CH_3CH_2OH \longrightarrow$

(f) $HOCH_2CH_2CH_2CO_2H \xrightarrow{H^+}$

(g) $HOCH_2CH_2CH_2\overset{O}{\overset{\|}{C}}CH_3 \xrightarrow{H^+}$

(h) $HOCH_2(CH_2)_7CO_2H \xrightarrow{H^+}$

20-16 How do condensation and addition polymers differ? Give an example of each.

20-17 8-Hydroxyoctanoic acid does not form a lactone when treated with acid or when heated. Instead, it forms a polymer. Write the structure of this polymer.

CARBOHYDRATES 21

During the nineteenth century, several compounds were discovered that have the empirical formula $C_n(H_2O)_n$. As a result, these compounds were designated as hydrates of carbon and were given the name *carbohydrates.* Examples of such compounds include corn starch ($C_6H_{12}O_6$) and cane sugar ($C_{12}H_{24}O_{12}$). One problem is that the name carbohydrate is not descriptive enough. The empirical formula $C_n(H_2O)_n$ is not unique to carbohydrates. For example, formaldehyde (CH_2O) and acetic acid ($C_2H_4O_2$) are certainly not carbohydrates. Furthermore, certain carbohydrates contain nitrogen and sulfur as well as carbon, hydrogen, and oxygen. Despite these problems, the name *carbohydrate* is used today to describe *the large class of compounds that are polyhydroxy aldehydes or polyhydroxy ketones or substances that yield such compounds on acid hydrolysis.* Carbohydrates are also called *sugars* or *saccharides* because of the sweet taste of the simpler members of the family.

Most plants contain 60 to 90 percent carbohydrates, whereas most animals contain only a small amount. Plants make carbohydrates by photosynthesis using carbon dioxide from air, water from the soil, and sunlight. They use carbohydrates as sources of energy and as their structural components. Animals cannot make carbohydrates and must obtain these vital compounds from plants.

The chemistry of carbohydrates usually involves only two functional groups: the carbonyl group of an aldehyde or ketone, and the alcohol hydroxy group. In addition, most carbohydrates contain one or more chiral centers. In this chapter, we will learn how these three features of carbohydrates contribute to their complexity. Let us start by classifying and naming carbohydrates.

21.1 CLASSIFYING CARBOHYDRATES

Carbohydrates are classified according to their acid hydrolysis product. We recognize three major categories. *Monosaccharides,* or simple sugars, are carbohydrates that cannot be hydrolyzed to smaller compounds. *Disaccharides* are carbohydrates that form 2 moles of monosaccharide per mole of carbohydrate when hydrolyzed. *Polysaccharides* are carbohydrates that form many moles of monosaccharides per mole of carbohydrate when hydrolyzed.

Monosaccharides can be further classified according to the number of carbon atoms they contain:

$C_3H_6O_3$ is a triose $\quad$ $C_4H_8O_4$ is a tetrose

$C_5H_{10}O_5$ is a pentose $\quad$ $C_6H_{12}O_6$ is a hexose

These terms contain a suffix indicating the number of carbon atoms in the compound and the ending *-ose,* which indicates that it is a sugar. Thus, triose is a carbohydrate containing three carbons.

We can also specify the kind of carbonyl group in a monosaccharide. A monosaccharide that is a ketone is called a *ketose,* whereas one that is an aldehyde is called an *aldose.* We can tell both how many carbon atoms are present and the kind of carbonyl group in a monosaccharide by combining these two terms into one word. For example, an aldohexose is a six-carbon sugar that has an aldehyde group. A ketopentose is a five-carbon sugar that has a ketone group. The following are additional examples:

```
     OH   OH                      OH
     |    |                       |
CH2CHCHCHCHCHO            CH2CHCHCCH2OH
 |     |    |              |  |  ||
 OH    OH   OH             OH OH O
   An aldohexose            A ketopentose
```

*EXERCISE 21-1 Write the structure of each of the following monosaccharides:
(a) an aldopentose (b) a ketotriose
(c) an aldotriose (d) a ketoheptose

*The answers for the exercises in this chapter begin on page 890.

EXERCISE 21-2 Classify each of the following monosaccharides:

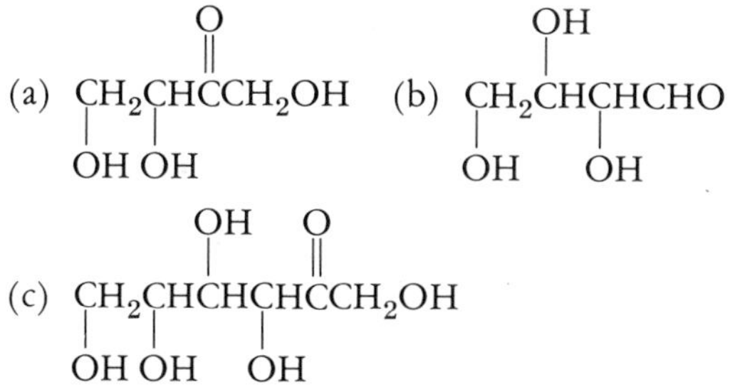

$$\text{(a)}\quad \begin{array}{c@{}c@{}c@{}c} & & O & \\ & & \| & \\ CH_2 & CH & C & CH_2OH \\ | & | & & \\ OH & OH & & \end{array} \qquad \text{(b)}\quad \begin{array}{c@{}c@{}c@{}c} & OH & & \\ & | & & \\ CH_2 & CH & CH & CHO \\ | & & | & \\ OH & & OH & \end{array}$$

$$\text{(c)}\quad \begin{array}{c@{}c@{}c@{}c@{}c@{}c} & & OH & & O & \\ & & | & & \| & \\ CH_2 & CH & CH & CH & C & CH_2OH \\ | & | & & | & & \\ OH & OH & & OH & & \end{array}$$

EXERCISE 21-3 How many chiral centers and how many stereoisomers are there in each of the following sugars? Place an asterisk beside each chiral carbon atom.

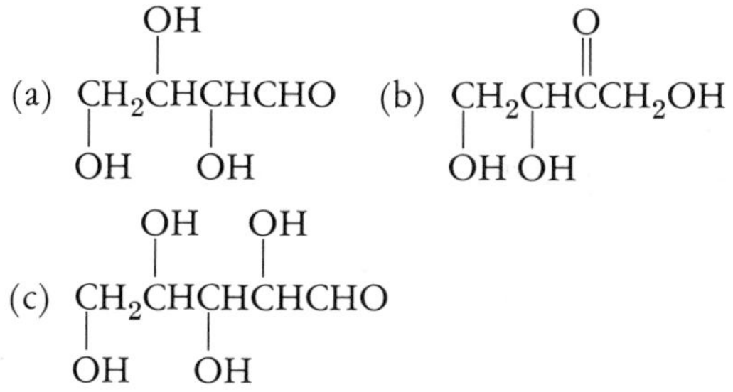

$$\text{(a)}\quad \begin{array}{c@{}c@{}c@{}c} & OH & & \\ & | & & \\ CH_2 & CH & CH & CHO \\ | & & | & \\ OH & & OH & \end{array} \qquad \text{(b)}\quad \begin{array}{c@{}c@{}c@{}c} & & O & \\ & & \| & \\ CH_2 & CH & C & CH_2OH \\ | & | & & \\ OH & OH & & \end{array}$$

$$\text{(c)}\quad \begin{array}{c@{}c@{}c@{}c@{}c} & OH & & OH & \\ & | & & | & \\ CH_2 & CH & CH & CH & CHO \\ | & & | & & \\ OH & & OH & & \end{array}$$

The only possible trioses are glyceraldehyde and dihydroxyacetone:

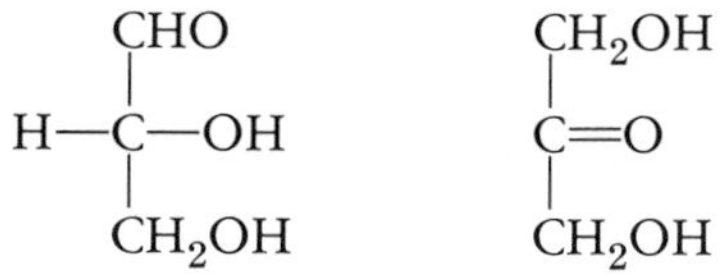

$$\begin{array}{ccc} & CHO & \\ & | & \\ H- & C & -OH \\ & | & \\ & CH_2OH & \end{array} \qquad\qquad \begin{array}{c} CH_2OH \\ | \\ C{=}O \\ | \\ CH_2OH \end{array}$$

Glyceraldehyde Dihydroxyacetone

In fact, these are the smallest molecules that can be classed as carbohydrates. Glyceraldehyde contains a chiral carbon atom, as we learned in Section 14.5, so it can exist as a pair of enantiomers. These enantiomers are especially important. They serve as reference points for designating the three-dimensional structures of all other monosaccharides.

21.2 THE THREE-DIMENSIONAL STRUCTURE OF MONOSACCHARIDES

One method used to represent the three-dimensional structure of molecules on a page was presented in Section 11.8. This method attempts to draw a

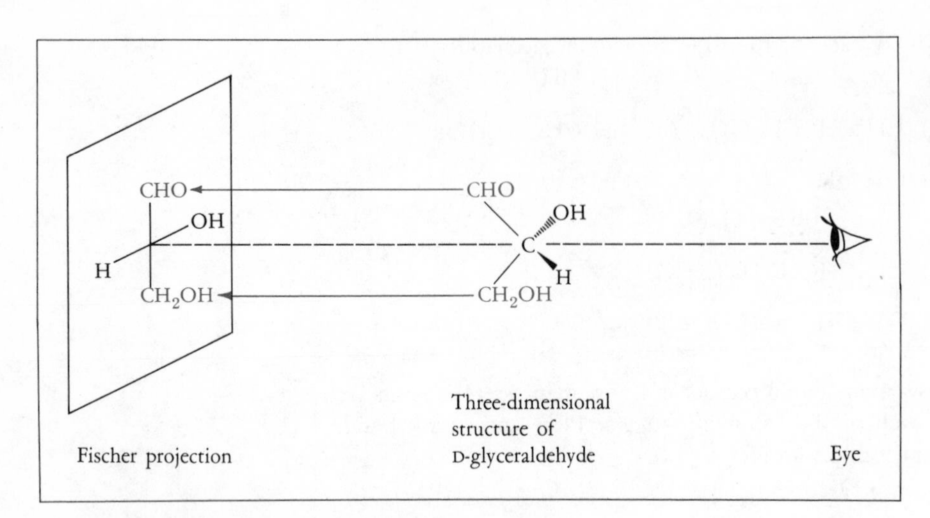

Fig. 21-1. The relationship between the three-dimensional representation of D-*glyceraldehyde and its Fischer projection.*

three-dimensional likeness of the molecule on the page. Another method is to project the three-dimensional structure onto the page. One such projection method was devised by Emil Fischer and is widely used in carbohydrate chemistry because of its simplicity. These *Fischer projections* are made by viewing the molecule as illustrated in Figure 21-1.
The projection is written on paper with horizontal and vertical lines as follows:

CHO
H—+—OH
CH_2OH

is equivalent to

CHO
H—C—OH
CH_2OH

The vertical line represents groups or atoms which project *away* from the viewer while the horizontal line represents groups or atoms which project *toward* the viewer. A carbon atom is located at the intersection of the two lines.

The two enantiomers of glyceraldehyde can be represented by these projection formulas as follows:

O=C–H (CHO)		H–C=O (CHO)
H—+—[OH]	¦	[HO]—+—H
CH_2OH	¦	CH_2OH
	Mirror	

The projection with the secondary alcohol hydroxy group that points to the right corresponds to the compound that was arbitrarily given the D-configuration. The other one, with the hydroxy group that points to the left, corresponds to the L-configuration of glyceraldehyde. The absolute configurations of these enantiomers have been verified by x-ray crystallography, as we learned in Section 14.5.

Monosaccharides are assigned to the D or L family according to the configuration of the chiral carbon *farthest from the aldehyde or ketone group.* If this carbon of a monosaccharide has the same configuration as D-glyceraldehyde, it is assigned to the D family. How do we make such a comparison? Fischer projection formulas are a convenient and simple way to do this. But first we must obtain the Fischer projection formula of the sugar. Usually, this is given. If not, it must be obtained by viewing a molecular model or other three-dimensional representation of the molecule according to the following rules:

1. The molecule is held so that the main carbon chain is vertical, with the lowest-numbered carbon atom at the top. For monosaccharides, this means that the carbonyl group is pointed to the top.
2. The molecule is arranged so that all the horizontal bonds to the main chain extend toward the viewer.

These rules are applied to the aldotetroses in Figure 21-2.

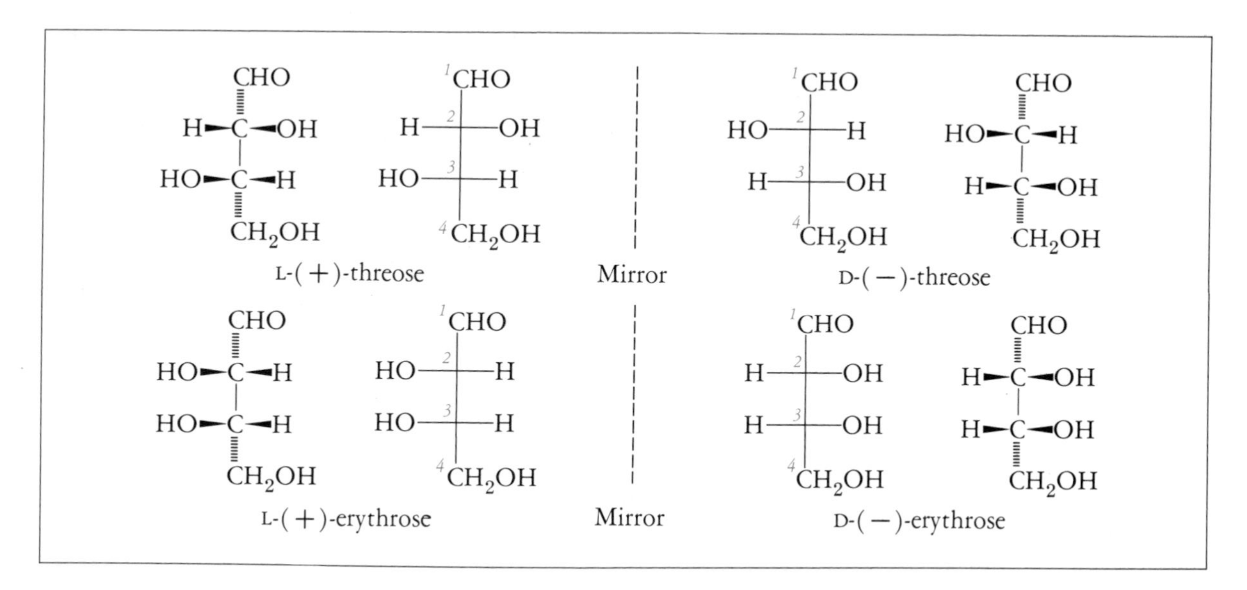

Fig. 21-2. The three-dimensional representations of the aldotetroses and their corresponding Fischer projections. An aldotetrose has two carbon centers, so it has four ($2^2 = 4$) *stereoisomers.*

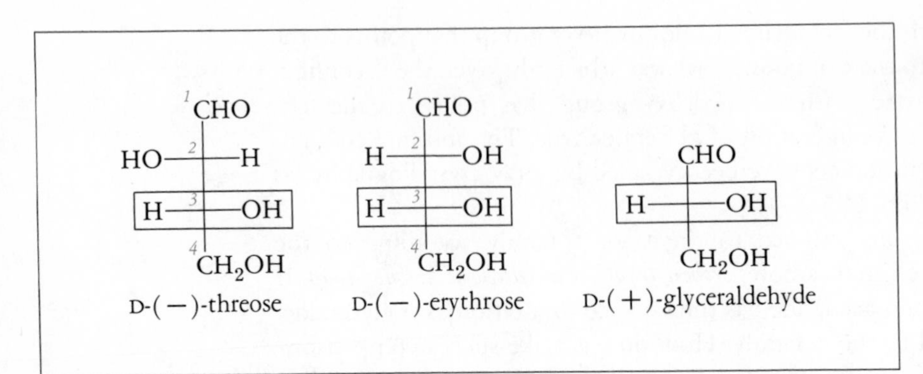

Fig. 21-3. The identical configuration of carbon 3 of D-(−)-*threose,* D-(−)-*erythrose, and the chiral center of* D-(+)-*glyceraldehyde.*

Now we can compare the configuration of carbon 3 of these four stereoisomers with the configuration of D- and L-glyceraldehyde. The hydroxy group of carbon 3 of the Fischer projection formulas of (−)-threose and (−)-erythrose and the hydroxy group of the chiral center of D-glyceraldehyde all point to the right. This is shown in Figure 21-3. Therefore, these three carbon atoms all have the same configuration and (−)-threose and (−)-erythrose are assigned to the D family.

EXERCISE 21-4 Assign the following monosaccharides to either the D or L family:

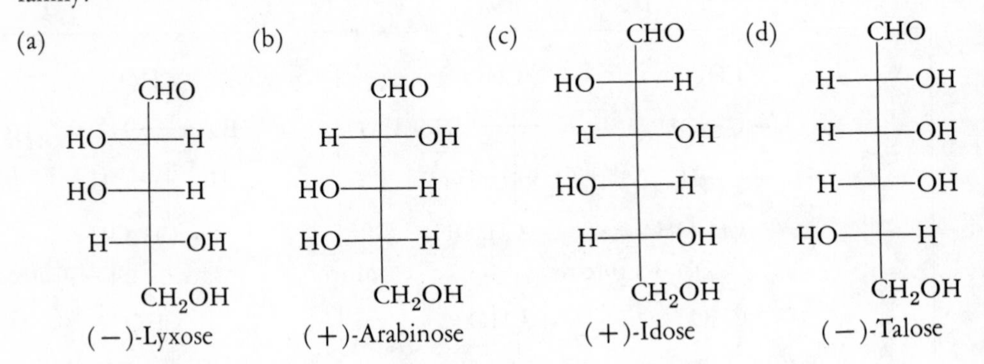

In a similar way, other aldoses and ketoses can be assigned to either the D or the L family. The names, Fischer projection formulas, and signs of optical rotation of all the aldoses up to and including the aldohexoses of the D family are given in Figure 21-4.

The aldopentoses contain three chiral centers, so eight ($2^3 = 8$) stereoisomers are possible (four pairs of enantiomers).

D-Ribose is a very important aldopentose because it is one of the parts of

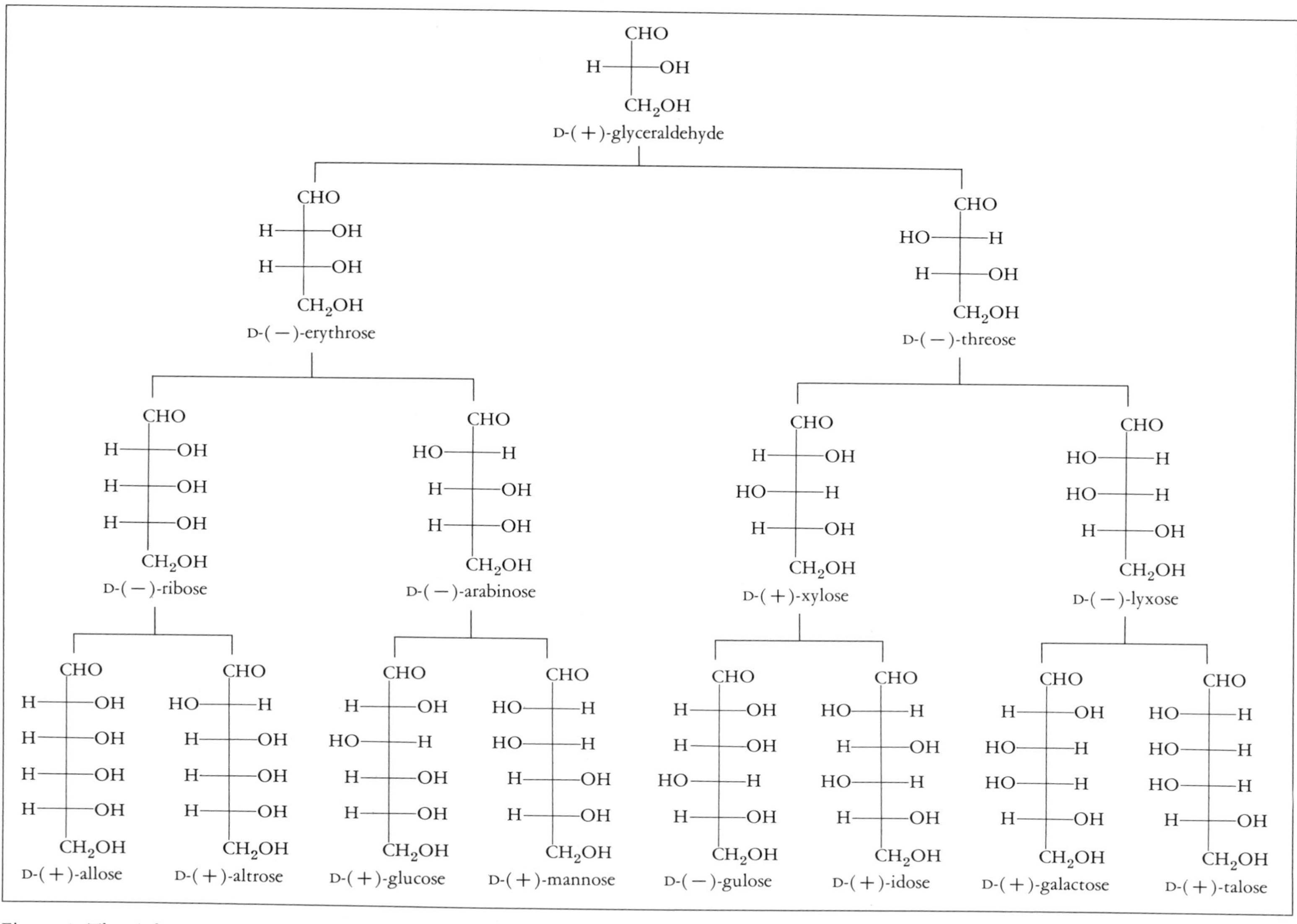

Fig. 21-4. The Fischer projection formulas of the D-family of the aldoses up to and including the aldohexoses. In all of these structures the hydroxy group on the chiral carbon farthest from the carbonyl group points to the right.

$$
\begin{array}{ccc}
 & \overset{1}{\text{CHO}} & \\
\text{H} & \overset{2}{—} & \text{OH} \\
\text{H} & \overset{3}{—} & \text{OH} \\
\text{H} & \overset{4}{—} & \text{OH} \\
 & \overset{5}{\text{CH}_2\text{OH}} & \\
\end{array}
\qquad
\begin{array}{ccc}
 & \overset{1}{\text{CHO}} & \\
\text{H} & \overset{2}{—} & \text{H} \\
\text{H} & \overset{3}{—} & \text{OH} \\
\text{H} & \overset{4}{—} & \text{OH} \\
 & \overset{5}{\text{CH}_2\text{OH}} & \\
\end{array}
$$

D-ribose   2-deoxy-D-ribose

certain nucleic acids. D-(−)-Ribose is found in *r*ibo*n*ucleic *a*cid (RNA). A compound related to D-ribose is 2-deoxy-D-ribose. Notice that the hydroxy group of carbon 2 of D-ribose has been replaced by a hydrogen atom in 2-deoxy-D-ribose. This compound is also a part of certain nucleic acids. 2-Deoxy-D-ribose is found in *d*eoxyribo*n*ucleic *a*cid (DNA). 2-Deoxy-D-ribose is not an aldopentose because its chemical formula, $C_5H_{10}O_4$, does not fit the definition for a carbohydrate or sugar. Despite this fact, both ribose and deoxyribose are referred to as the sugar parts of nucleic acids, as we will learn in Chapter 24.

The aldohexoses contain four chiral centers, so there are 16 ($2^4 = 16$) stereoisomers possible (eight pairs of enantiomers). Eight of these are in the D family (Figure 21-4), and eight are in the L family. Only three aldohexoses, all with the D configuration, are found in living systems. They are D-(+)-glucose, D-(+)-mannose, and D-(+)-galactose. The other 13 isomers have been prepared in the laboratory.

Derivatives of aldohexoses are known in which a hydroxy group is replaced by an amino group. These compounds are called amino sugars. There are three naturally occurring amino sugars: D-glucosamine, D-mannosamine, and D-galactosamine. In all three, the hydroxy group on carbon 2 of the parent monosaccharide is replaced by an amino group:

$$
\begin{array}{ccc}
 & \text{CHO} & \\
\text{H} & — & \text{NH}_2 \\
\text{HO} & — & \text{H} \\
\text{H} & — & \text{OH} \\
\text{H} & — & \text{OH} \\
 & \text{CH}_2\text{OH} & \\
\end{array}
\qquad
\begin{array}{ccc}
 & \text{CHO} & \\
\text{NH}_2 & — & \text{H} \\
\text{HO} & — & \text{H} \\
\text{H} & — & \text{OH} \\
\text{H} & — & \text{OH} \\
 & \text{CH}_2\text{OH} & \\
\end{array}
\qquad
\begin{array}{ccc}
 & \text{CHO} & \\
\text{H} & — & \text{NH}_2 \\
\text{HO} & — & \text{H} \\
\text{HO} & — & \text{H} \\
\text{H} & — & \text{OH} \\
 & \text{CH}_2\text{OH} & \\
\end{array}
$$

D-glucosamine   D-mannosamine   D-galactosamine

They all have the same three-dimensional structure as their parent carbohydrate. These amino sugars are widely found in nature as essential parts of many polysaccharides. They are important parts of the cell wall structure of many living organisms as well as the shells of crabs, lobsters, and shrimps.

So far, we have drawn the carbon structures of monosaccharides as a straight chain. This structure seems incorrect if we remember what we learned in Section 20.6 about predicting the reactions of polyfunctional compounds. We would expect that a hydroxy group and the aldehyde group of a monosaccharide would react to form a cyclic hemiacetal structure. This is exactly what happens.

21.3 THE CYCLIC STRUCTURE OF MONOSACCHARIDES

The straight-chain structure of monosaccharides explains many of their reactions. However, this structure does not agree with at least one experimental fact. This discrepancy can be illustrated by using D-(+)-glucose as an example.

It was discovered that D-(+)-glucose exists as two isomers. The α isomer has a specific rotation of +112 degrees and a melting point of 146° C. The β isomer has a specific rotation of +19 degrees and a melting point of 150° C. The existence of these two isomers is explained when we realize that all monosaccharides have a cyclic hemiacetal structure. The intramolecular reaction of the hydroxy group on carbon 5 of D-(+)-glucose (and most other monosaccharides) with the carbonyl group forms a cyclic hemiacetal with a six-member ring containing one oxygen and five carbon atoms. This is illustrated in Figure 21-5.

In Figure 21-5, we have used a Fischer projection to represent the cyclic hemiacetal form of D-(+)-glucose. This representation does not accurately represent the geometry of cyclic structures of monosaccharides. A better representation is the *Haworth formula.* The hemiacetal ring of D-(+)-glucose is represented as a regular hexagon with the ring oxygen away from the viewer.

This representation of D-(+)-glucose is given in Figure 21-6. In this projection the hydrogens attached to the ring carbons are not shown. Notice that a hydroxy group on carbon 2, 3, or 4 of the Fischer projection formula pointing to the right appears below the ring of the Haworth projection formula.

It is clear from the Haworth formulas in Figure 21-6 that D-(+)-glucose can exist as two isomers because a new chiral center is created at carbon 1

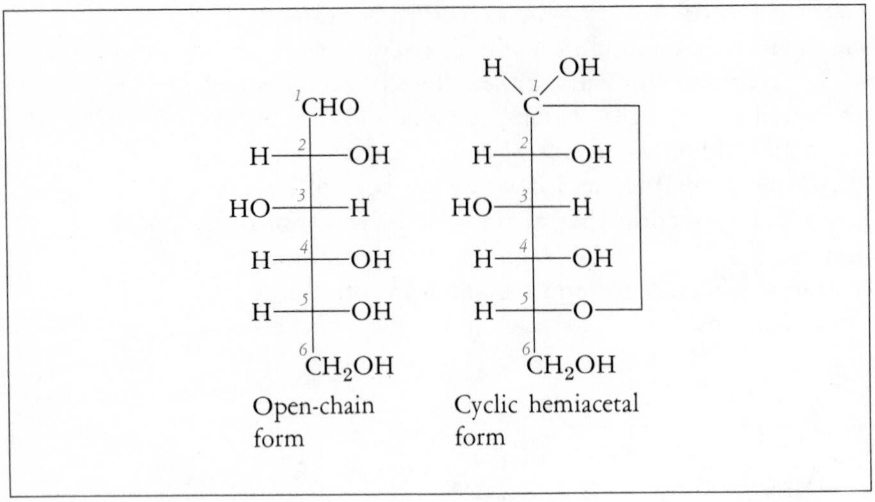

Open-chain form

Cyclic hemiacetal form

Fig. 21-5. Fischer projection formulas of the open and cyclic forms of D-(+)-*glucose.*

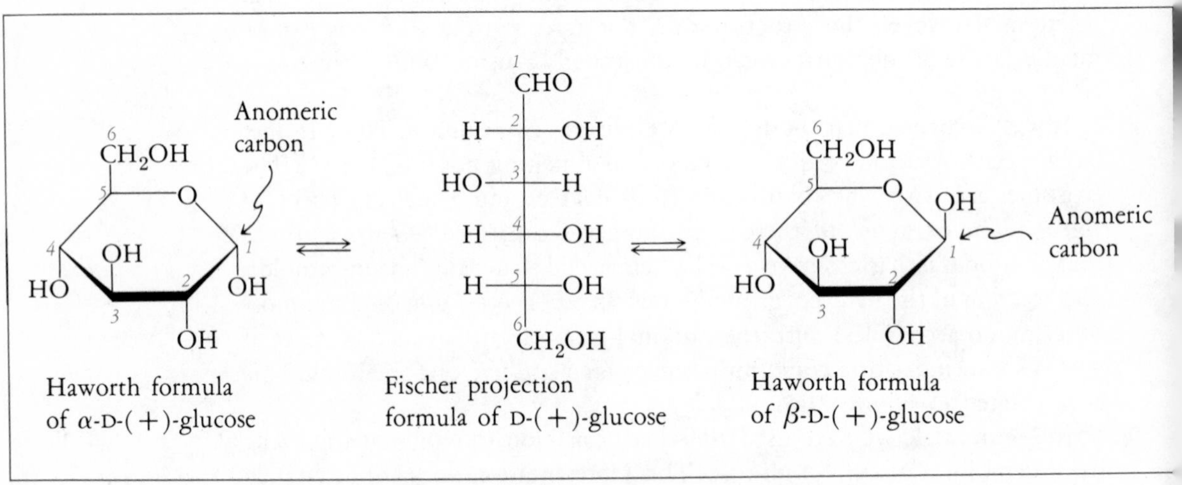

Anomeric carbon

Haworth formula of α-D-(+)-glucose

Fischer projection formula of D-(+)-glucose

Haworth formula of β-D-(+)-glucose

*Fig. 21-6. Haworth formulas of α- and β-*D-(+)-*glucose and their relationship to the Fischer projection formula of* D-(+)-*glucose.*

by formation of a cyclic hemiacetal. This new chiral carbon is called an *anomeric carbon.* As a result, there are now two isomers of D-(+)-glucose that differ only in their configuration at the hemiacetal, as shown in Figure 21-6. Isomers that differ only in this way are called *anomers.* By convention, the α isomer is defined as the one in which the hydroxy group at the anomeric carbon is below the plane of the ring. The isomer with the hydroxy group above the ring at the anomeric carbon is the β isomer.

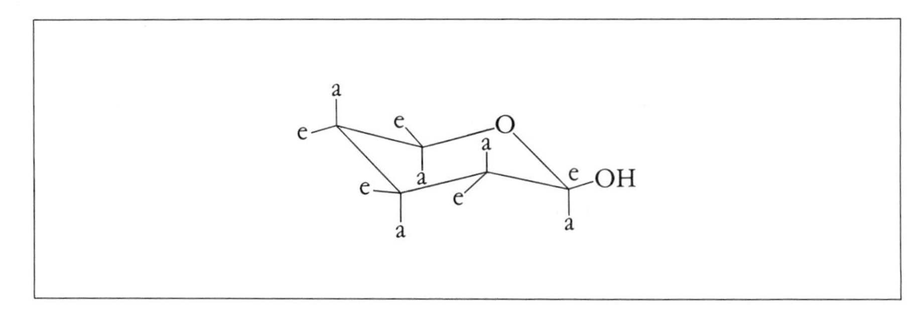

Fig. 21-7. The chair form of hemiacetal ring of monosaccharides. Equatorial bonds are designated e; *axial bonds are designated* a.

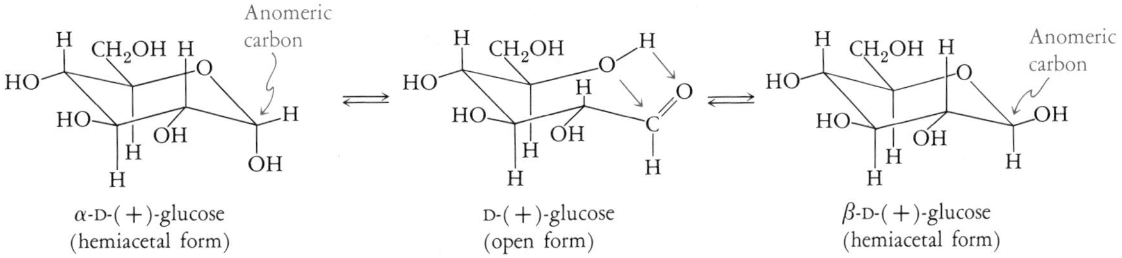

Anomeric carbon

α-D-(+)-glucose
(hemiacetal form)

D-(+)-glucose
(open form)

Anomeric carbon

β-D-(+)-glucose
(hemiacetal form)

Fig. 21-8. The open and cyclic structures of D-(+)-*glucose.*

The Haworth projection formulas show the six-member ring as planar. This is incorrect, because the ring adopts a chair form similar to that of cyclohexane (see Section 11.10). We can represent the ring as shown in Figure 21-7. The groups and atoms bonded to the ring carbons can occupy either axial or equatorial positions. These formulas, illustrated for D-(+)-glucose and its two anomers in Figure 21-8, give the best representation of true molecular shape and geometry. However, it is easier to see the three-dimensional structure of carbohydrates from the Haworth formulas. Therefore, we will use the Haworth formulas to represent the structure of carbohydrates in this book.

Intramolecular hemiacetal formation also occurs with other monosaccharides. D-(+)-Mannose, D-(+)-galactose, and D-(−)-fructose exist in cyclic form, as shown in Figure 21-9.

D-(+)-Galactose and D-(+)-mannose form six-member cyclic hemiacetals like glucose. In contrast, D-(−)-fructose is usually found in living systems as a five-member cyclic hemiketal.

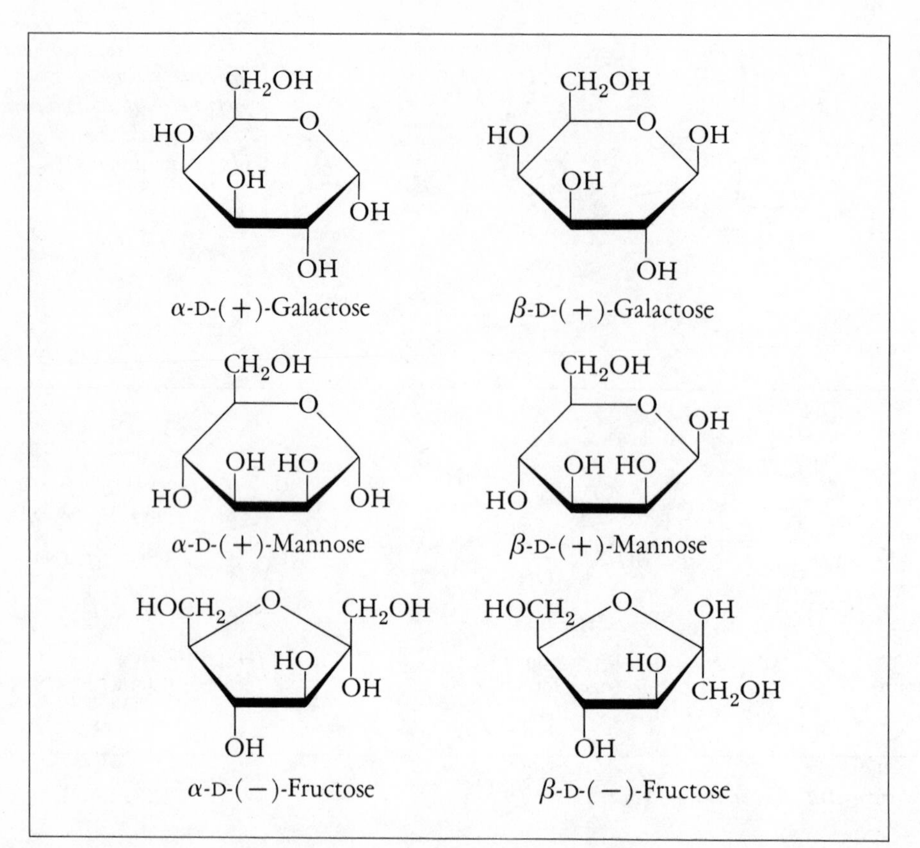

Fig. 21-9. The anomers of D-(+)-*galactose,* D-(+)-*mannose, and* D-(−)-*fructose.*

EXERCISE 21-5 For each of the following compounds, circle the hemiacetal or hemiketal functional group; decide whether it is the α or β isomer; and write the structure of its anomer.

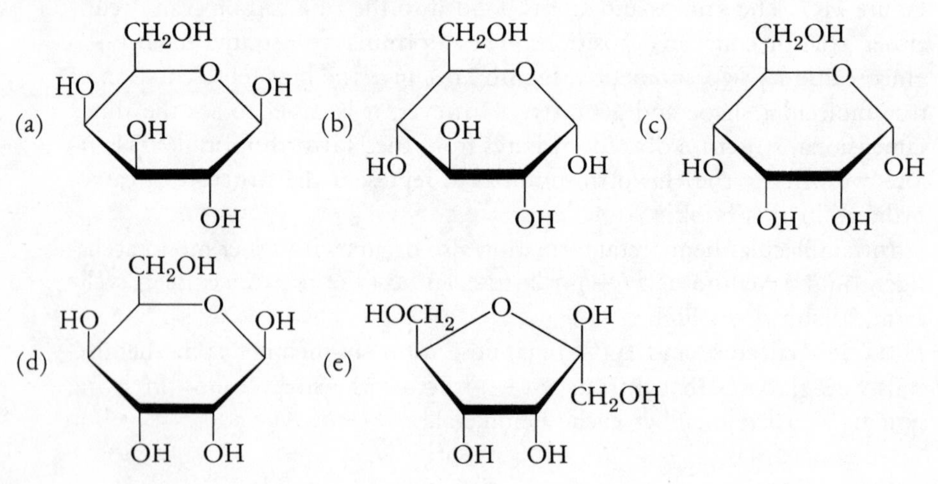

EXERCISE 21-6 Which of the following pairs of compounds are anomers?

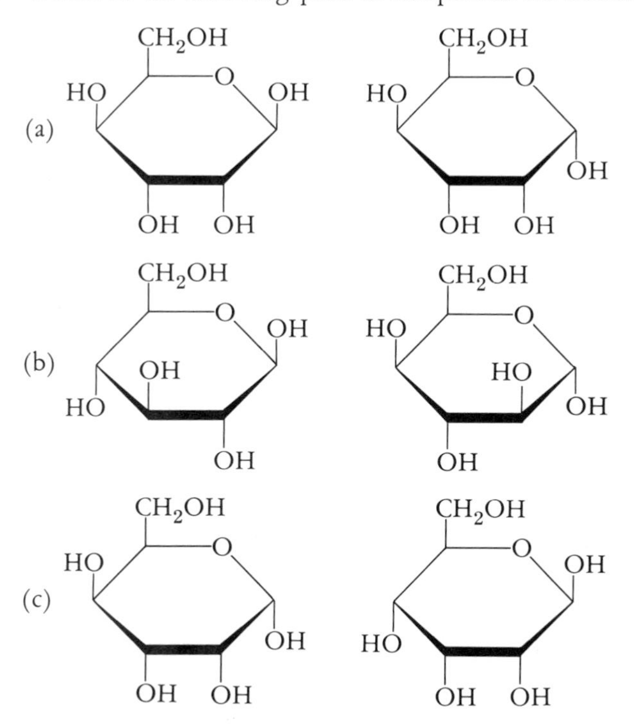

It is clear from the examples given in this section that monosaccharides exist in two anomeric cyclic forms. In solution, all three forms exist in equilibrium, as we will learn in the next section.

21.4 MUTAROTATION

Both α- and β-D-(+)-glucose are stable as solids. When either of these anomers is dissolved in water, the optical rotation of the solution slowly changes. For example, a freshly prepared solution of α-D-(+)-glucose in water has a specific rotation of +112 degrees. The rotation of this solution gradually changes with time until a constant value of +53 degrees is reached. This gradual change of optical rotation is called *mutarotation.* An aqueous solution of β-D-(+)-glucose also undergoes mutarotation. The initial value of +19 degrees of its optical rotation gradually changes to the same constant value of +53 degrees.

Mutarotation of glucose can be explained by the interconversion of the α and β anomers in solution through the open-chain aldehyde form as an intermediate. This equilibrium is shown in Figure 21-8. At equilibrium,

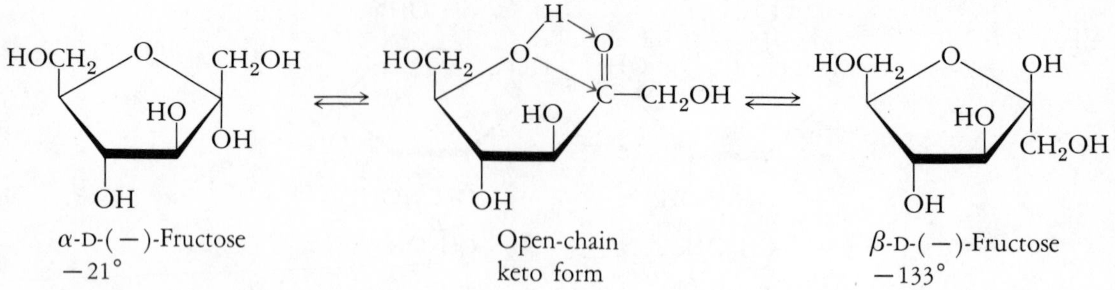

α-D-(−)-Fructose
−21°

Open-chain
keto form

β-D-(−)-Fructose
−133°

Fig. 21-10. Equilibrium mixture of the three forms of D-(−)-*fructose. At equilibrium, the specific rotation is −93 degrees, corresponding to a mixture of 37 percent of the α-isomer, 63 percent of the β-isomer, and a trace of the keto form.*

glucose consists of 63 percent of the β isomer, 37 percent of the α isomer, and a trace (less than 0.2 percent) of the open-chain aldehyde form.

Mutarotation is not a unique property of glucose. Solutions of all monosaccharides (and some disaccharides) undergo mutarotation. The mutarotation of D-(−)-fructose is shown in Figure 21-10.

Monosaccharides exist predominantly in the cyclic form. However, we must not forget that in solution the cyclic forms are in equilibrium with a small amount of the open-chain form. This equilibrium is important in the reactions of monosaccharides.

21.5 REACTIONS OF MONOSACCHARIDES

Based on the structure of monosaccharides described in the last section, we would predict that they would undergo the usual reactions of hydroxy, carbonyl, hemiacetal, and hemiketal functional groups. This prediction is correct. For example, all monosaccharides and some disaccharides react with Tollens' and Fehling's solutions (see Section 19.4). In these reactions, the small amount of aldehyde present at equilibrium is oxidized. This disturbs the equilibrium. In an attempt to restore equilibrium reaction occurs to convert the hemiacetal form to the open-chain form. Eventually, all of the hemiacetal form is converted to the aldehyde, which is oxidized. The oxidation of aldoses therefore occurs by reaction of the open-chain aldehyde form.

We learned in Section 19.4 that silver ions are the oxidizing reagent in the Tollens' test for aldehyde, and that cupric ions (Cu^{2+}) serve the same purpose in Fehling's solution. Cupric ion is reduced to cupric oxide (Cu_2O), and silver ion is reduced to silver metal by monosaccharides. Because of this property, monosaccharides (and some disaccharides) are called *reducing sugars.*

It is important to realize that D-(−)-fructose, a ketohexose, is also a reducing sugar: it reacts with Tollens' and Fehling's solutions just like an aldehyde. This is not expected of a ketone (see Section 19.4). The presence of a hydroxy group α to the ketone is responsible for this reaction. In general, α-hydroxy ketones are very easily oxidized (see Section 20.4). The same reagents that oxidize aldehydes also oxidize α-hydroxy ketones. Consequently, Tollens' and Fehling's solutions do not distinguish between ketoses and aldoses.

Like any aldehyde, the carbonyl group of an aldose can be reduced with reducing agents such as hydrogen (H_2) and a metal catalyst to form products called *alditols.* For example, D-(+)-glucose is reduced to D-(+)-glucitol, which is also known as sorbitol.

```
         CHO                            CH2OH
   H ────┼──── OH                 H ────┼──── OH
  HO ────┼──── H                 HO ────┼──── H
                     H2, Pt
   H ────┼──── OH   ───────→      H ────┼──── OH
   H ────┼──── OH                 H ────┼──── OH
        CH2OH                          CH2OH
  D-(+)-glucose                  D-(+)-glucitol
                                 (sorbitol)
```

The carbonyl groups of aldehydes and ketones react with phenylhydrazine to form phenylhydrazones (see Section 19.5), a specific example of a Schiff base. Monosaccharides undergo a similar reaction. However, the initial product reacts with more phenylhydrazine. The α-hydroxy group of the initial product is easily oxidized to a carbonyl group by a second mole of phenylhydrazine (see Section 20.4). This new carbonyl group then reacts with a third mole of phenylhydrazine to form an *osazone.* This is shown in Figure 21-11. Notice that phenylhydrazine reacts with carbons 1 and 2 of the monosaccharides.

Historically, osazones have been used to identify sugars. The reason for

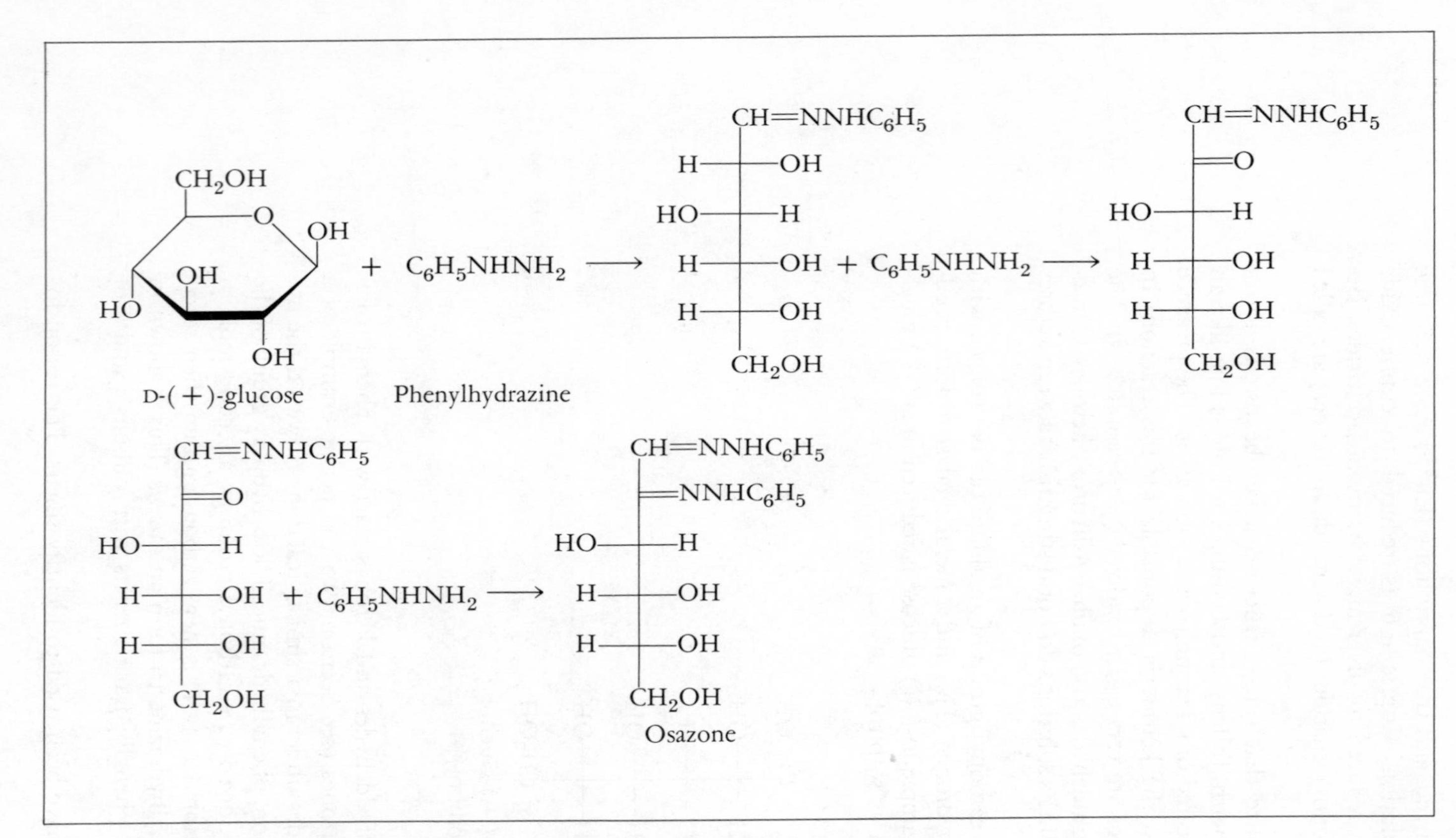

Fig. 21-11. The formation of an osazone by the reaction of D-(+)*-glucose and 3 moles of phenylhydrazine.*

this is that most sugars are very difficult to crystallize; they tend to remain as syrups. If the syrup is treated with excess phenylhydrazine, a crystalline osazone is formed. This solid has a characteristic melting point that is used to identify a particular sugar.

EXERCISE 21-7 Write the structure of the osazone formed from each of the following monosaccharides:
(a) L-(+)-threose (b) D-(+)-idose (c) D-(−)-arabinose

Monosaccharides contain alcohol groups that react to form esters (see Sections 16.4 and 18.5). Among the most important esters in living systems are the phosphate esters of monosaccharides (see Section 18.12). Two of these are D-(+)-glucose 1-phosphate and D-(+)-glucose 6-phosphate. In cells, D-(+)-glucose exists as a phosphate ester:

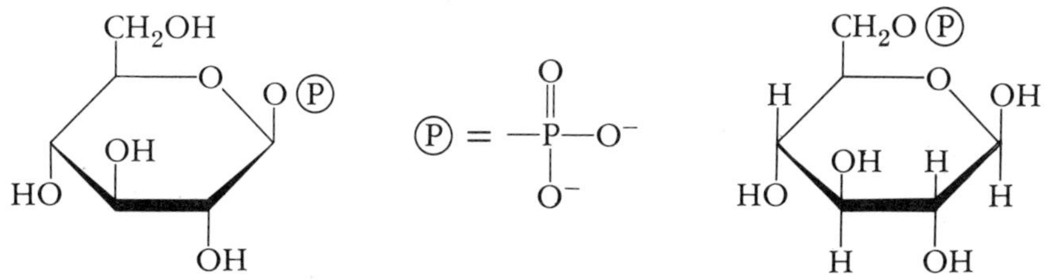

Ⓟ = $-PO_3^{2-}$

D-(+)-glucose 1-phosphate

D-(+)-glucose 6-phosphate

In Chapter 27, we will learn how these phosphate esters are involved in the formation and breakdown of D-(+)-glucose to produce energy for living systems.

Another important reaction of monosaccharides involves the hemiacetal or hemiketal functional group, which reacts with an alcohol and an acid to form an acetal or a ketal, respectively (see Section 19.5). For example, D-(+)-glucose undergoes an acid-catalyzed reaction with methanol to form two acetals, as shown in Figure 21-12. One is formed from the α anomer, and the other is formed from the β anomer of glucose.

The general term *glycoside* is given to the acetal or ketal formed by the reaction of a sugar with an alcohol or phenol. Individually, glycosides are named as derivatives of the parent monosaccharides by substituting the ending *-ide* for the *-e* ending of the parent name. For example, the glycosides in Figure 21-12 are glucosides (glucos + ide). We can further specify the alkyl group: they are α- and β-methyl glucosides.

These compounds are also called *O*-glycosides because the atom bonded

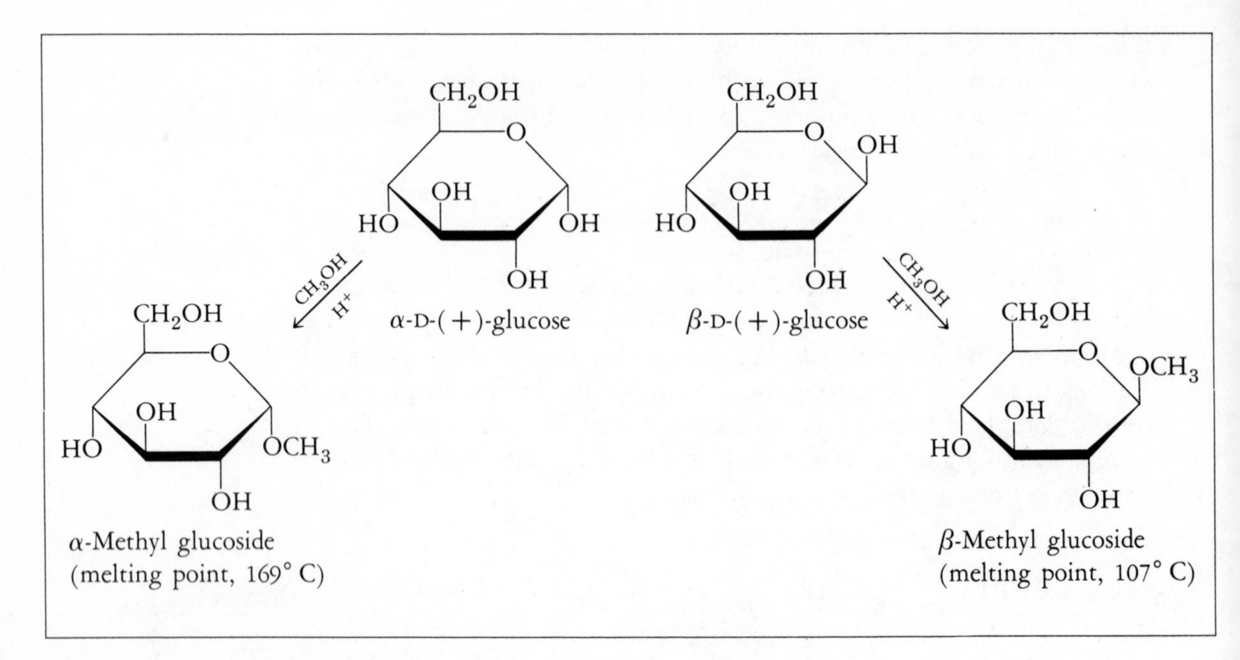

Fig. 21-12. The reaction of α- and β-D-(+)-glucose with methanol to form two isomeric acetals.

to the anomeric carbon is an oxygen. Glycosides with a nitrogen atom bonded to the anomeric carbon are also known. Such compounds are called *N*-glycosides. They are similar in structure to the *O*-glycosides, as shown in Figure 21-13.

The most common *N*-glycoside is adenosine, which is made up of adenine and D-(−)-ribose. Adenosine is part of the coenzyme *a*denosine *t*ri*p*hosphate (ATP), as we will learn in Chapter 24.

NH_2 N N N N — Adenine part

$HOCH_2$ O — D-ribose part

OH OH

Adenosine

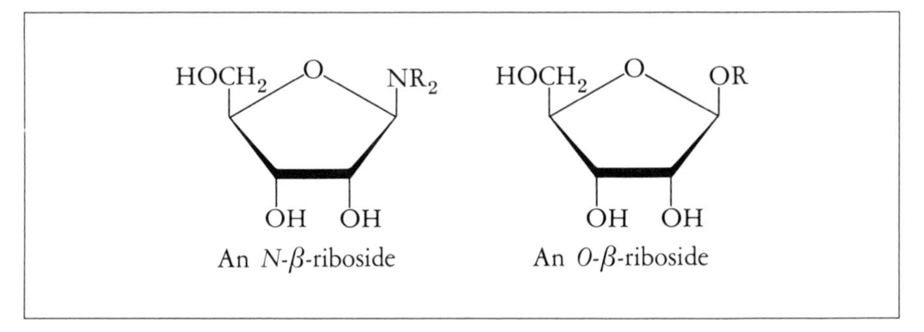

Fig. 21-13. Examples of structurally similar N- *and* O-*glycosides.*

In Chapter 26, we will learn that the primary role of this coenzyme is to furnish the energy for many reactions in living systems.

EXERCISE 21-8 Write the structures for both the α- and β-methyl glycosides of the following sugars:
(a) D-(+)-mannose (b) D-(+)-galactose
(c) D-(−)-fructose

In our example in Figure 21-12, we used methanol to react with D-(+)-glucose to form a glucoside. We could have used almost any alcohol. Suppose we use the hydroxy group from another monosaccharide? If we do this, we form a disaccharide.

21.6 DISACCHARIDES

Disaccharides are simply dimers made up of two molecules of monosaccharides that are the same or different. Monosaccharides are joined by glycoside bonds in two ways, as shown by structures *21-1* and *21-2* in Figure 21-14.

Notice carefully the difference between structures *21-1* and *21-2*. In *21-1* the hydroxy group on carbon 4 of monosaccharide B reacts with the hemiacetal hydroxy group of monosaccharide A to form an acetal. The bond is called a *1,4-glycoside bond.* Another way of looking at this is to assign a head and tail to each monosaccharide, as shown in Figure 21-14. Structure *21-1* is a dimer formed by the head of one monosaccharide bonding to the tail of another. The glycoside bond formed in this way is identical to the one in methyl D-(+)-glucoside. Both are acetals with the structure O—C—O—C—C. Most disaccharides and polysaccharides have this structure. Disaccharides of structure *21-1* still have a hemiacetal group. Consequently, they are reducing sugars, they can exist as α and β forms, and they

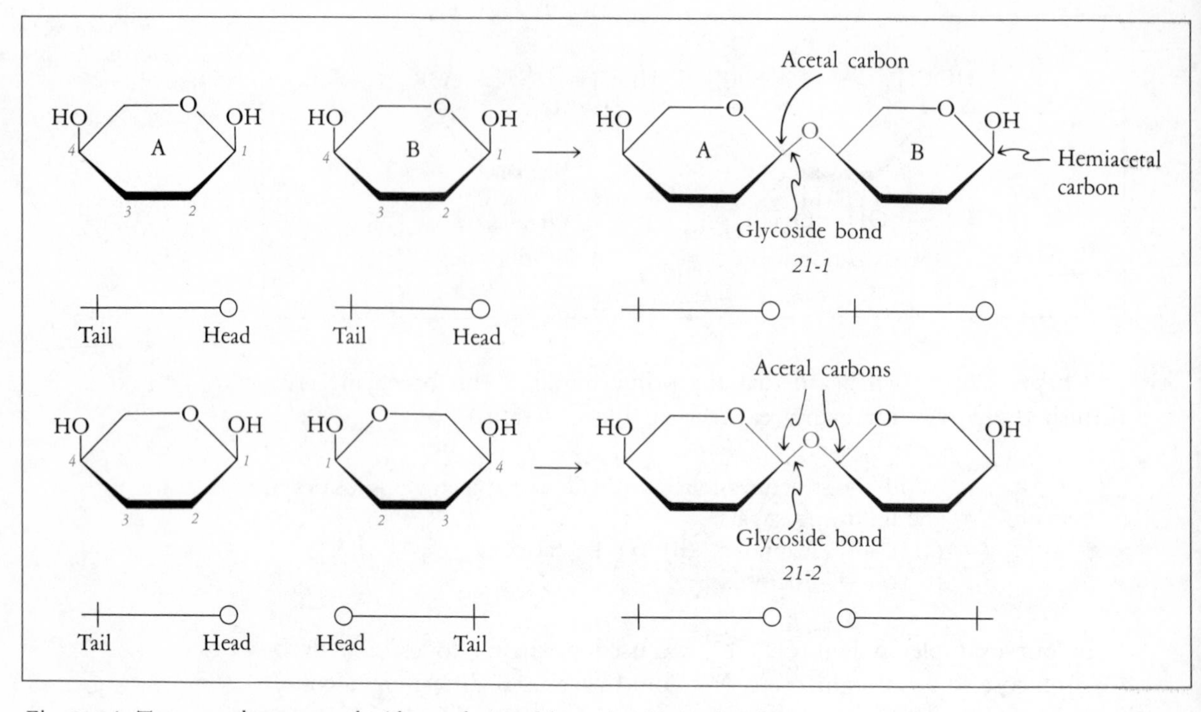

Fig. 21-14. Two ways that monosaccharides can be joined by a glycoside bond to form a disaccharide. All hydrogens and other —OH groups are omitted for clarity.

undergo mutarotation just like D-(+)-glucose, which contains a similar hemiacetal group.

Disaccharides with structure *21-2* are different because the glycoside bond is formed by joining the hemiacetal carbons of the two monosaccharides. It is a head-to-head dimer. As a result, the glycoside bond has the structure O—C—O—C—O. There is no hemiacetal group in a disaccharide with a structure such as *21-2*. Therefore, these are nonreducing sugars.

EXERCISE 21-9 Decide whether the glycoside bond in each of the following disaccharides resembles that of structure *21-1* or *21-2* in Figure 21-14.

(a)

CH_2OH CH_2OH

O O OH

OH OH

HO O

OH OH

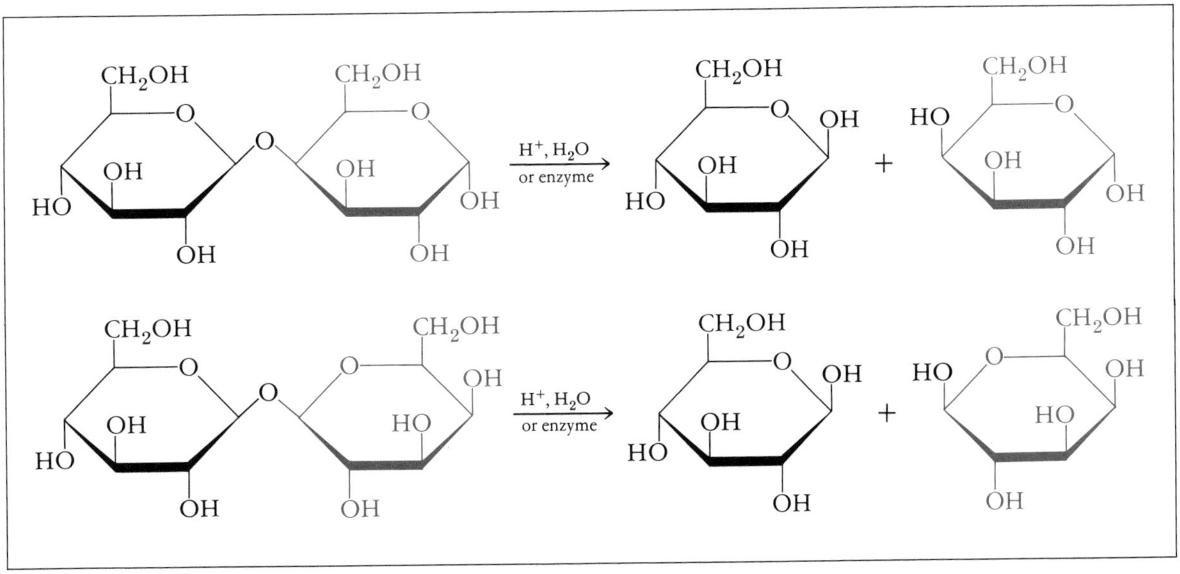

Fig. 21-15. The hydrolysis of disaccharides to their component monosaccharides.

(b)

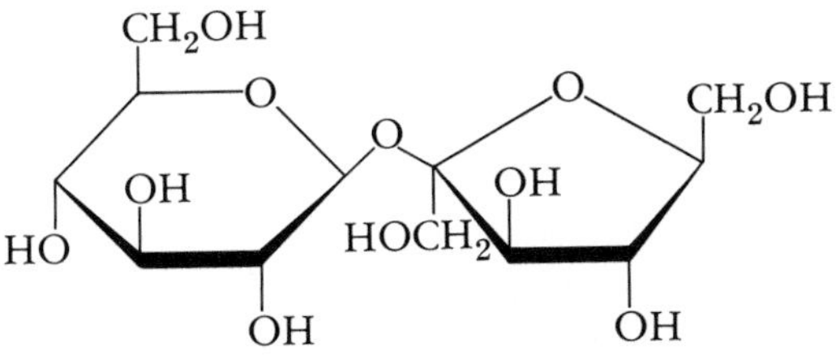

(c)

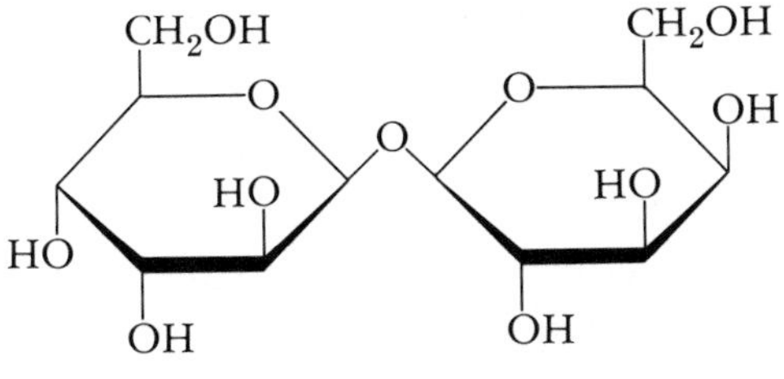

Disaccharides of structures *21-1* and *21-2* are hydrolyzed by dilute aqueous acid solutions or enzymes to their component monosaccharides, as shown in Figure 21-15. We can identify the two monosaccharides that make up a particular disaccharide by using this reaction.

EXERCISE 21-10 Draw the structures of the monosaccharides formed on hydrolysis of the following disaccharides:

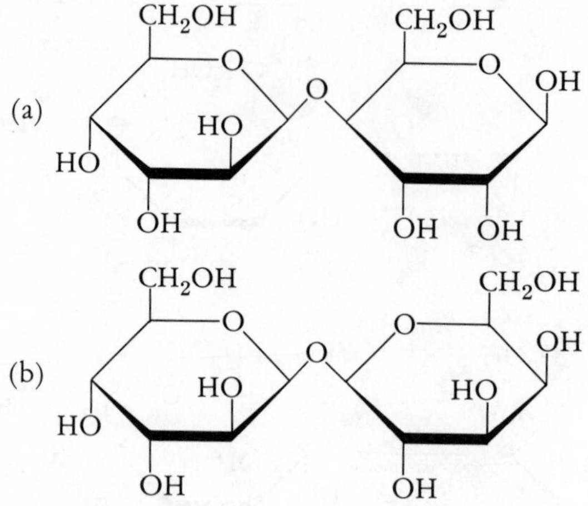

(a)

(b)

Lactose is an example of a disaccharide of structure *21-1*:

β-1,4-glycoside bond

β form

Hemiacetal carbon

D-(+)-galactose part

D-(+)-glucose part

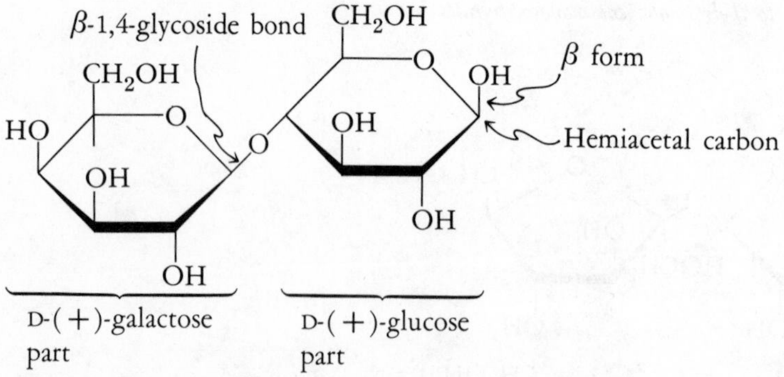

β-Lactose

Lactose is found in the milk of mammals. On hydrolysis, it yields equal amounts of D-(+)-glucose and D-(+)-galactose. These two sugars are joined by a β-1,4-glycoside bond. Lactose contains a hemiacetal group, so it exists in both α and β forms, undergoes mutarotation, and is a reducing sugar.

Maltose is another example of a disaccharide of structure *21-1*. This sugar is found mainly in malt liquors. On hydrolysis, maltose yields only D-(+)-glucose (2 moles per mole of maltose). Thus, two molecules are joined together by an α-1,4-glycoside bond. Like lactose, maltose exists in both the α and β forms, undergoes mutarotation, and is a reducing sugar.

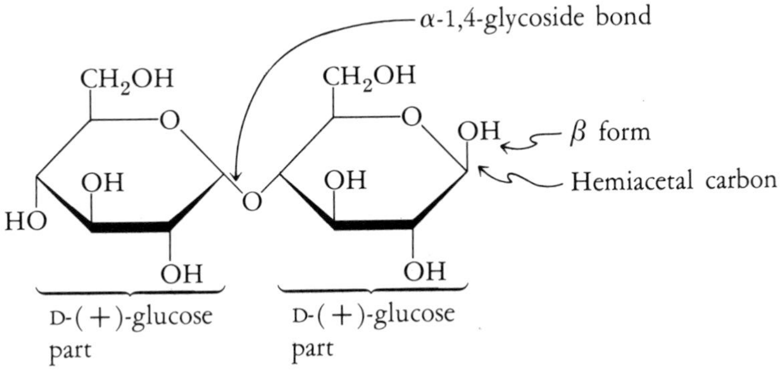

α-1,4-glycoside bond

CH_2OH CH_2OH OH ← β form

HO OH O OH ← Hemiacetal carbon

OH OH

D-(+)-glucose part D-(+)-glucose part

β-Maltose

EXERCISE 21-11 Draw the structures of α-lactose and α-maltose.

Sucrose, which we call table sugar, is an example of a disaccharide of structure *21-2*. On hydrolysis, it yields equal amounts of D-(+)-glucose and D-(−)-fructose. The two monosaccharides are joined by an α-glycoside bond from glucose. Sucrose does not contain a hemiacetal group, so it is a nonreducing sugar.

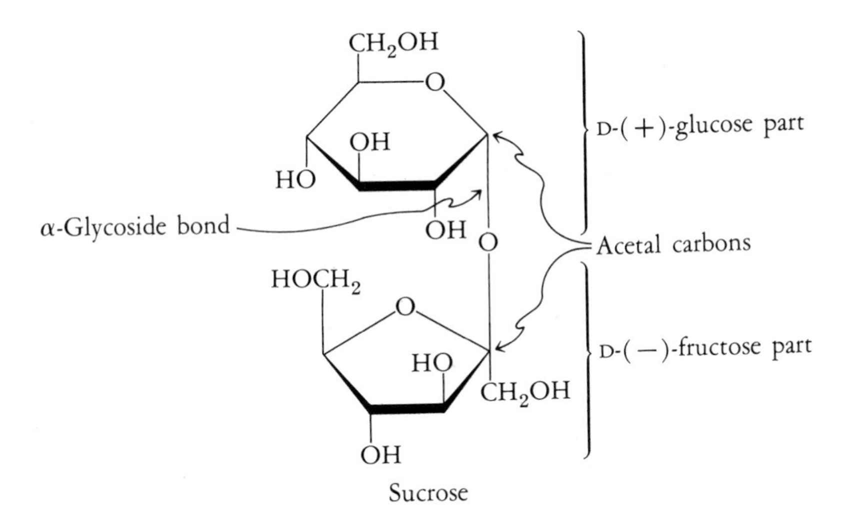

Sucrose

EXERCISE 21-12 Write the equations for and draw the structures of the products formed by hydrolysis of the following compounds:
(a) lactose (b) maltose (c) sucrose

We have learned in this section how monosaccharides combine to form disaccharides. Monosaccharides combine to form polysaccharides in a similar way.

21.7 POLYSACCHARIDES

Polysaccharides are polymers formed from monosaccharides. D-(+)-Glucose is the most common monosaccharide found in nature. Most of it is stored as food for plants and animals. It is not stored as glucose, which is too soluble in water for this purpose. Rather, it is converted into polymers of glucose, which are less water soluble. These polysaccharides have high molecular weights (25,000 to 10,000,000 amu), corresponding to 130 to 50,000 glucose molecules. The three most important polysaccharides in nature are starch, glycogen, and cellulose. These naturally occurring polymers are all made up of the same monomer, D-(+)-glucose. They differ in the way the individual D-(+)-glucose molecules are bonded together.

Starch is the major source of carbohydrates in the human diet. It is a mixture of about 20 percent *amylose* and 80 percent *amylopectin.* Amylose is a straight-chain polysaccharide made up entirely of D-(+)-glucose monomers joined by an α-1,4-glycoside bond, as shown in Figure 21-16. Thus, amylose is a head-to-tail polymer.

Amylopectin differs from amylose in that an occasional glucose molecule of the polymer forms an *additional* α-glycoside bond at carbon 6. This is shown in Figure 21-17. This results in a branched-chain structure for amylopectin. It is estimated that such branching occurs at one in every 20 to 25 glucose molecules in the chain.

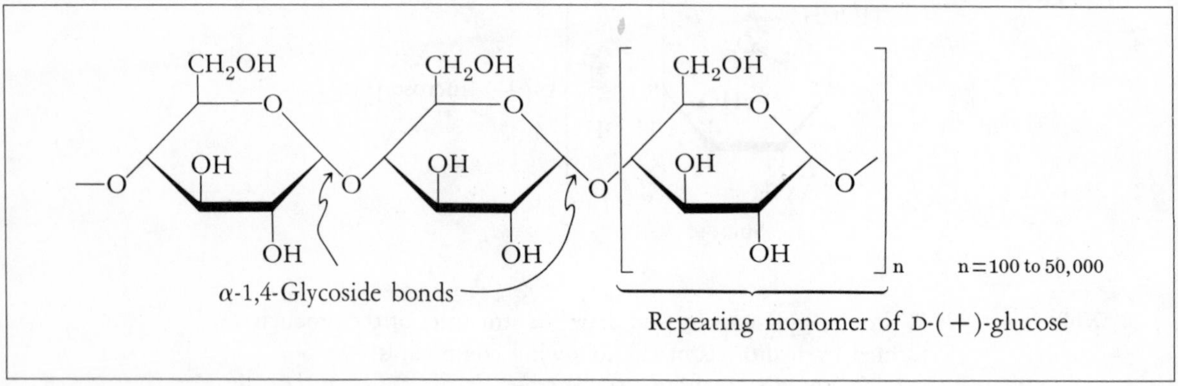

Fig. 21-16. The structure of amylose.

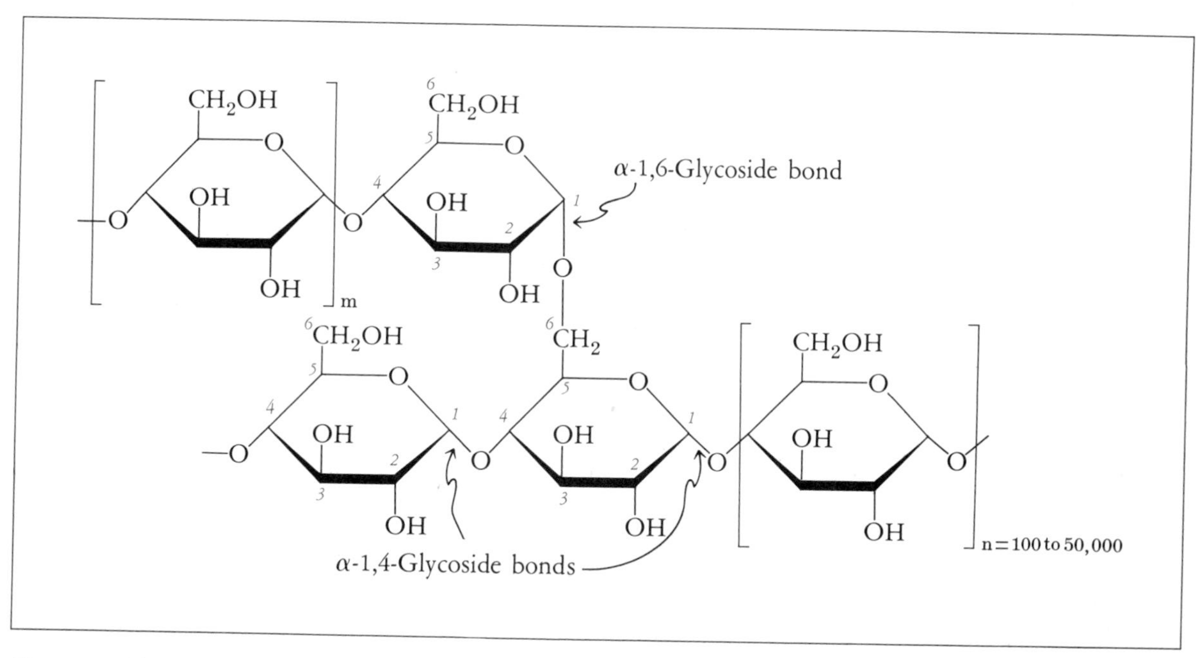

Fig. 21-17. The structure of amylopectin.

Glycogen is the reserve of carbohydrates in animals. Like amylopectin, glycogen is a branched polymer of glucose molecules joined by α-1,4- and α-1,6-glycoside bonds. The major source of glycogen is the starch in the food that we eat. Starch is hydrolyzed by enzymes in the intestines to glucose. The amount of glucose formed in this way is usually more than the body needs. The excess glucose is converted to glycogen and stored in the body. Between meals and during fasting, the glycogen is hydrolyzed to glucose as needed by the body.

Cellulose is found in all plants, so it is the most abundant of all carbohydrates. It is the material used to form cell walls and other structural features of the plants. Cotton fiber and filter paper are almost entirely cellulose. Wood is about 50 percent cellulose. Cellulose, like starch, is made up of D-(+)-glucose molecules joined head to tail, as shown in Figure 21-18. But it differs from starch in two important ways. First, the 1,4-glycoside bonds in cellulose are β rather than α. Second, the molecules of D-(+)-glucose are all arranged in long chains. There are very few branches. These two differences greatly affect their properties.

The chains of cellulose can be tightly packed because they contain few branches. The arrangement allows many hydrogen bonds to be formed between hydroxy groups of adjacent chains. As a result, these chains are

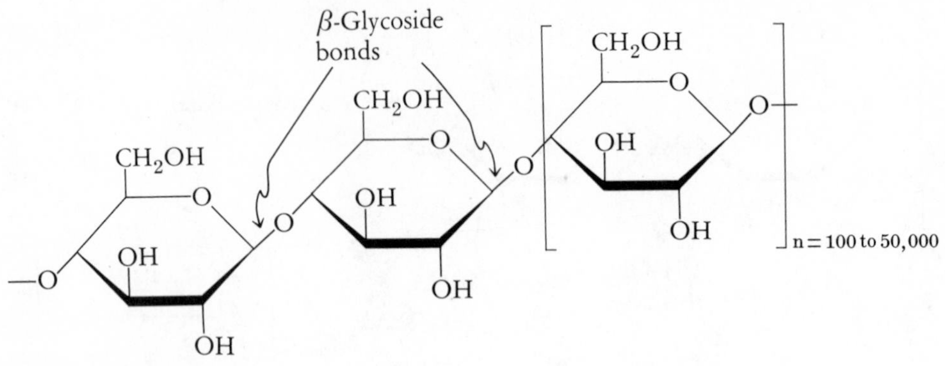

Fig. 21-18. The structure of cellulose.

quite strong and chemically stable. It is these properties of cotton that make it useful.

Humans and all other meat-eating animals are unable to use cellulose as a source of D-(+)-glucose. Our systems do not contain the necessary enzymes to hydrolyze the β-1,4-glycoside bonds of cellulose. On the other hand, many microorganisms do have these enzymes. As a result, they can digest cellulose. These microorganisms are often present in higher animals such as termites and cows, so that they too can digest cellulose.

The fact that an enzyme can recognize the difference between an α- and β-glycoside bond in a polysaccharide again shows the great stereospecificity of enzymatic reactions.

21.8 SUMMARY

Carbohydrates are polyhydroxy aldehydes or polyhydroxy ketones or substances that yield such compounds on hydrolysis. They are classified as monosaccharides, disaccharides, or polysaccharides. Monosaccharides can be further classified according to type of carbonyl group as aldoses or ketoses, or according to the number of carbon atoms as trioses, tetroses, pentoses, or hexoses. Combining these two terms classifies them according to both type of carbonyl group and number of carbon atoms (for example, aldohexoses and ketopentoses).

The open-chain form of monosaccharides is best represented by the Fischer projection formulas. Using these formulas, we can assign a particular monosaccharide to either the D or L family according to the configuration of the chiral carbon farthest from the carbonyl group. Monosaccharides exist predominantly in the cyclic acetal form. A new chiral center is created when this ring is formed, and monosaccharides can exist as α or β anomers. The optical rotation of a freshly prepared solution of a pure anomer slowly changes with time until a constant value is reached. This phenomenon is called mutarotation and is caused by the interconversion of one anomer to another through the open-chain aldehyde form as intermediate. In solution, all three forms of a monosaccharide—an open-chain form and two cyclic anomers—are in equilibrium.

The chemistry of monosaccharides involves the reactions of only two functional groups: the carbonyl group and the hydroxy group. Monosaccharides reduce Tollens' and Fehling's solutions and are therefore called reducing sugars. The carbonyl group of a monosaccharide is reduced by many reducing agents, including hydrogen (H_2) and a metal catalyst, and reacts with excess phenylhydrazine to form an osazone. The hydroxy groups of a monosaccharide react to form esters, and the hemiacetal hydroxy group reacts to form *O*-glycosides.

Disaccharides are dimers made up of two monosaccharides that are the same or different. Sucrose, maltose, and lactose are three important disaccharides found in nature. Sucrose and maltose are both α glycosides; lactose is a β glycoside. Both maltose and lactose are reducing sugars; sucrose is not.

Polysaccharides are polymers formed from monosaccharides. D-(+)-Glucose is the monomer that makes up most of the polysaccharides found in nature. It is found in starch, glycogen, and cellulose. These three polysaccharides differ in the type of glycoside bond that joins the D-(+)-glucose monomers and the amount of chain branching. Starch and glycogen are branched polymers in which the D-(+)-glucose molecules are joined by α-1,4- and α-1,6-glycoside bonds. Cellulose is a linear polymer that contains β-1,4-glycoside bonds.

Humans cannot use cellulose as a source of D-(+)-glucose because our systems do not contain the necessary enzymes to hydrolyze the β-1,4-glycoside bond. This fact illustrates the high stereospecificity of enzyme-catalyzed reactions.

REVIEW OF TERMS, CONCEPTS, AND REACTIONS

Terms

ALDITOLS The products formed by reducing the carbonyl group of a monosaccharide to an alcohol.

ALDOSE A monosaccharide that is an aldehyde.

ANOMERIC CARBON The chiral center formed when the hydroxy group at carbon 5 reacts with the carbonyl group of a monosaccharide to form a cyclic hemiacetal.

ANOMERS Isomers that differ only in the configuration at the hemiacetal or hemiketal carbon atom of the cyclic form of monosaccharides.

CARBOHYDRATES Compounds that are polyhydroxy aldehydes and polyhydroxy ketones or substances that yield such compounds on hydrolysis.

DISACCHARIDES Carbohydrates that form 2 moles of monosaccharide per mole of carbohydrate when hydrolyzed.

GLYCOSIDE The general name given to the acetal or ketal formed by the reaction of a saccharide with an alcohol or phenol.

GLYCOSIDE BOND The bond in a cyclic acetal or ketal between the anomeric carbon and the oxygen that is not part of the ring.

KETOSE A monosaccharide that is a ketone.

MONOSACCHARIDES Carbohydrates that cannot be hydrolyzed to smaller molecules.

MUTAROTATION The gradual change of the optical rotation with time of a solution of pure anomer until a constant value is reached.

N-GLYCOSIDE A glycoside in which the atom bonded to the anomeric carbon is a nitrogen.

O-GLYCOSIDE A glycoside in which the atom bonded to the anomeric carbon is an oxygen.

OSAZONE The product formed by the reaction of 3 moles of phenylhydrazine per mole of monosaccharide.

POLYSACCHARIDES A polymer made up of monosaccharides as monomers.

REDUCING SUGARS Monosaccharides and disaccharides that reduce Tollens' and Fehling's solutions.

SACCHARIDE Another name for carbohydrate.

Concepts

1. Carbohydrates contain hydroxy and carbonyl functional groups. They undergo the typical reactions of these groups. Particularly important is the intramolecular reaction between the hydroxy group at carbon 5 and the carbonyl group to form a cyclic hemiacetal or hemiketal.
2. Carbohydrates have numerous chiral centers and exist as stereoisomers.

3. Monosaccharides (and some disaccharides) exist in three forms: an open-chain form and two cyclic anomers. In solution, these three forms are in equilibrium.
4. A monosaccharide can be joined to another by a glycoside bond to form a disaccharide. The bond is usually between carbon 1 of one monosaccharide and carbon 4 of another monosaccharide.
5. D-(+)-Glucose is the monomer in most of the important naturally occurring polymers.
6. The glycoside bonds in starch and glycogen are all α; in cellulose, they are all β.

Reactions

1. *Mutarotation* (see Section 21.3)

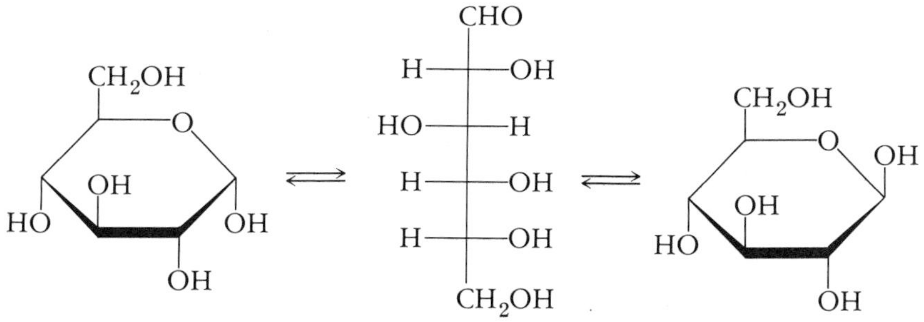

CH_2OH O OH HO OH OH ⇌ CHO H—OH HO—H H—OH H—OH CH_2OH ⇌ CH_2OH O OH OH HO OH

α-D-(+)-glucose   Open-chain form β-D-(+)-glucose

2. *Reactions of Monosaccharides* (see Section 21.5)
   a. Oxidation

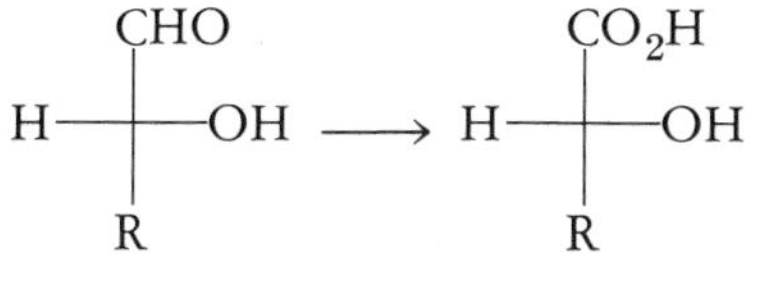

CHO, H—OH, R ⟶ CO_2H, H—OH, R   R = $HOCH_2(CHOH)_n$ n = 0, 1, 2, or 3

   b. Reduction

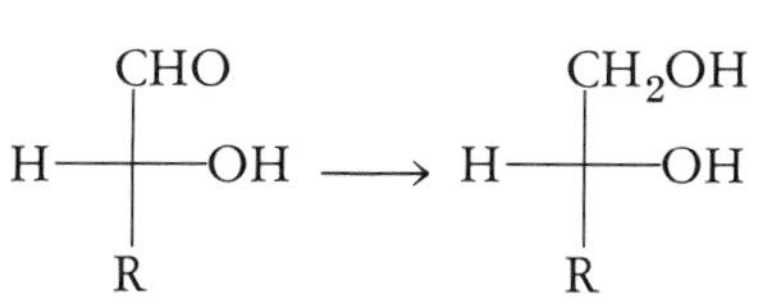

CHO, H—OH, R ⟶ CH_2OH, H—OH, R   R = $HOCH_2(CHOH)_n$ n = 0, 1, 2, or 3

c. Osazone formation

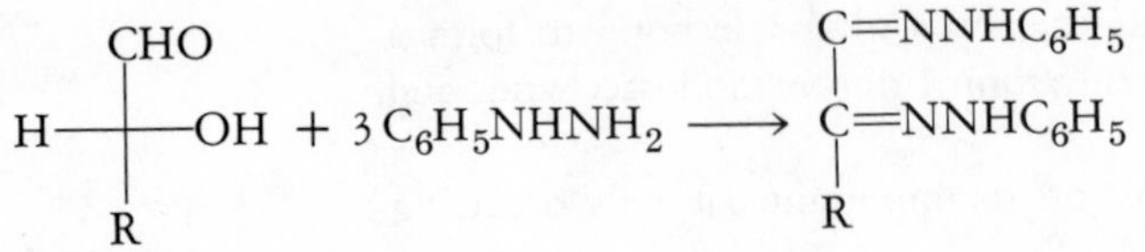

$$\begin{array}{c} CHO \\ | \\ H-C-OH \\ | \\ R \end{array} + 3\,C_6H_5NHNH_2 \longrightarrow \begin{array}{c} C{=}NNHC_6H_5 \\ | \\ C{=}NNHC_6H_5 \\ | \\ R \end{array} \qquad \begin{array}{l} R = HOCH_2(CHOH)_n \\ n = 0, 1, 2, \text{ or } 3 \end{array}$$

d. Glycoside formation

$$\text{(pyranose: } CH_2OH, OH, OH, HO, OH) + ROH \underset{}{\overset{H^+}{\rightleftharpoons}} \text{(pyranoside: } CH_2OH, OR, OH, HO, OH) + H_2O$$

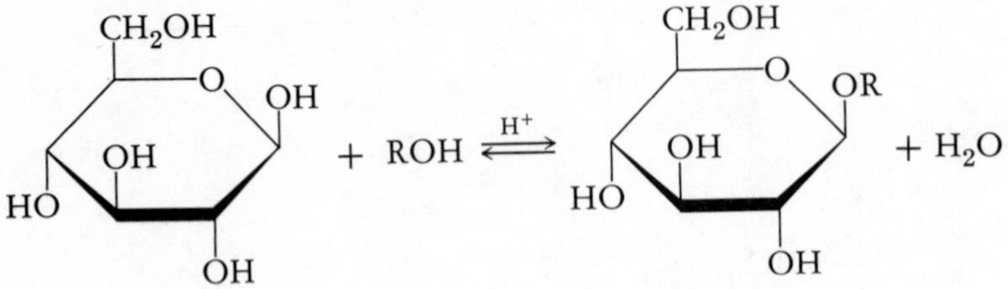

EXERCISES 21-13 Give an example of each of the following:

(a) a triose (b) an aldopentose
(c) an α anomer of a monosaccharide (d) a β-*O*-glycoside
(e) an α-*N*-glycoside (f) a disaccharide
(g) a reducing sugar (h) a nonreducing sugar

21-14 What is the difference between a glucoside and a glycoside?

21-15 Which of the following are reducing sugars?

(a) D-(+)-glyceraldehyde (b) D-(+)-glucose
(c) D-(−)-fructose (d) sucrose (e) lactose

21-16 Circle the glycoside bond in each of the following compounds:

(a) pyranose ring with substituents CH_2OH, HO, OC_2H_5, OH, OH

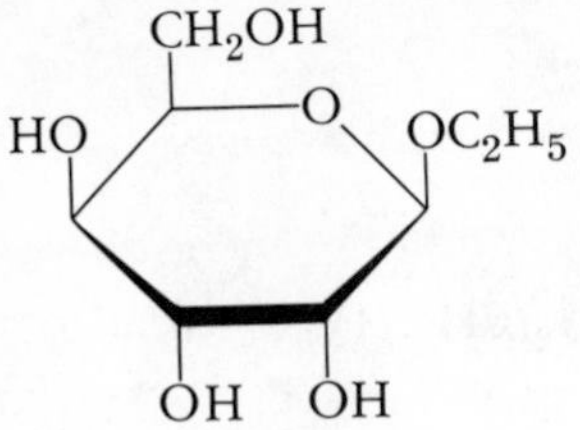

(b) pyranose ring (CH_2OH, OH, HO, OH) linked through O to a furanose ring (CH_2OH, OH, OH)

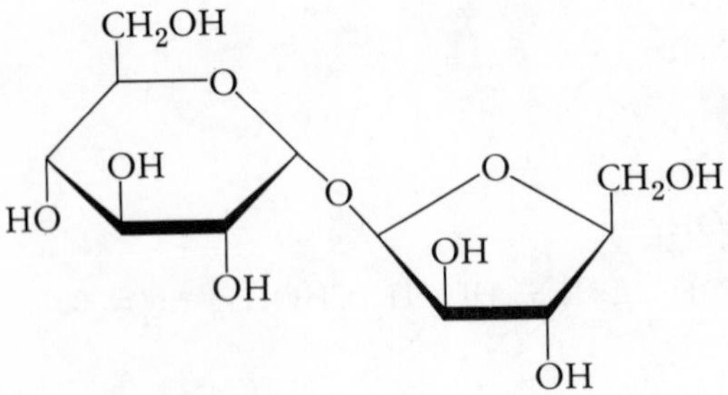

(c)

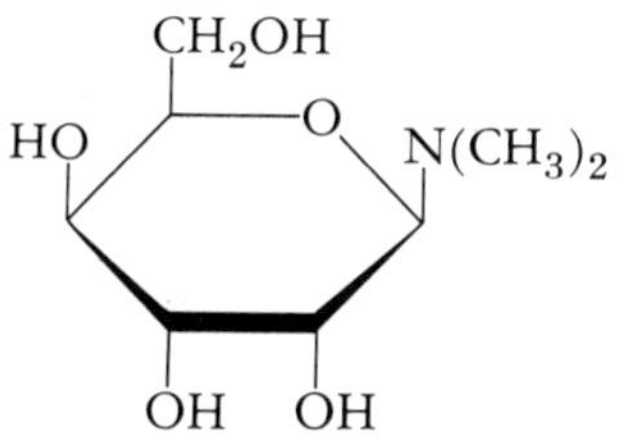

CH_2OH HO O $N(CH_3)_2$ OH OH

(d)

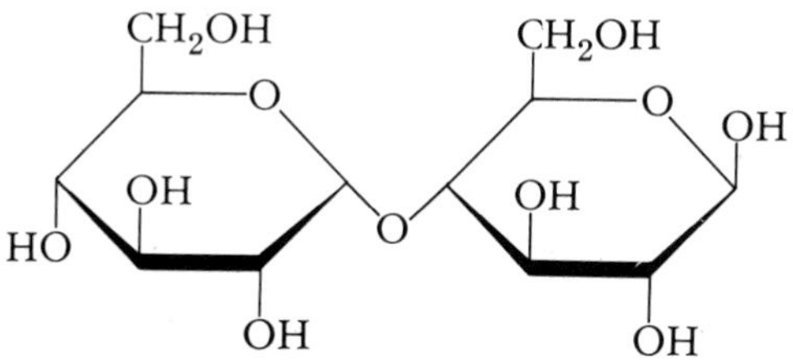

CH_2OH O OH HO OH O CH_2OH O OH OH OH

21-17 Write the structure of the osazone formed from the following compounds:
(a) D-(+)-glucose (b) D-(+)-mannose (c) D-(−)-fructose

21-18 Compare the configurations at carbons 3, 4, and 5 of the three osazones of exercise 21-17. Do they have the same or different configurations? What does this tell us about the configurations of carbons 3, 4, and 5 of D-(+)-glucose, D-(+)-mannose, and D-(−)-fructose?

21-19 Write the structure of the products formed by the reaction of D-(+)-mannose with each of the following reagents:
(a) phenylhydrazine (b) C_2H_5OH, anhydrous HCl
(c) Fehling's solution (d) acetic anhydride (e) H_2, Pt

21-20 How does starch differ from cellulose?

LIPIDS

22

The second major class of compounds of living systems that we will study are the lipids. We have classified compounds in previous chapters according to their functional groups. For example, compounds that contain a carbon-carbon double bond are classified as alkenes, carboxylic acids contain one or more $—CO_2H$ groups, and so forth. Lipids cannot be classified in this way because they do not have a characteristic functional group. Instead, lipids are classified according to their solubility. Lipids are readily soluble in nonpolar organic solvents such as diethyl ether, chloroform, or benzene but are insoluble in water. This means that when animal or plant material is crushed and mixed with these solvents, the compounds that dissolve are classified as lipids. This is one difference between lipids and the other two major classes of compounds of living systems, carbohydrates and proteins, which are insoluble in organic solvents because of their highly polar structure. Fats, vegetable oils, and waxes are examples of lipids that are familiar to everyone.

The structures of the compounds classified as lipids vary greatly, but they do have one common characteristic. A major portion of their structure is like a hydrocarbon. This is the reason that lipids are soluble in nonpolar organic solvents. Lipids have several important functions in living systems. They serve as the main energy reserve for living systems, form parts of cell membranes, and regulate the activities of cells and tissues.

Lipids have been classified in several ways. One way is to group them according to whether or not they undergo saponification. Lipids that undergo alkaline hydrolysis are called *saponifiable lipids.* Lipids that do not

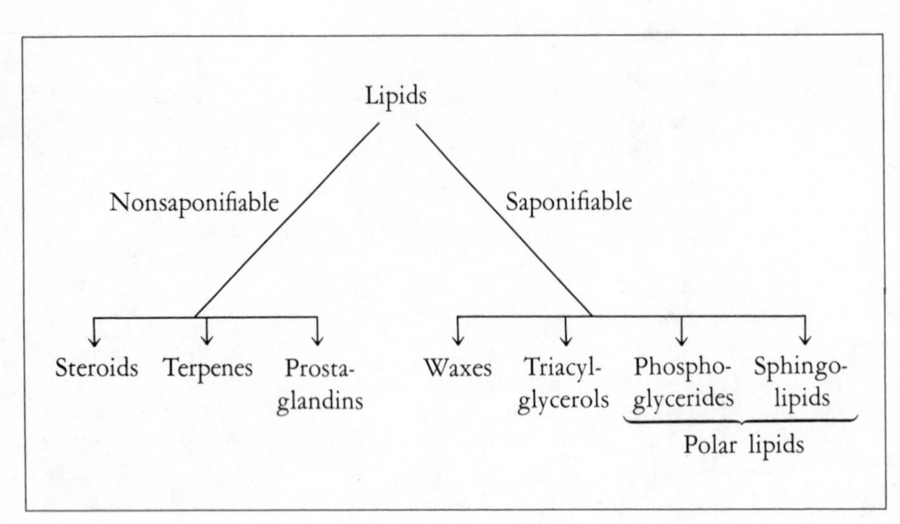

Fig. 22-1. Classification of lipids.

undergo such a reaction are called *nonsaponifiable lipids.* The lipids that we will study in this chapter are classified in this way in Figure 22-1.

Let us start our study of lipids by learning about fatty acids. They are one component of most of the saponifiable lipids.

22.1 FATTY ACIDS

Fatty acids are named for the fact that they are the major component of fats. They are carboxylic acids in which a long straight hydrocarbon chain is attached to the carboxyl group, as shown in Figure 22-2. More than 100 different fatty acids have been isolated from various lipids of animals and plants. Fatty acids differ from one another mainly in chain length and in the number and positions of double bonds in the hydrocarbon chain. Fatty acids whose hydrocarbon chain contains one or more double bonds are called *unsaturated fatty acids.* Those whose hydrocarbon chain contains no double bonds are called *saturated fatty acids.* Most fatty acids are straight-chain

$$\overbrace{CH_3CH_2CH_2CH_2CH_2CH_2CH_2CH_2CH_2CH_2CH_2CH_2CH_2CH_2CH_2}^{\text{Straight hydrocarbon chain}}\overbrace{CO_2H}^{\text{Carboxyl group}}$$

Fig. 22-2. The two parts of a fatty acid.

compounds and have an even number of carbon atoms between 14 and 22. Palmitic acid ($C_{16}H_{32}O_2$) and stearic acid ($C_{18}H_{36}O_2$) are the most common saturated fatty acids. Oleic acid ($C_{18}H_{34}O_2$) is the most common unsaturated fatty acid. Table 22-1 lists other common fatty acids. Notice that the double bonds of nearly all the naturally occurring fatty acids are in the *cis* configuration. Only a few are known that are in the *trans* configuration.

Mammals can make almost all the fatty acids they need. The major exceptions are linoleic and linolenic acids. Fatty acids that mammals require in their diets but cannot make are called *essential fatty acids.* Linoleic and linolenic acids are obtained from plants, in which they are abundant.

*EXERCISE 22-1 Write equations for two chemical reactions for each of the following fatty acids:
(a) palmitic acid (b) palmitoleic acid

Only traces of free fatty acids are found in cells. However, fatty acids are found in fats, vegetable oils, and waxes as the acyl part of esters. The simplest of these esters are the waxes. There is only one ester group in each wax molecule.

22.2 WAXES

Waxes are esters of fatty acids and long-chain monohydroxy alcohols. The acids and alcohols usually found in waxes contain an even number of carbon atoms between 26 and 34. Waxes are soft and pliable when warm, yet hard when cold. Waxes are found in both plants and animals. In plants they are found on the surfaces of stems and leaves, where they protect the plant from loss of moisture and attack by harmful insects. In animals, waxes also act as protective coatings. They are found on the surface of feathers, fur, and skin.

Carnauba wax is an example of a wax obtained from a plant. It is mostly myricyl cerotate and is widely used as floor wax and automobile wax.

$$CH_3(CH_2)_{24}C(=O)OCH_2(CH_2)_{28}CH_3$$

Myricyl cerotate
(major component of carnauba wax)

*The answers for the exercises in this chapter begin on page 892.

Table 22-1. Some Naturally Occurring Fatty Acids

Structure	No. of Carbon Atoms	IUPAC Name	Common Name	Melting Point (°C)
$CH_3(CH_2)_{10}CO_2H$	12	Dodecanoic acid	Lauric acid	44
$CH_3(CH_2)_{12}CO_2H$	14	Tetradecanoic acid	Myristic acid	54
$CH_3(CH_2)_{14}CO_2H$	16	Hexadecanoic acid	Palmitic acid	63
$CH_3(CH_2)_5CH{=}CH(CH_2)_7CO_2H$ (H, H cis)	16	*cis*-9-Hexadecenoic acid	Palmitoleic acid	−1
$CH_3(CH_2)_{16}CO_2H$	18	Octadecanoic acid	Stearic acid	70
$CH_3(CH_2)_7CH{=}CH(CH_2)_7CO_2H$ (H, H cis)	18	*cis*-9,12-Octadecenoic acid	Oleic acid	13
$CH_3(CH_2)_4CH{=}CHCH_2CH{=}CH(CH_2)_7CO_2H$ (H, H cis)	18	*cis,cis*-9,12-Octadecadienoic acid	Linoleic acid	−5
$CH_3CH_2CH{=}CHCH_2CH{=}CHCH_2CH{=}CH(CH_2)_7CO_2H$ (H, H cis)	18	*cis,cis,cis*-9,12,15-Octadecatrienoic acid	Linolenic acid	−11
$CH_3(CH_2)_{18}CO_2H$	20	Eicosanoic acid	Arachidic acid	77
$CH_3(CH_2)_{20}CO_2H$	22	Tetracosanoic acid	Lignoceric acid	86

Sometimes the alcohol part of a wax is a steroid (see Section 22.9). Lanolin is an example of such an ester that is widely used in cosmetics, creams, and ointments. It is a mixture of fatty acid esters of the steroid lanosterol.

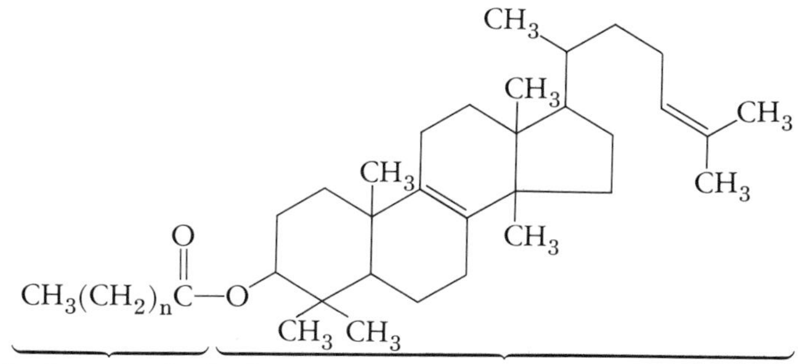

CH_3 CH_3 CH_3 CH_3 CH_3 CH_3

O

$CH_3(CH_2)_nC$—O

CH_3 CH_3

Acyl group obtained from fatty acids

Lanosterol part

Lanolin

EXERCISE 22-2 Write the structure of the wax formed by the acid-catalyzed reaction of $CH_3(CH_2)_{26}CO_2H$ and $CH_3(CH_2)_{22}CH_2OH$.

EXERCISE 22-3 Write the equations for three chemical reactions for each of the following compounds:
(a) myricyl cerotate (b) lanolin

Although waxes have only one ester group, other lipids are known that contain three ester groups in each molecule. These are the triacylglycerols.

22.3 TRIACYLGLYCEROLS (TRIGLYCERIDES*)

The name triacylglycerol accurately describes the structure of these compounds. They contain a glycerol backbone attached to three acyl groups, as shown in Figure 22-3. The acyl groups of the triacylglycerols are obtained from fatty acids. Thus, triacylglycerols are triesters formed from three fatty acids and glycerol.

* The name triglycerides was used until recently to identify these compounds. The new name, triacylglycerol, has been adopted on the recommendation of an international scientific committee.

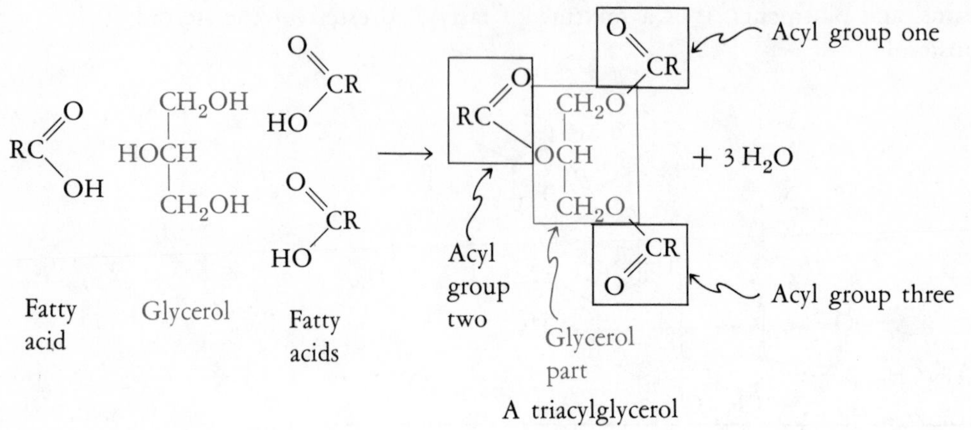

Fig. 22-3. The structure of triacylglycerols.

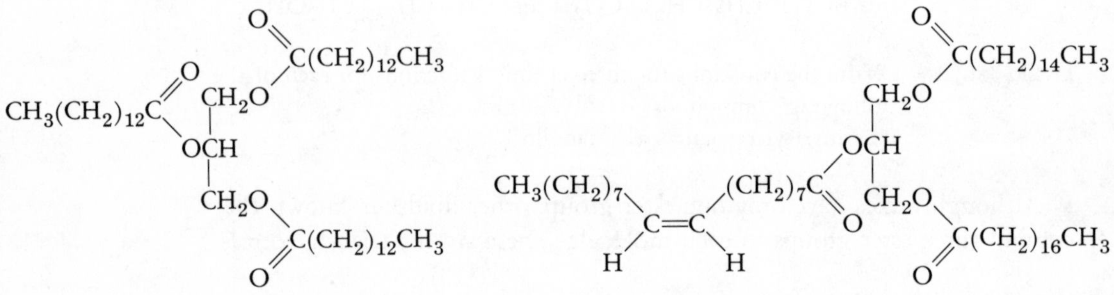

Fig. 22-4. Examples of triacylglycerols containing identical and different fatty acid parts.

Most of the lipids found in living systems are triacylglycerols. The main energy reserves are stored in this form in the cells and tissues of living systems. The acyl parts of triacylglycerols can be either identical or different, as shown in Figure 22-4. Triacylglycerols that are solid at room temperature are called fats; those that are liquid at room temperature are called oils. The major structural difference between fats and oils is that oils have acyl groups that contain double bonds. A natural fat or oil is not a pure compound, but a mixture of triacylglycerols. The difference is again in the type of acyl group in the triacylglycerols. This is illustrated in Table 22-2, where the type and percentage of each fatty acid making up a number of fats and oils are reported. It is still not understood what scientific principles determine the distribution of different fatty acids in natural triacylglycerols.

Triacylglycerols undergo the usual reactions of esters (see Section 18.5). A particularly important reaction is alkaline or enzyme-catalyzed hydrolysis. Alkaline hydrolysis (saponification) is important in the making of soaps. The digestion of fats that we eat involves an enzyme-catalyzed hydrolysis. These two reactions are shown in Figure 22-5.

Catalytic hydrogenation of the unsaturated acyl groups of the triacylglycerols in vegetable oils converts them to saturated triacylglycerols. This reaction is simply the catalytic hydrogenation of an alkene that we learned in Section 12.4. When the reaction is carried out on a vegetable oil, it is called hardening because it converts an oil into a solid vegetable fat. An example of this reaction is the formation of stearin from olein:

$$\begin{array}{l} CH_2O\text{–}C(=O)(CH_2)_7CH{=}CH(CH_2)_7CH_3 \\ |\\ CH\text{–}O\text{–}C(=O)(CH_2)_7CH{=}CH(CH_2)_7CH_3 \\ | \\ CH_2O\text{–}C(=O)(CH_2)_7CH{=}CH(CH_2)_7CH_3 \end{array} \qquad \text{Olein}$$

$$\Big\downarrow \; 3\,H_2,\ Pt$$

$$\begin{array}{l} CH_2O\text{–}C(=O)(CH_2)_{16}CH_3 \\ | \\ CH\text{–}O\text{–}C(=O)(CH_2)_{16}CH_3 \\ | \\ CH_2O\text{–}C(=O)(CH_2)_{16}CH_3 \end{array} \qquad \text{Stearin}$$

This reaction has been carried out industrially for many years to produce oleomargarine and cooking oils from inexpensive and abundant vegetable oils such as corn and soybean oils. In the industrial process, not all the double bonds are hydrogenated. The reason for this is that completely hydrogenated vegetable oils are brittle and unpalatable. Therefore, the

Table 22-2. Composition of Fats and Oils

	Composition (%)[a]						
	Saturated Fatty Acids				Unsaturated Fatty Acids		
	Lauric Acid C_{12}	Myristic Acid C_{14}	Palmitic Acid C_{16}	Stearic Acid C_{18}	Oleic Acid C_{18}	Linoleic Acid C_{18}	Linolenic Acid C_{18}
Fats							
Butter	1–3	8–14	25–29	8–13	18–29	1–2	3
Lard	—	1–2	25–30	12–17	41–60	3–8	—
Oils							
Cod Liver Oil	—	2–5	7–14	1	25–32	27–33	—
Corn Oil	—	1–2	8–12	3–4	20–50	35–60	—
Olive Oil	—	1	9	2	84	4	—
Peanut Oil	—	—	8	3	56	26	—
Coconut Oil	45–50	15–18	7–10	1–4	5–7	1	—

[a] Percentages vary with the source of the sample. Other fatty acids were not included in calculation of percentages.

amount of hydrogen added is controlled to leave some double bonds. In this way, a substance that is a creamy solid can be obtained.

The ester groups of triacylglycerols can be reduced by hydrogen to long-chain alcohols and glycerols under more vigorous conditions. This is a typical reaction of esters that we learned in Section 18.5:

$$CH_3(CH_2)_7CH{=}CH(CH_2)_7C(=O)OCH_2-CH[OC(=O)(CH_2)_{16}CH_3]-CH_2OC(=O)(CH_2)_{16}CH_3$$

(cis: both H on the same side of the C=C double bond; the upper CH_2O is drawn bonded to $C(=O)(CH_2)_{16}CH_3$ and the middle carbon drawn as OCH)

$$\xrightarrow[\text{Ni, } H_2]{\Delta \text{, Pressure}}$$

$$3\,CH_3(CH_2)_{16}CH_2OH + HOCH(CH_2OH)_2$$

(glycerol drawn as CH_2OH / HOCH / CH_2OH)

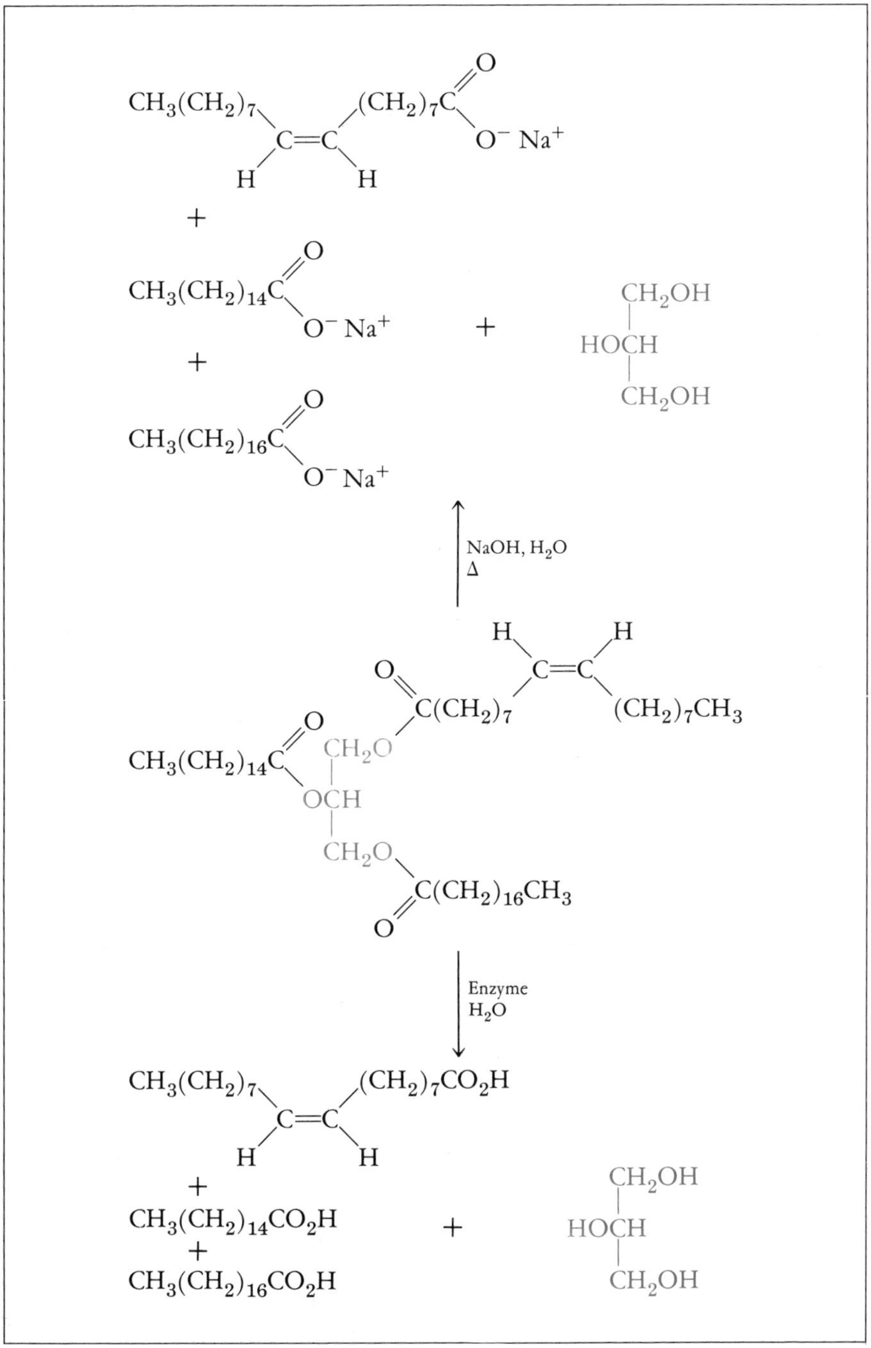

$$CH_3(CH_2)_7CH{=}CH(CH_2)_7CO_2^-\,Na^+$$

$$+$$

$$CH_3(CH_2)_{14}CO_2^-\,Na^+ \quad + \quad HOCH_2{-}CH(OH){-}CH_2OH$$

$$+$$

$$CH_3(CH_2)_{16}CO_2^-\,Na^+$$

↑ NaOH, H_2O, Δ

$$\begin{array}{l} CH_2O{-}C(O)(CH_2)_7CH{=}CH(CH_2)_7CH_3 \\ | \\ CH_3(CH_2)_{14}C(O){-}OCH \\ | \\ CH_2O{-}C(O)(CH_2)_{16}CH_3 \end{array}$$

↓ Enzyme, H_2O

$$CH_3(CH_2)_7CH{=}CH(CH_2)_7CO_2H$$

$$+$$

$$CH_3(CH_2)_{14}CO_2H \quad + \quad HOCH_2{-}CH(OH){-}CH_2OH$$

$$+$$

$$CH_3(CH_2)_{16}CO_2H$$

Fig. 22-5. The products of alkaline and enzyme-catalyzed hydrolysis of a triacylglycerol.

The long-chain alcohols are used in the manufacture of synthetic detergents.

EXERCISE 22-4 In each of the following structures, circle the glycerol part and the acyl part.

(a)

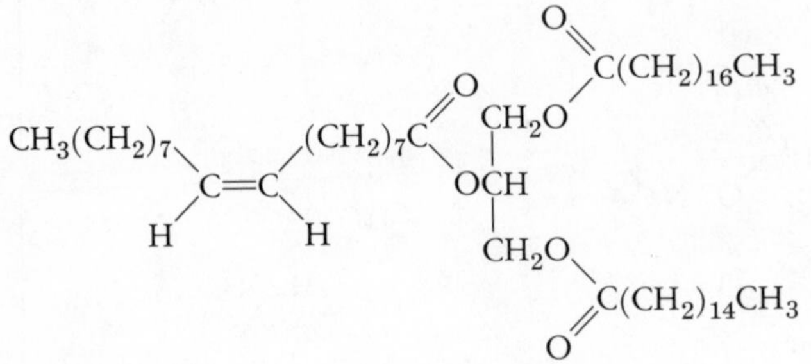

$$
\begin{array}{l}
CH_3(CH_2)_7CH{=}CH(CH_2)_7C({=}O)OCH \quad (cis;\ H, H) \\
CH_2OC({=}O)(CH_2)_{16}CH_3 \\
| \\
OCH \\
| \\
CH_2OC({=}O)(CH_2)_{14}CH_3
\end{array}
$$

(b)

$$
\begin{array}{l}
CH_2OC({=}O)(CH_2)_{12}CH_3 \\
| \\
CH_3(CH_2)_{14}C({=}O)OCH \\
| \\
CH_2OC({=}O)(CH_2)_{16}CH_3
\end{array}
$$

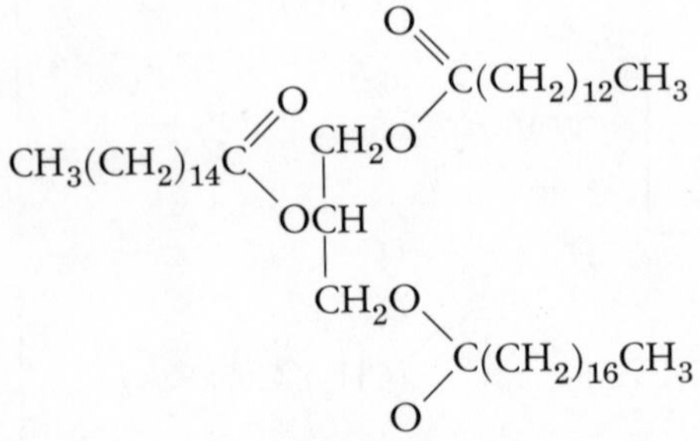

EXERCISE 22-5 Write the equation for the base- and enzyme-catalyzed hydrolysis of the compounds in exercise 22-4.

We learned in Section 16.4 that alcohols can form esters with inorganic acids such as phosphoric acid. Certain lipids have a structure in which one of the hydroxy groups of glycerol forms an ester with phosphoric acid rather than a fatty acid. This kind of compound is called a phosphoglyceride.

22.4 PHOSPHOGLYCERIDES

The general structure of phosphoglycerides is shown in Figure 22-6. Notice that phosphoglycerides are also esters of glycerol. However, they contain only two acyl groups. One of the primary hydroxy groups of glycerol forms an ester with phosphoric acid, which in turn is joined by a phosphate ester bond to an alcohol. Heating an acidic aqueous solution of phosphoglycerides hydrolyzes them to produce two fatty acids (the same or different), glycerol, phosphoric acid, and an alcohol.

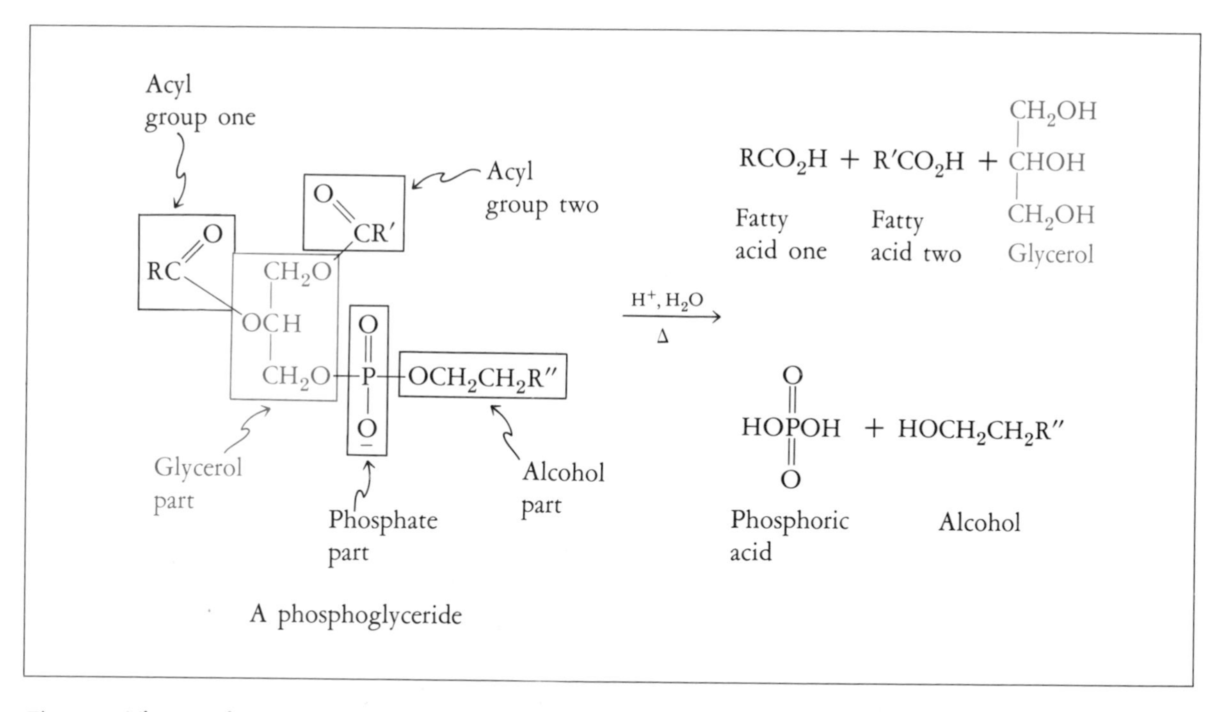

Fig. 22-6. The general structure of a phosphoglyceride and the products of its hydrolysis.

The principal phosphoglycerides in plants and animals are ethanolamine phosphoglyceride, *22-1,* choline phosphoglyceride, *22-2,* and serine phosphoglyceride, whose structures are shown in Figure 22-7. These three phosphoglycerides are a major part of most animal cell membranes.

The structures of these phosphoglycerides all have a polar head and a nonpolar hydrocarbon tail, as illustrated in Figure 22-8. This characteristic of their structure is very important to their role in forming cell membranes, as we will learn in Section 22.6. Phosphoglycerides usually have one saturated acyl group and one unsaturated acyl group bonded to the oxygen at carbons 1 and 2 of the glycerol, respectively.

EXERCISE 22-6 In each of the three structures *22-1, 22-2,* and *22-3* in Figure 22-7, circle the following:
(a) the glycerol part (b) the acyl group
(c) the phosphate part (d) the alcohol part

EXERCISE 22-7 Identify the polar and nonpolar parts of phosphoglycerides *22-1, 22-2,* and *22-3* (see Figure 22-7).

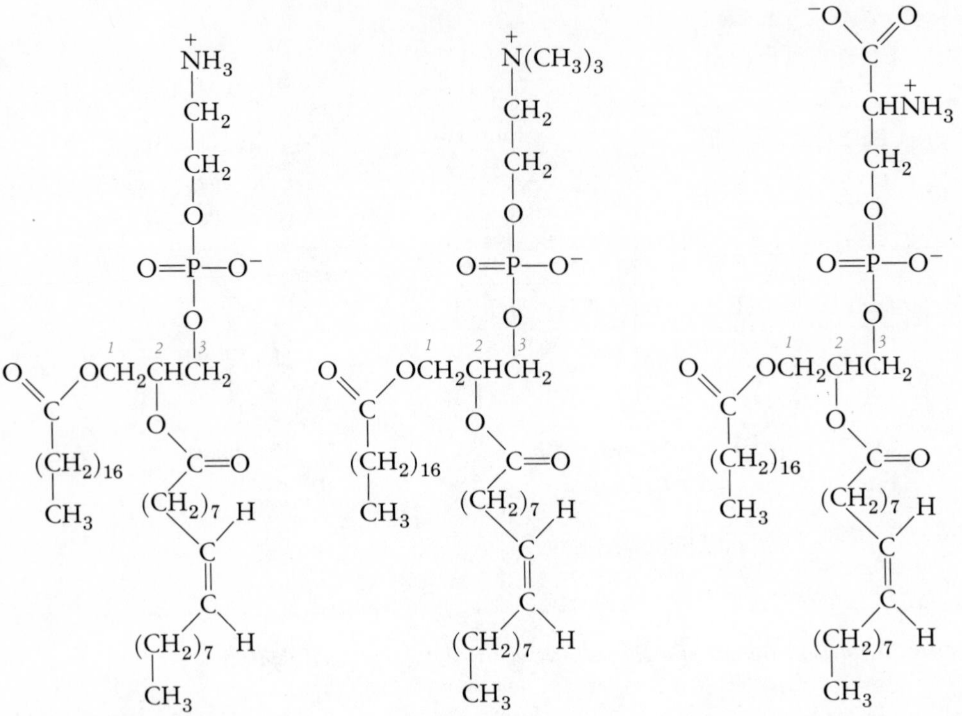

22-1
Ethanolamine phosphoglyceride

22-2
Choline phosphoglyceride

22-3
Serine phosphoglyceride

Fig. 22-7. The principal phosphoglycerides in plants and animals.

EXERCISE 22-8 Write the equation for the acid catalyzed hydrolysis of the phosphoglycerides *22-1*, *22-2*, and *22-3* (see Figure 22-7).

Other phosphorus-containing lipids that are important in cell membranes are the sphingolipids.

22.5 SPHINGOLIPIDS

Sphingolipids are complex lipids that contain sphingosine as one of their parts. These lipids are important components of the membranes of both animal and plant cells.

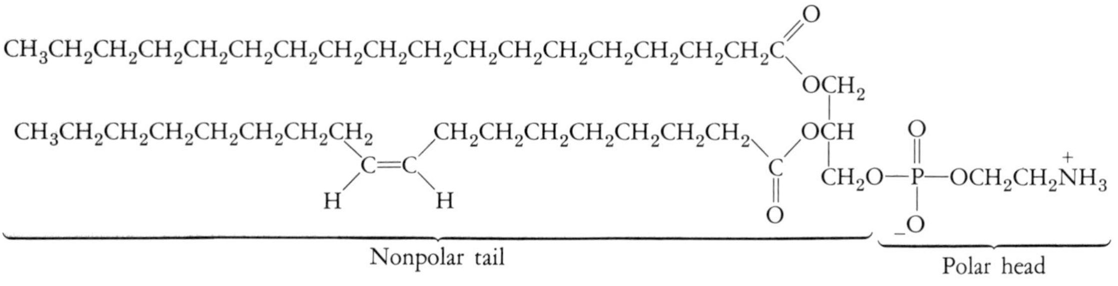

$$CH_3CH_2CH_2CH_2CH_2CH_2CH_2CH_2CH_2CH_2CH_2CH_2CH_2CH_2CH_2CH_2CH_2C(=O)\text{—}OCH_2$$

$$CH_3CH_2CH_2CH_2CH_2CH_2CH_2CH_2CH{=}CHCH_2CH_2CH_2CH_2CH_2CH_2CH_2C(=O)\text{—}OCH$$

$$CH_2O\text{—}P(=O)(O^-)\text{—}OCH_2CH_2\overset{+}{N}H_3$$

Nonpolar tail

Polar head

Fig. 22-8. The polar and nonpolar parts of phosphoglycerides.

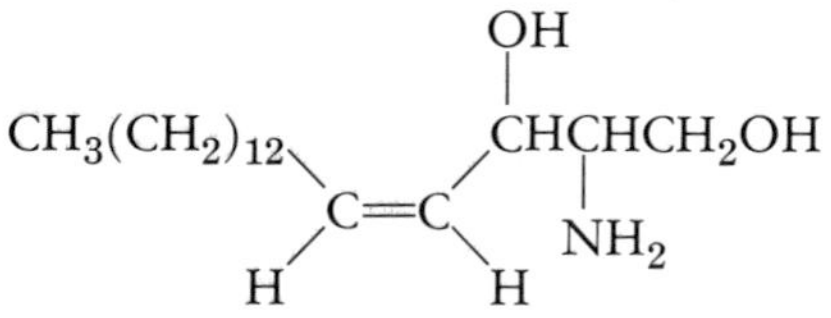

$$CH_3(CH_2)_{12}CH{=}CHCH(OH)CH(NH_2)CH_2OH$$

Sphingosine

Sphingolipids are present in large amounts in brain and nerve tissues of mammals, but only trace amounts are found in the lipid storage areas of tissues and cells.

The general structure of sphingolipids is shown in Figure 22-9. Notice that the acyl group is joined to sphingosine by an amide group. The primary

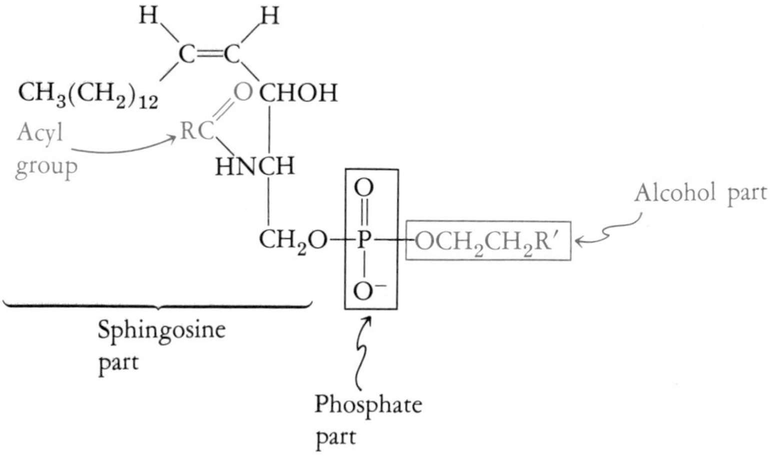

$$CH_3(CH_2)_{12}CH{=}CH\text{—}CHOH$$

$$RC(=O)\text{—}HNCH$$

$$CH_2O\text{—}P(=O)(O^-)\text{—}OCH_2CH_2R'$$

Acyl group

Alcohol part

Sphingosine part

Phosphate part

Fig. 22-9. The general structure of sphingolipids.

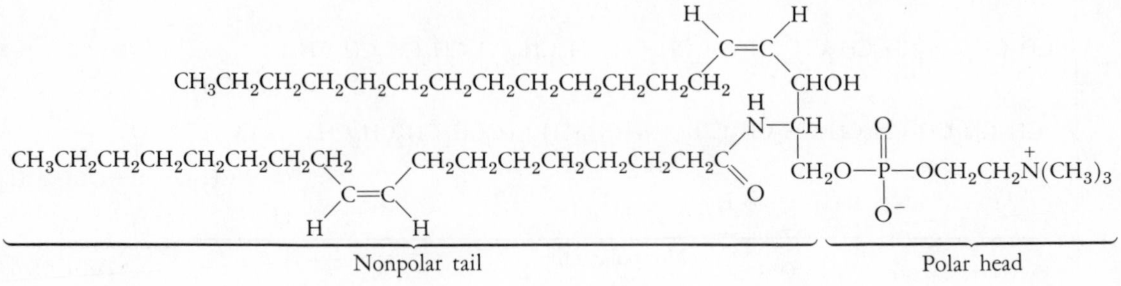

$$
\begin{array}{l}
\text{H} \quad \text{H} \\
\text{C}=\text{C} \\
CH_3CH_2CH_2CH_2CH_2CH_2CH_2CH_2CH_2CH_2CH_2CH_2CH_2 \qquad CHOH \\
\text{H} \\
\text{N}-CH \qquad \text{O} \\
CH_3CH_2CH_2CH_2CH_2CH_2CH_2CH_2 \quad CH_2CH_2CH_2CH_2CH_2CH_2CH_2C{=}O \quad CH_2O-\overset{\text{O}}{\overset{\|}{P}}-OCH_2CH_2\overset{+}{N}(CH_3)_3 \\
\text{C}=\text{C} \qquad\qquad\qquad\qquad\qquad\qquad\qquad\qquad\qquad\qquad\qquad\qquad \text{O}^- \\
\text{H} \quad \text{H}
\end{array}
$$

Nonpolar tail — Polar head

Fig. 22-10. Sphingomyelin.

hydroxy group of sphingosine forms an ester with phosphoric acid, which in turn is joined by a phosphate bond to an alcohol. Sphingomyelin, whose structure is shown in Figure 22-10, is a typical example of a sphingolipid.

Sphingolipids resemble phosphoglycerides in one important way: they both have a polar head and a nonpolar tail. As a result, they are also polar lipids. We will learn the importance of this structural feature in the next section.

22.6 BIOLOGICAL MEMBRANES

The major function of phosphorus-containing lipids in living systems is to form biological membranes. These membranes are indispensable for life. They are barriers that separate living systems into various regions where specific chemical reactions can occur that are important to the organism. For example, membranes separate cells from their surroundings and allow them to do their work. Most membranes contain about 40 percent lipids and 60 percent proteins, but there is considerable variation in these amounts, depending on the particular cell.

Although membranes form a barrier between different regions of an organism, certain molecules can pass through. For example, the nutrients of a cell and its waste products pass through the membranes. Other compounds are not allowed to pass. A membrane that is highly selective for the kinds of molecules or ions that it allows to pass is called a semipermeable membrane (see Section 8.6). In this way, membranes control the composition of the region within cells.

Membranes have other functions. They control the flow of information between cells and their surroundings. They do this by means of other compounds, called specific receptors, that are contained within the membranes. These compounds are usually proteins that undergo specific reactions. For example, the response of a particular cell to insulin, a hormone, is a chemical reaction with a specific receptor molecule in the membrane. Other chemical reactions can also occur in membranes. For example, photosynthesis, the series of chemical reactions that converts sunlight, carbon dioxide, and water into carbohydrates, occurs in the inner membranes of chloroplasts.

Phosphorus-containing lipids form membranes because they all possess an important common structural feature: a polar head and a nonpolar tail (see Sections 22.4 and 22.5). The nonpolar hydrocarbon tail is called the *hydrophobic* (water-repelling) part, and the polar head is called the *hydrophilic* (water-loving) part.

When these lipids are placed in water, only a small fraction dissolve to form a true solution. The rest form *micelles.* These are aggregates of lipids in which the hydrocarbon tails gather together to avoid the aqueous solution. In this way, they form a hydrophobic region in the center of the micelle. The hydrophilic part of the lipid forms the surface of the micelle that is in contact with aqueous surroundings. We can illustrate the structure of a micelle by using a shorthand notation for the structure of the polar lipids, as shown in Figure 22-11. The hydrophilic part is represented by a circle, and the two hydrophobic hydrocarbon tails are represented by two wavy lines.

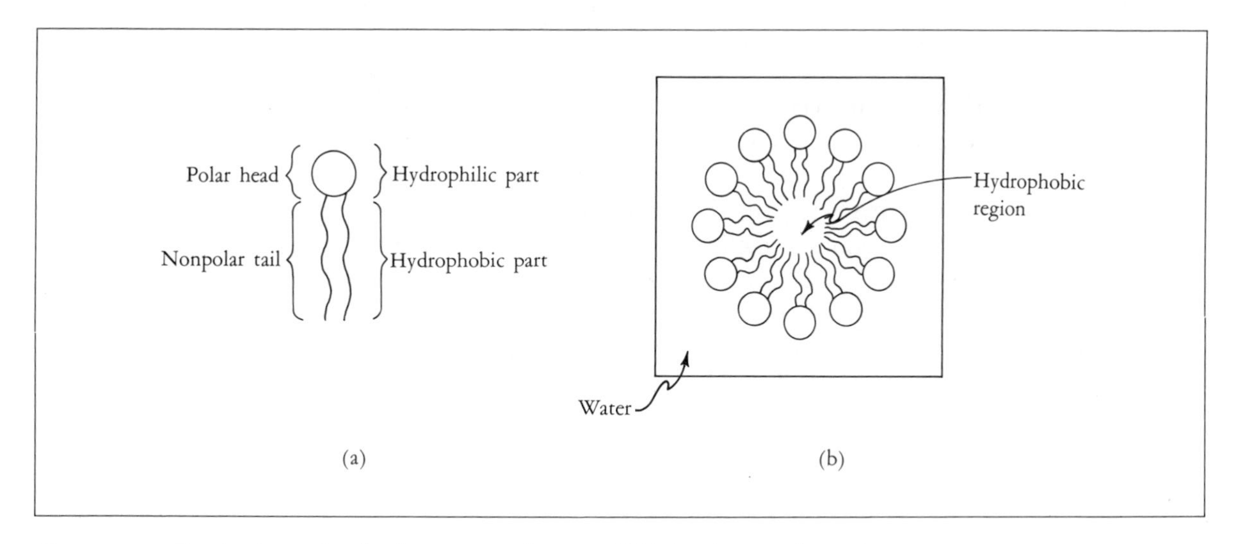

Fig. 22-11. a. Shorthand notation for the structure of a polar lipid. b. Diagram of a section of a micelle formed by polar lipid molecules.

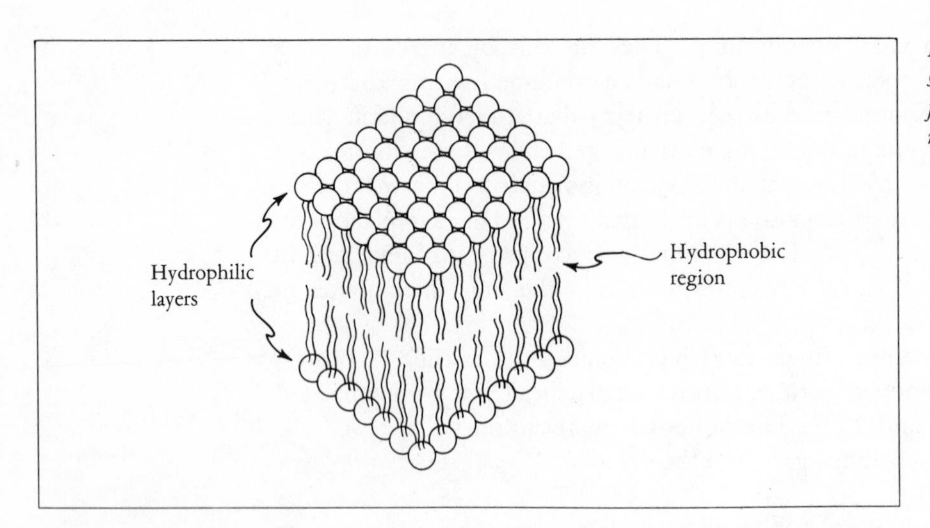

Fig. 22-12. A diagram of a section of a bimolecular sheet formed by polar lipid molecules.

Polar lipids can also form lipid bilayers, or bimolecular sheets, in addition to micelles, as shown in Figure 22-12. This arrangement is made up of two layers that are placed so that the hydrocarbon tails of each layer are adjacent to each other, forming a hydrophobic region. In this way the nonpolar tails avoid water and the polar heads interact with water. In fact, the bimolecular sheet is the favored structure for most polar lipids in aqueous solution. This preference is of great biological importance, because bimolecular sheets can grow to almost any size. Furthermore, they tend to close on themselves so that there are no ends with exposed hydrocarbon chains. In this way they enclose regions of living systems such as cells. This is the basic structure of membranes in living systems.

The formation of bimolecular sheets from polar lipids is a rapid and spontaneous process in water. The sheets form as a direct consequence of the hydrophobic and hydrophilic parts of the lipids. It is the specific interactions of these two parts with adjacent molecules that form bimolecular sheets. There are two major interactions. The first is the hydrogen bonding (see Section 6.3) between the polar heads and water molecules. The second is a van der Waals or hydrophobic attractive force (see Section 6.3) between the hydrocarbon tails in the hydrophobic region. These two interactions are cooperative. Both serve to stabilize the arrangement of polar lipid molecules in the bimolecular sheets.

Biological membranes also contain proteins that are located either completely or partially in the hydrophobic region of the bimolecular sheets, as shown in Figure 22-13. These proteins carry out the various functions of the membrane. It is believed that each type of protein has a specific function. Some transport ions or molecules across the membrane. Others act as recep-

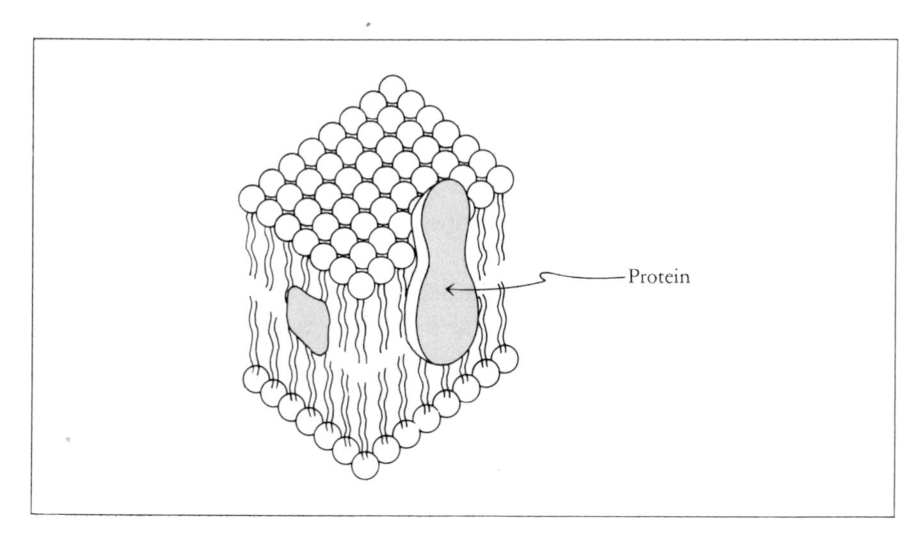

Fig. 22-13. The location of proteins in membranes.

tor sites for specific molecules that carry messages to the cell, and others are involved in the manufacture of particular compounds.

It is known that proteins carry out the functions of membranes, but scientists do not know in detail how they do it. This is one of the most active areas of research in biochemistry.

The same principle that governs the formation of micelles and bimolecular sheets is responsible for the cleansing ability of soaps.

22.7 SOAPS

Triacylglycerols are not polar lipids because they do not contain any polar groups. However, they form salts of fatty acids when they are saponified (see p. 613). These salts have a polar head and a long nonpolar hydrocarbon tail, as shown in Figure 22-14. Thus, they resemble the polar lipids. Sodium and

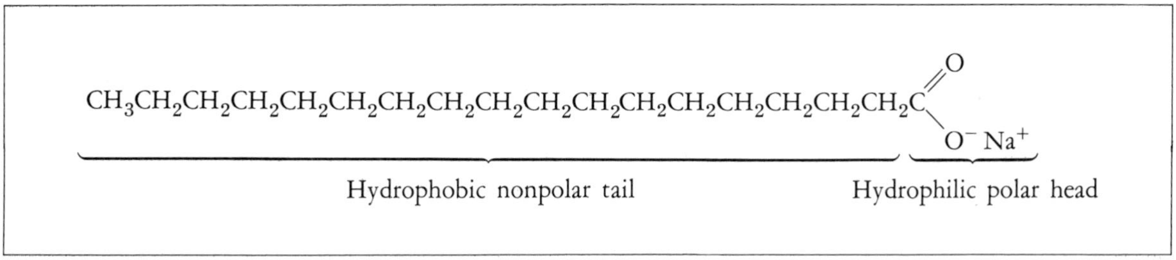

$$\underbrace{CH_3CH_2CH_2CH_2CH_2CH_2CH_2CH_2CH_2CH_2CH_2CH_2CH_2CH_2CH_2CH_2CH_2}_{\text{Hydrophobic nonpolar tail}}\underbrace{C\begin{matrix}\nearrow O \\ \searrow O^- Na^+\end{matrix}}_{\text{Hydrophilic polar head}}$$

Fig. 22-14. The polar head and nonpolar tail of a sodium salt of a fatty acid.

potassium salts of fatty acids have been used for centuries as soap. In fact, the word saponification means soap making.

Soaps are made by heating natural fats with an aqueous alkaline solution. The alkali can be sodium carbonate (Na_2CO_3), sodium hydroxide (NaOH), or even fire ashes, which are quite alkaline. After the saponification is completed, the soap is precipitated with salt. The water layer, which contains glycerol, salt, and excess alkali, is removed. The soap is then washed with water and purified. This process forms a simple soap that is used industrially. Perfumed toilet soaps and powdered and flaked laundry soaps are prepared by adding other chemicals.

A soap forms a micelle when added to water, just like polar lipids. The formation of micelles is responsible for the cleansing action of soaps. Dirt sticks to objects by a thin oil film. When the oil is removed, the solid dirt can be washed away. Soaps and other detergents remove the oil by forming a micelle around it. Their hydrocarbon tails are soluble in the oil, and the polar or hydrophilic ends compose the outer surface, as shown in Figure 22-15. The oil film forms an emulsion; that is, it exists as finely dispersed oil droplets in the aqueous solution. There is no tendency for the droplets to coagulate because of the ionic repulsion between the negative charges on the surfaces of the micelles. Consequently, the oil and the dirt are washed away in the water as small particles.

There are two disadvantages to the use of soaps for cleaning. First, they cannot be used in acidic water because the salts are converted to water-insoluble fatty acids. Second, soaps cannot be used in hard water because the calcium and magnesium ions present in hard water form precipitates. These precipitates are responsible for the insoluble soap scums that form bathtub rings and adhere to fabrics.

$$2\,C_{17}H_{35}C(=O)O^-Na^+ + Ca^{+2} \longrightarrow \left(C_{17}H_{35}C(=O)O^-\right)_2 Ca^{+2}\downarrow + 2\,Na^+$$

Water-soluble — Present in "hard" water — Water-insoluble

Water softeners, which replace calcium ions with sodium and potassium ions, and synthetic detergents were developed to solve these problems.

Synthetic detergents resemble soaps and polar lipids in that their molecules contain both polar and nonpolar regions. The most widely used syn-

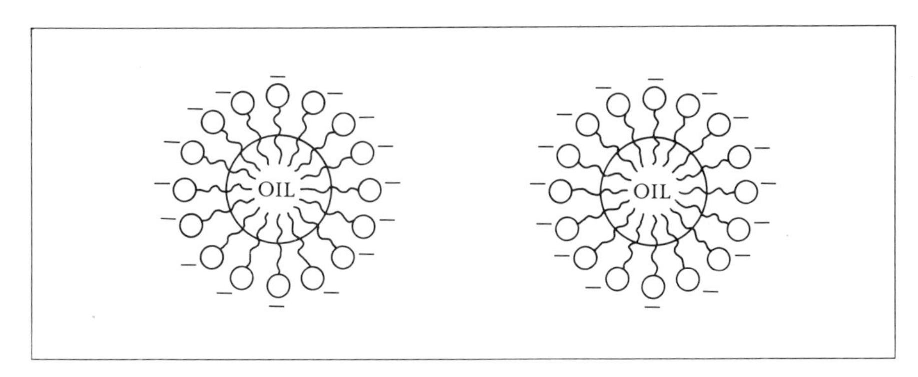

Fig. 22-15. Oil drop located in hydrophobic region of a micelle formed by soap molecules. Two micelles will not coalesce because of the repulsions between the negative charges on their surfaces.

thetic detergents are straight-chain alkyl sulfates and alkyl benzenesulfonates. The following are two examples:

$CH_3(CH_2)_{11}OSO_3^-\ Na^+$ — Sodium lauryl sulfate

$CH_3(CH_2)_{11}-C_6H_4-SO_3^-\ Na^+$ — Sodium 4-dodecyl benzenesulfonate

Their major advantage is that they do not form precipitates with the ions in hard water.

After a soap solution is used, it is thrown away. What happens to this dirty soapy water? In most cities, it eventually arrives at a sewage disposal plant, where the wastes are removed. Soaps can be removed from the water quite easily. Microorganisms present in the treatment process remove soaps and fatty acids by degrading them to smaller compounds. The first synthetic detergents used in the late 1950s presented a major problem. They passed through the sewage treatment plants unaffected. As a result, the water still contained these detergents and, when agitated, the water foamed. The problem was found to be in the structure of the synthetic detergents. The early ones contained branched alkyl chains. The microorganisms are unable to degrade such chains; they can only handle straight alkyl chains. Therefore, the early synthetic detergents were *not biodegradable.* The problem was solved by making synthetic detergents with only straight alkyl chains. Sodium lauryl sulfate and sodium 4-dodecyl benzenesulfonate are examples of *biodegradable* synthetic detergents that are degraded in the sewage treatment. Since 1966, only biodegradable synthetic detergents have been manufactured commercially.

EXERCISE 22-9 Circle the polar and nonpolar regions of the synthetic detergents sodium lauryl sulfate and sodium 4-dodecyl benzenesulfonate.

In this chapter, we have learned how nonpolar hydrophobic regions can exist in aqueous solutions. In living systems, many reactions are carried out in these hydrophobic regions. Let us now turn our attention to some of the nonpolar compounds that undergo important reactions in these regions.

22.8 TERPENES

Terpenes were introduced in Section 12.7. Terpenes contain carbon-carbon double bonds and are made up of two or more isoprene units. They are designated according to the number of isoprene units in the molecule, as shown in Table 22-3. The isoprene units are usually joined in a head-to-tail arrangement. In some cases, however, the tails of two isoprene units are joined. These two ways of joining isoprene units are illustrated in Figure 22-16. The dashed line indicates where the isoprene units are joined.

Many of the large number of monoterpenes identified in plants have characteristic odors or flavors. For example, limonene and α-pinene are the major components of lemon oil and turpentine, respectively:

CH_3 CH_3 CH_2 CH_3 CH_3 CH_3

Limonene α-Pinene

Table 22-3. Designation of Terpenes

No. of Carbon Atoms	No. of Isoprene Units	Designation
C_{10}	2	Monoterpenes
C_{15}	3	Sesquiterpenes
C_{20}	4	Diterpenes
C_{30}	6	Triterpenes
C_{40}	8	Tetraterpenes

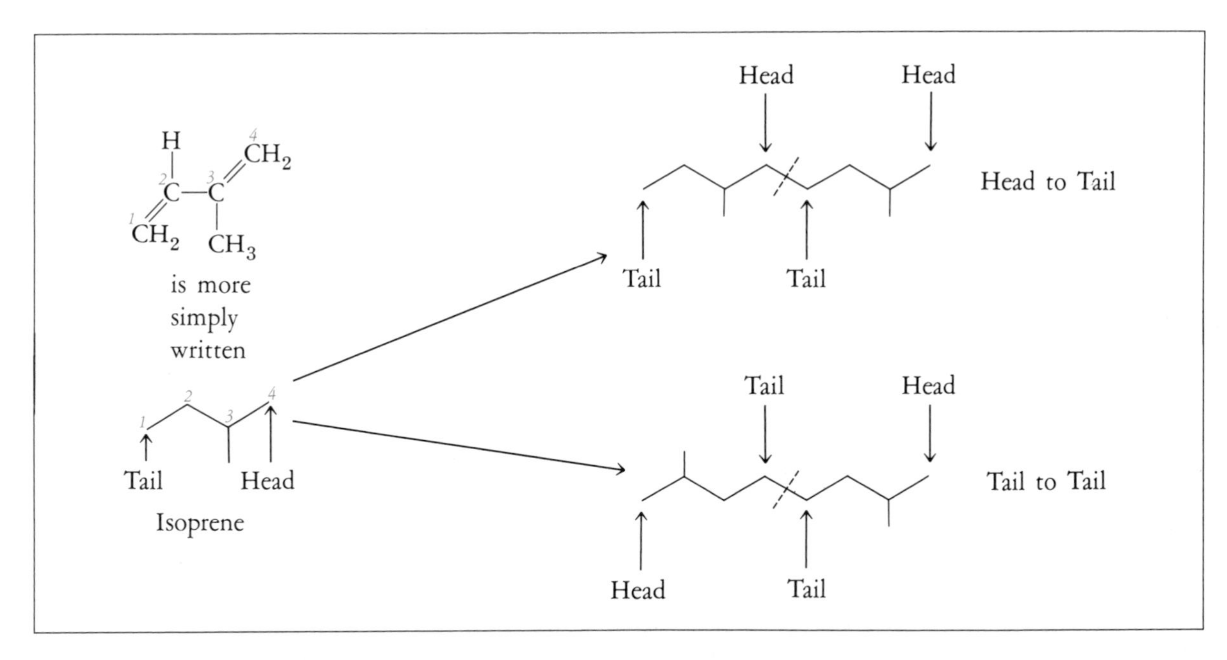

Fig. 22-16. Two ways of joining isoprene units to make terpene. The head-to-tail arrangement is the more common. The dashed line indicates where the isoprene units are joined.

Squalene (see Section 12.7) is an important triterpene, and the carotenoids are important tetraterpenes. In carotenoids, a tail-to-tail arrangement is present at the center of the molecule. An important carotenoid is β-carotene, the hydrocarbon precursor of vitamin A:

CH_3 CH_3 CH_3 CH_3 CH_3 CH_3 CH_3 CH_3 CH_3 CH_3

Tail-to-tail junction

β-Carotene

Carotenoids are the compounds that give the color to many yellow and green vegetables such as spinach, lettuce, and carrots. As we learned in Chapters 12 and 20, natural rubber is a polyterpene that contains hundreds of isoprene units in a linear head-to-tail arrangement.

Many compounds that are classified as terpenes also have functional groups containing oxygen. Among the most important of these compounds are the fat-soluble vitamins A, D, K, and E. Although they have been widely used in human nutrition and have been studied intensively, the way these compounds react at the molecular level in living systems is poorly understood.

Vitamin A, or retinol, is a tetraterpene that also contains a primary hydroxy group:

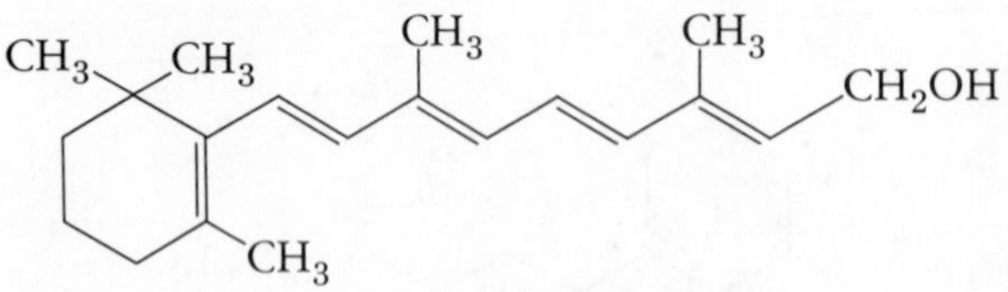

Vitamin A
(retinol)

Vitamin A is obtained from cod liver oil and animal liver. Vitamin A is formed in living systems from the carotenoids. Notice the similarity between the structures of vitamin A and β-carotene. Breaking the central double bond of β-carotene forms the correct carbon structure for two molecules of vitamin A. Transformation of the carbons of the initial double bond into alcohols forms vitamin A. This occurs in living systems by a sequence of enzyme-catalyzed chemical reactions.

The function of vitamin D is to promote the absorption of calcium ions from the intestines into the blood. Once in the bloodstream, calcium ions can travel to where they are needed—for example, for incorporation into the bone. Lack of vitamin D leads to the disease called *rickets.* The growth of bones is retarded in this disease as a result of lack of calcium ions. Several compounds are known to have vitamin D activity. The most important are vitamins D_2 (ergocalciferol) and vitamin D_3 (cholecalciferol):

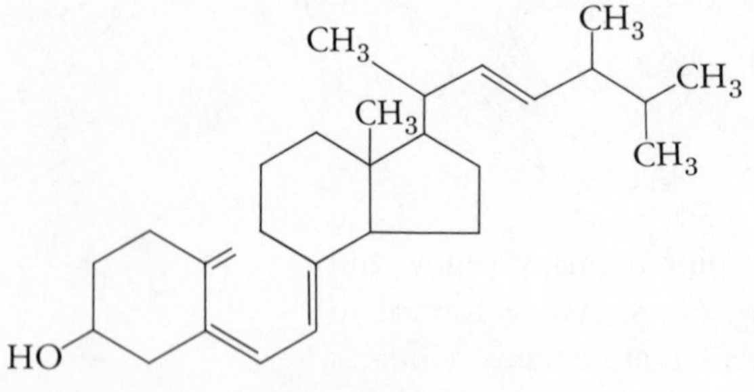

Vitamin D_2
(ergocalciferol)

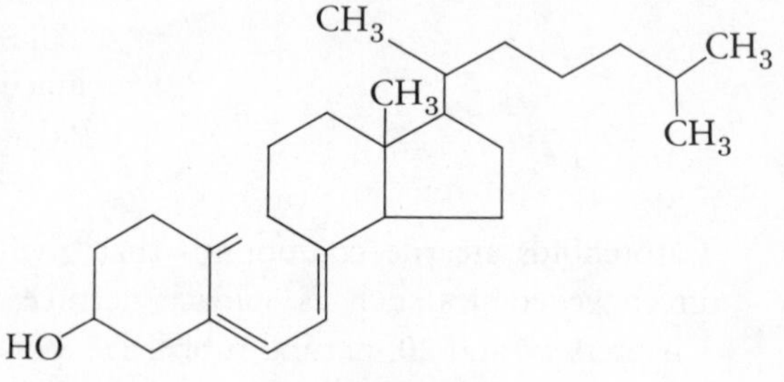

Vitamin D_3
(cholecalciferol)

Most natural foods contain little vitamin D, but it is easily prepared in humans by the action of sunlight on steroids present in the skin. People on a normal diet usually obtain vitamin D in this way.

Vitamin E is found in wheat germ. Several compounds have vitamin E properties, but the most abundant and active is α-tocopherol:

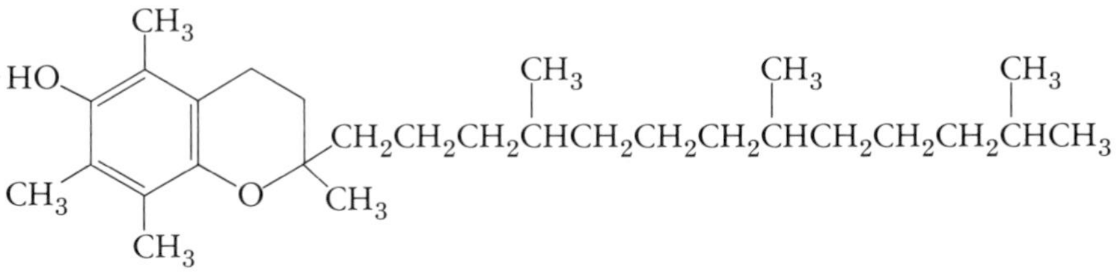

HO, CH_3, CH_3, CH_3 (ring substituents); $CH_2CH_2CH_2CH(CH_3)CH_2CH_2CH_2CH(CH_3)CH_2CH_2CH_2CH(CH_3)CH_3$; CH_3

Vitamin E
(α-Tocopherol)

The biological function of vitamin E is still obscure. One reason is that the lack of vitamin E produces several symptoms. It produces infertility in male and female rats and wasting of the skeletal muscles in guinea pigs. It is not known whether the lack of α-tocopherol causes infertility in humans.

Vitamin K is a quinone with a side chain made up of several isoprene units:

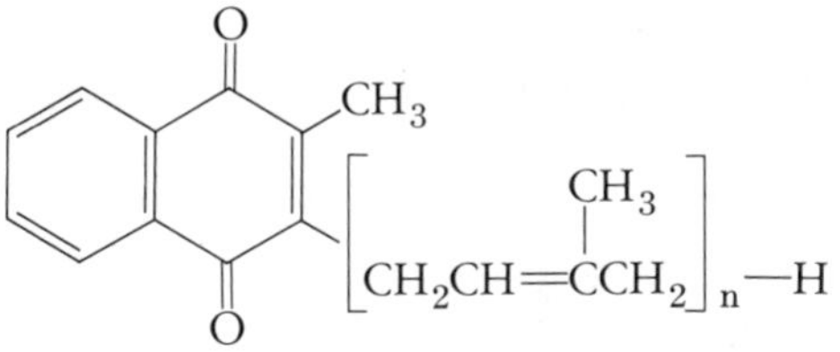

O, O, CH_3; $[CH_2CH{=}C(CH_3)CH_2]_n{-}H$

n = 6–10, depending on source

Vitamin K

The length of the side chain varies, depending on the source of the vitamin. Vitamin K is involved in some way in the blood-clotting mechanism. Vitamin K is formed by most plants and many microorganisms. Furthermore, it is found in the tissues of all organisms. For these reasons, it is believed that vitamin K has a more general biological activity than its role as a factor in blood clotting.

Terpenes are used in living systems as the starting materials for building many important compounds. Steroids are one such group of compounds. Let us turn our attention to these compounds.

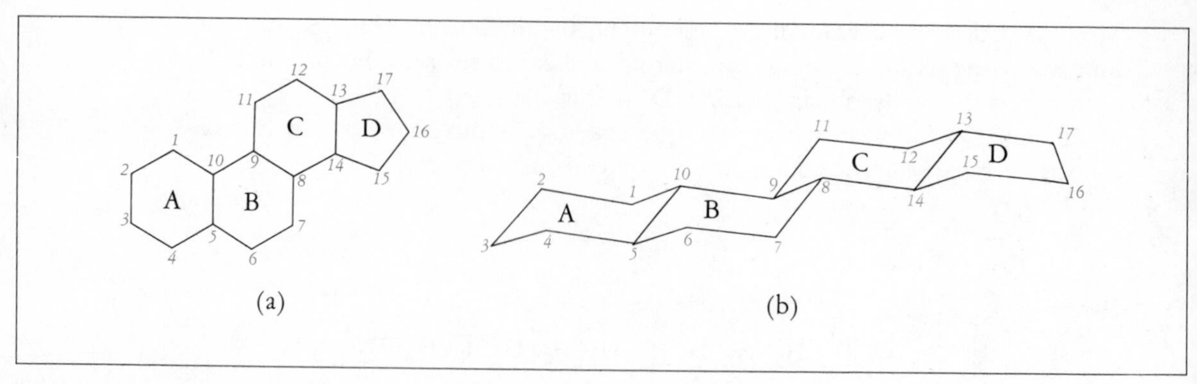

Fig. 22-17. Two representations of the steroid nucleus: (a) planar representation; (b) three-dimensional representation.

22.9 STEROIDS

A great many different steroids have been isolated from natural sources. They all have a common nucleus made up of four fused rings. Three of these contain six carbon atoms, and the other contains five carbon atoms. The rings are designated by capital letters, A, B, C, and D and the carbons are numbered as shown in Figure 22-17. Two representations of the steroid nucleus are given in Figure 22-17. One is the planar representation that is usually used because it is convenient. The other represents the three-dimensional structure of the four rings. Although individual steroids differ in the type and location of substitutions, we can make certain generalizations. Most steroids have methyl groups at carbons 10 and 13 and have substituents at carbons 3 and 17.

The steroids are formed in living systems by the cyclization of squalene 2,3-epoxide formed from the triterpene squalene, as shown in Figure 22-18. This extraordinary reaction clearly establishes the relationship between terpenes and steroids. Lanosterol is the first of the steroids formed, and it is converted in living systems to many important steroids. One of the most important is cholesterol, whose structure is shown in Figure 22-19.

Cholesterol is the best known and most abundant steroid in the human body. It is a part of biological membranes, and from it are made bile acids, steroid hormones, and vitamin D. Gallstones contain large amounts of cholesterol. Cholesterol is also found in the central nervous system.

Cholesterol is most commonly known because of the suspected link between its level in the blood of humans and certain types of heart disease such as *atherosclerosis.* This disease results from the build-up of excess cholesterol on the inner surfaces of the arteries, which decreases the diameter of

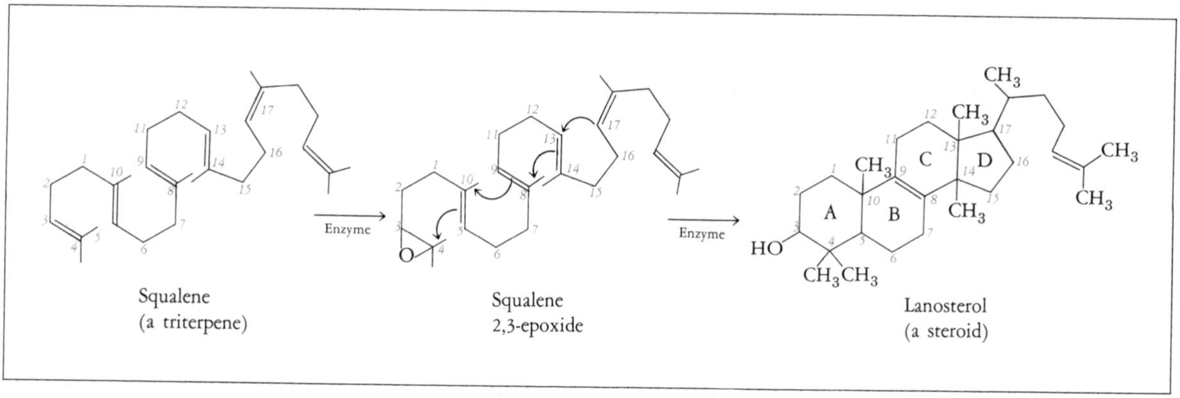

Squalene
(a triterpene)

Enzyme

Squalene
2,3-epoxide

Enzyme

CH_3 CH_3 CH_3 CH_3 A B C D HO CH_3CH_3 CH_3 CH_3

Lanosterol
(a steroid)

Fig. 22-18. The formation of the steroid lanosterol by the enzyme catalyzed cyclization of squalene 2,3-epoxide. The numbers of the carbon atoms of squalene and squalene 2,3-epoxide are based on the numbering system of lanosterol. They are given in this way to show where the carbons of squalene and squalene 2,3-epoxide end up in the lanosterol molecule. The numbering system of squalene and squalene 2,3-epoxide used in this figure are not correct based on the IUPAC rules given in Chapter 12.

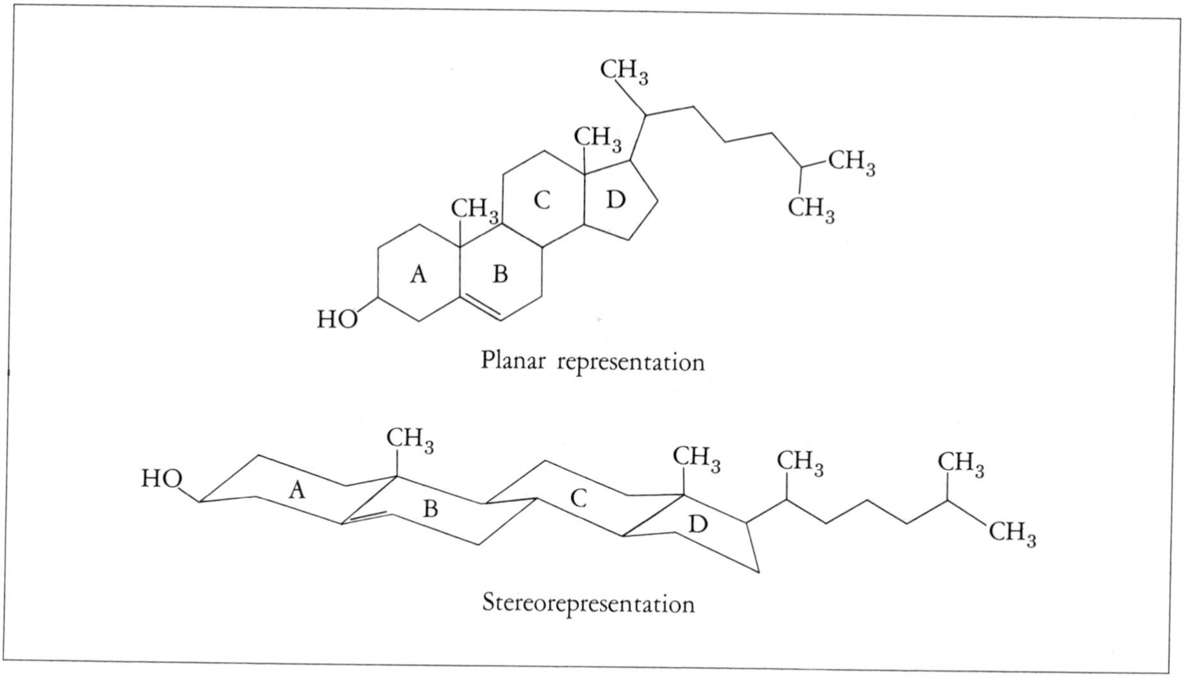

CH_3 CH_3 CH_3 CH_3 CH_3 A B C D HO

Planar representation

HO CH_3 A B C D CH_3 CH_3 CH_3 CH_3

Stereorepresentation

Fig. 22-19. The structure of cholesterol.

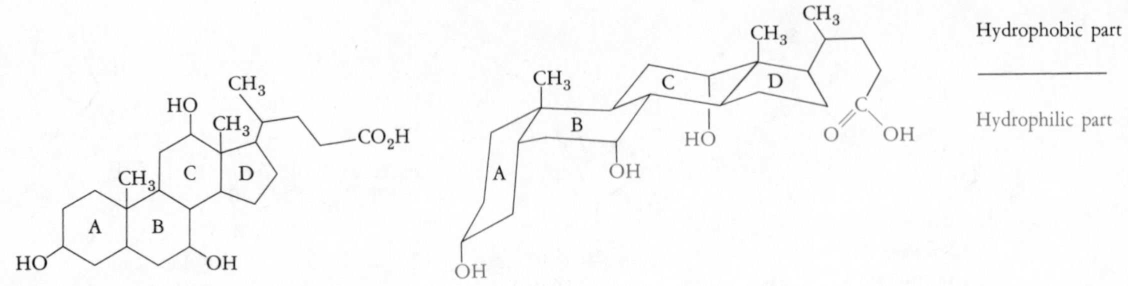

Fig. 22-20. Two representations of the structure of cholic acid.

the blood vessels, through which the blood must flow. The result is higher blood pressure and a greater risk of blood clot formation. To combat this disease, patients with high blood levels of cholesterol are placed on special low-fat diets to restrict their intake of cholesterol. Blood cholesterol levels do respond to diet. People on low-fat diets have lower blood cholesterol levels than do those on high-fat diets. Low-fat diets seem to help prevent this type of heart disease.

In the liver, cholesterol is transformed into bile acids. These compounds are stored in the gallbladder and are supplied to the small intestines on demand. The most abundant of the bile acids is cholic acid. Bile acids have two major functions. First, because they are made from cholesterol, they are a path for the elimination of cholesterol from the body. Second, they aid in the absorption of fats through the walls of the intestines. Cholic acid functions just like a detergent. It acts in this way because it has polar and nonpolar regions, as shown in Figure 22-20. The carboxylic acid group and the three hydroxy groups are all located on one side of the molecule, making that side hydrophilic. The other side of the molecule has only hydrocarbon groups and is therefore hydrophobic. Because of this structure, cholic acid can form micelles that aid in the absorption of lipids.

Many compounds with steroid nuclei are involved in the communication network between various parts of living systems. These compounds are called steroid hormones, and we will study a few examples in the next section.

22.10 STEROID HORMONES

The various parts of a multicellular organism must communicate with each other to survive. There are two methods of communication. One is the central nervous system. The other is by means of compounds called *hormones,* which act as messengers. Hormones are produced in one organ or gland and are transported by the blood to a specific location, where they exert their regulatory functions. These hormones are released only when needed and then in only very small amounts. They are very specific in their actions. They react only with specific receptors on target cells or target tissues. We will consider two major classes of steroid hormones in this section: the adrenocortical hormones and the sex hormones.

The cortex of the adrenal glands of mammals produces more than 30 steroid hormones that are essential to life. These are the adrenocortical hormones, and they have important roles in regulating physiological processes.

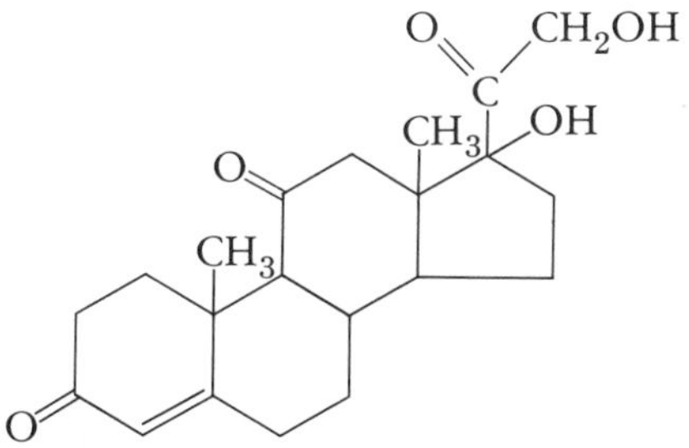

Cortisone

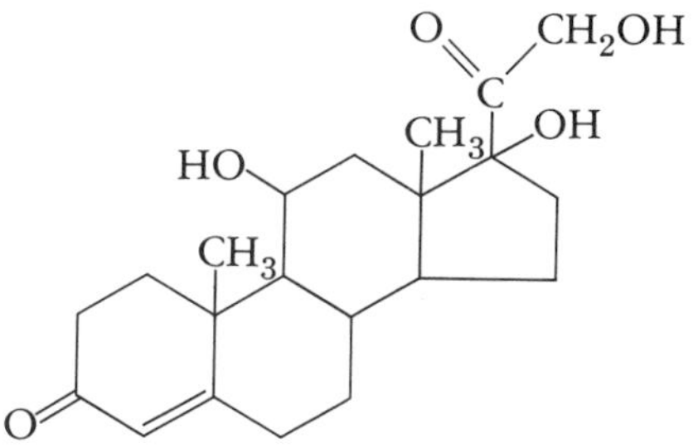

Cortisol

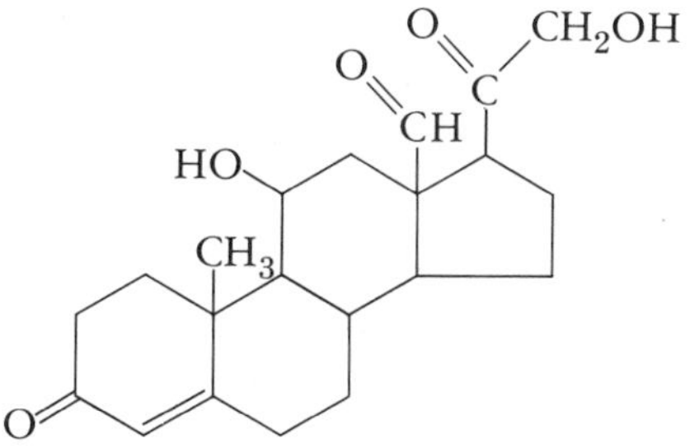

Aldosterone

For example, cortisol controls the glucose and carbohydrate balance in mammals. Aldosterone maintains the proper balance of ions (principally Na^+ and K^+) in body fluids. Cortisone is used in the treatment of rheumatic and other inflammatory diseases.

Sex hormones are compounds that are secreted by the testes of males and the ovaries of females. Their major functions are to promote the secondary sex characteristics. *Testosterone* is the male hormone. It is responsible for the facial hair and the deep voice of males.

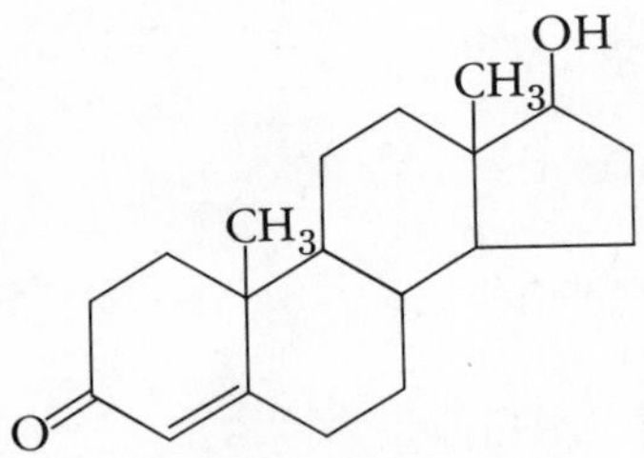

OH
CH_3
CH_3
O
Testosterone

Progesterone and the *estrogens* are the important sex hormones of females. The most important of the estrogens are estradiol and estrone. The levels of these hormones vary in the female body during the menstrual cycle.

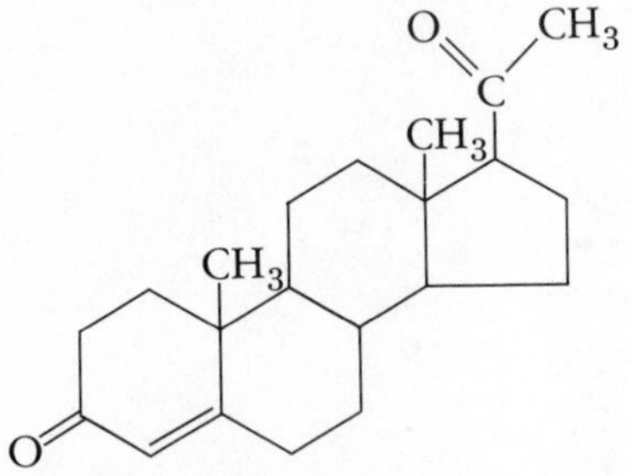

O CH_3
C
CH_3
CH_3
O
Progesterone

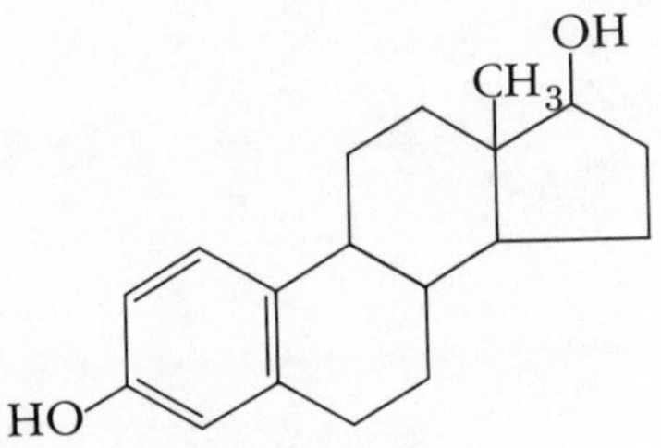

OH
CH_3
HO
Estradiol

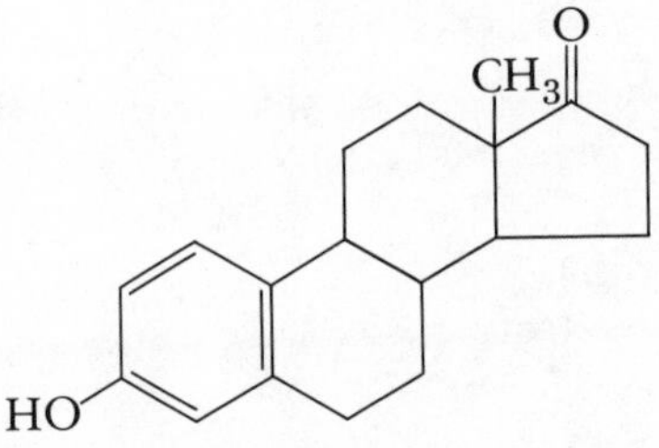

O
CH_3
HO
Estrone

During the last six months of pregnancy, the estrogens and progesterone are produced in large quantities. The high levels of these hormones insure that no additional eggs are released from the ovaries during pregnancy. The discovery of this fact led to the development of oral contraceptive pills. The idea is to maintain a high enough level of estrogens and progesterone to deceive the female body into thinking that it is pregnant. As a result, egg follicles are not developed and no egg is released into the fallopian tube.

However, progesterone is effective only when injected. Consequently, synthetic steroids have been prepared that can be taken orally and that function in the female body like progesterone. Among the numerous compounds used as oral contraceptives are mestranol, norethynodrel, and norethindrone:

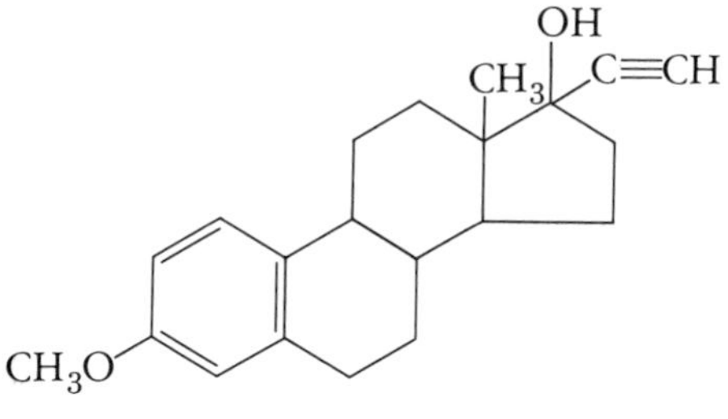

Mestranol

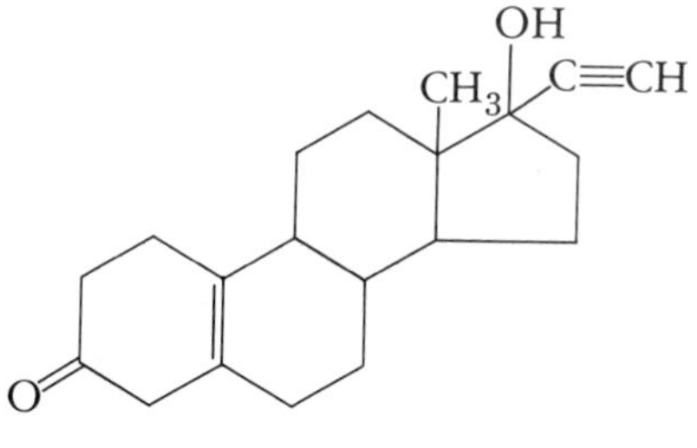

Norethynodrel

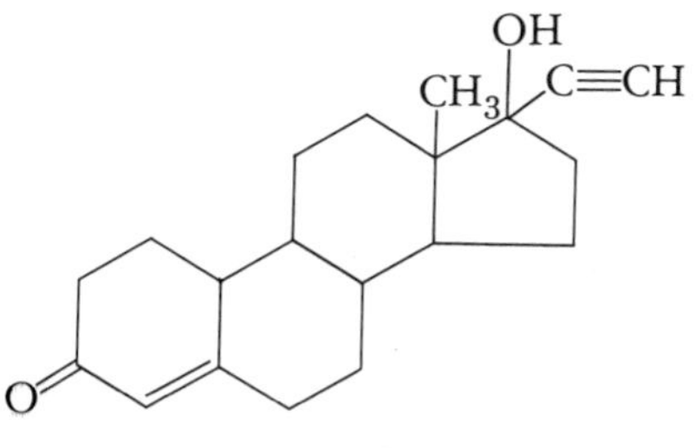

Norethindrone

Notice the similarity in structure among most of these hormones. This similarity is due to the fact that they are all obtained from cholesterol in living systems. Perhaps the most striking similarity is that between the structure of testosterone and estradiol. They have only slight structural differences, yet their physiological differences are enormous.

EXERCISE 22-10 Identify the steroid nucleus in the following compounds:
(a) testosterone (b) mestranol (c) estrone

EXERCISE 22-11 List the differences in chemical structure of each of the following pairs of compounds:
(a) testosterone and estradiol (b) cholesterol and progesterone
(c) norethynodrel and norethindrone

Recently, another group of compounds called prostaglandins have been discovered to have potential use as birth control agents.

22.11 PROSTAGLANDINS

Prostaglandins have very intense physiological activities. These compounds are found in extremely minute amounts in a variety of body tissues and fluids, including seminal and menstrual fluids. Despite their low concentrations, they have a wide range of functions. They affect blood pressure, heart rate, the menstrual cycle, and fertility; they relieve inflammation, asthma, and nasal congestion.

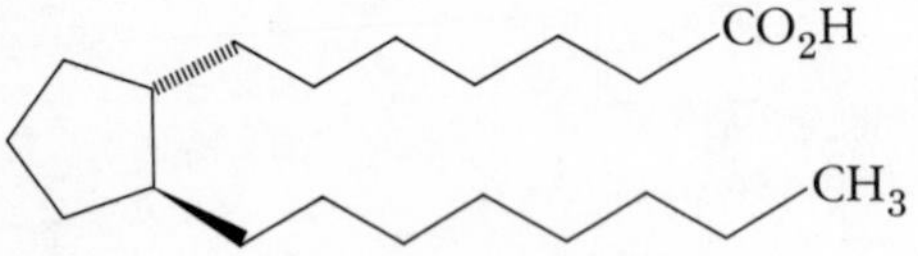

Prostanoic acid

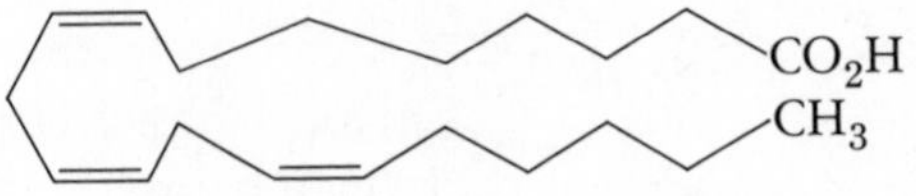

Homo-γ-linolenic acid

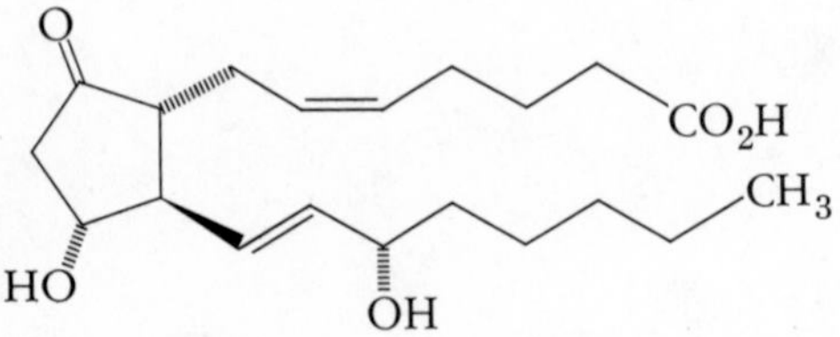

Prostaglandin E_2

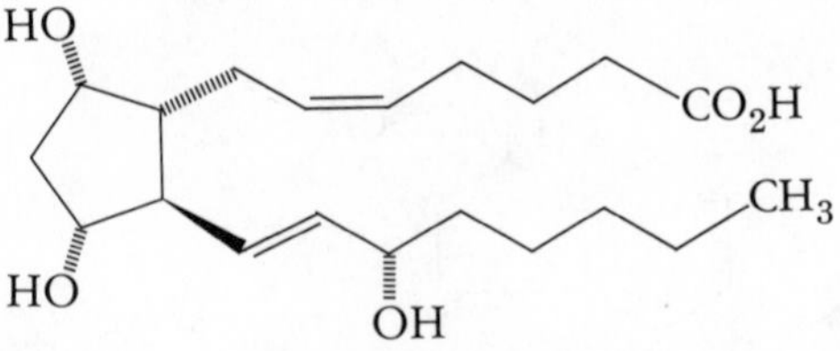

Prostaglandin $F_{2\alpha}$

Prostaglandins are made in cells from 20-carbon unsaturated fatty acids such as homo-γ-linolenic acid. Their close relationship can be seen by comparing their structures. Two prostaglandins, E_2 and $F_{2\alpha}$, have been found to induce abortions when given intravenously. This discovery has led to the possibility that these compounds can be used as once-a-month birth control agents.

Prostaglandins hold much promise as therapeutic drugs. The major problem is that the natural prostaglandins cannot be taken orally because they are rapidly degraded in humans; they do not survive long enough to carry out their functions. Much work is currently underway to develop active derivatives of prostaglandins that can be taken orally.

22.12 SUMMARY

Lipids are the parts of cells that are soluble in nonpolar organic solvents. Their structures vary widely, but in all cases a major portion of their structure is like a hydrocarbon. Lipids can be divided into two classes, saponifiable and nonsaponifiable.

Saponifiable lipids contain fatty acids, usually with an even number of carbon atoms (14 to 22). Triacylglycerols are esters made up of three fatty acids and the alcohol glycerol. Phosphoglycerides are also esters of glycerol made up of two fatty acids, but one of the hydroxy groups forms an ester with phosphoric acid rather than a fatty acid. Sphingolipids contain no glycerol, but they have two long hydrocarbon chains. One is a fatty acid, and the other is sphingosine, a long-chain aliphatic amino alcohol. Polar lipids, like phosphoglycerides and sphingolipids, spontaneously form micelles and bimolecular sheets in aqueous solutions.

The nonsaponifiable lipids include terpenes, steroids, and prostaglandins. Terpenes are linear or cyclic compounds that are made up of two or more isoprene units. Steroids are formed in living systems from the terpene squalene. The most abundant steroid in animal tissue is cholesterol. From it are made bile acids, adrenocortical hormones, and sex hormones.

Lipids are the main energy reserve of living systems, form parts of biological membranes, and regulate many activities of cells and tissues.

REVIEW OF TERMS, CONCEPTS, AND REACTIONS

Terms

FATTY ACIDS Long alkyl chain carboxylic acids.

HORMONES Compounds produced in one organ or gland and transported by the blood to specific sites where they exert their regulatory functions.

HYDROPHILIC The water-loving part of a molecule (usually, the polar part of the molecule).

HYDROPHOBIC The water-repelling part of a molecule (usually, the hydrocarbon part of the molecule).

LIPIDS Compounds of living systems that are soluble in nonpolar solvents.

MICELLES Aggregates of molecules in an aqueous solution in which the hydrocarbon tails of the molecules join together to form a hydrophobic region, and the polar heads form an outer region in contact with the water.

NONSAPONIFIABLE LIPIDS Lipids that do not contain fatty acids and do not undergo alkaline hydrolysis.

POLAR LIPIDS Lipids whose structures have polar and nonpolar regions.

SAPONIFIABLE LIPIDS Lipids that yield fatty acid salts on alkaline hydrolysis.

SATURATED FATTY ACID A fatty acid whose hydrocarbon chain contains no double bonds.

UNSATURATED FATTY ACID A fatty acid whose hydrocarbon chain contains one or more double bonds.

Concepts

1. The chemical structures of lipids vary greatly. Some are esters (e.g., triacylglycerols, phosphoglycerides, and sphingolipids), and others contain large hydrocarbon structures (e.g., terpenes, steroids, and prostaglandins).
2. Fats and vegetable oils are triacylglycerols. They are the major energy reserve of mammals.
3. Phosphoglycerides and sphingolipids are polar lipids. They spontaneously form micelles and bimolecular sheets. The latter form the basic structure of biological membranes.
4. The formation of micelles and bimolecular sheets by polar lipids provides hydrophobic regions in an aqueous solution. Many important biological reactions occur within these hydrophobic regions.
5. Soaps, whether natural or synthetic, have a polar head and a nonpolar tail and form micelles. These micelles are responsible for the cleansing action of soaps.
6. Steroids are made from terpenes in living systems. Cholesterol is the most abundant steroid in humans. Bile acids, adrenocortical hormones, and sex hormones are made from cholesterol in humans.
7. Prostaglandins regulate many activities of cells and tissues. They are formed from fatty acids in living systems.

Reactions

Triacylglycerols (see Section 22.3)

1. Saponification

$$\begin{array}{l} CH_2O_2CR \\ | \\ CHO_2CR' \\ | \\ CH_2O_2CR'' \end{array} \xrightarrow[H_2O]{NaOH} \begin{array}{l} CH_2OH \\ | \\ CHOH \\ | \\ CH_2OH \end{array} + \begin{array}{l} RCO_2^-Na^+ \\ R'CO_2^-Na^+ \\ R''CO_2^-Na^+ \end{array}$$

R, R′, and R″ are long alkyl chains that can be identical or different.

2. Enzyme-catalyzed hydrolysis

$$\begin{array}{l} CH_2O_2CR \\ | \\ CHO_2CR' \\ | \\ CH_2O_2CR'' \end{array} \xrightarrow[H_2O]{Enzyme} \begin{array}{l} CH_2OH \\ | \\ CHOH \\ | \\ CH_2OH \end{array} + \begin{array}{l} RCO_2H \\ R'CO_2H \\ R''CO_2H \end{array}$$

3. Addition of hydrogen

$$\begin{array}{l} CH_2O_2C(CH_2)_7CH{=}CH(CH_2)_7CH_3 \\ CHO_2C(CH_2)_7CH{=}CH(CH_2)_7CH_3 \\ CH_2O_2C(CH_2)_7CH{=}CH(CH_2)_7CH_3 \\ \quad\quad\quad\text{Olein} \end{array} \xrightarrow[H_2]{Pt} \begin{array}{l} CH_2O_2C(CH_2)_{16}CH_3 \\ CHO_2C(CH_2)_{16}CH_3 \\ CH_2O_2C(CH_2)_{16}CH_3 \\ \quad\quad\text{Stearin} \end{array}$$

4. Reduction

$$\begin{array}{l} CH_2O_2CR \\ CHO_2CR' \\ CH_2O_2CR'' \end{array} + H_2 \xrightarrow[\Delta]{Ni} \begin{array}{l} CH_2OH \\ CHOH \\ CH_2OH \end{array} + \begin{array}{l} RCH_2OH \\ R'CH_2OH \\ R''CH_2OH \end{array}$$

Formation of steroids

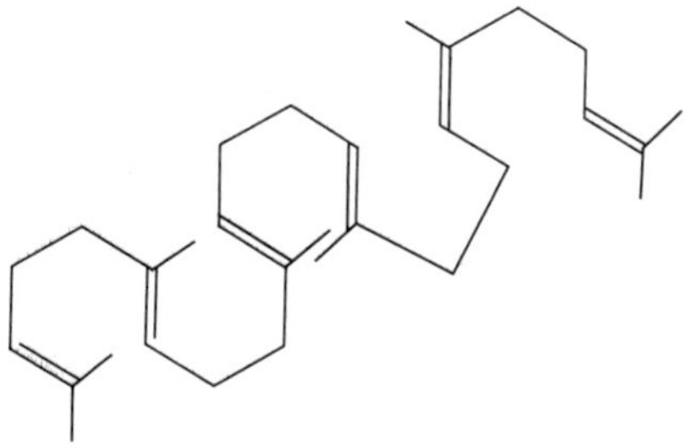

Squalene → Squalene 2,3-epoxide → Lanosterol

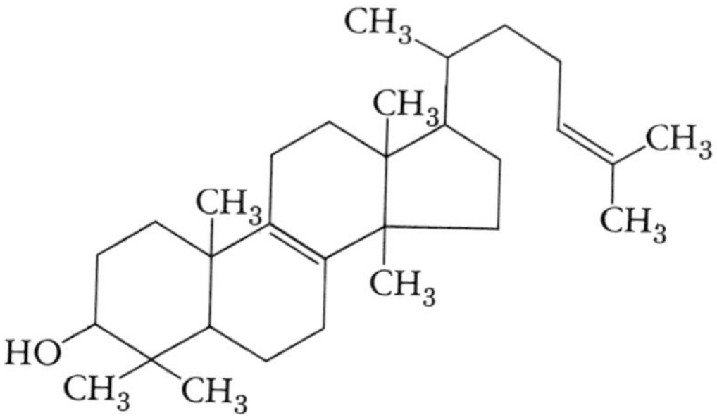

EXERCISES 22-12 Give an example of each of the following:
(a) a polar lipid (b) a nonsaponifiable lipid
(c) a triacylglycerol (d) a steroid (e) a terpene

22-13 For each of the following pairs of terms, what is the difference between the two terms?
(a) a fat and a vegetable oil (b) a fat and a wax
(c) a vegetable oil and a petroleum oil
(d) a soap and a synthetic detergent
(e) a micelle and a bimolecular sheet
(f) a terpene and a steroid

22-14 What are three functions of lipids in living systems?

22-15 Match the structure of each compound in column A with the term that best describes it in column B.

Column A

1.

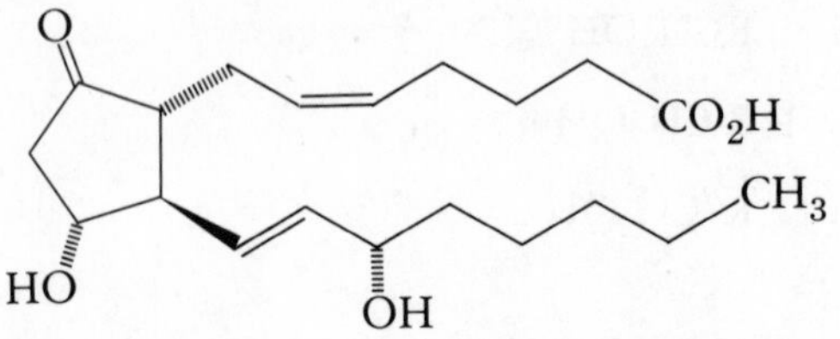

2.

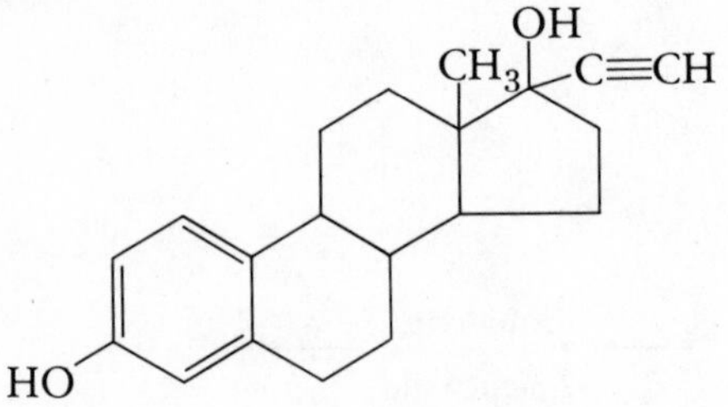

3.

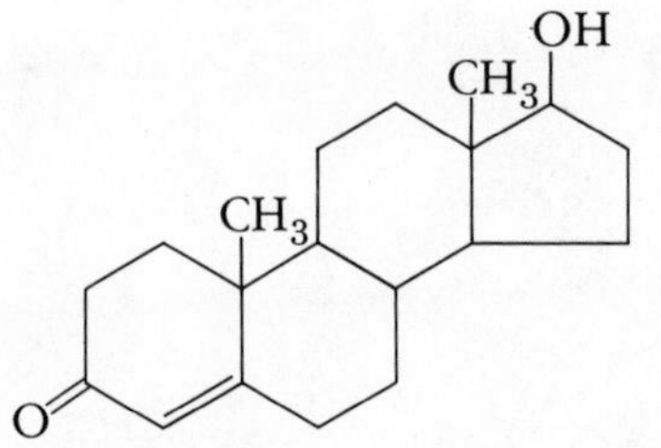

4. $CH_3(CH_2)_{20}C(=O)O(CH_2)_{17}CH_3$

5.

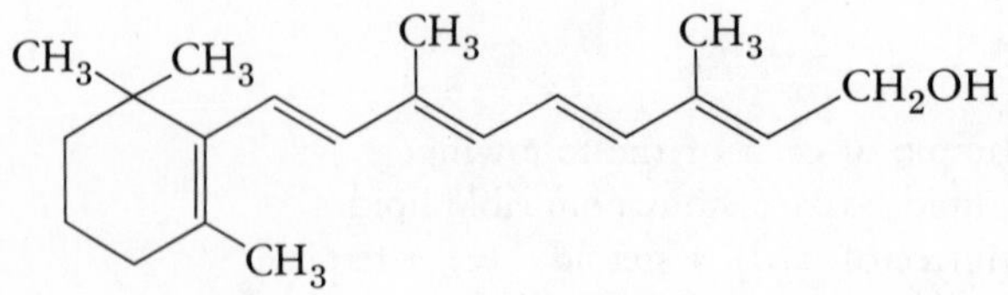

Column B

(a) soap
(b) sphingolipid
(c) hormone
(d) prostaglandin
(e) synthetic detergent
(f) wax
(g) a compound that shows progesterone activity
(h) a fat-soluble vitamin

6. $CH_3(CH_2)_{11}OSO_3^-Na^+$
7. $CH_3(CH_2)_{16}CO_2^-Na^+$
8.

$$\begin{array}{l} \text{H}\quad\ \text{H} \\ \text{C=C} \\ CH_3(CH_2)_{12}\quad \text{CHOH} \\ \quad\quad\quad | \\ \quad\quad\quad \text{CHNHC(=O)}(CH_2)_{16}CH_3 \\ \quad\quad\quad | \\ \quad\quad\quad CH_2\text{OP(=O)(O}^-\text{)O}CH_2CH_2\overset{+}{N}H_3 \end{array}$$

22-16 Write equations for the reactions of triacylglycerol *22-4* with
(a) NaOH, H_2O (b) H_2, Pt (c) Br_2 in CCl_4
(d) H_2, Ni, Δ, pressure

$$\begin{array}{l} CH_2O\text{–C(=O)}(CH_2)_7\text{CH=CH}CH_2\text{CH=CH}CH_2\text{CH=CH}CH_2CH_3 \quad (\text{all } cis) \\ | \\ CH_3(CH_2)_{16}\text{C(=O)–OCH} \\ | \\ CH_2O\text{–C(=O)}(CH_2)_7\text{CH=CH}(CH_2)_7CH_3 \quad (cis) \end{array}$$

22-4

22-17 The addition of HCl to an aqueous solution of sodium laurate forms a solid precipitate. Write the structure of this solid.

22-18 A lipid forms glycerol and oleic, stearic, and palmitic acids on saponification. Write a possible structure for this lipid.

PROTEINS 23

Proteins are the third major class of biologically important compounds that we will study. They make up about two thirds of the dry weight of cells and are crucial to virtually all processes in living systems. For example, proteins are the major components of muscles, skin, and bones. Specific proteins transport small molecules and ions within the system; other proteins called antibodies recognize and neutralize invading foreign substances such as viruses and bacteria. Proteins are also involved in the generation and transmission of nerve impulses. The primary function of proteins is to hold together, protect, and provide structure for the living organism. This is in contrast to carbohydrates, which are used primarily as energy sources, and lipids, which are used to form the structure of cell membranes and also are used as an energy source.

Despite the variety of their roles in living systems, the differences in their sizes (their molecular weights vary from several thousands to several million amu), and the differences in their physical properties (silk, skin, and bone are all protein), proteins are sufficiently similar in their molecular structure that we can classify them as a single family of compounds. In fact, proteins are an example of the enormous variety in physical properties and behavior that can be obtained by varying a simple structural theme.

When proteins are treated with a boiling acid or base solution, they are hydrolyzed to smaller compounds that have been identified as amino acids. Thus, proteins are polymers made up of amino acids as monomers. Twenty distinct amino acids have been obtained from the hydrolysis of proteins. Proteins of all living systems, from bacteria to humans, are constructed from this same set of 20 amino acids. In this chapter, we will study these amino acids and learn how they combine to form proteins.

23.1 AMINO ACIDS

As their name indicates, amino acids are compounds that contain an amino group and a carboxylic acid group. The amino acids in proteins have the amino group bonded to the α carbon of the carboxylic acid. As a result, they are called *α-amino acids.*

The amino group of an amino acid is sufficiently basic that it will react with the carboxylic acid group. This internal neutralization reaction forms a salt or *zwitter ion,* as shown in Figure 23-1. This is the structure of amino acids in a solid.

The 20 α-amino acids that make up proteins differ in the nature of the R group bonded to the α carbon. This R group is called the amino acid side chain. These side chains differ in size, shape, charge, hydrogen-bonding ability, and chemical reactivity. Consequently, each individual amino acid has unique properties. The structure of the 20 α-amino acids, their common names, and their three-letter abbreviations are given in Table 23-1. Common names are used exclusively for amino acids because their IUPAC names are too complicated and cumbersome. The abbreviations of the common names will be used as a shorthand to identify the amino acids in a protein. The amino acids in Table 23-1 are grouped according to the nature of their side chains. The simplest amino acid is glycine, which contains a hydrogen in place of the side chain. The other six amino acids in group A contain alkyl or aryl hydrocarbon side chains. Proline differs slightly from the other amino acids in Table 23-1. It contains a secondary rather than a primary amino group. Actually, proline is an imino acid. Its side chain is bonded to both the amino and the carboxylic acid groups and forms a ring. The side chains of all the compounds in group A are hydrophobic. The amino acids in group B have side chains that contain polar functional groups such as hydroxy, thiol, and amide groups. The side chains of the amino acids in

$$\overset{+}{N}H_3\underset{\underset{R}{|}}{C}HC\begin{matrix} \diagup\!\!\!\!\diagup O \\ \diagdown O^- \end{matrix}$$

Fig. 23-1. The zwitter ion structure of an α-amino acid.

Table 23-1. The Twenty Common Amino Acids

Structural Formula	Common Name	Three-Letter Abbreviation	pI
A. Amino acids with nonpolar side chains			
$H{-}CH(NH_3^+)CO_2^-$	Glycine	Gly	5.97
$CH_3{-}CH(NH_3^+)CO_2^-$	Alanine	Ala	6.00
$(CH_3)_2CH{-}CH(NH_3^+)CO_2^-$	Valine*	Val	5.96
$(CH_3)_2CHCH_2{-}CH(NH_3^+)CO_2^-$	Leucine*	Leu	5.98
$CH_3CH_2CH(CH_3){-}CH(NH_3^+)CO_2^-$	Isoleucine*	Ile	6.02
$C_6H_5{-}CH_2{-}CH(NH_3^+)CO_2^-$	Phenylalanine*	Phe	5.48
ring: $CH_2{-}CH_2{-}CH_2{-}NH_2^+{-}CHCO_2^-$	Proline	Pro	6.30
B. Amino acids with polar but neutral side chains			
indol-3-yl$\,{-}CH_2{-}CH(NH_3^+)CO_2^-$	Tryptophan*	Trp	5.89
$HOCH_2{-}CH(NH_3^+)CO_2^-$	Serine	Ser	5.68
$HOCH(CH_3){-}CH(NH_3^+)CO_2^-$	Threonine*	Thr	5.60
$HO{-}C_6H_4{-}CH_2{-}CH(NH_3^+)CO_2^-$	Tyrosine	Tyr	5.66

*These amino acids cannot be made by the body but must be obtained from the food we eat.

Table 23-1 (*Continued*)

Structural Formula	Common Name	Three-Letter Abbreviation	pI
$HSCH_2-\underset{+NH_3}{\underset{\vert}{C}}HCO_2^-$	Cysteine	Cys	5.07
$CH_3SCH_2CH_2-\underset{+NH_3}{\underset{\vert}{C}}HCO_2^-$	Methionine*	Met	5.74
$H_2N\overset{O}{\overset{\Vert}{C}}CH_2-\underset{+NH_3}{\underset{\vert}{C}}HCO_2^-$	Asparagine	Asn	5.41
$H_2N\overset{O}{\overset{\Vert}{C}}CH_2CH_2-\underset{+NH_3}{\underset{\vert}{C}}HCO_2^-$	Glutamine	Gln	5.65
C. Amino acids with acidic side chains			
$HO_2CCH_2-\underset{+NH_3}{\underset{\vert}{C}}HCO_2^-$	Aspartic acid	Asp	2.97
$HO_2CCH_2CH_2-\underset{+NH_3}{\underset{\vert}{C}}HCO_2^-$	Glutamic acid	Glu	3.22
D. Amino acids with basic side chains			
$NH_2CH_2CH_2CH_2CH_2-\underset{+NH_3}{\underset{\vert}{C}}HCO_2^-$	Lysine*	Lys	9.74
$NH_2\overset{NH}{\overset{\Vert}{C}}NHCH_2CH_2CH_2-\underset{+NH_3}{\underset{\vert}{C}}HCO_2^-$	Arginine*	Arg	10.76
$\overbrace{HC{=}N{-}CH{=}C{-}N(H){-}}^{\text{ring}}\;C-CH_2-\underset{+NH_3}{\underset{\vert}{C}}HCO_2^-$ (imidazole ring: N–H, HC, C, N—CH)	Histidine*	His	7.59

* These amino acids cannot be made by the body but must be obtained from the food we eat.

group C contain acidic functional groups, and those of amino acids in group D contain basic functional groups. We will learn in Section 23.7 that the nature of the side chains of the amino acids determines the structure and, ultimately, the function of proteins.

Half of the amino acids listed in Table 23-1 cannot be made by the human body. These ten are called *essential amino acids.* They are indicated by an asterisk in Table 23-1. These amino acids must be obtained from the food we eat.

In addition to the amino acids listed in Table 23-1, certain other special amino acids are present in protein. 4-Hydroxyproline and 5-hydroxylysine are two examples:

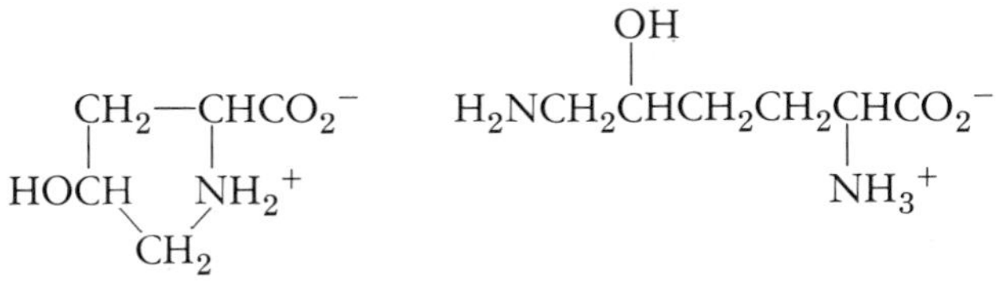

$$\begin{array}{c} CH_2{-}CHCO_2^- \\ |\qquad | \\ HOCH\quad NH_2^+ \\ \diagdown\ \diagup \\ CH_2 \end{array} \qquad\qquad \begin{array}{c} OH \\ | \\ H_2NCH_2CHCH_2CH_2CHCO_2^- \\ | \\ NH_3^+ \end{array}$$

4-Hydroxyproline — 5-Hydroxylysine

These amino acids are usually made from one of those listed in Table 23-1. They are formed by a specific enzyme-catalyzed reaction that modifies the common amino acid after it has been incorporated into a protein. 4-Hydroxyproline, for example, is made from the amino acid proline, which is already part of a protein.

Some amino acids are obtained from compounds that are not proteins. Three examples are given in Figure 23-2. The β-amino acid β-alanine is part of the structure of coenzyme A (see Section 26.5). γ-Aminobutyric acid is present in brain tissue and is involved in the transmission of nerve impulses. Ornithine is involved in the series of chemical reactions that converts nitrogen-containing compounds into urea (see Section 29.5). In all, more than 150 of these amino acids have been found.

Because of their structure, α-amino acids have some special properties.

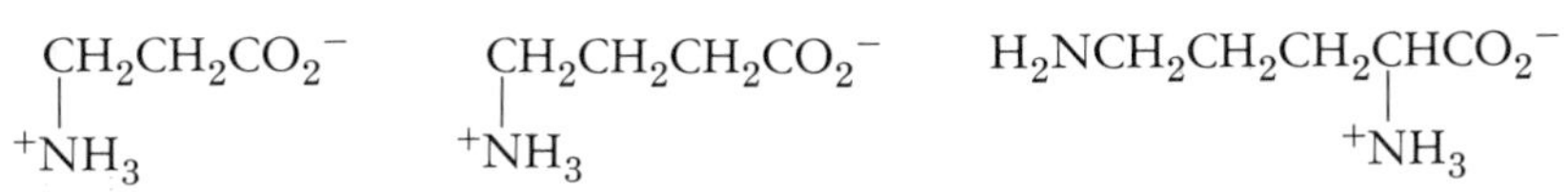

$$\begin{array}{c} CH_2CH_2CO_2^- \\ | \\ {}^+NH_3 \end{array} \qquad \begin{array}{c} CH_2CH_2CH_2CO_2^- \\ | \\ {}^+NH_3 \end{array} \qquad \begin{array}{c} H_2NCH_2CH_2CH_2CHCO_2^- \\ | \\ {}^+NH_3 \end{array}$$

β-Alanine — γ-Aminobutyric acid — Ornithine

Fig. 23-2. Amino acids obtained from nonprotein sources.

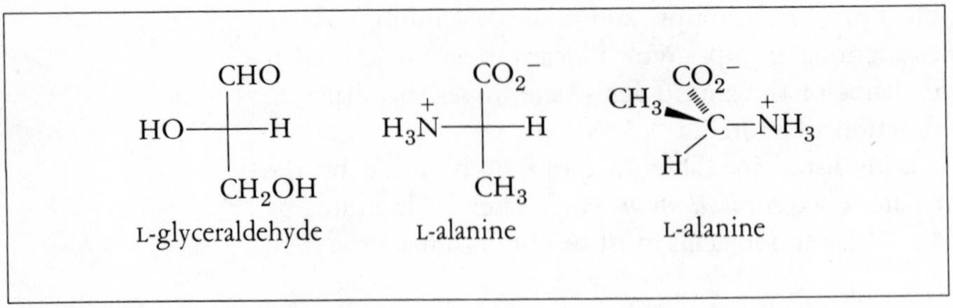

CHO	CO_2^-	CO_2^-
$HO-C-H$	$H_3\overset{+}{N}-C-H$	CH_3, H — $C-\overset{+}{N}H_3$
CH_2OH	CH_3	
L-glyceraldehyde	L-alanine	L-alanine

Fig. 23-3. The L*-configuration of α-amino acids obtained from proteins.*

23.2 PROPERTIES OF α-AMINO ACIDS

Except for that in glycine, the α carbons of all the other amino acids in Table 23-1 are chiral centers (see Section 14.1). Therefore, these α-amino acids may exist as a pair of enantiomers. However, in nearly all of the proteins obtained from plants and animals, the amino acids exist only in the L configuration relative to glyceraldehyde, as shown in Figure 23-3. This configuration is opposite that found for carbohydrates. Naturally occurring sugars belong to the D family (see Section 21.2).

D-amino acids are also found in nature, but not as parts of proteins. Some of them make up parts of the cell walls of bacteria; others are found in certain polypeptide chains of antibiotics. For example, the antibiotic gramicidin-S is a polypeptide chain made up of ten amino acid residues. One is D-phenylalanine and the other nine are L-amino acids.

The physical properties of amino acids resemble those of inorganic salts such as sodium chloride. The reason is that amino acids exist in solids as the salt or zwitter ion form. The attraction between opposite charges on adjacent molecules forms ionic bonds similar to those in salts (see Section 5.3). Consequently, crystalline amino acids are colorless, odorless, and melt with decomposition at temperatures greater than 200° C. They are much more soluble in water than in nonpolar organic solvents.

In neutral aqueous solution, an amino acid such as alanine also exists predominantly in the zwitter ion form. A compound with such a structure can act as either an acid or a base; that is, it can accept a proton or donate a proton. Substances that have this property are called *amphoteric.* Thus, if an amino acid is placed in a strong acid solution ($pH = 1$), it will accept a proton; in a strong basic solution ($pH = 11$), it will lose a proton. This property of amino acids is shown in Figure 23-4 for L-alanine. Notice that the zwitter ion structure of L-alanine has no net charge because it contains one positive and one negative charge, which cancel each other. In contrast, L-alanine has a net charge of $+1$ in strong acid solution and a net charge of -1 in strong base solution.

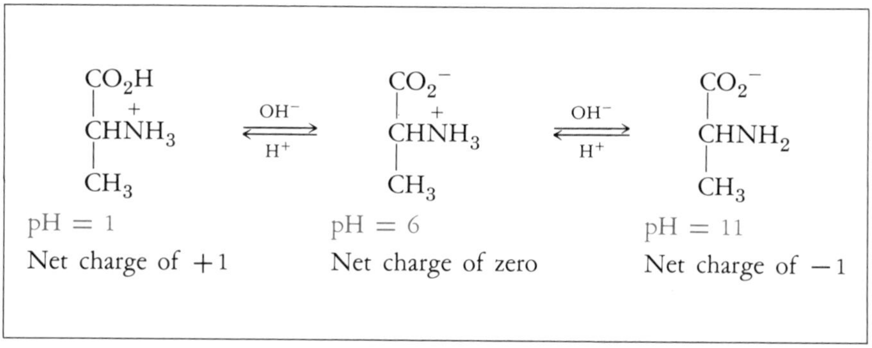

$$\underset{\text{pH} = 1 \text{, Net charge of } +1}{\overset{CO_2H}{\underset{CH_3}{C H\overset{+}{N}H_3}}} \underset{H^+}{\overset{OH^-}{\rightleftharpoons}} \underset{\text{pH} = 6 \text{, Net charge of zero}}{\overset{CO_2^-}{\underset{CH_3}{CH\overset{+}{N}H_3}}} \underset{H^+}{\overset{OH^-}{\rightleftharpoons}} \underset{\text{pH} = 11 \text{, Net charge of } -1}{\overset{CO_2^-}{\underset{CH_3}{CHNH_2}}}$$

Fig. 23-4. The predominant form of L-alanine at each of three pH values.

*EXERCISE 23-1 Give the net charge for each of the following structures:

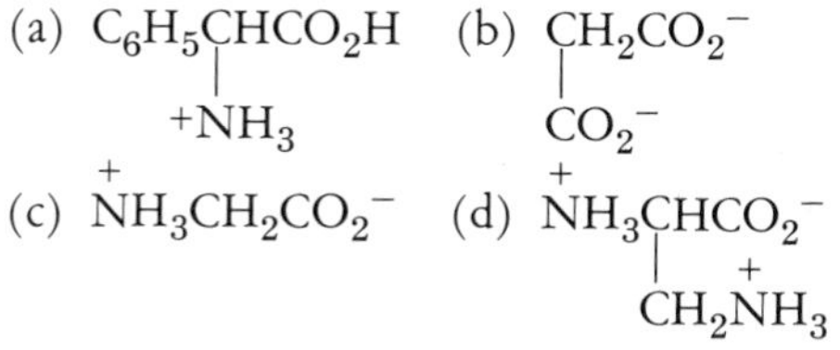

(a) $C_6H_5CH(\overset{+}{N}H_3)CO_2H$ (b) $CH_2CO_2^-$ bonded to CO_2^- ($^-O_2CCH_2CO_2^-$)

(c) $\overset{+}{N}H_3CH_2CO_2^-$ (d) $\overset{+}{N}H_3CH(CH_2\overset{+}{N}H_3)CO_2^-$

The pH at which the structure of an amino acid has no net charge is called its *isoelectric point.* It is given a special symbol, pI. The value of pI is a characteristic of each amino acid, as shown in Table 23-1, where the pI values of the 20 common amino acids are given. Several generalizations can be made about these pI values. The amino acids in groups A and B of Table 23-1, which contain only one amino group and one carboxylic acid group, have pI values in the range 5.0 to 6.3. Amino acids in group D of Table 23-1, which contain a basic group in their side chain, have pI values between 7.6 and 10.8. The two amino acids that contain an acidic group in their side chain (group C) have pI values of 2.97 and 3.22. Generally, at a pH 2 units or more above the pI value of an amino acid, the amino acid exists predominantly in the form that contains one or more negative charges. At a pH 2 units below its pI value, an amino acid exists predominantly in the form that contains one or more positive charges. This is illustrated for L-alanine in Figure 23-4 and for L-aspartic acid in Figure 23-5.

EXERCISE 23-2 Write the structures of the predominant forms of the following amino acids at pH values of 1, 7, and 11:
(a) L-valine (b) L-phenylalanine

*The answers for the exercises in this chapter begin on page 897.

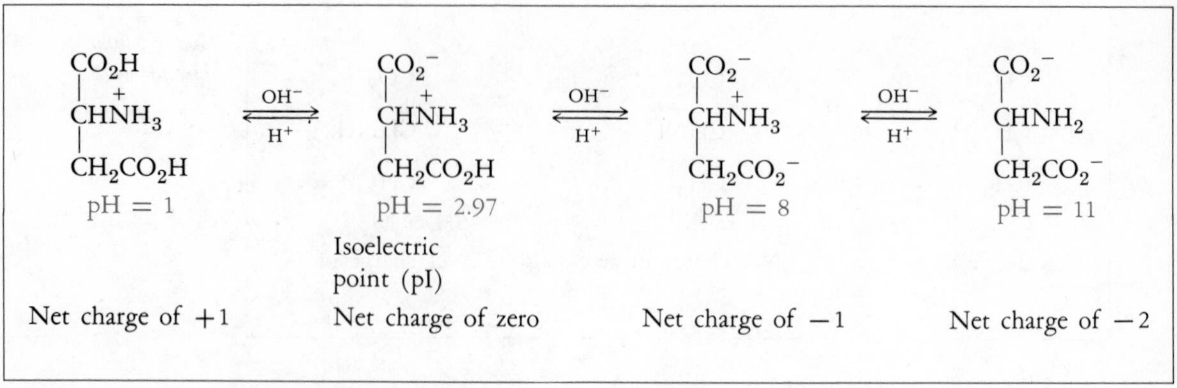

$$\underset{\text{pH} = 1}{\overset{\displaystyle CO_2H}{\underset{\displaystyle CH_2CO_2H}{\overset{|}{\underset{|}{CH\overset{+}{N}H_3}}}}} \underset{H^+}{\overset{OH^-}{\rightleftharpoons}} \underset{\text{pH} = 2.97}{\overset{\displaystyle CO_2^-}{\underset{\displaystyle CH_2CO_2H}{\overset{|}{\underset{|}{CH\overset{+}{N}H_3}}}}} \underset{H^+}{\overset{OH^-}{\rightleftharpoons}} \underset{\text{pH} = 8}{\overset{\displaystyle CO_2^-}{\underset{\displaystyle CH_2CO_2^-}{\overset{|}{\underset{|}{CH\overset{+}{N}H_3}}}}} \underset{H^+}{\overset{OH^-}{\rightleftharpoons}} \underset{\text{pH} = 11}{\overset{\displaystyle CO_2^-}{\underset{\displaystyle CH_2CO_2^-}{\overset{|}{\underset{|}{CHNH_2}}}}}$$

Isoelectric point (pI)

Net charge of +1 | Net charge of zero | Net charge of −1 | Net charge of −2

Fig. 23-5. The predominant form of L-aspartic acid at each of four pH values.

EXERCISE 23-3 The pH of blood is 7.4. Write the structure of the predominant form of each of the following amino acids at this pH:
(a) proline (b) tryptophan (c) glutamic acid
(d) histidine

EXERCISE 23-4 Using the pI values in Table 23-1, indicate the pH at which you would expect to find each of the following structures:

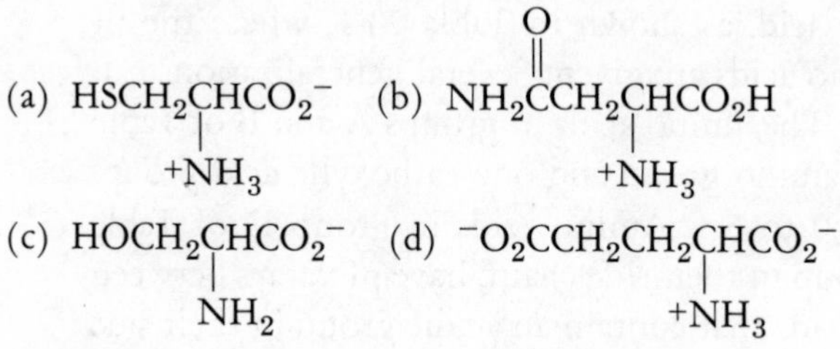

(a) $HSCH_2\underset{\displaystyle {}^+NH_3}{\underset{|}{C}}HCO_2^-$ (b) $NH_2\overset{\displaystyle O}{\overset{\|}{C}}CH_2\underset{\displaystyle {}^+NH_3}{\underset{|}{C}}HCO_2H$

(c) $HOCH_2\underset{\displaystyle NH_2}{\underset{|}{C}}HCO_2^-$ (d) ${}^-O_2CCH_2CH_2\underset{\displaystyle {}^+NH_3}{\underset{|}{C}}HCO_2^-$

Before we learn how amino acids are joined together to form proteins, let us briefly review their reactions and learn a few new ones.

23.3 REACTIONS OF AMINO ACIDS

An amino acid in its zwitter ion form cannot undergo the usual reactions of an amino group or a carboxylic acid group. It is necessary to make an aqueous solution of an amino acid sufficiently acidic to convert its carboxylate ion to a carboxylic acid group before it can undergo the usual reactions

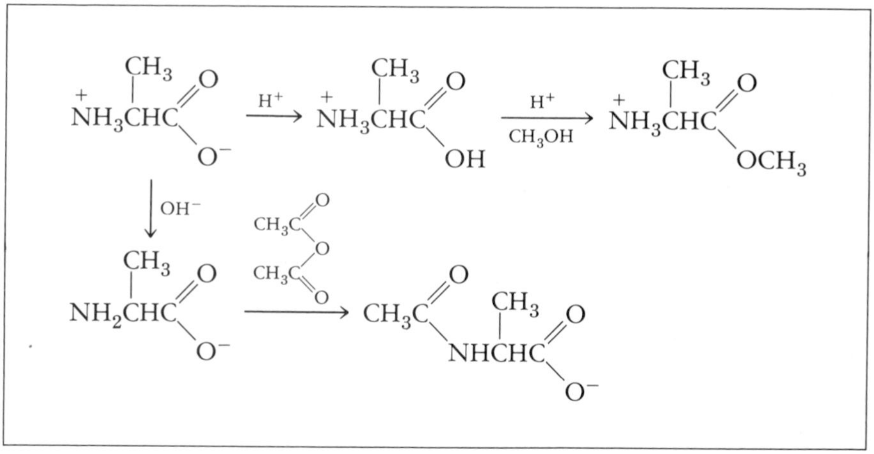

$$\overset{+}{N}H_3CH(CH_3)C(=O)O^- \xrightarrow{H^+} \overset{+}{N}H_3CH(CH_3)C(=O)OH \xrightarrow[CH_3OH]{H^+} \overset{+}{N}H_3CH(CH_3)C(=O)OCH_3$$

$$\overset{+}{N}H_3CH(CH_3)C(=O)O^- \xrightarrow{OH^-} NH_2CH(CH_3)C(=O)O^-$$

$$NH_2CH(CH_3)C(=O)O^- \xrightarrow{CH_3C(=O)OC(=O)CH_3} CH_3C(=O)NHCH(CH_3)C(=O)O^-$$

Fig. 23-6. The zwitter ion form of an amino acid must be converted to its amino or carboxylic group in order for them to undergo their usual reactions.

of a carboxylic acid. Similarly, the solution must be made basic to re-form the amino group before it can undergo its usual reactions. This is shown for L-alanine in Figure 23-6. In addition to their usual reactions, amino acids undergo several special reactions.

The side chains of amino acids have functional groups that undergo important reactions. The thiol group of cysteine is a typical example. This group is easily oxidized by the oxygen in air or by other mild oxidizing agents (see Section 16.8). The product is cystine, in which a covalent disulfide (S—S) bond is formed between the two cysteine molecules:

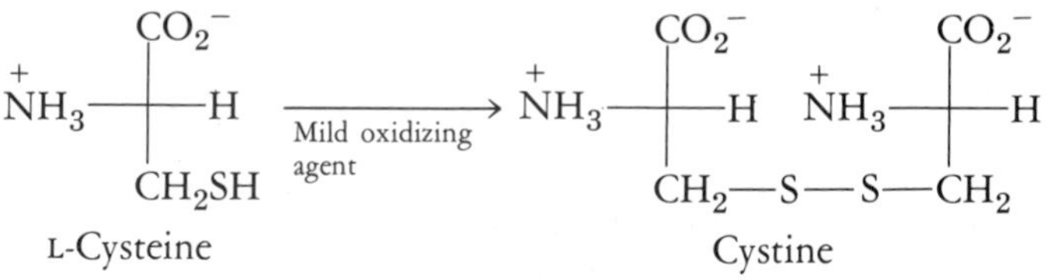

$$\underset{\text{L-Cysteine}}{\overset{+}{N}H_3-\underset{CH_2SH}{\overset{CO_2^-}{C}}-H} \xrightarrow[\text{agent}]{\text{Mild oxidizing}} \underset{\text{Cystine}}{\overset{+}{N}H_3-\underset{CH_2}{\overset{CO_2^-}{C}}-H \quad \overset{+}{N}H_3-\underset{CH_2}{\overset{CO_2^-}{C}}-H}$$

$$\text{(side chains joined: } CH_2-S-S-CH_2\text{)}$$

Cystine has a special role in protein structure because its disulfide bond is a covalent bond that links two protein chains or two points in a single chain (see Section 23.7). Such bonds are called *cross-links.*

EXERCISE 23-5 Identify the disulfide bond in cystine.

Another important and characteristic reaction of the thiol group in cysteine is its reaction with metal cations such as mercury (Hg^{+2}) and lead

(Pb^{+2}) to form mercaptides. The reaction, illustrated by mercuric ion, is as follows:

$$\overset{+}{N}H_3-\underset{CH_2SH}{\overset{CO_2^-}{C}}-H + Hg^{+2} \longrightarrow \left[\overset{+}{N}H_3-\underset{CH_2S}{\overset{CO_2^-}{C}}-H\right]_2 Hg + 2\,H^+$$

The mercaptide is insoluble in aqueous solutions. In living systems this reaction disrupts the disulfide bonds of proteins and causes the protein to precipitate out of solution. These metal cations also react with the carboxylate ion to form insoluble salts. As a result, these metal cations are toxic to most living systems.

EXERCISE 23-6 Write the equation for the reaction of L-alanine with each of the following reagents:
(a) CH_3OH, HCl (b) C_6H_5CHO at pH 8
(c) $(CH_3CO)_2O$ at pH 8

EXERCISE 23-7 Write the equation for the reaction of Ag^+ and Pb^{+2} with L-cysteine.

The most important reaction of amino acids is the formation of peptide bonds.

23.4 PEPTIDE BONDS

Proteins are formed by joining the carboxyl group of one amino acid to the α-amino group of another amino acid. The bond formed between the two amino acids is called a *peptide bond,* which is simply another name for an amide bond (see Section 18.8). When two amino acids are joined in this way, a dipeptide is formed, as shown in Figure 23-7. Joining three amino acids in this way forms a tripeptide; four, a tetrapeptide, and so forth. Proteinlike substances that have molecular weights less than 10,000 amu (about 100 amino acids) are called *polypeptides* or simply peptides. The name *protein* is given to compounds containing more than 100 amino acids. However, the distinction between a polypeptide and a protein is not clear cut. An amino acid unit in a polypeptide or protein is called a *residue.*

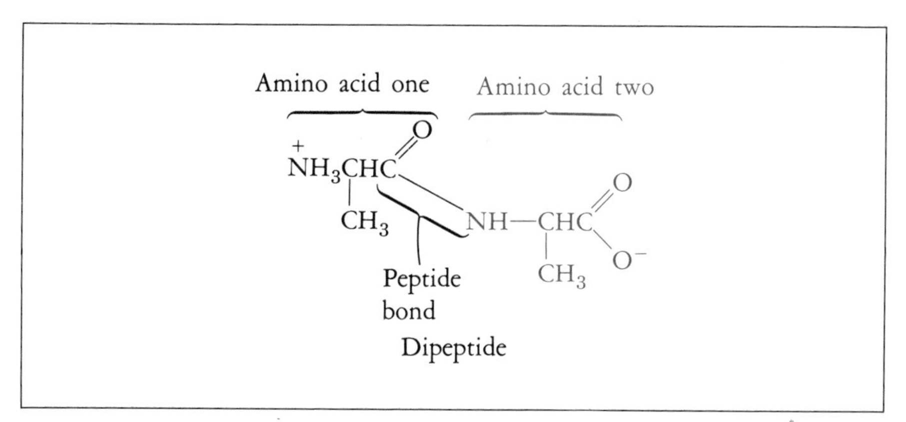

Fig. 23-7. Two amino acids joined by a peptide bond to form a dipeptide.

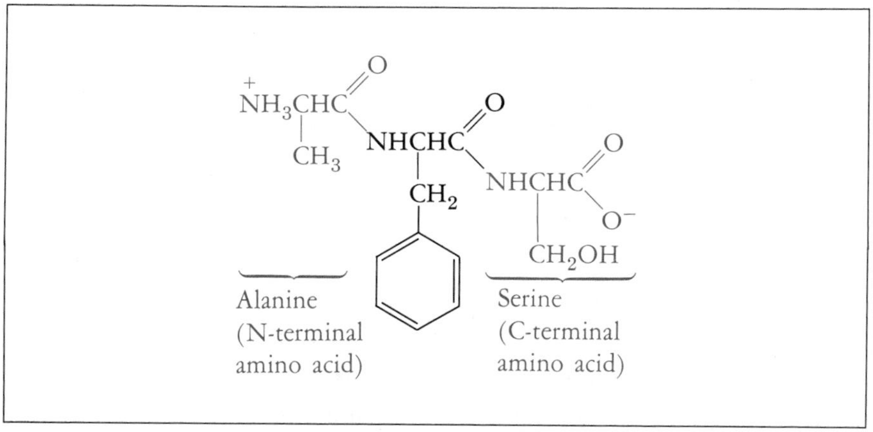

$\overset{+}{N}H_3CH(CH_3)C(=O)$—$NHCH(CH_2C_6H_5)C(=O)$—$NHCH(CH_2OH)C(=O)O^-$

Alanine (N-terminal amino acid)

Serine (C-terminal amino acid)

Fig. 23-8. The N- and C-terminal amino acids of a tripeptide.

By convention, peptides and proteins are written so that the free ammonium ion (NH_3^+) is on the left and the free carboxylate ion (CO_2^-) is on the right. Written in this way, the amino acid at the left end of the chain is called the *N-terminal amino acid* and the amino acid at the other end is called the *C-terminal amino acid.* Thus, alanine is the N-terminal amino acid and serine is the C-terminal amino acid in the tripeptide shown in Figure 23-8.

EXERCISE 23-8 Circle each peptide bond in the following peptides:

(a) 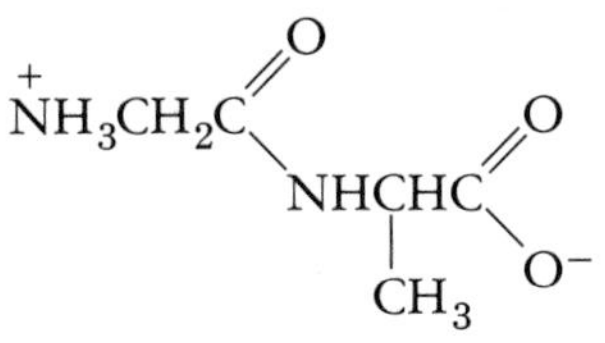 $\overset{+}{N}H_3CH_2C(=O)$—$NHCH(CH_3)C(=O)O^-$

(b) 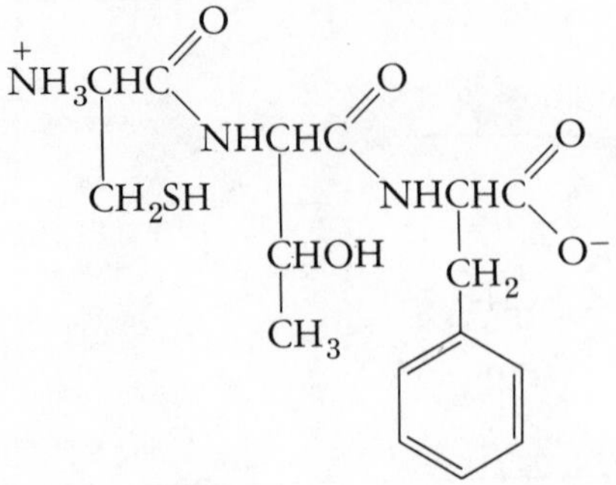 $\overset{+}{N}H_3CHC(=O)$—$NHCHC(=O)$—$NHCHC(=O)O^-$

CH_2SH $CHOH$ CH_2

CH_3 (phenyl ring)

(c) 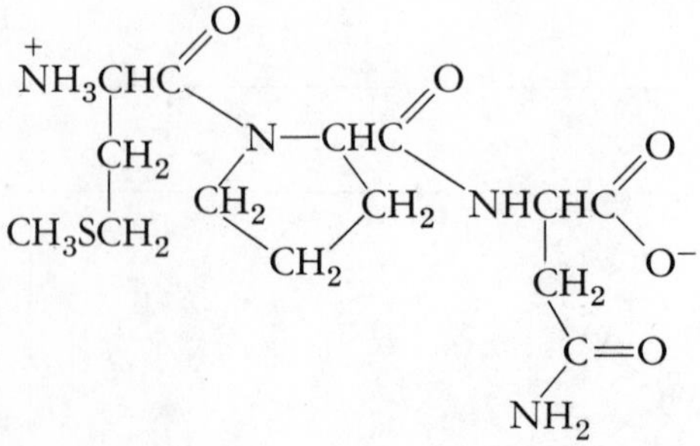 $\overset{+}{N}H_3CHC(=O)$—N—$CHC(=O)$—$NHCHC(=O)O^-$

CH_2 CH_2 CH_2 CH_2

CH_3SCH_2 CH_2 $C{=}O$

NH_2

EXERCISE 23-9 Identify the N-terminal and the C-terminal amino acids in the peptides in exercise 23-8.

Peptides are named by combining the names of each amino acid, named as an acyl group (see Section 18.1), in order starting from the N-terminal amino acid. The only exception is the C-terminal amino acid, which retains its name unchanged. The prefix L- is not included because it is understood that the amino acids have this configuration. Examples of the names of peptides are given in Figure 23-9.

A more convenient way of writing the structures of peptides and proteins is to use the three-letter abbreviations of the amino acids. These are placed in order and joined by dashes that represent the peptide bond. Examples are given in Figure 23-10.

EXERCISE 23-10 Write the abbreviated structural formula for the peptides in exercise 23-8.

EXERCISE 23-11 Write the complete structural formula for each of the following peptides:
(a) Gly-Ala-Ser (b) Ile-Tyr-Asn-Pro
(c) Gln-Met-Arg-Glu-Lys

EXERCISE 23-12 Give a name to each of the peptides in exercise 23-11.

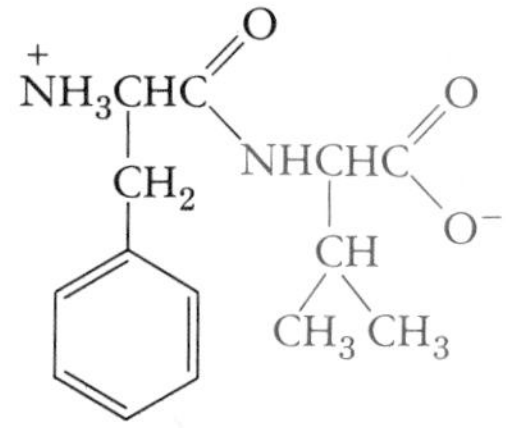

Fig. 23-9. Naming peptides.

$\overset{+}{N}H_3CH(CH_2C_6H_5)C(=O)–NHCH(CH(CH_3)_2)C(=O)O^-$

Phenylalanine	Valine	Names of individual amino acids
↓	↓	
Phenylalanyl	Valine	Names as acyl groups (except for C-terminal amino acid)
Phenylalanylvaline		Combined name

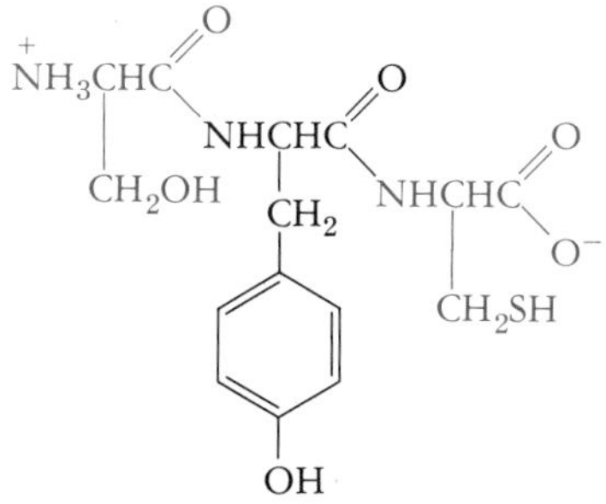

$\overset{+}{N}H_3CH(CH_2OH)C(=O)–NHCH(CH_2C_6H_4OH)C(=O)–NHCH(CH_2SH)C(=O)O^-$

Serine	Tyrosine	Cysteine	Names of individual amino acids
↓	↓	↓	
Seryl	Tyrosyl	Cysteine	Names as acyl groups
↘	↓	↙	
Seryltyrosylcysteine			Combined name

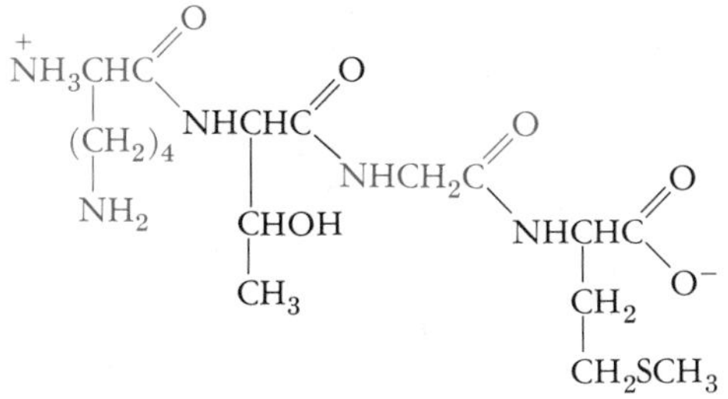

$\overset{+}{N}H_3CH((CH_2)_4NH_2)C(=O)–NHCH(CHOHCH_3)C(=O)–NHCH_2C(=O)–NHCH(CH_2CH_2SCH_3)C(=O)O^-$

Lysine	Threonine	Glycine	Methionine	Names of individual amino acids
↓	↓	↓	↓	
Lysyl	Threonyl	Glycyl	Methionine	Names as acyl groups
↘	↓	↙	↙	
Lysylthreonylglycylmethionine				Combined name

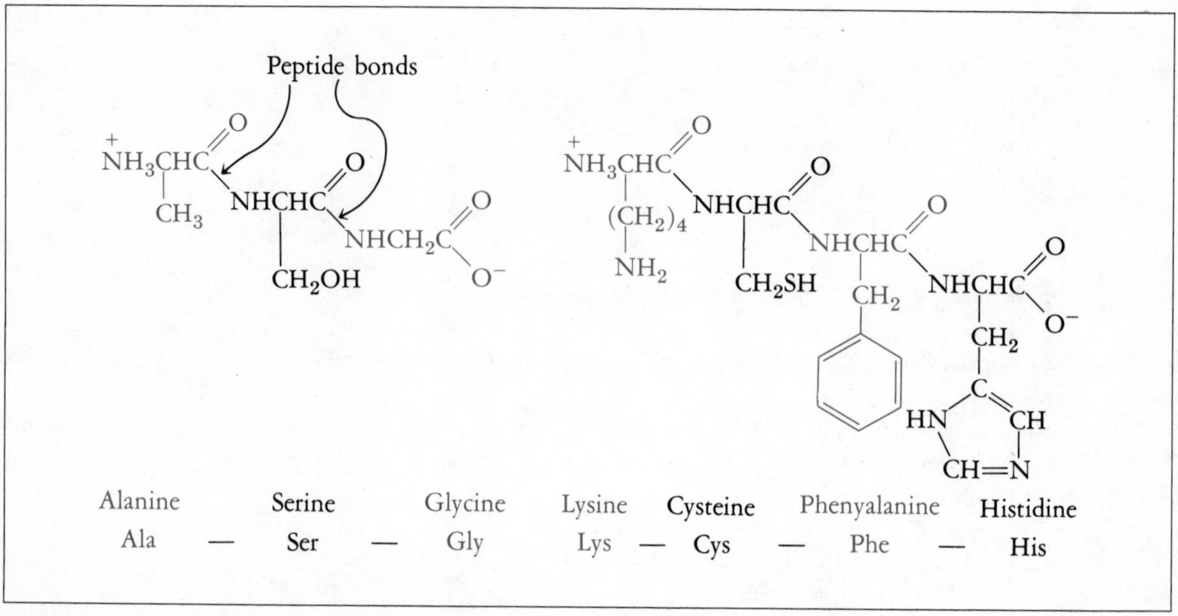

Peptide bonds

Alanine — Serine — Glycine

Ala — Ser — Gly

Lysine — Cysteine — Phenylalanine — Histidine

Lys — Cys — Phe — His

Fig. 23-10. The abbreviated structural formulas of peptides.

Polypeptides are linear sequences of amino acids joined together by peptide bonds. As a result, they are chain-like molecules. Many proteins contain a single chain of amino acids; others contain two or more chains. The sequence of amino acids is unique and precisely defined in each polypeptide or protein. It is this sequence that determines its biological function. If we are to understand how proteins function, we must learn their sequences of amino acids. The sequence of amino acids in a protein is called its primary structure.

23.5 PRIMARY STRUCTURE OF PROTEINS

A given number of amino acids can be joined by peptide bonds in many ways to form many isomeric peptides. For example, we can combine the three amino acids alanine, serine, and valine in the following six different ways:

Ala-Ser-Val	Ser-Ala-Val
Ala-Val-Ser	Ser-Val-Ala
Val-Ala-Ser	Val-Ser-Ala

Four different amino acids can form 24 isomeric combinations; 11 different amino acids can form 40 million isomeric structures! A similar relationship exists between the letters of the alphabet and the number of words they form. All of the words of the English language are made from combinations of only 26 letters.

Although many combinations of amino acids are possible, a particular protein of a living system has a unique sequence. The remarkable thing about living systems is that they produce a particular protein with the same amino acid sequence from generation to generation. How this sequence was originally formed or its exact meaning is still unknown. But there is no doubt that it is the basis of life.

The sequence of amino acids in a protein can be compared to the words in a sentence. The sequence of words gives a particular meaning to the sentence. Similarly, the sequence of amino acids in a protein carries a message that is vital to the living system. This sequence of amino acids is called its *primary structure.* It is responsible for the specific biological function of the protein. Even one amino acid incorrectly placed in a protein can alter its biological activity. An example is the disease sickle-cell anemia. It is caused by the fact that a glutamic acid molecule in normal hemoglobin is substituted by a valine molecule in the sickle cell hemoglobin. This seemingly minor change in the sequence of 146 amino acids in the protein alters its biological function and eventually results in the death of the person.

We must determine the primary structures of proteins if we are to understand the molecular basis of their biological activities. The method scientists use to do this can best be described by using an example. Suppose that we wish to determine the structure of a tripeptide that we have obtained from a living system. How do we do this? We must first learn the amino acid composition of the tripeptide. Then we must determine the sequence of these three amino acids. We establish the composition of a peptide or protein by heating it in strong aqueous hydrochloric acid solution for 12 hours. This hydrolyzes the peptide into its constituent amino acids, which are then separated and identified.* In this way, we find that our tripeptide is made up of alanine, serine, and histidine, as shown in Figure 23-11. We now know the composition but not the sequence of amino acids in our tripeptide.

We can learn the identity of the N-terminal amino acid of our tripeptide in the following way. The free amino group of the N-terminal amino acid is

* Many elegant methods for separating and identifying amino acids have been developed by scientists. Most methods are too complicated and involved to discuss at this time. It is enough to know that all the amino acids can be routinely separated from one another and identified.

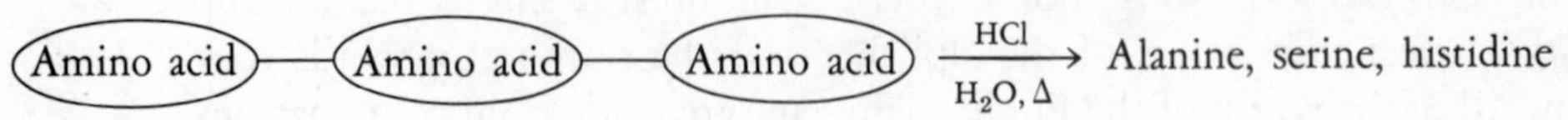

Amino acid — Amino acid — Amino acid $\xrightarrow[H_2O, \Delta]{HCl}$ Alanine, serine, histidine

Fig. 23-11. Establishing the composition of a tripeptide.

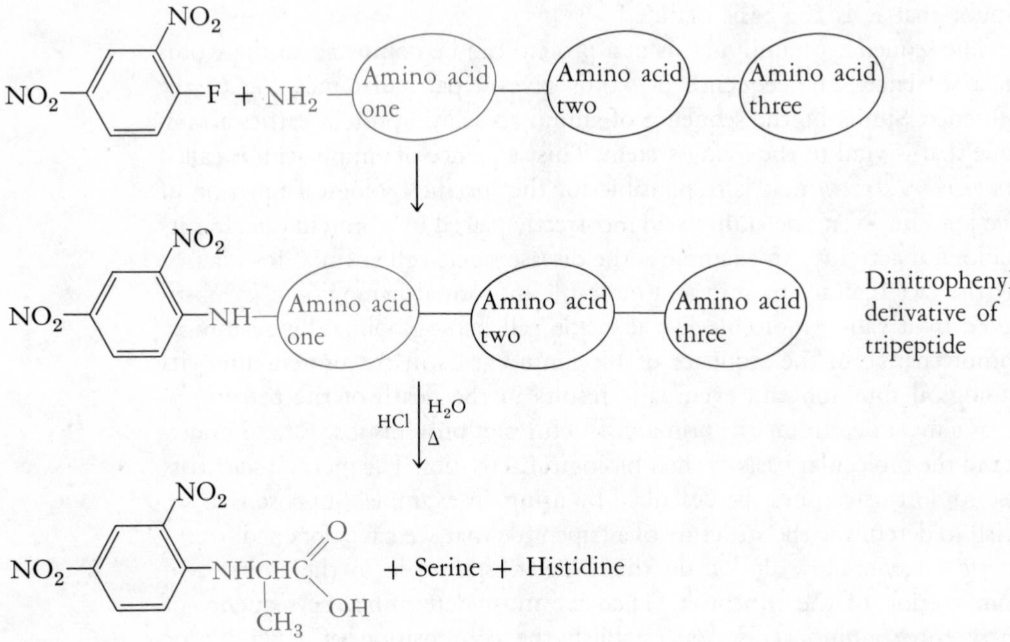

Fig. 23-12. Determining the identity of the N-terminal amino acid of a tripeptide.

reacted with a compound that will form a bond that is stable to acid hydrolysis. When we hydrolyze our tripeptide, one amino acid, the N-terminal one, will have a label attached to it. The compound that we use as a label is 2,4-dinitrofluorobenzene. It reacts with the uncharged terminal amino group to form a dinitrophenyl derivative (see Section 15.11), as shown in Figure 23-12. The C—N bond between the dinitrophenyl group and the N-terminal amino acid is stable to conditions that hydrolyze peptide bonds. Consequently, when the peptide or protein is hydrolyzed, the N-terminal amino acid has a dinitrophenyl group attached to its amino group. This serves to identify that amino acid. In our example, alanine is the N-terminal amino acid.

Although this method is useful for determining the N-terminal amino acid, it cannot be used more than once because the protein or polypeptide is completely hydrolyzed in the next step. The *Edman degradation* is a method that overcomes this problem. The terminal amino acid is labeled and removed from the peptide without disrupting the peptide bonds between the other amino acids of the peptide chain. In this way amino acids are removed one at a time from the amino end of the protein in their exact sequence. This method is illustrated for our tripeptide in Figure 23-13.

Phenyl isothiocyanate is the reagent used to react with the free amino group of the N-terminal amino acid. The product of this reaction is hydrolyzed under mild acidic conditions to remove only the labeled N-terminal amino acid. It is isolated as a cyclic compound, a phenylthiohydantoin (PTH) amino acid, and identified. In our example, this amino acid is alanine. The remaining intact dipeptide can react again with phenyl isothiocyanate. Amino acid number two is now at the N-terminal position. It can be removed, isolated, and identified. In our example, amino acid number two is histidine. The remaining amino acid, serine, must be the C-terminal amino acid of our tripeptide. In this way, we have succeeded in determining the amino acid sequence of our tripeptide.

We only needed to carry out two rounds of Edman degradation to reveal the complete sequence of the tripeptide in our example. For more complex proteins, many rounds are needed. This procedure has recently been automated, making the job much easier. Thus, it is possible to use a machine to carry out the Edman degradation. Each round takes about 2 hours to complete.

Scientists have established the amino acid sequence of more than 300 proteins using such methods. One of these proteins is lysozyme, an enzyme that destroys certain bacteria by breaking up the polysaccharide parts of their cell walls. Lysozyme is made up of a single chain of 129 amino acids and has a molecular weight of 14,600 amu. Four disulfide bonds acting as cross-links contribute to its stability. The amino acid sequence of lysozyme is shown in Figure 23-14.

In addition to its primary structure, a protein must have a well-defined three-dimensional structure to be biologically active. How proteins achieve such structures will be described in the next two sections.

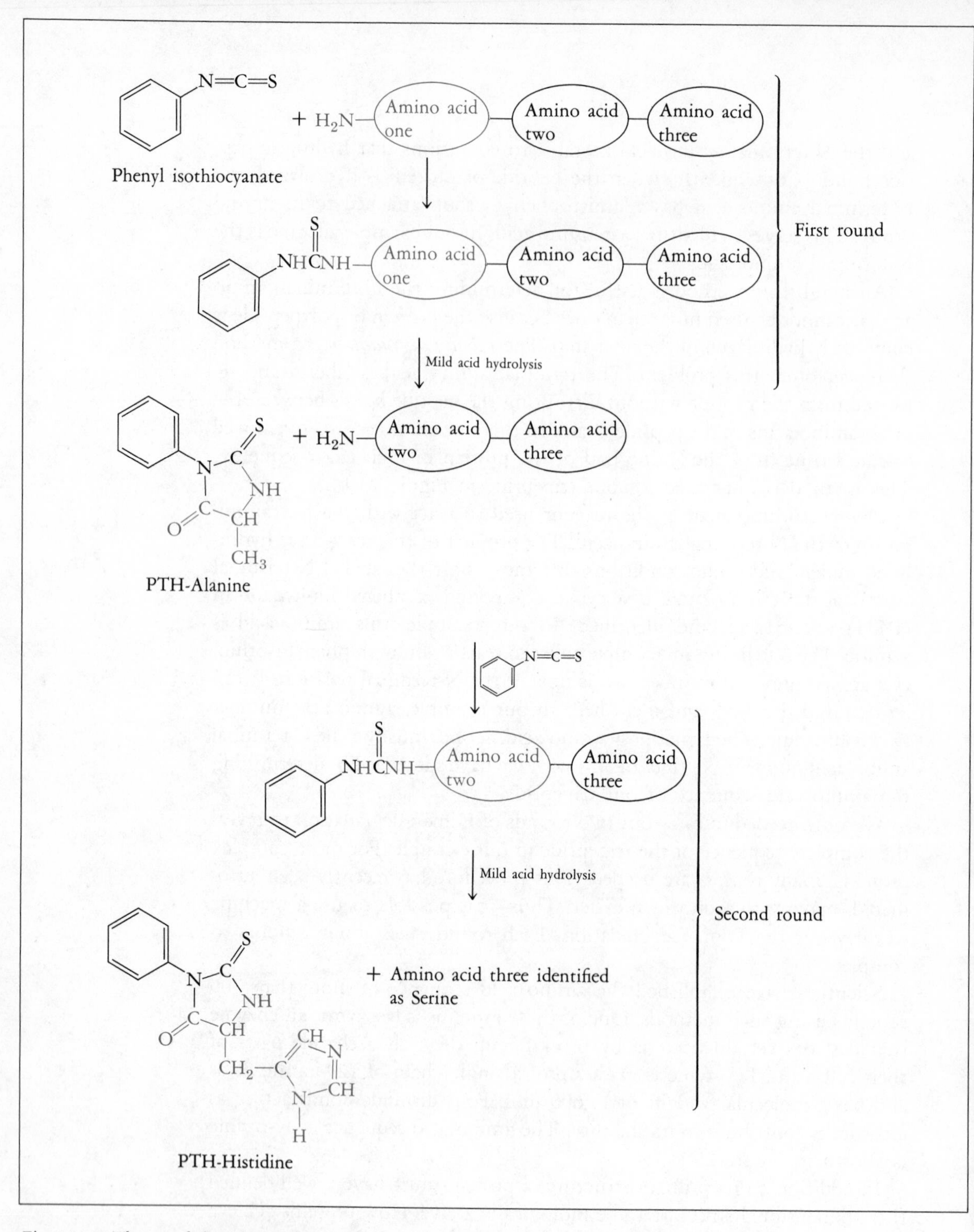

Fig. 23-13. The use of the Edman degradation to determine the amino acid sequence of a tripeptide. The sequence is Ala-His-Ser.

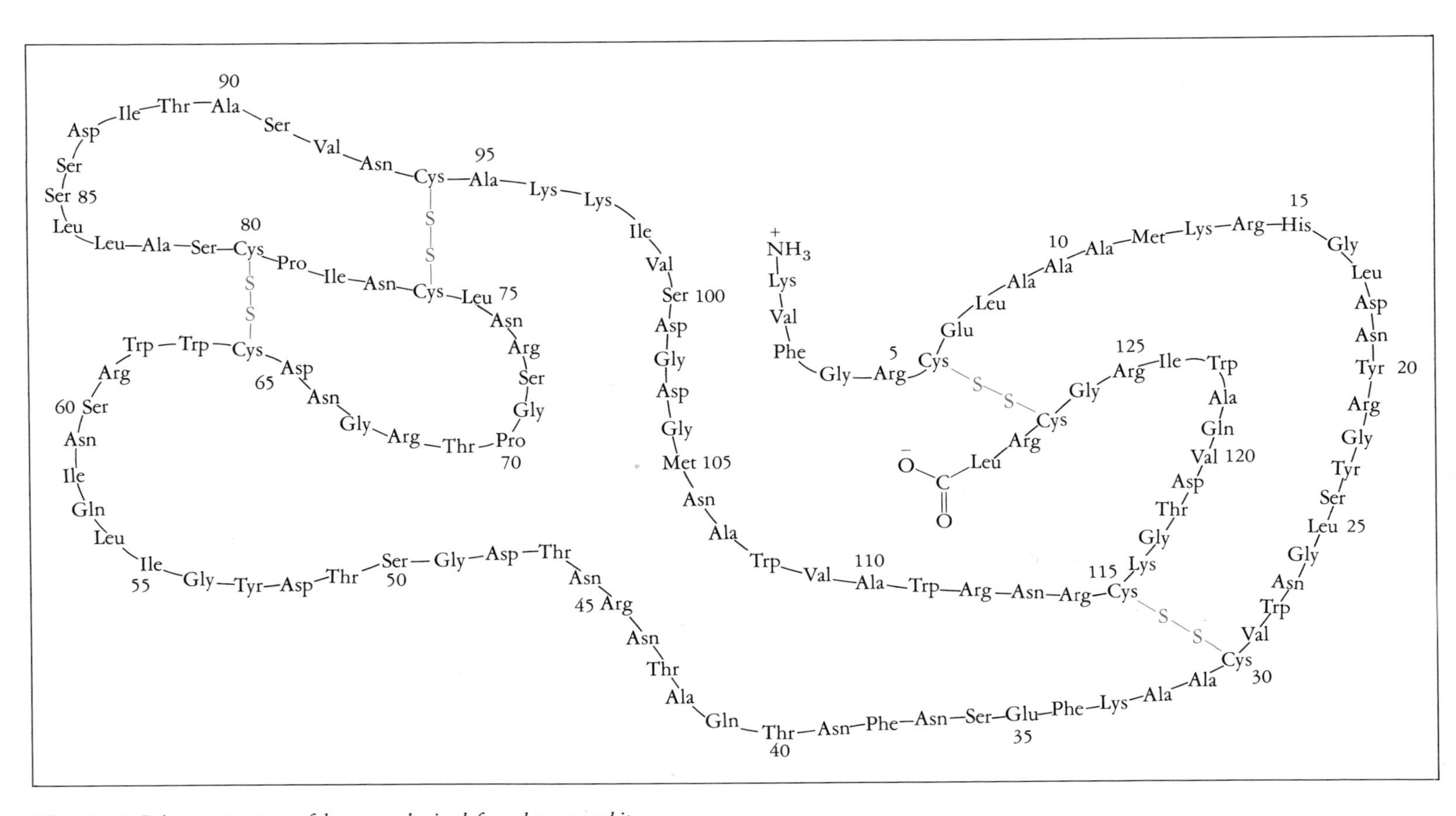

Fig. 23-14. Primary structure of lysozyme obtained from hen egg white.

23.6 SECONDARY STRUCTURES OF PROTEINS

The primary structure of lysozyme shown in Figure 23-14 does not represent the real three-dimensional structure of the protein. The chains of amino acids of this and other proteins do not extend aimlessly in space. Rather, they arrange themselves into definite periodic three-dimensional structures. These are called the *secondary structures* of proteins. Three such structures have been identified. They are the α helix, the β-pleated sheet, and the collagen triple helix.

The α-helical structure is shown in Figure 23-15b. In this structure, the chain of α-amino acids coils as a right-handed screw. This helix is formed and retains its shape because hydrogen bonds are formed between the amide

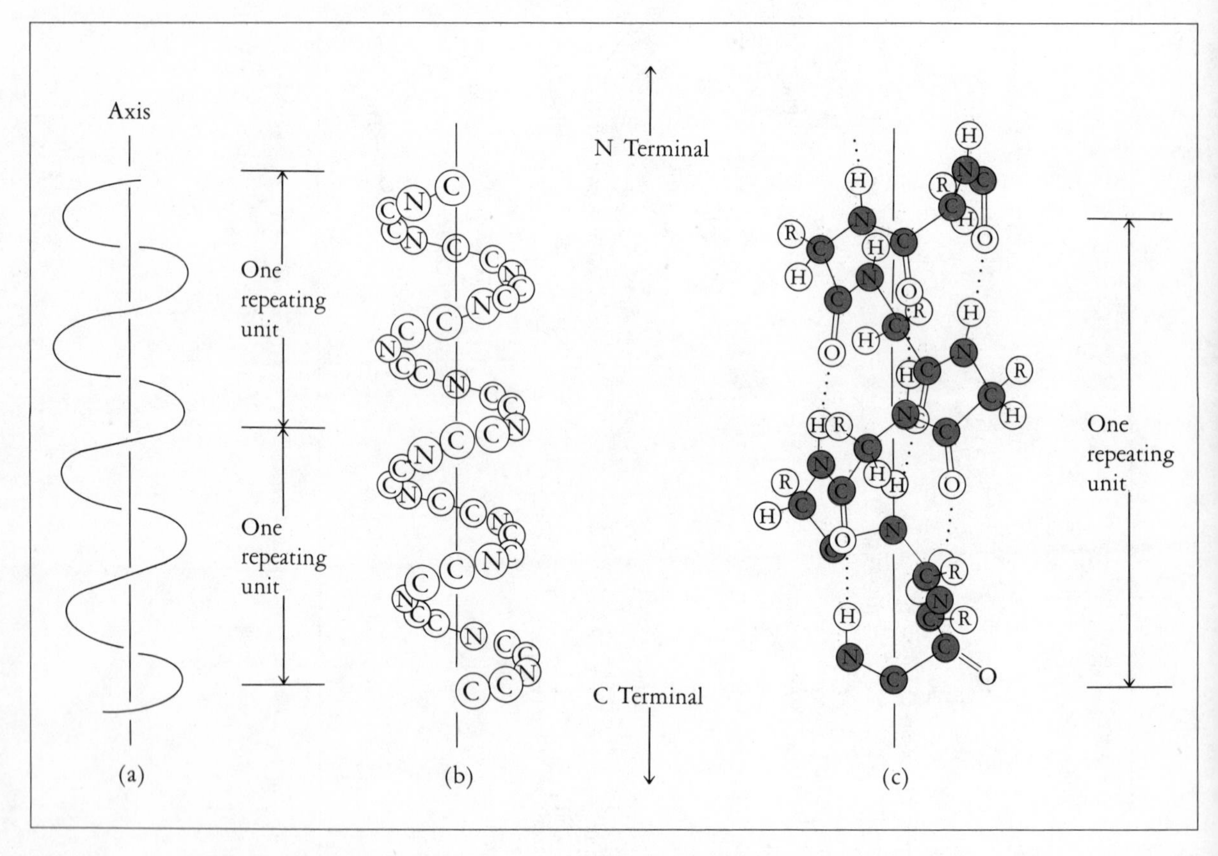

Fig. 23-15. (a) A right-handed helix. (b) The carbon and nitrogen atoms that form the backbone of a polypeptide chain arranged in an α-helix. (c) A ball-and-stick model of an α-helix. The dots represent intramolecular hydrogen bonds.

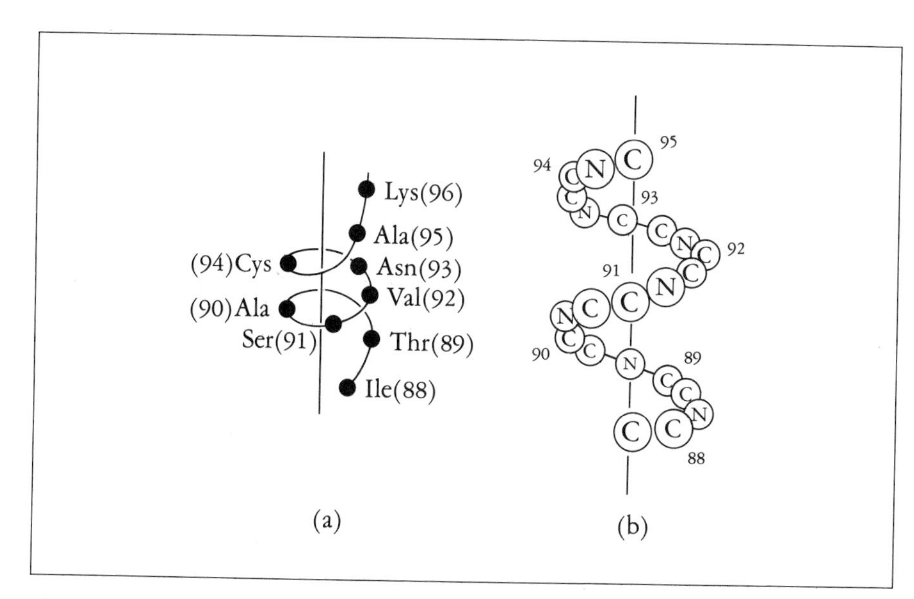

Fig. 23-16. Amino acid residues 88 to 96 of lysozyme form a short length of α-helix. (a) Schematic representation in which the dots represent the α-carbon atoms of the amino acid residue. (b) The carbon and nitrogen atoms that form the backbone of the α-helix portion of lysozyme.

hydrogen of one peptide bond with the carbonyl oxygen that is located above it on the next turn of the helix, as shown in Figure 23-15c. From x-ray studies, it is known that each turn of the helix contains about four amino acids. Notice also that side chains of the amino acid residues project outward from the helix.

The extent of the α-helical structure of proteins is highly variable. In some, the α helix is the major structural feature, whereas in others no α helix is present. For example, only a part of lysozyme exists as an α helix, as shown in Figure 23-16.

The β-pleated sheet structure is illustrated in Figure 23-17. In this structure, the polypeptide chains are arranged side by side and are held together by hydrogen bonds between the chains. Adjacent polypeptide chains run in opposite directions. The overall appearance of this structure is a pleated sheet in which the side chains of the amino acids project above and below the pleated surface.

Silk fiber is a protein that is known to exist predominantly as a β-pleated sheet. Short portions of the polypeptide chain of many proteins are arranged in a way similar to the β-pleated sheet. An example is lysozyme, as shown in Figure 23-18.

The triple helix is the unique structural feature of collagen, the protein that gives strength to bones, tendons, and skin. The molecules of collagen are made up of a polypeptide called tropocollagen. Each molecule of collagen is made up of three tropocollagen polypeptide chains wound around

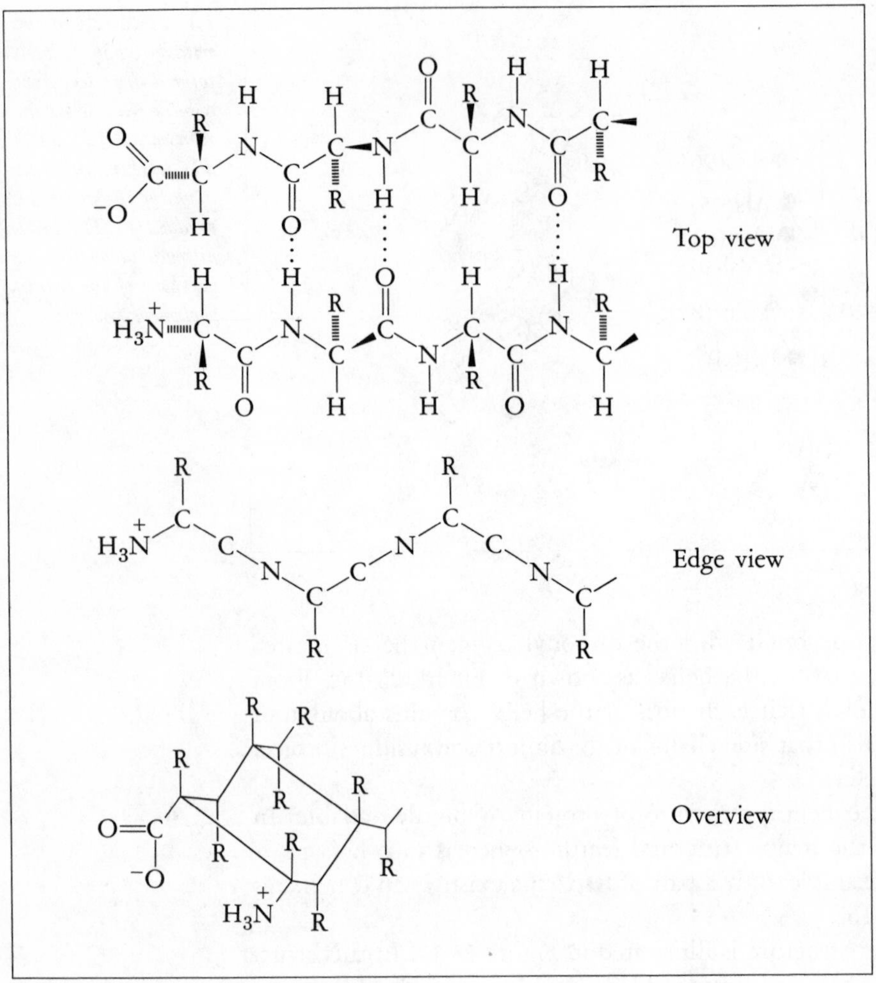

Top view

Edge view

Overview

Fig. 23-17. The top, edge, and overview of a β-pleated sheet.

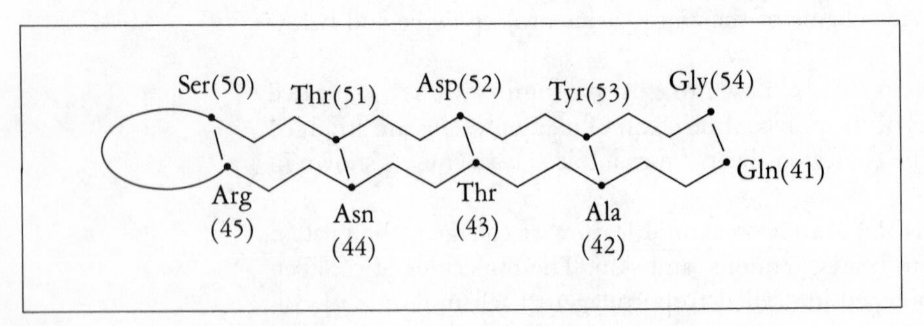

Fig. 23-18. Amino acid residues 41 through 45 and 50 through 54 of lysozyme fold back against each other to form a pleated sheet. A dot represents the α-carbon of the amino acid residue.

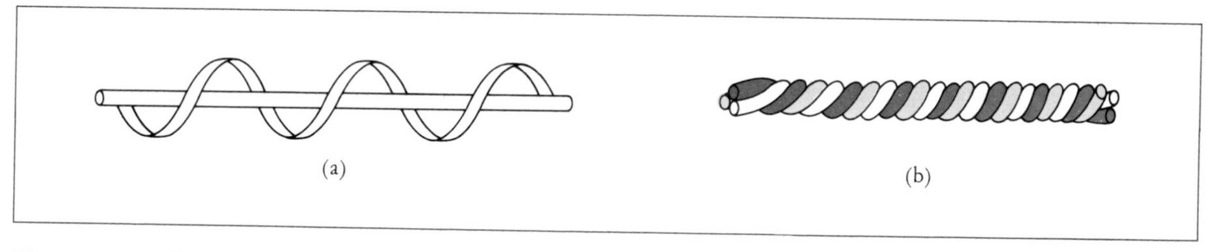

Fig. 23-19. (a) The helical structure of tropocollagen. (b) The triple helix formed by three tropocollagen helices winding about each other.

each other to form a triple helix. Such an arrangement looks like a three-strand rope, as shown in Figure 23-19. Each strand of the triple helix exists in the form of a helix. Thus, the triple helix is formed by winding three helices around each other. These helices are held together by hydrogen bonds. In this way, a stiff long fiber is formed that has remarkable strength. A load of 20 pounds is needed to break a fiber one sixteenth of an inch thick.

Hydrogen bonds are extremely important in establishing and maintaining the secondary structure of proteins. Individually, a hydrogen bond is weak; but several of them working together are a major force that holds the protein in its particular shape.

There are other attractive forces, in addition to hydrogen bonds, between parts of a protein that are important in determining their shape. We will examine these in the next section.

23.7 TERTIARY STRUCTURE OF PROTEINS

We have learned that proteins are made up of a particular sequence of amino acids. Because of this sequence, which is unique to each protein, they arrange themselves into particular structures (α helices, β-pleated sheets, or triple helices). It has been found that these structures can undergo further folding and bending, resulting in what is called the *tertiary structure* of proteins.

Attractive forces between the amino acid side chains making up the protein are responsible for the tertiary structure. These interactions and their effects on the structure of an α helix are shown in Figure 23-20. Among these attractive forces are disulfide bonds (see Sections 23.3 and 16.8), hydrophobic interactions (see Section 6.3), hydrogen bonds (see Sec-

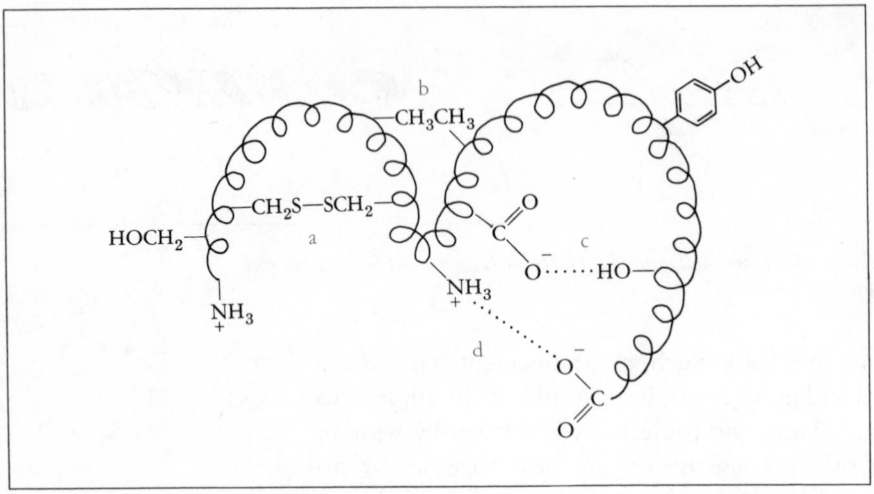

Fig. 23-20. The attractive interactions that stabilize the tertiary structure of a protein: (a) disulfide bond; (b) hydrophobic interactions; (c) hydrogen bond; (d) salt bridge.

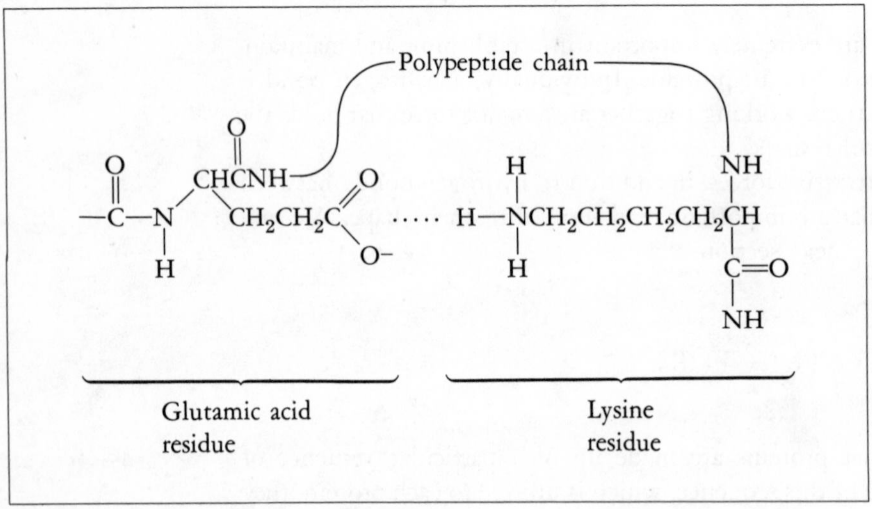

Fig. 23-21. A salt bridge between remotely situated glutamic acid and lysine residues.

tion 6.3), and salt bridges. We have encountered all of these interactions except salt bridges.

A salt bridge results from the attraction between two charged groups, one positive and the other negative, each located some distance apart on the same chain. For example, the attraction between the carboxylate ion on the side chain of a glutamic acid residue and the ammonium ion on the side chain of a remote lysine residue helps maintain a fold in the protein structure, as shown in Figure 23-21.

In the tertiary structure of a protein, the hydrophobic groups are usually placed inside the folds and the hydrophilic side chains remain on the outside, where they are exposed to the aqueous environment of the cell. This structure is very similar to that of micelles and bimolecular sheets (see Section 22.6). In all three, a hydrophobic environment is created in an aqueous solution. This is an important factor in the biological role of proteins, as we will learn in Chapter 25.

So far, we have learned about the structure of proteins made up of only one polypeptide chain. But many proteins contain more than one chain. This introduces another structural feature.

23.8 QUATERNARY STRUCTURE OF PROTEINS

Many proteins are made up of several identical or closely related polypeptide chains. The way that these polypeptide chains, called subunits, are joined together to form the protein is called its *quaternary structure.* Most proteins of molecular weight greater than 50,000 amu consist of a number of noncovalently joined subunits. Hemoglobin is an example of such a protein, as shown in Figure 23-22. It is made up of four separate subunits. Two, called the α chains, contain 141 amino acids each; the other two, called the β

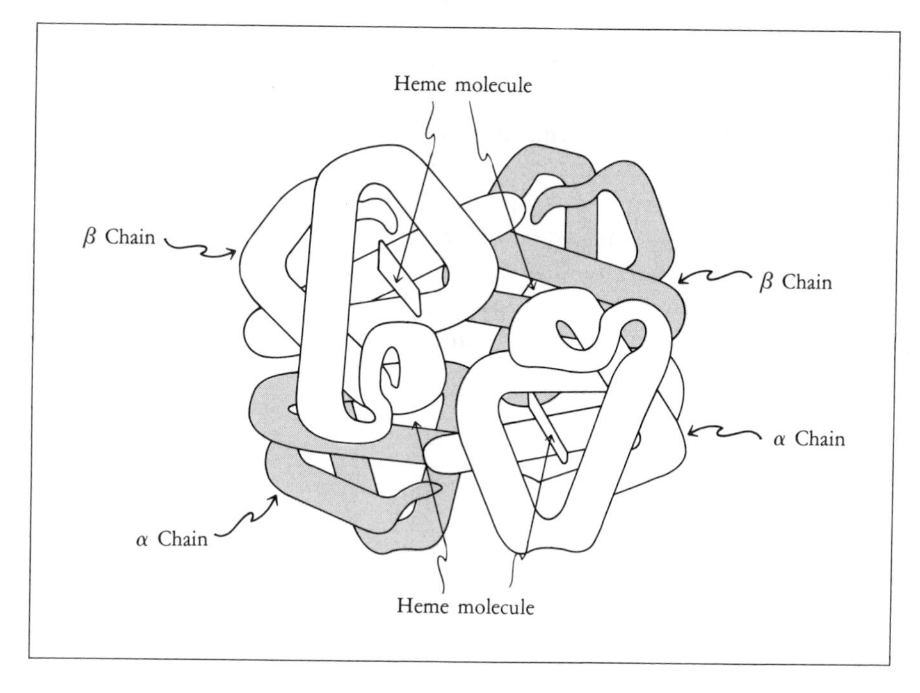

Fig. 23-22. The quaternary structure of hemoglobin. The hemoglobin molecule is made up of four polypeptide chains. Two identical chains are called α-chains, and the other two identical chains are called β-chains. Each chain encloses a heme molecule.

chains, contain 146 amino acids each. These subunits are held together by the attractive interactions discussed in Section 23.7. Each subunit also contains one molecule of heme, whose structure was given in Section 13.6.

EXERCISE 23-13 Adrenocorticotropin, a hormone of the anterior pituitary gland, consists of a single chain of 39 amino acid residues. Why doesn't this polypeptide have a quaternary structure?

In the previous three sections, we have concentrated on describing the three-dimensional structures that are found to a larger or smaller extent in all proteins. While such descriptions are vital to an understanding of their biological activity, it is not a convenient way of classifying proteins. Instead, proteins are usually classified according to their functions, overall shape, or composition.

23.9 CLASSIFYING PROTEINS

Proteins have been classified in three different ways. They can be broadly classified as *fibrous* or *globular* according to their overall shape. Fibrous proteins are made up of polypeptide chains that run parallel to an axis and are held together by hydrogen and disulfide bonds. These structures are strong and are generally insoluble in water. Examples of such proteins are collagen (Figure 23-19), silk, and the α-keratins of hair and wool. Globular proteins are made up of one or more α helices tightly rolled into a compact sphere, as shown in Figure 23-23. Globular proteins are usually soluble in water. Most enzymes, hormones, and antibodies are globular proteins.

A second method of classifying proteins is according to their composition. A *simple* protein forms only α-amino acids when hydrolyzed in strong acid solution. Albumins present in egg white (egg albumin) and blood (blood albumins) are the most common examples of simple proteins. They are soluble in water and dilute salt solutions. A *conjugated protein* yields

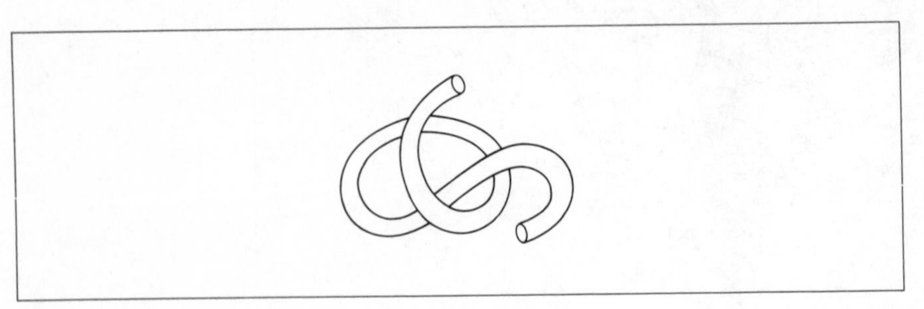

Fig. 23-23. A globular protein made up of a single α-helical coil.

nonpolypeptide material, called *prosthetic groups,* as well as amino acids when hydrolyzed. Hemoglobin (Figure 23-22) is an example of such a protein. The heme molecules are the prosthetic groups.

A third method of classifying proteins is by their function. The following are a few of the functions of proteins:

1. Enzymes	Biological catalysts that are vital to all living systems. They are all simple or conjugated proteins.
2. Structural proteins	Proteins that hold living systems together. The most common example is collagen (Figure 23-19).
3. Hormones	Proteins that act as messengers. The hormone insulin is an example.
4. Transport proteins	Proteins that carry molecules and ions from one place to another in the living system. Hemoglobin is an example; it carries oxygen from the lungs to cells.
5. Protective proteins	Proteins that destroy any foreign substance released into the living system by an infectious agent. An example is gamma globulin.
6. Toxins	Proteins that are poisons. Snake venom is an example.

No matter how proteins are classified, they all undergo denaturation under certain conditions.

23.10 DENATURATION OF PROTEINS

Soluble or globular proteins have widely different structures and functions, yet they all have one common property. They are extremely sensitive to small changes in their environments. When these changes occur, proteins lose all or part of their biological activity. That is, the proteins are *denatured.*

Denaturation occurs at the molecular level by disruption of the attractive forces (hydrogen bonds, disulfide bonds, hydrophobic attractions, and salt bridges) that maintain the unique secondary and tertiary structures of proteins. This is illustrated in Figure 23-24. Several chemicals or conditions are known to denature proteins. The following are the most common.

1. *Acids or bases:* The addition of large quantities of acids or bases causes changes in the state of ionization of the carboxyl and amino groups on the side chains of proteins. This interferes with the salt bridges and

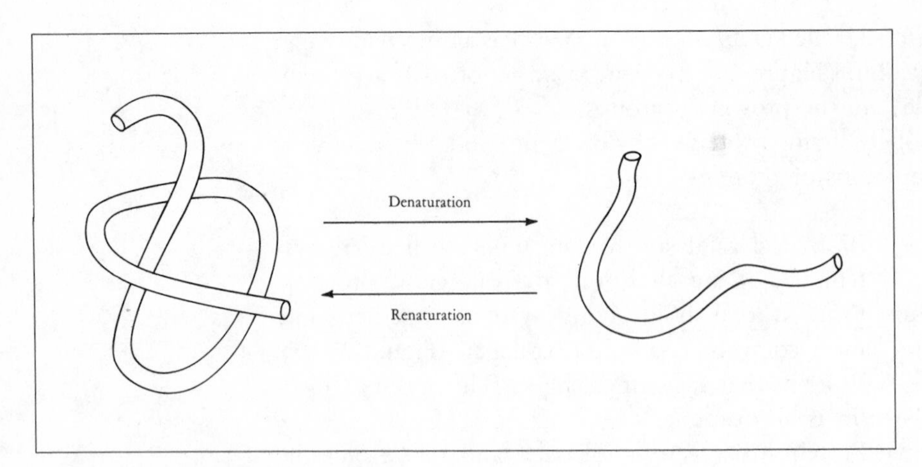

Fig. 23-24. The denaturation and renaturation of protein.

causes the protein to lose part of its structure. The amide bonds are eventually broken if the protein is left in a strong acid or base solution for a long time.

2. *Heat and ultraviolet light:* Heat and ultraviolet light cause proteins to coagulate, that is, to form an insoluble mass. For example, boiling or frying an egg causes the egg white protein to coagulate.
3. *Organic solvents:* Solvents like ethyl or isopropyl (rubbing) alcohol form intermolecular hydrogen bonds with proteins. As a result, they replace the intramolecular hydrogen bonds of the protein, resulting in loss of its three-dimensional structure.
4. *Heavy metal ions:* Cations such as Hg^{+2}, Ag^{+}, and Pb^{+2} react with the carboxylate ions and the thiol groups of the side chains of the proteins. This not only causes a disruption in the salt bridges and disulfide bonds, but also causes the protein to precipitate from solution.
5. *Vigorous agitation:* Denaturation of aqueous solutions of many proteins occurs by violently whipping or shaking them. An example is beating egg white into a meringue.
6. *Urea:* A solution of urea ($NH_2\overset{\overset{O}{\|}}{C}NH_2$) disrupts the intramolecular hydrogen bonds of proteins. Urea acts much like an alcohol because it also can form intermolecular hydrogen bonds with proteins.

The denaturation of a protein is sometimes reversible. Many examples are known in which a denatured protein molecule spontaneously regains its biological activity once it is returned to its natural environment. Such a process, called *renaturation,* returns the protein to its original structure, as shown schematically in Figure 23-24. From this fact we know that the amino acid sequence of the polypeptide chain (its primary structure) contains the necessary information to form its three-dimensional structure spontaneously. Furthermore, we know that this structure determines its biological activity.

23.11 SUMMARY

Proteins are involved in nearly all biological processes in living systems. Proteins are polymers with amino acids as monomers. The proteins of all living things from bacteria to humans are made from the same set of 20 amino acids. All but one of these amino acids are chiral and have the L configuration. The amino acids are joined together by peptide bonds to form polypeptides. A protein consists of one or more polypeptide chains. Each kind of protein has a unique sequence of amino acids that is genetically determined. This sequence of amino acids is called the primary structure of a protein.

The biological function of a protein is determined by the three-dimensional structure of its atoms in space. Three regular repeating three-dimensional structures of proteins are known: the α helix, the β-pleated sheet, and the collagen triple helix. These are the secondary structures of proteins. These secondary structures can be folded and bent to form the tertiary structure of a protein. Many proteins contain more than one polypeptide chain. The way that these chains are joined together is called the quaternary structure of a protein.

Attractive forces between various parts of the polypeptide chains are responsible for maintaining the secondary, tertiary, and quaternary structures of proteins. Among these forces are hydrogen bonds, disulfide bonds, hydrophobic attractions, and salt bridges. Any chemical or physical change that disrupts these attractive forces results in the loss of part or all of the biological activity of a protein. This disruption is called denaturation. Many actions such as heating or vigorous agitation can cause a protein to be denatured.

Denaturation can sometimes be reversed so that the protein regains all of its former biological activity. This fact indicates that the amino acid sequence of a protein contains all the information needed to specify its three-dimensional structure, and that this structure determines its biological activity.

REVIEW OF TERMS, CONCEPTS, AND REACTIONS

Terms

AMINO ACID A compound that contains both amino and carboxylic acid groups.

AMPHOTERIC SUBSTANCE A compound or ion that reacts with both acids and bases.

CONJUGATED PROTEIN A protein that yields nonprotein material as well as α-amino acids on hydrolysis.

C-TERMINAL AMINO ACID The amino acid at one end of a polypeptide chain that contains a free carboxylate ion.

DENATURATION Any chemical or physical change that causes a protein to lose part or all of its biological activity.

EDMAN DEGRADATION A method of determining the primary structure of proteins.

ESSENTIAL AMINO ACIDS Amino acids that cannot be made by a living system and must be obtained from its food.

FIBROUS PROTEIN A protein that is shaped like a fiber or rod.

GLOBULAR PROTEIN A protein that has a spherical shape.

ISOELECTRIC POINT The pH at which the structure of an amino acid has no net charge.

N-TERMINAL AMINO ACID The amino acid at one end of a polypeptide chain that contains a free amino group.

PEPTIDE BOND Another name for an amide bond ($-C(=O)NH-$).

POLYPEPTIDES A polymer consisting of 10 to 100 amino acids joined by peptide bonds.

PRIMARY STRUCTURE OF PROTEINS The unique sequence of amino acids in each kind of protein.

PROSTHETIC GROUP The nonprotein material that is a part of a conjugated protein.

PROTEIN A polymer made up of more than 100 amino acids joined by peptide bonds.

QUATERNARY STRUCTURE OF PROTEINS The way several polypeptide chains are joined together to form a protein.

RESIDUE An amino acid unit in a polypeptide.

SALT BRIDGE The attraction between two charged groups, one positive and the other negative, each located some distance apart on the same chain.

SECONDARY STRUCTURE OF PROTEINS The three-dimensional structure of polypeptides and proteins.

SIMPLE PROTEIN A protein that yields only α-amino acids on hydrolysis.

TERTIARY STRUCTURE OF PROTEINS The folding and bending of the secondary structures of polypeptides and proteins.

ZWITTER ION A compound that contains both positively charged and negatively charged functional groups.

Concepts

1. All polypeptides and proteins are made from the same set of 20 α-amino acids.
2. Every protein has a unique sequence of amino acids that determines its three-dimensional structure.
3. The three-dimensional structure of a protein determines its biological activity.
4. Attractive forces such as hydrogen bonds, hydrophobic attractions, salt bridges, and covalent bonds such as disulfide bonds hold polypeptides and protein molecules in their three-dimensional structures.
5. Denaturation occurs whenever any chemical or physical change disrupts these attractive forces or covalent disulfide bonds and destroys the three-dimensional structure of a protein.

Reactions

1. Hydrolysis (see Section 23.5).

$$\overset{+}{N}H_3CH(CH_2C_6H_5)C(=O)NHCH(CH_3)C(=O)NHCH(CH_2OH)C(=O)NHCH_2C(=O)O^- \xrightarrow[\text{concentrated HCl}]{H_2O,\ \Delta} \overset{+}{N}H_3CH(CH_2C_6H_5)C(=O)OH + \overset{+}{N}H_3CH(CH_3)CO_2H + \overset{+}{N}H_3CH(CH_2OH)CO_2H + \overset{+}{N}H_3CH_2CO_2H$$

2. Acid-base reactions of the zwitter-ion form of amino acids (see Section 23.2).

$$\overset{+}{N}H_3\underset{\underset{R}{|}}{C}HC\begin{matrix}\nearrow O \\ \searrow OH\end{matrix} \xleftarrow{H^+} \overset{+}{N}H_3\underset{\underset{R}{|}}{C}HC\begin{matrix}\nearrow O \\ \searrow O^-\end{matrix} \xrightarrow{OH^-} NH_2\underset{\underset{R}{|}}{C}HC\begin{matrix}\nearrow O \\ \searrow O^-\end{matrix}$$

3. Reactions with heavy metals (see Section 23.3).

$$2\,\overset{+}{N}H_3\underset{\underset{R}{|}}{C}HC\begin{matrix}\nearrow O \\ \searrow O^-\end{matrix} + M^{+2} \longrightarrow \left[\overset{+}{N}H_3\underset{\underset{R}{|}}{C}HC\begin{matrix}\nearrow O \\ \searrow O^-\end{matrix}\right]_2 M^{+2}\downarrow$$

$$2\,\overset{+}{N}H_3\underset{\underset{CH_2SH}{|}}{C}HC\begin{matrix}\nearrow O \\ \searrow O^-\end{matrix} + M^{+2} \longrightarrow \left[\overset{+}{N}H_3\underset{\underset{CH_2S}{|}}{C}HC\begin{matrix}\nearrow O \\ \searrow O^-\end{matrix}\right]_2 M + 2\,H^+$$

$M = Hg$ or Pb

4. Determining N-terminal amino acid (see Section 23.5).

$$\text{1-F-2,4-}(NO_2)_2C_6H_3 + NH_2\overset{\overset{R}{|}}{C}HC\begin{matrix}\nearrow O \\ \searrow NH\text{—(Polypeptide chain)}\end{matrix} \longrightarrow 2,4\text{-}(NO_2)_2C_6H_3\text{—}NH\overset{\overset{R}{|}}{C}HC\begin{matrix}\nearrow O \\ \searrow NH\text{—(Polypeptide chain)}\end{matrix}$$

$$\xrightarrow[\Delta]{HCl\;|\;H_2O} 2,4\text{-}(NO_2)_2C_6H_3\text{—}NH\overset{\overset{R}{|}}{C}HC\begin{matrix}\nearrow O \\ \searrow OH\end{matrix} + \text{Amino acids of polypeptide chain}$$

5. Edman degradation (see Section 23.5).

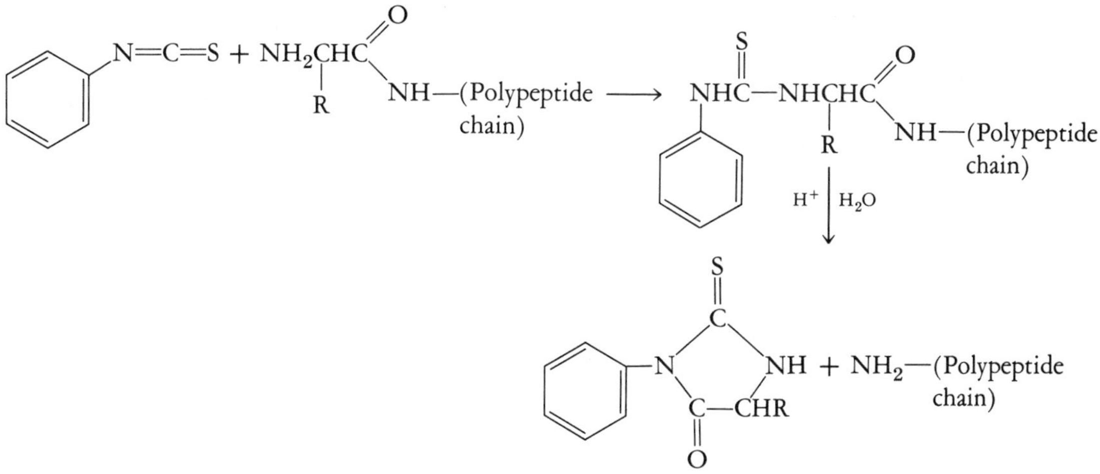

$$C_6H_5{-}N{=}C{=}S + NH_2CH(R)C({=}O){-}NH{-}\text{(Polypeptide chain)} \longrightarrow C_6H_5{-}NHC({=}S){-}NHCH(R)C({=}O){-}NH{-}\text{(Polypeptide chain)}$$

$$\xrightarrow{H^+,\ H_2O}\ \text{(ring: } C_6H_5{-}N{-}C({=}S){-}NH{-}CHR{-}C({=}O){-}\text{)} + NH_2{-}\text{(Polypeptide chain)}$$

EXERCISES 23-14 Write the three-dimensional structure for each of the following amino acids:
(a) L-alanine (b) L-serine (c) L-asparagine
(d) L-cysteine (e) L-lysine (f) L-glutamic acid

23-15 Write the structure of all possible tripeptides that contain one each of the following amino acids: threonine, proline, and arginine. Give a name to each structure.

23-16 Write the equation for the acid-catalyzed hydrolysis of each of the following peptides:

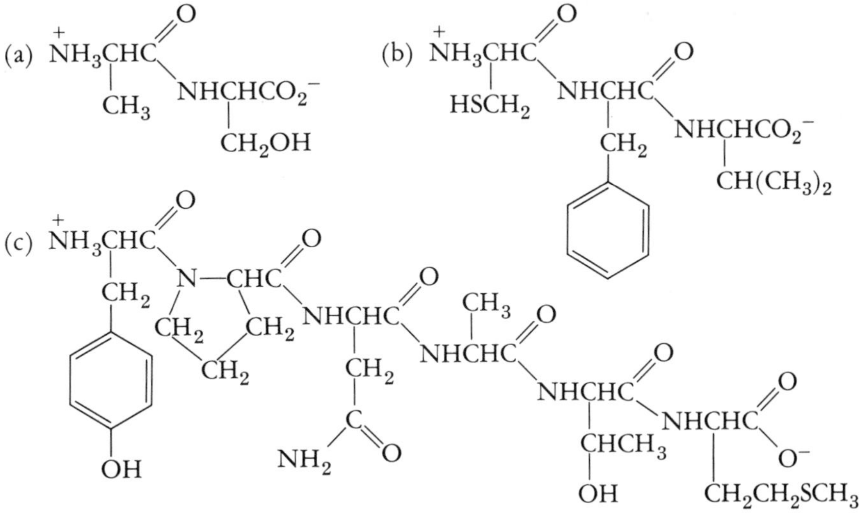

(a) $\overset{+}{N}H_3CH(CH_3)C({=}O){-}NHCH(CH_2OH)CO_2^-$

(b) $\overset{+}{N}H_3CH(CH_2SH)C({=}O){-}NHCH(CH_2C_6H_5)C({=}O){-}NHCH(CH(CH_3)_2)CO_2^-$

(c) $\overset{+}{N}H_3CH(CH_2C_6H_4OH)C({=}O){-}N\text{(pyrrolidine ring: }CH_2CH_2CH_2\text{)}{-}CHC({=}O){-}NHCH(CH_2C({=}O)NH_2)C({=}O){-}NHCH(CH_3)C({=}O){-}NHCH(CHCH_3OH)C({=}O){-}NHCH(CH_2CH_2SCH_3)C({=}O)O^-$

23-17 A pentapeptide is reacted with 2,4-dinitrofluorobenzene. After hydrolysis, the following compounds are isolated. Which is the N-terminal amino acid?

(a) $\overset{+}{N}H_3\underset{CH_3}{CH}C(=O)O^-$ (b) $\overset{+}{N}H_3\underset{CH_2OH}{CH}C(=O)O^-$

(c) $2,4\text{-}(NO_2)_2C_6H_3\text{-}NH\underset{CH_2\text{-}(3\text{-indolyl, HN})}{CH}C(=O)O^-$

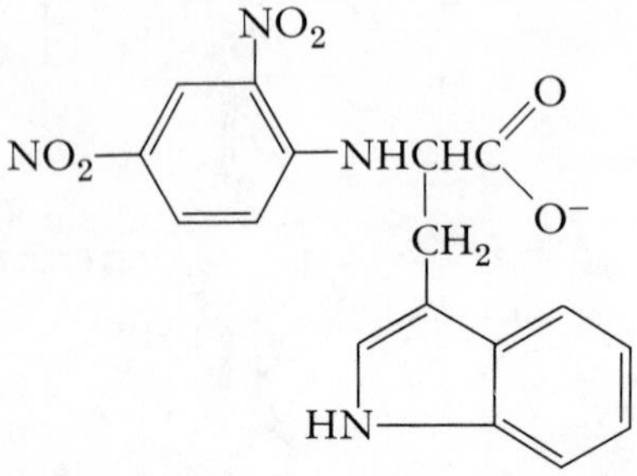

(d) $\overset{+}{N}H_3\underset{CH(CH_3)_2}{CH}C(=O)O^-$ (e) $\overset{+}{N}H_3\underset{CH_2C_6H_5}{CH}C(=O)O^-$

23-18 An Edman degradation is carried out on a hexapeptide. The results are given in the table:

	Round				
	1	2	3	4	5
Structure of PTH amino acid	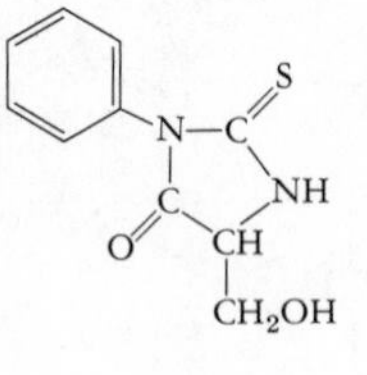 $C_6H_5\text{-}N\text{-}C(=S)\text{-}NH\text{-}\underset{CH_2OH}{CH}\text{-}C(=O)$ (ring)	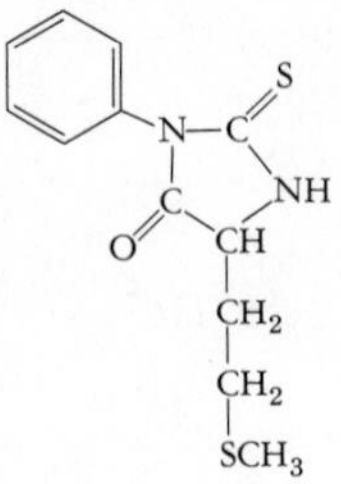 $C_6H_5\text{-}N\text{-}C(=S)\text{-}NH\text{-}\underset{CH_2CH_2SCH_3}{CH}\text{-}C(=O)$ (ring)	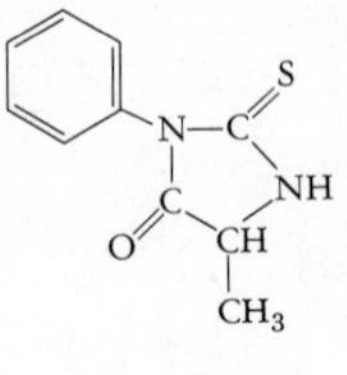 $C_6H_5\text{-}N\text{-}C(=S)\text{-}NH\text{-}\underset{CH_3}{CH}\text{-}C(=O)$ (ring)	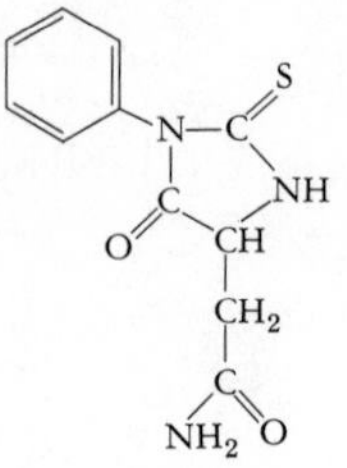 $C_6H_5\text{-}N\text{-}C(=S)\text{-}NH\text{-}\underset{CH_2C(=O)NH_2}{CH}\text{-}C(=O)$ (ring)	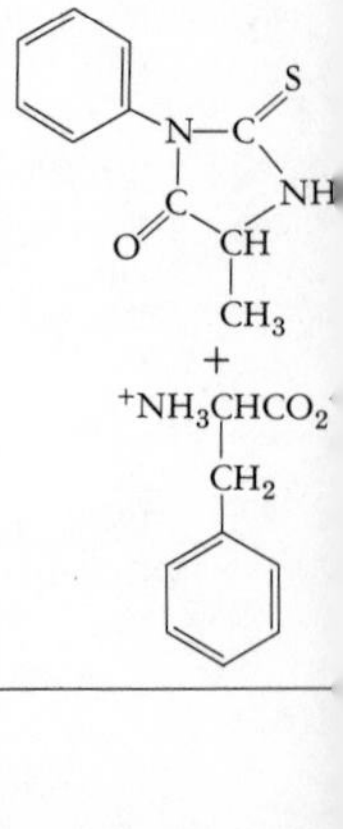 $C_6H_5\text{-}N\text{-}C(=S)\text{-}NH\text{-}\underset{CH_3}{CH}\text{-}C(=O)$ (ring) + $^+NH_3\underset{CH_2C_6H_5}{CH}CO_2^-$

(a) What is the amino acid composition of the hexapeptide?
(b) What is its amino acid sequence?

23-19 Covalent bonds are important in which structural feature of polypeptides and proteins?

23-20 Name and give an example of each of the noncovalent forces that are responsible for the three-dimensional structure of polypeptides and proteins.

23-21 What experimental evidence leads to the conclusion that the primary structure of a protein determines its biological activity?

NUCLEIC ACIDS 24

The unique feature of all living organisms is their ability to reproduce themselves. How this occurs at a molecular level has been a major scientific mystery until recently. Within the past 30 years, scientists have begun to understand the molecular basis of heredity and reproduction. They discovered that nucleic acids are the molecules responsible for transmitting the characteristics of a species from one generation to the next. Nucleic acids, like proteins and carbohydrates, are polymers made up of repeating monomers. It is the variety of ways in which these monomers are arranged in nucleic acids that is responsible, at the molecular level, for the characteristics that distinguish one species from another, transmit its characteristics to the next generation, and control its metabolism.

There are two types of nucleic acid: *r*ibo*n*ucleic *a*cids (RNA) and *d*eoxyribo*n*ucleic *a*cids (DNA). The master blueprint of an organism, in coded form, is DNA; RNA reads the DNA code and is involved in protein synthesis. We will learn in this chapter the structures of the compounds that make up nucleic acids and how they combine to form DNA and RNA. The structure of DNA will be examined, as will the way that DNA duplicates itself. Finally, we will learn how RNA is made from DNA, and how RNA acts as an intermediary in the synthesis of proteins. Let us start by examining the composition of nucleic acids.

24.1 COMPOSITION OF NUCLEIC ACIDS

The chemical composition of nucleic acids is best understood if we examine their stepwise hydrolysis products, given in Figure 24-1. Nucleic acids in

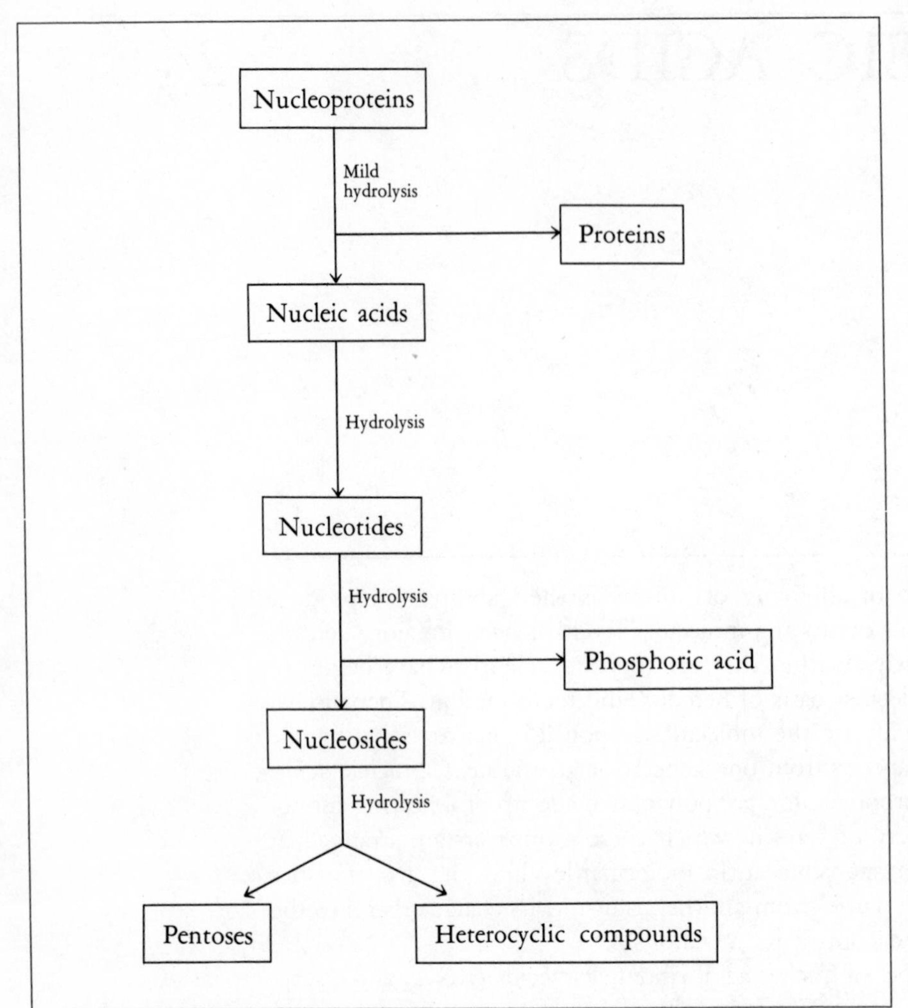

Fig. 24-1. Stages in the hydrolysis of nucleoproteins.

cells are usually found as *nucleoproteins,* that is, a complex of proteins and nucleic acids. The protein and nucleic acid parts of the complex can be separated by mild hydrolysis. The nucleic acid part can be hydrolyzed to a mixture of *nucleotides.* These nucleotides are composed of three parts. One part, phosphoric acid, can be removed by further hydrolysis. This leaves a smaller part called a *nucleoside* that yields a pentose sugar (see Section 21.1) and a heterocyclic compound (see Section 13.6) on hydrolysis.

Nucleic acids are composed of phosphoric acid, pentose sugars, and heterocyclic compounds. Only five different heterocyclic compounds and only two pentose sugars are needed to construct nucleic acids. Let us learn the structures of these seven compounds.

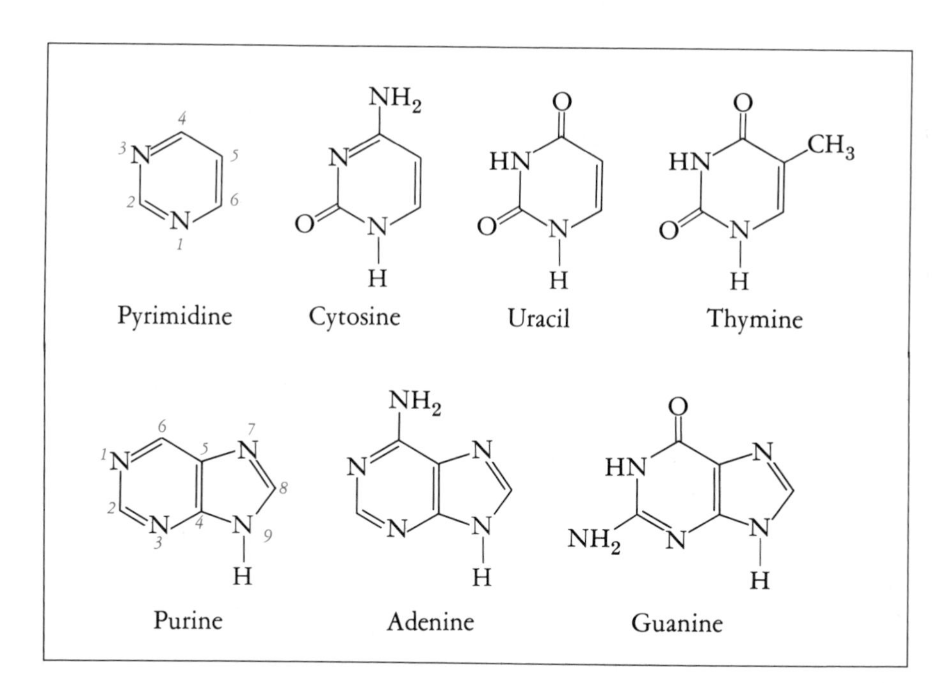

Fig. 24-2. The names and structures of the heterocyclic bases found in nucleic acids.

The Heterocyclic Compounds in Nucleic Acids

The structures and names of the heterocyclic compounds obtained from the hydrolysis of nucleic acids are given in Figure 24-2. These compounds are derivatives of either pyrimidine or purine. Three, *cytosine, uracil,* and *thymine,* are derivatives of pyrimidine. The other two, *adenine* and *guanine,* are derivatives of purine. When these compounds are dissolved in water, they form a basic solution (pH $>$ 7). For this reason, they are commonly called heterocyclic bases, or just bases.

Each atom of the rings of the five bases is given a number. These numbers are the same as the numbers given ring atoms of the corresponding purine or the pyrimidine, as shown in Figure 24-2. Remember these numbers, because we will refer to them later.

Both DNA and RNA contain adenine and guanine. However, they differ in the pyrimidine bases they contain: RNA contains cytosine and uracil, whereas DNA contains cytosine and thymine. DNA and RNA also contain different sugars.

The Sugars in Nucleic Acids

The Haworth projection formulas (see Section 21.3) of the two sugars *ribose* and *deoxyribose* are given in Figure 24-3. Both sugars are in the β form.

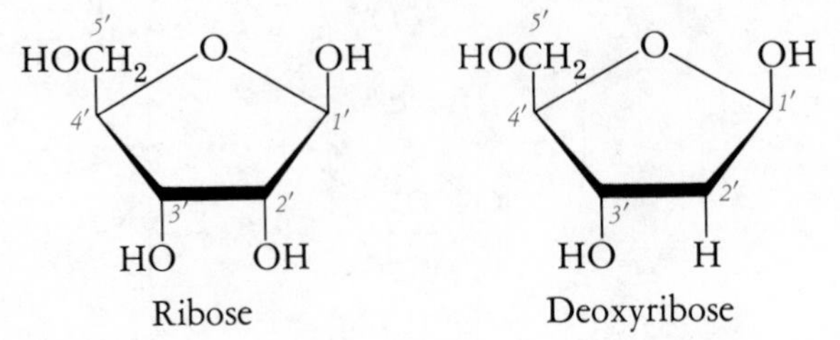

Ribose Deoxyribose

Fig. 24-3. The names and Haworth projection formulas of the two pentose sugars found in nucleic acids.

Table 24-1. The Composition of DNA and RNA

	Nucleic Acid	
Component	DNA	RNA
Purine bases	Adenine Guanine	Adenine Guanine
Pyrimidine bases	Cytosine Thymine	Cytosine Uracil
Pentose sugar	Deoxyribose	Ribose
Inorganic acid	Phosphoric acid	Phosphoric acid

Notice that each carbon atom of each sugar is given a number. When the sugar is part of a nucleic acid, the number has a prime symbol (1′, 2′, and so on) to distinguish the atoms in the sugar from those in the base.

A nucleic acid will contain either ribose or deoxyribose as its sugar, never both. The sugar in RNA is ribose, and that in DNA is deoxyribose. The differences in composition between DNA and RNA are summarized in Table 24-1.

Now that we have learned the structures of the various parts of nucleic acids, let us put them together.

24.2 STRUCTURE OF NUCLEOSIDES AND NUCLEOTIDES

Nucleosides and nucleotides are formed by combining the pentose sugars, the heterocyclic bases, and phosphoric acid in specific ways. Let us start by combining a pentose sugar and a base to form a nucleoside.

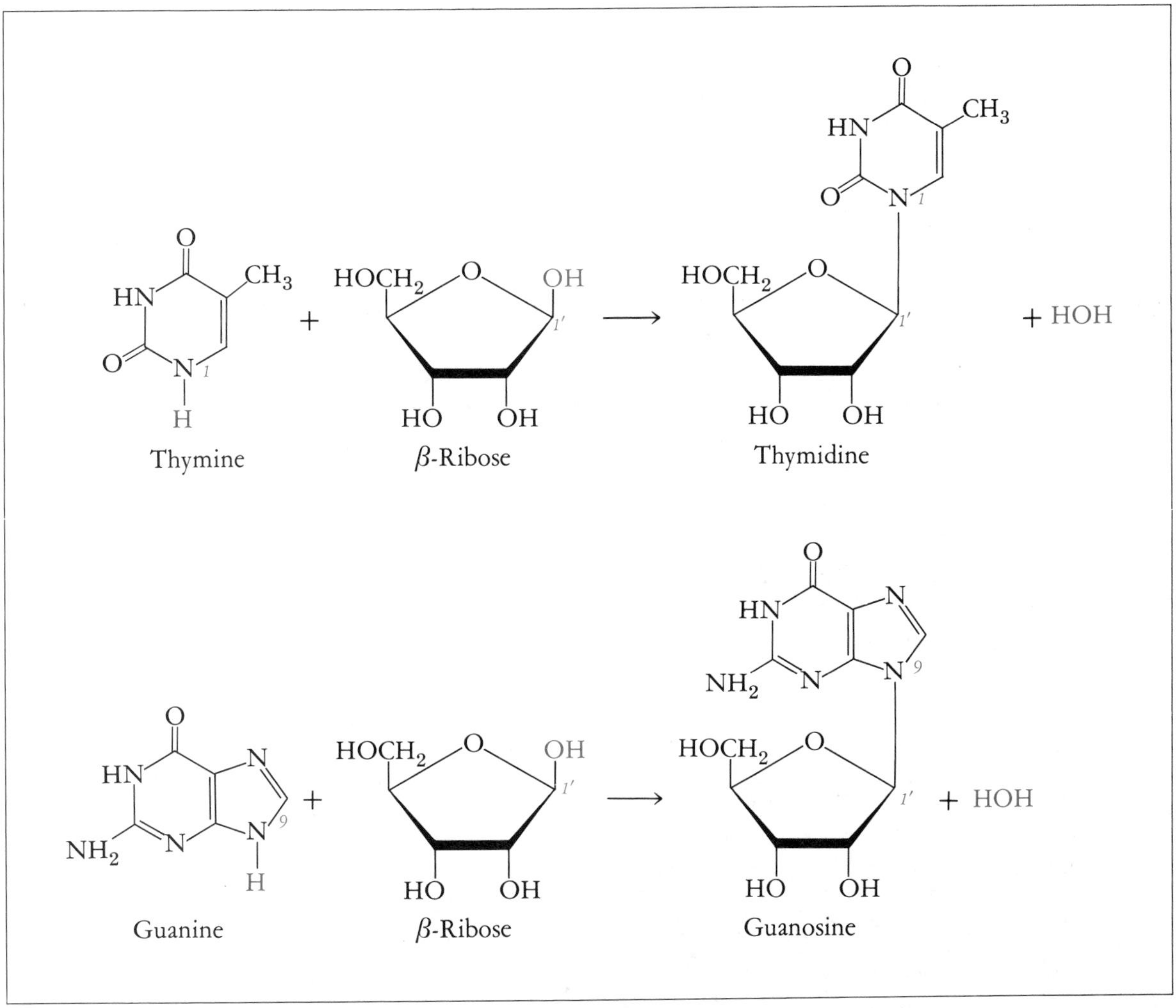

Fig. 24-4. Examples of the combination of a sugar and a base to form a nucleoside.

Nucleosides

Any one of the five heterocyclic bases can be combined with either β-ribose or β-deoxyribose to form a nucleoside. The two are joined by connecting carbon 1′ of the sugar and either nitrogen 1 of the pyrimidine base or nitrogen 9 of the purine base. Such a combination is shown in Figure 24-4. For such a combination to occur, a molecule of water must be eliminated. Notice that the N-glycoside bond (see Section 21.5) is always β in naturally occurring nucleosides.

*EXERCISE 24-1 Circle the N-glycoside bond in each nucleoside in Figure 24-4.

The common names of nucleosides indicate their structure. If the nucleoside contains a ribose unit and a purine base, its name is derived from the name of the purine base. This is done by changing the *-ine* ending of the base to *-osine.* For example, the nucleoside obtained from the reaction of adenine and ribose is called adenosine. The name of a nucleoside containing a pyrimidine base and a ribose sugar is also derived from the name of the base. But the ending of the pyrimidine base is changed to *-idine.* For example, the nucleoside formed from cytosine and ribose is called cytidine. If the bases are combined with deoxyribose, the prefix *deoxy-* is added to the name obtained, as explained before. For example, the nucleoside formed from cytosine and deoxyribose is called deoxycytidine. The name of the nucleoside for each sugar-base combination is given in Table 24-2.

Table 24-2. Names of Nucleosides

Base	Name When Combined with Ribose	Name When Combined with Deoxyribose
Cytosine	Cytidine	Deoxycytidine
Thymine	Thymidine	Deoxythymidine
Uracil	Uridine	Deoxyuridine
Adenine	Adenosine	Deoxyadenosine
Guanine	Guanosine	Deoxyguanosine

EXERCISE 24-2 Give the name of the nucleoside formed by combining the sugar and base in each of the following:
(a) ribose and guanine (b) thymine and deoxyribose
(c) deoxyribose and uracil

EXERCISE 24-3 Draw the structure of the base and the sugar that combine to form each of the following nucleosides:
(a) thymidine (b) deoxyguanosine (c) uridine

Combining a nucleoside with phosphoric acid forms nucleotides.

*The answers for the exercises in this chapter begin on page 899.

Nucleotides

Nucleotides are phosphate esters (see Section 18.12) of nucleosides. All naturally occurring nucleotides contain the phosphate group attached to carbon 5′ of the pentose in the nucleoside. The combination of phosphoric acid and two different nucleosides to form nucleotides are shown in Figure 24-5. Again, a molecule of water must be eliminated to form the phosphate ester.

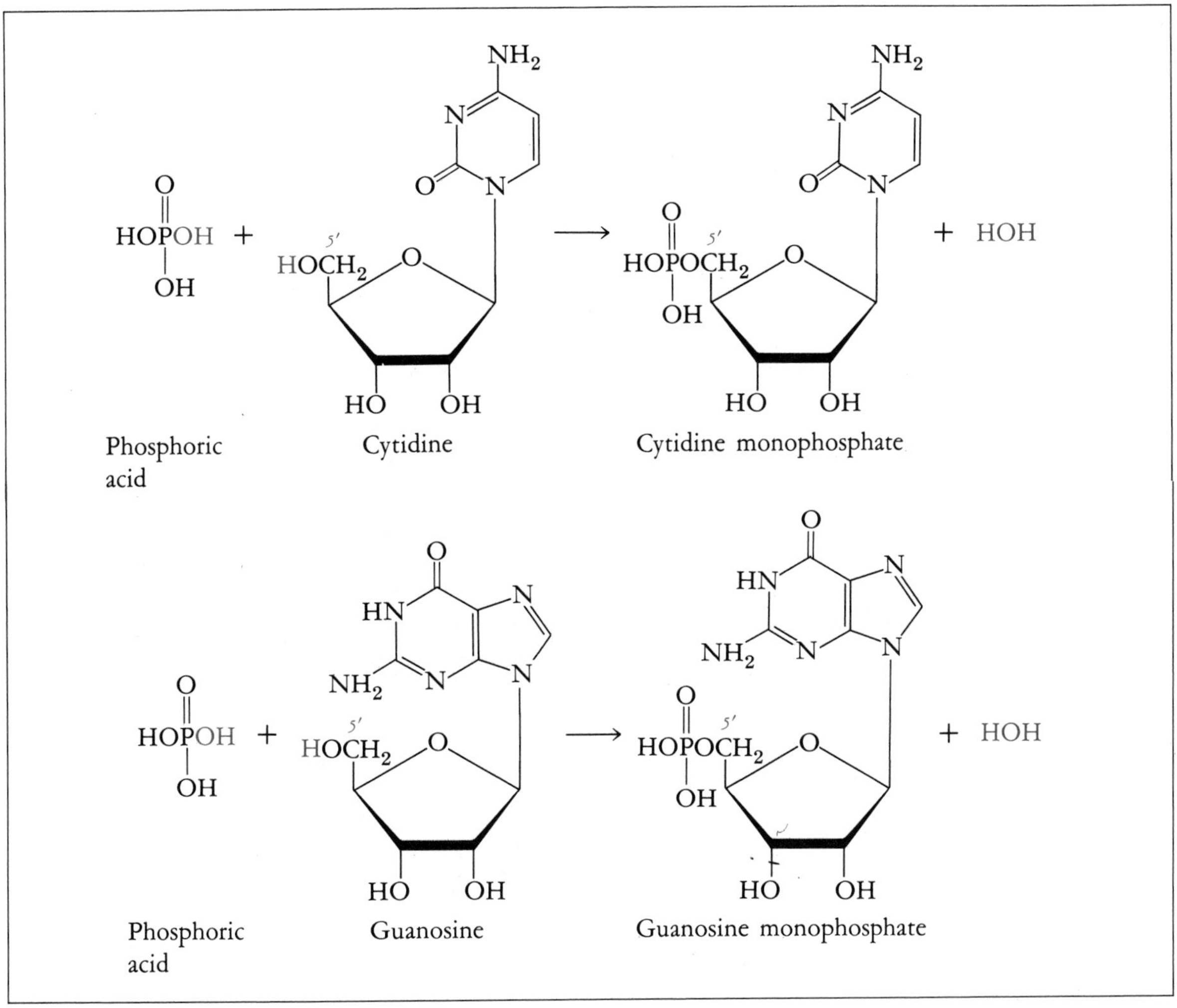

Fig. 24-5. Examples of the combination of phosphoric acid and a nucleoside to form a nucleotide. At the pH of the cell, the phosphate groups are ionized.

Table 24-3. Names of Nucleotides

Base	Name When Combined with Ribose	Abbreviation	Name When Combined with Deoxyribose	Abbreviation
Cytosine	Cytidine monophosphate	CMP	Deoxycytidine monophosphate	dCMP
Thymine	Thymidine monophosphate	TMP	Deoxythymidine monophosphate	dTMP
Adenine	Adenosine monophosphate	AMP	Deoxyadenosine monophosphate	dAMP
Guanine	Guanosine monophosphate	GMP	Deoxyguanosine monophosphate	dGMP
Uracil	Uridine monophosphate	UMP	Deoxyuridine monophosphate	dUMP

Nucleotides are named by combining the name of its nucleoside and the word monophosphate. For example, the name of the phosphate ester of adenosine is adenosine monophosphate. The names of other nucleotides are given in Table 24-3. Also included in Table 24-3 are the abbreviations used for the nucleotides. The letters A, T, U, C, and G stand for the ribonucleosides adenosine, thymidine, uridine, cytidine, and guanosine, respectively. The corresponding deoxyribonucleosides are indicated as dA, dT, dU, dC, and dG. The letter M stands for mono- and P stands for phosphate.

Individual nucleotides exist in cells. These free nucleotides usually exist as diphosphates or triphosphates. That is, they contain two and three phosphate groups attached to carbon 5′ of the pentose, as shown in Figure 24-6. The letters D and T in the abbreviation for the name of the nucleotide indicate that the nucleotide contains two or three phosphate groups, respectively. For example, CTP is the abbreviation for *c*ytidine *t*ri*p*hosphate. Di- and triphosphates are not a part of DNA or RNA. However, triphosphates are needed for the formation of polynucleotides, as we will learn.

24.3 POLYNUCLEOTIDES

Polynucleotides are polymers made up of repeating units of nucleotides. In this way, polynucleotides resemble proteins (see Chapter 23). In the case of proteins, the monomers are the 20 amino acids. However, only the four deoxyribose-containing nucleotides dAMP, dGMP, dCMP, and dTMP are

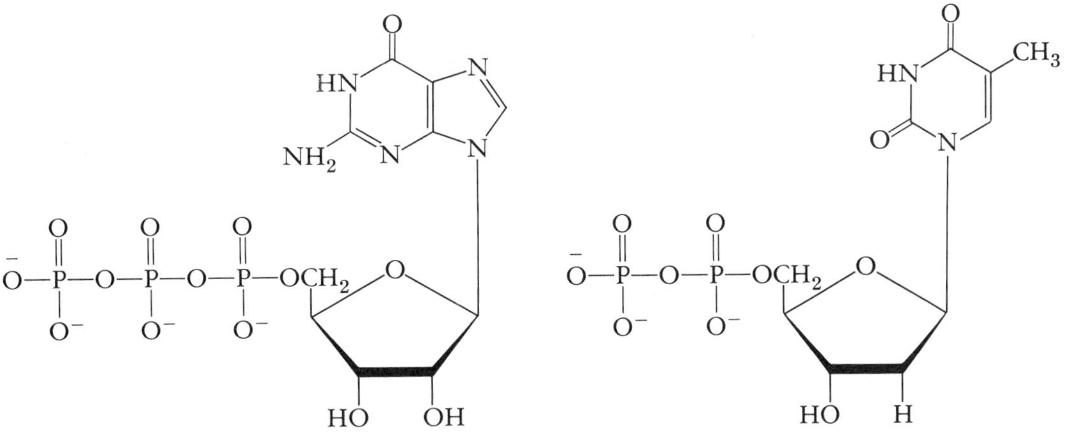

Guanosine triphosphate
(GTP)

Deoxythymidine diphosphate
(dTDP)

Fig. 24-6. Structures of guanosine triphosphate and deoxythymidine diphosphate.

needed to form the polynucleotide chain of DNA. The four ribose-containing nucleotides, AMP, GMP, CMP, and UMP, are the only ones needed to form the polynucleotide chain of RNA. The sequence of these nucleotides in the chain is called the *primary structure* of the polynucleotide or nucleic acid. This is similar to the primary structure of proteins (see Section 23.5).

The nucleotides in a polynucleotide chain are joined by a bond between the phosphate group of one nucleotide and the sugar of another nucleotide. This forms polymers that contain a chain or backbone of repeating sugar-phosphate units. The heterocyclic bases, joined to the pentose sugars, are off to one side. This basic structure of a polynucleotide is shown in Figure 24-7.

An examination of the detailed structure of a polynucleotide reveals that each nucleotide is joined to another nucleotide by a bond between its phosphate group (located on carbon 5′ of its sugar) and the carbon 3′ of the sugar of its neighboring nucleotide. A portion of the detailed primary structure of a polynucleotide is shown in Figure 24-8a. Notice that the phosphate group that joins the sugar parts of any two nucleotides is a diester (see Section 18.12). This group is called either a 3′,5′-phosphodiester bond or a 3′,5′-phosphodiester linkage.

The structural formulas of polynucleotides are written by convention so that the 5′-carbon of the sugar, located at one end of the chain, is always

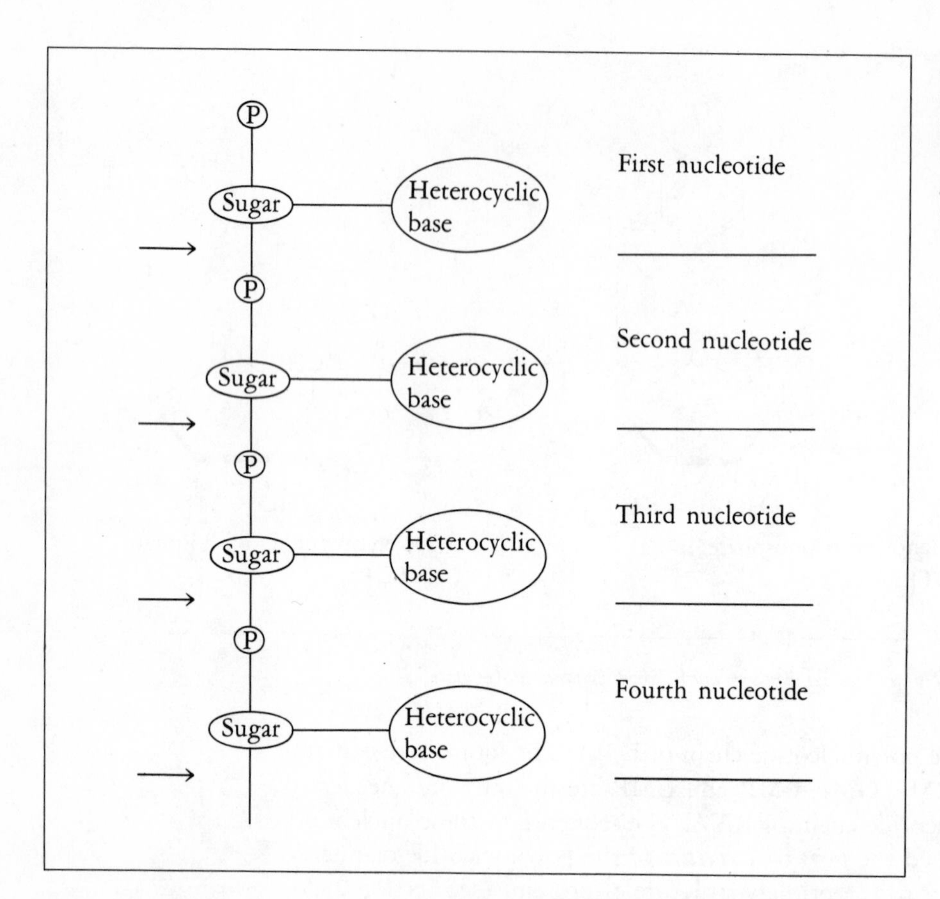

Fig. 24-7. The basic structure of polynucleotides. Each nucleotide is joined to another by a bond between a phosphate group and a sugar. The arrows indicate these bonds.

placed at the left end of the structure. The structural formula in Figure 24-8 is written in this way. For convenience, we can abbreviate the structure of a polynucleotide. This is done by using the letters for the nucleosides given in Section 24.2 and the letter p for phosphate. The location of the phosphate, either bonded to the 3′ or 5′ carbon of the sugar, is shown by the order in which the symbols are written. Thus, pA means that the phosphate is on carbon 5′ of the sugar of adenosine. The symbol Ap means that the phosphate is on carbon 3′ of the sugar of adenosine. This system can be extended. Thus, ApU means that adenosine is linked through a phosphate on carbon 3′ of its ribose to carbon 5′ of the ribose of uridine. This method of abbreviating structure of polynucleotides is applied to the structure in Figure 24-8.

EXERCISE 24-4 Circle and name each of the nucleotides in the structural formula in Figure 24-8.

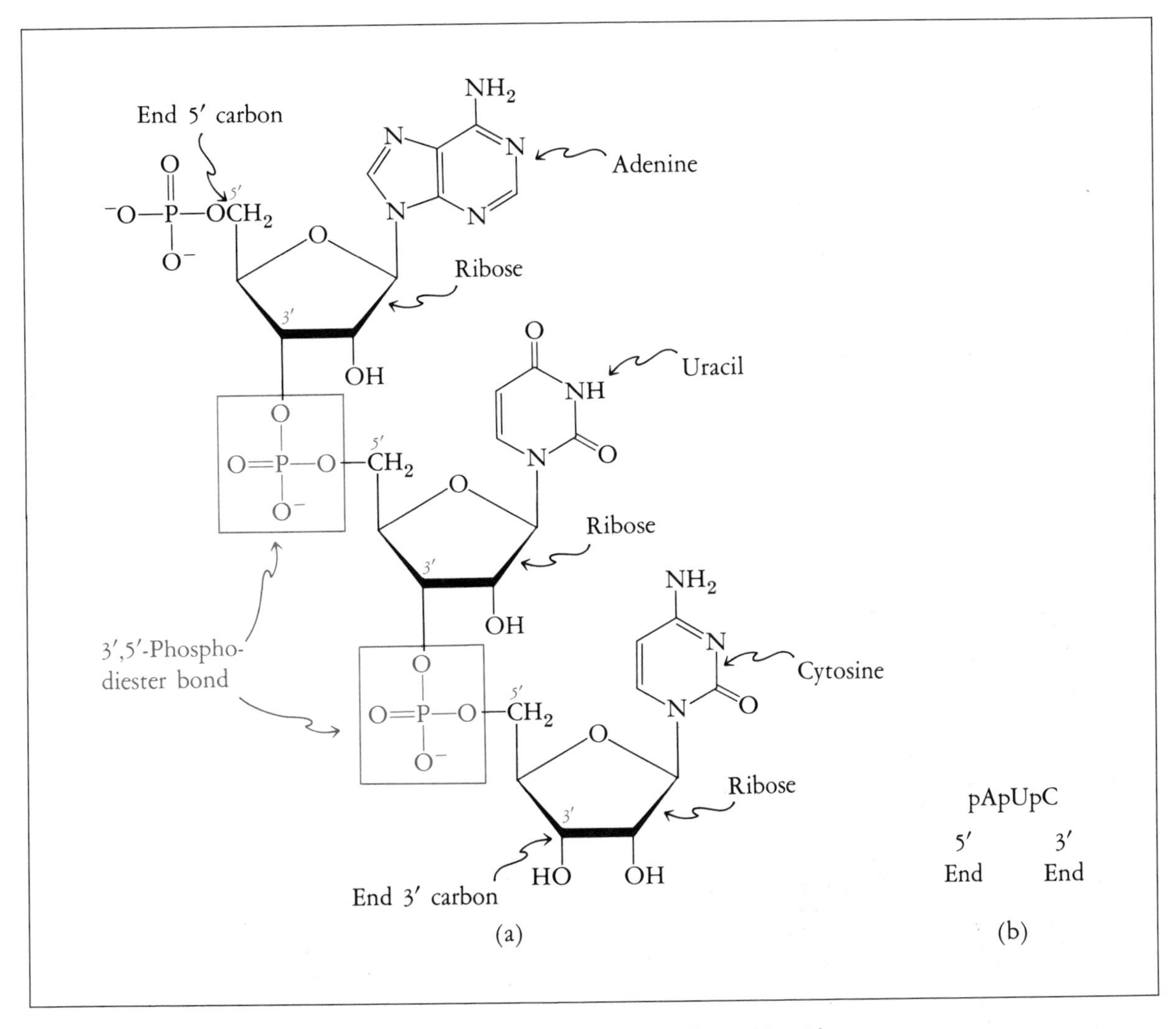

Fig. 24-8. A portion of the primary structure of the polynucleotide chain of a nucleic acid: (a) the complete structure; (b) an abbreviation of the structure.

EXERCISE 24-5 Which nucleotide in each of the following polynucleotides has a free 5′-hydroxy group? Which has a free 3′-hydroxy group?
(a) GpApTp (b) pApGpTp (c) ApGpTpCpA

EXERCISE 24-6 Write the structural formula for each of the following polynucleotides:
(a) TpApC (b) CpApTp (c) pApGpTpC

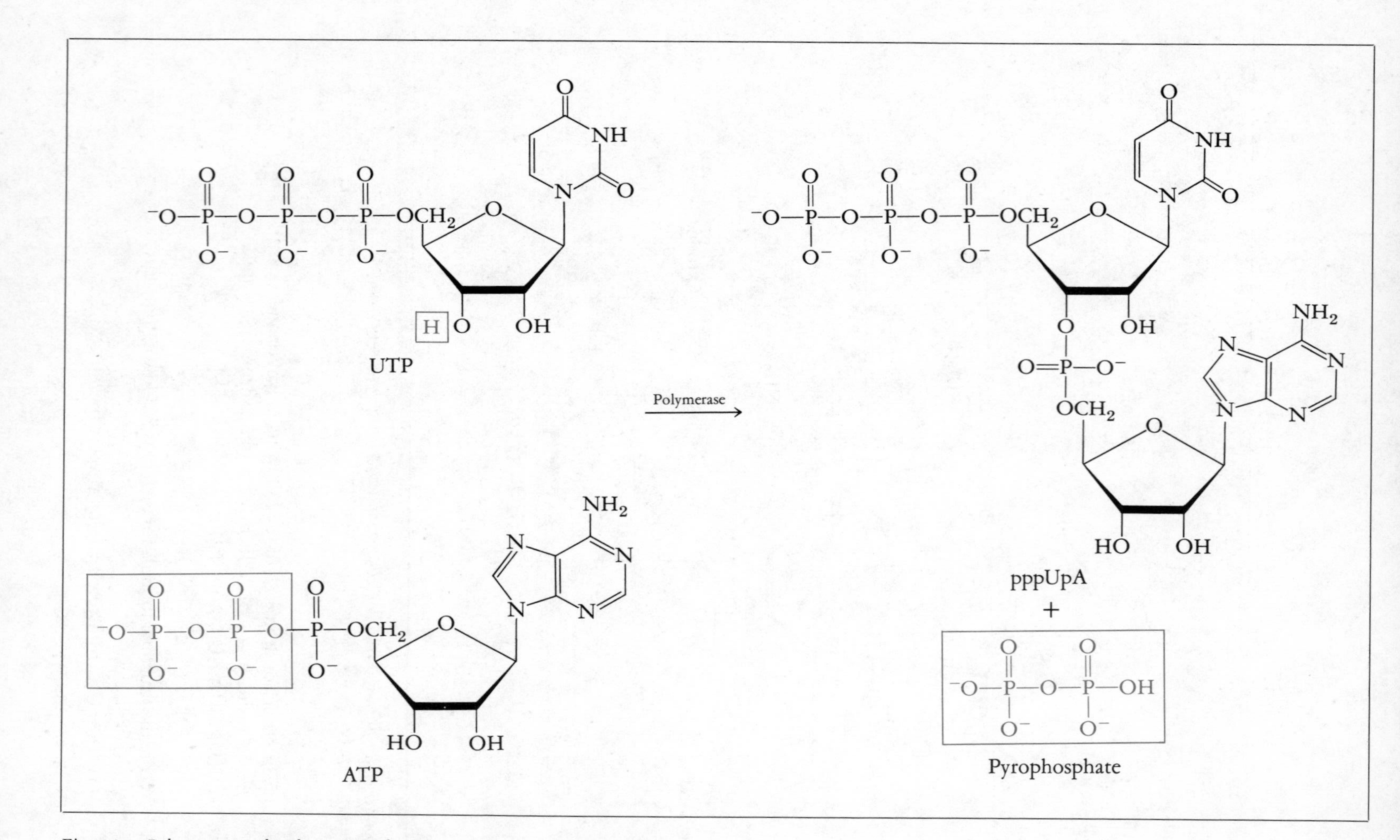

Fig. 24-9. Polymerase-catalyzed reaction of uridine triphosphate (UTP) and adenosine triphosphate (ATP) to form a dinucleotide.

Polynucleotides are formed in cells by the polymerization of nucleotide triphosphates. Two nucleotides are joined in the presence of an enzyme called polymerase with the elimination of pyrophosphate

$$^{-}O-\overset{\overset{\displaystyle O}{\|}}{\underset{\underset{\displaystyle O^-}{|}}{P}}-O-\overset{\overset{\displaystyle O}{\|}}{\underset{\underset{\displaystyle O^-}{|}}{P}}-OH$$

as shown in Figure 24-9. This is the reaction that allows DNA to make exact duplicates of itself, as we will learn in Section 24.5. But first let us learn the structure of DNA.

24.4 THE STRUCTURE OF DNA: THE DOUBLE HELIX

By 1953 a great deal was known about DNA. It was known that the genetic code was stored in DNA. From x-ray data, molecules of DNA were known to be long, fairly straight, and quite thin (only about a dozen atoms thick). Chemical analysis revealed the following facts about DNA. First, the base composition of DNA of a particular species is the same for all of its cells and is characteristic of that species. That is, DNAs from different species have different base compositions. Second, in the DNAs of all species, the molar amount of adenine is always equal to that of thymine. Similarly, the molar amounts of guanine and cytosine are equal. Finally, the total number of purine bases in DNA equals the number of pyrimidine bases.

James Watson and Francis Crick proposed a model of DNA based on these facts. The main feature of their model is that a molecule of DNA consists of *two chains of polynucleotides*. Each chain is coiled in a right-hand helix, and the two chains wind around each other to form a *double helix*. The deoxyribose sugars and the phosphate groups that are the backbone of the polynucleotides form the outside of the structure. The bases are located on the inside. A simplified representation of the double helix structure of DNA is shown in Figure 24-10.

The two chains of polynucleotides are held together by hydrogen bonds between specific bases on each chain. A guanine base on one chain always hydrogen bonds with a cytosine base opposite it on the other chain. Similarly, an adenine base on one chain is always paired by hydrogen bonding to a thymine base on the other chain. This hydrogen bonding between specific pairs of bases is shown by the dotted lines in Figure 24-10. Because of the great length of the DNA molecule, there are a large number of hydrogen bonds formed between the two chains. Although any one hydrogen bond is fairly weak (see Section 6.3), the sum of all these hydrogen bonds makes DNA a very stable molecule.

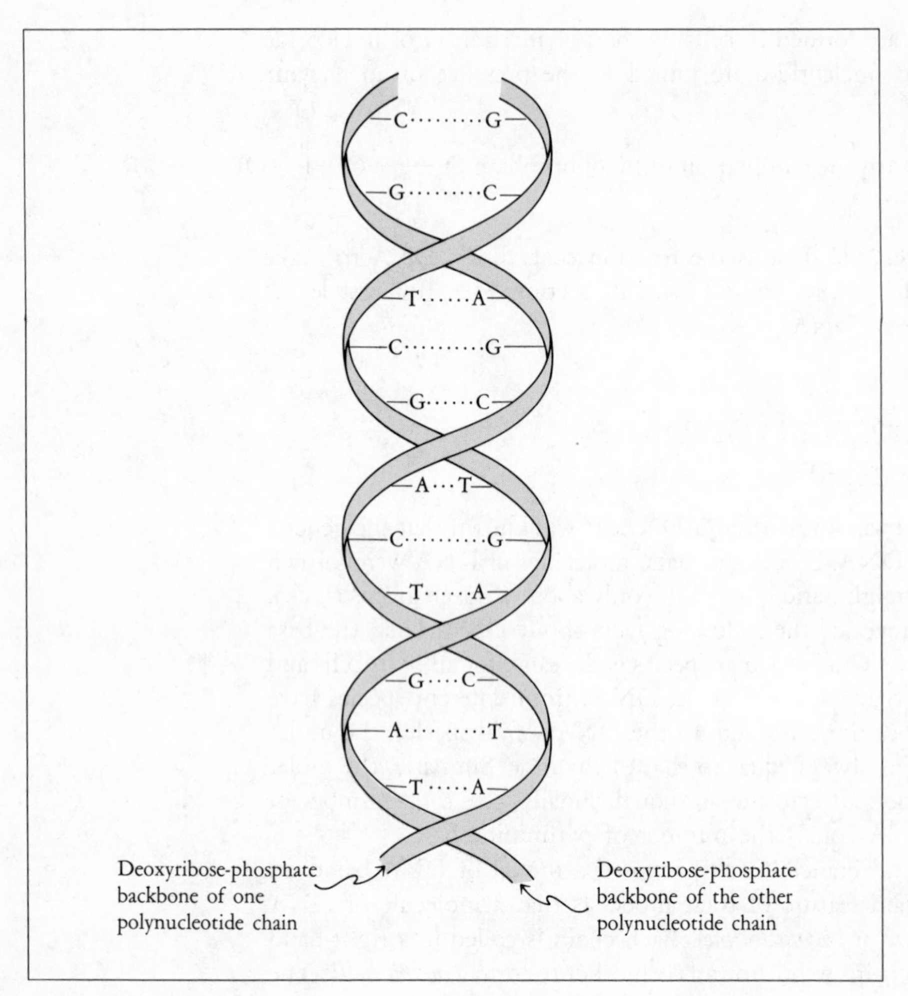

Fig. 24-10. A part of the double helix structure of DNA. The dotted lines (····) represent hydrogen bonds between pairs of bases.

Why do the bases on one chain of DNA always pair with a specific base on the other chain? The answer is that the two chains of the DNA double helix fit neatly together only when such pairing exists. The reason for this can be understood by examining the two sets of base pairs shown in Figure 24-11. In each set, one pyrimidine base is paired with a purine base. As a result, the distance between the two chains is the same (1.09 nm). If two pyrimidine bases were paired, the distance between two chains would need to be closer together. If two purine bases were paired, the distance would be farther apart. To maintain a constant distance between the two chains and consequently the most stable structure, a pyrimidine base always pairs with a purine base.

The two bases guanine and cytosine are called *complementary bases* of

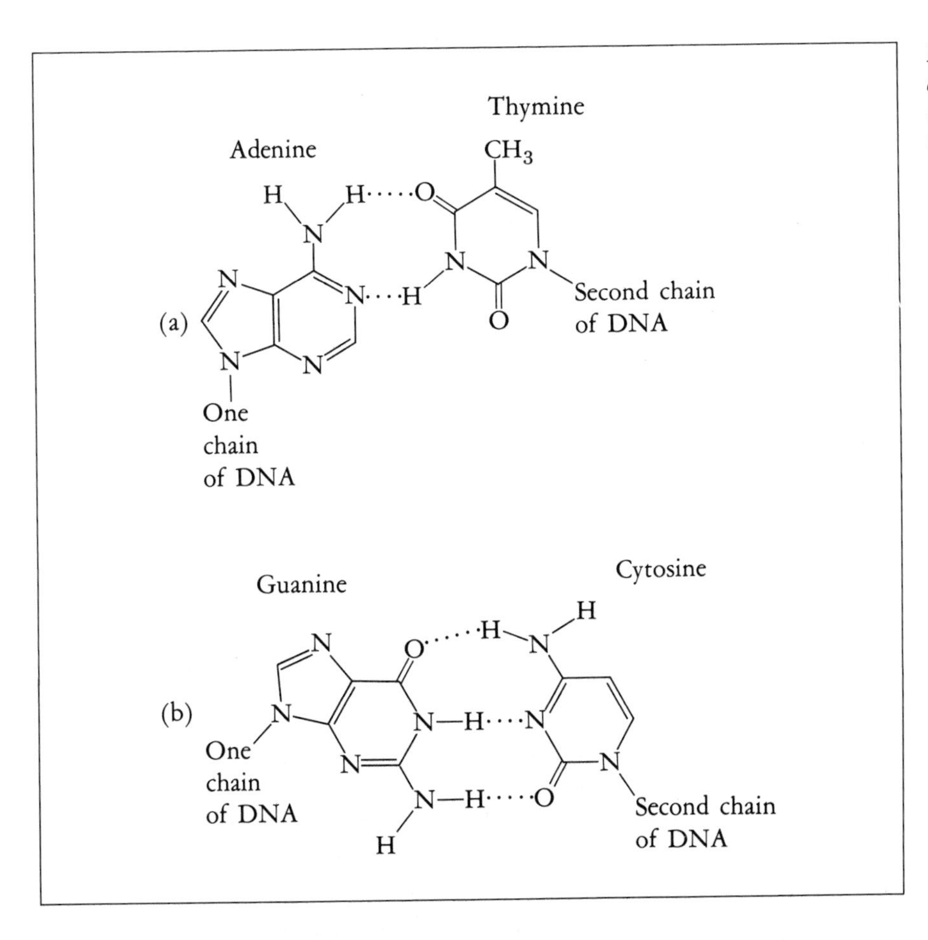

Fig. 24-11. The sets of complementary base pairs: (a) adenine and thymine; (b) guanine and cytosine.

DNA. The other two complementary bases are adenine and thymine. Whenever one of these bases, either a purine or pyrimidine base, is located on one chain of DNA, its complementary base is hydrogen bonded to it on the other chain.

The Watson-Crick double-helix structure of DNA provides us with something else, a model of how DNA produces an exact replica of itself.

24.5 HEREDITY AND DNA REPLICATION

Heredity is the process by which the physical and mental characteristics of parents are passed on to their children. The study of heredity is called *genetics.* At the molecular level, DNA contains the information needed to transmit the characteristics of a species from one generation to the next, as well as the information needed for the species to grow and live. Thus, DNA molecules are the chemical basis of heredity and govern protein synthesis.

Most DNA molecules are contained in chromosomes located in the nuclei of cells. The number of chromosomes depends on the species. Bacteria have a single chromosome, whereas human cells have 46 (23 from each parent). All DNA molecules have the same double-helix structure and the same deoxyribose-phosphate backbone, and all contain the same four bases. But that is the end of the similarity. The DNA molecules of different species differ in their length and, most important, in their sequence of four bases along the chains. *It is the sequence of bases in the structure of DNA that contains the genetic information of a particular species.* In humans, the chromosomes that combine to form the fertilized egg contain characteristic DNA molecules of humans. Thus, the fertilized egg of a human can only grow up to be another human. The particular sequence of bases in DNA molecules is different for every species.

Genetic information must be reproduced exactly each time a cell divides. This is done at the molecular level by making an exact copy of a DNA molecule. Thus, cells synthesize new and exact copies of DNA molecules that are passed to the new cell. The process by which DNA molecules reproduce themselves in the nucleus of cells is called *DNA replication.*

The Watson-Crick double-helix structure of DNA provides a model for replication, as shown in Figure 24-12. The process of replication begins when the two chains of a DNA molecule slowly unwind to expose each chain, as shown in Figure 24-12b. Each chain now serves as a template for the synthesis of two new molecules of DNA.

The cellular fluid surrounding DNA molecules in the nucleus is rich in the nucleotide triphosphates dATP, dCTP, dGTP, and dTTP needed to synthesize new DNA molecules. These nucleotide triphosphates move into position along the unwound part of the DNA chains. They pair with the proper complementary base (A to T or T to A, and C to G or G to C). As a result, each base on the unwound part of the DNA chains picks up only a nucleotide with a base identical to the one to which it was paired before unwinding (Figure 24-12c). Once the nucleotide triphosphate is in place, the enzyme DNA polymerase catalyzes the reaction that joins the nucleotide into the new chain with the elimination of pyrophosphate (Figure 24-12d). This reaction is similar to the reaction shown in Figure 24-9. As the original DNA molecule unwinds, additional nucleotide triphosphates pair with their complementary bases along the exposed chain. These are then joined to the lengthening new DNA chain. Eventually, a copy is made of each chain of the original DNA molecule and they separate to form two complete DNA molecules, as shown in Figure 24-13. Each contains a chain from the original DNA molecule and one new chain.

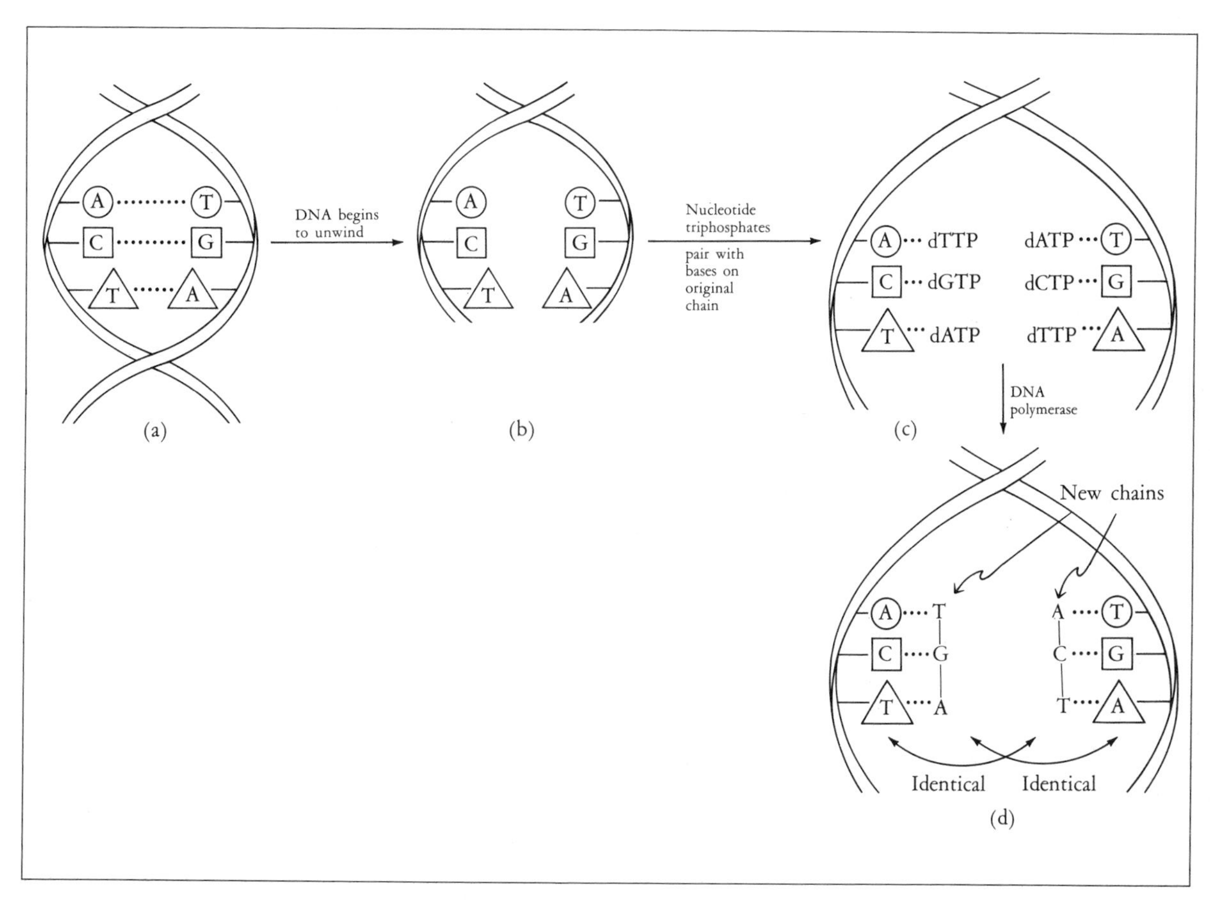

Fig. 24-12. The replication of DNA.

This is the way that each chain of the original DNA molecule forms a duplicate of its original partner. Whatever information was contained in the original DNA molecule is now contained in each of the replicates. During cell division, one DNA molecule is passed to the new cell, giving it the genetic information needed to survive. The other remains with the parent cell.

The replication of DNA molecules is the way genetic information is passed from one generation to the next. The base sequence is the important feature in DNA molecules. The information stored in this sequence also controls the specific amino acid sequence in proteins. As we will learn in Chapters 25, 27, 28, and 29, numerous proteins function as enzymes that control the metabolism and the synthesis of fundamental compounds in living organisms. Thus, by controlling the primary structure of proteins, DNA functions to specify the essential characteristics of the organism. But

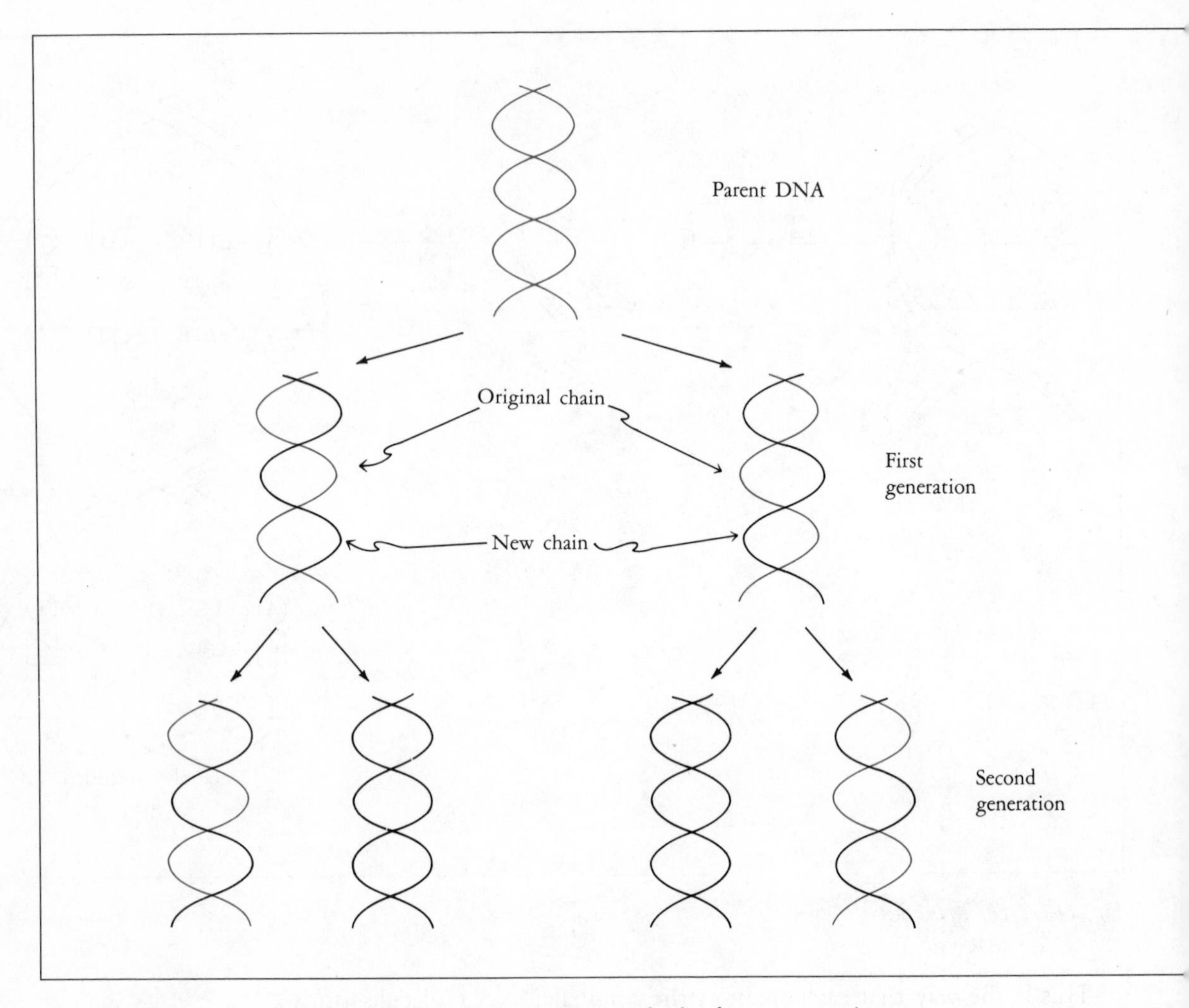

Fig. 24-13. The fate of the original chains in the parent DNA molecule after two generations.

the information carried by DNA is not transmitted directly to proteins. The information is first incorporated into the structure of molecules of RNA. There are actually three types of RNA molecule. Each has a specific function.

24.6 TRANSCRIPTION AND RNA

The genetic information carried by DNA is incorporated into the structure of RNA. The RNA molecules then use this information to specify and synthesize the amino acid sequence in proteins. As we learned in Section

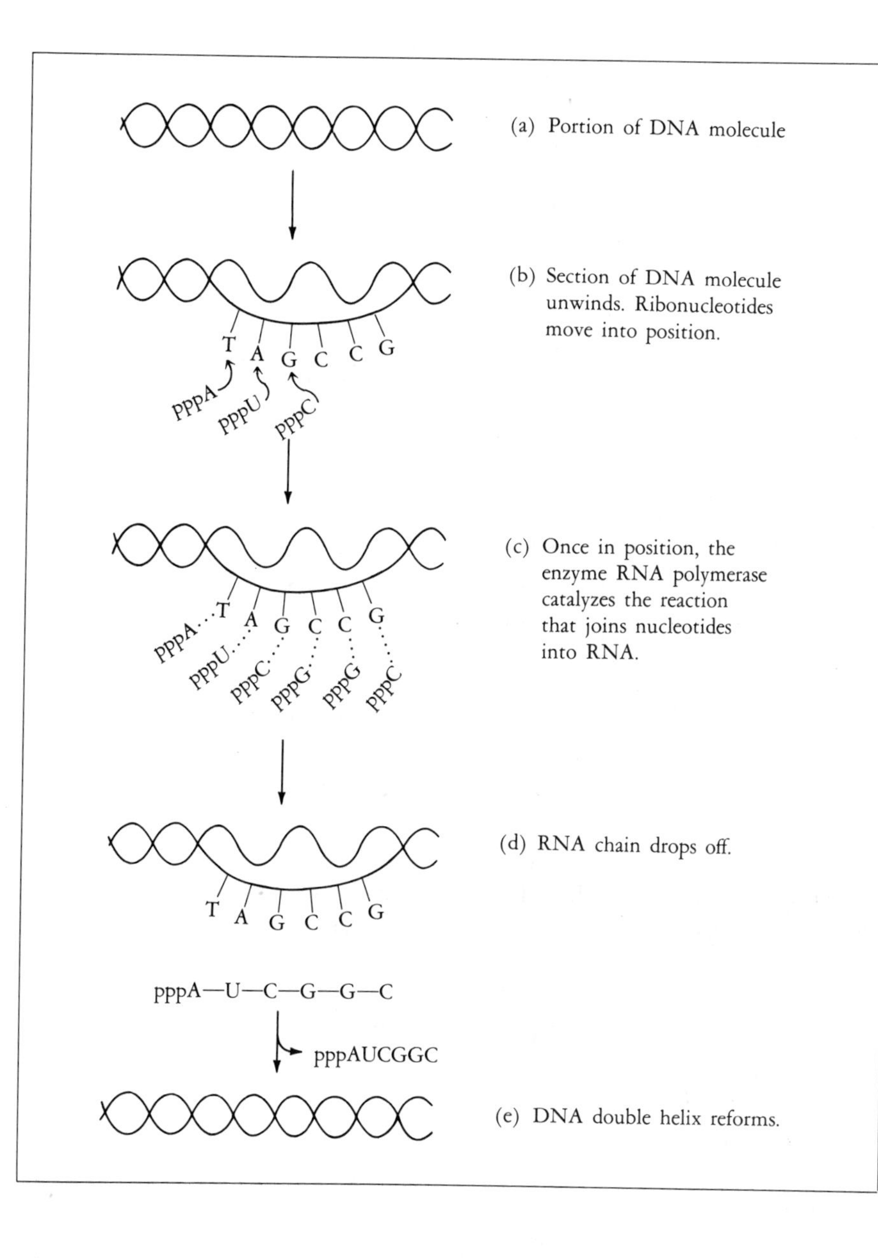
(a) Portion of DNA molecule
(b) Section of DNA molecule unwinds. Ribonucleotides move into position.
T A G C C G
pppA
pppU
pppC
(c) Once in position, the enzyme RNA polymerase catalyzes the reaction that joins nucleotides into RNA.
pppA
pppU
pppC
pppG
pppG
pppC
(d) RNA chain drops off.
pppA—U—C—G—G—C
pppAUCGGC
(e) DNA double helix reforms.

Fig. 24-14. Sequence of events in RNA synthesis (transcription).

24.1, RNA differs from DNA in three ways. First, RNA contains ribose instead of 2-deoxyribose as the sugar. Second, the base uracil is present in RNA instead of thymine. Finally, RNA is made up of a single polynucleotide chain rather than a double helix.

The process of making RNA from DNA is called *transcription.* The way that DNA is used as a template to make RNA is shown in Figure 24-14. A section of the double helix of DNA unwinds (Figure 24-14b) to expose a short segment that has the code for an RNA molecule. Only one of the complementary DNA chains is involved in directing the base sequence of RNA. Although the double helix of DNA is required for making RNA, one of the chains does not directly transmit information.

Ribonucleotide triphosphates, from the cellular fluids, move into position along the unwound part of the DNA double helix. Only one chain is used as a template, and transcription occurs in complementary fashion. Where an adenine base exists in DNA, a uracil will be incorporated into RNA (Figure 24-14c). Similarly, guanine from DNA is transcribed into cytosine, thymine into adenine, and cytosine into guanine. Once the nucleotides are in place, the enzyme RNA polymerase catalyzes the reaction that joins the nucleotides together into the RNA chain with the elimination of pyrophosphate (Figure 24-14c). Finally, the newly formed RNA chain drops off and the DNA re-forms the double helix (Figure 24-14d and e).

The DNA is used as a template to make three kinds of RNA: messenger RNA (abbreviated mRNA), transfer RNA (tRNA), and ribosomal RNA (rRNA). Each kind of RNA has a particular function.

The largest of the three, in both size and amount, is rRNA. Between 60 and 80 percent of the total RNA in cells is rRNA that has a molecular weight of several million amu. The rRNA combines with protein to form ribosomes, the intracellular substructures where proteins are synthesized. About 60 percent of the ribosome is rRNA, and the remaining 40 percent is protein.

Molecules of mRNA carry the genetic information from DNA to the ribosomes. When attached to the ribosomes, they direct protein synthesis. The size of an mRNA molecule depends on the size of the protein it is directed to make.

The smallest of the three kinds of RNA is tRNA. Each tRNA molecule consists of about 100 nucleotides in a single chain. The primary function of tRNA is to bring to the ribosomes the amino acids to be incorporated into the protein. Each of the 20 amino acids found in proteins has at least one particular tRNA molecule that carries it to the ribosome. The tRNA for each amino acid has a unique base sequence and fine structure, but all tRNA molecules have certain structural features in common. All have a similar shape that can be represented in two dimensions as the cloverleaf shown in Figure 24-15.

The shape of the tRNA molecule shown in Figure 24-15 is maintained by hydrogen bonds between base pairs. Other parts of the molecule that do not contain hydrogen-bonded base pairs exist as loops. Two parts of every tRNA molecule have important biological functions. The first is the site where the specific amino acid is joined. This is located at the 3′ end of the

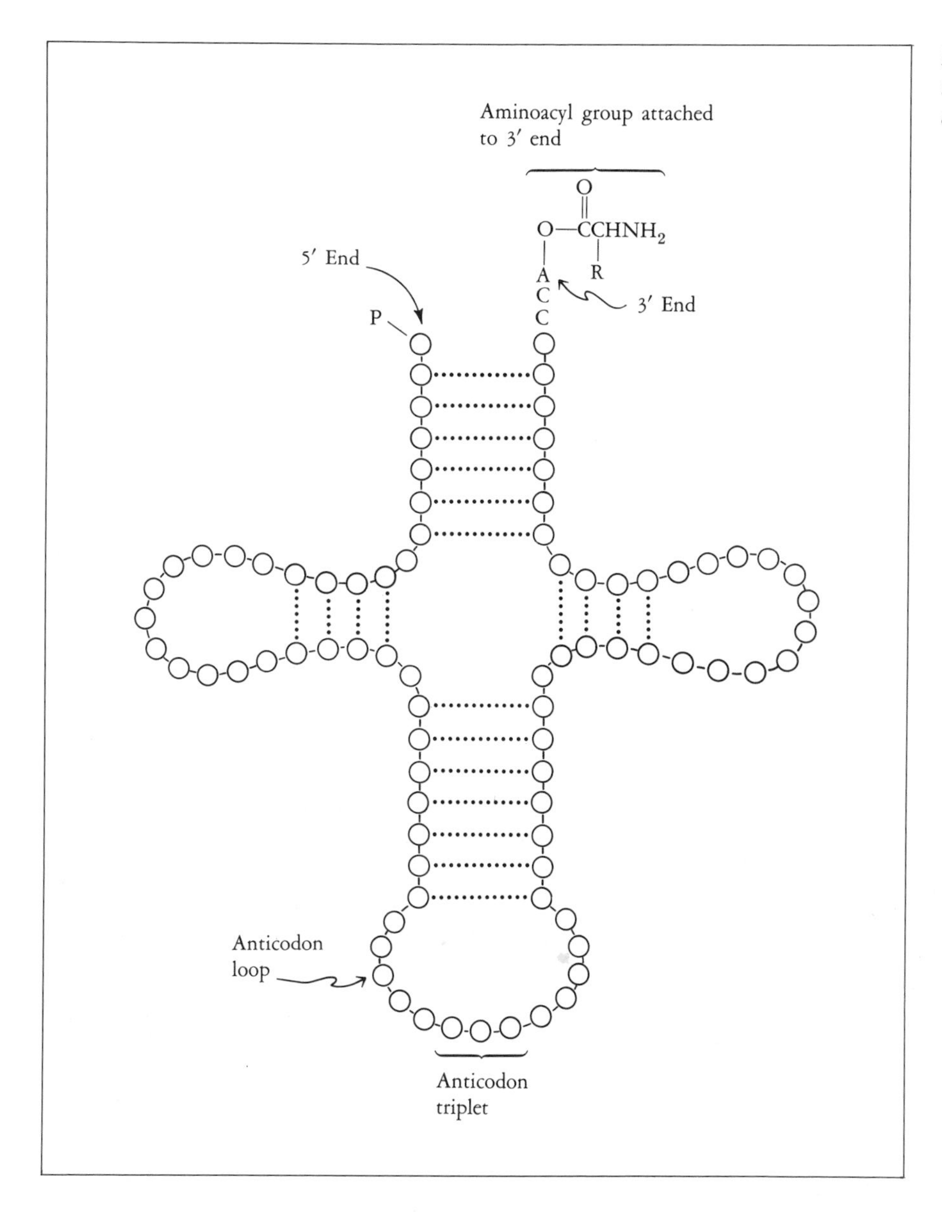

Fig. 24-15. Representation of tRNA. Each circle represents a ribonucleotide.

polynucleotide chain. The second important part is the loop at the other end of the molecule. This is the bottom of the molecule, as drawn in Figure 24-15. This loop contains a specific three-base sequence that represents a specific code for the amino acid carried by the tRNA. This three-base region is called the *anticodon* of the tRNA molecule. We will learn its importance and how it is used in the next two sections.

24.7 TRANSLATION AND THE GENETIC CODE

As we learned in the previous section, DNA acts as the template by which the genetic information needed for protein synthesis is transcribed to mRNA. Once formed, the mRNA molecule moves out of the nucleus and attaches itself to a ribosome. Here it is ready to begin protein synthesis.

The sequence of bases in an mRNA molecule contains the code needed to synthesize a protein. However, an mRNA molecule contains only four different bases: two purines and two pyrimidines. This four-unit code must be translated in some way into a 20-unit code that identifies each of the 20 amino acids used in protein synthesis. *Translation* is the process by which the four-base code in nucleic acids is turned into a 20-unit code needed to specify the amino acid sequence in proteins.

Translation occurs by using a specific sequence of three nucleotide bases on mRNA called a *base triplet* or *codon.* Each amino acid is specified by at least one set of base triplets. Using the four bases of mRNA in various combinations of three provides the code for each of the 20 amino acids. In fact, the four bases can form 64 combinations of three bases, many more than are needed for the 20 amino acids. These codons, which are the genetic code, have been identified. Only 61 of the possible 64 combinations are used, and they are given in Table 24-4.

The three letters in Table 24-4 that specify the codon for a particular amino acid represent the sequence of nucleotide bases, from the 5′ end to the 3′ end, on the mRNA molecule. Thus, the sequence AGU, which is the code for the introduction of the amino acid serine into a protein, is an abbreviation for the sequence of adenosine, guanosine, and uridine. The codons are strung consecutively along the mRNA chain. They are read without interruption; that is, there are no punctuation marks. The only exceptions are the codons UAA, UAG, and UGA, which signal the end of a protein chain. Thus, the sequence UCACUACACGAUUAUUAA codes for the synthesis of a serylleucylhistidylaspartyltyrosine polypeptide.

EXERCISE 24-7 Find in Table 24-4 the abbreviations for the six codons that are the codes for the introduction of serine into a protein.

EXERCISE 24-8 Using Table 24-4, name the amino acid specified by each of the following codons:
(a) AUU (b) GAU (c) GCU (d) GGG (e) CGA (f) CAU

EXERCISE 24-9 The following sequence is the code for the synthesis of what polypeptide?
UCAUUGACCGAG

Table 24-4. The Genetic Code

First Base (5′ End)	Center Base				Last Base (3′ End)
	U	C	A	G	
U	Phe	Ser	Tyr	Cys	U
U	Phe	Ser	Tyr	Cys	C
U	Leu	Ser	End	End	A
U	Leu	Ser	End	Trp	G
C	Leu	Pro	His	Arg	U
C	Leu	Pro	His	Arg	C
C	Leu	Pro	Gln	Arg	A
C	Leu	Pro	Gln	Arg	G
A	Ile	Thr	Asn	Ser	U
A	Ile	Thr	Asn	Ser	C
A	Ile	Thr	Lys	Arg	A
A	Met*	Thr	Lys	Arg	G
G	Val	Ala	Asp	Gly	U
G	Val	Ala	Asp	Gly	C
G	Val	Ala	Glu	Gly	A
G	Val	Ala	Glu	Gly	G

* The codon AUG serves both as a code for methionine and as a code for starting protein synthesis.

The codon that is the code for a specific amino acid is the same in all living systems. For example, UCG specifies serine for any species. Differences between species are due to the fact that the sequences of bases on their mRNA molecules are different. Consequently, the sequences of codons are different, resulting in the synthesis of proteins with different sequences of amino acids.

Let us now learn how the codons on mRNA actually direct the synthesis of proteins.

24.8 PROTEIN SYNTHESIS

Protein synthesis begins when the codons on an mRNA molecule combine with the complementary bases of the anticodons of tRNA, as shown in Figure 24-16. Only a part of the mRNA nucleotide base sequence is shown in Figure 24-16. The first codon, AUG, is not only the code for the amino

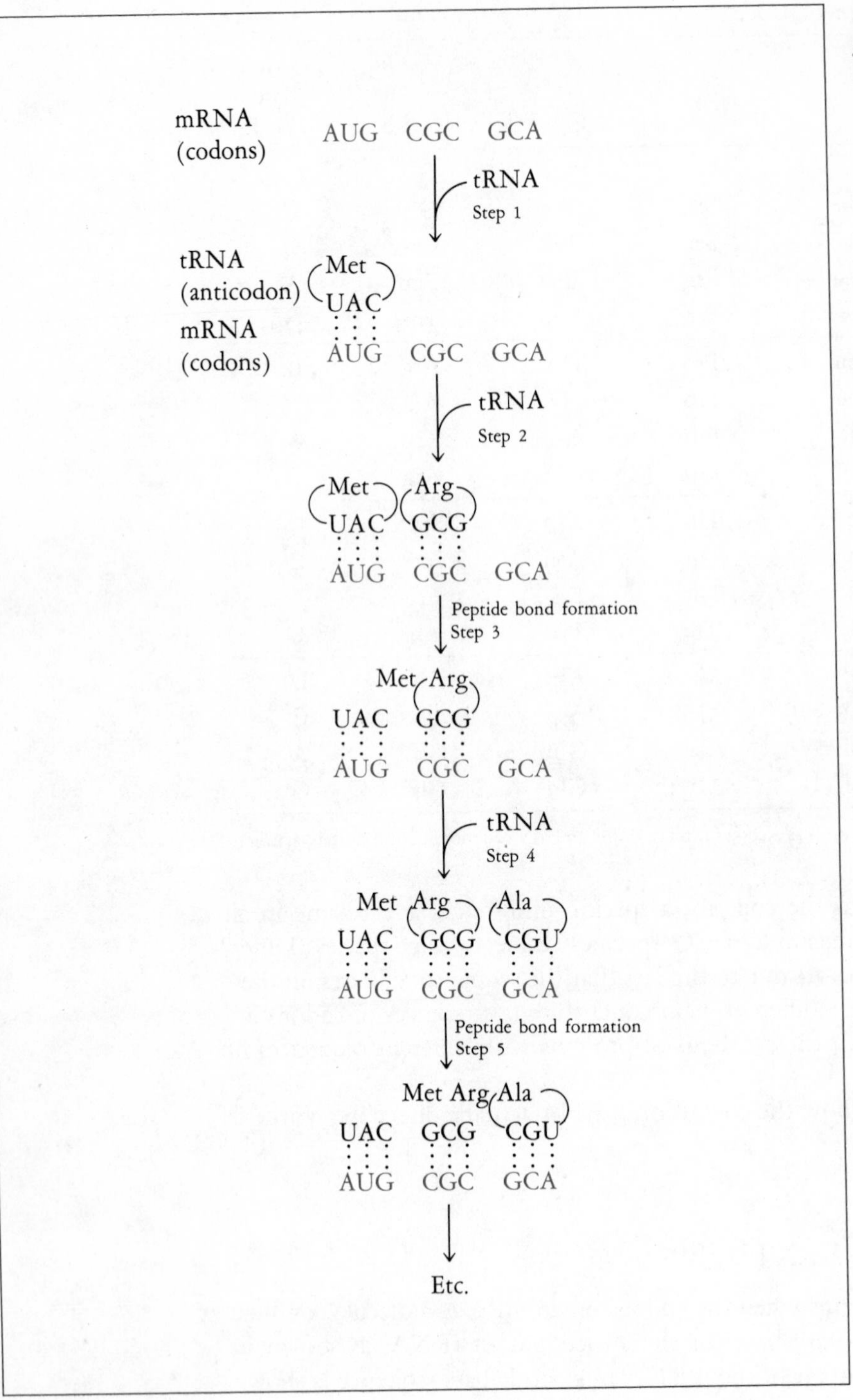

mRNA (codons) AUG CGC GCA

tRNA
Step 1

tRNA (anticodon) Met UAC
mRNA (codons) AUG CGC GCA

tRNA
Step 2

Met Arg
UAC GCG
AUG CGC GCA

Peptide bond formation
Step 3

Met Arg
UAC GCG
AUG CGC GCA

tRNA
Step 4

Met Arg Ala
UAC GCG CGU
AUG CGC GCA

Peptide bond formation
Step 5

Met Arg Ala
UAC GCG CGU
AUG CGC GCA

Etc.

Fig. 24-16. Translation of the genetic code for protein synthesis.

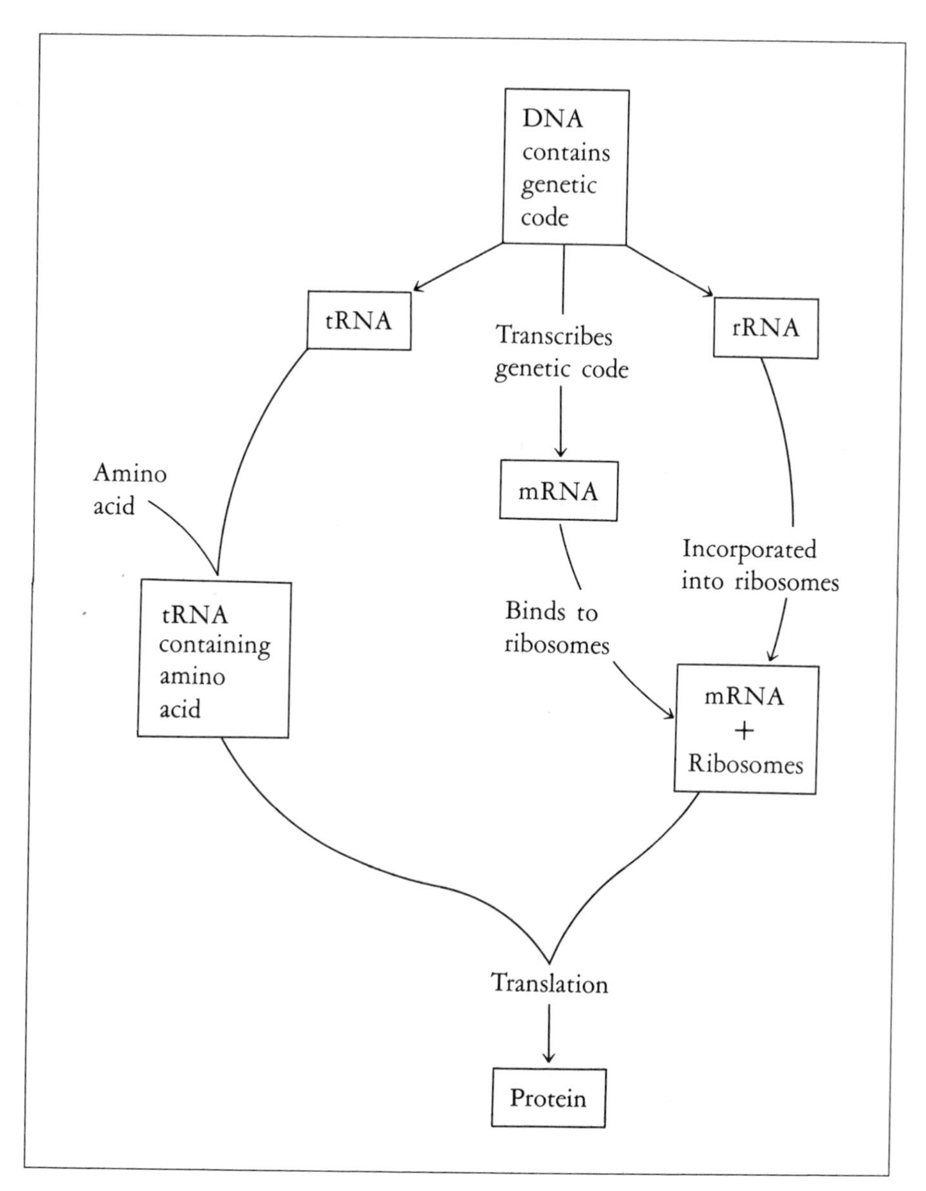

Fig. 24-17. The role of RNA as an intermediary in the synthesis of proteins from the genetic code of DNA.

acid methionine, but is also the code for starting protein synthesis. The codon AUG can pair by hydrogen bonding with a tRNA molecule that contains the complementary anticodon UAC, as shown in step 1, Figure 24-16. Such a tRNA molecule is a carrier of the amino acid methionine. The next codon, CGC, pairs with a tRNA molecule that has the complementary anticodon GCG and carries the amino acid arginine (step 2, Figure 24-16).

When the two tRNA molecules are correctly placed on the mRNA chain, a peptide bond is formed between the two amino acids (step 3, Figure 24-16). The sequence of steps is repeated for the next codon, GCA. This codon pairs with a tRNA molecule that has the complementary codon CGU and carries the amino acid alanine (step 4, Figure 24-16). This amino acid is then joined by a peptide bond to the dipeptide previously formed (step 5, Figure 24-16). A protein is synthesized with a specific sequence of amino acids in this way. This sequence of steps continues until one of the three codons UAA, UAG, or UGA is reached. These are the stop signals that terminate protein synthesis on the mRNA chain.

The way that DNA directs protein synthesis through RNA as an intermediary is summarized in Figure 24-17. The DNA contains the genetic code in its sequence of bases. This code is transcribed to mRNA (see Section 24.6). Molecules of mRNA bind to ribosomes (made up partly of rRNA), and the genetic code is translated by means of base triplets, or codons (see Section 24.7). Matching codons of mRNA with their complementary anticodons of tRNA results in a particular sequence of amino acids in a protein (see Section 24.8).

Occasionally something happens to one or more of the bases in a DNA molecule, and this leads to mutations.

24.9 MUTATIONS

A *mutation* is any chemical or physical change that alters the sequence of bases in a DNA molecule. When the change involves a single base in DNA, it is called a *point mutation.* Some point mutations may occur spontaneously; others result from ultraviolet light, ionizing radiation (see Section 4.2), or a variety of chemical compounds. Anything that causes mutation is called a *mutagen.* Point mutations cause the synthesis of proteins with defects in their amino acid sequence. Let us examine three examples of point mutations and learn how they affect the amino acid sequence in proteins.

One point mutation, called insertion, involves the addition of one or more extra nucleotides besides those already present in a DNA chain. Another point mutation, called deletion, is just the opposite of insertion. One or more nucleotides is missing from the normal sequence in DNA. The change in amino acid sequence, caused by insertion or deletion, occurs during the translation of mRNA. Because the codons of mRNA are read without interruption, deletion or insertion causes a shift in the code. This results in a frame-shift mutation, as shown in Figure 24-18. Frame-shift mutations usually lead to a large number of changes in the amino acid

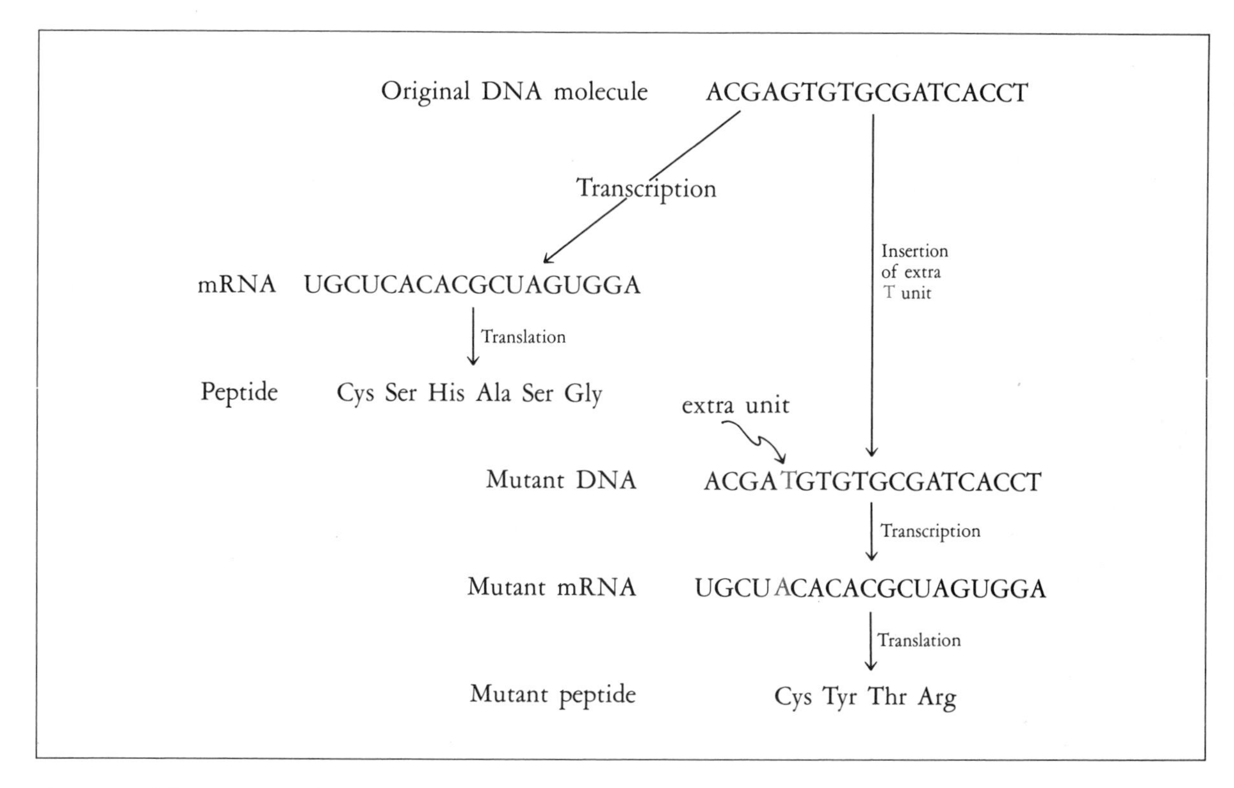

Fig. 24-18. Effect of a frame-shift mutation on protein synthesis. Notice that the mutant peptide not only has the incorrect order of amino acids, but is also shorter. This is because the UAG codon on the mutant mRNA signals the end of the synthesis. (Figure adapted from I. Danishefsky, Biochemistry for Medical Sciences. *Boston: Little, Brown, 1980.)*

sequence of a protein. Consequently, the protein no longer carries out its biological function. This is usually fatal to the organism.

Another type of point mutation involves the substitution of one base for another. For example, a guanosine may replace an adenosine in the DNA chain. As a result, the complementary chain of the DNA molecule would then contain a cytidine instead of the usual thymidine. In other words, the original A-T pair of nucleosides in DNA is replaced by a G-C pair. These mutations occur during DNA replication. In general, they cause an error of only one amino acid in the sequence of amino acids of a protein. As a result, this substitution only rarely affects the function of the protein.

Recently scientists have made major changes in the structure of DNA molecules by joining segments of DNA from one organism to pieces of DNA from another organism. The new DNA molecules formed in this way are called recombinant DNA.

24.10 RECOMBINANT DNA AND GENETIC ENGINEERING

The term *recombinant DNA,* or gene splicing, as it is often called, refers to new DNA molecules formed by joining, or splicing, parts of DNA molecules from one organism into parts of DNA molecules from another organism. Although the DNA of any organism (plant, animal, or bacteria) can be used, DNA molecules from the bacteria *Escherichia coli* (*E. coli*) have been used most frequently. There are two reasons for the use of *E. coli.* First, its genetic identity has been thoroughly studied. Second, a particular strain has been developed that can survive only under the carefully controlled conditions of the laboratory. This minimizes the chance of producing a new and deadly bacteria that would survive in the world. The steps in the production of recombinant DNA are shown in Figure 24-19.

The *E. coli* contains plasmids that are small circular units of DNA molecules separate from the principal chromosome units. These plasmids serve as the vehicle for the recombinant technique. The *E. coli* bacteria are placed in a detergent solution to break open the cells. The plasmids are isolated from the solution (step 1, Figure 24-19) and then cleaved by a specific enzyme called a restriction enzyme to form a linear DNA unit (step 2, Figure 24-19). Some DNA molecules from another organism (foreign DNA) are combined with the same restriction enzyme to form segments of DNA that have ends complementary to those of the DNA from the plasmids (host DNA) (step 3, Figure 24.19). These complementary ends are called "sticky ends." The foreign DNA segments and the host DNA are combined with the enzyme DNA ligase, which joins the two into the recombinant DNA (step 4, Figure 24-19). This step usually recloses the plasmids, and they are then put back into the cells of *E. coli* (step 5, Figure 24-19). Whenever these cells of *E. coli* divide, new plasmids are synthesized that are copies of the plasmids containing recombinant DNA. These new plasmids have new characteristics dictated by the combination of the original genes and the genes transplanted from a different species.

Much scientific and public controversy has surrounded recombinant DNA research. This research has both benefits and risks. Opponents of this research fear the synthesis of new and deadly bacteria that would proliferate uncontrollably, causing mass disease and death. As a result of these fears, controls on the types of bacteria used in this research and strict laboratory containment procedures were introduced in 1976. Some people also fear that the results of this research could be used for political purposes. For example, these techniques might be used for genetic engineering to produce

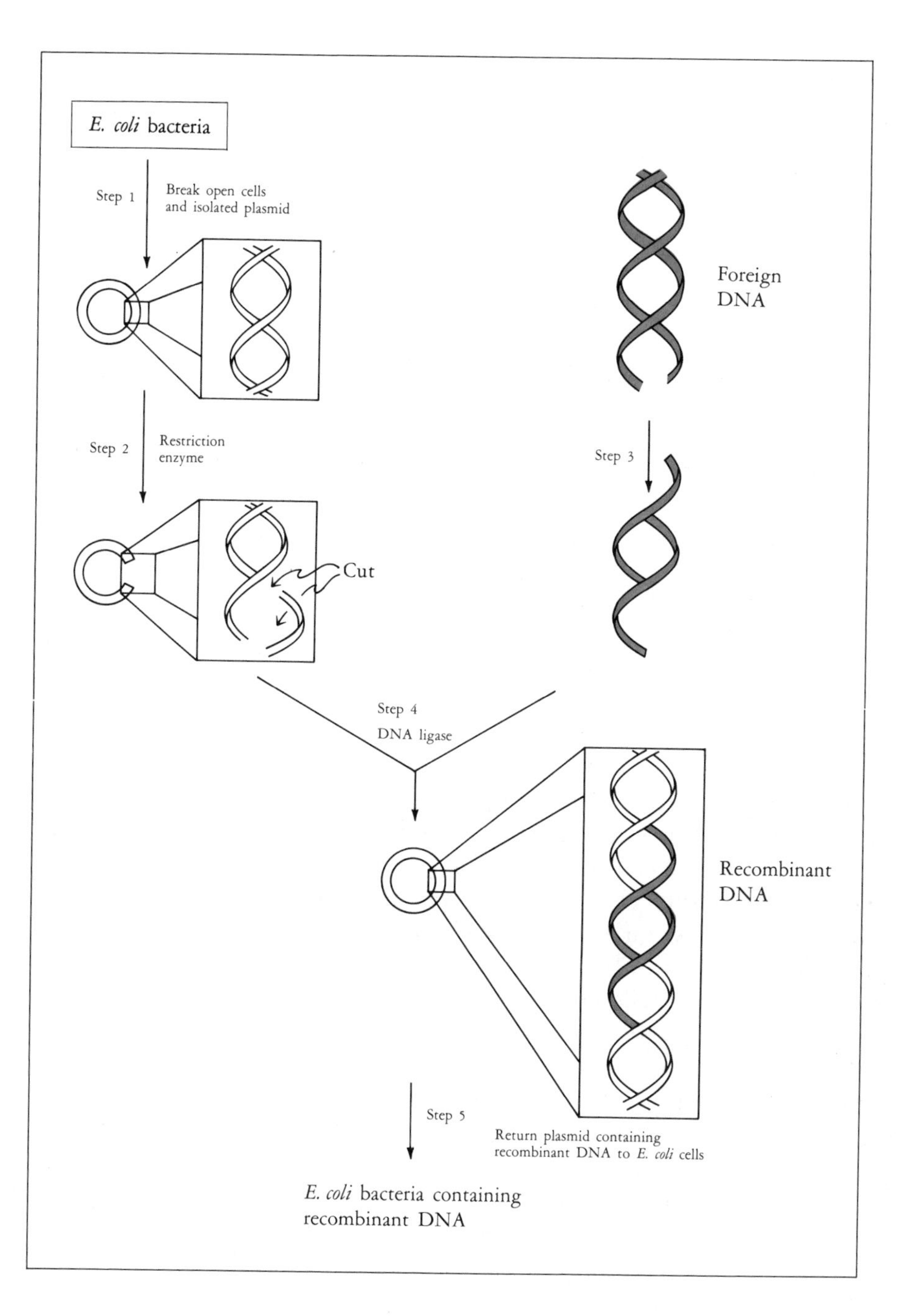

Fig. 24-19. Production of recombinant DNA.

humans of a predetermined intelligence and social behavior. Finally, some people believe that we should not meddle in natural evolution.

The proponents of this research are excited about its potential benefits. It is conceivable that recombinant DNA research could lead to cures for genetic diseases such as sickle cell anemia (see Section 23.5). The introduction of appropriate segments of DNA into *E. coli* has already been used to produce bacteria that are essentially miniature pharmaceutical factories. For example, a synthetic insulin gene has been made that produces insulin faster and cheaper than insulin obtained from animals. Research is under way to use recombinant DNA in the manufacture of vaccines against viral diseases such as influenza and hepatitis. Other potential uses of recombinant DNA include modifying important cereal crops so that they can fix their own nitrogen, thereby eliminating the need for most fertilizers; producing a pesticide that is specific for a particular destructive insect; and designing a bacterium that can extract oil from tar sands. Such wide potential application means that recombinant DNA will surely have profound effects on our future.

24.11 SUMMARY

Nucleic acids are composed of only four of five heterocyclic bases, one of two pentose sugars, and phosphate groups. Two important examples of nucleic acids are DNA and RNA. Molecules of DNA contain the heterocyclic bases adenine, guanine, cytosine, and thymine, and have deoxyribose as the pentose sugar. Molecules of RNA differ from DNA in that they contain uracil rather than thymine and ribose rather than deoxyribose.

Both DNA and RNA are polynucleotides. The nucleotides in a polynucleotide chain are joined by a 3′,5′-phosphodiester bond between the sugar of one nucleotide and the sugar of another nucleotide. This forms polymers that contain a chain or backbone of repeating sugar-phosphate units. The heterocyclic bases, joined to the pentose sugars, are off to one side.

Molecules of DNA consist of two chains of polynucleotides. Each chain is coiled in a right-handed helix, and the two chains wind around each other to form a double helix. The deoxyribose sugars and the phosphate groups form the outside of the structure, whereas the bases are located on the inside. The sequence of bases in the structure of DNA contains the genetic information of a particular species. This information is passed along to new and exact copies of DNA by a process called replication.

In addition to serving as a template for making new molecules of itself, DNA is also the template for the synthesis of RNA molecules by a process called transcription. The RNA molecules use the information passed on to them from DNA to synthesize proteins. Three kinds of RNA, mRNA, tRNA, and rRNA, are synthesized from DNA. Each has a specific purpose. Molecules of rRNA combine with proteins to form ribosomes, the sites of protein synthesis. Molecules of mRNA carry the genetic code from DNA to the ribosomes, and molecules of tRNA bring specific amino acids to the ribosomes for incorporation into the protein.

The four-base code in nucleic acids is turned into a 20-unit code needed to specify the sequence of amino acids in proteins by a process called translation. Each amino acid is specified on an mRNA molecule by at least one set of base triplets, or codons. To synthesize a protein, complementary sequences of bases, called anticodons, on tRNA combine with the codons of mRNA to place the amino acids in their proper sequence in the protein molecule.

An error or change in the sequence of bases in DNA molecules affects the amino acid sequence in a protein. These changes cause mutations. Some mutations occur naturally. Others can be caused by scientists using a technique called recombinant DNA. Considerable controversy surrounds this research.

REVIEW OF TERMS AND CONCEPTS

Terms

ANTICODON The three-base region on a tRNA molecule that indicates which amino acid is carried by the tRNA molecule.

CODON The three-base sequence on a mRNA molecule that specifies the order of amino acids in a protein.

DNA Deoxyribonucleic acid: the macromolecule in living systems that contains the genetic information.

DNA REPLICATION The mechanism by which DNA duplicates itself.

DOUBLE HELIX The structure of molecules of DNA. Two chains, each a helix, wind around each other to form a double helix.

GENETIC CODE The heredity message stored in the arrangement of the bases in DNA molecules.

mRNA Messenger ribonucleic acid. The nucleic acid that carries the directions for protein synthesis from DNA molecules to the ribosomes.

MUTAGEN Any agent that causes mutation.

MUTATION Any change, chemical or physical, in the base sequence of DNA molecules.

NUCLEIC ACIDS Polymers of nucleotides.

NUCLEOSIDE Compounds formed by joining carbon 1′ of a pentose and either nitrogen 1 of a pyrimidine base or nitrogen 9 of a purine base.

NUCLEOTIDE Phosphate esters of nucleosides. The phosphate group is attached to carbon 5′ of the pentose in the nucleoside.

RECOMBINANT DNA Molecules of DNA formed by inserting portions of DNA from one organism into DNA of another.

rRNA Ribosomal ribonucleic acid. The nucleic acid that combines with proteins to form ribosomes.

TRANSCRIPTION The way the code on DNA is copied to give the complementary code on RNA.

TRANSLATION The way the four-base code in nucleic acids is turned into a 20-unit code needed to specify the amino acid sequence in proteins.

tRNA Transfer ribonucleic acid. The nucleic acid that carries amino acids to the ribosomes.

Concepts

1. Nucleic acids are made up of relatively few kinds of molecules. Only two kinds of pentose sugars, five heterocyclic bases, and phosphoric acid serve as the building blocks of nucleic acids.
2. DNA contains the genetic information that is responsible for the characteristics that distinguish one species from another, transmit its characteristics to the next generation, and control its metabolism.
3. The structure of DNA is a double helix. This structure allows DNA to reproduce itself and also serve as a template for making mRNA, rRNA, and tRNA.
4. Genetic information on DNA is transcribed to RNA; translation of this code occurs to enable the synthesis of proteins.

EXERCISES

24-10 Distinguish between nucleosides and nucleotides.

24-11 Nucleic acids, proteins, and carbohydrates are all polymers. What is the monomer in each of these polymers?

24-12 Give the name and draw the structure of the pentose sugar in DNA and the pentose sugar in RNA.

24-13 How do DNA and RNA differ in their chemical composition and their function?

24-14 Give the name of the nucleotide represented by each of the following abbreviations:
(a) ATP (b) UTP (c) dADP (d) dGMP

24-15 Give the names and structures of the complementary base pairs in DNA and in RNA.

24-16 What holds the complementary base pairs in DNA together?

24-17 Name and briefly describe the function of each of the three types of RNA.

24-18 Distinguish between transcription and translation.

24-19 What is the relationship between codons and anticodons?

24-20 Part of a DNA chain consists of the following sequence of nucleosides: AAGCCGACAGATAGC.
(a) What is the order of nucleotides in the complementary mRNA chain formed from this part of DNA?
(b) What is the sequence of amino acids coded by this part of the DNA chain?

ENZYMES 25

In previous chapters, we have learned that many of the reactions of living systems are similar in most respects to those found in the laboratory. For example, hydration of alkenes, aldol condensations, hydrolysis of esters and amides, the oxidation of alcohols, and many others occur both in the laboratory and in living systems. However, reactions in living systems do differ in one important respect. They always occur with the aid of *enzymes.* Enzymes are proteins produced by living systems; they catalyze specific biological reactions. They are among the most remarkable molecules because of their extraordinary ability to catalyze reactions of living systems.

The idea that some kind of catalyst is involved in biological reactions has been known since 1835, when Johann Berzelius discovered that starch is hydrolyzed faster by a solution of diastase of malt (now known to be an enzyme) than by water alone. For many years, it was believed that enzymes were a fundamental part of the structure and life of a cell. In 1897 Hans Büchner and Eduard Büchner succeeded in extracting from yeast cells the enzymes that catalyze alcohol fermentation. This result clearly shows that enzymes can act without the cell. However, it was not until 1926 that James B. Sumner isolated the first enzyme in pure crystalline form. He was also the first to present evidence that enzymes are proteins. Today nearly 2,000 different enzymes are known, and at least 200 of them have been obtained in crystalline form. All known enzymes are proteins.

In previous chapters, we have referred briefly to enzymes and their roles as catalysts. Because they were presented without explanation, their abilities to catalyze specific reactions may seem to be almost magical. In this chapter, we want to remove some of the mystery of enzymes. To do this, we must learn how the molecular structure of an enzyme is responsible for its catalytic action. Let us start by learning more about the role of enzymes as catalysts in living systems.

25.1 ENZYMES AS CATALYSTS

The principal function of enzymes is to act as catalysts for the reactions of living systems. As catalysts, enzymes have a number of unique characteristics. In this section, we will examine two of them: their ability to speed up reactions, and their specificity.

Enzymes have the ability to increase the speed (or rate) of chemical reactions. For example, the hydrolysis of urea occurs in water at 20° C in a slightly acidic solution according to the the following equation:

$$\underset{\text{Urea}}{NH_2\overset{\overset{\displaystyle O}{\|}}{C}NH_2} + 2\,H_2O + H^+ \longrightarrow 2\,NH_4^+ + HCO_3^-$$

This reaction can occur with or without the enzyme urease. However, in the presence of the enzyme, the reaction occurs 100 trillion times faster than the corresponding uncatalyzed reaction! This is not an isolated example. In general, enzyme-catalyzed reactions occur anywhere from 1 million to 1000 trillion times faster than the corresponding uncatalyzed reaction. Few man-made catalysts can approach this catalytic activity of enzymes.

The ability of an enzyme to catalyze a particular reaction of a compound or class of compounds is called its *specificity.* The specificity of enzymes can vary. Some have nearly absolute specificity for a particular compound and will not catalyze the reactions of even very closely related compounds. Other enzymes will catalyze the reactions of a whole class of compounds that have a common structural feature.

Aspartase is an example of an enzyme that has nearly absolute specificity. It catalyzes only the addition of ammonia to fumarate and the deamination of L-aspartate according to the following equation:

$$\underset{\text{Fumarate}}{(H)(^-O_2C)C{=}C(CO_2^-)(H)} + \overset{+}{N}H_4 \underset{}{\overset{\text{Aspartase}}{\rightleftharpoons}} \underset{\text{L-aspartate}}{^-O(O{=})CCH_2\text{–}C(H)(CO_2^-)\text{–}\overset{+}{N}H_3}$$

Aspartase is unable to catalyze the addition of ammonia to maleate (the *cis* isomer of fumarate) or to esters or amides of fumaric acid. Aspartase also has strict stereospecificity. It will catalyze the deamination only of L-aspartate. It is ineffective toward the deamination of D-aspartate. It is this absolute stereospecificity of many enzymes, which we first learned about in Section 14.9, that is especially remarkable. Another stereospecific enzyme is lactate dehydrogenase (see Section 27.4), which is specific for the L enantiomer of lactate.

Other enzymes have relatively broad specificities. They are capable of catalyzing the reactions of a number of different but structurally related compounds. One example is chymotrypsin, an enzyme that catalyzes the hydrolysis of proteins in the gastrointestinal tract. It catalyzes not only the hydrolysis of peptide bonds, but also the hydrolysis of a wide variety of esters and amides. A few of the compounds are shown in Figure 25-1. Carboxyesterase is an enzyme that also has broad specificity. It catalyzes the hydrolysis of the esters of various carboxylic acids. Carboxypeptidase is yet another example. It catalyzes the hydrolysis of the peptide bond of the C-terminal end of a peptide, regardless of the identity of the amino acid or the length of the peptide.

The specificity of enzymes is important to living cells. A cell contains several thousand different compounds, and there are many combinations of chemical reactions that these compounds can undergo. Because of the specificity of enzymes, however, only certain reactions occur: those that are essential to the continued life of the cell. In this way, enzymes are vital to living cells.

Each of the thousands of reactions in a cell is controlled by an enzyme. To identify each of these enzymes, we need a way to name and classify them.

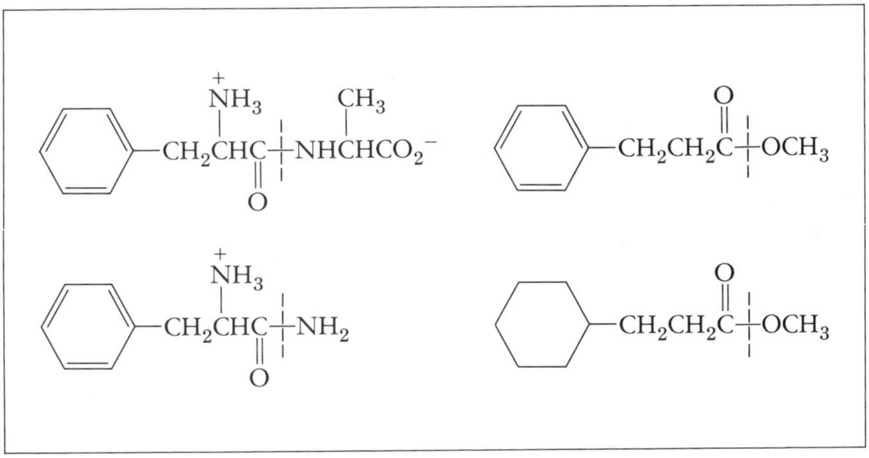

$C_6H_5-CH_2CH(\overset{+}{N}H_3)C(=O)\,|\,NHCH(CH_3)CO_2^-$

$C_6H_5-CH_2CH_2C(=O)\,|\,OCH_3$

$C_6H_5-CH_2CH(\overset{+}{N}H_3)C(=O)\,|\,NH_2$

$C_6H_{11}-CH_2CH_2C(=O)\,|\,OCH_3$

Fig. 25-1. Four compounds hydrolyzed by chymotrypsin. The dashed line passes through the bond that is broken.

25.2 NAMING AND CLASSIFYING ENZYMES

The compound or class of compounds on which an enzyme acts is called its *substrate.* Many enzymes are named by adding the ending *-ase* to the name of the substrate. For example, urease is the enzyme that catalyzes the hydrolysis of urea (see Section 25.1), and phosphatase catalyzes the following hydrolysis of phosphate esters:

$$\mathrm{RO\overset{\overset{\displaystyle O}{\|}}{\underset{\underset{\displaystyle OH}{|}}{P}}OH} + H_2O \xrightarrow{\text{Phosphatase}} \mathrm{ROH} + \mathrm{HO\overset{\overset{\displaystyle O}{\|}}{\underset{\underset{\displaystyle OH}{|}}{P}}OH}$$

Other enzymes have been named for the type of reaction they catalyze. For example, the enzyme lactate dehydrogenase catalyzes the following oxidation of L-(+)-lactate to pyruvate:

$$\underset{\text{L-(+)-Lactate}}{\mathrm{CH_3\overset{\overset{\displaystyle OH}{|}}{C}HCO_2^-}} + NAD^+ \xrightarrow[\text{dehydrogenase}]{\text{Lactate}} \underset{\text{Pyruvate}}{\mathrm{CH_3\overset{\overset{\displaystyle O}{\|}}{C}CO_2^-}} + NADH + H^+$$

Several enzymes that have been known for a long time carry names that were given by their discoverers. Trypsin and pepsin are two examples. These names do not tell us anything about the chemical reaction that the enzyme catalyzes.

Until 1961, enzymes were named in a rather haphazard manner. At that time, a systematic classification of enzymes was adopted on the recommendation of the International Union of Biochemistry. This system divides enzymes into six major classes according to the type of reaction catalyzed. These six classes and an example of each are summarized in Table 25-1. Each enzyme is assigned a recommended name, usually a short one for everyday use, and a systematic name that identifies the reaction it catalyzes. The systematic name is obtained from the classification in Table 25-1. For example, the recommended name of the enzyme that catalyzes the addition of ammonia to fumarate (see Section 25.1) is aspartase. Its systematic name is aspartate ammonia-lyase.

Table 25-1. International Classification of Enzymes

Classification of Enzyme	Type of Reaction Catalyzed	Typical Reaction	Specific Examples
Oxidoreductase	Oxidation-reduction	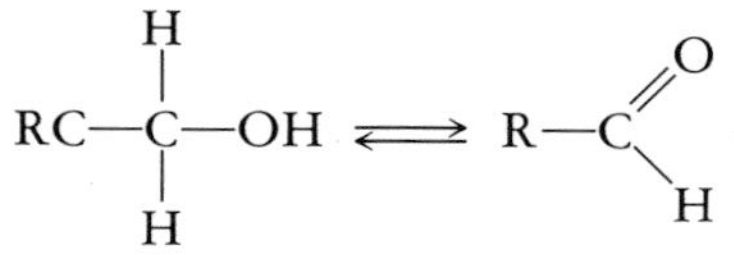 $RCH_2OH \rightleftharpoons RCHO$ (RC—C(H)(H)—OH ⇌ R—C(=O)H)	L-malate ⟶ oxaloacetate (Section 16.5)
Transferase	Transfer of groups or atoms from one compound to another	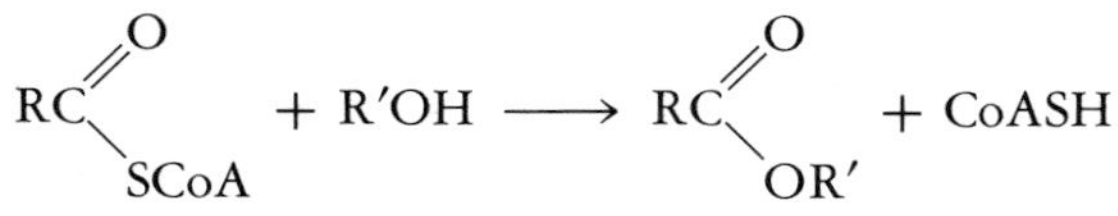 $RC(=O)SCoA + R'OH \longrightarrow RC(=O)OR' + CoASH$	Acetylation of choline (Section 18.7) Transamination (Section 29.3)
Hydrolyase	Hydrolysis of a variety of compounds	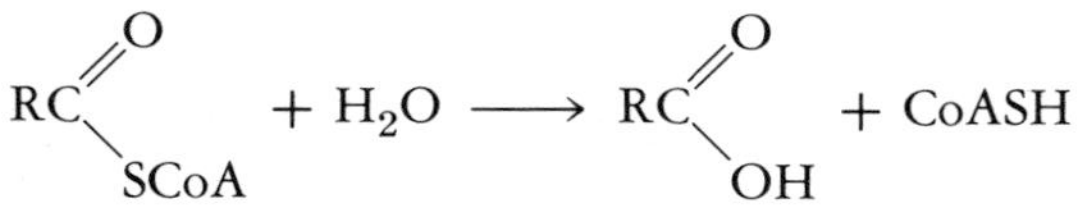 $RC(=O)SCoA + H_2O \longrightarrow RC(=O)OH + CoASH$	Citryl CoA ⟶ citrate + CoASH (Section 27.5)
Lyase	Additions to >C=C<, >C=O, and >C=N—	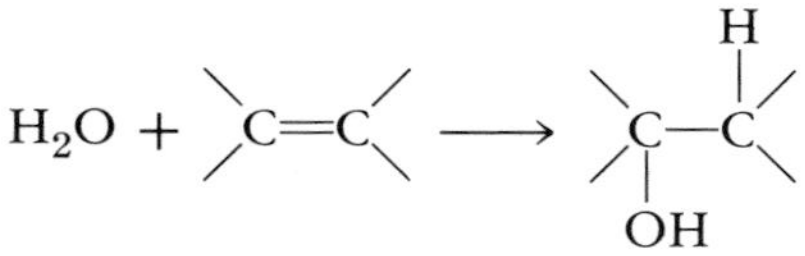 H_2O + >C=C< ⟶ >C(OH)—C(H)<	Fumarate ⟶ L-malate (Sections 12.4 and 27.5)
Isomerases	Isomerization reactions	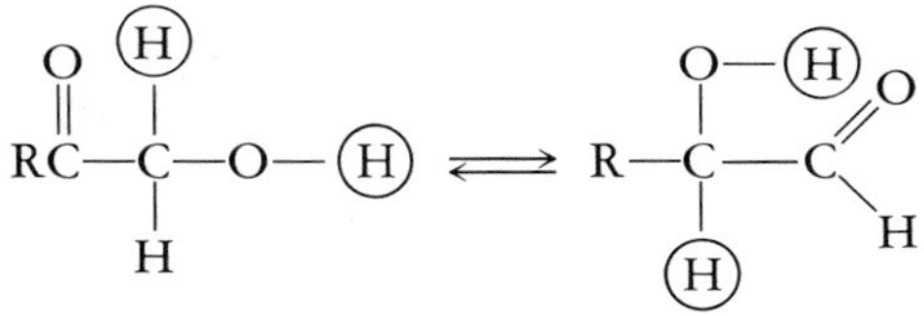 RC(=O)—C(Ⓗ)(H)—O—Ⓗ ⇌ R—C(O—Ⓗ)(Ⓗ)—C(=O)H	Dihydroxyacetone phosphate ⇅ Glyceraldehyde 3-phosphate (Section 27.3)
Ligase	Formation of bonds with breaking of phosphate bond of ATP	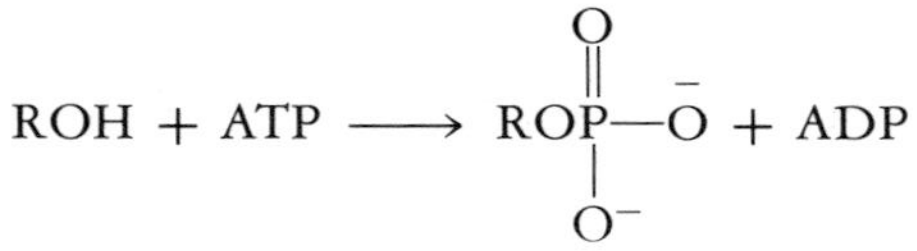 ROH + ATP ⟶ ROP(=O)(O⁻)—O⁻ + ADP	Glucose ⟶ glucose 6-phosphate (Section 27.3)

*EXERCISE 25-1 What is the substrate for each of the following enzymes?
(a) peptidase (b) lipase (c) sucrase
(d) cellulase (e) lactase

EXERCISE 25-2 What kind of reaction does each of the following enzymes catalyze?
(a) reductase (b) hydrolyase (c) isomerase
(d) transferase

The names of enzymes do not tell us anything about their structures. The reason for this is simple. It is much easier to determine the chemical function of an enzyme than its complicated structure. Yet a great deal is known about the structures of enzymes. For example, it is known that many enzymes contain nonprotein parts that have important effects on their catalytic activity. We will examine this feature of enzymes in the next section.

25.3 ENZYME COFACTORS

Enzymes vary widely in their structures. Some are simple proteins (see Section 23.9) and are biologically active by themselves. Others require the presence of a nonprotein part called a *cofactor* to be biologically active. The cofactor may be either a metal ion or an organic molecule that is called a *coenzyme.* The protein with its cofactor is called a *holoenzyme.* The protein without its cofactor is called an *apoenzyme.* The relationship between these parts of an enzyme and its biological activity is shown schematically in Figure 25-2.

It has been found that the structures of many coenzymes are directly related to certain water-soluble vitamins. Vitamins are organic molecules that are needed in small amounts in the diets of animals. The function of these vitamins is to serve as structural units that can be modified by the animal into the coenzymes it needs. Most of the biochemical roles of the water-soluble vitamins are known. These vitamins and the coenzymes obtained from them are given in Table 25-2. The structures of these coenzymes will be presented in later chapters, when we will examine their biological roles in more detail.

From the information in Table 25-2, it is clear that, rather than synthesize these coenzymes from the simpler compounds already in cells, many organisms, including humans, obtain the preformed basic structural skeleton

*The answers for the exercises in this chapter begin on page 902.

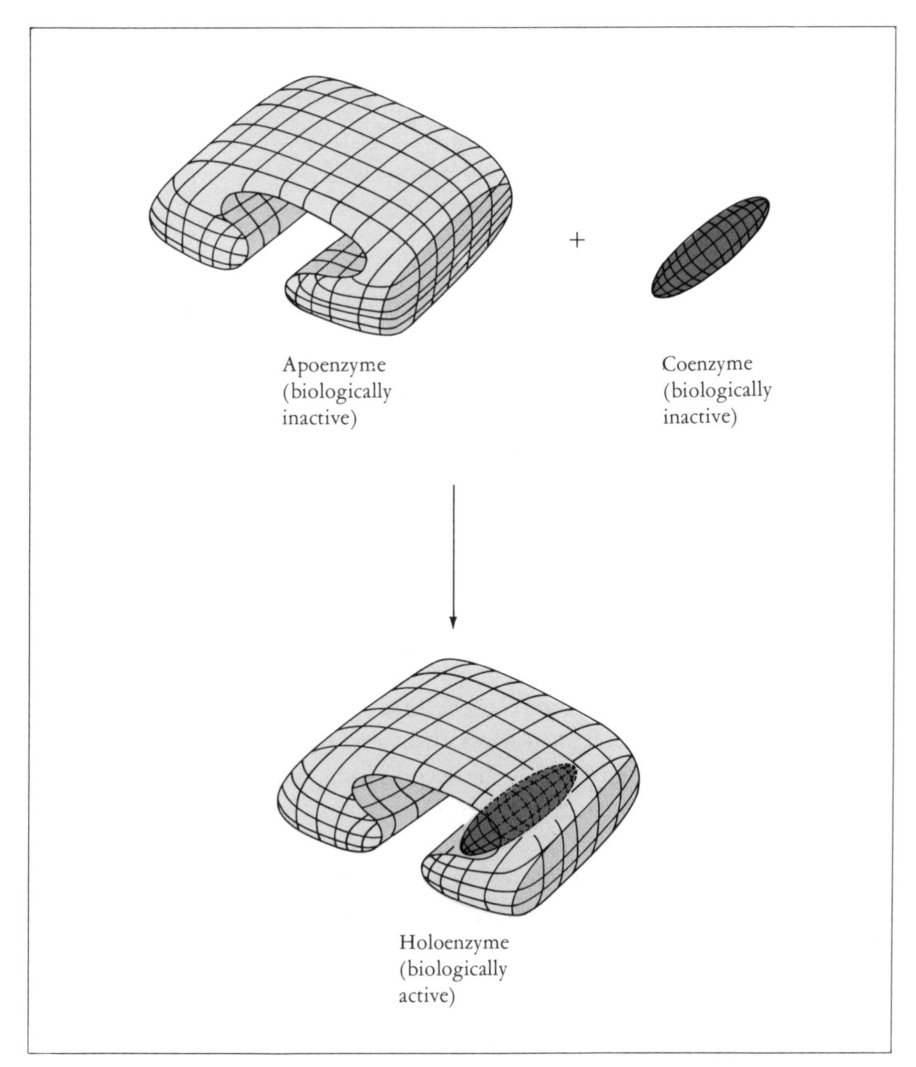

Fig. 25-2. The relationship between apoenzyme, coenzyme, and holoenzyme.

from the food they eat. For this reason these vitamins are essential for living systems.

Some coenzymes are tightly bound to the apoenzyme and are sometimes called prosthetic groups (see Section 23.9). Thus, coenzymes can be regarded as specific examples of prosthetic groups. In other cases, the coenzyme is only loosely attached to the apoenzyme and essentially acts as one of the specific substrates of that enzyme.

Now that we have identified various parts of enzymes, let us turn our attention to how enzymes catalyze reactions.

Table 25-2. Coenzymes Obtained from Some Essential Water-Soluble Vitamins and Their Biological Functions

Vitamin	Coenzyme	Biological Function
B_1 (Thiamine)	Thiamine pyrophosphate	Decarboxylation reactions
B_2 (Riboflavin)	FAD (*F*lavin *A*denine *D*inucleotide) and FMN (*F*lavin *M*ono*n*ucleotide)	Oxidation-reduction reactions
B_6 (Pyridoxine)	Pyridoxal phosphate	Group transfer reactions
Nicotinate (Niacin)	NAD (*N*icotinamide *A*denine *D*inucleotide)	Oxidation-reduction reactions (Section 7.6)
Pantothenate	Coenzyme A	Transfer of acyl groups (Section 18.10)

25.4 HOW ENZYMES CATALYZE REACTIONS

To explain the catalytic activity of enzymes, the general two-step cyclic mechanism shown in Figure 25-3 has been proposed. In the first step, the enzyme (E) joins with a molecule of substrate (S) to form an intermediate complex called an enzyme-substrate complex (ES). In the second step, the reaction occurs in this complex, which breaks apart to form the product (P) and the enzyme. The enzyme is now ready to react with another molecule of substrate. In very few cases, enzyme-substrate complexes have been isolated and their three-dimensional structures determined. These structures have provided us with many details of how enzymes catalyze reactions. From them, it is concluded that the enzyme and the substrate are held together in the complex by the same forces that maintain the structure of proteins (Chapter 23): covalent bonds, hydrogen bonds, salt bridges, and hydropho-

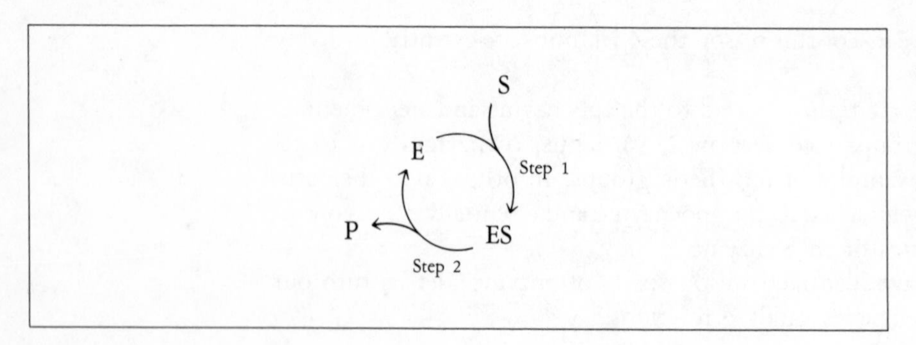

Fig. 25-3. A two-step cyclic mechanism for the catalytic action of an enzyme (E), where S is the substrate, P is the product, and ES is the enzyme-substrate complex.

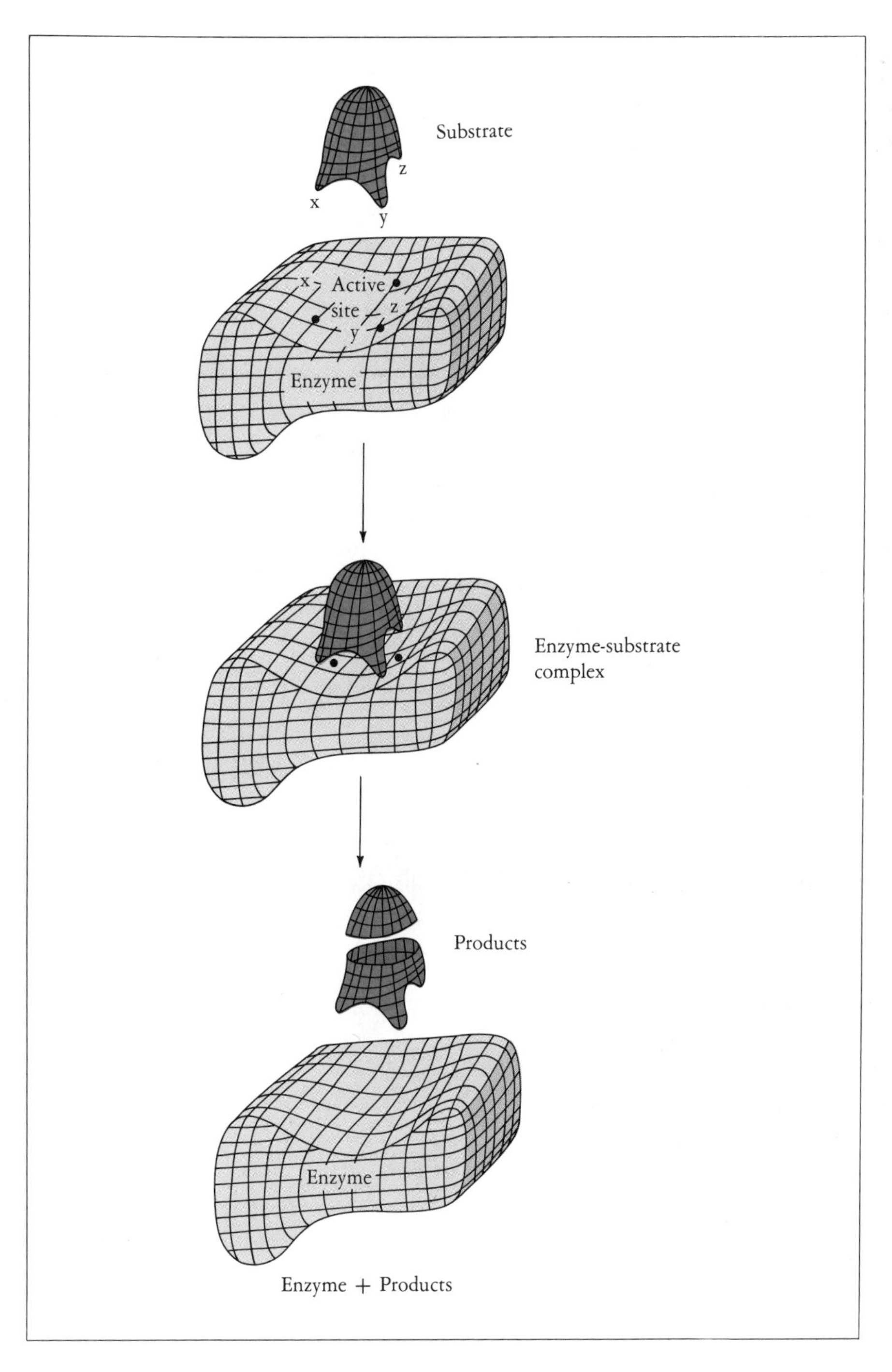

Fig. 25-4. The lock and key model of enzyme action. The symbol ● represents catalytic groups.

bic interactions. Furthermore, the enzyme molecule is much larger than the substrate. As a result, only a small part of the enzyme is involved in forming the complex. This region where the substrate binds to the enzyme and then undergoes reaction is called the *active site* of the enzyme. The active site is a specific three-dimensional region that has a unique arrangement of groups that enable the enzyme and its substrate to fit together properly.

A simple picture of this interaction between the enzyme and substrate is provided by the lock and key model. According to this model, the active site and the substrate have complementary structures and they fit together like a key in a lock. This model is shown schematically in Figure 25-4. We learned in Section 14.9 that the ability of enzymes to distinguish between enantiomers can be explained by proposing that an enzyme and its substrate must be joined by at least three points. These sites are labeled x, y, and z in Figure 25-4.

Once a substrate is bound to the active site, certain functional groups contained in the side chains of the amino acid residues of the enzyme assist in the reaction. These are called *catalytic groups* and they are shown schematically in Figure 25-4. One example of the assistance of such groups was given in Section 16.5.

25.5 LYSOZYME: A SPECIFIC EXAMPLE OF ENZYME CATALYSIS

OPTIONAL SECTION

Lysozyme, as we learned in Section 23.5, is an enzyme that catalyzes the hydrolysis of the polysaccharide component of the cell walls of certain bacteria. It also catalyzes the hydrolysis of chitin, a polysaccharide found in the shells of crabs and lobsters. Chitin is made up of repeating units of

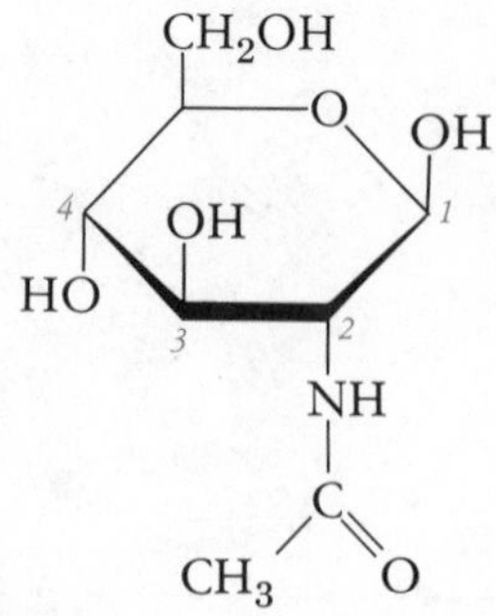

Fig. 25-5. N-*acetylglucosamine is the amino sugar that makes up the polysaccharide chitin.*

N-acetylglucosamine. This monosaccharide is simply the amino sugar glucosamine (see Section 21.2) in which the amino group is acetylated. Its structure is shown in Figure 25-5. Chitin is a polymer made up of repeating units of *N*-acetylglucosamine joined in the 1,4 positions by β-glycoside bonds (see Section 21.7). Lysozyme catalyzes the hydrolysis of this glycoside bond, as shown in Figure 25-6.

To learn how lysozyme catalyzes this reaction, it is necessary first to determine its three-dimensional structure. We know its primary structure

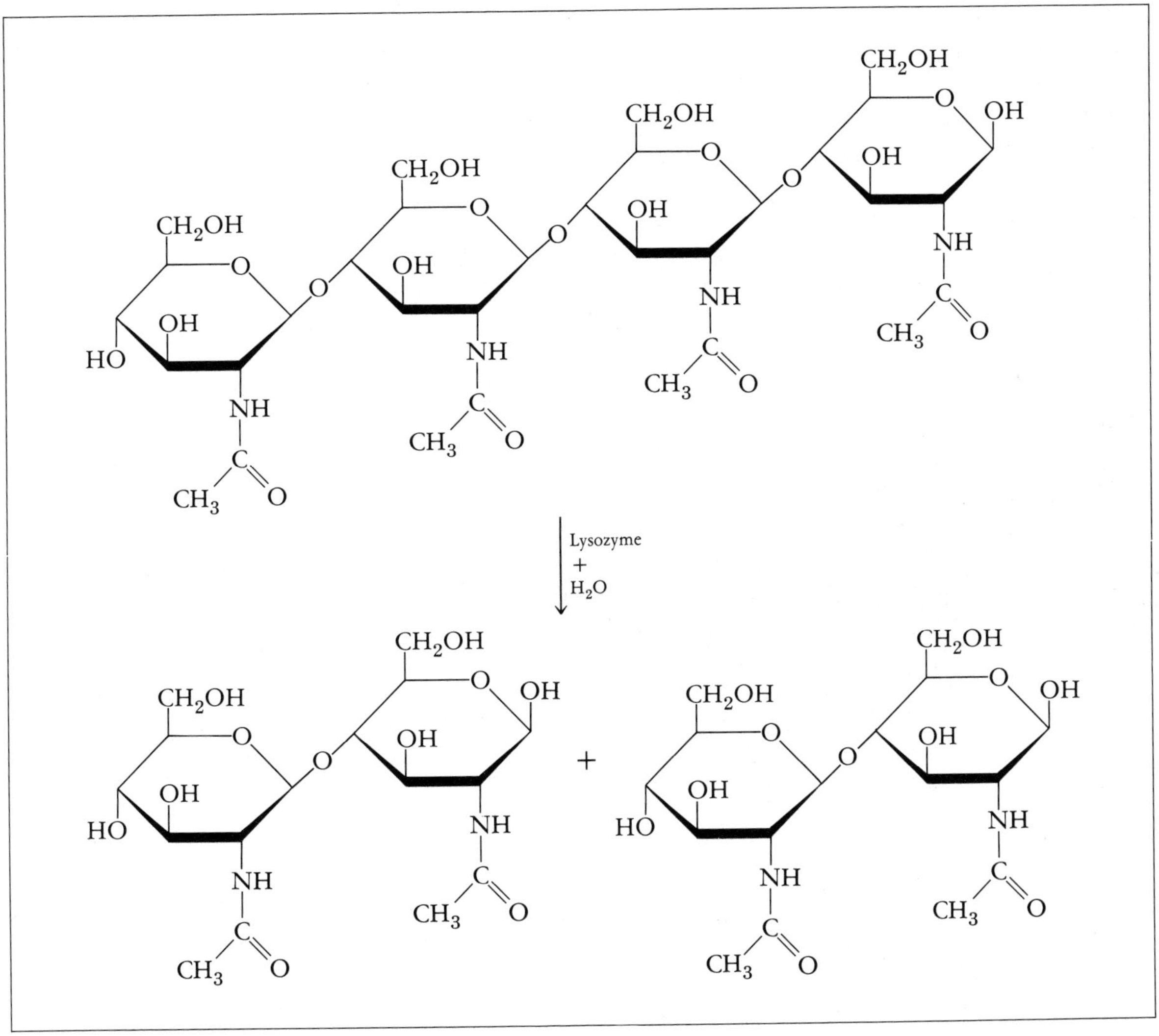

Fig. 25-6. Lysozyme catalyzes the hydrolysis of the β-glycoside bond of chitin.

(see Chapter 23, Figure 23-14). By x-ray crystallography, David Phillips and his colleagues in 1965 determined the three-dimensional structure of lysozyme. However, this still did not identify the active site.

The active site of lysozyme was identified by determining the three-dimensional structure of a complex formed between lysozyme and the trimer of *N*-acetylglucosamine. This trimer is not hydrolyzed by lysozyme, but it does bind to the active site of the enzyme. From the x-ray crystallographic study of this complex, it was possible to identify not only the active site, but also the interactions responsible for the specific binding of the substrate. Once this information was obtained, a detailed mechanism of the enzyme's action was proposed.

The trimer of *N*-acetylglucosamine binds to lysozyme in a cleft on the surface of the enzyme. The substrate is held in position by the hydrogen bonds shown in Figure 25-7. We must use the correct three-dimensional representations of *N*-acetylglucosamine to show these interactions accurately. In addition, a large number of hydrophobic interactions between the enzyme and the substrate assist in binding the substrate to the active site.

Once the position of the substrate on the active site had been established,

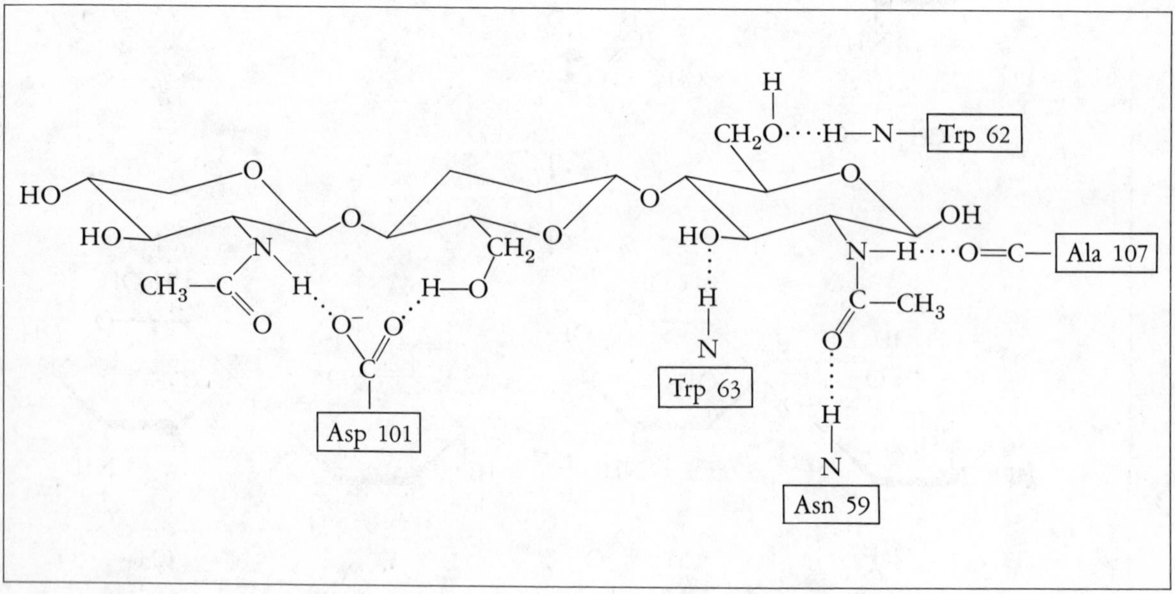

*Fig. 25-7. Hydrogen bonds at the active site between specific amino acid residues of lysozyme and the substrate, tri-*N*-acetylglucosamine. Hydrogens and other nonessential atoms have been omitted for clarity.*

a search was made for possible catalytic groups close to the glycoside bond that is broken. Most catalytic groups serve as hydrogen ion donors or acceptors. The donation or removal of a proton is a critical step in most enzyme-catalyzed reactions. Phillips and his colleagues have proposed that the two amino acid residues aspartic 52 and glutamic 35 are the catalytic groups. The two groups are on opposite sides of the glycoside bond. Furthermore, the two acid side chains are located in quite different environments. Aspartic acid is in a polar environment, whereas glutamic acid is in a nonpolar region. Thus, it is likely that the carboxylic acid group of aspartic 52 exists in the ionized form $\left(-C(=O)O^{-}\right)$, whereas that of glutamic 35 exists in the un-ionized form $\left(-C(=O)OH\right)$.

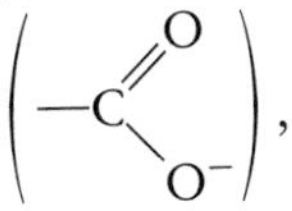

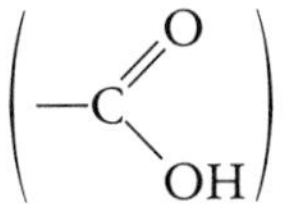

The role of glutamic 35 in the proposed mechanism is to donate a proton to the oxygen atom of the glycoside bond. This occurs as the first step shown in Figure 25-8. The glycoside bond breaks in the second step to form a carbonium ion that is stabilized by the carboxylate ion of the aspartic 52. The aspartic 52 residue is located in exactly the right spot to provide this stabilization. The carbonium ion reacts with water in the third step to form a protonated hemiacetal. Loss of a proton to glutamic 35 and diffusion of the products from the enzyme completes the reaction.

The details of the mechanism given in Figure 25-8 are very similar to the mechanism of the acid-catalyzed preparation and hydrolysis of acetals given in Section 19.7. The same type of carbonium ions are involved in both mechanisms. The major difference between the two is that enzyme-catalyzed reactions have catalytic groups located in close proximity to the reaction site that assist in making and breaking bonds. Thus, we can conclude that, not only are many of the reactions of living systems similar to those in the laboratory, but they also occur by a similar mechanism.

EXERCISE 25-3 Write the mechanism for the acid-catalyzed hydrolysis of di-*N*-acetylglucosamine. Compare this mechanism with the one given in Figure 25-8.

Sometimes the active site of an enzyme can bind a molecule that does not undergo any further reaction. This is called inhibition of enzyme activity.

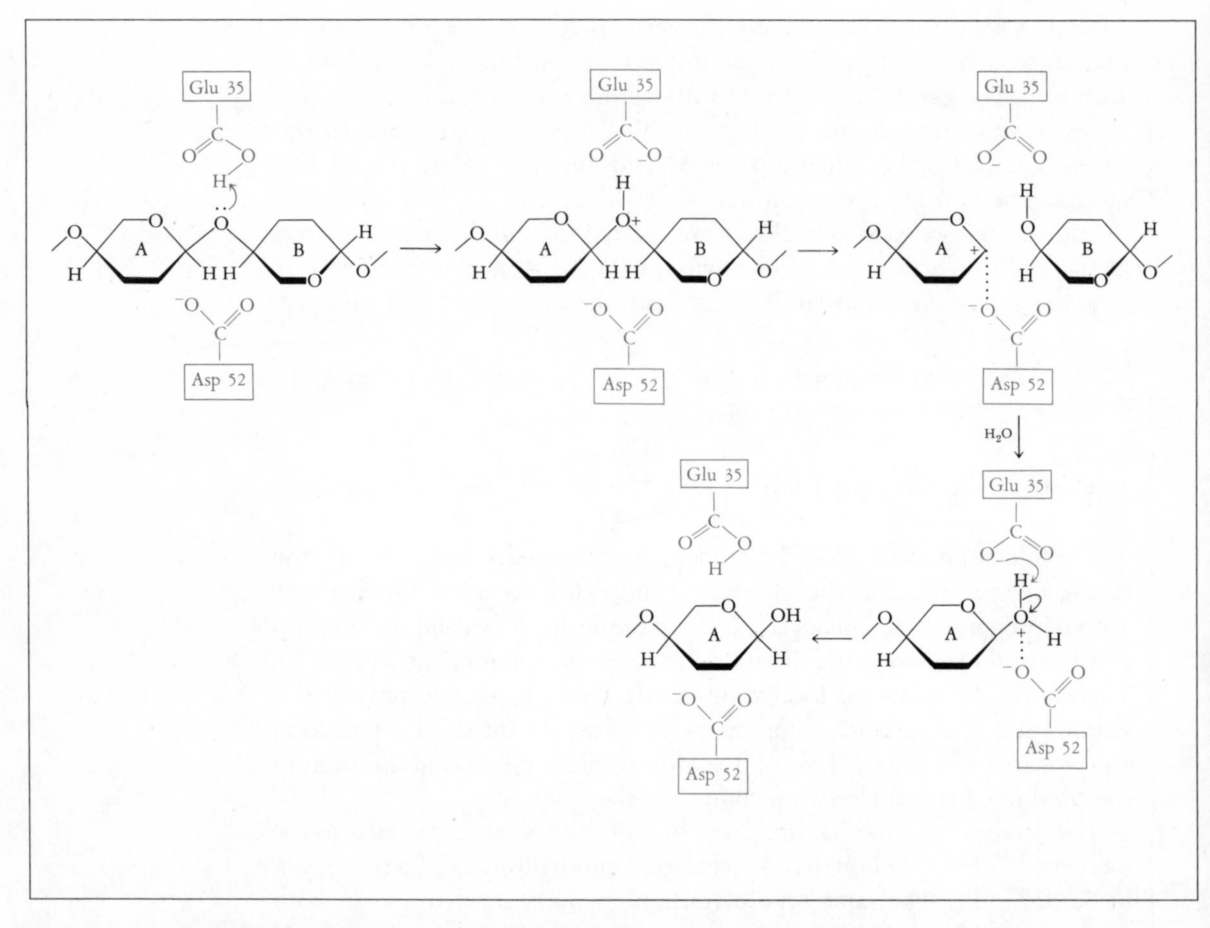

Fig. 25-8. The proposed mechanism of the enzyme-catalyzed hydrolysis of chitin. All nonessential atoms on the polysaccharide chain have been omitted for clarity.

25.6 ENZYME INHIBITION

Sometimes an enzyme can be fooled. It binds a compound that has a close structural resemblance to its normal substrate. Because it is structurally different, this compound cannot react and it remains bound to the enzyme. As a result, it prevents the normal substrate from occupying the active site. Compounds of this kind are called *inhibitors.* They inhibit the enzyme from carrying out its normal function. There are two general classes of inhibitors, competitive and noncompetitive.

A competitive inhibitor competes with the normal substrate for the active site of an enzyme. The effectiveness of a competitive inhibitor depends on its concentration relative to that of the normal substrate. If the inhibitor is present in large excess, it will block all the active sites on all the enzyme molecules; the result is complete inhibition. But the formation of the enzyme-inhibitor complex is reversible. That is, the inhibitor can leave the active site, giving the normal substrate a chance to occupy the site. If the concentration of substrate is increased, it can compete more effectively with the inhibitor. In fact, competitive inhibition can be completely suppressed by the addition of a large excess of the normal substrate.

A classic example of competitive inhibition is the action of malonate on succinate dehydrogenase. The enzyme catalyzes the removal of one hydrogen atom from each of the two methylene carbon atoms of succinate (dehydrogenated), as shown in Figure 25-9. Malonate will also bind to the active

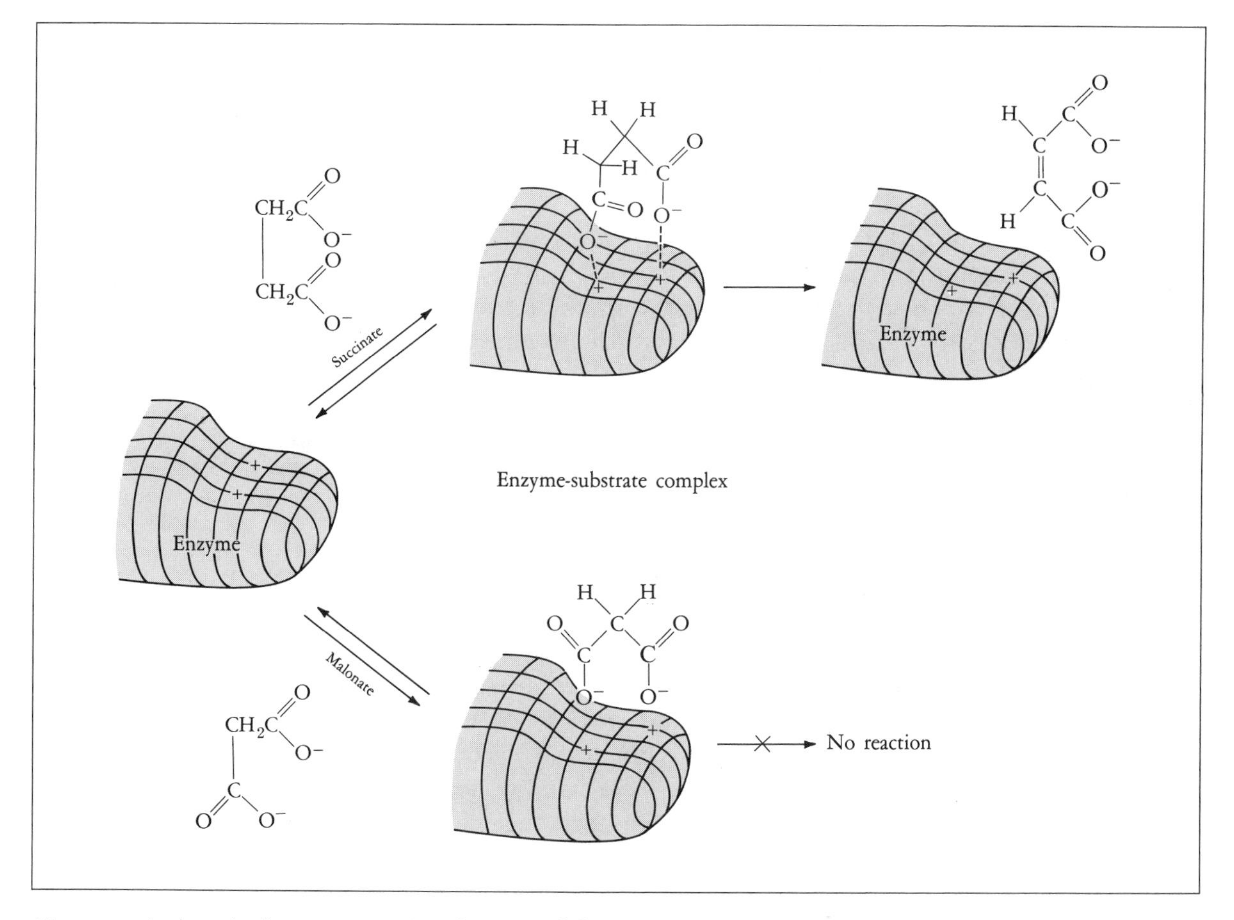

Fig. 25-9. A schematic diagram representing the action of the competitive inhibitor malonate on the enzyme succinate dehydrogenase.

site, because the distance between its two carboxylate groups is similar to that of the carboxylate groups of succinate. But malonate cannot be dehydrogenated, and it remains bound to the active site.

EXERCISE 25-4 Explain why pyrophosphate,

$$^{-}O-\overset{\overset{O}{\|}}{\underset{\underset{O^{-}}{|}}{P}}-O-\overset{\overset{O}{\|}}{\underset{\underset{O^{-}}{|}}{P}}-O^{-}$$

, oxaloacetate $^{-}O_2CCH_2\overset{\overset{O}{\|}}{C}CO_2^{-}$, and oxalate $^{-}O_2CCO_2^{-}$ are also inhibitors of succinate dehydrogenase.

Noncompetitive inhibitors react with an enzyme at a site other than the active site. The most common type of noncompetitive inhibitor is a reagent that reacts with a functional group of the enzyme outside of the active site that is essential to maintaining its catalytically active three-dimensional structure. In this way, the inhibitor deforms the enzyme so that it cannot form an enzyme-substrate complex.

We learned in Chapter 23 that thiol and carboxylate groups help maintain the three-dimensional structures of proteins. Heavy metal ions such as lead (Pb^{+2}) and mercury (Hg^{+2}) react with these groups (see Section 23.10) and destroy their ability to maintain the structure of proteins and enzymes. Thus, these heavy metal ions are noncompetitive inhibitors.

Many enzymes contain a metal ion as a cofactor. These enzymes are inhibited noncompetitively by compounds that can remove the metal ion from the enzyme. For example, the compound ethylenediamine tetra-acetate (EDTA) removes magnesium ions (Mg^{+2}) from enzymes. This changes the shape of the enzyme and destroys its biological activity, as shown schematically in Figure 25-10.

The action of inhibitors is not all bad. Scientists have used inhibitors to aid them in determining the location and structure of the active sites of several enzymes. Most enzyme-substrate complexes cannot be isolated because they react too quickly to form products. Consequently, their structures cannot be determined by x-ray crystallography. A competitive inhibitor forms a complex by occupying the same site as the substrate. This enzyme-inhibitor complex does not undergo further reaction. Therefore, it can be isolated, and its structure can be determined by x-ray crystallography. Because of the close structural similarity between the substrate and its competitive inhibitor, we can learn the location of the active site and the location of the catalytic groups of the enzyme from the structure of the

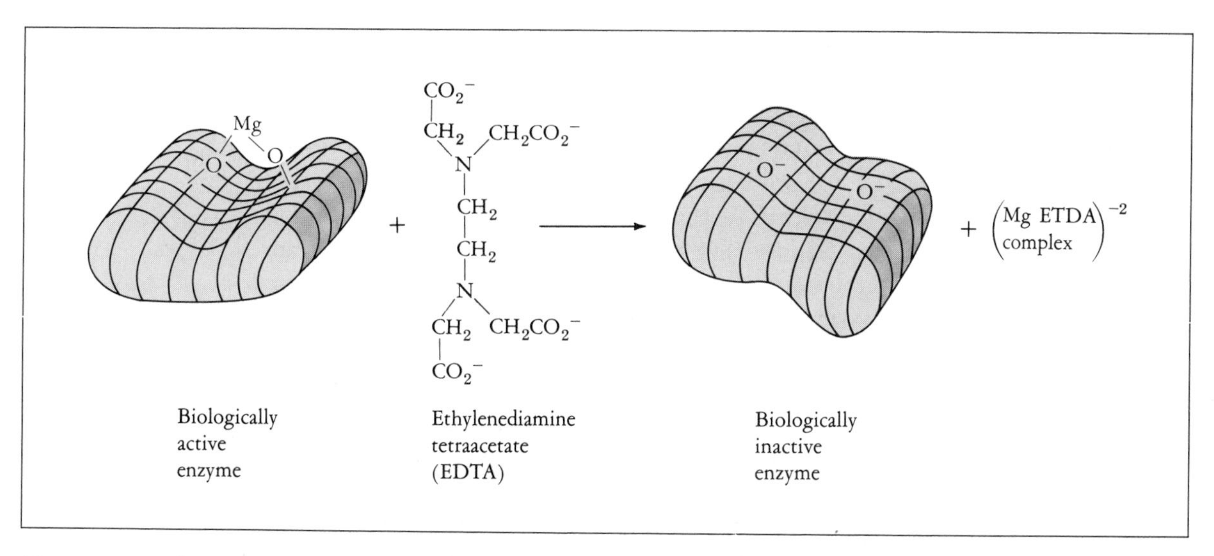

Fig. 25-10. A schematic representation of the action of noncompetitive inhibitor EDTA.

enzyme-inhibitor complex. This is how the active site and the catalytic groups of lysozyme were identified (see Section 25.5). Tri-*N*-acetylglucosamine is actually a competitive inhibitor of lysozyme.

An important function of certain inhibitors is the control of enzyme activity in cells.

25.7 CONTROL OF ENZYME ACTIVITY

We have learned that enzymes control the kinds of chemical reactions that take place in a cell. But enzymes also control when certain reactions occur and how fast they occur. The chemical reactions that occur in a cell depend on the demand placed on the cell. For example, when you pick up a book, certain of your muscles do work. To carry out this action, the brain signals the cells in your muscles to work. On command, the enzymes responsible for the chemical reactions involved in the muscle action must act immediately. When the signal to stop work comes, these same enzymes must stop. Therefore, the activities of enzymes in a cell are under rigid control. This control of the activity of an enzyme is really a control of the speed of the enzyme-catalyzed reactions of the cell, which is accomplished in several ways.

Enzymes are proteins and are sensitive to changes in the pH of their environment. We learned in Section 23.10 that changes in pH cause denaturation of proteins. Denaturation is a disruption of the three-dimensional structure of a protein. For an enzyme, such a change in structure results in loss of catalytic activity. All enzymes show maximal activity within a narrow pH range. The median value of this range is called the *optimal pH.* With the exception of enzymes in the gastric juices, the optimal pH of most enzymes is between 6 and 8. The pH of the environment of a cell can be used to control the activity of an enzyme. Thus, a substance may be released in the cell that changes slightly the intracellular pH. As a result, the activity of the enzyme changes and the speed of the enzyme-catalyzed reaction also changes.

The activity of enzymes can be controlled by limiting the amount of the compounds in the cell that form the enzyme-substrate complex. Thus, controlling the amount of substrate, the amount of enzyme, or the amount of cofactor needed by a particular apoenzyme can slow down or speed up the enzyme-catalyzed reaction.

As the amount of substrate is increased, the activity of the enzyme also increases. This is true up to a point. That point is reached when all the enzyme molecules are bound to substrate. Extra substrate molecules must wait until the enzyme-substrate complex has formed products and free enzyme before they can react. Thus, there is a maximal activity determined by the amount of enzyme present. But increasing the amount of enzyme will increase the maximal speed of the catalyzed reaction. Thus, increasing or decreasing the amount of either substrate or enzyme controls the speed of the chemical reactions in the cell.

Most cells can control the amount of cofactor produced for a particular apoenzyme. As long as the cofactor is absent, the apoenzyme is inactive. On command, the cell can release the cofactor that reacts with the apoenzyme to form the biologically active holoenzyme. In this way, specific chemical reactions in a cell can be started or stopped.

Another way of controlling enzyme activity is by *end-product inhibition.* This control occurs when the end product of a series of enzyme-catalyzed reactions is a specific inhibitor of the enzyme that catalyzes a reaction at or near the beginning of the sequence. The classic example of end-product inhibition is the series of five enzyme-catalyzed reactions that convert L-threonine to L-isoleucine shown in Figure 25-11. The first enzyme of the sequence, L-threonine dehydratase, is strongly inhibited by L-isoleucine, the final product. As the amount of L-isoleucine formed in the cell gradually increases, it begins to inhibit the first reaction. As a result, less L-threonine reacts and less L-isoleucine is formed. When there is enough L-isoleucine for the needs of the cell, the reaction is completely inhibited and no more

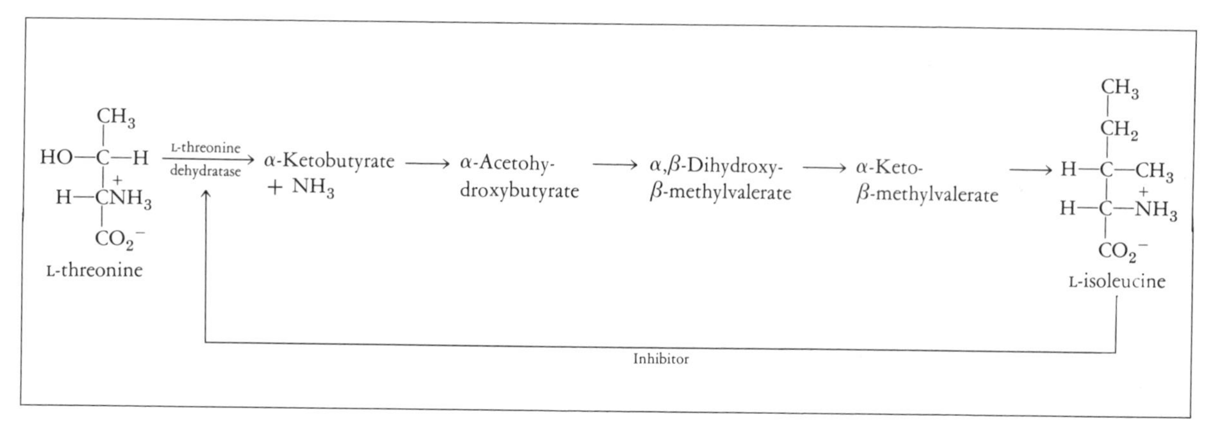

Fig. 25-11. L-isoleucine is an inhibitor of L-threonine dehydratase, the enzyme of the first of five enzyme-catalyzed reactions that form L-isoleucine from L-threonine.

L-isoleucine is formed. As the cell uses L-isoleucine, its concentration decreases. When its concentration decreases to the point where it cannot completely inhibit the first step, the reaction of L-threonine begins again to form L-isoleucine. In this way, the amount of L-isoleucine in the cell is controlled so that neither too little nor too much is present at any time.

The enzymes that show end-product inhibition are known to have at least two binding sites. One is the active site that binds the substrate. The other site binds the end product and is called the *allosteric site.* When the end product binds to this site, it changes the three-dimensional structure of the enzyme. This distorts the active site sufficiently that it is rendered inactive. This inhibition is shown schematically in Figure 25-12.

A different way of controlling enzyme activity is to change an inactive form of an enzyme, called a *zymogen,* into its catalytically active form. This change usually requires an enzyme-catalyzed breaking of chemical bonds. The classic examples of this method of control are the digestive enzymes pepsin, trypsin, and chymotrypsin. These enzymes catalyze the hydrolysis of proteins in the gastrointestinal tract. They are made from the zymogens (pepsinogen, trypsinogen, and chymotrypsinogen) that are secreted into the gastrointestinal tract, where they are changed into their active forms by the hydrolysis of one or more specific peptide bonds. For example, the zymogen trypsinogen is changed into active trypsin by the action of the enzyme enterokinase. Enterokinase catalyzes the removal of a hexapeptide from the N-terminal end of trypsinogen to form trypsin.

Finally, hormones can control the activity of enzymes. These substances are formed in trace amounts by various glands and act as chemical messen-

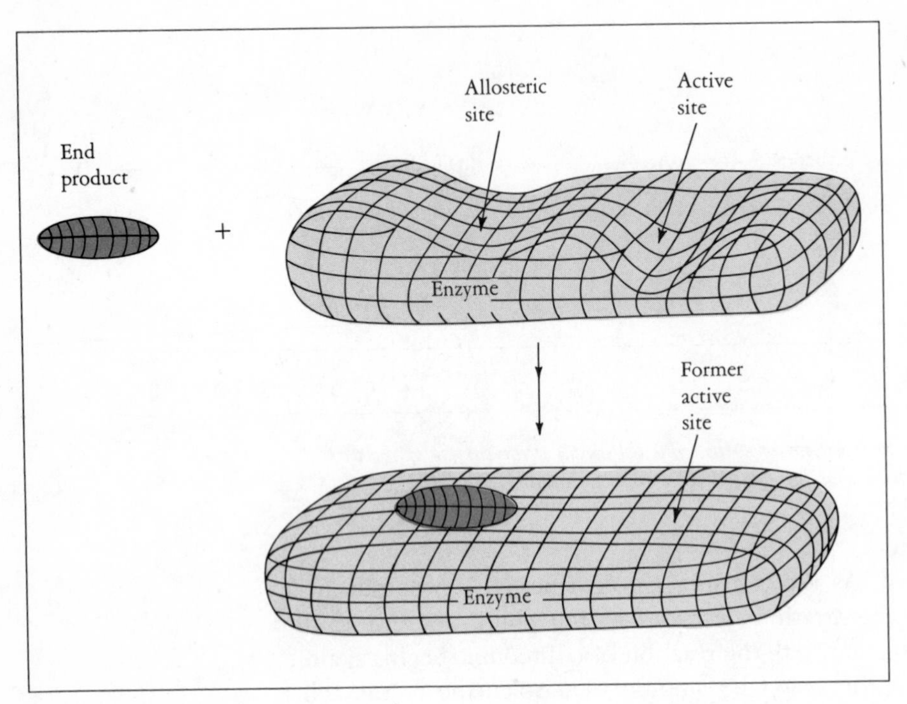

Fig. 25-12. A schematic representation of allosteric inhibition.

gers. They are carried by the blood to specific hormone receptors, usually located in the cell membrane (see Section 22.6). The binding of the hormone to its receptor causes the formation of a messenger within the cell that stimulates or depresses some characteristic chemical activity within the cell.

Feeding a foreign substance to an organism can interfere with many of these controls. Sometimes this interference can destroy the organism.

25.8 POISONS

Poisons are compounds that cause death when fed to an organism. They act by inhibiting the activity of key enzymes in the ways described in Section 25.6. For example, the cyanide ion (CN^-) of sodium or potassium cyanide forms strong bonds with the metal ions essential to the enzymes involved in the transport of oxygen to the cells. Removal of these metal ions changes the three-dimensional structure of the enzyme, and it cannot carry out its function. As a result, the organism dies of suffocation.

We learned in Section 23.10 that heavy metal ions such as Pb^{+2} and Hg^{+2} react with essential thiol groups of proteins. This causes denaturation and results in loss of biological activity. A particularly dangerous derivative of mercury is dimethylmercury, $(CH_3)_2Hg$. Elemental mercury, dumped into the environment, finds its way to the bottom of lakes and rivers, where it is changed into dimethyl mercury by a series of chemical reactions. This compound is particularly dangerous because the methyl group allows the mercury to enter many hydrophobic regions of living systems. In humans, it attacks the central nervous system and causes permanent damage. A horrible example of such poisoning has been documented in the inhabitants of the Japanese fishing village of Minamata. This village is located near a chemical plant that discharged mercury into the surrounding waters. The mercury was gradually changed into dimethyl mercury. It found its way into algae that were eaten by fish. The villagers ate the fish and became seriously ill from mercury poisoning. This serious incident focused worldwide attention on the problem of industrial pollution of the environment.

A number of poisons are known that are classed as "nerve gases." These poisons interfere with the signals transmitted by nerve cells. The nerve cells transmit signals to other nerve cells at junctions called synapses, or synaptic clefts. The nerve signals are carried across most synaptic clefts by means of chemical transmitters. One of these is acetylcholine (see Section 18.7). The arrival of a nerve impulse leads to release of acetylcholine into the synaptic

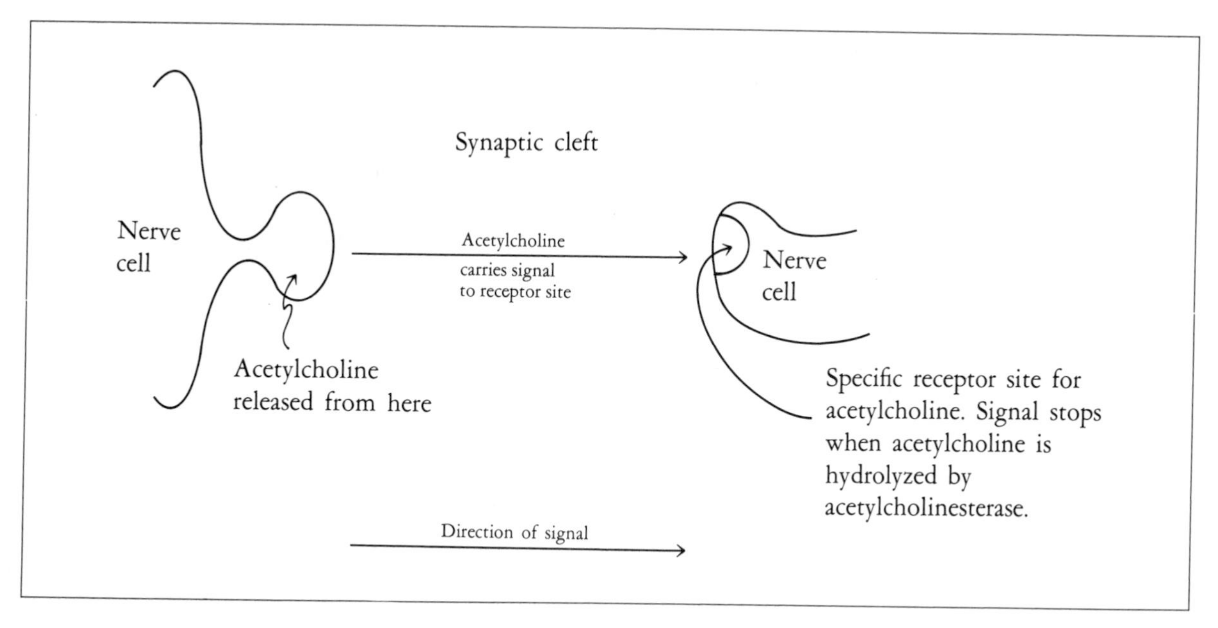

Fig. 25-13. Schematic representation of the transmission of signals across the synaptic cleft by acetylcholine.

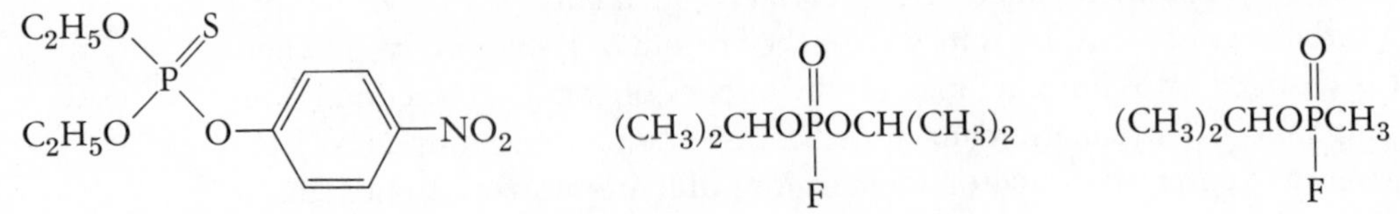

$(C_2H_5O)_2P(=S)-O-C_6H_4-NO_2$

Parathion

$(CH_3)_2CHOP(=O)(F)OCH(CH_3)_2$

Diisopropyl phosphofluoridate

$(CH_3)_2CHOP(=O)(F)CH_3$

Sarin

Fig. 25-14. Organophosphorus compounds that interfere with the functioning of the nervous system.

cleft. The acetylcholine molecules travel a short distance to the specific receptor site on the next nerve cell, where they react with receptor molecules. This is shown in Figure 25-13.

After acetylcholine interacts with the receptor, it is quickly hydrolyzed by acetylcholinesterase. This stops the signal. Some poisons deactivate the enzyme acetylcholinesterase and prevent the hydrolysis of acetylcholine.

A particularly powerful inhibitor of acetylcholinesterase is diisopropyl phosphofluoridate. This compound reacts with the active site of acetylcholinesterase to form a very stable phosphorus-enzyme complex. Other organophosphorus compounds that have been prepared for use as pesticides or nerve gases are shown in Figure 25-14. Parathion has been widely used as an insecticide. These compounds all cause respiratory paralysis.

From these examples, it is clear that a poison inhibits key enzymes, with the result that the organism usually dies. If this organism is one that is infectious to humans, we now have a method of using chemical compounds to destroy it without harming other organisms. This is the idea behind many drugs, as we will learn in the next section.

25.9 CHEMICAL COMPOUNDS THAT FIGHT INFECTIOUS DISEASES*

One of the major advances in medicine during the last 50 years has been the use of chemical compounds (drugs) to destroy infectious organisms without harming the human host. One of the major problems in finding such compounds is that the chemical reactions in all living organisms, from

* This use of chemical compounds was originally called chemotherapy. However, the word chemotherapy today means the use of chemical compounds to fight cancer.

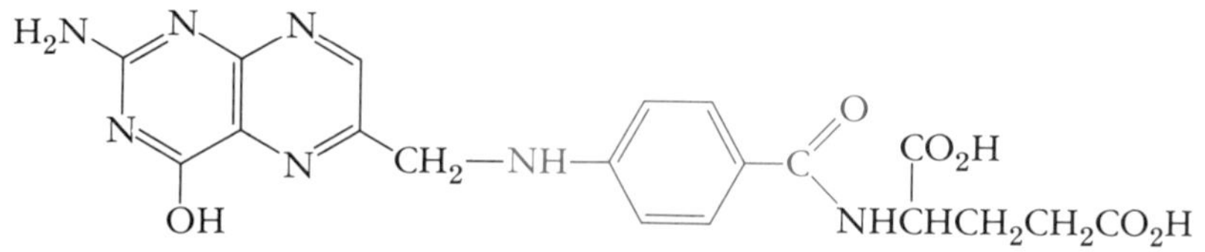

H_2N N N N N OH CH_2—NH— —C(=O) CO_2H $NHCHCH_2CH_2CO_2H$

Part of folic acid obtained from 4-aminobenzoic acid

Folic acid, an important enzyme

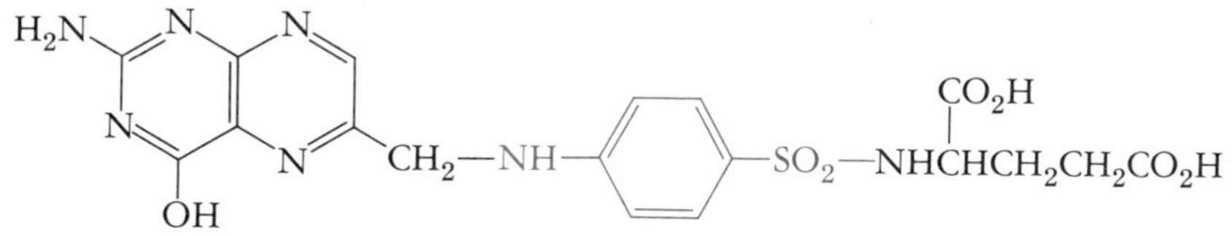

H_2N N N N N OH CH_2—NH— —SO_2—$NHCHCH_2CH_2CO_2H$ CO_2H

Part of altered folic acid in which sulfanilimide replaces 4-aminobenzoic acid

Altered folic acid (biologically inactive)

Fig. 25-15. The structures of folic acid and altered folic acid.

bacteria to humans, are similar. As a result, it is not easy to find chemical compounds that attack infectious organisms without also attacking the host. However, great progress has been made in discovering safe and effective drugs for a variety of diseases. Examples of two types of drugs, antimetabolites and antibiotics, will be discussed in this section.

An *antimetabolite* is a compound that is a competitive inhibitor of an important enzyme-catalyzed reaction of the invading organism. The sulfa drugs (Section 17.6) are the classic example of antimetabolites. A sulfa drug such as sulfanilamide is effective because it resembles 4-aminobenzoic acid, a compound vital to the formation of the coenzyme folic acid in many bacteria. Sulfanilamide fools the enzyme into using it in place of 4-aminobenzoic acid, as shown in Figure 25-15. However, the altered folic acid will not work as a coenzyme in the bacteria and the organism dies. The effect of the sulfa drug is to inhibit an important chemical reaction in the life cycle of the bacteria. Humans are not affected by sulfa drugs because we cannot make folic acid (we must get it from the food we eat).

Antibiotics are compounds made by one organism that are poisonous to another organism. The classic example is penicillin, first isolated by Alexander Fleming in 1928 from a mold called *Penicillium notatum.* Several forms of penicillin are known that differ only in the structure of the side chain, as

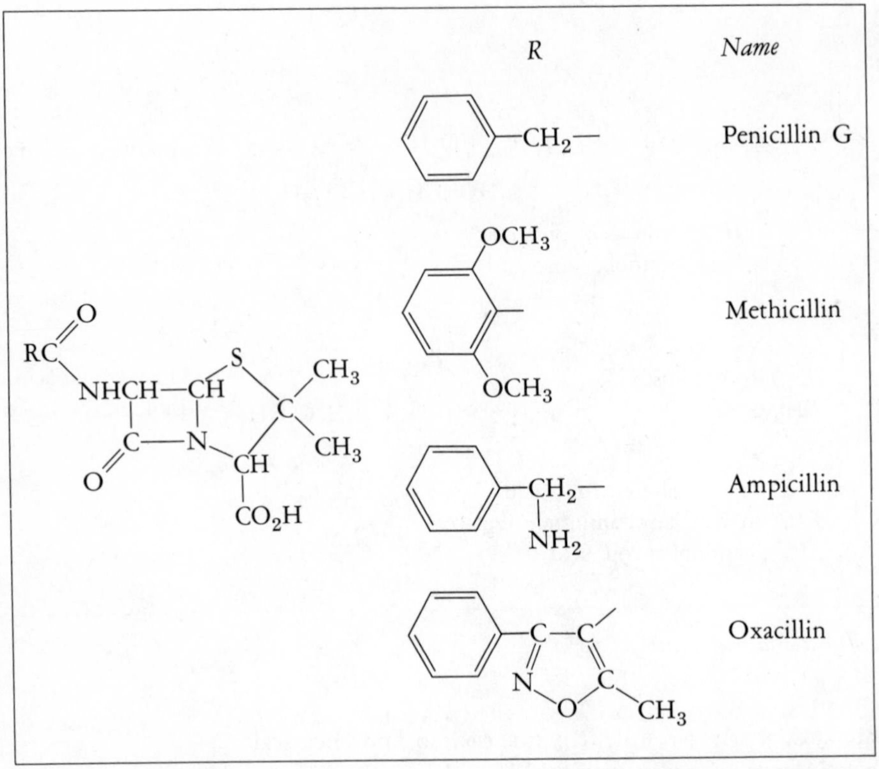

R	Name
C_6H_5—CH_2—	Penicillin G
2,6-(OCH_3)$_2C_6H_3$—	Methicillin
C_6H_5—$CH(NH_2)$—	Ampicillin
C_6H_5—C=C(—)—C(CH_3)=N—O (isoxazolyl)	Oxacillin

General structure: $RC(=O)NHCH-CH$ (β-lactam ring fused to thiazolidine ring: S, $C(CH_3)_2$, $CH-CO_2H$, N, $C=O$)

Fig. 25-16. The structural formulas of four penicillins.

shown in Figure 25-16. The first penicillin to be used on a large scale was penicillin G.

The penicillins are believed to stop the growth of bacteria by inhibiting the enzyme transpeptidase, which catalyzes the last step in the construction of its cell membrane. The cell must have this barrier to survive. When formation of the cell membrane is stopped, the cell cannot grow and the spread of the bacterial infection is stopped.

Many organisms have developed a resistance to penicillin G by developing an enzyme, penicillinase, that breaks it down as follows:

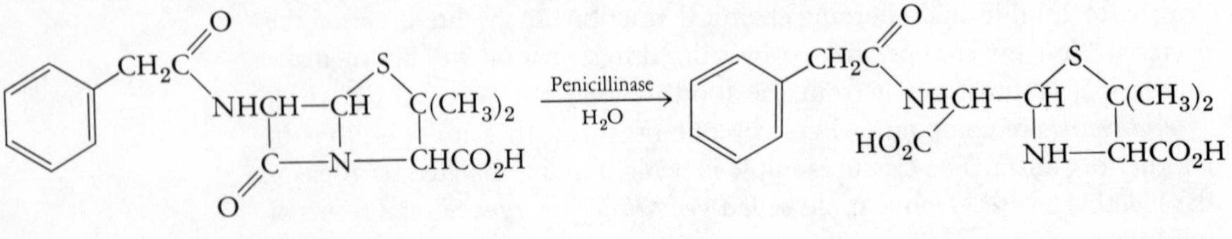

$$C_6H_5CH_2C(=O)-NHCH-CH(S)C(CH_3)_2-C(=O)-N-CHCO_2H \xrightarrow[H_2O]{\text{Penicillinase}} C_6H_5CH_2C(=O)-NHCH(CO_2H)-CH(S)C(CH_3)_2-NH-CHCO_2H$$

Penicillin G

To combat this problem, scientists have made many derivatives of penicillin in the laboratory that are resistant to penicillinase. Among these are methicillin, ampicillin, and oxacillin, whose structures are given in Figure 25-16.

EXERCISE 25-5 What kind of reaction is catalyzed by the enzyme penicillinase?

The tetracyclines, originally obtained from microbes in soil samples, are another group of important antibiotics. The chemical structures of two tetracyclines are as follows:

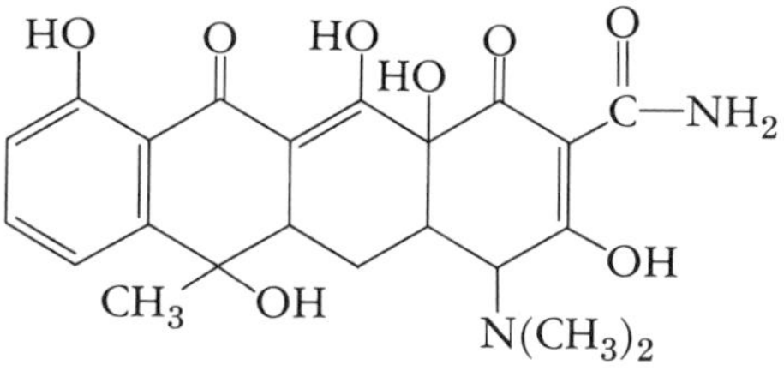

Tetracycline

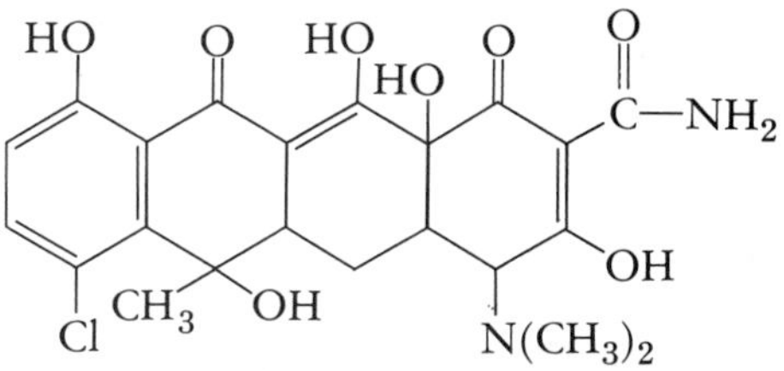

Chlortetracycline

These antibiotics act by inhibiting protein synthesis in certain organisms. This stops their growth. The chemical structures of the compounds used in protein synthesis in bacteria and animal cells are sufficiently different that the tetracyclines act only on the bacteria.

From these examples, it is clear that antibiotics and antimetabolites are simply specific poisons for infectious organisms. They act by inhibiting the catalytic role of enzymes in key chemical reactions of the organism.

25.10 SUMMARY

Enzymes are the catalysts of biological reactions. All known enzymes are proteins. Enzymes have two unique characteristics. They greatly increase the speed of chemical reactions, and they show specificity toward particular compounds. Some enzymes, such as lysozyme, are simple proteins that are biologically active by themselves. Others need cofactors to be biologically active. Cofactors may be either metal ions or organic compounds. An enzyme acts by binding substrate to a small specific region on its surface, called the active site, to form an enzyme-substrate complex. The chemical reaction takes place in this complex. Once the reaction is over, the enzyme and product separate, and the enzyme is ready to accept another molecule of substrate.

Enzymes can be inhibited by small molecules or ions. A competitive inhibitor is one that occupies the active site because it is structurally similar to the usual substrate. Noncompetitive inhibitors react with the enzyme at a site other than the active site. Inhibition of an enzyme is one way of controlling enzyme activity. Poisons are compounds that cause death by inhibiting the activity of key enzymes in an organism. Certain chemical compounds are poisons specifically for infectious organisms. These compounds destroy the infectious organisms without harming the human host.

REVIEW OF TERMS AND CONCEPTS

Terms

ACTIVE SITE A specific three-dimensional region on the enzyme surface that has a unique arrangement of groups enabling the enzyme and its substrate to fit together properly.

ANTIBIOTIC A compound made by one organism that is a poison for another organism.

ANTIMETABOLITE A compound that is a competitive inhibitor of an important enzyme-catalyzed reaction of an invading organism.

APOENZYME The enzyme protein, without its cofactor.

CATALYTIC GROUPS Functional groups located near the active site that assist in the reaction.

COENZYME A cofactor that is an organic molecule.

COFACTOR The nonprotein part of an enzyme. Its presence is required for the enzyme to be biologically active.

END-PRODUCT INHIBITION The end product of a series of enzyme-catalyzed reactions is a specific inhibitor of the enzyme that catalyzes a reaction at or near the beginning of the series of reactions.

ENZYME Proteins that are produced by living systems to catalyze specific biological reactions.

HOLOENZYME The apoenzyme and its cofactor.

INHIBITORS Compounds or ions that block the catalytic activity of enzymes.

OPTIMAL pH The pH at which an enzyme has its greatest activity.

SPECIFICITY The ability of an enzyme to catalyze a particular reaction of a compound or class of compounds.

SUBSTRATE The compound or class of compounds on which an enzyme acts.

ZYMOGEN A protein that has no catalytic activity but can be changed into an enzyme by a specific chemical reaction within the organism.

Concepts

1. Enzymes are proteins made by a living organism that catalyze specific reactions.
2. An enzyme acts by combining with its substrate to form an enzyme-substrate complex. The chemical reaction occurs within this complex.
3. The active site is the place on the enzyme where the substrate binds. It contains a unique three-dimensional arrangement of groups that hold the substrate and enzyme together, mainly by hydrogen bonds and hydrophobic interactions.
4. The catalytic activity of enzymes can be destroyed by inhibitors.
5. Some inhibitors act by occupying the active site. This prevents the substrate from binding to the site. Others act by changing the three-dimensional shape of enzymes. This, in effect, denatures the enzyme.

EXERCISES 25-6 Give two functions of enzymes in living systems.

25-7 How are vitamins and certain coenzymes related?

25-8 How does the lock-and-key model explain the way an enzyme works?

25-9 What is the difference between competitive and noncompetitive inhibitors?

25-10 Give four ways of controlling enzyme activity.

25-11 How is a zymogen changed into an enzyme?

25-12 What is the difference between an antimetabolite and an antibiotic?

ENERGY TRANSFER AND METABOLISM

26

The events of the past ten years have dramatically shown how dependent we are on energy. Without it, our society would collapse. We would be unable to heat our homes, drive our cars, or use any of the modern machines of industry.

Living organisms also need energy to function. Without it, our bodies could not perform mechanical work such as muscle contraction, could not transport molecules and ions between various parts of the body, could not synthesize macromolecules from simple molecules. In short, without energy, a living organism cannot carry out the chemical reactions that are responsible for its existence.

Where do we get energy? Our society gets it from petroleum, coal, or natural gas. Living organisms get energy from their food, as we learned in Section 2.6. But all of this energy is really obtained from the same place, the sun. Food, petroleum, and coal are simply energy of the sun that has been transferred into chemical compounds and stored for later use. In this chapter, we will trace the transfer of energy from the sun into chemical compounds that can be used by humans. We will also trace the major paths by which cell components are synthesized and degraded, and we will identify several compounds that play important roles in the use of energy within living organisms. Let us begin by reviewing the concept of energy. (You may wish to review Sections 2.6 and 2.7.)

26.1 CHEMICAL ENERGY

We learned in Section 2.6 that energy exists in many forms, such as electrical energy, potential energy, kinetic energy, chemical energy, heat, work,

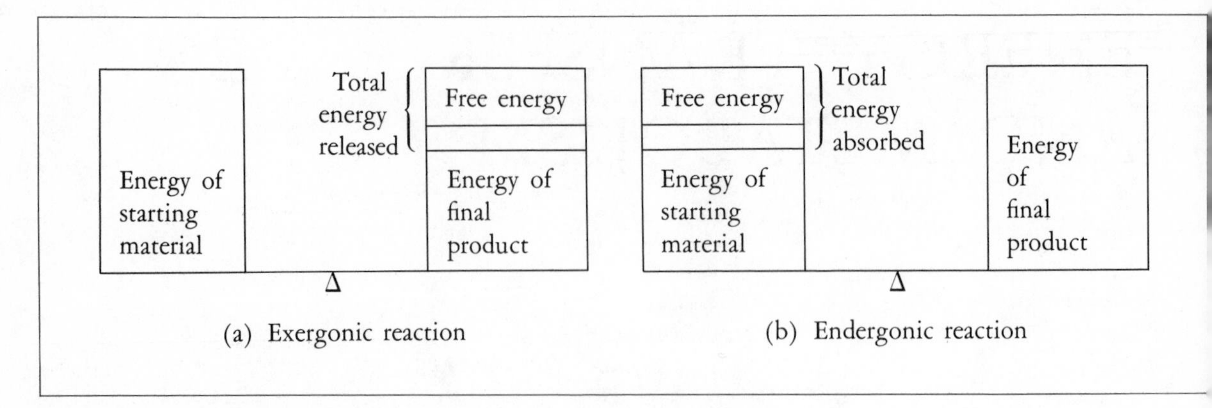

Fig. 26-1. A schematic representation of (a) exergonic and (b) endergonic chemical reactions.

and light. Whenever matter undergoes a chemical reaction or a physical change, its energy usually changes. This change occurs by transferring energy. Work is one of the ways to transfer energy. Scientifically, we are interested in only one kind of work: the kind called useful work, which is beneficial to us. The energy that gives us useful work is given a special name. It is called *free energy.* Because the free energy transferred in living systems is obtained from compounds, let us examine the source of chemical energy.

Energy is contained in the chemical bonds of the molecules of a compound. When a compound undergoes a chemical reaction, bonds are broken and new ones are formed. As a result, there is a change in energy. If the amount of energy in the products is less than that in the starting materials, energy is released. Not all of this energy is free energy. But if some free energy is released in such a change, the reaction is called an *exergonic reaction.* When the energy of the products is greater than that of the starting material, energy must be absorbed for the reaction to occur. When free energy is absorbed in a change, the reaction is called an *endergonic reaction.* These two energy changes are shown in Figure 26-1.

It is important to realize that, although a compound contains energy, it never contains heat or work. Only if a reaction occurs and a transfer of energy becomes necessary do heat and work appear. Often when the term heat or work is used, what is actually meant is that the change in energy of a substance will be visible as heat or work.

Many examples of the interconvertibility of the various forms of energy are available to us everyday. In the engine of a car, the chemical energy of the fuel not only does work to make the engine run, but is also converted into heat and sound. A steam engine converts heat from burning fuel to work. An object exposed to sunlight becomes warm because the energy of the sunlight is converted to heat.

Chemical reactions involved in our body also demonstrate the interconvertibility of energy. From the food we eat, our body gets the energy to carry out its normal work and gets the heat to maintain a constant body temperature. Thus, from a quantity of energy available in food, the human body produces both heat and work.

During an exergonic reaction, the excess free energy can be transferred to another compound rather than released as heat or work. Such transfers of free energy occur frequently in living systems. In fact, this is the way that the energy of the sun is converted to a form of energy that can be used by living organisms. This conversion involves several stages, as we will learn in the next section.

26.2 THE FLOW OF ENERGY IN NATURE

Think for a moment about the food you eat. Where does it come from? Some of it, meat, comes from animals. The rest of it, cereals, vegetables, and fruits, comes from plants. If you investigate the food source of animals, you will soon realize that they too get their food from plants. Thus, plants are the source of food and, consequently, energy of the entire animal world.

What is the source of food and energy of plants? Green plants require nitrogen-containing compounds, minerals, and water from the soil, carbon dioxide (CO_2) from the atmosphere, and, most important, sunlight. The energy of the sun is absorbed by the chlorophyll of the green plant cells and is used to carry out the synthesis of glucose from carbon dioxide and water. The chemical equation is as follows:

$$6\,CO_2 + 6\,H_2O + \text{sunlight} \longrightarrow C_6H_{12}O_6 + 6\,O_2$$

This equation represents the process called *photosynthesis.* This process is extremely complex and involves more than 100 sequential chemical reactions, each catalyzed by a specific enzyme. The result of all these chemical reactions is the *transfer of part of the energy of the sunlight into the molecules of glucose.* This energy is stored in the glucose and is released when glucose is broken down into smaller compounds. Not only is energy released, but these small molecules are used by the plant to form other important carbon-containing compounds such as lipids and proteins. Animals that can-

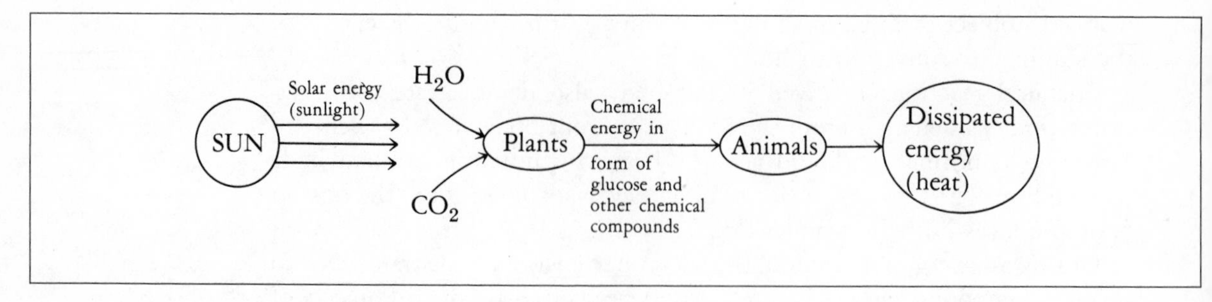

Fig. 26-2. The flow of energy in nature.

not convert the energy of sunlight into chemical energy eat plants to obtain their energy and certain essential chemical compounds. This sequence is shown in Figure 26-2.

The fact that only certain organisms can use the energy of sunlight to make organic compounds from carbon dioxide and water allows us to divide cells into two large classes, depending on their food. *Autotrophic* cells use carbon dioxide as the sole source of carbon. From carbon dioxide, water, and minerals, they build all the chemical compounds they need. *Heterotrophic* cells on the other hand, cannot use carbon dioxide but must use relatively complex carbon-containing compounds as their food.

It is important to realize that not all the cells of an organism are of the same class. For example, the green leaves of plants contain photosynthetic autotrophic cells. But the roots contain heterotrophic cells. Furthermore, autotrophic cells in plants act as heterotrophic cells in the dark. This means that at night these cells feed on some of the glucose they made during the day. These cells have a remarkable flexibility.

Living organisms depend on each other in many ways. In the next section, we will learn one way in which autotrophic and heterotrophic cells help each other.

26.3 THE CARBON AND OXYGEN CYCLES

Living organisms in nature depend on each other for food in a number of ways. The most important are the cycles of carbon and oxygen in which photosynthetic autotrophic cells and heterotrophic cells feed each other. Photosynthetic autotrophic cells use carbon dioxide, water, minerals, and solar energy to form organic compounds such as glucose. Heterotrophic cells use these organic compounds as fuel and building blocks to make the

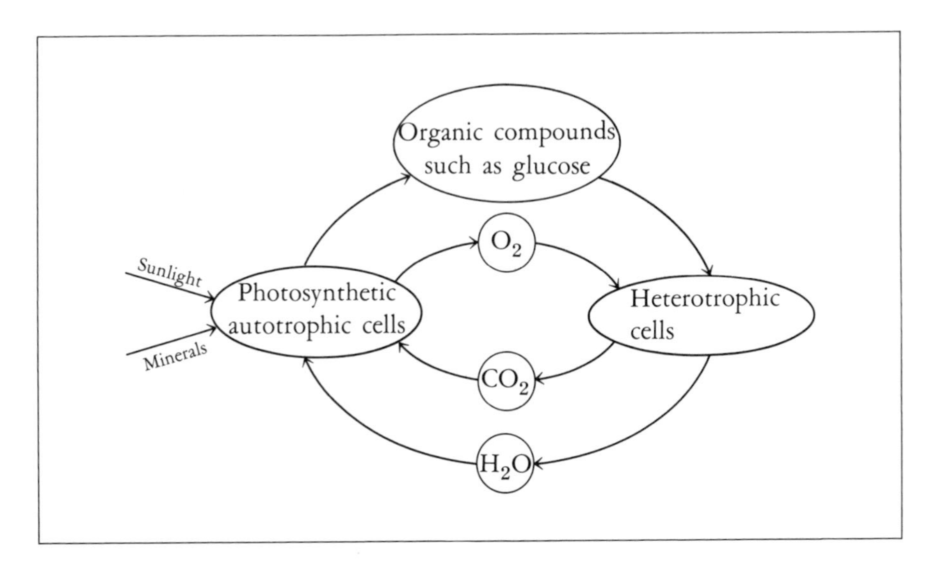

Fig. 26-3. The carbon and oxygen cycles in nature. Photosynthetic autotrophic and heterotrophic cells each produce compounds that can be used by the other as food.

compounds they need. In the process, carbon dioxide is formed as an end product. This carbon dioxide is returned to the atmosphere, where it is used by photosynthetic autotrophic cells. This carbon cycle is shown in Figure 26-3.

Along with the carbon cycle, there is an exchange of oxygen between the photosynthetic autotrophic and heterotrophic cells. Most synthetic cells form oxygen, which is used by the heterotrophic cells to oxidize organic compounds. The oxidation of organic compounds by heterotrophic cells forms carbon dioxide and water, which are used in photosynthesis. This oxygen cycle is also shown in Figure 26-3.

In addition to the carbon and oxygen cycles, the element nitrogen also cycles through living organisms.

26.4 THE NITROGEN CYCLE

The element nitrogen is found in proteins, nucleic acids, and many other important compounds in nature. As with the carbon and oxygen cycles, there exists an interdependence among various organisms in the way they use nitrogen. Nitrogen gas (N_2) occurs in large amounts in the atmosphere. However, it is chemically inert; most forms of life cannot use it as a source of nitrogen. Instead, they must get it from some compound of nitrogen such as ammonia, nitrate salts (such as $NaNO_3$ or NH_4NO_3), or amino acids.

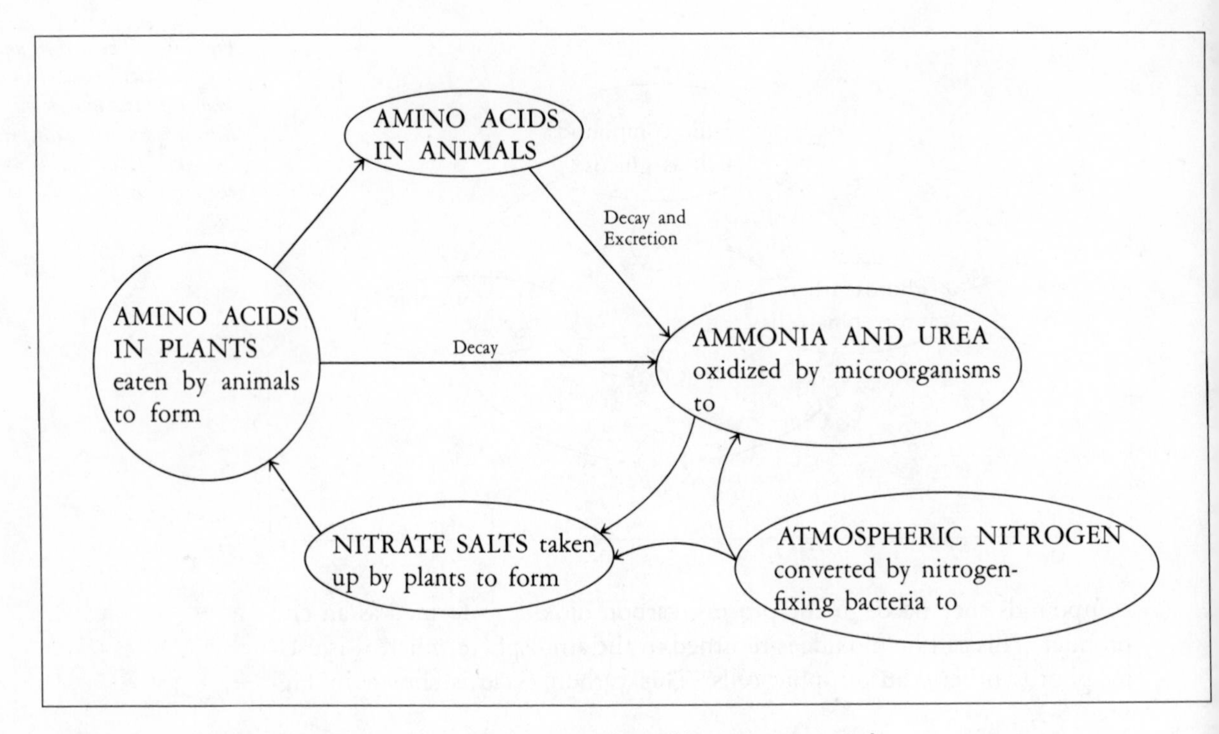

Fig. 26-4. The nitrogen cycle. The main forms of nitrogen in the world are given in capital letters.

Most plants get their nitrogen from the soil in the form of nitrate salts. The plants then transform these nitrates into amino acids and other important nitrogen-containing compounds that are used to build proteins. Animals get their nitrogen from the plant proteins they eat. The heterotrophic cells of both plants and animals eventually return their nitrogen to the soil, usually in the forms of urea or ammonia. Animals return their nitrogen in products of excretion; plants return it as products of decay after their death. In the soil, microorganisms oxidize ammonia to nitrate salts, which can be used again by plants. This nitrogen cycle is shown in Figure 26-4.

A few forms of life, such as the nitrogen-fixing bacteria, can convert the nitrogen in the atmosphere to either ammonia or nitrate salts. This increases the supply of nitrogen-containing compounds that can be used by living organisms.

So far, we have traced the paths by which energy, carbon, oxygen, and nitrogen are used in nature. In the next section, we will examine some of these paths in more detail.

26.5 METABOLISM

Metabolism is the general name given to all the enzyme-catalyzed chemical reactions that are responsible for the life of an organism. These reactions include obtaining energy for the organism from its food, synthesizing macromolecules from smaller molecules, performing muscular motion, and transporting molecules from place to place in the organism. Metabolism involves several thousand different enzyme-catalyzed chemical reactions. You may well wonder how it is possible to learn anything about such a large number of reactions. It turns out that, although the total number of individual reactions is very large, the number of different kinds of reactions is small. For example, the way that cells extract energy from their food and use this energy to synthesize other molecules is remarkably similar in most forms of life. Because of these similarities, scientists can study the chemical reactions involved in the metabolism of simple organisms and then use these results to learn about the corresponding processes in humans and other complex organisms. In this section, we will outline the common pathways by which the major classes of food are degraded and how other compounds are synthesized.

Metabolism is divided into two parts, catabolism and anabolism. *Catabolism* refers to the series of enzyme-catalyzed reactions in which macromolecules such as carbohydrates, lipids, and proteins are broken up into smaller and simpler molecules such as lactic acid, carbon dioxide, ammonia, or urea. During catabolism, part of the energy of these macromolecules is released to the organism for its use.

Anabolism is the series of enzyme-catalyzed reactions in which macromolecules such as nucleic acids, proteins, polysaccharides, and lipids, are made from simpler molecules. Energy is needed to do this. Consequently, during anabolism the living organism must furnish energy to carry out these reactions. As we will see, the energy released in catabolism is stored and then used in anabolism.

Anabolism and catabolism occur together and at the same time in cells. However, they are controlled independently. Metabolism occurs in a series of chemical reactions forming mainly intermediate compounds. As a result, the term *intermediary metabolism* is used to indicate this feature. The intermediate compounds of metabolism are called *metabolites.*

Anabolism and catabolism occur by means of a number of consecutive enzyme-catalyzed chemical reactions organized into three major stages. These stages are shown in Figure 26-5.

Consider first catabolism. In stage 1, the three important classes of or-

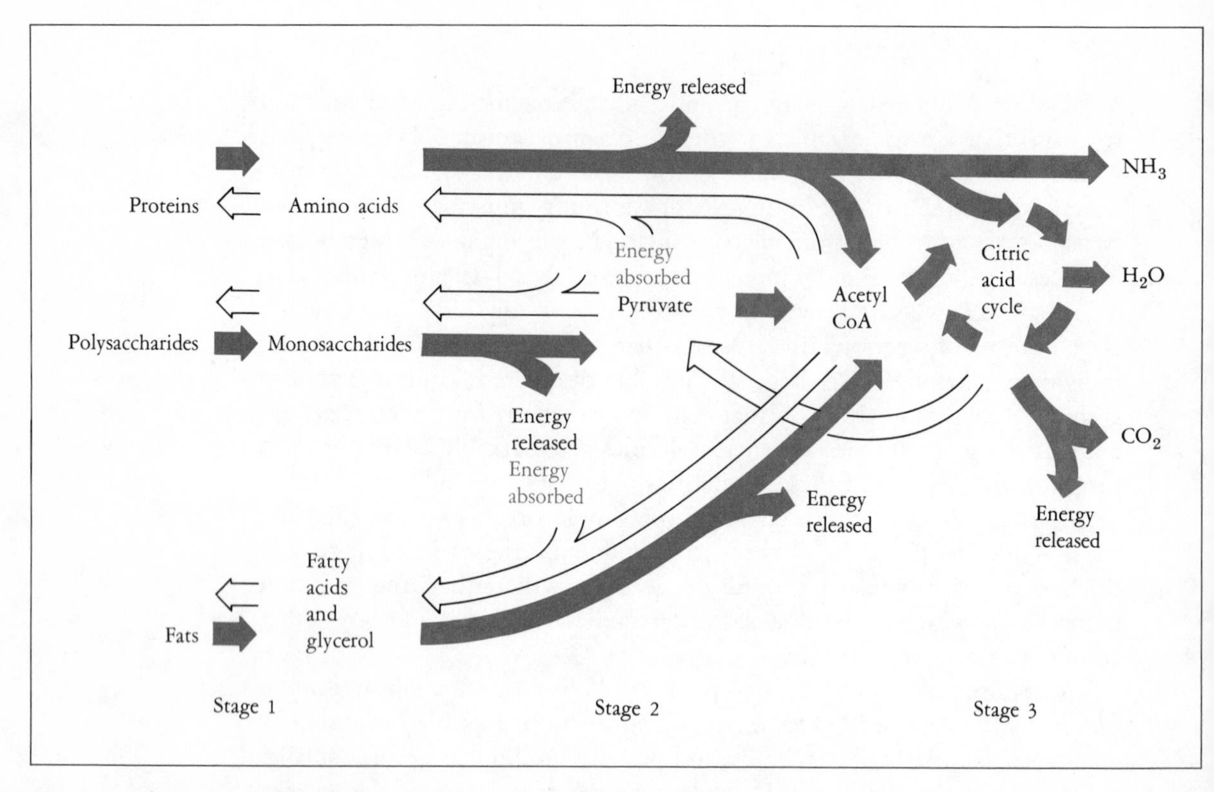

Fig. 26-5. The three stages of metabolism. The blue arrows indicate the catabolic pathways, and the white arrows indicate the anabolic pathways.

ganic compounds in living systems are degraded to their monomers or component parts. Thus, proteins yield amino acids, fats yield fatty acids and glycerol, and polysaccharides yield monosaccharides. In stage 2, the many different products of stage 1 are degraded to fewer, simpler molecules. The amino acids, fatty acids, monosaccharides, and glycerol are degraded to the two-carbon acetate unit in *acetyl coenzyme A*.

Acetyl coenzyme A has a central role in metabolism, as can be seen from Figure 26-5. Not only is it formed in the catabolic pathway, but it is also the starting point for the anabolic pathway and the starting material for the synthesis of steroids and terpenes. The structure of acetyl coenzyme A is shown in Figure 26-6. We learned some of the reactions of acetyl coenzyme A in Sections 17.6 and 18.7. We will encounter this compound many times in the next two chapters.

Finally, the acetyl group of acetyl coenzyme A as well as other products of stage 2 are completely oxidized to carbon dioxide and water. This oxida-

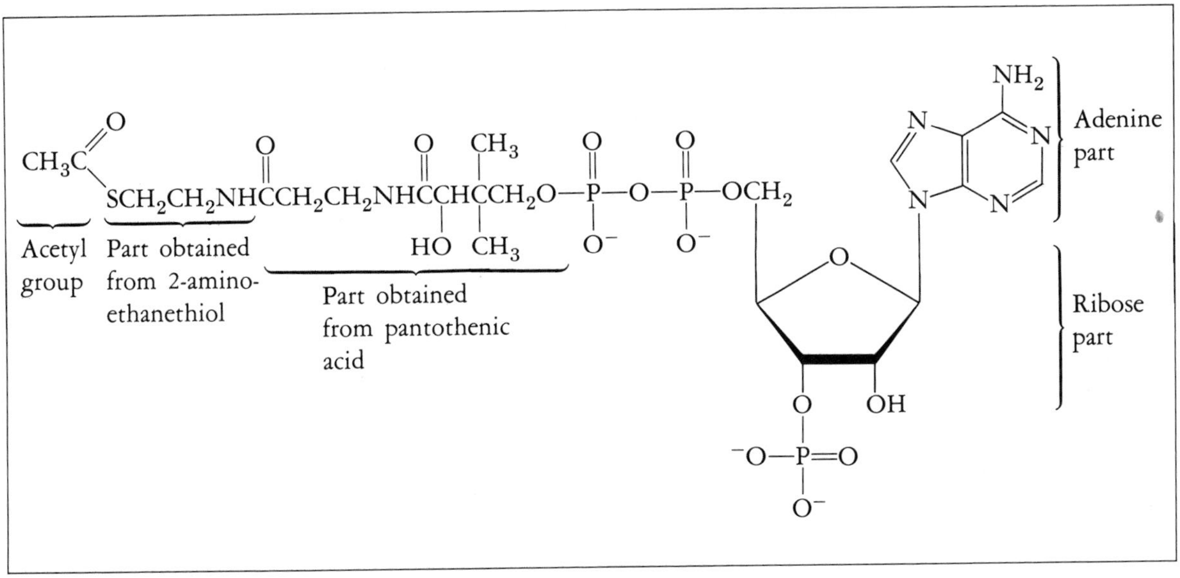

$CH_3C(=O)SCH_2CH_2NHC(=O)CH_2CH_2NHC(=O)CH(OH)C(CH_3)_2CH_2O-P(=O)(O^-)-O-P(=O)(O^-)-OCH_2$

Acetyl group

Part obtained from 2-amino-ethanethiol

Part obtained from pantothenic acid

Adenine part

Ribose part

$^-O-P(=O)-O^-$

OH

Fig. 26-6. Acetyl coenzyme A.

tion is carried out by a series of reactions known by any one of three names: the Krebs cycle, the citric acid cycle, or the tricarboxylic acid cycle.

The synthesis of organic molecules in living systems also takes place in three stages. Small molecules of stage 3 are joined together to form the monomers of the macromolecules in stage 2. These monomers are finally assembled into macromolecules in stage 1. For example, the synthesis of protein starts in stage 3. Here, certain α-keto acids are formed that are transformed to amino acids by amino group donors in stage 2. Finally, the amino acids are joined together into polypeptide chains in stage 1. This is shown in Figure 26-7.

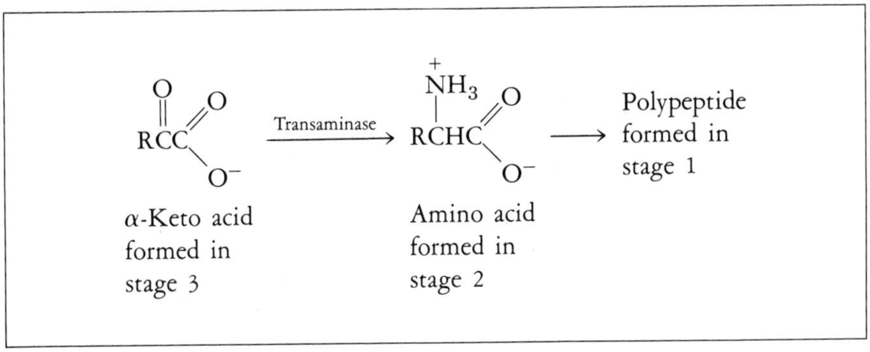

$RC(=O)CO_2^-$ $\xrightarrow{\text{Transaminase}}$ $RCH(\overset{+}{N}H_3)CO_2^-$ $\longrightarrow$ Polypeptide formed in stage 1

α-Keto acid formed in stage 3

Amino acid formed in stage 2

Fig. 26-7. The three stages of the synthesis of proteins.

As shown in Figure 26-5, the catabolic and anabolic pathways for a particular substrate are *not* the reverse of each other. For example, a series of enzyme-catalyzed reactions are responsible for the degradation of glycogen to pyruvate. It might seem logical and economical to use the reverse of these reactions to synthesize glycogen from pyruvate. Unfortunately, this is not possible because of the different energy requirements of each pathway. Catabolic pathways release energy, whereas anabolic pathways require energy. Thus, degrading complex organic molecules releases energy; their synthesis requires energy. This means that the two pathways cannot be the exact reverse of each other.

An advantage of independent catabolic and anabolic pathways is that they can be independently regulated. For example, the degradation of glycogen to pyruvate is controlled by enzymes different from those controlling the reverse process, the synthesis of glycogen from pyruvate. As a result, the reactions between glycogen and pyruvate can be regulated in both directions. Such dual regulation is a general characteristic of parallel catabolic and anabolic pathways.

Catabolic and anabolic pathways often differ in another respect. They usually occur in different places in the cells. For example, the degradation of fatty acids to acetyl coenzyme A occurs in the mitochondria. The reverse set of reactions that synthesize fatty acids from acetyl coenzyme A occur outside of the mitochondria. These two sets of reactions require two different sets of enzymes. Physically separating the catabolic and anabolic pathways allows both of them to occur independently yet simultaneously.

There is one central meeting place of all catabolic and anabolic pathways. This place is stage 3, which has a dual role. It can act catabolically to bring about the final degradation of complex molecules, or it can act anabolically to furnish the small molecules needed to synthesize the complex molecules required by the living organism.

In each enzyme-catalyzed reaction of metabolism, a characteristic energy change occurs. Specific reactions in the catabolic pathways conserve a portion of the energy of the metabolite. This energy is then used in specific reactions in the anabolic pathway. As we examine several metabolic pathways in the next few chapters, we must be concerned with two parts of each individual enzyme-catalyzed chemical reaction. The first part is the structural change that occurs in the reactants as they are transformed into products. The second part is the change in the chemical energy as the reaction occurs. Is energy released or is it absorbed? Does the product conserve part of the energy for furture use by the system?

It turns out that one compound in particular, adenosine triphosphate

(ATP), is usually involved in the storage or conversion of chemical energy in living organisms.

26.6 ATP: THE UNIVERSAL ENERGY TRANSFER AGENT

The energy released by the degradation of large molecules is transformed into a special form before it is used in living organisms. It is transferred to a special carrier of energy called *a*denosine *trip*hosphate (ATP). This ATP is made up of three parts: an adenine unit, a ribose unit, and a triphosphate unit arranged as shown in Figure 26-8. At the pH of the cell, the triphosphate unit exists in the anion form.

The ATP molecule is an energy carrier because it is energy rich. A large amount of its energy is released when ATP is converted to *a*denosine *di*-*p*hosphate (ADP) or *a*denosine *mono**p*hosphate (AMP). These reactions and the structures of ADP and AMP are given in Figure 26-9. The energy released in these reactions is used to drive reactions in living organisms that require energy. The energy of ATP is renewed by reversing the reactions in Figure 26-9. That is, ADP or AMP, phosphate, and energy obtained from certain steps in the catabolic pathway combine to reform ATP. This ATP-ADP cycle, shown in Figure 26-10, *is the fundamental process of energy exchange in living organisms.*

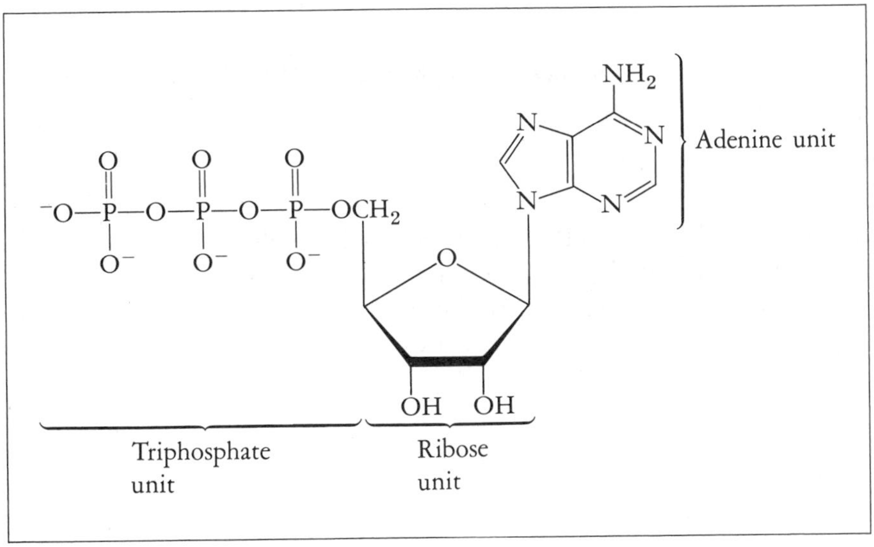

Adenine unit

Triphosphate unit

Ribose unit

Fig. 26-8. The structure of adenosine triphosphate (ATP).

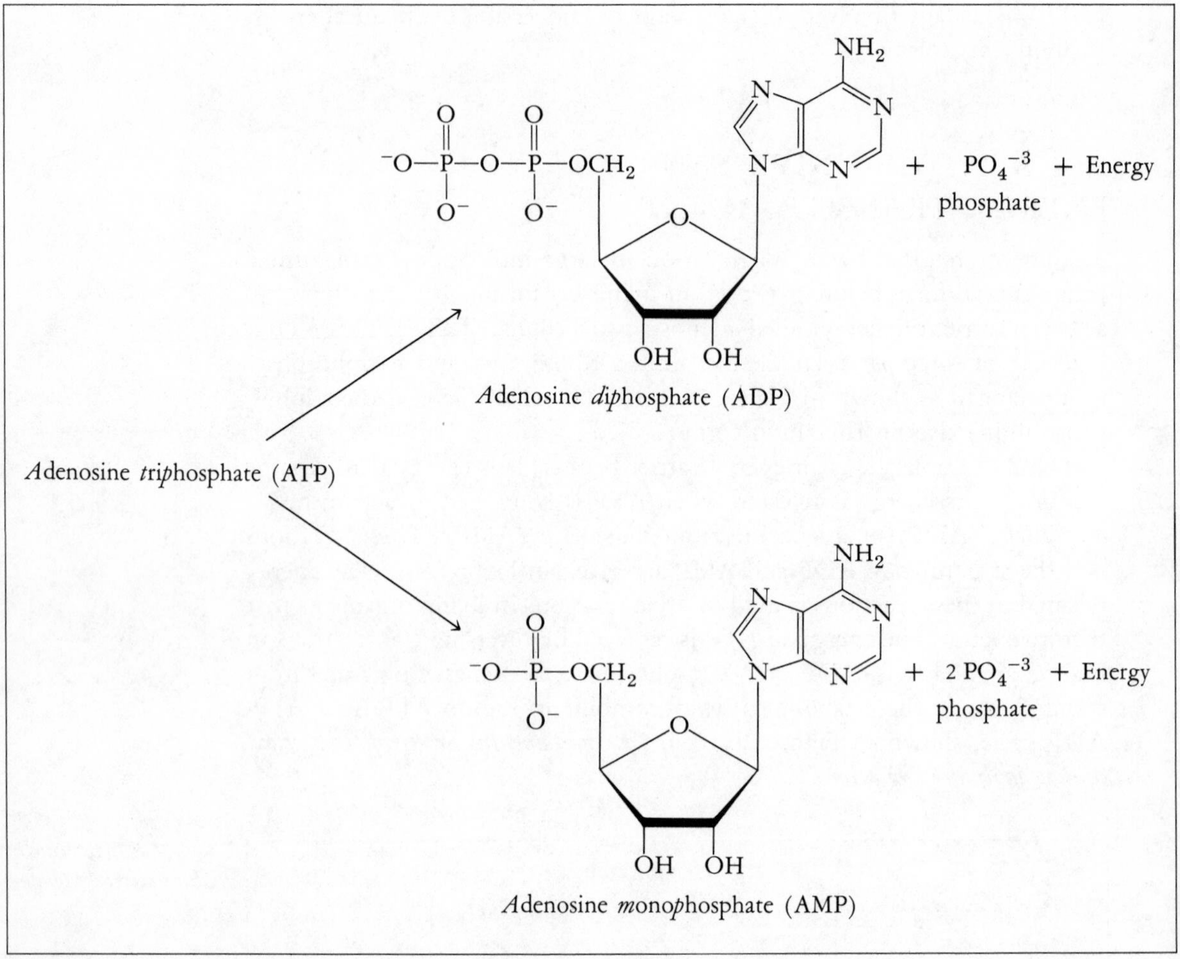

Adenosine triphosphate (ATP) → *Adenosine diphosphate* (ADP) + PO_4^{-3} (phosphate) + Energy

Adenosine triphosphate (ATP) → *Adenosine monophosphate* (AMP) + 2 PO_4^{-3} (phosphate) + Energy

Fig. 26-9. The conversion of ATP to ADP or AMP with the release of energy.

Because much energy is released when ATP reacts, it is called a high-energy phosphate compound. Its phosphate bonds are often referred to as high-energy bonds. To emphasize this, the high energy phosphate bonds are sometimes designated by a squiggle (∼). Thus, the structure of ATP can be written as follows:

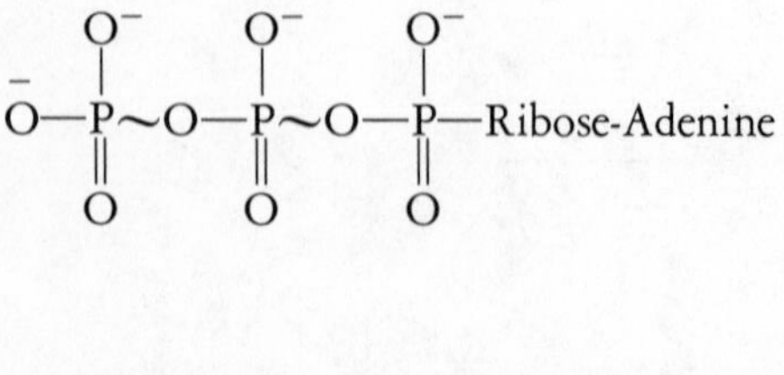

$$^{-}O{-}\overset{O^{-}}{\underset{O}{P}}{\sim}O{-}\overset{O^{-}}{\underset{O}{P}}{\sim}O{-}\overset{O^{-}}{\underset{O}{P}}{-}\text{Ribose-Adenine}$$

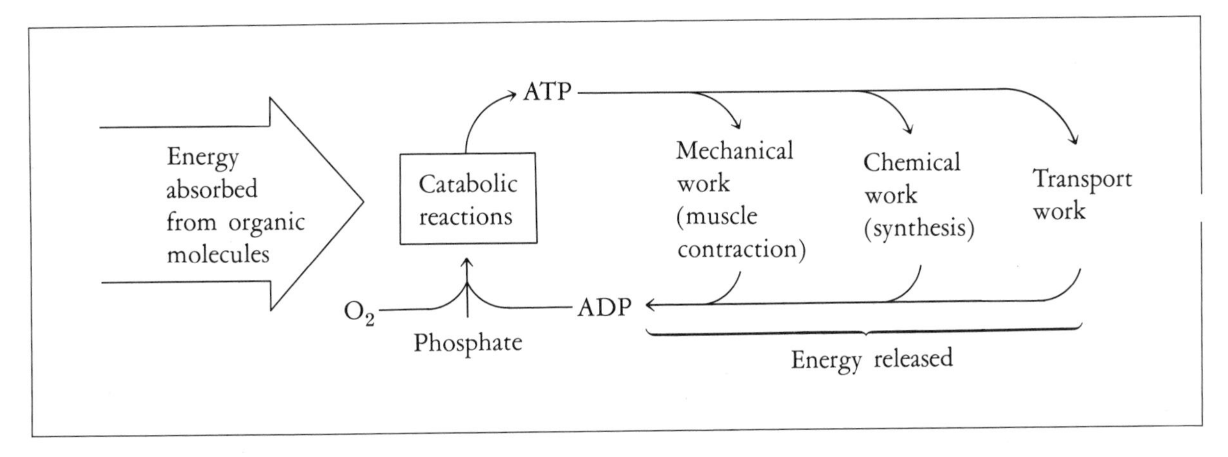

Fig. 26-10. The ATP-ADP cycle. The energy obtained from catabolic reactions is transferred to ATP and is carried to various places in the cell, where it is released to do work.

Notice that ATP contains *two* high-energy bonds. The energy in ATP is transferred when ATP acts as an intermediate in coupled reactions.

26.7 COUPLED REACTIONS

We learned in Section 26.5 that metabolism occurs by means of several consecutive reactions. The only way that chemical energy can be transferred from one reaction to another in these sequences is for the two reactions to have a common reaction intermediate; ATP is such an intermediate. In the consecutive reactions responsible for energy transfer by ATP, chemical energy is transferred to ADP in one reaction to form ATP as a reaction product. In the next reaction, ATP is the substrate. Its phosphate group is transferred to another molecule. In this way, ATP transfers its energy and changes a compound into one with a higher energy content. In these two reactions, ATP is a common intermediate. It couples enzyme-catalyzed reactions involving the transfer of phosphate groups and is thus the carrier for the transfer of chemical energy. This role is illustrated schematically in Figure 26-11.

In the next chapter we will examine in more detail many reactions in which energy is transferred with ATP as the carrier.

Another way to make ATP is by oxidation reactions in living organisms. In the next section, we will learn the structure of compounds that are involved in many oxidation and reduction reactions.

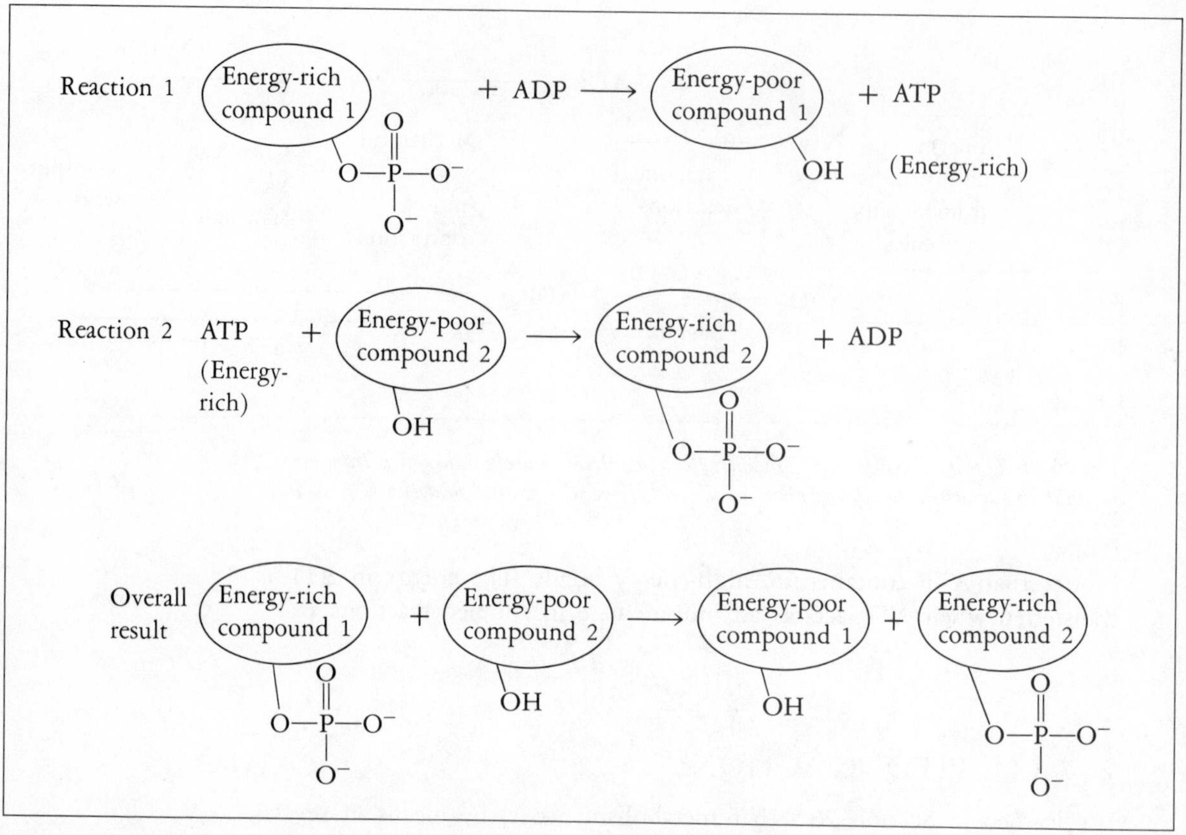

Reaction 1: Energy-rich compound 1 (O—P(=O)(O⁻)—O⁻) + ADP ⟶ Energy-poor compound 1 (OH) + ATP (Energy-rich)

Reaction 2: ATP (Energy-rich) + Energy-poor compound 2 (OH) ⟶ Energy-rich compound 2 (O—P(=O)(O⁻)—O⁻) + ADP

Overall result: Energy-rich compound 1 (O—P(=O)(O⁻)—O⁻) + Energy-poor compound 2 (OH) ⟶ Energy-poor compound 1 (OH) + Energy-rich compound 2 (O—P(=O)(O⁻)—O⁻)

Fig. 26-11. A schematic representation of the role of ATP as an intermediate in energy transfer. In reaction 1, the energy of compound 1 is transferred to ATP. In reaction 2, ATP transfers part of its energy to compound 2. The overall result is the transfer of energy from compound 1 to compound 2 using ATP as the carrier.

26.8 IMPORTANT OXIDATIVE COENZYMES

Most organisms need oxygen. They use it for *respiration.* This word means not only breathing, but also the use of oxygen by the cell to oxidize organic compounds. In the reactions that do this, oxygen is reduced to water, and ATP is made by a series of reactions called oxidative phosphorylation. The enzymes that catalyze all these reactions are complex in structure, and their specific actions are not understood. Nevertheless, several compounds are known to be involved as coenzymes in these oxidation reactions.

The structures of the coenzymes *n*icotinamide *a*denine *d*inucleotide (NAD^+) and *n*icotinamide *a*denine *d*inucleotide *p*hosphate ($NADP^+$) are given in Figure 26-12. The reactive part of these coenzymes is the pyridine ring, as we learned in Section 16.5. It accepts a hydride ion to form the reduced coenzyme NADH (or NADPH), according to the following equation:

Alcohol (hydride ion donor) NAD+ (hydride ion acceptor) → NADH (reduced form) $+ \; >C{=}O + H^+$

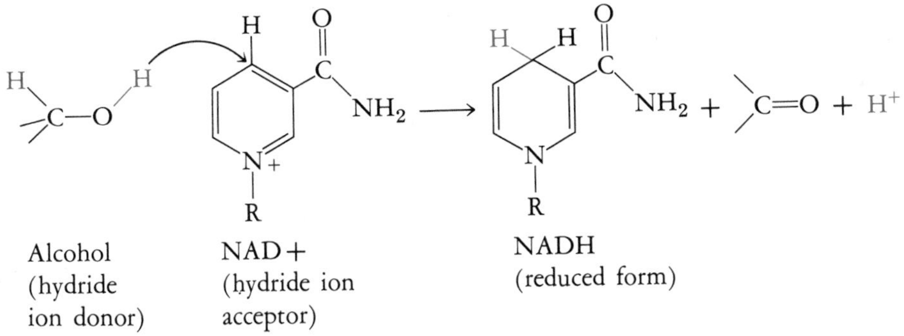

Another coenzyme involved in the oxidation of organic compounds is *f*lavin *a*denine *d*inucleotide (FAD). A related coenzyme is *f*lavin *m*ono*n*ucleotide (FMN). The structures of these two enzymes are given in Figure 26-13. The reactive portion of these coenzymes is the flavin group. It accepts a hydride ion to form the reduced coenzymes $FADH_2$ and $FMNH_2$ according to the following equation:

FAD or FMN (hydride ion acceptor) + Hydride ion donor → $FADH_2$ or $FMNH_2$ (reduced form)

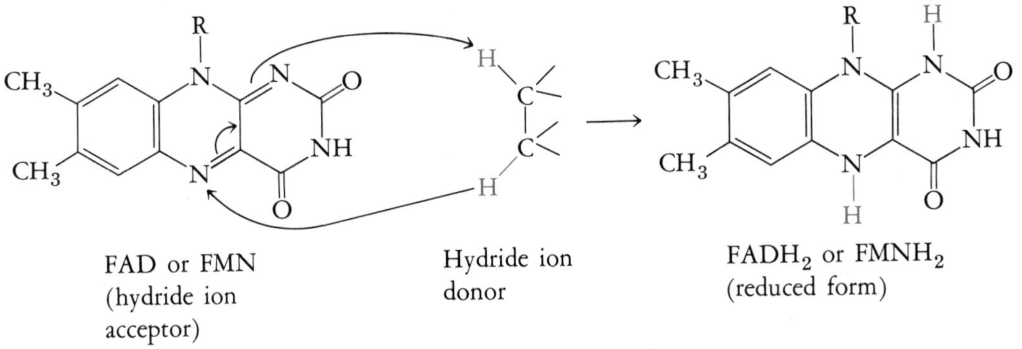

Notice that part of the structure of all three of these coenzymes is obtained from a water-soluble B vitamin. The vitamin nicotinamide is part of NAD^+; the vitamin riboflavin is part of both FAD and FMN.

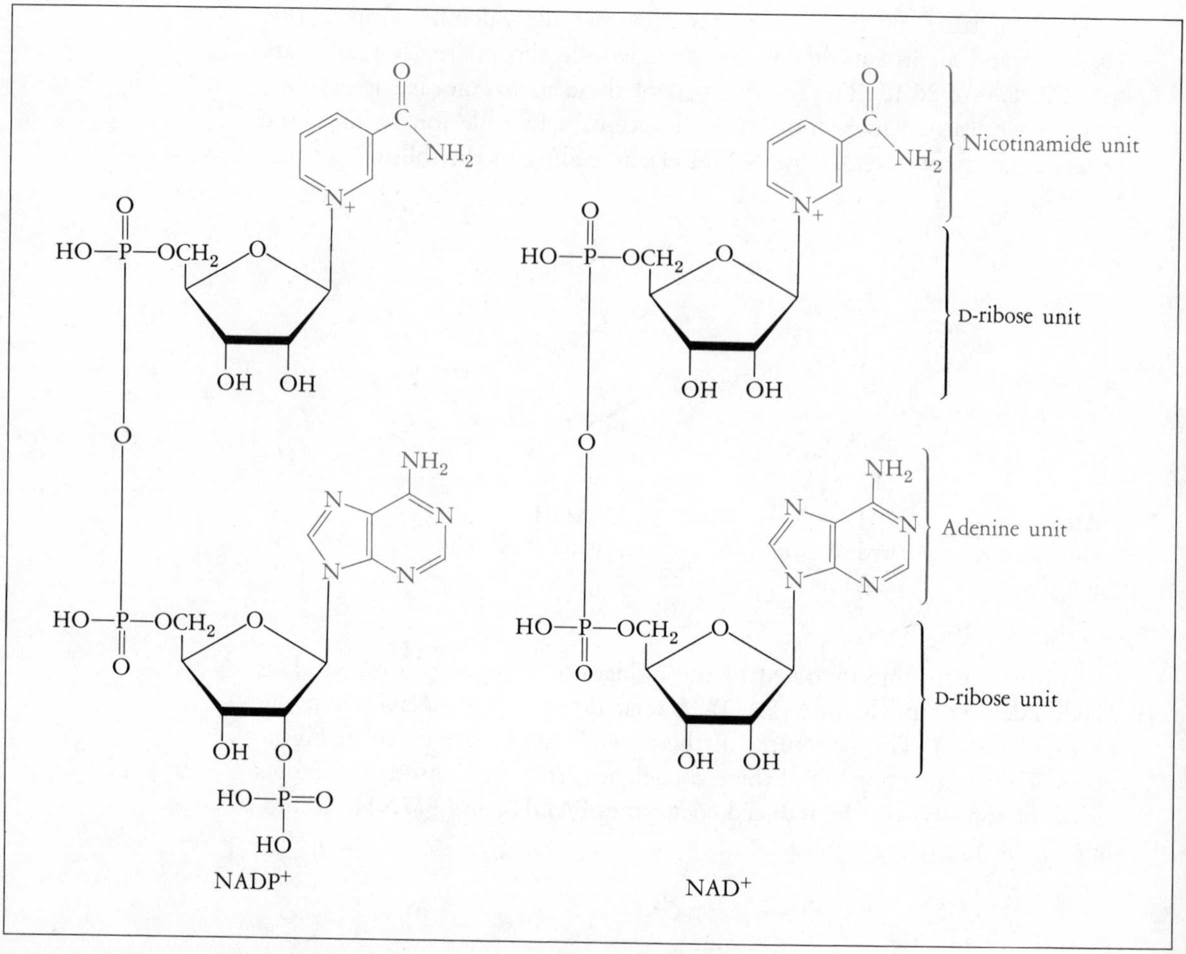

Fig. 26-12. Structures of nicotinamide adenine dinucleotide phosphate ($NADP^+$) *and nicotinamide adenine dinucleotide* (NAD^+).

Coenzyme Q, or ubiquinone, is a quinone (see Section 16.6). Its structure is given in Figure 26-14. Actually, there are several ubiquinones differing only in the length of the side chain. These coenzymes accept a hydride ion to form the reduced form of the coenzymes according to the equation given in Figure 26-14.

The reduced form of all of these enzymes is reoxidized by oxygen in the cell. Coupled with this reoxidation is the formation of ATP from ADP by oxidative phosphorylation. This sequence of reactions is a major source of ATP for all living organisms.

In the next few chapters, we will study the details of the metabolism of

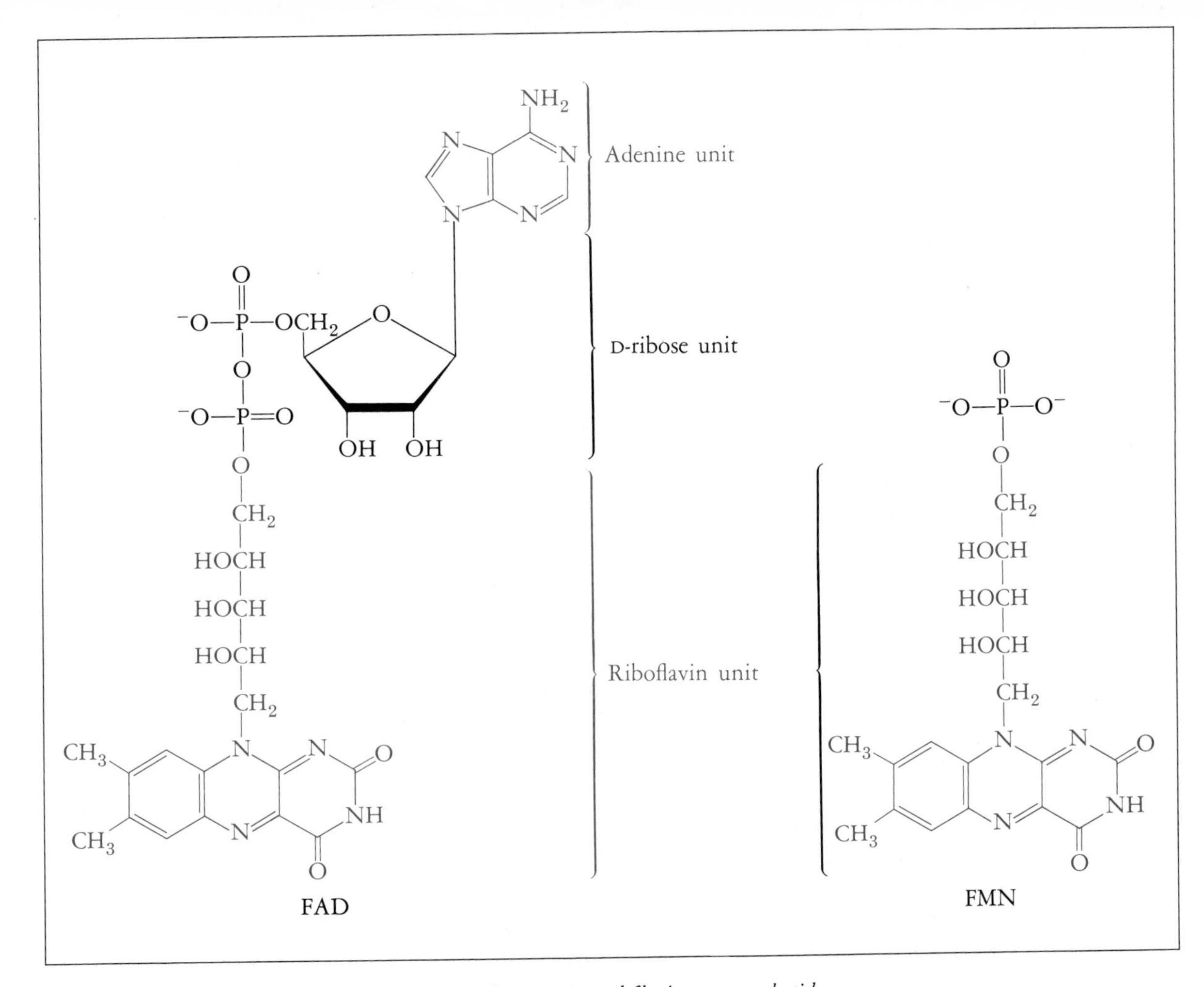

Fig. 26-13. Structures of flavin adenine dinucleotide (FAD) and flavin mononucleotide (FMN).

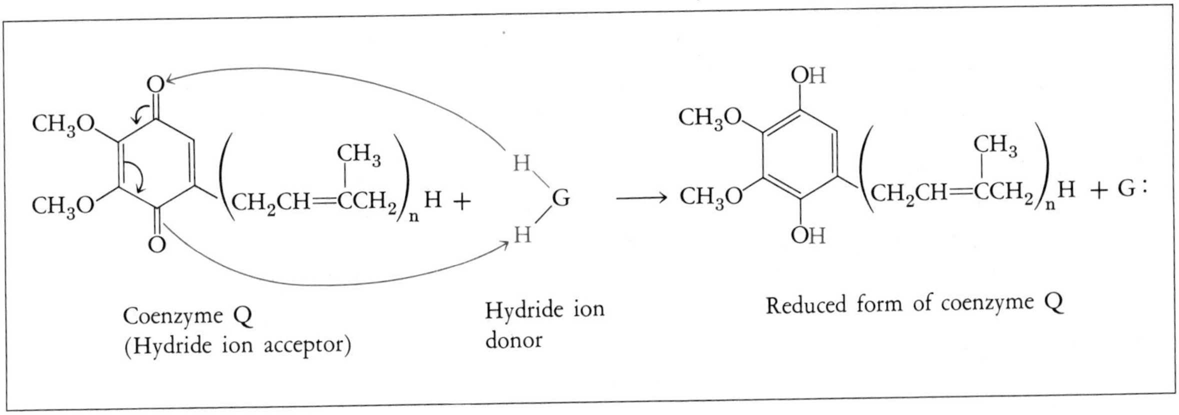

CH_3O CH_3O $\left(CH_2CH{=}\overset{CH_3}{|}{C}CH_2\right)_n H$ + H–G–H ⟶ CH_3O CH_3O OH OH $\left(CH_2CH{=}\overset{CH_3}{|}{C}CH_2\right)_n H$ + G:

Coenzyme Q
(Hydride ion acceptor)

Hydride ion
donor

Reduced form of coenzyme Q

Fig. 26-14. The reduction of coenzyme Q by a hydride ion donor.

the three major classes of bioorganic molecules. At that time, we will learn where these coenzymes act in the various metabolic pathways.

26.9 SUMMARY

All living organisms obtain their energy from the sun. Most organisms cannot use solar energy directly, so it must be converted to chemical energy. This is done by photosynthesis in plants. Chemical energy is contained in the chemical bonds of all organic compounds. Certain of these organic compounds are used as food by living organisms. They are food for two reasons. First, they provide the source of carbon, hydrogen, nitrogen, oxygen, and other elements needed by the organism. Second, when they undergo reactions, they release energy that is used immediately by the living organism or is stored for future use. During this process which involves thousands of chemical reactions, carbon, oxygen, and nitrogen are continually recycled.

Energy is obtained from food in three stages by a process called catabolism. In the first stage, large compounds are broken down into smaller ones such as amino acids, monosaccharides, and fatty acids. In the second stage, these compounds are degraded to a very few smaller ones that have important roles in metabolism. The third and final stage is the citric acid cycle, in which foods are completely oxidized to carbon dioxide and water.

Energy is used by living organisms to build compounds that it needs by a process called anabolism. Small compounds, obtained from the degradation of large, complex organic ones, are first used to make the monomers, which are then joined together to form macromolecules such as proteins, polysaccharides, and fats. The catabolic and anabolic pathways are not the exact reverse of each other. The reason for this is that the catabolic pathways release energy, whereas the anabolic pathways take energy from the organism.

The universal energy transfer agent is ATP. It is formed in the catabolic pathways from ADP. The ATP supplies the energy for the reactions of the anabolic pathways. Thus, ATP stores the energy that the organism gets from food and then releases it when it is needed for such things as building complex compounds, contracting muscles, and transporting molecules or ions. The ATP transfers its energy by means of coupled reactions. The process of respiration also forms ATP. This occurs when certain coenzymes involved in oxidation-reduction reactions are reoxidized by oxygen.

REVIEW OF TERMS AND CONCEPTS

Terms

ANABOLISM The series of enzyme-catalyzed chemical reactions in which simple organic compounds are used to build large and complex ones.

AUTOTROPHIC CELLS Cells that use carbon dioxide as their sole source of food.

CATABOLISM The series of enzyme-catalyzed chemical reactions in which complex organic compounds are broken down into simpler and smaller ones.

ENDERGONIC REACTION A chemical reaction in which free energy is absorbed.

EXERGONIC REACTION A chemical reaction in which free energy is released.

FREE ENERGY A measure of the ability of an object to do useful work.

HETEROTROPHIC CELLS Cells that use complex organic molecules as their food.

METABOLISM All the enzyme-catalyzed chemical reactions responsible for the life of an organism.

METABOLITES The intermediate compounds of metabolism.

PHOTOSYNTHESIS The conversion of solar energy into the chemical energy of glucose.

RESPIRATION The use of oxygen by cells to oxidize organic compounds.

Concepts

1. The energy of the sun is transferred by photosynthesis to organic compounds that can be used as foods by plants and animals.
2. Living organisms get their energy by breaking up complex organic molecules. This process is called catabolism.
3. The energy obtained in catabolism is transferred to ATP. The ATP then carries the energy to various parts of the cell, where it is released to do work.
4. Living organisms make the organic compounds they need by anabolism. The energy needed for this process is supplied by ATP.

EXERCISES

26-1 Natural gas is often used as a fuel in home heating. Explain what happens to the energy of the natural gas as it is burned.

26-2 Chemicals are used as fuels in rockets. Into what form of energy is the chemical energy transformed at liftoff?

26-3 Why do living organisms need food?

26-4 What three classes of organic compounds are used as food by living organisms?

26-5 What are the three stages of catabolism?

26-6 Where in metabolism are the reactions of catabolism and anabolism identical?

26-7 What are the final products of catabolism of each of the following?
(a) fatty acids (b) amino acids (c) glucose

26-8 What is the difference between ATP, ADP, and AMP?

26-9 Why is ATP called the universal energy transfer agent?

26-10 Write the equation for the oxidation of glucose in living systems.

26-11 What vitamins are needed to make the following conenzymes?
(a) FAD (b) NAD^+ (c) FMN

26-12 Write the equation for the photosynthesis of glucose from carbon dioxide and water.

CARBOHYDRATE METABOLISM

27

We learned in a general way in Chapter 26 how a living organism obtains its energy by degrading the complex organic compounds that are its food. Now it is time to be specific. We want to learn the exact sequence of enzyme-catalyzed chemical reactions that recover part of the chemical energy of food. We will start with carbohydrates. We will first examine *glycolysis,* a catabolic pathway by which many organisms obtain energy from glucose in the absence of oxygen. Then we will examine the citric acid cycle. This is a sequence of enzyme-catalyzed chemical reactions of almost universal occurrence by which organisms use oxygen to obtain more energy from their food. Finally, we will discuss the specific steps in the anabolism of glucose. Let us start by learning how sugars are handled by the digestive system.

27.1 DIGESTION OF CARBOHYDRATES

The process of catabolism of carbohydrates starts as soon as they are placed in your mouth. It is here that the process called *digestion* starts. Enzymes in the saliva begin the hydrolysis of carbohydrates into monosaccharides that can be absorbed by the body. This process continues through the stomach into the small intestine, where the monosaccharides are absorbed.

We can use starch to trace the digestion of carbohydrates. Starch is a polysaccharide made up of glucose* units, as we learned in Section 21.7.

* All naturally occurring monosaccharides have the D configuration, as we learned in Chapter 21.

Digestion of starch starts in the mouth, where the enzyme ptyalin catalyzes its hydrolysis to a mixture of polysaccharides of intermediate chain length called dextrins. Hydrolysis continues as starch and dextrins pass through the esophagus. But in the stomach, hydrolysis almost stops because the acidity of the stomach destroys the enzyme and the hydrolysis reaction is too slow without it. Starch and dextrins then pass on to the small intestine, where digestion continues. Here a number of enzymes catalyze the hydrolysis of dextrin and any remaining starch into glucose, which is used directly by the body.

Many polysaccharides are made from glucose units. However, not all of them can be used as food by humans. For example, we cannot digest paper and wood, which are mostly cellulose, to obtain glucose. The reason for this is that humans do not have the enzyme that hydrolyzes β-glycoside bonds. This is a major difference between humans and other organisms. For example, termites have enzymes in their digestive tract that can catalyze the hydrolysis of the cellulose in wood. Humans, of course, cannot use wood as food. Yet the structural difference between cellulose and glucose is slight (see Section 21.7) and certain enzymes can recognize this difference. This is another example of the stereospecificity of enzyme-catalyzed reactions.

The complete hydrolysis of the carbohydrates we eat produces only three monosaccharides: glucose, fructose, and galactose. These monosaccharides are absorbed through the walls of the small intestine into the blood. Fructose and galactose are carried to the liver, where galactose is converted to glucose 1-phosphate and fructose is converted to fructose 1-phosphate. Glucose 1-phosphate is isomerized to glucose 6-phosphate and fructose 1-phosphate is cleaved into glyceraldehyde and dihydroxyacetone phosphate. Glyceraldehyde is phosphorylated to glyceraldehyde 3-phosphate. Glucose 6-phosphate, glyceraldehyde 3-phosphate, and dihydroxyacetone phosphate can all be metabolized, as we will learn in Section 27-3. Glucose is the monosaccharide that is used in all cells. But what happens to the glucose when it gets into the blood? We will answer this question in the next section.

27.2 WHAT HAPPENS TO GLUCOSE IN LIVING ORGANISMS?

The glucose formed by digesting carbohydrates quickly enters the blood, where it becomes available to cells in the organism. However, the amount of glucose permitted in the blood is closely regulated. If the blood contains too much glucose, a condition called *hyperglycemia* develops; too little glucose in the blood causes a condition called *hypoglycemia*.

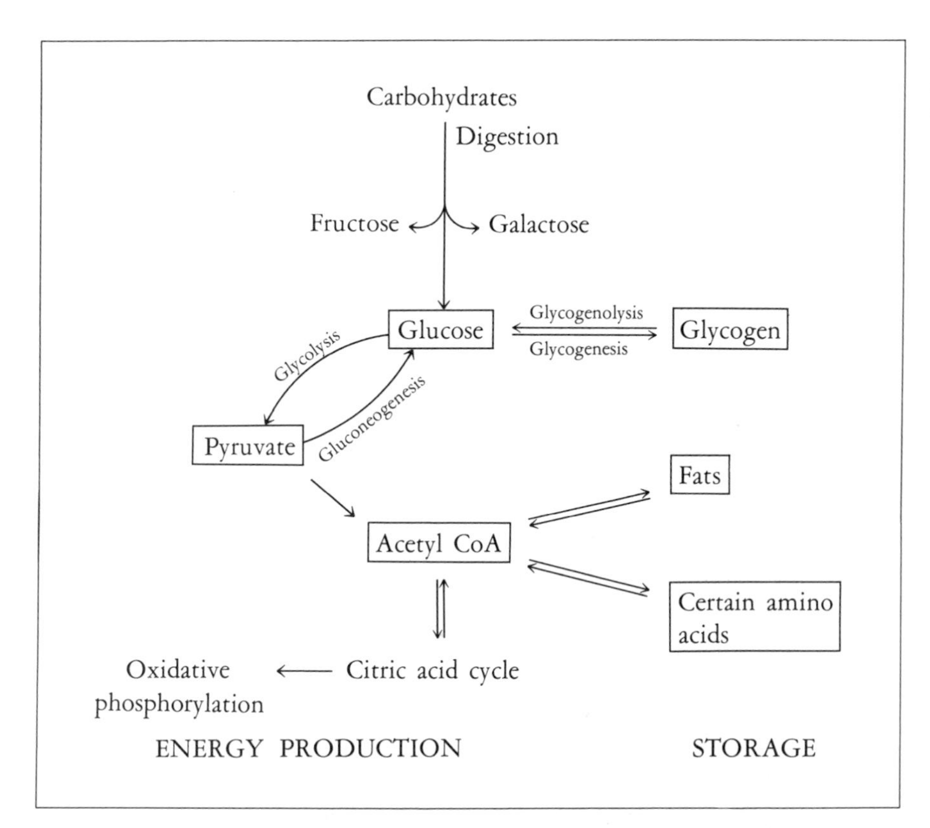

Fig. 27-1. The formation and utilization of glucose in living organisms.

The amount of glucose is regulated by a series of interrelated metabolic pathways that insure a continuous and adequate supply of glucose in the blood. Some of these pathways oxidize glucose to carbon dioxide and water to produce energy for the organism. Other pathways store glucose for later use. These various pathways are outlined in Figure 27-1.

If too much glucose is present in the blood for immediate use, a pathway called glycogenesis exists through which it can be stored as glycogen. Between meals, when the glucose level in the blood decreases, the organism draws on this reserve of glycogen. The glycogen is hydrolyzed to glucose by a sequence of reactions called glycogenolysis, and the glucose is released into the blood to maintain the blood glucose level.

Glucose in cells is oxidized to provide energy and smaller compounds that are used to build fats and certain amino acids. When necessary, this process can be reversed to make glucose from amino acids. This is another way that the blood level of glucose can be maintained.

In all cells, the most important role of glucose is the release of energy by catabolism. Let us turn our attention to glycolysis, the first series of reactions in the catabolism of glucose.

27.3 GLYCOLYSIS

The series of reactions that change glucose into pyruvate with the formation of ATP is called *glycolysis.* The overall equation for this process is the following:

$$\underset{\text{Glucose}}{C_6H_{12}O_6} + 2\,ADP + 2\,P_i + 2\,NAD^+ \longrightarrow \underset{\text{Pyruvate}}{2\,CH_3\overset{\overset{\displaystyle O}{\|}}{C}CO_2^-} + 2\,ATP + 4\,H^+ + 2\,NADH$$

The symbol P_i is used to designate inorganic phosphate (PO_4^{-3}). Glycolysis occurs without oxygen and is therefore called an *anaerobic catabolic process.* The individual reactions of glycolysis are summarized in Figure 27-2.

At first glance the scheme in Figure 27-2 may seem like a complicated series of new reactions. But this is not so! We have learned each type of reaction in this scheme in previous chapters. As we examine each reaction in turn, we will point out its relationship to previous reactions. In this way, it will be clear that the principles of organic chemistry apply to these reactions of living organisms. Let us now start our journey along the glycolysis trail.

Reaction ①* is the *phosphorylation of glucose by adenosine triphosphate (ATP) to form glucose 6-phosphate,* a phosphate monoester. We learned about such esters in Section 16.4. Phosphate monoesters are prepared more easily in living systems than in the laboratory. They are readily formed in living systems by enzymes called kinases. These enzymes transfer a phosphate group from ATP to an acceptor. Because a phosphate group is transferred to a hexose, this enzyme is called hexokinase. Phosphate group transfer is one of the fundamental reactions in living organisms. We will encounter it repeatedly.

Reaction ② is the *isomerization of glucose 6-phosphate to fructose 6-phosphate,* catalyzed by the enzyme phosphoglucose isomerase. In this reaction, the six-member ring of glucose 6-phosphate is converted to the five-member ring of fructose 6-phosphate. The isomerization, which may appear complicated, can be explained by a sequence of two simple reactions that we have already learned. They involve the tautomerism that we learned about in Section 20.2. The two reactions are shown in Figure 27-3. The isomerization starts from the open form of glucose 6-phosphate. The ene-diol intermediate, *27-1,* is formed by a tautomeric shift of the hydrogen indicated in the structure of glucose 6-phosphate. Another tautomeric shift converts *27-1* into the open form of fructose 6-phosphate. Thus, this remarkable isomeri-

* These numbers correspond to the numbers given the reactions in Figure 27-2.

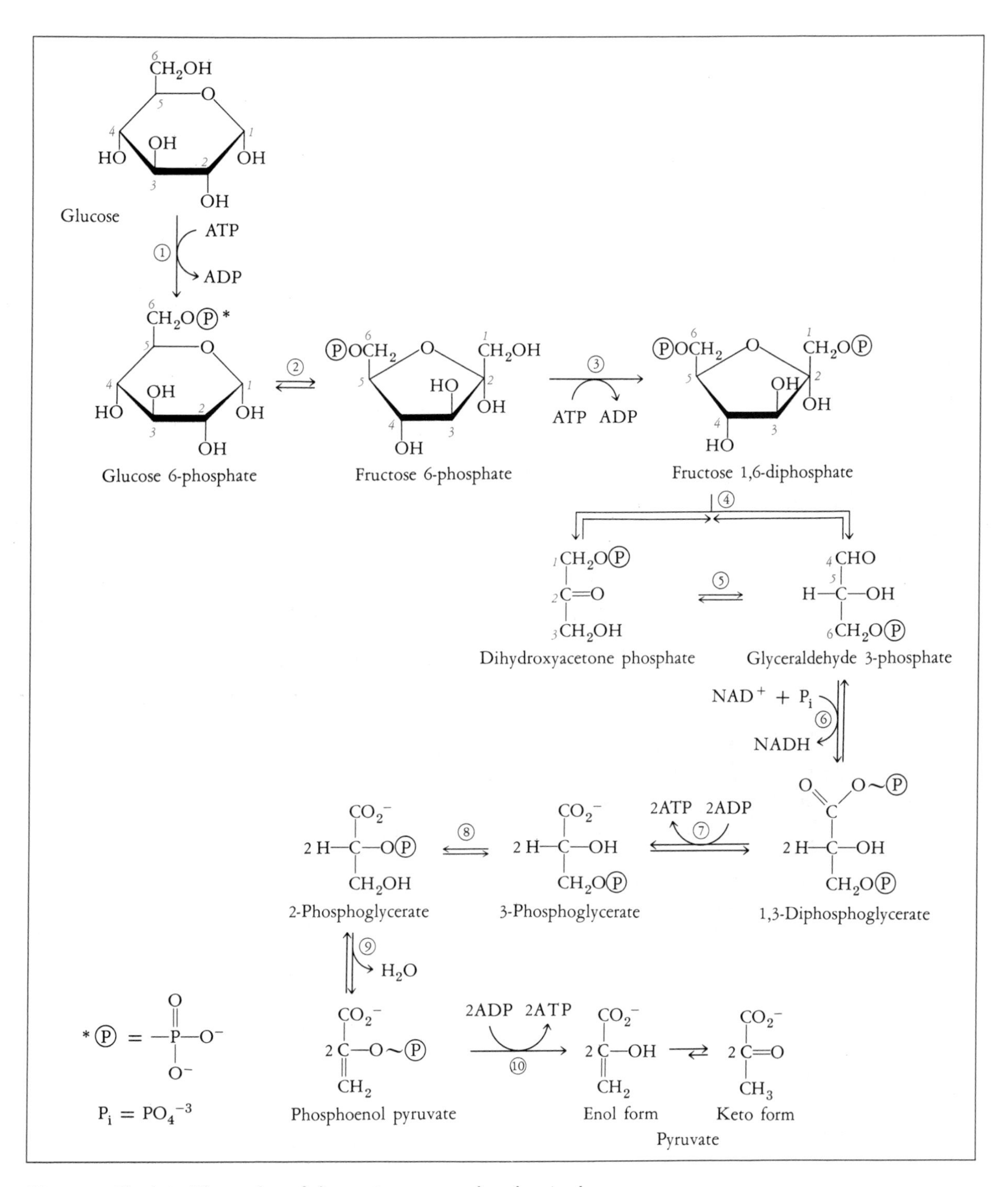

Fig. 27-2. Glycolysis. The numbers of the reactions correspond to those in the text.

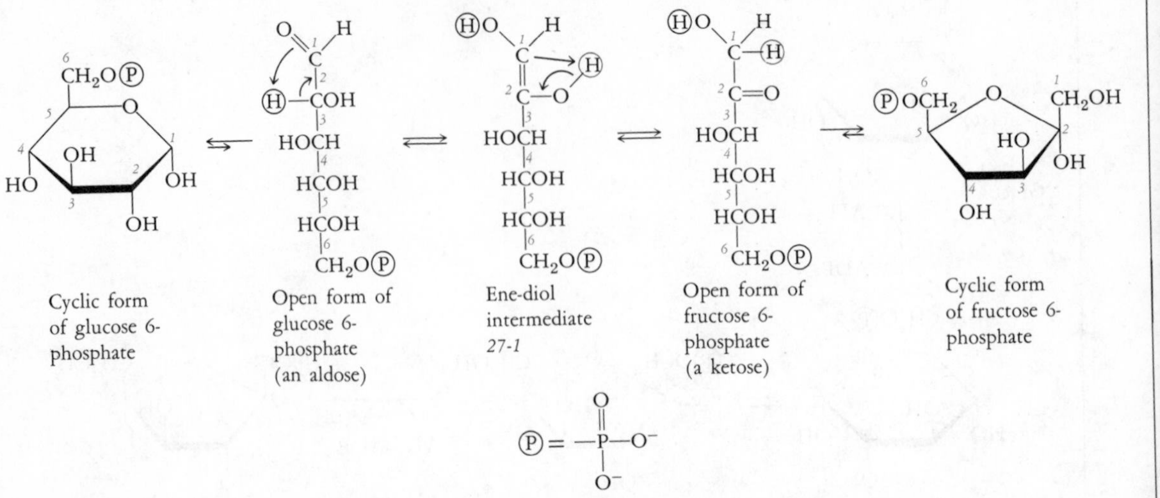

Fig. 27-3. The tautomeric shifts involved in the isomerization of glucose 6-phosphate to fructose 6-phosphate.

zation can be understood in terms of tautomerism. Two such equilibria convert the aldose into a ketose.

Reaction (3) is a second phosphorylation step catalyzed by the enzyme phosphofructokinase. This enzyme transfers a phosphate group from ATP to fructose 6-phosphate to form fructose 1,6-diphosphate.

Reaction (4) involves the breaking of the six-carbon molecule fructose 1,6-diphosphate into two three-carbon molecules, glyceraldehyde 3-phosphate and dihydroxyacetone phosphate. This reaction is catalyzed by the enzyme aldolase. The enzyme is given this name because the reverse of reaction (4) is an aldol condensation, a reaction that we learned in Section 19.8. In fact, reaction (4) is an example of a retroaldol condensation.

In reaction (5), dihydroxyacetone phosphate is isomerized to glyceraldehyde 3-phosphate by the enzyme triose phosphate isomerase. The reason for this reaction is that glyceraldehyde 3-phosphate is on the direct path of glycolysis, but dihydroxyacetone phosphate is not. By the action of aldolase in reaction (4) and triose phosphate isomerase in reaction (5), two molecules of glyceraldehyde 3-phosphate are formed for each molecule of fructose 1,6-diphosphate.

This isomerization can be explained by the same sequence of two tautomeric hydrogen shifts used to explain the isomerization in reaction (2). This is shown in Figure 27-4.

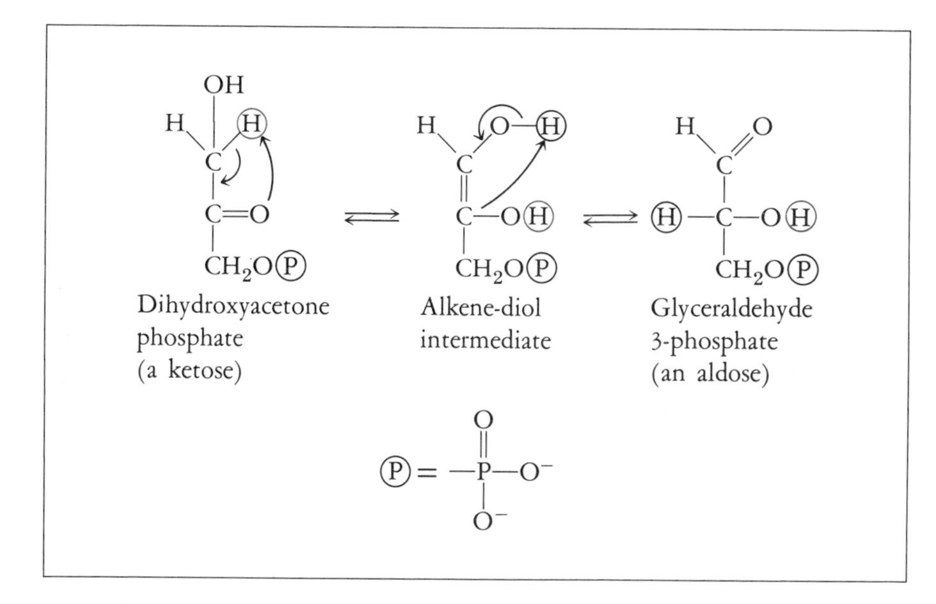

Fig. 27-4. The tautomeric hydrogen shifts involved in the isomerization of dihydroxacetone phosphate to glyceraldehyde 3-phosphate.

In the first five steps of glycolysis, one molecule of glucose has been split into two molecules of glyceraldehyde 3-phosphate. So far, we have not obtained any of the energy of glucose. Instead, it has cost us energy because we have used two molecules of ATP, one in reaction (1) and another in reaction (3). But in the following reactions, we will recover these two molecules of ATP and get two more.

In reaction (6), glyceraldehyde 3-phosphate is converted to 1,3-diphosphoglycerate. This reaction is catalyzed by the enzyme glyceraldehyde 3-phosphate dehydrogenase, which contains nicotinamide adenine dinucleotide (NAD^+) as the prosthetic group. This conversion is another example of an addition reaction of an aldehyde group, which we learned in Section 19.5. The steps involved in this reaction are shown in Figure 27-5. In the first step, a thiol group of the enzyme adds to the carbonyl group of glyceraldehyde 3-phosphate. In the second step, a hydride ion is transferred to the pyridine ring of NAD^+ to form NADH. In the final step, a phosphate ion (inorganic phosphate, PO_4^{-3} or P_i) attacks the carbonyl carbon to form 1,3-diphosphoglycerate as product.

In reaction (7), 1,3-diphosphoglycerate reacts to form 3-phosphoglycerate and ATP. This reaction is highly exergonic, and part of this free energy is transferred to ATP. This is the first reaction we have encountered that forms ATP. The enzyme phosphoglycerate kinase catalyzes the transfer of a phosphate group from the acyl phosphate of 1,3-diphosphoglycerate to ADP to form ATP and 3-phosphoglycerate as products.

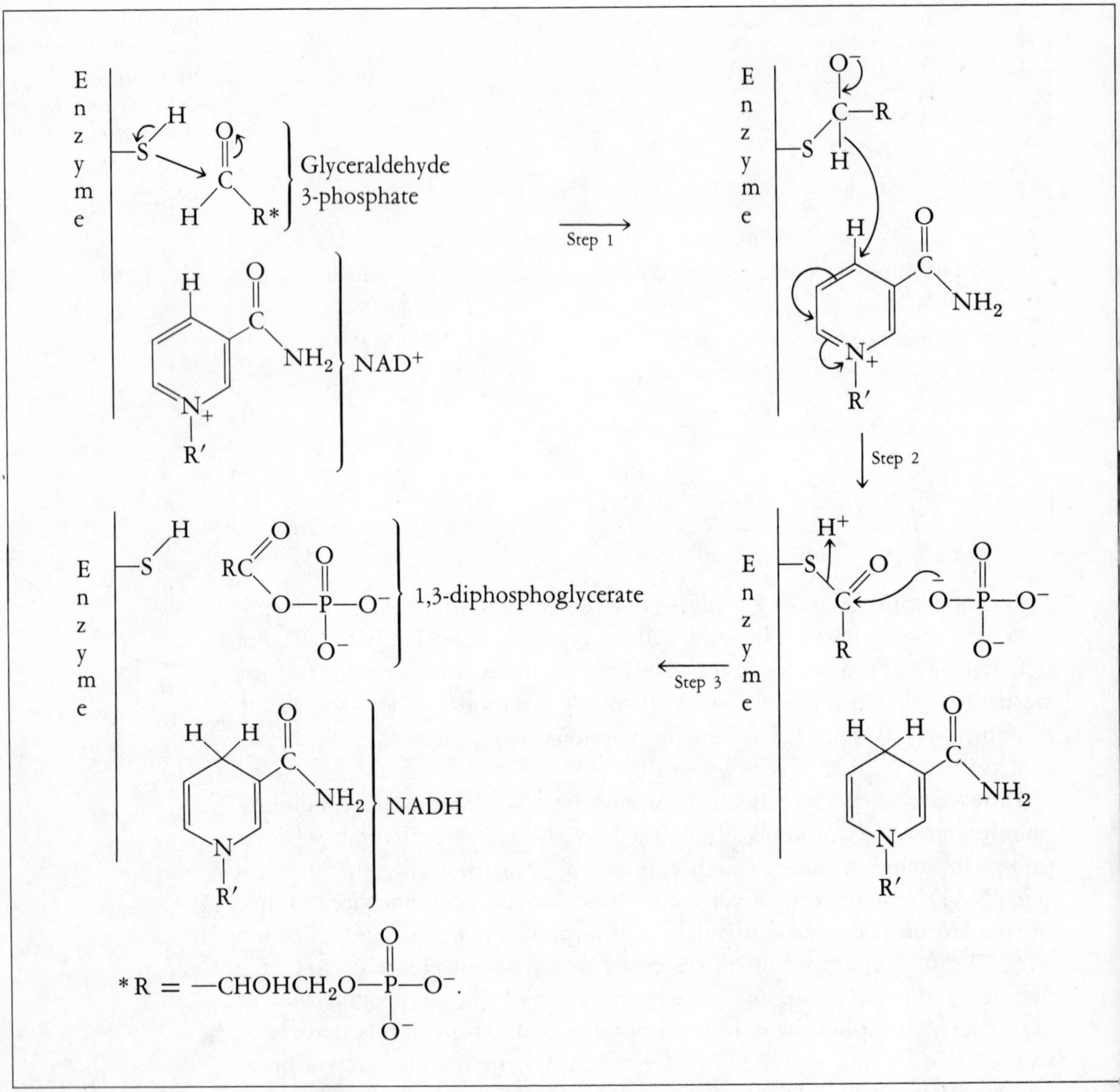

$$^{*}R = -CHOHCH_2O-\overset{\overset{O}{\|}}{\underset{\underset{O^-}{|}}{P}}-O^-.$$

Fig. 27-5. The three steps in the conversion of glyceraldehyde 3-phosphate to 1,3-diphosphoglycerate.

Reaction ⑧ involves a shift in the position of the phosphate group from the 3 to the 2 position. This reaction is catalyzed by the enzyme phosphoglyceromutase.

In reaction ⑨, an enol phosphate is formed by the dehydration of 2-phosphoglycerate. This reaction is catalyzed by the enzyme enolase. Enol phosphates are high-energy compounds. Thus, in this simple reaction, a low-energy compound is transformed into a higher-energy one that can be used in the next step.

The last reaction, ⑩, is the formation of pyruvate and ATP. The transfer of a phosphate group from phosphoenolpyruvate to ADP is catalyzed by the enzyme pyruvate kinase. The loss of a phosphate group forms the enol form of pyruvate first, which then quickly tautomerizes to the keto form (see Section 20.2).

Let us summarize the first ten steps in glycolysis. The intermediates have either six or three carbon atoms. The three carbon compounds are phosphate esters of dihydroxyacetone, glyceraldehyde, glycerate, or pyruvate. There are only five types of reaction involved in the entire sequence:

1. Phosphate group transfer. Reactions ① and ③ involve transfer from ATP; reactions ⑦ and ⑩ involve transfer to form ATP; reaction ⑥ involves transfer to form a high-energy acyl phosphate anhydride.
2. Isomerization. Reaction ② involves the isomerization of an aldose to a ketose; reaction ⑤ involves the reverse.
3. Phosphate shift. Reaction ⑧. Shift in the position of the phosphate group from the 3 to the 2 position of glycerate.
4. Dehydration. Reaction ⑨. Phosphoenol pyruvate is formed by the dehydration of 2-phosphoglycerate.
5. Retroaldol condensation. Reaction ④. Fructose 1,6-diphosphate forms two three-carbon molecules.

In terms of energy, two ATP molecules are used in the early reactions, ① and ③, but we get these two back in reaction ⑦, plus two more in reaction ⑩. Thus, we have a net gain of two ATP molecules after the first ten steps.

Notice that all of the intermediates in the glycolytic pathway between glucose and pyruvate are phosphorylated. At the pH of the cell (pH $\approx$ 7), the phosphate group is fully ionized. As a result, these compounds cannot escape from the cell, because the cell membrane generally does not allow highly polar molecules to pass. This prevents the intermediates from leaking out of the cell. These phosphate groups have a second function. They serve as binding groups in the formation of enzyme-substrate complexes. Finally, these phosphate groups serve to conserve part of the energy of glucose, because they eventually become the terminal groups of ATP during glycolysis.

This is not the end, however. The product pyruvate can undergo any of several reactions, depending on the organism.

27.4 REACTIONS OF PYRUVATE

The sequence of ten reactions of glycolysis that we learned in the preceding section is very similar in all organisms. In contrast, the fate of pyruvate varies. Pyruvate can be converted to ethanol, lactate, or the acetyl group of acetyl coenzyme A, depending on the type of cell and the amount of oxygen present.

Ethanol is formed from pyruvate in several microorganisms such as yeast. The two reactions involved in this conversion are given in Figure 27-6.

The first step is the decarboxylation of pyruvate to acetaldehyde, catalyzed by the enzyme pyruvate decarboxylase, which contains thiamine pyrophosphate as a coenzyme. This coenzyme is synthesized in humans from vitamin B (thiamine), one of the essential water-soluble vitamins (see Section 25.3). The structure of thiamine pyrophosphate is as follows:

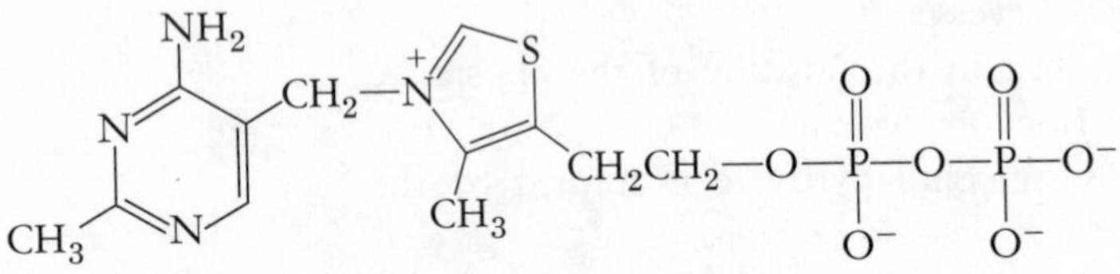

Thiamine pyrophosphate

Pyruvate $\xrightarrow{H^+ \quad CO_2}$ Acetaldehyde $\xrightarrow{NADH + H^+ \quad NAD^+}$ Ethanol

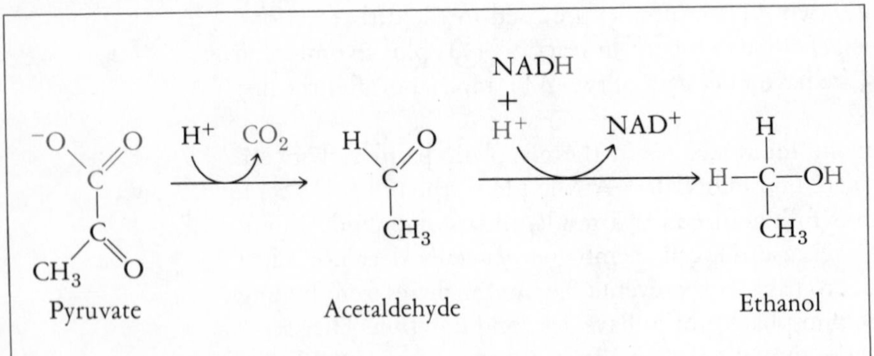

Fig. 27-6. The conversion of pyruvate to ethanol.

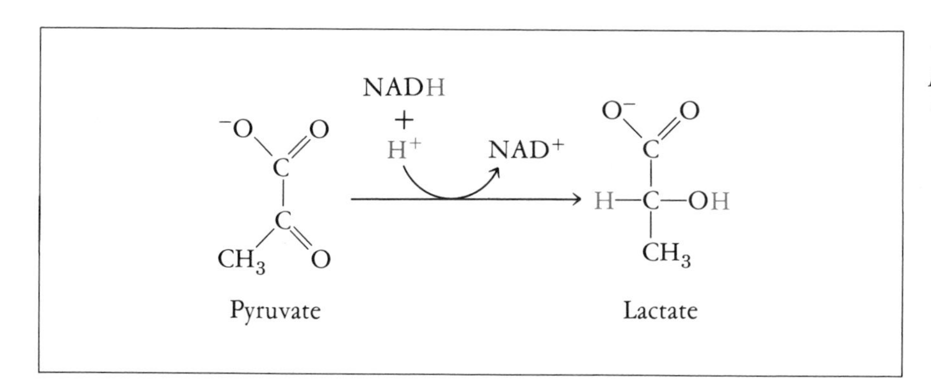

Fig. 27-7. Reduction of pyruvate to lactate by NADH.

This coenzyme catalyzes the decarboxylation of an α-keto acid. This is not a typical laboratory reaction of this class of compound, but it is a typical reaction in living systems.

The second step is the reduction of acetaldehyde to ethanol, which is catalyzed by the enzyme alcohol dehydrogenase. This reaction is a typical reduction of an aldehyde to an alcohol (see Section 19.6). Here the reducing agent is NADH, as we learned in Section 26.8. The NAD^+ is used in reaction ⑥ of Figure 27-2.

The formation of ethanol from pyruvate occurs under anaerobic conditions. The two reactions in Figure 27-6 represent the last stages of the process called *alcoholic fermentation.* In this process, starches in malt or other cereals and the sugar in grapes are converted into alcoholic beverages such as beer and wine.

In the cells of the skeletal muscles, pyruvate is converted into lactate. This reaction is catalyzed by the enzyme lactate dehydrogenase. The equation for this reaction is given in Figure 27-7. This reaction is a typical reduction of a ketone to a secondary alcohol (see Section 19.6). Again, NADH is the reducing reagent. The NAD^+ formed is also used in reaction ⑥ of Figure 27-2.

These two reactions of pyruvate are very important to the continued function of the glycolytic pathway shown in Figure 27-2. Glycolysis involves the partial oxidation of glucose to pyruvate. In the process, NAD^+ is reduced to NADH in reaction ⑥ (Figure 27-2). If there were no way to reoxidize the accumulated NADH to NAD^+, glycolysis would stop. One way to reoxidize NADH is by using oxygen, as we will learn in the next section. But certain cells are often forced to work under anaerobic conditions. For example, during strenuous muscle activity, skeletal muscles operate under anaerobic conditions. The way these cells meet their need for NAD^+ is to reduce pyruvate to lactate. In the process, NADH is oxidized

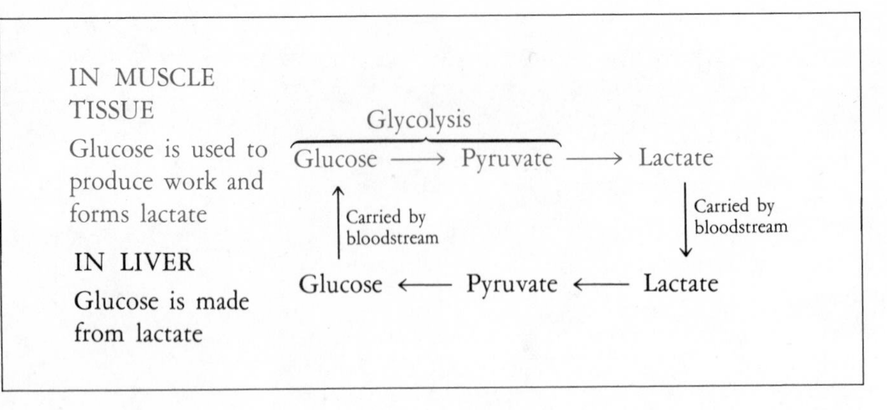

Fig. 27-8. The Cori cycle.

to NAD^+, which is used to continue the oxidation of glucose by the glycolytic pathway.

But there is one problem. Lactate builds up in the muscle cells. It must be removed before it increases the acidity in the cell. An increase in acidity decreases muscle performance and leads to fatigue. Lactate is a dead end in metabolism. To be eliminated, it must be reoxidized to pyruvate. This is done in the following way.

Most of the lactate diffuses into the bloodstream and is carried to the liver. In the liver it is oxidized to pyruvate by NAD^+ (the reverse of the reduction reaction in muscles). Pyruvate is converted to glucose by the gluconeogenesis pathway (see Section 27.2) in the liver. Glucose diffuses into the bloodstream and is taken back to the skeletal muscles for further anaerobic catabolism and generation of ATP. In this way, lactate is converted to glucose in the liver. This glucose is furnished to skeletal muscles, which get ATP from the glycolysis of glucose to lactate. This cycle shifts part of the burden of running the skeletal muscles to the liver. This conversion, called the Cori cycle, is shown in Figure 27-8.

The anaerobic oxidation of glucose (glycolysis) does not produce much energy. We only get two molecules of ATP per molecule of glucose. But we still have left pyruvate, a three-carbon molecule. More energy can be obtained by the oxidation of pyruvate and several other compounds in the citric acid cycle.

27.5 THE CITRIC ACID CYCLE

A sequence of reactions that completes the oxidation of glucose to carbon dioxide and water was first proposed by Hans Krebs, who received a Nobel Prize for his work in 1953. In the first step, pyruvate is decarboxylated to the two-carbon acetyl group of acetyl coenzyme A, which then enters the citric acid cycle. This cycle is the final common series of reactions for the oxidation of *all* molecules used as foods. Proteins, fats, and carbohydrates all end up here. Most enter the cycle as acetyl coenzyme A, although other points of entry are also possible. This cycle has another purpose. It provides the small molecules for the synthesis of all the compounds needed by a living organism. The sequence of reactions in the citric acid cycle is shown in Figure 27-9. Each reaction is numbered, and the individual reactions will be discussed in some detail.

Reaction [1] is the oxidative decarboxylation of pyruvate to form acetyl coenzyme A. This reaction, which is irreversible, funnels the product of glycolysis into the citric acid cycle. It requires six cofactors: coenzyme A, lipoic acid, thiamine pyrophosphate, NAD^+, flavin adenine dinucleotide (FAD), and Mg^{+2}. The reaction is believed to occur in four rapid steps catalyzed by a multienzyme system called pyruvate dehydrogenase complex.

Reaction [2] is actually the start of the cycle. It joins the two-carbon acetate group of acetyl coenzyme A to the four-carbon oxaloacetate to form citrate and coenzyme A. This reaction occurs by means of the two following steps:

Step 1. (Aldol-type condensation)

$$\underset{\text{Oxaloacetate}}{{}^-O_2CCH_2C(=O)CO_2^-} + \underset{\text{Acetyl CoA}}{H{-}CH_2C(=O)SCoA} \longrightarrow \underset{\text{Citryl CoA}}{{}^-O_2CCH_2C(OH)(CO_2^-)CH_2C(=O)SCoA}$$

Step 2. (Hydrolysis)

$$\underset{\text{Citryl CoA}}{{}^-O_2CCH_2C(OH)(CO_2^-)CH_2C(=O)SCoA} + H_2O \longrightarrow \underset{\text{Citrate}}{{}^-O_2CCH_2C(OH)(CO_2^-)CH_2C(=O)O^-}$$

The first step is an aldol-type condensation, a reaction that we learned in Section 19.8. It is catalyzed by the enzyme citrate synthetase. The second step is the hydrolysis of a thioester, a reaction that we learned in Section 18.6.

Reaction [3] is the isomerization of citrate to isocitrate. This is accomplished by a dehydration step followed by a hydration step. One enzyme called aconitase catalyzes both reactions, which have been discussed in Section 15.9.

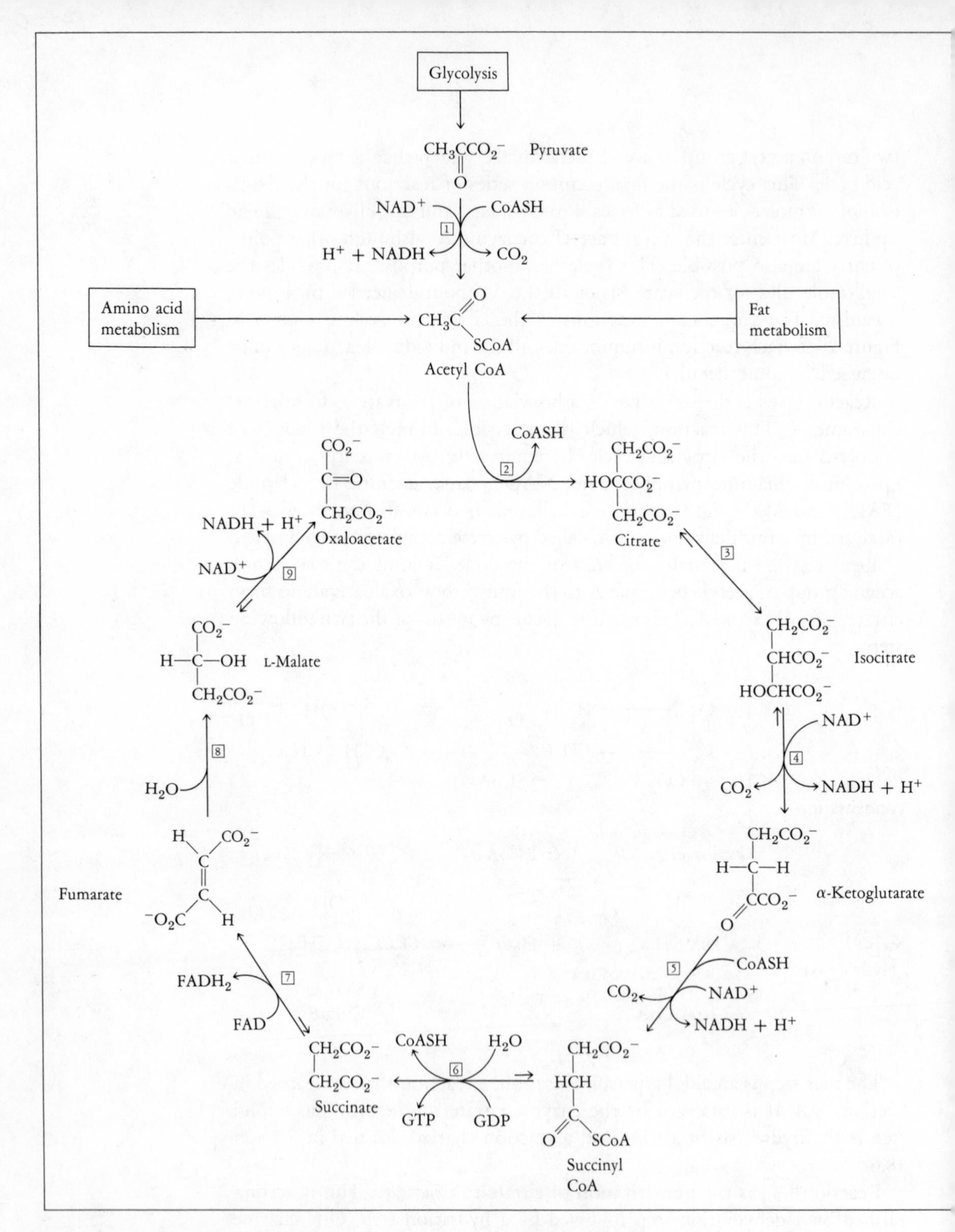

Fig. 27-9. The citric acid cycle. The numbers of the reactions correspond to those in the text.

Reaction $\boxed{4}$ is the oxidative decarboxylation of isocitrate to α-ketoglutarate catalyzed by the enzyme isocitrate dehydrogenase. This reaction occurs by means of the following two steps:

$$
\begin{array}{c}
CO_2^- \\
| \\
CH_2 \\
| \\
H-C-CO_2^- \\
| \\
H-C-O-H \\
| \\
CO_2^- \\
\text{Isocitrate}
\end{array}
\xrightarrow{NAD^+ \quad NADH + H^+}
\begin{array}{c}
CO_2^- \\
| \\
CH_2 \\
| \\
H-C-CO_2^- \\
| \\
C=O \\
| \\
CO_2^- \\
\text{Oxalosuccinate}
\end{array}
\xrightarrow{H^+ \quad CO_2}
\begin{array}{c}
CO_2^- \\
| \\
CH_2 \\
| \\
H-C-H \\
| \\
C=O \\
| \\
CO_2^- \\
\alpha\text{-Ketoglutarate}
\end{array}
$$

The first step is a typical oxidation reaction of a secondary alcohol to form a ketone (see Section 19.3). The product, oxalosuccinate, is a β-ketocarboxylic acid. As we know from Section 20.2, compounds that contain such a functional group easily lose carbon dioxide. Oxalosuccinate is no exception. It loses carbon dioxide to form α-ketoglutarate. This is the first oxidation reaction of the cycle, and it produces the first molecule of carbon dioxide.

Reaction $\boxed{5}$ involves a second oxidative decarboxylation reaction, in which another molecule of carbon dioxide is produced. This reaction is almost identical to reaction $\boxed{1}$. It is catalyzed by another multienzyme system called α-ketoglutarate dehydrogenase complex, which requires the same six cofactors as the enzymes in reaction $\boxed{1}$. Reaction $\boxed{5}$ is also irreversible, and this prevents the citric acid cycle from operating in the reverse direction.

In reaction $\boxed{6}$, the high-energy molecule succinyl coenzyme A, formed in reaction $\boxed{5}$, is hydrolyzed to succinate and coenzyme A. This reaction is the same as the second step of reaction $\boxed{2}$. Part of the energy of succinyl coenzyme A is transferred to guanosine triphosphate (GTP). The terminal phosphate group of GTP can be readily transferred to adenosine diphosphate (ADP) to form ATP. In reaction $\boxed{6}$, one molecule of the high-energy ATP is formed for each acetate entering the cycle. This is the only reaction in the cycle that conserves its energy as ATP. Reaction $\boxed{7}$ involves the oxidation of succinate to fumarate, catalyzed by the enzyme succinate dehydrogenase. The oxidizing agent is FAD. There is no simple example of such an oxidation reaction in the laboratory. Reaction $\boxed{8}$ is the hydration of fumarate to form L-malate. Fumarase catalyzes this highly stereospecific reaction, which we used as an example of the hydration of an alkene in

Section 12.4. Reaction [9] is the oxidation of L-malate to oxaloacetate to complete the cycle. Again, this reaction is a typical oxidation of a secondary alcohol to a ketone. Reaction [9] is catalyzed by malate dehydrogenase, and NAD^+ is again the oxidizing agent.

Let us summarize the reactions involved in the citric acid cycle. Two carbon atoms enter the cycle as the acetyl group of acetyl coenzyme A. Two carbon atoms leave the cycle in the form of carbon dioxide. There are only six types of organic reaction involved in the cycle. All but one of them have been presented in previous chapters. They are:

1. Aldol condensation. First step of reaction [2].
2. Hydrolysis. Second step of reaction [2], and reaction [6].
3. Decarboxylation. Second step of reaction [4], reactions [5] and [1].
4. Dehydration. First step of reaction [3].
5. Hydration. Second step of reaction [3] and reaction [8].
6. Oxidation of a secondary alcohol to a ketone. First step of reaction [4] and reaction [9].

Reaction [7] is the only new reaction to learn to understand the organic chemistry of the functional groups involved in the citric acid cycle.

In terms of energy, we have found that only one molecule of ATP is produced for each acetyl group of acetyl coenzyme A. But we have not completed the story. The three molecules of NADH and one molecule of $FADH_2$ that are formed in the cycle must be reoxidized to keep the cycle operating. When this happens, ATP is formed. This occurs by a process called oxidative phosphorylation.

27.6 ATP FROM OXIDATIVE PHOSPHORYLATION

The coenzymes NADH and $FADH_2$ are oxidized to NAD^+ and FAD by a series of reactions that result in the eventual reduction of oxygen to water. The overall equations for these two reactions are as follows:

$$2\,H^+ + 2\,NADH + O_2 \longrightarrow 2\,NAD^+ + 2\,H_2O$$

$$2\,FADH_2 + O_2 \longrightarrow 2\,FAD + 2\,H_2O$$

The oxidation of NADH and $FADH_2$ is far more complicated than is indicated by these two equations. The series of oxidation-reduction reactions involved is known as the *respiratory chain.* Because oxidation-reduction reactions require transfer of electrons, this series of reactions is also known as the *electron transport system.* This series of reactions is the common pathway by which most of the food we eat is oxidized by the oxygen we breathe.

Tightly coupled to the respiratory chain are the phosphorylation reactions. In this series of reactions, the energy released by the respiratory chain is transferred to ADP and inorganic phosphate (P_i) to form ATP. This series of reactions is called *oxidative phosphorylation.* These are the reactions that transfer most of the energy of glucose to ATP. In these coupled reactions (see Section 26.7), each molecule of NADH oxidized forms three molecules of ATP; each molecule of $FADH_2$ oxidized forms two molecules of ATP.

Oxidative phosphorylation occurs in the mitochondria of cells. The mitochondria contain the enzymes of the citric acid cycle, the enzymes of fatty acid oxidation, and the molecules involved in the respiratory chain. These molecules are placed in the mitochondria in such a way that they form an assembly line that operates quickly and efficiently. Thus, the NADH and $FADH_2$ formed in the citric acid cycle and from fatty acid metabolism (Chapter 28) are efficiently oxidized to form the energy-rich molecule ATP, which can then be used by the cell. The mitochondrion is truly the power plant of the cell. We will learn how much energy living systems get from glucose in the next section.

27.7 TOTAL ENERGY CONSERVED IN CARBOHYDRATE CATABOLISM

The amount of ATP formed in each stage of carbohydrate metabolism is given in Table 27-1. A grand total of 36 molecules of ATP are formed from the oxidation of each molecule of glucose to carbon dioxide and water. The overall equation is as follows:

$$C_6H_{12}O_6 + 36\,ADP + 36\,P_i + 6\,O_2 \longrightarrow 6\,CO_2 + 36\,ATP + 6\,H_2O$$

Notice that the two molecules of NADH formed in reaction ⑥ yield only two molecules of ATP. The reason for this is that glycolysis and oxidative phosphorylation occur in different parts of the cell. The NADH

Table 27-1. Summary of the Reactions That Form ATP in the Oxidation of Glucose

Pathway	Reaction[a]	Molecules of ATP Formed/ Molecule of Glucose	Total
Glycolysis	Reaction (1): phosphorylation	−1	
	Reaction (2): phosphorylation	−1	
	Reaction (7): 2 molecules of ATP formed	+2	
	Reaction (10): 2 molecules of ATP formed	+2	+2
Citric acid cycle	Reaction [6]: 2 molecules of ATP formed	+2	+2
Oxidative phosphorylation	Reaction (6): 2 molecules of NADH formed	+4[b]	
	Reactions [1], [4], [6], and [9]: each forms 2 molecules of NADH	+24[c]	
	Reaction [7]: 2 molecules of $FADH_2$ formed	+4	+32
	Grand total		+36

[a] The circled numbers refer to the numbered reactions in Figure 27-2. The numbers in boxes refer to the numbered reactions in Figure 27-9.
[b] Two ATP molecules are used in glycolysis to transport a three-carbon compound. See text for full explanation.
[c] Each reaction forms 2 NADH molecules, equivalent to 6 molecules of ATP. Therefore, four reactions ×6 ATP per reaction = 24 ATP molecules formed.

formed during glycolysis must be transported to another part of the cell. This costs energy. Thus, each NADH formed in glycolysis yields only two ATP molecules rather than three.

It is clear from the information in Table 27-1 that the vast majority of the ATP molecules (32 of 36) are formed by oxidative phosphorylation. Living organisms conserve 38 percent of the free energy available from glucose in the ATP formed in glycolysis, the citric acid cycle, and oxidative phosphorylation.

Now that we have learned how a living organism gets its energy, let us briefly look at the series of anabolic reactions in which it uses a part of this energy to make glucose.

27.8 GLUCONEOGENESIS

The series of reactions that forms glucose from pyruvate is called *gluconeogenesis,* the creation of "new" glucose.

In glycolysis, glucose is converted to pyruvate, whereas in gluconeogenesis, glucose is made from pyruvate. But gluconeogenesis is not the exact reversal of glycolysis. Most of the reactions in the glycolytic pathways are reversible, and the same enzymes catalyze both the forward and reverse reactions. However, three reactions in the glycolytic pathway are highly exergonic. As a result, they are virtually irreversible. The living organism must overcome these free energy barriers by providing alternate pathways. But these pathways require energy, and it is here that some of the energy of ATP is used to make glucose.

The first free energy barrier encountered in gluconeogenesis is the conversion of pyruvate to phosphoenol pyruvate (the reverse of reaction ⑩ in Figure 27-2). This barrier is bypassed by the two new reactions given in Figure 27-10. The reaction of pyruvate with carbon dioxide to form oxaloacetate is catalyzed by the enzyme pyruvate carboxylase. This reaction requires the free energy from one molecule of ATP. This reaction is an aldol condensation between pyruvate and carbon dioxide. Then oxaloacetate is decarboxylated and phosphorylated to form phosphoenol pyruvate. This reaction is catalyzed by the enzyme phosphoenol pyruvate carboxykinase and requires the free energy of GTP, another high-energy phosphorylating reagent.

$$\underset{\text{Pyruvate}}{{}^{-}O_2C{-}C({=}O){-}CH_3} \xrightarrow[\text{ATP} \;\to\; \text{ADP} + P_i]{CO_2} \underset{\text{Oxaloacetate}}{{}^{-}O_2C{-}C({=}O){-}CH_2CO_2^-} \xrightarrow[\text{GTP} \;\to\; \text{GDP} + P_i]{-CO_2} \underset{\text{Phosphoenol pyruvate}}{{}^{-}O_2C{-}C({-}O{\sim}Ⓟ){=}CH_2}$$

$$Ⓟ = {-}P({=}O)(O^-){-}O^- \qquad P_i = PO_4^{-3}$$

Fig. 27-10. The conversion of pyruvate to phosphoenol pyruvate in gluconeogenesis.

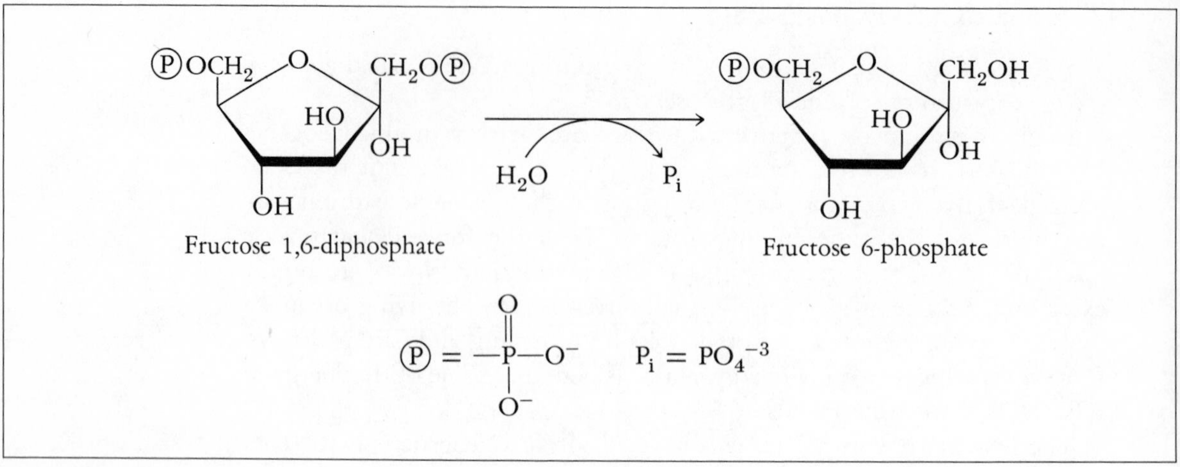

ⓅOCH_2 O CH_2OⓅ HO OH OH

Fructose 1,6-diphosphate

H_2O P_i

ⓅOCH_2 O CH_2OH HO OH OH

Fructose 6-phosphate

Ⓟ = $-\overset{O}{\overset{\|}{P}}-O^-$ with O^- below; $P_i = PO_4^{-3}$

Fig. 27-11. The hydrolysis of fructose 1,6-diphosphate to fructose 6-phosphate in gluconeogenesis.

The second free energy barrier is the formation of fructose 6-phosphate from fructose 1,6-diphosphate (the reverse of reaction ③ in Figure 27-2). Here the bypass is carried out by the enzyme fructose 1,6-diphosphatase, which catalyzes the specific hydrolysis of the 1-phosphate group according to the reaction given in Figure 27-11. This reaction is an example of the hydrolysis of a monoalkyl phosphate ester, a reaction that we learned in Section 16.4.

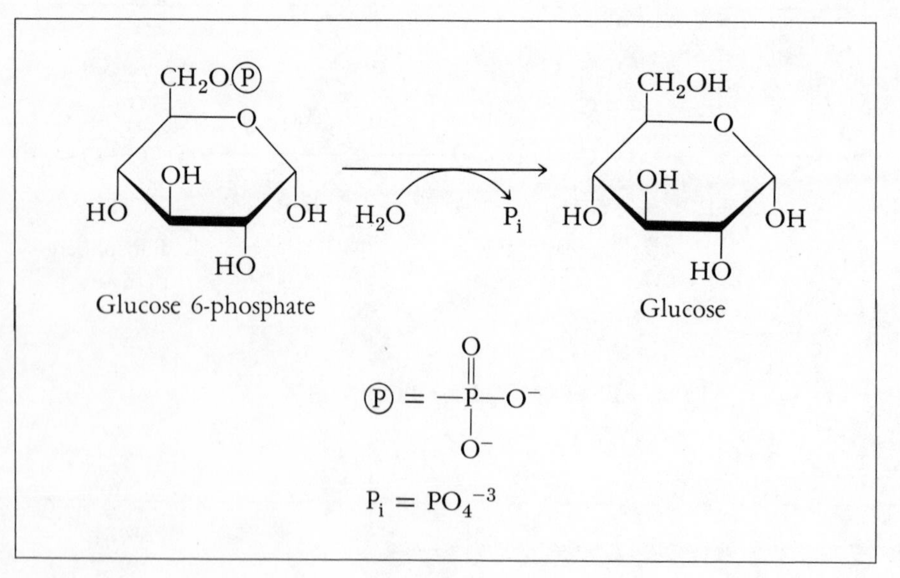

Fig. 27-12. The hydrolysis of glucose 6-phosphate to glucose in gluconeogenesis.

The final barrier is the formation of glucose from glucose 6-phosphate (the reverse of reaction ① in Figure 27-2). Again, the bypass is carried out by using a different enzyme, called glucose 6-phosphatase, to catalyze the hydrolysis of the phosphate group, as given by the equation in Figure 27-12. This, again, is simply the hydrolysis of a monophosphate ester. The pathways of glycolysis and gluconeogenesis are summarized in Figure 27-13.

Gluconeogenesis occurs mostly in the liver. A small amount of glucose is also made in the cortex of the kidney. Very little gluconeogenesis takes place in muscles or in the brain. In fact, the brain and muscles have a high demand for glucose, which they obtain from the blood. Gluconeogenesis in the liver and kidneys maintains the glucose level in the blood so that the brain and muscles will have sufficient glucose for their needs.

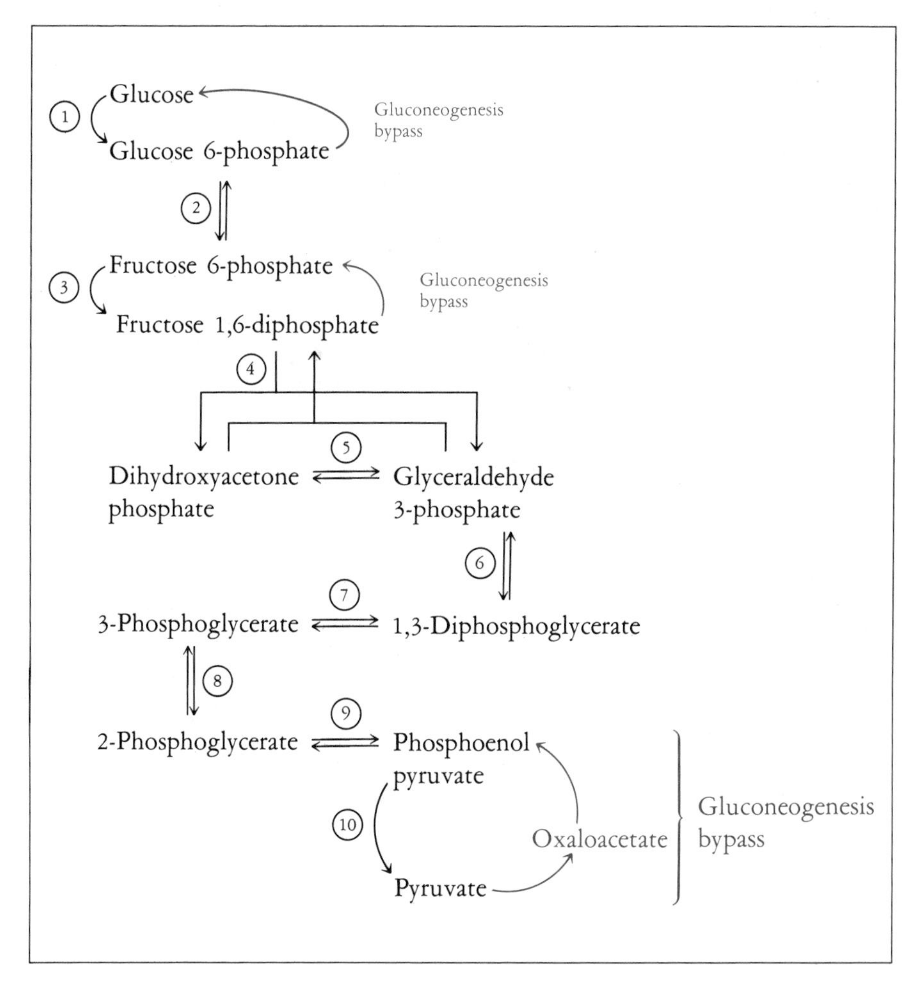

Fig. 27-13. A comparison of the pathways for glycolysis and gluconeogenesis. The circled numbers refer to the reactions in Figure 27-2.

27.9 SUMMARY

Glucose is a major source of energy for all living organisms. Humans obtain glucose by eating carbohydrates, which are degraded by enzymes in the digestive tract. Once in the blood, glucose is transported to the cells. Several reaction sequences common to most living organisms transfer part of the free energy of glucose to ATP. The first of these pathways is glycolysis, a series of ten reactions that converts glucose to pyruvate. In aerobic organisms, pyruvate is the entry into the citric acid cycle and oxidative phosphorylation. Most of the free energy of glucose is converted to ATP in oxidative phosphorylation. Many of the chemical reactions involved in glycolysis and the citric acid cycle are similar to the chemical reactions that we learned in previous chapters.

Glucose is also one of the sources of the small molecules that are used by living organisms to build special compounds they need. If the need arises, even glucose itself can be made, starting with pyruvate, by a series of reactions called gluconeogenesis. Several of the reactions that convert pyruvate to glucose are common to glycolysis. However, four new reactions are needed. These new reactions use part of the free energy of ATP to overcome three exergonic reactions of the glycolytic pathway. It is in these reactions where the free energy is needed to convert pyruvate to glucose.

These two pathways demonstrate how living organisms use energy. Part of the free energy (38 percent) of glucose is transferred to ATP, where it can be stored temporarily. It can then be used to make new compounds when called on. In the next two chapters, we will learn how the energy of ATP is used to make fats and certain amino acids.

REVIEW OF TERMS, CONCEPTS, AND REACTIONS

Terms

AEROBIC With oxygen.

ALCOHOLIC FERMENTATION The formation of ethanol from pyruvate by organisms under anaerobic conditions.

ANAEROBIC Without oxygen.

ELECTRON TRANSPORT SYSTEM Another name for the respiratory chain.

GLYCOGENESIS The chemical reaction that converts glucose into glycogen.

GLYCOGENOLYSIS The chemical reaction that converts glycogen into glucose (the reverse of glycogenesis).

GLYCOLYSIS A series of reactions that change glucose into pyruvate with the formation of ATP.

GLUCONEOGENESIS A series of reactions that form glucose from pyruvate.

HYPERGLYCEMIA An excess of sugar in the blood.

HYPOGLYCEMIA Not enough sugar in the blood.

OXIDATIVE PHOSPHORYLATION A series of reactions that transfer the energy of the respiratory chain to ATP.

PHOSPHORYLATION The transfer of a phosphate group from one molecule to another.

RESPIRATORY CHAIN A series of oxidization-reduction reactions that oxidize the coenzymes NADH and $FADH_2$ and reduce oxygen to water.

Concepts

1. Glucose is one of the major foods of all living organisms.
2. The amount of glucose in the blood is regulated by several pathways that make or use it as needed. When too much glucose is in the blood, it is removed by storing it as glycogen in the liver or muscles or by oxidizing it to form energy. The amount of glucose in the blood can be increased by removing it from storage or by making it from simpler molecules.
3. Glucose serves two purposes in living organisms. It provides energy and compounds to the citric acid cycle, from which large molecules needed by the organism can be made.
4. Part of the chemical energy of glucose is converted to ATP by the glycolytic pathway, the citric acid cycle, and oxidative phosphorylation.

Reactions

1. Phosphorylation (see Section 27.3)
   Glucose + ATP $\longrightarrow$ ADP + glucose 6-phosphate
2. Aldose-ketose isomerization (see Sections 27.3 and 20.2)
   Glucose 6-phosphate $\longrightarrow$ fructose 6-phosphate
3. Retroaldol condensation (see Sections 27.3 and 19.8)
   Fructose 6-phosphate $\longrightarrow$ dihydroxyacetone phosphate + glyceraldehyde 3-phosphate
4. Dehydration (see Sections 27.3, 12.4, and 15.9)
   2-Phosphoglycerate $\longrightarrow$ phosphoenol pyruvate
5. Reduction (see Sections 27.4, 16.3, 19.3, and 26.8)
   Acetaldehyde + NADH $\longrightarrow$ ethanol + NAD^+

6. Aldol condensation (see Sections 27.5 and 19.8)
   Oxaloacetate + acetyl CoA $\longrightarrow$ citryl CoA
7. Hydrolysis (see Sections 27.5, 18.5, and 18.6)
   Citryl CoA + H_2O $\longrightarrow$ citrate + CoASH
8. Decarboxylation (see Sections 27.5 and 20.2)
   α-Ketoglutarate + CoASH $\longrightarrow$ succinyl CoA + CO_2
9. Oxidation (see Sections 27.5, 16.3, 19.3, and 26.8)
   L-Malate + NAD^+ $\longrightarrow$ oxaloacetate + NADH
10. Dehydrogenation (see Section 27.5)
    Succinate + FAD $\longrightarrow$ fumarate + $FADH_2$
11. Addition of carbon dioxide (see Section 27.8)
    Pyruvate + CO_2 + ATP $\longrightarrow$ oxaloacetate + ADP + P_i

EXERCISES

27-1 Table sugar (sucrose) is made up of two monosaccharides, glucose and fructose. How is sucrose metabolized?

27-2 Why does the digestion of carbohydrates stop in the stomach?

27-3 The body maintains a constant level of glucose in the blood despite the fact that carbohydrates are usually eaten only at meals. Explain how the body can do this.

27-4 If the glucose level in the blood decreases, how can the liver help to increase the glucose supply?

27-5 Name four compounds that can be obtained from the metabolism of pyruvate.

27-6 Why does the fermentation process continue to form ethanol instead of stopping at pyruvate?

27-7 The sour taste of vinegar is due to acetic acid. When wine is left standing open to the air, it slowly acquires a sour taste. Explain chemically what is happening.

27-8 What happens to the lactate formed by the activity of muscles?

27-9 Why is the citric acid cycle so important to the metabolic reactions of living organisms?

27-10 Label the two carbon atoms of the acetyl group of acetyl coenzyme A. Are these the same two carbon atoms that are lost as carbon dioxide in the citric acid cycle?

TRIACYLGLYCEROL METABOLISM

28

The catabolism of carbohydrates and fats provides nearly all the energy needed by humans and the higher animals and plants. Carbohydrates are the readily available source of energy, and fats are the major energy reserve. More energy can be obtained from fats than from glycogen, which is the reason why more energy is stored in the body as fat than as glycogen. For example, a person who weighs 70 kg (154 lb) has a reserve of 13 kg (28.6 lb) of fat, 5 kg (11 lb) of protein (mostly in muscle), 1 kg (2.20 lb) of glycogen and 10 g (0.02 lb) of glucose.

In this chapter we will learn the chemical reactions of the catabolism of the most abundant type of compound making up fats, triacylglycerols (see Section 22.3). We will also learn how part of their free energy is transferred to adenosine triphosphate (ATP). In addition, we will examine the anabolism of fatty acids and triacylglycerols. Finally, we will learn how important acetyl coenzyme A is to the metabolism of both carbohydrates and triacylglycerols. Let us start by learning how triacylglycerols are digested.

28.1 DIGESTION OF TRIACYLGLYCEROLS

The digestion of triacylglycerols does not begin until they pass through the stomach and enter the small intestine. Here they are hydrolyzed to fatty acids and glycerol (see Section 22.3) according to the following equation:

$$\underset{\text{Triacylglycerol}}{\begin{array}{l} \ \ \text{CH}_2\text{OC(=O)R} \\ \text{R'C(=O)OCH} \\ \ \ \text{CH}_2\text{OC(=O)R''} \end{array}} + 3\,H_2O \xrightarrow{\text{Lipases}} \underset{\text{Glycerol}}{\begin{array}{l} \text{CH}_2\text{OH} \\ \text{HOCH} \\ \text{CH}_2\text{OH} \end{array}} + \underset{\text{Fatty acids}}{\begin{array}{l} \text{RCO}_2\text{H} \\ \text{R'CO}_2\text{H} \\ \text{R''CO}_2\text{H} \end{array}}$$

This process requires several steps. First, the large fat globules must be broken up into smaller globules so that they can be more easily attacked by water-soluble enzymes. This is accomplished by the bile acids (see Section 22.9). These compounds are secreted by the gall bladder and act as emulsifying agents. Then a hormone stimulates the pancreas to secrete the pancreatic juices that contain the enzymes, called lipases, that catalyze the hydrolysis of triacylglycerols.

Glycerol and fatty acids are then absorbed through the intestinal membrane. The products of hydrolysis can follow any of several pathways after they are absorbed. The fatty acids can be made into triacylglycerols and phospholipids. The triacylglycerol can be stored in cells or carried to the liver, where it can be oxidized to form energy. Catabolism of glycerol occurs by the glycolytic pathway.

28.2 GLYCEROL CATABOLISM

The glycerol formed in the hydrolysis of triacylglycerols is converted into dihydroxyacetone phosphate by the following two reactions:

$$\underset{\text{L-glycerol}}{\begin{array}{l} \text{CH}_2\text{OH} \\ \text{CHOH} \\ \text{CH}_2\text{OH} \end{array}} \xrightarrow[\text{ATP} \to \text{ADP}]{} \underset{\text{L-glycerol 3-phosphate}}{\begin{array}{l} \text{CH}_2\text{OH} \\ \text{CHOH} \\ \text{CH}_2\text{O—Ⓟ} \end{array}} \xrightarrow[\text{NAD}^+ \to \text{NADH} + \text{H}^+]{} \underset{\text{Dihydroxyacetone phosphate}}{\begin{array}{l} \text{CH}_2\text{OH} \\ \text{C=O} \\ \text{CH}_2\text{O—Ⓟ} \end{array}}$$

$$\text{Ⓟ} = -\text{P(=O)(O}^-\text{)—O}^-$$

We are familiar with both of these types of reaction. The first is a phosphorylation of a primary hydroxy group similar to reaction (1) in Figure 27-2. The second reaction is the oxidation of a secondary alcohol to a ketone (see Section 27.3 and reaction [9] in Figure 27-9).

Dihydroxyacetone phosphate is one of the compounds in the glycolytic pathway (Figure 27-2). Thus, glycerol enters the glycolytic pathway and is oxidized to form energy, carbon dioxide, and water. The fatty acids have a different catabolic pathway.

28.3 FATTY ACID CATABOLISM

Fatty acids contain the major part of the energy of triacylglycerols. Their catabolism occurs in a series of four reactions, repeated several times, that removes two carbon atoms at a time from the long chain of carbon atoms. These reactions occur with a transfer of a part of the energy to ATP. The enzymes that catalyze these reactions are located in the mitochondria along with the enzymes of the citric acid cycle, the respiratory chain, and oxidative phosphorylation. All of the major energy-producing systems are concentrated in one place. As noted in Section 27.6, this makes the mitochondria the power plants of the cell.

As a result of much experimental work, the details of the catabolism of fatty acids are quite well understood. The four reactions that occur repeatedly are given in Figure 28-1. Let us examine each reaction in detail. Each reaction in Figure 28-1 is numbered to correspond to the reaction numbers in the following discussion.

To begin the series of reactions, the fatty acids must be activated. This occurs by the reaction of the fatty acid with coenzyme A to form a thioester called an *acyl coenzyme A*. This reaction actually occurs in two steps, according to the equations given in Figure 28-2. In the first step, the fatty acid reacts with ATP to form an acyl adenylate, a mixed anhydride. It then reacts with coenzyme A to form an acyl coenzyme A. We have encountered both of these reactions before. The first is the formation of a mixed anhydride of a carboxylic acid and phosphoric acid (see Section 18.12). The second reaction is a typical substitution reaction of carboxylic acid derivatives (see Section 18.5). In living organisms, both of these reactions are catalyzed by the enzyme acyl coenzyme A synthetase. Both reactions occur at the active site. That is, after the acyl adenylate is formed, it remains bound to the enzyme, where it undergoes the second reaction to form acyl coenzyme A.

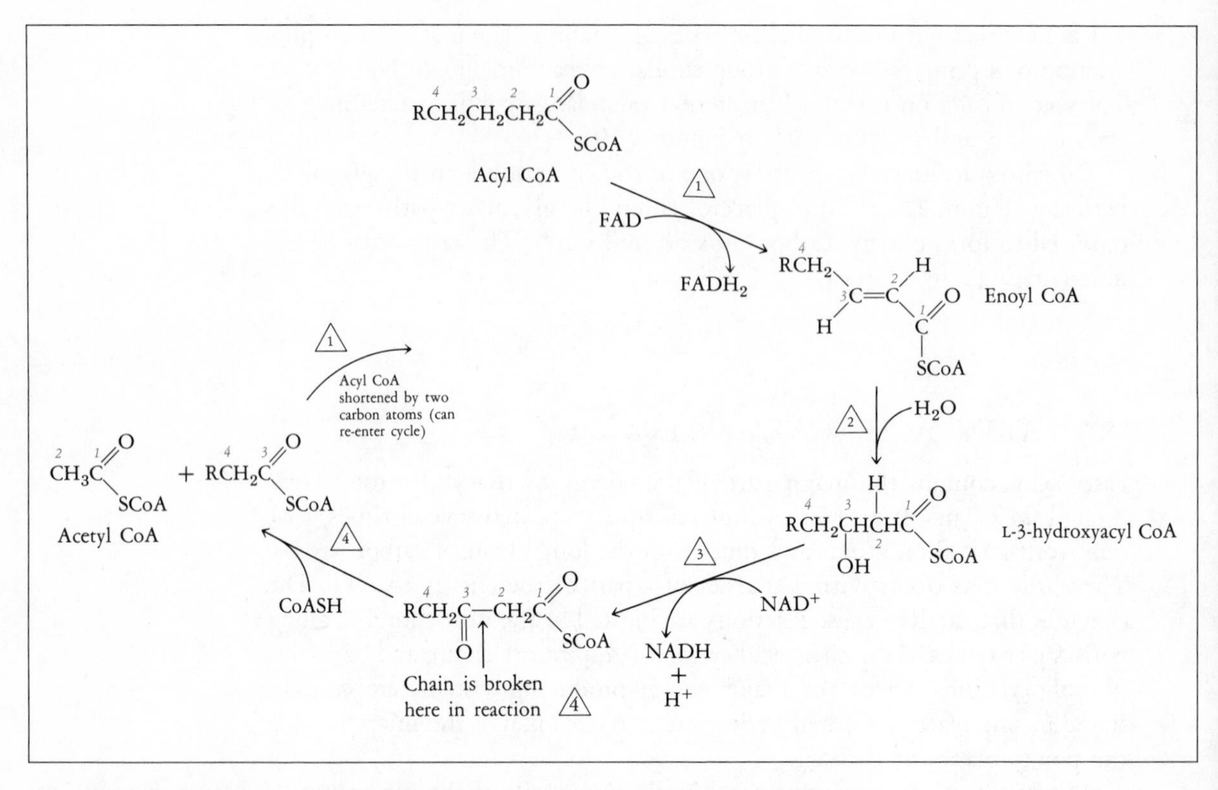

Fig. 28-1. The four-reaction sequence by which two carbon atoms are removed from a fatty acid carbon chain.

*EXERCISE 28-1	Circle the mixed anhydride functional group in acyl adenylate.
EXERCISE 28-2	Circle the thioester functional group in acyl coenzyme A.
EXERCISE 28-3	Circle the group that is replaced by coenzyme A in its reactions with acyl adenylate.

Once the fatty acid is activated, its catabolism can begin. Reaction 1 in Figure 28-1 is the oxidation of the acyl coenzyme A catalyzed by the enzyme acyl coenzyme A dehydrogenase. The coenzyme flavin adenine dinucleotide (FAD) removes two hydrogens, one each from carbons 2 and 3 to form a double bond with the *trans* configuration. A similar reaction occurs in the citric acid cycle (succinate to fumarate), as we learned in Section 27.5.

Reaction 2 is the stereospecific hydration of the *trans* double bond to form L-3-hydroxylacyl coenzyme A. This reaction, which is catalyzed by the

* The answers for the exercises in this chapter begin on page 903.

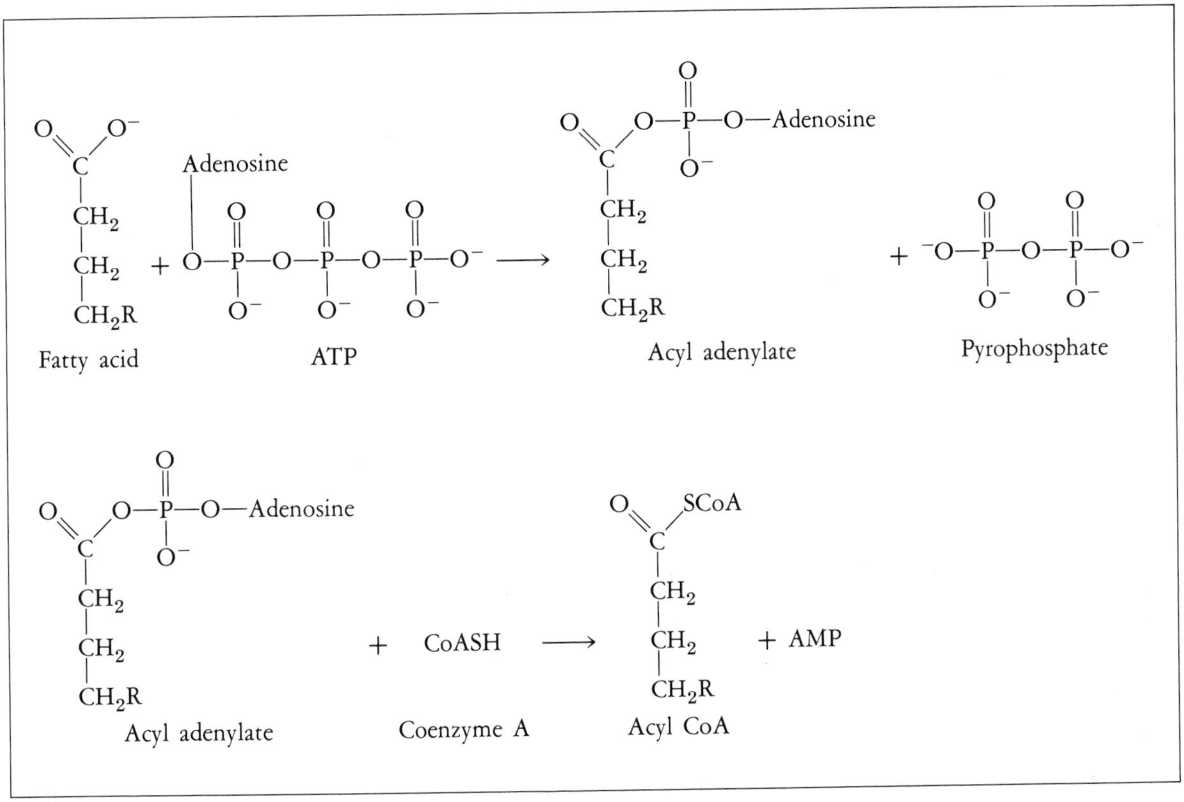

$$\underset{\text{Fatty acid}}{RCH_2-CH_2-CH_2-C(=O)-O^-} + \underset{\text{ATP}}{\text{Adenosine}-O-P(=O)(O^-)-O-P(=O)(O^-)-O-P(=O)(O^-)-O^-} \longrightarrow \underset{\text{Acyl adenylate}}{RCH_2-CH_2-CH_2-C(=O)-O-P(=O)(O^-)-O-\text{Adenosine}} + \underset{\text{Pyrophosphate}}{{}^-O-P(=O)(O^-)-O-P(=O)(O^-)-O^-}$$

$$\underset{\text{Acyl adenylate}}{RCH_2-CH_2-CH_2-C(=O)-O-P(=O)(O^-)-O-\text{Adenosine}} + \underset{\text{Coenzyme A}}{\text{CoASH}} \longrightarrow \underset{\text{Acyl CoA}}{RCH_2-CH_2-CH_2-C(=O)-\text{SCoA}} + \text{AMP}$$

Fig. 28-2. The formation of acyl coenzyme A.

enzyme enoyl coenzyme A hydratase, is similar to the hydration of fumarate in the citric acid cycle.

In reaction △3, a secondary alcohol is oxidized by nicotinamide adenine dinucleotide (NAD^+) to a keto group. This typical oxidation reaction is catalyzed by the enzyme L-3-hydroxyacyl coenzyme A dehydrogenase. This enzyme is absolutely specific for the L-isomer of the hydroxyacyl substrate. Again, a similar reaction occurs in the citric acid cycle (L-malate to oxalo-acetate).

The first three reactions of this sequence have oxidized carbon 3 of the fatty acid chain to a keto group. Reaction △4 is the breaking of the carbon bond between carbons 2 and 3 to form acetyl coenzyme A and an acyl coenzyme A shortened by two carbon atoms. This reaction, which is cata-lyzed by the enzyme β-keto thiolase, is the reverse of the Claisen reaction that we learned in Section 19.8. The shortened acyl coenzyme A can un-dergo another sequence of four reactions to lose another two carbons. This

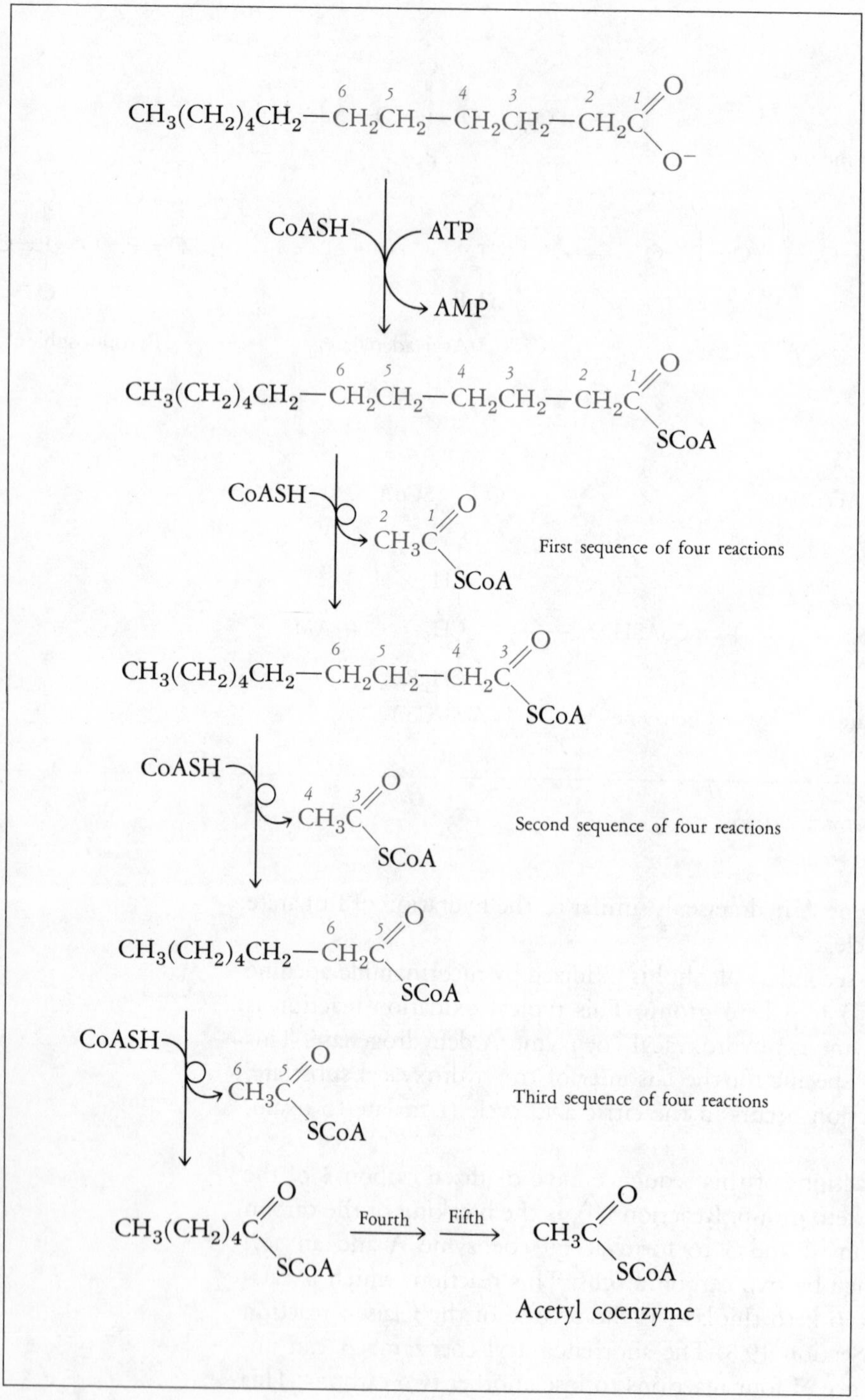

$$\overset{6}{} \quad \overset{5}{} \quad \overset{4}{} \quad \overset{3}{} \quad \overset{2}{} \quad \overset{1}{}$$
$$CH_3(CH_2)_4CH_2{-}CH_2CH_2{-}CH_2CH_2{-}CH_2C(=O)O^-$$

CoASH, ATP → AMP

$$CH_3(CH_2)_4CH_2{-}CH_2CH_2{-}CH_2CH_2{-}CH_2C(=O)SCoA$$

CoASH → $\overset{2}{C}H_3\overset{1}{C}(=O)SCoA$ First sequence of four reactions

$$CH_3(CH_2)_4CH_2{-}\overset{6}{C}H_2\overset{5}{C}H_2{-}\overset{4}{C}H_2\overset{3}{C}(=O)SCoA$$

CoASH → $\overset{4}{C}H_3\overset{3}{C}(=O)SCoA$ Second sequence of four reactions

$$CH_3(CH_2)_4CH_2{-}\overset{6}{C}H_2\overset{5}{C}(=O)SCoA$$

CoASH → $\overset{6}{C}H_3\overset{5}{C}(=O)SCoA$ Third sequence of four reactions

$$CH_3(CH_2)_4C(=O)SCoA \xrightarrow{\text{Fourth}} \xrightarrow{\text{Fifth}} CH_3C(=O)SCoA$$

Acetyl coenzyme

Fig. 28-3. First three sequences of four reactions in the catabolism of laurate. Two carbon units are removed in each sequence of four reactions as acetyl coenzyme A. After five sequences, the entire chain is broken down to six molecules of acetyl coenzyme A.

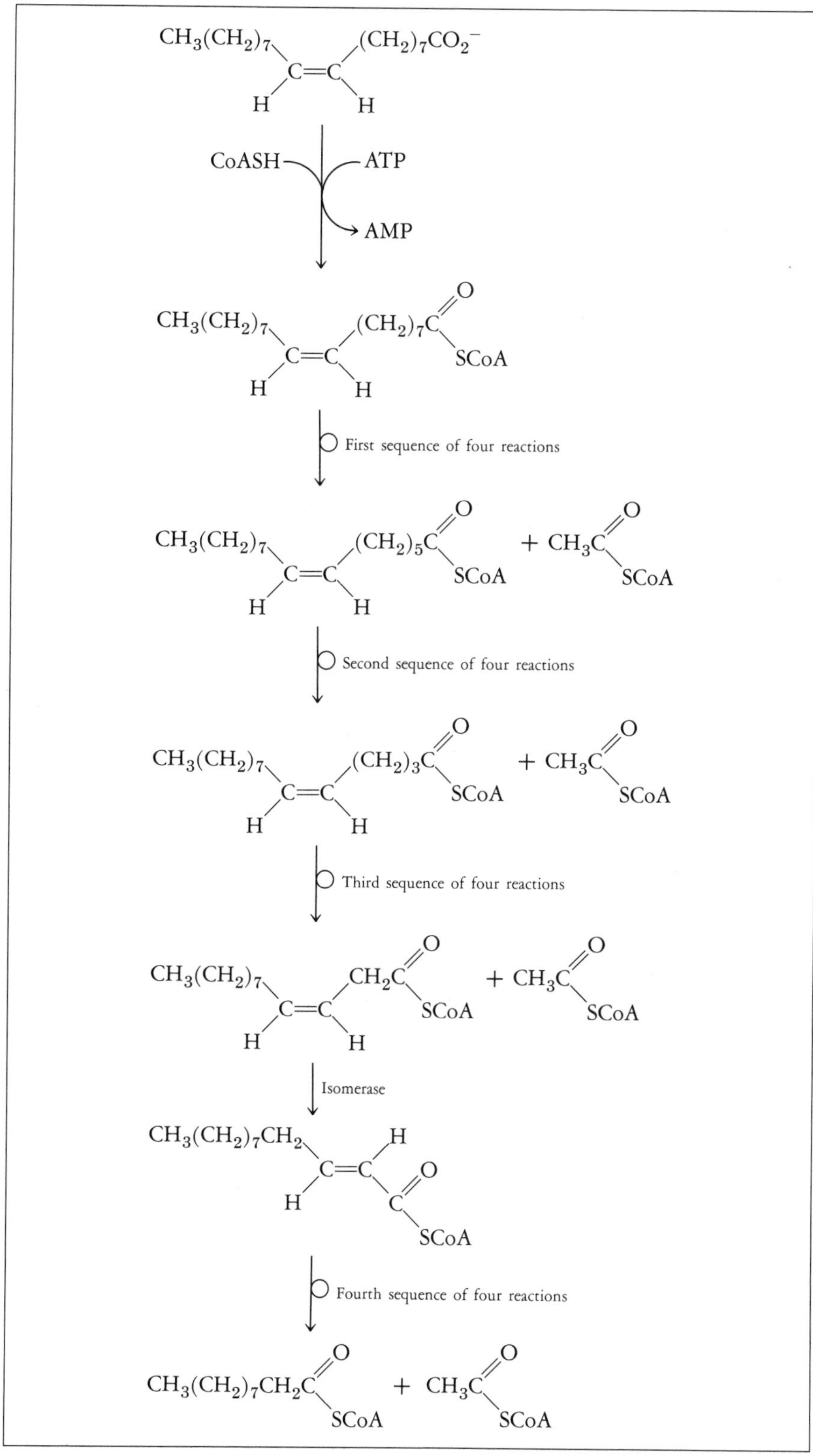

$CH_3(CH_2)_7CH{=}CH(CH_2)_7CO_2^-$ (cis: H, H on the same side of C=C)

CoASH, ATP → AMP

$CH_3(CH_2)_7CH{=}CH(CH_2)_7C({=}O)SCoA$

First sequence of four reactions

$CH_3(CH_2)_7CH{=}CH(CH_2)_5C({=}O)SCoA + CH_3C({=}O)SCoA$

Second sequence of four reactions

$CH_3(CH_2)_7CH{=}CH(CH_2)_3C({=}O)SCoA + CH_3C({=}O)SCoA$

Third sequence of four reactions

$CH_3(CH_2)_7CH{=}CHCH_2C({=}O)SCoA + CH_3C({=}O)SCoA$

Isomerase

$CH_3(CH_2)_7CH_2CH{=}CHC({=}O)SCoA$ (trans)

Fourth sequence of four reactions

$CH_3(CH_2)_7CH_2C({=}O)SCoA + CH_3C({=}O)SCoA$

Fig. 28-4. The catabolism of unsaturated fatty acids.

is shown in Figure 28-3, where the first three sequences of four reactions in the catabolism of laurate are given. The reactions continue until only the last two carbons of the chain remain as acetyl coenzyme A.

The catabolism of unsaturated fatty acids occurs by a similar series of reactions. But there is a problem in the catabolism of unsaturated fatty acids, as shown in Figure 28-4, for the catabolism of oleic acid. This fatty acid has a double bond between carbons 9 and 10. Oleic acid is activated just like the saturated fatty acids, and it undergoes the same sequence of four reactions as laurate. However, the product after the third sequence of four reactions is not a substrate for the enzyme of reaction △1. This product has a double bond between carbons 3 and 4, which prevents the formation of a double bond between carbons 2 and 3. This problem is solved by a new reaction that shifts both the position and the configuration of the double bond. An enzyme called isomerase isomerizes the *cis* double bond to a *trans* double bond between carbons 2 and 3. This compound is now a regular substrate for reaction △2, and the remaining reactions are the same as those of saturated fatty acids. Thus, only one more enzyme is needed to use the same series of chemical reactions for the catabolism of both saturated and unsaturated fatty acids.

EXERCISE 28-4 Write each reaction involved in the conversion of

$$CH_3(CH_2)_7CH{=}CHCH_2C(=O)SCoA \quad \text{to} \quad CH_3(CH_2)_7CH_2C(=O)SCoA$$

EXERCISE 28-5 How many molecules of acetyl coenzyme A will be obtained from the catabolism of palmitate ($C_{15}H_{31}CO_2^-$)?

So far, we have concentrated on the chemical fate of the fatty acids. In the next section we will learn how much energy the organism gets from these chemical reactions.

28.4 TOTAL ENERGY CONSERVED IN FATTY ACID CATABOLISM

An enormous amount of energy is obtained by an organism from the oxidation of a fatty acid. Let us find out how much energy by using laurate, $C_{12}H_{23}O_2^-$, as an example. We will trace its metabolism and determine the gain or loss of ATP at each step.

The first step is activation. It costs two of the phosphate bonds of ATP to activate the fatty acid: ATP yields adenosine monophosphate (AMP) plus

Table 28-1. Summary of the Amount of ATP Formed by the Catabolism of Laurate

	Amount of ATP
Activation of laurate costs 2 ATP	−2
6 acetyl CoA molecules formed; each produces 12 ATP (6 × 12)	+72
5 $FADH_2$ molecules formed (1 per sequence); each produces 2 ATP (5 × 2)	+10
5 NADH molecules formed (1 per sequence); each produces 3 ATP (5 × 3)	+15
Grand total	+95

two molecules of inorganic phosphate (P_i). This is the same as two molecules of ATP in terms of energy used. So the first step costs energy. But as laurate is degraded, six molecules of acetyl coenzyme A are formed. Each of these molecules can be oxidized by the citric acid cycle and its coupled respiratory chain to form three molecules of NADH, one molecule of $FADH_2$, and one molecule of guanosine triphosphate (GTP) (see Section 27.5). Moreover, each sequence of four reactions of fatty acid catabolism forms one molecule of NADH and one molecule of $FADH_2$. Reoxidation of these compounds in the respiratory chain forms three and two molecules of ATP, respectively (see Section 27.6), whereas GTP forms one molecule of ATP (see Section 27.5). The amount of ATP formed in each of these reactions is summarized in Table 28-1.

The oxidation of one molecule of laurate forms a grand total of 95 ATP molecules. It is instructive to compare the amounts of ATP formed from the oxidation of a fatty acid and a sugar containing the same number of carbon atoms. Two molecules of glucose contain the same number of carbon atoms (12) as laurate. Oxidation of one molecule of glucose forms 36 molecules of ATP, as we learned in Section 27.7. Therefore, two molecules of glucose will form 72 molecules of ATP on oxidation. This is less than the 95 ATP molecules formed from oxidation of one laurate molecule.

But there is another fact to consider. The molecular weight of two molecules of glucose (360 amu) is greater than that of one molecule of laurate (199 amu). If we take this difference into account, the oxidation of laurate forms 0.48 ATP molecule per unit of molecular weight (95 ATP/199 amu), whereas the oxidation of two glucose molecules forms only 0.20 ATP per unit of molecular weight (72 ATP/360 amu). This confirms the earlier statement that organisms get more energy from fats than from carbohydrates.

EXERCISE 28-6 Calculate the amount of ATP formed by the oxidation of palmitate ($C_{15}H_{31}CO_2^-$).

EXERCISE 28-7 The oxidation of one molecule of laurate forms six acetyl coenzyme A molecules, yet only five NADH molecules and five $FADH_2$ molecules are formed. Explain.

Each sequence of four reactions of the catabolism of fatty acids produces one molecule of acetyl coenzyme A. What do living organisms do with all this acetyl coenzyme A? We will learn the answer to this question in the next section.

28.5 THE IMPORTANCE OF ACETYL COENZYME A

Acetyl coenzyme A is a very important intermediate in living organisms because it is involved in many metabolic pathways. The most important pathways are given in Figure 28-5. Acetyl coenzyme A has two main roles. Its acetyl group can be oxidized to provide energy by the citric acid cycle, as we learned in Section 27.5. Its other role is to serve as the starting point for the synthesis of many compounds needed by living organisms.

We have learned how acetyl coenzyme A is formed by the catabolism of glucose and fatty acids, and we have traced its path through the citric acid cycle. Now it is time to learn how it is used to synthesize other compounds.

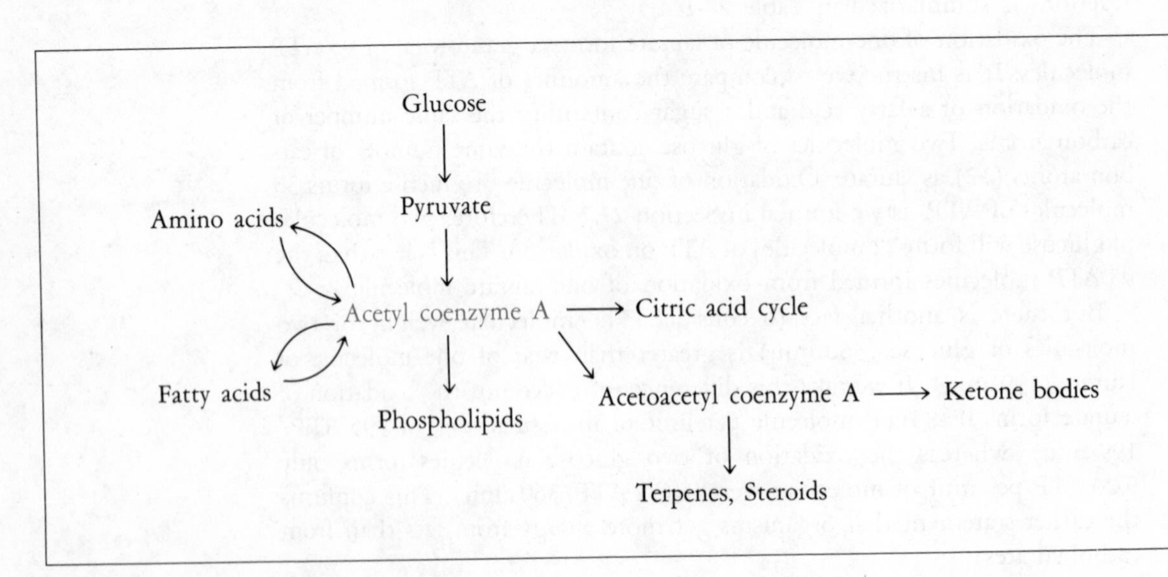

Fig. 28-5. The central role of acetyl coenzyme A in metabolism.

In this chapter, we will learn how it is transformed into fatty acids and ketone bodies; in the next chapter, we will learn how it is used to form amino acids. Let us start with the anabolism of fatty acids.

28.6 FATTY ACID ANABOLISM

The synthesis of fatty acids does not occur by the reverse of the catabolic pathway. Instead, a new series of reactions is involved. The pathways of catabolism and anabolism are completely different. In fact, the two pathways are located in different regions of the cell. Furthermore, the enzymes involved in the anabolism are organized into a multienzyme complex called fatty acid synthetase. The various intermediates formed during the synthesis are bound to the enzyme complex. The intermediates never become free, and the sequence of reactions is determined by the position of each enzyme in the complex.

The reactions involved in the synthesis of fatty acids are given in Figure 28-6. The first thing to notice about the reaction in Figure 28-6 is that *the fatty acid chain is built up two carbon atoms at a time* by a repeating sequence of four reactions. But before this cycle can start, acetyl coenzyme A must be converted into acetyl acyl carrier protein (ACP) and malonyl ACP. Let us examine these reactions and those in the anabolic sequence in some detail. As before, each reaction is given a number that corresponds to the number in the following discussion.

An important reaction must take place before the acetyl group of acetyl coenzyme A can be used by the fatty acid synthetase complex. It must first be converted to malonyl coenzyme A. This occurs in reaction ①. The enzyme acetyl coenzyme A carboxylase catalyzes the reaction of acetyl coenzyme A and bicarbonate ion (HCO_3^-) to form malonyl coenzyme A. The carbon atom of the bicarbonate ion becomes the carbon of the free carboxylate group of malonyl coenzyme A, as shown in Figure 28-6. This reaction is an example of a Claisen condensation (see Section 19.8).

Reactions ② and ③ are quite similar. Both involve the reaction of an acyl coenzyme A with an acyl carrier protein (ACP). The ACP is a single polypeptide chain of 77 amino acid residues. Its function is to anchor acyl groups (in this case, acetyl and malonyl groups) to the enzyme complex and carry them through the reaction sequence. The ACP contains a free thiol group. In reactions ② and ③ the acyl groups are transferred from the sulfur of coenzyme A to the sulfur of ACP (see Section 18.10) to form acetyl ACP and malonyl ACP. These reactions are catalyzed by the enzyme transacylase. In this way, the substrates are bound to the enzymes, and chain building can start.

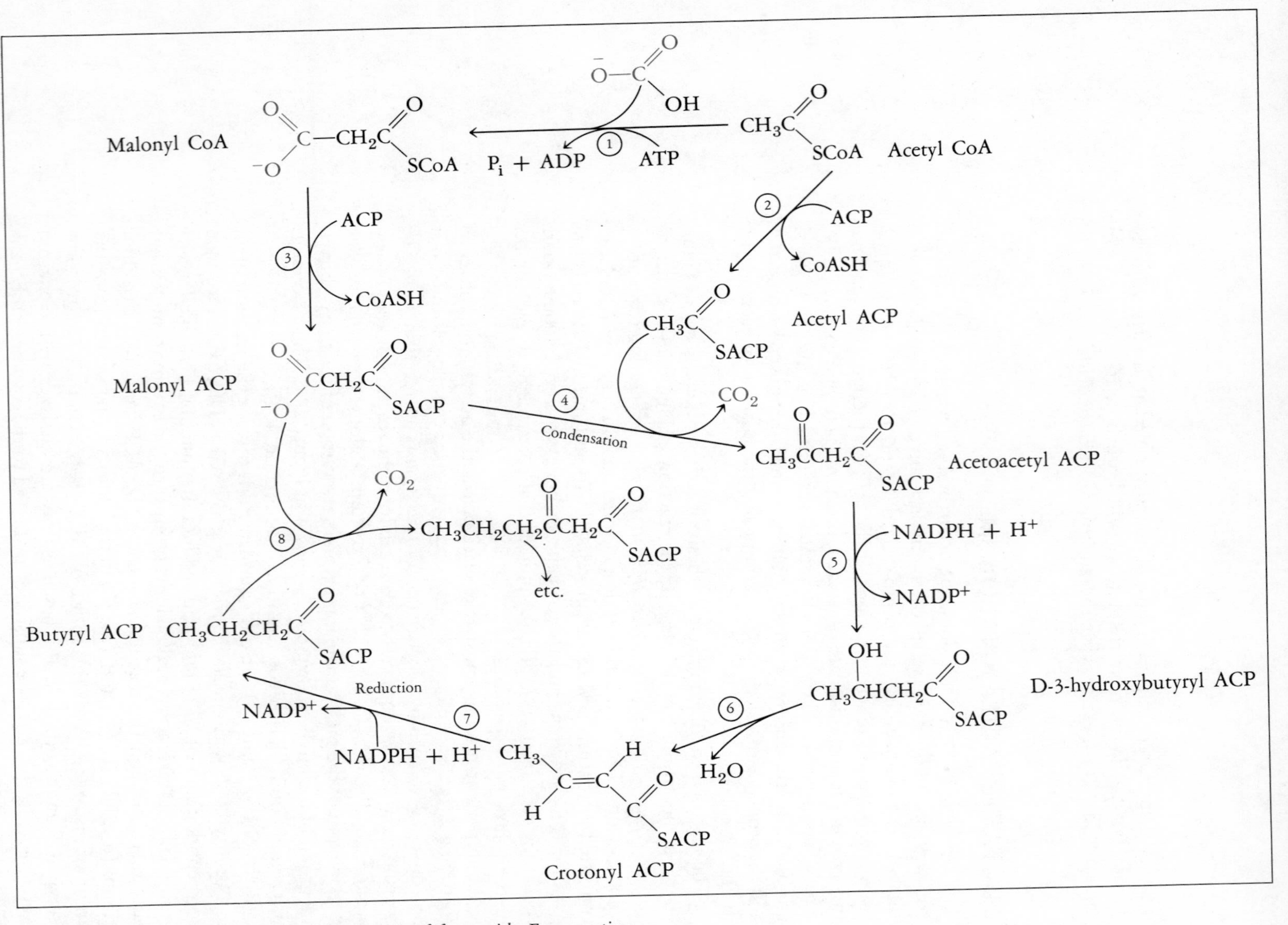

Fig. 28-6. The sequence of reactions in the synthesis of fatty acids. Four reactions (condensation, reduction, dehydration, and reduction) are repeated each time, adding two carbons to the acid chain. ACP = acyl carrier protein.

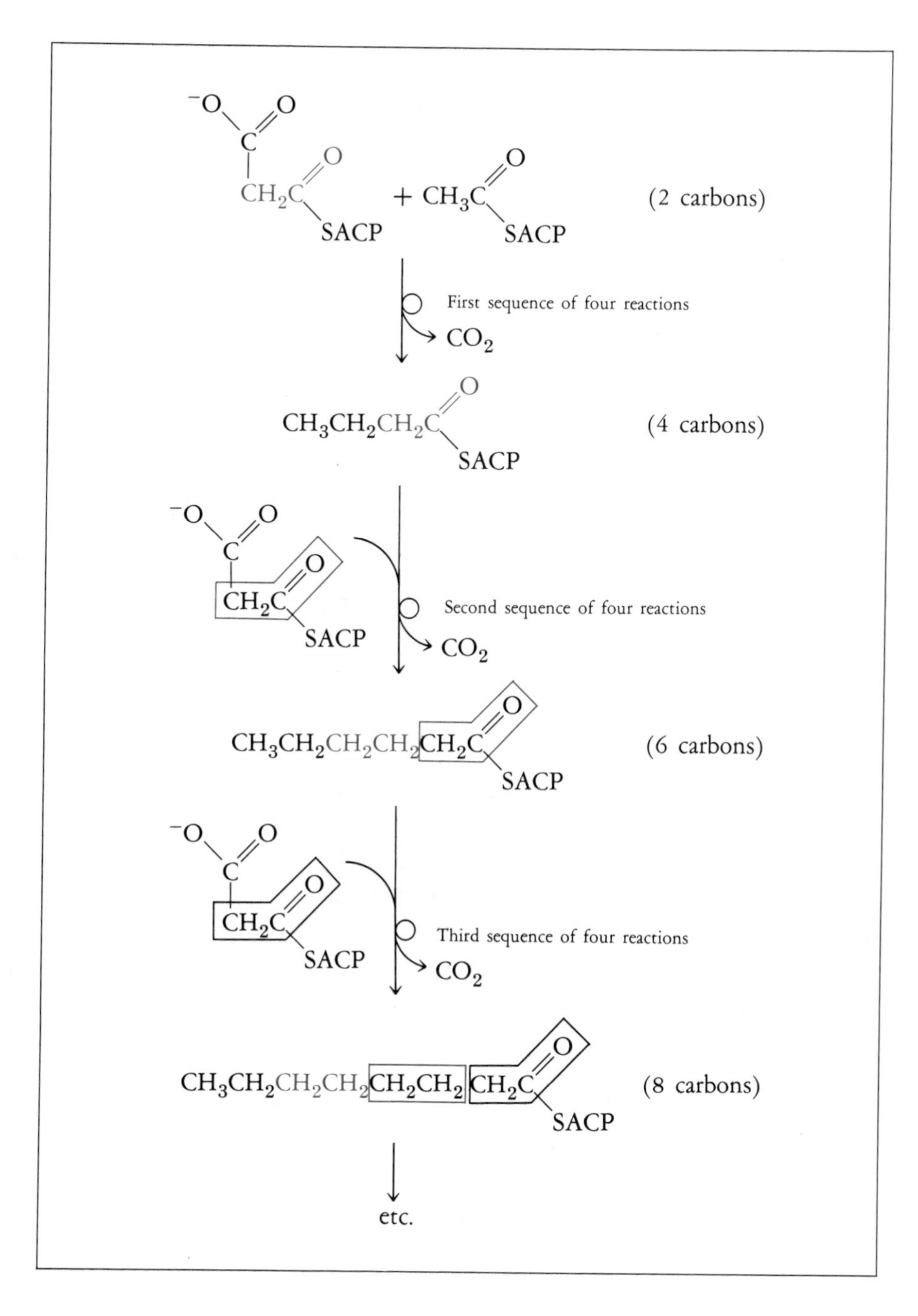

Fig. 28-7. The first three sequences of four reactions in the synthesis of a fatty acid. Each sequence of four reactions adds two carbons to the chain.

Malonyl ACP and acetyl ACP react to form acetoacetyl ACP in reaction ④. This reaction is an example of a malonic ester condensation reaction (see Section 20.2) in a living organism. Here it is catalyzed by the enzyme β-ketoacyl ACP synthetase. This is the reaction that increases the length of the carbon chain by two carbons. Notice that the carbon dioxide released is the same one that was added as bicarbonate ion in the formation of malonyl coenzyme A in reaction ①. Thus, carbon dioxide (as bicarbonate ion in reaction ①) is actually a catalyst in this sequence; that is, the carbon of carbon dioxide does not become part of the fatty acid carbon chain. All the carbon atoms in the even-number fatty acids come from acetyl coenzyme A.

The next three reactions, ⑤, ⑥, ⑦, are the reverse of reactions △1, △2, and △3 of the catabolism of fatty acids given in Figure 28-1. But they use different coenzymes. Reactions ⑤ and ⑦ require the phosphorylated coenzymes nicotinamide adenine dinucleotide phosphate ($NADP^+$) and NADPH instead of the NAD^+ and NADH used in reactions △1 and △3.

The formation of butyryl ACP completes one sequence of four reactions. The next sequence of four reactions starts with the condensation of butyryl ACP and malonyl ACP in reaction ⑧ to form an acyl ACP containing six carbon atoms. In this way, the fatty acid chain is synthesized by adding two carbon atoms at a time. This is shown in Figure 28-7 for three sequences of four reactions. We now understand why most naturally occurring fatty acids contain an even number of carbon atoms.

The reactions continue until the chain length reaches 16 carbon atoms. In most organisms, fatty acid synthesis then stops. This happens because the enzyme that catalyzes reaction ④, β-ketoacyl ACP synthetase, cannot accept an acyl ACP containing 16 carbons. This acyl group is too long to fit into the active site. Instead, it is hydrolyzed to the 16-carbon fatty acid (palmitic acid) and ACP. The synthesis of longer fatty acids occurs elsewhere in the organism by a different pathway.

Notice that separate pathways exist for the anabolism and catabolism of both fatty acids and carbohydrates. These pathways are located in different parts of the cell and they occur independently. As a result, the activity of each pathway is determined by the needs of the organism.

Once fatty acids are synthesized, they can be converted into triacylglycerols by a series of reactions given in the next section.

28.7 TRIACYLGLYCEROL ANABOLISM

Triacylglycerols are made in most organisms by the reaction of L-glycerol 3-phosphate and fatty acid coenzyme A. L-Glycerol 3-phosphate is obtained

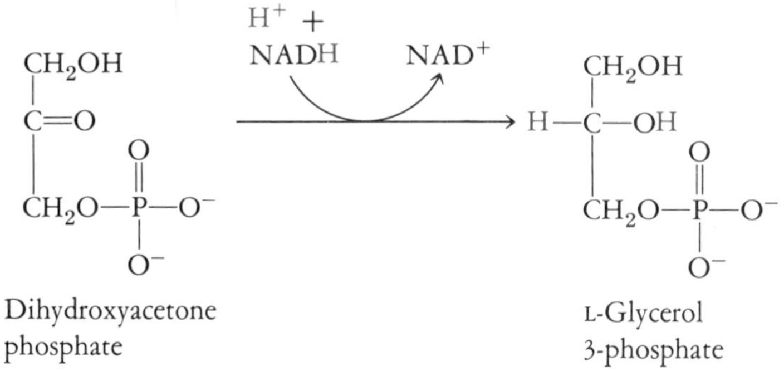

$$\underset{\text{Dihydroxyacetone phosphate}}{\begin{array}{l} CH_2OH \\ | \\ C{=}O \\ | \\ CH_2O{-}\overset{\overset{O}{\|}}{\underset{\underset{O^-}{|}}{P}}{-}O^- \end{array}} \xrightarrow[\quad]{H^+ + NADH \quad NAD^+} \underset{\text{L-Glycerol 3-phosphate}}{\begin{array}{l} \quad\; CH_2OH \\ \quad\;\; | \\ H{-}C{-}OH \\ \quad\;\; | \\ \quad\; CH_2O{-}\overset{\overset{O}{\|}}{\underset{\underset{O^-}{|}}{P}}{-}O^- \end{array}}$$

Fig. 28-8. Preparation of L-glycerol 3-phosphate from dihydroxyacetone phosphate.

from two different sources. One is from dihydroxyacetone phosphate, produced in the glycolysis of glucose (see Section 27.3), by means of the reaction given in Figure 28-8. This reaction is the reverse of the second step of the catabolism of glycerol given in Section 28.2. L-Glycerol 3-phosphate is also obtained by the reaction of glycerol, from the hydrolysis of triacylglycerols, and ATP. This is the first step of the catabolism of glycerol given in Section 28.2.

The sequence of reactions that form triacylglycerols from fatty acid coenzyme A and L-glycerol 3-phosphate is shown in Figure 28-9.

The first two reactions are acyl transfer reactions (see Section 18.10). The acyl groups are transferred from the sulfur of coenzyme A to the hydroxy groups of L-glycerol 3-phosphate to form phosphatidic acid. Hydrolysis occurs in the next step to form a diacylglycerol that reacts with a third fatty acid coenzyme A to form a triacylglycerol in the final step. In the fats of animals and plant tissues, the triacylglycerols usually contain two or more different fatty acids. It is not known what determines the identity of the fatty acids of a particular triacylglycerol.

We have learned that acetyl coenzyme A is the starting material for the synthesis of fatty acids. Acetyl coenzyme A is also the starting point for the formation of ketone bodies.

28.8 KETONE BODIES

A large part of the acetyl coenzyme A in cells enters into the citric acid cycle by condensing with oxaloacetate to form citrate (see Section 27.5). During certain illnesses such as diabetes or during fasting, the catabolism of fats is

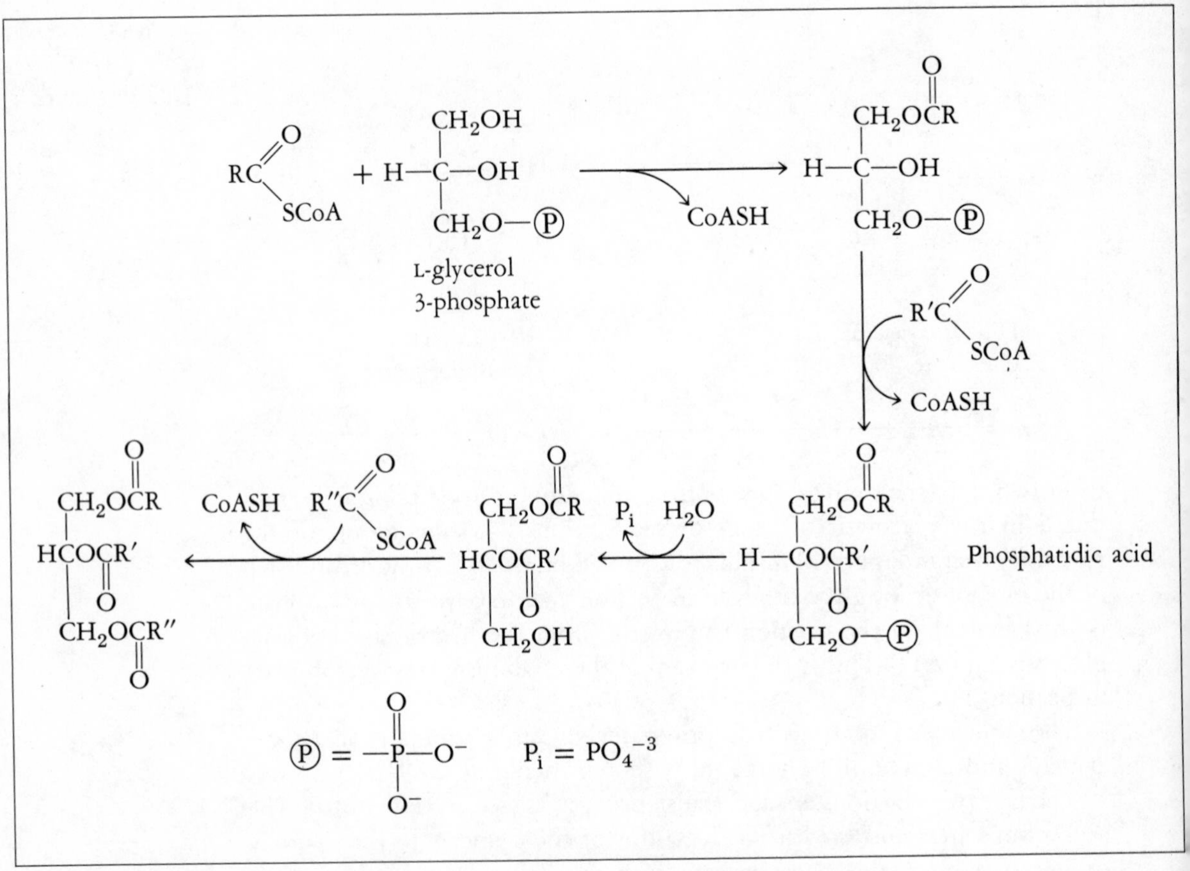

$$RC(=O)SCoA + \text{L-glycerol 3-phosphate} \longrightarrow CH_2OC(=O)R{-}CH(OH){-}CH_2O{-}Ⓟ + CoASH$$

L-glycerol
3-phosphate

$R'C(=O)SCoA$ → CoASH

Phosphatidic acid

P_i H_2O

CoASH $R''C(=O)SCoA$

$$Ⓟ = -P(=O)(O^-)-O^- \qquad P_i = PO_4^{-3}$$

Fig. 28-9. The reactions that form triacylglycerols from fatty acid coenzyme A and L-glycerol 3-phosphate.

greater than the catabolism of carbohydrates. When this happens, there is not enough oxaloacetate to react with acetyl coenzyme A. Then acetyl coenzyme A reacts to form acetoacetate and D-3-hydroxybutanoate by the pathway given in Figure 28-10.

The first reaction is a Claisen-type condensation reaction to form acetoacetyl coenzyme A. This reaction was discussed in Section 19.9. This reaction is also the reverse of reaction △4 in Figure 28-1. The second reaction is an aldol-type condensation reaction (see Section 19.8) between a third molecule of acetyl coenzyme A and acetoacetyl coenzyme A to form 3-hydroxy-3-methylglutaryl coenzyme A. This compound can undergo several reactions. One pathway leads to the formation of terpenes and steroids. The reaction that interests us and that leads to the formation of acetoacetate is a retroaldol

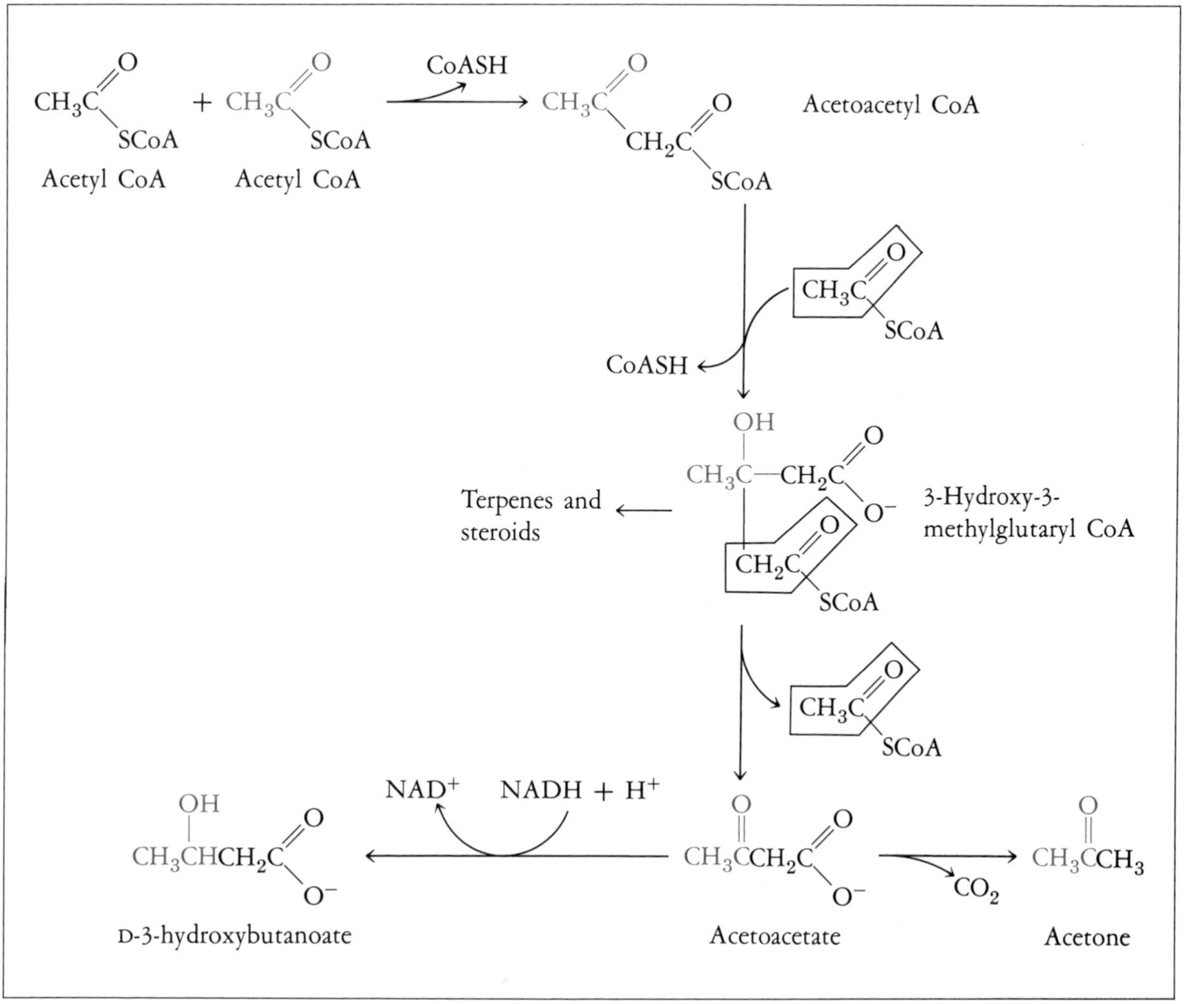

Fig. 28-10. *The sequence of reactions that form acetoacetate, acetone, and D-3-hydroxybutanoate from acetyl coenzyme A.*

condensation. Notice again that these reactions are examples in living organisms of well-known condensation reactions used by chemists in the laboratory.

The acetoacetate formed by the reactions in Figure 28-10 undergoes two further reactions. The first is an enzyme-catalyzed reduction to D-3-hydroxybutanoate. Because acetoacetate contains a β-ketocarboxylate group, it can be decarboxylated to form acetone (see Section 20.2). These three compounds, acetoacetate, D-3-hydroxybutanoate, and acetone, are called *ketone bodies.* Notice, however, that D-3-hydroxybutanoate is *not* a ketone. We will learn in Chapter 30 how the formation of these compounds affects living organisms.

28.9 SUMMARY

Fats are the major energy reserve of humans. Fat consists mostly of triacylglycerols. When digested, triacylglycerols are hydrolyzed to fatty acids and glycerol. The glycerol is converted to dihydroxyacetone phosphate, which enters the glycolytic pathway. The first step of the catabolism of fatty acids is a reaction that forms an acyl coenzyme A. The acyl coenzyme A enters a cycle of four reactions that removes two carbons from the fatty acid chain as acetyl coenzyme A. This cycle is repeated until the entire fatty acid chain is converted into acetyl coenzyme A. Much of the energy of the fatty acid is transferred to ATP as a result of these reactions.

Acetyl coenzyme A is an important intermediate in the metabolism of carbohydrates and fatty acids. It has two roles. It is oxidized to form energy, and it is the starting material for the synthesis of fatty acids, amino acids, terpenes, steroids, and ketone bodies.

The anabolism and catabolism of fatty acids occur by two different pathways. Although these pathways are different, they do have certain similarities. In particular, both pathways involve adding or removing a two-carbon portion in each sequence of four reactions. In the anabolism, a cycle of four reactions is involved that builds a chain up to 16 carbon atoms, at which point fatty acid synthesis stops.

If the amount of acetyl coenzyme A builds up in the cell, another way that it can be removed is by forming ketone bodies. They are formed by a series of reactions in which two molecules of acetyl coenzyme A react to form acetoacetate, which reacts further to form D-3-hydroxybutanoate and acetone. This series of reactions becomes important when the catabolism of fats is greater than that of carbohydrates. This occurs during starvation and in certain illnesses such as diabetes.

REVIEW OF TERMS, CONCEPTS, AND REACTIONS

Terms

ACP An acyl carrier protein. A polypeptide chain of 77 amino acid residues that serves to anchor acyl groups to the enzyme complex.

ACYL COA A thiol ester made up of a fatty acid and coenzyme A.

KETONE BODIES The three compounds, acetoacetate, D-3-hydroxybutanoate, and acetone, formed by a series of reactions starting from acetyl coenzyme A.

Concepts

1. Catabolism of fatty acids occurs by a sequence of four reactions (oxidation, hydration, oxidation, and cleavage) that removes two carbons from the chain as acetyl coenzyme A. This sequence of four reactions is repeated until the entire fatty acid chain is converted to acetyl coenzyme A.
2. Acetyl coenzyme A has two roles in metabolism. It is oxidized to generate energy, and it is the starting material for the synthesis of many compounds.
3. Catabolism and anabolism of fatty acids occur by two different pathways.
4. The anabolism of fatty acids occurs by a sequence of four reactions (condensation, reduction, dehydration, and reduction) that adds two carbon atoms to the fatty acid chain.

Reactions

1. Hydrolysis of triacylglycerols (see Sections 22.3 and 28.1)
   $$\text{Triacylglycerol} \xrightarrow[\text{lipases}]{H_2O,\ NaOH\ \text{or}} \text{glycerol} + 3 \text{ fatty acids}$$
2. Phosphorylation (see Sections 16.4, 27.3, and 28.2)
   Glycerol + ATP ⟶ L-glycerol 3-phosphate + ADP
3. Oxidation (see Sections 16.4, 19.3, 26.8, 27.3, and 28.2)
   (a) L-Glycerol 3-phosphate + NAD^+ ⟶ dihydroxyacetone phosphate + NADH + H^+
   (b) $$RCH(OH)CH_2C(=O)SCoA + NAD^+ \longrightarrow RC(=O)CH_2C(=O)SCoA + NADH + H^+$$
4. Formation of mixed anhydride (see Sections 18.12 and 28.3)
   Fatty acid + ATP ⟶ acyl adenylate + 2 P_i
5. Acyl transfer (see Sections 18.10 and 28.3)
   (a) Acyl adenylate + CoASH ⟶ acyl CoA + AMP
   (b) Acyl CoA + ACPSH ⟶ acyl SACP + CoASH
   (c) $$\text{Acyl CoA} + \text{L-glycerol 3-phosphate} \longrightarrow H{-}C(CH_2OC(=O)R)(OH){-}CH_2O{-}P(=O)(O^-){-}O^- + \text{CoASH}$$

EXERCISES

28-8 Write the equations for three examples of acyl transfer reactions given in this chapter.

28-9 List the similarities and differences in the anabolic and catabolic pathways of fatty acids.

28-10 Write the equations for the reactions that convert glucose to triacylglycerol.

28-11 Why does a lack of carbohydrates in the diet of a human stimulate the formation of ketone bodies?

28-12 Write the equation for an example of each of the following types of reaction given in this chapter:
(a) dehydrogenation (b) aldol-type condensation
(c) hydration (d) hydrolysis (e) hydrogenation
(f) alcohol oxidation (g) Claisen-type condensation
(h) retroaldol-type condensation (i) dehydration
(j) reduction of a ketone

AMINO ACID METABOLISM

29

Amino acids are used in three ways by living organisms: (*1*) as the starting material for the synthesis of proteins; (*2*) as the source of carbon and nitrogen atoms for the synthesis of other compounds needed by the organism; and (*3*) as energy for the organism.

The most important of these functions is building proteins and polypeptides. About 75 percent of the amino acids obtained from food are used in this way. The proteins of cells must be continuously resynthesized to replace the proteins lost or damaged during the normal life of the organism. It is estimated that half of the proteins in vital organs such as heart, liver, and kidneys are replaced in this way. Enzymes are replaced more quickly, and half of the muscle protein is replaced about every 180 days. In a normal healthy adult weighing 70 kg (154 lb), about 450 g (1 lb) of protein is replaced every day. Anywhere from 6 to 20 g (0.01 to 0.04 lb) of nitrogen is excreted daily in the form of nitrogen-containing compounds, principally urea.

Under normal conditions, a person's intake of nitrogen in food is equal to the nitrogen lost by excretion; that is, there is a *nitrogen balance.* Sometimes an organism can have a *positive nitrogen balance.* In this case, nitrogen intake is greater than loss of nitrogen. This occurs whenever tissue is being synthesized, for example, during periods of growth. A *negative nitrogen balance* (nitrogen intake less than loss of nitrogen) occurs during fasting or with a diet that lacks or is deficient in protein.

Certain organisms are capable of synthesizing all 20 amino acids from water, carbon dioxide, and nitrates from the soil. Humans and higher ani-

Table 29-1. Essential and Nonessential Amino Acids

Essential	Nonessential
Arginine	Alanine
Histidine	Asparagine
Isoleucine	Aspartate
Leucine	Cysteine
Lysine	Glutamate
Methionine	Glutamine
Phenylalanine	Glycine
Threonine	Proline
Tryptophan	Serine
Valine	Tyrosine

mals are not such organisms, because they can synthesize only about half of the amino acids they need. The amino acids that cannot be synthesized by humans are called *essential amino acids* and they are listed in Table 29-1.

Amino acids provide a source of carbon and nitrogen atoms that are used in the synthesis of nonprotein nitrogen-containing molecules. For example, amino sugars (see Section 21.2), the porphyrin ring of hemoglobin (see Section 13.6), the choline and ethanolamine parts of phosphoglycerides (see Section 22.4), and many others are synthesized using the carbon and nitrogen atoms of amino acids.

An organism with a positive nitrogen balance cannot store the excess amino acids. In this way, amino acids differ from carbohydrates and fats. The carbon skeletons of the excess amino acids are degraded to one of the intermediates of the citric acid cycle. Thus, they enter the pool of compounds that can be either directly oxidized to provide energy or converted to fat or glycogen for storage. Oxidation of the carbon skeleton of amino acids either directly or indirectly from fat and glycogen supplies about 20 percent of the total energy needed by an adult.

Although the carbon skeleton of an amino acid can be degraded and stored in the body, there is no such permanent storage for the nitrogen atoms obtained from amino acids. Most of the nitrogen atoms not needed immediately by the organism are converted to urea and excreted in the urine.

In this chapter, we will trace the fate of the carbon skeletons of amino acids. We will also learn how α-amino groups are lost and what happens to them. Finally, we will examine briefly how certain amino acids are synthesized in living organisms. Let us start by learning how amino acids are formed by the digestion of proteins.

29.1 DIGESTION OF PROTEINS

The digestion of proteins begins in the stomach. Here the hydrolysis of proteins is started by the gastric juices. The high acidity of the gastric juices denatures the proteins and makes it easier for the *proteolytic* (protein-cleaving) enzymes to hydrolyze them. About 10 percent of the peptide bonds are hydrolyzed at this stage.

The major proteolytic enzyme in the stomach is pepsin. It enters the stomach as the zymogen pepsinogen (see Section 25.7). Pepsinogen is hydrolyzed to pepsin by stomach acid. Once some pepsin is formed, it catalyzes the conversion of pepsinogen to more pepsin. *This is an example of autocatalysis.* That is, the product of a chemical reaction catalyzes its formation. Pepsin catalyzes the hydrolysis of peptide bonds within the protein molecule. Few free amino acids are formed at this stage. Instead, the proteins are broken into polypeptides.

Protein digestion continues in the small intestine. Here the digestive juices of the pancreas contain the zymogens chymotrypsinogen, trypsinogen, and procarboxypeptidases A and B. Trypsinogen is converted to active trypsin by the enzyme enterokinase. Trypsin then converts chymotrypsinogen to chymotrypsin. Once formed, these enzymes catalyze the hydrolysis of the polypeptides. Each enzyme has a sufficiently broad range of specificity that the polypeptides are converted into short peptide chains and free amino acids. The remaining peptide chains are hydrolyzed to free amino acids by enzymes called peptidases.

The hydrolysis of proteins to free amino acids is the result of the combined action of the proteolytic enzymes in the gastrointestinal tract. Once formed, the amino acids are absorbed into the blood and carried to the liver. Here the metabolism of amino acids takes place.

Unlike carbohydrates and fatty acids, amino acids are not catabolized through just one pathway. Each amino acid has a different sequence of reactions for its degradation. Despite the complex nature of the catabolism, there is one common feature: all amino acids are degraded into two parts. One part contains the α-amino groups. The other part contains compounds that have the carbon skeleton of the original amino acids. The nitrogen-containing compounds are finally degraded to urea, and the carbon skeletons are finally degraded to carbon dioxide and water. These paths are outlined in Figure 29-1.

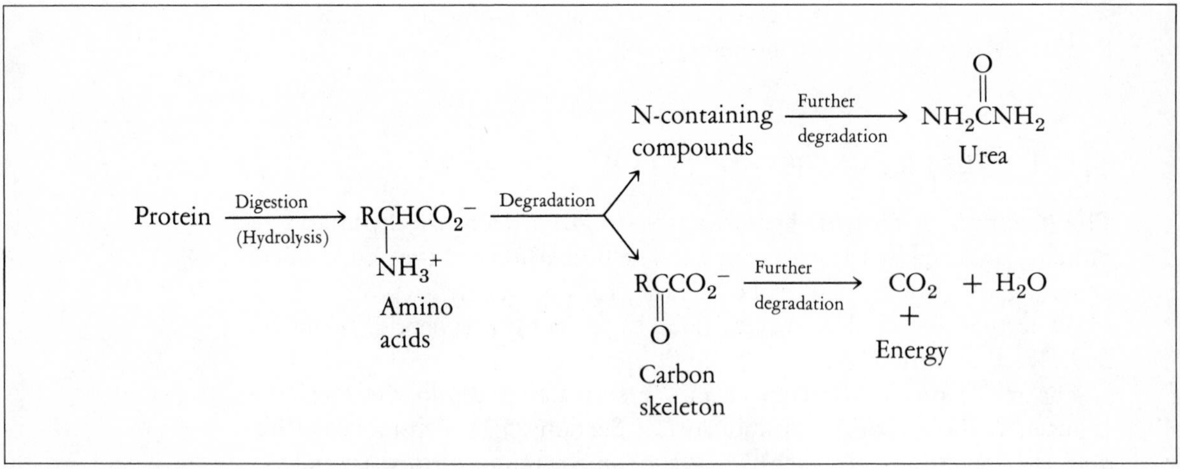

$$\text{Protein} \xrightarrow[\text{(Hydrolysis)}]{\text{Digestion}} \underset{\substack{\text{Amino}\\ \text{acids}}}{\underset{|}{\underset{NH_3^+}{RCHCO_2^-}}} \xrightarrow{\text{Degradation}} \begin{cases} \text{N-containing compounds} \xrightarrow[\text{degradation}]{\text{Further}} \underset{\text{Urea}}{NH_2\overset{\overset{O}{\|}}{C}NH_2} \\ \underset{\text{Carbon skeleton}}{R\underset{\underset{O}{\|}}{C}CO_2^-} \xrightarrow[\text{degradation}]{\text{Further}} CO_2 + H_2O + \text{Energy} \end{cases}$$

Fig. 29-1. The catabolism of amino acids involves separating the α-amino group from the carbon skeleton. Once this happens, the parts containing nitrogen and carbon are catabolized by separate paths.

It turns out that the carbon-containing parts of the amino acids are degraded to only five compounds that are part of the citric acid cycle.

29.2 THE FATES OF THE CARBON PORTIONS OF THE AMINO ACIDS

There are 20 different multienzyme-catalyzed reaction sequences for the catabolism of the carbon parts of the 20 different amino acids found in proteins in humans and higher animals. However, all of these pathways finally lead to only five compounds. All of them are part of the metabolic pathways that we learned in previous chapters. These five products, acetyl coenzyme A, oxaloacetate, fumarate, succinyl coenzyme A, and α-ketoglutarate, and the amino acids from which they are formed are shown in Figure 29-2.

The carbon atoms of 11 amino acids eventually form acetyl coenzyme A, either directly or through pyruvate or acetoacetyl coenzyme A. Two amino acids are degraded to oxaloacetate, three are degraded to succinyl coenzyme A, and five are degraded to α-ketoglutarate. Three amino acids, phenylalanine, tyrosine, and isoleucine, are degraded so that one part of the carbon skeleton forms acetyl coenzyme A and the other part forms either succinyl

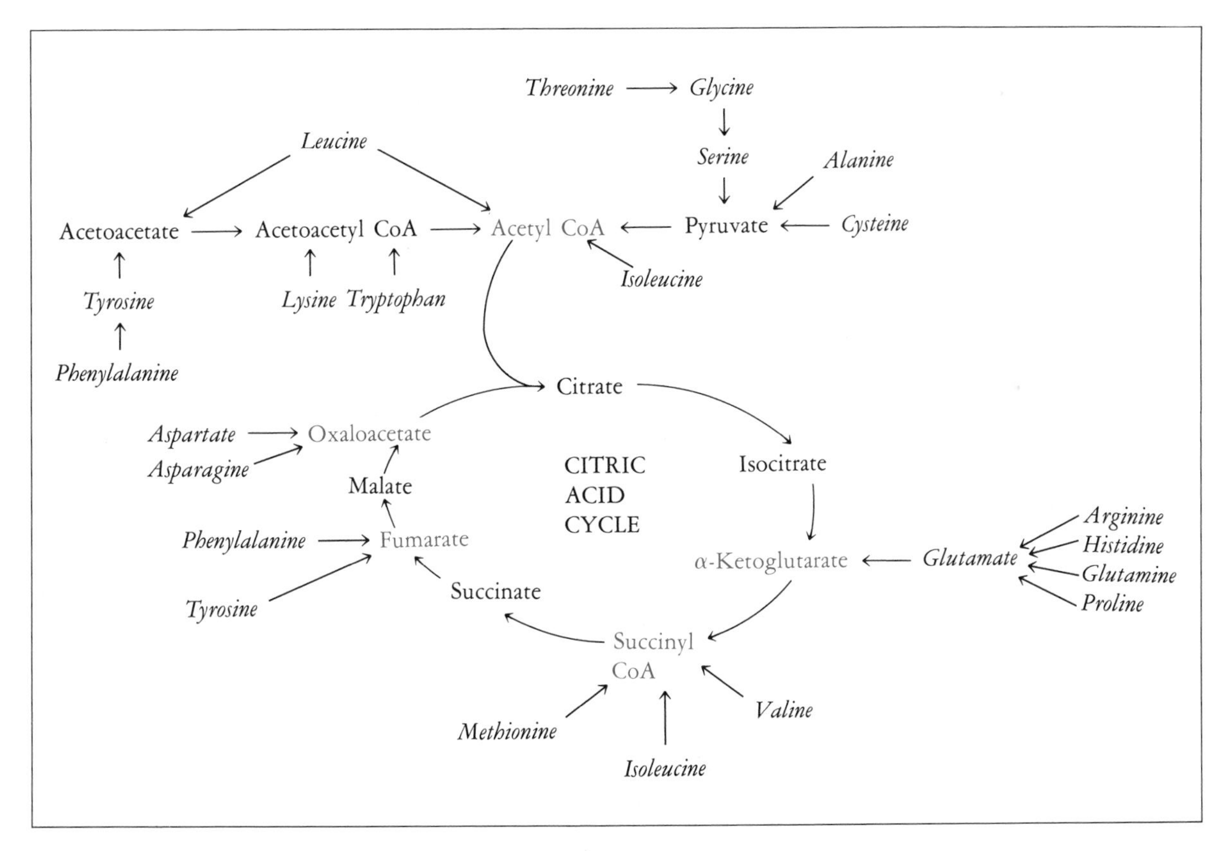

Fig. 29-2. The five products of the degradation of the carbon part of amino acids.

coenzyme A (for isoleucine) or fumarate (for phenylalanine and tyrosine). Not all the carbon atoms of each of the 20 amino acids end up in these five compounds. Some are lost as carbon dioxide in decarboxylation reactions.

The catabolic pathways of amino acids are usually long and complex and contain numerous intermediates. Many of these intermediates have additional functions in the cell. They are often essential to the synthesis of other cell components. In this way, the organism makes use of the carbon atoms in the skeleton of a particular amino acid for its own purposes.

*EXERCISE 29-1 List the 11 amino acids that are eventually degraded to acetyl coenzyme A.

*The answers for the exercises in this chapter begin on page 904.

To form the five compounds given in Figure 29-2, the α-amino group of each amino acid must be removed. The reactions that do this are the same in most organisms. We will learn these reactions in the next two sections.

29.3 TRANSAMINATION: REMOVAL OF α-AMINO GROUPS

Transamination is a reaction that enzymatically removes the α-amino group in the catabolism of at least 12 amino acids (alanine, arginine, asparagine, aspartate, cysteine, isoleucine, leucine, lysine, phenylalanine, tryptophan, tyrosine, and valine). In this reaction, the α-amino group of the amino acid is transferred to the α carbon of α-ketoglutarate according to the following equation:

$$\underset{\text{Amino acid}}{RCH(NH_2)CO_2^-} + \underset{\alpha\text{-Ketoglutarate}}{{}^-O_2CC(=O)CH_2CH_2CO_2^-} \longrightarrow \underset{\text{L-Glutamate}}{{}^-O_2CCH(NH_2)CH_2CH_2CO_2^-} + \underset{\alpha\text{-Ketocarboxylate}}{RC(=O)CO_2^-}$$

The products of this reaction are L-glutamate and an α-ketocarboxylate, with the same carbon skeleton as the original amino acid.

Transamination is catalyzed by enzymes called transaminases, a large number of which are known. Most are specific for both α-ketoglutarate and the amino acid. For example, the reactions given in Figure 29-3 are each catalyzed by a specific enzyme.

Transamination appears to be a new chemical reaction. However, the equations given in Figure 29-3 are the net reactions. That is, the reaction actually involves several steps, and their sum total is given in the net equation. If we examine each of these steps, we will find that each is a reaction we have learned. We can make some generalizations about these reactions. First, all of the transaminases have a coenzyme. In fact, the coenzyme is the same for all transaminases; it is pyridoxal phosphate, which has the following structure:

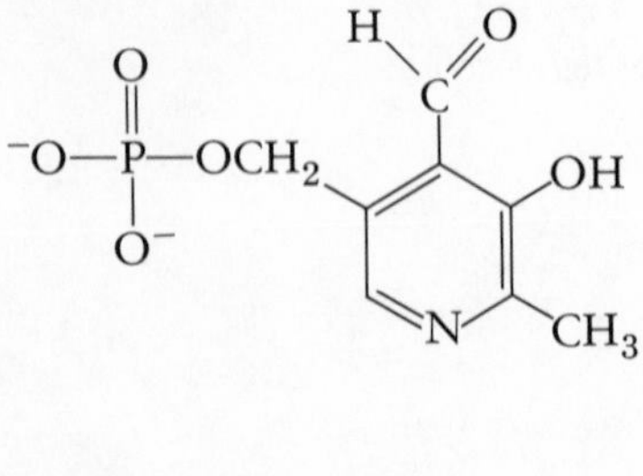

$$^-O-P(=O)(O^-)-OCH_2-\text{(pyridine ring bearing } -CHO\text{, } -OH\text{, } -CH_3\text{; ring N)}$$

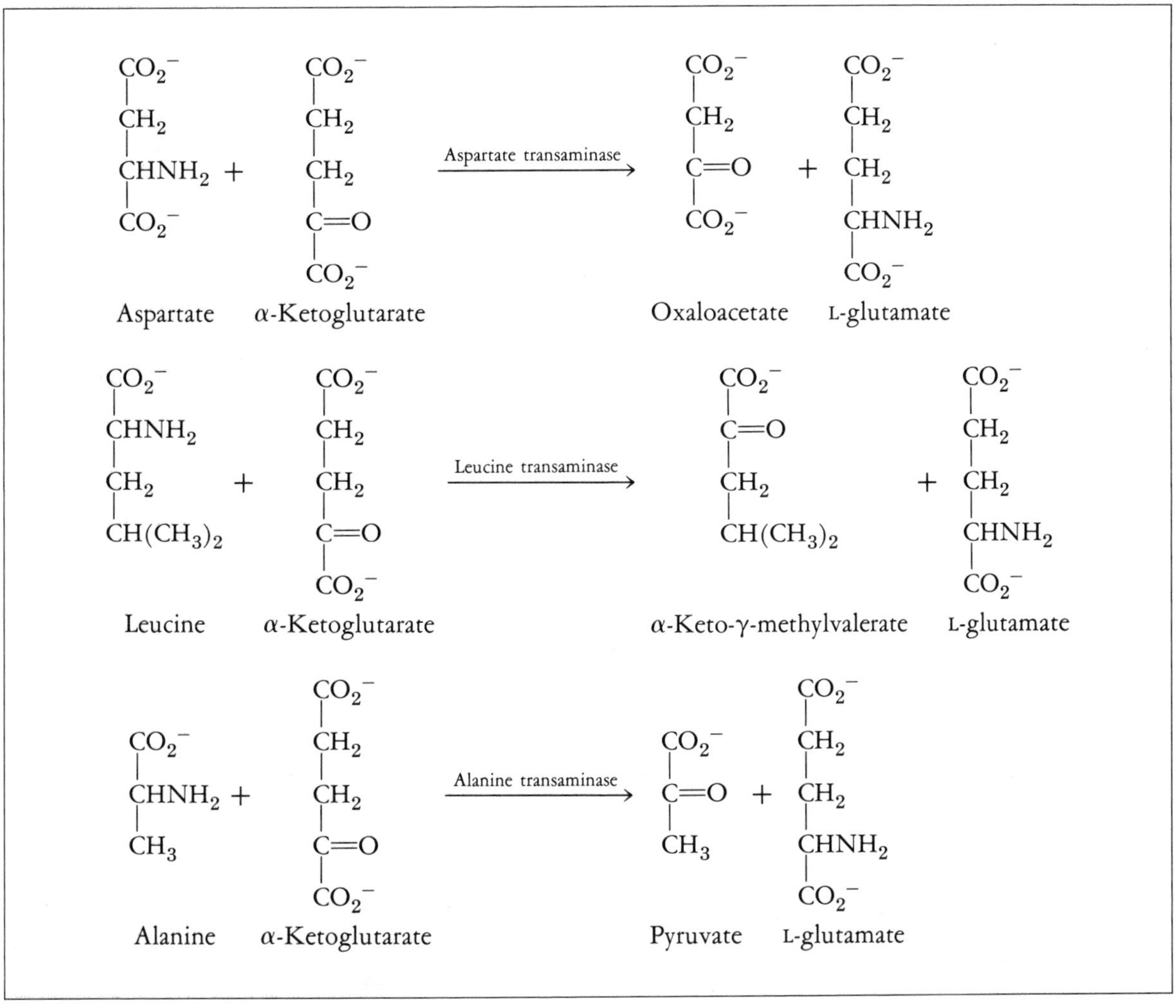

$$\underset{\text{Aspartate}}{CO_2^- - CH_2 - CHNH_2 - CO_2^-} + \underset{\alpha\text{-Ketoglutarate}}{CO_2^- - CH_2 - CH_2 - C{=}O - CO_2^-} \xrightarrow{\text{Aspartate transaminase}} \underset{\text{Oxaloacetate}}{CO_2^- - CH_2 - C{=}O - CO_2^-} + \underset{\text{L-glutamate}}{CO_2^- - CH_2 - CH_2 - CHNH_2 - CO_2^-}$$

$$\underset{\text{Leucine}}{CO_2^- - CHNH_2 - CH_2 - CH(CH_3)_2} + \underset{\alpha\text{-Ketoglutarate}}{CO_2^- - CH_2 - CH_2 - C{=}O - CO_2^-} \xrightarrow{\text{Leucine transaminase}} \underset{\alpha\text{-Keto-}\gamma\text{-methylvalerate}}{CO_2^- - C{=}O - CH_2 - CH(CH_3)_2} + \underset{\text{L-glutamate}}{CO_2^- - CH_2 - CH_2 - CHNH_2 - CO_2^-}$$

$$\underset{\text{Alanine}}{CO_2^- - CHNH_2 - CH_3} + \underset{\alpha\text{-Ketoglutarate}}{CO_2^- - CH_2 - CH_2 - C{=}O - CO_2^-} \xrightarrow{\text{Alanine transaminase}} \underset{\text{Pyruvate}}{CO_2^- - C{=}O - CH_3} + \underset{\text{L-glutamate}}{CO_2^- - CH_2 - CH_2 - CHNH_2 - CO_2^-}$$

Fig. 29-3. Specific transamination reactions.

Pyridoxal phosphate is synthesized from vitamin B_6 (pyridoxine), one of the water soluble vitamins (see Section 25.3). Pyridoxal phosphate has the key role in all transaminase-catalyzed reactions. Consequently, the pathways of all the transaminases are the same, and we can show this pathway using only one amino acid as an example. Alanine is our example, and the reaction sequence for its transamination is shown in Figure 29-4.

The pyridoxal phosphate is tightly bound to the enzyme. Its role in the reaction is to act as an amino group carrier. In the first step in Figure 29-4, pyridoxal phosphate forms a Schiff base with the unprotonated amino

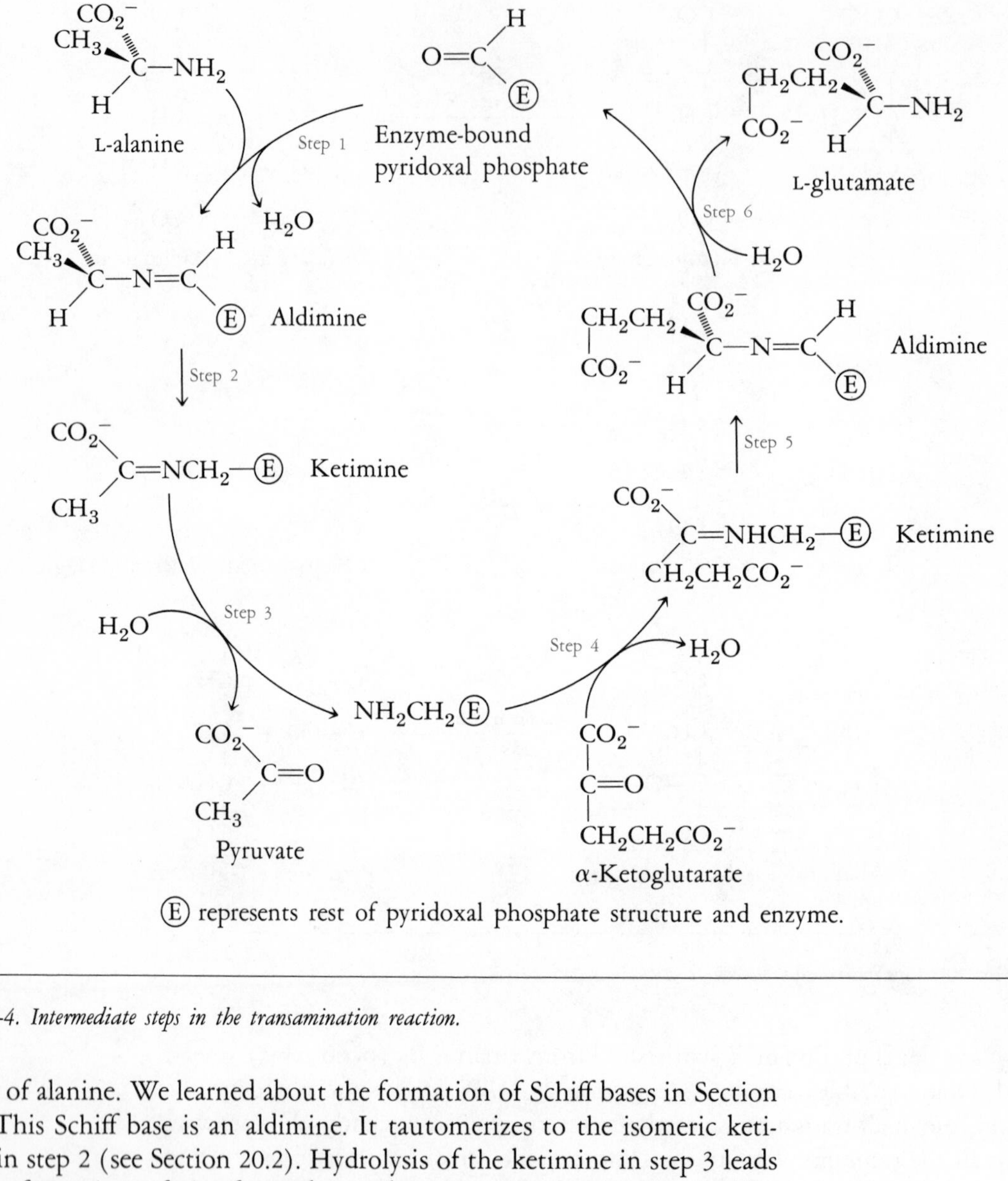

Ⓔ represents rest of pyridoxal phosphate structure and enzyme.

Fig. 29-4. Intermediate steps in the transamination reaction.

group of alanine. We learned about the formation of Schiff bases in Section 19.5. This Schiff base is an aldimine. It tautomerizes to the isomeric ketimine in step 2 (see Section 20.2). Hydrolysis of the ketimine in step 3 leads to the formation of the free α-ketocarboxylate (see Section 19.5). The pyridoxamine phosphate enzyme complex then forms a Schiff base with α-ketoglutarate. This starts the reverse of steps 1, 2, and 3 to form L-glutamate and pyridoxal phosphate. Once the cycle is completed, the enzyme complex is ready to start the cycle over again.

The conversion of the coenzyme from its aldehyde to its amino form and back again occurs by a series of chemical reactions that we have learned in previous chapters. All are known to occur in the laboratory. By means of these reactions, pyridoxal phosphate acts as a carrier of an amino group from an amino acid to α-ketoglutarate.

Notice that this reaction of the catabolism of 12 amino acids forms one common α-amino acid, L-glutamate. In this way, all the amino groups are collected into one compound that enters a series of reactions that eventually converts the amino group into urea, ammonia, or uric acid. We will learn the first reaction in this series in the next section.

29.4 OXIDATIVE DEAMINATION

The amino group of L-glutamate is converted into ammonium ions by the following oxidative deamination reaction:

$$\underset{\text{L-Glutamate}}{\begin{matrix} CO_2^- \\ | \\ CHNH_2 \\ | \\ CH_2 \\ | \\ CH_2CO_2^- \end{matrix}} + \begin{matrix} NAD^+ \\ (NADP^+) \end{matrix} + H_2O \longrightarrow \overset{+}{N}H_4 + \underset{\alpha\text{-Ketoglutarate}}{\begin{matrix} O \diagdown \quad \diagup CO_2^- \\ C \\ | \\ CH_2 \\ | \\ CH_2CO_2^- \end{matrix}} + \begin{matrix} NADH \\ (NADPH) \end{matrix}$$

This oxidization requires either nicotinamide adenine dinucleotide (NAD^+) or nicotinamide adenine dinucleotide phosphate ($NADP^+$) and is catalyzed by the enzyme glutamate dehydrogenase. The other product formed in the reaction, α-ketoglutarate, can undergo transamination with any of several amino acids to form more L-glutamate. Thus, a cycle is set up as shown in Figure 29-5.

Transamination transfers the amino group of an amino acid to α-ketoglutarate to form L-glutamate. Oxidative deamination of L-glutamate converts its α-amino group into ammonium ion and reforms α-ketoglutarate, which can start the cycle again. In this way, α-ketoglutarate and L-glutamate convert the amino groups of α-amino acids into ammonium ions.

Ammonium ions are extremely toxic to humans and higher animals. Therefore, they must be removed from the organism. The major pathway for this is the urea cycle.

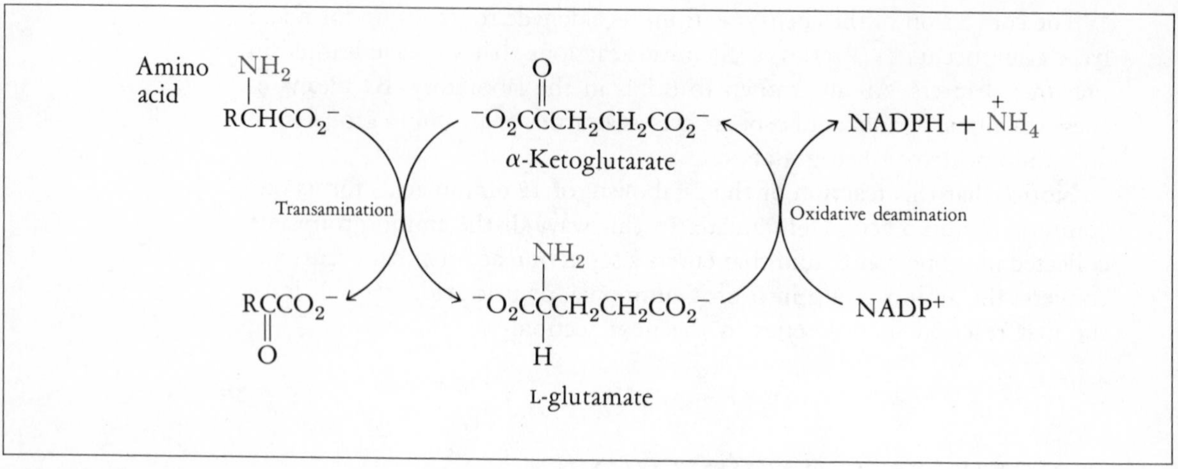

$$\text{Amino acid } RCH(NH_2)CO_2^- + {}^-O_2CC(=O)CH_2CH_2CO_2^- \ (\alpha\text{-Ketoglutarate}) \xrightarrow{\text{Transamination}} RC(=O)CO_2^- + {}^-O_2CCH(NH_2)CH_2CH_2CO_2^- \ (\text{L-glutamate})$$

$$\text{L-glutamate} + NADP^+ \xrightarrow{\text{Oxidative deamination}} \alpha\text{-Ketoglutarate} + NADPH + NH_4^+$$

Fig. 29-5. The role of α-ketoglutarate and L-glutamate in converting the α-amino groups of amino acids into NH_4^+.

29.5 UREA CYCLE

The ammonium ions formed by the oxidative deamination of L-glutamate are converted into urea by means of a series of five reactions called the urea cycle. These reactions convert a toxic ion into a neutral, nontoxic compound that is transported by the blood from the liver to the kidneys, where it is excreted in the urine. The sequence of reactions in the urea cycle is shown in Figure 29-6. Let us examine each of these reactions in more detail. As in previous chapters, the numbers given the reactions in the text correspond to those given in Figure 29-6.

Reaction ▽1 is the formation of carbamoyl phosphate from carbon dioxide, ammonium ion, adenosine triphosphate (ATP), and water in a reaction that is catalyzed by the enzyme carbamoyl phosphate synthetase. This reaction can be viewed as taking place in two steps. The first step is the addition of ammonia to one carbon-oxygen double bond of carbon dioxide (see Section 19.6), as follows:

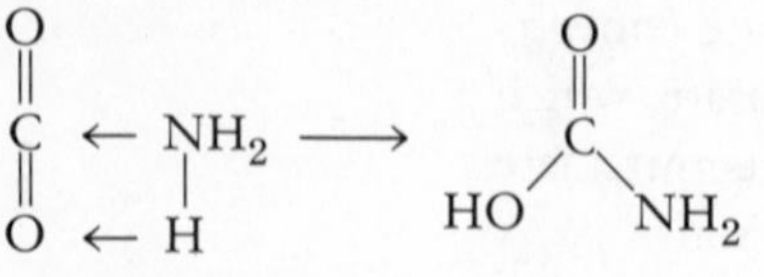

$$O{=}C{=}O + H{-}NH_2 \longrightarrow HO{-}C({=}O){-}NH_2$$

The second step is the formation of an acyl phosphate, a mixed anhydride (see Section 18.12), and adenosine diphosphate (ADP):

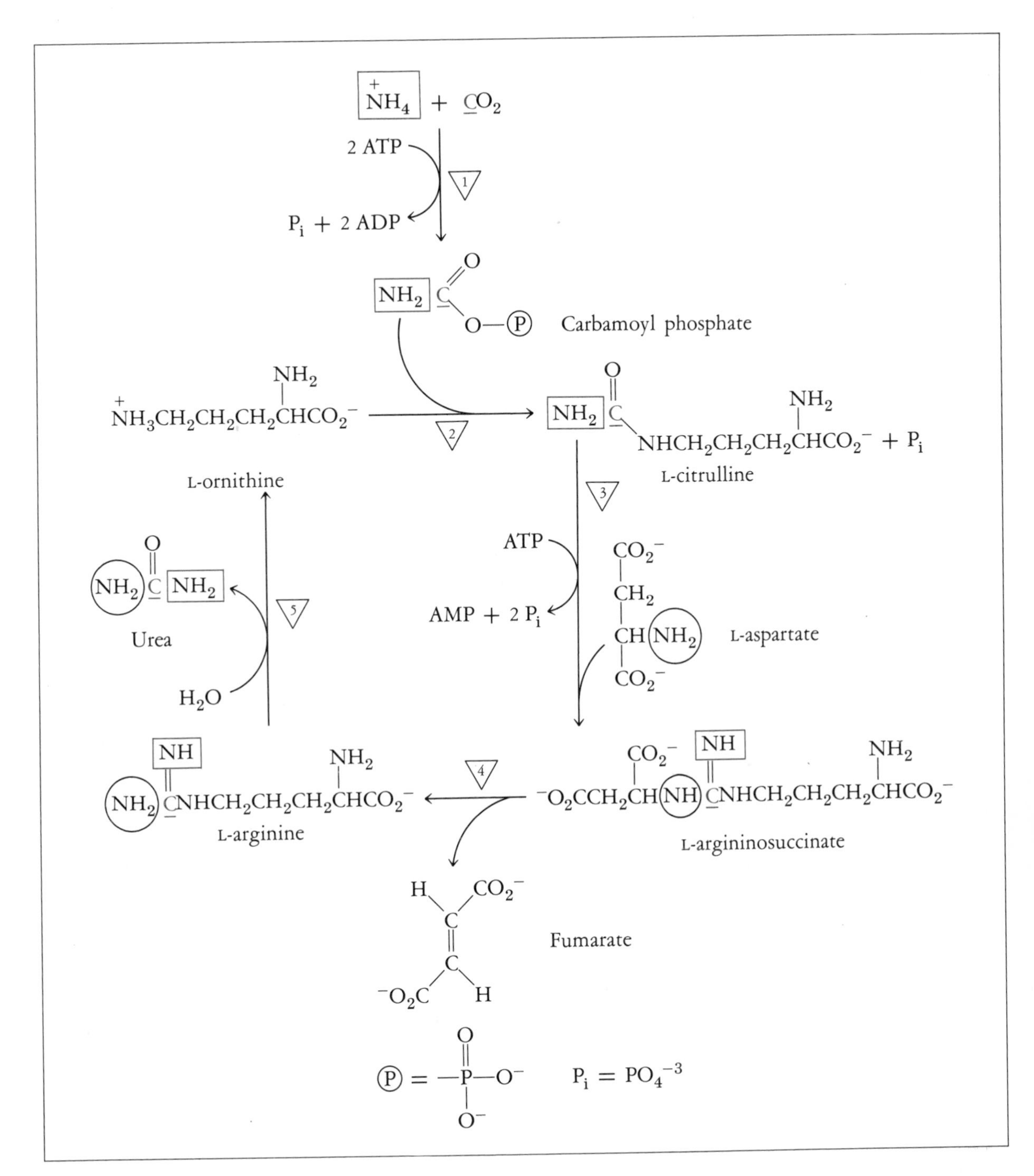

Fig. 29-6. The urea cycle. The reaction numbers correspond to those given in the text.

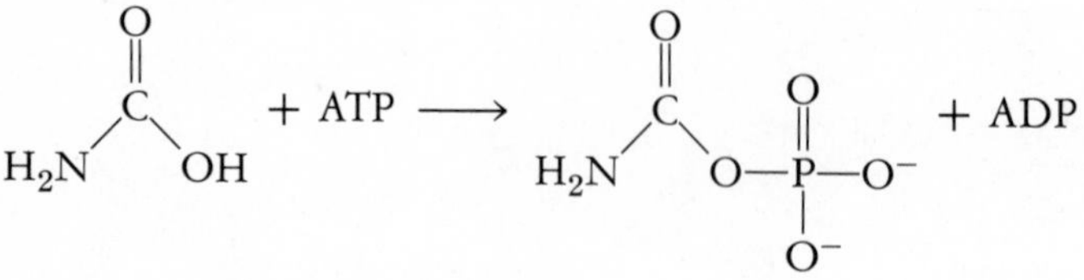

$$H_2N{-}C({=}O){-}OH + ATP \longrightarrow H_2N{-}C({=}O){-}O{-}P({=}O)(O^-){-}O^- + ADP$$

Thus, this complex-looking reaction can be understood in terms of reactions that we have already learned.

In reaction ▽2, citrulline is formed from ornithine and carbamoyl phosphate. This reaction is catalyzed by the enzyme ornithine transcarbamoylase. In this reaction, the amino group on carbon 5 of ornithine replaces the phosphate group of carbamoyl phosphate. This is a typical reaction of anhydrides (see Section 18.9). Another way of looking at this reaction is that a carbamoyl group is transferred to ornithine to form citrulline, as shown in Figure 29-7. This is similar to an acyl transfer reaction (see Section 18.10).

In reaction ▽3, aspartate provides the second amino group needed for forming urea. Aspartate gets this amino group from L-glutamate by means of transamination. The amino group of aspartate adds to the carbamoyl carbon of L-citrulline to form L-argininosuccinate. This reaction requires ATP and is catalyzed by argininosuccinate synthetase.

Reaction ▽4 is an elimination reaction (see Section 15.9) to form fumarate and L-arginine. We can imagine that the reaction occurs as follows:

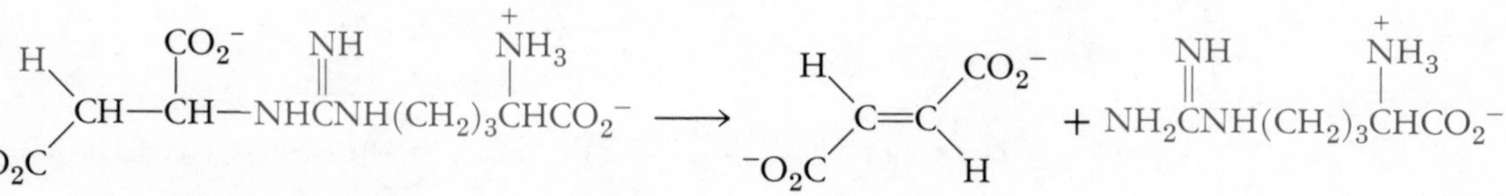

$$\underset{\text{L-Argininosuccinate}}{{}^-O_2C(H)CH{-}CH(CO_2^-){-}NHC({=}NH)NH(CH_2)_3CH(\overset{+}{N}H_3)CO_2^-} \longrightarrow \underset{\text{Fumarate}}{{}^-O_2C(H)C{=}C(H)CO_2^-} + \underset{\text{L-Arginine}}{NH_2C({=}NH)NH(CH_2)_3CH(\overset{+}{N}H_3)CO_2^-}$$

Finally, reaction ▽5 is the hydrolysis of L-arginine to form urea and L-ornithine, which begins the cycle again. This reaction is catalyzed by the enzyme arginase.

Let us summarize the reactions in the urea cycle. One of the nitrogen atoms of the urea is obtained from ammonium ion, and the second is obtained from aspartate. Carbon dioxide furnishes the carbon atom of urea. The sources of these atoms and their location in each compound of the cycle are indicated in Figure 29-6. We are familiar with the chemical reactions involved in the urea cycle:

1. Acyl-type transfer reaction. Reaction ▽2.
2. Carbonyl addition reaction. Reaction ▽3.

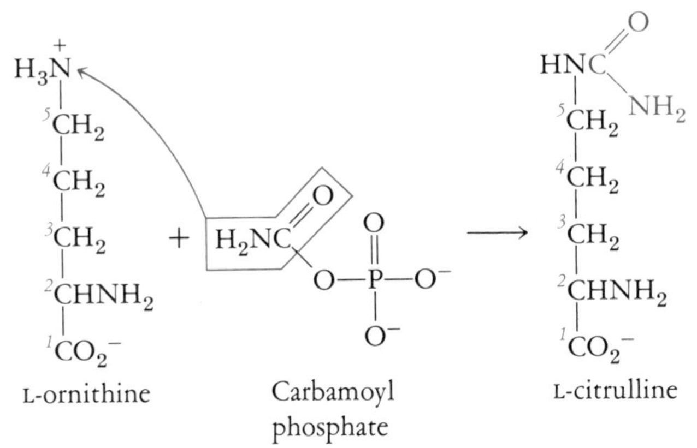

Fig. 29-7. *The transfer of a carbamoyl group from carbamoyl phosphate to ornithine to form citrulline.*

3. Elimination reaction. Reaction 4.
4. Hydrolysis reaction. Reaction 5.

In addition, reaction 1 can be viewed as a carbonyl addition reaction followed by the formation of a mixed anhydride. In terms of energy, four molecules of ATP are spent to form one molecule of urea. Thus, forming urea from water, carbon dioxide, and nitrogen-containing ions and compounds is an endergonic process.

Not all of the nitrogen obtained from the amino groups of amino acids is converted to urea. Some is used in the anabolism of the nonessential amino acids.

29.6 ANABOLISM OF SEVERAL NONESSENTIAL AMINO ACIDS

Sources of nitrogen that can be used by living organisms are relatively scarce in nonliving environments. Consequently, most living organisms are economical in the use of nitrogen-containing compounds formed by catabolism of amino acids. These compounds are often reused to synthesize the nonessential amino acids and other nitrogen compounds needed by the organism.

The 20 amino acids are synthesized by 20 different multienzyme-catalyzed reaction sequences. As in the case of carbohydrates and fats, the pathways of amino acid anabolism are, for the most part, different from those used in their catabolism. We shall consider only the anabolism of several nonessential amino acids. There are three reasons for this. First, they are made from simple compounds. Second, their synthetic pathways are relatively short. Third, the anabolism of essential amino acids does not occur in humans. Let us start with one of the amino acids that we have already encountered.

Glutamate

L-Glutamate plays an important role in the synthesis of many amino acids, as we will learn. L-Glutamate is the source of the α-amino group of many amino acids. It is formed from α-ketoglutarate by transamination, as we learned in Section 29.3.

Alanine and Aspartate

L-Alanine and L-aspartate are formed from the corresponding α-ketocarboxylic acids by transamination according to the reactions given in Figure 29-8. Remember that pyruvate is an intermediate in the glycolytic pathway and that oxaloacetate is a compound in the citric acid cycle. Notice that L-glutamate is the α-amino group donor. As discussed in Section 29.3, pyridoxal phosphate is the coenzyme in these transamination reactions.

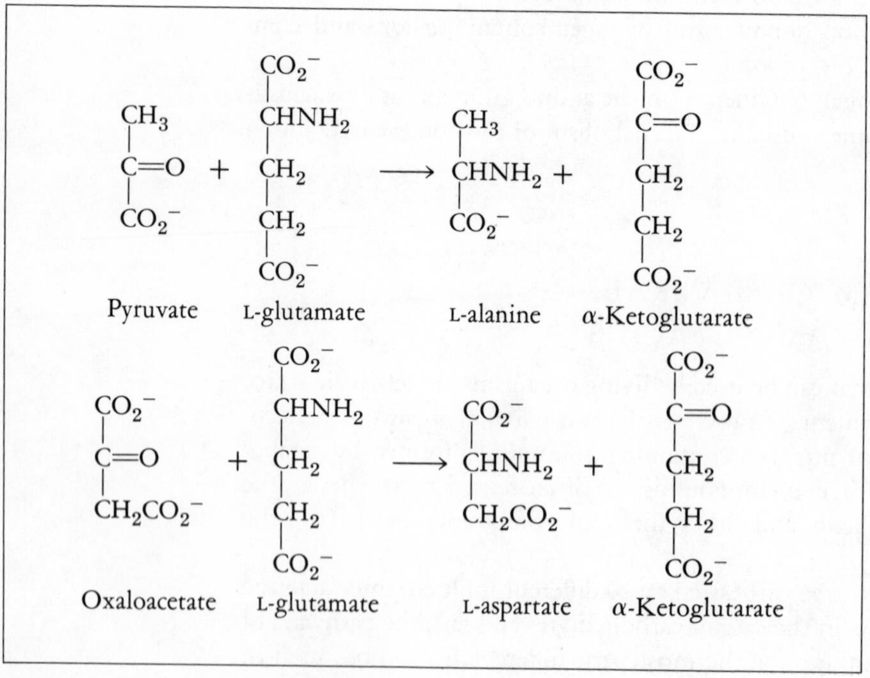

$$\begin{array}{ccccccc}
 & & CO_2^- & & & & CO_2^- \\
 & & | & & & & | \\
CH_3 & & CHNH_2 & & CH_3 & & C{=}O \\
| & & | & & | & & | \\
C{=}O & + & CH_2 & \longrightarrow & CHNH_2 & + & CH_2 \\
| & & | & & | & & | \\
CO_2^- & & CH_2 & & CO_2^- & & CH_2 \\
 & & | & & & & | \\
 & & CO_2^- & & & & CO_2^- \\
\text{Pyruvate} & & \text{L-glutamate} & & \text{L-alanine} & & \alpha\text{-Ketoglutarate}
\end{array}$$

$$\begin{array}{ccccccc}
 & & CO_2^- & & & & CO_2^- \\
 & & | & & & & | \\
CO_2^- & & CHNH_2 & & CO_2^- & & C{=}O \\
| & & | & & | & & | \\
C{=}O & + & CH_2 & \longrightarrow & CHNH_2 & + & CH_2 \\
| & & | & & | & & | \\
CH_2CO_2^- & & CH_2 & & CH_2CO_2^- & & CH_2 \\
 & & | & & & & | \\
 & & CO_2^- & & & & CO_2^- \\
\text{Oxaloacetate} & & \text{L-glutamate} & & \text{L-aspartate} & & \alpha\text{-Ketoglutarate}
\end{array}$$

Fig. 29-8. The formation of L-alanine and L-aspartate from the corresponding α-ketocarboxylic acids.

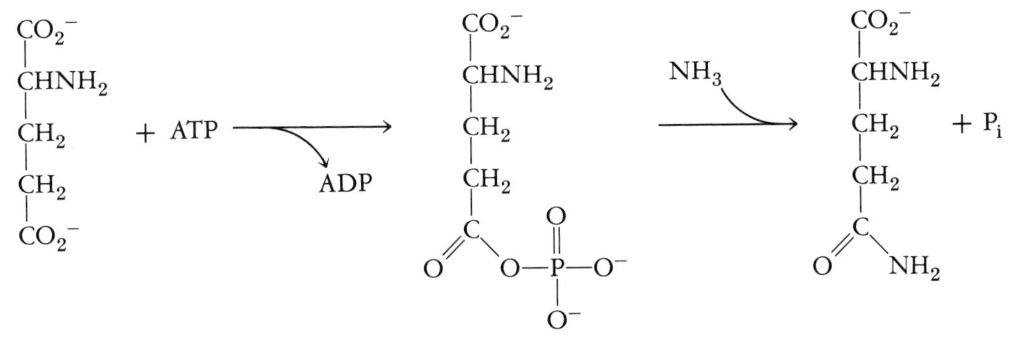

$$\underset{\text{L-glutamate}}{CO_2^- - CHNH_2 - CH_2 - CH_2 - CO_2^-} + ATP \xrightarrow{\quad} ADP + \underset{\gamma\text{-Glutamyl phosphate}}{CO_2^- - CHNH_2 - CH_2 - CH_2 - C(=O) - O - PO_3^{2-}} \xrightarrow{NH_3} \underset{\text{L-glutamine}}{CO_2^- - CHNH_2 - CH_2 - CH_2 - C(=O)NH_2} + P_i$$

Fig. 29-9. Formation of L-glutamine from L-glutamate.

Glutamine and Asparagine

L-Glutamine is another amino acid that is formed from L-glutamate. This reaction, catalyzed by the enzyme glutamine synthetase, occurs in two steps involving γ-glutamyl phosphate as an enzyme-bound intermediate, as shown in Figure 29-9.

The first step is the formation of a mixed anhydride, a reaction that we have seen a number of times in metabolism. In the second step, ammonia replaces the phosphate group on the carbonyl carbon, a reaction typical of all mixed anhydrides. The overall reaction is the conversion of a carboxylate ion to an amide group. This is an example of a reaction in living organisms that is well known to occur in nonliving systems.

L-Asparagine is formed from L-aspartate in a similar way.

EXERCISE 29-2 Write both the overall equation and the two steps involved in the formation of L-asparagine from L-aspartate.

Proline and 4-Hydroxyproline

L-Proline is formed from L-glutamate by the series of chemical reactions shown in Figure 29-10.

The first step is the reduction of L-glutamate to γ-glutamate semialdehyde. A similar reduction occurs in the glycolytic pathway when 3-phosphoglycerate is reduced to glyceraldehyde 3-phosphate (see Section 27.3). In the next step, the free α-amino group reacts with the aldehyde group to form an

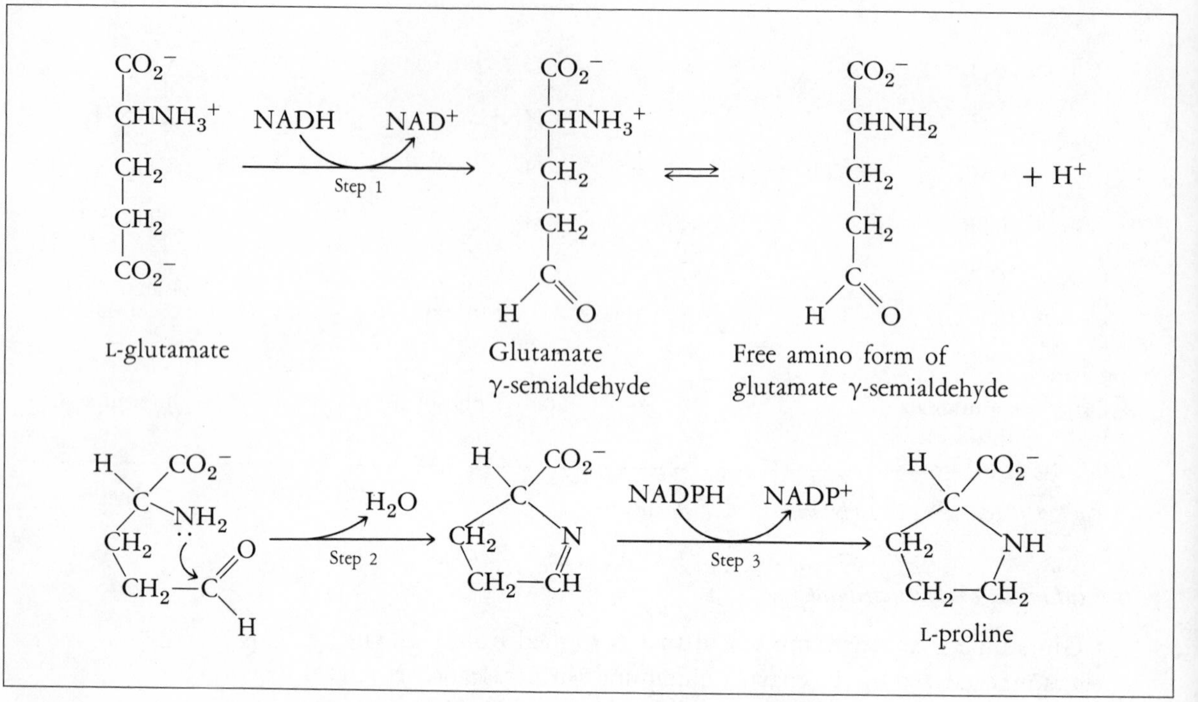

L-glutamate $\xrightarrow[\text{Step 1}]{NADH \rightarrow NAD^+}$ Glutamate γ-semialdehyde $\rightleftharpoons$ Free amino form of glutamate γ-semialdehyde $+ H^+$

$\xrightarrow[\text{Step 2}]{H_2O}$ $\xrightarrow[\text{Step 3}]{NADPH \rightarrow NADP^+}$ L-proline

Fig. 29-10. The series of chemical reactions involved in the formation of L-proline from L-glutamate.

internal Schiff base. The last step is reduction of the carbon-nitrogen double bond by NADPH to form proline.

The 4-hydroxyproline residues in collagen and other fibrous proteins are formed from certain protein-bound proline residues. This reaction is catalyzed by the enzyme proline 4-monooxygenase, which acts *only* on proline residues that are part of polypeptide chains. This complex enzyme system requires ferric ion (Fe^{+3}) and ascorbic acid (Vitamin C) as cofactors and uses α-ketoglutarate as a co-reducing agent. The overall reaction is the following:

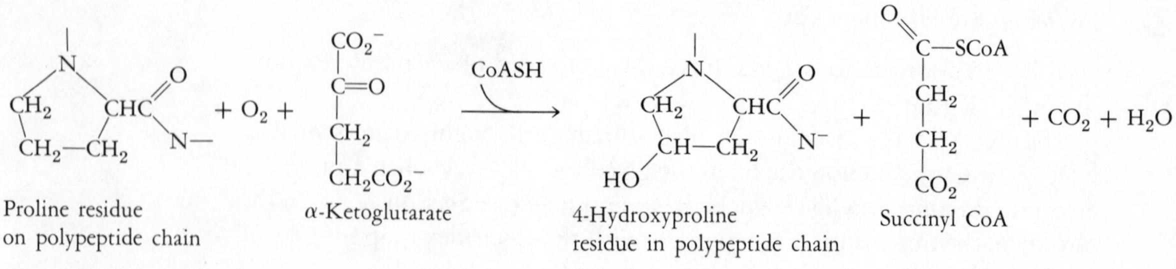

Proline residue on polypeptide chain $+ O_2 +$ α-Ketoglutarate $\xrightarrow{CoASH}$ 4-Hydroxyproline residue in polypeptide chain $+$ Succinyl CoA $+ CO_2 + H_2O$

Tyrosine

L-Tyrosine is formed from the essential amino acid phenylalanine by a reaction called hydroxylation (see Section 13.6). This reaction, which is catalyzed by the enzyme phenylalanine 4-monooxygenase, replaces the *para*-H of the phenyl ring by a hydroxy group. The overall reaction is as follows:

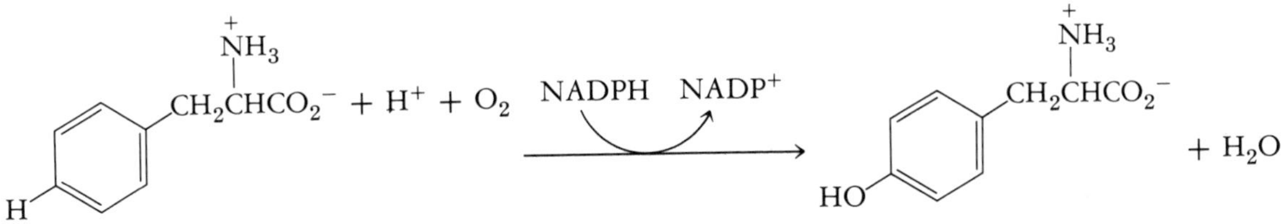

$$\text{L-phenylalanine} + H^+ + O_2 \xrightarrow{\text{NADPH} \;\to\; \text{NADP}^+} \text{L-tyrosine} + H_2O$$

L-phenylalanine: $H{-}C_6H_4{-}CH_2CH(\overset{+}{N}H_3)CO_2^-$

L-tyrosine: $HO{-}C_6H_4{-}CH_2CH(\overset{+}{N}H_3)CO_2^-$

Serine

L-Serine is formed from 3-phosphoglycerate by the series of reactions shown in Figure 29-11. The first step is the oxidation of a secondary alcohol to a

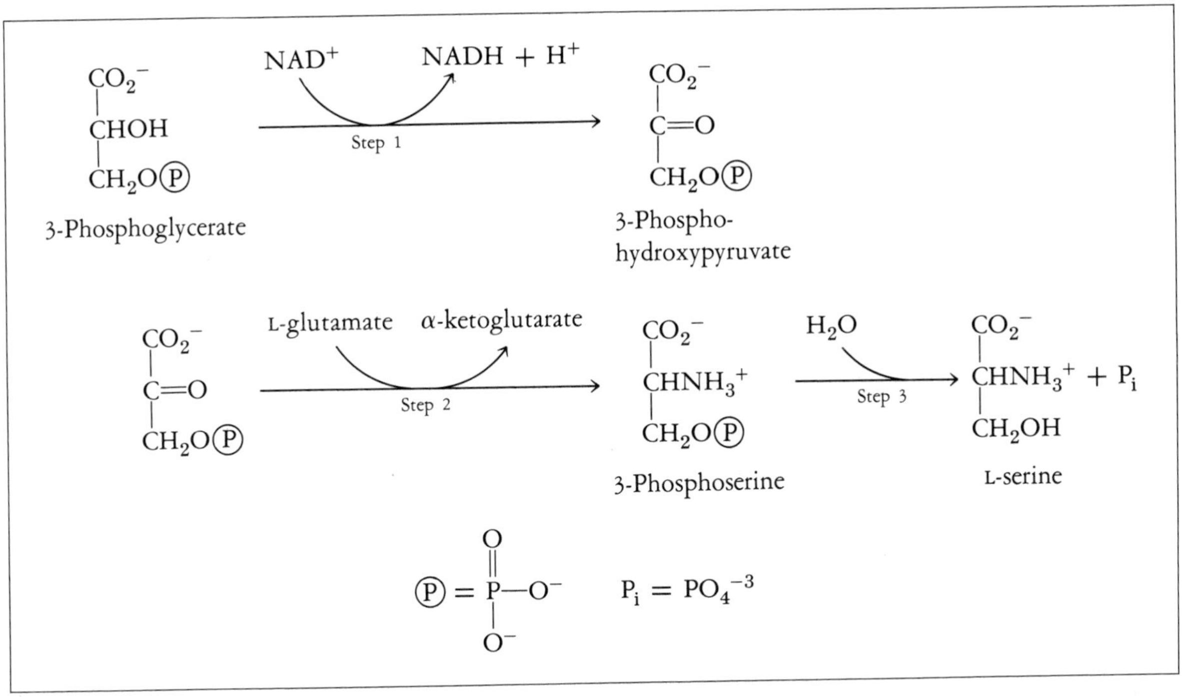

Step 1: 3-Phosphoglycerate (CO_2^- / CHOH / CH_2OⓅ) $\xrightarrow{NAD^+ \;\to\; NADH + H^+}$ 3-Phospho-hydroxypyruvate (CO_2^- / C=O / CH_2OⓅ)

Step 2: 3-Phosphohydroxypyruvate (CO_2^- / C=O / CH_2OⓅ) $\xrightarrow{\text{L-glutamate} \;\to\; \alpha\text{-ketoglutarate}}$ 3-Phosphoserine (CO_2^- / $CHNH_3^+$ / CH_2OⓅ)

Step 3: 3-Phosphoserine $\xrightarrow{H_2O}$ L-serine (CO_2^- / $CHNH_3^+$ / CH_2OH) + P_i

Ⓟ = $-P(=O)(O^-){-}O^-$ $\qquad P_i = PO_4^{-3}$

Fig. 29-11. The series of chemical reactions involved in the formation of L-serine from 3-phosphoglycerate.

keto group by NAD^+ followed by transamination in the second step. Finally, hydrolysis of the phosphate group forms serine. We have seen examples of these three reactions before in other metabolic pathways.

From these examples, we can see that the nonessential amino acids are made from compounds that already have the correct carbon skeleton. Most of these compounds are the products of the various metabolic pathways. It seems that if the organism cannot manufacture the correct carbon skeleton, it cannot synthesize that particular amino acid. This seems to be the reason why humans cannot synthesize certain amino acids. We simply do not have the enzyme-catalyzed reactions needed to make the carbon skeletons of the essential amino acids. Consequently, we must get these amino acids from our food.

29.7 SUMMARY

The proteolytic enzymes pepsin, trypsin, chymotrypsin, and carboxypeptidases carry out the complete hydrolysis of proteins to amino acids in the digestive tract. The free amino acids are then absorbed into the blood, which transports them to the liver, where their catabolism takes place.

The catabolism of amino acids starts by removal of their α-amino groups by transamination to an α-keto acid. As a result, the catabolic pathways of amino acids have two branches. One involves the degradation of the carbon skeleton of the amino acids. The other involves the degradation of the amino groups to ammonium ions.

The carbon skeletons of all 20 amino acids are eventually degraded to only five compounds: acetyl coenzyme A, α-ketoglutarate, succinyl coenzyme A, fumarate, and oxaloacetate. These five compounds are either oxidized to carbon dioxide, water, and energy, or are used as starting materials for the synthesis of fats, glycogen, amino acids, and other compounds needed by the organism.

The α-amino groups of many amino acids are transferred to α-ketoglutarate to form L-glutamate. Oxidative deamination of L-glutamate forms ammonium ion, which in humans and higher animals is converted to urea. The urea is carried by the blood to the kidneys, where it is excreted in the urine.

Not all the α-amino groups are converted to ammonium ions. Some of the nitrogen is saved to use in the synthesis of amino acids. There are 20 different pathways for the synthesis of the 20 amino acids. The pathways for the synthesis and degradation of amino acids are different. The pathways for synthesis of nonessential amino acids are much shorter and less complex than are those for synthesis of essential amino acids. L-Glutamate furnishes the amino group in the synthesis of most of the nonessential amino acids.

REVIEW OF TERMS, CONCEPTS, AND REACTIONS

Terms

AUTOCATALYSIS The product of a chemical reaction catalyzes the reaction by which it is formed.

ESSENTIAL AMINO ACIDS The amino acids that an organism cannot synthesize.

NEGATIVE NITROGEN BALANCE The intake of nitrogen by an organism is less than its loss of nitrogen.

NITROGEN BALANCE The intake of nitrogen by an organism is equal to its loss of nitrogen.

POSITIVE NITROGEN BALANCE The intake of nitrogen by an organism is greater than its loss of nitrogen.

PROTEOLYTIC ENZYMES Enzymes that catalyze the hydrolysis of proteins in the gastrointestinal tract.

TRANSAMINATION The enzyme-catalyzed reaction in which the α-amino group of an amino acid is transferred to the α carbon of an α-ketoacid.

Concepts

1. Amino acids are used in three ways by living organisms.
2. The catabolism of amino acids breaks them into two parts; one part contains the carbon skeleton, and the other part contains the amino group.
3. The carbon skeletons of all 20 amino acids are degraded to only five compounds that are part of or can enter the citric acid cycle.
4. The nitrogen parts of all 20 amino acids are degraded to ammonium ions, which can be converted to urea in the urea cycle or used to synthesize other nitrogen-containing compounds.
5. The anabolic and catabolic pathways of amino acids are different.

Reactions

1. Transamination (see Section 29.3)
   L-aspartate + α-ketoglutarate $\longrightarrow$ α-ketosuccinate + L-glutamate
2. Oxidative deamination (see Section 29.4)
   L-glutamate + NAD^+ (or $NADP^+$) + H_2O $\longrightarrow$
   NH_4^+ + α-ketoglutarate + NADH (or NADPH)

EXERCISES

29-3 What are the three ways that amino acids are used by living organisms?

29-4 Why does a baby have a positive nitrogen balance?

29-5 Why do humans need a continual supply of amino acids?

29-6 Write the equation for the transamination reaction of each of the following amino acids with α-ketoglutarate:
(a) alanine (b) cysteine (c) isoleucine (d) tryptophan

29-7 What is the function of pyridoxal phosphate in the transamination reaction?

29-8 If phenylpyruvate, $C_6H_5CH_2\overset{O}{\overset{\|}{C}}CO_2^-$, is fed to an organism, the organism can synthesize phenylalanine. Write the equation for the chemical reaction by which this occurs.

29-9 From the reactions of amino acids described in this chapter, give an example of each of the following types of reaction:
(a) Schiff base formation (b) hydrolysis of a peptide bond
(c) hydrolysis of an aldimine (d) oxidative deamination
(e) elimination (f) transamination
(g) carbonyl addition reaction (h) hydrolysis of a ketimine

29-10 Why is the urea cycle important to the metabolism of amino acids?

29-11 What compounds are common to the metabolism of glucose, triacylglycerols, and amino acids?

29-12 The catabolic pathway by which valine is converted to CO_2 and propionyl CoA is given below. We have seen examples of all of these reactions in this and previous chapters. For each reaction, specify the type of reaction.

$$(CH_3)_2CH\overset{\overset{+}{N}H_3}{\overset{|}{C}}HCO_2^- \xrightarrow{(a)} (CH_3)_2CH\overset{O}{\overset{\|}{C}}CO_2^- \xrightarrow{(b)} (CH_3)_2CHC(=O)SCoA \xrightarrow{(c)} CH_2{=}\underset{CH_3}{\underset{|}{C}}C(=O)SCoA$$

$$\xrightarrow{(d)} HOCH_2\underset{CH_3}{\underset{|}{C}}HC(=O)SCoA \xrightarrow{(e)} HOCH_2\underset{CH_3}{\underset{|}{C}}HC(=O)OH \xrightarrow{(f)} H\overset{O}{\overset{\|}{C}}\underset{CH_3}{\underset{|}{C}}HC(=O)OH \xrightarrow{(g)} CH_3CH_2C(=O)SCoA + CO_2$$

THE INTERRELATIONSHIPS AND CONTROL OF METABOLISM: SUMMARY AND CONCLUSIONS

30

The theme of Chapters 27, 28, and 29 has been to demonstrate the similarities between the reactions of certain functional groups in the laboratory and their reactions in the metabolism of carbohydrates, triacylglycerols, and amino acids. We have paid little attention to where these reactions occur in living organisms, particularly the human body. We will turn our attention to this subject in this chapter. Another important aspect of metabolism and one that we have hinted at in previous chapters is the set of interrelationships among the three major metabolic pathways. We shall examine these interrelationships. Finally, we shall examine in some detail how these metabolic pathways are controlled and how they respond as various demands are made on them. Let us start by interrelating the three metabolic pathways that we learned in Chapters 27, 28, and 29.

30.1 INTERRELATIONSHIPS AMONG METABOLIC PATHWAYS

The interrelationships among the metabolic pathways of carbohydrates, lipids, and proteins are given in Figure 30-1. The metabolic pathways are not given in detail; only compounds common to all three pathways are shown.

Notice that the three different foods (carbohydrates, lipids, and proteins) are all degraded to only a few common compounds: pyruvate, acetyl coenzyme A, and the compounds of the citric acid cycle. The latter compounds

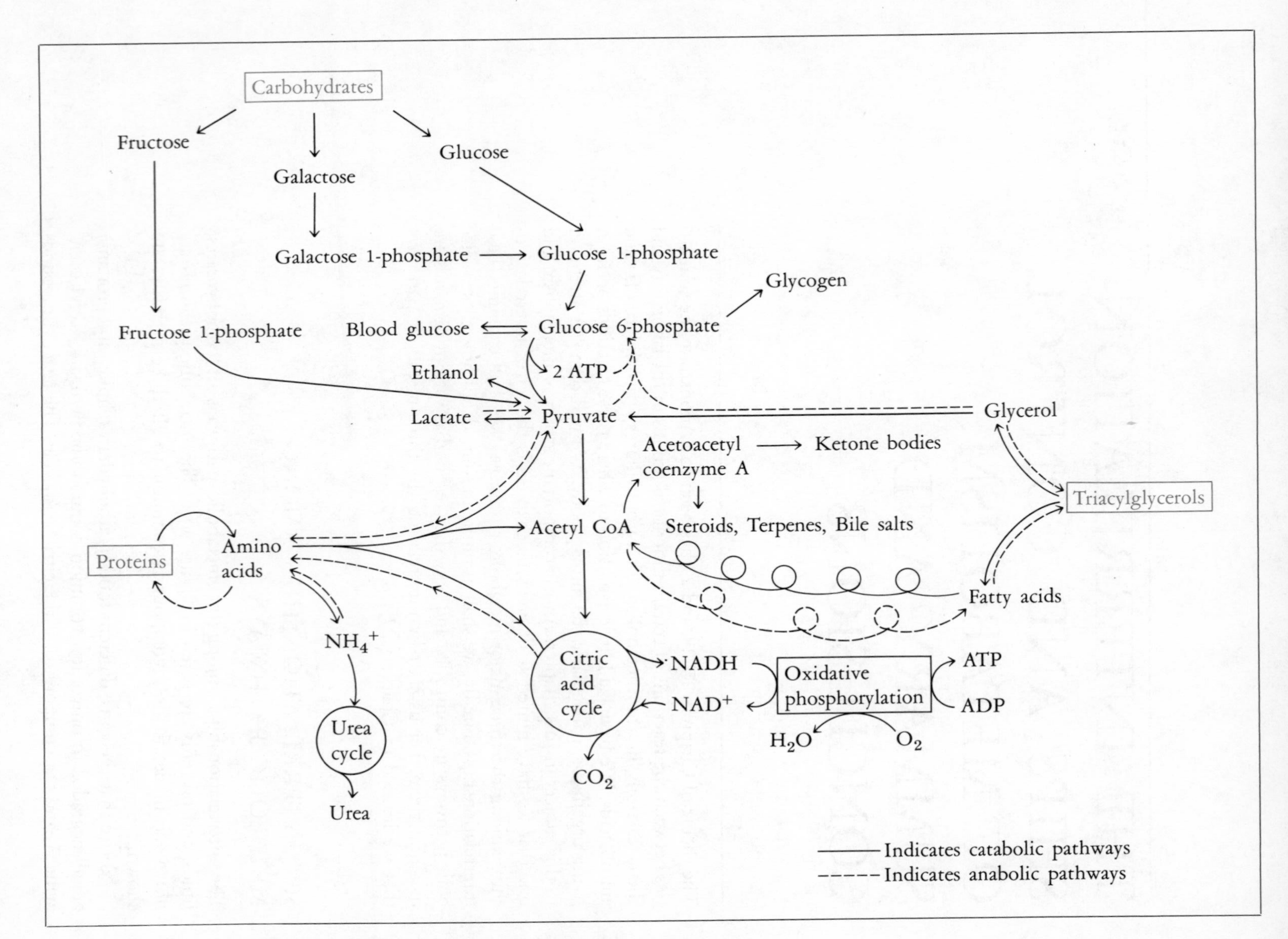

Fig. 30-1. Interrelating the metabolism of carbohydrates, proteins, and triacylglycerols.

are particularly important because they are coupled to oxidative phosphorylation. This series of reactions is the final stage of the aerobic oxidation that stores most of the energy of food in the form of adenosine triphosphate (ATP). In addition, these molecules are the starting materials for the synthesis of amino acids, fats, and other compounds needed by the organism.

An important fact to notice about the scheme in Figure 30-1 is that there is no reaction that forms pyruvate from acetyl coenzyme A. This means that acetyl coenzyme A formed by the catabolism of fatty acids *cannot* be converted into glucose or glycogen. In humans and higher animals, pyruvate obtained from other sources is the starting point for gluconeogenesis (see Section 27.8). Most plants can synthesize carbohydrates from acetyl coenzyme A because they have an additional series of enzyme-catalyzed reactions for this purpose that is not present in humans and higher animals.

There is an interrelationship not only between the metabolic pathways but also between the various parts of an organism. Communication and a supply of nutrients between the various regions are vital for its existence. One of the ways the human body does this is by means of the blood.

30.2 BLOOD

Blood is the transport system of the body. It carries the major nutrients from the intestine to the liver, where they are processed, and then carries them to other organs. The waste products of cells are carried by the blood to the kidneys for excretion. The blood carries oxygen from the lungs to cells and returns carbon dioxide from cells to the lungs. In addition, hormones and other chemical messengers are carried from various glands to their target molecules.

The body contains 5 to 6 L of blood. The cellular components make up about 55 percent of the volume of blood. Red blood cells, which carry oxygen and some carbon dioxide, are the most numerous. White blood cells, which attack invading bacteria, are the other type of cells in blood. The rest of the blood consists of noncellular blood plasma and blood platelets, which are needed for blood clotting. Blood plasma is a solution made up of 90 percent water and 10 percent dissolved compounds. The chemical composition of blood plasma is given in Table 30-1. Plasma proteins are the most numerous and one of the most important types of compound in the blood plasma. *Albumins* and *globulins* are two plasma proteins that have important metabolic functions. Albumins carry hydrophobic nutrients such as fatty acids, lipids, vitamins, and hormones. Such globulins carry ions that otherwise would not be soluble at the pH ($\sim$7) of the blood. The remain-

Table 30-1. Major Components of Normal Human Blood Plasma

Component	%	Function
Serum albumin	60	Carriers of fatty acids and lipids
α, β, and γ-Globulins	30	Antibodies, carriers of lipids and ions
Lipids	7	Fuels that are carried to sites of storage or oxidation
Glucose	1	Source of glucose in blood
Amino acids	0.6	For synthesis of proteins
Urea	0.3	Waste products of amino acid metabolism
Lactate	0.2	Product of glycolysis in muscles
Pyruvate	0.02	Product of glycolysis in muscles
Acetoacetate, β-Hydroxybutyrate	0.08	Ketone bodies

der of the blood plasma is made up of other metabolites and dissolved wastes.

The composition of blood plasma remains nearly constant at all times. This is quite remarkable given that the body suddenly receives a large quantity of nutrients whenever food is eaten. The body has a number of controls to maintain the constant composition of blood plasma. We will learn a few of them in Section 30.4. Because the composition of the blood is nearly constant, measuring the amounts of specific components in the blood (and urine) is extemely important in medicine. A change in the amount of one or more components can often indicate the metabolic nature of an illness and the effectivenss of medication to cure it.

The blood supplies specific compounds and nutrients to all organs of the body. Many of these organs have specific metabolic roles.

30.3 MAJOR METABOLIC ACTIVITIES OF THE VARIOUS ORGANS IN HUMANS

Most of the chemical reactions of the major metabolic pathways can be carried out by all the major organs of the body. However, each organ does have a characteristic pattern of metabolism in keeping with its specialized function. Let us summarize the special metabolic features of the major organs of the human body. Because the first organ to receive the nutrients is the liver, it is appropriate that we start with this organ.

The Liver

The liver receives most of the nutrients immediately after they are absorbed from the intestinal tract. Because of the wide variety of food that we eat, the liver must have a great metabolic flexibility. It must be able to handle the three major classes of food plus other chemicals that we consume. For example, ethanol from alcoholic beverages, caffeine from coffee and tea, and many other compounds are metabolized by the liver. The liver must be able to handle the sudden increase of food when we eat and then the period of fasting between meals. Thus, the liver not only metabolizes the food, but also maintains the constant level of nutrients in the blood.

The liver handles changes in the type of food it receives simply by making more of the enzymes needed to metabolize that particular food. For example, the liver of a person on a high-protein diet has a large quantity of the enzymes that catalyze amino acid catabolism and gluconeogenesis. If the person changes to a diet high in carbohydrates, these enzymes almost completely disappear from the liver and are replaced within a short time by large quantities of enzymes that catalyze carbohydrate metabolism. Thus, the liver can regulate the synthesis of the enzymes it needs to metabolize the foods that it receives.

About 60 percent of the glucose that the liver receives enters the liver cells and is phosphorylated to glucose 6-phosphate. The rest passes through the liver and becomes blood glucose. On a normal diet, most of the glucose 6-phosphate is converted to glycogen or fatty acids. Very little is oxidized because the oxidation of fatty acids and amino acids provides most of the ATP needed by the liver.

The amino acids that enter the liver have numerous metabolic fates. Some of them pass through the liver and into the blood, where they are carried to other organs for use in protein synthesis. Some of the amino acids are used by the liver to synthesize compounds that it needs such as enzymes, plasma proteins, and liver tissue. Certain amino acids are converted to specific nitrogen-containing compounds. Excess amino acids are deaminated and degraded. Some of the carbon skeletons are used in gluconeogenesis, and others enter the citric acid cycle to form ATP, carbon dioxide, and water. The amino groups are eventually converted to urea by the urea cycle. The liver is the only organ that can synthesize urea.

After hydrolysis, fatty acids are oxidized to yield acetyl coenzyme A, which either forms ketone bodies, cholesterol, and bile salts or enters the citric acid cycle to form ATP. The ATP formed in the liver is used for the synthesis of glycogen, fatty acids, triacylglycerols, liver enzymes, plasma proteins, urea, and other small molecules such as nonessential amino acids.

Most of the triacylglycerol formed in the liver is not stored there. Instead, it is stored in adipose tissue.

Adipose Tissue

Adipose tissue is made up of specialized fat cells that contain large globules of triacylglycerol in nearly pure form. These globules make up about 90 percent of the weight of these cells. Adipose tissue is found under the skin (particularly around the waist) and in mammary glands. These cells are not inert storage depots. Rather, the triacylglycerols in these cells are continuously being synthesized and degraded in response to the energy needs of the body.

Adipose tissue makes triacylglycerol from glucose, which provides not only acetyl coenzyme A but also glycerol 3-phosphate. Triacylglycerol can also be made starting with the fatty acids that arrive as lipoproteins in the blood. Adipose tissue also degrades glucose and fatty acids to form ATP via the citric acid cycle. This ATP is then used to synthesize triacylglycerols.

The major function of adipose tissue is to act as a readily available source of triacylglycerol. The stored triacylglycerol is released from the cell after it is hydrolyzed to fatty acids and glycerol within the cell. The fatty acids are then transferred to the blood, where they are bound to serum albumin. In this form they are carried to the heart and muscles, where they are oxidized to form the ATP needed by these organs.

The liver and adipose tissue are the major regions of storage, metabolism, and distribution of the fuel molecules needed by the body. Between them, they maintain a constant level of energy-rich compounds in the blood, making them available to all cells in the body. They maintain this constant level at all times, even after feeding and during times of fasting or starvation.

The fatty acids bound to serum albumin are the major source of energy of the heart.

The Heart

While at rest or during moderate exercise, the heart obtains the ATP it needs by aerobic metabolism. It uses glycolysis as a source of ATP only under emergency conditions. Fatty acids bound to serum albumin and carried by the blood are the major energy-rich compounds used by the heart. Some ketone bodies, blood glucose, and lactate can also be used as a source of energy in emergencies. All of these compounds form the ATP needed by the heart by oxidation via the citric acid cycle and oxidative phosphorylation.

The heart has an active amino acid metabolism that is used to synthesize and degrade heart protein continuously. Skeletal muscles are similar to the heart in a number of ways.

The Skeletal Muscles

Contracting the skeletal muscles allows the body to move. During normal activity, the energy (in the form of ATP) needed to contract these muscles is obtained from the aerobic oxidation of the fatty acids bound to serum albumin. In this way, the skeletal muscles resemble the heart. During maximal exercise, such as running, the skeletal muscles, like the heart, increase their metabolism of glucose. Most of the glucose used under these conditions is from glycogen stored in the muscles. It is converted by glycolysis to lactate, which is reconverted to glucose by the Cori cycle (see Section 27.4). In this way, the liver helps the muscles during periods of extreme activity. This is an example of the interdependence of organs.

A muscle contains only about 1 percent of its weight in glycogen. During continuous maximal contraction, this is quickly used up. For example, all of this glycogen is used within 20 seconds under anaerobic oxidation or within 3.5 minutes under aerobic oxidation. The muscle must then obtain glucose from other sources such as the blood or the liver. More glucose is available from the liver, where it is synthesized by degrading amino acids. Glucose is synthesized in this way because fatty acids cannot be converted to glucose. If the demand for glucose becomes great enough, the body begins to degrade its own protein to synthesize the glucose it needs.

The level of glucose in the blood must also be maintained for the brain to function.

The Brain

The brain of a normal person uses only glucose as its source of energy; it has no reserve of glycogen or triacylglycerol and consequently depends on the glucose in the blood. The metabolism of the brain is critically dependent on the level of blood glucose. If the level drops, symptoms of brain dysfunction appear. At very low levels of blood glucose, a person falls into a coma. Glucose metabolism occurs in the brain entirely by glycolysis and the citric acid cycle.

The ATP formed by the catabolism of glucose is used in the brain to synthesize acetylcholine and other nerve impulse transmitters. The brain uses energy at a constant rate. Intense concentration or active thought causes no apparent increase in the metabolism of glucose in the brain.

Most of the energy needed by the body is obtained by aerobic metabolism. We get the oxygen for these processes through the lungs.

The Lungs

The body needs a large amount of oxygen to make ATP by the oxidative phosphorylation reactions. The body gets this oxygen through the lungs. Here the rich supply of oxygen comes into direct contact with the hemoglobin, a conjugated protein contained in the red blood cells. Hemoglobin reacts with oxygen to form oxyhemoglobin, which is carried to the cells by the blood. Oxyhemoglobin gives up its oxygen to the cells and re-forms hemoglobin, which then picks up some of the carbon dioxide and carries it back to the lungs to be expelled. The hemoglobin can now react with more oxygen and restart the cycle.

Not all of the oxygen of oxyhemoglobin is given to the cells. About 60 percent of it remains in the blood as an oxygen reserve. A person exercising strenuously has a high oxygen need and uses this reserve as a ready source of oxygen. With prolonged heavy exercise, the need for more oxygen is satisfied by heavy and deep breathing.

Finally, the waste products of metabolism are carried to the kidneys, where they are processed and excreted.

The Kidneys

The kidneys, like the liver, have a high degree of metabolic flexibility because they, too, receive a wide variety of compounds. They metabolize almost all of the compounds formed in the body. Blood glucose, fatty acids, ketone bodies, and amino acids are all degraded to acetyl coenzyme A, which then enters the citric acid cycle oxidative phosphorylation reaction to form ATP. About 75 percent of the energy obtained from these reactions is used by the kidneys to form urine. The remainder is used to transport and regulate the amount of cations (mostly Na^+ and K^+) in the blood.

In a healthy person, all these organs carry out their functions at the same time. However, the metabolism of one organ is not independent of the others; there is an interdependence that is the result of many controls. We will learn a few of these controls in the next section.

30.4 CONTROL OF METABOLISM

Nothing within a cell is static. All of the compounds that make up the parts of the cell are constantly being built up and broken down. Anything that affects the speed at which the synthesis and degradation occur will affect the overall metabolic picture in some way. Thus, any chemical reaction that plays a significant role in metabolism may have a controlling influence. Because molecules can react with each other in so many different ways, there are countless ways to control metabolism. Despite the complexities, only a few reactions are responsible for controlling several of the major metabolic pathways. We will examine these reactions, starting with glycolysis and gluconeogenesis.

Glycolysis and Gluconeogenesis

The glycolytic pathway, which degrades glucose to pyruvate (see Section 27.3), is primarily controlled by the enzyme phosphofructokinase. This enzyme catalyzes the formation of fructose 1,6-diphosphate from fructose 6-phosphate (Reaction ③, Figure 27-2). This enzyme is stimulated by adenosine monophosphate (AMP) and adenosine diphosphate (ADP) but is inhibited by ATP, citrate, and reduced nicotinamide adenine dinucleotide (NADH). Hexokinase, which catalyzes the formation of glucose 6-phosphate from glucose (Reaction ①, Figure 27-2) is a secondary point of control. This enzyme is inhibited by the product, glucose 6-phosphate.

Notice how the stimulators and inhibitors control glycolysis. As ATP is used by the cell, the amounts of ADP and AMP increase. This stimulates the enzyme phosphofructokinase to start the glycolytic pathway to produce more ATP. When too much ATP is formed, it acts as an inhibitor that stops glycolysis, and no more ATP is formed. These are examples of end-product inhibition of enzyme activity (see Section 25.7).

Gluconeogenesis is primarily controlled by the enzyme pyruvate carboxylase, which catalyzes the formation of oxaloacetate from pyruvate (see Section 27.8). This enzyme is stimulated by acetyl coenzyme A. Consequently, whenever the amount of acetyl coenzyme A in the cell is greater than it needs to be, glucose synthesis is stimulated. This diverts the metabolism of pyruvate from acetyl coenzyme A formation to glucose formation. The secondary control point of gluconeogenesis is the enzyme fructose 1,6-diphosphatase, which catalyzes the hydrolysis of fructose 1,6-diphosphate. This enzyme is stimulated by citrate and 3-phosphoglycerate, which are both needed to synthesize glucose. It is inhibited by AMP.

The control points for glycolysis and gluconeogenesis are placed in rela-

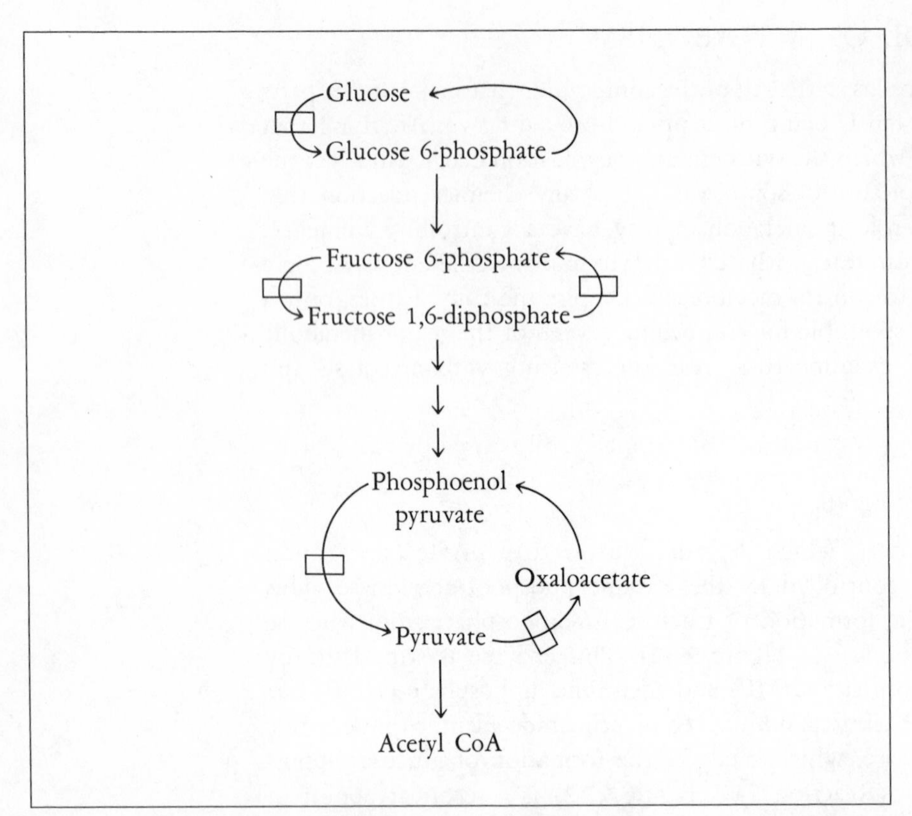

Fig. 30-2. Control of the opposing pathways of glycolysis and gluconeogenesis. The symbol ▭ indicates reactions that control the metabolic pathways.

tion to the other reactions in Figure 30-2. *Notice that the enzymes controlling glycolysis and gluconeogenesis are located at points where the two pathways are different.* This allows each pathway to be controlled independently of the other. Thus, when the cell has sufficient ATP and an ample supply of compounds such as acetyl coenzyme A and citrate that can be used in the citric acid cycle as fuel, gluconeogenesis is stimulated and glycolysis is inhibited. On the other hand, when the cell needs ATP or when compounds such as acetyl coenzyme A and citrate are in short supply, glycolysis is promoted and gluconeogenesis is inhibited. Thus, controlling a few reactions regulates the pathways for glycolysis and gluconeogenesis. In a similar way, the citric acid cycle is controlled.

Citric Acid Cycle

In most cells, the aldol condensation of acetyl coenzyme A and oxaloacetate (Reaction [2], Figure 27-9) is the primary control reaction of the citric acid cycle. The enzyme citrate synthetase, which catalyzes this reaction, is inhibited by succinyl coenzyme A, one of the compounds of the citric acid cycle. Another reaction that appears to be under control is succinate dehydrogenation (Reaction 7, Figure 27-9). This reaction is stimulated by large amounts of succinate in the cell and is inhibited by oxaloacetate.

The citric acid cycle can also be controlled by limiting the amount of acetyl coenzyme A that enters the cycle (Reaction 1, Figure 27-9). This is done by controlling the activity of the pyruvate dehydrogenase complex that catalyzes the formation of acetyl coenzyme A from pyruvate. ATP is the energy-rich product of the citric acid cycle and oxidative phosphorylation. When the amount of ATP becomes too high, the pyruvate dehydrogenase complex is inhibited. This slows down the formation of acetyl coenzyme A and the subsequent formation of ATP. When ATP is used by the cell and the amount of ADP becomes too high and enough pyruvate is available, the pyruvate dehydrogenase complex is stimulated and acetyl coenzyme A and ATP are formed.

The control of amino acid metabolism is more complicated because of the many independent pathways for their catabolism and anabolism. However, two general types of control have been identified.

Amino Acids

The synthesis of amino acids is known to be controlled in at least two ways. One is end-product inhibition. This control is achieved by the inhibition by the final product of the first of a series of enzyme-catalyzed reactions. An example of this control was given in Section 25.7. End-product inhibition is a delicate control of synthesis because it is capable of second-by-second adjustment of the amount of final product formed.

The second type of control is to inhibit the synthesis of one or more enzymes that catalyze reactions of the synthetic pathway. This method of controlling amino acids is slower to respond to changing metabolic conditions than is end-product inhibition. Most of the pathways leading to the synthesis of amino acids are controlled by a combination of end-product inhibition and inhibition of enzyme synthesis. These pathways are often complex and vary from one amino acid to another.

Not all methods of controlling the metabolism of glucose and amino acids have been discovered. However, as we will learn next, there is a network of interrelated controls within the cell that allows all of the metabolic cycles to respond in different ways to different conditions.

30.5 ADJUSTING THE BODY'S METABOLISM TO CHANGING CONDITIONS

Let us examine two important situations in which a change in metabolism occurs: eating a meal with large amounts of carbohydrates, and fasting.

When we eat a meal heavy in carbohydrates, the blood glucose level quickly increases. Part of the excess glucose is converted into glycogen until these reserves are full. More excess glucose is metabolized to carbon dioxide and water. This leads to an increase in the amount of ATP in the cell. When the level of ATP becomes too high, it inhibits the citric acid cycle. As a result, the amount of acetyl coenzyme A increases in the cell. Acetyl coenzyme A can be removed by converting it to malonyl coenzyme A, which starts the synthesis of fatty acids and eventually the formation of triacylglycerols. Thus, excess glucose is converted to fat in our bodies. These reactions occur to some extent in all cells, but they are most important in adipose tissue (see Section 30.3).

The body uses all of its reserves to maintain its energy needs during fasting. The reserves of energy-rich compounds in an average adult are given in Table 30-2. Notice that the reserves of glucose and glycogen are very small—so small, in fact, that they can provide the normal energy needs of the body for only about one day. In contrast, the reserves of triacylglycerols and proteins are quite large. In particular, the triacylglycerols can provide the normal energy needs for more than two months. Let us trace what happens to the metabolism of the body during fasting.

During the first few days of a fast, the reserve of easily metabolized glycogen is used. This is mostly the glycogen stored in the liver. Once this reserve is used, the metabolism of triacylglycerols increases. But surprisingly, the amount of nitrogen excreted in the urine as urea also increases. This means that the body proteins are being degraded. Why should the

Table 30-2. Energy Reserves in an Average 70-kg Person

Compound	Weight (kg)	Percent
Triacylglycerols in adipose tissue	13	19
Proteins in muscles and other tissues	5	7.1
Glycogen in liver and muscles	1	1.4
Glucose, fatty acids, and triacylglycerols in blood	0.02	0.03

body begin to use protein for its energy when it has such a large supply of triacylglycerols?

The answer to this question is that the brain needs a large amount of glucose to function (see Section 30.3). Because the liver glycogen is almost all used, glucose must be synthesized from other compounds in the body to meet the needs of the brain. The body is forced to degrade body proteins and use their carbon skeletons to synthesize glucose *because it cannot synthesize glucose from fatty acids.* Using body proteins in this way causes the loss of certain biological functions that need proteins.

While the glycerol part of triacylglycerols can form some glucose, no glucose can be obtained from fatty acids in the body. Remember that, in the body, acetyl coenzyme A cannot be converted to pyruvate, the starting point of gluconeogenesis. Consequently, the blood glucose needed by fasting persons is obtained from body protein. The carbon skeleton of the amino acids form pyruvate and compounds of the citric acid cycle. These compounds synthesize glucose by gluconeogenesis. The degradation of the amino acids in the liver converts the α-amino group to urea, which is then excreted in the urine. This accounts for the increase in the amount of urea in the urine after the first few days of fasting.

The demands of the brain dominate the entire metabolism of a fasting person. Body proteins that perform important functions are sacrificed to maintain the blood sugar level needed by the brain. The body proteins are sacrificed in a definite order. The first to be lost are the digestive enzymes. This is not serious, because they are not used while a person is fasting. Next to be lost are the enzymes in the liver that process the incoming nutrients from the intestines. Then the muscle proteins are used. When this occurs, the fasting person becomes physically inactive. Clearly, the body chooses which proteins to degrade to keep the central nervous system working.

The body protein is degraded rapidly during the first week of fasting. If this continued, the body could not last much more than a month. After several weeks, another metabolic adjustment occurs. The brain develops the ability to use other compounds in addition to glucose as fuel. In particular, the ketone bodies (see Section 28.8) such as acetoacetate that are produced during fatty acid oxidation in the liver are used as a source of energy. The brain now obtains a large part of its energy from the oxidation of acetoacetate which is obtained from the relatively large reserve of fat. This change in metabolism allows the remainder of the body protein to be conserved. A fasting person will still degrade some body protein, but much more slowly. This continues until all the triacylglycerol reserve has been completely used.

Once the reserve of triacylglycerols has been consumed, the entire energy needs of the body must be met by the degradation of body protein. The muscles, which contain the largest amounts of protein, are used first. The body gradually consumes itself to an inevitable end.

Clearly, a normal healthy body can adjust its metabolism to meet a variety of conditions. Sometimes there is an error in the metabolism that the body cannot overcome. A number of these metabolic disorders have been identified.

30.6 DISEASES DUE TO ERRORS IN METABOLISM

One of the most common human diseases is diabetes (diabetes mellitus). About 400 out of every million people develop juvenile diabetes between the ages of 8 and 12 years. Another 30,000 of the million, or 3 percent, develop the disease by 40 to 50 years of age. More than 70 percent of persons in their late seventies are affected. Diabetes is a complex disease. It is partly hereditary, yet diet and exercise have a great effect on its occurrence.

The main symptom of diabetes is a high blood glucose level (hyperglycemia). In acute diabetes, hunger, thirst, weight loss, excretion of large volumes of urine, and ketonemia (large amounts of ketone bodies in the blood) are also symptoms. The metabolic cause of diabetes is a disruption of the normal glucose metabolism. There is an increase in gluconeogenesis, and glucose is no longer converted into fatty acids via acetyl coenzyme A.

Normally, as much as a third of the carbohydrates eaten by a person are converted into fatty acids and then into triacylglycerols to be stored in adipose tissue. This does not occur in people with diabetes. In fact, their metabolism tries to maintain as high a level of glucose in the blood as possible. This is accomplished by increasing gluconeogenesis from amino acids and inhibiting the conversion of glucose into fatty acids. The reason for this change in metabolism is not known. It may be due to an inability of the body to transport blood glucose across the membranes of skeletal muscles and other tissues. The body may try to overcome this transport problem by increasing the level of glucose in the blood. Whatever the reason, the metabolic change is evident in the blood and urine. Large amounts of glucose and nitrogen-containing compounds are found in the urine. Much water is lost in the urine, which accounts for the thirst of diabetics. The constant loss of glucose in the urine occurs at the expense of amino acids and body protein. This accounts for the constant hunger yet loss of weight in diabetics.

While much glucose is synthesized, very little of it is used for energy by diabetics. Most of the cells, except those in the brain, use triacylglycerols as their energy source. As a result, the amount of acetyl coenzyme A in the cell increases to the point where the citric acid cycle cannot oxidize all of it. The excess acetyl coenzyme A is converted into ketone bodies. These appear in the blood (ketonemia) and the urine (ketonuria). When excess amounts of ketone bodies are present in the body, a condition called *ketosis* results.

All these metabolic disorders can be reversed by giving insulin to the diabetic. Insulin is believed to increase the metabolism of glucose by increasing the amounts of the enzymes glucokinase, phosphofructokinase and pyruvate kinase in the liver. The formation of enzymes needed in gluconeogenesis is suppressed, which decreases the degradation of amino acids to form glucose. Insulin also restores the normal conversion of glucose to fatty acids. As a result, glucose blood levels and the amount of ketone bodies in the blood and urine return to normal.

The exact chemical basis of the action of insulin is unknown despite decades of study. There are two current theories to explain its action. The first proposes that this hormone acts on the membranes of all tissues and cells, making them more permeable. This allows glucose as well as other compounds and ions to pass more easily into the cells and allow their normal metabolism to occur. Insulin, according to this view of its action, is a general hormone that helps all nutrients to enter cells so they can be used or stored. The second theory proposes that the increased gluconeogenesis and lower triacylglycerol synthesis occur to compensate for the lack of insulin. According to this view, insulin is involved in the transport of glucose. To overcome the lack of insulin, the body increases the amount of glucose in the blood so that the cells and tissues at the extremities of the body will get enough glucose.

Another disease that is caused by the inability of certain people to metabolize sugar is called *lactose intolerance.* People with this disease cannot eat any food that contains lactose because they do not have the enzyme β-galactosidase. This enzyme catalyzes the hydrolysis of lactose (a dissacharide) in the gastrointestinal tract. Without β-galactosidase, these people cannot absorb lactose normally. This is not a serious illness, and it can be controlled by maintaining a lactose-free diet.

Some diseases are known to be caused by disorders in the storage of glycogen in the liver or muscles. The first of these was discovered in 1929 by Edgar von Gierke. People with this disease have an enlarged abdomen caused by a massive increase in the size of the liver. Another symptom of this disease is hypoglycemia between meals. People with this disease do not have the enzyme glucose 6-phosphatase. As a result, glucose 6-phosphate cannot be hydrolyzed to glucose. The problem is that only glucose, not

$$H\!-\!\left(CH_2\overset{CH_3}{\overset{|}{C}}HCH_2CH_2\right)_3CH_2\underset{\beta}{\overset{CH_3}{\overset{|}{C}}H}\underset{\alpha}{CH_2}C\!\!\begin{array}{l}\nearrow O\\ \searrow SCoA\end{array} \longrightarrow H\!-\!\left(CH_2\overset{CH_3}{\overset{|}{C}}HCH_2CH_2\right)_3CH_2\underset{\beta}{\overset{CH_3}{\overset{|}{C}}H}\underset{\alpha}{C}\!\!\begin{array}{l}\nearrow O\\ \searrow SCoA\end{array} + CO_2$$

β Oxidation is blocked

α Oxidation

Phytanate

Degradation continues by usual β oxidation

Fig. 30-3. The α-oxidation in the first step of the catabolism of the fatty acid phytanate.

glucose 6-phosphate, can pass through the cell membranes into the blood. Glucose 6-phosphate is trapped in the liver. However, glucose can still be converted to glycogen in these people. Consequently, the liver becomes enlarged as the amounts of glycogen and glucose 6-phosphate stored there continually increase. To get rid of some of the glucose 6-phosphate, there is an increase in glycolysis in the liver. This leads to high levels of lactate and pyruvate in the blood. Because the amount of glucose available to the cells is limited, most of the energy needed is obtained from triacylglycerol metabolism.

A disease called Refsum's disease is known to be caused by a disorder of fatty acid catabolism. The 20-carbon branched-chain fatty acid phytanate accumulates in the brain. This fatty acid is formed in the body by oxidation of the alcohol phytol, which is obtained from the hydrolysis of chlorophyll. Phytanate has a methyl group on the β carbon. Consequently, the usual β-oxidation of fatty acids is blocked. Instead, the first step in the catabolism of phytanate involves an α-oxidation, as shown in Figure 30-3. People with Refsum's disease are unable to metabolize phytanate because they do not have the necessary enzymes to carry out this first step.

It is not known whether the symptoms of this disease are due to the large amount of phytanate present in the brain or are due to the lack of some product of its degradation.

L-Tyrosine is synthesized in humans from the essential amino acid L-phenylalanine, as we learned in Section 29.6. This reaction is catalyzed by the enzyme phenylalanine 4-monooxygenase. This enzyme is absent from about 1 of every 10,000 persons. Such people cannot metabolize phenylalanine by the usual pathway. Instead, phenylalanine undergoes transamination with α-ketoglutarate to form phenylpyruvate, as shown in Figure 30-4. Phenylpyruvate accumulates in the blood and is eventually excreted in the urine.

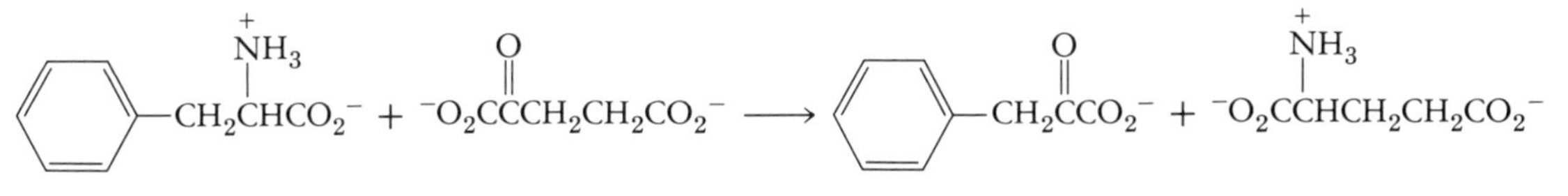

$$\text{C}_6\text{H}_5\text{-CH}_2\overset{\overset{+}{\text{NH}_3}}{\text{CH}}\text{CO}_2^- + {}^-\text{O}_2\text{C}\overset{\text{O}}{\overset{\|}{\text{C}}}\text{CH}_2\text{CH}_2\text{CO}_2^- \longrightarrow \text{C}_6\text{H}_5\text{-CH}_2\overset{\text{O}}{\overset{\|}{\text{C}}}\text{CO}_2^- + {}^-\text{O}_2\text{C}\overset{\overset{+}{\text{NH}_3}}{\text{CH}}\text{CH}_2\text{CH}_2\text{CO}_2^-$$

L-phenylalanine | α-Ketoglutarate | Phenylpyruvate | L-glutamate

Fig. 30-4. The formation of phenylpyruvate by a transamination reaction.

Excess phenylpyruvate is dangerous to children. It interferes with normal brain development and causes mental retardation. This disorder is called *phenylketonuria,* or PKU for short. It was one of the first metabolic disorders recognized in humans. Mental retardation of children with this disease can be prevented by limiting the amount of L-phenylalanine in their diet.

30.7 SUMMARY AND CONCLUSIONS

The metabolic pathways of the three major foods of humans are closely interrelated. In fact, they are all degraded to only a few common compounds that either are nutrients or are used as a source of energy by oxidation via the citric acid cycle and oxidative phosphorylation. The nutrients are carried to the various parts of the body by the blood. Most of the major organs of the body can carry out the chemical reactions of the major metabolic pathways. However, each organ does have a characteristic pattern of metabolism in keeping with its specific function. The liver has a great metabolic flexibility because it receives all the nutrients immediately after they are absorbed from the intestinal tract. Adipose tissue stores triacylglycerols that are used by the body as a readily available source of energy. The heart and skeletal muscles metabolize fatty acids to get their energy, whereas the brain gets its energy from glucose metabolism. Because the brain has no reserves of glycogen, it is very dependent on the level of blood glucose. The oxygen needed by the body for aerobic oxidation is obtained through the lungs. Finally, the kidneys, like the liver, have a high degree of metabolic flexibility because they receive a wide variety of compounds that are transformed into waste products for excretion.

The major metabolic pathways are controlled by only a few reactions. This control usually occurs at points where the catabolic and anabolic pathways are different. This allows each pathway to be controlled independently of the other. These controls are important when the body is placed under stress, such as after eating a meal heavy in carbohydrates or when one is fasting. Many diseases due to errors in metabolism have been identified. Among them are diabetes, lactose intolerance, von Gierke's disease, Refsum's disease, and phenylketonuria. These diseases are caused either by the lack of a particular enzyme, or by a breakdown in the control of one or more of the metabolic pathways.

It should be clear by now that the life of all organisms, from bacteria to humans, can be described as series of interrelated chemical reactions. It is true that there are countless chemical reactions involved in numerous complex series in a living organism. But most of these chemical reactions are similar, at least in type, to those found with simpler compounds in the laboratory, as we have learned in the previous chapters. This clearly shows that the principles learned by studying science in general and organic chemistry in particular can be directly applied to living systems. This is the justification for the statement, "no new laws of science are needed to account for the process of life," made in Chapter 1.

REVIEW OF TERMS AND CONCEPTS

Terms

ADIPOSE TISSUE Specialized fat cells that contain large globules of triacylglycerols.

ALBUMINS Plasma proteins that carry hydrophobic nutrients such as fatty acids, lipids, vitamins, and hormones in the blood.

BLOOD PLASMA The noncellular part of blood.

GLOBULINS Plasma proteins that carry ions in the blood.

KETOSIS A condition caused by excessive amounts of ketone bodies in cells and blood.

LACTOSE INTOLERANCE A disease caused by the lack of the enzyme β-galactosidase. Persons with this disease cannot eat any product containing lactose.

PHENYLKETONURIA (PKU) A disease caused by the lack of the enzyme phenylalanine monooxygenase. Persons with this disease have large amounts of phenylpyruvate in their blood.

Concepts

1. Metabolic pathways for all three major types of food are interrelated and lead to only a few common compounds.

2. In humans, there is no chemical reaction that converts acetyl coenzyme A to pyruvate. As a result, fatty acids cannot be converted to glucose.
3. Most of the organs of the human body can carry out the chemical reactions of the major metabolic pathways.
4. The major metabolic pathways are controlled by only a few chemical reactions.

EXERCISES

30-1 What two compounds are formed from the catabolism of glucose, triacylglycerols, and certain amino acids?

30-2 What compound or compounds provide the major source of energy for each of the following organs?
(a) heart (b) skeletal muscles (c) brain (d) liver
(e) kidneys (f) adipose tissue

30-3 Explain how the levels of ATP, ADP, and AMP are maintained by the metabolism of glucose.

30-4 During the fasting state, why does the body use its protein as a source of energy before using its store of triacylglycerol?

30-5 During the advanced fasting state, why does the brain use ketone bodies for energy rather than glucose?

30-6 What abnormality in metabolism causes ketosis?

30-7 Explain how the lack of the enzyme glucose 6-phosphatase causes an increase in the size of a person's liver.

30-8 Trace the pathway by which carbohydrates are converted to triacylglycerols.

30-9 Trace the pathway by which L-alanine is converted to glucose.

30-10 Trace the pathway by which triacylglycerols are converted to L-alanine.

APPENDIX

A BRIEF MATHEMATICAL REVIEW

The ability to use mathematics is important to the study of chemistry. This short review is intended for students who have doubts about their ability to handle mathematical calculations.

A.1 REARRANGING ALGEBRAIC EQUATIONS

There are many algebraic equations in science that relate two or more quantities. For example, the following algebraic equation relates temperatures on the Kelvin scale to those on the Celsius scale (Section 1.5).

$$K = {}^\circ C + 273.15^\circ$$

This equation as written is convenient to convert a temperature from °C to K. However, to convert a temperature from °C to K, it is necessary to rearrange the equation so that °C stands alone on one side. The key to rearranging any algebraic equation is to remember that the *equality between the two sides of the equation will remain as long as the same mathematical operation is carried out on both sides of the equation.* A mathematical operation means to multiply, divide, add, or subtract any quantity. Thus the equation relating °C and K can be rearranged as follows:

Original equation	$K = °C + 273.15°$
Subtract 273.15° from both sides	$K - 273.15° = °C + 273.15° - 273.15°$
Rearranged equation	$K - 273.15° = °C$

This rearranged equation has the desired quantity °C alone on one side.

* EXERCISE A-1 Rearrange each of the following equations so that the quantity x stands alone on one side of the equation.
(a) $y = x - 25$ (b) $20 + x = 15 - 17$
(c) $0.20 = -0.50 - x$

Many algebraic equations contain products of terms on each side. For example, the relationship between pressure and volume of a gas is given by the following algebraic equation (Section 6.8).

$$P_iV_i = P_fV_f$$

To rearrange the equation so that only one quantity, for example, V_i, is on the left side, divide both sides of the equation by P_i.

Original equation $$P_iV_i = P_fV_f$$

Divide both sides by P_i $$\frac{\cancel{P_i}V_i}{\cancel{P_i}} = \frac{P_fV_f}{P_i}$$

Rearranged equation $$V_i = \frac{P_fV_f}{P_i}$$

To rearrange some equations, it is necessary to multiply both sides of the equation by a quantity. This technique is used to rearrange the following equation, which relates the pressure and temperature of a gas (Section 6.8) so that only P_i is on the left side.

*The answers for the exercises in this chapter begin on page 904.

Original equation $\frac{P_i}{T_i} = \frac{P_f}{T_f}$

Multiply both sides by T_i $\frac{T_i P_i}{\cancel{T_i}} = \frac{P_f T_i}{T_f}$

Rearranged equation $P_i = \frac{P_f T_i}{T_f}$

EXERCISE A-2 Rearrange each of the following equations so that the quantity x stands alone on one side of the equation.

(a) $5x = 15$ (b) $\frac{x}{2} = 2y$ (c) $\frac{3x}{b} = \frac{5}{3}$

It may be necessary to use more than one operation to rearrange some equations. Consider the following equation.

$$\frac{x}{2} - y = 7$$

This equation can be rearranged so that x stands alone on one side by first adding y to both sides.

$$\frac{x}{2} - y + y = 7 + y$$

Then multiply both sides by 2.

$$2\frac{x}{2} = 2(7 + y)$$

$$x = 14 + 2y$$

EXERCISE A-3 Rearrange each of the following equations so that the quantity x stands alone on one side of the equation.

(a) $4x + 16 = y$ (b) $\frac{y}{17} = x - 5$ (c) $x(3 + y) - 4 = 7$

A.2 PROPORTIONS

Proportions are useful in many chemical calculations. A proportion is an extension of a ratio. A ratio can be set up between any two related quantities. For example, if apples cost \$1.00 for 12 apples, the ratio is:

$$\frac{\$1.00}{12 \text{ apples}}$$

This ratio can be used to determine the cost of any number of apples. For example, what is the cost of 15 apples? The answer is obtained by setting up the following proportion.

$$\frac{\$1.00}{12 \text{ apples}} = \frac{x}{15 \text{ apples}}$$

Rearranging the equation (Section A.1):

$$\frac{\$1.00}{12 \cancel{\text{apples}}}(15 \cancel{\text{apples}}) = x = \$1.25 \qquad \text{(A-1)}$$

This proportion has been set up so that the same unit (apples) is in the denominator on both sides of the equation. As a result, the unit apples cancels and the correct unit, \$, is obtained. The proportion can be set up just as well with the same unit in the numerator on both sides of the equation. Thus the following proportion is equally correct.

$$\frac{12 \text{ apples}}{\$1.00} = \frac{15 \text{ apples}}{x}$$

$$x = \frac{\$1.00}{12 \cancel{\text{apples}}}(15 \cancel{\text{apples}}) = \$1.25$$

The chance of setting up a proportion incorrectly is greatly reduced if the units of the quantities are included.

Notice that the ratio \$1.00/12 apples is also a conversion factor (Section 1.4). Thus, setting up a proportion or using a conversion factor is really the same thing. To show that this is true, let us calculate the cost of 15 apples using the conversion factor method.

15 apples × conversion factor = cost

The conversion factor must have the unit of apples in the denominator to cancel apples. Thus the conversion factor is \$1.00/12 apples.

$$15 \cancel{\text{apples}} \frac{(\$1.00)}{12 \cancel{\text{apples}}} = \$1.25 \qquad \text{(A-2)}$$

Equations A-1 and A-2 are identical. So both methods are the same.

EXERCISE A-4 (a) A patient must take three tablets per day. How many days will one hundred tablets last the patient?
(b) A factory chimney emits 2.5 g of pollutant per hour. How many grams of pollutant are emitted in 24 hours?

A.3 PERCENTAGE

Any number can be changed into a *percentage* term by multiplying it by 100%. Numbers expressed as a percent are often used to report how much of a certain thing is contained in a sample. For example, a box contains 15 red balls and 32 white balls. The percentage of red balls in the box is obtained as follows.

$$\begin{matrix}\text{Percent} \\ \text{Red balls}\end{matrix} = \frac{\text{Number of red balls}}{\text{Total number of balls in the box}} \times 100\%$$

$$\begin{matrix}\text{Percent} \\ \text{Red balls}\end{matrix} = \frac{15}{47} \times 100\% = 0.32 \times 100\% = 32\%$$

EXERCISE A-5 (a) There are 100 women and 75 men in a class. What is the percent of men and the percent of women in the class?
(b) A box contains 700 red marbles, 300 blue marbles, and 75 white marbles. What is the percent of each type of marble in the box?

A.4 ACCURACY, PRECISION, AND SIGNIFICANT FIGURES

Accuracy and precision are two terms that are used frequently in science. Their use can be illustrated by the following example.

A physician prescribes both the type and the amount of medication for a patient. Usually the medication is given in the form of a pill or capsule that contains a specific amount of the drug. For example, the tranquilizer chlordiazepoxide is available in capsules that contain 10 mg of this drug. If we were to determine the amount of drug in one of these capsules, would it be exactly 10 mg? The answer to this question depends upon how accurately the manufacturer can make the capsule. By this is meant how close the agreement is between the actual amount of drug in the capsule and 10 mg. *Accuracy* then means the degree of agreement between the measured value (in the example, the amount of drug in the capsule) and the true value, 10 mg in this case.

Do all these capsules contain 10 mg? Or do some of them contain more or less? In other words, how large a variation is there in the amount of tranquilizer in each of a number of capsules? The degree of agreement between successive measurements is called *precision.* The smaller the variation between successive measurements, the better the precision. Good precision does not necessarily mean good accuracy. For example, if only 5 mg of tranquilizer is placed in each capsule due to a consistent error in manufacturing, the precision could be good but the accuracy would be very poor. In such a case, the patient taking these capsules would receive only half as much medication as prescribed. For this reason pharmaceutical companies continually check the amount and purity of medication in pills or capsules.

EXERCISE A-6 According to the label on three different brands of cereal, each contains 375 g of cereal. The following are the weights of cereal found in 5 boxes of each brand. Decide if the precision and accuracy of the weight of cereal are good or bad in each brand.

Brand A	*Brand B*	*Brand C*
374 g	360 g	360 g
373 g	355 g	380 g
376 g	365 g	390 g
377 g	360 g	330 g
375 g	360 g	342 g

The number of significant figures in a measurement is an indication of its precision. The *significant figures* in a number are all the digits whose values are known with certainty plus the first digit whose value is uncertain. For example, the number 125 contains three significant figures. The 1 hundred and the 2 tens are known with certainty while the 5 units are uncertain. The digit whose value is uncertain is always the last digit on the right.

The digit 0 can be a significant figure or not. If it is used to place the decimal point, 0 is not a significant figure. The digit 0 is always a significant figure whenever a nonzero digit is on both its left and right sides in a number. For example, the number 46075 contains five significant figures no matter where the decimal point is located. Thus 46075 μm, 4.6075 cm, and .046075 m all have the same number of significant figures. Each of these is the same number expressed in different units. The digit 0 between the decimal point and the digit 4 in the number .046075 is not a significant figure because it is used only to place the decimal point.

Sometimes it is impossible to determine if the digit zero is used to place the decimal point or whether it is part of the measurement. This is the case in the number 4200. This number contains four digits but it is impossible to tell how many significant figures it contains. The two zeroes may be significant or they may be included just to indicate the decimal point. It is best to write such a number in exponential notation to avoid confusion (Section A.5).

EXERCISE A-7 How many significant figures are there in each of the following numbers?
(a) 12 (b) 1.23 (c) .0402 (d) 4003 (e) 43000
(f) .00043

Frequently, multiplications and divisions are carried out with numbers that have different numbers of significant figures. How many significant figures are there in the result? This is an important question because hand-held calculators give results that have up to 10 digits. How many of these digits are really significant? The answer to this question is that *there are no more significant figures in the answer than there are in the number with the least number of significant figures involved in the calculation.* This is shown in the following example.

$$\frac{34.23 \times 14.5}{23.235} = 21.36152356 \text{ (answer from a hand-held calculator)}$$

The number 14.5 has the least number of significant figures, three, of all the numbers involved in the calculation. Therefore the answer cannot have more than three significant figures. To arrive at this number, the answer must be rounded off to three significant figures. To round off means to increase the last significant figure by 1 if the digit to its right is 5 or greater or to leave it as it is if the digit is 4 or less. Since the digit to the right of the third significant figure is 6, the answer is rounded off to 21.4.

Notice that the numbers involved in the calculation are never rounded off before the calculation is carried out. Rounding off is done only to the answer of a calculation.

EXERCISE A-8 Carry out the following calculations and give the answer to the correct number of significant figures.

(a) 5.3×10.7 (b) $\frac{75.43}{2.67}$ (c) $\frac{43.2 \times 7.23}{6.743}$

(d) $\frac{7.54 \times 6.710}{5.32 \times 4.320}$

The number of significant figures in the answers to addition and subtraction problems is determined in a different way. The placing of the decimal point is important in determining how many digits are significant. This is shown in the following example.

$$\begin{array}{r|l} 15.99 & \\ 1.73 & 5 \\ .04 & 37 \\ \hline 17.76 & 87 \end{array}$$

The last digit in each of the numbers being added is uncertain. The number 15.99 has an uncertainty in the one-hundredth place. The other two numbers have uncertainties in the one-thousandth and one-ten-thousandth place, respectively. The uncertainty in the one-hundredth place of the number 15.99 means that the answer must also have an uncertainty in the one-hundredth place. Consequently, any digit beyond the hundredth place is uncertain and is not significant. Thus the answer after rounding off is 17.77.

EXERCISE A-9 Carry out the following additions or subtractions and give the answer to the correct number of significant figures.

(a) $17.3 + 5.25$ (b) $7.552 - 1.34$

(c) $4.73 + .0125 + 1.376$

Certain numbers, called *pure numbers,* are defined exactly. For example, if we count six people, the number is exactly six, no more and no less. We can write this number as simply 6 or 6.00 or with as many zeroes after the decimal as we want or need in a calculation. Some conversion factors are pure numbers because they are defined. The important point to remember about pure numbers is that they never limit the number of significant figures an answer can have.

A.5 EXPONENTIAL NOTATION

Writing Numbers in Exponential Notation

Extremely large or small numbers are most commonly written in exponential notation. In this method numbers are written as the product of two terms:

Coefficient → $A \times \underbrace{10^{b}}_{\text{Exponential}}$ ← Exponent

A, called the coefficient, is a number between one and ten; b, called the exponent, is either a positive or negative whole number. The exponent is written as a superscript on the number 10. When the exponent is a positive whole number, it indicates how many times the number ten will be multiplied by itself. For example:

$$10^1 = 10$$
$$10^2 = 10 \times 10 = 100$$
$$10^3 = 10 \times 10 \times 10 = 1000$$
$$10^4 = 10 \times 10 \times 10 \times 10 = 10000$$
$$10^8 = 10 \times 10 \times 10 \times 10 \times 10 \times 10 \times 10 \times 10 = 100000000$$

When the exponent is a negative whole number, it indicates how many times the number 1/10 will be multiplied by itself. For example:

$$10^{-1} = \frac{1}{10} = .1$$

$$10^{-2} = \frac{1}{10} \times \frac{1}{10} = \frac{1}{100} = .01$$

$$10^{-3} = \frac{1}{10} \times \frac{1}{10} \times \frac{1}{10} = \frac{1}{1000} = .001$$

$$10^{-4} = \frac{1}{10} \times \frac{1}{10} \times \frac{1}{10} \times \frac{1}{10} = \frac{1}{10000} = .0001$$

$$10^{-8} = \frac{1}{10} \times \frac{1}{10} \times \frac{1}{10} \times \frac{1}{10} \times \frac{1}{10} \times \frac{1}{10} \times \frac{1}{10} \times \frac{1}{10} = \frac{1}{100000000} = .00000001$$

Notice that positive exponents represent large numbers while negative exponents represent small numbers.

Any number can be written in exponential notation. For example, the number 2300 can be written 2.3×10^3. To determine the value of the exponent, count the number of places to the right or left that the decimal must be moved in order to place it after the first significant figure. If the decimal point must be moved to the right, the exponent is a negative number. In the example 2300 the decimal point must be moved three places to the left to get the coefficient 2.3. The exponent is 3 and the number is 2.3×10^3 in exponential notation. The following are additional examples.

$123 = 1.23 \times 10^2$	$.0123 = 1.23 \times 10^{-2}$	$654000 = 6.54 \times 10^5$
21	12	54321
left ←	right →	left ←

EXERCISE A-10 Express the following numbers in exponential notation.
(a) 265 (b) 173000 (c) .00573 (d) .126 (e) .0543

To convert a number from exponential notation to its nonexponential form, simply multiply the coefficient by the exponential. Thus:

$$5.75 \times 10^{-3} = 5.75 \times \frac{1}{10} \times \frac{1}{10} \times \frac{1}{10} = 5.75 \times \frac{1}{1000} = .00575$$

$$6.25 \times 10^2 = 6.25 \times 10 \times 10 = 625$$

EXERCISE A-11 Express the following numbers in nonexponential form.
(a) 1.25×10^2 (b) 2.5×10^{-2} (c) 3.75×10^3
(d) 5.23×10^5 (e) 6.2×10^{-8}

Exponential Notation and Significant Figures

Numbers are written in exponential notation so that all the digits in the coefficient are significant. This allows us to easily show the significant figures in some numbers that we could not specify any other way. Suppose, for example, that the number 30000 were known to two significant figures. Writing it as 30000 does not indicate this. But writing it 3.0×10^4 clearly indicates that the number has two significant figures. The following are additional examples.

Exponential Form	*Nonexponential Form*	*No. of Significant Figures*
6.02×10^2	602	Three: 6, 0, and 2
2.14×10^{-2}	.0214	Three: 2, 1, and 4
4.5×10^3	4500	Two: 4 and 5
6.000×10^5	600000	Four: 6, 0, 0, and 0

Multiplying Numbers in Exponential Notation

To multiply numbers in exponential notation, multiply the coefficients but *add the exponents.* For example:

$$(2 \times 10^2)(4 \times 10^3) = \underbrace{8}_{4 \times 2} \times 10^{5 \leftarrow 2+3}$$

$$(1.3 \times 10^{-2})(1.7 \times 10^4) = \underbrace{2.2}_{1.3 \times 1.7} \times 10^{2 \leftarrow -2+4}$$

$$(1.53 \times 10^{-3})(2.4 \times 10^{-2}) = \underbrace{3.7}_{1.53 \times 2.4} \times 10^{-5 \leftarrow -3+(-2)}$$

EXERCISE A-12 Multiply the following numbers.
(a) $(3 \times 10^2)(2 \times 10^3)$ (b) $(1.6 \times 10^{-4})(2.3 \times 10^5)$
(c) $(2.75 \times 10^{-4})(6.94 \times 10^{-3})$

Dividing Numbers in Exponential Notation

To divide numbers written in exponential notation, divide the coefficient in the numerator by the coefficient in the denominator but *subtract the exponent in the denominator from the exponent in the numerator.* For example:

$$\frac{6 \times 10^4}{3 \times 10^2} = 2 \times 10^2 \leftarrow 4 - 2$$

$$\frac{4.5 \times 10^{-3}}{2.5 \times 10^{-2}} = 1.8 \times 10^{-1} \leftarrow -3 - (-2)$$

$$\frac{8.23 \times 10^4}{2.4 \times 10^{-2}} = 3.4 \times 10^6 \leftarrow 4 - (-2)$$

EXERCISE A-13 Divide the following numbers.

(a) $\frac{4 \times 10^6}{2 \times 10^3}$ (b) $\frac{3.5 \times 10^{-3}}{1.2 \times 10^2}$ (c) $\frac{7.54 \times 10^2}{1.65 \times 10^{-2}}$

(d) $\frac{8.54 \times 10^{-2}}{6.15 \times 10^{-3}}$

A.6 LOGARITHMS

A logarithm is an exponent. A logarithm to the base 10, usually written simply as log, is an exponent of the number 10. The logarithm of any number, N, is the exponent (or the power) to which 10 must be raised to equal N. For example, the number 100 is equal to 10^2. The logarithm of 100 is therefore 2. These statements can be written as equations. Thus, if

$$N = 10^x$$

then

$$\log N = x.$$

The quantity x is the logarithm of N to the base 10. The following are additional examples.

10	$\log 10 = 1$	$10 = 10^1$
100	$\log 100 = 2$	$100 = 10^2$
1000	$\log 1000 = 3$	$1000 = 10^3$
100000	$\log 100000 = 5$	$100000 = 10^5$
0.1	$\log 0.1 = -1$	$.1 = \frac{1}{10} = 10^{-1}$
.001	$\log .001 = -3$	$.001 = \frac{1}{10^3} = 10^{-3}$

Most of the numbers whose logarithms we want to determine are written in exponential notation. The logarithm of a number such as 3.5×10^{-3} is calculated by adding the logs of the coefficient (3.5) and the exponential (10^{-3}). Thus:

$$\log(3.5 \times 10^{-3}) = \log 3.5 + \log 10^{-3}$$

The value of log 3.5 (obtained from a log table) is 0.54 while the value of $\log 10^{-3}$ is -3. Adding these together gives the value of the log of 3.5×10^{-3}.

$$\log 3.5 \times 10^{-3} = 0.54 - 3.00 = -2.45$$

EXERCISE A-14 Determine the logarithm of each of the following numbers.
(a) 1.54 (b) 1.76×10^{2} (c) 5.76×10^{-3}
(d) 8.56×10^{-10} (e) 3.25×10^{7}

To find the number corresponding to a logarithm, we reverse the procedure given in the preceding paragraph. For example, what is the number whose logarithm is -2.28? Since the logarithm is an exponent, the number is $10^{-2.28}$. But this is not the usual way that we write numbers in exponential notation. Such numbers always have a coefficient followed by an exponential that has a whole number as an exponent (Section A.5). In order to convert $10^{-2.28}$ into exponential notation, we rewrite the exponent -2.28 as $-3 + 0.72$. Thus:

$$10^{-2.28} = 10^{-3} \times 10^{.72}$$ (Remember that adding exponents means to multiply exponentials.)

Now we want to find the number whose log is 0.72. From a log table, we find that this number is 5.24. Therefore the number whose log is -2.28 is 5.24×10^{-3}.

EXERCISE A-15 Determine the number corresponding to each of the following logarithms.
(a) -1.64 (b) -5.73 (c) 7.84 (d) -8.45 (e) -11.12

ANSWERS TO EXERCISES

CHAPTER 1

1-1

Unit	Symbol	Relationship to the Gram
teragram	Tg	10^{12} g
gigagram	Gg	10^{9} g
megagram (metric ton)	Mg	10^{6} g
kilogram	kg	10^{3} g
gram	g	g
decigram	dg	10^{-1} g
centigram	cg	10^{-2} g
milligram	mg	10^{-3} g
microgram	μg	10^{-6} g
nanogram	ng	10^{-9} g
picogram	pg	10^{-12} g

1-2

Unit	Symbol	Relationship to the Liter
teraliter	TL	10^{12} L
gigaliter	GL	10^{9} L
megaliter	ML	10^{6} L
kiloliter	kL	10^{3} L
liter	L	L
deciliter	dL	10^{-1} L
centiliter	cL	10^{-2} L
milliliter	mL	10^{-3} L
microliter	μL	10^{-6} L
nanoliter	nL	10^{-9} L
picoliter	pL	10^{-12} L

1-3 (a) $1 = \frac{454\ g}{1\ lb}; \frac{1\ lb}{454\ g} = 1$ (d) $1 = \frac{28.4\ g}{1\ oz}; \frac{1\ oz}{28.4\ g} = 1$ (g) $1 = \frac{0.0648\ g}{1\ grain}; \frac{1\ grain}{0.0648\ g} = 1$

(b) $1 = \frac{3.78\ L}{1\ gal}; \frac{1\ gal}{3.78\ L} = 1$ (e) $1 = \frac{0.473\ L}{1\ pt}; \frac{1\ pt}{0.473\ L} = 1$

(c) $1 = \frac{1.60\ km}{1\ mi}; \frac{1\ mi}{1.60\ km} = 1$ (f) $1 = \frac{0.914\ m}{1\ yd}; \frac{1\ yd}{0.914\ m} = 1$

1-4 (a) 265 gal (b) 29.5 in. (c) 710 g (d) 297 m (e) 5.01 L (f) 40.0 km (g) 11.0 lb (h) 7.57 L

1-5 (a) 373 K (b) 253 K (c) 310 K (d) 623 K

1-6 (a) 24.9° C (b) −263° C (c) −123° C (d) 1227° C

1-7 (a) 35° C (b) 14° F (c) 302° F (d) −18° C (e) 228 K (f) −325° F

1-8 (a) 1.13×10^7 g (b) 216 g (c) 3.42×10^3 g

1-9 (a) 1.84 mL (b) 556 mL (c) 2.89×10^3 mL

1-10 (a) 0.790 (b) 1.025 (c) 10.5 (d) 1.026

CHAPTER 2

2-1 (a) clear, colorless, odorless liquid at room temperature
(b) odorless, yellow solid at room temperature
(c) odorless, white solid at room temperature
(d) gas at room temperature, supports combustion
(e) solid at room temperature, transparent

2-2 (a) homogeneous (b) heterogeneous (c) homogeneous (d) homogeneous

2-3 (a) liquid to gas (evaporation) (b) solid to liquid (melting) (c) solid to gas (sublimation)

2-4 (a) physical change (b) chemical reaction (c) physical change (d) chemical reaction

2-5 (a) kinetic (a motion is involved)
(b) chemical (the burning of the fuel is a chemical reaction)
(c) potential (if the stone moves, an energy change will result)

2-6 (a) Some of the chemical energy of the coal is converted to heat, to light (the fire glows), and to sound (the fire crackles). The remainder of the energy is contained in the products of combustion.
(b) The potential energy of the book is converted into kinetic energy, which is converted into sound when the book hits the floor.
(c) The kinetic energy of the hammer is converted into sound and is transferred into the kinetic energy of the nail.
(d) The chemical energy of the food is converted into heat (you perspire when jogging) and the mechanical action of the muscles.

2-7 2.4×10^4 cal, 24 kcal

2-8 (a) 54.0 kcal (b) 15.9 kcal (c) 8.4 kcal (d) 5.9 kcal

2-9 (a) 1.00 kcal (b) 0.0949 kcal (c) 0.581 kcal (d) 0.217 kcal

CHAPTER 3

3-1

Atom	A	D	G	H
Number of protons	5	8	3	10
Number of electrons	5	8	3	10
Atomic number	5	8	3	10

3-2 (a) 6 (b) 17 (c) 26 (d) 50 (e) 79 (f) 47 (g) 92

3-3 Atoms M and Q are isotopes.

3-4

	Element	Mass Number	Atomic Number
(a)	Cl	35	17
(b)	Cl	37	17
(c)	N	14	7
(d)	N	15	7

3-5 (a) ${}^{35}_{17}Cl$ (b) ${}^{37}_{17}Cl$ (c) ${}^{14}_{7}N$ (d) ${}^{15}_{7}N$

3-6 (a) Be $1s^2 2s^2$ (b) C $1s^2 2s^2 2p_x 2p_y$ (c) F $1s^2 2s^2 2p_x{}^2 2p_y{}^2 2p_z$ (d) Na $1s^2 2s^2 2p_x{}^2 2p_y{}^2 2p_z{}^2 3s$

3-7 (a) B:2:3 (b) N:2:5 (c) Al:2:8:3 (d) P:2:8:5 (e) Cl:2:8:7

3-8 shell 1, 2 electrons; shell 2, 8 electrons; shell 3, 18 electrons; shell 4, 32 electrons

3-9 (a) 5 (b) 4 (c) 7 (d) 1 (e) 8 (f) 4

CHAPTER 4

4-1 (a) ${}^{40}_{20}Ca$ (b) ${}^{220}_{86}Rn$ (c) ${}^{233}_{90}Th$ (d) ${}^{234}_{90}Th$

4-2 (a) ${}^{24}_{12}Mg$ (b) ${}^{12}_{6}C$ (c) ${}^{206}_{82}Pb$ (d) ${}^{1}_{1}H$

4-3 22.8 days

4-4 31,250 atoms

CHAPTER 5

5-1 (a) Ca^{+2}:2:8:8 (b) S^{-2}:2:8:8 (c) Li^{+}:2 (d) F^{-}:2:8 (e) Al^{+3}:2:8

5-2 (a) and (d) are elements. (b), (c), (e), (f), and (g) are ions.

5-3 (a) K· (b) Mg: (c) :S̈: (d) :F̤̈· (e) Al· (f) Li·

5-4 (a) Li^+Cl^- 1:1 (b) $Mg^{+2}F^-$ 1:2 (c) Li^+S^{-2} 2:1 (d) $Mg^{+2}O^{-2}$ 1:1 (e) $Al^{+3}O^{-2}$ 2:3 (f) $Ca^{+2}Cl^-$ 1:2

5-5 (a) K_2O (b) $CaCl_2$ (c) $FeBr_2$ (d) CuS (e) Fe_2O_3

5-6 (a) lithium bromide (b) sodium iodide (c) calcium fluoride (d) ferric chloride (e) aluminum oxide (f) cupric sulfide (g) mercuric chloride

5-7 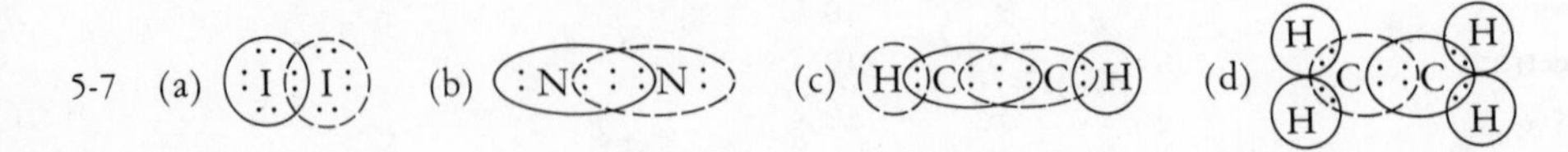

5-8 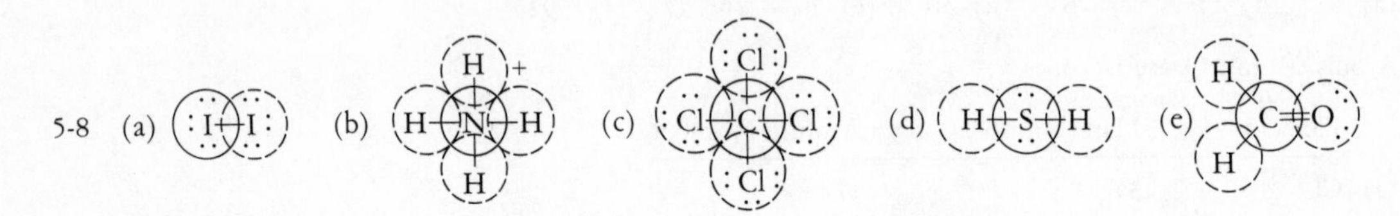

5-9 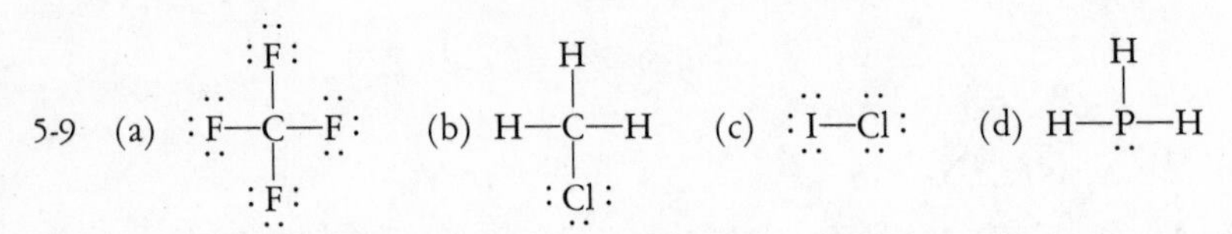

5-10 (a) LiOH (b) $NaHCO_3$ (c) $CaCO_3$ (d) $(NH_4)_2HPO_4$ (e) $Ba_3(PO_4)_2$

5-11 (a) sodium nitrite (b) barium phosphate (c) potassium hydrogen phosphate (d) lithium hydrogen sulfate (e) ammonium sulfate

5-12

Ion	Total number of protons	Total positive charge	Total number of electrons	Total negative charge	Charge of ion
(a) NH_4^+	11	+11	10	−10	+1
(b) CO_3^{-2}	30	+30	32	−32	−2

5-13 $O{=}N^+(O^-)(O^-)$ ⟷ $O^-{-}N^+({=}O)(O^-)$ ⟷ $O^-{-}N^+(O^-)({=}O)$

CHAPTER 6

6-1 The nitrogen atom is located at the center of a tetrahedron and the four hydrogens occupy the corners. The H—N—H bond angle is close to 109.5 degrees.

6-2 HF and NH_3 are the only two polar molecules.

6-3 (a) van der Waals or hydrophobic attractions
(b) and (c) dipolar attractions
(d) and (e) hydrogen bonds

6-4 Both are difficult to compress. Solids remain rigid while liquids take the shape of their container.

6-5 60.8 L

6-6 2 atm

6-7 127 mL

6-8 149 K or −124° C

6-9 2 atm

6-10 596 K or 323° C

6-11 2.6 L

6-12 373 K or 100° C

6-13 P_{N_2} 593 mm Hg; P_{O_2} 160 mm Hg; P_{Ar} 7.1 mm Hg; P_{CO_2} 0.53 mm Hg

CHAPTER 7

7-1 (a) 1 N, 3 H (b) 1 C, 4 Cl (c) 2 Cl (d) 2 C, 6 H, 1 O (e) 12 C, 26 H (f) 2 Na, 1 H, 1 P, 4 O

7-2 (a) 28.02 amu (b) 70.90 amu (c) 17.04 amu (d) 153.81 amu (e) 84.93 amu (f) 180.18 amu (g) 342.34 amu (h) 110.98 amu (i) 119.98 amu (j) 105.99 amu

7-3 (a) %C 27.3%, %O 72.7%
(b) %N 82.2%, %H 17.8%
(c) %C 14.1%, %H 2.38, %Cl 83.5%
(d) %C 40.0%, %H 6.73%, %O 53.3%
(e) %C 40.7%, %H 8.55%, %N 23.7%, %O 27.1%

7-4 (a) $H_2 + Cl_2 \longrightarrow 2\,HCl$
(b) $C + 2\,H_2 \longrightarrow CH_4$
(c) $4\,Al + 3\,O_2 \longrightarrow 2\,Al_2O_3$
(d) $2\,PtO \longrightarrow 2\,Pt + O_2$
(e) $C_2H_2 + 2\,H_2 \longrightarrow C_2H_6$

7-5 (a) 70.90 g (b) 84.93 g (c) 101.96 g (d) 58.44 g (e) 60.06 g

7-6 (a) 0.0670 mol (b) 0.428 mol (c) 1.28 mol (d) 5.55 mol

7-7 (a) 21.0 g (b) 43.8 g (c) 183 g (d) 164 g

7-8 (a) 118 L (b) 44.8 L (c) 12.9 L (d) 72.8 L

7-9 (a) 0.416 mol (b) 2.00 mol (c) 0.901 mol (d) 3.37 mol

7-10 (a) 22.0 g (b) 508 g (c) 149 g (d) 1.01×10^3 g

7-11 245 g; 1.25×10^5 mL

7-12 35.8 g

7-13 2.81 g

7-14 14.7 g

7-15 (a) exothermic (b) endothermic (c) exothermic (d) endothermic

7-16 266 kcal

7-17 Li is oxidized; Br_2 is reduced; $Li \longrightarrow Li^+ + e^-$; $Br_2 + 2e^- \longrightarrow 2\,Br^-$; $2\,Li + Br_2 \longrightarrow 2\,LiBr$

7-18 (a) C +2; O −2 (c) O −2; C +4 (e) Fe +3; O −2 (g) H +1; O −2; C −2
(b) O −2; H +1 (d) Mg +2; Cl −1 (f) C −4; H +1

7-19 (a) Oxidation numbers: Al 0; O_2 0; in Al_2O_3 Al +3, O −2. Al is oxidized. O_2 is reduced. Al is reducing agent. O_2 is oxidizing agent. Oxidation half-reaction $Al \longrightarrow Al^{+3} + 3e^-$. Reduction half-reaction $O_2 + 4e^- \longrightarrow 2\,O^{-2}$.
(b) Oxidation numbers: H_2 0; Cu 0; in H_2O, H +1, O −2; in CuO, Cu +2, O −2. H_2 is oxidized. Cu is reduced. H_2 is reducing agent. Cu is oxidizing agent. Oxidation half-reaction $H_2 \longrightarrow 2\,H^+ + 2e^-$. Reduction half-reaction $Cu^{+2} + 2e^- \longrightarrow Cu$.
(c) Oxidation numbers: Zn 0; Cu 0; in $CuSO_4$, Cu +2, S +6, O −2; in $ZnSO_4$, Zn +2, S +6, O −2. Zn is oxidized. Cu is reduced. Zn is reducing agent. Cu is oxidizing agent. Oxidation half-reaction $Zn \longrightarrow Zn^{+2} + 2e^-$. Reduction half-reaction $Cu^{+2} + 2e^- \longrightarrow Cu$.
(d) Oxidation numbers: H_2 0; in CH_2O, C 0, H +1, O −2; in CH_4O, C −2, H +1, O −2. H_2 is oxidized. C is reduced. C is oxidizing agent. H_2 is reducing agent. Oxidation half-reaction $H_2 \longrightarrow 2\,H^+ + 2e^-$. Reduction half-reaction $CH_2O + 2\,H^+ + 2e^- \longrightarrow CH_4O$.

CHAPTER 8

8-1 (a) 1.48% (b) 1.38%

8-2 (a) 3.3% (b) 3.0%

8-3 (a) 1.00% (b) 2.50%

8-4 (a) 320 mg/100 mL (b) 2.0×10^4 mg/100 mL

8-5 0.50 ppm

8-6 (a) 1.00 M (b) 0.370 M

8-7 5.84 g (b) 67.6 g (c) 25.7 g

8-8 (a) 2.17 g (b) 2.00 g (c) 2.10 g

8-9 (a) 1.00 Eq (b) 0.300 Eq (c) 2.00 Eq (d) 1.05 Eq (e) 3.00 Eq (f) 1.50 Eq

8-10 (a) 1 mEq/L Li^+; 1 mEq/L Br^- (b) 100 mEq/L Na^+; 100 mEq/L CO_3^{-2}

8-11 59.5 mEq/L

8-12 (a) 1 mol H^+, 1 mol Cl^- (c) 1 mol Ca^{+2}, 2 mol Cl^- (e) 3 mol Li^+, 1 mol PO_4^{-3}
(b) 1 mol Na^+, 1 mol OH^- (d) 2 mol Na^+, 1 mol CO_3^{-2}

8-13 (a) toward 1 M NaCl solution (b) toward 1 M glucose solution (c) toward 1.0 M NaCl solution

8-14 (a) 1 M LiBr (b) 1 M Na_2CO_3 (c) 0.6 M NaCl

8-15 (a) 1 M LiBr is hypertonic. Glucose solution is hypotonic.
(b) Na_2CO_3 solution is hypertonic. NaCl solution is hypotonic.
(c) NaCl solution is hypertonic. Glucose solution is hypotonic.

CHAPTER 9

9-1 $Ag^+ + Cl^- \longrightarrow AgCl\downarrow$

9-2 (a) $Fe^{+2} + S^{-2} \longrightarrow FeS\downarrow$
(b) no reaction
(c) no reaction
(d) $Cu^{+2} + S^{-2} \longrightarrow CuS\downarrow$
(e) no reaction
(f) no reaction
(g) $Ag^+ + Cl^- \longrightarrow AgCl\downarrow$
(h) $Ba^{+2} + SO_4^{-2} \longrightarrow BaSO_4\downarrow$
(i) no reaction

9-3 (a) $K = \dfrac{[CH_3OH]}{[CO][H_2]^2}$
(b) $K = \dfrac{[Br_2][HCl]^2}{[Cl_2][HBr]^2}$
(c) $K = \dfrac{[NO]^2}{[N_2][O_2]}$
(d) $K = \dfrac{[NO_2]^2}{[N_2O_4]}$
(e) $K = \dfrac{[H_2O][C_4H_8O_2]}{[C_2H_4O_2][C_2H_6O]}$
(f) $K = \dfrac{[H_2O][C_2H_5Cl]}{[C_2H_6O][HCl]}$
(g) $K = \dfrac{[Ag(NH_3)_2{}^+]}{[Ag^+][NH_3]^2}$
(h) $K = \dfrac{[CN_2H_4O][H_2O]}{[NH_3]^2[CO_2]}$

9-4 (a) both reactants and products
(b) both reactants and products
(c) mostly products
(d) mostly reactants

9-5 (a) Reaction occurs to (i) left (ii) right (iii) right
(b) Reaction occurs to (i) left (ii) right (iii) left

9-6 (a) 1×10^{-10} M (b) 1×10^{-7} M (c) 1×10^{-4} M (d) 1×10^{-2} M

9-7 (a) acidic (b) acidic (c) basic (d) basic (e) neutral (f) neutral

9-8 (a) $HBr + H_2O \longrightarrow H_3O^+ + Br^-$
(b) $HI + H_2O \longrightarrow H_3O^+ + I^-$
(c) $HNO_3 + H_2O \longrightarrow H_3O^+ + NO_3^-$
(d) $HClO_4 + H_2O \longrightarrow H_3O^+ + ClO_4^-$
(e) $C_2Cl_3O_2H + H_2O \longrightarrow H_3O^+ + C_2Cl_3O_2^-$

9-9 $H_3PO_4 + H_2O \rightleftharpoons H_2PO_4^- + H_3O^+$; $H_2PO_4^- + H_2O \rightleftharpoons HPO_4^{-2} + H_3O^+$; $HPO_4^{-2} + H_2O \rightleftharpoons PO_4^{-3} + H_3O^+$

9-10 $KOH + HBr \longrightarrow H_2O + KBr$

9-11

	Full Equation	Net Ionic Equation
(a)	$KOH + HCl \longrightarrow H_2O + KCl$	$H^+ + OH^- \longrightarrow H_2O$
(b)	$NaOH + HBr \longrightarrow H_2O + NaBr$	$H^+ + OH^- \longrightarrow H_2O$
(c)	$KOH + HNO_3 \longrightarrow H_2O + KNO_3$	$H^+ + OH^- \longrightarrow H_2O$
(d)	$NaOH + H_2SO_4 \longrightarrow NaHSO_4 + H_2O$	$H^+ + OH^- \longrightarrow H_2O$
(e)	$KOH + HClO_4 \longrightarrow KClO_4 + H_2O$	$H^+ + OH^- \longrightarrow H_2O$
(f)	$2\,KOH + H_2SO_4 \longrightarrow K_2SO_4 + H_2O$	$H^+ + OH^- \longrightarrow H_2O$

9-12 Equation 9-11: $H_2CO_3 + OH^- \longrightarrow HCO_3^- + H_2O$; Equation 9-12: $HCO_3^- + OH^- \longrightarrow CO_3^{-2} + H_2O$.

9-13 (a) strong acid, strong base (b) strong acid, strong base (c) weak acid, strong base (d) weak acid, strong base (e) strong acid, strong base (f) strong acid, strong base (g) strong acid, strong base (h) strong acid, strong base

9-14 (a) neutral (b) neutral (c) basic (d) basic (e) neutral (f) neutral (g) neutral (h) neutral

CHAPTER 10

10-1 (a) 4 (b) 2 (c) 6 (d) 8 (e) 12

10-2 (a) 3.24 (b) 4.91 (c) 6.10 (d) 7.46 (e) 9.06

10-3 1.3×10^{-5} M to 5.0×10^{-9} M

10-4 Solution A, 0.310 M; Solution B, 0.376 M; Solution C, 0.625 M.

10-5 25.0 mL

10-6 (a) 36.5 g (b) 40.0 g (c) 85.7 g (d) 49.0 g (e) 29.2 g (f) 20.6 g

10-7 (a) 1 (b) 1 (c) 2 (d) 2 (e) 2 (f) 3

10-8 (a) 0.198 Eq (b) 3.03 Eq (c) 0.145 Eq (d) 1.71 Eq

10-9 2.75 Eq for (a), (b), (c), and (d).

10-10 (a) 0.802 N (b) 2.51 N (c) 3.44 N (d) 6.09 N

10-11 (a) 0.218 N (b) 0.112 N (c) 0.176 N

10-12 (a) acid (b) base (c) acid (d) base (e) base (f) base (g) base

10-13 (a) HOH (b) ClO_4^- (c) HBr (d) I^- (e) HNO_3 (f) H_2CO_3 conjugate acid; CO_3^{-2} conjugate base

10-14 (a) H_2O, H_3O^+, HBr, Br^- (b) H_2SO_4, HSO_4^-; H_2O, H_3O^+ (c) H_3O^+, H_2O; OH^-, H_2O (d) OH^-, H_2O; $C_2H_3O_2H$, $C_2H_3O_2^-$

10-15 (a) $HSO_4^- + H_2O \rightleftarrows SO_4^{-2} + H_3O^+$ $\quad HSO_4^- + H_3O^+ \rightleftarrows H_2SO_4 + H_2O$

(b) $HSO_3^- + H_2O \rightleftarrows SO_3^{-2} + H_3O^+$ $\quad HSO_3^- + H_3O^+ \rightleftarrows H_2SO_3 + H_2O$

(c) $H_2PO_4^- + H_2O \rightleftharpoons HPO_4^{-2} + H_3O^+$ $\quad H_2PO_4^- + H_3O^+ \rightleftharpoons H_3PO_4 + H_2O$

10-16 (a) to the left (b) to the left (c) to the right (d) to the left

10-17 (a) $HCN + H_2O \rightleftharpoons H_3O^+ + CN^- \quad K_a = \frac{[H^+][CN^-]}{[HCN]}$

(b) $HF + H_2O \rightleftharpoons H_3O^+ + F^- \quad K_a = \frac{[H^+][F^-]}{[HF]}$

(c) $H_2SO_4 + H_2O \rightleftharpoons HSO_4^- + H_3O^+ \quad K_{a_1} = \frac{[H^+][HSO_4^-]}{[H_2SO_4]}$

$HSO_4^- + H_2O \rightleftharpoons H_3O^+ + SO_4^{-2} \quad K_{a_2} = \frac{[H^+][SO_4^{-2}]}{[HSO_4^-]}$

(d) $H_2SO_3 + H_2O \rightleftharpoons HSO_3^- + H_3O^+ \quad K_{a_1} = \frac{[H^+][HSO_3^-]}{[H_2SO_3]}$

$HSO_3 + H_2O \rightleftharpoons SO_3^{-2} + H_3O^+ \quad K_{a_2} = \frac{[H^+][SO_3^{-2}]}{[HSO_3^-]}$

(e) $H_3PO_4 + H_2O \rightleftharpoons H_3O^+ + H_2PO_4^- \quad K_{a_1} = \frac{[H^+][H_2PO_4^-]}{[H_3PO_4]}$

$H_2PO_4^- + H_2O \rightleftharpoons H_3O^+ + HPO_4^{-2} \quad K_{a_2} = \frac{[H^+][HPO_4^{-2}]}{[H_2PO_4^-]}$

$HPO_4^{-2} + H_2O \rightleftharpoons H_3O^+ + PO_4^{-3} \quad K_{a_3} = \frac{[H^+][PO_4^{-3}]}{[HPO_4^{-2}]}$

10-18 (a) HCl (b) H_2SO_4 (c) $C_2H_3O_2H$

10-19 (a) $K_b = \frac{[Ag^+][OH^-]}{[AgOH]}$ (c) $K_{b_1} = \frac{[BaOH^+][OH^-]}{[Ba(OH)_2]} \quad K_{b_2} = \frac{[Ba^{+2}][OH^-]}{[BaOH^+]}$

(b) $K_b = \frac{[H_2CO_3][OH^-]}{[HCO_3^-]}$

10-20 (a) 2.81 (b) 5.56 (c) 7.26 (d) 9.03

CHAPTER 11

11-1 $CH_3CH_2CH_2CH_3$; $CH_3CH(CH_3)CH_3$

11-2

Functional Group		Compounds
$>C{=}C<$	alkene	$CH_3CH{=}CH_2$, $(CH_3)_2C{=}CH_2$, $CH_3CH_2CH{=}CH_2$
$-C{\equiv}C-$	alkyne	$CH_3C{\equiv}CCH_3$, $CH_3C{\equiv}CH$, $CH_3CH_2C{\equiv}CH$
(benzene ring)	aromatic	$C_6H_5CH_2CH_3$, $C_6H_5CH_3$, $C_6H_5CH(CH_3)_2$
$-\overset{\vert}{\underset{\vert}{C}}-X$	halide	$CH_3CH_2CH_2Br$, $CH_3\underset{Cl}{\underset{\vert}{C}}HCH_3$
$-\overset{\vert}{\underset{\vert}{C}}-OH$	alcohol	$CH_3\overset{OH}{\overset{\vert}{C}}HCH_3$, $CH_3(CH_2)_4CH_2OH$, CH_3CH_2OH
$-\overset{\vert}{\underset{\vert}{C}}-NH_2$	amine	$CH_3\overset{CH_3}{\underset{CH_3}{C}}NH_2$, $CH_3CH_2NH_2$
$-\overset{\vert}{\underset{\vert}{C}}-C(=O)OH$	$(-\overset{\vert}{\underset{\vert}{C}}-CO_2H)$ carboxylic acid	$CH_3CH_2CO_2H$, CH_3CO_2H, $(CH_3)_2CHCO_2H$
$-\overset{\vert}{\underset{\vert}{C}}-C(=O)O\overset{\vert}{\underset{\vert}{C}}-$	$(-\overset{\vert}{\underset{\vert}{C}}-CO_2\overset{\vert}{\underset{\vert}{C}}-)$ ester	$CH_3CO_2CH_3$, $CH_3CH_2CO_2CH_3$, $CH_3CH_2CO_2CH_2CH_3$, $CH_3CO_2CH(CH_3)_2$
$-\overset{\vert}{\underset{\vert}{C}}-C(=O)NH_2$	$(-\overset{\vert}{\underset{\vert}{C}}CONH_2)$ amide	CH_3CONH_2, $CH_2CH_3CONH_2$, $(CH_3)_3CCONH_2$
$-\overset{\vert}{\underset{\vert}{C}}-C(=O)X$	$(-\overset{\vert}{\underset{\vert}{C}}-COX)$ acid halide	CH_3COCl, CH_3CH_2COCl, $(CH_3)_2CHCOCl$
$-\overset{\vert}{\underset{\vert}{C}}-C(=O)H$	$(-\overset{\vert}{\underset{\vert}{C}}-CHO)$ aldehyde	CH_3CH_2CHO, CH_3CHO
$-\overset{\vert}{\underset{\vert}{C}}-\overset{O}{\overset{\Vert}{C}}-\overset{\vert}{\underset{\vert}{C}}-$	ketone	$CH_3CH_2\overset{O}{\overset{\Vert}{C}}CH_2CH_3$, $CH_3\overset{O}{\overset{\Vert}{C}}CH_2CH_3$, $CH_3\overset{O}{\overset{\Vert}{C}}CH_3$
$-CN$	nitrile	CH_3CH_2CN, $(CH_3)_2CHCN$

11-3 $C_{15}H_{32}$; $C_{20}H_{42}$; $C_{25}H_{52}$; $C_{50}H_{102}$

11-4 (a) 3,4-diethylhexane (b) 4-ethyl-2,3-dimethylheptane (c) 5-isopropyl-3-methyloctane

11-5 (a) $(CH_3)_2CHCH_2CH_3$

(b) 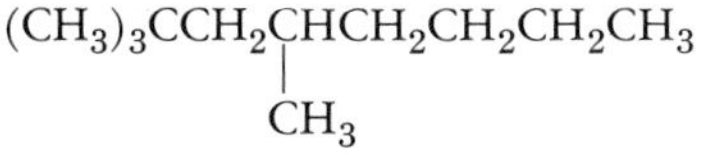 $(CH_3)_3CCH_2CHCH_2CH_2CH_2CH_3$ with CH_3 on the CH

(c) 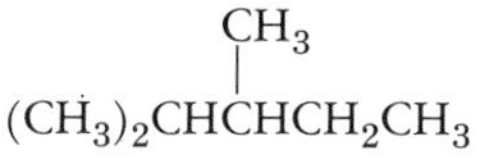 $(CH_3)_2CHCHCH_2CH_3$ with CH_3 on the second CH

(d) 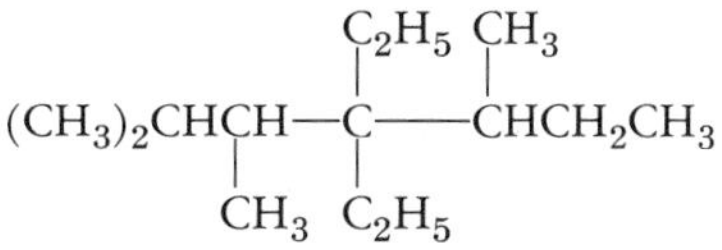 $(CH_3)_2CHCH{-}C{-}CHCH_2CH_3$ with CH_3 on the CH, C_2H_5 above and C_2H_5 below the C, and CH_3 on the CH

11-6 (a) The longest chain contains 5 carbons. The correct name is pentane.
(b) Substituents have not been given lowest number. The correct name is 2,2-dimethypentane.
(c) The longest chain contains 7 carbons. The correct name is 3-methylheptane.
(d) Substituents have not been given lowest possible number. The correct name is 4-ethyl-3-methyl-5-propyloctane.

11-7 (a) 1,2-dimethylcyclopropane
(b) 1,1-dimethylcyclopropane
(c) 1-ethyl-4-isopropylcycloheptane
(d) neopentylcyclohexane
(e) 1-cyclohexyl-2-methylpentane

11-8 (a) 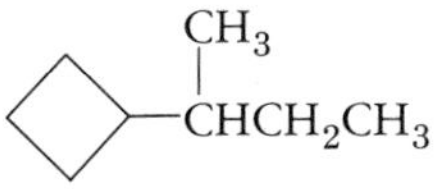 cyclobutane ring bearing $-CHCH_2CH_3$ with CH_3 on the CH

(b) 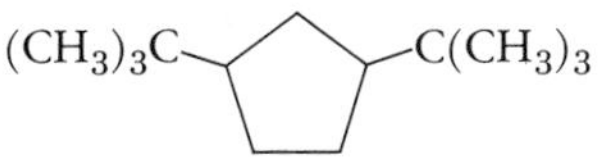 $(CH_3)_3C$– cyclopentane ring –$C(CH_3)_3$

(c) cyclooctane ring bearing $-CH_2CH(CH_3)_2$ and $-CH_2CH_2CH_3$

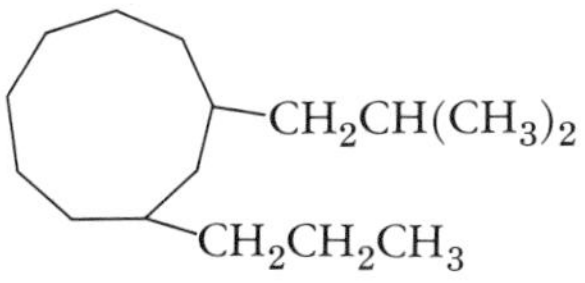

(d) 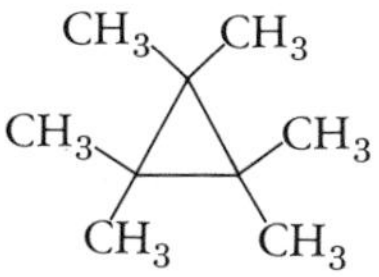 cyclopropane ring bearing six CH_3 groups

11-9 (a) 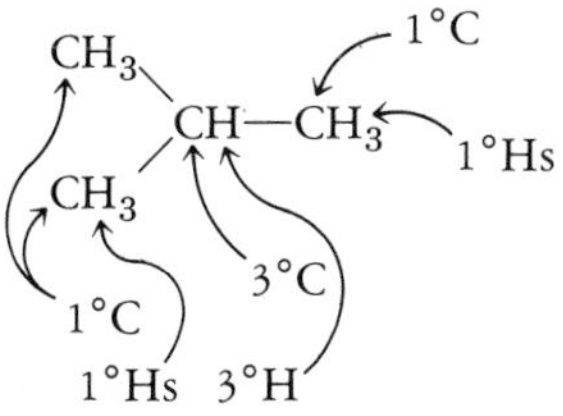 $(CH_3)_2CH{-}CH_3$: 1°C, 1°Hs on each CH_3; 3°C, 3°H on the CH

(b) $CH_3{-}CH_2CH{-}(CH_3)_2$: 1°C, 1°Hs; 2°C, 2°Hs; 3°C, 3°Hs; 1°C, 1°Hs

(c) $CH_3{-}CH_2CH_2{-}CH_3$: 1°C, 1°Hs; 2°C, 2°Hs; 1°C, 1°Hs

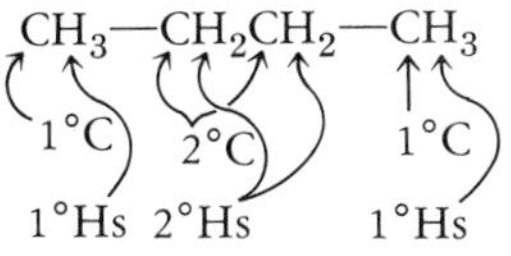

(d) 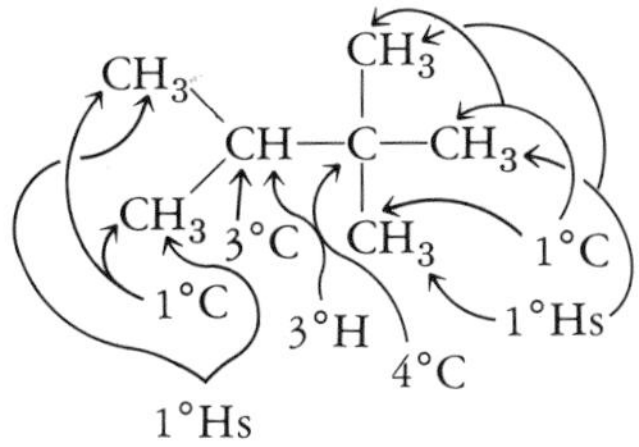 $(CH_3)_2CH{-}C(CH_3)_3$: 1°C, 1°Hs; 3°C, 3°H; 4°C; 1°C, 1°Hs

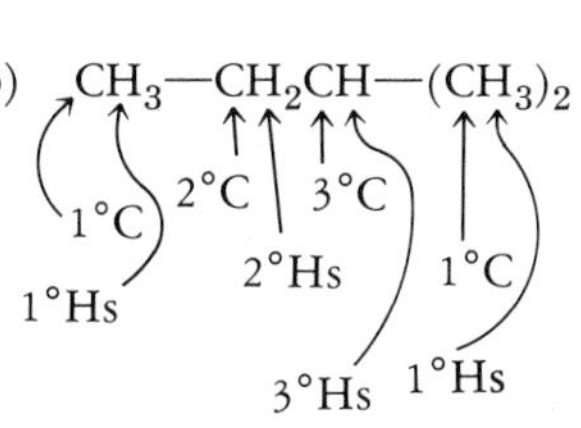

11-10 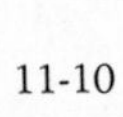

Cl Cl Cl Cl Cl Cl — Staggered conformation

Cl Cl Cl Cl Cl Cl — Eclipsed conformation

11-11

	Alkane-like part	Carbon adjacent to functional group	Functional group
(a)	$CH_3CH_2CH_2$	CH_2	OH
(b)	$(CH_3)_2$	CH	CO_2H
(c)	$(CH_3)_3CCH_2$	CH_2	NH_2

(d) 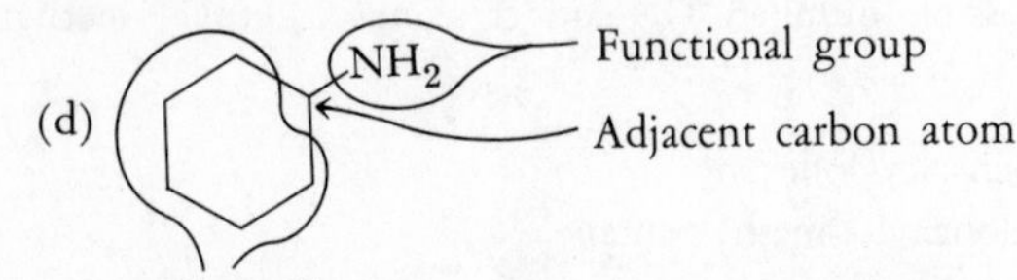

(e) 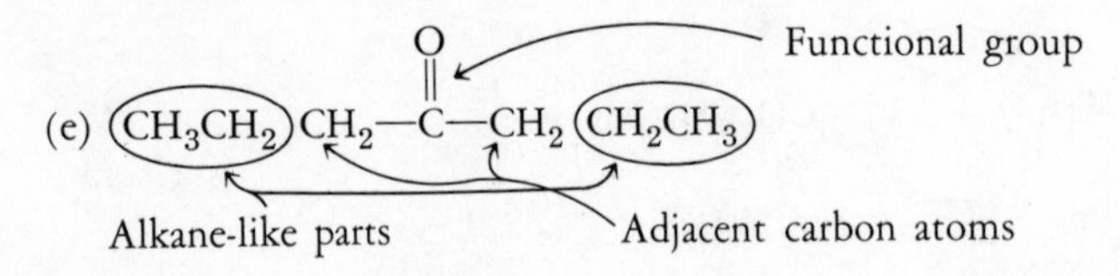

CHAPTER 12

12-1 Ethylene is ethene; propylene is propene; isobutylene is methylpropene.

12-2 (a) 4-methyl-1-pentene
(b) 2-isobutyl-4-methyl-1,4-pentadiene
(c) 3-isopropyl-1,2-dimethylcyclopropene
(d) 3,5-dimethylcyclohexene
(e) 1,3,5-cycloheptatriene

12-3 (a) [1,3-cyclohexadiene structure]

(b) cyclopropyl–$CH{=}CHCH_3$

(c) 1,2-dimethylcyclopentene: CH_3, CH_3

(d) $CH_2{=}CHCHC(CH_3)_3$ with $(CH_3)_3C$ on the third carbon

(e) $(CH_3)_2C{=}CHCH_2CH(CH_3)_2$

(f) $(CH_3)_2C{=}CCH_2CH_3$ with CH_3 on the third carbon

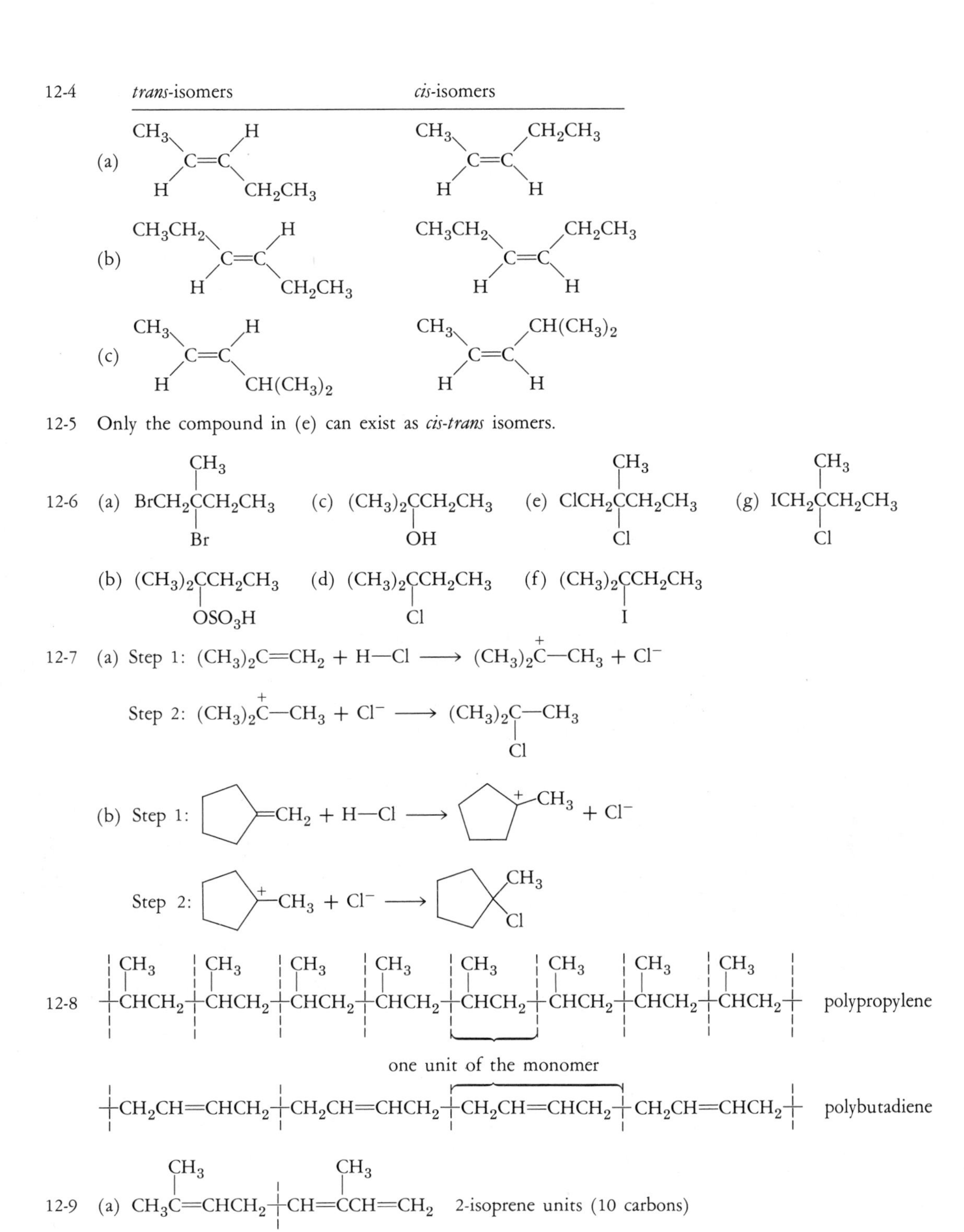
12-4
trans-isomers
cis-isomers
(a)
(b)
(c)
12-5 Only the compound in (e) can exist as cis-trans isomers.
12-6
12-7 (a) Step 1:
Step 2:
(b) Step 1:
Step 2:
12-8
polypropylene
one unit of the monomer
polybutadiene
12-9 (a)
2-isoprene units (10 carbons)

(b) [structure: bicyclic sesquiterpene with CH_3, $(CH_3)_2CH$, and CH_3 substituents] — 3 isoprene units (15 carbons)

12-10 (a) $CH_3CH(OH)CH_2OH$ (b) cyclopentane-1,2-diol (OH, OH) (c) 1-methylcyclohexane-1,2-diol (CH_3, OH, OH) (d) $HOCH_2CH(OH)CH_2C(CH_3)_3$

12-11 (a) CH_3CHCH_2 (epoxide, O) $+ C_6H_5CO_2H$ (c) 1-methylcyclohexene oxide (CH_3, O) $+ C_6H_5CO_2H$

(b) cyclopentene oxide (O) $+ C_6H_5CO_2H$ (d) $CH_2CHCH_2C(CH_3)_3$ (epoxide, O) $+ C_6H_5CO_2H$

12-12 (a) $CH_3CO_2H + CO_2$ (c) $CH_3CCH_2CH_2CH_2CH_2CO_2H$ (C=O)
(b) $HO_2CCH_2CH_2CH_2CO_2H$ (d) $CO_2 + HO_2CCH_2C(CH_3)_3$

12-13 (a) 2-butyne (c) 4,5-dimethyl-2-heptyne (e) 3,3-dimethyl-1-butyne
(b) 3-hexyne (d) 3-cyclopentyl-1-butyne (f) 1-butyne

12-14 (a) cyclohexyl–$C{\equiv}CH$ (c) cyclopropyl–$CH_2C{\equiv}CH$

(b) $CH_3C{\equiv}CC(CH_3)_3$ (d) $CH_3C{\equiv}CCH(C(CH_3)_3)CH(CH(CH_3)_2)CH(CH_3)CH_2CH_3$

12-15 (a) 1-butyne (b) 2-pentyne (c) 2-butyne (d) 3-methyl-1-pentyne (e) 2,5-dimethyl-3-hexyne

12-16 (a) $CH_3CH_2C(Cl){=}CHCl$ (c) $CH_3CH_2C(Cl){=}CH_2$ (e) $CH_3CH_2CH{=}CH_2$

(b) $CH_3CH_2C(=O)CH_3$ (d) $CH_3CH_2CCl_2CH_3$

CHAPTER 13

13-1 (a) fluorobenzene
(b) *t*-butylbenzene
(c) 3-phenyl-1-butyne
(d) phenyliodomethane
(e) 3-chloro-2-phenylpentane

13-2 (a) 1,3-dichlorobenzene
(b) 3-*t*-butylphenol
(c) 4-nitrobenzenesulfonic acid
(d) 3-nitrotoluene
(e) 2-chloroaniline
(f) 4-isobutylbenzenesulfonic acid
(g) 3-ethylbenzoic acid
(h) 2-bromoacetophenone
(i) 3-nitrobenzonitrile
(j) 3-propylbenzaldehyde

13-3 (a)  F, Cl
(e) 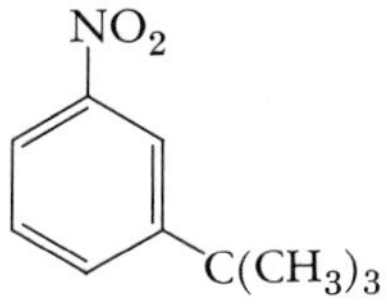 NO_2, $C(CH_3)_3$
(i)

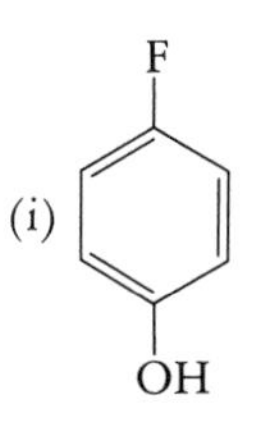

(m) $COCH_3$, CH_2CH_3
(b) CO_2H, $(CH_2)_2CH_3$
(f) NH_2, CHC_2H_5, CH_3
(j) NO_2, CH_2CH_3
(c) CH_3, F 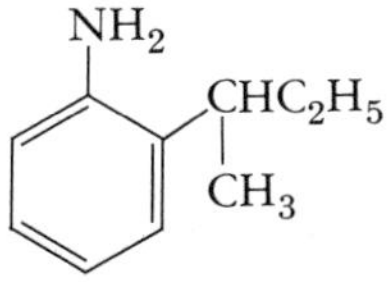
(g) 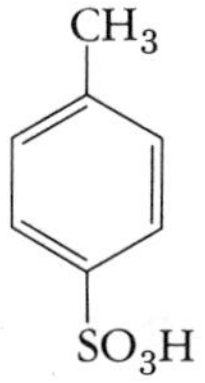 CH_3, SO_3H
(k) 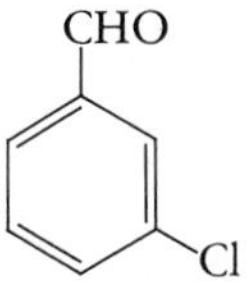 CHO, Cl
(d) NH_2, NO_2
(h) CH_3
(l) 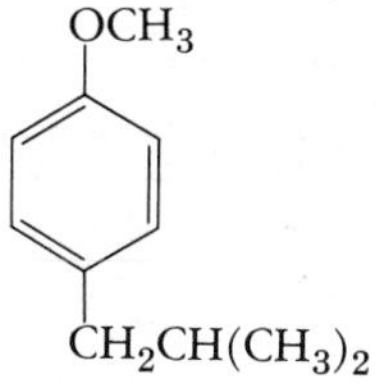 OCH_3, $CH_2CH(CH_3)_2$

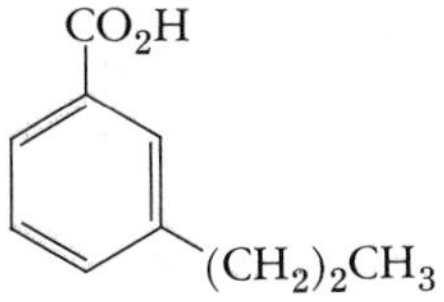

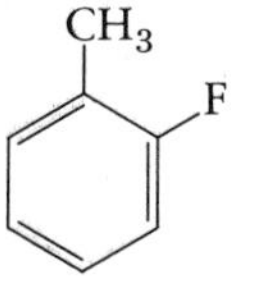

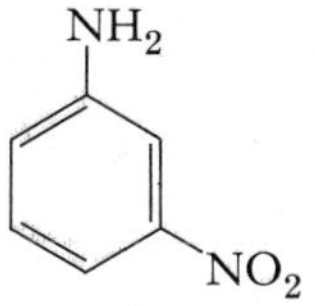

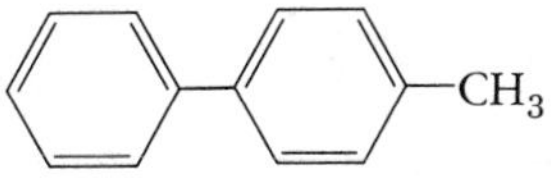

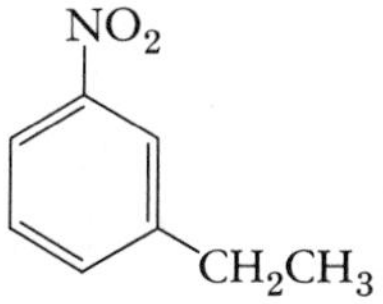

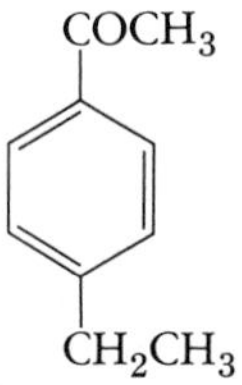

13-4 (a) 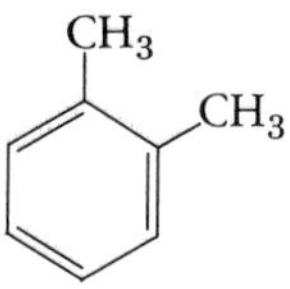 CH_3, CH_3
(b)  CH_3, CH_3
(c) 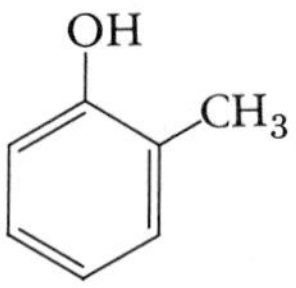 OH, CH_3
(d) 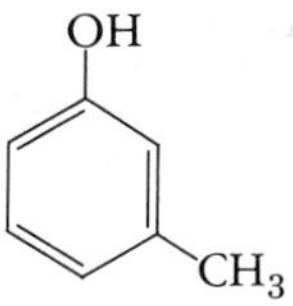 OH, CH_3

13-5 (a) o-$ClC_6H_4NO_2$ + p-$ClC_6H_4NO_2$
(b) o-$C_2H_5OC_6H_4NO_2$ + p-$C_2H_5OC_6H_4NO_2$
(c) m-$NO_2C_6H_4CO_2CH_3$
(d) m-$NO_2C_6H_4COCH_3$
(e) o-$NO_2C_6H_4C(CH_3)_3$ + p-$NO_2C_6H_4C(CH_3)_3$

13-6 (a) o-ClC_6H_4Cl + p-ClC_6H_4Cl
(b) o-$ClC_6H_4OC_2H_5$ + p-$ClC_6H_4OC_2H_5$
(c) m-$ClC_6H_4CO_2CH_3$
(d) m-$ClC_6H_4COCH_3$
(e) o-$ClC_6H_4C(CH_3)_3$ + p-$ClC_6H_4C(CH_3)_3$

13-7 (a) slower (b) faster (c) slower (d) slower (e) faster

13-8 $FeCl_3 + Cl{-}Cl \longrightarrow FeCl_4^-Cl^+$; $FeCl_4^-Cl^+ + C_6H_6 \longrightarrow$ [arenium ion: ring carbon bearing Cl and H, + charge] $+ FeCl_4^-$;

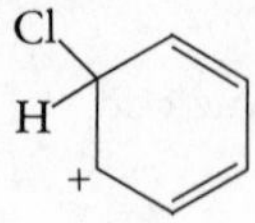

[arenium ion: ring carbon bearing Cl and H, + charge] $+ FeCl_4^- \longrightarrow C_6H_5Cl + HCl + FeCl_3$

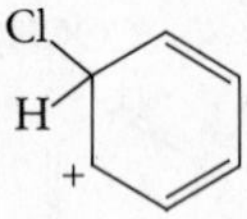

13-9 $CH_3Br + AlBr_3 \longrightarrow \overset{\delta+}{CH_3}{-}\overset{\delta-}{Br}AlBr_3$; $C_6H_6 + \overset{\delta+}{CH_3}{-}BrAlBr_3 \longrightarrow$ [arenium ion: ring carbon bearing CH_3 and H, + charge] $+ AlBr_4^-$;

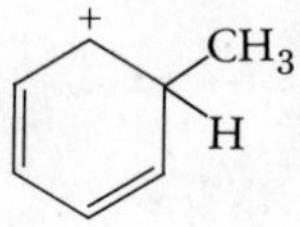

[arenium ion: ring carbon bearing CH_3 and H, + charge] $+ AlBr_4^- \longrightarrow C_6H_5CH_3 + AlBr_3 + HBr$

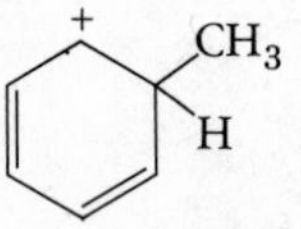

13-10 [Indole ring systems circled in four structures:]

CH_2CHCO_2H with NH_2 (indole N–H) ; HO-substituted indole with $CH_2CH_2NH_2$ (indole N–H) ; CH_3O-substituted indole fused to N-containing ring system with CH_2CH_3, H, H (indole N–H) ; OH-substituted indole with $CH_2CH_2N(CH_3)_2$ (indole N–H)

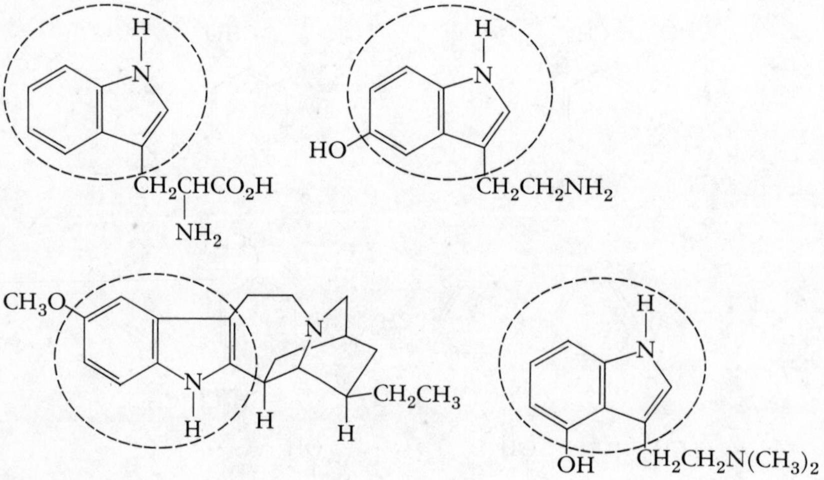

CHAPTER 14

14-1 (c) and (d)

14-2 (a) H—C(H)(H)H   H(H)(H)C—H (b) H(H)(H)C—Cl   Cl—C(H)(H)H (c) Cl—C(F)(H)H   H(F)(H)C—Cl

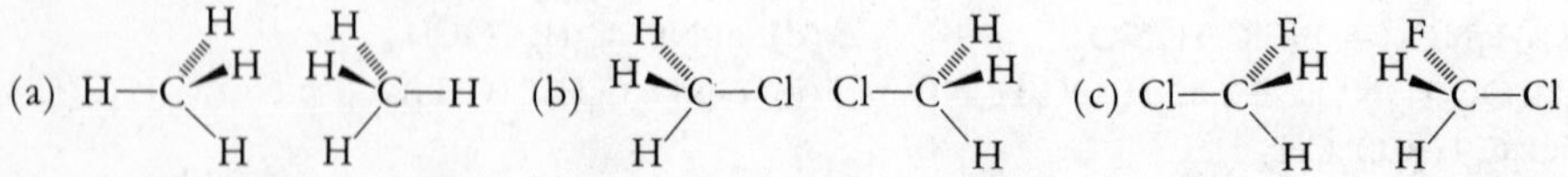

(d) 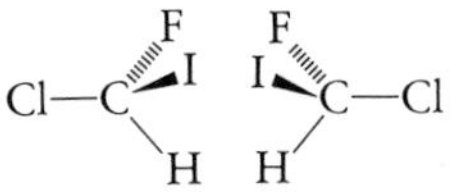 Cl—C(F)(I)(H) I(F)(H)C—Cl (e) 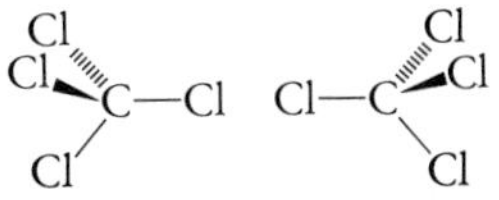 Cl(Cl)(Cl)C—Cl Cl—C(Cl)(Cl)(Cl) Only (d) is chiral.

14-3 The industrial preparation produces equal amounts of each enantiomer. Therefore the compounds are optically inactive.

14-4 (a) 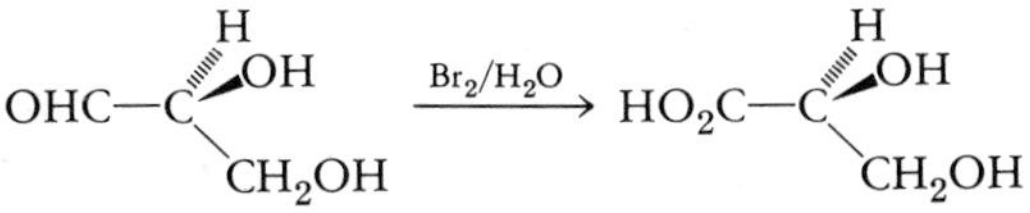 OHC—C(H)(OH)(CH_2OH) $\xrightarrow{Br_2/H_2O}$ HO_2C—C(H)(OH)(CH_2OH)

L family

(b) 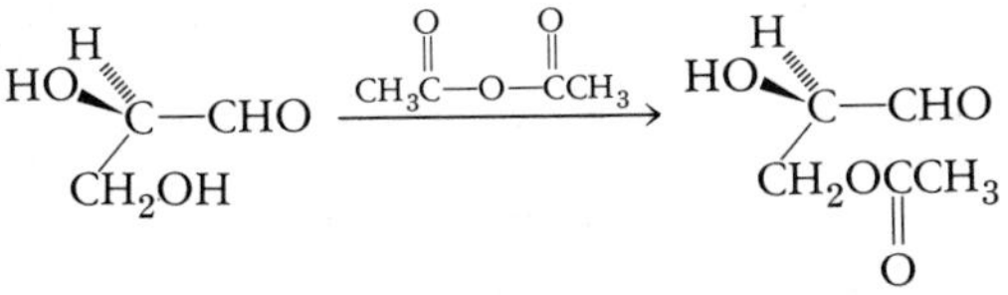 HO(H)C—CHO (CH_2OH) $\xrightarrow{CH_3\overset{O}{\overset{\|}{C}}—O—\overset{O}{\overset{\|}{C}}CH_3}$ HO(H)C—CHO ($CH_2O\underset{\underset{O}{\|}}{C}CH_3$)

D family

(c) 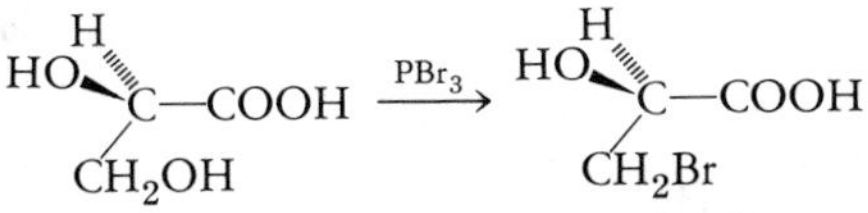 HO(H)C—COOH (CH_2OH) $\xrightarrow{PBr_3}$ HO(H)C—COOH (CH_2Br)

D family

In all three reactions, no bonds to the chiral carbon are broken.

14-5 Glucose has 4 chiral carbons: $HOCH_2\overset{*}{C}H(OH)\overset{*}{C}H(OH)\overset{*}{C}H(OH)\overset{*}{C}H(OH)CHO$. There are 2^4 or 16 possible stereoisomers.

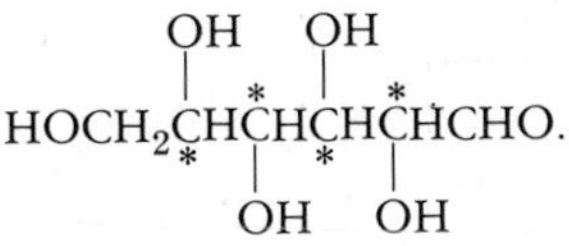

14-6

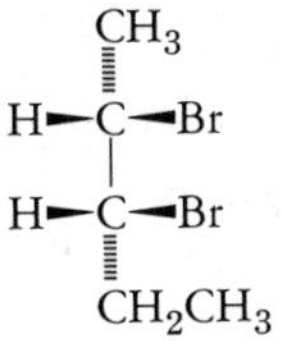

14-15	*14-16*	*14-17*	*14-18*
CH_3	CH_3	CH_3	CH_3
H—C—Br	Br—C—H	Br—C—H	H—C—Br
H—C—Br	Br—C—H	H—C—Br	Br—C—H
CH_2CH_3	CH_2CH_3	CH_2CH_3	CH_2CH_3

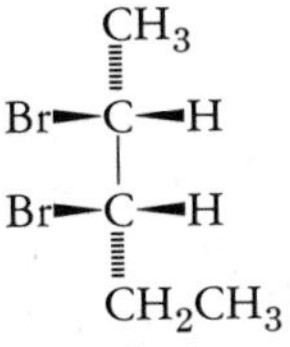

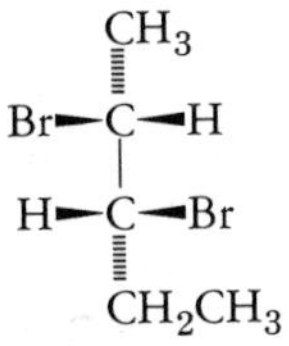

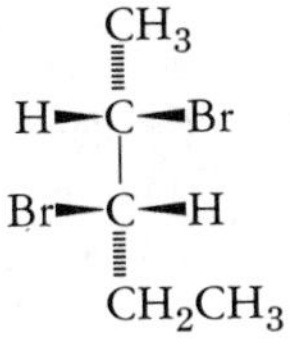

Enantiomers: *14-15* and *14-16*, *14-17* and *14-18*.

Diastereomers: *14-15* and *14-17*, *14-15* and *14-18*, *14-16* and *14-17*, *14-16* and *14-18*.

14-7 (a) and (c)

14-8

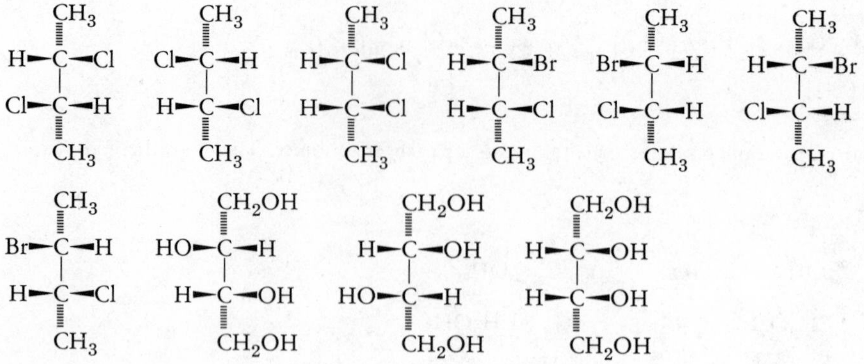

CH_3	CH_3	CH_3	CH_3	CH_3	CH_3
H–C–Cl	Cl–C–H	H–C–Cl	H–C–Br	Br–C–H	H–C–Br
Cl–C–H	H–C–Cl	H–C–Cl	H–C–Cl	Cl–C–H	Cl–C–H
CH_3	CH_3	CH_3	CH_3	CH_3	CH_3

CH_3	CH_2OH	CH_2OH	CH_2OH
Br–C–H	HO–C–H	H–C–OH	H–C–OH
H–C–Cl	H–C–OH	HO–C–H	H–C–OH
CH_3	CH_2OH	CH_2OH	CH_2OH

CHAPTER 15

15-1
(a) 2-bromo-3-methylbutane
(b) 3-bromo-2-chloro-2-methylpentane
(c) 3,5-diiodo-3,7-dimethylnonane
(d) 1-chloro-1-phenylpropane
(e) 1,4-dichloro-2,5-difluoro-3,6-dimethylcyclohexane
(f) *trans*-1-chloro-3,3-dimethyl-1-butene
(g) *cis*-4-chloro-2-pentene
(h) 2-isobutyl-4-chloro-1-hexene
(i) 3-chlorocyclohexene

15-2
(a) $CH_3CH(Br)CH_2CH_2CH_2CH_3$
(b) 1-chlorocyclohexene (ring structure with Cl on the double-bond carbon)
(c) cyclopentyl–Br
(d) $CH_3CH(I)CH_2CH_2CH(I)CH_2CH_3$
(e) $ClCH_2CH{=}CH_2$
(f) $(CH_3)_2CHCH(CH_3)CH_2CH(Cl)CH_3$
(g) $CH_2(Br)CH(CH_3)CH_2C_6H_5$
(h) cyclobutane ring bearing CH_3 and Cl on the same carbon
(i) cyclohexene ring with Br on the double-bond carbon and $C(CH_3)_3$
(j) $CH_2{=}CHCH_2CHBrCH_3$

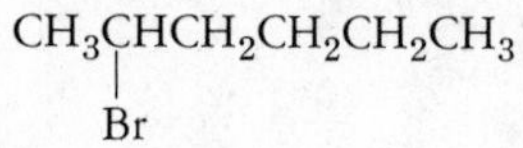

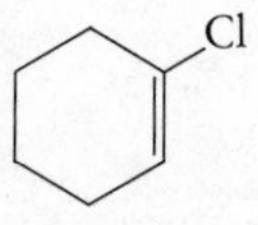

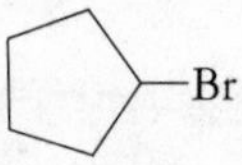

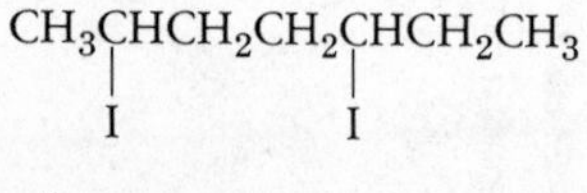

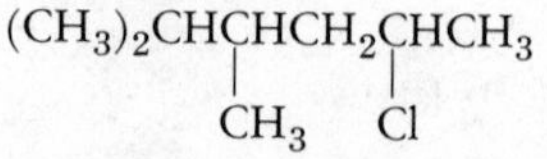

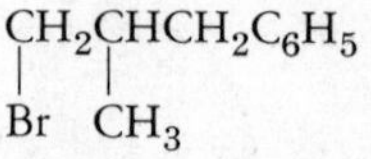

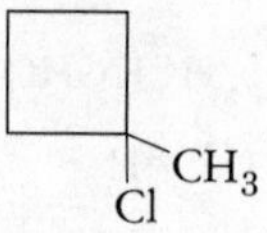

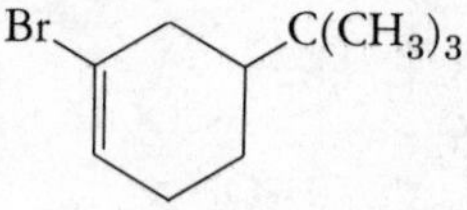

15-3
(a) 1-bromopropane
(b) 2-iodobutane
(c) 2-chloropropane
(d) 1-fluoro-2,2-dimethylpropane
(e) chloroethane
(f) 1-bromo-2-methylpropane
(g) trichloromethane
(h) 2-iodo-2-methylpropane
(i) tetrachloromethane
(j) 1,2-dichloroethane
(k) chloroethene
(l) dichloromethane

15-4

Exercise 15-1	Exercise 15-2	Exercise 15-3
(a) 2°	(a) 2°	(a) 1°
(b) chloro 3°, bromo 2°	(b) vinyl	(b) 2°
(c) 3-iodo 3°, 5-iodo 2°	(c) 2°	(c) 2°

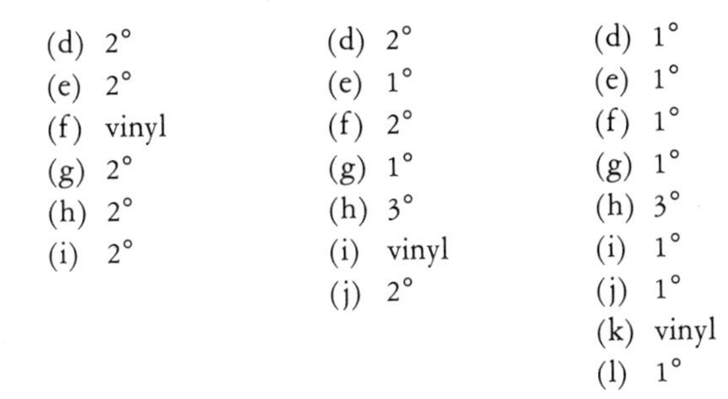

(d) 2°	(d) 2°	(d) 1°
(e) 2°	(e) 1°	(e) 1°
(f) vinyl	(f) 2°	(f) 1°
(g) 2°	(g) 1°	(g) 1°
(h) 2°	(h) 3°	(h) 3°
(i) 2°	(i) vinyl	(i) 1°
	(j) 2°	(j) 1°
		(k) vinyl
		(l) 1°

15-5 (a), (c), (e), and (f) are nucleophiles

15-6 Nucleophile is underlined. Leaving group is circled. The site of nucleophilic attack is indicated by ↓.

(a) $\underline{CH_3O^-} + C_6H_5\overset{\downarrow}{C}H_2\text{(Br)} \longrightarrow C_6H_5CH_2OCH_3 + Br^-$

(b) $\underline{CH_3S^-} + CH_3\overset{\downarrow}{C}H_2\text{(Cl)} \longrightarrow CH_3CH_2SCH_3 + Cl^-$

(c) $\underline{I^-} + CH_3\overset{\downarrow}{C}H_2\text{(Br)} \longrightarrow CH_3CH_2I + Br^-$

(d) $\underline{N_3^-} + CH_3CH_2\overset{\downarrow}{C}H_2\text{(}OSO_2C_6H_5\text{)} \longrightarrow CH_3CH_2CH_2N_3 + {}^-OSO_2C_6H_5$

15-7 (a) CH_3CH_2CN (b) CH_3OCH_3 (c) $CH_3CO_2CH_2CH_2CH_2CH_3$ (d) $(CH_3)_2CHCH_2OC_6H_5$

15-8 One of the circled hydrogens could be lost in an elimination reaction.

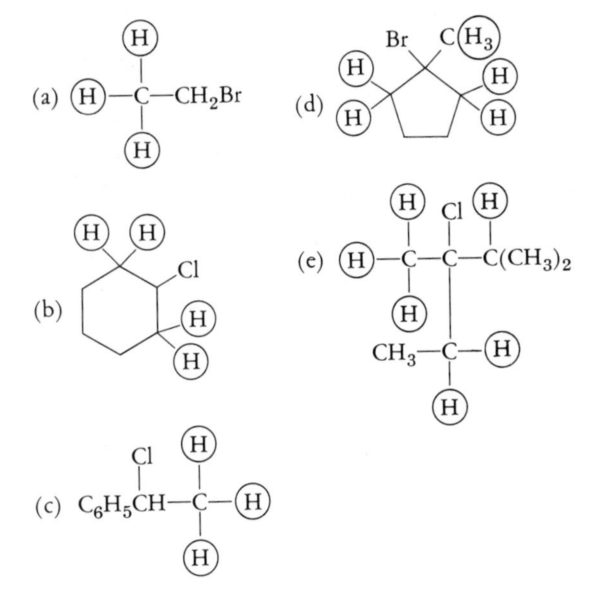

15-9 Equivalent hydrogens in any one compound are indicated by the same symbol.

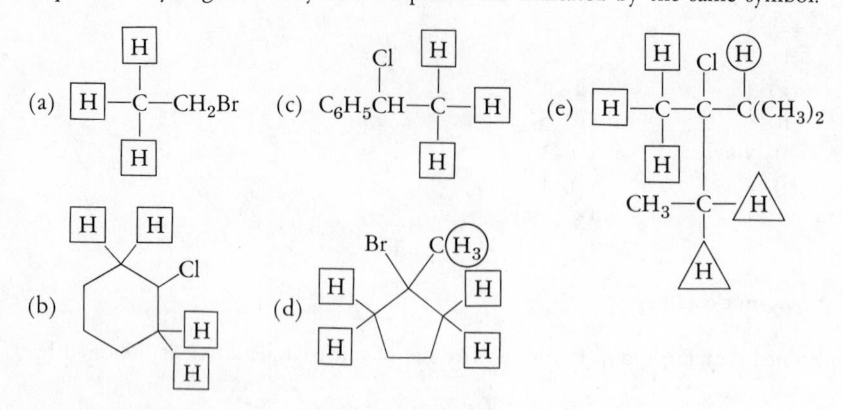

15-10 Six: three pairs of *cis-trans* isomers.

15-11 (a) 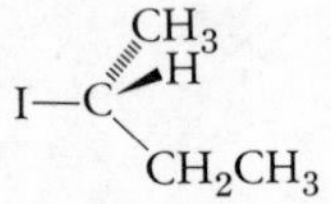 $I—C(CH_3)(H)(CH_2CH_3)$ (b) 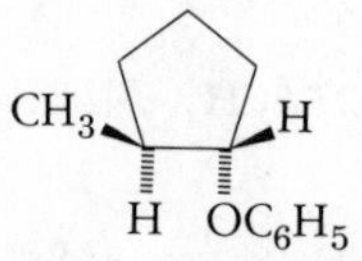 CH_3, H, H, OC_6H_5 on cyclopentane ring

CHAPTER 16

16-1 (a) propyl alcohol (b) isobutyl alcohol (c) isopropyl alcohol (d) cyclopentyl alcohol (e) *t*-butyl alcohol (f) *sec*-butyl alcohol (g) pentyl alcohol (h) benzyl alcohol

16-2 (a) 1° (b) 1° (c) 2° (d) 2° (e) 3° (f) 2° (g) 1° (h) 1°

16-3 (a) 1-propanol (b) 2-methyl-1-propanol (c) 2-propanol (d) cyclopentanol (e) 2-methyl-2-propanol (f) 2-butanol (g) 1-pentanol (h) phenylmethanol

16-4 (a) $(CH_3)_3CCH_2OH$
(b) $CH_3CH_2CH(CH_3)CH_2OH$
(c) 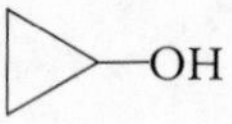 cyclopropyl—OH
(d) $CH_3CH(Cl)CH_2OH$
(e) 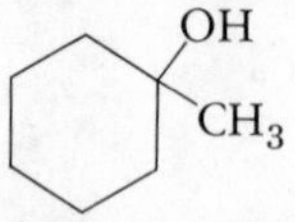 1-methylcyclohexanol (ring with OH and CH_3)
(f) $HOCH_2CH_2OH$
(g) $CH_3CH(OH)C(CH_3){=}CH_2$
(h) $CH{\equiv}CCH_2OH$
(i) $CH_2(OH)CH(OH)—CH_2(OH)$
(j) $CH_3CH_2CH(C_6H_5)OH$
(k) cyclohex-2-en-1-ol (ring with OH)

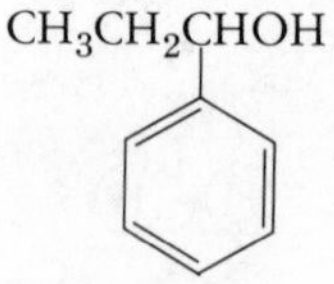

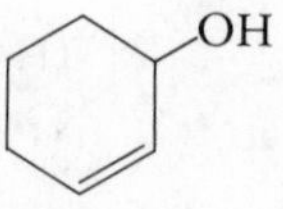

16-5 (a) methanol (b) cyclobutanol (c) 1,2-propanediol (d) 2-chloroethanol (e) 1-phenyl-2-propen-1-ol (f) 4-chlorocyclohexanol (g) 3-bromo-1-cyclopentyl-1-propanol (h) 3-cyclopentenol (i) *cis*-4-chloro-2-buten-1-ol (j) 4-iodo-2,2-dimethyl-1-pentanol (k) 1,2,3,4-butantetraol

16-6 No. Hydrogen bonds can form between molecules of ethanol but not between molecules of dimethyl ether.

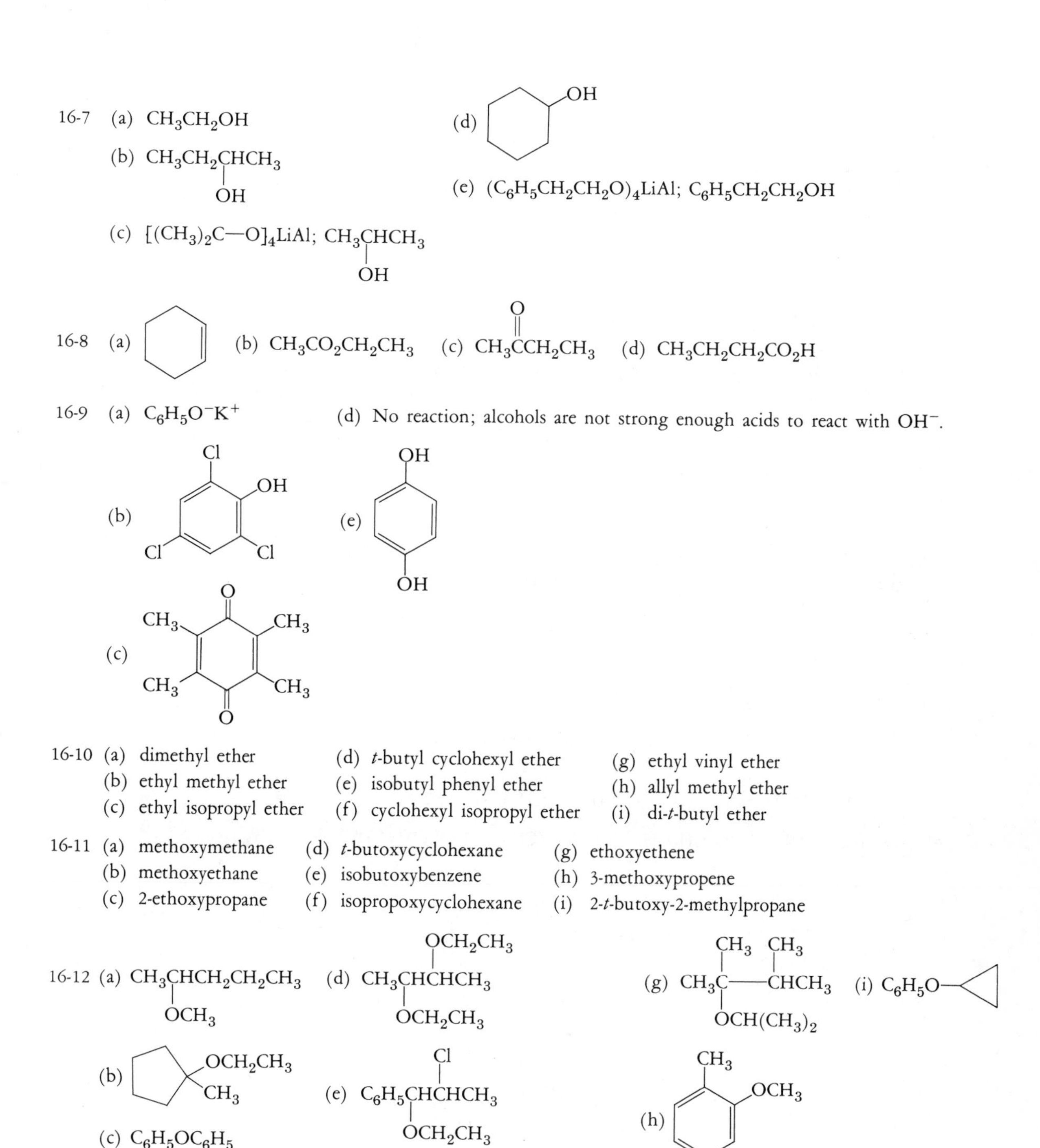

16-7 (a) CH_3CH_2OH
(b) $CH_3CH_2CH(OH)CH_3$
(c) $[(CH_3)_2C{-}O]_4LiAl$; $CH_3CH(OH)CH_3$
(d) cyclohexanol (OH)
(e) $(C_6H_5CH_2CH_2O)_4LiAl$; $C_6H_5CH_2CH_2OH$

16-8 (a) cyclohexene (b) $CH_3CO_2CH_2CH_3$ (c) $CH_3COCH_2CH_3$ (d) $CH_3CH_2CH_2CO_2H$

16-9 (a) $C_6H_5O^-K^+$
(b) Cl, OH, Cl, Cl (trichlorophenol)
(c) CH_3, CH_3, CH_3, CH_3, O, O (tetramethylbenzoquinone)
(d) No reaction; alcohols are not strong enough acids to react with OH^-.
(e) OH, OH (hydroquinone)

16-10 (a) dimethyl ether
(b) ethyl methyl ether
(c) ethyl isopropyl ether
(d) *t*-butyl cyclohexyl ether
(e) isobutyl phenyl ether
(f) cyclohexyl isopropyl ether
(g) ethyl vinyl ether
(h) allyl methyl ether
(i) di-*t*-butyl ether

16-11 (a) methoxymethane
(b) methoxyethane
(c) 2-ethoxypropane
(d) *t*-butoxycyclohexane
(e) isobutoxybenzene
(f) isopropoxycyclohexane
(g) ethoxyethene
(h) 3-methoxypropene
(i) 2-*t*-butoxy-2-methylpropane

16-12 (a) $CH_3CH(OCH_3)CH_2CH_2CH_3$
(b) 1-ethoxy-1-methylcyclopentane (OCH_2CH_3, CH_3)
(c) $C_6H_5OC_6H_5$
(d) $CH_3CH(OCH_2CH_3)CH(OCH_2CH_3)CH_3$
(e) $C_6H_5CH(OCH_2CH_3)CH(Cl)CH_3$
(f) $CH_3CH_2CH_2OC{\equiv}CCH_2CH_2Cl$
(g) $CH_3C(CH_3)(OCH(CH_3)_2){-}CH(CH_3)CH_3$
(h) CH_3, OCH_3, OCH_3 (substituted benzene)
(i) C_6H_5O–cyclopropyl

16-13 (a) $CH_3OH + CH_3Br$ (b) $HOCH_2CH_2Br$ (c) $HOCH_2CH_2SH$

CHAPTER 17

17-1 (a) *t*-butylamine
(b) diethylamine
(c) N-methylpropylamine
(d) trimethylamine
(e) N-methyl-N-vinylbutylamine
(f) cyclohexylamine
(g) N-methylcyclopropylamine
(h) isobutylamine
(i) allylamine
(j) N-methylaniline
(k) N,N-dimethylisopropylamine
(l) N-isopropylbutylamine

17-2 (a) 1° (b) 2° (c) 2° (d) 3° (e) 3° (f) 1° (g) 2° (h) 1° (i) 1° (j) 2° (k) 3° (l) 2°

17-3 (a) 2-methyl-2-propanamine
(b) N-ethylethanamine
(c) N-methyl-1-propanamine
(d) N,N-dimethylmethanamine
(e) N-methyl-N-vinyl-1-butanamine
(f) cyclohexanamine
(g) N-methylcyclopropanamine
(h) 2-methyl-1-propanamine
(i) 2-propen-1-amine
(j) N-methylaniline
(k) N,N-dimethyl-2-propanamine
(l) N-isopropylbutanamine

17-4 (a) $CH_3CH_2CH_2CH_2NH_2$
(b) $CH_3CH_2\underset{\displaystyle NHC_2H_5}{\underset{|}{C}}HCH_2CH_2CH_2CH_3$
(c) $[(CH_3)_3C]_3N$
(d) $C_6H_5N(C_2H_5)_2$
(e) $CH_3(CH_2)_4CH_2NHCH_2CH_2CH_3$

17-5 (a) butylammonium bromide
(b) N-ethylanilinium chloride
(c) tetrapropylammonium iodide
(d) isopropylmethylammonium bromide

17-6 (a) $C_6H_5CH{=}CH_2 + N(CH_3)_3 + H_2O$

(b) $C_6H_5(H)C{=}C(H)CO_2^- + NH_3$ (trans)

(c) $CH_3CH_2CH{=}CH_2 + N(CH_2CH_2CH_2CH_3)_3 + H_2O$

17-7 (a) $(CH_3)_4\overset{+}{N}I^-$ (b) $CH_3CH_2CH_2CH_2\overset{+}{N}(CH_3)_3\ I^-$ (c) $(CH_3)_3\overset{+}{N}CH_2CH_3\ I^-$

17-8 (a) phthalimide anion N^- + CH_3CH_2Br ⟶ N-ethylphthalimide (NCH_2CH_3) + Br^-;

N-ethylphthalimide (NCH_2CH_3) + HCl $\xrightarrow{2\,H_2O}$ phthalic acid ($C_6H_4(CO_2H)_2$) + $CH_3CH_2\overset{+}{N}H_3Cl^-$;

$CH_3CH_2\overset{+}{N}H_3Cl^- + OH^- \longrightarrow CH_3CH_2NH_2 + H_2O + Cl^-$

(b) 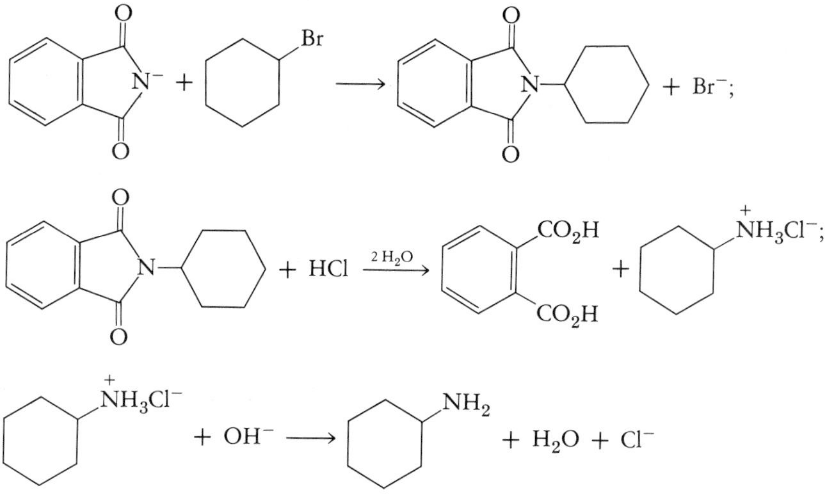 O, N^-, O + Br $\longrightarrow$ O, N—, O + Br^-;

O, N—, O + HCl $\xrightarrow{2\,H_2O}$ CO_2H, CO_2H + $\overset{+}{N}H_3Cl^-$;

$\overset{+}{N}H_3Cl^-$ + OH^- $\longrightarrow$ NH_2 + H_2O + Cl^-

17-9 (a) H_2O (c) H_2O (e) NO_3^- (g) Cl^- (i) HSO_4^- (k) $CH_3\underset{\underset{O}{\|}}{C}CH_2^-$

(b) CH_3CO_2H (d) HCl (f) CH_3NH_2 (h) $C_6H_5O^-$ (j) $CF_3CO_2^-$

17-10 (a) $HI—I^-$; $H_2O—H_3O^+$ (c) $HNO_3—NO_3^-$; $CH_3NH_2—CH_3\overset{+}{N}H_3$

(b) $OH^-—H_2O$; $\overset{-}{N}H_2—NH_3$ (d) $CH_3CHO—\overset{-}{C}H_2CHO$; $H_2O—OH^-$

17-11 Order of increasing acid strength $NH_3 < \overset{+}{N}H_4 < CH_3—C_6H_4—\overset{+}{N}H_3 < CH_3CO\overset{+}{N}H_3$

17-12 Order of increasing base strength $NH_2—C_6H_5 <$ N (pyridine) $< NH_3 < CH_3NH_2$

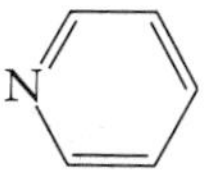

17-13 (a) $CH_3C_6H_4NH_2$ (b) CH_3NH_2 (c) NH_3 (d) $CH_3C_6H_4NH_2$ (e) $(CH_3)_2NH$

17-14 (a) $CH_3CONHCH_3 + CH_3\overset{+}{N}H_3Cl^-$ (c) $C_6H_5SO_2NH(CH_2)_2CH_3 + NaCl + H_2O$

(b) $CH_3CON(CH_3)_2 + CH_3CO_2H$

(d) $CH_3CONHC_6H_4SO_2NH—$ (N, S thiazole ring)

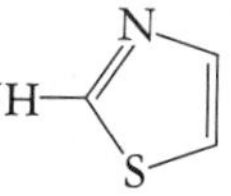

CHAPTER 18

18-1 (a) propionic acid or propanoic acid
(b) 3-methylbutanoic acid
(c) 2-bromo-2-methylpropanoic acid
(d) cyclopentanecarboxylic acid
(e) sodium cyclohexanecarboxylate
(f) *trans*-4-chloro-2-butenoic acid
(g) 5-methyl-3-phenyl-4-hexenoic acid
(h) 2,2-dimethylpropanedioic acid
(i) 2-cyclohexenecarboxylic acid

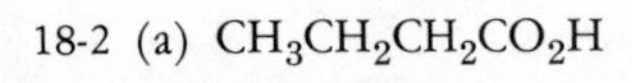

18-2 (a) $CH_3CH_2CH_2CO_2H$

(b) $CH_2{=}CHCO_2H$

(c) cycloheptane ring bearing CO_2H

(d) $CH_3CH_2CO_2^-Na^+$

(e) $\mathrm{H}(\mathrm{HO_2C})C{=}C(\mathrm{H})\mathrm{CO_2H}$ (H, H on the same side; HO_2C, CO_2H on the same side)

(f) $HO_2C\underset{\displaystyle CH_3}{\underset{|}{C}}HCH_2\underset{\displaystyle CH_3}{\underset{|}{C}}HCO_2H$

(g) $CH_3\underset{\displaystyle CH_3}{\underset{|}{C}}HCH_2CH_2\underset{\displaystyle CH_3}{\underset{|}{C}}HCO_2H$

(h) $CH_2{=}CH(CH_2)_6CO_2H$

(i) cyclobutane ring bearing CO_2H and Cl on opposite corners

18-3 (a) Propanoyl (propionyl) (b) 2-methylpropanoyl (c) methanoyl (formyl) (d) 2-bromoethanoyl (α-bromoacetyl) (e) propenoyl (f) octanoyl

18-4 Molecules of alcohols, 1° amides, and carboxylic acids are capable of forming intermolecular hydrogen bonds, while molecules of ketones, 3° amides, and esters are not.

18-5 (a) $CH_3CH_2CO_2H$ propanoic acid
(b) $HO_2CCH_2CH_2CO_2H$ butanedioic acid
(c) $CH_3CH_2CO_2H$ (2 molecules/molecule alkene) propanoic acid
(d) $CH_3CH_2CH_2CO_2H + HO_2CCH_2CH_3$ butanoic acid + propanoic acid
(e) $HO_2C(CH_2)_4CO_2H$ hexanedioic acid
(f) $C_6H_5CO_2H$ benzoic acid

18-6 (a) $(CH_3)_2CHCO_2^-K^+$ potassium 2-methylpropanoate
(b) $CH_3CH_2CH_2CO_2^-(CH_3)_2\overset{+}{N}H_2$ dimethylammonium butanoate
(c) $CH_3CH_2CO_2H$ propanoic acid

18-7 (a) propyl methanoate (b) isopropyl ethanoate (c) ethyl propanoate (d) cyclohexyl ethanoate (e) methyl *trans*-2-butenoate (f) phenyl 3-methylbutanoate

18-8 (a) $CH_3CH_2CH_2CO_2CH_3$
(b) $CH_3(CH_2)_3CO_2(CH_2)_5CH_3$
(c) $C_6H_5CO_2CH_2CH_2CH_3$
(d) $CH_3CH_2CO_2C_6H_5$
(e) $CH_3CH_2CH_2\underset{\displaystyle CH_3}{\underset{|}{C}}HCO_2C_6H_4Cl$
(f) $CH_2{=}CHCO_2CH_2CH(CH_3)_2$
(g) $CH_3CO_2CH_2C_6H_5$

18-9 (a) $CH_3CO_2H + CH_3CH_2OH \overset{H^+}{\rightleftharpoons} CH_3CO_2CH_2CH_3 + H_2O$

(b) $C_6H_5CO_2H + (CH_3)_2CHOH \overset{H^+}{\rightleftharpoons} C_6H_5CO_2CH(CH_3)_2 + H_2O$

(c) $C_6H_5CH_2CH_2CO_2H + CH_3CH_2OH \overset{H^+}{\rightleftharpoons} C_6H_5CH_2CH_2CO_2CH_2CH_3 + H_2O$

18-10 (a) $CH_3CO_2^-Na^+ + HOCH_2CH_3$ sodium acetate + ethyl alcohol
(b) $(CH_3)_2CHCO_2^-Na^+ + HOCH_3$ sodium 2-methylpropanoate + methanol
(c) $CH_3CO_2^-Na^+ + HOCH_2CH_2OH + CH_3CH_2CO_2^-Na^+$ sodium ethanoate + 1,2-ethanediol + sodium propanoate
(d) $C_6H_5CO_2^-Na^+ + HOC(CH_3)_3$ sodium benzoate + 2-methyl-2-propanol

18-11 (a) $CH_3CH_2CH_2CONH_2 + CH_3OH$

(b) $C_6H_5CO_2CH(CH_3)_2 + CH_3CH_2OH$

(c) (H)(CH$_3$)C=C(H)(CONHCH(CH$_3$)$_2$) $+ (CH_3)_2CHOH$

18-12 (a) $(CH_3)_2CHCO_2H + HSCH_2CH_2CH_3$

(b) $(CH_3)_2CHCONHC_2H_5 + HSCH_2CH_2CH_3$

(c) $(CH_3)_2CHCOSCH_3 + HSCH_2CH_2CH_3$

(d) $(CH_3)_2CHCO_2CH(CH_3)_2 + HSCH_2CH_2CH_3$

18-13 (a) $CH_3CO_2(CH_2)_2CH_3 + HSCoA$

(b) $CH_3CH_2CH_2CONHCH(CH_3)C_6H_5 + HSCoA$

18-14 (a) propanamide
(b) N-methylmethanamide
(c) N-cyclopentylbenzamide
(d) N-isobutylbutanamide
(e) N-methyl-2-chloropropanamide
(f) N,N-dipropyl-2-methylpropanamide

18-15 (a) $CH_3CH_2CH_2CONH_2$

(b) $CH_3CH_2CH(CH_3)CH_2CONH_2$

(c) $CH_3CH_2CH_2CH_2CONHCH_3$

(d) $C_6H_5CON(CH_3)_2$

(e) $CH_2{=}CHCONH(CH_2)_2CH_3$

18-16 Exercise 18-14 (a) $CH_3CH_2CO_2H + \overset{+}{N}H_4$

(b) $HCO_2H + CH_3\overset{+}{N}H_3$

(c) $C_6H_5CO_2H +$ cyclopentyl$-\overset{+}{N}H_3$

(d) $CH_3CH_2CH_2CO_2H + (CH_3)_2CHCH_2\overset{+}{N}H_3$

(e) $CH_3CH(Cl)CO_2H + CH_3\overset{+}{N}H_3$

(f) $(CH_3)_2CHCO_2H + (CH_3CH_2CH_2)_2\overset{+}{N}H_2$;

Exercise 18-15 (a) $CH_3CH_2CH_2CO_2H + \overset{+}{N}H_4$

(b) $CH_3CH_2CH(CH_3)CH_2CO_2H + \overset{+}{N}H_4$

(c) $CH_3(CH_2)_3CO_2H + CH_3\overset{+}{N}H_3$

(d) $C_6H_5CO_2H + (CH_3)_2\overset{+}{N}H_2$

(e) $CH_2{=}CHCO_2H + CH_3CH_2CH_2\overset{+}{N}H_3$

18-17 (a) $CH_3CONH_2 + H_2O \xrightarrow{H^+} CH_3CO_2H + \overset{+}{N}H_4$

(b) $CH_3CONHCH_3 + H_2O \xrightarrow{OH^-} CH_3CO_2^- + CH_3NH_2$

(c) $CH_3CON(CH_3)_2 + CH_3CH_2OH \longrightarrow CH_3CO_2CH_2CH_3 + (CH_3)_2NH$

18-18 (a) Step 1: $CH_3C(=O)OC(=O)CH_3 + H_2O \longrightarrow CH_3C(O^-)(\overset{+}{O}H_2)OC(=O)CH_3$

Step 2: $CH_3C(O^-)(\overset{+}{O}H_2)OC(=O)CH_3 \longrightarrow CH_3C(=O)\overset{+}{O}H_2 + CH_3C(=O)O^-$

Step 3: $CH_3C(=O)\overset{+}{O}H_2 \longrightarrow CH_3C(=O)OH + H^+$

(b) Step 1: $CH_3C(=O)OC(=O)CH_3 + \ddot{N}H(CH_3)_2 \longrightarrow CH_3C(O^-)(OC(=O)CH_3)\!-\!\overset{+}{N}H(CH_3)_2$

Step 2: $CH_3C(O^-)(OC(=O)CH_3)\!-\!\overset{+}{N}H(CH_3)_2 \longrightarrow CH_3C(=O)\overset{+}{N}H(CH_3)_2 + CH_3C(=O)O^-$

Step 3: $CH_3C(=O)\overset{+}{N}H(CH_3)_2 \longrightarrow CH_3C(=O)N(CH_3)_2 + H^+$

18-19 (a) Step 1: $CH_3CH_2C(=O)OCH_3 + H^+ \rightleftharpoons CH_3CH_2\overset{+}{C}(OH)\!-\!OCH_3$ (common to all 3 mechanisms)

Step 2: $CH_3CH_2\overset{+}{C}(OH)OCH_3 + H_2\ddot{O} \rightleftharpoons CH_3CH_2C(OH)(\overset{+}{O}H_2)\!-\!OCH_3$

Step 3: $CH_3CH_2C(OH)(\overset{+}{O}H_2)\!-\!OCH_3 \rightleftharpoons CH_3CH_2C(OH)_2\!-\!\overset{+}{O}(H)CH_3$

Step 4: $CH_3CH_2C(\ddot{O}H)(OH)\!-\!\overset{+}{O}(H)CH_3 \rightleftharpoons CH_3CH_2C(=\overset{+}{O}H)OH + CH_3OH$

Step 5: $CH_3CH_2C(=\overset{+}{O}H)OH \rightleftharpoons CH_3CH_2C(=O)OH + H^+$

(b) Step 2: $CH_3CH_2\overset{OH}{\underset{+}{C}}OCH_3 + H\ddot{O}CH_2CH_3 \rightleftharpoons CH_3CH_2\overset{OH}{C}(-OCH_3)(-\overset{+}{O}(H)CH_2CH_3)$

Step 3: $CH_3CH_2C(OH)(-OCH_3)(-\overset{+}{O}(H)CH_2CH_3) \rightleftharpoons CH_3CH_2C(OH)(-\overset{+}{O}(H)CH_3)(OCH_2CH_3)$

Step 4: $CH_3CH_2C(\ddot{O}H)(-\overset{+}{O}(H)CH_3)(OCH_2CH_3) \rightleftharpoons CH_3CH_2C(=\overset{+}{O}H)OCH_2CH_3 + CH_3OH$

Step 5: $CH_3CH_2C(=\overset{+}{O}H)OCH_2CH_3 \rightleftharpoons CH_3CH_2C(=O)OCH_2CH_3 + H^+$

(c) Step 2: $CH_3CH_2\overset{OH}{\underset{+}{C}}OCH_3 + :NH_3 \rightleftharpoons CH_3CH_2C(OH)(-OCH_3)(\overset{+}{N}H_3)$

Step 3: $CH_3CH_2C(OH)(-OCH_3)(\overset{+}{N}H_3) \rightleftharpoons CH_3CH_2C(OH)(-\overset{+}{O}(H)CH_3)(NH_2)$

Step 4: $CH_3CH_2C(\ddot{O}H)(-\overset{+}{O}(H)CH_3)(NH_2) \rightleftharpoons CH_3CH_2C(=\overset{+}{O}H)NH_2 + CH_3OH$

Step 5: $CH_3CH_2C(=\overset{+}{O}H)NH_2 \rightleftharpoons CH_3C(=O)NH_2 + H^+$

CHAPTER 19

19-1 (a) formaldehyde (b) valeraldehyde (c) isobutyraldehyde (d) phenylacetaldehyde (e) α-chloropropionaldehyde (f) β-methylbutyraldehyde

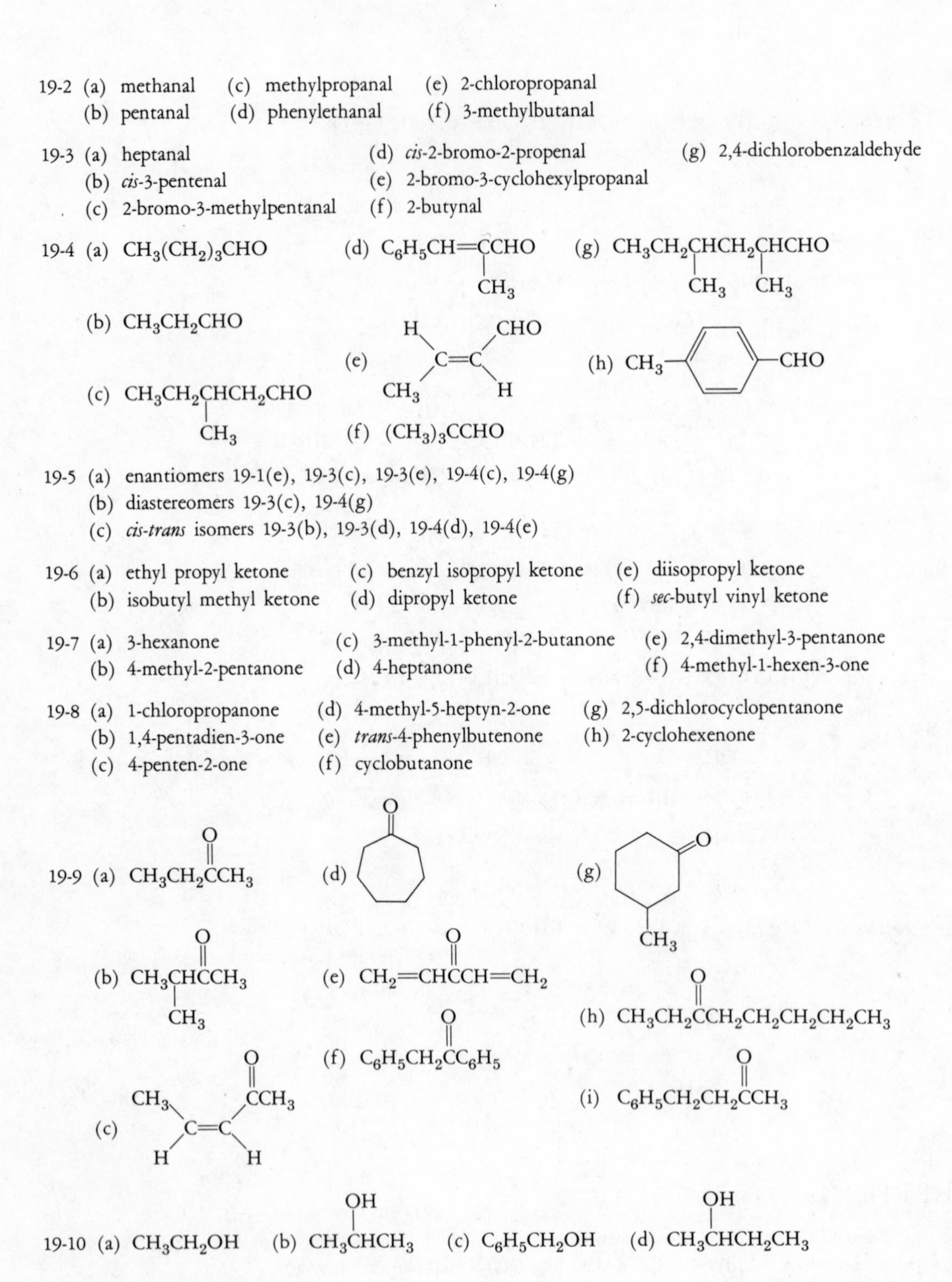

19-2 (a) methanal (b) pentanal (c) methylpropanal (d) phenylethanal (e) 2-chloropropanal (f) 3-methylbutanal

19-3 (a) heptanal (b) *cis*-3-pentenal (c) 2-bromo-3-methylpentanal (d) *cis*-2-bromo-2-propenal (e) 2-bromo-3-cyclohexylpropanal (f) 2-butynal (g) 2,4-dichlorobenzaldehyde

19-4 (a) $CH_3(CH_2)_3CHO$
(b) CH_3CH_2CHO
(c) $CH_3CH_2CHCH_2CHO$ (with CH_3 on the CH)
(d) $C_6H_5CH{=}CCHO$ (with CH_3 on the C)
(e) H and CH_3 on one carbon, CHO and H on the other: $C{=}C$
(f) $(CH_3)_3CCHO$
(g) $CH_3CH_2CHCH_2CHCHO$ (with CH_3 on each CH)
(h) CH_3–C_6H_4–CHO (para)

19-5 (a) enantiomers 19-1(e), 19-3(c), 19-3(e), 19-4(c), 19-4(g)
(b) diastereomers 19-3(c), 19-4(g)
(c) *cis-trans* isomers 19-3(b), 19-3(d), 19-4(d), 19-4(e)

19-6 (a) ethyl propyl ketone (b) isobutyl methyl ketone (c) benzyl isopropyl ketone (d) dipropyl ketone (e) diisopropyl ketone (f) *sec*-butyl vinyl ketone

19-7 (a) 3-hexanone (b) 4-methyl-2-pentanone (c) 3-methyl-1-phenyl-2-butanone (d) 4-heptanone (e) 2,4-dimethyl-3-pentanone (f) 4-methyl-1-hexen-3-one

19-8 (a) 1-chloropropanone (b) 1,4-pentadien-3-one (c) 4-penten-2-one (d) 4-methyl-5-heptyn-2-one (e) *trans*-4-phenylbutenone (f) cyclobutanone (g) 2,5-dichlorocyclopentanone (h) 2-cyclohexenone

19-9 (a) $CH_3CH_2\overset{O}{\overset{\|}{C}}CH_3$
(b) $CH_3CH\overset{O}{\overset{\|}{C}}CH_3$ (with CH_3 on the CH)
(c) CH_3 and H on one carbon, $\overset{O}{\overset{\|}{C}}CH_3$ and H on the other: $C{=}C$
(d) cycloheptanone
(e) $CH_2{=}CH\overset{O}{\overset{\|}{C}}CH{=}CH_2$
(f) $C_6H_5CH_2\overset{O}{\overset{\|}{C}}C_6H_5$
(g) 3-methylcyclohexanone (ring with O and CH_3)
(h) $CH_3CH_2\overset{O}{\overset{\|}{C}}CH_2CH_2CH_2CH_2CH_3$
(i) $C_6H_5CH_2CH_2\overset{O}{\overset{\|}{C}}CH_3$

19-10 (a) CH_3CH_2OH (b) $CH_3\overset{OH}{\overset{|}{C}}HCH_3$ (c) $C_6H_5CH_2OH$ (d) $CH_3\overset{OH}{\overset{|}{C}}HCH_2CH_3$

19-11 (a) $CH_3\underset{\underset{OH}{|}}{C}HOCH_2CH_3$; $CH_3CH(OCH_2CH_3)_2$

(b) $C_6H_5\overset{\overset{OH}{|}}{C}HOC_2H_5$; $C_6H_5CH(OC_2H_5)_2$

(c) $CH_2{=}CH\overset{\overset{OH}{|}}{C}HOC_2H_5$; $CH_2{=}CHCH(OC_2H_5)_2$

(d) $CH_3(CH_2)_2\overset{\overset{OH}{|}}{C}HOC_2H_5$; $CH_3(CH_2)_2CH(OC_2H_5)_2$

19-12 (a) $(CH_3)_2\overset{\overset{OH}{|}}{C}OC_2H_5$; $(CH_3)_2C(OC_2H_5)_2$

(b) 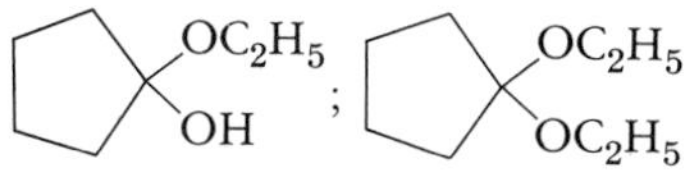 cyclopentane C with OC_2H_5 and OH ; cyclopentane C with OC_2H_5 and OC_2H_5

(c) CH_3, C_6H_5 C OC_2H_5, OH ; CH_3, C_6H_5 $C(OC_2H_5)_2$

(d) 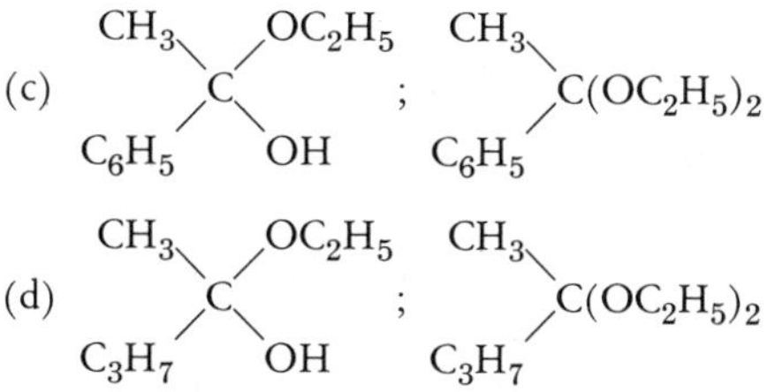 CH_3, C_3H_7 C OC_2H_5, OH ; CH_3, C_3H_7 $C(OC_2H_5)_2$

19-13 (a) $CH_3\underset{\underset{OH}{|}}{C}HOCH_2CH_2OH$; CH_3CH (O—CH_2, O—CH_2)

(b) $C_6H_5\underset{\underset{OH}{|}}{C}HOCH_2CH_2OH$; C_6H_5CH (O—CH_2, O—CH_2)

(c) $CH_2{=}CH\overset{\overset{OH}{|}}{C}HOCH_2CH_2OH$; $CH_2{=}CHCH$ (O—CH_2, O—CH_2)

(d) $C_3H_7\overset{\overset{OH}{|}}{C}HOCH_2CH_2OH$; C_3H_7CH (O—CH_2, O—CH_2)

(e) $(CH_3)_2\underset{\underset{OH}{|}}{C}OCH_2CH_2OH$; $(CH_3)_2C$ (O—CH_2, O—CH_2)

(f) cyclopentane C with OH and OCH_2CH_2OH ; cyclopentane C (O—CH_2, O—CH_2)

(g) CH_3, C_6H_5 C OH, OCH_2CH_2OH ; CH_3, C_6H_5 C (O—CH_2, O—CH_2)

(h) CH_3, C_3H_7 C OH, OCH_2CH_2OH ; CH_3, C_3H_7 C (O—CH_2, O—CH_2)

19-14 (a) $(CH_3)(C_3H_7)C{=}O + H_2NNHC_6H_5 \longrightarrow (CH_3)(C_3H_7)C{=}NNHC_6H_5 + H_2O$

(b) $(CH_3)(C_3H_7)C{=}O + NH_2OH \longrightarrow (CH_3)(C_3H_7)C{=}N{-}OH + H_2O$

(c) $(CH_3)(C_3H_7)C{=}O + H_2NNH{-}C_6H_3(NO_2){-}NO_2 \longrightarrow (CH_3)(C_3H_7)C{=}NNH{-}C_6H_3(NO_2){-}NO_2 + H_2O$

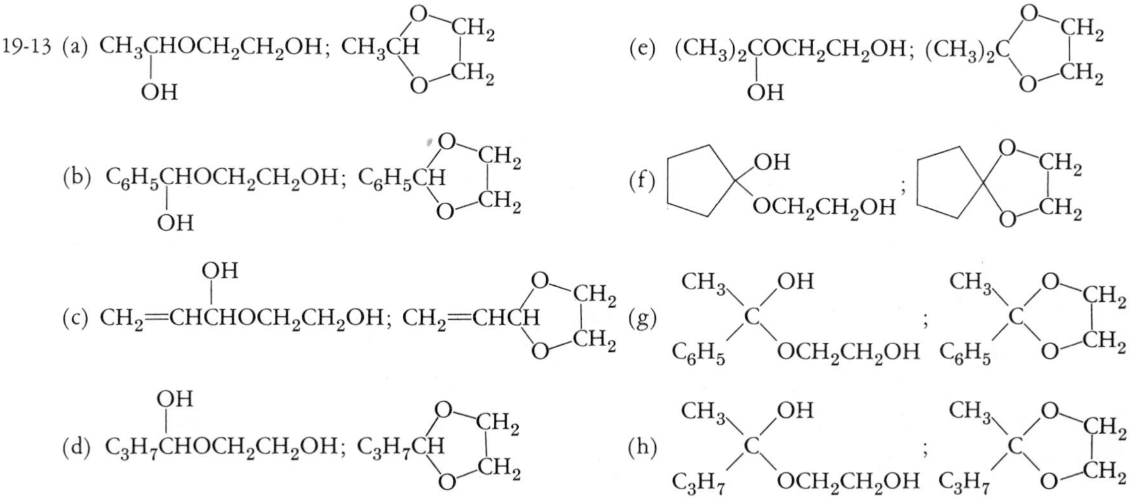

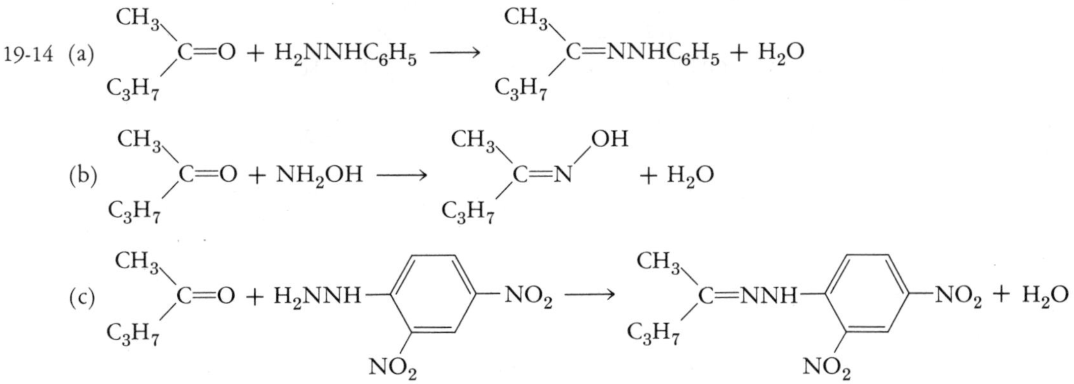

(d)
$$(CH_3)(C_3H_7)C{=}O + H_2NNHCNH_2 \longrightarrow (CH_3)(C_3H_7)C{=}NNHCNH_2 + H_2O$$
(in both formulas the C of $NHCNH_2$ carries $=O$)

19-15 (a) Step 1:
$$CH_3(H)C{=}O + H^+ \rightleftharpoons CH_3(H)\overset{+}{C}{-}OH$$

Step 2:
$$CH_3(H)\overset{+}{C}{-}OH + H_2\ddot{O}: \rightleftharpoons CH_3(H)C(OH)(\overset{+}{O}H_2)$$

Step 3:
$$CH_3(H)C(OH)(\overset{+}{O}H_2) \rightleftharpoons CH_3(H)C(OH)(OH) + H^+$$

(b) Step 1:
$$CH_3(H)C{=}O + H^+ \rightleftharpoons CH_3(H)\overset{+}{C}{-}OH$$

Step 2:
$$CH_3(H)\overset{+}{C}{-}OH + CH_2CH_2OH\,(:OH) \rightleftharpoons CH_3(H)C(OH)(\overset{+}{O}(H)CH_2CH_2OH)$$

Step 3:
$$CH_3(H)C(OH)(\overset{+}{O}(H)CH_2CH_2OH) \rightleftharpoons CH_3(H)C(OH)(OCH_2CH_2OH) + H^+;$$

$$CH_3(H)C(\ddot{O}H)(OCH_2CH_2OH) + H^+ \rightleftharpoons CH_3(H)C(\overset{+}{O}H_2)(OCH_2CH_2OH);$$

$$CH_3(H)C(\overset{+}{O}H_2)(OCH_2CH_2OH) \rightleftharpoons CH_3(H)C^+(OCH_2CH_2OH) + H_2O;$$

$$CH_3(H)C^+(OCH_2CH_2OH) \rightleftharpoons CH_3(H)C\langle \overset{+}{O}(H){-}CH_2{-}CH_2{-}O \rangle;$$

$$\mathrm{CH_3CH(\overset{+}{O}H\!-\!CH_2\!-\!CH_2\!-\!O)} \rightleftharpoons \mathrm{CH_3CH(O\!-\!CH_2\!-\!CH_2\!-\!O)} + \mathrm{H^+}$$

(c) Step 1: $\mathrm{CH_3(H)C{=}O} + \mathrm{H^-} \longrightarrow \mathrm{CH_3(H)C(H)O^-}$

Step 2: $\mathrm{CH_3(H)C(H)O^-} + \mathrm{H{-}OH} \longrightarrow \mathrm{CH_3(H)C(H)OH} + \mathrm{OH^-}$

(d) Step 1: $\mathrm{CH_3(H)C{=}O} + \mathrm{H_2\ddot{N}CH_3} \rightleftharpoons \mathrm{CH_3(H)C(\overset{+}{N}H_2CH_3)O^-}$

Step 2: $\mathrm{CH_3(H)C(\overset{+}{N}H_2CH_3)O^-} \rightleftharpoons \mathrm{CH_3(H)C(NHCH_3)OH}$;

$\mathrm{CH_3(H)C(NHCH_3)OH} + \mathrm{H^+} \rightleftharpoons \mathrm{CH_3(H)C(NHCH_3)\overset{+}{O}H_2}$;

$\mathrm{CH_3(H)C(N(H)CH_3)\overset{+}{O}H_2} \longrightarrow \mathrm{CH_3(H)C{=}NCH_3} + \mathrm{H_2O} + \mathrm{H^+}$

19-16 (a) $\mathrm{C_4H_9CH(OH)CH(C_3H_7)CHO}$ (b) $\mathrm{(CH_3)_2CHCH(OH)C(CH_3)_2CHO}$ (c) $\mathrm{C_6H_5CH_2CH(OH)CH(C_6H_5)CHO}$ (d) $\mathrm{H_2C(OH)CH_2C({=}O)CH_3}$

19-17 (a) $\mathrm{CH_3C({=}O)CH_2C({=}O)OCH_3}$ (b) $\mathrm{CH_3CH_2CH_2C({=}O)CH(CH_2CH_3)C({=}O)OCH_2CH_3}$

19-18 Ⓟ$\mathrm{OCH_2{-}C(H)(HO){-}C(H)(OH){-}CH(OH)C({=}O)CH_2O}$Ⓟ

19-19 (a) $CH_3\underline{CH_2}CHO$ (b) $CH_3\underline{CH_2}C(=O)\underline{CH_3}$ (c) $\underline{CH_3}CO_2CH_2CH_3$ (d) $(CH_3)_2\underline{CH}CHO$

19-20 (a) $CH_3CH_2CHO + OH^- \rightleftharpoons \left[CH_3\bar{C}HC(=O)H \longleftrightarrow CH_3CH{=}CH(O^-)\right] + H_2O$

(b) $CH_3CH_2C(=O)CH_3 + OH^- \rightleftharpoons \left[CH_3\bar{C}H{-}C(=O)CH_3 \longleftrightarrow CH_3CH{=}C(O^-)CH_3\right] + H_2O$;

$CH_3CH_2C(=O)CH_3 + OH^- \rightleftharpoons \left[CH_3CH_2C(=O)\bar{C}H_2 \longleftrightarrow CH_3CH_2C(O^-){=}CH_2\right] + H_2O$

(c) $CH_3CO_2CH_2CH_3 + OH^- \rightleftharpoons \left[\bar{C}H_2C(=O){-}OCH_2CH_3 \longleftrightarrow CH_2{=}C(O^-)OCH_2CH_3\right] + H_2O$

(d) $(CH_3)_2CHCHO + OH^- \rightleftharpoons \left[(CH_3)_2\bar{C}C(=O)H \longleftrightarrow (CH_3)_2C{=}CH(O^-)\right] + H_2O$

19-21 Step 1: $CH_3CH(H)C(=O)H + {}^-OH \rightleftharpoons CH_3\bar{C}HC(=O)H + H_2O$

Step 2: $CH_3CH_2C(=O)H + CH_3\bar{C}HC(=O)H \rightleftharpoons CH_3CH_2C(O^-)(H){-}CH(CH_3)CHO$

Step 3: $CH_3CH_2C(O^-)(H){-}CH(CH_3)CHO + H{-}OH \rightleftharpoons CH_3CH_2C(OH)(H){-}CH(CH_3)CHO + OH^-$

CHAPTER 20

20-1 (a) $CH_2{=}CHC{\equiv}CH$ (d) $CH_3CH{=}CHCH_2CH_2SH$ (g) $H_2NCH_2CH_2C(=O)CH_3$

(b) $CH_3CH(OH)CH_2CH(NH_2)CH_3$ (e) $CH_3CH_2CH_2C(=O)CH_2CN$ (h) epoxide ring bearing CH_3 and $COCl$ (2-methyloxirane-3-carbonyl chloride)

(c) $CH_3CH(OH)CH_2CONH_2$ (f) $ClCH_2CH_2C(=O)CH(CH_3)CH_2CO_2H$ (i) $CH_3OCH_2CH{=}CHCHO$

20-2 (a) 1,4-dihydroxybutanone (b) *trans*-4-hexen-3-one (c) 3-methyl-2-butenal (d) *cis*-4-bromo-2-buten-1-ol (e) 5-hydroxy-3-pentynamide (f) 5-methoxy-2-cyclohexenone

20-3 (a) $CH_3\overset{\overset{\displaystyle OH}{|}}{C}{=}CH_2$ (b) $CH_3O\overset{\overset{\displaystyle OH}{|}}{C}{=}CH\overset{\overset{\displaystyle O}{\|}}{C}OCH_3$

20-4 (a) $CH_3CH_2\overset{\overset{\displaystyle O}{\|}}{C}CH_3$ (b) $CH_3CH_2CO_2H$ (c) CF_3H (d) $CH_3CO_2CH_3$

20-5 (a) $CH_3\overset{\overset{\displaystyle O}{\|}}{C}\overset{-}{C}H\overset{\overset{\displaystyle O}{\|}}{C}OC_2H_5 \longleftrightarrow CH_3\overset{\overset{\displaystyle O^-}{|}}{C}{=}CH\overset{\overset{\displaystyle O}{\|}}{C}OC_2H_5 \longleftrightarrow CH_3\overset{\overset{\displaystyle O}{\|}}{C}CH{=}\overset{\overset{\displaystyle O^-}{|}}{C}OC_2H_5$

(b) $CH_3\overset{\overset{\displaystyle O}{\|}}{C}\overset{-}{C}H\overset{\overset{\displaystyle O}{\|}}{C}CH_3 \longleftrightarrow CH_3\overset{\overset{\displaystyle O^-}{|}}{C}{=}CH\overset{\overset{\displaystyle O}{\|}}{C}CH_3 \longleftrightarrow CH_3\overset{\overset{\displaystyle O}{\|}}{C}CH{=}\overset{\overset{\displaystyle O^-}{|}}{C}CH_3$

20-6 (a) $\overset{-}{C}H(CO_2C_2H_5)_2$

(b) $CH_3(CH_2)_6CH_2CH(CO_2C_2H_5)_2$

(c) $CH_3(CH_2)_6CH_2CH_2CO_2H + 2\,C_2H_5OH + CO_2$

(d) $CH_3CH{=}CHCH_2CH_2\overset{\overset{\displaystyle O}{\|}}{C}CH(CO_2C_2H_5)_2$

(e) $CH_3CH{=}CHCH_2CH_2\overset{\overset{\displaystyle O}{\|}}{C}CH_3 + 2\,C_2H_5OH + 2\,CO_2$

20-7 (a) (b)

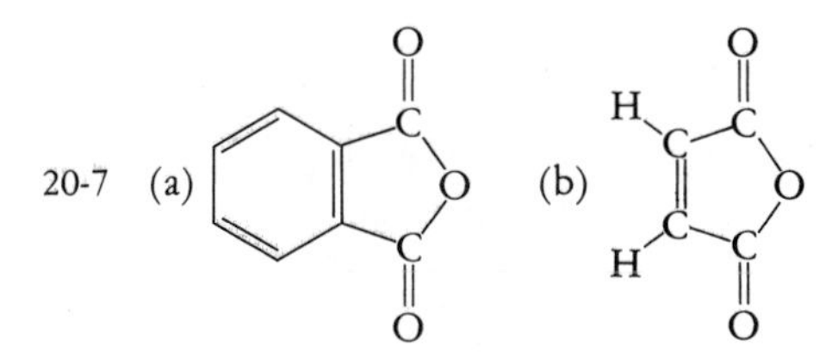

20-8 (a) lactic acid, glyceric acid, β-hydroxybutyric acid, malic acid, and mevalonic acid
(b) tartaric acid

20-9 $\left[-\overset{\overset{\displaystyle O}{\|}}{C}(CH_2)_5NH-\right]_n$

20-10 (a) addition polymer $CH_2{=}CCl_2$
(b) condensation polymer $HO_2C(CH_2)_5NH_2$
(c) addition polymer $CH_2{=}\overset{\overset{\displaystyle Cl}{|}}{C}CH{=}CH_2$
(d) addition polymer $CF_2{=}CF_2$

20-11 Glutaric acid forms an anhydride instead of reacting with 1,6-diaminohexane to form a polyamide.

CHAPTER 21

21-1 (a) 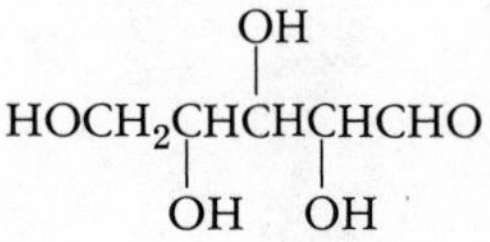 $HOCH_2CH(OH)CH(OH)CH(OH)CHO$ (c) 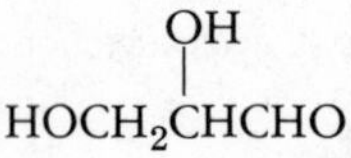 $HOCH_2CH(OH)CHO$

(b) 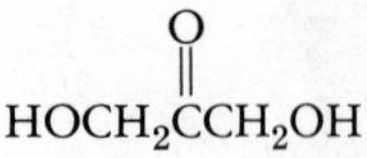 $HOCH_2C(=O)CH_2OH$ (d) 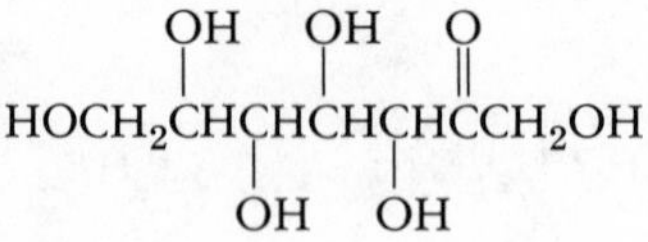 $HOCH_2CH(OH)CH(OH)CH(OH)CH(OH)C(=O)CH_2OH$

21-2 (a) ketotetrose (b) aldotetrose (c) ketohexose

21-3 (a) 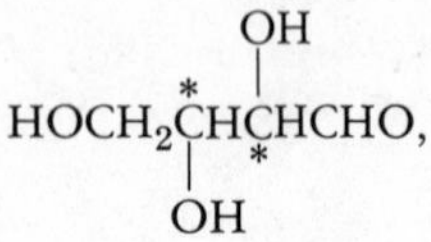 $HOCH_2\overset{*}{C}H(OH)\overset{*}{C}H(OH)CHO$, 2 chiral centers, 4 stereoisomers

(b) 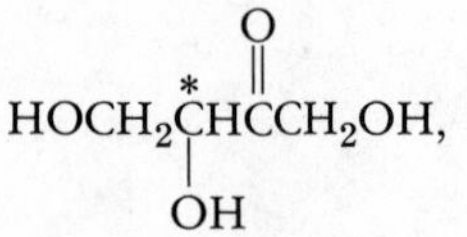 $HOCH_2\overset{*}{C}H(OH)C(=O)CH_2OH$, 1 chiral center, 2 stereoisomers

(c) 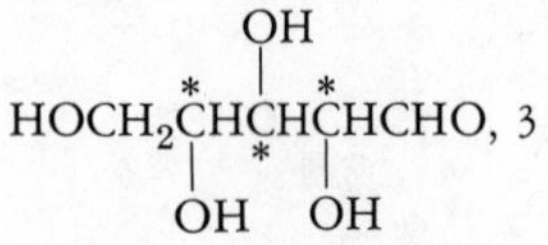 $HOCH_2\overset{*}{C}H(OH)\overset{*}{C}H(OH)\overset{*}{C}H(OH)CHO$, 3 chiral centers, 8 enantiomers

21-4 (a) D-family (b) L-family (c) D-family (d) L-family

21-5

Compound	hemiacetal or hemiketal	α- or β-isomers	anomer
(a) [ring structure: CH_2OH, HO, OH, O, OH, OH]	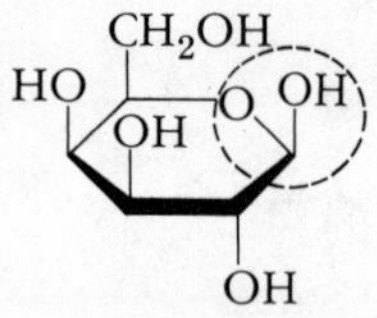 hemiacetal	β-isomer	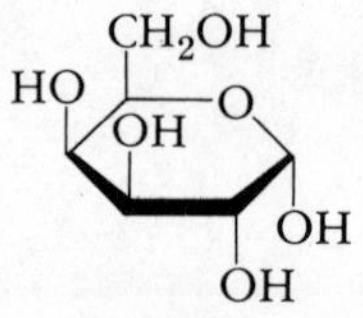 [ring structure: CH_2OH, HO, OH, O, OH, OH]
(b) [ring structure: CH_2OH, OH, O, HO, OH, OH]	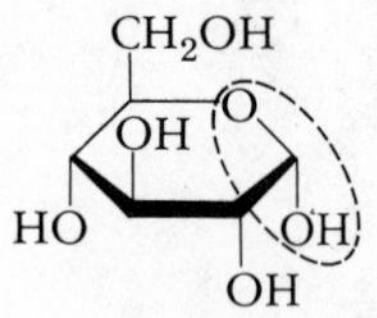 hemiacetal	α-isomer	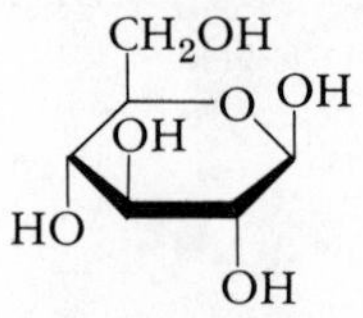 [ring structure: CH_2OH, OH, O, OH, HO, OH]
(c) [ring structure: CH_2OH, O, HO, OH, OH, OH]	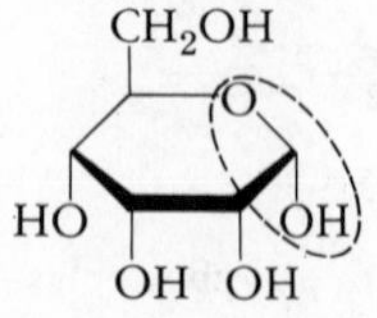 hemiacetal	α-isomer	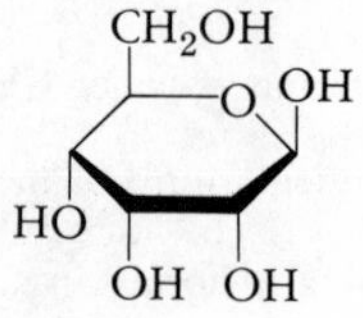 [ring structure: CH_2OH, O, OH, HO, OH, OH]

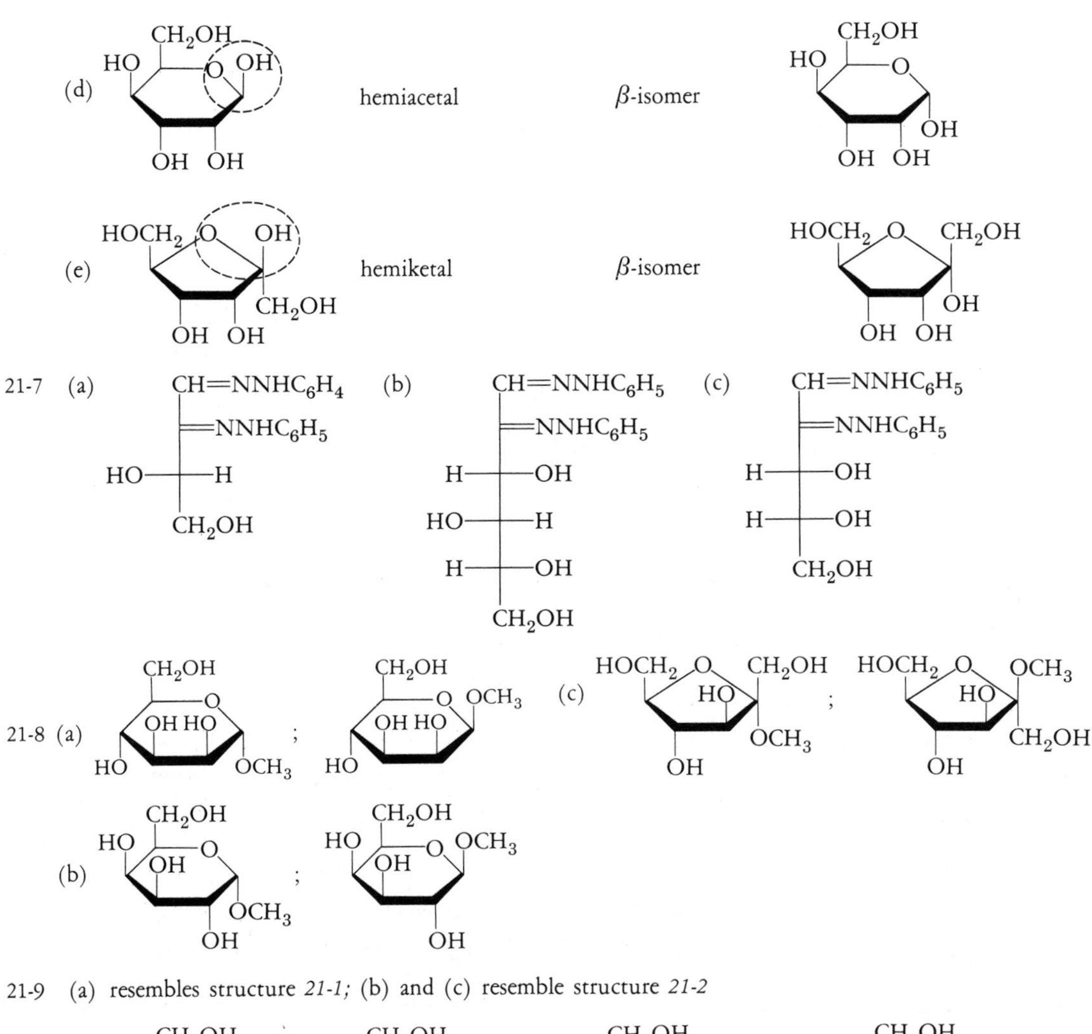

21-9 (a) resembles structure *21-1;* (b) and (c) resemble structure *21-2*

21-10 (a) + (b) +

21-11

α-lactose

α-maltose

21-12 (a) lactose $\xrightarrow[H^+]{H_2O}$ CH_2OH, HO, O, OH, OH, OH + CH_2OH, O, OH, HO, OH, OH

(b) maltose $\xrightarrow[H^+]{H_2O}$ CH_2OH, O, OH, HO, OH, OH + CH_2OH, O, OH, HO, OH, OH

(c) sucrose $\xrightarrow[H^+]{H_2O}$ CH_2OH, O, OH, HO, OH, OH + $HOCH_2$, O, OH, OH, CH_2OH, OH

CHAPTER 22

22-1 (a) $CH_3(CH_2)_{14}CO_2H$ $\xrightarrow{CH_3OH,\ H^+}$ $CH_3(CH_2)_{14}CO_2CH_3$

$CH_3(CH_2)_{14}CO_2H$ $\xrightarrow{aq.\ NaOH}$ $CH_3(CH_2)_{14}CO_2Na^+$

(b) $CH_3(CH_2)_5(H)C{=}C(H)(CH_2)_7CO_2H$ $\xrightarrow{O_3}$ $CH_3(CH_2)_5C({=}O)OH$ (HO) + $O{=}C(OH)(CH_2)_7CO_2H$

$\xrightarrow{KMnO_4,\ KOH}$ $CH_3(CH_2)_5CH(OH){-}CH(OH)(CH_2)_7CO_2^-K^+$

22-2 $CH_3(CH_2)_{26}CO_2CH_2(CH_2)_{22}CH_3$

22-3 (a) $CH_3(CH_2)_{24}CO_2CH_2(CH_2)_{28}CH_3$ $\xrightarrow{CH_3OH,\ H^+}$ $CH_3(CH_2)_{24}CO_2CH_3 + CH_3(CH_2)_{28}CH_2OH$

$\xrightleftharpoons{H_2O,\ H^+}$ $CH_3(CH_2)_{24}CO_2H + CH_3(CH_2)_{28}CH_2OH$

$\xrightarrow{CH_3NH_2}$ $CH_3(CH_2)_{24}CONHCH_3 + CH_3(CH_2)_{28}CH_2OH$

(b) Lanolin $\xrightarrow{NH_3}$ $CH_3(CH_2)_nCONH_2$ +

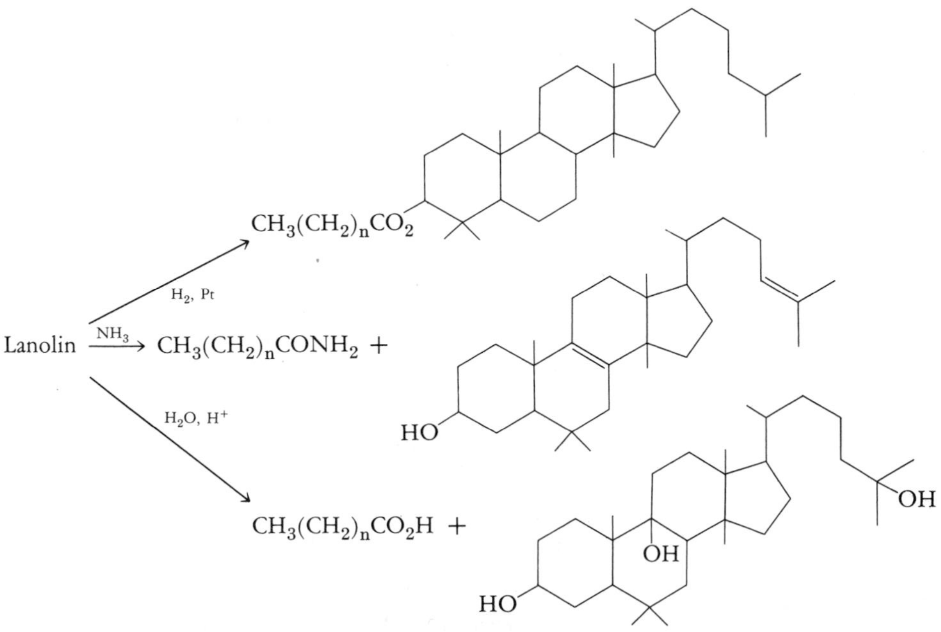

$\xrightarrow{H_2,\ Pt}$ $CH_3(CH_2)_nCO_2$

$\xrightarrow{H_2O,\ H^+}$ $CH_3(CH_2)_nCO_2H$ +

HO

HO OH OH

22-4 (a) $CH_3(CH_2)_7CH{=}CH(CH_2)_7C(=O)$—O—CH($CH_2OC(=O)(CH_2)_{16}CH_3$)($CH_2OC(=O)(CH_2)_{14}CH_3$)

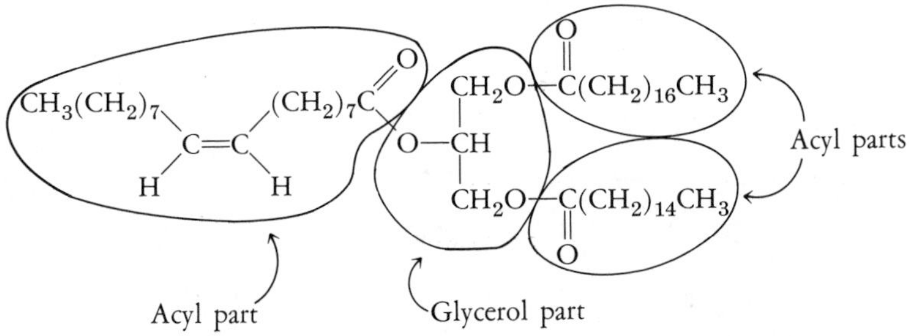

Acyl part   Glycerol part   Acyl parts

(b) $CH_3(CH_2)_{14}C(=O)$—OCH($CH_2OC(=O)(CH_2)_{12}CH_3$)($CH_2OC(=O)(CH_2)_{16}CH_3$)

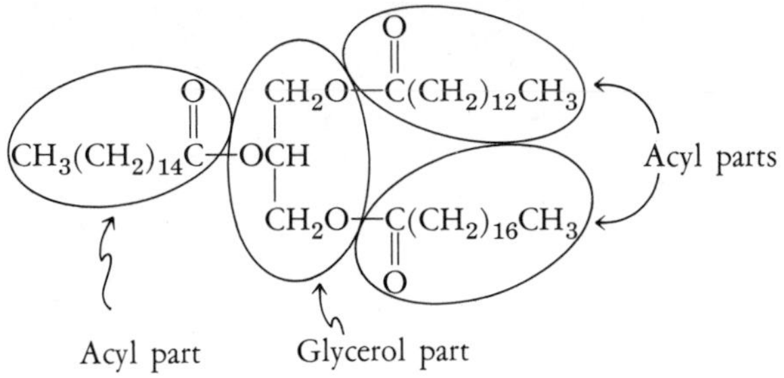

Acyl part   Glycerol part   Acyl parts

22-5 (a)

$$\underset{H}{\overset{CH_3(CH_2)_7}{}}\!\!\!\!\!\!>C{=}C<\!\!\!\!\!\!\underset{H}{\overset{(CH_2)_7CO_2^-\,Na^+}{}} + \begin{array}{l} CH_2OH \\ | \\ CHOH \\ | \\ CH_2OH \end{array} + \begin{array}{l} CH_3(CH_2)_{16}CO_2^-\,Na^+ \\ \\ CH_3(CH_2)_{14}CO_2^-\,Na^+ \end{array}$$

$$\uparrow \text{aq. NaOH, } \Delta$$

$$\underset{H}{\overset{CH_3(CH_2)_7}{}}\!\!\!\!\!\!>C{=}C<\!\!\!\!\!\!\underset{H}{\overset{(CH_2)_7\overset{O}{\overset{\|}{C}}{-}OCH}{}} \quad \begin{array}{l} CH_2O\overset{O}{\overset{\|}{C}}(CH_2)_{16}CH_3 \\ | \\ \text{(CH)} \\ | \\ CH_2O\underset{O}{\underset{\|}{C}}(CH_2)_{14}CH_3 \end{array}$$

$$\downarrow \text{Enzyme}$$

$$\underset{H}{\overset{CH_3(CH_2)_7}{}}\!\!\!\!\!\!>C{=}C<\!\!\!\!\!\!\underset{H}{\overset{(CH_2)_7CO_2H}{}} + \begin{array}{l} CH_2OH \\ | \\ CHOH \\ | \\ CH_2OH \end{array} + \begin{array}{l} CH_3(CH_2)_{16}CO_2H \\ \\ CH_3(CH_2)_{14}CC_2H \end{array}$$

(b)

$$CH_3(CH_2)_{14}CO_2^-\,Na^+ + \begin{array}{l} CH_2OH \\ | \\ CHOH \\ | \\ CH_2OH \end{array} + \begin{array}{l} CH_3(CH_2)_{12}CO_2^-\,Na^+ \\ \\ CH_3(CH_2)_{16}CO_2^-\,Na^+ \end{array}$$

$$\uparrow \text{aq. NaOH, } \Delta$$

$$CH_3(CH_2)_{14}\overset{O}{\overset{\|}{C}}OCH \quad \begin{array}{l} CH_2O\overset{O}{\overset{\|}{C}}(CH_2)_{12}CH_3 \\ | \\ \text{(CH)} \\ | \\ CH_2O\underset{O}{\underset{\|}{C}}(CH_2)_{16}CH_3 \end{array}$$

$$\downarrow \text{Enzyme}$$

$$\begin{array}{l} CH_3(CH_2)_{12}CO_2H \\ CH_3(CH_2)_{14}CO_2H \\ CH_3(CH_2)_{16}CO_2H \end{array} + \begin{array}{l} CH_2OH \\ | \\ CHOH \\ | \\ CH_2OH \end{array}$$

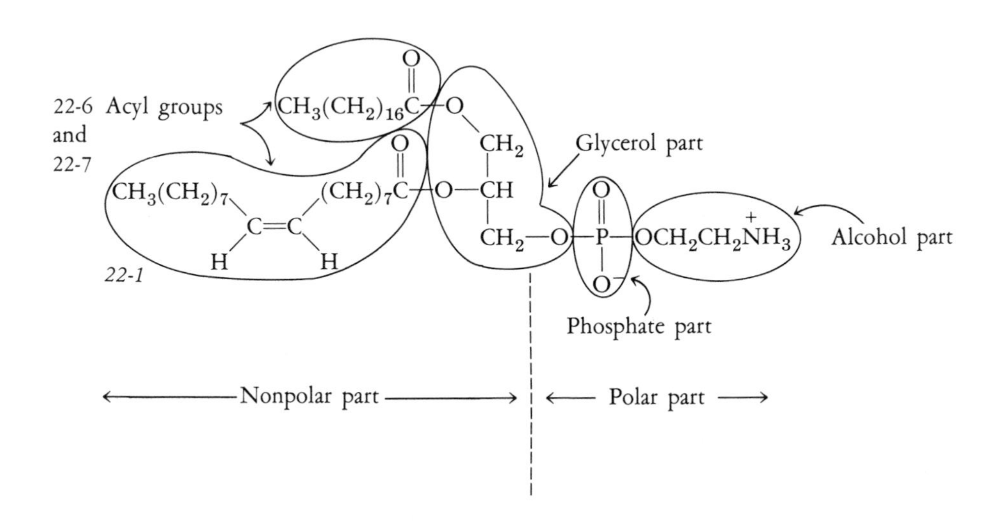

22-6 and 22-7
Acyl groups
$CH_3(CH_2)_{16}C$
O
CH_2
Glycerol part
$CH_3(CH_2)_7$
$(CH_2)_7C$
O—CH
C=C
H
H
CH_2—O
P
$OCH_2CH_2\overset{+}{N}H_3$
Alcohol part
O^-
22-1
Phosphate part
Nonpolar part
Polar part

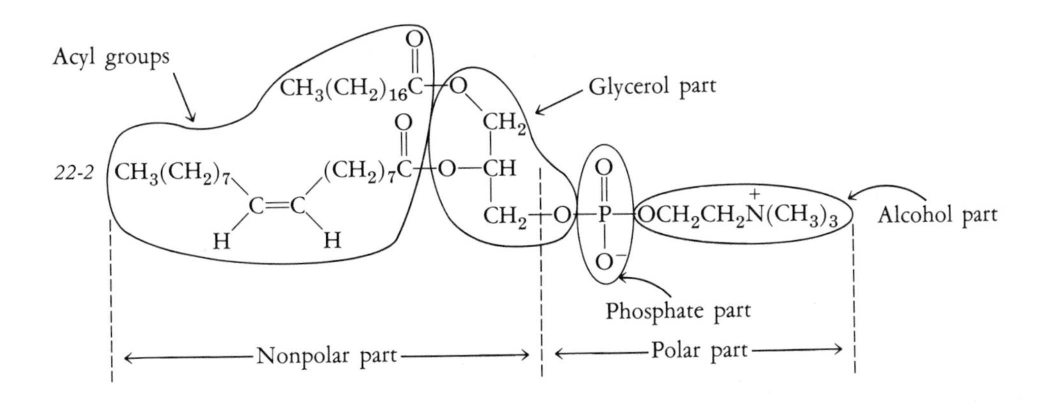

Acyl groups
$CH_3(CH_2)_{16}C$
O
Glycerol part
CH_2
22-2
$CH_3(CH_2)_7$
$(CH_2)_7C$
O—CH
C=C
H
H
CH_2—O
P
$OCH_2CH_2\overset{+}{N}(CH_3)_3$
Alcohol part
O^-
Phosphate part
Nonpolar part
Polar part

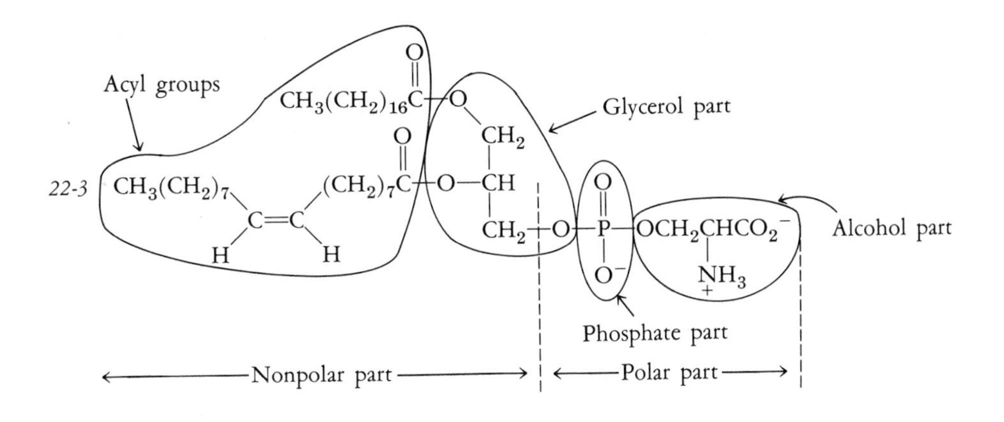

Acyl groups
$CH_3(CH_2)_{16}C$
O
Glycerol part
CH_2
22-3
$CH_3(CH_2)_7$
$(CH_2)_7C$
O—CH
C=C
H
H
CH_2—O
P
$OCH_2CHCO_2^-$
$\overset{+}{N}H_3$
Alcohol part
O^-
Phosphate part
Nonpolar part
Polar part

22-8 *22-1* $\xrightarrow[\Delta]{H_2O,\ H^+}$ $CH_3(CH_2)_{16}CO_2H$ + $CH_3(CH_2)_7CH{=}CH(CH_2)_7CO_2H$ (cis; H, H) + $CH_2OH{-}CHOH{-}CH_2OH$ + H_3PO_4 + $HOCH_2CH_2\overset{+}{N}H_3$

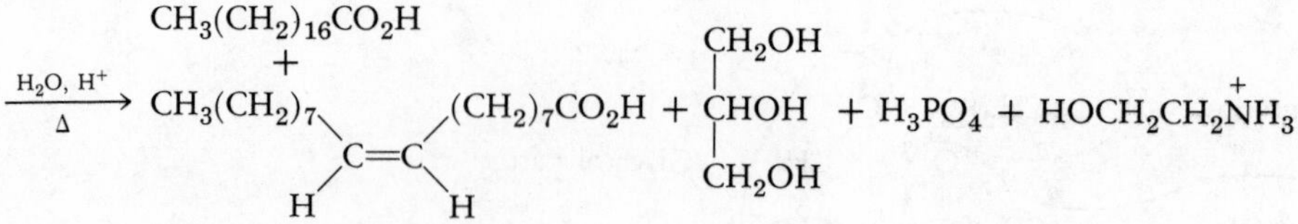

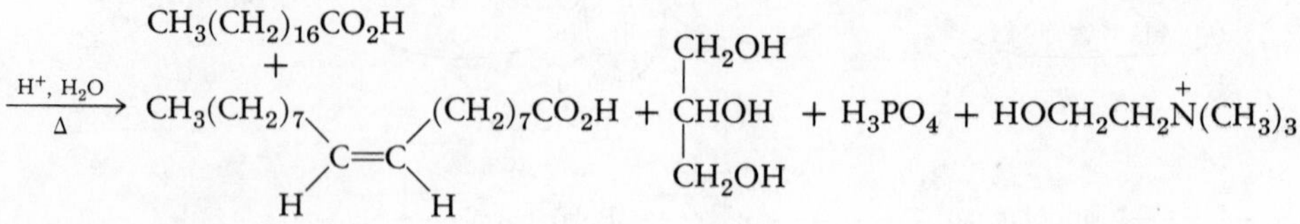

22-2 $\xrightarrow[\Delta]{H^+,\ H_2O}$ $CH_3(CH_2)_{16}CO_2H$ + $CH_3(CH_2)_7CH{=}CH(CH_2)_7CO_2H$ (cis; H, H) + $CH_2OH{-}CHOH{-}CH_2OH$ + H_3PO_4 + $HOCH_2CH_2\overset{+}{N}(CH_3)_3$

22-3 $\xrightarrow[\Delta]{H^+,\ H_2O}$ $CH_3(CH_2)_{16}CO_2H$ + $CH_3(CH_2)_7CH{=}CH(CH_2)_7CO_2H$ (cis; H, H) + $CH_2OH{-}CHOH{-}CH_2OH$ + H_3PO_4 + $HOCH_2CH(\overset{+}{N}H_3)CO_2H$

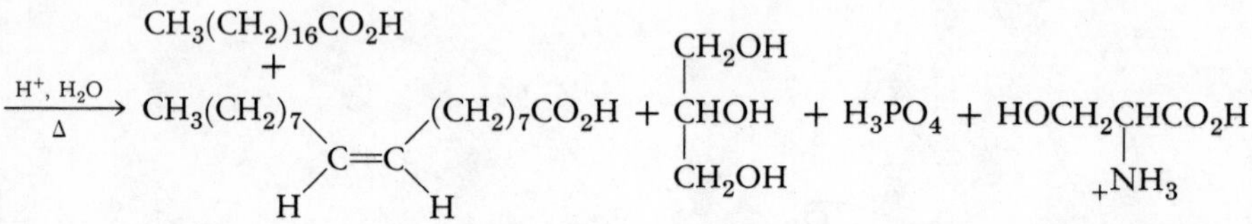

22-9 $CH_3(CH_2)_{11}$ | $OSO_3^-\ Na^+$

Nonpolar part | Polar part

$CH_3(CH_2)_{11}{-}C_6H_4{-}$ | $SO_3^-\ Na^+$

Nonpolar part | Polar part

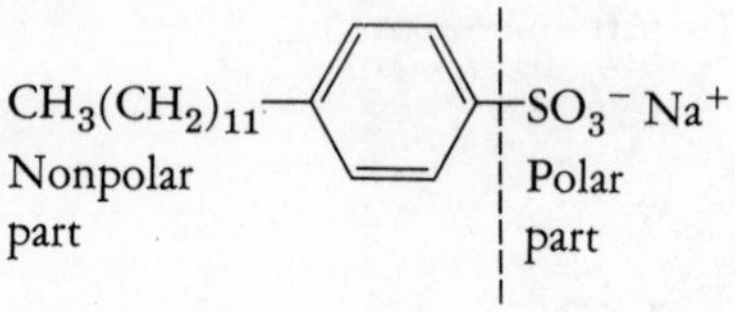

22-10 The steroid nucleus in all three compounds is

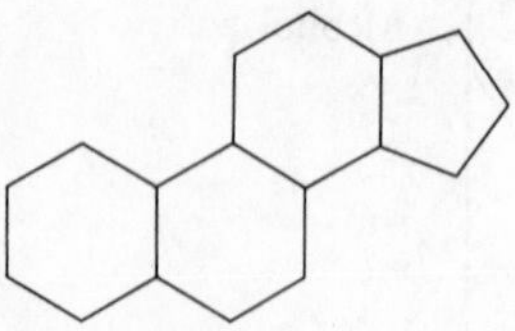

[steroid nucleus: three fused six-membered rings and a five-membered ring]

22-11 (a)

Structural Feature	Testosterone	Estradiol
Ring A	Contains C=O at C 3 and a double bond between C 4 and C 5	Phenol group at C 3
C 10	CH_3 group	No CH_3 group

(b)

Structural Feature	Progesterone	Cholesterol
C 17	Acetyl group	Long saturated hydrocarbon chain
C 3	C=O	C—OH
C 4 and C 5	C=C	no C=C
C 5 and C 6	no C=C	C=C

(c) Norethylnodrel and norethindrone differ only in the position of the C=C.

CHAPTER 23

23-1 (a) +1 (b) −2 (c) 0 (d) +1

23-2 (a) pH 1 $(CH_3)_2CHCH(\overset{+}{N}H_3)CO_2H$; pH 7 $(CH_3)_2CHCH(\overset{+}{N}H_3)CO_2^-$; pH 11 $(CH_3)_2CHCH(NH_2)CO_2^-$

(b) pH 1 $C_6H_5CH_2CH(\overset{+}{N}H_3)CO_2H$; pH 7 $C_6H_5CH_2CH(\overset{+}{N}H_3)CO_2^-$; pH 11 $C_6H_5CH_2CH(NH_2)CO_2^-$

23-3 (a) 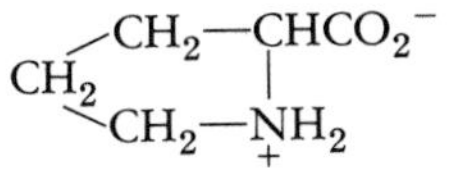 CH_2 ring: $CH_2—CHCO_2^-$ / $CH_2—\overset{+}{N}H_2$ (cyclic)

(c) $^-O_2CCH_2CH_2CH(NH_2)CO_2^-$

(b) 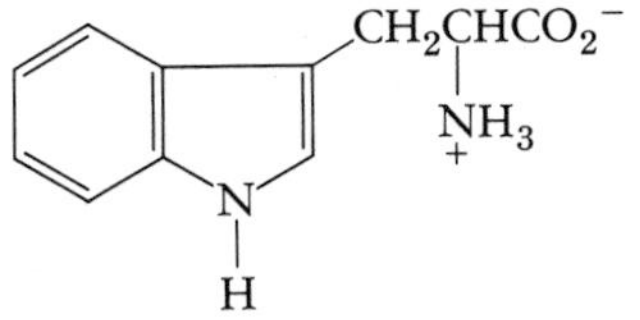 indole ring (N—H) with $CH_2CH(\overset{+}{N}H_3)CO_2^-$

(d) 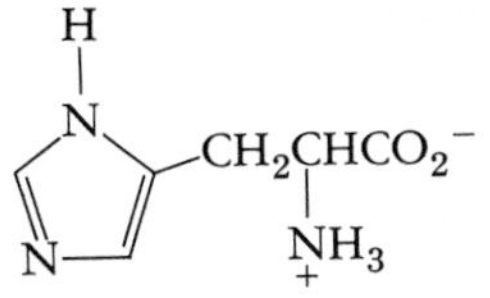 imidazole ring (N—H, N) with $CH_2CH(\overset{+}{N}H_3)CO_2^-$

23-4 (a) pH 3.07 to 7.07 (c) pH 3.68 to 7.68
(b) pH below 3.65 (d) pH above 5.2

23-5

$\overset{+}{N}H_3$—C(H)(CO_2^-)—CH_2—S—S—CH_2—C(H)(CO_2^-)—$\overset{+}{N}H_3$

Disulfide bond (S—S)

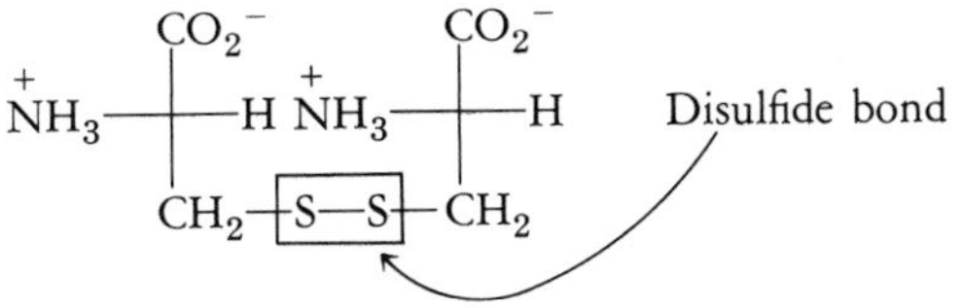

23-6

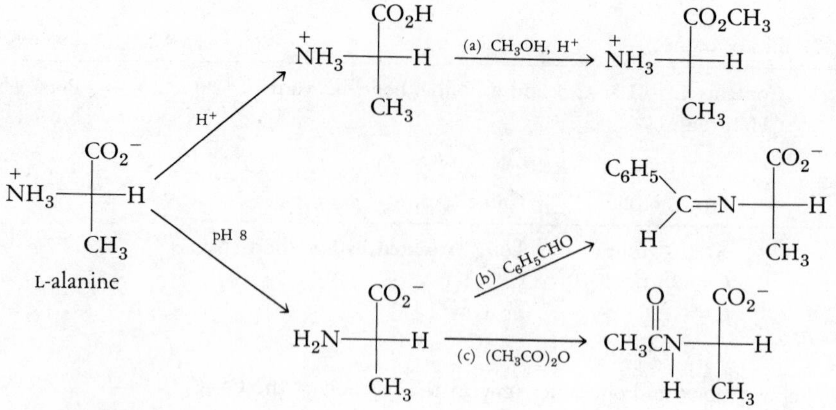

L-alanine: $\overset{+}{N}H_3$–C(H)(CO_2^-)(CH_3)

$\xrightarrow{H^+}$ $\overset{+}{N}H_3$–C(H)(CO_2H)(CH_3) $\xrightarrow{\text{(a) } CH_3OH,\ H^+}$ $\overset{+}{N}H_3$–C(H)(CO_2CH_3)(CH_3)

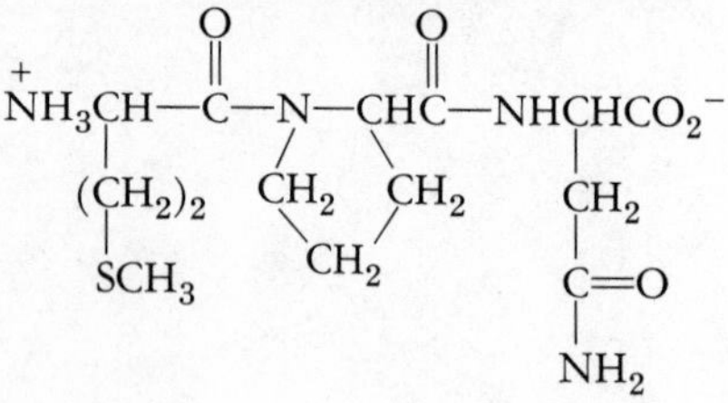

$\xrightarrow{\text{pH 8}}$ H_2N–C(H)(CO_2^-)(CH_3)

H_2N–C(H)(CO_2^-)(CH_3) $\xrightarrow{\text{(b) } C_6H_5CHO}$ $C_6H_5(H)C{=}N$–C(H)(CO_2^-)(CH_3)

H_2N–C(H)(CO_2^-)(CH_3) $\xrightarrow{\text{(c) } (CH_3CO)_2O}$ $CH_3C(=O)N(H)$–C(H)(CO_2^-)(CH_3)

23-7 $\overset{+}{N}H_3$–C(H)(CO_2^-)(CH_2SAg) $+ 2\,H^+$

$[\overset{+}{N}H_3$–C(H)(CO_2^-)(CH_2S)$]_2$Pb $+ 2\,H^+$

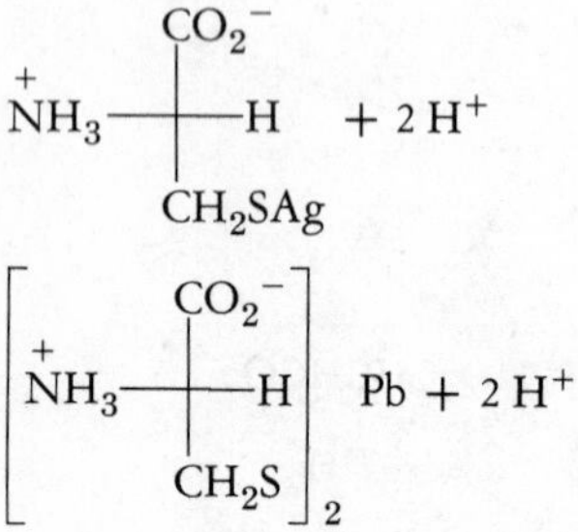

23-8

(a) 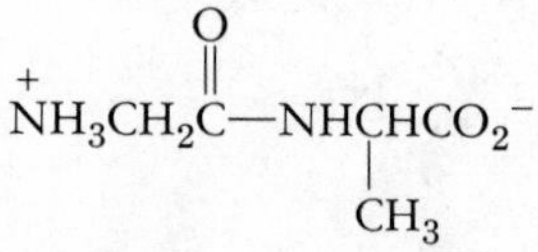 $\overset{+}{N}H_3CH_2C(=O)$—$NHCH(CH_3)CO_2^-$

(b) 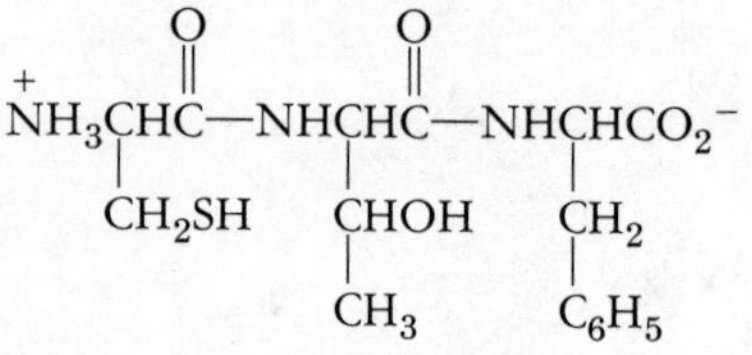 $\overset{+}{N}H_3CH(CH_2SH)C(=O)$—$NHCH(CHOHCH_3)C(=O)$—$NHCH(CH_2C_6H_5)CO_2^-$

(c) $\overset{+}{N}H_3CH((CH_2)_2SCH_3)$—$C(=O)$—N—CHC(=O)—$NHCH(CH_2C(=O)NH_2)CO_2^-$ (proline ring: N, CH, CH_2, CH_2, CH_2)

23-9		23-10
glycine	alanine	Gly-Ala
cysteine	phenylalanine	Cys-Thr-Phe
methionine	asparagine	Met-Pro-Asn

23-11

(a) $\overset{+}{N}H_3CH_2\overset{O}{\overset{\|}{C}}-NH\underset{CH_3}{CH}\overset{O}{\overset{\|}{C}}-NH\underset{CH_2OH}{CH}CO_2^-$

(b) $\overset{+}{N}H_3CH(CH(CH_3)CH_2CH_3)\overset{O}{\overset{\|}{C}}-NHCH(CH_2C_6H_4OH)\overset{O}{\overset{\|}{C}}-NHCH(CH_2C(=O)NH_2)\overset{O}{\overset{\|}{C}}-N-CHCO_2^-$ (proline ring: $N-CH_2-CH_2-CH_2-CH$)

(c) $\overset{+}{N}H_3CH((CH_2)_2C(=O)NH_2)\overset{O}{\overset{\|}{C}}-NHCH((CH_2)_2SCH_3)\overset{O}{\overset{\|}{C}}-NHCH((CH_2)_3NHC(=NH)NH_2)\overset{O}{\overset{\|}{C}}-NHCH(CH_2CH_2CO_2H)\overset{O}{\overset{\|}{C}}-NHCH((CH_2)_4NH_2)CO_2^-$

23-12

glycylalanylserine

isoleucyltyrosylasparagylproline

glutaminylmethionylarginylglutamyllysine

23-13 Adrenocortitrophin has only one polypeptide chain; therefore it cannot have a quaternary structure.

CHAPTER 24

24-1

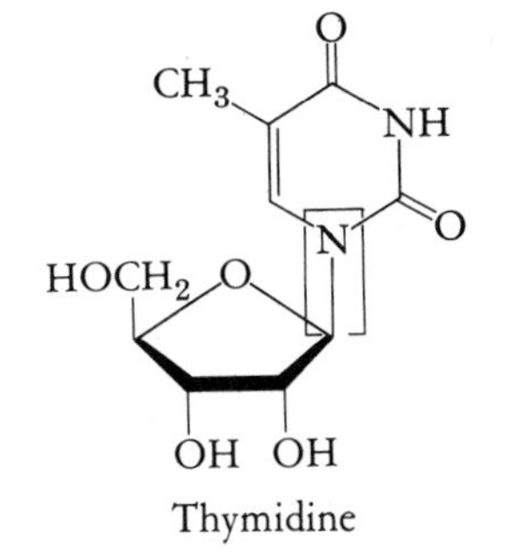

Thymidine

$HOCH_2$ O N N NH O NH_2 N OH OH

Guanosine

24-2 (a) guanosine (b) deoxythymidine (c) deoxyuridine

24-3 (a) 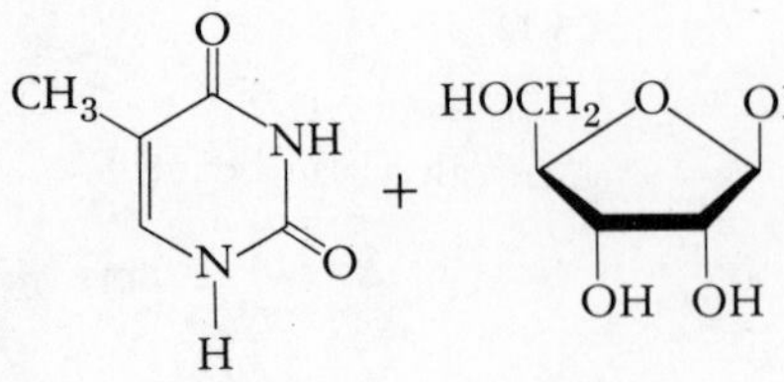 CH_3 O NH N O H + $HOCH_2$ O OH OH OH

(b) 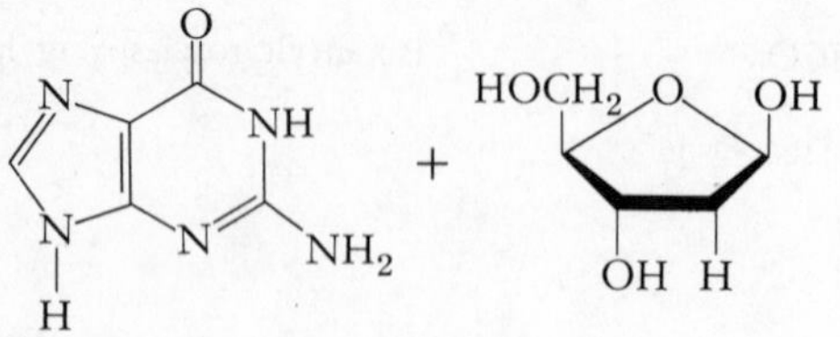 N O NH N N NH_2 H + $HOCH_2$ O OH OH H

(c) 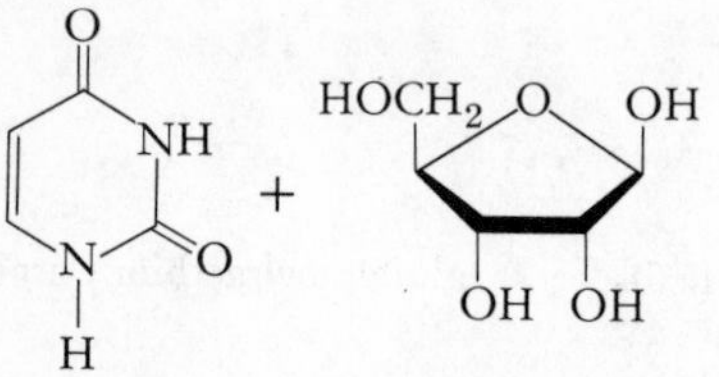 O NH N O H + $HOCH_2$ O OH OH OH

24-4

NH_2 N N N N O $^{-}O-P-OCH_2$ O^- O

Adenosine monophosphate

O OH O NH

$O=P-OCH_2$ O^- N O O

Uridine monophosphate

O OH NH_2 N

$O=P-OCH_2$ O^- N O O

Cystidine monophosphate

OH OH

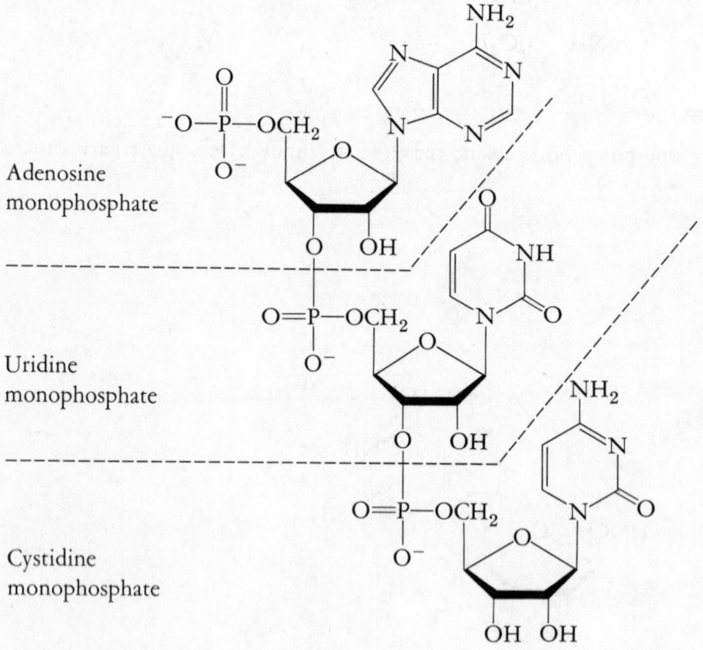

24-5 (a) and (c) have a free 5′-hydroxy group; (c) has a free 3′-hydroxy group

24-6 (a)

O
CH_3
NH
N
O
$HOCH_2$ O
NH_2
O OH
N
N
N
N
O=P—OCH_2
O^-
O
NH_2
N
O OH
O=P—OCH_2
N
O
O^-
O
OH OH

(b)

NH_2
N
N
O
$HOCH_2$ O
NH_2
N
N
O OH
N
N
O=P—OCH_2
O^-
O
O
CH_3
NH
O OH
O=P—OCH_2
N
O
O^-
O
O OH
O=P—O^-
O^-

(c)

NH_2
N
N
O
N
N
$^-$O—P—OCH_2
O
O^-
O
O OH
N
NH
N
N
O=P—OCH_2
O
O^-
O
CH_3
NH
O OH
O=P—OCH_2
N
O
O^-
O
NH_2
O OH
N
O=P—OCH_2
N
O
O^-
O
OH OH

24-7 UCU, UCC, UCA, UCG, AGU, and AGC

24-8 (a) isoleucine (b) aspartic acid (c) alanine (d) glycine (e) arginine (f) histidine

24-9 serylleucylthreonylglutamic acid

CHAPTER 25

25-1 (a) peptide (b) lipid (c) sucrose (d) cellulose (e) lactose

25-2 (a) reduction (c) isomerization
(b) hydrolysis (d) transfer of group or atom from one compound to another

25-3

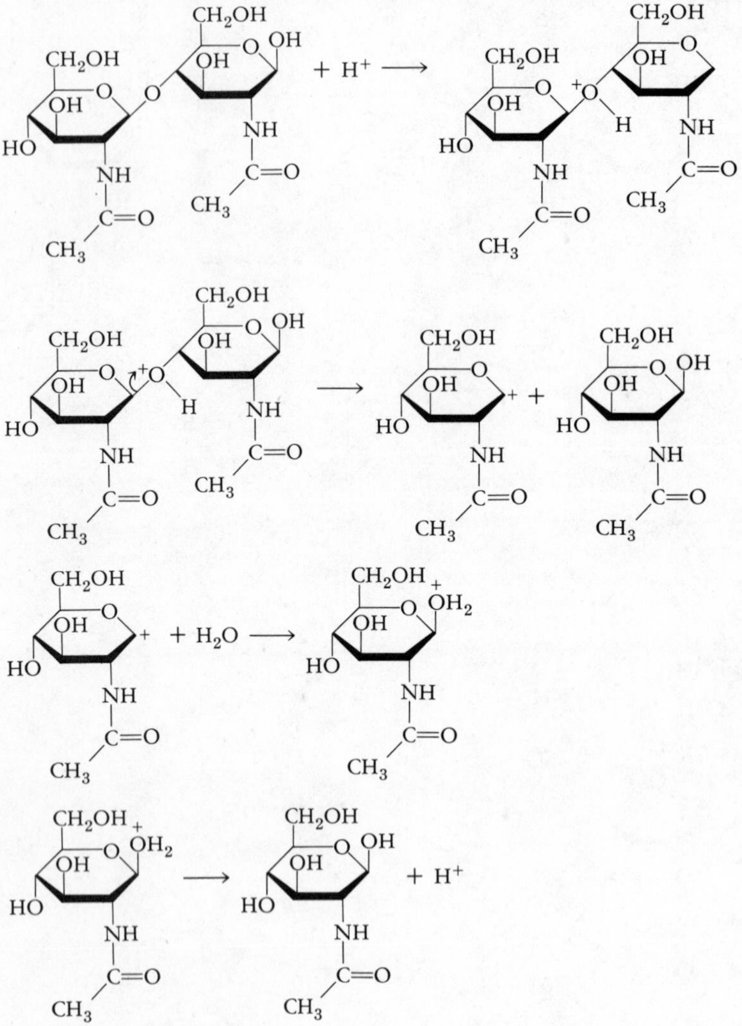

25-4 The distance between the two $—O^-$ atoms in each molecule is about the same. They all bind to the active sites on succinate dehydrogenase.

25-5 It catalyzes the hydrolysis of the amide bond of penicillin G.

CHAPTER 28

28-1 $RCH_2CH_2CH_2C(=O)-O-P(=O)(O^-)-O$—ribose-adenine (the $C(=O)-O-P(=O)(O^-)-$ portion is circled)

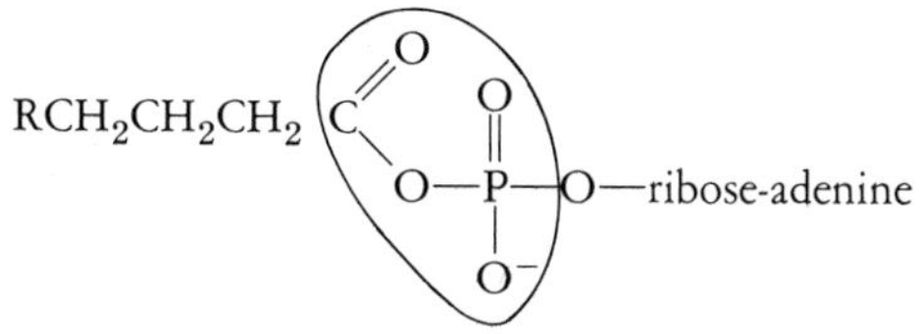

28-2 $RCH_2CH_2CH_2C(=O)S$ CoA (the $C(=O)S$ portion is circled)

28-3 $RCH_2CH_2CH_2C(=O)-O-P(=O)(O^-)-O$—ribose adenine ⟵ replaced by CoA (the $O-P(=O)(O^-)-O$—ribose adenine portion is circled)

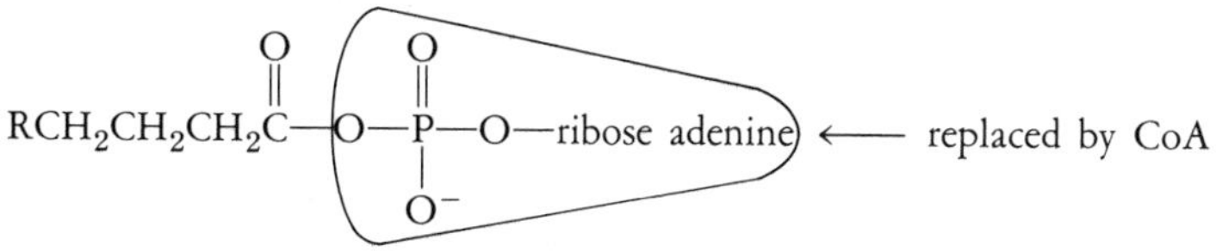

28-4

$CH_3(CH_2)_7CH=CHCH_2C(=O)-SCoA$ (cis: H, H on same side)

↓ Isomerase

$CH_3(CH_2)_7CH_2CH=CHC(=O)SCoA$ (trans)

↓ H_2O

$CH_3(CH_2)_7CH_2CH(OH)CH_2C(=O)-SCoA$

↓ NAD^+ → NADH

$CH_3(CH_2)_7CH_2C(=O)CH_2C(=O)-SCoA$

↓ CoASH

$CH_3(CH_2)_7CH_2C(=O)-SCoA + CH_3C(=O)-SCoA$

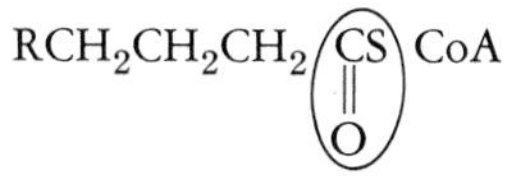

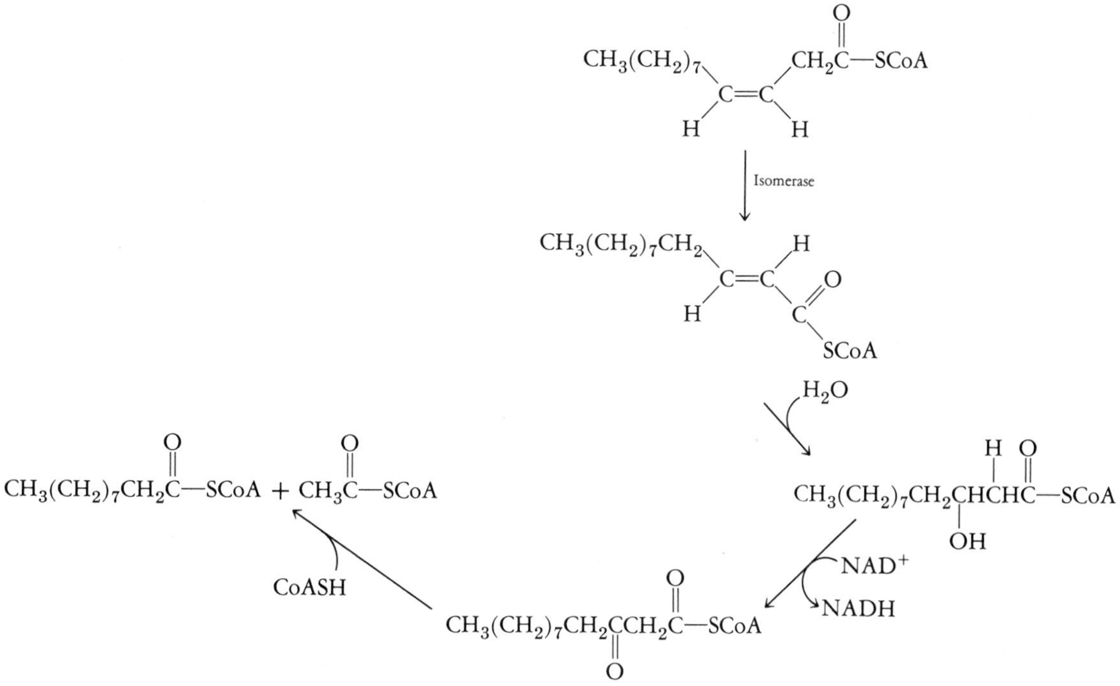

28-5 eight

28-6 129 ATP

28-7 It takes five rounds of the fatty acid degradation cycle to catabolize the 12-carbon compound laurate. Each round produces 1 molecule of $FADH_2$, 1 molecule of NADH, and 1 molecule of acetyl coenzyme A. The last molecule of acetyl coenzyme A is the end product of the last cycle.

CHAPTER 29

29-1 leucine, tyrosine, phenylalanine, lysine, tryptophan, threonine, glycine, serine, alanine, cysteine, and isoleucine

29-2

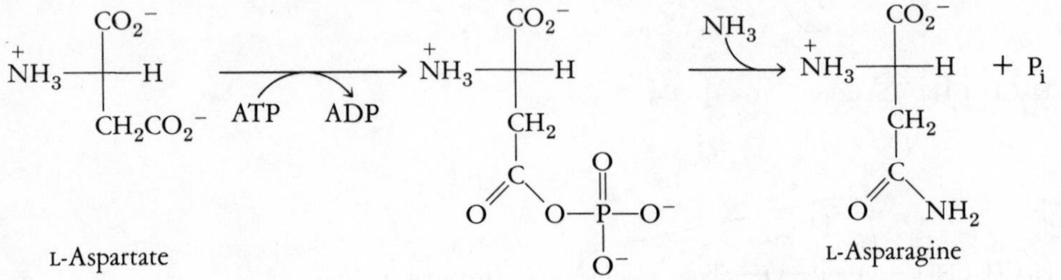

CO_2^- | $\overset{+}{N}H_3$—C—H | $CH_2CO_2^-$ (L-Aspartate) $\xrightarrow[\text{ATP} \to \text{ADP}]{}$ CO_2^- | $\overset{+}{N}H_3$—C—H | CH_2 | C(=O)—O—P(=O)(O^-)—O^- $\xrightarrow{NH_3}$ CO_2^- | $\overset{+}{N}H_3$—C—H | CH_2 | C(=O)—NH_2 (L-Asparagine) + P_i

APPENDIX

A-1 (a) $25 + y = x$ (b) $x = -22$ (c) $x = -0.70$

A-2 (a) $x = 3$ (b) $x = 4y$ (c) $x = \frac{5b}{9}$

A-3 (a) $x = \frac{y - 16}{4}$ (b) $\frac{y}{17} + 5 = x$ (c) $x = \frac{11}{3 + y}$

A-4 (a) 33 days (b) 60 g

A-5 (a) men 42.9%; women 57.1% (b) red 65.1%; blue 27.9%; white 7.0%

A-6 Brand A, good precision, good accuracy; Brand B, good precision, poor accuracy; Brand C, poor precision, poor accuracy.

A-7 (a) 2 (b) 3 (c) 3 (d) 4 (e) 2 (f) 2

A-8 (a) 57 (b) 28.3 (c) 46.3 (d) 2.20

A-9 (a) 22.6 (b) 6.21 (c) 6.12

A-10 (a) 2.65×10^2 (b) 1.73×10^5 (c) 5.73×10^{-3} (d) 1.26×10^{-1} (e) 5.43×10^{-2}

A-11 (a) 125 (b) 0.025 (c) 3750 (d) 523000 (e) 0.000000062

A-12 (a) 6×10^5 (b) 37 (c) 1.91×10^{-6}

A-13 (a) 2×10^3 (b) 2.9×10^{-5} (c) 4.57×10^4 (d) 13.9

A-14 (a) 0.187 (b) 2.241 (c) -2.240 (d) -9.068 (e) 7.512

A-15 (a) 2.29×10^{-2} (b) 1.86×10^{-6} (c) 6.92×10^7 (d) 3.55×10^{-9} (e) 7.59×10^{-12}

INDEX

Page numbers in *italics* indicate illustrations. Page numbers followed by *t* indicate tables.